Space Sciences Series of ISSI

Volume 96

The Space Sciences Series of ISSI books are coherent reports of the findings, discussions, and ideas that result from international scientific workshops regularly held at the International Space Science Institute (ISSI) in Bern, Switzerland. ISSI's main task is to contribute to the achievement of a deeper understanding of the results from space-research missions, adding value to those results through multi-disciplinary research in an atmosphere of international cooperation. The books are reprints of special issues in the Space Science Reviews journal and occasionally of special issues in the Surveys in Geophysics journal.

Dominik C. Hezel • Herbert Palme •
Klaus Mezger • Alessandro Morbidelli •
Jutta Zipfel
Editors

Evolution of the Solar System: Constraints from Meteorites

Previously published in *Space Science Reviews, Topical Collection: Evolution of the Solar System: Constraints from Meteorites*

Springer

Editors
Dominik C. Hezel
Institut für Geowissenschaften
Goethe Universität Frankfurt am Main
Frankfurt am Main, Hessen, Germany

Herbert Palme
Senckenberg Forschungsinstitut und
Naturmuseum Frankfurt
Frankfurt am Main, Hessen, Germany

Klaus Mezger
Institut für Geologie
Universität Bern
Bern, Switzerland

Alessandro Morbidelli
Collège de France
Paris, France

Jutta Zipfel
Senckenberg Forschungsinstitut und
Naturmuseum Frankfurt
Frankfurt am Main, Hessen, Germany

ISSN 1385-7525
Space Sciences Series of ISSI
ISBN 978-94-024-2368-6

Cover illustration: A compound chondrule from the Allende meteorite. The left side is a barred olivine chondrule, the right side represents porphyritic chondrule. The diameter of the chondrule is about 1mm. Photo by J. Zipfel, Senckenberg, Frankfurt

This Springer imprint is published by the registered company Springer Nature B.V.
The registered company address is: Van Godewijckstraat 30, 3311 GX Dordrecht, The Netherlands

Contents

Space Science Reviews (2025) 221:125
https://doi.org/10.1007/s11214-025-01254-7

Evolution of the Solar System: Constraints from Meteorites – Editorial

Dominik C. Hezel[1] · Herbert Palme[2] · Jutta Zipfel[2] · Alessandro Morbidelli[3,4] · Klaus Mezger[5]

Received: 25 November 2025 / Accepted: 26 November 2025 / Published online: 8 December 2025

The early Solar System must have been an exceptional place, where habitable planets formed within a few tens of million years, starting from nm- to μm sized dust in a gaseous disk around a newly forming star. Today, meteorites and materials collected from asteroids are used together with samples from the Moon, Mars, but also from our Earth to decipher the chronology of the various physical and chemical processes occurring in the protoplanetary disk.

The quantity and quality of chemical, mineralogical and isotopic data on samples of the various planetary material building blocks increased substantially over the past two decades. In particular, the ability to measure isotope ratios on very small samples with exquisite precision led to the identification of intriguing anomalies that suggest the existence of processes in the formation and evolution of the disk never contemplated before. As a consequence, new and more detailed ideas on the mechanics, dynamics and chronology of the early Solar System were formulated. Interestingly enough, the foundation for some of the ideas are rooted in Immanuel Kant's *nebular hypothesis* from 1755 that invoked the formation of rings in a spinning cloud as precursors to planet formation. The many spectacular images taken by the ALMA telescope over the past 10 years depict protoplanetary disks with ring structures that support our current ideas driven by constraints derived from meteorites. In addition to analytical studies on meteorites, primarily performed by meteoriticists, astrophysicists have, in great detail, simulated the evolution of the early Solar System in computer models. These cover the complete evolution from the interstellar medium to the Solar System as we know it today. Such computer models include detailed evidence from meteorites to constrain input parameters and test their predictions, notably concerning the distribution of

✉ D.C. Hezel
dominik.hezel@em.uni-frankfurt.de

1 Institut für Geowissenschaften, Goethe-Universität Frankfurt am Main, Altenhöferallee 1, 60438 Frankfurt am Main, Germany

2 Senckenberg Forschungsinstitut und Naturmuseum, Senckenberganlage 25, 60325 Frankfurt am Main, Germany

3 Collège de France, Centre National de la Recherche Scientifique, Université Paris Sciences et Lettres, Sorbonne Université, 75014 Paris, France

4 Laboratoire Lagrange, Centre National de la Recherche Scientifique, Observatoire de la Côte d'Azur, Université Côte d'Azur, 06304 Nice, France

5 Institut für Geologie, Universität Bern, 3012 Bern, Switzerland

isotope anomalies and the abundances of short-lived radioactive elements. The streaming instability model has brought an entirely new perspective on the formation of planetesimals starting from a disk of gas and chondrule-sized grains, which seems consistent with many constraints derived from meteorites and their different components. In summary, the two communities, meteoriticists and astrophysicists, have been increasingly working together over the past two decades to produce coherent models and ideas describing the origin and evolution of the early Solar System. However, despite important progress and new findings, a robust foundational framework comparable to natural selection in biology or plate tectonics in geology is still missing.

The aim of this workshop, held from 5–9 June 2023 at ISSI in Bern, was to highlight the progress in early Solar System research during the past two decades and from this provide an update on the current knowledge of protoplanetary disk processes, as well as planetesimal and planet formation and their early evolution. Leading experts from the fields of meteoritics and astrophysics were invited and presented the latest progress in their research fields. For the resulting publications, the authors were encouraged to present controversial views and interpretations to illustrate the current limits of our understanding, and, where possible, formulate questions or make suggestions for future research.

The following contributions are the result of the presentations and extensive discussions at the workshop. The present collection starts with five articles (Jacquet et al., Desch & Miret-Roig, Liu et al., Tissot et al., Lodders, Bergemann & Palme) setting the framework in which the early Solar System formed and developed. The first article introduces basic meteoritical and astrophysical terms, parameters, and concepts. The next four contributions describe the initial chemical and isotopic composition of our Solar System, the environment in which it formed, interactions and material exchange with its surroundings – both from the interstellar medium to the disk and within the disk – as well as material growth and dynamical processes. The following seven articles (Marrocchi et al., van Kooten et al., Schönbächler et al., Krot et al., Lee et al., Nakamura et al., Zolensky et al.) reconstruct the chronology and processes that formed and altered the chondritic components, starting with their formation from dust in the protoplanetary disk all the way to their agglomeration and reprocessing in planetesimals. The results of the remarkable samples of asteroids and comets collected during space missions underline the importance of this ground-truth material, given that the returned material was often sufficiently different to initial expectations. The concluding two articles (Sossi et al., Lodders et al.) illustrate the connection of protoplanetary disk processes to the bulk composition of the terrestrial planets and, particularly, Earth.

Acknowledgements We are grateful to the International Space Science Institute and all their organizers, in particular Andrea Fischer, for the opportunity to hold this workshop in their wonderful location in Bern. We had a highly inspiring atmosphere and a delightful social dinner. We thank Frank Schulz from Springer for his always prompt and helpful responses, which aided negotiating the bumps of the sometimes enigmatic submission system. Finally, and foremost, we thank all the authors and workshop attendees who made this collection possible and provided such a timely and valuable overview of early Solar System sciences research.

Declarations

Competing Interests The authors declare no competing interests.

Space Science Reviews (2024) 220:78
https://doi.org/10.1007/s11214-024-01112-y

The Early Solar System and Its Meteoritical Witnesses

Emmanuel Jacquet[1] · Cornelis Dullemond[2] · Joanna Drążkowska[3] · Steven Desch[4]

Received: 8 April 2024 / Accepted: 11 September 2024 / Published online: 30 September 2024

Abstract

Meteorites, and in particular primitive meteorites (chondrites), are irreplaceable probes of the solar protoplanetary disk. We review their essential properties and endeavour to place them in astrophysical context. The earliest solar system solids, refractory inclusions, may have formed over the innermost au of the disk and have been transported outward by its expansion or turbulent diffusion. The age spread of chondrite components may be reconciled with the tendency of drag-induced radial drift if they were captured in pressure maxima, which may account for the non-carbonaceous/carbonaceous meteorite isotopic dichotomy. The solid/gas ratio around unity witnessed by chondrules, if interpreted as nebular (non-impact) products, suggests efficient radial concentration and settling at such locations, conducive to planetesimal formation by the streaming instability. The cause of the pressure bumps, e.g. Jupiter or condensation lines, remains to be ascertained.

1 Introduction

The early history of the Solar System is still largely a (space-time) *Terra Incognita.* To be sure, the relevant physics is essentially classical, so were we to know the initial conditions of the protosun, its protoplanetary disk and its environment, and had we infinite computational power, the evolution would be predetermined. Yet we are far from being such Laplace

✉ E. Jacquet
emmanuel.jacquet@mnhn.fr

✉ C. Dullemond
dullemond@uni-heidelberg.de

J. Drążkowska
drazkowska@mps.mpg.de

S. Desch
steven.desch@asu.edu

1 IMPMC, Muséum national d'Histoire naturelle, 57 rue Cuvier, 75005, Paris, France

2 Institute for Theoretical Astrophysics, Center for Astronomy, Heidelberg University, Albert-Ueberle-Str. 2, Heidelberg, 69120, Germany

3 Max Planck Institute for Solar System Research, Justus-von-Liebig-Weg 3, Göttingen, 37077, Germany

4 School of Earth and Space Exploration, Arizona State University, PO Box 876004, Tempe, 85287-6004, AZ, USA

daemons. Even assuming present-day protoplanetary disks are a suitable proxy of the solar nebula, telescopic observations could not resolve any substructure in these disks until about a decade ago. The Very Large Telescope (VLT) at Paranal in Chile now takes detailed images of protoplanetary disks around young stars about 140 parsec away, but cannot probe the disk midplanes owing to the optical thickness of the disks at optical and infrared wavelengths. Still, the Atacama Large Millimeter Array at Llano de Chajnantor in Chile (ALMA) can observe these disks at millimeter wavelengths, which are dominated by direct thermal emission from large (approximately millimeter size) dust grains mostly settled toward the midplane, but the resolution remains limited to 10 s of au at best. On the theoretical side, while computational power has greatly increased over recent decades, simulations can be carried out over a disk lifetime only if resolution is kept coarser than the spatial scales relevant to turbulence etc., so require assumptions on sub-grid physics. Some may be tested by more local simulations over shorter timescales (typically hundreds of orbits), others calculated *ab initio* but the knowledge of effective physical laws and appropriate approximation regimes remains fragmentary.

In this context, meteorites provide key constraints. Multiple lines of evidence indicate that they record the first few million years of existence of the Solar System, when the proto-Sun was still surrounded by its gaseous protoplanetary disk, also known as the "solar nebula". Save for rare Martian and lunar samples, their radiometric absolute ages tightly cluster around 4.56 Ga, consistent with the helioseismological dating at 4600 ± 40 Ma (Houdek and Gough 2011), or 4569 ± 6 Ma or 4587 ± 7 Ma (Bonanno and Fröhlich 2015). The oldest Solar System-made meteoritic components, the refractory inclusions, set a Time Zero at 4567 or 4568 Ma (Bouvier and Wadhwa 2010; Connelly et al. 2017; Piralla et al. 2023; Desch et al. 2023), notwithstanding a systematic error due to the uncertainty on the half-lives of uranium isotopes (e.g. Tissot et al. 2017). The evidence of short-lived radionuclides such as aluminum-26 suggest a recent (more or less averaged) input from massive stars in the parental molecular cloud (e.g. Desch et al. 2022). The composition of chondrite components, in particular refractory inclusions, bear witness to exchanges with a gas of solar (H_2-dominated) composition (Davis and Richter 2014), that is a protoplanetary disk environment, before disk accretion and photoevaporation dissipated it. Paleomagnetic data suggest the disappearance of nebular magnetic fields, and of the solar nebula altogether, by $\sim$4 Ma (Weiss et al. 2021), consistent with the few-Ma disk half-lives in neighbouring molecular clouds (Williams and Cieza 2011). Meteorites are thus invaluable time probes of disk interiors at optical depths and microscopic scales that will always elude telescopic capabilities.

Still, meteorites are largely rocks without geological context. Even in those cases where a pre-encounter orbit can be calculated, the chaotic orbital evolutions undergone since ejection from their parent bodies preclude the ready identification of the latter. Even when the parent bodies are identified, where they orbit now is not necessarily a good proxy of where they originally accreted, depending on the orbital instabilities that reshuffled planet trajectories in the past (e.g. Walsh et al. 2011). As to chronology, radiometric age dating at sub-Ma resolution necessary to distinguish disk evolutionary periods remains challenging and somewhat model-dependent, if at all possible for the events of interest. Planetesimal accretion, for example, cannot be directly dated and is a free parameter of thermal evolution models. Also, what part of meteorite petrology relates to the "background" protoplanetary disk in contradistinction to more local (e.g. planetary) environments is not necessarily clear beforehand. Meteorites are thus not ideal sensors of nebular conditions at known positions and epochs. This is somewhat like trying to unravel the history of a destroyed Egyptian pyramid based on debris randomly scattered around the globe. Yet this is the beauty of it. The very astrophysics which meteorites are called to help is thus also needed to provide template contexts to make sense of them. Astrophysics constrain meteoritics as much as meteoritics constrain astrophysics (Jacquet 2014b).

The purpose of this article is thus to provide some basic astrophysical context on the early Solar System for the more specific meteorite-based discussions of the other contributions in this collection. We will focus on the evolution of the solids in the disk. Yet, however astronomically-minded, these introductory sections could not, of necessity, be blind to meteoritical constraints to guide their very content, and we will thus start with cosmochemical prerequisites.

2 Meteoritical Prolegomena

2.1 Overview of Meteorite Classification

2.1.1 Petrology

A first-order division among meteorites separates *primitive meteorites*, or *chondrites*, from *differentiated meteorites*. Chondrites represent about 86% of falls (Grady 2000). As supported by their close match with the solar photosphere for nonvolatile element abundances (especially for CI chondrites, see Lodders et al. 2024), chondrites represent material largely unmodified since the accretion of their parent bodies. They are conglomerates of various solids (chondrules, refractory inclusions etc.) which were once freely floating in the solar nebula and/or in a planetesimal-forming environment and whose presentation is deferred until the next subsection.

Not all planetesimals retained a chondritic structure after accretion. The decay of short-lived radionuclides, chiefly aluminum-26, caused heating and partial melting of metal-sulfides (or magnetite for the most oxidized protoliths; McCoy et al. 2018) and silicates in many of them. Metal-sulfide melts, driven by negative buoyancy, segregated into cores, while silicate partial melts migrated toward the surface to form a crust enriched in incompatible elements (that is, not easily accomodated in the crystal lattice of major rock-forming minerals), with or without a magma ocean stage. The intervening mantle comprises residues left over after melt extraction and/or early crystallizations from the said primary melts. The whole process leading to these compositionally distinctive layers in the parent bodies is called *differentiation* (McCoy et al. 2006). Among differentiated meteorites, *irons* are iron-nickel-dominated fragments arising from the core and *achondrites* are the stony fragments of the mantle or crust. The latter comprise *evolved achondrites*, which are igneous or magmatic rocks (i.e. rocks crystallized out of magmas) conformable to terrestrial terminology (basalts, orthopyroxenites, etc.; Jacquet 2022) and rarer *primitive achondrites*, which are partial melting residues (or restites) and make a transition with chondrites. Some irons and achondrites may however arise from localized impact melting on chondritic targets.

Although the focus of most of this book will be on chondrites, it is worth noting that the parent bodies of differentiated meteorites must have accreted earlier than those of present-day chondrites, so as to secure enough radioactive heat for partial melting (even though the differentiated meteorites themselves may have crystallized and reached isotopic closure later). Hf-W dating of irons indeed suggest accretion within $\sim$1 Ma after refractory inclusions (Spitzer et al. 2021) (see Schönbächler et al. in prep). Differentiated parent bodies represent thus an early generation of planetesimals unsampled by chondrites that must be taken into account in models of disk-wide compositional evolution.

While heating was weaker for them, chondrites were not wholly spared by secondary processes (within the parent bodies). Melting of co-accreted ice led to aqueous alteration of some originally anhydrous minerals, especially in carbonaceous chondrites. Further temperature increases (e.g. in the drier ordinary chondrites) caused metamorphism, with various degrees of recrystallization and gradual disappearance of the original chondritic texture. van Schmus and Wood (1967) devised a scale of petrographic (or petrologic) types from 1 to 6 to quantify these secondary effects. As reinterpreted in the 1970s, type 3 corresponds to the most primitive chondrites, while types 2 and 1 mark increasing degrees of aqueous alteration and types 4 to 6 increasing metamorphism. The later introduced types 7 (Tomkins et al. 2020) and 8 (Jacquet 2022) already correspond to primitive achondrites. Type 3 chondrites are not free of secondary effects and have been subdivided in decimal subtypes (e.g. 3.0, 3.1, 3.2 etc.; Sears et al. 1980; Grossman and Brearley 2005; Sears 2016). Disentangling secondary from primary (preaccretionary) features is thus not always unequivocal (see Krot et al. 2024; Lee et al. 2024).

2.1.2 Genetics

Geological processes within asteroids are not the only cause of variation among meteorites. It had long been known that primitive meteorites differ in redox state (Prior 1920), but Urey and Craig (1953) showed that they did in bulk composition too ("bulk", or "whole-rock" refers to the global composition of a rock, irrespective of the distribution of the elements among its petrographic components). The Urey-Craig work evinced metal-silicate (Fe/Si) fractionation (*fractionation* meaning the partial or total separation of compositionally distinct fractions), but other fractionations were later identified, e.g. varying degrees of refractory element enrichment and/or depletions of volatile elements (see Sossi et al. in prep) as well as Mg/Si fractionation (Anders 1964; Larimer and Anders 1970; Wasson and Chou 1974; Braukmüller et al. 2018). Isotopic ratios, starting with oxygen (Clayton et al. 1973, 1976), were also later found to vary away from terrestrial values (Dauphas and Schauble 2016; Kleine et al. 2020). Last not least, these whole-rock compositional variations correlate with variations in the textures and abundances of the chondrite components directly visible under the microscope, and it is an open question to what extent chondrite fractionations reflect varying proportions of a small number of component types common to all the disk (Anders 1964; Zanda et al. 2006; Alexander 2019a,b; Bryson and Brennecka 2021) or not (Goldberg et al. 2015; Hezel et al. 2018). At any rate, obviously, the solar nebula was not homogeneous. It may have inherited heterogeneity from the parental molecular cloud but it also developed some *in situ* (see Tissot et al. 2024).

Present-day chondrites do not show a continuum in compositional space, but discrete clusters known as *chemical groups* (e.g. Scott and Krot 2014). Chondrite groups are given letter designations, such as the H, L, and LL ordinary chondrite groups (originally meaning "high iron", "low iron", "low iron low metal"), which can be combined with a petrologic type in compact classifications such as "H5". Chemical groups are generally interpreted as representatives of distinct primary parent bodies (that is, prior to any collisional disruption; Greenwood et al. 2020), as sometimes supported by peaks in cosmic ray exposure ages suggesting common ejection events (Herzog and Caffee 2014). The lack of a compositional continuum is not simply a matter of incomplete sampling of the main belt asteroids by meteorites. In fact, the intermediate compositions, which likely existed in the protoplanetary disk, must have mostly disappeared, e.g. by incorporation into planets, since the present-day main belt, not exceeding a twentieth of lunar mass, is but a very partial sampling of the original population of planetesimals (Jacquet 2022). Taking into account ungrouped meteorites (as the recognition of a group requires a minimum of five unpaired members; Wasson

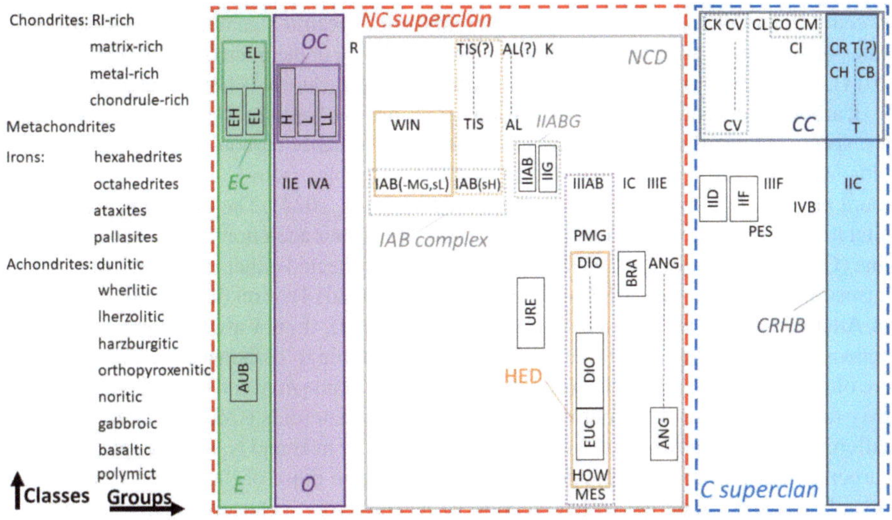

Fig. 1 Asteroidal meteorite taxonomy. The *y* axis displays a petrologic classification of meteorites (that is, based on the nature of the rock; chondrites on top and differentiated meteorites at bottom) whereas the arbitrary *x* axis allows the representation of genetic affinities (made explicit by boxing in high-level groups; dashed lines also link different portions of low-level groups unrepresented in the intervening petrologic classes). For example, aubrites (AUB), which are orthopyroxenitic achondrites, and EH and EL chondrites sample a common isotopic reservoir termed the E reservoir (colored in green), itself a subset of the NC superclan. Abbreviations: AL= Acapulco-Lodran group (acapulcoites and lodranites), ANG = angrite, AUB = aubrite, BRA = brachinite, C = carbonaceous, CB = carbonaceous Bencubbin-type, CH = carbonaceous ALH 85085-type, CI = carbonaceous Ivuna-type, CK = carbonaceous Karoonda-type, CL = carbonaceous Loongana 001-type, CM = carbonaceous Mighei-type, CO = carbonaceous Ornans-type, CR = carbonaceous Renazzo-type, CV = carbonaceous Vigarano-type, DIO = diogenite, E = enstatite (EH, EL), H = high iron, HED = howardite-eucrite-diogenite, HOW = howardite, K = Kakangari-type, L = low iron, LL = low iron low metal, MES = mesosiderite, NC = noncarbonaceous, NCD = normal NC Differentiated meteorite signature, O = ordinary, PES = pallasite Eagle Station group, PMG = pallasite main group, R = Rumuruti-type, RI-rich = refractory inclusion-rich, T = tafassite (Bouvier et al. 2021), TIS = tissemouminite (Stephant et al. 2023), URE = ureilite, WIN = winonaite. "Metachondrites" represent type 7 primitive achondrites and chondritic impact melt rocks. Modified after Jacquet (2022)

1974), meteorites may represent 95-148 primary parent bodies (Greenwood et al. 2020). Yet the distinction between primary parent bodies of origin may be difficult if some happened to accrete close-by in the disk and thus in compositional space, so the empirical concept of groups needs not be given a parent body interpretation (Jacquet 2022), although it certainly has genetic significance at least in pointing to a restricted space-time section of origin in the disk. Groups may be further combined in higher-level taxa, sometimes called "clans", such as the ordinary chondrites, encompassing more inclusive regions of the disk ("reservoirs" in cosmochemical parlance). The current taxonomy is depicted in Fig. 1.

At the highest level, there is a clear isotopic dichotomy between carbonaceous (C) and non-carbonaceous (NC) chondrites (Warren 2011). C chondrites are richer in many neutron-rich isotopes (but also in oxygen-16) than NC meteorites which are closer to terrestrial values. NC chondrites chiefly comprise ordinary chondrites (chiefly composed of olivine, pyroxene, iron-nickel metal and iron sulfides), and the very reduced enstatite chondrites (dominated by magnesian pyroxene (enstatite), iron-nickel metal and multifarious sulfides), which are isotopically closest to the Earth. Samples returned from S type asteroid Itokawa, which resembled LL4-6 chondrites Nakamura et al. (2011), confirmed the genetic link between ordinary chondrites and S type asteroids, with spectral signatures of Fe-bearing silicates olivine and pyroxene, which dominate inside 2.5 au in the main belt (DeMeo and Carry 2014), the region primarily sampled by Earth-crossing meteoroids (Binzel et al. 2015).

Carbonaceous chondrites (where group names always begin by "C", e.g. CI, CV, CM...), are spectroscopically similar to the low-albedo, presumably carbon-rich C complex asteroids, as confirmed by the recent return of CI-like samples from C asteroid Ryugu (Nakamura et al. 2023) and B asteroid Bennu (Lauretta et al. 2024). The prevalence of aqueous alteration effects among C chondrites may result from their accretion beyond the water snow line (Grimm and McSween 1989). Micrometeoroids collected either in the stratosphere (and termed *Interplanetary Dust Particles* or IDPs; Bradley 2014) or on the ground, in particular in Antarctic snow (*micrometeorites*; Genge et al. 2020), do not always show evidence of aqueous alteration but are on the whole most comparable to carbonaceous chondrites, and are often even more carbon-rich (Engrand et al. 2023). Some may derive from comets, which may form a compositional continuum with C complex asteroids (Gounelle et al. 2008). Yet differences exist. Known comets are isotopically heavier in H and N, for example, than most carbonaceous chondrites (e.g. Aléon 2010), even though those of the CRHB clan, a subset of carbonaceous chondrites (often metal-rich, but refractory inclusion-depleted) show some isotopic affinities. So the C superclan has some structure. The CI may make another division, drastically underrepresented in the meteoritical record because of their friability judging from the Ryugu and Bennu sample returns, but it may be premature though to elevate them on a par with NC and other C in some disk-scale "trichotomy" (Hopp et al. 2022), as they may have contributed to both CRHB and "normal" refractory inclusion-rich carbonaceous chondrites ("CC-RI"; Marrocchi et al. 2022), and are not more remarkable than the CRHB/CC-RI divide.

Overall then, NC chondrites may originate from the inner disk and C chondrites from the outer disk (where the boundary, mobile or not, may have been at a few au (astronomical units), perhaps near Jupiter). In isotopic space, differentiated meteorites also plot in either the C or NC superclan (Warren 2011; Kleine et al. 2020), suggesting that their protoliths, whose isotopic composition would not change upon differentiation, resembled present-day C or NC chondrites. Most differentiated meteorites belong to the NC superclan (if at more ^{16}O-rich compositions than most NC chondrites) in line with more rapid accretion (and thus higher ^{26}Al abundances) at shorter heliocentric distances. Most iron meteorite groups do not seem to have achondrite, let alone mantle, counterparts (Scott et al. 2015). This suggests early stripping and loss of stony layers, perhaps in inner disk regions more collisional than the environment in which Vesta, the presumed parent body of howardites-eucrites-diogenites (HED), retained its largely intact basaltic crust.

2.2 Chondrite Components

Thus far, we have been mainly interested in the bulk composition of chondrites, but their individual components are the witnesses of protoplanetary disk processes prior to their accretion. Chondrites may indeed be described as "cosmic sediments", welding together varous millimeter or submillimeter-size components such as *chondrules* and *refractory inclusions* in a finer-grained *matrix*. These are illustrated in the micrographs of two chondrites, carbonaceous and non-carbonaceous (Figs. 2 and 3) and will now be introduced individually.

2.2.1 Refractory Inclusions

Refractory inclusions are the oldest dated components formed in the Solar System (~4567-4568 Ma). The tight clustering of their initial ^{26}Al/^{27}Al ratios around the canonical value

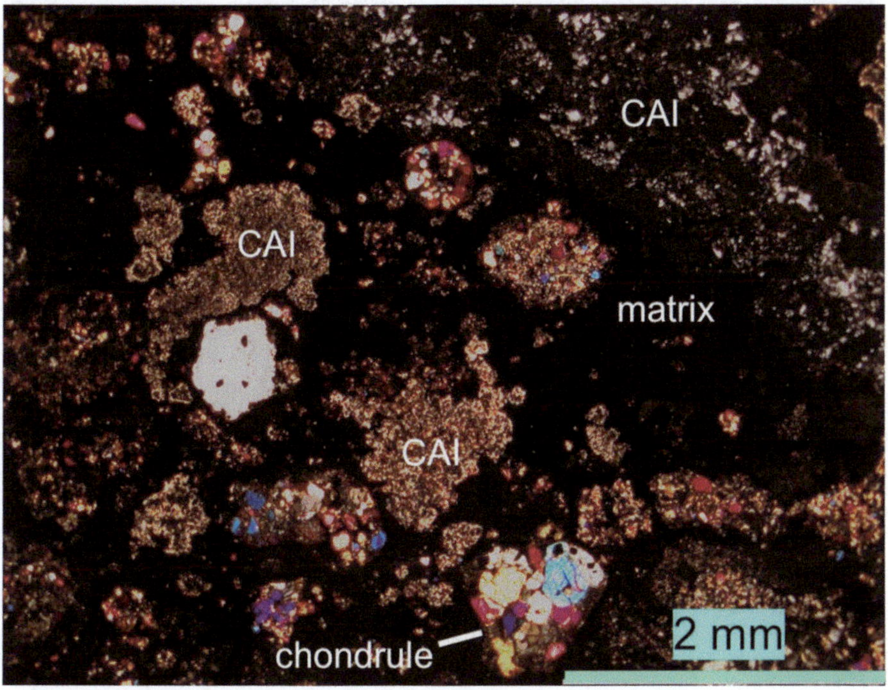

Fig. 2 Transmitted polarized and analyzed light microscopy of the Allende (CV3) carbonaceous chondrites. The polarization/analysis of the light yields interference colors which contribute to the identification of minerals. The CAI on the upper right mainly consists of coarse-grained melilite and has presumably undergone at least partial melting (unlike the two other fine-grained CAIs indicated). The matrix appears opaque because it is finer-grained than the thickness ($\sim 30\ \mu m$) of the thin section (3181lm2 of the Muséum national d'Histoire naturelle in Paris)

5.2×10^{-5} (e.g. Lee et al. 1976; Huss et al. 2001; Sugiura and Krot 2007; MacPherson et al. 2012) suggests a formation epoch shorter than $\sim 10^5$ years. Refractory inclusions comprise the *Calcium-Aluminum-rich Inclusions* (CAIs), which may be fine-grained aggregates of mineralogically zoned nodules or coarser-grained igneous objects, and the less refractory *Amoeboid Olivine Aggregates* (AOAs) which are more or less annealed (forsteritic) olivine-rich and may contain CAI-like inclusions. Although some "lappige Chondren" (lobate "chondrules") were reported by Tschermak (1885) in the Renazzo meteorite, the first extensive description of a CAI, which also made the connection with condensation models, was that of Christophe Michel-Lévy (1968) in the Vigarano meteorite. This was just one year before the massive fall of Allende, also a Vigarano-type carbonaceous chondrite (CV) laden with CAIs, definitively secured the attention of the meteoritical community for them (Clarke et al. 1971). The CAI mineralogy corresponds indeed to the first phases expected to condense out of a cooling gas of solar composition (or the last residues to evaporate during heating), assuming thermodynamic equilibrium, with AOAs corresponding to the onset of the condensation of "main component" elements, in particular in the form of olivine and enstatite (Hutchison 2004; Davis and Richter 2014). Mass-dependent light isotope enrichment in Si and Mg in fine-grained refractory inclusions is consistent with a kinetic condensation signature, perhaps within cooling timescales of weeks (Marrocchi et al. 2019b; Fukuda et al.

Fig. 3 Reflected light microscopy of the Hallingeberg (L3.4) ordinary chondrite (thin section 3576lm1 of the Muséum national d'Histoire naturelle in Paris)

2021), but many of them have undergone reheating, up to melting and partial evaporation (MacPherson 2005).

Despite intergroup mineralogical variation, the common average isotopic signature of refractory inclusions in different chondrite groups (e.g. as ^{16}O-rich as the solar wind; McKeegan et al. 2011), indicate a common reservoir of origin. Although the high temperatures required for refractory inclusion condensation ($\gtrsim$1500 K) have long suggested an origin in the vicinity of the Sun, they are strikingly rare in non-carbonaceous chondrites (~0.01 vol%; Dunham et al. 2023) whereas they may make up to ~10 vol% of carbonaceous chondrites (in particular CVs; Ebel et al. 2016), despite their presumed remoter region of provenance. Although C chondrites are isotopically intermediate between NC meteorites and refractory inclusions, mass balance calculations preclude making the former by simple addition of refractory inclusions to a NC-like endmember (Nanne et al. 2019). However, an isotopically CAI-like component with *chondritic* chemistry could do (Nanne et al. 2019; Schneider et al. 2020); this may indicate a nonrefractory isotopically CAI-like complement mixed in the matrix and chondrule precursors.

2.2.2 Chondrules

Chondrules make up ~20-90 vol% of chondrites (excepting the nearly chondrule-free CIs which have also undergone extensive aqueous alteration; Morin et al. 2022), hence their names, ultimately derived from the Greek "khondros", "small grain". Chondrules are igneous spheroids, generally 0.1-1 mm in diameter, mostly made of the ferromagnesian silicates olivine and pyroxene set in a (more or less) glassy mesostasis (Russell et al. 2018).

Opaque phases (that is, opaque in transmitted light microscopy) such as metal and sulfides also occur, and isolated opaque or silicate grains found in the interchondrule matrix may have been expelled from chondrules (Zanda et al. 1994; Jacquet et al. 2013; Jacquet 2021). Most chondrules show porphyritic textures indicative of incomplete melting (with some identifiable relict grains from the precursor material) but the more spectacular nonporphyritic chondrules solidified from undercooled droplets. Ferromagnesian chondrules present a redox dichotomy between *type I* chondrules, which are reduced (with very magnesian olivine and pyroxene and most Fe as metal) and more oxidized *type II* chondrules (with Fe mostly in sulfides or silicates). Chondrule textures are consistent with formation by melting (incomplete for the most common porphyritic ones) and cooling within timescales variously estimated from minutes to days (Jones et al. 2018). Al-Mg and Pb-Pb dating suggest chondrules are ~1-3 Ma younger than CAIs (Kita and Ushikubo 2012). Isotopic and mineralogical properties indicate chondrule provinciality in that chondrules from different groups were not produced as a single population, but were produced in several regions both in the C and NC reservoirs (Jones 2012; Schneider et al. 2020).

Despite 222 years of research since their original description by Jacques-Louis de Bournon (in Howard 1802), the nature of the chondrule-forming mechanism(s) remains elusive. Proposed chondrule-forming scenarios are generally classified as either "nebular" or "planetary". While this is essentially a disciplinary division between astrophysics and geology, one may define these categories after the assumed nature of the chondrule precursors (that is, the material which was melted to produce chondrules), with nebular scenarios assuming free-floating mm-size aggregates while planetary scenarios derive them from already accreted planetesimals. Examples of nebular scenarios include shock waves passing through the disk (Desch et al. 2005), e.g. because of planetary embryos in eccentric orbits (Boley et al. 2013) or the expansion of an impact plume (Stewart et al. 2024), melting at the disk inner edge (X point; Shu et al. 1996) or lightning discharges (Desch and Cuzzi 2000; Johansen and Okuzumi 2018) while planetary scenarios chiefly mean impact splashing (Asphaug et al. 2011; Sanders and Scott 2012) or jetting (Johnson et al. 2015), depending on whether the target material was already molten or not. Nebular scenarios are favored by evidence of refractory inclusion admixtures in chondrule bulk compositions (Misawa and Nakamura 1988; Schneider et al. 2020; Jacquet 2022) and relict grains (Krot et al. 2006; Marrocchi et al. 2019a; Piralla et al. 2021). However, planetary scenarios fare better with evidence for large concentrations of condensable materials in chondrule-forming environments, at least two orders of magnitude above solar (Tenner et al. 2015; Cuzzi and Alexander 2006; Jacquet 2021; Alexander et al. 2008). A "planetary" (impact plume) origin is arguably consensual for the unusual (and mostly nonporphyritic) CB and many CH chondrite chondrules (e.g. Krot et al. 2005), but this needs not apply to "mainsteam" chondrules. Thus, as yet, it is not known what significance chondrule formation had in the planet-forming processes, in particular whether it affected most planetesimal building blocks or only a late epiphenomenal minority (Herbst and Greenwood 2019). The article by Marrocchi et al. (2024) is devoted to this chondrule mystery.

2.2.3 Matrix

The matrix of the most pristine chondrites comprises submicron-size grains. These include amorphous silicate phases, crystalline silicates and sulfides (Brearley and Jones 1998). Organic matter is also present, diffuse or as discrete globules, ranging from insoluble macromolecular, partly aromatic, material to soluble species such as aminoacids which have attracted interest regarding the genesis of life on Earth (Gilmour 2014). Since matrix is most abundant in carbonaceous chondrites, their carbon content may be high (a few wt%), hence their name historically, but this is not always the case. Water (or, strictly, hydroxyl ions) is also present in the amorphous silicates (Le Guillou et al. 2015) or phyllosilicates and may result from melting of ice grains on the parent body (i.e. aqueous alteration), although preaccretionary hydration is possible (Marrocchi et al. 2023). The porosity of CI chondrites, which are nearly pure matrix, reaches 34.9 vol% (Flynn et al. 2018) but matrix porosity may have been higher before lithification as a coherent rock, up to 70-80 vol% judging from fabric in fine-grained rims around Allende chondrules (Bland et al. 2011).

Some essentially monomineralic grains present very anomalous isotopic compositions, sometimes requiring log-scales in plots (Zinner 2014). These grains, now termed *presolar grains*, are thought to have condensed in the envelopes of dying stars (red giants, supernovae...) before the formation of the Sun, and as such they inherited the isotopic signatures of their characteristic nucleosynthetic processes. They predate CAIs (the oldest solids *formed* in the Solar System) by millions to hundreds of millions of years according to cosmic ray exposure estimates (Heck et al. 2020). Although they now represent only a minority (less than a thousandth; Haenecour et al. 2018) of chondrite matter, the original dust of the Solar System was evidently entirely presolar. Most likely, the isotopic variations for many elements among bulk meteorites are ultimately traceable to various proportions of different populations of presolar grains, whether by differential destruction during "thermal processing" (e.g. Niemeyer 1988; Van Kooten et al. 2016) or uneven original distribution in the parental molecular cloud and/or the disk (Jacquet et al. 2019). Specifically, C chondrites may have incorporated a greater proportion of grains enriched in r process isotopes (i.e. formed through *r*apid neutron additions, in neutron-rich environments, e.g. supernovae) in contradistinction to s(low) process carriers (see Liu et al. 2024). We note that bulk isotopic variations in oxygen do not seem to relate to varying nucleosynthetic contributions as originally proposed by Clayton et al. (1973), but rather to one as yet elusive mass-independent process, such as CO self-shielding, which may have produced ^{16}O-poor water exchanging isotopes with originally isotopically CAI-like nebular rocks (Young et al. 2008).

We do not know what proportion of matrix (or IDPs) is still presolar (Alexander et al. 2017; Bradley et al. 2022), for some grains may be too small to be resolvable by NanoSIMS, and grains formed in the interstellar medium itself, as opposed to stellar atmospheres/ejecta, would not be readily identifiable isotopically (although this is suggested for D- or ^{15}N-rich "hotspots" in organic matter; Gilmour 2014). Part of the matrix may relate to the condensation events which formed refractory inclusions, and another still may be related to chondrules, whether as small chondrule fragments or disequilibrium condensates. The contention that matrix derives from the same solar (in some respect) reservoir as chondrules and have exchanged elements with them in an overall closed system is termed chondrule-matrix complementarity and will be discussed in the article by van Kooten et al. (2024). Matrix may form distinctive fine-grained rims around chondrules, which some consider as dust accreted immediately after chondrule formation (Zanetta et al. 2021; Pinto et al. 2022).

3 Dust in the Early Solar System

3.1 Early Condensates in the Solar System

What did the solar protoplanetary disk look like? One practical reference is the Minimum Mass Solar Nebula (MMSN; Hayashi 1981) where present-day planetary masses are mentally supplemented by solar gas to restore solar abundances. With some smoothing, one obtains a surface density Σ_g profile (Hayashi 1981):

$$\Sigma_g = 1.7 \times 10^4\ \mathrm{kg/m^2} \left(\frac{1\ \mathrm{au}}{r}\right)^{1.5} \tag{1}$$

with r the heliocentric distance. The MMSN would be about 0.01 $\mathrm{M}_\odot$ in mass. This of course ignores any movement and loss of disk solids and gas to the Sun. In fact, simulations of disk building would predict an initial disk mass one order of magnitude higher (e.g. Yang and Ciesla 2012; Pignatale et al. 2018).

The MMSN is typically associated a temperature T profile corresponding to passive heating by unattenuated sunlight (Hayashi 1981):

$$T = 280\ \mathrm{K} \left(\frac{1\ \mathrm{au}}{r}\right)^{0.5} \tag{2}$$

This is however an overestimate, because of absorption of direct sunlight by the intervening dust would make the actual midplane temperature a factor of $\sim$2 lower at a few au (e.g. Chiang and Goldreich 1997). Even then, only the inner boundary of the disk would be hot enough for refractory inclusion formation, as in the X wind model (Shu et al. 1996). However, the dissipation of turbulence can heat the disk to higher temperatures. For a steady disk of mass accretion rate $\dot{M}$, the corresponding contribution is (Jacquet et al. 2012):

$$\begin{aligned} T &= \left(\frac{3}{128\pi^2}\frac{\kappa m}{\sigma k_B \alpha}\dot{M}^2\Omega_K^3\right)^{1/5} \\ &= 500\ \mathrm{K}\left(\frac{\kappa}{0.5\ \mathrm{m^2/kg}}\right)^{1/5}\left(\frac{10^{-3}}{\alpha}\right)^{1/5}\left(\frac{\dot{M}}{10^{-8}\mathrm{M}_\odot/\mathrm{a}}\right)^{2/5}\left(\frac{1\ \mathrm{au}}{r}\right)^{9/10} \end{aligned} \tag{3}$$

with κ the (Rosseland mean) specific opacity, k_B the Boltzmann constant, m the mean molecular mass (2.3 amu for a solar gas), $\Omega_K = \sqrt{GM_*/r^3}$ the Keplerian angular velocity (with G the gravitational constant and M_* the stellar mass) and α the Shakura and Sunyaev (1973) turbulence parameter.[1] Such "viscous heating" (termed such because, ideally, the large-scale effects of turbulence may be equated with that of an enhanced viscosity) must dominate at short heliocentric distances, and for early (Class 0 or I) disks (with high $\dot{M}$), may ensure CAI-forming temperature up to 1 au (e.g. Yang and Ciesla 2012; Pignatale et al. 2018; Owen and Jacquet 2015). In fact, evidence of rapid condensation of refractory inclusions (Marrocchi et al. 2019b) suggests that they formed in localized events under a lower ambient temperature, which thus may have been beyond their nominal condensation isotherms.

While refractory inclusion formation may thus have occurred over a wider extent of the early disk than previously believed, their presence at the large heliocentric distances (several au) housing the C superclan still requires outward transport. Disk winds have difficulties in lifting particles larger than a few tens of microns (e.g. Marrocchi et al. 2019b; Desch et al. 2021) in a MMSN disk, but early disks in formation may have replenished the envelope with mm-size dust that way (Tsukamoto et al. 2021). Despite difficulties inherent to the survival of solids up to and from the disk inner edge (Desch et al. 2010), the X-wind model,

[1]The definition of α is more general than the case of turbulence (see e.g. Jacquet 2013) but the equations involved herein will assume this type of flow, with a Schmidt number (the ratio of kinematic effective viscosity to turbulent diffusivity) of unity for simplicity.

or something in this vein, is still occasionnally referred to in the cosmochemical litterature as a way to leapfrog the (NC) inner disk but this could hardly be an expected effect given the sensitivity of the transport range to size (Hu 2010). Turbulent diffusion alone may be insufficient (e.g. Ciesla 2010), but if the solar protoplanetary disk started off sufficiently compact, a viscous spreading could have incurred outward mean velocities in the vicinity of the CAI-forming region (Jacquet et al. 2011b; Yang and Ciesla 2012; Pignatale et al. 2018; Marschall and Morbidelli 2023).

3.2 Dust Growth

Whether condensed in the disk or inherited from the parental molecular cloud, the dust first grows by coagulation, which is a very complex process (Birnstiel et al. 2016). Small grains stick to each other by van der Waals forces. As a rule of thumb the time scale of growth is (cf. Birnstiel et al. 2012)

$$\tau_{\mathrm{grow}} \equiv \frac{a}{\dot{a}} \sim \frac{1}{\Omega_K \epsilon} \tag{4}$$

where a is the radius of the dust grain, $\dot{a}$ its time derivative, and ϵ the dust-to-gas mass ratio, about 0.01 for a solar gas. Given that the growth time scale τ_{grow} does not depend on the grain size, the growth process (at least in this simplification) is exponential. The growth from micron size to millimeter size, for $\epsilon = 0.01$, thus takes about a hundred orbits. Around this size, however, collision velocities, due to turbulence or differential drift (Birnstiel et al. 2012), are too high for sticking, and may rather lead to bouncing, erosion or fragmentation, so further growth by coagulation alone is frustrated (e.g. Blum 2018).

The typical size of chondrules or refractory inclusions may thus be accounted by this coagulation bottleneck (Jacquet 2014a). In principle, fluffy aggregates of submicron monomers might grow fractally to planetesimal masses and beyond (Okuzumi et al. 2012), but the sticking behaviour beyond cm size is untested experimentally (Simon et al. 2022). The very existence of refractory inclusions and chondrules also implies that chondrite parent bodies did not simply grow from such ultraporous aggregates, but have always had nonnegligible filling factors ($\gtrsim 0.1$; e.g. Bland et al. 2011), but the question may be left open for cometesimals.

3.3 Drift

In the same way snowflakes drift downward (toward higher pressures) relative to a wind carrying them, once dust aggregates reach millimeter size, their drift becomes significant. Their radial velocity is (Birnstiel et al. 2012):

$$v_{r,d} = \left(u_r + \frac{t_f}{\rho_g}\frac{\partial p}{\partial r}\right)\left(1 + \mathrm{St}^2\right)^{-1} \tag{5}$$

where $p = \rho_g c_s^2$ is the midplane gas pressure, ρ_g the gas density, u_r the gas radial velocity and $c_s = \sqrt{k_B T/m}$ the isothermal sound speed of the gas. Finally, $\mathrm{St} \equiv \Omega_K t_f$ is the Stokes number of the dust particle, with t_f the friction time scale, also often called the stopping time. The limit $\mathrm{St} \gg 1$ corresponds to planetesimals, which essentially follow Keplerian motion.

For particles smaller than the gas molecular mean free path, as appropriate for mm-size objects, the Epstein drag law applies and, at the midplane (Birnstiel et al. 2010)):

$$\mathrm{St} = \frac{\pi}{2}\frac{\rho_\bullet}{\Sigma_g} a \tag{6}$$

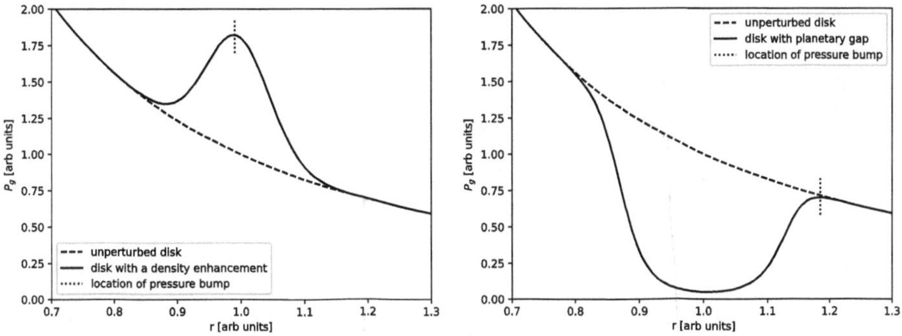

Fig. 4 Sketches of how an annular density enhancement (left) or a planetary gap (right) can lead to a local pressure maximum ('pressure bump'), which can trap dust

where $\rho_\bullet$ is the effective material density of the dust aggregate, with porosity accounted for. The Stokes number is therefore a dimensionless measure of the grain radius a. For St$\ll 1$ and negligible u_r, we have:

$$v_{r,d} \approx \frac{\partial \ln p}{\partial \ln r}\frac{c_s^2}{v_K}\,\mathrm{St} = -0.1\mathrm{m/s}\left(\frac{a}{1\,\mathrm{mm}}\right)\left(\frac{\rho_\bullet}{2\,\mathrm{g/cm^3}}\right)\left(\frac{r}{3\,\mathrm{au}}\right)^{1.5} \tag{7}$$

with $v_K = \sqrt{GM_*/r}$ the Kepler velocity. The numerical evaluation assumes the MMSN. The radial drift time scale of a 1 mm dust particle at $r = 3$ au is thus about 10^5 years.

This short time scale shows that radial drift is a major problem in the planet-forming region of the protoplanetary disk. It has long been known from telescopic observations that mm-size "pebbles" do survive in the outer regions ($r \gg 30$ au) of disks of millions of years of age. Within our own solar system, single chondrites can host chondrules and refractory inclusions with age differences much larger than 0.1 Ma (Villeneuve et al. 2009; Connelly et al. 2017). This suggests that the material that eventually forms an asteroid must have been able to withstand the radial drift for prolonged stretches of time.

3.4 Dust Trapping

One solution is the presence of one or more local maxima in the radial gas pressure profile (Fig. 4). At the peak of one such local maximum, one has $\partial \ln p/\partial \ln r = 0$ (no drift), and inside of this peak one has even $\partial \ln p/\partial \ln r > 0$ (outward drift). Such a local "pressure bump", which manifests itself generally as a ring in a Keplerian disk, tends to accumulate dust particles. They drift toward the peak from both sides, and get trapped there (Pinilla et al. 2012). This allows these particles to stay there indefinitely, as long as the local pressure bump is long-lived.

Indeed, observations suggest smooth disks are the exception rather than the rule. VLT images of protoplanetary disks around young stars about 140 parsec away (Benisty et al. 2023, and Fig. 5) reveal a rich variety of substructures in these disks. Many disks display several concentric rings in their surface layers, while others show complex spiral patterns. Some of these disks show evidence of being warped, i.e. non-planar, and some show signatures of ongoing infall of matter from farther away (e.g., Ginski et al. 2021). The ALMA images (Fig. 6) show that the large mm-size dust grains are arranged mostly in concentric rings. These rings are most likely "dust traps", in which the dust grains are trapped at circular local pressure maxima (Dullemond et al. 2018). Indeed, the larger the dust grain, the more it drifts toward pressure maxima.

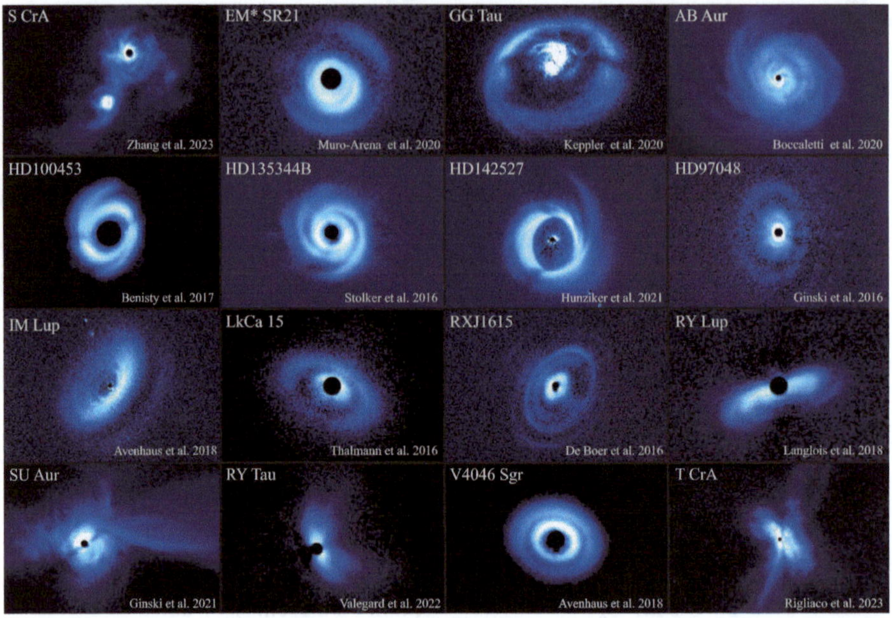

Fig. 5 Protoplanetary disk images taken with the SPHERE instrument on the VLT at optical wavelengths. Image credit: C. Ginski

Possibly then, such dust traps may have isolated several reservoirs in the solar nebula, in particular the NC and C reservoirs. The CAI enrichment in the C reservoir could be due to such trapping, while most inner disk CAIs would have been lost to the protosun (Kruijer et al. 2020; Desch et al. 2018). However, the similar level of Mo isotopic anomalies shown by NC irons and chondrites suggest that CAIs (or at least CAIs isotopically similar to their CC counterparts) cannot have been abundant early on there, perhaps as a result of their removal by an early outward flow (Jacquet et al. 2011b; Marschall and Morbidelli 2023). A pressure bump would still allow fine dust (tightly coupled to the gas) to go through (Haugbølle et al. 2019; Stammler et al. 2023), in the direction of the flow relative to it, so some transfer of material would remain possible, and may possibly account for an apparent secular isotopic evolution of the NC superclan in the direction of C meteorites (e.g. Schiller et al. 2018), though the putative apex would lie outside the range of sampled C meteorites (Burkhardt et al. 2021). In the C superclan, an evolution from chondrule- and refractory inclusion-rich CCs toward inclusion-poor CI may have been incurred by differential drift (Hellmann et al. 2023). Within each dust trap, so long it is open to influx of drifting or infalling dust, isotopic variation might then be essentially a matter of accretion time. Unfortunately, ALMA does not have the spatial resolution to see whether dust rings occur at distances between 1 and 5 au from a young star.

The cause of such local ring-shaped pressure maxima in a protoplanetary disk remains a topic of debate. The most natural explanation could be the existence of a planet (Dipierro et al. 2015; Dong et al. 2015). If a planet is massive enough to open a gap, it will produce a local pressure maximum at the outer boundary of the gap (Fig. 4-right). This will halt the radial drift of millimeter size dust aggregates. In planet formation theories, the minimum planetary mass to produce such a dust trap is called the "pebble isolation mass" (Lambrechts

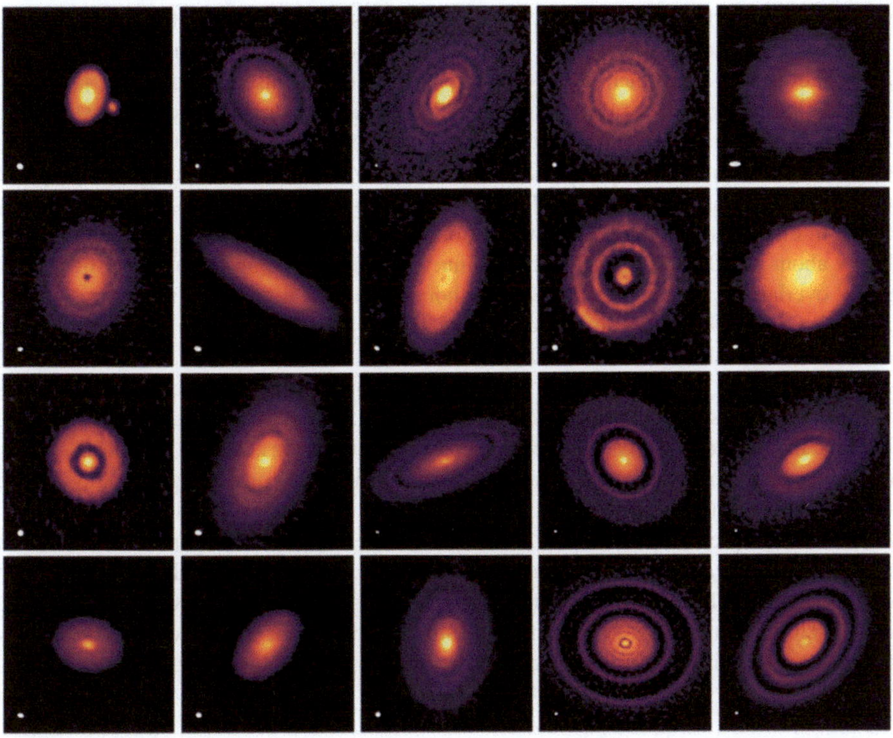

Fig. 6 Protoplanetary disk images taken with ALMA at millimeter wavelength. The images shown here are from the DSHARP campaign (Andrews et al. 2018)

et al. 2014; Bitsch et al. 2018), which can be roughly written as

$$M_{\rm iso} = 20 \left(\frac{h_p/r}{0.05} \right)^3 M_\oplus \tag{8}$$

where $M_\oplus$ is the Earth mass and $h_p = c_s/\Omega_K$ the pressure scale height of the disk. For the MMSN, h_p/r is between 0.03 (near 1 au) and 0.06 (near 10 au). This implies that as Jupiter was still forming, it would already early on (at less than 10% of its final mass) produce a pressure bump just outside of its orbit that stops radial drift of pebbles. As Jupiter grew more massive, it might even have caused additional pressure bumps inward of its orbit, thus trapping dust there as well (Bae et al. 2017). However, whether this multi-trap phenomenon is viable depends on the details of the thermodynamics of the disk (Ziampras et al. 2020).

There are, however, still some problems with the planetary origin of dust traps. While massive planets certainly produce dust traps, not all dust traps can be explained well with planets. For instance, Uranus and Neptune, according to the above formula, have masses well below the pebble isolation mass, and cannot thus produce a dust trap. This means that they cannot have prevented radial drift of material that produced the Trans Neptunian Objects. Also, since dust particles grow within about a hundred orbits (see above), radial drift may already set in before the first planets have grown massive enough to stop it.

There are however numerous alternative planet-free models being suggested that could lead to local pressure bumps. Should infall proceed along streamers e.g., pressure bumps may be created where they land on the disk (Kuznetsova et al. 2022), but these may diffuse away once the streamers cease. Radial gas transport mechanisms inside the disk, if of unequal efficiency as a function of heliocentric distance, may also lead to "traffic jams".

Radial transport close to the Sun is generally attributed to turbulence driven by the magnetorotational instability (MRI). If the turbulent viscosity is a sufficiently strong function of the magnetic pressure, weakly magnetized zones will accumulate matter at the expense of the more magnetized ones (e.g. Iwasaki et al. 2024). The MRI, however, cannot operate at all distances, for the required ionization levels are not ensured by thermal ionization below about ~1000 K, so the MRI-active inner region must give way to a MRI-free "dead zone" overlain by active layers ionized by nonthermal radiations from the Sun or cosmic rays (Gammie 1996). Since the effective viscosity drops, a pressure bump may arise at the dead zone inner edge (Flock et al. 2015).

A sublimation line, such as the water ice line, could also incur short-scale variations in the effective viscosity. The viscosity $\nu = \alpha k_B T/(m\Omega_K)$ may drop inside because of the increase in molecular mass m (assuming a constant α; Charnoz et al. 2021). Outside, it may also increase because the temperature T will be enhanced by condensing grains increasing the local opacity (Ciesla and Cuzzi 2006), but those grains may also lower the effective α by scavenging electrons available in the active layers (Kretke and Lin 2007). However, transport in the dead zone may be driven by magnetized winds leaving the disk rather than disk turbulence. Those winds may also produce large-scale ring-like pressure bumps (Riols et al. 2020; Suriano et al. 2018). Since the simulations do not extend to more than a few hundred orbits, it is unknown whether such bumps can be long-lived.

4 Planet(esimal) Formation

4.1 Dust Concentration

Stopping the drift of dust is not enough to accrete planetesimals. We saw earlier that simple growth by sticking reached a bottleneck around mm-cm size. The mutual gravity of dust grains would be helpful, but to overcome competing effects such as shear of the differentially rotating disk, one needs to reach a critical density called the Roche density, defined, for a fluid, as (Chandrasekhar 1967)

$$\rho_{\rm R} = 3.5\frac{M_\star}{r^3} = 2\cdot 10^{-6}\frac{M_\star}{M_\odot}\left(\frac{r}{\rm au}\right)^{-3}\ {\rm g/cm^3}, \tag{9}$$

(The exact threshold applicable in a dusty disk is somewhat controversial, and could be a factor of 5 lower for an axisymmetric ring of tightly coupled dust and gas, see e.g. Chiang and Youdin 2010). In the MMSN, the Roche density translates into a midplane dust-to-gas ratio of

$$\epsilon = 1530\cdot\left(\frac{r}{\rm au}\right)^{-1/4}, \tag{10}$$

five orders of magnitude above the solar value. To reach such a high concentration, we may count on dust settling toward the midplane, since pressure is maximum there, for a given radial location. For settling to be significant, dust must grow to larger aggregates with $S \equiv {\rm St}/\alpha > 1$ (Jacquet et al. 2012). Indeed, the competition between settling and turbulence results in a midplane density enhancement factor of $(1+S)^{1/2}$. In astrophysical literature

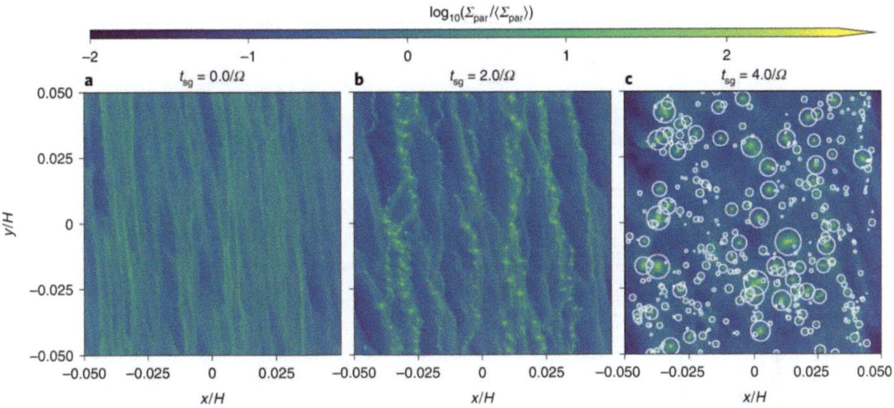

Fig. 7 Streaming instability simulation from Nesvorný et al. (2019). The panels show the time evolution, in the units of local orbital time counted after particle self-gravity was switched on, of the vertically integrated density of pebbles relative to the initially uniform surface density. The x and y coordinates show the dimensions of the simulated domain in units of the gas disk scale height. The filaments already present at the time of switching on self-gravity fragment into pebble clumps. The circles in the last panel show the Hill spheres of gravitationally-bound clumps

(and quite irrespective of geological usage), such aggregates are called "pebbles", with a Stokes number (see equation (6)) falling in between 0.001 and 1. The corresponding size range depends on gas density but at the distance of the asteroid belt, pebbles are typically at least millimeter-sizes and this minimum size increases as we move closer to the star or consider a disk more massive than the MMSN. Thus, it is unclear whether individual chondrules can be such pebbles or not. Most likely, aggregates of chondrules and matrix dust may be needed to reach the pebble sizes. In this case, an early coagulation of the latter may be a way to preserve chemical complementarity until parent body accretion (Jacquet 2014a), but motion of independent chondrules and dust grains would not have destroyed complementarity if the S of the chondrules was < 1 (Goldberg et al. 2015), so matrix-chondrule complementarity might not be a stringent constraint on the dynamics of chondrite components.

For typical values of $\mathrm{St} = 10^{-2}$ and $\alpha = 10^{-4}$, settling causes the dust-to-gas ratio to increase by one order of magnitude and Kelvin-Helmholtz instabilities may then prevent further settling (Weidenschilling 1980). However, further enhancement is needed to proceed with planetesimal formation, which must be in radial and/or azimuthal direction. There are a range of processes discussed in the literature that lead to such enhancements. We have already mentioned axisymmetric pressure bumps for radial pileup. Vortices invoked for explaining non-axisymmetric features in observed disk may concentrate dust in the azimuthal direction (Raettig et al. 2015; Li et al. 2020). In a turbulent cascade, dust aggregates may get concentrated by the gas eddies (Hartlep and Cuzzi 2020). However, it is unlikely that any of these processes alone leads to dust concentration up to the Roche density.

4.2 Streaming Instability

Now, if only a solid/gas ratio of order unity (i.e. two orders of magnitude above solar) could be achieved, the *streaming instability* could take over to bring the pebble clumps to the Roche density, depending on the value of the Stokes number. The streaming instability is the instability of the dust drift in the sub-Keplerian gas disk to local overdensities when the

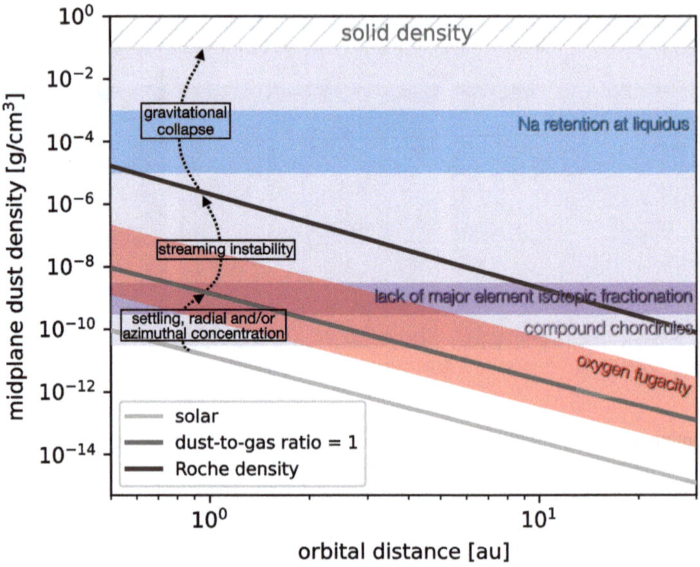

Fig. 8 Comparison of typical midplane dust densities in the Minimum Mass Solar Nebula model reached during various stages of planetesimal formation (see the text boxes) and the constraints on chondrule formation derived from: the oxygen fugacity estimated from olivine Fa contents (assuming a CI anhydrous dust±ice composition for the solids; Tenner et al. 2015), lack of major element isotopic fractionation (Cuzzi and Alexander 2006), compound chondrules (assuming a 1 h collision time; Jacquet 2021), Na retention at liquidus in type II chondrules (Alexander et al. 2008). We note that the oxygen fugacity constraint is a constraint on the solid/gas ratio so is proportional to the assumed density profile; in showing the total range from type I to type II chondrules irrespective of the heliocentric distances, we have also not tried to refine the constraints as a function of their presumed provenance, e.g. with respect to the snow line (both chondrule types occur in the two meteorite superclans anyway). We also note that the compound chondrule constraint is only a lower limit, because we only bound the chondrule-chondrule relative velocity by a *maximum* of 1 km/s (Arakawa and Nakamoto 2019)

momentum feedback from dust to gas is taken into account (Youdin and Goodman 2005). Schematically, in the linear phase, dust is dragged toward pressure perturbation maxima, and gas is dragged along, with the help of the Coriolis force, strengthening the pressure maxima, hence a positive feedback (Jacquet et al. 2011a) (see also Squire and Hopkins 2020). In the nonlinear phase, an overdensity of pebbles drags the surrounding gas along, collectively moving like a much larger body and collecting isolated pebbles which move with different radial and azimuthal velocities. This way, the initial overdensity grows. Numerical models show that, under some conditions, the dust structures produced by the instability fragment into very dense clumps which reach the Roche density and gravitationally collapse (Johansen et al. 2007; Simon et al. 2016; Yang et al. 2017, and many others), as presented in Fig. 7.

Since the streaming instability would operate rapidly (within a few hundred or thousand orbits, depending on St; Yang et al. 2017), any excess solid abundance above the threshold for its operation would rapidly be locked up in new planetesimals and the background disk dust density would return to marginal stability. Hence, one might expect a streaming instability-regulated solid/gas ratio around unity. This compares favorably with many chondrule constraints plotted in Fig. 8, which are consistent for the inner part of the solar nebula, up to a few au. Still, the incompatibly much higher densities (above the Roche density) estimated by Alexander et al. (2008) from Na retention at liquidus temperatures for type

II chondrules, would need to be explained away somehow (e.g. if the Na-bearing olivine cores, which are normally associated with the onset of olivine crystallization, are in fact relict grains). So chondrules may have formed in the settled dust disk, but not during the individual collapse events therein, to which we now turn.

4.3 Collapse of the Pebble Cloud

Once the streaming instability leads to very dense, gravitationally-bound pebble clumps, the following collapse to planetesimals is thought to be very quick. In this sense, the planetesimals produced this way may be seen as a snapshot of the local disk composition. This seems to be consistent with the fact that single chondrite groups appear homogeneous in bulk isotopic composition even though their members may come from different depths within their parent body (Pedersen et al. 2019; Przylibski et al. 2023).

Wahlberg Jansson and Johansen (2017) and Visser et al. (2021) studied the collapse of pebble clouds with one-dimensional models, assuming they are spherical and not rotating but including collisions between the constituent pebbles. They found that in the presence of protoplanetary gas, the collapse only takes about one-third of the local orbital time. Both models agree that there should be some degree of dynamical sorting of the infalling pebbles, such that the largest pebbles end up closest to the final planetesimal center and the smallest pebbles at its surface. This is qualitatively consistent with the sizes of CO chondrite chondrules increasing with petrological subtype and thus depth inside of the parent body (Pinto et al. 2021). However, in the models cited above, only one type of material was included and the pebbles might be dusty agglomerates rather than individual chondrules or CAIs, whose admixtures are yet to be studied in this context.

Nesvorný et al. (2021) and Polak and Klahr (2023) followed the pebble cloud collapse in three-dimensional models, including possible initial rotation of the pebble cloud. In these models, the collapse takes slightly longer than in the one-dimensional models but the timescales are still comparable with the local orbital time. Remarkably, a significant amount of binary planetesimals are formed with predominantly prograde rotation, which is consistent with the observations of Trans-Neptunian Objects (Fraser et al. 2017; Nesvorný et al. 2019). However, these three-dimensional models assumed monodisperse pebbles, thus they could not study the possible size sorting during the collapse.

4.4 Planetesimal Formation Sites

Although streaming instability could, indeed, take over for solid/gas ratios of order unity at the midplane, this is not an altogether easy threshold to reach. The hydrodynamic models, such as presented in Fig. 7, are done in very local domains due to the very high resolution needed to resolve the streaming instability, and thus assuming arbitrary initial conditions. Global models are needed to study the interplay of dust coagulation and radial drift potentially leading to streaming instability conditions at some locations in the disk. However, often such global models of dust evolution do not lead to any planetesimal formation at all (Estrada and Umurhan 2023). The very pressure bumps mentioned above, which concentrate solids radially, may be the necessary havens for accretion (Stammler et al. 2019; Jiang and Ormel 2021; Lau et al. 2022). Several models have shown that planetesimal formation is favored in the vicinity of the water snowline (Schoonenberg and Ormel 2017; Drążkowska and Alibert 2017; Charnoz et al. 2019; Hyodo et al. 2019, 2021) or the silicate sublimation line (Morbidelli et al. 2022; Izidoro et al. 2022). Independently of pressure bumps, their solid abundances could be indeed increased there because evaporation would slow down the inward drift of condensable matter inside them, by zeroing the drift relative to gas (which may have originally flowed outward; Morbidelli et al. 2022) for the evaporated species, and also reducing the size of the remaining, more refractory dust grains (e.g. Ciesla and Cuzzi 2006). So mass would accumulate more rapidly from the outer disk than it could be removed.

From the meteorite record, we also know that planetesimal accretion started early after CAI formation, which possibly corresponds to the protoplanetary disk buildup stage, and proceeded for several Ma. This can be generally reproduced in models following disk formation and modeling the planetesimal formation at the snow line (Drążkowska and Dullemond 2018; Lichtenberg et al. 2021). However, fulfilling all the requirements at once: that is starting early, forming planetesimals at more than one location, and keeping pebble supply for continuing planetesimal formation over several Ma, is very challenging (Marschall and Morbidelli 2023). One possibility is photoevaporation increasing the solid/gas ratio at the midplane (Carrera et al. 2017; Ercolano et al. 2017), which may explain a new surge in planetesimal accretion after a relative dearth between 1 and 2 Ma after CAIs (Sugiura and Fujiya 2014), as proposed by A. Morbidelli in the workshop. Another would be to include the full planet formation model in the planetesimal formation models. To date, these models typically only consider dust growth to pebbles and planetesimal formation but without their further evolution. Nevertheless, it is known that a giant planet, if formed early, may trigger another burst of planetesimal formation outside of its orbit (Eriksson et al. 2020; Shibaike and Alibert 2020; Miller et al. 2021), consistent with the concept of Jupiter splitting the NC and C reservoirs.

4.5 Rapid Formation of Giant Planets?

Certainly, as we saw in the previous section, the rapid formation of a giant planet may prevent pebbles from drifting inward too quickly and to keep planetesimal formation going for several Ma, consistent with the Solar System record. However, such a fast core growth is not typically seen in planet formation models. The timescale of Jupiter formation is a longstanding problem in planet formation theory, first recognized by Pollack et al. (1996). In the classical models, where planet growth is driven by the accretion of planetesimals, already forming Jupiter rocky core before the gas disk dissipates is a challenge.

An alternative could be top-down formation of the giant planet by fragmentation of gravitationally unstable disk (Boss 1997; Mayer et al. 2002). However, many problems have been pointed out regarding this mode of planet formation. The protoplanetary disk would need to be atypically cold to allow for its fragmentation (Kratter et al. 2010). Planets forming this way would be much more massive than Jupiter and on wider orbits but migrate quickly in those massive disks and fall onto the central star (Kratter and Lodato 2016). What is most important for us, the early planetesimal formation is likely hampered in massive gravitationally unstable, turbulent disks, as streaming instability requires a calm midplane pebble layer (Drążkowska and Dullemond 2018) and the spiral waves present in gravitationally unstable disks are not helping to enhance the pebble-to-gas ratio (Vorobyov et al. 2018). Further studies are needed on the perspective of planetesimal formation in gravitationally unstable disks and its interplay with planet formation in this scenario.

In recent years, the *pebble accretion* paradigm was introduced, where planetary core growth is aided by the accretion of the pebbles rather than planetesimals. Pebble accretion may be faster as pebbles are not scattered by the gravity of massive core as easily as planetesimals (Ormel and Klahr 2010; Lambrechts and Johansen 2012). However, even with pebble accretion, the formation of Jupiter's core in a smooth disk with the initial dust-to-gas ratio of 0.01 takes over 1 Ma (Lambrechts and Johansen 2014). This result is very sensitive to the exact model parameters such as the turbulence level and disk mass (Drążkowska et al.

2021). It is worth noting that in order to accrete pebbles efficiently, the planetary core must already be quite massive with the minimum mass sensitively dependent on the disk pressure gradient and pebble Stokes number (see, e.g., Liu and Ji 2020). At 1 au in a typical disk model, this mass would correspond to 100 km-sized body and it increases with the orbital distance.

It seems that, within the core accretion paradigm, the only way to produce massive planetary cores quickly is to start with an initial pressure bump helping to enhance the solid abundance up to the streaming instability threshold and at the same time providing an abundance of pebbles for fast pebble accretion (Morbidelli 2020; Jiang and Ormel 2023). Lau et al. (2022) presented a model in which a core of several Earth masses forms from a pebble ring at 10 au on the timescale of just 0.1 Ma. This is possible thanks to considering pebble growth and drift, planetesimal formation, and planetesimal growth by both planetesimal and pebble accretion simultaneously. This way, a few massive bodies emerging out of the very first planetesimals grow quickly and stir the planetesimals formed later preventing the emergence of further planetary cores, which otherwise would compete for accreting pebbles present in the ring. So the pressure bump invoked for the NC/C dichotomy (Brasser and Mojzsis 2020) would be a cause rather than an effect of the growth of Jupiter. Now, while the observation of the NC/C dichotomy for early accreted iron meteorites is generally attributed to an early growth of Jupiter (Kruijer et al. 2017), the forming Jupiter may have efficiently accreted bodies with intermediate isotopic compositions (Jacquet et al. 2019), or that intermediate region of the disk may have been unfavorable to planetesimal formation, if devoid of pressure bumps (Morbidelli et al. 2022).

5 Conclusion

As resolution of astronomical observations improves, our picture of the solar nebula may become less nebulous. Rings such as those observed in neighbouring protoplanetary disks may have been the safe havens for preventing drift loss of solids and producing age-diverse collections of them as observed in chondrites. These would have been concentrated to the level recorded by chondrules, conducive to the streaming instability and thence planetesimal accretion. The rings could interrupt, or limit, communications between different disk regions (e.g. inner and outer disk) accounting for the NC/C isotopic dichotomy among meteorites. Presumably, these rings trace pressure maxima, however the origin of such pressure maxima remains contentious. Giant planets such as Jupiter might be responsible, but such planets may also be consequences of such pressure maxima. In early times, more or less contemporaneously with infall, large-scale transport may have been unimpeded, as suggested by refractory inclusions, now chiefly found far from the Sun. None of this, of course, makes a standard model of meteorite formation "geography" yet, but this at least offers tantalizing hints to explore further, whether in the microscope or the telescope, or the minds above them.

Acknowledgements The authors are grateful to the organizers of the International Space Science Institute workshop "Evolution of the Solar System: Constraints from Meteorites" (June 5-9, 2023, in Bern). JD was funded by the European Union under the European Union's Horizon Europe Research and Innovation Programme 101040037 (PLANETOIDS). This article is dedicated to the memory of the lead author's grandmother Marguerite Jacquet (1923-2023). *Nous aurons bien ri de l'éternité.*

Funding Open Access funding enabled and organized by Projekt DEAL.

Data Availability Not applicable.

Declarations

Competing Interests The authors have no competing interests to declare that are relevant to the content of this article.

References

Aléon J (2010) Meteorites and the physico-chemical conditions in the early solar nebula. In: Montmerle T, Ehrenreich D, Lagrange AM (eds) EAS Publications Series, pp 253–300. https://doi.org/10.1051/eas/1041020

Alexander CMO (2019a) Quantitative models for the elemental and isotopic fractionations in chondrites: the carbonaceous chondrites. Geochim Cosmochim Acta 254:277–309. https://doi.org/10.1016/j.gca.2019.02.008

Alexander CMO (2019b) Quantitative models for the elemental and isotopic fractionations in the chondrites: the non-carbonaceous chondrites. Geochim Cosmochim Acta 254:246–276. https://doi.org/10.1016/j.gca.2019.01.026

Alexander CMOD, Grossman JN, Ebel DS, et al (2008) The formation conditions of chondrules and chondrites. Science 320(5883):1617. https://doi.org/10.1126/science.1156561

Alexander CMO, Nittler LR, Davidson J, et al (2017) Measuring the level of interstellar inheritance in the solar protoplanetary disk. Meteorit Planet Sci 52(9):1797–1821. https://doi.org/10.1111/maps.12891

Anders E (1964) Origin, age, and composition of meteorites. Space Sci Rev 3:583–714. https://doi.org/10.1007/BF00177954

Andrews SM, Huang J, Pérez LM, et al (2018) The disk substructures at high angular resolution project (DSHARP). I. Motivation, sample, calibration, and overview. Astrophys J Lett 869(2):L41. https://doi.org/10.3847/2041-8213/aaf741. arXiv:1812.04040 [astro-ph.SR]

Arakawa S, Nakamoto T (2019) Compound chondrule formation in optically thin shock waves. Astrophys J 877(2):84. https://doi.org/10.3847/1538-4357/ab1b3e. arXiv:1904.09580 [astro-ph.EP]

Asphaug E, Jutzi M, Movshovitz N (2011) Chondrule formation during planetesimal accretion. Earth Planet Sci Lett 308:369–379. https://doi.org/10.1016/j.epsl.2011.06.007

Bae J, Zhu Z, Hartmann L (2017) On the formation of multiple concentric rings and gaps in protoplanetary disks. Astrophys J 850(2):201. https://doi.org/10.3847/1538-4357/aa9705. arXiv:1706.03066 [astro-ph.EP]

Benisty M, Dominik C, Follette K, et al (2023) Optical and near-infrared view of planet-forming disks and protoplanets. In: Inutsuka S, Aikawa Y, Muto T, et al (eds) Protostars and planets VII, p 605. https://doi.org/10.48550/arXiv.2203.09991. arXiv:2203.09991

Binzel RP, Reddy V, Dunn TL (2015) The near-Earth object population: connections to comets, main-belt asteroids, and meteorites. In: Asteroids IV, pp 243–256. https://doi.org/10.2458/azu_uapress_9780816532131-ch013

Birnstiel T, Dullemond CP, Brauer F (2010) Gas- and dust evolution in protoplanetary disks. Astron Astrophys 513:A79+. https://doi.org/10.1051/0004-6361/200913731. arXiv:1002.0335 [astro-ph.EP]

Birnstiel T, Klahr H, Ercolano B (2012) A simple model for the evolution of the dust population in protoplanetary disks. Astron Astrophys 539:A148. https://doi.org/10.1051/0004-6361/201118136. arXiv:1201.5781 [astro-ph.EP]

Birnstiel T, Fang M, Johansen A (2016) Dust evolution and the formation of planetesimals. Space Sci Rev 205(1–4):41–75. https://doi.org/10.1007/s11214-016-0256-1. arXiv:1604.02952 [astro-ph.SR]

Bitsch B, Morbidelli A, Johansen A, et al (2018) Pebble-isolation mass: scaling law and implications for the formation of super-earths and gas giants. Astron Astrophys 612:A30. https://doi.org/10.1051/0004-6361/201731931. arXiv:1801.02341 [astro-ph.EP]

Bland PA, Howard LE, Prior DJ, et al (2011) Earliest rock fabric formed in the Solar System preserved in a chondrule rim. Nat Geosci 4:244–247. https://doi.org/10.1038/ngeo1120

Blum J (2018) Dust evolution in protoplanetary discs and the formation of planetesimals. What have we learned from laboratory experiments? Space Sci Rev 214(2):52. https://doi.org/10.1007/s11214-018-0486-5. arXiv:1802.00221 [astro-ph.EP]

Boley AC, Morris MA, Desch SJ (2013) High-temperature processing of solids through solar nebular bow shocks: 3D radiation hydrodynamics simulations with particles. Astrophys J 776:101. https://doi.org/10.1088/0004-637X/776/2/101. arXiv:1308.2968 [astro-ph.EP]

Bonanno A, Fröhlich HE (2015) A Bayesian estimation of the helioseismic solar age. Astron Astrophys 580:A130. https://doi.org/10.1051/0004-6361/201526419. arXiv:1507.05847 [astro-ph.SR]

Boss AP (1997) Giant planet formation by gravitational instability. Science 276:1836–1839. https://doi.org/10.1126/science.276.5320.1836

Bouvier A, Wadhwa M (2010) The age of the Solar System redefined by the oldest pb-pb age of a meteoritic inclusion. Nat Geosci 3:637–641

Bouvier A, Ma N, Neumann W, et al (2021) Early and sequential formation of planetesimals in an outer region of the protoplanetary disc. In: European Planetary Science Congress EPSC2021–846. https://doi.org/10.5194/epsc2021-846

Bradley JP (2014) Early solar nebula grains - interplanetary dust particles. In: Davis AM (ed) Meteorites and cosmochemical processes, vol 1, pp 287–308

Bradley JP, Ishii HA, Bustillo K, et al (2022) On the provenance of GEMS, a quarter century post discovery. Geochim Cosmochim Acta 335:323–338. https://doi.org/10.1016/j.gca.2022.06.036

Brasser R, Mojzsis SJ (2020) The partitioning of the inner and outer Solar System by a structured protoplanetary disk. Nat Astron 4:492–499. https://doi.org/10.1038/s41550-019-0978-6

Braukmüller N, Wombacher F, Hezel DC, et al (2018) The chemical composition of carbonaceous chondrites: implications for volatile element depletion, complementarity and alteration. Geochim Cosmochim Acta 239:17–48. https://doi.org/10.1016/j.gca.2018.07.023

Brearley AJ, Jones RH (1998) Chondritic meteorites. In: Papike JJ (ed) Planetary materials. Mineralogical society of America, chap 3, pp 3–1–3–398

Bryson JFJ, Brennecka GA (2021) Constraints on chondrule generation, disk dynamics, and asteroid accretion from the compositions of carbonaceous meteorites. Astrophys J 912(2):163. https://doi.org/10.3847/1538-4357/abea12

Burkhardt C, Spitzer F, Morbidelli A, et al (2021) Terrestrial planet formation from lost inner Solar System material. Sci Adv 7(52):eabj7601. https://doi.org/10.1126/sciadv.abj7601

Carrera D, Gorti U, Johansen A, et al (2017) Planetesimal formation by the streaming instability in a photoevaporating disk. Astrophys J 839(1):16. https://doi.org/10.3847/1538-4357/aa6932. arXiv:1703.07895 [astro-ph.EP]

Chandrasekhar S (1967) Ellipsoidal figures of equilibrium–an historical account. Commun Pure Appl Math 20(2):251–265. https://doi.org/10.1002/cpa.3160200203. https://onlinelibrary.wiley.com/doi/pdf/10.1002/cpa.3160200203

Charnoz S, Pignatale FC, Hyodo R, et al (2019) Planetesimal formation in an evolving protoplanetary disk with a dead zone. Astron Astrophys 627:A50. https://doi.org/10.1051/0004-6361/201833216

Charnoz S, Avice G, Hyodo R, et al (2021) Forming pressure traps at the snow line to isolate isotopic reservoirs in the absence of a planet. Astron Astrophys 652:A35. https://doi.org/10.1051/0004-6361/202038797. arXiv:2105.00456 [astro-ph.EP]

Chiang EI, Goldreich P (1997) Spectral energy distributions of T Tauri stars with passive circumstellar disks. Astrophys J 490:368–376. https://doi.org/10.1086/304869. arXiv:astro-ph/9706042

Chiang E, Youdin AN (2010) Forming planetesimals in solar and extrasolar nebulae. Annu Rev Earth Planet Sci 38:493–522. https://doi.org/10.1146/annurev-earth-040809-152513. arXiv:0909.2652

Christophe Michel-Lévy M (1968) Un chondre exceptionnel dans la météorite de vigarano. Bull Soc Fr Minéral Cristallogr 91:212–214

Ciesla FJ (2010) The distributions and ages of refractory objects in the solar nebula. Icarus 208:455–467. https://doi.org/10.1016/j.icarus.2010.02.010

Ciesla FJ, Cuzzi JN (2006) The evolution of the water distribution in a viscous protoplanetary disk. Icarus 181:178–204. https://doi.org/10.1016/j.icarus.2005.11.009. arXiv:astro-ph/0511372

Clarke JRS, Jarosewich E, Mason B, et al (1971) In: The Allende, Mexico, meteorite shower. Smithsonian contributions to the Earth sciences, vol 5, pp 1–53

Clayton RN, Grossman L, Mayeda TK (1973) A component of primitive nuclear composition in carbonaceous meteorites. Science 182:485–488. https://doi.org/10.1126/science.182.4111.485

Clayton RN, Onuma N, Mayeda TK (1976) A classification of meteorites based on oxygen isotopes. Earth Planet Sci Lett 30(1):10–18. https://doi.org/10.1016/0012-821X(76)90003-0

Connelly JN, Bollard J, Bizzarro M (2017) Pb-Pb chronometry and the early Solar System. Geochim Cosmochim Acta 201:345–363. https://doi.org/10.1016/j.gca.2016.10.044

Cuzzi JN, Alexander CMO (2006) Chondrule formation in particle-rich nebular regions at least hundreds of kilometres across. Nature 441:483–485. https://doi.org/10.1038/nature04834

Dauphas N, Schauble EA (2016) Mass fractionation laws, mass-independent effects, and isotopic anomalies. Annu Rev Earth Planet Sci 44:709–783. https://doi.org/10.1146/annurev-earth-060115-012157

Davis AM, Richter FM (2014) Condensation and evaporation of Solar System materials Elsevier, Amsterdam, pp 335–360

DeMeo FE, Carry B (2014) Solar System evolution from compositional mapping of the asteroid belt. Nature 505:629–634. https://doi.org/10.1038/nature12908

Desch SJ, Cuzzi JN (2000) The generation of lightning in the solar nebula. Icarus 143:87–105. https://doi.org/10.1006/icar.1999.6245

Desch SJ, Ciesla FJ, Hood LL, et al (2005) Heating of chondritic materials in solar nebula shocks. In: Krot AN, Scott ERD, Reipurth B (eds) Chondrites and the protoplanetary disk, pp 849–871

Desch SJ, Morris MA, Connolly HC, et al (2010) A critical examination of the X-wind model for chondrule and calcium-rich, aluminum-rich inclusion formation and radionuclide production. Astrophys J 725:692. https://doi.org/10.1088/0004-637X/725/1/692. arXiv:1011.3483 [astro-ph.EP]

Desch SJ, Kalyaan A, O'D. Alexande CM (2018) The effect of Jupiter's formation on the distribution of refractory elements and inclusions in meteorites. Astrophys J Suppl Ser 238:11. https://doi.org/10.3847/1538-4365/aad95f. arXiv:1710.03809 [astro-ph.EP]

Desch SJ, Herbst AK, Williams CD, et al (2021) Radial distribution of CAIs. In: 52nd Lunar and Planetary Science Conference, Lunar and Planetary Science Conference, p 2663

Desch SJ, Young ED, Dunham ET, et al (2022) Short-Lived Radionuclides in Meteorites and the Sun's Birth Environment. https://doi.org/10.48550/arXiv.2203.11169. arXiv:2203.11169

Desch SJ, Dunlap DR, Dunham ET, et al (2023) Statistical chronometry of meteorites. I. A test of ^{26}Al homogeneity and the Pb-Pb age of the Solar System's t = 0. Icarus 402:115607. https://doi.org/10.1016/j.icarus.2023.115607. arXiv:2212.00390 [astro-ph.EP]

Dipierro G, Price D, Laibe G, et al (2015) On planet formation in HL Tau. Mon Not R Astron Soc 453(1):L73–L77. https://doi.org/10.1093/mnrasl/slv105. arXiv:1507.06719 [astro-ph.EP]

Dong R, Zhu Z, Whitney B (2015) Observational signatures of planets in protoplanetary disks I. gaps opened by single and multiple young planets in disks. Astrophys J 809(1):93. https://doi.org/10.1088/0004-637X/809/1/93. arXiv:1411.6063 [astro-ph.EP]

Drążkowska J, Alibert Y (2017) Planetesimal formation starts at the snow line. Astron Astrophys 608:A92. https://doi.org/10.1051/0004-6361/201731491. arXiv:1710.00009 [astro-ph.EP]

Drążkowska J, Dullemond CP (2018) Planetesimal formation during protoplanetary disk buildup. Astron Astrophys 614:A62. https://doi.org/10.1051/0004-6361/201732221. arXiv:1803.00575 [astro-ph.EP]

Drążkowska J, Stammler SM, Birnstiel T (2021) How dust fragmentation may be beneficial to planetary growth by pebble accretion. Astron Astrophys 647:A15. https://doi.org/10.1051/0004-6361/202039925. arXiv:2101.01728 [astro-ph.EP]

Dullemond CP, Birnstiel T, Huang J, et al (2018) The disk substructures at high angular resolution project (DSHARP). VI. Dust trapping in thin-ringed protoplanetary disks. Astrophys J Lett 869(2):L46. https://doi.org/10.3847/2041-8213/aaf742. arXiv:1812.04044 [astro-ph.EP]

Dunham ET, Sheikh A, Opara D, et al (2023) Calcium-aluminum-rich inclusions in non-carbonaceous chondrites: abundances, sizes, and mineralogy. Meteorit Planet Sci 58(5):643–671. https://doi.org/10.1111/maps.13975

Ebel DS, Brunner C, Konrad K, et al (2016) Abundance, major element composition and size of components and matrix in CV, CO and Acfer 094 chondrites. Geochim Cosmochim Acta 172:322–356. https://doi.org/10.1016/j.gca.2015.10.007

Engrand C, Lasue J, Wooden DH, et al (2023) Chemical and physical properties of cometary dust. https://doi.org/10.48550/arXiv.2305.03417. arXiv:2305.03417

Ercolano B, Jennings J, Rosotti G, et al (2017) X-ray photoevaporation's limited success in the formation of planetesimals by the streaming instability. Mon Not R Astron Soc 472(4):4117–4125. https://doi.org/10.1093/mnras/stx2294. arXiv:1709.00361 [astro-ph.EP]

Eriksson LEJ, Johansen A, Liu B (2020) Pebble drift and planetesimal formation in protoplanetary discs with embedded planets. Astron Astrophys 635:A110. https://doi.org/10.1051/0004-6361/201937037. arXiv:2001.11042 [astro-ph.EP]

Estrada PR, Umurhan OM (2023) Formation of the first planetesimals via the streaming instability in globally turbulent protoplanetary disks? Astrophys J 946(1):15. https://doi.org/10.3847/1538-4357/acb7db. arXiv:2302.03163 [astro-ph.EP]

Flock M, Ruge JP, Dzyurkevich N, et al (2015) Gaps, rings, and non-axisymmetric structures in protoplanetary disks. From simulations to ALMA observations. Astron Astrophys 574:A68. https://doi.org/10.1051/0004-6361/201424693. arXiv:1411.2736 [astro-ph.EP]

Flynn GJ, Consolmagno GJ, Brown P, et al (2018) Physical properties of the stone meteorites: implications for the properties of their parent bodies. Chem Erde 78(3):269–298. https://doi.org/10.1016/j.chemer.2017.04.002

Fraser WC, Bannister MT, Pike RE, et al (2017) All planetesimals born near the Kuiper Belt formed as binaries. Nat Astron 1:0088. https://doi.org/10.1038/s41550-017-0088. arXiv:1705.00683 [astro-ph.EP]

Fukuda K, Brownlee DE, Joswiak DJ, et al (2021) Correlated isotopic and chemical evidence for condensation origins of olivine in comet 81P/Wild 2 and in AOAs from CV and CO chondrites. Geochim Cosmochim Acta 293:544–574. https://doi.org/10.1016/j.gca.2020.09.036

Gammie CF (1996) Layered accretion in T tauri disks. Astrophys J 457:355–362. https://doi.org/10.1086/176735

Genge MJ, Van Ginneken M, Suttle MD (2020) Micrometeorites: insights into the flux, sources and atmospheric entry of extraterrestrial dust at Earth. Planet Space Sci 187:104900. https://doi.org/10.1016/j.pss.2020.104900

Gilmour I (2014) Structural and isotopic analysis of organic matter in carbonaceous chondrites. In: Davis AM (ed) Meteorites and cosmochemical processes, vol 1, pp 215–233

Ginski C, Facchini S, Huang J, et al (2021) Disk evolution study through imaging of nearby young stars (DESTINYS): late infall causing disk misalignment and dynamic structures in SU Aur. Astrophys J Lett 908(2):L25. https://doi.org/10.3847/2041-8213/abdf57. arXiv:2102.08781 [astro-ph.EP]

Goldberg AZ, Owen JE, Jacquet E (2015) Chondrule transport in protoplanetary discs. Mon Not R Astron Soc 452:4054–4069. https://doi.org/10.1093/mnras/stv1610

Gounelle M, Morbidelli A, Bland PA, et al (2008) Meteorites from the outer Solar System? In: Barucci MA, Boehnhardt H, Cruikshank DP, Morbidelli A (eds) The Solar System beyond Neptune. University of Arizona Press, Tucson, pp 525–541.

Grady MM (2000) Catalogue of meteorites. Cambridge University Press, Cambridge

Greenwood RC, Burbine TH, Franchi IA (2020) Linking asteroids and meteorites to the primordial planetesimal population. Geochim Cosmochim Acta 277:377–406. https://doi.org/10.1016/j.gca.2020.02.004

Grimm RE, McSween HY (1989) Water and the thermal evolution of carbonaceous chondrite parent bodies. Icarus 82:244–280. https://doi.org/10.1016/0019-1035(89)90038-9

Grossman JN, Brearley AJ (2005) The onset of metamorphism in ordinary and carbonaceous chondrites. Meteorit Planet Sci 40:87. https://doi.org/10.1111/j.1945-5100.2005.tb00366.x

Haenecour P, Floss C, Zega TJ, et al (2018) Presolar silicates in the matrix and fine-grained rims around chondrules in primitive CO3.0 chondrites: evidence for pre-accretionary aqueous alteration of the rims in the solar nebula. Geochim Cosmochim Acta 221:379–405. https://doi.org/10.1016/j.gca.2017.06.004

Hartlep T, Cuzzi JN (2020) Cascade model for planetesimal formation by turbulent clustering. Astrophys J 892(2):120. https://doi.org/10.3847/1538-4357/ab76c3. arXiv:2002.06321 [astro-ph.EP]

Haugbølle T, Weber P, Wielandt DP, et al (2019) Probing the protosolar disk using dust filtering at gaps in the early Solar System. arXiv:1903.12274 [astro-ph.EP]

Hayashi C (1981) Structure of the solar nebula, growth and decay of magnetic fields and effects of magnetic and turbulent viscosities on the nebula. Prog Theor Phys Suppl 70:35–53. https://doi.org/10.1143/PTPS.70.35

Heck PR, Greer J, Kööp L, et al (2020) Lifetimes of interstellar dust from cosmic ray exposure ages of presolar silicon carbide. Proc Natl Acad Sci 117(4):1884–1889. https://doi.org/10.1073/pnas.1904573117

Hellmann JL, Schneider JM, Wölfer E, et al (2023) Origin of isotopic diversity among carbonaceous chondrites. Astrophys J Lett 946(2):L34. https://doi.org/10.3847/2041-8213/acc102. arXiv:2303.04173 [astro-ph.EP]

Herbst W, Greenwood JP (2019) A radiative heating model for chondrule and chondrite formation. Icarus 329:166–181. https://doi.org/10.1016/j.icarus.2019.03.039. arXiv:1903.12224 [astro-ph.EP]

Herzog GF, Caffee MW (2014) Cosmic-ray exposure ages of meteorites. In: Davis AM (ed) Meteorites and cosmochemical processes, pp 419–454

Hezel DC, Bland PA, Palme H, et al (2018) Composition of chondrules and matrix and their complementary relationship in chondrites. In: Russell SR, Connolly HCJ, Krot AN (eds) Chondrules: records of protoplanetary disk processes. Cambridge planetary science, pp 91–121

Hopp T, Dauphas N, Abe Y, et al (2022) Ryugu's nucleosynthetic heritage from the outskirts of the Solar System. Sci Adv 8(46):eadd8141. https://doi.org/10.1126/sciadv.add8141

Houdek G, Gough DO (2011) On the seismic age and heavy-element abundance of the Sun. Mon Not R Astron Soc 418(2):1217–1230. https://doi.org/10.1111/j.1365-2966.2011.19572.x. arXiv:1108.0802 [astro-ph.SR]

Howard E (1802) Experiments and observations on certain stony and metalline substances which at different times are said to have fallen on the Earth; also on various kinds of native irons. Philos Trans 92:168–212

Hu R (2010) Transport of the first rocks of the Solar System by X-winds. Astrophys J 725:1421–1428. https://doi.org/10.1088/0004-637X/725/2/1421. arXiv:1010.2532 [astro-ph.EP]

Huss GR, MacPherson GJ, Wasserburg GJ, et al (2001) 26Al in CAIs and chondrules from unequilibrated ordinary chondrites. Meteorit Planet Sci 36:975–997. https://doi.org/10.1111/j.1945-5100.2001.tb01934.x

Hutchison R (2004) Meteorites. Cambridge University Press, Cambridge

Hyodo R, Ida S, Charnoz S (2019) Formation of rocky and icy planetesimals inside and outside the snow line: effects of diffusion, sublimation, and back-reaction. Astron Astrophys 629:A90. https://doi.org/10.1051/0004-6361/201935935. arXiv:1907.04621 [astro-ph.EP]

Hyodo R, Guillot T, Ida S, et al (2021) Planetesimal formation around the snow line. II. Dust or pebbles? Astron Astrophys 646:A14. https://doi.org/10.1051/0004-6361/202039894. arXiv:2012.06700 [astro-ph.EP]

Iwasaki K, Tomida K, Takasao S, et al (2024) Dynamics Near the Inner Dead-Zone Edges in a Proprotoplanetary Disk. https://doi.org/10.48550/arXiv.2401.03733. arXiv:2401.03733 [astro-ph.EP]

Izidoro A, Dasgupta R, Raymond SN, et al (2022) Planetesimal rings as the cause of the Solar System's planetary architecture. Nat Astron 6:357–366. https://doi.org/10.1038/s41550-021-01557-z. arXiv:2112.15558 [astro-ph.EP]

Jacquet E (2013) On vertical variations of gas flow in protoplanetary disks and their impact on the transport of solids. Astron Astrophys 551:A75. https://doi.org/10.1051/0004-6361/201220812. arXiv:1301.5817 [astro-ph.EP]

Jacquet E (2014a) The quasi-universality of chondrule size as a constraint for chondrule formation models. Icarus 232:176–186

Jacquet E (2014b) Transport of solids in protoplanetary disks: comparing meteorites and astrophysical models. C R Geosci 346:3–12. arXiv:1402.0533 [astro-ph.EP]

Jacquet E (2021) Collisions and compositional variability in chondrule-forming events. Geochim Cosmochim Acta 296:18–37. https://doi.org/10.1016/j.gca.2020.12.025. arXiv:2101.10083 [astro-ph.EP]

Jacquet E (2022) Meteorite petrology versus genetics: toward a unified binominal classification. Meteorit Planet Sci 57(9):1774–1794. https://doi.org/10.1111/maps.13896. arXiv:2209.07377 [astro-ph.EP]

Jacquet E, Balbus S, Latter H (2011a) On linear dust-gas streaming instabilities in protoplanetary discs. Mon Not R Astron Soc 415:3591–3598. https://doi.org/10.1111/j.1365-2966.2011.18971.x. arXiv:1104.5396 [astro-ph.EP]

Jacquet E, Fromang S, Gounelle M (2011b) Radial transport of refractory inclusions and their preservation in the dead zone. Astron Astrophys 526:L8. https://doi.org/10.1051/0004-6361/201016118. arXiv:1101.0699 [astro-ph.EP]

Jacquet E, Gounelle M, Fromang S (2012) On the aerodynamic redistribution of chondrite components in protoplanetary disks. Icarus 220:162–173. https://doi.org/10.1016/j.icarus.2012.04.022. arXiv:1204.4337 [astro-ph.EP]

Jacquet E, Paulhiac-Pison M, Alard O, et al (2013) Trace element geochemistry of CR chondrite metal. Meteorit Planet Sci 48:1981–1999

Jacquet E, Pignatale FC, Chaussidon M, et al (2019) Fingerprints of the protosolar cloud collapse in the Solar System. II. Nucleosynthetic anomalies in meteorites. Astrophys J 884(1):32. https://doi.org/10.3847/1538-4357/ab38c1. arXiv:1910.09299 [astro-ph.EP]

Jiang H, Ormel CW (2021) Survival of ALMA rings in the absence of pressure maxima. Mon Not R Astron Soc 505(1):1162–1179. https://doi.org/10.1093/mnras/stab1278. arXiv:2103.01236 [astro-ph.EP]

Jiang H, Ormel CW (2023) Efficient planet formation by pebble accretion in ALMA rings. Mon Not R Astron Soc 518(3):3877–3900. https://doi.org/10.1093/mnras/stac3275. arXiv:2207.13002 [astro-ph.EP]

Johansen A, Okuzumi S (2018) Harvesting the decay energy of ^{26}Al to drive lightning discharge in protoplanetary discs. Astron Astrophys 609:A31. https://doi.org/10.1051/0004-6361/201630047. arXiv:1708.05157 [astro-ph.EP]

Johansen A, Oishi JS, Mac Low M, et al (2007) Rapid planetesimal formation in turbulent circumstellar disks. Nature 448:1022–1025. https://doi.org/10.1038/nature06086

Johnson BC, Minton DA, Melosh HJ, et al (2015) Impact jetting as the origin of chondrules. Nature 517:339–341. https://doi.org/10.1038/nature14105

Jones RH (2012) Petrographic constraints on the diversity of chondrule reservoirs in the protoplanetary disk. Meteorit Planet Sci 47:1176–1190. https://doi.org/10.1111/j.1945-5100.2011.01327.x

Jones RH, Villeneuve J, Libourel G (2018) Thermal histories of chondrules. In: Russell SS, Connolly JHC, Krot AN (eds) Chondrules: records of protoplanetary disk processes, pp 57–90. https://doi.org/10.1017/9781108284073.003

Kita NT, Ushikubo T (2012) Evolution of protoplanetary disk inferred from ^{26}Al chronology of individual chondrules. Meteorit Planet Sci 47:1108–1119. https://doi.org/10.1111/j.1945-5100.2011.01264.x

Kleine T, Budde G, Burkhardt C, et al (2020) The non-carbonaceous-carbonaceous meteorite dichotomy. Space Sci Rev 216(4):55. https://doi.org/10.1007/s11214-020-00675-w

Kratter K, Lodato G (2016) Gravitational instabilities in circumstellar disks. Annu Rev Astron Astrophys 54:271–311. https://doi.org/10.1146/annurev-astro-081915-023307. arXiv:1603.01280 [astro-ph.SR]
Kratter KM, Murray-Clay RA, Youdin AN (2010) The runts of the litter: why planets formed through gravitational instability can only be failed binary stars. Astrophys J 710(2):1375–1386. https://doi.org/10.1088/0004-637X/710/2/1375. arXiv:0909.2644 [astro-ph.EP]
Kretke KA, Lin DNC (2007) Grain retention and formation of planetesimals near the snow line in MRI-driven turbulent protoplanetary disks. Astrophys J Lett 664(1):L55–L58. https://doi.org/10.1086/520718 arXiv:0706.1272 [astro-ph]
Krot AN, Amelin Y, Cassen P, et al (2005) Young chondrules in CB chondrites from a giant impact in the early Solar System. Nature 436:989–992. https://doi.org/10.1038/nature03830
Krot AN, McKeegan KD, Huss GR, et al (2006) Aluminum-magnesium and oxygen isotope study of relict Ca-Al-rich inclusions in chondrules. Astrophys J 639:1227–1237. https://doi.org/10.1086/498610
Krot AN, Petaev MI, Piani L, et al (2024) Metasomatic alteration of type 3 ordinary and carbonaceous chondrites. Space Sci Rev 220
Kruijer TS, Burkhardt C, Budde G, et al (2017) Age of Jupiter inferred from the distinct genetics and formation times of meteorites. Proc Natl Acad Sci 114:6712–6716. https://doi.org/10.1073/pnas.1704461114
Kruijer TS, Kleine T, Borg LE (2020) The great isotopic dichotomy of the early Solar System. Nat Astron 4:32–40. https://doi.org/10.1038/s41550-019-0959-9
Kuznetsova A, Bae J, Hartmann L, et al (2022) Anisotropic infall and substructure formation in embedded disks. Astrophys J 928(1):92. https://doi.org/10.3847/1538-4357/ac54a8. arXiv:2202.05301 [astro-ph.EP]
Lambrechts M, Johansen A (2012) Rapid growth of gas-giant cores by pebble accretion. Astron Astrophys 544:A32. https://doi.org/10.1051/0004-6361/201219127. arXiv:1205.3030 [astro-ph.EP]
Lambrechts M, Johansen A (2014) Forming the cores of giant planets from the radial pebble flux in protoplanetary discs. Astron Astrophys 572:A107. https://doi.org/10.1051/0004-6361/201424343. arXiv:1408.6094 [astro-ph.EP]
Lambrechts M, Johansen A, Morbidelli A (2014) Separating gas-giant and ice-giant planets by halting pebble accretion. Astron Astrophys 572:A35. https://doi.org/10.1051/0004-6361/201423814. arXiv:1408.6087 [astro-ph.EP]
Larimer JW, Anders E (1970) Chemical fractionations in meteorites–III. Major element fractionations in chondrites. Geochim Cosmochim Acta 34:367–387. https://doi.org/10.1016/0016-7037(70)90112-2
Lau TCH, Drążkowska J, Stammler SM, et al (2022) Rapid formation of massive planetary cores in a pressure bump. Astron Astrophys 668:A170. https://doi.org/10.1051/0004-6361/202244864. arXiv:2211.04497 [astro-ph.EP]
Lauretta DS, Connolly JHC, Aebersold JE, et al (2024) Asteroid (101955) Bennu in the Laboratory: Properties of the Sample Collected by OSIRIS-REx. https://doi.org/10.48550/arXiv.2404.12536. arXiv:2404.12536 [astro-ph.EP]
Le Guillou C, Changela HG, Brearley AJ (2015) Widespread oxidized and hydrated amorphous silicates in CR chondrites matrices: implications for alteration conditions and H_2 degassing of asteroids. Earth Planet Sci Lett 420:162–173. https://doi.org/10.1016/j.epsl.2015.02.031
Lee T, Papanastassiou DA, Wasserburg GJ (1976) Demonstration of Mg-26 excess in Allende and evidence for Al-26. Geophys Res Lett 3:41–44. https://doi.org/10.1029/GL003i001p00041
Lee MR, O'D Alexander CM, Bischoff A, et al (2024) Low-temperature aqueous alteration of chondrites. Space Sci Rev 220
Li YP, Li H, Li S, et al (2020) Planet-induced vortices with dust coagulation in protoplanetary disks. Astrophys J Lett 892(2):L19. https://doi.org/10.3847/2041-8213/ab7fb2. arXiv:2003.06522 [astro-ph.EP]
Lichtenberg T, Drążkowska J, Schönbächler M, et al (2021) Bifurcation of planetary building blocks during Solar System formation. Science 371(6527):365–370. https://doi.org/10.1126/science.abb3091. arXiv:2101.08571 [astro-ph.EP]
Liu B, Ji J (2020) A tale of planet formation: from dust to planets. Res Astron Astrophys 20(10):164. https://doi.org/10.1088/1674-4527/20/10/164. arXiv:2009.02321 [astro-ph.EP]
Liu N, Lugaro M, Leitner J, et al (2024) Presolar grains as probes of supernova nucleosynthesis. Space Sci Rev 220
Lodders K, Bergemann M, Palme H (2024) Solar system elemental abundances from the solar photosphere and CI-chondrites. Space Sci Rev 220
MacPherson GJ (2005) Calcium-aluminum-rich inclusions in chondritic meteorites. In: Davis AM, Holland HD, Turekian KK (eds) Meteorites, comets and planets: treatise on geochemistry, Chap. 1.08, vol 1. Elsevier B, pp 201–246
MacPherson GJ, Kita NT, Ushikubo T, et al (2012) Well-resolved variations in the formation ages for Ca-Al-rich inclusions in the early Solar System. Earth Planet Sci Lett 331:43–54. https://doi.org/10.1016/j.epsl.2012.03.010

Marrocchi Y, Euverte R, Villeneuve J, et al (2019a) Formation of CV chondrules by recycling of amoeboid olivine aggregate-like precursors. Geochim Cosmochim Acta 247:121–141. https://doi.org/10.1016/j.gca.2018.12.038

Marrocchi Y, Villeneuve J, Jacquet E, et al (2019b) Rapid condensation of the first Solar System solids. Proc Natl Acad Sci 116(47):23461–23466. https://doi.org/10.1073/pnas.1912479116

Marrocchi Y, Piralla M, Regnault M, et al (2022) Isotopic evidence for two chondrule generations in CR chondrites and their relationships to other carbonaceous chondrites. Earth Planet Sci Lett 593:117683. https://doi.org/10.1016/j.epsl.2022.117683

Marrocchi Y, Rigaudier T, Piralla M, et al (2023) Hydrogen isotopic evidence for nebular pre-hydration and the limited role of parent-body processes in CM chondrites. Earth Planet Sci Lett 611:118151. https://doi.org/10.1016/j.epsl.2023.118151

Marrocchi Y, Jones RH, Russell SS, et al (2024) Chondrule properties and formation conditions. Space Sci Rev 220:69 https://doi.org/10.1007/s11214-024-01102-0

Marschall R, Morbidelli A (2023) An inflationary disk phase to explain extended protoplanetary dust disks. Astron Astrophys 677:A136. https://doi.org/10.1051/0004-6361/202346616. arXiv:2307.01249 [astro-ph.EP]

Mayer L, Quinn T, Wadsley J, et al (2002) Formation of giant planets by fragmentation of protoplanetary disks. Science 298(5599):1756–1759. https://doi.org/10.1126/science.1077635. arXiv:astro-ph/0301088 [astro-ph]

McCoy TJ, Mittlefehldt DW, Wilson L (2006) Asteroid differentiation. In: Lauretta DS, McSween HY (eds) Meteorites and the Early Solar System II. University of Arizona Press, Tucson, pp 733–745

McCoy T, Lunning N, Corrigan CM (2018) Experimental investigation of core formation in the CV-CK chondrite parent body. In: AGU Fall Meeting abstracts, P52B–06

McKeegan KD, Kallio APA, Heber VS, et al (2011) The oxygen isotopic composition of the Sun inferred from captured solar wind. Science 332(6037):1528. https://doi.org/10.1126/science.1204636

Miller E, Marino S, Stammler SM, et al (2021) The formation of wide exoKuiper belts from migrating dust traps. Mon Not R Astron Soc 508(4):5638–5656. https://doi.org/10.1093/mnras/stab2935. arXiv:2110.04007 [astro-ph.EP]

Misawa K, Nakamura N (1988) Demonstration of REE fractionation among individual chondrules from the Allende (CV3) chondrite. Geochim Cosmochim Acta 52:1699–1710. https://doi.org/10.1016/0016-7037(88)90238-4

Morbidelli A (2020) Planet formation by pebble accretion in ringed disks. Astron Astrophys 638:A1. https://doi.org/10.1051/0004-6361/202037983. arXiv:2004.04942 [astro-ph.EP]

Morbidelli A, Baillié K, Batygin K, et al (2022) Contemporary formation of early Solar System planetesimals at two distinct radial locations. Nat Astron 6:72–79. https://doi.org/10.1038/s41550-021-01517-7. arXiv:2112.15413 [astro-ph.EP]

Morin GLF, Marrocchi Y, Villeneuve J, et al (2022) ^{16}O-rich anhydrous silicates in CI chondrites: implications for the nature and dynamics of dust in the solar accretion disk. Geochim Cosmochim Acta 332:203–219. https://doi.org/10.1016/j.gca.2022.06.017

Nakamura T, Noguchi T, Tanaka M, et al (2011) Itokawa dust particles: a direct link between S-type asteroids and ordinary chondrites. Science 333:1113–1116. https://doi.org/10.1126/science.1207758

Nakamura T, Matsumoto M, Amano K, et al (2023) Formation and evolution of carbonaceous asteroid Ryugu: direct evidence from returned samples. Science 379(6634):abn8671. https://doi.org/10.1126/science.abn8671

Nanne JAM, Nimmo F, Cuzzi JN, et al (2019) Origin of the non-carbonaceous-carbonaceous meteorite dichotomy. Earth Planet Sci Lett 511:44–54. https://doi.org/10.1016/j.epsl.2019.01.027

Nesvorný D, Li R, Youdin AN, et al (2019) Trans-Neptunian binaries as evidence for planetesimal formation by the streaming instability. Nat Astron 3:808–812. https://doi.org/10.1038/s41550-019-0806-z. arXiv:1906.11344 [astro-ph.EP]

Nesvorný D, Li R, Simon JB, et al (2021) Binary planetesimal formation from gravitationally collapsing pebble clouds. Planet Sci J 2(1):27. https://doi.org/10.3847/PSJ/abd858. arXiv:2011.07042 [astro-ph.EP]

Niemeyer S (1988) Isotopic diversity in nebular dust - the distribution of TI isotopic anomalies in carbonaceous chondrites. Geochim Cosmochim Acta 52:2941–2954. https://doi.org/10.1016/0016-7037(88)90159-7

Okuzumi S, Tanaka H, Kobayashi H, et al (2012) Rapid coagulation of porous dust aggregates outside the snow line: a pathway to successful icy planetesimal formation. Astrophys J 752:106. https://doi.org/10.1088/0004-637X/752/2/106. arXiv:1204.5035 [astro-ph.EP]

Ormel CW, Klahr HH (2010) The effect of gas drag on the growth of protoplanets. Analytical expressions for the accretion of small bodies in laminar disks. Astron Astrophys 520:A43. https://doi.org/10.1051/0004-6361/201014903. arXiv:1007.0916 [astro-ph.EP]

Owen JE, Jacquet E (2015) Astro- and cosmochemical consequences of accretion bursts - I. The D/H ratio of water. Mon Not R Astron Soc 446:3285–3296. https://doi.org/10.1093/mnras/stu2254. arXiv:1410.6844 [astro-ph.SR]

Pedersen SG, Schiller M, Connelly JN, et al (2019) Testing accretion mechanisms of the H chondrite parent body utilizing nucleosynthetic anomalies. Meteorit Planet Sci 54(6):1215–1227. https://doi.org/10.1111/maps.13269

Pignatale FC, Charnoz S, Chaussidon M, et al (2018) Making the planetary material diversity during the early assembling of the Solar System. Astrophys J Lett 867:L23. https://doi.org/10.3847/2041-8213/aaeb22. arXiv:1810.10827 [astro-ph.EP]

Pinilla P, Birnstiel T, Ricci L, et al (2012) Trapping dust particles in the outer regions of protoplanetary disks. Astron Astrophys 538:A114. https://doi.org/10.1051/0004-6361/201118204. arXiv:1112.2349 [astro-ph.EP]

Pinto GA, Marrocchi Y, Morbidelli A, et al (2021) Constraints on planetesimal accretion inferred from particle-size distribution in CO chondrites. Astrophys J Lett 917(2):L25. https://doi.org/10.3847/2041-8213/ac17f2. arXiv:2107.12658 [astro-ph.EP]

Pinto GA, Marrocchi Y, Jacquet E, et al (2022) Formation of chondrule fine-grained rims from local nebular reservoirs. Meteorit Planet Sci 57(5):1004–1017. https://doi.org/10.1111/maps.13812

Piralla M, Villeneuve J, Batanova V, et al (2021) Conditions of chondrule formation in ordinary chondrites. Geochim Cosmochim Acta 313:295–312. https://doi.org/10.1016/j.gca.2021.08.007

Piralla M, Villeneuve J, Schnuriger N, et al (2023) A unified chronology of dust formation in the early Solar System. Icarus 394:115427. https://doi.org/10.1016/j.icarus.2023.115427

Polak B, Klahr H (2023) High-resolution study of planetesimal formation by gravitational collapse of pebble clouds. Astrophys J 943(2):125. https://doi.org/10.3847/1538-4357/aca58f. arXiv:2211.13318 [astro-ph.EP]

Pollack JB, Hubickyj O, Bodenheimer P, et al (1996) Formation of the giant planets by concurrent accretion of solids and gas. Icarus 124:62–85. https://doi.org/10.1006/icar.1996.0190

Prior GT (1920) The classification of meteorites. Mineral Mag 19:51–63

Przylibski TA, Długosz-Lisiecka M, Łuszczek K, et al (2023) Results of statistical analysis of primary components concentrations in bulk chemical composition of ordinary chondrite groups. Meteorit Planet Sci 58(9):1246–1259. https://doi.org/10.1111/maps.14050

Raettig N, Klahr H, Lyra W (2015) Particle trapping and streaming instability in vortices in protoplanetary disks. Astrophys J 804(1):35. https://doi.org/10.1088/0004-637X/804/1/35. arXiv:1501.05364 [astro-ph.EP]

Riols A, Lesur G, Menard F (2020) Ring formation and dust dynamics in wind-driven protoplanetary discs: global simulations. Astron Astrophys 639:A95. https://doi.org/10.1051/0004-6361/201937418. arXiv:2006.01194 [astro-ph.EP]

Russell SS, Connolly HCJ, Krot AN (eds) (2018) Chondrules: record of protoplanetary disk processes Cambridge University Press, Cambridge

Sanders IS, Scott ERD (2012) The origin of chondrules and chondrites: debris from low-velocity impacts between Molten planetesimals? Meteorit Planet Sci 47:2170–2192. https://doi.org/10.1111/maps.12002

Schiller M, Bizzarro M, Fernandes VA (2018) Isotopic evolution of the protoplanetary disk and the building blocks of Earth and the Moon. Nature 555:507–510. https://doi.org/10.1038/nature25990

Schneider JM, Burkhardt C, Marrocchi Y, et al (2020) Early evolution of the solar accretion disk inferred from Cr-Ti-O isotopes in individual chondrules. Earth Planet Sci Lett 551:116585. https://doi.org/10.1016/j.epsl.2020.116585. arXiv:2009.08684 [astro-ph.EP]

Schönbächler M, et al (in prep) The timing of planetesimal formation. Space Sci Rev

Schoonenberg D, Ormel CW (2017) Planetesimal formation near the snowline: in or out? Astron Astrophys 602:A21. https://doi.org/10.1051/0004-6361/201630013. arXiv:1702.02151 [astro-ph.EP]

Scott ERD, Krot AN (2014) Chondrites and their components. In: Davis AM (ed) Meteorites and cosmochemical processes, pp 65–137

Scott ERD, Keil K, Goldstein JI, et al (2015) In: Early impact history and dynamical origin of differentiated meteorites and asteroids, pp 573–595. https://doi.org/10.2458/azu_uapress_9780816532131-ch030

Sears DWG (2016) The CO chondrites: major recent Antarctic finds, their thermal and radiation history, and describing the metamorphic history of members of the class. Geochim Cosmochim Acta 188:106–124. https://doi.org/10.1016/j.gca.2016.05.033

Sears DW, Grossman JN, Melcher CL, et al (1980) Measuring metamorphic history of unequilibrated ordinary chondrites. Nature 287:791–795. https://doi.org/10.1038/287791a0

Shakura NI, Sunyaev RA (1973) Black holes in binary systems. Observational appearance. Astron Astrophys 24:337–355

Shibaike Y, Alibert Y (2020) Planetesimal formation at the gas pressure bump following a migrating planet. I. Basic characteristics of the new formation model. Astron Astrophys 644:A81. https://doi.org/10.1051/0004-6361/202039086. arXiv:2010.10594 [astro-ph.EP]

Shu FH, Shang H, Lee T (1996) Toward an astrophysical theory of chondrites. Science 271:1545–1552. https://doi.org/10.1126/science.271.5255.1545

Simon JB, Armitage PJ, Li R, et al (2016) The mass and size distribution of planetesimals formed by the streaming instability. I. The role of self-gravity. Astrophys J 822(1):55. https://doi.org/10.3847/0004-637X/822/1/55. arXiv:1512.00009 [astro-ph.SR]

Simon JB, Blum J, Birnstiel T, et al (2022) Comets and Planetesimal Formation. https://doi.org/10.48550/arXiv.2212.04509. arXiv:2212.04509 [astro-ph.EP]

Sossi PA Hin R Kleine T, et al (in prep) The history of volatile and moderately volatile elements in the Earth and terrestrial planets. Space Sci Rev

Spitzer F, Burkhardt C, Nimmo F, et al (2021) Nucleosynthetic Pt isotope anomalies and the Hf-W chronology of core formation in inner and outer Solar System planetesimals. Earth Planet Sci Lett 576:117211. https://doi.org/10.1016/j.epsl.2021.117211. arXiv:2109.15250 [astro-ph.EP]

Squire J, Hopkins PF (2020) Physical models of streaming instabilities in protoplanetary discs. Mon Not R Astron Soc 498(1):1239–1251. https://doi.org/10.1093/mnras/staa2311. arXiv:2003.01738 [astro-ph.EP]

Stammler SM, Drążkowska J, Birnstiel T, et al (2019) The DSHARP rings: evidence of ongoing planetesimal formation? Astrophys J Lett 884(1):L5. https://doi.org/10.3847/2041-8213/ab4423. arXiv:1909.04674 [astro-ph.EP]

Stammler SM, Lichtenberg T, Drążkowska J, et al (2023) Leaky dust traps: how fragmentation impacts dust filtering by planets. Astron Astrophys 670:L5. https://doi.org/10.1051/0004-6361/202245512. arXiv:2301.05505 [astro-ph.EP]

Stephant A, Carli C, Anand M, et al (2023) Tissemouminites: a new group of primitive achondrites spanning the transition between acapulcoites and winonaites. Meteorit Planet Sci 58(1):111–134. https://doi.org/10.1111/maps.13944

Stewart ST, Carter PJ, Lock SJ, et al (2024) The formation of chondritic mixtures by vaporizing collisions between planetesimals. In: LPI contributions, p 2449

Sugiura N, Fujiya W (2014) Correlated accretion ages and ϵ^{54}Cr of meteorite parent bodies and the evolution of the solar nebula. Meteorit Planet Sci 49(5):772–787. https://doi.org/10.1111/maps.12292

Sugiura N, Krot AN (2007) 26Al-26Mg systematics of Ca-Al-rich inclusions, amoeboid olivine aggregates, and chondrules from the ungrouped carbonaceous chondrite Acfer 094. Meteorit Planet Sci 42:1183–1195. https://doi.org/10.1111/j.1945-5100.2007.tb00568.x

Suriano SS, Li ZY, Krasnopolsky R, et al (2018) The formation of rings and gaps in magnetically coupled disc-wind systems: ambipolar diffusion and reconnection. Mon Not R Astron Soc 477(1):1239–1257. https://doi.org/10.1093/mnras/sty717. arXiv:1712.06217 [astro-ph.SR]

Tenner TJ, Nakashima D, Ushikubo T, et al (2015) Oxygen isotope ratios of FeO-poor chondrules in CR3 chondrites: influence of dust enrichment and H_2O during chondrule formation. Geochim Cosmochim Acta 148:228–250. https://doi.org/10.1016/j.gca.2014.09.025

Tissot FLH, Dauphas N, Grove TL (2017) Distinct $^{238}U/^{235}U$ ratios and REE patterns in plutonic and volcanic angrites: geochronologic implications and evidence for U isotope fractionation during magmatic processes. Geochim Cosmochim Acta 213:593–617. https://doi.org/10.1016/j.gca.2017.06.045

Tissot FLH, Burkhardt C, Kuznetsova A, et al (2024) Infall and disk processes: the message from meteorites. Space Sci Rev 220

Tomkins AG, Johnson TE, Mitchell JT (2020) A review of the chondrite-achondrite transition, and a metamorphic facies series for equilibrated primitive stony meteorites. Meteorit Planet Sci 55(4):857–885. https://doi.org/10.1111/maps.13472

Tschermak G (1885) Die mikroskopische Beschaffenheit der Meteoriten, erlaütert durch photographische Abbildungen. Schweitzerbart, E

Tsukamoto Y, Machida MN, Si I (2021) “Ashfall” induced by molecular outflow in protostar evolution. Astrophys J Lett 920(2):L35. https://doi.org/10.3847/2041-8213/ac2b2f. arXiv:2109.12779 [astro-ph.SR]

Urey HC, Craig H (1953) The composition of the stone meteorites and the origin of the meteorites. Geochim Cosmochim Acta 4:36–82. https://doi.org/10.1016/0016-7037(53)90064-7

Van Kooten EMME, Wielandt D, Schiller M, et al (2016) Isotopic evidence for primordial molecular cloud material in metal-rich carbonaceous chondrites. Proc Natl Acad Sci 113(8):2011–2016

van Kooten EMME, Brearley A, Ebel DS, et al (2024) Is there a genetic relationship between chondrules and matrix? Space Sci Rev 220

van Schmus WR, Wood JA (1967) A chemical-petrologic classification for the chondritic meteorites. Geochim Cosmochim Acta 31:747–747. https://doi.org/10.1016/S0016-7037(67)80030-9

Villeneuve J, Chaussidon M, Libourel G (2009) Homogeneous distribution of aluminum-26 in the Solar System from the magnesium isotopic composition of chondrules. Science 325:985–988

Visser RG, Drążkowska J, Dominik C (2021) The radial structure of planetary bodies formed by the streaming instability. Astron Astrophys 647:A126. https://doi.org/10.1051/0004-6361/202039769. arXiv:2101.09209 [astro-ph.EP]

Vorobyov EI, Akimkin V, Stoyanovskaya O, et al (2018) Early evolution of viscous and self-gravitating circumstellar disks with a dust component. Astron Astrophys 614:A98. https://doi.org/10.1051/0004-6361/201731690. arXiv:1801.06898 [astro-ph.EP]
Wahlberg Jansson K, Johansen A (2017) Radially resolved simulations of collapsing pebble clouds in protoplanetary discs. Mon Not R Astron Soc 469:S149–S157. https://doi.org/10.1093/mnras/stx1470. arXiv:1706.03655 [astro-ph.EP]
Walsh KJ, Morbidelli A, Raymond SN, et al (2011) A low mass for Mars from Jupiter's early gas-driven migration. Nature 475:206–209. https://doi.org/10.1038/nature10201
Warren PH (2011) Stable-isotopic anomalies and the accretionary assemblage of the Earth and Mars: a subordinate role for carbonaceous chondrites. Earth Planet Sci Lett 311:93–100. https://doi.org/10.1016/j.epsl.2011.08.047
Wasson JT (1974). Meteorites: classification and properties. Minerals and rocks, vol 10. Springer, Berlin. https://doi.org/10.1007/978-3-642-65863-1
Wasson JT, Chou C (1974) Fractionation of moderately volatile elements in ordinary chondrites. Meteoritics 9:69–84
Weidenschilling SJ (1980) Dust to planetesimals - settling and coagulation in the solar nebula. Icarus 44:172–189. https://doi.org/10.1016/0019-1035(80)90064-0
Weiss BP, Bai XN, Fu RR (2021) History of the solar nebula from meteorite paleomagnetism. Sci Adv 7(1):eaba5967. https://doi.org/10.1126/sciadv.aba5967. arXiv:2103.02011 [astro-ph.EP]
Williams JP, Cieza LA (2011) Protoplanetary disks and their evolution. Annu Rev Astron Astrophys 49:67–117. https://doi.org/10.1146/annurev-astro-081710-102548. arXiv:1103.0556 [astro-ph.GA]
Yang L, Ciesla FJ (2012) The effects of disk building on the distributions of refractory materials in the solar nebula. Meteorit Planet Sci 47:99–119. https://doi.org/10.1111/j.1945-5100.2011.01315.x
Yang CC, Johansen A, Carrera D (2017) Concentrating small particles in protoplanetary disks through the streaming instability. Astron Astrophys 606:A80. https://doi.org/10.1051/0004-6361/201630106. arXiv:1611.07014 [astro-ph.EP]
Youdin AN, Goodman J (2005) Streaming instabilities in protoplanetary disks. Astrophys J 620:459–469. https://doi.org/10.1086/426895. arXiv:astro-ph/0409263
Young ED, Kuramoto K, Marcus RA, et al (2008) Mass-independent oxygen isotope variation in the solar nebula. In: MacPherson GJ (ed) Oxygen in the Solar System. Reviews in Mineralogy & Geochemistry, vol 68. Mineralogical Society of America, pp 187–218. https://doi.org/10.1515/9781501508509-009
Zanda B, Bourot-Denise M, Perron C, et al (1994) Origin and metamorphic redistribution of silicon, chromium, and phosphorus in the metal of chondrites. Science 265:1846–1849. https://doi.org/10.1126/science.265.5180.1846
Zanda B, Hewins RH, Bourot-Denise M, et al (2006) Formation of solar nebula reservoirs by mixing chondritic components. Earth Planet Sci Lett 248:650–660. https://doi.org/10.1016/j.epsl.2006.05.016
Zanetta PM, Leroux H, Le Guillou C, et al (2021) Nebular thermal processing of accretionary fine-grained rims in the Paris CM chondrite. Geochim Cosmochim Acta 295:135–154. https://doi.org/10.1016/j.gca.2020.12.015
Ziampras A, Kley W, Dullemond CP (2020) Importance of radiative effects in gap opening by planets in protoplanetary disks. Astron Astrophys 637:A50. https://doi.org/10.1051/0004-6361/201937048. arXiv:2003.02298 [astro-ph.EP]
Zinner E (2014) Presolar grains. In: Davis AM (ed) Treatise on geochemistry. Meteorites and cosmochemical processes vol 1. Elsevier, Oxford, pp 181–213. https://doi.org/10.1016/B978-0-08-095975-7.00101-7

Space Science Reviews (2024) 220:76
https://doi.org/10.1007/s11214-024-01113-x

The Sun's Birth Environment: Context for Meteoritics

Steve Desch[1] · **Núria Miret-Roig**[2]

Received: 30 April 2024 / Accepted: 12 September 2024 / Published online: 26 September 2024

Abstract
Meteorites trace planet formation in the Sun's protoplanetary disk, but they also record the influence of the Sun's birth environment. Whether the Sun formed in a region like Taurus-Auriga with $\sim 10^2$ stars, or a region like the Carina Nebula with $\sim 10^6$ stars, matters for how large the Sun's disk was, for how long and from how far away it accreted gas from the molecular cloud, and how it acquired radionuclides like ^{26}Al. To provide context for the interpretation of meteoritic data, we review what is known about the Sun's birth environment. Based on an inferred gas disk outer radius $\approx 50-90$ AU, radial transport in the disk, and the abundances of noble gases in Jupiter's atmosphere, the Sun's molecular cloud and protoplanetary disk were exposed to an ultraviolet flux $G_0 \sim 30-3000$ during its birth and first ≈ 10 Myr of evolution. Based on the orbits of Kuiper Belt objects, the Solar System was subsequently exposed to a stellar density $\approx 100\,M_\odot\,\mathrm{pc}^{-3}$ for ≈ 100 Myr, strongly implying formation in a bound cluster. These facts suggest formation in a region like the outskirts of the Orion Nebula, perhaps 2 pc from the center. The protoplanetary disk might have accreted gas for many Myr, but a few $\times 10^5$ yr seems more likely. It probably inherited radionuclides from its molecular cloud, enriched by inputs from supernovae and especially Wolf-Rayet star winds, and acquired a typical amount of ^{26}Al.

Keywords Stars: formation · Stars: Wolf-Rayet · Photon-dominated region (PDR) · Protoplanetary disks

1 Introduction

Today the Sun and the planets in their orbits evolve slowly, due to the internal workings of the Solar System, isolated from the goings-on in the Galaxy. Not so at its birth, when the effects of the early Sun's surroundings determined how the Solar System would evolve for the next billions of years. All new solar systems are greatly affected by their birth environments,

✉ S. Desch
steve.desch@asu.edu

✉ N. Miret-Roig
nuria.miret.roig@univie.ac.at

1 School of Earth and Space Exploration, Arizona State University, PO Box 876004, Tempe, AZ 85287-6004, USA

2 Department of Astrophysics, University of Vienna, Türkenschanzstrasse 17 (Sternwarte), 1180 Wien, Austria

especially the number of stars, N, in their birth clusters, which can range from $N \sim 10^2$ stars in low-mass star-forming regions like Taurus-Auriga and Lupus, to $\sim 10^3$ stars in high-mass star-forming regions like Upper Scorpius (Upper Sco), to $\sim 10^4$ stars in regions like the Orion Nebula Cluster (ONC), to $\sim 10^6$ stars in regions like the Carina Nebula. A major goal of Solar System studies is to determine the number of stars in the Sun's birth environment, and the effects those stars had on its future evolution.

About as many stars form in regions with $N \sim 10^2$ as in clusters with $N \sim 10^4$ or $N \sim 10^6$, but their formation and evolution will proceed very differently, because of the different stellar densities and especially ultraviolet (UV) fluxes (Lada and Lada 2003; Hester and Desch 2005; Adams 2010). Clusters with large N have a greater likelihood of randomly generating high-mass O and B stars, which will emit tremendous amounts of UV that can ionize and heat the gas, creating a so-called H II region. A single O6 star like θ^1 Ori C in the ONC emits 5×10^5 times the luminosity of the Sun, half of that energy in the UV. The ratio of the UV flux to the background UV flux in the Galaxy ($1.6 \times 10^{-3}\,\mathrm{erg\,cm^{-2}\,s^{-1}}$) is denoted G_0. Regions like Taurus-Auriga have $N \sim 10^2$, lack OB stars and are characterized by $G_0 \sim 1$ and stellar densities $\rho_\star \sim 1\,M_\odot\,\mathrm{pc}^{-3}$ (Güdel et al. 2007). Regions like Upper Sco have $N \sim 10^3$, dozens of B stars and $G_0 \sim 10^3$, and $\rho_\star \sim 1\,M_\odot\,\mathrm{pc}^{-3}$ (Luhman 2020). The adjacent ρ Ophiuchi (ρ Oph) region has $N \sim 10^3$ stars, including 2 B stars, $G_0 = 57^{+202}_{-31}$ and $\rho_\star \sim 13^{+25}_{-6}\,M_\odot\,\mathrm{pc}^{-3}$ (Parker et al. 2012). The ONC, with θ^1 Ori C, a few O stars, and a dozen B stars, sees $G_0 \sim 5000\,(r/1\,\mathrm{pc})^{-2}$, where r is the distance from the cluster center (θ^1 Ori C) (Störzer and Hollenbach 1999). The stellar density is $3.3 \times 10^4\,M_\odot\,\mathrm{pc}^{-3}$ at the cluster center, and the average density within 2.06 pc is $49\,M_\odot\,\mathrm{pc}^{-3}$ (Hartmann et al. 1998). And the most extreme locales like the Trumpler 14 subregion of the $N \sim 10^6$ Carina Nebula by themselves have $N \sim 10^3 - 10^4$, dozens of O stars, $G_0 \sim 10^6$, and $\rho_\star > 10^3\,M_\odot\,\mathrm{pc}^{-3}$ (Mesa-Delgado et al. 2016). A priori, the Sun is equally likely to have formed in any of these environments (Adams 2010).

The outcomes of formation in these different environments are quite divergent, starting with the sizes of protoplanetary disks. As reviewed by Marchington and Parker (2022), these are greatly affected by external photoevaporation by the UV flux of nearby massive stars. In low-mass star-forming regions like the ~ 6 Myr-old Lupus cloud, the median radius of dust emission is about 100 AU. In the similarly old (3–10 Myr; Ratzenböck et al. 2023; Miret-Roig et al. 2022) Upper Sco region, the median dust radius is 40 AU; but it is only 15 AU for disks ('proplyds') in the 2.5 Myr-old ONC. In the extreme environment of Trumpler 14, no disks > 3 AU in radius are observed, despite its young age (few $\times 10^5$ yr); planet formation is effectively shut off in these systems (Mesa-Delgado et al. 2016).

The same UV flux photoevaporates the molecular gas that would feed protostars and their disks. In low-mass star-forming regions, even older Class II (few Myr old) protostars and their disks remain embedded in their natal molecular clouds and may continue to be fed by streamers of gas, as are AB Aur, GM Aur, SU Aur, and DG Tau (Pineda et al. 2023). The proplyds in the ONC or Trumpler 14 H II regions are completely isolated from their molecular clouds in $< 10^5$ yr (Mesa-Delgado et al. 2016).

It is not clear how quickly accretion is shut off to disks in regions of intermediate mass like Upper Sco or in the Orion Nebula far from the cluster center. In all H II regions, massive stars launch ionization fronts that propagate outward leaving in their wake ionized and heated gas that then leaves the cluster. A very massive O star can clear out an entire parsecs-wide cluster in as little as $\sim 10^5$ yr after its formation (Wall et al. 2020). What follows after that is a D-type ionization front in which a shock front propagates into the molecular cloud at speeds of a few km/s (few pc/Myr), compressing it; several tenths of a parsec behind the shock front, an ionization eats into the compressed gas, heating and ionizing it. The duration of accretion onto a protostar depends on the relative timing between its formation and when these ionization fronts pass it. If the shock preceding a D-type ionization front triggered the formation of the protostar, the relative timing is short, $\sim 10^5$ yr (Hester and Desch 2005). Theoretical models (Dale et al. 2007; Wall et al. 2020) and observations (Snider et al. 2009; Karr and Martin 2003) suggest a substantial fraction ($\approx 25\%$) of stars in H II regions were triggered to form in this way. But the more-abundant, spontaneously-formed stars several pc from the cluster center may have to wait many Myr to be uncovered.

Models of planet formation and the interpretation of meteoritic data require pinning down which of these birth environments the Sun formed in. For example, it has been suggested that planet-forming pebbles drifted in from > 100 AU (Lambrechts and Johansen 2014), which is only possible in large disks. Likewise, it has been suggested that the isotopic dichotomy in the Solar System (Trinquier et al. 2009; Warren 2011; Kruijer et al. 2017) is due to late infall of material from the molecular cloud (Liu et al. 2022), possibly along streamers (Arzoumanian et al. 2023), which would require small G_0. The Solar System started with live ^{26}Al that was the cause of asteroidal melting (Grimm and McSween 1993; Lichtenberg et al. 2019). It is commonly (but not universally, see e.g., Jura et al. 2013; Young 2014; Desch et al. 2023) assumed that solar systems born with sufficient ^{26}Al to melt asteroids might be rare (Ciesla et al. 2015; Lichtenberg et al. 2019) unless they formed in an especially large birth cluster (Williams and Gaidos 2007). Until recently, our intuition of how stars form was informed by studies of low-mass star-forming regions like Taurus-Auriga, because this is the closest (140 pc), most easily-observed star-forming region (cf. Hester and Desch 2005), and these starting assumptions would be reasonable for such an environment. But it has been increasingly recognized that stars like the Sun have a high probability of forming in high-mass star-forming regions (Lada and Lada 2003; Adams 2010), and the signatures of such an origin have been sought in the Solar System.

To provide the proper context for disk and planet formation models and interpretation of meteoritic data, we are compelled to ask:

- How large was the Sun's protoplanetary disk?
- How long did infall from the molecular cloud last?
- Was all this material local to the Sun's cloud core (and thus homogeneous), or brought in from a distance?
- How did the Solar System acquire ^{26}Al and other radionuclides, and is the amount it acquired common?
- **Did the Sun form in a high-mass or low-mass star-forming region?**

To answer these questions, below we review astronomical observations, astrophysical modeling, and planetary and meteoritic data. We conclude that the Sun formed in a high-mass star-forming region with $N \sim 10^3 - 10^4$ stars. Its disk of solids late (> 5 Myr) in its evolution is inferred to have been ≈ 50 AU, and its gas disk $\approx 50 - 90$ AU in radius. More likely than not, it was cut off from accretion from the molecular cloud within a few $\times 10^5$ yr of its birth, but it cannot be ruled out that it accreted gas for a few Myr, although it is not clear that gas would be compositionally different. The amount of ^{26}Al it received appears typical for stars formed in a high-mass star-forming region, acquired from a molecular cloud enriched over the previous tens of Myr by massive stars. The outskirts of the Orion Nebula, more than 2 pc from the center, appear to be a good analog environment.

2 Evidence the Sun Formed Near Massive Stars

2.1 Astronomical Evidence

Planetary and astronomical observations suggest that the Sun itself formed in a region with high stellar density and high UV flux, both signposts of formation in a high-mass star-forming region.

2.1.1 Structure of the Kuiper Belt

The strongest evidence comes from the architecture of the Kuiper Belt, especially the population of 'detached' Kuiper Belt Objects (KBOs). One of the major populations of KBOs is scattered objects, with large semi-major axes a and perihelia $q \approx 30$ AU. These were evidently scattered by Neptune as it migrated outward. The timing of Neptune's migration is constrained to be ≈ 60 Myr, by the sequence of events that led to the formation of the dwarf planet Haumea and its dynamical family (Noviello et al. 2022). It is also very likely that Neptune's migration was due to the outer Solar System dynamical ('Nice') instability (Tsiganis et al. 2005), which simulations of Solar System analogs suggest took place most likely at 37-62 Myr (de Sousa et al. 2020). It is also likely that the Nice instability led to the Giant Impact that formed the Earth-Moon system, at about 60 Myr (Barboni et al. 2017). A subset of about 200 of these objects scattered by Neptune have $a > 50$ AU and detached perihelia 35 AU $< q <$ 50 AU that require them to have been gravitationally perturbed after they were scattered (Kenyon and Bromley 2004; Morbidelli and Levison 2004; Brasser et al. 2006). Sedna, with $a = 506$ AU and $q = 76$ AU, is a similar but more extreme example. Their radial distribution has been successfully modeled only by a stellar encounter; the one simulation that matched observations involved a $0.17\,M_\odot$ star passing 175 AU from the Sun, after Neptune's migration. This would be improbable unless the Sun formed in a dense cluster.

For the stellar encounter creating the detached KBOs to be probable requires the product of stellar density, η, and time exposed to this density, T, to be $\eta T > 1 \times 10^4\,\mathrm{pc}^{-3}$ Myr. Assuming the encounter took place in the first 100 Myr of Solar System evolution, but after Neptune's migration at ≈ 50 Myr, a lower limit $\eta > 200\,\mathrm{pc}^{-3}$ is inferred. (An interval of tens of Myr, comparable to the timescale of Neptune's migration, also is necessary to create sufficient numbers of detached scattered KBOs; Nesvorný et al. 2023). An upper limit $\eta T < 3 \times 10^4\,\mathrm{pc}^{-3}$ Myr can be placed from the relatively low inclinations of dynamically cold KBOs (Batygin et al. 2020). Assuming here $T \approx 100$ Myr, $\eta < 300\,\mathrm{pc}^{-3}$ is inferred. These limits are loose, based on stochastic probabilities, but suggest a stellar density $\approx 250\,\mathrm{pc}^{-3}$, or $\approx 100\,M_\odot\,\mathrm{pc}^{-3}$ (assuming an average stellar mass $0.4\,M_\odot$; Kroupa 2001). Just as importantly, this high stellar density must have lasted for up to 100 Myr (tens of Myr after Neptune's migration) to explain the Kuiper Belt.

Therefore, it is essential that the Sun formed in a cluster with a minimum number of sibling stars and a sufficient density to remain gravitationally bound for many tens of Myr, which is quite diagnostic. For example, Forbes et al. (2021) have suggested the ρ Ophiuchi is a good analog to the Solar System. We also find that the UV flux and stellar density there are good matches to what we infer for the early Solar System, and stars in that region are likely to acquire live ^{26}Al; but the cluster is not likely to stay bound. The ρ Ophiuchi cluster has about 500 stars and a stellar density $\approx 90\,M_\odot\,\mathrm{pc}^{-3}$ (Ratzenböck et al. 2023; Miret-Roig et al. 2022; Evans et al. 2009); but integrating the trajectories of the stars in a Galactic potential, these will disperse to densities $< 12\,M_\odot\,\mathrm{pc}^{-3}$ in 10 Myr, and $< 1\,M_\odot\,\mathrm{pc}^{-3}$ in 100

Myr (Miret-Roig et al., in prep.). The vast majority of stars similarly disperse from their birth clusters in < 10 Myr, and only a small fraction $\approx 10\%$ remain gravitationally bound to their siblings for longer times (Lada and Lada 2003; Adams 2010). What determines whether a bound, open cluster forms is not clear, but observational evidence suggests the mechanism works most efficiently in clusters with roughly $N > 2000$ stars (Lada and Lada 2003; Hao et al. 2023) to $N > 10^4$ stars (Adams 2010). The Orion Nebula, with 2200 stars within 2 pc (Megeath et al. 2012), is generally considered likely to form a bound cluster, but just marginally so (Hartmann et al. 1998; Kroupa et al. 2001).

2.1.2 Truncation of the Disk

A second line of evidence for the Sun's formation in a high-mass star-forming region comes from the inferred size of the protoplanetary disk. The orbits of cold classical KBOs, with low orbital inclinations and eccentricities that reflect formation in the disk, extend only out to semi-major axes ≈ 47 AU (Jewitt et al. 1998; Trujillo and Brown 2001), strongly suggesting KBOs did not form beyond this distance. Because the growth of KBOs is attributed to streaming instability (McKinnon et al. 2020) and streaming instability effectively only concentrates particles with Stokes numbers $> 10^{-3}$ (Li and Youdin 2021), this indicates a probable lack of St $> 10^{-3}$ particles (i.e., particles larger than tens of μm in size) beyond about 47 AU.

Two ideas exist to explain why protoplanetary disks might be truncated. One idea is tidal truncation of both gas and solids in the disk, due to the gravitational effects of a star passing at about 150 AU (Ida et al. 2000). The second is that external photoevaporation by UV radiation with $G_0 \sim 10^3$ also could truncate the disk at 50-100 AU (Hollenbach and Adams 2004). Both mechanisms would demand formation in a high-mass star-forming region. For a given stellar density, studies show that disks are much more likely to be truncated by UV photoevaporation from massive stars than by dynamical interaction with passing stars (Winter et al. 2018).

Although KBOs trace the primordial distribution of solid matter only, the gas density also can be inferred, with the recognition that gas and particles are decoupled. In recent years, it has been understood that the gas disk extended beyond 50 AU, but that large particles had drifted radially inward due to aerodynamic drag by the time KBOs formed. For example, using standard formulas (Desch et al. 2017), and assuming a sound speed $c_s \approx 0.3\,\mathrm{km\,s^{-1}}$ and typical disk properties, only particles with Stokes numbers St $< 10^{-3}$ would have drift speeds $v_{r,\mathrm{d}} < 5\,\mathrm{cm\,s^{-1}}$ and avoid spiralling inward tens of AU within 5 Myr, the typical timescale for formation of KBOs (Bierson and Nimmo 2019). This implies much reduced masses of submillimeter particles beyond a radius very close to the observed 47 AU. This would make the solar nebula consistent with other evolved protoplanetary disks, which are commonly observed to have gas emission that extends up to a factor of 2 larger radii than millimeter emission from solids (Ansdell et al. 2018). The Sun's gas disk probably extended out to at least 50 AU, and possibly closer to 90 AU.

The size of the Sun's protoplanetary disk argues for formation in an intermediate-mass star-forming region. Marchington and Parker (2022) compiled literature data about disk sizes and masses, and G_0, across various star-forming regions, finding disks are generally smaller in more massive star-forming regions, largely due to external photoevaporation. In low-mass star-forming regions, G_0 is low and disks are large: in the several Myr-old Taurus-Auriga region, $G_0 \sim 1$ and the median radius of dust emission, r_d, is about 50 AU; and in the 6 Myr-old (Luhman 2020) Lupus region, $G_0 \sim 4$ (Galli et al. 2020; Cleeves et al. 2016), and the median $r_d \approx 100$ AU (Tazzari et al. 2017; Ansdell et al. 2018). In the 3–10 Myr-old

(Ratzenböck et al. 2023; Miret-Roig et al. 2022) Upper Sco region, and in the adjacent ρ Oph cloud, $G_0 \approx 50$ (Trapman et al. 2020; Luhman 2020; Parker et al. 2012) and the median $r_d \approx 60$ AU. Within the central parsec or so of the 2.5 Myr-old ONC (Jeffries et al. 2011; Da Rio et al. 2010), $G_0 \sim 10^4$ (Hartmann et al. 1998) and the median $r_d \approx 15$ AU. Finally, the majority of disks in the Trumpler 14 cluster in the Carina Nebula, which see $G_0 \sim 10^6$, have $r_d < 3$ AU (Mesa-Delgado et al. 2016). Formation in the Carina Nebula is probably excluded, as is formation in the central parts of the Orion Nebula, where $< 5\%$ of dust disks are as large as 47 AU (Marchington and Parker 2022). The Sun's disk size is more compatible with environments with lower G_0, like Lupus or Taurus-Auriga, or Upper Sco, where disks see a range of G_0 from 5 to 500 (Trapman et al. 2020), or parts of the Orion Nebula > 2 pc from the center. The upper limit on G_0 is $> 10^3$ but $< 10^4$.

One last aspect of externally photoevaporated disks is more compatible with intermediate G_0 rather than a Taurus-like $G_0 \approx 1$. The *Stardust* samples of the Wild 2 comet include refractory inclusions like *Inti* thought to have formed inside 1 AU, then transported outward in the disk to the comet-forming region (Zolensky et al. 2006). Because the gas flow in disks is inward inside the transition radius, r_T, and outward for $r > r_T$, this requires r_T to remain at a few AU even late in disk evolution, in contrast to self-similar models of isolated, viscously evolving disks, which predict outward movement of r_T to > 10 AU after a few Myr (Hartmann et al. 1998). Even modest amounts of external photoevaporation ($G_0 \sim 30$) can keep r_T fixed at about 3 AU, but not $G_0 \sim 1$ (Desch 2007; Kalyaan et al. 2015; Desch et al. 2018).

2.1.3 Jupiter's Noble Gas Abundances

A third line of evidence for high UV fluxes comes from Jupiter's noble gas abundances. The abundances of various species X relative to H and solar abundances, $(X/H)/(X/H)_\odot$, have been measured in Jupiter's atmosphere: He, Ne, Ar, Kr, Xe, C, N, O, and S by the *Galileo* entry probe Neutral Mass Spectrometer; P by the *Galileo* orbiter Infrared Mapping Spectrometer; and O by the *Juno* microwave radiometer (Li et al. 2020). Compared to solar abundances (Asplund et al. 2009), He and Ne are slightly depleted, certainly due to selective sequestration into liquid helium droplets that sink to Jupiter's core (Wilson and Militzer 2010), but the other species are all enhanced by the same factor, ≈ 3 (Guillot and Hueso 2006; Monga and Desch 2015; Li et al. 2020).

Enhancements of the noble gases, especially Ar, are difficult to explain, as they fractionate from H_2/He gas only when Ar enters ice, which requires deposition of water vapor as amorphous ice, at temperatures < 25 K (Bar-Nun et al. 1988). The only successful model for these uniform enhancements is that H (and He and Ne), but not Ar and other elements, were lost from the solar nebula before Jupiter accreted most of its gas (i.e., by about 5 Myr; Desch et al. 2018). If two-thirds of the just the H_2, He and Ne were lost, then all other species would appear uniformly enhanced relative to H, by the same factor of 3. Monga and Desch (2015), building on the model of Guillot and Hueso (2006), showed that if photoevaporation removes H_2 gas from the outer edge of the disk at a location where the disk temperature is < 25 K (i.e., beyond 50 AU), then Ar could potentially be trapped in large amorphous ice grains that could spiral back in the planet-forming region.

The Monga and Desch (2015) model is consistent with $G_0 \approx 300$. A lower limit of roughly $G_0 > 30$ can be placed by the need in the model for UV to photodesorb enough water vapor to recondense and trap Ar and noble gases. An upper limit of about $G_0 < 3000$ can be placed by the need for gas loss at the outer edge of the disk to take place where temperatures are cold enough (roughly 25 K) to trap Ar, which requires the outer disk edge to remain at 50-100 AU.

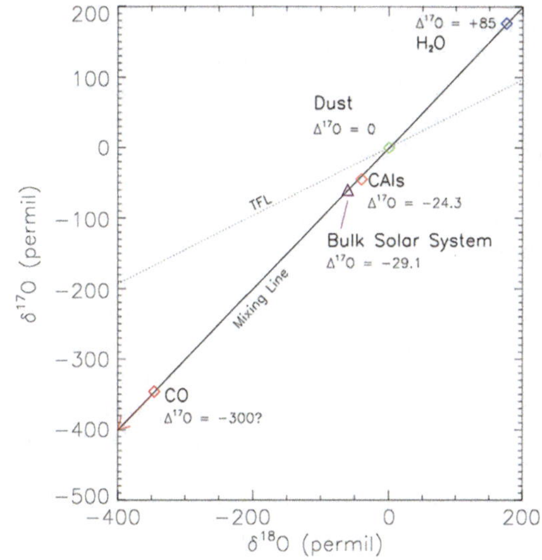

Fig. 1 Three-isotope plot on which terrestrial samples are typically found at $\delta^{17}O = \delta^{18}O = 0‰$ or along the Terrestrial Fractionation Line (TFL) with slope 0.52. Planetary materials are typically found near the terrestrial value, but the Sun is found on a mixing line with slope 1.0, at $\delta^{17}O = \delta^{18}O \approx -60‰$ and $\Delta^{17}O = -29.1‰$; CAIs are found on the same mixing line, at $\delta^{17}O = \delta^{18}O \approx -50‰$ and $\Delta^{17}O = -24.3‰$. The mixed components are hypothesized also to fall on this line: isotopically heavy water ($\Delta^{17}O \approx +85‰$), normal dust ($\Delta^{17} \approx 0‰$) and light CO ($\Delta^{17}O < -300‰$)

2.2 Meteoritic Evidence

Further tests of whether or not the Sun formed in a high-mass star-forming region could come from astrophysical modeling combined with meteoritic data, especially data pertaining to the mass-independent fractionation of oxygen and sulfur isotopes, and short-lived radionuclides.

2.2.1 Oxygen Isotopes

A very useful tool in planetary science is the "three-isotope" plot of oxygen isotopic compositions, like the one displayed in Fig. 1. The vertical axis is $\delta^{17}O = (^{17}O/^{16}O)/(^{17}O/^{16}O)_{SMOW} - 1$ and the horizontal axis is $\delta^{18}O = (^{18}O/^{16}O)/(^{18}O/^{16}O)_{SMOW} - 1$, where the terrestrial standard, Standard Mean Ocean Water, has $\delta^{17}O = \delta^{18}O = 0‰$ (where ‰ is parts per thousand). Almost all chemical processes fractionate oxygen isotopes in a mass-dependent way, along a line with slope 0.52. The vertical displacement of another sample from this line is $\Delta^{17}O = \delta^{17}O - 0.52\,\delta^{18}O$. Essentially all samples on Earth fall along a single line of slope 0.52 and $\Delta^{17}O = 0‰$. Samples from another asteroid will also array along a parallel line of slope 0.52, with a common, distinct $\Delta^{17}O$; for example, $\Delta^{17}O = -0.3‰$ for all meteorites from the asteroid 4 Vesta (Ireland et al. 2020).

Surprisingly, calcium-rich, aluminum-rich inclusions (CAIs) contain minerals arraying along a line of slope 1.0 in a three-isotope plot, some extending to planetary-like values $\Delta^{17}O \approx 0‰$, while most CAIs plot at $\Delta^{17}O < -24‰$ (Ireland et al. 2020). This is interpreted as a mixing line between planetary (silicate) materials and some other, ^{16}O-rich (or $^{17,18}O$-poor) reservoir. The Sun lies at $\Delta^{17}O \approx -29‰$, close to CAIs in the three-isotope plot, as inferred from solar wind measurements by the *Genesis* mission (McKeegan et al. 2011). The process that created that other reservoir is distinct from normal chemical processes, and is called "mass-independent fractionation" (MIF).

Practically the only widely accepted model for the MIF is isotopically selective photodissociation of CO gas in the protoplanetary disk or molecular cloud, by UV radiation from either the Sun or a nearby massive star (Yurimoto and Kuramoto 2004; Lyons and

Young 2005; Young 2007). Because CO is photodissociated by first exciting the molecule to a distinct (but unbound) energy level, $C^{16}O$, $C^{17}O$, and $C^{18}O$ are photodissociated by photons of different UV wavelengths. Because ^{16}O is so much more abundant, the $C^{16}O$-dissociating photons are absorbed close to the edge of the disk or molecular cloud, an effect called "self-shielding". Gas near the edge also is likely to escape in a photoevaporative flow. For practical purposes, the material incorporated into meteorites comes from deep within the disk or cloud, in which $C^{17}O$ and $C^{18}O$ are equally photodissociated. This leaves the remaining CO richer in ^{16}O, with $\delta^{17}O \approx \delta^{18}O$ and very negative $\Delta^{17}O$. The liberated O reacts with hydrogen to form H_2O uniformly enriched in ^{17}O and ^{18}O, with $\delta^{17}O \approx \delta^{18}O$ and very positive $\Delta^{17}O$. Any mixture sampling oxygen from the starting composition of silicates, CO gas and H_2O will array along a line of slope 1.0 in a three-isotope plot. This model is strongly supported by the "cosmic symplectite" material in the Acfer 094 chondrite, containing minerals like magnetite that are products of aqueous alteration, that clearly exchanged O isotopes with H_2O; it has $\delta^{17}O \approx \delta^{18}O \approx +180‰$ and $\Delta^{17}O \approx +85‰$ (Sakamoto et al. 2007). The solar nebula appears to have had isotopically light CO, isotopically heavy water, and silicates with intermediate isotopic composition near $\delta^{17}O \approx \delta^{18}O \approx 0‰$.

Strong support for self-shielded photodissociation of CO comes from astronomical observations of CO absorption in some protostellar systems. Smith et al. (2009) measured the values of $\delta^{17}O$ and $\delta^{18}O$ in CO gas in the protoplanetary disk in the VV CrA system and found an excess of ^{16}O on the order of $\Delta^{17}O \approx -200‰$, as expected if CO gas was photodissociated in an isotopically selective way. Unfortunately, such observations leave open the question of whether the CO was photodissociated by UV radiation from the central protostar or from nearby OB stars, and whether the CO was photodissociated in the disk or during an earlier molecular cloud stage.

If irradiation in the disk could be ruled out, then the irradiation would have to have taken place in the molecular cloud, due to nearby OB stars. Models of self-shielding in the disk have been constructed (e.g., Young 2007; Lyons et al. 2009), although the need to create distinct reservoirs, well-mixed across the entire disk, before the formation of many mineral phases in CAIs, is problematic (see Alexander et al. 2017; Krot et al. 2020). Such models rely on large values of the turbulence parameter, $\alpha \sim 10^{-2}$, so that the isotopic signature produced in the outer disk can be transmitted to the inner disk in $< 3 \times 10^4$ years (Young 2007; Lyons et al. 2009). Whether such a high α is reasonable or not is debated. It is difficult to argue from first principles what the value should be because the physical mechanism of transport in disks has not been identified definitively. While many observations argue for $\alpha \lesssim 10^{-4}$ in the outer portions of many protoplanetary disks (Youdin and Lithwick 2007; Pinte et al. 2016; Flaherty et al. 2020), there is observational support for high values in some disks (Sengupta et al. 2024). Models of the outer regions of the Sun's disk in particular argue for a low value $\alpha \sim 10^{-5}$ or else there would be rapid mixing between the "NC" inner-disk reservoir with the "CC" outer disk reservoir (Desch et al. 2018). As well, the values of α leading to mixing of an outer disk $\Delta^{17}O$ signature with the inner disk would also lead to rapid transport of angular momentum and high mass accretion rates. The coefficient of turbulent mixing, $\mathcal{D} \sim r^2/t_{\rm mix}$ is comparable ($\mathcal{D} \sim \nu$) to the turbulent viscosity $\nu = \alpha C^2/\Omega$. Because the mass accretion rate is $\dot{M} \approx 3\pi\,\Sigma\nu$, the mass-loss timescale $(\pi r^2\,\Sigma)/\dot{M}$ therefore is necessarily comparable to the mixing timescale $t_{\rm mix}$. Rapid mixing of the $\Delta^{17}O$ signature from the outer disk to the inner disk before CAI formation would also imply loss of the disk on comparable timescales $\sim 10^5$ years, not several Myr. Then again, models for the NC/CC dichotomy exist that invoke a rapid evolution of the solar nebula and high values of α that might allow rapid mixing (Liu et al. 2022). There are significant challenges to models in which self-shielding exists in the disk, but such models have not been ruled out.

In contrast, isotopically selective photodissociation of CO in the molecular cloud is a more established idea (Bally and Langer 1982) and the basis for models of MIF of oxygen isotopes in the Solar System in particular (Yurimoto and Kuramoto 2004). Unfortunately, such models do not make firm predictions of G_0, but values $\gg 1$ presumably are needed to create δ^{16}O-poor H_2O deep in the molecular cloud.

2.2.2 Sulfur Isotopes

Recently Vacher et al. (2021) analyzed magnetite-pentlandite minerals in the same cosmic symplectite material in Acfer 094, for S isotopes. Like oxygen, S isotopes should fractionate according to a slope-1/2 line in $\delta^{33}S = (^{33}S/^{32}S)/(^{33}S/^{32}S)_{CDT} - 1$ vs. $\delta^{34}S$, or a slope-2 line in $\delta^{36}S$ vs. $\delta^{34}S$ space (here CDT is the terrestrial standard, Canyon Diablo Troilite). The deviations of a sample from these lines can be labeled $\Delta^{33}S$ or $\Delta^{36}S$. Almost all samples on Earth—and indeed, throughout most of the Solar System—have $\Delta^{33}S \approx 0‰$ and $\Delta^{36}S \approx 0‰$; but the cosmic symplectite is characterized by $\Delta^{33}S \approx +4‰$ and $\Delta^{36}S \approx -6‰$, yielding a ratio $\Delta^{36}S/\Delta^{33}S = -1.58 \pm 0.66$.

Vacher et al. (2021) interpreted this as arising from photodissociation of H_2S in the molecular cloud. The details of photodissociation of H_2S differ from those of CO, but lead to the same result: the different isotopes of S are liberated from gas-phase H_2S, as elemental S^0, at different rates. As long as the temperature is < 370 K, the S^0 recondenses on dust grains before converting back to H_2S, imparting a MIF isotopic signature in S in solids. The resultant ratio $\Delta^{36}S/\Delta^{33}S$ in the solids is sensitive to the spectrum of the UV radiation. UV radiation from T Tauri stars like the early Sun is dominated by Lyα photons, and photodissociation by this spectrum results in S^0 with $\Delta^{36}S/\Delta^{33}S \approx -2.73 \pm 0.73$ (Chakraborty et al. 2013), incompatible with the isotopic signature. Photodissociation by UV radiation from OB stars, or the interstellar radiation field, result in S^0 with $\Delta^{36}S/\Delta^{33}S \approx -1.53 \pm 0.67$ or -1.75 ± 0.71, repectively, matching the cosmic symplectite.

In similarity to self-shielding of CO, the isotopic anomalies themselves ($\Delta^{17}O$, $\Delta^{34}S$, $\Delta^{36}S$) do not directly constrain the strength G_0 of the UV field. Self-shielding of H_2S in the protoplanetary disk rather than molecular cloud also probably cannot be ruled out at this time. However, unlike the case of CO, photodissociation of H_2S seems to demand a UV flux from nearby OB stars, which therefore constrains the Sun's birth environment.

The photodissociation of CO and H_2S together also do place some indirect limits on the UV flux and G_0, through the gas temperatures that are implied. CO must be in the gas phase where irradiation takes place, implying temperatures > 50 K; but H_2O must condense, suggesting temperatures < 170 K, roughly. Likewise, H_2S must be in the gas phase, implying temperatures > 70 K; but S^0 must condense before back-reacting to H_2S, implying temperatures < 370 K (Vacher et al. 2021). These temperatures, $\approx 70 - 170$ K, are quite characteristic of photodissociation regions (PDRs) with intermediate UV flux. In the PDR models of Adams et al. (2004), these temperatures arise at various optical depths into gas with G_0 in the range of 300 to 3×10^4. They did not present results for lower G_0, but a lower limit to $G_0 > 10$ or so also must exist: most molecular clouds and filaments exposed to smaller G_0 are overwhelmingly associated with very cold temperatures < 20 K (e.g., Hacar et al. 2023), at which CO and H_2S would be in the solid phase (although desorption of molecules from grains by GCR bombardment complicates this picture).

2.2.3 Short-Lived Radionuclides

For a long time, the existence of short-lived ($t_{1/2} <$ tens of Myr) radionuclides (SLRs) like ^{26}Al in the solar nebula was taken as demanding formation very near a supernova (Cameron and Truran 1977; Adams 2010). The supernova was presumed to inject ^{26}Al and other SLRs either into the Sun's protoplanetary disk or into its molecular cloud, just prior to collapse of the Sun's cloud core. Models requiring such proximity of a Wolf-Rayet star or supernova to the Sun tend to require higher stellar densities $\gg 10\,M_\odot\,\mathrm{pc}^{-3}$ (Pfalzner and Vincke 2020; Parker et al. 2023) and larger number of stars (e.g., $N > 2500$; Portegies Zwart 2019) in the Sun's birth cluster, but even then the relative timing of events would be improbable (Ouellette et al. 2010). It seems more likely instead that the Sun and its protoplanetary disk inherited their SLRs from the molecular cloud, which was itself enriched by Wolf-Rayet winds and supernovae, by virtue of being in a high-mass star-forming region.

Strong evidence for the inheritance of SLRs from the molecular cloud is provided by their homogeneity in meteoritic samples, especially bulk abundances of achondrites. Desch et al. (2023a) and Desch et al. (2023b) compiled and analyzed the times of formation of 14 bulk achondrites according to radiometric dating using the Pb-Pb, Al-Mg, Mn-Cr, and Hf-W chronometers. Across these systems, 38 times of formation are made concordant in a statistically rigorous sense ($\chi_\nu^2 = 1.1$), assuming values for only four parameters (initial abundances of ^{53}Mn and ^{182}Hf, a half-life of 3.8 Myr for ^{53}Mn, and a Pb-Pb age of the Solar System of 4568.4 Myr). This strongly supports homogeneity in the solar nebula of the SLRs ^{26}Al, ^{53}Mn, and ^{182}Hf, which therefore must have been inherited from the molecular cloud.

The abundances of *all* the known SLRs in the early Solar System of nucleosynthetic origin (^{41}Ca, ^{36}Cl, ^{26}Al, ^{60}Fe, ^{53}Mn, ^{107}Pd, ^{182}Hf, ^{247}Cm, ^{129}I), as well as longer-lived radionuclides (^{92}Nb, ^{244}Pu, ^{146}Sm, ^{235}U, ^{238}U) appear remarkably consistent with the predictions of a simple box model in which the Sun formed in a spiral arm of the Galaxy, from molecular clouds continuously enriched by inputs from massive stars (Jacobsen 2005, Huss et al. 2009, Young 2014, Young 2020; see review by Desch et al. 2023). In essence, massive stars forming in the spiral arms rapidly evolve and eject material as Wolf-Rayet winds and supernovae before leaving the spiral arm, contaminating molecular clouds, which are the sites of subsequent star formation. Because molecular clouds are gravitationally drawn into the spiral arms, the SLR enrichments do not readily leave the spiral arms, allowing some steady-state buildup over many $\times 10^8$ yr within these structures (Fujimoto et al. 2018, 2020). While very short-lived SLRs (^{41}Ca, ^{26}Al) must be much less abundant in steady state than longer-lived ones, their abundances, like the abundances of all 14 radionuclides above, are exactly consistent with the predictions of the model, assuming the material entering molecular clouds is overwhelmingly from Wolf-Rayet winds and not from supernovae. This is surprising but justifiable: massive stars evolve to eject Wolf-Rayet winds within < 4 Myr, while they are still near their birth clouds, whereas most stars explode as supernovae only after $> 10 - 20$ Myr, by which time they have dispersed far from their birth clouds.

At one point, a high abundance $^{60}\mathrm{Fe}/^{56}\mathrm{Fe} \sim 10^{-6}$ was inferred to have existed in the solar nebula, which would have practically demanded injection of material from a nearby supernova (Tachibana et al. 2006). Subsequent analyses have shown how these values were surely overestimated (Ogliore et al. 2011; Telus et al. 2016). It is now understood that ^{60}Fe did exist in the solar nebula, but at levels $^{60}\mathrm{Fe}/^{56}\mathrm{Fe} \sim 10^{-8}$ (Tang and Dauphas 2012). These levels are consistent with the levels inferred from gamma ray observations to exist in the Galaxy and in molecular clouds (Fujimoto et al. 2018; Desch et al. 2023). While the Sun probably formed in a region with supernovae, the evidence from ^{60}Fe does not demand it formed near (< 1 pc) a contemporaneous supernova.

Molecular clouds in spiral arms also should contain ^{10}Be at levels consistent with the values in the solar nebula. The abundance of ^{10}Be in the solar nebula was a remarkably uniform $^{10}\mathrm{Be}/^{9}\mathrm{Be} \approx 7.1 \times 10^{-4}$, indicating inheritance of ^{10}Be from the Sun's natal molecular cloud (Dunham et al. 2022), as with the other SLRs. Live ^{10}Be is produced by spallation of oxygen

Table 1 Constraints on the Sun's star-forming environment

Parameter	Constraint	Reason
Cluster size	$N \gtrsim 2000$ stars	Must form bound cluster.
Cluster lifetime	$T \approx 100$ Myr	Scattered KBOs' perihelia must be raised after scattering by Neptune migrating at ≈ 60 Myr.
Stellar density	$\eta \approx 200-300\,\mathrm{pc}^{-3}$	A stellar encounter must raise perihelia of scattered KBOs, but dynamically cold KBOs must not be over-excited.
UV Flux	$30 \lesssim G_0 \lesssim 3000$	Jupiter's noble gas abundances; disk truncated at ≈ 100 AU; outward gas flow beyond a few AU.
Galactic star-forming environment	Molecular cloud in a spiral arm, enriched by Wolf-Rayet stars	Abundances of short-lived radionuclides like ^{26}Al, ^{10}Be, ^{53}Mn, ^{60}Fe, etc.

and other nuclei, by Galactic cosmic rays (GCRs). The solar nebula abundance was thought to be too high to have been inherited from the Sun's molecular cloud (Tatischeff et al. 2014), assuming a GCR flux based on observations of the Sun's current neighborhood, which is relatively quiescent. But molecular clouds in spiral arms see higher GCR flux, as GCRs are accelerated in supernova shocks, and supernovae are concentrated in spiral arms. Allowing for the higher GCR fluxes in spiral arms 4.57 Gyr ago, a ratio $^{10}\mathrm{Be}/^{9}\mathrm{Be} = 7.1 \times 10^{-4}$ is a very probable outcome (Dunham et al. 2022; Desch et al. 2023).

While SLRs like ^{26}Al were once taken as the smoking gun for birth of the Sun near a massive star, the story is probably more subtle. The abundances of the SLRs are all explained by inheritance from a molecular cloud in a spiral arm, enriched in material from supernovae and especially Wolf-Rayet winds. This places the Sun more definitively than ever in the general vicinity of a high-mass star-forming region, but not necessarily less than a parsec from a supernova. Acquisition of live ^{26}Al from previous generations of supernovae and massive stars would seem readily achieved in regions like Upper Scorpius (Krause et al. 2018; Ratzenböck et al. 2023) or the Orion Nebula (Foley et al. 2023).

3 The Sun's Formation

Based on the arguments presented here, summarized in Table 1, we place the Sun's formation in a molecular cloud in a spiral arm, contaminated by previous generations of star formation. The Sun formed in a bound stellar cluster that maintained densities of several hundred stars per cubic parsec, for about 100 Myr. It experienced a UV flux much higher than is found in the interstellar medium outside star-forming regions.

The weight of this astronomical evidence places the Sun's formation in an environment like the outskirts of the Orion Nebula, perhaps 2 pc from the central Orion Nebula cluster and θ^1 Ori C. Figure 2 shows an optical image of the Orion Nebula taken by the *Hubble Space Telescope*, superimposed with a circle 15 arcminutes (2 pc at 450 pc) in radius, centered on θ^1 Ori C. We consider the Sun to have formed in a region like those depicted in the left or upper corners of the figure, at the distance of the yellow circle. Although the density of stars there is not as great as in the center of the cluster, it is still substantial. For reference, the Orion Nebula half-mass radius is $\lesssim 1$ pc (Megeath et al. 2016). In Fig. 3 we also show a recent near-infrared image taken by the NIRCam instrument on the *James Webb Space*

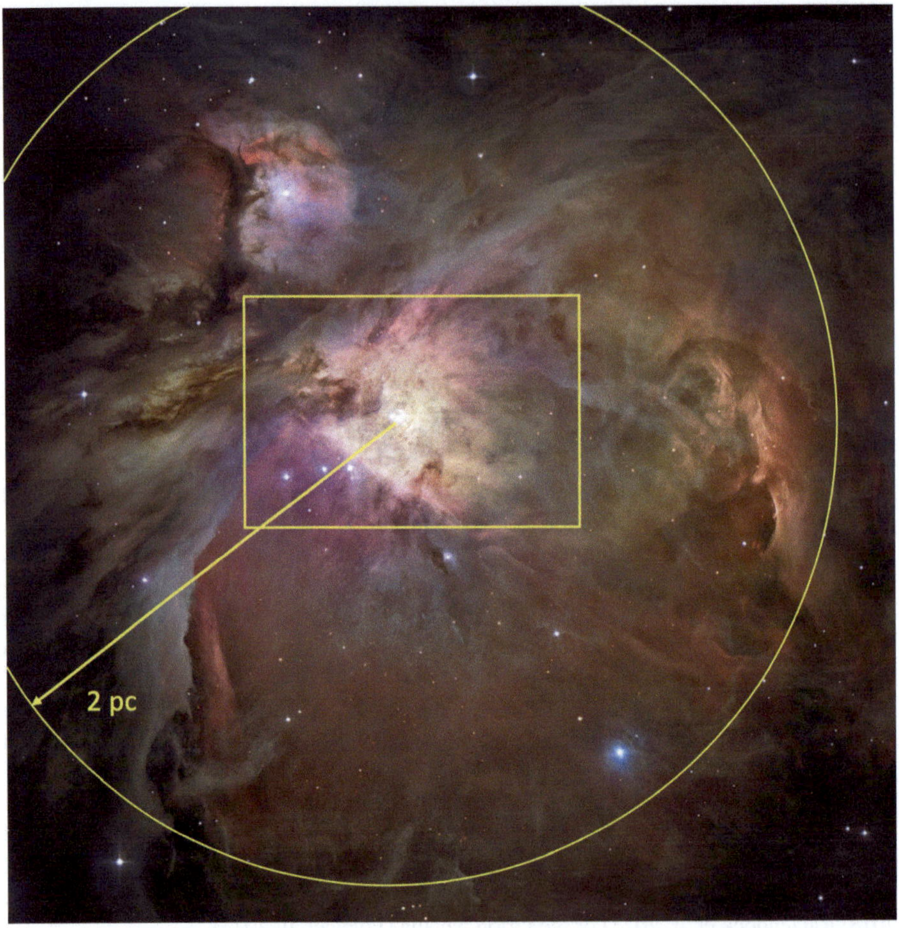

Fig. 2 *Hubble Space Telescope* optical (*Advanced Camera for Surveys*) image of the Orion Nebula. The field of view is 30.0 by 30.0 arcminutes, roughly 4.0 pc by 4.0 pc. The Trapezium cluster and θ^1 Ori C are apparent at the center of the superimposed circle, which has a radius of 15 arcminutes, or 2 pc. The yellow rectangle shows the approximate field of view of the image in Fig. 3. (Credit: NASA, ESA, M. Robberto (Space Telescope Science Institute / ESA) and the Hubble Space Telescope Orion Treasury Project Team)

Telescope (McCaughrean and Pearson 2023). This image only depicts the central region (< 2 pc from the cluster center), but better conveys the structure of the nebulosity.

Many aspects of the Orion Nebula match those inferred for the Solar System's early evolution. As reviewed above, the Sun's molecular cloud and protoplanetary disk must have been exposed to a UV flux with $G_0 \approx 30 - 3000$, based on the need to create isotopically fractionated reservoirs of oxygen and sulfur in the molecular cloud, based on the truncation of the Sun's solid disk at ≈ 47 AU (edge of the classical Kuiper Belt), based on the need for photoevaporation to shape the dynamics of the gas disk, and based on the need to remove hydrogen by photoevaporation to explain the abundances of Jupiter's atmosphere. The UV flux in the Orion Nebula scales with distance r from the center of the cluster (θ^1 Ori C) roughly as $G_0 \approx 5000\,(r/1\,\mathrm{pc})^{-2}$ (Störzer and Hollenbach 1999). Therefore, the Sun's birth place is consistent with environments at distances larger than about 1.3 pc from the Orion Nebula

Fig. 3 *James Webb Space Telesope NIRCam* near-infrared image of the Orion Nebula. The field of view is 7.5 by 10.9 arcminutes, roughly 0.85 pc by 1.25 pc, and corresponds to the rectangle in Fig. 2. Emission from ionized gas at the center is colored purple, while greens and browns denote emission from molecular gas and dust. Below and to the left of the very bright BN-KL object (seen in reddish pink) is the Trapezium cluster including θ^1 Ori C. (Credit: NASA/ESA/CSA/STScI, K. Pontoppidan, A. Pagan)

center, but not beyond 13 pc. Formation in the central 1 pc of a region like the Orion Nebula can be ruled out, as the UV flux and stellar density would be too high. The Sun's protoplanetary disk must have been exposed to a stellar density $\approx 100\,M_\odot\,\mathrm{pc}^{-3}$ for ≈ 100 Myr. The average stellar density in the Orion Nebula within 2 pc of the center is $\approx 100\,M_\odot\,\mathrm{pc}^{-3}$, and it is likely to form a bound cluster, maintaining these stellar densities for 100 Myr (Hartmann et al. 1998). The Orion Molecular Cloud is expected to have already experienced 10-20 supernovae (some preceded by Wolf-Rayet winds) in the last 12 Myr (Bally 2008), and the region is associated with gamma-ray emission from the decay of gas-phase ^{26}Al at a level similar to the solar nebula abundance of ^{26}Al (Diehl 2002). This highlights the likelihood of stars in regions like this incorporating live radionuclides like ^{26}Al.

Formation of the Sun in such an environment has important implications for the early evolution of the solar nebula. If the Sun formed in a region like the Orion Nebula, its formation may have been triggered by the nearby massive stars, although the mechanism is probably not as simplistic as a cloud core triggered to collapse by a supernova shock. A detailed history of triggered star formation has recently been worked out for the Scorpius-Centaurus region (Krause et al. 2018; Ratzenböck et al. 2023). There, star formation is sequential, has been ongoing for the past 20 Myr, and has propagated inside-out towards multiple directions. Krause et al. (2018) advocate that the Upper Sco molecular cloud was triggered to collapse by the surrounding superbubble of hot, X-ray-emitting gas created by previous generations of supernovae and massive star (Wolf-Rayet) winds. Ratzenböck et al. (2023) refined this idea to suggest that the superbubble triggered collapse and star formation in gas arranged along filaments, resulting in chains of star clusters. Similarly, in the Orion

Nebula, the center of ^{26}Al gamma-ray emission appears coincident with a center of expansion of stars, implying one or more stellar explosions at that point about 6 Myr ago, which have triggered star formation (Foley et al. 2023). Like the Sco-Cen region, the Orion Nebula contains a superbubble of X-ray emitting, ^{26}Al-bearing gas, the Orion-Eridanus superbubble (Reynolds and Ogden 1979) that also may have triggered star formation across this region of space.

The question of how normal the Sun's ^{26}Al abundance was can be answered in this context. Contamination by Wolf-Rayet stars resembles the model advocated by Dwarkadas et al. (2017), formation at the edge of a bubble launched by Wolf-Rayet stars, except that in that model the Sun would be rare, corresponding only to material in the swept-up shell. In our interpretation, like others, the Wolf-Rayet stars contaminate the molecular clouds before star formation, making ^{26}Al more common. In their model of how stars form at the periphery of a massive star-forming region, Gaidos et al. (2009) found that the modal abundance would be $^{26}\mathrm{Al}/^{27}\mathrm{Al} \sim 10^{-6} - 10^{-5}$, compared to the initial value in the Solar System, 5×10^{-5}. In their whole-galaxy simulations of ^{26}Al enrichment in clouds, Fujimoto et al. (2018) found similar values, $^{26}\mathrm{Al}/^{27}\mathrm{Al} \sim 10^{-5}$. The box model of Young (2020) predicts a modal abundance very similar to the Solar System value, but with a range of values from factor-of-5 lower to a factor-of-5 larger. In all models, stars like the Sun are predicted to have been born with an initial $^{26}\mathrm{Al}/^{27}\mathrm{Al}$ value similar to, or perhaps an order of magnitude smaller than, the Solar System initial value, with a range of probable values extending perhaps an order of magnitude both smaller and larger. Of course, some fraction ($< 20\%$; Lada and Lada 2003) of stars form in low-mass star-forming regions ($N \sim 10^2$ or lower) and may acquire relatively little ^{26}Al.

If the Sun formed in a region like the Orion Nebula, late infall might be possible, although it is not at all clear this would happen. Late infall onto the Sun's protoplanetary disk would be a likely outcome if the Sun had formed in a low-mass star-forming region like Taurus-Auriga (Pineda et al. 2023). It would be very unlikely if it had formed in a high-mass star-forming region like the Carina Nebula; whether the Sun was triggered to form or not, proximity to gas experiencing high G_0 implies the Sun's disk would have been shut off from accreting gas within $\sim 10^5$ yr timescales. For an intermediate-mass region like the Orion Nebula, especially 2 pc from the center of the Orion Nebula Cluster, late infall along streamers has not been observed, but cannot entirely be ruled out. However, the fact that the Sun apparently formed at the edge of a photodissociation region (PDR) suggests that it quickly (within a few $\times 10^5$ yr) would be uncovered by the ionization front, as in other star-forming regions (Snider et al. 2009). Even presuming the Sun accreted late material from its molecular cloud, it is not at all clear that it would be compositionally distinct from earlier-accreted material. In 1 Myr, diffusion due to hydromagnetic turbulence should homogenize material over lengthscales > 0.3 pc (Pan et al. 2012) and is further homogenized during accretion (Kuffmeier et al. 2016).

Based on the evidence presented here, we conclude the Sun formed in a region like the outskirts (> 2 pc from the center) of the Orion Nebula, a region characterized by $G_0 \sim 10^3$, stellar densities $\approx 100\, M_\odot\, \mathrm{pc}^{-3}$, in a cluster that will remain gravitationally bound for ≈ 100 Myr. Such a region would contain $N \sim 10^3 - 10^4$ stars. Some estimates of the Sun's birth cluster size are slightly larger (e.g., Pfalzner and Vincke 2020; see review by Bergin et al. 2023), mostly because they do not account for the fact that the gas disk could be much larger than the dust disk outer edge at 47 AU (allowing for smaller G_0); and especially because they demand proximity to a Wolf-Rayet star or supernova to acquire sufficient ^{26}Al, rather than considering previous contamination of the molecular cloud as the source. Other estimates (Forbes et al. 2021) are smaller, because they do not account for the need for the

Sun to remain in a bound cluster for 100 Myr. Formation in a region like the Orion Nebula is consistent with the Sun's disk size, disk dynamics, Jupiter's abundances, the existence of detached Kuiper belt objects yet the relative non-disturbance of cold classical Kuiper belt objects, the generation of isotopically fractionated reservoirs of oxygen and sulfur in the solar nebula, and the abundances of short-lived radionuclides. Late ($>$ a few $\times 10^5$ yr) infall cannot be ruled out, but seems unlikely, and should be judged in this context.

Acknowledgements The authors are grateful to the organizers of the International Space Science Institute workshop "Evolution of the Solar System: Constraints from Meteorites (June 5-9, 2023, Bern, Switzerland).

Funding Open access funding provided by University of Vienna.

Declarations

Competing Interests The authors declare no competing interests.

References

Adams FC (2010) The birth environment of the Solar System. Annu Rev Astron Astrophys 48:47–85. https://doi.org/10.1146/annurev-astro-081309-130830. arXiv:1001.5444 [astro-ph.SR]

Adams FC, Hollenbach D, Laughlin G, et al (2004) Photoevaporation of circumstellar disks due to external far-ultraviolet radiation in stellar aggregates. Astrophys J 611(1):360–379. https://doi.org/10.1086/421989. arXiv:astro-ph/0404383 [astro-ph]

Alexander CMO, Nittler LR, Davidson J, et al (2017) Measuring the level of interstellar inheritance in the solar protoplanetary disk. Meteorit Planet Sci 52(9):1797–1821. https://doi.org/10.1111/maps.12891

Ansdell M, Williams JP, Trapman L, et al (2018) ALMA survey of Lupus protoplanetary disks. II. Gas disk radii. Astrophys J 859(1):21. https://doi.org/10.3847/1538-4357/aab890. arXiv:1803.05923 [astro-ph.EP]

Arzoumanian D, Arakawa S, Kobayashi MIN, et al (2023) Insights on the Sun birth environment in the context of star cluster formation in hub-filament systems. Astrophys J Lett 947(2):L29. https://doi.org/10.3847/2041-8213/acc849. arXiv:2303.15695 [astro-ph.GA]

Asplund M, Grevesse N, Sauval AJ, et al (2009) The chemical composition of the Sun. Annu Rev Astron Astrophys 47(1):481–522. https://doi.org/10.1146/annurev.astro.46.060407.145222. arXiv:0909.0948 [astro-ph.SR]

Bally J (2008) Star formation in Orion and beyond. In: Star formation across the Milky Way galaxy, p 7

Bally J, Langer WD (1982) Isotope-selective photodestruction of carbon monoxide. Astrophys J 255:143–148. https://doi.org/10.1086/159812

Bar-Nun A, Kleinfeld I, Kochavi E (1988) Trapping of gas mixtures by amorphous water ice. Phys Rev B 38(11):7749–7754. https://doi.org/10.1103/PhysRevB.38.7749

Barboni M, Boehnke P, Keller B, et al (2017) Early formation of the Moon 4.51 billion years ago. Sci Adv 3(1):e1602365. https://doi.org/10.1126/sciadv.1602365

Batygin K, Adams FC, Batygin YK, et al (2020) Dynamics of planetary systems within star clusters: aspects of the Solar System's early evolution. Astron J 159(3):101. https://doi.org/10.3847/1538-3881/ab665d. arXiv:2002.05656 [astro-ph.EP]

Bergin EA, Alexander C, Drozdovskaya M, et al (2023) Interstellar Heritage and the Birth Environment of the Solar System. E-prints arXiv:2301.05212 [astro-ph.EP]. https://doi.org/10.48550/arXiv.2301.05212

Bierson CJ, Nimmo F (2019) Using the density of Kuiper Belt objects to constrain their composition and formation history. Icarus 326:10–17. https://doi.org/10.1016/j.icarus.2019.01.027

Brasser R, Duncan MJ, Levison HF (2006) Embedded star clusters and the formation of the Oort cloud. Icarus 184(1):59–82. https://doi.org/10.1016/j.icarus.2006.04.010

Cameron AGW, Truran JW (1977) The supernova trigger for formation of the Solar System. Icarus 30(3):447–461. https://doi.org/10.1016/0019-1035(77)90101-4

Chakraborty S, Yanchulova P, Thiemens MH (2013) Mass-independent oxygen isotopic partitioning during gas-phase SiO_2 formation. Science 342(6157):463–466. https://doi.org/10.1126/science.1242237

Ciesla FJ, Mulders GD, Pascucci I, et al (2015) Volatile delivery to planets from water-rich planetesimals around low mass stars. Astrophys J 804(1):9. https://doi.org/10.1088/0004-637X/804/1/9. arXiv:1502.07412 [astro-ph.EP]

Cleeves LI, Öberg KI, Wilner DJ, et al (2016) The coupled physical structure of gas and dust in the IM Lup protoplanetary disk. Astrophys J 832(2):110. https://doi.org/10.3847/0004-637X/832/2/110. arXiv:1610.00715 [astro-ph.SR]

Da Rio N, Robberto M, Soderblom DR, et al (2010) A multi-color optical survey of the Orion Nebula cluster. II. The H-R diagram. Astrophys J 722(2):1092–1114. https://doi.org/10.1088/0004-637X/722/2/1092. arXiv:1008.1265 [astro-ph.GA]

Dale JE, Bonnell IA, Whitworth AP (2007) Ionization-induced star formation - I. The collect-and-collapse model. Mon Not R Astron Soc 375(4):1291–1298. https://doi.org/10.1111/j.1365-2966.2006.11368.x. arXiv:astro-ph/0612128 [astro-ph]

de Sousa RR, Morbidelli A, Raymond SN, et al (2020) Dynamical evidence for an early giant planet instability. Icarus 339:113605. https://doi.org/10.1016/j.icarus.2019.113605

Desch SJ (2007) Mass distribution and planet formation in the solar nebula. Astrophys J 671(1):878–893. https://doi.org/10.1086/522825

Desch SJ, Estrada PR, Kalyaan A, et al (2017) Formulas for radial transport in protoplanetary disks. Astrophys J 840(2):86. https://doi.org/10.3847/1538-4357/aa6bfb. arXiv:1704.01267 [astro-ph.EP]

Desch SJ, Kalyaan A, Alexander CMOD (2018) The effect of Jupiter's formation on the distribution of refractory elements and inclusions in meteorites. Astrophys J Suppl Ser 238(1):11. https://doi.org/10.3847/1538-4365/aad95f. arXiv:1710.03809 [astro-ph.EP]

Desch SJ, Dunlap DR, Dunham ET, et al (2023a) Statistical chronometry of meteorites. I. A test of ^{26}Al homogeneity and the Pb-Pb age of the Solar System's $t = 0$. Icarus 402:115607. arXiv:2212.00390 [astro-ph.EP]. https://doi.org/10.1016/j.icarus.2023.115607

Desch SJ, Dunlap DR, Williams CD, et al (2023b) Statistical chronometry of meteorites: II. Initial abundances and homogeneity of short-lived radionuclides. Icarus 402:115611. arXiv:2212.00145 [astro-ph.EP]. https://doi.org/10.1016/j.icarus.2023.115611

Desch SJ, Young ED, Dunham ET, et al (2023) Short-lived radionuclides in meteorites and the Sun's birth environment. In: Inutsuka S et al (eds) Protostars and Planets VII. Astronomical Society of the Pacific, San Francisco, pp 759–798

Diehl R (2002) ^{26}Al production in the Vela and Orion regions. New Astron Rev 46(8–10):547–552. https://doi.org/10.1016/S1387-6473(02)00199-9

Dunham ET, Wadhwa M, Desch SJ, et al (2022) Uniform initial ^{10}Be/^{9}Be inferred from refractory inclusions in CV3, CO3, CR2, and CH/CB chondrites. Geochim Cosmochim Acta 324:194–220. https://doi.org/10.1016/j.gca.2022.02.002

Dwarkadas VV, Dauphas N, Meyer B, et al (2017) Triggered star formation inside the shell of a Wolf-Rayet bubble as the origin of the Solar System. Astrophys J 851(2):147. https://doi.org/10.3847/1538-4357/aa992e. arXiv:1712.10053 [astro-ph.SR]

Evans NJ, Dunham MM, Jørgensen JK, et al (2009) The Spitzer c2d legacy results: star-formation rates and efficiencies; evolution and lifetimes. Astrophys J Suppl Ser 181(2):321–350. https://doi.org/10.1088/0067-0049/181/2/321. arXiv:0811.1059 [astro-ph]

Flaherty K, Hughes AM, Simon JB, et al (2020) Measuring turbulent motion in planet-forming disks with ALMA: a detection around DM Tau and nondetections around MWC 480 and V4046 Sgr. Astrophys J 895(2):109. https://doi.org/10.3847/1538-4357/ab8cc5. arXiv:2004.12176 [astro-ph.SR]

Foley MM, Goodman A, Zucker C, et al (2023) A 3D view of Orion. I. Barnard's loop. Astrophys J 947(2):66. https://doi.org/10.3847/1538-4357/acb5f4. arXiv:2212.01405 [astro-ph.GA]

Forbes JC, Alves J, Lin DNC (2021) A Solar System formation analogue in the ophiuchus star-forming complex. Nat Astron 5:1009–1016. https://doi.org/10.1038/s41550-021-01442-9. arXiv:2108.09326 [astro-ph.EP]

Fujimoto Y, Krumholz MR, Tachibana S (2018) Short-lived radioisotopes in meteorites from Galactic-scale correlated star formation. Mon Not R Astron Soc 480(3):4025–4039. https://doi.org/10.1093/mnras/sty2132. arXiv:1802.08695 [astro-ph.GA]

Fujimoto Y, Krumholz MR, Si I (2020) Distribution and kinematics of ^{26}Al in the Galactic disc. Mon Not R Astron Soc 497(2):2442–2454. https://doi.org/10.1093/mnras/staa2125. arXiv:2006.03057 [astro-ph.GA]

Gaidos E, Krot AN, Williams JP, et al (2009) ^{26}Al and the formation of the Solar System from a molecular cloud contaminated by Wolf-Rayet winds. Astrophys J 696(2):1854–1863. https://doi.org/10.1088/0004-637X/696/2/1854. arXiv:0901.3364 [astro-ph.EP]
Galli PAB, Bouy H, Olivares J, et al (2020) Lupus DANCe. Census of stars and 6D structure with Gaia-DR2 data. Astron Astrophys 643:A148. https://doi.org/10.1051/0004-6361/202038717. arXiv:2010.00233 [astro-ph.SR]
Grimm RE, McSween HY (1993) Heliocentric zoning of the asteroid belt by aluminum-26 heating. Science 259(5095):653–655
Güdel M, Briggs KR, Arzner K, et al (2007) The XMM-Newton extended survey of the Taurus molecular cloud (XEST). Astron Astrophys 468(2):353–377. https://doi.org/10.1051/0004-6361:20065724. arXiv: astro-ph/0609160 [astro-ph]
Guillot T, Hueso R (2006) The composition of Jupiter: sign of a (relatively) late formation in a chemically evolved protosolar disc. Mon Not R Astron Soc 367(1):L47–L51. https://doi.org/10.1111/j.1745-3933.2006.00137.x. arXiv:astro-ph/0601043 [astro-ph]
Hacar A, Clark SE, Heitsch F, et al (2023) Initial conditions for star formation: a physical description of the filamentary ISM. In: Inutsuka S et al (eds) Protostars and Planets VII. Astronomical Society of the Pacific, San Francisco, pp 153–192. arXiv:2203.09562
Hao CJ, Xu Y, Hou LG, et al (2023) Unveiling the initial conditions of open star cluster formation. Res Astron Astrophys 23(7):075023. https://doi.org/10.1088/1674-4527/acd58d. arXiv:2305.04415 [astro-ph.GA]
Hartmann L, Calvet N, Gullbring E, et al (1998) Accretion and the evolution of T tauri disks. Astrophys J 495(1):385–400. https://doi.org/10.1086/305277
Hester JJ, Desch SJ (2005) Understanding our origins: star formation in HII region environments. In: Krot AN, Scott ERD, Reipurth B (eds) Chondrites and the protoplanetary disk. ASP Conference Series, vol 341. Astronomical Society of the Pacific, San Francisco, pp 107–130. arXiv:astro-ph/0506190
Hollenbach D, Adams FC (2004) Dispersal of disks around young stars: constraints on Kuiper Belt formation. In: Caroff L, Moon LJ, Backman D, et al (eds) Debris disks and the formation of planets, p 168
Huss GR, Meyer BS, Srinivasan G, et al (2009) Stellar sources of the short-lived radionuclides in the early Solar System. Geochim Cosmochim Acta 73(17):4922–4945. https://doi.org/10.1016/j.gca.2009.01.039
Ida S, Larwood J, Burkert A (2000) Evidence for early stellar encounters in the orbital distribution of Edgeworth-Kuiper Belt objects. Astrophys J 528(1):351–356. https://doi.org/10.1086/308179. arXiv: astro-ph/9907217 [astro-ph]
Ireland TR, Avila J, Greenwood RC, et al (2020) Oxygen isotopes and sampling of the Solar System. Space Sci Rev 216(2):25. https://doi.org/10.1007/s11214-020-0645-3
Jacobsen SB (2005) The birth of the Solar System in a molecular cloud: evidence from the isotopic pattern of short-lived nuclides in the early Solar System. In: Krot AN, Scott ERD, Reipurth B (eds) Chondrites and the protoplanetary disk, p 548
Jeffries RD, Littlefair SP, Naylor T, et al (2011) No wide spread of stellar ages in the Orion Nebula cluster. Mon Not R Astron Soc 418(3):1948–1958. https://doi.org/10.1111/j.1365-2966.2011.19613.x. arXiv: 1108.2052 [astro-ph.SR]
Jewitt D, Luu J, Trujillo C (1998) Large Kuiper Belt objects: the Mauna Kea 8K CCD survey. Astron J 115(5):2125–2135. https://doi.org/10.1086/300335
Jura M, Xu S, Young ED (2013) ^{26}Al in the early Solar System: not so unusual after all. Astrophys J Lett 775(2):L41. https://doi.org/10.1088/2041-8205/775/2/L41. arXiv:1308.6325 [astro-ph.EP]
Kalyaan A, Desch SJ, Monga N (2015) External photoevaporation of the solar nebula. II. Effects on disk structure and evolution with non-uniform turbulent viscosity due to the magnetorotational instability. Astrophys J 815(2):112. https://doi.org/10.1088/0004-637X/815/2/112. arXiv:1511.05620 [astro-ph.EP]
Karr JL, Martin PG (2003) Triggered star formation in the W5 H II region. Astrophys J 595(2):900–912. https://doi.org/10.1086/376590
Kenyon SJ, Bromley BC (2004) Stellar encounters as the origin of distant Solar System objects in highly eccentric orbits. Nature 432(7017):598–602. https://doi.org/10.1038/nature03136. arXiv:astro-ph/0412030 [astro-ph]
Krause MGH, Burkert A, Diehl R, et al (2018) Surround and Squash: the impact of superbubbles on the interstellar medium in Scorpius-Centaurus OB2. Astron Astrophys 619:A120. https://doi.org/10.1051/0004-6361/201732416. arXiv:1808.04788 [astro-ph.GA]
Krot AN, Nagashima K, Lyons JR, et al (2020) Oxygen isotopic heterogeneity in the early Solar System inherited from the protosolar molecular cloud. Sci Adv 6(42):eaay2724. https://doi.org/10.1126/sciadv.aay2724
Kroupa P (2001) On the variation of the initial mass function. Mon Not R Astron Soc 322(2):231–246. https://doi.org/10.1046/j.1365-8711.2001.04022.x. arXiv:astro-ph/0009005 [astro-ph]
Kroupa P, Aarseth S, Hurley J (2001) The formation of a bound star cluster: from the Orion Nebula cluster to the Pleiades. Mon Not R Astron Soc 321(4):699–712. https://doi.org/10.1046/j.1365-8711.2001.04050.x. arXiv:astro-ph/0009470 [astro-ph]

Kruijer TS, Burkhardt C, Budde G, et al (2017) Age of Jupiter inferred from the distinct genetics and formation times of meteorites. Proc Natl Acad Sci 114(26):6712–6716. https://doi.org/10.1073/pnas.1704461114

Kuffmeier M, Frostholm Mogensen T, Haugbølle T, et al (2016) Tracking the distribution of 26Al and 60Fe during the early phases of star and disk evolution. Astrophys J 826(1):22. https://doi.org/10.3847/0004-637X/826/1/22. arXiv:1605.05008 [astro-ph.SR]

Lada CJ, Lada EA (2003) Embedded clusters in molecular clouds. Annu Rev Astron Astrophys 41:57–115. https://doi.org/10.1146/annurev.astro.41.011802.094844. arXiv:astro-ph/0301540 [astro-ph]

Lambrechts M, Johansen A (2014) Forming the cores of giant planets from the radial pebble flux in protoplanetary discs. Astron Astrophys 572:A107. https://doi.org/10.1051/0004-6361/201424343. arXiv:1408.6094 [astro-ph.EP]

Li R, Youdin AN (2021) Thresholds for particle clumping by the streaming instability. Astrophys J 919(2):107. https://doi.org/10.3847/1538-4357/ac0e9f. arXiv:2105.06042 [astro-ph.EP]

Li C, Ingersoll A, Bolton S, et al (2020) The water abundance in Jupiter's equatorial zone. Nat Astron 4:609–616. https://doi.org/10.1038/s41550-020-1009-3. arXiv:2012.10305 [astro-ph.EP]

Lichtenberg T, Golabek GJ, Burn R, et al (2019) A water budget dichotomy of rocky protoplanets from ^{26}Al-heating. Nat Astron 3:307–313. https://doi.org/10.1038/s41550-018-0688-5. arXiv:1902.04026 [astro-ph.EP]

Liu B, Johansen A, Lambrechts M, et al (2022) Natural separation of two primordial planetary reservoirs in an expanding solar protoplanetary disk. Sci Adv 8(16):eabm3045. https://doi.org/10.1126/sciadv.abm3045. arXiv:2204.10651 [astro-ph.EP]

Luhman KL (2020) A Gaia survey for young stars associated with the Lupus clouds. Astron J 160(4):186. https://doi.org/10.3847/1538-3881/abb12f. arXiv:2009.05123 [astro-ph.SR]

Lyons JR, Young ED (2005) CO self-shielding as the origin of oxygen isotope anomalies in the early solar nebula. Nature 435(7040):317–320. https://doi.org/10.1038/nature03557

Lyons JR, Bergin EA, Ciesla FJ, et al (2009) Timescales for the evolution of oxygen isotope compositions in the solar nebula. Geochim Cosmochim Acta 73(17):4998–5017. https://doi.org/10.1016/j.gca.2009.01.041

Marchington B, Parker RJ (2022) The evolution of protoplanetary disc radii and disc masses in star-forming regions. Mon Not R Astron Soc 515(4):5449–5466. https://doi.org/10.1093/mnras/stac2145. arXiv:2208.04330 [astro-ph.EP]

McCaughrean MJ, Pearson SG (2023) A JWST survey of the Trapezium Cluster & inner Orion Nebula. I. Observations & overview. E-prints arXiv:2310.03552 [astro-ph.SR]. https://doi.org/10.48550/arXiv.2310.03552

McKeegan KD, Kallio APA, Heber VS, et al (2011) The oxygen isotopic composition of the Sun inferred from captured solar wind. Science 332(6037):1528. https://doi.org/10.1126/science.1204636

McKinnon WB, Richardson DC, Marohnic JC, et al (2020) The solar nebula origin of (486958) Arrokoth, a primordial contact binary in the Kuiper Belt. Science 367(6481):aay6620. https://doi.org/10.1126/science.aay6620. arXiv:2003.05576 [astro-ph.EP]

Megeath ST, Gutermuth R, Muzerolle J, et al (2012) The Spitzer Space Telescope survey of the Orion A and B molecular clouds. I. A census of dusty young stellar objects and a study of their mid-infrared variability. Astron J 144(6):192. https://doi.org/10.1088/0004-6256/144/6/192. arXiv:1209.3826 [astro-ph.GA]

Megeath ST, Gutermuth R, Muzerolle J, et al (2016) The Spitzer Space Telescope survey of the Orion A and B molecular clouds. II. The spatial distribution and demographics of dusty young stellar objects. Astron J 151(1):5. https://doi.org/10.3847/0004-6256/151/1/5. arXiv:1511.01202 [astro-ph.GA]

Mesa-Delgado A, Zapata L, Henney WJ, et al (2016) Protoplanetary disks in the hostile environment of Carina. Astrophys J Lett 825(1):L16. https://doi.org/10.3847/2041-8205/825/1/L16. arXiv:1605.08809 [astro-ph.EP]

Miret-Roig N, Galli PAB, Olivares J, et al (2022) The star formation history of upper Scorpius and Ophiuchus. A 7D picture: positions, kinematics, and dynamical traceback ages. Astron Astrophys 667:A163. https://doi.org/10.1051/0004-6361/202244709. arXiv:2209.12938 [astro-ph.GA]

Monga N, Desch S (2015) External photoevaporation of the solar nebula: Jupiter's noble gas enrichments. Astrophys J 798(1):9. https://doi.org/10.1088/0004-637X/798/1/9. arXiv:1410.4870 [astro-ph.EP]

Morbidelli A, Levison HF (2004) Scenarios for the origin of the orbits of the trans-Neptunian objects 2000 CR_{105} and 2003 VB_{12} (Sedna). Astron J 128(5):2564–2576. https://doi.org/10.1086/424617. arXiv:astro-ph/0403358 [astro-ph]

Nesvorný D, Bernardinelli P, Vokrouhlický D, et al (2023) Radial distribution of distant trans-Neptunian objects points to Sun's formation in a stellar cluster. Icarus 406:115738. https://doi.org/10.1016/j.icarus.2023.115738. arXiv:2308.11059 [astro-ph.EP]

Noviello JL, Desch SJ, Neveu M, et al (2022) Let it go: geophysically driven ejection of the Haumea family members. Planet Sci J 3(9):225. https://doi.org/10.3847/PSJ/ac8e03

Ogliore RC, Huss GR, Nagashima K (2011) Ratio estimation in SIMS analysis. Nucl Instrum Methods Phys Res B 269(17):1910–1918. https://doi.org/10.1016/j.nimb.2011.04.120. arXiv:1106.0797 [astro-ph.IM]

Ouellette N, Desch SJ, Hester JJ (2010) Injection of supernova dust in nearby protoplanetary disks. Astrophys J 711(2):597–612. https://doi.org/10.1088/0004-637X/711/2/597

Pan L, Desch SJ, Scannapieco E, et al (2012) Mixing of clumpy supernova ejecta into molecular clouds. Astrophys J 756(1):102. https://doi.org/10.1088/0004-637X/756/1/102. arXiv:1206.6516 [astro-ph.SR]

Parker RJ, Maschberger T, Alves de Oliveira C (2012) A search for mass segregation of stars and brown dwarfs in ρ Ophiuchi. Mon Not R Astron Soc 426(4):3079–3085. https://doi.org/10.1111/j.1365-2966.2012.21790.x. arXiv:1208.0005 [astro-ph.GA]

Parker RJ, Lichtenberg T, Patel M, et al (2023) Short-lived radioisotope enrichment in star-forming regions from stellar winds and supernovae. Mon Not R Astron Soc 521(4):4838–4851. https://doi.org/10.1093/mnras/stad871. arXiv:2303.11393 [astro-ph.EP]

Pfalzner S, Vincke K (2020) Cradle(s) of the Sun. Astrophys J 897(1):60. https://doi.org/10.3847/1538-4357/ab9533. arXiv:2005.11260 [astro-ph.EP]

Pineda JE, Arzoumanian D, Andre P, et al (2023) From bubbles and filaments to cores and disks: gas gathering and growth of structure leading to the formation of stellar systems. In: Inutsuka S et al (eds) Protostars and Planets VII. Astronomical Society of the Pacific, San Francisco, pp 233–273. arXiv:2205.03935

Pinte C, Dent WRF, Ménard F, et al (2016) Dust and gas in the disk of HL tauri: surface density, dust settling, and dust-to-gas ratio. Astrophys J 816(1):25. https://doi.org/10.3847/0004-637X/816/1/25. arXiv:1508.00584 [astro-ph.SR]

Portegies Zwart S (2019) The formation of solar-system analogs in young star clusters. Astron Astrophys 622:A69. https://doi.org/10.1051/0004-6361/201833974. arXiv:1810.12934 [astro-ph.EP]

Ratzenböck S, Großschedl JE, Alves J, et al (2023) The star formation history of the Sco-Cen association. Coherent star formation patterns in space and time. Astron Astrophys 678:A71. https://doi.org/10.1051/0004-6361/202346901. arXiv:2302.07853 [astro-ph.SR]

Reynolds RJ, Ogden PM (1979) Optical evidence for a very large, expanding shell associated with the I Orion OB association, Barnard's loop, and the high galactic latitude Halpha filaments in Eridanus. Astrophys J 229:942–953. https://doi.org/10.1086/157028

Sakamoto N, Seto Y, Itoh S, et al (2007) Remnants of the early Solar System water enriched in heavy oxygen isotopes. Science 317(5835):231. https://doi.org/10.1126/science.1142021

Sengupta D, Cuzzi JN, Umurhan OM, et al (2024) Length and velocity scales in protoplanetary disk turbulence. Astrophys J 966(1):90. https://doi.org/10.3847/1538-4357/ad2c89. arXiv:2402.15475 [astro-ph.EP]

Smith RL, Pontoppidan KM, Young ED, et al (2009) High-precision $C^{17}O$, $C^{18}O$, and $C^{16}O$ measurements in young stellar objects: analogues for co self-shielding in the early Solar System. Astrophys J 701(1):163–175. https://doi.org/10.1088/0004-637X/701/1/163. arXiv:0906.1024 [astro-ph.EP]

Snider KD, Hester JJ, Desch SJ, et al (2009) Spitzer observations of the H II region NGC 2467: an analysis of triggered star formation. Astrophys J 700(1):506–522. https://doi.org/10.1088/0004-637X/700/1/506. arXiv:0711.1515 [astro-ph]

Störzer H, Hollenbach D (1999) Photodissociation region models of photoevaporating circumstellar disks and application to the proplyds in Orion. Astrophys J 515(2):669–684. https://doi.org/10.1086/307055

Tachibana S, Huss GR, Kita NT, et al (2006) ^{60}Fe in chondrites: debris from a nearby supernova in the early Solar System? Astrophys J Lett 639(2):L87–L90. https://doi.org/10.1086/503201

Tang H, Dauphas N (2012) Abundance, distribution, and origin of ^{60}Fe in the solar protoplanetary disk. Earth Planet Sci Lett 359:248–263. https://doi.org/10.1016/j.epsl.2012.10.011. arXiv:1212.1490 [astro-ph.EP]

Tatischeff V, Duprat J, de Séréville N (2014) Light-element nucleosynthesis in a molecular cloud interacting with a supernova remnant and the origin of Beryllium-10 in the protosolar nebula. Astrophys J 796(2):124. https://doi.org/10.1088/0004-637X/796/2/124. arXiv:1410.1455 [astro-ph.HE]

Tazzari M, Testi L, Natta A, et al (2017) Physical properties of dusty protoplanetary disks in Lupus: evidence for viscous evolution? Astron Astrophys 606:A88. https://doi.org/10.1051/0004-6361/201730890. arXiv:1707.01499 [astro-ph.EP]

Telus M, Huss GR, Ogliore RC, et al (2016) Mobility of iron and nickel at low temperatures: implications for ^{60}Fe-^{60}Ni systematics of chondrules from unequilibrated ordinary chondrites. Geochim Cosmochim Acta 178:87–105. https://doi.org/10.1016/j.gca.2015.11.046

Trapman L, Rosotti G, Bosman AD, et al (2020) Observed sizes of planet-forming disks trace viscous spreading. Astron Astrophys 640:A5. https://doi.org/10.1051/0004-6361/202037673. arXiv:2005.11330 [astro-ph.SR]

Trinquier A, Elliott T, Ulfbeck D, et al (2009) Origin of nucleosynthetic isotope heterogeneity in the solar protoplanetary disk. Science 324(5925):374. https://doi.org/10.1126/science.1168221

Trujillo CA, Brown ME (2001) The radial distribution of the Kuiper Belt. Astrophys J Lett 554(1):L95–L98. https://doi.org/10.1086/320917

Tsiganis K, Gomes R, Morbidelli A, et al (2005) Origin of the orbital architecture of the giant planets of the Solar System. Nature 435(7041):459–461. https://doi.org/10.1038/nature03539

Vacher LG, Ogliore RC, Jones C, et al (2021) Cosmic symplectite recorded irradiation by nearby massive stars in the Solar System's parent molecular cloud. Geochim Cosmochim Acta 309:135–150. https://doi.org/10.1016/j.gca.2021.06.026

Wall JE, Mac Low MM, McMillan SLW, et al (2020) Modeling of the effects of stellar feedback during star cluster formation using a hybrid gas and N-body method. Astrophys J 904(2):192. https://doi.org/10.3847/1538-4357/abc011. arXiv:2003.09011 [astro-ph.SR]

Warren PH (2011) Stable isotopes and the noncarbonaceous derivation of ureilites, in common with nearly all differentiated planetary materials. Geochim Cosmochim Acta 75(22):6912–6926. https://doi.org/10.1016/j.gca.2011.09.011

Williams JP, Gaidos E (2007) On the likelihood of supernova enrichment of protoplanetary disks. Astrophys J Lett 663(1):L33–L36. https://doi.org/10.1086/519972. arXiv:0705.3459 [astro-ph]

Wilson HF, Militzer B (2010) Sequestration of noble gases in giant planet interiors. Phys Rev Lett 104(12):121101. https://doi.org/10.1103/PhysRevLett.104.121101. arXiv:1003.5940 [astro-ph.EP]

Winter AJ, Clarke CJ, Rosotti G, et al (2018) Protoplanetary disc truncation mechanisms in stellar clusters: comparing external photoevaporation and tidal encounters. Mon Not R Astron Soc 478(2):2700–2722. https://doi.org/10.1093/mnras/sty984. arXiv:1804.00013 [astro-ph.SR]

Youdin AN, Lithwick Y (2007) Particle stirring in turbulent gas disks: including orbital oscillations. Icarus 192(2):588–604. https://doi.org/10.1016/j.icarus.2007.07.012. arXiv:0707.2975 [astro-ph]

Young ED (2007) Time-dependent oxygen isotopic effects of CO self shielding across the solar protoplanetary disk. Earth Planet Sci Lett 262(3–4):468–483. https://doi.org/10.1016/j.epsl.2007.08.011

Young ED (2014) Inheritance of solar short- and long-lived radionuclides from molecular clouds and the unexceptional nature of the Solar System. Earth Planet Sci Lett 392:16–27. https://doi.org/10.1016/j.epsl.2014.02.014. arXiv:1403.0832 [astro-ph.EP]

Young ED (2020) The birth environment of the Solar System constrained by the relative abundances of the solar radionuclides. In: Elmegreen BG, Tóth LV, Güdel M (eds) Origins: from the protosun to the first steps of life. Proceedings IAU Symposium, vol 345. Cambridge University Press, Cambridge, pp 70–77. https://doi.org/10.1017/S1743921319001777. arXiv:1909.06361

Yurimoto H, Kuramoto K (2004) Molecular cloud origin for the oxygen isotope heterogeneity in the Solar System. Science 305(5691):1763–1766. https://doi.org/10.1126/science.1100989

Zolensky ME, Zega TJ, Yano H, et al (2006) Mineralogy and petrology of comet 81P/Wild 2 nucleus samples. Science 314(5806):1735. https://doi.org/10.1126/science.1135842

Space Science Reviews (2024) 220:88
https://doi.org/10.1007/s11214-024-01122-w

Presolar Grains as Probes of Supernova Nucleosynthesis

Nan Liu[1] · Maria Lugaro[2,3,4,5] · Jan Leitner[6,7] · Bradley S. Meyer[8] · Maria Schönbächler[9]

Received: 11 July 2024 / Accepted: 29 October 2024 / Published online: 12 November 2024

Abstract
We provide an overview of the isotopic signatures of presolar supernova grains, specifically focusing on ^{44}Ti-containing grains with robustly inferred supernova origins and their implications for nucleosynthesis and mixing mechanisms in supernovae. Recent technique advancements have enabled the differentiation between radiogenic (from ^{44}Ti decay) and nonradiogenic ^{44}Ca excesses in presolar grains, made possible by enhanced spatial resolution of Ca-Ti isotope analyses with the Cameca NanoSIMS (Nano-scale Secondary Ion Mass Spectrometer) instrument. Within the context of presolar supernova grain data, we discuss (*i*) the production of ^{44}Ti in supernovae and the impact of interstellar medium heterogeneities on the galactic chemical evolution of ^{44}Ca/^{40}Ca, (*ii*) the nucleosynthesis processes of neutron bursts and explosive H-burning in Type II supernovae, and (*iii*) challenges in identifying the progenitor supernovae for ^{54}Cr-rich presolar nanospinel grains. Drawing on constraints and insights derived from presolar supernova grain data, we also provide an overview of our current understanding of the roles played by various supernova types – including Type II, Type Ia, and electron capture supernovae – in accounting for the diverse array of nucleosynthetic isotopic variations identified in bulk meteorites and meteoritic components. We briefly overview the potential mechanisms that have been proposed to explain these nucleosynthetic variations by describing the transport and distribution of presolar dust carriers in the protoplanetary disk. We highlight existing controversies in the interpretation of presolar grain data and meteoritic nucleosynthetic isotopic variations, while also outlining potential directions for future research.

Keywords Circumstellar matter · Stellar astronomy · Circumstellar dust · Nucleosynthesis · Type II supernovae · Type Ia supernovae · Electron capture supernovae · Meteorites

1 Introduction

Supernovae, extraordinary cosmic events, are pivotal in the nucleosynthesis of the common elements that are essential for forming star-planet systems, including our solar system. In cosmochemistry, the term "supernova" frequently appears in literature discussions on nucleosynthetic isotope variations and the initial presence of short-lived nuclides, which have been identified in diverse meteoritic components and bulk meteorites (e.g., Leya et al. 2009; Burkhardt et al. 2011; Akram et al. 2013; Cook and Schönbächler 2017; Ek et al.

Extended author information available on the last page of the article

2020; Schiller et al. 2020; Hopp et al. 2020; Nie et al. 2023). Some of these isotopic signatures have been suggested to originate from the heterogenous distribution of "supernova" nucleosynthesis products across the protoplanetary disk (e.g., Hutchison et al. 2022). "Supernova", in fact, is a broad term and encompasses various types, such as Type I, Type II, and electron-capture supernovae (ECSNe). Each type of supernovae contributes uniquely to element and isotope synthesis during galactic chemical evolution (GCE) – the process by which the abundances and isotopic compositions of the chemical elements in a galaxy, for example, the Milky Way Galaxy, change over time (Timmes et al. 1995; Prantzos et al. 2018; Kobayashi et al. 2020; Arcones and Thielemann 2023). Differentiating between nucleosynthesis products from the different supernova types and linking isotopic variations in meteoritic components to a specific supernova type are, therefore, crucial in understanding the origin and formation of the solar system.

Type II, Type Ib, and Type Ic supernovae, as classified spectroscopically, are all theoretically considered as core-collapse supernovae (CCSNe) originating from the collapse of the Fe core of massive stars (Modjaz et al. 2019; Burrows and Vartanyan 2021). The key similarity among them is the shared mechanism of Fe-core collapse. The core collapse occurs when the central Fe core of a massive star, which is produced by previous burning phases (from He burning up to Si burning), cannot sustain gravity anymore via energy released by nuclear burning and undergoes a gravitational collapse, leading to the infalling matter to bounce back from the forming neutron star and, consequently, generating a cataclysmic explosion. Type II supernovae are commonly associated with red supergiants with extended H envelopes, and Type Ib supernovae with progenitors such as Wolf-Rayet (WR) stars, where the outer layers of H were previously stripped by strong stellar winds. Type Ic supernovae are also associated with WR stars, except that in this case the WR stars lost both their H and He layers. Compared to Type Ib and Type Ic supernovae, Type II supernovae are more common and thus provide a more significant contribution to the GCE (Kobayashi et al. 2020). Type II supernovae were historically believed to be the astrophysical sites for rapid neutron-capture process (r-process) that produces roughly half of the cosmic abundances of the elements that are heavier than Fe and up to Th and U (e.g., Hillebrandt et al. 1976; Meyer et al. 1992; Woosley and Hoffman 1992). Recent simulations and observations, however, have changed this view. Model simulations encountered difficulties in producing, in typical Type II supernovae, the necessary neutron-rich environment for the r-process (e.g., Fischer et al. 2010). In addition, several pieces of evidence imply the infrequent occurrence of r-process events and challenge the suggestion that the frequently occurring Type II supernovae are the primary sites for the r-process, including large scatter in r-process elements detected among metal-poor stars (Sneden et al. 2008) and the initial ^{244}Pu abundances inferred for the early solar system and the present-day interstellar medium (ISM) based on meteorite and deep-sea measurements, respectively (Hotokezaka et al. 2015). The current consensus posits that the r-process could occur in rare Type II supernovae such as magnetorotational supernovae, alongside the mergers of compact objects such as neutron stars (NS) and black holes (Thielemann et al. 2017). Indeed, the gravitational wave event GW170817 that was associated to a short γ-ray burst and caused by a NS-NS merger event, was observed as a kilonova r-process event (Kilpatrick et al. 2017; Pian et al. 2017).

Type Ia supernovae instead originate from the thermonuclear explosion of white dwarf stars (see Ruiter 2020 for a review). In a binary star system, where two stars orbit each other, if one or both of the stars are white dwarfs, they can accrete mass from the companion or merge, eventually undergoing a cataclysmic event and leading to a Type Ia supernova. Type Ia supernovae, alongside Type II supernovae, contribute significantly to the Fe-group elements present in the ISM but at a later time compared to Type II supernovae (Tinsley

1979; Greggio and Renzini 1983). This temporal variation in the contributions of Type II and Type Ia supernovae to the galactic Fe-group element inventory is caused by the longer evolutionary time of Type Ia supernovae, given the time it takes for their progenitor stars to transition to the white dwarf phase and the following accretion and merger.

Electron capture supernovae represent another intriguing class, which is triggered by electron capture in the core and could occur as thermonuclear or collapsing ECSNe (Jones et al. 2019a). These events were predicted to occur in $8-10\ M_\odot$ stars with degenerate O/Ne/Mg cores (Nomoto 1984, 1987). Model predictions of the stellar evolution within this mass regime, however, remain uncertain (Jones et al. 2013), and observational evidence of their existence was lacking. Thus, ECSNe had long faced skepticism until a recent breakthrough detection that confirmed the occurrence of an ECSN (Hiramatsu et al. 2021). Wanajo et al. (2011) suggested the potential for a weak r-process, i.e., reaching only up to atomic mass $\approx$ 100, in ECSNe, albeit with some debate surrounding this proposal (Jones et al. 2016).

Accurate nucleosynthesis model predictions for all these supernova types are critical for distinguishing between their nucleosynthesis products and linking isotopic variations in meteoritic components to the specific supernova type, thus allowing for a fundamental understanding of the origin of the solar system. Many challenges, however, persist due to uncertainties in the structure and evolution of the progenitor star, the explosion hydrodynamics, and the nuclear reaction rates, as well as neutrino transport related to the explosion mechanism of Type II supernovae, the still debated evolution channel(s) for Type Ia supernovae, and the unknown likelihood of 8–10 $M_\odot$ stars evolving to ECSNe. Direct observations of isotope abundances within these supernovae offer ideal tests for nucleosynthesis models. Such measurements, however, are currently limited to a few radioactive isotopes (*e.g.*, ^{56}Ni and ^{44}Ti; Diehl 2017).

In the aftermath of a supernova explosion, the condensation of dust becomes plausible when the temperature undergoes a significant drop (below ~2500 K) while maintaining a high density in the ejecta (Sarangi et al. 2018). Current spectroscopic observations suggest that (*i*) dust forms efficiently in Type II supernova remnants (Wesson and Bevan 2021), e.g., SN 1987A (Fig. 1b), (*ii*) Type Ia supernovae show no definitive evidence of dust, potentially due to low formation efficiencies and/or elevated destruction rates (Douvion et al. 2001), and (*iii*) dust formation in ECSNe remains unknown because of the only recent first detection of such an event (Hiramatsu et al. 2021). Considering the observed efficient dust formation around and high occurrence frequencies of Type II supernovae, dust from supernovae in the ISM is anticipated to come predominantly from Type II supernovae. Supernova dust is expected to have served as one of the foundational constituents of the ISM material that formed the solar system at about 4.6 Ga ago and then became incorporated and preserved in primitive extraterrestrial materials – the leftover building blocks of the planets.

Consistent with this expectation, ancient stardust grains, which include contributions predominantly from asymptotic giant branch (AGB) stars, Type II supernovae, and possibly novae, have been identified in primitive extraterrestrial materials, offering a direct probe into the conditions for the birth of the solar system (Nittler and Ciesla 2016; Liu 2024). These grains range from submicron up to tens of microns in size and are diverse in chemical composition, including phases such as silicon carbide (SiC), silicon nitride (Si_3N_4), graphite, oxides, and silicates. Certain nanospinel grains, showing large ^{54}Cr excesses, have been suggested to originate from Type Ia supernovae or ECSNe (Nittler et al. 2018), albeit a potential Type II supernova origin was also proposed (Dauphas et al. 2010; Qin et al. 2011; den Hartogh et al. 2022). Collectively termed as "presolar grains", these minute cosmic fossils recorded the nucleosynthetic legacy of their parent stars. In the literature, presolar SiC

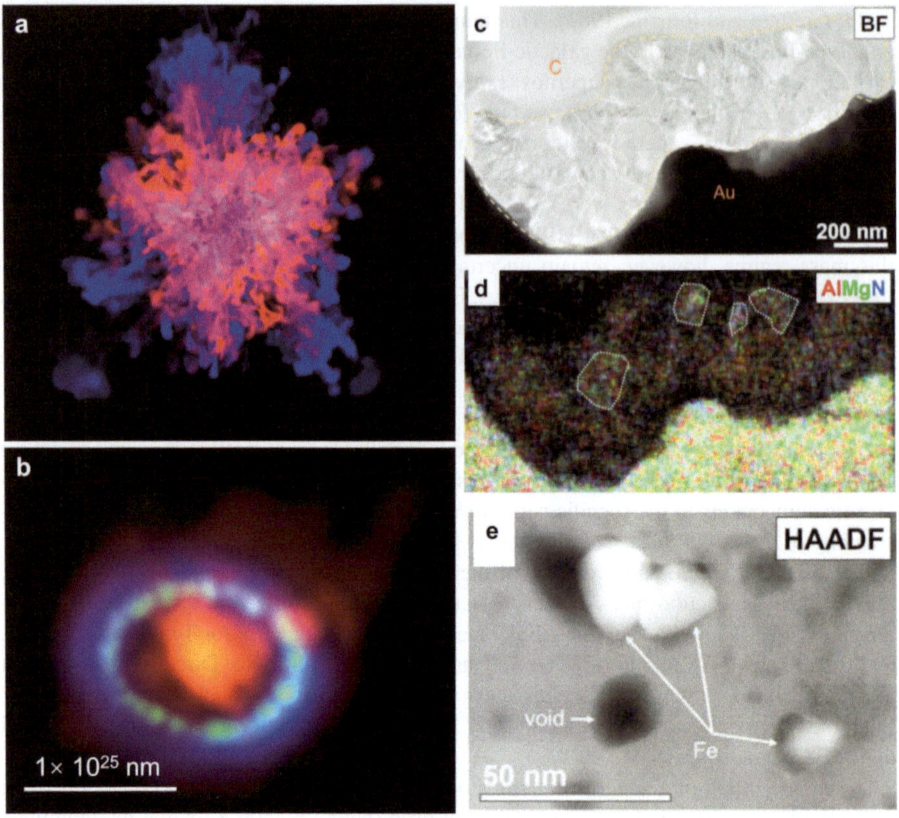

Fig. 1 Panel (a) presents the 3D simulation of a CCSN at 333 seconds post-ignition with elements represented in different colors: Ni (blue), O (red), C (green), and Ni-O mixture (pink), calculated by Dr. H.-T. Janka (Max Planck Institute for Astrophysics). Panel (b) shows the Atacama Large Millimeter/submillimeter Array (ALMA) telescope multiwavelength observation of SN 1987A (Indebetouw et al. 2014), highlighting dust (red), the circumstellar gas ring (green), and the supernova blast-induced shock wave (purple) (image credit: Alexandra Angelich). Panels (c) through (e) feature the bright field (BF) image, energy-dispersive X-ray composite image, and high-angle annular dark field (HAADF) image, respectively, of a Type-X presolar SiC grain, which formed in the remnant of a Type II supernova explosion that occurred before the solar system formation (Singerling et al. 2021). The X grain is denoted by the contour line in panel (c), in which the letters "c" and "Au" denote carbon deposited on top of the grain during sample preparation and the gold foil on which the grain was initially dispersed, respectively (see Singerling et al. 2021 for details). In panel (d), the high Mg concentration (in green) of the X grain results from the radiogenic decay of ^{26}Al ($t_{1/2} = 0.72$ Ma), since X grains are enriched in Al with high initial ^{26}Al/^{27}Al ratios (Hoppe et al. 2023; Liu et al. 2024a)

and O-rich grains are divided into different groups based on their isotopic compositions, and presolar graphite grains are divided into high-density and low-density grains based on their density separation during the acid extraction procedure (Amari et al. 1994; Stephan et al. 2024). The adopted classification schemes for various presolar phases are useful in identifying their stellar origins (see Liu 2024 for a recent review).

Figure 1 exemplifies the interconnection between the laboratory studies of these grains and the broader astronomical and astrophysical research in understanding Type II supernovae. Meteorite investigations point to the role of Type II supernovae as prolific dust factories (Hoppe et al. 2000; Nittler and Ciesla 2016; Liu 2024), in line with the implication from the multiwavelength observations of SN 1987A remnant via ALMA (Fig. 1b; Indebetouw et al. 2014). X grains are a group of presolar SiC grains with robust Type II supernova origins inferred from their multielement isotope data (Nittler et al. 1996; Lin et al. 2010; Xu et al. 2015; Liu et al. 2018a, 2024a). The presence of Fe subgrains within an X grain shown in Fig. 1e offers a granular view of the mixing process that ensues following a supernova explosion (Singerling et al. 2021), phenomena that can now be juxtaposed with three-dimensional (3D) simulations, as exemplified in Fig. 1a (Wongwathanarat et al. 2015, 2017). By piecing together insights from large-scale computational models, telescope observations, and microscopic-scale meteoritic data, we can significantly enhance our capability to comprehend the evolution of supernovae, their explosive nucleosynthesis, and dust formation in these spectacular celestial events.

This review will delve into recent advancements in expanding the inventory and multielement isotope data of presolar supernova grains identified within primitive meteorites and the interpretation of such data, aimed at unraveling new implications for supernova nucleosynthesis. Guided by presolar supernova grain data, we will focus on a comparative analysis of the stellar nucleosynthesis of the three supernova types — Type II, Type Ia, and ECSNe. Finally, we will contextualize these implications within the realm of nucleosynthetic isotope variations in bulk meteorites. This comprehensive examination will guide our future exploration of (*i*) the supernova type that is responsible for the observed nucleosynthetic variations in meteorites and (*ii*) the early solar system processes that led to the widespread distribution of isotopic variations across the protoplanetary disk, from the perspective of existing presolar grain data.

2 Presolar Supernova Grains and Their Isotopic Signatures

2.1 Titanium-44: A Smoking Gun of Supernova Nucleosynthesis

2.1.1 Production of ^{44}Ti in Supernovae

In the context of supernova nucleosynthesis, the radioactive isotope ^{44}Ti ($t_{1/2} = 60$ a) emerges as a compelling indicator, particularly for Type II supernovae. This unequivocal status stems from two key observations: (*i*) ^{44}Ti is predicted to be produced primarily during explosive O burning and α-rich freeze outs from nuclear statistical equilibrium (NSE) in the deep layers of Type II supernovae (The et al. 2006; Sieverding et al. 2023), and (*ii*) unambiguous detection of γ-ray emissions from the decay of ^{44}Ti was reported in the Type II supernova remnants SN 1987a and Cassiopeia A (Cas A; Grefenstette et al. 2014, 2017; Boggs et al. 2015). In contrast, Type Ia supernovae and ECSNe are anticipated to yield ^{44}Ti in quantities one order of magnitude lower than Type II supernovae (Nomoto 1984, 1987; Nomoto et al. 1997; The et al. 2006). This is because, in the former two types of supernovae, ^{44}Ti is produced only through explosive He burning since explosive O burning and the conditions for α-rich freeze outs are typically not attained (Woosley et al. 1986). Notably, the detection of γ-ray emissions from ^{44}Ti in Type Ia supernovae remains elusive (Weinberger et al. 2020), and no detection of ^{44}Ti γ-ray emissions in ECSNe has been reported. *Integrating these pieces of evidence suggests that the initial presence of ^{44}Ti in a presolar grain provides a direct piece of evidence to support its origin in supernovae, although further examination based on other isotope signatures is always required to pinpoint the specific supernova type.*

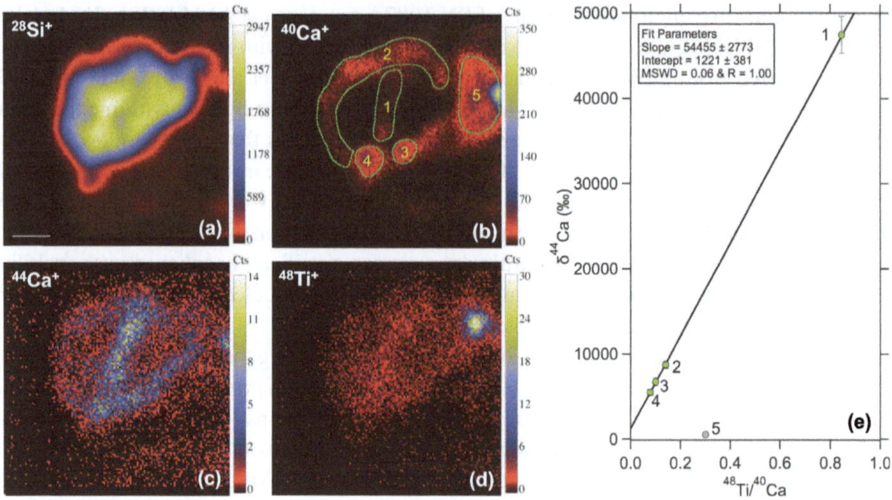

Fig. 2 NanoSIMS Ca-Ti isotope data of a Type-X SiC grain from Liu et al. (2023a). Panels (a) to (d) show the ^{28}Si, ^{40}Ca, ^{44}Ca, and ^{48}Ti ion images, respectively, collected from the grain at a spatial resolution of ~150 nm. The white scalebar in panel (a) denotes 500 nm. In panel (b), five different regions of interest (ROIs) were defined for reducing the Ca-Ti isotope data shown in panel (e), where δ^{44}Ca was calculated as $[(^{44}Ca/^{40}Ca)_{grain} / (^{44}Ca/^{40}Ca)_{standard} - 1] \times 1000$, and the linear fit was obtained using the CEREs-Fit.xlsm tool of Stephan and Trappitsch (2023). MSWD stands for mean squared weighted deviation, and R is the Pearson correlation coefficient. Errors are all 1σ. The data for ROI #5 was excluded from the linear fit because of its significant deviation. This deviation is likely predominantly due to some Ti contaminant present at the grain edge, resulting in an increase in the $^{48}Ti/^{40}Ca$ ratio

The absence of initial ^{44}Ti in a presolar grain, however, does not conclusively exclude a supernova origin, considering its low productions in Type Ia supernovae and ECSNe and the potential lack of ^{44}Ti (despite its production in the deep region) incorporated into the parent supernova ejecta of the grain. Nevertheless, here we focus our discussions primarily on presolar grains with inferred initial presence of ^{44}Ti because of their robustly determined supernova origins.

2.1.2 Detection of ^{44}Ti in Presolar Grains

Due to the short half-life of ^{44}Ti compared to the residency of presolar grains in the ISM (up to ~1 Ga; Gail et al. 2009; Heck et al. 2020) and within the solar system (4.6 Ga), any ^{44}Ti atoms initially present within a presolar grain have already fully decayed to ^{44}Ca. This decay process can potentially yield a detectable $^{44}Ca/^{40}Ca$ excess in the grain, depending on its Ca/Ti ratio (i.e., $^{40}Ca/^{48}Ti$ in Fig. 2e). The analysis of Ca-Ti isotope systematics in presolar grains is routinely conducted using the Cameca Nanoscale Secondary Ion Mass Spectrometer (NanoSIMS; Hoppe 2006). In the past, direct, quantitative distinction between radiogenic (from ^{44}Ti decay) and nonradiogenic ^{44}Ca excess was challenging, leading to the derivation of the initial $^{44}Ti/^{48}Ti$ ratio based on the assumption that the detected ^{44}Ca excess in a presolar grain is fully due to the ^{44}Ti decay (Nittler et al. 1996). Recent advancements in spatial resolution, enabled by the Hyperion radio-frequency plasma O^- ion source manufactured by Oregon Physics (Malherbe et al. 2016), have empowered NanoSIMS analyses to directly discriminate between radiogenic and nonradiogenic ^{44}Ca excess for μm-sized and/or Ti-rich (and Ca-poor) presolar grains (Liu et al. 2023b; Nittler et al. 2023; Leitner et al. 2024).

Figure 2 provides an illustrative example of high-resolution Ca and Ti imaging data. The observed correlation of Ca and Ti ions at high spatial resolution enables the differentiation between radiogenic and nonradiogenic ^{44}Ca excess in a Type-X SiC grain. The evident correlation between the distributions of $^{44}Ca^+$ and $^{48}Ti^+$ (rather than $^{40}Ca^+$) within the grain directly implies that $^{44}Ca^+$ was initially present in the grain as $^{44}Ti^+$. In other words, the predominant radiogenic origin of $^{44}Ca^+$ as the decay product of $^{44}Ti^+$ is demonstrated by the fact that $^{44}Ca^+$ is not spatially correlated with $^{40}Ca^+$ that shares the same chemical property with $^{44}Ca^+$ but instead with $^{48}Ti^+$ that is chemically the same as $^{44}Ti^+$. The variable $^{48}Ti/^{40}Ca$ ratios across the grain further allow for the derivation of the initial $^{44}Ti/^{48}Ti$ ratio based on the linear fit, known as an internal isochron, shown in Fig. 2(e). The slope of the isochron yields an initial $^{44}Ti/^{48}Ti$ ratio of 1.18 ± 0.06, and the y-axis intercept corresponds to a nonradiogenic ^{44}Ca excess of $1221 \pm 381‰$, revealing the production of both ^{44}Ca and ^{44}Ti during Type II supernova nucleosynthesis. Thus, high-resolution NanoSIMS imaging of Ca and Ti isotopes provides a powerful tool for evaluating the origin of ^{44}Ca excesses in presolar grains, thereby allowing for a more accurate determination of both the initial $^{44}Ti/^{48}Ti$ and nonradiogenic $^{44}Ca/^{40}Ca$ ratios in these grains.

To date, ^{44}Ca excesses[1] have been detected in various presolar phases and groups, including a substantial fraction of Type X SiC grains (Nittler et al. 1996; Lin et al. 2010; Xu et al. 2015), a few Type C and ungrouped SiC grains (Hoppe et al. 2010; Xu et al. 2015), a single Type AB SiC grain (high-resolution imaging; Liu et al. 2023b), certain presolar graphite grains (Nittler et al. 1996; Amari et al. 2014), a few Group 1 and Group 4 silicates (Leitner and Hoppe 2022; Leitner et al. 2024), a few Group 4 oxides (Nittler et al. 2008), one Group 3 oxide (Nittler et al. 2011), and a ungrouped ^{16}O-rich oxide (Gyngard et al. 2010). Among the phases analyzed, SiC is one of the most suitable phases for inferring the initial $^{44}Ti/^{48}Ti$ ratio because of its Ti enrichment, low intrinsic Ca consent, and typically large size. In contrast, oxides and silicates[2] typically show higher Ca concentrations compared to Ti, resulting in their overall smaller radiogenic ^{44}Ca excesses (e.g., Nittler et al. 2008). The presence of intrinsic Ca in these phases, coupled with small ^{44}Ca excesses, introduce ambiguities in the interpretation of their ^{44}Ca excesses. This is because (*i*) both ^{44}Ca and ^{44}Ti can be produced by stellar nucleosynthesis in a star, and, additionally, (*ii*) the initial composition of the star may differ from the solar composition because of the effects of GCE and heterogeneities in the ISM. Heterogenous ISM compositions could mean local enrichments of ^{44}Ti and/or ^{44}Ca in the ISM (The et al. 2006), and stars that subsequently form in such zones may exhibit ^{44}Ca excesses in their initial composition relative to the broader Galactic baseline at the time. This results in their stardust grains to display GCE-related nonradiogenic ^{44}Ca excesses.

Finally, the determination of the initial $^{44}Ti/^{48}Ti$ ratio from SIMS Ca-Ti isotope data of presolar SiC and potentially graphite grains, faces additional uncertainties related to the SIMS relative sensitivity factor (RSF), which is needed for determining the Ca/Ti elemental ratio. The RSF, which corrects for different ionization efficiencies of different elements in the SIMS instrument, can be different for different materials and must be determined through

[1] The ^{44}Ca excesses detected in presolar grains so far have mostly been collected using NanoSIMS instruments equipped with the previous generation of O^- plasma ion source at poorer spatial resolutions, compared to the data shown in Fig. 2. These previous studies, in most cases, could not quantitively distinguish between radiogenic and nonradiogenic contributions to their observed ^{44}Ca excesses and assumed that the observed ^{44}Ca excesses were solely due to ^{44}Ti decay.

[2] In addition, presolar silicates are measured *in situ* in meteoritic matrices, which leads to unavoidably sampling adjacent Ca-rich matrix materials if present and, in turn, diluted isotopic anomalies.

measurements of standards. Currently, this factor is calibrated using NIST O-rich glass standards due to the absence of viable alternatives. Recent findings by Liu et al. (2024a) reported that the utilization of O-rich standards for calibrating the SIMS RSF value for the elemental Mg/Al ratio results in a factor of two reduction in the derived initial $^{26}Al/^{27}Al$ ratio for presolar SiC grains. This is caused by different sample versus standard oxidation states. This discrepancy may have similarly impacted the inferred initial $^{44}Ti/^{48}Ti$ ratios of presolar SiC and graphite grains. Consequently, it is essential to bear in mind a factor of two potential, systematic uncertainty when comparing presolar SiC and graphite grain data to model calculations.

In summary, high-resolution NanoSIMS imaging has enabled quantitative distinction between nonradiogenic and radiogenic ^{44}Ca excesses in μm-sized presolar grains. The inferred initial $^{44}Ti/^{48}Ti$ ratios of presolar SiC and graphite grains, however, still suffer from uncertainties in the SIMS RSF value, which needs to be addressed in future work.

2.1.3 The Initial Presence of ^{44}Ti in Type-X SiC Grains

Hypothesis: Production of ^{44}Ti in Rare Type II Supernovae The production of ^{44}Ti is recognized as one of the contributing sources of the cosmic abundance of ^{44}Ca. It is, however, debated whether typical Type II supernovae alone govern the production of ^{44}Ti in the Galaxy. Two key observations challenge this view (The et al. 2006): (*i*) given the short half-life of ^{44}Ti and the Galactic Type II supernova occurrence rate of approximately three per 100 a, several ^{44}Ti γ-ray sources are expected, yet only one, the 360-year-old Cas A, was definitively identified in the Galaxy; and (*ii*) Cas A is located far from the inner Galaxy, contrary to the expectation that the inner region, rich in young massive stars, should be the primary location for ^{44}Ti γ-ray sources if typical Type II supernovae are the main contributors of ^{44}Ti. In light of these inconsistencies, The et al. (2006) proposed that either (*i*) Type II supernovae have been improbably rare in the Galaxy during the past few centuries, or (*ii*) ^{44}Ti is not predominantly produced by typical supernovae but by rare events.

Therefore, the question arises whether the lack of inferred initial ^{44}Ti in some X grains may provide evidence for the idea that ^{44}Ti is not predominantly produced in typical Type II supernovae or reflects reduced incorporation of ^{44}Ti-containing supernova materials into some of the grains. X grains are characterized by large ^{28}Si excesses (Nittler et al. 1996), likely due to explosive O burning in the Si/S zone (see Fig. 3) (Lin et al. 2010) and/or α-rich freeze outs in the Fe/Ni zone (see Fig. 3) (Hoppe et al. 2000; Liu et al. 2024a). Explosive O burning in the Si/S zone significantly overproduces ^{28}Si, while α-rich freeze outs in the Fe/Ni zone can overproduce both ^{28}Si and ^{29}Si with ^{29}Si being produced by proton capture onto ^{28}Si because of the free protons released by neutrino-nucleus reactions there (Liu et al. 2024a). The latter explains the fact that, in the plot of δ^{29}Si versus δ^{30}Si, X grains predominantly fall along a slope-2/3 line instead of the slope-1 line that is predicted if the Si/S zone with pure ^{28}Si dominantly controls the ^{28}Si excesses of X grains (Fig. 4a). In a Type II supernova, predominant production of ^{44}Ti is predicted by α-rich freeze outs and explosive O burning (The et al. 1998). This leads to the expectation that if the parent supernova of an X grain produced ^{44}Ti in these zones, ^{28}Si and ^{44}Ti could be incorporated into the grain together. In turn, the absence of inferred initial ^{44}Ti in some X grains may imply the lack of ^{44}Ti production in their parent supernovae. Let us examine this scenario below in detail based on X grain data compiled in the Presolar Grain Database (PGD; Stephan et al. 2024).

X Grain Formation: Large-Scale Versus Local Mixing Scenarios To compare the isotopic composition of an X grain with nucleosynthesis models for Type II supernova explosions,

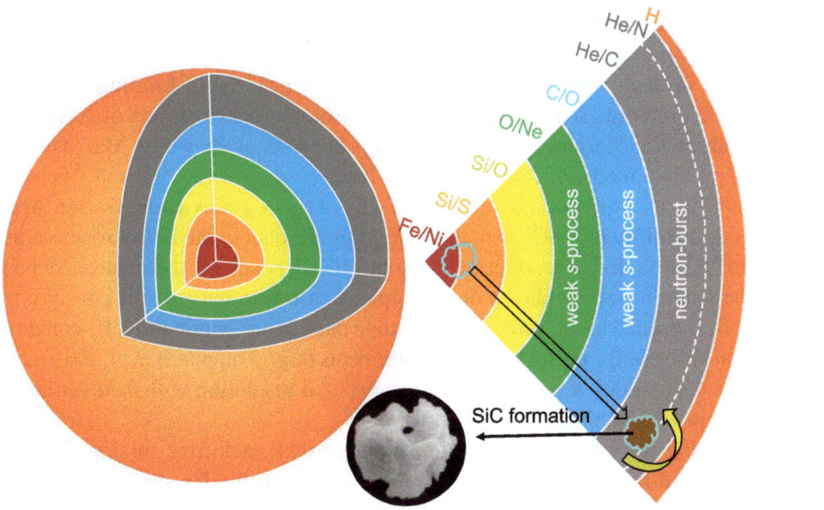

Fig. 3 Illustration of the onion shell structure of a presupernova massive star. Each zone (not to scale) is labeled by the two most abundant elements it contains (Meyer et al. 1995). On the right, illustrated is the formation process of X grains after the explosion when assuming that the grains incorporated material from the Fe/Ni zone adjacent to the central neutron star, the inner Si/S zone, and the He (and possibly H) shells

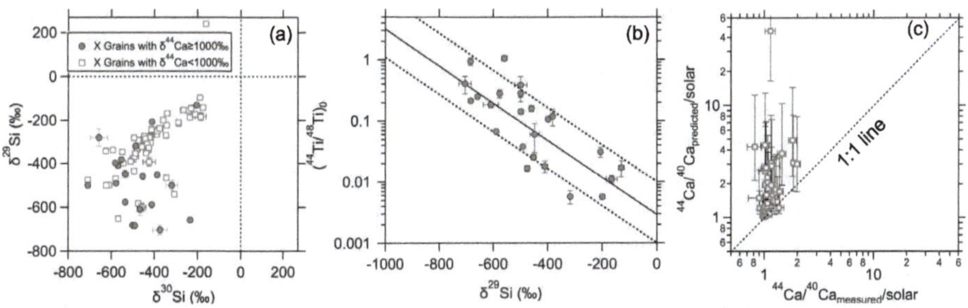

Fig. 4 Panel (a) shows Si isotope data for 26 SiC X grains with ^{44}Ca excesses $\gtrsim$ 1000‰ (Amari et al. 1992, 2003; Nittler et al. 1996; Besmehn and Hoppe 2003; Lin et al. 2010; Xu et al. 2015; Liu et al. 2023a) and 53 SiC X grains from Lin et al. (2010) and Liu et al. (2018a, 2023a) with smaller ^{44}Ca excesses ($<$ 500‰) and available Ca and Ti concentration data. Panel (b) illustrates isotope data for the 26 SiC X grains with large ^{44}Ca excesses. The solid line is an exponential fit to the grain data, and 90% of the grains lie within the dashed lines. Panel (c) shows predicted versus measured values of the 53 SiC X grains with smaller ^{44}Ca excesses. See the main text for the calculation of ^{44}Ca/^{40}Ca$_{\mathrm{predicted}}$/solar

parametrized mixing of materials ejected from varying zones of the parent star has been traditionally invoked (Hoppe et al. 1996b; Nittler et al. 1996). This is because mixing processes in state-of-the-art 3D supernova simulations are quite uncertain and not well constrained (Müller 2020). Thus, comparing unmixed one-dimensional (1D) supernova nucleosynthesis model predictions for the different shells shown in Fig. 3 with the multielement isotopic compositions of X grains, offers a feasible approach for constraining the involved nucleosynthesis processes (Lin et al. 2010).

Traditionally, it was proposed that X grains incorporated materials predominantly from the Si/S and He/C zones (but not the shells inbetween), with the former contributing ^{28}Si-rich and the latter contributing C-rich material (C > O) that is essential for the formation of SiC grains (Amari et al. 1992; Nittler et al. 1996). Additional evidence from Ca-Ti isotope systematics further implicated the mixing of Fe/Ni material in forming X grains (Hoppe et al. 2000; Lin et al. 2010), although the Fe isotopic compositions of X grains seem to suggest otherwise (Marhas et al. 2008). This is because, as shown by Marhas et al. (2008), the 25 $M_{\odot}$ CCSN model of Rauscher et al. (2002) predicts large ^{54}Fe and ^{56}Fe enrichments in the Fe/Ni zone, which are not observed in any of the X grains studied so far. According to Woosley and Hoffman (1992), whether α-rich freeze outs in the Fe/Ni zone predominantly produce ^{56}Fe (the decay product of ^{56}Ni) or ^{54}Fe depends on the neutron richness there. Since the neutron-rich condition during alpha-rich freeze outs is still poorly constrained, more work is needed to interpret the Fe isotope data. More importantly, Fe contamination is a common problem for presolar SiC measurements (e.g., Trappitsch et al. 2018). NanoSIMS Fe isotope analyses of X grains at improved spatial resolution will shed further light onto the mixing process in their parent supernovae.

Challenging the conventional notion of large-scale, selective mixing, Pignatari et al. (2013b) posited that X grains could derive solely from the He shell and nearby regions, attributing their ^{28}Si enrichments to high-temperature α-capture in a newly identified zone, the so-called C/Si zone, produced in a higher-than-standard energetic 1D supernova model. The energy required for forming the C/Si zone in the model of Pignatari et al. (2013b) corresponds to hypernovae,[3] which are more massive and rarer than typical CCSNe. Several subsequent studies supported this local mixing theory to explain the multielement isotopic compositions of X grains (Xu et al. 2015; Hoppe et al. 2018, 2023, 2024; Pignatari et al. 2018; den Hartogh et al. 2022; Schofield et al. 2022). However, these data-model comparisons relied on an ad hoc mixing calculation method with intrinsic uncertainties because of its random selection and weighting of different shell materials.

Addressing these uncertainties, several recent investigations have reported correlations between different isotope ratios to deconvolute the isotopic signatures of different contributing supernova zones, thereby enhancing the accuracy in constraining supernova nucleosynthesis (Liu et al. 2018a, 2024a). For example, Liu et al. (2024a) highlighted a shortfall of the local mixing hypothesis: although Hoppe et al. (2023) reproduced their X grain isotopic compositions by invoking the C/Si zone and explosive H burning (i.e., the localized mixing scenario) based on the 1D, parametric models of Pignatari et al. (2015), the explosive H-burning zone (^{30}Si-rich) produces much higher ^{26}Al/^{27}Al ratios than the C/Si zone (^{28}Si-rich), which consequently implies a positive trend between ^{26}Al/^{27}Al and δ^{30}Si. In contrast to this expectation, a negative trend was observed among X grains (Liu et al. 2024a). The observed positive correlation between ^{26}Al and ^{28}Si excesses in X grains may instead support a large-scale mixing scenario (Fig. 3), in which neutrino-nucleus interactions within the Fe/Ni zone foster a proton-rich condition that enables substantial ^{26}Al production, alongside ^{28}Si synthesis via α-rich freeze outs in the same region. More detailed calculations of neutrino-nucleus reactions following a supernova explosion are currently being pursued to investigate modelling uncertainties, based on which constraints can be derived on relevant parameters based on the X grain data (Walls and Meyer 2024).

Another example is related to the abundance pattern of α elements. The α-rich freeze outs in the Fe/Ni zone and α-capture in the C/Si zone can produce similar isotopic signatures

[3] A hypernova is an extremely energetic type of supernovae resulting from the Fe-core collapse in massive stars (>30 $M_{\odot}$), leading to the formation of black holes. It ejects material with exceptionally high kinetic energy, around ten times greater than typical supernovae (Moriya et al. 2018). Alternatively, if asymmetric explosions are typical of CCSNe, such an energetic explosive condition may be achieved in certain directions during a typical CCSN explosion.

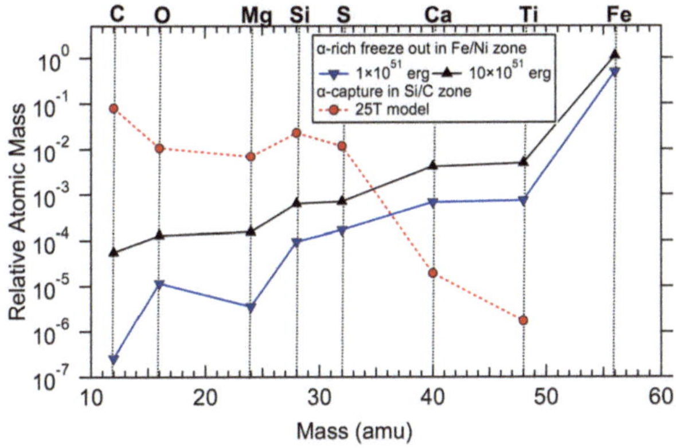

Fig. 5 Plot comparing the productions of α elements by (i) α-capture in the C/Si zone of the 25T CCSN model of Pignatari et al. (2013b) (in red) and by α-rich freeze outs in the 25 $M_{\odot}$ model from Liu et al. (2024a) at two different explosion energies (in black and blue). These latter calculations (Rauscher-Meyer models hereafter) explored explosive nucleosynthesis based on the 25 $M_{\odot}$ presupernova stellar model of Rauscher et al. (2002) and the explosion model and reaction network of Bojazi and Meyer (2014). Iron-56 is the decay product of ^{56}Ni made by NSE and the most abundant Fe isotope in this region. Nickel-56 decays to ^{56}Co with $t_{1/2} = 6.1$ d, which then decays to ^{56}Fe with $t_{1/2} = 77$ d

(Liu et al. 2018a; Schofield et al. 2022), e.g., enrichments in ^{44}Ti, ^{40}Ca, ^{48}Ti. Current models, however, predict that the two processes lead to distinct abundance patterns for α elements for the following reasons (Fig. 5). Alpha-rich freeze outs can modify the NSE products that are characterized by a peak at ^{56}Ni because of its highest binding energy (Woosley and Hoffman 1992). In comparison, the seeds for α-capture in the C/Si zone are the ^{12}C produced by He burning. While α-capture onto ^{12}C at high temperatures can lead to efficient production of heavier α elements such as Si and S, the efficiency of the α-capture, however, drops significantly in synthesizing heavier elements in the C/Si zone.

The isotopic compositions of X grains argue against the low Ti/Si and Ti/Ca ratios that characterize α-capture in the C/Si zone in the models of Pignatari et al. (2013b, 2015) and thus prefer the large-scale mixing scenario. The negative correlation between the radiogenic ^{49}Ti (from ^{49}V decay) and ^{28}Si excesses of X grains (Liu et al. 2018a, 2023a) can be reproduced by mixing between materials from the Si/S and He/C zones (Liu et al. 2023a). In addition, with respect to ^{49,50}Ti excesses in X grains, much higher than expected ^{46}Ca excesses have been observed in Si_3N_4 grains (Liu et al. 2024b), which likely condensed out of the same supernova ejecta with X grains but at lower temperatures according to their isotopic similarities (Nittler et al. 1995; Hoppe et al. 1996a; Lin et al. 2010). This observation can be explained by mixing materials from the Fe/Ni zone and the He/C zone because ^{48}Ti is produced in a larger quantity than ^{40}Ca in the Fe/Ni zone so that its mixing with the He/C zone leads to larger reduction in ^{49,50}Ti/^{48}Ti than in ^{46}Ca/^{40}Ca in the ejecta (Liu et al. 2024b). Thus, both the observations argue against the low Ti/Si and Ti/Ca ratios that characterize α-capture in the C/Si zone (Fig. 5) in the models of Pignatari et al. (2013b, 2015).

Further information can be derived from current 3D CCSN explosion model simulations and astronomical observations. Observations of CCSN remnants show that ^{56}Ni and ^{44}Ti knots from the innermost region lie outside of the central region of the remnants, supporting large-scale macroscopic mixing (Grefenstette et al. 2017). The microscopic mixing required to form dust with material from the inner and the outer layers (as required in the large-scale

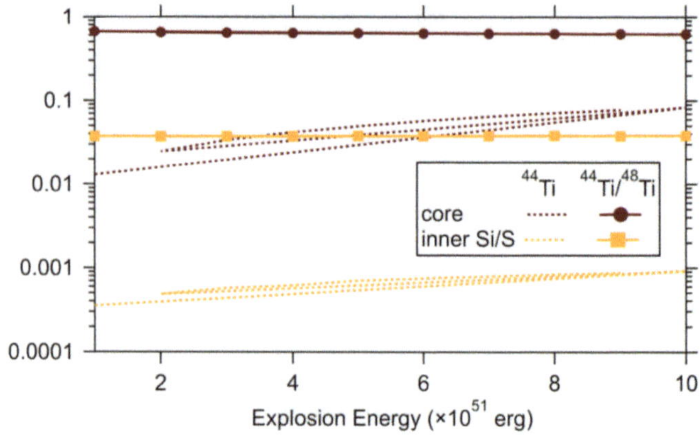

Fig. 6 The ^{44}Ti/^{48}Ti ratio and ^{44}Ti abundance in the Fe/Ni and inner Si/S zones predicted by the Rauscher-Meyer model calculations for a 25 $M_{\odot}$ progenitor massive star as a function of explosion energy. The corresponding model calculations at two of the energies (1×10^{51} erg and 10×10^{51} erg) are shown in Fig. 5. The Fe/Ni and inner Si/S zones are shown here for illustration of the large-scale mixing scenario that explains the isotopic compositions of X grains (see Fig. 3; Lin et al. 2010; Liu et al. 2023a, 2024b)

mixing scenario) is further supported by some phenomenon observed in 3D hydrodynamic simulations. The 3D models predict that the outward-expanding shock can become decelerated due to a Rayleigh–Taylor-unstable density inversion near the He/H interface (Hammer et al. 2010; Wongwathanarat et al. 2017), thus allowing for sufficient mixing of material from the innermost region with He-shell material. *In conclusion, the isotopic compositions of X grains are in better agreement with the large-scale mixing scenario based on existing stellar nucleosynthesis models and observations for CCSNe.*

In the following, we will, therefore, focus on the large-scale mixing scenario in which X grains sampled materials from the Fe/Ni zone (Lin et al. 2010; Liu et al. 2024a,b), the inner Si/S zone (Liu et al. 2018a, 2023a), and the He (and possibly H) shells (Lodders and Fegley 1995; Nittler et al. 1996; Fedkin et al. 2010) but without any contributions from the Si/O, O/Ne, and C/O layers inbetween, as illustrated in Fig. 3. Finally, before moving onto the next section, we would like to stress that future investigations are required to explore whether α-capture in the outer region of a CCSN can lead to significantly enhanced Ti/Si and Ti/Ca ratios as required by the X and Si_3N_4 grain data. This is because the models of Pignatari et al. (2015) did not include the main physical mechanisms that could shape the He abundance profile in C/O zone in the pre-explosive O/C zone, for example, secular instabilities such as rotational mixing to gravity waves (see discussion in Pignatari et al. 2013a), which could affect the effective extension and α-capture

Varying Initial ^{44}Ti/^{48}Ti Ratios of X Grains Given that He- and H-shell materials lack ^{44}Ti but contain ^{48}Ti from the initial stellar composition, the proportion of He- and H-shell materials in the supernova ejecta from which X grains formed inversely affects the ^{44}Ti/^{48}Ti ratio (Lin et al. 2010): a higher percentage of He- and H-shell (^{29}Si-rich) materials corresponds to a lower ^{44}Ti/^{48}Ti ratio and a higher δ^{29}Si value in the mixed supernova ejecta. In addition, the inner (^{28}Si-rich) Si/S zone exhibits ^{44}Ti/^{48}Ti ratios that are more than an order of magnitude lower than the Fe/Ni zone (Fig. 6), which implies that a higher percentage of inner Si/S material (with respect to Fe/Ni material) in the mixed ejecta leads to reduction in both ^{44}Ti/^{48}Ti and δ^{29}Si in the mixed supernova ejecta. Consequently, the absence of inferred initial presence of ^{44}Ti in an X grain may be attributed to various factors: (*i*) lower-than-predicted ^{44}Ti/^{48}Ti ratios produced by α-rich freeze outs in the deepest supernova region, (*ii*) substantial dilution of Fe/Ni material with He- and H-shell material and/or inner Si/S material, with the effect of the Fe/Ni and Si/S material being more pronounced in ^{28}Si-rich X grains and that of the He- and H-shell material in ^{29}Si-rich X grains, and/or (*iii*) significant terrestrial Ca contamination, resulting in an undetectable ^{44}Ca excess from the grain.

We examine the three possibilities based on the X grains that exhibit significant ^{44}Ca excesses ($\gtrsim$ 1000‰) shown in Fig. 4. Their large excesses minimize the effects of analytical uncertainties and potential GCE-influenced initial composition[4] on the derivation of initial ^{44}Ti/^{48}Ti ratios. These X grains collectively display a compelling negative trend in Fig. 4b, despite being analyzed across different laboratories under varying conditions and using different ion microprobe techniques (including Cameca 3f-, 6f-, NanoSIMS ion microprobes). The trend in Fig. 4b points to the coproduction of ^{44}Ti (relative to ^{48}Ti) and ^{28}Si (relative to ^{29}Si), in line with the theoretical expectation for nucleosynthesis in deep supernova regions as discussed earlier. The imperfect correlation between ^{44}Ti/^{48}Ti and δ^{29}Si does not necessarily point to production of variable ^{44}Ti/^{48}Ti ratios because some small variability in the isotopic and/or elemental compositions of the two endmembers inferred from the linear mixing line – the inner ^{28}Si-rich endmember from the Fe/Ni and Si/S region, and the outer ^{29,30}Si-rich endmember from the He and H shells – could all contribute to the scatter in the X grain data.

In contrast to the subset of X grains (~35%) exhibiting significant ^{44}Ca excesses shown in Fig. 4b, the majority of X grains (~65%) reported in the literature have no or much smaller ^{44}Ca excesses (Stephan et al. 2024). If all X grains originated from the same supernova ejecta, then we could hypothesize that the large ^{44}Ca enrichments of the X grains in Fig. 4b are purely caused by their relatively low Ca/Ti ratios, and vice versa, i.e., the Scenario (*iii*) above. This hypothesis is further scrutinized based on X grains for which elemental concentration data are available (Fig. 4c). The influence of variable mixing ratios between the two endmembers can be corrected by calculating the value of ^{44}Ca/^{40}Ca$_{\text{predicted}}$/solar. The first step of this calculation is based on the established correlation in Fig. 4b (solid line), which shows the ^{44}Ti/^{48}Ti ratio for a given δ^{29}Si value. From the predicted ^{44}Ti/^{48}Ti ratio, the second step in the calculation is to use the Ca and Ti concentration data to predict the ^{44}Ca excess for a grain, i.e., ^{44}Ca/^{40}Ca$_{\text{predicted}}$/solar. This method implicitly assumes that the nonradiogenic δ^{44}Ca of an X grain equates to zero, a basis for the derivation of most initial ^{44}Ti/^{48}Ti data in Fig. 4b. Estimations for the lower and upper boundaries of δ^{44}Ca$_{\text{predicted}}$ are anchored to the dashed lines shown in Fig. 4b. The resultant trend in Fig. 4c shows that the measured ^{44}Ca excesses in the majority of X grains fall beneath the values that we would expect if the X grains in Fig. 4c were sampled from the same supernova ejecta as those in Fig. 4b. Although potential Ti contamination, if sampled during the analyses, could lead to inflated ^{44}Ca/^{40}Ca$_{\text{predicted}}$/solar values, NanoSIMS analysis of X grains at high-spatial resolution suggests that Ti contamination is rare (Liu et al. 2023a).

The scenario (*i*), i.e., reduced ^{44}Ti production in the deep regions of their parent supernovae, can be further excluded, given the invariable ^{44}Ti/^{48}Ti production ratio as a function of varying explosion energies shown in Fig. 6. The lack of any energy dependence is caused by the fact that the productions of ^{44}Ti and ^{48}Ti both increase with increasing explosion energy in the deep region of a Type II supernova. This is because ^{44}Ti is an α nuclide and ^{48}Ti

[4]The GCE effect could lead to temporal and spatial ^{44}Ti and ^{44}Ca variations in the ISM and thus in the initial stellar composition (see Section "Potential effect of heterogeneities in the ISM on ^{44}Ca" for discussion in detail). In this case, the ^{44}Ca excess detected in an X grain would be irrelevant to the ^{44}Ti production in its parent supernova.

is copiously produced as the α nuclide ^{48}Cr, which subsequently decays to ^{48}V ($t_{1/2} = 22$ h) and then to the stable isotope ^{48}Ti ($t_{1/2} = 16$ d).

By excluding the Scenarios (i) and (iii), we thus infer that the two sets of X grains with large and small ^{44}Ca excesses likely sampled varying amounts of materials from the Fe/Ni zone (^{28}Si, ^{44}Ti-rich), the inner Si/S zone (^{28}Si-rich, ^{44}Ti-poor), and the outer shells (^{29}Si-rich, ^{44}Ti-free), i.e., the Scenario (ii). Variation in the relative mixing ratios between these zones explains the varying initial ^{44}Ti/^{48}Ti ratios observed among X grains. In other words, the X grains with small ^{44}Ca excesses sampled more materials from both the Si/S zone and outer shells (i.e., accounting for their comparable ^{28}Si excesses) than the X grains with large ^{44}Ca excesses. *The absence of inferred initial presence of ^{44}Ti in the majority of X grains thus cannot be used as evidence for the idea that ^{44}Ti is predominantly produced in rare supernova events.*

Potential Effect of Heterogeneities in the ISM on ^{44}Ca Two pieces of indirect evidence from presolar grain studies may still suggest that ^{44}Ti is predominantly produced in rare supernova events, based on which a heterogenous distribution of ^{44}Ti (and thus its decay product ^{44}Ca) is expected (spatially and temporally) in the ISM: (*i*) the broader-than-anticipated scatter in the ^{44}Ca/^{40}Ca ratios of presolar hibonite grains (Nittler et al. 2008) and (*ii*) the lack of observable correlation between the ^{44}Ca excesses of presolar silicates and their Ti-rich subgrains (Nittler et al. 2023; Leitner et al. 2024). Nittler et al. (2008) observed varying ^{44}Ca/^{40}Ca ratios in presolar hibonite grains of AGB stellar origin with similar ^{18}O/^{16}O ratios, which infer a similarity in the initial stellar metallicities of their parent stars (Nittler 2009). Since AGB nucleosynthesis barely produces any ^{44}Ca, this observation suggests that AGB stars that were born with similar metallicities, i.e., with similar ^{18}O/^{16}O ratios, had relatively variable initial ^{44}Ca/^{40}Ca ratios, thus corroborating the expected heterogenous distribution of ^{44}Ti (and thus its decay product ^{44}Ca) in the ISM due to the dominant production of ^{44}Ti in rare supernova events. In addition, Nittler et al. (2023) and Leitner et al. (2024) observed that, in two presolar silicate grains, their ^{44}Ca excesses were not spatially correlated with ^{48}Ti signals. One of the silicate examples is shown in Fig. 7. In Fig. 7, we see that the small ^{44}Ca excess of the silicate grain remains constant with varying ^{48}Ti/^{40}Ca ratios, which is in great contrast to the linear correlation observed in the X grain shown in Fig. 2e and thus points to a nonradiogenic origin for the observed small ^{44}Ca excess. Together with the hibonite grain data from Nittler et al. (2008), the presolar silicate data provide further evidence for the heterogenous distributions of ^{44}Ti and thus ^{44}Ca in the ISM. Furthermore, the grain data in Fig. 7 deviate from the homogenous GCE trend reported by Nittler et al. (2008) based on the GCE calculation of Timmes et al. (1995), and its ^{44}Ca excess suggests local ^{44}Ca enrichments in the ISM, which could have resulted from rare supernova events. A caveat is that the evidence abovementioned directly reflects the effect of heterogeneous ISM composition on the ^{44}Ca/^{40}Ca ratio, which could have been caused by many different astrophysical processes, e.g., insufficient mixing of supernova ejecta with the ISM material (e.g., Lugaro et al. 1999; Nittler 2005). *In conclusion, presolar grain data provide no direct support for the suggestion of predominant ^{44}Ti production in rare types of supernova events.*

2.2 Further Insights into Supernova Nucleosynthesis

As discussed so far, presolar supernova grains are direct condensates from supernova ejecta and thus have recorded their isotopic compositions, offering a direct window into the intricacies of supernova nucleosynthesis. Advanced isotopic analyses harnessing both modern NanoSIMS and resonance ionization mass spectrometric (RIMS) techniques are now able to

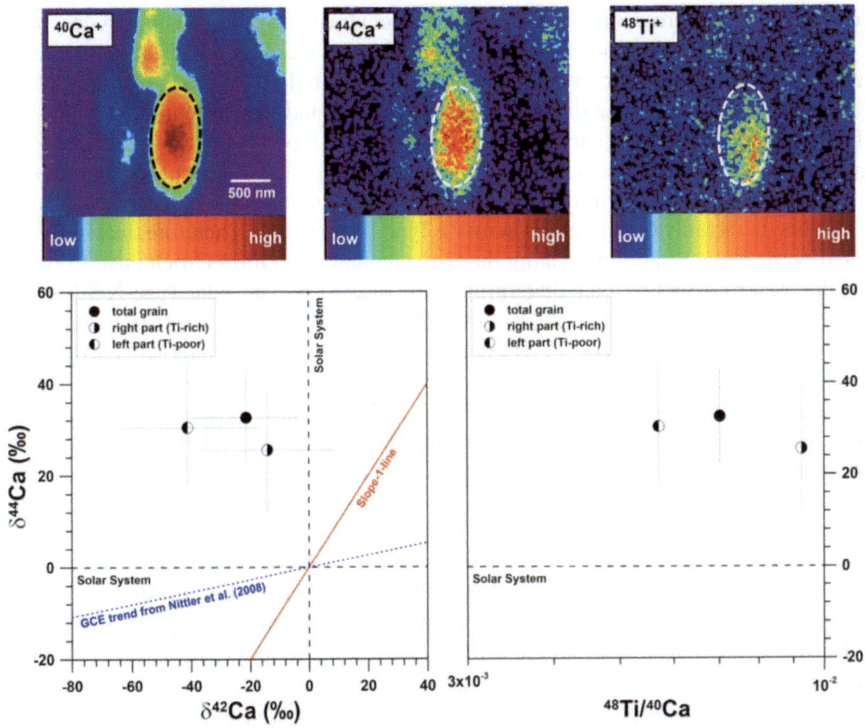

Fig. 7 NanoSIMS Ca-Ti isotope data of a ^{17}O-rich silicate grain of low-mass AGB stellar origin from Leitner et al. (2024). The upper panel shows ^{40}Ca, ^{44}Ca, and ^{48}Ti ion images (with color bars shown at the bottom) collected from the silicate grain (denoted by dashed ellipses) at a spatial resolution of ~100 nm. The white scalebar in the ^{40}Ca image denotes 500 nm. Three different ROIs were defined for reducing the Ca-Ti isotope data shown in the upper panel, as denoted by the three data points shown in the lower panel

extract accurate isotope ratios across elements from C to Nd in micron-sized presolar grains at permil- to percent-level precision (Liu et al. 2022a). These multielement isotope datasets are instrumental to shed new light on nucleosynthesis processes, particularly in Type II supernovae (Hoppe et al. 2018, 2023, 2024; Kodolányi et al. 2018; Liu et al. 2018a, 2024a; Pignatari et al. 2018; Stephan et al. 2018; Ott et al. 2019), given the dominance of their dust in the presolar grain inventory. The reader is referred to the abovementioned literature for recent advancements in this field. In this section, we will primarily focus on the production of neutron-rich isotopes in supernovae and their signatures in supernova grains, setting a foundational context for the subsequent discussion on nucleosynthetic isotope variations in bulk meteorites in Sect. 2.3.

2.2.1 Neutron Burst in Type II Supernovae

Multiple lines of evidence suggest that the *r*-process does not occur in typical Type II supernovae, nevertheless, these stellar events are suggested to be the sites for other neutron-capture processes, including the weak slow neutron capture process (*s*-process) during the presupernova stellar evolution (Pignatari et al. 2010) and the neutron-burst process during the explosion (Meyer et al. 2000). Specifically, the weak *s*-process is believed to occur during the hydrostatic convective core He-burning phase and the subsequent convective shell

C-burning phase of a presupernova massive star. In contrast, the neutron-burst process takes place during a Type II supernova explosion in the He/C zone (Fig. 3). These two neutron-capture processes are similar in that they are both powered by neutrons released by the $^{22}Ne(\alpha, n)^{25}Mg$ reaction but at different neutron densities and on different timescales.

The degree to which ejecta carrying the imprint of weak *s*-process nucleosynthesis are incorporated into supernova grains – and the extent of any superimposed explosive nucleosynthesis signatures – is not yet fully explored. In contrast, the He/C zone, where the neutron burst occurs, is the only C-rich region across a Type II supernova. Therefore, material from the He/C zone is necessary for creating the C-rich condition to form C-rich dust. Corroborating this, neutron-burst isotopic fingerprints were discerned in presolar Type-X SiC grains (Pellin et al. 2006; Stephan et al. 2018), strengthening the argument that C-rich conditions are required for the formation of SiC (Lodders and Fegley 1995). An alternative hypothesis by Bliss et al. (2018) postulated that the neutrino-driven winds following a CCSN explosion could create a neutron-rich environment that is suitable to produce the Mo and Ru isotopic patterns observed in X grains. However, Bliss et al. (2018) calculated the neutrino-driven-wind condition based on an analytical model and focused solely on heavy-element isotopic signatures, thus making it challenging to explore the collateral consequences for the other isotopic signatures (e.g., C, Ti) of X grains and the required C-rich formation condition. In contrast, CCSN stellar nucleosynthesis models (Woosley and Heger 2007; Heger and Woosley 2010) predict the neutron-burst process to occur within the He/C zone, and this satisfies the prerequisite of C-rich conditions for the formation of X grains. Therefore, the discussion below will focus on the neutron-burst scenario to interpret the neutron-rich isotopic compositions of X grains.

The neutron-burst process, characterized by a neutron density up to 10^{17-20} neutrons/cm^3 at temperatures around 10^9 K (Meyer et al. 2000), predominantly produces intermediate-mass Mo isotopes like ^{95}Mo and ^{97}Mo (Fig. 8), and, therefore, differentiates itself from both the *s*- ($\sim 10^7$–10^{10} neutrons/cm^3) and *r*-processes ($>10^{20}$ neutrons/cm^3) (Cowan and Thielemann 2004; Liu et al. 2019). Such ^{95,97}Mo-enhanced isotopic patterns, accompanied by excesses of ^{88}Sr, ^{96}Zr, and ^{138}Ba (Fig. 8b), were detected in X grains (Pellin et al. 2000, 2006; Stephan et al. 2018), in line with neutron-burst nucleosynthesis model predictions (Meyer et al. 2000). The preservation of neutron-burst isotopic imprints for heavy elements in X grains can be ascribed to the following two aspects. (*i*) The production of these isotopes is orders of magnitude lower in the Fe/Ni and Si/S zones than in the He shell, minimizing the influence of inner supernova materials on the isotope ratios of an X grain for these elements (Stephan et al. 2018). And (*ii*) nucleosynthesis in the H envelope barely alters the initial stellar composition, which is inferred to lie close to the solar composition in the parent supernovae of X grains based on their ^{46,47}Ti/^{48}Ti ratios (Liu et al. 2024b); consequently, mixing with the H envelope reduces isotope variations produced by neutron bursts but barely alters the neutron-burst isotopic patterns. Discussions on X grain data in the context of Type II supernova models that delve into these isotope ratios are reported in the literature (Pignatari et al. 2018; Stephan et al. 2018; Ott et al. 2019). Going forward, below we investigate whether the neutron-burst isotopic compositions of X grains for different elements can be explained coherently. X grains are divided into several subtypes based on Si isotopes (Stephan et al. 2024). Below, we will mention the subtypes X1 and X2 in the discussion: X1 grains are the dominant subtype of X grains and lie along the slope-2/3 line in the 3-Si-isotope plot, while X2 grains are much lower in abundance and more enriched in ^{30}Si than X1 grains (Lin et al. 2010; Stephan et al. 2024).

Figure 8a showcases that the neutron-burst process results in the overproduction of isotopes, such as ^{46}Ca, ^{50}Ti, ^{88}Sr, ^{96}Zr, ^{95}Mo, ^{97}Mo, and ^{138}Ba, with the degree of overproduction generally increasing with increasing energy, except for ^{96}Zr. These predicted variations

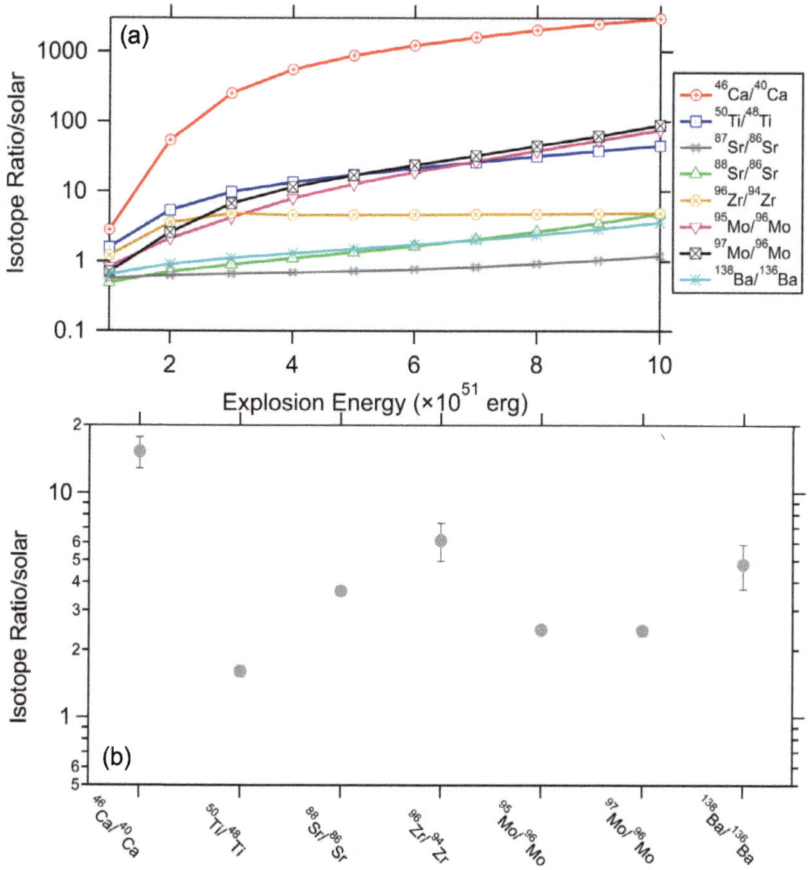

Fig. 8 Panel (a) displays the dependences of varying isotope ratios on explosion energy in the He/C zone predicted by the same Rauscher-Meyer models from Fig. 5. The radioactive isotopes involved in the production of the isotopes of interest all have half-lives less than tens of days and should have fully decayed to their daughter isotopes by the time of X grain condensation in the supernova ejecta (Liu et al. 2018a; Ott et al. 2019; Niculescu-Duvaz et al. 2022). The isotope ratios are normalized to solar composition, i.e., unity in y-axis. Panel (b) summarizes the maximum excesses in the neutron-rich isotopes of Ca, Ti, Sr, Zr, Mo, and Ba observed in X grains so far (Amari et al. 2003; Pellin et al. 2006; Liu et al. 2024b). Plotted are 1σ errors

generally align with observations from X grains: (*i*) The comparable ^{95,97}Mo/^{96}Mo ratios predicted across different energies, explain the similar levels of ^{95}Mo and ^{97}Mo enrichments in X grains (Fig. 8b; Pellin et al. 2000), despite that they could have originated from diverse parent supernovae whose He/C zones were heated to varying temperatures during the explosions; and (*ii*) the prominent ^{86}Sr enhancements (suggested by the large, negative $\delta^{87,88}$Sr values that are normalized to ^{86}Sr) observed in two X2 grains (Stephan et al. 2018) well correspond to neutron bursts at lower explosion energies (e.g., $\delta^{87,88}\text{Sr} \approx -500‰$ at 1×10^{51} erg) as compared to X1 grains (Pellin et al. 2000; Stephan et al. 2018).

However, a challenge arises in reconciling these model predictions with observed isotope data in a quantitative way. The models suggest that the neutron-burst process should yield ^{95,97}Mo excesses at levels an order of magnitude greater than those of ^{88}Sr and ^{138}Ba and at least as much as ^{96}Zr (Fig. 8a). Yet, this expected excess pattern has not been identified in the multielement isotopic composition of any known X grains (Fig. 8b; Pellin et al. 2000): the maximum excesses in ^{95,97}Mo observed in X grains so far are not higher than those in ^{88}Sr, ^{96}Zr, and ^{138}Ba. A possible explanation for the data-model discrepancy might be asteroidal and/or terrestrial Mo contamination, which can lower the observed excesses in ^{95,97}Mo without affecting other isotope ratios (Liu et al. 2019, 2022c). Under this premise, the upper limits of observed enrichments in ^{88}Sr, ^{96}Zr, and ^{138}Ba, approximately 2000 − 3000‰ (Fig. 8b; Pellin et al. 2000; Stephan et al. 2018), could correspond to neutron-burst processes occurring at $(9-10) \times 10^{51}$ erg in the 25 $M_\odot$ supernova model shown in Fig. 6. Compared to astronomical observations of supernova remnants, which suggest a mean energy per unit mass of 0.85×10^{51} erg /10 $M_\odot$ (Rubin et al. 2016), the inferred explosion energies of up to $(9-10) \times 10^{51}$ erg / 25 $M_\odot$ from the presolar grain data are significantly higher. However, these values still fall within the range observed among Type II supernova remnants $(0.2-20) \times 10^{51}$ erg /10 $M_\odot$ and may be related to explosion asymmetries or progenitor massive stars with lower initial stellar masses. On the other hand, the observed data-model discrepancies could imply uncertainties in current neutron-burst model calculations for the following reasons. The model calculations suffer from uncertainties in the adopted nuclear reaction rates, especially given that the neutron-capture and beta-decay rates of many short-lived isotopes involved in the neutron-capture path are based on theoretical calculations. Recently, efforts on providing experimental constraints have been pursued (e.g., Spyrou et al. 2024), making it promising to have more reliable neutron-burst model predictions in the foreseen future. In addition, the accuracy of neutron-burst model calculations is also affected by the adopted isotope abundances produced by presupernova nucleosynthesis, and neutron-capture processes in presupernova massive stars need to be fully explored.

Another problem is that at such high energies the neutron-burst nucleosynthesis is predicted to produce substantial excesses in ^{46}Ca (2×10^6‰) and ^{50}Ti (4×10^4‰). Yet, the observed ^{50}Ti excesses in X grains peak at roughly 700‰, which is substantially lower than expected, with many grains showing smaller excesses (Fig. 8b; Liu et al. 2024b). While ^{46}Ca/^{40}Ca ratios for X grains remain unknown due to their general lack of intrinsic Ca, recent analyses of Murchison supernova Si_3N_4 grains – isotopic twins of X grains (Nittler et al. 1995; Lin et al. 2010) – have uncovered ^{46}Ca excesses up to 10,000‰ (Fig. 8b; Liu et al. 2024b). These observed excesses in both ^{46}Ca and ^{50}Ti (Fig. 8b) fall nearly two orders of magnitude below the model predictions. As discussed earlier, this discrepancy can be reconciled by incorporating material from the Si/S and/or Fe/Ni zones, where significant amounts of ^{40}Ca and ^{48}Ti are produced.

In conclusion, the isotopic compositions of Sr, Zr, and Ba in X grains can be explained by the neutron-burst process occurring, for example, at explosion energies up to (9–10) × 10^{51} erg in a 25 $M_\odot$ supernova. The observed lower ^{46}Ca and ^{50}Ti excesses in supernova grains are primarily due to the incorporation of Fe/Ni and/or Si/S material, characterized by significant production of ^{40}Ca and ^{48}Ti, resulting in smaller ^{46}Ca/^{40}Ca and ^{50}Ti/^{48}Ti ratios in the mixed ejecta. The lower-than-predicted levels of ^{95,97}Mo excesses in X grains could suggest the presence of significant Mo contamination sampled in the study of Pellin et al. (2000).

To test the contamination hypothesis, further Mo isotope analyses using new generation RIMS instruments at higher spatial resolution (Stephan et al. 2016; Pal et al. 2022) are needed, which will help minimize the effect of potential extrinsic contamination. Moreover, uncertainties in neutron-burst model calculations also need to be examined in the future by adopting experimentally constrained reaction rates and by considering uncertainties in the isotope abundances that are produced by presupernova nucleosynthesis. Future modeling work also needs to extend the initial mass and metallicity ranges, especially given that the

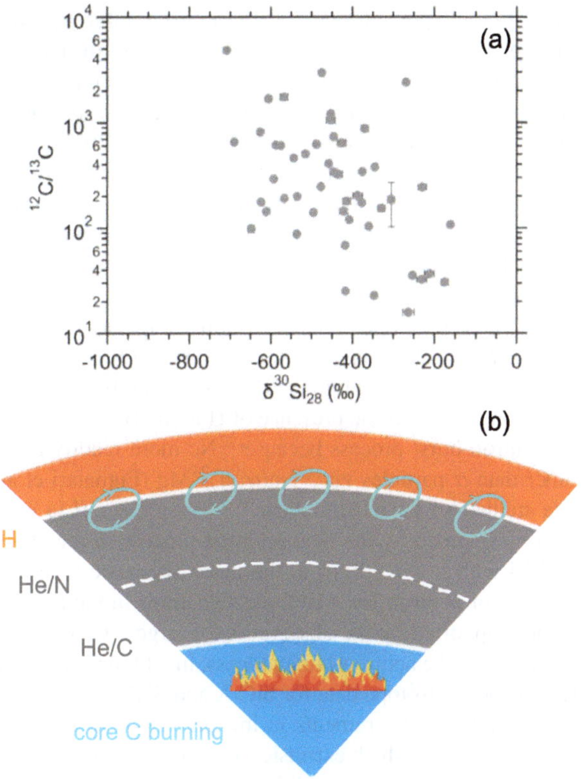

Fig. 9 Panel (a) presents literature C and Si isotope data for X grains from Hoppe et al. (2018) and Liu et al. (2024a). In panel (b), two possible mechanisms for introducing H into the He shell (the grey region) are illustrated. (i) Mixing of H from the bottom of the envelope into the upper layers of the He shell may occur through overshoot enabled by the convective motions of eddies in the envelope. This process results in an exponentially decaying distribution of H within the He/N zone (Liu et al. 2018b). (ii) Alternatively, mixing can be driven by the rapid release of a significant amount of energy during core C-burning, leading to an unstable convective He shell above. The instability may enable mixing of H across the entire He shell at levels reaching up to one percent (Pignatari et al. 2015)

progenitor stars of the vast majority of CCSNe are expected to have initial masses lower than 25 $M_\odot$ according to the initial stellar mass function. Given the complex dependencies of isotope ratios on various parameters in supernova nucleosynthesis models, it is crucial to determine the properties of the parent supernovae of X grains to refine the parameter space. This refined understanding will enable us to better constrain the role of Type II supernovae in accounting for nucleosynthetic isotopic variations observed in bulk meteorites, given the confirmed contribution of the parent Type II supernovae of X grains to the seeds for the formation of the solar system.

2.2.2 Explosive H Burning in Type II Supernovae

Effect on Neutron Bursts In this section, we will be focusing on the process of explosive H burning during Type II supernova explosions. This phenomenon specifically refers to the presence of H in the He shell during the explosions, which results in efficient production of proton-rich isotopes such as ^{13}C, ^{15}N, ^{25}Mg, and ^{26}Al in the He shell (Pignatari et al. 2015; Schofield et al. 2022). In the absence of H, the He shell is predicted to consist predominantly of ^{12}C produced by triple-α reaction (Woosley and Heger 2007). The observed positive correlation between ^{13}C and ^{30}Si enrichments among X grains (Fig. 9a) suggests that the ^{30}Si-rich endmember ($\delta^{30}Si \geq 0$‰, representing material from the He and H shells) has $^{12}C/^{13}C$ ratios $\lesssim$100, which could result from the efficient production of ^{13}C (*i*) in the He/N zone and H envelope by presupernova nucleosynthesis and/or (*ii*) in the He/C and He/N zones by explosive H burning during the explosion. The possibility (*ii*) is favored by the fact that X grains with subsolar $^{12}C/^{13}C$ ratios exhibit a positive correlation between their ^{13}C and ^{15}N excesses (Liu et al. 2024a), which are in line with the signatures of explosive H burning in the He shell. In comparison, stellar temperatures during presupernova evolution and in the H shell during the explosion are both too low to enable hot CNO cycle for the production of ^{15}N (Lin et al. 2010; Hoppe et al. 2018).

The exact region within the He shell where explosive H burning takes place is still debated due to the still unknown mechanism by which H is introduced into the He shell. Although the two processes illustrated in Fig. 9b differ in the quantity and the distribution of H within the He shell, modeling outcomes from both predict remarkably similar isotopic imprints (Pignatari et al. 2015; Liu et al. 2018b), and both provide a satisfying explanation to the C and N isotope ratios of X grains (Fig. 9a).

The two mechanisms, however, diverge in their impact on the neutron-burst process: H added via overshooting into the He/N zone barely influences the neutron-burst process in the He/C zone, whereas the presence of H in the He/C zone can suppress the neutron source for the neutron-burst process because ^{22}Ne more readily captures protons via $^{22}Ne(p,\gamma)^{23}Na$ rather than α particles via $^{22}Ne(\alpha, n)^{25}Mg$ (Pignatari et al. 2015). In comparison, X grain data studied so far mostly show $^{49,50}Ti$ excesses (Liu et al. 2018a, 2023a, 2024b), which can be explained by the neutron-burst process in the He/C zone, in addition to ^{49}V decay (to ^{49}Ti with $t_{1/2} = 330$ d) in the inner Si/S zone (Liu et al. 2018a, 2023a). The common presence of neutron-burst isotopic signatures in the parent supernova ejecta of X grains, is further supported by the observation of large ^{46}Ca excesses, which are distinctive isotopic signatures of the neutron-burst process, in all four Si_3N_4 grains studied by Liu et al. (2024b). The Ti and Ca isotope data for SiC X and Si_3N_4 grains, respectively, were mostly collected with NanoSIMS instruments in imaging mode at high spatial resolution (~100 nm, e.g., Fig. 2), based on which extrinsic Ti and Ca contamination would have been recognized and suppressed if existed. In addition, presolar SiC grains are rich in Ti and Si_3N_4 in Ca (Liu et al. 2024b) so that the effects of extrinsic contamination were mostly small. In contrast, heavy-element isotope data for SiC X grains were collected with RIMS instruments in spot mode at 1–5 μm (Pellin et al. 2000, 2006; Stephan et al. 2018), and it was shown that the RIMS data for presolar SiC grains were affected by Sr, Ba, and Mo contamination (Barzyk et al. 2007; Liu et al. 2022c). Thus, the lack of neutron-burst isotopic signatures in some of the X grains studied by Pellin et al. (2000, 2006) is possibly caused by contamination and thus do not argue directly against the common presence of neutron-burst isotopic signatures in X grains. *The suggested common presence of neutron-burst isotopic patterns in X and Si_3N_4 grains could rule out the mixing scenario on the basis of core C-burning, although it has been argued that non-uniform, localized mixing of H induced by core C-burning could permit neutron bursts in less affected regions of the He/C zone (Hoppe et al. 2018; Pignatari et al. 2018).*

In general, the deactivation of neutron burst in the He/C zone significantly reduces the production of certain neutron-rich isotopes like ^{60}Fe (e.g., Limongi and Chieffi 2006). Future studies need to investigate the effect of H ingestion on affecting the ^{60}Fe production based on the multiple pieces of astronomical and geological evidence for the amounts of fresh ^{60}Fe produced from nearby Type II supernovae (Binns et al. 2016; Fimiani et al. 2016; Wallner et al. 2016).

Effect on Inferring Stellar Origins for Presolar Grains Independent of the mechanism for inducing explosive H burning into the He shell, the probable occurrence of this process in the parent supernovae of X grains initiated a reevaluation of the stellar origins of several other types of presolar grains. This includes grains previously attributed to nova origins (putative nova grains hereafter) and Type AB grains, both of which exhibit large ^{13}C enrichments but different ranges of $^{14}N/^{15}N$ ratios and distinct Si isotopic compositions (Pignatari et al. 2015; Liu et al. 2016, 2017, 2018b; Hoppe et al. 2019), as well as ^{25}Mg-rich Group 1 presolar silicates that are characterized by ^{17}O enhancements (Leitner and Hoppe 2019). The detected enrichments of p-process isotopes in one such grain (Savina et al. 2007) supports a potential supernova provenance for AB grains as well as the initial ^{44}Ti presence in another grain (Liu et al. 2023b). The hypothesized supernova origins of putative nova SiC grains and ^{25}Mg-rich Group 1 silicates, however, should be considered cautiously. This is because their characteristic isotopic signatures – excesses in ^{13}C, ^{15}N, ^{17}O, ^{25}Mg, or ^{26}Al – point solely to high-temperature H-burning nucleosynthesis, which is not exclusive to Type II supernovae but could occur in various stellar environments such as novae (José and Hernanz 2007), intermediate-mass AGB stars (Karakas and Lugaro 2016), and super-AGB stars (Doherty et al. 2014). Interestingly, due to its fragile nature, ^{18}O can be easily destroyed during CNO cycle, leading to large ^{18}O depletions that characterize explosive H burning in the He shell during a Type II supernova explosion, although these depletions could be erased if mixing occurred with adjacent He-and H-shell material and the pre-supernova material (Pignatari et al. 2015; Leitner and Hoppe 2019). In comparison, Type II supernova oxides (i.e., Group 4 grains) and graphite grains with the inferred initial presence of ^{44}Ti are characterized by ^{18}O enrichments (Nittler et al. 2008; Groopman et al. 2012), pointing to the occurrence of the reactions $^{14}N(\alpha,\gamma)^{18}F(\beta^+)^{18}O$ in the absence of protons in the He/C zone.

Despite their ambiguous stellar origins, putative nova grains, Type AB grains, and ^{25}Mg-rich Group 1 silicates provide a unique opportunity to specifically investigate high-temperature H-burning processes, and intercomparison of their isotope data between these grain types may shed new lights onto their stellar origin(s). Existing model predictions for novae, intermediate-mass AGB stars, and super-AGB stars face a number of problems in explaining the multielement isotope data of putative nova SiC grains and ^{25}Mg-rich Group 1 silicates (Amari et al. 2001; Nittler and Hoppe 2005; Liu et al. 2016; Leitner and Hoppe 2019). It remains to be seen whether these data-model discrepancies can be used to firmly exclude these stellar sources or reflect modeling uncertainties, e.g., missing physics due to the simplified 1D modeling approach, largely uncertain mass-loss rates for intermediate-mass AGB stars and super AGB stars.

2.2.3 Production of ^{54}Cr in Supernovae

Chromium-54 isotopic variations in bulk meteorites led Dauphas et al. (2010) and Qin et al. (2011) to independently discover presolar Cr-rich spinel grains, tens to hundreds of nanometers in size, with up to thousands of permil enrichments in ^{54}Cr. Initial assessments of Cr isotopes in nanospinel grains using NanoSIMS instruments equipped with a DuoPlasmatron O^- ion source were constrained by a spatial resolution of $\sim$0.5–1.0 μm, leading to significantly diluted isotopic anomalies due to the sampling of Cr contamination from neighboring grains (Dauphas et al. 2010; Qin et al. 2011). Subsequent advancements in the reduced beam size of the Hyperion O^- plasma ion source enabled in situ Cr isotope analyses at refined resolution ($\sim$100 nm), and this led to more accurate determinations of intrinsic ^{54}Cr enrichments of up to 50 times its solar abundance (Nittler et al. 2018).

Substantial ^{54}Cr can be produced by (*i*) neutron-burst processes in the O/Ne, C/O, and He/C zones in Type II supernovae (den Hartogh et al. 2022) and (*ii*) quasi-equilibrium in high-density Type Ia supernovae (Woosley 1997) and (*iii*) quasi-equilibrium in ECSNe (Wanajo et al. 2013). The pronounced ^{54}Cr enrichments in nanospinel grains, therefore, reveal inconclusive insights into their exact stellar origin. Nittler et al. (2018) suggested an

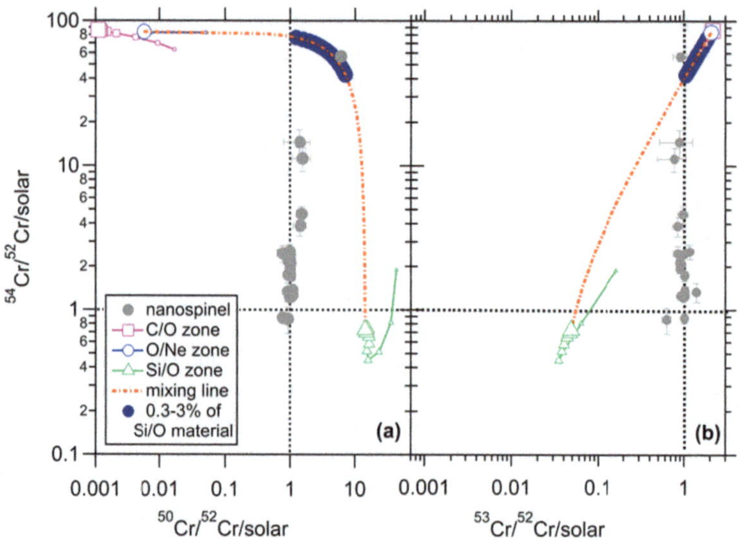

Fig. 10 Presolar nanospinel grain data from Nittler et al. (2018) are compared to the Rauscher-Meyer CCSN models shown in Figs. 6 and 8. The black dashed lines represent the solar ratios. Predictions for different supernova zones are shown as symbols with lines (as indicated in the legend), with the symbol size increasing with increasing explosion energy from 1×10^{51} to 10×10^{51} erg. The orange dashed line represents mixing of materials from the O/Ne and underlying Si/O zones at 10×10^{51} erg, and the blue shaded region along the orange line highlights the compositions predicted by mixing $0.3-3\%$ of material from the Si/O zone into the O/Ne zone. The decay of ^{53}Mn to ^{53}Cr was not included in the model calculations, given its relatively long half-life ($t_{1/2} = 3.7$ Ma) and unknown Mn/Cr ratios in the nanospinel grains

ECSN or Type Ia supernova origin for these grains, but detailed comparisons with Type II nucleosynthesis models by den Hartogh et al. (2022) suggested that the nanospinel data from Nittler et al. (2018) could also be consistently explained by nucleosynthesis in the O/Ne and C/O zones of a Type II supernova, underscoring the ambiguity in determining the stellar source of the grains solely based on the Cr isotope ratios. This complexity reflects the similarities of nucleosynthesis products across the supernova types, despite their distinct evolutionary paths and explosion mechanisms. In particular, given the diversity in nucleosynthesis across various zones of Type II supernovae and the potential mixing of materials from these regions during the explosion, categorically ruling out a Type II supernova as the progenitor of certain presolar grains or as a contributor to certain meteoritic isotopic anomalies is challenging.

Figure 10 provides an illustrative example for Cr isotopes and reveals that supernova zones possess distinct Cr isotopic signatures, delineating that only a narrow range of mixing ratios between zones rich in ^{54}Cr (such as He/C, C/O, O/Ne) and those rich in ^{50}Cr (such as Si/O, Si/S) can approximate the most extreme Cr isotopic composition observed in nanospinel grains (as indicated by the blue shaded region in Fig. 10).[5] The models predict substantial ^{52}Cr production in the Fe/Ni zone via α-rich freeze outs and unaltered initial Cr isotopic composition in the He/N and H zones. The information above allows us to infer that (*i*) the isotopic compositions within the blue shaded area could also be reproduced by

[5]For the data-model comparison in Fig. 10, we assumed that the ion signals at mass 50 were fully from ^{50}Cr and not ^{50}Ti. The assumption is needed because the isobaric interreference at mass 50, if existed, could not be resolved in the NanoSIMS measurements of Nittler et al. (2018).

incorporating material from the C/O and Si/S zones, which exhibit Cr isotopic ratios akin to O/Ne and Si/O zones, respectively, and (*ii*) the less extreme Cr isotopic compositions of the nanospinel grains could emerge from dispersing the described ejecta with contributions from the He/N zone and H envelope.

In comparison to Type II supernova nucleosynthesis (in high-entropy conditions), the quasi-equilibrium nucleosynthesis process, occurring in low-entropy conditions predicted for ECSNe and for rare types of high-density Type Ia supernovae, preferentially produces ^{54}*Cr with minimal impact on the other Cr isotopes (Woosley 1997; Wanajo et al. 2013; Jones et al. 2019b), thus providing a direct explanation to the observed isotopic compositions of nanospinel grains.* It is commonly believed based on spectroscopic observations that Type Ia supernova explosions produce significantly less dust than Type II supernova explosions (Gomez et al. 2012). Recent mid-infrared observations of the Type Ia supernova 2018evt, revealed the presence of about 1 $M_{\odot}$ of newly formed dust after the supernova explosion (Wang et al. 2024), which is comparable to the maximum amount of dust observed in Type II supernova remnants (Niculescu-Duvaz et al. 2022). Furthermore, the simulations of Jones et al. (2019b) predict efficient productions of O, Mg, Si in and large mass ejections from thermonuclear ECSNe, thus providing a preferential condition for dust formation. Information on the existence of ECSNe and their occurrence rate and dust formation efficiency based on observations, is urgently needed to evaluate their possible contribution to the presolar grain inventory in the solar system. Future detailed isotopic analyses encompassing elements such as O, Ti, Fe, and Cr with both NanoSIMS and RIMS instruments are expected to enhance our ability to discriminate among the proposed formation scenarios for these grains.

2.3 Supernova Nucleosynthesis and Nucleosynthetic Variability in Meteorites

2.3.1 Bulk Meteoritic Isotopic Variations

Nucleosynthetic isotope variations in bulk meteorites have been observed for a variety of elements, including major elements like Fe (e.g., Cook and Schönbächler 2017; Schiller et al. 2020; Hopp et al. 2022), trace elements like Mo and Pd (e.g., Ek et al. 2020), refractory elements like Ca and Ti (e.g., Dauphas et al. 2014), and moderately volatile elements like K and Zn (e.g., Martins et al. 2023; Nie et al. 2023). In addition, intra- and inter-variations have been observed in the nucleosynthetic isotope compositions of non-carbonaceous (NC) and carbonaceous (CC) meteorites, which are assumed to have formed in the inner and outer parts of the disk, respectively (Leya et al. 2008; Desch et al. 2018). The NC-CC isotopic differences may indicate distinct nucleosynthetic isotopic signatures in these reservoirs, although the exact origin of these differences is still debated. For instance, when considering Mo isotopes, the variations within NC and CC meteorites clearly point to s-process[6] variations (e.g., Dauphas et al. 2002; Burkhardt et al. 2011). However, the cause of inter-variations in Mo isotopic compositions between NC and CC meteorites remains ambiguous. It was proposed that the offset between the trends in NC and CC meteorites was due to an additional r-process component in the CC isotopic reservoir (e.g., Budde et al. 2016). However, Stephan and Davis (2021) showed that s-process variability could alternatively explain the observed offsets in all Mo isotopes. In this scenario, the CC isotopic reservoir

[6]The s-process nucleosynthesis hereafter refers to the main s-process specifically, occurring in AGB stars (Lugaro et al. 2023). It is noteworthy that there is also the weak s-process operating in massive stars (see discussion in Sect. 2.2.1), which can efficiently produce heavy elements before the first s-process peak at Sr (see Pignatari et al. 2010 and Lugaro et al. 2023 for details).

sampled more s-process material formed at higher-than-average neutron densities than the NC isotopic reservoir, given that this material was homogenously distributed within the two reservoirs. It is unclear, however, how such two reservoirs remained separated in the ISM from the time the dust was ejected by their parent AGB star(s) to the formation of the solar system, or if they separated instead within the protoplanetary disk due to, e.g., size differences (see Sect. 2.3.3).

In addition, enrichments in neutron-rich isotopes like ^{48}Ca, ^{50}Ti, ^{54}Cr (e.g., Niederer et al. 1980; Papanastassiou 1986; Leya et al. 2009; Chen et al. 2015; Kööp et al. 2018; Akram et al. 2013) as well as r-process isotopes of elements with A $<$ 140 (e.g., Brennecka et al. 2013; Akram et al. 2013) and p-process isotope enrichments (e.g., Charlier et al. 2021), were identified in calcium, aluminum-rich inclusions (CAIs). These inclusions are, so far, the oldest dated material in the solar system and represent the earliest condensates in the solar nebula (MacPherson 2014). Given that bulk meteorites generally exhibit variations in the neutron-rich isotopes of refractory elements like ^{48}Ca and ^{50}Ti, it has been proposed that CAIs (and potentially CAI-like dust that displays isotopic similarities to CAIs but differs chemically) are largely responsible for these observed anomalies (Leya et al. 2008; Trinquier et al. 2009; Akram et al. 2015; Burkhardt et al. 2019; Yap and Tissot 2023).

In summary, the pattern of nucleosynthetic variations in bulk meteorites suggests at least two distinct nucleosynthetic components at play: s-process nucleosynthetic products most likely carried by presolar AGB stardust grains and another unknown component with its isotopic compositions recorded by CAIs. It is also important to note that most of the dust – from which our Sun and planets formed – had an averaged isotope composition of the local ISM and only $\sim$3% of the ISM dust (i.e., presolar grains) kept the extreme isotopic compositions inherited from their parent stars (Hoppe et al. 2017; Ek et al. 2020). These presolar grain components are then responsible for the nucleosynthetic signatures found in bulk meteorites.

2.3.2 Linking Meteoritic Nucleosynthetic Variations to Astrophysical Sites

Nucleosynthetic Variations in CAIs and Their Astrophysical Sites The stellar sites of the r- and p-processes remain a subject of ongoing debate (Arnould and Goriely 2003; Tanvir et al. 2017; Kajino et al. 2019). The proposed astrophysical sites for the r-process include NS-NS mergers, black hole and NS mergers, and rare types of Type II supernovae, i.e., special classes of Type II supernovae with fast rotation and high magnetic fields (Cowan and Thielemann 2004; Cowan et al. 2021), while those for the p-process include Type Ia and Type II supernovae (Travaglio et al. 2011; Sasaki et al. 2022; Roberti et al. 2023). In addition, a weak r-process, i.e., reaching up to atomic mass $\approx$ 100 only, could potentially occur in ECSNe (Wanajo et al. 2011), and several processes producing isotopes up to this mass may occur during α-rich freeze outs and in the neutrino winds from the nascent neutron star during CCSNe (e.g., Akram et al. 2013).

While neutron-burst processes in Type II supernovae can account for the ^{50}Ti and ^{54}Cr enrichments of CAIs, neutron-burst models predict orders of magnitude higher production of ^{46}Ca than ^{48}Ca (see discussion in Sect. 2.2.1), inconsistent with the Ca isotope data of CAIs that shows distinct ^{48}Ca enrichments (Chen et al. 2015; Kööp et al. 2018). In addition, deep supernova regions like the Fe/Ni zone are characterized by prolific production of the α-isotope ^{40}Ca with no significant production of ^{48}Ca, i.e., $\delta^{48}\text{Ca} = -1000‰$. This is because ^{48}Ca cannot be produced in high-entropy conditions that are characteristic of Type II supernovae (Meyer et al. 1996). Meyer et al. (1996) suggested that low-entropy conditions like those that may be achieved in rare Type Ia supernovae are ideal for efficient ^{48}Ca production. This was later confirmed based on detailed nucleosynthesis model calculations of high-density Type Ia supernovae and ECSNe (Woosley 1997; Wanajo et al. 2013; Jones et al. 2019b). In addition, Ni and Fe isotope data of bulk meteorites suggest that both Type II and Type Ia supernova nucleosynthesis can explain the observed isotopic signatures (Nanne et al. 2019; Cook et al. 2021; Hopp et al. 2022). Moreover, it has been observed that variations in ^{46}Ti and ^{50}Ti, the most proton- and neutron-rich Ti isotopes, respectively, are well correlated among CAIs (Davis et al. 2018).[7] Therefore, it appeared that multiple stellar sources, e.g., Type Ia and Type II supernovae, contributed materials to the CAI-forming region.

More recently, Meyer and Bermingham (2023) showed that isotopes produced by different nucleosynthetic processes, including the neutron-rich isotopes ^{48}Ca and ^{96}Zr, the proton-rich isotope ^{46}Ti, the *p*-process isotope ^{84}Sr, could be produced together in an exploding white dwarf, corresponding to a rare high-density Type Ia supernova or an ECSN event. This is because the required distinct nucleosynthetic processes can occur at varying temperatures and densities at different stellar radii in these events. Detailed modeling of the stellar evolution of white dwarf explosions, coupled with nucleosynthesis calculations, is needed to test the proposal of Meyer and Bermingham (2023). GCE calculations are also needed to evaluate the likelihood of such events in contributing to the observed meteoritic isotopic variations in the early solar system.

In conclusion, existing isotopic evidence suggests that the nucleosynthetic variations of CAIs require the contribution of materials to the CAI-forming region from (i) multiple stellar sources, e.g., Type Ia and Type II supernovae or (ii) a rare high-density Type Ia supernova or an ECSN event as suggested by Meyer and Bermingham (2023).

***s*-Process Nucleosynthetic Variations** In contrast to ambiguities in the stellar sites of the neutron-rich isotopes of the intermediate-mass elements Ca, Ti, and Cr, and the *r*- and *p*-process isotopes, the discovery of Tc absorption lines in Mira-type variable stars by astronomer Paul Merrill (Merrill 1952) provided concrete evidence that the short-lived radioactive isotope ^{99}Tc ($t_{1/2} = 0.21$ Ma) is freshly synthesized in stellar interiors through the *s*-process along the beta stability valley. This discovery established a direct connection between *s*-process nucleosynthesis and low-mass AGB stars. Various research groups have developed sophisticated stellar models to elucidate the evolution of AGB stars and the associated nucleosynthesis processes (Cristallo et al. 2009; Karakas and Lugaro 2016; Battino et al. 2019; Vescovi et al. 2020). These AGB stellar models confirmed that the *s*-process predominantly operates in the He-intershell regions of low-mass AGB stars ($1.5\ M_\odot \lesssim M \lesssim 4\ M_\odot$) (Gallino et al. 1988). The presence of *s*-process isotopic signatures in presolar SiC and graphite grains for heavy elements like Zr and Mo provides valuable insights that have led to stringent constraints on AGB modeling parameters (Lugaro et al. 2003; Vescovi et al. 2020; Liu et al. 2018c, 2022a). With a reasonably comprehensive understanding of the *s*-process in AGB stars, Ek et al. (2020), based on nucleosynthetic isotope data of Pd, Mo, Ru and Zr from in bulk meteorites, identified presolar dust from high-metallicity AGB stars as the primary source that is responsible for the *s*-process isotopic variabilities of the bulk meteorites. Smaller isotope effects in heavier element (e.g., Nd, Hf) are predicted as a collateral consequence of high-metallicity AGB stars being sources for the *s*-process variations, and this was indeed observed in the meteorite record (e.g., Ek et al. 2020; Akram et al. 2013).

[7] Interestingly, variations in ^{46}Ti and ^{50}Ti are not correlated in chondritic matrices according to leaching experiments (Trinquier et al. 2009). This could be supporting evidence that small CAI dust originating from different stellar production sites was fused into forming the large CAIs present in chondrites.

2.3.3 Thermal Processing Versus Size Sorting of Dust Carriers

Considering that Pd is more volatile than Mo and Ru, Ek et al. (2020) further proposed that the diminished Pd isotopic effects stem from incomplete elemental condensation around AGB stars. They also proposed that the observed *s*-process variations are due to potential destruction of more fragile ISM dust, relative to AGB dust in the inner protoplanetary disk in the early solar system. This scenario predicts the lack of nucleosynthetic variations observed in the more volatile *s*-process elements such as Cd (Toth et al. 2020). On the other hand, recent findings have revealed substantial isotopic variations in the neutron-rich isotopes of the moderately volatile elements K and Zn (^{40}K and ^{70}Zn) within and between CC and NC meteorites (Martins et al. 2023; Nie et al. 2023). The discrepancy between Cd and K, Zn isotopes, highlights the role of dust carrier in controlling the preservation of its carried isotopic anomaly in meteorites. It is conceivable that ^{40}K and ^{70}Zn were enclosed in a thermally more robust carrier phase than, e.g., Cd, and thus their signature survived from the thermal processing. Given the still unclear nucleosynthetic origins of Zn and K isotopes combined with the unknown nature of their carriers, the observed K and Zn isotopic anomalies currently provide no direct clues to test the scenario of thermal processing.

Besides thermal processing, size sorting has been postulated as an alternative mechanism for introducing nucleosynthetic isotope variations that are assumed to be carried in diverse presolar phases within bulk meteorites (Pignatale et al. 2017, 2019). The gas-dust model simulations of Hutchison et al. (2022) used the two dust populations with their grain size distributions defined by the meteorite record: the nano-um sized presolar grain fraction and solar system silicates that grew up to mm-cm size (i.e. single grains, chondrules). Their results show that, if presolar grains remain small and do not form aggregates, they can become enriched in the outer disk through dust transport by viscous expansion within the protoplanetary disk. This enrichment mechanism may potentially explain the amplified isotopic signatures, such as those seen in ^{54}Cr, ^{50}Ti, and ^{96}Zr, within the CC reservoir located at greater heliocentric distances in comparison to the NC reservoir. Moreover, nucleosynthetic effects of the presolar grain fraction can be reduced in pressure bumps, e.g, produced by Jupiter in the disk, because the larger solar system dust is trapped in the bumps, while the smaller presolar grains are coupled to the gas and can move more freely through the bump.

Finally, investigations into presolar grains suggest an interesting trend. Specifically, studies on presolar grains revealed that the abundances of presolar grains from low-metallicity AGB stars (e.g., Type-Y and -Z SiC grains) and supernovae (e.g., Type-X SiC grains, Group 3 and 4 silicates, and Group 4 oxides) both increase with decreasing grain size (Hoppe et al. 2010, 2015; Liu et al. 2022b). It is important to note that the size disparities between grains from distinct stellar sources may be subtle, with most grains falling within the submicron (and maybe even smaller) range. Low-metallicity AGB dust is expected to carry stronger *s*-process isotopic signatures for the following reasons. Stellar observations of *s*-process element enhancements at the surface of C-rich stars suggest increasing *s*-process efficiency with decreasing stellar metallicity (e.g., Lugaro et al. 2020). This is because the *s*-process efficiency is linearly correlated with the number of neutrons per Fe atom – the seed for the *s*-process. In low-metallicity AGB stars, the amount of Fe is reduced because of the lowered metallicity, and the number of neutrons remains similar. Thus, the *s*-process occurs at enhanced neutron densities in low-metallicity AGB stars (see Käppeler et al. 2011 and Liu et al. 2022a for reviews of the *s*-process in AGB stars).

To explore the effect of size disparities further, future simulations are warranted to assess the feasibility of enriching relatively smaller presolar grains from lower-metallicity AGB stars and supernovae in the outer protoplanetary disk. If such enrichment processes existed, it could lead to (*i*) higher concentrations of low-metallicity AGB dust carrying *s*-process isotopic signatures at higher-than-average neutron densities (Liu et al. 2019, 2022c), providing a mechanism for the scenario proposed by Stephan and Davis (2021) to explain the NC-CC dichotomy for Mo isotopes, and (*ii*) supernova dust hosting neutron-rich isotopic signatures like ^{54}Cr within the outer protoplanetary disk. While these potential enrichments in low-metallicity AGB and supernova dust within the outer protoplanetary disk may offer an explanation to the observed variations in *s*-process and neutron-rich isotopic signatures among bulk meteorites, it is noteworthy that SiC grains of Type Z and Y from low-metallicity AGB stars have been observed to show heavy-element isotopic signatures that differ from low-metallicity AGB model predictions (Liu et al. 2019) and more investigations are needed.

3 Conlusions

In conclusion, the presence of presolar supernova grains in extraterrestrial materials and their distinctive isotopic signatures have provided ground-truth information to investigate a variety of nucleosynthesis processes in supernovae, especially in Type II supernovae. Below, we summarize the key implications and conclusions drawn from our overview.

1. Type II supernovae are known for their significant dust production based on spectroscopic observations, which is further corroborated by the dominance of Type II supernova dust among presolar supernova grains. Specifically, evidence for the Type II supernova origin of Type-X SiC grains includes the initial presence of ^{44}Ti in them, demonstrated by recent high-resolution NanoSIMS imaging data that allowed direct distinguishing between radiogenic and nonradiogenic ^{44}Ca excesses. Our analysis of the literature Ca-Ti isotope data from X grains suggested that the varying ^{44}Ti/^{48}Ti ratios observed among X grains resulted from sampling materials from the ^{44}Ti-producing Fe/Ni and Si/S zones and outer ^{44}Ti-free He and H shells with varying mixing ratios. In addition, nonradiogenic origins of small ^{44}Ca excesses were observed in two presolar silicate grains. The detected nonradiogenic ^{44}Ca excesses imply a spatially and/or temporally heterogenous distribution of the ^{44}Ca/^{40}Ca ratio in the ISM and possibly, in general, contributed to the ^{44}Ca/^{40}Ca variations observed among presolar oxides and silicates of low-mass AGB stellar origin.
2. In addition to the presence of ^{44}Ti, the Type II supernova origin of X grains is further corroborated by their other distinct isotopic signatures such as large excesses in ^{28}Si and ^{26}Al. The heavy-element isotopic compositions of X grains are characterized by the signatures of neutron bursts occurring in the He/C zone during the explosions of their parent supernovae. Current supernova model calculations face challenges in simultaneously explaining the neutron-burst isotopic signatures of X grains across multiple elements. Further studies are required to explore potential terrestrial/asteroidal contamination effects on grain measurements for varying elements and uncertainties in model predictions due to, e.g., uncertain nuclear reaction rates.
3. The light-element isotopic compositions of X grains indicate explosive H burning occurring in the He shell during the explosions of their parent supernovae, thus requiring mixing of H from the upper H shell into the underlying He shell. Two proposed mechanisms include H overshooting into the underlying He/N zone and H mixing throughout the entire He shell driven by rapid energy release during core C-burning (i.e., H ingestion). Model simulations based on the two mechanisms yield similar isotopic signatures and can both explain the light-element isotopic signatures of X grains. The two mechanisms, however, differ in their effects on the neutron-burst process. While mixing of H into the H/N zone has no effect on this process, the H-ingestion process is anticipated

to deactivate the neutron source for the neutron-burst process in the He/C zone, which seemingly contradicts the common presence of neutron-burst isotopic signatures in X grains. Arguments, however, have been made for the H-ingestion scenario to be in line with the X grain data that non-uniform, localized mixing of H induced by core C-burning could allow for neutron bursts in less affected regions of the He/C zone. Independent of the mechanism for inducing explosive H burning into the He shell, the probable occurrence of this process in Type II supernovae has led the proposal that presolar grains that carry explosive H burning isotopic signatures could all have come from Type II supernovae. Intercomparison of the isotope data of these potential supernova grains may shed new lights onto their ambiguous stellar origin(s).

4. Presolar nanospinel grains exhibit substantial ^{54}Cr excesses that point to their supernova origin, although the specific supernova type remains ambiguous, with high-density Type Ia, ECSNe, or Type II supernovae as potential candidates. Further investigations are essential to understand the specific mixing process required for the observed Cr isotopic signatures based on Type II supernova nucleosynthesis, quantify the relative contributions of these distinct supernova sources to the presolar dust reservoirs in the solar system, and acquire multielement isotopic compositions of presolar nanospinel grains to better distinguish between these distinct supernova sources.
5. The presence of nucleosynthetic isotope variations in bulk meteorites implies the heterogeneous distribution of various nucleosynthetic components that at least include *s*-process nucleosynthesis component carried in presolar AGB stardust grains and an unresolved supernova component with its isotopic compositions recorded by CAIs. Multiple supernova sources, including Type Ia supernovae, Type II supernovae, and ECSNe, have been suggested to explain the isotopic composition of the CAI component. In addition, systematic differences in the isotopic compositions of NC and CC meteorites point to a separation of the two isotopic reservoirs in the protoplanetary disk in the early solar system, based on which two mechanisms, thermal processing and size sorting, have been proposed. Future studies are required to investigate (i) the nucleosynthetic origins of observed K and Zn isotopic anomalies with the absence of Cd isotopic variations in bulk meteorites for testing the mechanism of thermal processing and (ii) the feasibility of enriching the relatively smaller presolar grains from low-metallicity AGB stars and supernovae in the outer protoplanetary disk based on presolar grain data.

These findings above underscore the complex interplay of stellar evolution, nucleosynthesis, and mixing, and the temporal and spatial effects of GCE on the ISM composition, highlighting the need for continued research to unravel the intricate origins of isotopic anomalies in various meteoritic components and their relations.

Acknowledgements This paper was instigated at the International Space Science Institute (ISSI) workshop "Evolution of the solar system: constraints from meteorites." The authors would like to express their gratitude to ISSI for the travel support. The authors also would like to thank Dr. Maria Bergemann, Dr. Larry Nittler, and an anonymous referee for a careful reading of the manuscript and helpful feedback and Drs. Marco Pignatari and Carolyn Doherty for insightful discussions and inputs. We thank the European Union's Horizon 2020 research and innovation program (ChETEC-INFRA – Project no. 101008324).

Funding Open Access funding enabled and organized by Projekt DEAL. This work was supported by NASA through grants 80NSSC23K1034 and 80NSSC24K0132 to N.L. and 80NSSC20K0338 to B.S.M. M.L. acknowledges the support of the Lendület Program (LP2023-10) of the Hungarian Academy of Sciences and the NKFIH excellence grant TKP2021-NKTA-64. J.L. acknowledges support by the German Research Foundation (DFG) through grant LE 3279/3-1.

Declarations

Competing Interests The authors declare that they have no conflict of interest.

References

Akram W, Schönbächler M, Sprung P, Vogel N (2013) Zirconium-hafnium isotope evidence from meteorites for the decoupled synthesis of light and heavy neutron-rich nuclei. Astrophys J 777:169

Akram W, Schönbächler M, Bisterzo S, Gallino R (2015) Zirconium isotope evidence for the heterogeneous distribution of s-process materials in the Solar System. Geochim Cosmochim Acta 165:484–500

Amari S, Hoppe P, Zinner E, Lewis RS (1992) Interstellar SiC with unusual isotopic compositions: grains from a supernova? Astrophys J 394:L43

Amari S, Lewis RS, Anders E (1994) Interstellar grains in meteorites: I. Isolation of SiC, graphite and diamond; size distributions of SiC and graphite. Geochim Cosmochim Acta 58:459–470

Amari S, Gao X, Nittler LR, Zinner E, José J, Hernanz M, Lewis RS (2001) Presolar grains from novae. Astrophys J 551:1065–1072

Amari S, Zinner E, Gallino R, Lewis RS (2003) Silicon and titanium isotopic analysis of silicon carbide grains of Type X and Z. Meteor Planet Sci Suppl 38:5118

Amari S, Zinner E, Gallino R (2014) Presolar graphite from the Murchison meteorite: an isotopic study. Geochim Cosmochim Acta 133:479–522

Arcones A, Thielemann F-K (2023) Origin of the elements. Astron Astrophys Rev 31:1

Arnould M, Goriely S (2003) The p-process of stellar nucleosynthesis: astrophysics and nuclear physics status. Phys Rep 384:1–84

Barzyk JG, Savina MR, Davis AM, Gallino R, Gyngard F, Amari S, Zinner E, Pellin MJ, Lewis RS, Clayton RN (2007) Constraining the ^{13}C neutron source in AGB stars through isotopic analysis of trace elements in presolar SiC. Meteor Planet Sci 42:1103–1119

Battino U, Tattersall A, Lederer-Woods C, Herwig F, Denissenkov P, Hirschi R, Trappitsch R, den Hartogh JW, Pignatari M, NuGrid Collaboration (2019) NuGrid stellar data set - III. Updated low-mass AGB models and s-process nucleosynthesis with metallicities $Z = 0.01$, $Z = 0.02$, and $Z = 0.03$. Mon Not R Astron Soc 489:1082–1098

Besmehn A, Hoppe P (2003) A NanoSIMS study of Si- and Ca-Ti-isotopic compositions of presolar silicon carbide grains from supernovae. Geochim Cosmochim Acta 67:4693–4703

Binns WR, Israel MH, Christian ER, Cummings AC, de Nolfo GA, Lave KA, Leske RA, Mewaldt RA, Stone EC, von Rosenvinge TT, Wiedenbeck ME (2016) Observation of the ^{60}Fe nucleosynthesis-clock isotope in galactic cosmic rays. Science 352:677–680

Boggs SE, Harrison FA, Miyasaka H, Grefenstette BW, Zoglauer A, Fryer CL, Reynolds SP, Alexander DM, An H, Barret D, Christensen FE, Craig WW, Forster K, Giommi P, Hailey CJ, Hornstrup A, Kitaguchi T, Koglin JE, Madsen KK, Mao PH, Mori K, Perri M, Pivovaroff MJ, Puccetti S, Rana V, Stern D, Westergaard NJ, Zhang WW (2015) ^{44}Ti gamma-ray emission lines from SN1987A reveal an asymmetric explosion. Science 348:670–671

Bojazi MJ, Meyer BS (2014) Explosive nucleosynthesis of N15 in a massive-star model. Phys Rev C 89:025807

Brennecka GA, Borg LE, Wadhwa M (2013) Evidence for supernova injection into the solar nebula and the decoupling of r-process nucleosynthesis. Proc Natl Acad Sci 110:17241–17246

Budde G, Burkhardt C, Brennecka GA, Fischer-Gödde M, Kruijer TS, Kleine T (2016) Molybdenum isotopic evidence for the origin of chondrules and a distinct genetic heritage of carbonaceous and non-carbonaceous meteorites. Earth Planet Sci Lett 454:293–303

Burkhardt C, Kleine T, Oberli F, Pack A, Bourdon B, Wieler R (2011) Molybdenum isotope anomalies in meteorites: constraints on solar nebula evolution and origin of the Earth. Earth Planet Sci Lett 312:390–400

Burkhardt C, Dauphas N, Hans U, Bourdon B, Kleine T (2019) Elemental and isotopic variability in Solar System materials by mixing and processing of primordial disk reservoirs. Geochim Cosmochim Acta 261:145–170

Burrows A, Vartanyan D (2021) Core-collapse supernova explosion theory. Nature 589:29–39

Charlier BLA, Tissot FLH, Vollstaedt H, Dauphas N, Wilson CJN, Charlier BLA, Marquez RT (2021) Survival of presolar p-nuclide carriers in the nebula revealed by stepwise leaching of Allende refractory inclusions. Sci Adv 7:eabf6222

Chen H-W, Lee T, Lee D-C, Chen J-C (2015) Correlation of ^{48}Ca, ^{50}Ti, and ^{138}La heterogeneity in the Allende refractory inclusions. Astrophys J 806:L21

Cook DL, Schönbächler M (2017) Iron isotopic compositions of troilite (FeS) inclusions from iron meteorites. Astron J 154:172

Cook DL, Meyer BS, Schönbächler M (2021) Iron and nickel isotopes in IID and IVB iron meteorites: evidence for admixture of an SN II component and implications for the initial abundance of ^{60}Fe. Astrophys J 917:59

Cowan JJ, Thielemann F-K (2004) R-process nucleosynthesis in supernovae. Phys Today 57(10):47–53

Cowan JJ, Sneden C, Lawler JE, Aprahamian A, Wiescher M, Langanke K, Martínez-Pinedo G, Thielemann F-K (2021) Origin of the heaviest elements: the rapid neutron-capture process. Rev Mod Phys 93:015002

Cristallo S, Straniero O, Gallino R, Piersanti L, Domínguez I, Lederer MT (2009) Evolution, nucleosynthesis, and yields of low-mass asymptotic giant branch stars at different metallicities. Astrophys J 696:797–820

Dauphas N, Marty B, Reisberg L (2002) Molybdenum nucleosynthetic dichotomy revealed in primitive meteorites. Astrophys J 569:L139–L142

Dauphas N, Remusat L, Chen JH, Roskosz M, Papanastassiou DA, Stodolna J, Guan Y, Ma C, Eiler JM (2010) Neutron-rich chromium isotope anomalies in supernova nanoparticles. Astrophys J 720:1577–1591

Dauphas N, Chen JH, Zhang J, Papanastassiou DA, Davis AM, Travaglio C (2014) Calcium-48 isotopic anomalies in bulk chondrites and achondrites: evidence for a uniform isotopic reservoir in the inner protoplanetary disk. Earth Planet Sci Lett 407:96–108

Davis AM, Zhang J, Greber ND, Hu J, Tissot FLH, Dauphas N (2018) Titanium isotopes and rare Earth patterns in CAIs: evidence for thermal processing and gas-dust decoupling in the protoplanetary disk. Geochim Cosmochim Acta 221:275–295

den Hartogh J, Petö MK, Lawson T, Sieverding A, Brinkman H, Pignatari M, Lugaro M (2022) Comparison between core-collapse supernova nucleosynthesis and meteoric stardust grains: investigating magnesium, aluminium, and chromium. Astrophys J 927:220

Desch SJ, Kalyaan A, Alexander CMO'D (2018) The effect of Jupiter's formation on the distribution of refractory elements and inclusions in meteorites. Astrophys J Suppl Ser 238:11

Diehl R (2017) Gamma-ray line measurements from supernova explosions. In: Marcowith A, et al (eds) IAU symp. 331, Supernova 1987A: 30 years later—cosmic rays and nuclei from supernovae and their aftermaths. Cambridge University Press, Cambridge, p 157

Doherty CL, Gil-Pons P, Lau HHB, Lattanzio JC, Siess L (2014) Super and massive AGB stars - II. Nucleosynthesis and yields - $Z = 0.02$, 0.008 and 0.004. Mon Not R Astron Soc 437:195–214

Douvion T, Lagage PO, Cesarsky CJ, Dwek E (2001) Dust in the Tycho, Kepler and Crab supernova remnants. Astron Astrophys 373:281–291

Ek M, Hunt AC, Lugaro M, Schönbächler M (2020) The origin of s-process isotope heterogeneity in the solar protoplanetary disk. Nat Astron 4:273–281

Fedkin AV, Meyer BS, Grossman L (2010) Condensation and mixing in supernova ejecta. Geochim Cosmochim Acta 74:3642–3658

Fimiani L, Cook DL, Faestermann T, Gómez-Guzmán JM, Hain K, Herzog G, Knie K, Korschinek G, Ludwig P, Park J, Reedy RC, Rugel G (2016) Interstellar ^{60}Fe on the surface of the Moon. Phys Rev Lett 116:151104

Fischer T, Whitehouse SC, Mezzacappa A, Thielemann F-K, Liebendörfer M (2010) Protoneutron star evolution and the neutrino-driven wind in general relativistic neutrino radiation hydrodynamics simulations. Astron Astrophys 517:A80

Gail H-P, Zhukovska SV, Hoppe P, Trieloff M (2009) Stardust from asymptotic giant branch stars. Astrophys J 698:1136–1154

Gallino R, Busso M, Picchio G, Raiteri CM, Renzini A (1988) On the role of low-mass asymptotic giant branch stars in producing a Solar System distribution of s-process isotopes. Astrophys J 334:L45

Gomez HL, Clark CJR, Nozawa T, Krause O, Gomez EL, Matsuura M, Barlow MJ, Besel M-A, Dunne L, Gear WK, Hargrave P, Henning T, Ivison RJ, Sibthorpe B, Swinyard BM, Wesson R (2012) Dust in historical galactic Type Ia supernova remnants with Herschel. Mon Not R Astron Soc 420:3557–3573

Grefenstette BW, Harrison FA, Boggs SE, Reynolds SP, Fryer CL, Madsen KK, Wik DR, Zoglauer A, Ellinger CI, Alexander DM, An H, Barret D, Christensen FE, Craig WW, Forster K, Giommi P, Hailey CJ, Hornstrup A, Kaspi VM, Kitaguchi T, Koglin JE, Mao PH, Miyasaka H, Mori K, Perri M, Pivovaroff MJ, Puccetti S, Rana V, Stern D, Westergaard NJ, Zhang WW (2014) Asymmetries in core-collapse supernovae from maps of radioactive ^{44}Ti in Cassiopeia A. Nature 506:339–342

Grefenstette BW, Fryer CL, Harrison FA, Boggs SE, DeLaney T, Laming JM, Reynolds SP, Alexander DM, Barret D, Christensen FE, Craig WW, Forster K, Giommi P, Hailey CJ, Hornstrup A, Kitaguchi T, Koglin JE, Lopez L, Mao PH, Madsen KK, Miyasaka H, Mori K, Perri M, Pivovaroff MJ, Puccetti S, Rana V, Stern D, Westergaard NJ, Wik DR, Zhang WW, Zoglauer A (2017) The distribution of radioactive ^{44}Ti in Cassiopeia A. Astrophys J 834:19

Greggio L, Renzini A (1983) The binary model for Type I supernovae - theoretical rates. Astron Astrophys 118:217–222

Groopman E, Bernatowicz, Zinner E (2012) C, N, and O isotopic heterogeneities in low-density supernova graphite grains from Orgueil. Astrophys J Lett 754:L8. 6pp

Gyngard F, Zinner E, Nittler LR, Morgand A, Stadermann FJ, Mairin Hynes K (2010) Automated NanoSIMS measurements of spinel stardust from the Murray meteorite. Astrophys J 717:107–120

Hammer NJ, Janka H-T, Müller E (2010) Three-dimensional simulations of mixing instabilities in supernova explosions. Astrophys J 714:1371–1385

Heck PR, Greer J, Kööp L, Trappitsch R, Gyngard F, Busemann H, Maden C, Ávila JN, Davis AM, Wieler R (2020) Lifetimes of interstellar dust from cosmic ray exposure ages of presolar silicon carbide. Proc Natl Acad Sci 117:1884–1889

Heger A, Woosley SE (2010) Nucleosynthesis and evolution of massive metal-free stars. Astrophys J 724:341–373

Hillebrandt W, Takahashi K, Kodama T (1976) r-process nculesoynthesis: a dynamical model. Astron Astrophys 52:63–68

Hiramatsu D, Howell DA, Van Dyk SD, Goldberg JA, Maeda K, Moriya TJ, Tominaga N, Nomoto KI, Hosseinzadeh G, Arcavi I, McCully C, Burke J, Bostroem KA, Valenti S, Dong Y, Brown PJ, Andrews JE, Bilinski C, Williams GG, Smith PS, Smith N, Sand DJ, Anand GS, Xu C, Filippenko AV, Bersten MC, Folatelli G, Kelly PL, Noguchi T, Itagaki K (2021) The electron-capture origin of supernova 2018zd. Nat Astron 5:903–910

Hopp T, Budde G, Kleine T (2020) Heterogeneous accretion of Earth inferred from Mo-Ru isotope systematics. Earth Planet Sci Lett 534:116065

Hopp T, Dauphas N, Spitzer F, Burkhardt C, Kleine T (2022) Earth's accretion inferred from iron isotopic anomalies of supernova nuclear statistical equilibrium origin. Earth Planet Sci Lett 577:117245

Hoppe P (2006) NanoSIMS: a new tool in cosmochemistry. Appl Surf Sci 252:7102–7106

Hoppe P, Strebel R, Eberhardt P, Amari S, Lewis RS (1996a) Small SiC grains and a nitride grain of circumstellar origin from the Murchison meteorite: implications for stellar evolution and nucleosynthesis. Geochim Cosmochim Acta 60:883–907

Hoppe P, Strebel R, Eberhardt P, Amari S, Lewis RS (1996b) Type II supernova matter in a silicon carbide grain from the Murchison meteorite. Science 272:1314–1316

Hoppe P, Strebel R, Eberhardt P, Amari S, Lewis RS (2000) Isotopic properties of silicon carbide X grains from the Murchison meteorite in the size range 0.5-1.5 μm. Meteorit Planet Sci 35:1157–1176

Hoppe P, Leitner J, Gröner E, Marhas KK, Meyer BS, Amari S (2010) NanoSIMS studies of small presolar SiC grains: new insights into supernova nucleosynthesis, chemistry, and dust formation. Astrophys J 719:1370–1384

Hoppe P, Leitner J, Kodolányi J (2015) New constraints on the abundances of silicate and oxide stardust from supernovae in the Acfer 094 meteorite. Astrophys J Lett 808:L9

Hoppe P, Leitner J, Kodolányi J (2017) The stardust abundance in the local interstellar cloud at the birth of the Solar System. Nat Astron 1:617–620

Hoppe P, Pignatari M, Kodolányi J, Gröner E, Amari S (2018) NanoSIMS isotope studies of rare types of presolar silicon carbide grains from the Murchison meteorite: implications for supernova models and the role of ^{14}C. Geochim Cosmochim Acta 221:182–199

Hoppe P, Stancliffe RJ, Pignatari M, Amari S (2019) Isotopic signatures of supernova nucleosynthesis in presolar silicon carbide grains of Type AB with supersolar ^{14}N/^{15}N ratios. Astrophys J 887:8

Hoppe P, Leitner J, Pignatari M, Amari S (2023) New constraints for supernova models from presolar silicon carbide X grains with very high ^{26}Al/^{27}Al ratios. Astrophys J Lett 943:L22

Hoppe P, Leitner J, Pignatari M, Amari S (2024) Isotope studies of presolar silicon carbide grains from supernovae: new constraints for hydrogen-ingestion supernova models. Mon Not R Astron Soc 532:211–222

Hotokezaka K, Piran T, Paul M (2015) Short-lived ^{244}Pu points to compact binary mergers as sites for heavy r-process nucleosynthesis. Nat Phys 11:1042

Hutchison MA, Bodénan J-D, Mayer L, Schönbächler M (2022) Presolar grain dynamics: creating nucleosynthetic variations through a combination of drag and viscous evolution. Mon Not R Astron Soc 512:5874–5894

Indebetouw R, Matsuura M, Dwek E, Zanardo G, Barlow MJ, Baes M, Bouchet P, Burrows DN, Chevalier R, Clayton GC, Fransson C, Gaensler B, Kirshner R, Lakićević M, Long KS, Lundqvist P, Martí-Vidal I, Marcaide J, McCray R, Meixner M, Ng C-Y, Park S, Sonneborn G, Staveley-Smith L, Vlahakis C,

van Loon J (2014) Dust production and particle acceleration in supernova 1987A revealed with ALMA. Astrophys J Lett 782:L2
Jones S, Hirschi R, Nomoto K, Fischer T, Timmes FX, Herwig F, Paxton B, Toki H, Suzuki T, Martínez-Pinedo G, Lam YH, Bertolli MG (2013) Advanced burning stages and fate of 8-10 $M_{\odot}$ stars. Astrophys J 772:150
Jones S, Röpke FK, Pakmor R, Seitenzahl IR, Ohlmann ST, Edelmann PVF (2016) Do electron-capture supernovae make neutron stars? First multidimensional hydrodynamic simulations of the oxygen deflagration. Astron Astrophys 593:A72
Jones S, Côté B, Röpke FK, Wanajo S (2019a) A new model for electron-capture supernovae in galactic chemical evolution. Astrophys J 882:170
Jones S, Röpke FK, Fryer C, Ruiter AJ, Seitenzahl IR, Nittler LR, Ohlmann ST, Reifarth R, Pignatari M, Belczynski K (2019b) Remnants and ejecta of thermonuclear electron-capture supernovae. Constraining oxygen-neon deflagrations in high-density white dwarfs. Astron Astrophys 622:A74
José J, Hernanz M (2007) The origin of presolar nova grains. Meteor Planet Sci 42:1135–1143
Käppeler F, Gallino R, Bisterzo S, Wako A (2011) The s process: Nuclear physics, stellar models, and observations. Rev Mod Phys 83:157–194
Kajino T, Aoki W, Balantekin AB, Diehl R, Famiano MA, Mathews GJ (2019) Current status of r-process nucleosynthesis. Prog Part Nucl Phys 107:109–166
Karakas AI, Lugaro M (2016) Stellar yields from metal-rich asymptotic giant branch models. Astrophys J 825:26
Kilpatrick CD, Foley RJ, Kasen D, Murguia-Berthier A, Ramirez-Ruiz E, Coulter DA, Drout MR, Piro AL, Shappee BJ, Boutsia K, Contreras C, Di Mille F, Madore BF, Morrell N, Pan Y-C, Prochaska JX, Rest A, Rojas-Bravo C, Siebert MR, Simon JD, Ulloa N (2017) Electromagnetic evidence that SSS17a is the result of a binary neutron star merger. Science 358:1583–1587
Kobayashi C, Karakas AI, Lugaro M (2020) The origin of elements from carbon to uranium. Astrophys J 900:179
Kodolányi J, Stephan T, Trappitsch R, Hoppe P, Pignatari M, Davis AM, Pellin MJ (2018) Iron and nickel isotope compositions of presolar silicon carbide grains from supernovae. Geochim Cosmochim Acta 221:127–144
Kööp L, Nakashima D, Heck PR, Kita NT, Tenner TJ, Krot AN, Nagashima K, Park C, Davis AM (2018) A multielement isotopic study of refractory FUN and F CAIs: mass-dependent and mass-independent isotope effects. Geochim Cosmochim Acta 221:296–317
Leitner J, Hoppe P (2019) A new population of dust from stellar explosions among meteoritic stardust. Nat Astron 3:725–729
Leitner J, Hoppe P (2022) Ca-isotope investigation of silicate stardust. In: 53rd Lunar and Planetary Science Conference. Abstract #1949
Leitner J, Hoppe P, Trieloff M (2024) Calcium and potassium isotope studies of presolar silicates. In: 55th Lunar and Planetary Science Conference. Abstract #1443
Leya I, Schönbächler M, Wiechert U, Krähenbühl U, Halliday AN (2008) Titanium isotopes and the radial heterogeneity of the Solar System. Earth Planet Sci Lett 266:233–244
Leya I, Schönbächler M, Krähenbühl U, Halliday AN (2009) New titanium isotope data for Allende and Efremovka CAIs. Astrophys J 702:1118
Limongi M, Chieffi A (2006) The nucleosynthesis of ^{26}Al and ^{60}Fe in solar metallicity stars extending in mass from 11 to 120 M_{solar}: the hydrostatic and explosive contributions. Astrophys J 647:483–500
Lin Y, Gyngard F, Zinner E (2010) Isotopic analysis of supernova SiC and Si_3N_4 grains from the Qingzhen (EH3) chondrite. Astrophys J 709:1157–1173
Liu N (2024) Presolar grains. In: Treatise on geochemistry, 3rd edn., vol 7, p 113
Liu N, Nittler LR, Alexander CMO'D, Wang J, Pignatari M, José J, Nguyen A (2016) Stellar origins of extremely ^{13}C- and ^{15}N-enriched presolar SiC grains: novae or supernovae? Astrophys J 820:140
Liu N, Nittler LR, Pignatari M, Alexander CMO'D, Wang J (2017) Stellar origin of ^{15}N-rich presolar SiC grains of Type AB: supernovae with explosive hydrogen burning. Astrophys J Lett 842:L1
Liu N, Gallino R, Cristallo S, Bisterzo S, Davis AM, Trappitsch R, Nittler LR (2018c) New constraints on the major neutron source in low-mass AGB stars. Astrophys J 865:112
Liu N, Nittler LR, Alexander CMO'D, Wang J (2018a) Late formation of silicon carbide in Type II supernovae. Sci Adv 4:eaao1054
Liu N, Stephan T, Boehnke P, Nittler LR, Meyer BS, Alexander CMO'D, Davis AM, Trappitsch R, Pellin MJ (2018b) Common occurrence of explosive hydrogen burning in Type II supernovae. Astrophys J 855:144
Liu N, Stephan T, Cristallo S, Gallino R, Boehnke P, Nittler LR, Alexander CMO'D, Davis AM, Trappitsch R, Pellin MJ, Dillmann I (2019) Presolar silicon carbide grains of types Y and Z: their molybdenum isotopic compositions and stellar origins. Astrophys J 881:28

Liu N, Cristallo S, Vescovi D (2022a) Slow neutron-capture process: low-mass asymptotic giant branch stars and presolar silicon carbide grains. Universe 8:362
Liu N, Dauphas N, Cristallo S, Palmerini S, Busso M (2022b) Oxygen and aluminum-magnesium isotopic systematics of presolar nanospinel grains from CI chondrite Orgueil. Geochim Cosmochim Acta 319:296–317
Liu N, Stephan T, Cristallo S, Vescovi D, Gallino R, Nittler LR, Alexander CMO'D, Davis AM (2022c) Presolar silicon carbide grains of types Y and Z: their strontium and barium isotopic compositions and stellar origins. Eur Phys J A 58:216
Liu N, Alexander CMO'D, Nittler LR, Wang J (2023a) Late formation of silicon carbide in supernova remnants confirmed by new vanadium-titanium isotope data. In: 54th Lunar and Planetary Science Conference. Abstract #2496
Liu N, Alexander CMO'D, Wang J, Nittler LR, Meyer BS (2023b) Radioactive ^{44}Ti found in ^{15}N-rich AB presolar SiC grains. In: 86th annual meeting of the meteoritical society. Abstract #6230
Liu N, Alexander CMO'D, Meyer BS, Nittler LR, Wang J, Stroud RM (2024a) Explosive nucleosynthesis in core-collapse Type II supernovae: insights from new C, N, Si, and Al-Mg isotopic compositions of presolar grains. Astrophys J Lett. 961:L22
Liu N, Alexander CMO'D, Wang J, Nittler LR, Meyer B (2024b) Potassium-calcium isotope systematics of supernova silicon nitride grains. In: 55th Lunar and Planetary Science Conference. Abstract #1478
Lodders K, Fegley B Jr (1995) The origin of circumstellar silicon carbide grains found in meteorites. Meteoritics 30:661
Lugaro M, Zinner E, Gallino R, Amari S (1999) Si isotopic ratios in mainstream presolar SiC grains revisited. Astrophys J 527:369–394
Lugaro M, Davis AM, Gallino R, Pellin MJ, Straniero O, Käppeler F (2003) Isotopic compositions of strontium, zirconium, molybdenum, and barium in single presolar SiC grains and asymptotic giant branch stars. Astrophys J 593:486–508
Lugaro M, Cseh B, Világos B, Karakas A, Ventura P, Dell'Agli F, Trappitsch R, Hampel M, D'Orazi V, Pereira CB, Tagliente G, Szabó GM, Pignatari M, Battino U, Tattersall A, Ek M, Schönbächler M, Hron J, Nittler LR (2020) Origin of large meteoritic SiC stardust grains in metal-rich AGB stars. Astrophys J 898:96
Lugaro M, Pignatari M, Reifarth R, Wiescher M (2023) The *s* process and beyond. Annu Rev Nucl Part Sci 73:315–340
MacPherson GJ (2014) Calcium-aluminum-rich inclusions in chondritic meteorites. In: Davis AM (ed) Meteorites and cosmochemical processes. Elsevier, Amsterdam, pp 139–179
Malherbe J, Penen F, Isaure M-P, Frank J, Hause G, Dobritzsch D, Gontier E, Horréard F, Hillion F, Schaumlöffel D (2016) A new radio frequency plasma oxygen primary ion source on nano secondary ion mass spectrometry for improved lateral resolution and detection of electropositive elements at single cell level. Anal Chem 88:7130–7136
Marhas KK, Amari S, Gyngard F, Zinner E, Gallino R (2008) Iron and nickel isotopic ratios in presolar SiC grains. Astrophys J 689:622–645
Martins R, Kuthning S, Coles BJ, Kreissig K, Rehkämper M (2023) Nucleosynthetic isotope anomalies of zinc in meteorites constrain the origin of Earth's volatiles. Science 379:369–372
Merrill PW (1952) Emission lines in the spectra of long-period variable stars. J R Astron Soc Can 46:181
Meyer BS, Bermingham KR (2023) Galactic chemical evolution of NRLEE dust. In: 54th Lunar and Planetary Science Conference. Abstract #2762
Meyer BS, Mathews GJ, Howard WM, Woosley SE, Hoffman RD (1992) R-process nucleosynthesis in the high-entropy supernova bubble. Astrophys J 399:656
Meyer BS, Weaver TA, Woosley SE (1995) Isotope source table for a 25 $M_\odot$ supernova. Meteoritics 30:325–334
Meyer BS, Krishnan TD, Clayton DD (1996) ^{48}Ca production in matter expanding from high temperature and density. Astrophys J 462:825
Meyer BS, Clayton DD, The L-S (2000) Molybdenum and zirconium isotopes from a supernova neutron burst. Astrophys J Lett 540:L49–L52
Modjaz M, Gutiérrez CP, Arcavi I (2019) New regimes in the observation of core-collapse supernovae. Nat Astron 3:717–724
Moriya TJ, Sorokina EI, Chevalier RA (2018) Superluminous supernovae. Space Sci Rev 214:59
Müller B (2020) Hydrodynamics of core-collapse supernovae and their progenitors. Living Rev Comput Astrophys 6:3
Nanne JAM, Nimmo F, Cuzzi JN, Kleine T (2019) Origin of the non-carbonaceous-carbonaceous meteorite dichotomy. Earth Planet Sci Lett 511:44–54
Niculescu-Duvaz M, Barlow MJ, Bevan A, Wesson R, Milisavljevic D, De Looze I, Clayton GC, Krafton K, Matsuura M, Brady R (2022) Dust masses for a large sample of core-collapse supernovae from

optical emission line asymmetries: dust formation on 30-year time-scales. Mon Not R Astron Soc 515:4302–4343

Nie NX, Wang D, Torrano ZA, Carlson RW, Alexander CMO'D, Shahar A (2023) Meteorites have inherited nucleosynthetic anomalies of potassium-40 produced in supernovae. Science 379:372

Niederer FR, Papanastassiou DA, Wasserburg GJ (1980) Endemic isotopic anomalies in titanium. Astrophys J 240:L73–L77

Nittler LR (2005) Constraints on heterogeneous galactic chemical evolution from meteoritic stardust. Astrophys J 618:281–296

Nittler LR (2009) On the mass and metallicity distributions of the parent AGB stars of O-rich presolar stardust grains. Publ Astron Soc Aust 26:271–277

Nittler LR, Ciesla F (2016) Astrophysics with extraterrestrial materials. Annu Rev Astron Astrophys 54:53–93

Nittler LR, Hoppe P (2005) Are presolar silicon carbide grains from novae actually from supernovae? Astrophys J Lett 631:L89–L92

Nittler LR, Hoppe P, Alexander CMO'D, Amari S, Eberhardt P, Gao X, Lewis RS, Strebel R, Walker RM, Zinner E (1995) Silicon nitride from supernovae. Astrophys J Lett 453:L25

Nittler LR, Amari S, Zinner E, Woosley SE, Lewis RS (1996) Extinct ^{44}Ti in presolar graphite and SiC: proof of a supernova origin. Astrophys J Lett 462:L31

Nittler LR, Alexander CMO'D, Gallino R, Hoppe P, Nguyen AN, Stadermann FJ, Zinner EK (2008) Aluminum-, calcium- and titanium-rich oxide stardust in ordinary chondrite meteorites. Astrophys J 682:1450–1478

Nittler LR, Gyngard F, Zinner E, Stroud RM (2011) Mg and Ca isotopic anomalies in presolar oxides: large anomalies in a group 3 hibonite grain. In: 42nd Lunar and Planetary Science Conference. Abstract #1872

Nittler LR, Alexander CMOD, Liu N, Wang J (2018) Extremely ^{54}Cr- and ^{50}Ti-rich presolar oxide grains in a primitive meteorite: formation in rare types of supernovae and implications for the astrophysical context of Solar System birth. Astrophys J Lett 856:L24

Nittler LR, Barosch J, Alexander CMOD, Wang J, Brearley A, Dobrică E (2023) Presoalr grain aggregates: a composite oxide-silicate from a low-metallicity AGB star and a possible circumstellar-interstellar grain aggregate. In: 86th annual meeting of the meteoritical society. Abstract #6079

Nomoto K (1984) Evolution of 8-10 solar mass stars toward electron capture supernovae. I - formation of electron-degenerate O + NE + MG cores. Astrophys J 277:791–805

Nomoto K (1987) Evolution of 8-10 M_{sun} stars toward electron capture supernovae. II. Collapse of an O + NE + MG core. Astrophys J 322:206

Nomoto K, Iwamoto K, Nakasato N, Thielemann F-K, Brachwitz F, Tsujimoto T, Kubo Y, Kishimoto N (1997) Nucleosynthesis in Type Ia supernovae. Nucl Phys A 621:467–476

Ott U, Stephan T, Hoppe P, Savina MR (2019) Isotopes of barium as a chronometer for supernova dust formation. Astrophys J 885:128

Pal I, Jadhav M, Savina MR, Shulaker DZ, Dory CJ, Gyngard F, Kita N, Amari S (2022) Heavy element isotopic measurements of high density presolar graphite grains with LION. In: 53rd Lunar and Planetary Science Conference. Abstract #1262

Papanastassiou DA (1986) Chromium isotopic anomalies in the Allende meteorites. Astrophys J 308:L27–L30

Pellin MJ, Calaway WF, Davis AM, Lewis RS, Amari S, Clayton RN (2000) Toward complete isotopic analysis of individual presolar silicon carbide grains: C, N, Si, Sr, Zr, Mo, and Ba in single grains of Type X. In: 31st Lunar and Planetary Science Conference. Abstract #1917

Pellin MJ, Savina MR, Calaway WF, Tripa CE, Barzyk JG, Davis AM, Gyngard F, Amari S, Zinner E, Lewis RS, Clayton RN (2006) Heavy metal isotopic anomalies in supernovae presolar grains. In: 37th Lunar and Planetary Science Conference. Abstract #2041

Pian E, D'Avanzo P, Benetti S, Branchesi M, Brocato E, Campana S, Cappellaro E, Covino S, D'Elia V, Fynbo JPU, Getman F, Ghirlanda G, Ghisellini G, Grado A, Greco G, Hjorth J, Kouveliotou C, Levan A, Limatola L, Malesani D, Mazzali PA, Melandri A, Møller P, Nicastro L, Palazzi E, Piranomonte S, Rossi A, Salafia OS, Selsing J, Stratta G, Tanaka M, Tanvir NR, Tomasella L, Watson D, Yang S, Amati L, Antonelli LA, Ascenzi S, Bernardini MG, Boër M, Bufano F, Bulgarelli A, Capaccioli M, Casella P, Castro-Tirado AJ, Chassande-Mottin E, Ciolfi R, Copperwheat CM, Dadina M, De Cesare G, di Paola A, Fan YZ, Gendre B, Giuffrida G, Giunta A, Hunt LK, Israel GL, Jin Z-P, Kasliwal MM, Klose S, Lisi M, Longo F, Maiorano E, Mapelli M, Masetti N, Nava L, Patricelli B, Perley D, Pescalli A, Piran T, Possenti A, Pulone L, Razzano M, Salvaterra R, Schipani P, Spera M, Stamerra A, Stella L, Tagliaferri G, Testa V, Troja E, Turatto M, Vergani SD, Vergani D (2017) Spectroscopic identification of r-process nucleosynthesis in a double neutron-star merger. Nature 551:67–70

Pignatale FC, Gonzalez J-F, Cuello N, Bourdon B, Fitoussi C (2017) Size and density sorting of dust grains in SPH simulations of protoplanetary discs. Mon Not R Astron Soc 469:237–254

Pignatale FC, Gonzalez J-F, Bourdon B, Fitoussi C (2019) Size and density sorting of dust grains in SPH simulations of protoplanetary discs - II. Fragmentation. Mon Not R Astron Soc 490:4428–4446

Pignatari M, Gallino R, Heil M, Wiescher M, Käppeler F, Herwig F, Bisterzo S (2010) The weak s-process in massive stars and its dependence on the neutron capture cross sections. Astrophys J 710:1557–1577

Pignatari M, Hirschi R, Wiescher M, Gallino R, Bennett M, Beard M, Fryer C, Herwig F, Rockefeller G, Timmes FX (2013a) The $^{12}C + ^{12}C$ reaction and the impact on nucleosynthesis in massive stars. Astrophys J 762:31

Pignatari M, Wiescher M, Timmes FX, de Boer RJ, Thielemann F-K, Fryer C, Heger A, Herwig F, Hirschi R (2013b) Production of carbon-rich presolar grains from massive stars. Astrophys J Lett 767:L22

Pignatari M, Zinner E, Hoppe P, Jordan CJ, Gibson BK, Trappitsch R, Herwig F, Fryer C, Hirschi R, Timmes FX (2015) Carbon-rich presolar grains from massive stars: subsolar $^{12}C/^{13}C$ and $^{14}N/^{15}N$ ratios and the mystery of ^{15}N. Astrophys J 808:L43

Pignatari M, Hoppe P, Trappitsch R, Fryer C, Timmes FX, Herwig F, Hirschi R (2018) The neutron capture process in the He shell in core-collapse supernovae: presolar silicon carbide grains as a diagnostic tool for nuclear astrophysics. Geochim Cosmochim Acta 221:37–46

Prantzos N, Abia C, Limongi M, Chieffi A, Cristallo S (2018) Chemical evolution with rotating massive star yields - I. The solar neighbourhood and the s-process elements. Mon Not R Astron Soc 476:3432–3459

Qin L, Nittler LR, Alexander CMO'D, Wang J, Stadermann FJ, Carlson RW (2011) Extreme ^{54}Cr-rich nano-oxides in the CI chondrite Orgueil - implication for a late supernova injection into the Solar System. Geochim Cosmochim Acta 75:629–644

Rauscher T, Heger A, Hoffman RD, Woosley SE (2002) Nucleosynthesis in massive stars with improved nuclear and stellar physics. Astrophys J 576:323–348

Roberti L, Pignatari M, Psaltis A, Sieverding A, Mohr P, Fülöp Z, Lugaro M (2023) The γ-process nucleosynthesis in core-collapse supernovae. I. A novel analysis of γ-process yields in massive stars. Astron Astrophys 677:A22

Rubin A, Gal-Yam A, De Cia A, Horesh A, Khazov D, Ofek EO, Kulkarni SR, Arcavi I, Manulis I, Yaron O, Vreeswijk P, Kasliwal MM, Ben-Ami S, Perley DA, Cao Y, Cenko SB, Rebbapragada UD, Woźniak PR, Filippenko AV, Clubb KI, Nugent PE, Pan Y-C, Badenes C, Howell DA, Valenti S, Sand D, Sollerman J, Johansson J, Leonard DC, Horst JC, Armen SF, Fedrow JM, Quimby RM, Mazzali P, Pian E, Sternberg A, Matheson T, Sullivan M, Maguire K, Lazarevic S (2016) Type II supernova energetics and comparison of light curves to shock-cooling models. Astrophys J 820:33

Ruiter AJ (2020) Type Ia supernova sub-classes and progenitor origin. Proc Int Astron Union 15(S357):1–15

Sarangi A, Matsuura M, Micelotta ER (2018) Dust in supernovae and supernova remnants I: formation scenarios. Space Sci Rev 214:63

Sasaki H, Yamazaki Y, Kajino T, Kusakabe M, Hayakawa T, Cheoun M-K, Ko H, Mathews GJ (2022) Impact of hypernova νp-process nucleosynthesis on the galactic chemical evolution of Mo and Ru. Astrophys J 924:29

Savina MR, Pellin MJ, Davis AM, Lewis RS, Amari S (2007) p-Process signature in a unique presolar silicon carbide grain. In: 38th Lunar and Planetary Science Conference. Abstract #2231

Schiller M, Bizzarro M, Siebert J (2020) Iron isotope evidence for very rapid accretion and differentiation of the proto-Earth. Sci Adv 6(7):eaay7604

Schofield J, Pignatari M, Stancliffe RJ, Hoppe P (2022) Isotopic ratios for C, N, Si, Al, and Ti in C-rich presolar grains from massive stars. Mon Not R Astron Soc 517:1803–1820

Sieverding A, Kresse D, Janka H-T (2023) Production of ^{44}Ti and iron-group nuclei in the ejecta of 3D neutrino-driven supernovae. Astrophys J Lett 957:L25

Singerling SA, Liu N, Nittler LR, Alexander CMO'D, Stroud RM (2021) TEM analyses of unusual presolar silicon carbide: insights into the range of circumstellar dust condensation conditions. Astrophys J 913:90

Sneden C, Cowan JJ, Gallino R (2008) Neutron-capture elements in the early galaxy. Annu Rev Astron Astrophys 46:241–288

Spyrou A, Mücher D, Denissenkov PA, Herwig F, Good EC, Balk G, Berg HC, Bleuel DL, Clark JA, Dembski C, DeYoung PA, Greaves B, Guttormsen M, Harris C, Larsen AC, Liddick SN, Lyons S, Markova M, Mogannam MJ, Nikas S, Owens-Fryar J, Palmisano-Kyle A, Perdikakis G, Pogliano F, Quintieri M, Richard AL, Santiago-Gonzalez D, Savard G, Smith MK, Sweet A, Tsantiri A, Wiedeking M (2024) First study of the ^{139}Ba (n,γ)^{140}Ba reaction to constrain the conditions for the astrophysical I process. Phys Rep Lett 132:202701

Stephan T, Davis AM (2021) Molybdenum isotope dichotomy in meteorites caused by s-process variability. Astrophys J 909:8

Stephan T, Trappitsch R (2023) Reliable uncertainties: error correlation, rotated error bars, and linear regressions in three-isotope plots and beyond. Int J Mass Spectrom 491:117053
Stephan T, Trappitsch R, Davis AM, Pellin MJ, Rost D, Savina MR, Yokochi R, Liu N (2016) CHILI - the Chicago Instrument for Laser Ionization - a new tool for isotope measurements in cosmochemistry. Int J Mass Spectrom 407:1–15
Stephan T, Trappitsch R, Davis AM, Pellin MJ, Rost D, Savina MR, Jadhav M, Kelly CH, Gyngard F, Hoppe P, Dauphas N (2018) Strontium and barium isotopes in presolar silicon carbide grains measured with CHILI - two types of X grains. Geochim Cosmochim Acta 221:109–126
Stephan T, Trappitsch R, Hoppe P, Davis AM, Bose M, Boujibar A, Gyngard F, Hynes KM, Liu N, Nittler LR, Ogliore RC (2024) The presolar grain database. I. Silicon carbide. Astrophys J, Suppl Ser 270:27
Tanvir NR, Levan AJ, González-Fernández C, Korobkin O, Mandel I, Rosswog S, Hjorth J, D'Avanzo P, Fruchter AS, Fryer CL, Kangas T, Milvang-Jensen B, Rosetti S, Steeghs D, Wollaeger RT, Cano Z, Copperwheat CM, Covino S, D'Elia V, de Ugarte Postigo A, Evans PA, Even WP, Fairhurst S, Figuera Jaimes R, Fontes CJ, Fujii YI, Fynbo JPU, Gompertz BP, Greiner J, Hodosan G, Irwin MJ, Jakobsson P, Jørgensen UG, Kann DA, Lyman JD, Malesani D, McMahon RG, Melandri A, O'Brien PT, Osborne JP, Palazzi E, Perley DA, Pian E, Piranomonte S, Rabus M, Rol E, Rowlinson A, Schulze S, Sutton P, Thöne CC, Ulaczyk K, Watson D, Wiersema K, Wijers RAMJ (2017) The emergence of a lanthanide-rich kilonova following the merger of two neutron stars. Astrophys J Lett 848:L27
The L-S, Clayton DD, Diehl R, Hartmann DH, Iyudin AF, Leising MD, Meyer BS, Motizuki Y, Schönfelder V (2006) Are ^{44}Ti-producing supernovae exceptional? Astron Astrophys 450:1037–1050
Thielemann F-K, Eichler M, Panov IV, Wehmeyer B (2017) Neutron star mergers and nucleosynthesis of heavy elements. Annu Rev Nucl Part Sci 67:253–274
Timmes FX, Woosley SE, Weaver TA (1995) Galactic chemical evolution: hydrogen through zinc. Astrophys J, Suppl Ser 98:617
Tinsley BM (1979) Stellar lifetimes and abundance ratios in chemical evolution. Astrophys J 229:1046–1056
Toth ER, Fehr MA, Friebel M, Schönbächler M (2020) Cadmium isotopes in chondrites and acid leachates: nucleosynthetic homogeneity and a monitor for thermal neutron capture effects. Geochim Cosmochim Acta 274:286–301
Trappitsch R, Stephan T, Savina MR, Davis AM, Pellin MJ, Rost D, Gyngard F, Gallino R, Bisterzo S, Cristallo S, Dauphas N (2018) Simultaneous iron and nickel isotopic analyses of presolar silicon carbide grains. Geochim Cosmochim Acta 221:87–108
Travaglio C, Röpke FK, Gallino R, Hillebrandt W (2011) Type Ia supernovae as sites of the p-process: two-dimensional models coupled to nucleosynthesis. Astrophys J 739:93
Trinquier A, Elliott T, Ulfbeck D, Coath C, Krot AN, Bizzarro M (2009) Origin of nucleosynthetic isotope heterogeneity in the solar protoplanetary disk. Science 324:374–376
Vescovi D, Cristallo S, Busso M, Liu N (2020) Magnetic-buoyancy-induced mixing in AGB stars: presolar SiC grains. Astrophys J Lett 897:L25
Wallner A, Feige J, Kinoshita N, Paul M, Fifield LK, Golser R, Honda M, Linnemann U, Matsuzaki H, Merchel S, Rugel G, Tims SG, Steier P, Yamagata T, Winkler SR (2016) Recent near-Earth supernovae probed by global deposition of interstellar radioactive ^{60}Fe. Nature 532:69–72
Walls LS, Meyer BS (2024) Neutrino-induced production of carbon and aluminum and relevance for SiC X grains. In: 55th Lunar and Planetary Science Conference. Abstract #1578
Wanajo S, Janka H-T, Müller B (2011) Electron-capture supernovae as the origin of elements beyond iron. Astrophys J Lett 726:L15
Wanajo S, Janka H-T, Müller B (2013) Electron-capture supernovae as origin of ^{48}Ca. Astrophys J Lett 767:L26
Wang L, Hu M, Wang L, Yang YY, Jiawen GH, Chen S, Hu L, Chen T-W, Mo J, Wang X, Baade D, Hoeflich P, Wheeler JC, Pignata G, Burke J, Hiramatsu D, Howell DA, McCully C, Pellegrino C, Galbany L, Hsiao EY, Sand DJ, Zhang J, Uddin SA, Anderson JP, Ashall C, Cheng C, Gromadzki M, Inserra C, Lin H, Morrell N, Morales-Garoffolo A, Müller-Bravo TE, Nicholl M, Gonzalez EP, Phillips MM, Pineda-García J, Sai H, Smith M, Shahbandeh M, Srivastav S, Stritzinger MD, Yang S, Young DR, Yu L, Zhang X (2024) Newly formed dust within the circumstellar environment of SN Ia-CSM 2018evt. Nat Astron 8:504–519
Weinberger C, Diehl R, Pleintinger MMM, Siegert T, Greiner J (2020) ^{44}Ti ejecta in young supernova remnants. Astron Astrophys 638:A83
Wesson R, Bevan A (2021) Observational limits on the early-time dust mass in SN 1987A. Astrophys J 923:148
Wongwathanarat A, Müller E, Janka H-T (2015) Three-dimensional simulations of core-collapse supernovae: from shock revival to shock breakout. Astron Astrophys 577:A48
Wongwathanarat A, Janka H-T, Müller E, Pllumbi E, Wanajo S (2017) Production and distribution of ^{44}Ti and ^{56}Ni in a three-dimensional supernova model resembling Cassiopeia A. Astrophys J 842:13

Woosley SE (1997) Neutron-rich nucleosynthesis in carbon deflagration supernovae. Astrophys J 476:801–810
Woosley SE, Heger A (2007) Nucleosynthesis and remnants in massive stars of solar metallicity. Phys Rep 442:269–283
Woosley SE, Hoffman RD (1992) The alpha-process and the r-process. Astrophys J 395:202
Woosley SE, Taam RE, Weaver TA (1986) Models for Type I supernova. I. Detonations in white dwarfs. Astrophys J 301:601
Xu Y, Zinner E, Gallino R, Heger A, Pignatari M, Lin Y (2015) Sulfur isotopic compositions of submicrometer SiC grains from the Murchison meteorite. Astrophys J 799:156
Yap TE, Tissot FLH (2023) The NC-CC dichotomy explained by significant addition of CAI-like dust to the Bulk Molecular Cloud (BMC) composition. Icarus 405:115680

Authors and Affiliations

Nan Liu[1] · Maria Lugaro[2,3,4,5] · Jan Leitner[6,7] · Bradley S. Meyer[8] · Maria Schönbächler[9]

✉ N. Liu
nanliu@bu.edu

✉ J. Leitner
jan.leitner@mpic.de

1 Institute for Astrophysical Research, Boston University, Boston, MA 02215, USA

2 Konkoly Observatory, HUN-REN Research Centre for Astronomy and Earth Sciences, Konkoly Thege Miklós út 15-17., H-1121, Budapest, Hungary

3 CSFK, MTA Centre of Excellence, Budapest, Konkoly Thege Miklós út 15-17, H-1121, Hungary

4 ELTE Eötvös Loránd University, Institute of Physics and Astronomy, Budapest 1117, Pázmány Péter sétány 1/A, Hungary

5 School of Physics and Astronomy, Monash University, VIC 3800, Melbourne, Australia

6 Max Planck Institute for Chemistry, D-55128 Mainz, Germany

7 Institute of Earth Sciences, Heidelberg University, D-69120 Heidelberg, Germany

8 Department of Physics and Astronomy, Clemson University, Clemson, SC 29634, USA

9 Institute for Geochemistry and Petrology, ETH Zürich, 8092 Zurich, Switzerland

Space Science Reviews (2025) 221:85
https://doi.org/10.1007/s11214-025-01207-0

Infall and Disk Processes – the Message from Meteorites

François L.H. Tissot[1] · Christoph Burkhardt[2] · Aleksandra Kuznetsova[3,4,5] · Andreas Pack[6] · Martin Schiller[7] · Fridolin Spitzer[2] · Elishevah M.M.E. Van Kooten[7] · Teng Ee Yap[1]

Received: 10 September 2024 / Accepted: 6 August 2025 / Published online: 17 September 2025

Abstract
How do planetary systems, in general, and our own Solar System (SS), in particular, form? In conjunction, Astronomy and Isotope Cosmochemistry provide us with powerful tools to answer this age-old question. In this contribution, we review recent advances in our understanding of circumstellar disk evolution, including infall and disk processes, as explored through astrophysical models and nucleosynthetic isotope anomalies of SS materials. Astronomically, filamentary structures and anisotropy are observed across the dynamic range of star formation and disk substructures are found to be ubiquitous, highlighting how star- and planet-forming environments are far more complex and dynamic than previously thought. Isotopically, two decades of investigation of nucleosynthetic anomalies in bulk meteorites and refractory inclusions have produced a rich dataset, revealing the existence of pervasive heterogeneity in the early SS, both at the large- (i.e., NC-CC dichotomy) and fine-scale (i.e., trends within the NC group). Using an updated data compilation, we review the systematics and emerging structures of these anomalies as a function of their nucleosynthetic origin. We present the two main families of models – *inheritance* vs *unmixing* – that have been proposed to explain the origin of the observed isotope heterogeneities, and discuss their respective implications for cloud infall and thermal processing in the disk. We also discuss how the extension of nucleosynthetic anomaly analyses to other chondritic components (Ameboid Olivine Aggregates, chondrules, matrix) has started to yield insights into transport, processing, and mixing of dust in the disk. Limitations, open questions, and key avenues for future work are presented in closing.

Keywords Protoplanetary disk · Streamers · Meteorites · Isotope anomalies · Nucleosynthesis

1 Introduction

From primitive mythological musings to more sophisticated Ptolemaic and Copernican theories, the origin of the Solar System (SS) and its planets has been amongst the most fundamental questions facing humankind. The discovery of exoplanetary systems has only made this question more relevant, offering a wealth of new samples to probe for comparison. Since the turn of the century, astronomical surveys have unveiled the remarkable diversity of planetary systems and broadly contextualized our SS within the galactic planetary census (Pinte

Extended author information available on the last page of the article

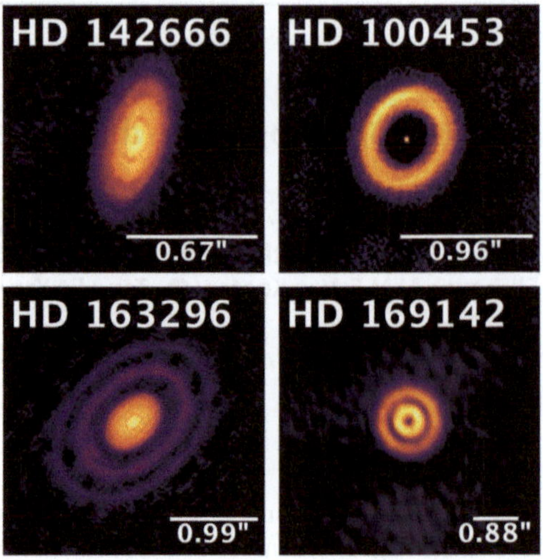

Fig. 1 A selection of ALMA images (Band 6 and 7 continuum) illustrating the diversity of sizes and structures of circumstellar disks. Scale bar is 100 AU (value denotes the angular scale in arcseconds). Modified from Stapper et al. (2022)

et al. 2020; Thompson et al. 2018). From these observations emerges the view that our SS is statistically anomalous in its architecture (Rosenthal et al. 2022), suggesting the inadequacy of Astronomy alone in solving the essential mystery: how *did* our SS form? In this article, we review recent theoretical and experimental insights into the dynamics of SS formation, with a focus on those brought on by the study of meteorites.

To leading order, the astrophysics of planetary system formation is well understood. Indeed, despite limitations in the spatial resolution of astronomical observations, the plethora of systems available for study have shed light on the governing mechanisms that drive disk evolution. The onset of gravitational fragmentation within giant molecular clouds (GMC) initiates the collapse of dense cores, in which conservation of angular momentum leads to the formation of a disk that grows outwards from the central proto-star by viscous spreading. Within the forming disk, micron-sized dust particles grow to the mm- and cm-scale by hit-and-stick collisions (e.g., Armitage 2020; Blum and Wurm 2008). Hydrodynamical processes, such as resonant drag instabilities (e.g., Squire and Hopkins 2018; Youdin and Goodman 2005; see Sect. 2), then proceed to concentrate these particles into dust clumps, which will themselves collapse into $\sim$100-km-scale planetesimals. Astronomical observations of star-forming regions and their protoplanetary systems provide snapshots of the structures that emerge from these phenomena, from GMC collapse and subsequent proto-cluster cloud formation (Friesen et al. 2016; Svoboda et al. 2019), to disk formation and evolution (Manara et al. 2022; van der Marel et al. 2021) (Fig. 1), and planet-disk interaction (e.g., Keppler et al. 2018; Law et al. 2023).

Over the past decade, our understanding of the dynamics of star and planet formation has rapidly evolved due to (i) high spectral and spatial resolution studies in the (sub)-millimeter domain over a range of physical scales, as well as (ii) the increased dynamic range achieved by magneto-hydrodynamic (MHD) simulations of star formation from cluster to disk scales. The emerging astrophysical picture is that of a hierarchy of heterogeneous structures, present from star-forming filaments on molecular cloud scales (10 s of parsecs) down to non-axisymmetric infalling flows called 'streamers' extending from cloud core to planet-forming disk scales (< 0.1 pc – 10 s of AU) (see review in Pineda et al. 2022). In addition, interferometric surveys of planet-forming disks at various ages with multiple tracers, via the Atacama Large Millimeter/submillimeter Array (ALMA) and the Very Large Array

(VLA) have revealed that protoplanetary disks can be found with structured distributions of millimeter-sized dust concentrated in rings, gaps, spirals, or crescents, often differentially distributed to molecular gas (Andrews et al. 2018; Öberg et al. 2021; Ohashi et al. 2023; Tobin et al. 2020). These advances have raised exciting new questions, also relevant for the early SS, regarding the dynamical evolution of planet-forming disks, the transport and delivery of planetary ingredients, and the timing of planet formation.

In essence, our SS provides a test case for models of planetary system formation. While it is only one of a myriad of planetary systems in the Cosmos, the processes that shaped it can be mapped with exquisite detail through the study of meteorites. Indeed, meteorites act as witnesses to SS formation, preserving imprints of the physico-chemical processes at play in the earliest stages of its history. By inspecting the elemental and isotopic composition of meteorites and their components, one can probe the dynamic dialogue between the collapsing cloud and forming disk, constrain the physical, thermal, and chemical history of meteorite parent bodies, and trace the transport and mixing of materials between different disk reservoirs. The latter processes are most clearly revealed through the investigation of nucleosynthetic isotope anomalies. These anomalies reflect the heterogeneous distribution of products from stellar nucleosynthesis in the protosolar disk and, as elegantly stated by Clayton (1982), carry a *cosmic chemical memory*. Much like DNA – which made robust reconstructions of human migration patterns possible – nucleosynthetic anomalies hold the key to retracing the migration patterns of solids in the early SS, offering the opportunity to expose its primordial architecture and evolution through time.

This contribution centers on recent developments concerning circumstellar disk evolution as explored through astrophysical models and nucleosynthetic isotope anomalies of SS materials. The former is discussed in Sect. 2, with emphasis on stellar birth environments, streamers, and the formation of disk sub-structures. In Sect. 3, we review the fundamentals and history of isotopic anomalies, and present their correlations in multi-element space. We then discuss how these anomalies can trace compositional changes in cloud infall in Sect. 4.1, and in Sect. 4.2, discuss how they may instead reflect transport and thermal processes in the disk. In Sect. 5, we detail what can be learned about transport, processing, and mixing of dust in the disk from isotopic anomalies in cometary materials and components of chondritic meteorites. Finally, we review the current limitations to our understanding of disk evolution and end with a discussion of open questions and an outlook in Sect. 6.

2 Astrophysical Context

2.1 Structures of Star-Forming Regions and Protoplanetary Disks

Detailed observations of star-forming regions in the Solar neighborhood have revealed an array of potential templates on which to base our understanding of the birth environment of the SS (e.g., Alves et al. 2020; Barclay et al. 2018), variable in evolutionary age (< 1 Myr in Ophiuchus to 5–10 Myr in Upper Sco), stellar density (10–1000 $M_{\odot} pc^{-3}$), and high energy radiation (UV to X-ray). These regions comprise a collection of young, protostellar systems, many of which host dust-rich planet-forming disks, inferred to have been born within a few Myr of each other and still embedded within their natal filamentary cloud. Studies of the dynamics and morphology of star-forming regions reveal that the natal gas distribution is quite heterogeneous: the larger scale filaments in which stars form were found to be composed of many different sub-filaments dubbed 'fibers', velocity coherent on relatively small

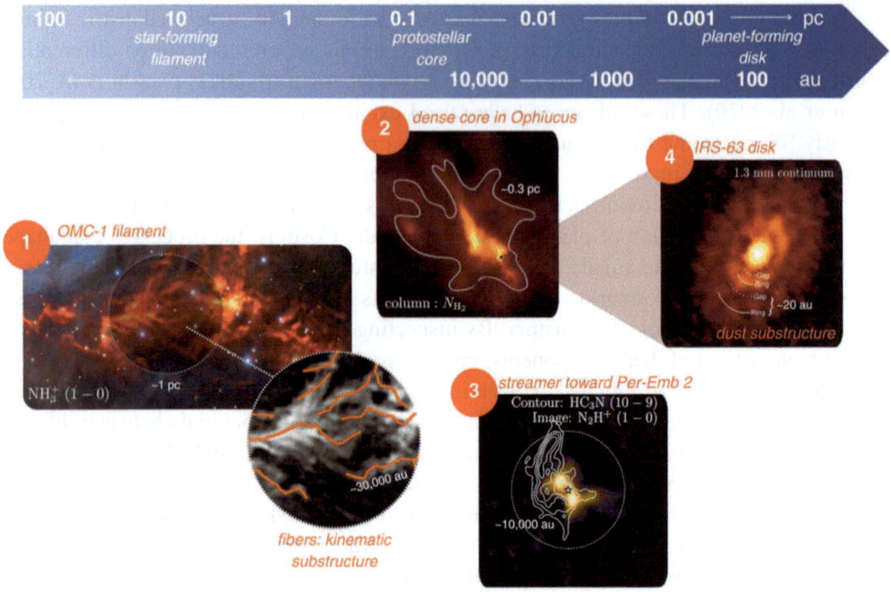

Fig. 2 Filamentary structures and anisotropy are observed across the dynamic range of star formation. Starting at the star-forming scale, (1) shows the OMC-1 star-forming filament observed by ALMA+IRAM 30 m in the form of 'fibers' identified as velocity coherent on 30,000 AU scales (Hacar et al. 2018). Such filamentary behavior and anisotropy is associated with the presence of (2) dense protostellar cores, as in the heterogeneous column density of L1709, inferred from Herschel observations (Arzoumanian et al. 2019) and fit with ballistic free-fall trajectories as in the case of (3) the 10,500 AU 'streamer' structure around the protostar Per-Emb 2 seen in HC_3N by ALMA (Pineda et al. 2020; Pinte et al. 2020). The structured infalling envelopes of cores like L1709 can be associated with the presence of substructure in the (4) thermal dust continuum (1.3 mm) emission from ALMA of its young planet-forming disk (< 500 kyr old) IRS-63 (Segura-Cox et al. 2020)

scales ($\sim$ 0.1–0.3 pc) (Hacar et al. 2018) (Fig. 2). In numerical simulations that track star-formation from cluster to protostellar core scales, such structures are a natural consequence of the gravitational collapse of turbulent clouds. Recent studies find that assembly and accretion of protostellar material should be episodic and directional (Kuffmeier et al. 2017; Kuznetsova et al. 2019), reflecting the inherent heterogeneity of the star-forming environment. In these simulations (which have variable approaches to the treatment of feedback from massive stars), episodes of accretion onto nascent disks have been found to correspond to distinct spatial cloud reservoirs, suggesting a picture in which protostellar systems are supplied with fresh material from surrounding fibers. Such filamentary flows of infalling cloud material, from 10^3 to 10^4 AU in length and extending down to disk scales, dubbed 'streamers' (Pineda et al. 2020), have been found around a number of sources in low-mass star-forming regions at various evolutionary stages. Recent observations of streamer sources in the NGC 1333 star-forming region connect the smaller-scale inflow onto protostellar sources to large-scale flows originating outside of the host filament (Valdivia-Mena et al. 2024), supporting a picture in which early phase disks can be expected to experience later-stage accretion of compositionally distinct material. The dynamical timescales of observed streamers are relatively short (10 s–100 s of kyr) compared to the typical disk lifetime (3–10 Myr, Mamajek et al. 2004), suggesting that multiple episodes of infall during disk assembly and evolution are probable. For instance, the infalling streamer in Per-Emb 2 (Pineda et al.

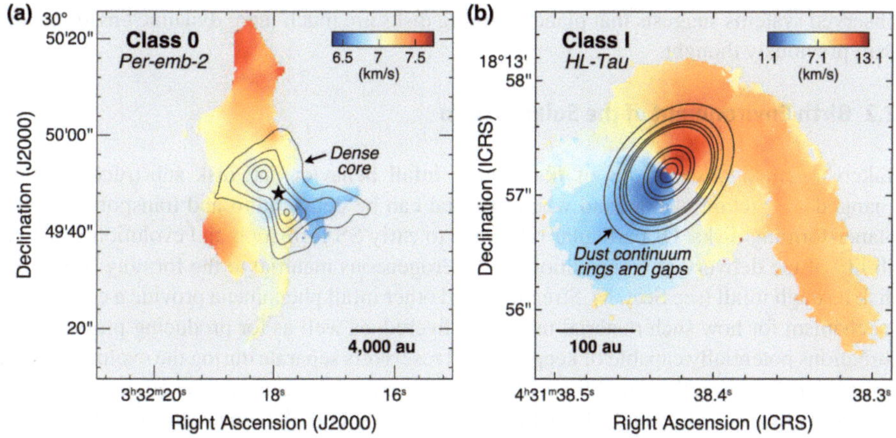

Fig. 3 Examples of streamers at two different scales corresponding to two classes of young stellar objects (YSO) (modified from Pineda et al. 2022). (a) Large scale streamer and dense molecular cloud core around the Per-emb-2 Class 0 YSO. The dense core was traced using N_2H^+ (1–0), and the streamer centroid velocity (color bar) using HC_3N (10–9) from NOEMA observations. (b) Streamer funneling material onto the disk of the Class 1 YSO HL-Tau. The centroid velocity is derived from HCO^+ (3–2), and the disk structure (rings and gaps) is as identified by ALMA Partnership et al. (2015). Streamers could facilitate the delivery of compositionally variable materials into protoplanetary disks

2020) traced by HC_3N emission, is kinematically distinct from a remnant envelope traced by NH_3^+ (Fig. 2 & 3). Notably, this type of anisotropic streaming infall in sources assumed to be at the earliest phases of disk evolution has been found to coincide with annular structures, rings and gaps, in the millimeter dust emission profile of their planet-forming disks (Segura-Cox et al. 2020; Yen et al. 2019) (Fig. 2 & 3). Additionally, episodes of filamentary infall do not appear to be restricted to the early stages of disk formation, as several sources at later evolutionary ages (Class II) have been suggested to host remnant infalling envelopes or infalling streams (GM Aur: Huang et al. 2021, SU Aur: Akiyama et al. 2019).

While disk substructures have been identified in interferometric surveys targeting the thermal millimeter dust emission of nearby sources in younger embedded systems (Kido et al. 2023; Sai Insa Choi et al. 2023; Sheehan et al. 2020), they are particularly ubiquitous at later phases, when the disk is no longer as heavily enshrouded within an infalling envelope (Andrews et al. 2018). Dust concentration within substructures has been shown to be robustly produced when grains, drifting inward as they lose energy to gas drag, are arrested in their drift by a perturbation of the disk's pressure gradient (Pinilla et al. 2012). The presence of such perturbations, often referred to as dust traps, can have consequences for the long-term evolution of the disk structure and transport. Dust traps serve as both barriers between solid reservoirs, restricting the flow of solids, and as potential sites for efficient planet formation, locally enhancing the dust concentration and speeding up growth through coagulation and clumping mechanisms. Such perturbations, i.e., pressure bumps, can be directly produced by the spiral wakes from the tidal forces of planets (Bae and Zhu 2018; Dong and Fung 2017; Zhu et al. 2015), at the edges of gaps opened by larger planets (Hammer et al. 2019; Zhu et al. 2012), at so-called 'dead zones' wherein turbulence is quenched (e.g., Gammie 1996; Okuzumi and Hirose 2011; Pinilla et al. 2016; Dullemond and Penzlin 2018), or even due to hydrodynamic instabilities induced by infall itself (Kuznetsova et al. 2022). While there is no single agreed upon physical interpretation for the cause of observed disk substructures (see Bae et al. 2022; Morbidelli et al. 2024), their ubiquity among observed systems suggests that planet-forming disks are much more dynamic environments than previously thought.

2.2 Birth Environment of the Solar System

Taken together, observations of filamentary infall behavior and disk substructures have changed the picture of how and when material can be delivered to and transported within planet-forming disks. Of particular relevance to early SS formation and evolution is the viability of the delivery of compositionally heterogeneous material to the forming protosolar disk through infall (see Sect. 4). Streamers and other infall phenomena provide a compelling mechanism for how such material may be delivered, as well as for producing pressure perturbations potentially capable of keeping solid reservoirs separate during the evolution of the protosolar disk. Whether the SS was likely to have experienced such heterogeneous infall events in its early history and whether the composition of infalling material could adequately reproduce observed meteoritic trends may depend, in part, on the properties of the SS birth environment. We need to emphasize that the discovery and characterization of streamers is quite recent (i.e., within the past 5 years), and most of the known sources at this point have been discovered serendipitously in existing ALMA data, not in any dedicated searches. As such, we keep in mind the aphorism "absence of evidence is not evidence of absence" when discussing the likelihood of streamers existing in specific types of environments.

The best-studied astrophysical systems are in the low-mass star-forming regions in the solar neighborhood. Because ALMA campaigns with adequate sensitivity and observational configurations to detect streamers have primarily focused on disks in such regions, most streamer detections come from low-mass star-forming regions. If the SS was born in such an environment, the available data already suggests that it would be likely to have experienced multiple episodes of infall during disk assembly and evolution. However, more than half of low-mass stars like our Sun are expected to form in high-mass star-forming regions (N>1000), and a number of works argue in favor of the Sun being born in such an environment, in close proximity to a massive star (e.g., Adams 2010; Desch and Miret-Roig 2024). External irradiation from a nearby massive star resulting in $\sim 10^4$ times the standard interstellar radiation field in the solar neighborhood today would result in a large degree of photoevaporation. If the protosolar disk evolved in a gas-poor environment cleared by external photoevaporation, akin to that of the proplyds (i.e., photoionized protoplanetary disks) in Orion, accretion of fresh material through streamers or late-stage infall would be highly unlikely. Recent works, however, have used cosmochemical signatures such as the O and S isotopic ratios and the relative abundance of H, He, to Ar, Xe in Jupiter's atmosphere to constrain external irradiation levels from a nearby massive star to more moderate levels between 300 and 3,000 times the standard interstellar radiation field in the solar neighborhood today (Desch and Miret-Roig 2024; Monga and Desch 2014). The size of the SS is consistent with such far-UV (FUV) fluxes, and the question is whether infall of compositionally heterogeneous material in such environment is likely. Given the bias of existing data towards low-mass star-forming regions, a proper discussion of likelihood is not possible. However, the recent detection of a streamer introducing compositionally variable material onto the Class I protostar Oph IRS 63 (Flores et al. 2023; Podio et al. 2024) shows that streamers do occur under moderate photoevaporation. Below, we examine whether enrichment of the early protosolar disk through infall is possible in scenarios in which the protosun is initially born in a more embedded environment or in proximity to massive stars.

Massive (OB) stars are core-collapse supernova progenitors, born in more densely packed massive star-forming regions such as Orion, rather than the nearby low-mass star forming regions inhabited by the most well-studied planet-forming disks. If the SS came from a massive star-forming region, it may have shared its birth environment with a supernova (SN) progenitor responsible for injection of fresh material to the molecular cloud, including the Fe-peak elements and their nucleosynthetic anomalies. However, proximity to a SN progenitor or a SN explosion before and during SS formation is thought to result in hostile conditions for star formation and disk survival. If enrichment occurred early in the protosolar disk's lifetime and in close proximity to the progenitor (< 1 pc), as in the 'injection scenario', special conditions would be needed for the protosolar disk to survive photoevaporation from the strong FUV/EUV radiation emitted from the nearby massive star prior to explosion. Specifically, the protosolar system must have either (i) spent the majority of its time in the outskirts of a massive star-forming region (Adams 2010; Desch and Miret-Roig 2024) or (ii) been formed as a compact and massive enough system to withstand high amounts of external photoevaporation (Parker 2020). Considering direct injection through a more clumpy ejecta geometry loosens the constraints on the proximity of the SN progenitor to a few parsecs (Ouellette et al. 2010). However, without additional shielding, at a few parsecs, elevated levels of FUV still place significant limits on the disk lifetime, and current models find that disk destruction occurs within 0.5–1 Myr for all but the most compact systems (Gárate et al. 2024).

Recent work argues that in an embedded environment, not only could the nascent molecular cloud filament effectively shield the early SS from the effects of stellar feedback, but that streamers could themselves aid in the delivery of fresh materials (including short lived radionuclides) (Arzoumanian et al. 2023). Indeed, numerical simulations of star cluster formation that take into account shielding from high energy radiation by the surrounding gas and dust find much greater probabilities of disk survival for systems in massive star forming regions (Wilhelm et al. 2023). Recent observations of OMC-1, a region in close proximity (< 1 pc) to the massive star photoionizing the Trapezium region in Orion, find that the filamentary substructures in the protostellar environs (Hacar et al. 2018) are of such high density ($n_H > 10^7\ cm^{-3}$), as to be largely unaffected by high-energy radiation from nearby stars (Hacar et al. 2020).

Membership within a massive star-forming region is, however, not necessarily required for delivery of SN material. Indeed, not all massive stars remain within their natal environments, and an estimated 10–15% of O stars are observed to be unbound from any star cluster (Maíz Apellániz et al. 2018). These 'runaways' are thought to be ejected from their host clusters through dynamical interactions. Estimates based on expulsion from binary systems due to the "kick" from the explosion of a more massive companion predict that O stars could travel hundreds of parsecs from their birthplaces before their own SN event (Renzo et al. 2019). Furthermore, several works contend that there may exist some base level of enrichment in the natal gas due to self-pollution from massive (e.g., Wolf-Rayet) stars and/or supernovae (Desch et al. 2022; Dwarkadas et al. 2017; Jacobsen 2005; Young 2014). If the SS formed in a scenario akin to that proposed for the nearby low-mass star-forming regions of the Local Bubble – in which star formation was triggered by the expansion of SN ejecta into the interstellar medium (Zucker et al. 2022) – enrichment and formation of the SS birth cluster may very well go hand in hand (Gounelle and Meynet 2012). Compression from the pressure-driven expansion of SN ejecta can trigger star formation in marginally stable clouds, but enriched ejecta from the SN explosion must cool before it can be incorporated into star-forming material. Numerical simulations tracking SN material within turbulent star-forming molecular clouds find that hot ejecta do not mix efficiently with the existing cool molecular material. Enriched material is thus primarily incorporated after it has cooled sufficiently to be accreted by protostellar systems following their initial phase of disk formation (t > 100 kyr) (Kuffmeier et al. 2016).

These recent insights suggest that it is important to consider that the time at which compositionally distinct material is accreted onto the protosolar disk does not necessarily correspond to the timing of an enrichment event like a SN explosion. The composition of accreted material can also change over time when protostars move through and accrete from spatial reservoirs of varying composition. Numerical simulations find that protostars accrete material from well outside the bounds of the initial cores, in some cases traveling on the order of 0.3–1.0 pc from their initial formation location within < 1 Myr (Kuznetsova et al. 2015). Observed SN remnants show spatial compositional heterogeneity (e.g., Holland-Ashford et al. 2020; Winkler et al. 2014), suggesting anisotropies in the ejecta are imprinted on their structure. Recent 3D simulations of SN explosions have even been able to capture the development of these anisotropies across scales (Mandal et al. 2023), even accounting for variable yields of Ti/Fe in ejecta compared to spherically symmetric models (Sieverding et al. 2023).

Two important constraints when considering the birth environment of the protosolar disk are (i) the source(s) and (ii) the timing of the enrichment of the natal gas in short-lived radionuclides (SLR) and other r-/s-process elements through stellar feedback: both through winds driven by massive stars and the ejecta of core-collapse supernovae. Short-lived radionuclides such as ^{26}Al ($t_{1/2} = 0.7$ Myr) inform on the relative timing of enrichment, as their very presence in the early SS requires that they were produced and injected into the parental cloud shortly before SS formation (e.g., Dauphas and Chaussidon 2011; Tissot et al. 2016; Young 2014). If present, isotopic heterogeneity for SLR would additionally require that the enrichment itself is spatially or temporally heterogeneous. Although there is some debate about the (small-scale) homogeneity of ^{26}Al (e.g., Desch et al. 2023c; Krestianinov et al. 2023) and ^{92}Nb (Hibiya et al. 2023), the remarkable concordance of various SLR chronometers among different planetary bodies points to a homogeneous distribution of SLRs in the early SS to within a level of a few per cent (e.g., Dauphas and Chaussidon 2011; Davis and McKeegan 2014; Desch et al. 2023c,b; Nyquist et al. 2009). By contrast, isotopic anomalies of non-radiogenic, nucleosynthetic origin are observed among different early SS materials in the parts per ten thousand to parts per million range (ε to μ levels). Even though these isotopic variations are tiny, they provide essential information about the formation and evolution of the SS.

In summary, star formation is a multi-scale and multi-physics problem of immense dynamic range. Our theoretical understanding has evolved in order to account for how the large-scale environment affects the formation and evolution of planetary systems, resulting in a much more dynamic and complex picture than previously thought (Fig. 2). While the birth environment of the SS remains uncertain, constraints on the timing and relative amounts of compositional variation traced by meteoritic evidence are available. These could help constrain the likelihood of various enrichment scenarios and/or environmental conditions relevant to the early SS, and establish its context within the greater Galactic environment.

3 Isotope Anomalies in Meteorites

3.1 Basics and History

3.1.1 What are isotope anomalies?

As samples of planetary bodies forming in the solar accretion disk, meteorites are key for constraining the origin, makeup, and physico-chemical evolution of matter in the Sun's birth environment. Investigating isotope variations in meteorites and their components is particularly useful in this respect. These variations are classified into two broad categories: *mass-dependent* and *mass-independent*. As the name indicates, *mass-dependent* isotope effects are those in which the degree of isotopic fractionation scales proportionally with isotopic mass differences. These effects arise during equilibrium processes in (iso-)chemical exchange reactions due to differences in the zero-point energy of isotopically substituted molecules, as well as during kinetic (i.e., unidirectional) processes as a result of the faster reaction and/or diffusion of the lighter isotopes relative to the heavier ones. Mass-dependent isotope effects are generally large (‰ level or more) and can provide insights into the conditions of phase equilibria, or processes like evaporation, condensation, and diffusion. Extensive description of mass-dependent effects, including their quantum mechanical origin, can be found in the literature (e.g., Albarède et al. 2004; Bigeleisen and Goeppert-Mayer 1947; Dauphas and Schauble 2016; Ibañez-Mejia and Tissot 2021; Maréchal et al. 1999; Urey 1947; Young et al. 2002).

In contrast, *mass-independent* effects are those that *do not* scale proportionally with the mass difference between isotopes but relate to other nuclide properties like radioactivity, neutron capture cross section, or radius. These effects present themselves as deviations from mass-dependent behavior (Fig. 4). In order to isolate them, the measured data need to be corrected for natural and analytical mass-dependent effects. Such corrections can only be done for elements with at least three stable isotopes, which unfortunately means that for elements like H, C and N (three key volatile elements), differentiating between mixing and material processing is extremely arduous. For elements with 3 isotopes or more, correction of mass-dependent effects is typically performed using an internal normalization approach, whereby the degree of mass-dependent fractionation (often denoted β) experienced by the sample is determined by (i) fixing the "normalizing ratio" to a reference value (most often the terrestrial composition), and (ii) assuming a particular mass-dependent law (typically, the exponential law) (e.g., Maréchal et al. 1999; Russell et al. 1978). The β value thus obtained is then used to remove the effect of mass-dependent fractionation from the isotope ratios of interest. Residual departures from mass-dependency after internal normalization are mass-independent effects.

Not all mass-independent effects are considered "isotope anomalies", a term reserved here for isotopic variations that cannot be explained by radiogenic ingrowth, cosmo- and nucleogenic production, or Nuclear Field Shift effects (Fig. 4). These isotope anomalies are typically small, and most often reported in epsilon (ε) and mu (μ) notations, which are, respectively, part-per-ten-thousand and part-per-million (ppm) deviations from the terrestrial composition. At high precision (i.e., ppm level), spurious effects can arise from a host of analytical effects, requiring that great care be taken to measure samples and standards under identical conditions. Similarly, for samples that have experienced large degrees of mass-dependent fractionation, correction with an inappropriate mass-fractionation law can result in apparent anomalies, which can mislead data interpretation (e.g., Budde et al. 2023; Davis et al. 2015; Olsen et al. 2013; Zhang et al. 2014). In this contribution, we will focus on two types of isotopic anomalies: (i) those affecting oxygen isotopes, the origin of which remains debated but is most often understood as a result of isotope-selective photodissociation of CO in the solar nebula (Clayton 2002), and (ii) those of nucleosynthetic origin, which reflect the heterogeneous distribution of isotopically anomalous presolar grains in the solar nebula (e.g., Dauphas and Schauble 2016).

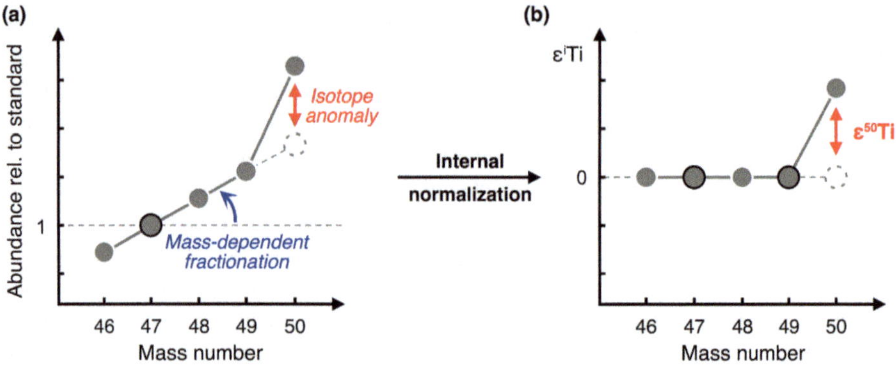

Fig. 4 Illustration of the difference between mass-dependent vs mass-independent isotope effects. In panel (a) the y-axis represents the abundance of the various titanium isotopes in a hypothetical sample (normalized to the abundance of ^{47}Ti) divided by the same ratios in a terrestrial standard. Values on the y-axis are thus the ratio of the ^{x}Ti/^{47}Ti in the sample divided by the same ratio in the standard, with x being any given isotope. The linear array defined by the isotopic abundances of most Ti isotopes (46 to 49) indicates that the sample experienced some degree of mass-dependent fractionation compared to the standard. The departure from this array of the abundance of ^{50}Ti (which is highly exaggerated for clarity), represents the influence of a mass-independent process. Panel (b) shows how, after internal normalization (see for instance Russell et al. 1978 and Maréchal et al. 1999), the impact of mass-dependent effects has been removed. The y-axis in panel (b) represents the deviations of the internally normalized composition of the sample relative to that of the standard, in epsilon notation (or parts per ten thousands)

3.1.2 Initial discoveries to present golden age

Historically, it was assumed that all SS materials had the same initial isotopic composition as a result of homogenization in a hot solar nebula (Cameron 1962). A string of discoveries spanning the last five decades have since falsified this notion, revealing the existence of isotopic heterogeneity in SS materials from the micron- to planetary-scale. The history of this paradigm shift regarding the chemistry of the SS is well summarized in Zinner (2014) and Dauphas and Schauble (2016). In brief, the first evidence for isotopic heterogeneity of the solar nebula was discovered in the noble gases Xe (Reynolds and Turner 1964) and Ne (Black 1972; Black and Pepin 1969) extracted from chondrites (chondrites are meteorites from asteroidal parent bodies that did not experience sufficient heating to melt, preserving original disk materials). Subsequent work involving the sequential digestion of chondrites eventually led to the identification of nanodiamonds (Lewis et al. 1987), SiC grains (Bernatowicz et al. 1987; Tang and Anders 1988), and graphite grains (Amari et al. 1990) as the presolar carriers of these isotope anomalies. This revealed that grains forming in the outflows of dying stars survived the formation of the SS, and provided a benchmark to test models of stellar nucleosynthesis on natural samples in the lab. Since then, the use of advanced microanalytical techniques (e.g., Nanoscale Secondary Ion Mass Spectrometry, or Nano-SIMS; Resonance Ionization Mass Spectrometry, or RIMS), has led to the identification of a variety of additional presolar phases, including oxides and silicates, and yielded isotope anomaly data for an ever increasing number of elements and stellar sources (Stephan et al. 2024).

The discovery of isotope anomalies in major rock-forming elements came only shortly after the finding of variations in the noble gases and began with O (Clayton et al. 1973), before rapidly expanding to Mg, Ca, Ti, Cr, and Ba (e.g., Clayton et al. 1988 and references therein).These anomalies were found in Ca-Al-rich inclusions (CAIs; sub-mm to cm-sized

Table 1 Cosmochemical classification of the elements for which nucleosynthetic isotope anomalies have been identified in bulk meteorites

Elements	Lithophile (silicate + oxides)	Siderophile/ chalcophile (metal + sulfides)
Refractory (T_C 1850–1355 K)	Zr, Nd, Sm, **Ti, Ca,** Sr, Ba	Os, W, Mo, Ru, Pt
Moderately refractory (T_C 1355–1250 K)	**Mg, Si,** Cr	**Ni, Fe,** Pd
Moderately volatile (T_C 1250–252 K)	K, Zn	
Highly volatile (T_C < 252 K)	(O), Ne, Xe	

Elements in order of decreasing condensation temperatures (T_C) at a pressure of 10^{-4} bar (Lodders 2003). Major elements are shown in boldface. Modified after (McDonough et al. 2016)

highly refractory ceramic-like aggregates present in chondrites; their mineralogy is consistent with that predicted for the first solids condensing from a cooling gas of solar composition, Grossman 1975, 1972) and are smaller than those in the presolar grains. This is an effect of averaging and dilution, which also makes it challenging to link the anomalies in CAIs to specific carriers and stellar sources. Indeed, not all presolar carriers are refractory phases that can survive chemical digestion with strong acids like SiC, nanodiamonds, or graphite, and painstaking workflows involving physical and chemical separation protocols are needed to identify and study these less-refractory carriers. A notable success in the application of such an approach has been the identification of nano-spinels as the carrier of ^{54}Cr anomalies (Dauphas et al. 2010; Qin et al. 2010).

With advancements in mass spectrometry, more subtle anomalies became resolvable. In particular, the development of Multi-Collector Inductively Coupled Plasma Mass Spectrometry (MC-ICPMS) instruments has revealed that isotopic anomalies of refractory elements are not confined to presolar grains and meteorite inclusions, but also present in bulk meteorites and samples of planetary-sized bodies (e.g., the Moon, Mars, Earth). Dauphas et al. (2002a) first showed that nucleosynthetic Mo anomalies persisted in magmatic iron meteorites, which are thought to sample the metallic cores of differentiated planetary bodies (10 s to 100 s of km in diameter). By picking up on earlier work (Niemeyer 1988a; Rotaru et al. 1992), widespread heterogeneity was then also described in the Cr, Ti, and Ni isotope composition of bulk chondrites and achondrites (e.g., Leya et al. 2008; Regelous et al. 2008; Trinquier et al. 2009, 2007). These results triggered a zealous search for nucleosynthetic anomalies in numerous refractory elements (see Dauphas and Schauble 2016), and have ushered in a golden age for the field (Fig. 5). Today, nucleosynthetic isotope anomalies have been documented in elements spanning a wide range of geo- and cosmochemical behavior: from lithophile (e.g., Ca, Ti, Cr, Ba) to siderophile (e.g., Fe, Mo, Ni, W), and ranging in volatility from highly refractory (e.g., Ca, Ti, Sr, Zr) to moderately volatile (K, Zn) (see Table 1). While the carrier phases of almost all these anomalies remain enigmatic, their heterogeneous distribution in SS materials can be leveraged to understand the evolution of the early SS.

3.1.3 Isotope anomalies as tracers of SS formation

Forged under the extreme temperatures, pressures, and neutron fluxes of stellar environments, nucleosynthetic signatures are essentially immune to modification by physicochemical processes taking place in interstellar media and circumstellar disks. As such, the presolar grains present in the SS parent molecular cloud are *de facto* inert and robust tracers of its collapse and infall onto the nascent protoplanetary disk, as well as the transport

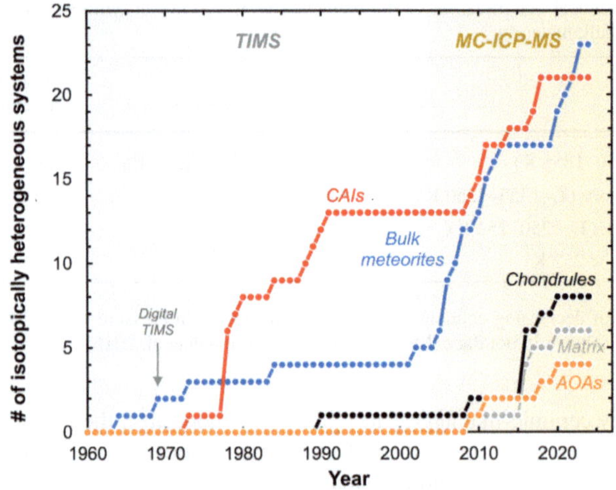

Fig. 5 Temporal evolution of the number of systems in which isotopic heterogeneity has been documented for bulk meteorites (blue), and their components: CAIs (red), AOAs (orange), chondrules (black), and matrix (grey). Gas-source mass spectrometry is key for work on noble gases, O, and S, and early work on Si. Following the advent of digital TIMS instruments, isotopic anomalies were rapidly discovered for numerous refractory elements in CAIs (~1980s). Two-decades later, the introduction of MC-ICPMS instruments led to a similar rise in the discovery of isotope anomalies in bulk meteorites. This temporal delay reflects the much larger magnitude of isotope anomalies in CAIs (typically on the order of a few to 10 ε in normal CAIs, and up to 100 s of ε in FUN CAIs) compared to bulk meteorites and planets (typically on the order of 0.1 to a couple of ε). Investigation of anomalies in chondrules, matrix, and AOAs has begun only recently. See Supplementary Materials for data source

and mixing of materials therein. The end product of these complex processes is the current distribution of isotope anomalies in SS materials, which reflect different contributions of nuclides produced in different stages of stellar evolution and astrophysical settings (e.g., Burbidge et al. 1957; Cameron 1957; Clayton 1983). Most relevant to the subject of nucleosynthetic anomalies discussed here are the so-called Fe-peak elements (e.g., Ti, Cr, Fe, Ni, Zn) and the elements with atomic masses beyond ~80 (e.g., Sr, Zr, Mo, Ru, Nd). Iron-peak elements are produced in neutron-rich environments either at or near nuclear statistical equilibrium (NSE) in massive stars ($\geq 8\ M_\odot$) prior and during core-collapse (Type-II) supernovae, as well as in rare Type Ia supernovae, which are thought to result from binary stellar evolution (Heger et al. 2014). For some Fe-peak elements, anomalies are attributed to excesses in neutron-rich isotopes (e.g., ^{54}Cr, ^{50}Ti, ^{48}Ca) (Dauphas et al. 2014; Trinquier et al. 2009, 2008). For others, anomalies reflect excesses in the neutron-poor isotopes (^{58}Ni, ^{54}Fe) (Hopp et al. 2022a,b; Steele et al. 2011). All elements beyond the Fe-peak are produced via either the s-, r-, and/or p-process, which respectively stand for 'slow neutron capture', 'rapid neutron capture', and 'proton-capture' or 'photodisintegration' (Burbidge et al. 1957; Cameron 1957). The s-process occurs in intermediate-mass stars (0.5–8 $M_\odot$) during the asymptotic giant branch (AGB) phase of their evolution (Gallino et al. 1998). For instance, variation in s-process contributions to SS materials is thought to be reflected in isotope anomalies in the s-peak elements Zr, Mo, and Ru (Akram et al. 2013; Burkhardt et al. 2011; Chen et al. 2010). The r-process occurs in environments characterized by extremely high neutron fluxes, such as supernovae and neutron star mergers. The p-process (more often named γ-process) is less well understood, and a multiplicity of sites have been proposed for its operation, including the O/Ne-shell of exploding massive stars, white dwarf explosions, or thermonuclear burning on the surface of neutron stars (Rauscher et al. 2013).

By quantifying the minute differences in isotope anomalies between meteorites and planetary bodies at high precision, isotope cosmochemists have been able to place powerful constraints on early SS dynamics and planetary formation. Most notably, these anomalies have revealed that for virtually all elements, bulk planetary materials can be classified into two isotopically distinct supergroups formed early in, and preserved across, SS history. These two supergroups, termed non-carbonaceous (NC) and carbonaceous (CC), are thought to broadly correspond to inner and outer SS materials, respectively, with their separation conventionally attributed to either (i) the early formation of proto-Jupiter close to the water-ice sublimation front (Burkhardt et al. 2019; Kleine et al. 2020; Kruijer et al. 2017; Warren 2011), or (ii) preferential planetesimal formation at the silicate and/or water sublimation fronts, with Jupiter playing a secondary role (Lichtenberg et al. 2021; Morbidelli et al. 2022). The existence of such isotopic heterogeneity has provided unprecedented opportunities to probe the history of the SS and its materials. For instance, leveraging differences in the siderophilic affinities of elements, nucleosynthetic anomalies have been shown to track the provenance of planetary building blocks across accretion of the terrestrial planets (e.g., Dauphas 2017). Recently, anomalies in the more volatile elements K and Zn have also been used to investigate the origin of terrestrial and Martian volatiles (Kleine et al. 2023; Martins et al. 2023; Nie et al. 2023; Savage et al. 2022; Steller et al. 2022).

Key to the interpretation of the isotopic heterogeneity observed in SS materials is the question of its origin. Were the presolar carriers themselves heterogeneously distributed in the parent molecular cloud, such that an isotopically heterogeneous disk is merely a reflection of compositional changes in infalling material (Burkhardt et al. 2019; Nanne et al. 2019; Yap and Tissot 2023) (see Sect. 4.1)? Or on the contrary, were the carriers homogeneously distributed in the cloud, with the observed heterogeneity arising from physical (e.g., dust grain size-sorting) and/or thermal processing during disk evolution (e.g., Burkhardt et al. 2012; Dauphas et al. 2010; Trinquier et al. 2009) (see Sect. 4.2)? Similarly, in the search for planetary building blocks, recent work has hinted at the existence of a "missing" building block for the Earth that is not sampled by bulk meteorites (Burkhardt et al. 2021). Where is this material and why has it eluded detection? These questions have sparked lively debate in the field. As an illustrative example, in the past two years there have been studies challenging the existence of an NC-CC dichotomy between SS materials (Onyett et al. 2023), supporting it (e.g., Yap and Tissot 2023), and positing the existence of an NC-CC-CI *trichotomy* instead (Dauphas et al. 2024; Hopp et al. 2022a). In this contribution, we aim to present the available data and the opposing views they have kindled, focusing on the overarching question: *What do isotope anomalies tell us about infall and/or disk processes?*

3.2 Oxygen Isotope Anomalies

A key rock forming element, oxygen has three stable isotopes (^{16}O, ^{17}O, ^{18}O). These display some of the largest, non-radiogenic, anomalies observed in SS bodies and have been historically critical in the identification of genetic links between SS materials (see also review by Ireland et al. 2020, and references therein). The processes at the origin of O isotope anomalies, however, are distinct from those envisioned for other elements, e.g., Cr or Ti (see below). Furthermore, apart from mass-independent fractionation, mass-dependent fractionation and exchange between reservoirs (e.g., gas and dust) play a role in setting the O isotopic composition of meteoritic materials. Here we use the classical $\delta^{17}O$ and $\delta^{18}O$ notations along with the $\Delta'^{17}O$ notation as introduced by Miller (2002). The $\Delta'^{17}O$, which is defined as:

$$\Delta'^{17}O = ln\left(1 + \frac{\delta^{17}O_{SMOW}}{1000}\right) - 0.528\, ln\left(1 + \frac{\delta^{18}O_{SMOW}}{1000}\right), \quad (1)$$

is synonymous with the "^{17}O-excess" parameter used by the water community.

3.2.1 Discovery and proposed origins/mechanisms

Clayton et al. (1973) first discovered O isotope anomalies in CAIs in carbonaceous chondrites and observed that the data fell on a slope ~1 line in $\delta^{17}O$ vs. $\delta^{18}O$ space. Back then, the O isotope composition of silicates was analyzed by fluorinating rocks and converting the liberated O_2 to CO_2 for mass spectrometric analysis. Clayton et al. (1973) observed a correlation between the apparent $\delta^{13}C$ and the $\delta^{18}O$ of the analyzed CO_2 and correctly attributed the shift in $\delta^{13}C$ to mass-independent fractionation in $\delta^{17}O$. Later, direct analysis of the liberated O_2 became common and allowed measurements of $\Delta'^{17}O$ with much higher precision.

Different processes have been presented to explain O isotope anomalies. Originally, Clayton et al. (1973) proposed that the bulk SS has a composition similar to that of the Earth ($\delta^{18}O$ = +5.5‰, $\Delta'^{17}O$ = –0.05‰, Pack et al. 2016) and suggested a nucleosynthetic origin of O isotope anomalies in ^{16}O-rich CAIs (see also Clayton et al. 1977). The large, differentiated inner SS bodies from which we have samples (Earth, Moon, Mars, Vesta), all have O isotope compositions defining a tight cluster around $-0.3 \leq \Delta'^{17}O \leq +0.3$‰. Without any measurements and following Occam's razor, one would conclude that the bulk SS is at $\delta^{18}O \approx +5$‰ and $\Delta'^{17}O \approx 0$‰. The question of the bulk O isotope composition of the SS has since been addressed by measuring solar wind particles implanted in lunar regolith. However, this returned ambiguous estimates of ^{16}O-rich (Hashizume and Chaussidon 2005) and ^{16}O-poor compositions (Ireland et al. 2006). The dispute was resolved when the results from the GENESIS mission became available (McKeegan et al. 2011) and, independently by spectroscopic measurements of the solar photosphere (Lyons et al. 2018). The data from GENESIS were interpreted to show mass-dependent fractionation and projection of the data on the slope-1 line defined by CAIs yielded $\delta^{17}O = \delta^{18}O = -58$‰ ($\Delta'^{17}O = -28$‰). The direct measurements of the solar photosphere gave $\delta^{18}O = -50 \pm 11$‰ (Lyons et al. 2018).

The spectroscopic data support the correction of the GENESIS data and confirm that the SS is ^{16}O-rich relative to asteroids and planets. These observations contradict the interpretation of CAIs having incorporated anomalous ^{16}O-rich material and that the bulk SS is at $\Delta'^{17}O \approx 0$‰. If the CAIs had incorporated material with nucleosynthetic anomalies for O, one would also expect large, correlated isotope anomalies with other elements. In one early study, Clayton and Mayeda (1977) reported on two CAIs with anomalous oxygen, for which the anomaly correlated with the anomaly in Mg isotopes (Wasserburg et al. 1977), which has been traced back to the presence of short-lived ^{26}Al (Lee et al. 1977). However, correlated anomalies for a larger set of samples and elements have not been observed (Clayton 1993, 1978) and the nucleosynthetic origin of the O isotope anomalies in bulk meteorites has since largely been dismissed.

Thiemens and Heidenreich (1983) and Heidenreich and Thiemens (1983) experimentally observed a novel chemical process that leads to O isotope fractionation similar to that observed in meteorite components. Their experiments on ozone formation led to the formation of O_3 with large positive $\Delta'^{17}O$ and O_2 with negative $\Delta'^{17}O$. Educt O_2 and reaction product O_3 fall on a slope-1 line in a $\ln(\delta^{17}O + 1)$ vs. $\ln(\delta^{18}O + 1)$ space, just like refractory inclusions in carbonaceous chondrites do. The mass-independent fractionation is due to symmetry effects in the O-O-O molecule and differences in reaction rates. An $^{16}O_3$ reaction product tends to decompose faster from the excited state back to O_2 whereas asymmetric O_3 with substituted ^{17}O or ^{18}O has a longer half-life making it more likely to react to stable O_3. The authors suggest that the formation of symmetric molecules containing oxygen (e.g., SiO_2, TiO_2) could be responsible for the fractionation along a slope-1 line observed in meteorites (see also volume η [Greek "eta"] effect, Marcus 2004).

A different type of chemical reaction that produced isotope anomalies was suggested by Marcus (2004). He proposed that the equilibration between adsorbed monoxides (XO_{ads}) and oxygen atoms (O_{ads}) and excited dioxides ($XO^*_{2,ads}$, Eq. (2)) can lead to mass-independent effects, e.g., during condensation of dust in the early solar nebula:

$$XO_{ads} + O_{ads} \rightleftharpoons XO^*_{2,ads} \tag{2}$$

The element "X" in Eq. (2) could be Si, Al, Ti, or other metals. Such surface η effect yields $XO^*_{2,ads}$ that is mass-independently enriched in ^{17}O and ^{18}O and XO_{ads} and O_{ads} that are depleted in ^{17}O and ^{18}O (Eq. (2)). The excited $XO^*_{2,ads}$ evaporates and reacts with H_2 to form ^{16}O-poor H_2O, whereas the adsorbed XO and O react with adsorbed metal atoms (e.g., Ca) or monoxides into ^{16}O-rich oxides and silicates of the CAIs. Chakraborty et al. (2013) experimentally demonstrated that the oxidation of gaseous SiO, which is the major Si-bearing component in a gas of solar composition at high temperatures, by OH leads to formation of solid SiO_2 (Eq. (3)), which is selectively depleted in ^{16}O. This could explain the ^{16}O-poor composition of silicates (e.g., chondrules) relative to the ^{16}O-rich composition of the solar nebula without a nucleosynthetic ^{16}O source (and CO self-shielding).

$$SiO_g + OH_g \rightarrow SiO_{2,s} + H_g \tag{3}$$

The reaction shown in Eq. (3) may go through a vibrationally excited transition state (SiOOH*). As in the case of O_3 formation, the deactivation of the intermediate state would then be the place of mass-independent fractionation (Chakraborty et al. 2013).

A different perspective on the origin of the mass-independent fractionation in meteorites and their components has been published by Clayton (2002). He suggested that predissociation and self-shielding of CO leads to formation of ^{16}O-poor H_2O. Isotope separation by self-shielding of CO had been observed in molecular clouds (Bally and Langer 1982). Those photons that predissociate $C^{16}O$ are quickly consumed and, at greater penetration depths, only $C^{17}O$ and $C^{18}O$ are pre-dissociated. The O radical reacts with ambient H_2 to H_2O (Eq. (4)). This H_2O is then selectively, and by the same factor, enriched in ^{17}O and ^{18}O.

$$CO + H_2 + \text{UV-light} \rightarrow CO^* + H_2 \rightarrow C + O + H_2 \rightarrow C + H_2O \tag{4}$$

Silicates and oxides that react with such H_2O would then fall along a slope-1 line in the three-oxygen isotope space. The different O isotope fractionation regimes are illustrated in Fig. 6 (modified after Thiemens and Lin 2021). Originally, Clayton (2002) suggested that the source of the UV radiation is the young Sun and that self-shielding occurred in the inner SS. High ambient temperatures, however, would lead to quick back-equilibration and disappearance of the anomaly. Lyons and Young (2005) modeled self-shielding with an external UV source (O or B star within 1 pc) and location of self-shielding at the disk surface. The environment in the outer, cool parts of the disk may resemble the conditions in cold molecular clouds. Yurimoto and Kuramoto (2004) placed the origin of the mass-independent O isotope fractionation in the cold molecular cloud from which the SS formed. In such an environment, dissociation of CO is followed by surface-mediated reaction of O

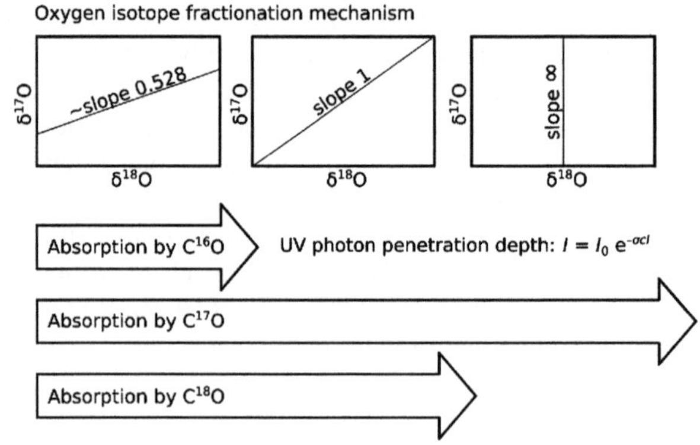

Fig. 6 Sketch illustrating the effect of self-shielding of CO (modified after Thiemens and Lin 2021). The penetration depth of the isotopologue-specific photons is a function of the concentration of the isotopologues (Lambert-Beer law). In the region of dissociation of all three CO isotopologues, no or only mass-dependent fractionation occurs. At greater penetration depth, only $C^{17}O$ and $C^{18}O$ are dissociated and the fractionation has a slope of one. At even greater depths, where only $C^{17}O$ is absorbed, fractionation would follow a vertical slope

with H_2 on the cold surfaces of grains. The ice is ^{17}O and ^{18}O rich, whereas the CO is ^{16}O rich. These studies (Clayton 2002; Lyons and Young 2005; Yurimoto and Kuramoto 2004) have in common that they predict that the SS (i.e., the Sun) has a ^{16}O-rich composition with $\delta^{17}O \approx \delta^{18}O \approx -50‰$, reflected in the composition of CAIs and confirmed by results of the GENESIS mission.

The existence of a ^{17}O- and ^{18}O-rich H_2O reservoir is suggested by the observation of high $\Delta'^{17}O$ in the most oxidized phases (magnetite) in Semarkona (Choi et al. 1998). These are suggested to have formed by reaction of Fe metal or troilite with ^{16}O-poor water. Sakamoto et al. (2007) reported oxidized phases in chondrites with $\Delta'^{17}O$ as high as +80‰. This is within the range of predicted H_2O (ice) in the model of Lyons and Young (2005) with an enhanced UV flux of only 5× (relative to the local interstellar medium). Organic material with even higher $\Delta'^{17}O$ has been analyzed by Hashizume et al. 2011. All these observations are in agreement with the existence of ^{16}O-poor H_2O. On the other end, CO formed the counterbalancing ^{16}O-rich reservoir that had been lost. If the bulk SS is at $\delta^{17}O = \delta^{18}O = -50‰$, CO is expected to have had lower $\delta^{17}O$ and $\delta^{18}O$ values. So far, only three objects have been analyzed that may reflect the composition of the lost ^{16}O-rich CO reservoir: a chondrule from the CH3 Acfer 214 carbonaceous chondrite (Kobayashi et al. 2003), and two CAIs from the CH/CB-like Isheyevo carbonaceous chondrite (Gounelle et al. 2009).

Despite the success of these models, debate remains about the exact origin of O isotope anomalies in SS materials. For instance, a line of arguments was presented by Thiemens and Lin (2021) that criticized the self-shielding model in the form suggested by Clayton (2002) and Lyons and Young (2005). Also, experiments were not able to reproduce the effects predicted by the self-shielding model (Chakraborty et al. 2008). Some observations are also surprising when seen through the lens of the self-shielding hypothesis. For instance, apart from magnetite in some unequilibrated OC chondrites, no chondrite silicates and oxides have been observed that plot towards the ^{16}O-poor self-shielding-produced H_2O with $\Delta'^{17}O > 7‰$ (Choi et al. 1998; Doyle et al. 2015). While the high $\Delta'^{17}O$ of magnetite in unequilibrated OCs is thought to be inherited from the aqueous fluids on the parent bodies, the corresponding alteration products (fayalite, magnetite) from some CCs plot below the terrestrial fractionation line (Doyle et al. 2015). This suggests that the water on these CC

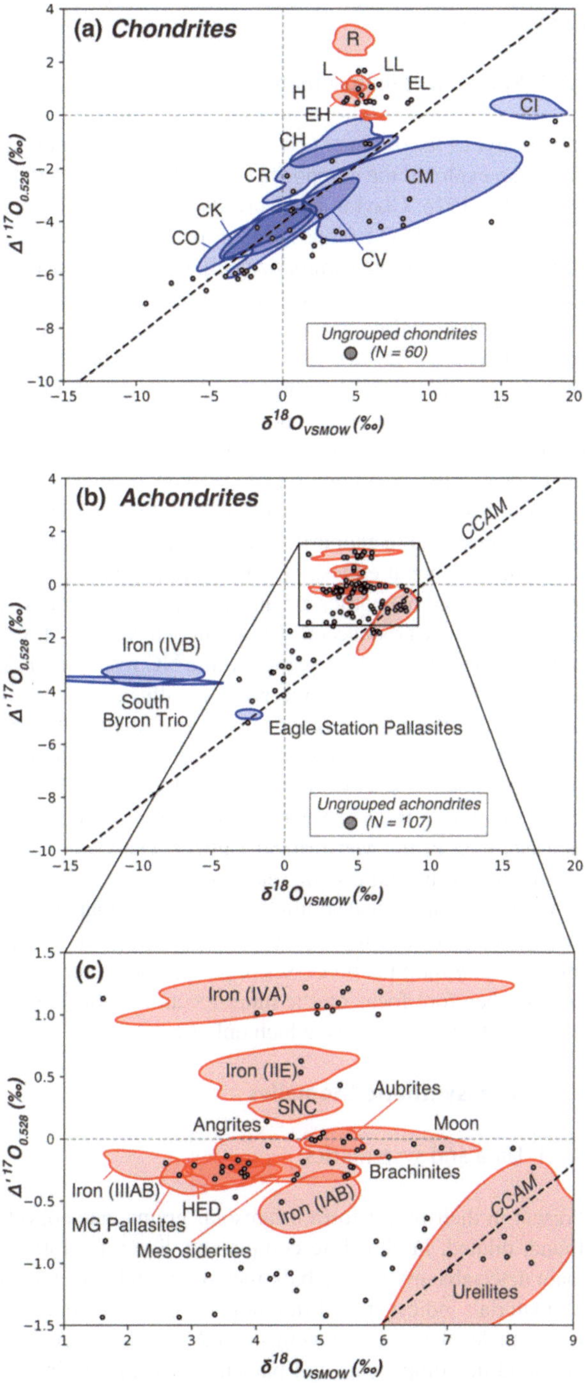

Fig. 7 Diagrams of mass-independent ($\Delta'^{17}O$) against mass-dependent ($\delta^{18}O$) oxygen isotope signatures of bulk meteorites. Compilation of chondrites (a) and achondrites (b, c) bulk rock O isotope data extracted from the Meteoritical Bulletin Database (https://www.lpi.usra.edu/meteor/, accessed March 2024, N = 1749). Data from ungrouped achondrites (grey points in panel b and c) suggest (i) distinct parent bodies in the fields typical of CC and NC achondrites, and (ii) underexplored links between meteorites sampling different reservoirs within the same parent body

parent bodies had lower $\Delta'^{17}O$ than the water in the unequilibrated OC parent body, which is opposite to what would be expected since CC parent bodies are thought to form at greater heliocentric distances.

3.2.2 Systematics in bulk meteorite data

Irrespective of the origin of O isotope anomalies, in a number of seminal papers, Clayton and his group explored the composition of bulk meteorites (Clayton 2008, 2003, 1993; Clayton et al. 1991, 1976; Clayton and Mayeda 1999, 1996, 1984, 1983). A number of laboratories now analyze triple O isotopes at high-precision and the amount of data has increased substantially. We therefore summarize bulk rock O isotope data compiled from the Meteoritical Bulletin database (https://www.lpi.usra.edu/meteor/) and display the data in Fig. 7.

The main chondrite groups span a large range in $\Delta'^{17}O$ from -5 to +3‰, i.e. spanning some 8000 ppm (Fig. 7a). The observed bulk-rock variations are large in comparison to the analytical resolution in the range of <10 ppm. With exception of the aqueously altered CI chondrites, all carbonaceous chondrites have in common that they are enriched in ^{16}O compared to the non-carbonaceous chondrites, the Earth and most differentiated meteorites, most likely due to the incorporation of ^{16}O-rich refractory inclusions. The achondrites, which originate from parent bodies that have experienced melting and lost the primary phases and textures preserved in chondrites, mostly fall in a narrower range of $-0.5 \leq \Delta'^{17}O \leq +1.5$‰ (Fig. 7b). Only the ureilites, the Eagle Station group pallasites, the IVB iron meteorites, and some ungrouped meteorites are an exception here and fall in the field of carbonaceous chondrites along the Carbonaceous Chondrite Anhydrous Mineral (CCAM) line. Despite their O isotope similarity, ureilites belong to the NC reservoir (Fig. 7c), while the other meteorites have been linked to the CC reservoir. Finally, a large set of meteorites ($N = 107$) from the Meteoritic Bulletin Database have been classified as "ungrouped achondrites". The data are shown in Fig. 7c. They suggest additional distinct parent bodies in the fields typical of CC and NC achondrites, as well as underexplored links between meteorites sampling different reservoirs within the same parent body. The data demonstrate that the collection of large datasets, e.g., in the Meteoritical Bulletin Database, can be of high value for identifying distinct reservoirs and genetic relationships.

Despite the historical significance of, and fundamental discoveries brought on by, O isotope anomalies in meteorites, their interpretation is complicated by the fact that they are influenced by multiple processes (e.g., gas-melt interactions, water interactions). Hereafter, we will therefore focus on the more recent insights gained into genetic relationships from nucleosynthetic anomalies, which only trace mixing.

3.3 Nucleosynthetic Anomalies

3.3.1 Data and systematics

Present in their purest form in presolar grains, nucleosynthetic isotope anomalies are also found, diluted, in chondrite components like refractory inclusions or chondrules, and remain detectable even at the bulk meteorite- and planetary-scale. The anomalies observed in SS materials indicate that different types of stellar sources contributed material to the Sun's parent molecular cloud. How this cloud-scale heterogeneity relates to the isotope anomalies of chondrite components and planetary bodies forming in the protosolar disk, and hence, what we can learn from these anomalies about early SS evolution, remains a matter of intense debate. To guide this debate, we present here an overview of the most up-to-date dataset of isotope anomalies in SS materials. These anomalies exhibit systematic trends in multi-element isotope space, reflecting a handful of key nucleosynthetic pathways in stars and supernovae, and forming the basis for any spatial and temporal interpretations regarding the dynamics of SS formation (Sects. 4 and 5).

Figure 8 summarizes the collective effort of the community over the past two decades and shows the distribution of isotope anomalies among bulk meteoritic and planetary materials in multi-element isotope space. The elements shown in Fig. 8 were chosen for illustrative purposes and cover a range of nucleosynthetic origins and geo-/cosmochemical behaviors. As is evident, the most salient feature of this dataset is a fundamental isotopic difference among bulk SS materials (red vs. blue symbols), referred to as the NC-CC dichotomy (Sect. 3.1). This clustering of the data is most pronounced in spaces were the anomalies are large (e.g., ε^{54}Cr vs ε^{50}Ti), but is also visible when the anomalies are small (e.g., ε^{54}Fe vs ε^{64}Ni). The NC group contains enstatite (EC), ordinary (OC), Rumuruti, and Kakangari chondrites, while the CC group contains the carbonaceous chondrites (e.g., CI, CV, CO, CM, CR). Importantly, both groups contain iron meteorites and stony achondrites, sampling differentiated planetesimals that accreted early in the protosolar disk relative to chondrites. The Earth, Moon, and Mars belong to the NC group, with the isotopic composition of the Bulk Silicate Earth (BSE) being closest to that of ECs, and that of the Bulk Silicate Mars (BSM) being typically intermediate to ECs and OCs. As such, it is commonly invoked that the NC group represents relatively volatile-poor, inner SS materials, while the CC group reflects materials richer in volatiles accreted at greater heliocentric distance (e.g., Hilton et al. 2022; Kleine et al. 2020; Spitzer et al. 2021).

The separation of materials into two nebular reservoirs was first proposed based on the ε^{54}Cr–Δ^{17}O systematics of differentiated NC bodies, OCs, and CC chondrites (Trinquier et al. 2007). A few years later, Warren (2011) compiled available isotope anomaly data for Cr, Ti, Ni, and O and showed that meteoritic materials defined two distinct clusters in isotope-isotope plots for these elements. Importantly, all meteorites consistently fall into one of the two isotopic clusters regardless of the combination of elements considered, indicating this bimodality is a ubiquitous feature among SS bodies. The NC–CC dichotomy has since been observed for numerous other elements with different geochemical affinities (Table 1). These include the siderophiles Mo, Ru, Fe, W, and lithophiles Ca, Si, Sr, Zr, Ba, Sm, and Nd. All these elements are refractory to moderately refractory, meaning that they condense at relatively high temperatures (>1250 K) from a gas of solar composition (Lodders 2003). In contrast, nucleosynthetic isotope anomalies in volatile elements – with half-mass condensation temperatures (T_C) below those of the major rock-forming elements Mg, Si, and Fe (<1250 K) – were long believed to be absent due to extensive homogenization within the disk (Vollstaedt et al. 2020). Recent studies have, however, identified nucleosynthetic isotope anomalies for the lithophile and moderately volatile elements Zn ($T_C \sim 704$ K) and K ($T_C \sim 993$ K) (Martins et al. 2023; Nie et al. 2023; Savage et al. 2022; Steller et al. 2022) providing an additional valuable tool for constraining the provenance of Earth's (and Mars') volatiles. The fact that the dichotomy persists across elements with different cosmochemical and geochemical behaviors is an important clue to understanding its origin (Sect. 4).

Returning to Fig. 8, the *inter*-reservoir trend between NC and CC bodies (i.e., the dichotomy) can be described as an enrichment of supernova produced (NSE and r-process) isotopes in CC bodies relative to NC bodies. The NC-CC dichotomy, phrased more specifically, is the systematic offset of CC bodies from NC bodies towards CAIs in isotope-isotope space. Two families of models have been proposed to explain the origin of the dichotomy, which are discussed in detail in Sect. 4. Before diving into this interpretative discussion, we continue with a description of the structure of the data. Closer examination of Fig. 8 also

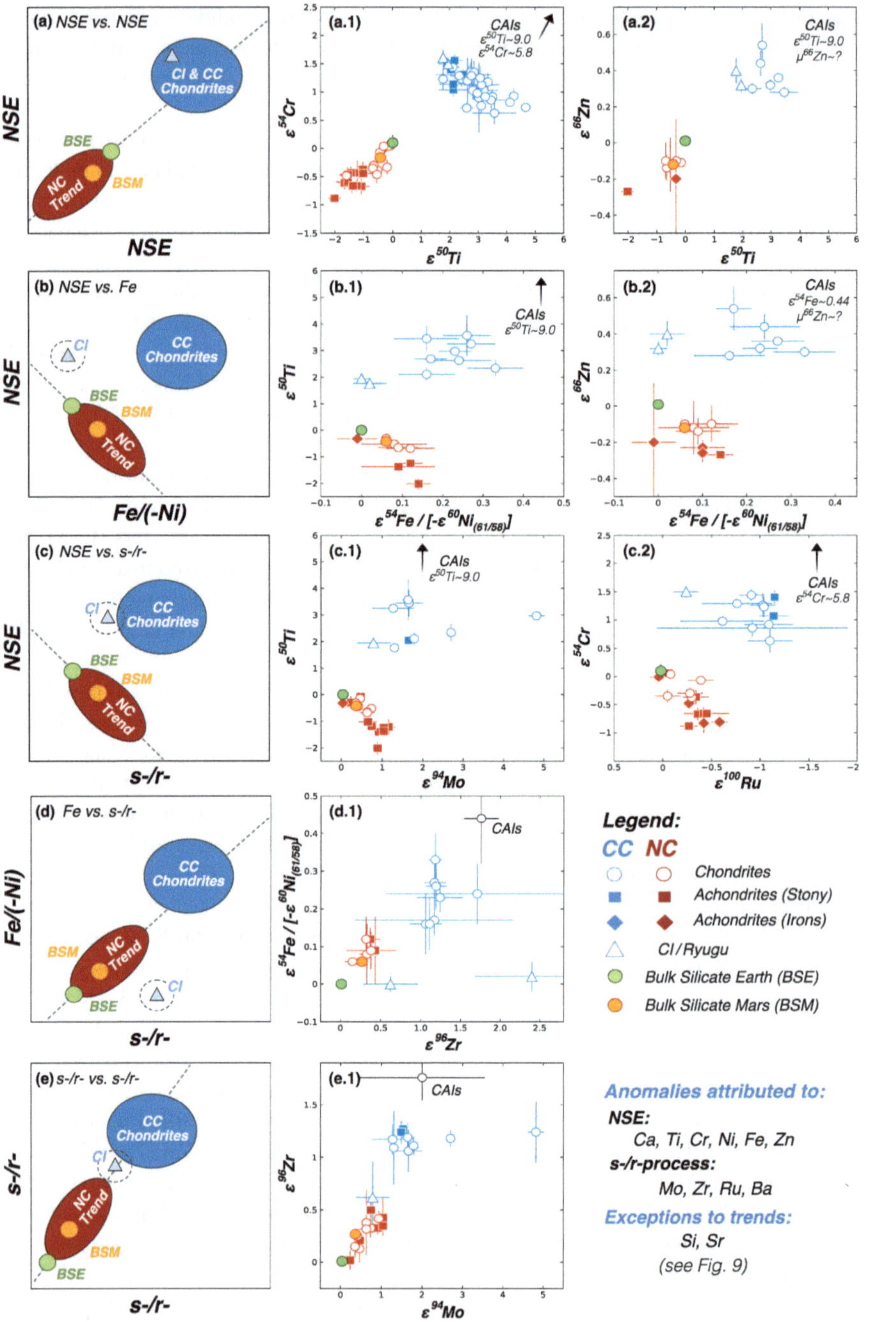

Fig. 8 Isotope anomaly systematics of NC and CC meteorites and silicate portions of Earth (BSE) and Mars (BSM) in multielement space. NC meteorites representing the inner SS are shown in red, and CC meteorites representing the outer SS are shown in blue. Note that in plots with ε^{100}Ru, we inverted the axis such that enrichment in supernova-derived nuclides (towards CAIs) is directed to the top right, as in all other plots. Owing to the conventionally adopted internal normalization scheme for Ru anomalies, higher ε^{100}Ru reflects greater *s*-process contributions. See Supplementary Materials for data source

reveals finer structure beyond the NC-CC dichotomy: i.e., *intra*-reservoir trends. In particular, NC bodies define a linear array consistent across multi-element isotope space, with the BSE as an endmember. Depending on the elements considered, the NC array points either towards the CC group (and CAIs), or away from them. In the former case, the BSE is positioned either closest to, or farthest away from, the CC bodies. An *intra*-reservoir trend in the CC group is only clearly apparent in some isotope spaces (e.g., ε^{50}Ti-ε^{54}Cr), and the spread in the data mostly correlates with the relative abundance of CAIs, chondrules, and matrix in CC chondrites (Alexander 2019; Hellmann et al. 2023, 2020).

The exact morphology of the ε-ε plots (i.e., 2D cross sections of multi-element isotope space) reflects the nucleosynthetic origin of the isotope anomalies being considered. These can be separated into 5 sub-categories, corresponding to the combinations of the different nucleosynthetic pathways responsible for these anomalies:

(a) When two isotopes whose anomalies result from NSE contributions (e.g., ^{50}Ti, ^{48}Ca, ^{54}Cr, ^{66}Zn; Fe and Ni excluded, see below) are plotted against each other (Fig. 8a), the NC trend points towards the CCs, with the BSE endmember closest to the CCs.

(b) When anomalies attributed to NSE are plotted against those of Fe or Ni (Fig. 8b), the BSE remains closest to the CCs, but the NC trend points away from them, and the CI chondrites separate from the CC cluster.

(c) When anomalies attributed to NSE are plotted against those of elements beyond the Fe-peak (i.e., reflecting excesses in s- and r-process nuclides; Mo, Zr, Ru, Ba) (Fig. 8c), the data distribution is akin to (b) but CI chondrites are not clearly distinct from the other CCs. Instead, CI chondrites appear to be an endmember of the CC group (as it does for anomalies of all elements except Fe and Ni).

(d) When plotting s- and r-process anomalies against those in Fe or Ni (Fig. 8d), the NC trend points towards the CCs as in (a), but the BSE sits farthest from the CCs. The CI chondrites are separated from the rest of the CCs.

(e) Finally, when s- and r-process anomalies are plotted against each other, a data distribution akin to that in (d) is retrieved, but with CI chondrites seemingly in the CC group (Fig. 8e).

At first glance, two elements do not seem to conform to the general trends described above (Fig. 9). We consider them in more detail below.

Silicon: Silicon isotope anomalies (μ^{30}Si) were recently reported in bulk SS materials by Onyett et al. (2023). In contrast to other elements, NC and CC chondrites have overlapping anomalies, and a dichotomy is seen between chondrites and NC achondrites instead. Taken at face value, these observations suggest a temporal evolution of the inner disk's Si isotopic composition from NC achondrite- to CI-like. However, Dauphas et al. (2024) pointed out that the elevated μ^{30}Si values in enstatite chondrites may be arising from an improper correction of mass-dependent fractionation. Using the equilibrium law (relevant to fractionation at high temperatures) instead of the exponential law would result in a ^{30}Si anomaly of enstatite chondrites within error of the BSE. This would align the Si isotope data more closely with the general trends described in the above section (Fig. 9a). With only one study to date, future studies are needed to resolve this debate.

Strontium: With only four isotopes (^{84}Sr, ^{86}Sr, ^{87}Sr, and ^{88}Sr), including one (^{87}Sr) typically produced overwhelmingly by decay of ^{87}Rb, it has been famously difficult to identify the nucleosynthetic origin of Sr isotope anomalies. After normalization to ^{86}Sr and ^{88}Sr, anomalies are only apparent in ^{84}Sr. Whether these represent true variations in ^{84}Sr abundance (i.e., a result of the p-process) or variations in abundances of the normalizing isotopes (^{86}Sr, s-process; ^{88}Sr: s- and r-process) has been a long standing question (Burkhardt et al. 2019; Charlier et al. 2021, 2019; Fukai and Yokoyama 2019; Hans et al. 2013; Moynier

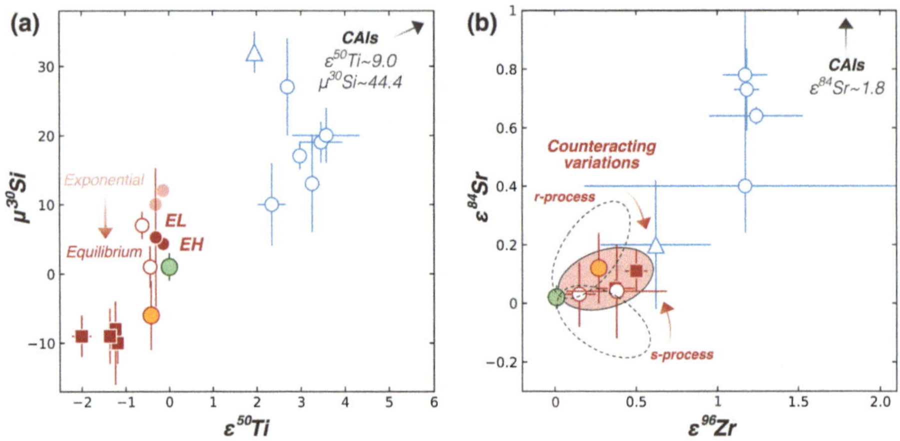

Fig. 9 Isotope anomalies of (a) Si vs Ti, and (b) Sr vs Zr. At first glance, both Si and Sr do not seem to conform to the general trends observed for other elements (see Fig. 8). (a) The elevated Si isotope anomalies observed in enstatite chondrites may be due to improper use of the exponential law to correct for MDF, which is expected to follow the equilibrium law at the high-T applicable to the early solar nebula. (b) The undistinguishable Sr isotope anomalies in NC materials, which contrast with the clear trends in NC materials observed for other elements, is thought to testify to correlated *s*- and *r*-/*p*-anomalies, whose effect cancel out. See main text for further information. Symbols as in Fig. 8

et al. 2012; Papanastassiou and Wasserburg 1978; Paton et al. 2013; Schneider et al. 2023; Yokoyama et al. 2015). Most relevant to our discussion is the systematics observed in bulk meteorites and planetary materials. As for other elements, CC meteorites have Sr anomalies that are generally offset from NC materials towards those of CAIs. Unlike most other elements, however, NC materials have indistinguishable compositions, and some CC chondrites have ε^{84}Sr anomalies similar to NC materials (Charlier et al. 2017; Schneider et al. 2023) (Fig. 9b). These apparent departures from the general trends described above (Fig. 8) were recently proposed to reflect variable *s*- and (*r*, *p*)-process contributions (Schneider et al. 2023). In NC meteorites, these variations are correlated, resulting in ε^{84}Sr excesses and deficits of similar magnitude that effectively cancel each other out. In CC meteorites, the presence of CAIs (which have elevated ε^{84}Sr values) leads to the apparent ^{84}Sr excess. In the absence of additional Sr isotopes to unambiguously quantify *s*-, *r*- and *p*-process variations in SS materials, the proposal of Schneider et al. (2023) leverages the isotope-isotope trends seen in other elements to explain the seemingly different Sr systematics.

3.3.2 Constraints from inferred accretion ages

Armed solely with the dataset of nucleosynthetic anomalies (Fig. 8), it is impossible to tell if the structure in the data described above reflects a temporal and/or spatial evolution of the early SS. In particular, the NC-CC dichotomy could reflect (i) the simultaneous existence of two spatially separated reservoirs, or (ii) a temporal evolution in the isotopic composition of the disk as whole (Warren 2011). Distinguishing between these two hypotheses requires coupling the genetic information from nucleosynthetic isotope anomalies in NC and CC meteorites to the chronological information of the accretion age of their parent bodies.

Accretion ages of meteorite parent bodies are inferred through a combination of isotopic analyses and thermal models of internal heating from ^{26}Al decay. For stony achondrites

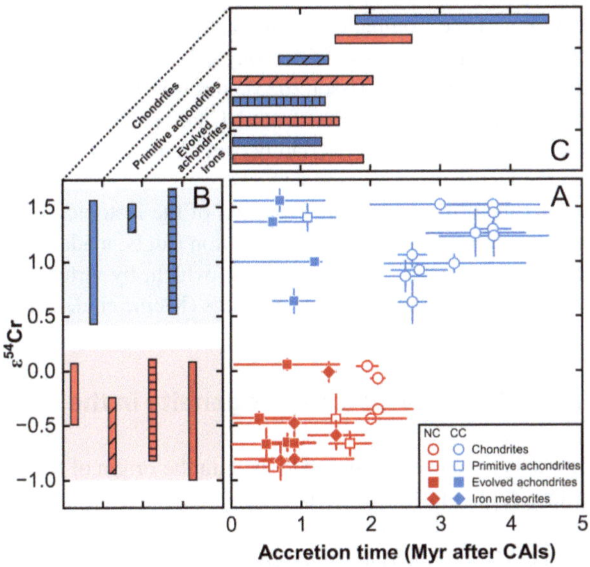

Fig. 10 Diagram of chromium isotopic composition versus inferred accretion ages of various meteorite groups (a) and bar diagrams of ε^{54}Cr (b) and inferred accretion ages (c) of NC and CC meteorites individually. Panel c) indicates that iron meteorite and stony achondrite parent body accretion started rapidly and coevally within the first ~2 Myr after CAIs in both reservoirs followed by the formation of chondrites at ~2–4 Myr after CAIs. Importantly, the isotopic dichotomy was already established within ~1 Myr after CAIs (a). Furthermore, iron meteorites, stony achondrites, and chondrites appear to cover almost the entire isotopic range of both reservoirs (b) arguing against a temporal evolution for the overall isotopic evolution of the NC and CC reservoirs. Red and blue shaded areas in b) indicate the full isotopic range of NC and CC meteorites, respectively. The size of the bars in b) and c) include the respective uncertainties on the sample values of a certain group. Data as reported in Tables S2–3

and iron meteorites, thermal modeling is anchored by Hf-W model ages of core segregation: ~1–2 Myr and ~3 Myr after CAI condensation for NC and CC iron meteorite parent bodies, respectively (Spitzer et al. 2021). For chondrites, thermal models are informed by either their cooling histories following thermal metamorphism, as deduced from Pb-Pb (e.g., Blackburn et al. 2017) and Hf-W systematics (e.g., Hellmann et al. 2019), or radiometric dating of aqueous alteration (i.e., via carbonates) by the Mn-Cr system (e.g., Jogo et al. 2017). Combined, these techniques indicate that planetesimal formation took place as early as several 10^5 years after CAIs, and lasted for ~4 Myr (Budde et al. 2018; Kruijer et al. 2020; Neumann et al. 2023; Spitzer et al. 2021; Sugiura and Fujiya 2014). This is roughly consistent with the timing of disk dissipation as inferred from paleomagnetic studies (Borlina et al. 2022; Wang et al. 2017), and the extinct radionuclide (e.g., Al-Mg) systematics of chondrules (Dauphas and Chaussidon 2011). These records imply an inside-out gradual disk dispersal that limited planetesimal accretion in the inner disk to about 2 Myr after CAIs, while in the outer disk parent-body formation continued until about 4.5 Myr after CAIs. The bodies forming within ~1.5 Myr after CAIs differentiated due to the presence of ^{26}Al and are sampled by iron meteorites and stony achondrites, while chondrites sample bodies assembled from about 2 Myr onwards, when ^{26}Al decayed to levels insufficient to induce melting and differentiation (e.g., Spitzer et al. 2021).

Based on an apparent linear relationship between ε^{54}Cr and the time of accretion for NC stony meteorites and CC chondrites, Sugiura and Fujiya (2014) supported the hypothesis of a temporal evolution of the isotopic composition of the disk. This conclusion, however, was based on a limited dataset, devoid of CC stony achondrites (only a few tens of which have been identified so far; Yap and Tissot 2023), and iron meteorites. In Fig. 10, we plot the ε^{54}Cr of both stony and iron meteorites from NC and CC groups against their inferred time of accretion. The data reveals an overlap between NC and CC parent body accretion ages, testifying to the fact that the NC and CC reservoirs co-existed in separate regions of the early SS, and that there was no simple evolution of the isotopic composition within or between the reservoirs over time. The same observation can be made using nucleosynthetic anomalies of other elements (e.g., Fe, Ni, Mo, Ru), which, by virtue of their siderophilic affinities, are present in both stony and iron meteorites (Kleine et al. 2020).

4 Origin of Large-Scale Isotopic Heterogeneity in the Early SS

Two families of models have been proposed to explain the origin of the NC-CC dichotomy (Fig. 11). In the first (Fig. 11a-b), the dichotomy reflects the inheritance of isotopic heterogeneity initially present in the molecular cloud, and CC bodies constitute mixtures of isotopically NC and CAI-like materials, representing late and early infall, respectively (e.g., Burkhardt et al. 2019; Nanne et al. 2019; Yap and Tissot 2023). In the second (Fig. 11c-d), the presolar carriers were homogeneously distributed within the cloud, and isotopic heterogeneity arose from sorting of presolar carriers through thermal and physical processing within the disk (e.g., Burkhardt et al. 2012; Dauphas et al. 2010; Davis et al. 2018; Trinquier et al. 2009).

4.1 Tracing the Molecular Cloud Infall

The *intra*- and *inter*-reservoir trends presented in Sect. 3.3 may reflect a plethora of superimposed nebular processes. Nonetheless, their persistence across elements with a wide range of geo- and cosmo-chemical behavior suggests they are primarily governed by mixing between several isotopically distinct nebular components inherited from the parent molecular cloud (Fig. 11a-b). Magnetohydrodynamic models of protostellar accretion indicate significant variance in the source of infalling material within collapsing molecular clouds, and associated infall timescales (e.g., Kuffmeier et al. 2023). These astrophysical considerations lend credence to models that suggest the spatial order of isotopic heterogeneity in the nascent disk derives from a chronological order of distinct infalling materials (Burkhardt et al. 2021, 2019; Jacquet et al. 2019; Kleine et al. 2020; Nanne et al. 2019; Van Kooten et al. 2016; Yap and Tissot 2023). Here, we review the key aspects and recent developments of such models, as well as the enticing avenues for future work they bring forth.

4.1.1 Inheritance models

What we henceforth denote as *inheritance models* rely on the idea that the apparent trend in multi-element isotope space from NC to CC to CAI compositions is the result of mixing of isotopically distinct endmembers (Fig. 8). This trend was initially observed for Ti isotopes, and the anomalous Ti isotope composition of carbonaceous chondrites was interpreted as reflecting variable mixtures of their isotopically distinct, major constituents: matrix, chondrules, and CAIs (Niemeyer 1988a,b; Niemeyer and Lugmair 1984), which themselves formed from precursor materials that retained a "cosmic chemical memory",

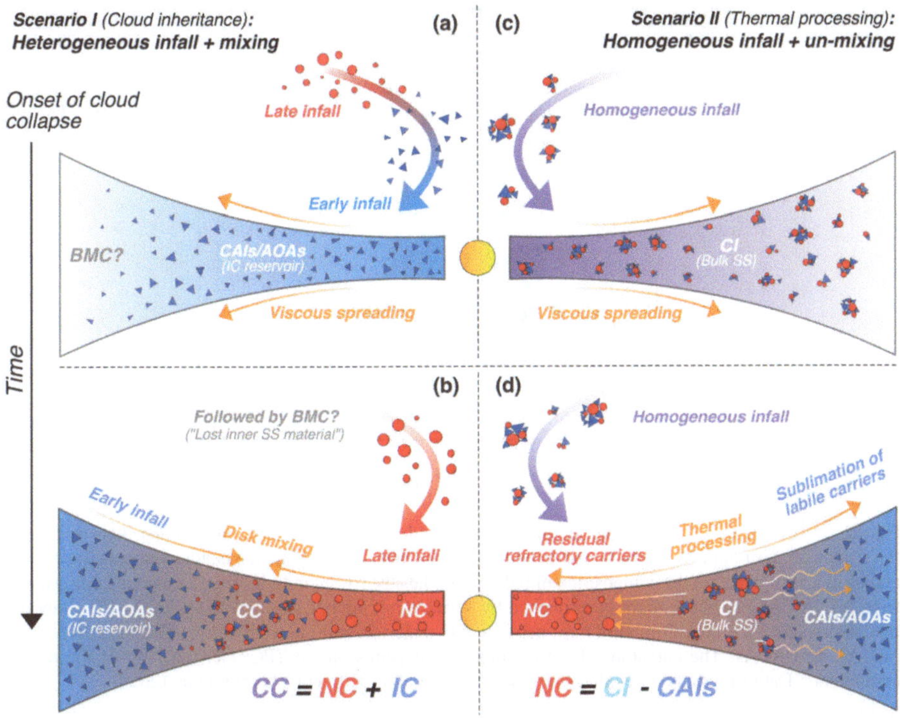

Fig. 11 Schematic representation of the two families of models that have been proposed to explain the isotopic heterogeneity observed in SS materials. In (a,b) the NC-CC dichotomy reflects the inheritance of isotopic heterogeneity initially present in the molecular cloud, and CC bodies constitute mixtures of isotopically NC and Inclusion-like Chondritic (IC) materials, where IC denotes an hypothetical reservoir that is isotopically CAI-like but chemically chondritic (e.g., Nanne et al. 2019; Burkhardt et al. 2019; Yap and Tissot 2023). In (c,d) the presolar carriers were homogeneously distributed within the protosolar nebula (CI-like bulk SS), and isotopic heterogeneity arose from sorting of presolar carriers through thermal and physical processing within the disk (e.g., Trinquier et al. 2009; Dauphas et al. 2010; Burkhardt et al. 2012; Davis et al. 2018)

i.e., isotopic heterogeneities inherited from the molecular cloud (Clayton 1982). In his seminal work, Niemeyer (1988a) further deduced that CI chondrites, which are taken as a proxy for the chemical composition of the Sun for all but the volatile elements, may not be representative of the *isotopic* composition of the bulk SS, and that mass-balance dictates that "*our present collection of* [*bulk*] *SS samples is missing a large reservoir that is more enhanced in* ^{50}Ti (e.g., *the Sun*)". All isotope anomaly data obtained since then for carbonaceous chondrites and their components may be interpreted along the premises laid out by Niemeyer (1988a). By now, it is evident that the offset of CC meteorites from the NC meteorites towards the isotopic composition of CAIs is a global feature seen for all elements showing isotope anomalies (Fig. 8). Importantly, this includes not only elements that are enriched in CAIs (e.g., Ca, Ti, Sr), but also elements that are depleted in CAIs (e.g., Cr, Ni). Hence, while CAIs *sensu stricto* play a role in setting the isotopic composition of carbonaceous chondrites for refractory elements, they do not contribute significantly to the anomalies in less-refractory elements. This observation led to the postulation that material with an isotopic composition like CAIs, but a less-refractory chemistry, must be present within carbonaceous chondrites. In other words, CAIs are the refractory part of an isotopically distinct reservoir in the disk, and both refractory and non-refractory products forming from this

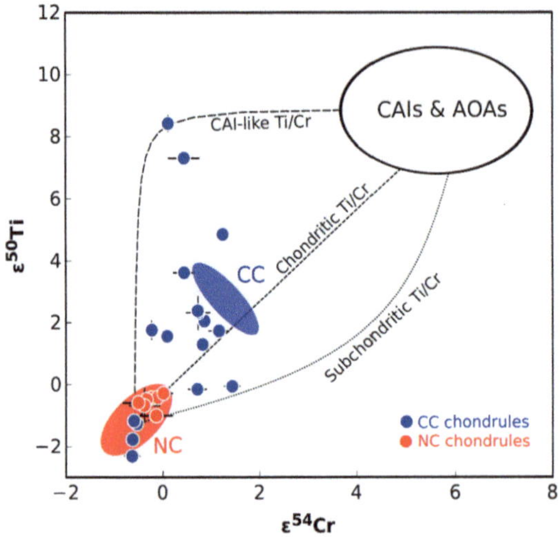

Fig. 12 ε^{54}Cr versus ε^{50}Ti isotopic anomalies of individual NC and CC chondrules and their relation to the composition of bulk NC and CC bodies and refractory inclusions. While the isotopic composition of NC chondrules is in line with the composition of bulk NC bodies the CC chondrules span a range of compositions from NC to CAI-like. This might be explained by variable mixing of CAIs and AOAs in the CC chondrule precursors. Curves represent mixing lines between NC materials and material with an isotopic composition of CAIs and AOAs. The curvature of the mixing lines depends on the Ti/Cr element ratio of the mixing endmember. Data from Schiller et al. (2020), Schneider et al. (2020) and Williams et al. (2020)

reservoir must have contributed to the makeup and isotopic composition of carbonaceous chondrites (Burkhardt et al. 2019; Nanne et al. 2019; Yap and Tissot 2023). The existence of such an Inclusion-like Chondritic (IC) reservoir (Burkhardt et al. 2019) is supported by (i) the finding of common Ti, Cr, and O isotope anomalies in CAIs and the less-refractory AOAs from CV chondrites (Jansen et al. 2024; Torrano et al. 2024), and (ii) the variable Ti and Cr isotopic composition of individual chondrules from CV and CK carbonaceous chondrites, ranging from NC to the hypothetical IC compositions (Fig. 12) (Gerber et al. 2017; Olsen et al. 2016; Schneider et al. 2020; Williams et al. 2020; Zhu et al. 2019). Hence, the CC reservoir can be understood as a mixture of NC and IC materials (CC=NC+IC), whose distinct isotopic compositions may be directly inherited from the molecular cloud through infall processes.

CAIs are the SS's oldest solids and formed at high temperatures, most likely close to the young Sun, but are now most abundant in bodies forming later in the outer disk (i.e., carbonaceous chondrites and comets; Joswiak et al. 2017; Krot et al. 2019; McKeegan et al. 2006). This requires the transport of CAIs to the outer disk, a process that is naturally explained by (i) radial disk expansion accompanying material infall close to the protostar (i.e., viscous spreading; expansion of the centrifugal radius), and/or (ii) ballistic mechanisms (i.e., disk winds and/or stellar outflows) (e.g., Desch et al. 2018; Hueso and Guillot 2005; Morbidelli et al. 2024; Scott et al. 2018; Yang and Ciesla 2012 & references therein). The virtual absence of CAIs in NC meteorites suggests CAI transport (either to the outer disk, or into the Sun due to radial drift) was quantitative, and implies that material falling in early and close to the protosun became enriched in the outer disk, while late infalling material dominated the inner disk (Nanne et al. 2019).

By combining infall and disk evolution models with the isotopic signatures of meteorites, a framework describing the early evolution of the SS emerged that, despite its simplicity, can

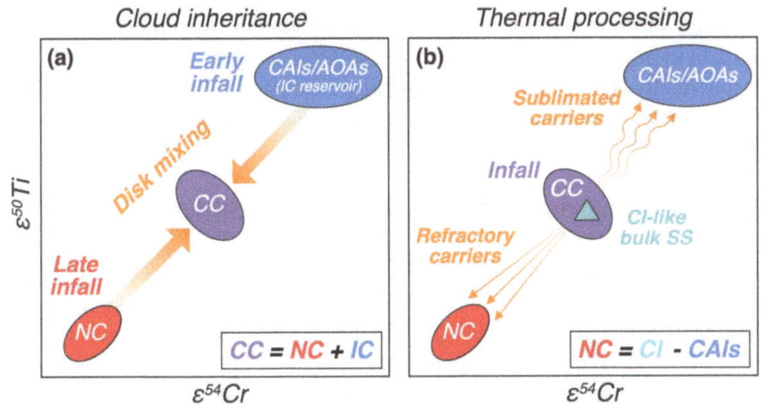

Fig. 13 Schematic representation in isotope-isotope space of the two families of models that have been proposed to explain the isotopic heterogeneity observed in SS materials. For a description of each model, see Fig. 11

explain many observables of the meteoritic record (Burkhardt et al. 2019; Jacquet et al. 2019; Nanne et al. 2019; Yap and Tissot 2023). In essence, this model equates the IC reservoir with early infalling cloud material, and the NC reservoir with late infalling cloud material. Hence, it implies that many of the overarching isotopic variations among planetary materials in the SS ultimately date back to the infall and disk building phase and are inherited from an isotopically heterogeneous molecular cloud core. A cartoon of the model in its simplest form is provided in Fig. 11a-b, and its isotopic implications are shown on Fig. 13.

During the initial stages of collapse, infalling material is IC in composition and limited to the vicinity of the protosun, where it was thermally processed before being transported outwards by viscous spreading. As such, the nascent disk and a significant portion of our Sun would have been CAI-like in isotopic composition: i.e., enriched in NSE and *r*-process nuclides (see quote from Niemeyer (1988a) above). With time, as the disk expanded, the composition of the infalling material transitioned to be more NC-like. Since most of the infalling mass is assumed to be deposited close to the Sun, the later NC dust dilutes and overprints IC isotopic signatures in the inner disk. As infall dwindles, the disk thus formed is characterized by a gradient from NC material in the inner disk to IC material in the outer disk, with a mixture of NC and IC materials (i.e., CC material) in-between. To a certain extent, this primordial disk gradient must have been retained during planetesimal formation (see Sect. 4.1.3). Most likely, this happened via pressure bumps in the disk (e.g., induced by Jupiter's formation at the water-ice sublimation line), which blocked the passage of inward drifting IC materials and triggered NC and CC planetesimal accretion (e.g., Hellmann et al. 2023; Kruijer et al. 2017; Lichtenberg et al. 2021; Morbidelli et al. 2022).

A number of studies explored the permittable parameter space of the infall models with respect to the meteoritic constraints in more detail. They indicate that the angular momentum of the infalling material must have been rather low in order to be constrained to centrifugal radius close to the star, while the initial viscosity of the disk must have been high (α_0 >0.01), and the infall time short (<5×10^5 yr) to allow for high rates of viscous spreading with CAI outward transport and massive outer disk buildup (Jongejan et al. 2023; Marschall and Morbidelli 2023; Woitke et al. 2024). Magnetic braking and magnetohydrodynamic effects might allow for these constraints to be met, but thus far these effects have not been directly included in parameter studies. Nevertheless, these studies all agree that infall and viscous spreading likely played a key role in setting the initial conditions for the SS. Hence, while the details remain to be ironed out, and the model setup is certainly oversimplified (e.g., the IC material spherically contained in the surrounding NC material layer), it must be highlighted that the very existence of a self-consistent model explaining the meteoritic evidence within an astronomical framework is a success in itself.

4.1.2 CI as chemical but not isotopic proxies of primitive SS dust

In the inheritance model, the CC meteorites, including the CI chondrites, represent variable mixtures of NC and IC materials. This is of interest since the chemical composition of CI chondrites is taken as representative of the non-volatile element composition of the Sun (Asplund et al. 2009; Lodders 2003; Palme et al. 2014), and in chemical and isotopic mixing models of carbonaceous chondrites, CIs are used as representative of a matrix endmember component (e.g., Alexander 2019; Hellmann et al. 2020; Nie et al. 2021). Thus, CI chondrites are generally thought to be good proxies for the unprocessed primitive dust in the protosolar disk, and also are used as starting compositions in models assigning the origin of nucleosynthetic anomalies in CAIs and bulk meteorites to differential dust processing in the disk (see Sect. 4.2). However, there is growing evidence that CI chondrites themselves are mixtures, lending credence to the idea that they formed from NC and IC materials. For instance, it was found that compared to NC bodies, CI chondrites like other CC chondrites exhibit an excess in thulium (Tm) relative to neighboring rare earth elements, indicating the presence of CAIs in their precursor materials (Barrat et al. 2016; Dauphas and Pourmand 2015). Likewise, based on a trend between ε^{50}Ti anomalies and Ti/Si elemental ratios among chondrites, it was inferred that the precursor materials of CI chondrites might have contained up to 3% of refractory CAI-like dust (Bryson and Brennecka 2021; Burkhardt et al. 2019). The presence of IC material is further supported by reports of relict CAIs (up to ~ 0.5 atom. %; Frank et al. 2023) and AOA-like olivines (Kawasaki et al. 2022; Morin et al. 2022) in CI chondrites. Together with the finding of chondrule-like olivines (Kawasaki et al. 2022), these observations point towards a bimodal origin of refractory dust in CI chondrites: some being ^{16}O-rich like CAIs, AOAs, and the Sun, and some being ^{16}O-depleted like the NC materials. We note that AOAs have about chondritic relative abundances of refractory elements (Jansen et al. 2024; Torrano et al. 2024), such that a significant amount of AOAs could be accommodated in CI chondrites without changing their overall solar-like bulk composition. Furthermore, it has been suggested that the uniquely distinct Fe and Ni isotopic compositions of CI compared to other CC chondrites reflect physical fractionation of a metal component among their precursor materials, possibly because their formation was triggered by photoevaporation of the gas towards the end of the disk's lifetime (Spitzer et al. 2024). Hence, while CI chondrites have a chemical composition similar to the Sun, they nonetheless appear to be mixtures of isotopically distinct source materials. It is therefore possible that the *isotope* composition of CIs is not representative of the average unprocessed primitive dust from which our SS and the protosolar disk formed. In fact, a growing number of studies have documented the existence of various chemically CI-like materials with different nucleosynthetic compositions, such as CI-like dark clasts in different host chondrites (Goodrich et al. 2021; Van Kooten et al. 2024, 2017b).

4.1.3 Origin of the intra-reservoir NC trend: missing end-member

Aside from the NC-CC-IC offset, the most salient feature in the isotope anomaly data of the bulk planetesimal and planetary bodies is the global correlation of the anomalies in the

NC reservoir (Fig. 8). These correlations are present amongst elements with different cosmochemical (refractory to moderately volatile) and geochemical (lithophile to siderophile) behavior, and among isotopes of different nucleosynthetic origin (stellar and explosive nucleosynthesis in different types of low- and high-mass stars). Furthermore, the anomalies of different nucleosynthetic heritage are not only correlated among different elements, but also among different isotopes of the same element. For instance, correlated variations in ε^{46}Ti and ε^{50}Ti are associated with anomalies originating in Type Ia and Type II supernovae (e.g., Davis et al. 2018; Zhang et al. 2011). Likewise, variations in ε^{94}Mo and ε^{95}Mo among NC meteorites were found to reflect correlated contributions of s- and r-process nuclides (Spitzer et al. 2020), as does the ε^{84}Sr signature of NC bodies (Schneider et al. 2023). Moreover, no trend is apparent between the formation age of NC bodies and their isotopic composition (Fig. 10). Together, these features of the NC trend imply that the global isotopic variations are the result of a two component mixing, where one mixing endmember is enriched in s-process and NSE-associated nuclides relative to the composition of the BSE (e.g., positive ε^{30}Si, ε^{48}Ca, ε^{50}Ti, ε^{54}Cr, ε^{64}Ni, ε^{66}Zn, ε^{100}Ru, and negative ε^{54}Fe, ε^{96}Zr, ε^{94}Mo) and the other is depleted in these nuclides. The exact nature and composition of the mixing endmembers remain vague, but the mixing trend must have been established early, and is unlikely to be caused by the differential processing of individual carriers, or by fractionation due to specific element properties, as these processes cannot easily explain the *global* nature of the observed correlations. It has thus been suggested that the NC isotopic anomaly trend, like the NC-CC-IC offset, is related to heterogeneous infall and represents a gradient in the inner disk (e.g., Spitzer et al. 2020). In such a scenario the very end of infall would represent the "lost" inner SS mixing endmember that balances any contribution of known meteoritic material to the forming Earth, and which is expected to dominate the composition of the disk sunwards of Earth's orbit (Burkhardt et al. 2021; Yap and Tissot 2023).

In summary, the heterogeneous accretion and selective processing of NC and IC materials in different nebular environments, combined with the heterogeneous distribution of the resulting nebular products in the disk can readily account for the observed isotopic and elemental variations between, and to a large extent also within, the NC and CC reservoirs (Burkhardt et al. 2019; Hellmann et al. 2023; Jacquet et al. 2019; Nanne et al. 2019; Yap and Tissot 2023). As such, much of the compositional diversity of meteorites seems to have its origin in the very initial stages of disk building and evolution.

4.1.4 Implications for short-lived radionuclides

The interpretation of the isotopic anomaly record as an inherited signature of heterogeneous infall is in line with recent astronomical observations of filamentary infall and disk substructures, and the change they have initiated in the perception of how and when material can be delivered to planet-forming disks (Sect. 2). Streamers and other observed infall phenomena provide compelling mechanisms for delivering isotopically heterogeneous cloud material to the evolving protosolar disk. Such heterogeneous infall not only can explain the overarching NC/CC dichotomy, but also a range of other meteoritic anomaly signatures. For instance, some rare CAIs seem to have formed with no or much lower initial abundance of the short-lived nuclide ^{26}Al than normal CAIs. These include the so-called PLACs (platy hibonite crystals, Ireland 1988), and hibonite or corundum dominated FUN CAIs (Fractionation and Unidentified Nuclear effects; Desch et al. 2023a,b,c; Krot et al. 2014 and references therein). These ^{26}Al-poor CAIs often also have variable and large nucleosynthetic anomalies and are generally interpreted to have formed before the addition of ^{26}Al to the SS (e.g., Kööp et al.

2018, however also see Desch et al. 2023a,b,c for an alternative interpretation). Within the IC-NC inheritance scenario this would imply that the IC infall forming the normal CAIs was preceded by an even earlier infall of ^{26}Al-free cloud material (e.g., Brennecka et al. 2020). However, given the observation of streamers at different stages of disk evolution, an alternative interpretation of the ^{26}Al-free CAIs could be their formation at a later time from material introduced into the disk by a short-lived streamer sampling cloud material not polluted by freshly synthesized ^{26}Al. To what extent such transient heterogeneities in the accreting materials affected the bulk isotopic evolution of the disk, and in particular the debated distribution of ^{26}Al among forming planetary bodies (e.g., Bollard et al. 2019; Budde et al. 2018; Desch et al. 2023a,b,c; Jacobsen et al. 2008; Kita et al. 2013; Krestianinov et al. 2023; Larsen et al. 2020, 2011; Schiller et al. 2015; Van Kooten et al. 2016; Wasserburg et al. 2012) remains to be investigated. In any case, the possibility of short-lived streamers should be taken into account in any future assessment of isotopic variability among SS materials.

4.2 Thermal Processing

A second family of models, which we henceforth denote as *unmixing models* (Fig. 11c-d), have been proposed to explain the isotopic heterogeneity of SS materials. These models rely on the idea that presolar carriers were initially homogeneously distributed within the cloud of CI chondritic chemical and isotopic composition, and that the sorting of the carriers through thermal and physical processes in the disk led to the isotopic structure seen today (e.g., Burkhardt et al. 2012; Dauphas et al. 2010; Davis et al. 2018; Trinquier et al. 2009).

Recent sample return missions from the asteroids Ryugu and Bennu have shown that CI chondrites likely represent an abundant reservoir in the outer SS (Lauretta et al. 2024). A similar idea had previously been suggested by models of CC matrix compositions (Alexander 2019; Hellmann et al. 2020; Nie et al. 2021), as well as analyses of CC matrices (Van Kooten et al. 2021, 2019, Van Kooten and Moynier 2019; Zanda et al. 2018). In contrast to inheritance models that invoke the hypothetical IC component as a large missing reservoir of the outer SS, unmixing models build on the proposed ubiquity of CI-like material in the outer SS. It is then the thermal processing of this CI-like material in the early SS that leads to its unmixing into two dust populations.

Physico-chemical gradients are inherent to accretion disks and the thermal processing of dust is thought to be a generic process in at least the inner disk regions (e.g., Colmenares et al. 2024; Henning and Semenov 2013). Given the variable properties of different (presolar) minerals, and the variable cosmo- and geochemical behavior of different elements, it is thus conceivable that the selective processing of presolar solids in the protosolar disk induced a significant fraction of the isotopic anomalies seen among planetary materials. Indeed, many of the isotope anomalies at the bulk meteorite scale have been assigned to disk processing in their discovery publications. For instance, variable s-process Mo deficits among meteorites were assigned to the selective processing of presolar SiC (Dauphas et al. 2002b), as was the correlation between Mo and Ru isotope anomalies (Dauphas et al. 2004), and Nd variations among carbonaceous chondrites (Carlson et al. 2007). Likewise, the observation of Ni isotope anomalies among meteorites was suggested to be caused by size-sorting of presolar dust grains during transport within the disk (Regelous et al. 2008).

Based on correlated anomalies in ε^{46}Ti, ε^{50}Ti, and ε^{54}Cr among planetary materials Trinquier et al. (2009) proposed a thermal unmixing model of CI-like dust that has been frequently adopted in later works (e.g., Ek et al. 2020; Larsen et al. 2011; Paton et al. 2013; Poole et al. 2017; Schiller et al. 2015). In essence, this model suggests CI chondrites represent the precursor dust from which the protosun and the surrounding disk accreted, and that the inter-reservoir isotopic anomaly trend of NC, CC, and CAI materials can be explained by unmixing of this dust by thermal processing (Fig. 11 & 13). This would imply that even though ^{46}Ti, and ^{50}Ti and ^{54}Cr have different nucleosynthetic origins and their carriers may not have a similar mineralogy, they must have similar physicochemical characteristics that allow them to be thermally processed together. Trinquier et al. (2009) suggested that, since chondritic matrix shows evidence of thermal processing by loss of moderately volatile elements (Bland et al. 2005), a similar process must have destroyed the thermally labile ^{54}Cr, ^{46}Ti, and ^{50}Ti carriers. In more detail, the thermal processing model invokes the sublimation of thermally labile isotopically anomalous material. During the thermal processing of labile carriers, the surrounding gas is progressively enriched in supernova-derived nuclides (^{54}Cr, ^{50}Ti, ^{26}Al, ^{48}Ca, ^{88}Sr) and eventually forms CAIs and AOAs with this signature by condensation, whereas the residual material is depleted in these isotopes relative to the CI starting composition and ends up in NC bodies. Alternatively, it has been proposed that the thermally labile carrier is represented by ISM-derived dust or icy and organic-rich mantles around silicate dust (Ek et al. 2020). In this way, the isotopic divide may also relate to the petrological divide between water-poor NC and water-rich CC bodies.

In any case, although thermal processing of dust in protoplanetary disks seems unavoidable, a general concern of the thermal processing models is that it remains unclear how they work in detail. In particular, they have to overcome the challenge that the carriers of different types of anomalies and different elements likely respond differently to thermal processing; yet, at the bulk meteorites scale, there are consistent isotope anomaly correlations between elements with different nucleosynthetic heritage and different cosmochemical and geochemical properties (Fig. 8). A profound hurdle on the way to a better understanding of the relevance of thermal processing for the generation of bulk parent body anomalies is the lack of knowledge about the properties of the anomaly carrier phases. Although isotope studies on sequentially digested primitive chondrites can provide some hints (e.g., Burkhardt et al. 2019, 2012; Frossard et al. 2024; Paton et al. 2013; Schiller et al. 2015), identification and detailed nano-characterization of the presolar anomaly carriers will be needed to understand how these grains respond to thermal processing (Nittler et al. 2020). This is not an easy task, since presolar grains are rare. Within chondritic matrices, presolar grains constitute up to a few hundreds of ppm of the total matrix. As a result, to date, only the carriers of ^{54}Cr anomalies have been successfully identified in the form of presolar spinel (Dauphas et al. 2010; Qin et al. 2010).

Thermal processing of dust has, so far, mainly been considered to drive the nucleosynthetic diversity recorded by meteorites and their components but it may also operate during planetary accretion. Considering this process may allow to resolve the apparent lack of a nucleosynthetic endmember component in the meteoritic record that appears to be required to explain the composition of the Earth in multi-element-multi-isotope plots such as ε^{96}Zr vs. ε^{50}Ti and ε^{94}Mo vs ε^{54}Cr (Fig. 8). In detail, Johansen et al. (2023) presented a model of pebble accretion during which gas-drag assisted accretion of mm-to-cm sized particles onto a growing planet would lead to the inevitable formation of a hot envelope around the accreting planet. Processing of solids that accrete onto a planet with a hot envelope may modify the planet's bulk chemical composition, whereby sublimated solids can be lost to the surrounding protoplanetary disk via recycling flows (e.g., Steinmeyer et al. 2023). Onyett et al. (2023) proposed that the same process may also allow modification of isotope signatures such as those controlled by carriers with extreme nucleosynthetic isotope compositions. Building on the proposal of an inward flux of CI-like dust to accreting planets (Schiller et al. 2020, 2018), Onyett et al. (2023) proposed that a modest loss, via recycling flows through the envelope, of isotopically "normal" Zr, Nd, and Mo relative to refractory and highly s-enriched isotope carriers (e.g., SiC grains) could alleviate the need for a hidden reservoir of these isotope signatures in the innermost disk and explain the isotope composition of the Earth. As such, the apparent "missing end-member" needed to explain Earth's isotope signature in Zr, Nd, and Mo in inheritance models may instead be indeed "missing" in the form of isotopically normal material that was lost during the Earth's accretion. The interpretation of Onyett et al. (2023) has been challenged, however, as it remains unclear how this process makes the Earth an isotopic endmember for multiple geochemically and cosmochemically distinct elements, and why various differentiated and undifferentiated objects form common NC isotope anomaly trends in multi-elemental isotope space in the first place (Morbidelli et al. 2025). Nevertheless, if interpreted in the framework of dust processing during terrestrial planet formation, the larger implication of the isotope signatures would be that terrestrial planet formation did not follow the classical model of stochastic planetary growth of ever larger collisions, but instead was mainly driven by pebble accretion (Johansen et al. 2021, 2015).

4.3 Dichotomy vs Trichotomy

The unique nucleosynthetic anomalies of Fe and Ni in CI chondrites (and Ryugu) are at the origin of an ongoing debate. At the current limits of precision, CI chondrites and Ryugu have identical Fe and Ni isotope anomalies, which are distinct from that of other CC chondrites (Fig. 14): i.e., in any ε-ε plot involving either Fe or Ni isotopes, CIs and Ryugu are positioned well outside the CC field. The chemical affinity between Fe and Ni suggests their anomalies share a common carrier. These anomalies primarily reflect variations in the abundances of nuclides ^{54}Fe, ^{58}Ni, and ^{60}Ni. Relative to other CCs, CIs/Ryugu appear depleted in ^{54}Fe, plotting close to the BSE in μ^{54}Fe (see Sect. 3.3; also Fig. 8), but enriched in ^{58}Ni and ^{60}Ni. This departure from a coupled excess in the neutron-poor nuclides ^{54}Fe and ^{58}Ni, which defines the CC offset from NCs, suggests the parent bodies of CIs/Ryugu sampled a distinct carrier during their formation, deriving from a region and/or time in the protosolar disk wherein (i) the carrier was uniquely present (i.e., a trichotomy in disk isotopic reservoirs; Hopp et al. 2022a), and/or (ii) astrophysical conditions (i.e., the mode of planetesimal formation; Spitzer et al. 2024) were conducive to its incorporation.

An isotopic trichotomy was initially proposed on the basis of Fe anomalies in Ryugu (Dauphas et al. 2024; Hopp et al. 2022a). In particular, the composition of CIs/Ryugu was posited to reflect that of a rarely sampled reservoir lying beyond the CC formation zone, around the birthplaces of Uranus and Neptune. Some subsequent works (Marrocchi et al. 2023; Yap and Tissot 2023) suggested the appearance of a trichotomy resulted from a lack of Fe isotope data, predicting that CC achondrites would bridge the gap between CIs/Ryugu and the other CCs, thereby reaffirming CIs as an endmember in the CC intra-reservoir trend. Analyses of Fe anomalies in those achondrites have since falsified this prediction (Spitzer et al. 2024; Rego et al. 2024), establishing the distinct genetic heritage of CIs/Ryugu as a true feature. Subsequent Ni anomalies measured in Ryugu, CIs, and CCs corroborated the special status of CIs/Ryugu (Spitzer et al. 2024), and inspired the suggestion that micron-sized Fe-Ni grains (distinct from the bulk inventory of chondritic iron such as that hosted in chondrules) constitute the carrier of Fe and Ni anomalies. As planetesimal formation by hydrodynamic instabilities (e.g., the streaming instability) tend to feature the largest (typically mm-cm in size; Yap and Batygin 2024), and thus the most aerodynamically coupled, dust grains in the disk locality (e.g., Squire and Hopkins 2018; Youdin and Goodman 2005), these micron-sized Fe-Ni grains are envisioned to have contributed minimally to the building blocks of most planetesimals, such as the parent bodies of the CC chondrites. Towards

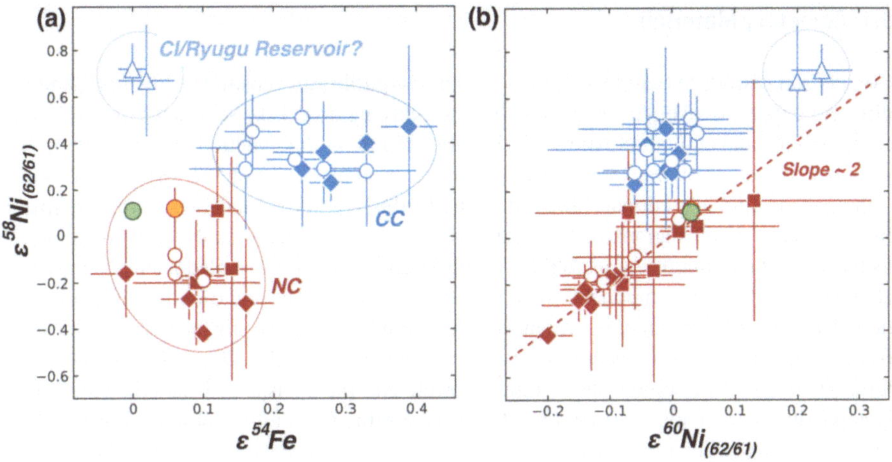

Fig. 14 Isotope anomalies of (a) ε^{58}Ni vs ε^{54}Fe, and (b) ε^{58}Ni vs ε^{60}Ni. Nickel anomalies are internally normalized to the terrestrial ^{62}Ni/^{61}Ni ratio (see subscript). At current limits of precision, CI chondrites and Ryugu (blue triangles) exhibit identical Fe and Ni anomalies, distinct from those of other CC chondrites. In particular, CIs/Ryugu are characterized by a lower ε^{54}Fe, and higher ε^{58}Ni and ε^{60}Ni, relative to the other CCs. This observation suggests the parent bodies of CIs and Ryugu incorporated a unique carrier (e.g., micron-sized Fe-Ni grains; Spitzer et al. 2024) during their formation, and underlies the (presently debated) proposition that they derive from a third isotopic reservoir in the early SS (Hopp et al. 2022a; see Sect. 4.3). Symbols as in Fig. 8

the end of the protosolar disk lifetime (~ 4 Myr from CAIs; Borlina et al. 2022; Wang et al. 2017), photoevaporation facilitates gas dissipation and with it, a rise in metallicity across the disk. This is thought to trigger a final wave of planetesimal formation, not by said hydrodynamic instabilities, but through direct gravitational collapse (e.g., Goldreich and Ward 1973). Such collapse does not involve grain size-sorting, thus incorporating the anomalous micron-sized Fe-Ni grains in abundance. CI chondrites and Ryugu are suggested to originate from this last generation of planetesimals (Spitzer et al. 2024).

Physical interpretations for the observed systematics, while the ultimate goal, remain largely unsubstantiated. Indeed, the suggested anomalous Fe-Ni grains have yet to be identified (i.e., shown to exist), isolated, and studied. At present, all that can be inferred with certainty is that CIs and Ryugu accreted in a different place and/or time than the other CCs.

5 Evidence for Disk Transport and Mixing: Insights from Isotope Anomalies in Nebular Dust Components

Observations of young stellar objects reveal that the amount of crystalline to amorphous silicate dust grains in outer disk regions is higher than in the ISM. This is interpreted as evidence for the outward transport of crystalline materials formed by condensation or thermal annealing of amorphous dust close to the protostar (Watson et al. 2009). Similar transport and mixing of primitive and processed dust also happened in our SS, and the investigation of cometary materials and chondritic meteorites provides clues for reconstructing these processes in some detail.

5.1 Cometary Materials

Comets and chondrites can be understood as non-equilibrium "cosmic sediments" sampling different depositional environments within the evolving disk. Like sediments on Earth, their composition, makeup, and texture provide constraints on the genetic heritage of their constituents, as well as on transport, sorting, and alteration processes on the way to – and within – their respective accretionary environments. In this regard, comets are cosmic sediments from the cold distal regions of the outer SS. Yet, the laboratory analysis of material collected from the coma of comet Wild-2 in the Stardust mission revealed a surprisingly high abundance of high-T components such as refractory inclusions and olivine. These observations have been interpreted as evidence for the transport of early-formed crystalline solids from the inner disk to the accretion region of comets (Brownlee 2014). Furthermore, isotopic analyses of the comet dust grains showed that the transported materials stem from isotopically distinct source reservoirs. In particular, the O-isotopic compositions of refractory crystalline solids imply that the outward-transported cometary dust formed at different radii in the disk, ^{16}O-enriched grains close to the protosun, and ^{16}O-depleted grains further out, possibly near the accretion region of chondrites (McKeegan et al. 2006). Together, these results led to the realization that the outward transport of dust is a natural outcome of early disk evolution (see Sect. 4.1), and that even highly primitive SS objects like comets are mixtures of processed and unprocessed disk materials of diverse heritage.

5.2 Chondrite Components

Stardust samples aside, the main source of information for investigating transport, mixing, and processing of dust in the solar accretion disk are chondrite meteorites. They come in much larger sample sizes compared to the few returned cometary grains, and, importantly, in a range of different flavors, each thought to represent a specific snapshot of the materials present in the protosolar disk at the time and place of their parent bodies' accretion. The NC chondrites are thought to have formed $\sim$2 Myr after CAIs in the inner regions of the SS, close to the formation region of the terrestrial planets, while the CC chondrites formed $\sim$2–4 Myr after CAIs in further out regions, with proposals ranging from beyond Jupiter's orbit to beyond Neptune's orbit (Fig. 11). The main constituents of chondrites are the eponymous chondrules (sub-mm to mm-sized once-molten, fast-cooled silicate spherules formed by unclear processes in the presence of nebular gas; Russell et al. 2018), opaque phases (e.g., metal and sulfides), refractory inclusions (CAIs and AOAs), and a fine-grained matrix consisting of crystalline and amorphous solids, fragments of other chondrite components, organics, and presolar grains. Variations in the relative abundances and physico-chemical characteristics (e.g., size, mineralogy) of the main chondrite components form the basis for the phenomenological classification of chondrites into different groups and types, and testify to the distinct/changing accretion environments in the disk (e.g., Scott and Krot 2014). Understanding the origin and relation of the different components within and among chondrites is key for understanding the mechanisms and extent of disk transport and processing. To date, much of the cosmochemical debate related to this point has centered around the question of whether the main components of a given chondrite formed mostly by fractionation from an isolated local disk volume, or whether they were transported over wide distances from separate disk regions to mix into the accreting parent body. As conservative source tracers, isotope anomalies in meteorite components provide constraints for teasing apart these scenarios and better understand transport, mixing, and accretion processes in the disk.

Refractory inclusions: Clear evidence for the transport of material from different disk source regions to the accretion region of chondrites comes from the isotopic analysis of refractory inclusions. For many elements, CAIs and AOAs have relatively uniform, but highly anomalous isotopic compositions compared to the composition of their host chondrites (e.g., Jansen et al. 2024; Torrano et al. 2024). This necessitates a different source reservoir of refractory inclusions compared to other chondrite components, and – like for the CAIs found in the Stardust samples – implies an outward transport of early-formed phases. Moreover, within the different classes and types of chondrites, the abundance of refractory inclusions is highly variable. While CAIs are very rare in NC chondrites (<0.01 area %, Dunham et al. 2023), they constitute up to ~5 area % of CC chondrites (Hezel et al. 2008). This heterogeneity implies variability in the distribution of refractory inclusions in space and time, which has been linked to the formation of chondrite parent bodies within pressure bumps in the disk (Desch et al. 2018; Hellmann et al. 2023). More specifically, these models suggest that the CC chondrites formed through the pile-up of inward-drifting particles at different times and/or different distances in pressure bumps outside of proto-Jupiter. These particles include CAIs transported to the outer disk (see Sect. 4.1). In this scenario, the near absence of CAIs in NC chondrites is the result of both trapping of CAIs at pressure bumps, and radial drift of inner SS CAIs into the protosun early on.

An important implication of the variable abundance of refractory inclusions in chondrites is that it exerts a significant control on the bulk isotope anomalies and chemical compositions of CC chondrites. This is evidenced by trends between the abundances of CAIs, the abundances of refractory elements like Ca, Ti, Sr, Zr, or Nd, and the bulk CC chondrite isotope anomalies (e.g., Burkhardt et al. 2019; Yap and Tissot 2023). Moreover, these trends pass through the compositions of CC chondrules and end at the composition of NC chondrules (Fig. 2 in Gerber et al. 2017). This implies that CAIs and AOAs are not only present as entities in carbonaceous chondrites, but also have been part of the precursor materials of CC chondrules.

Chondrules: As the dominating component in most chondrites, chondrules likely played a critical role in mediating the accretion of dust into planetesimals (e.g., Budde et al. 2016; Scott and Krot 2014). Determining the origin of chondrules within a given chondrite, as well as among different chondrite groups and types is thus of prime interest for investigating dust transport in the disk, and ultimately provides clues about the nature and site of chondrule formation and planetesimal accretion. Isotopic variability among and within chondrules has been observed for a number of elements with different cosmochemical properties, including O, Ti, and Cr (e.g., Marrocchi et al. 2024b; Olsen et al. 2016; Schneider et al. 2020; Van Kooten et al. 2021, 2020, 2016; Williams et al. 2020).

Oxygen is nominally a volatile element, but in oxides and silicates also part of refractory and main component minerals of nebular dust. Given these attributes, the O isotopic composition of individual chondrules is the result of a combination of precursor heterogeneity, open system gas-melt interaction during chondrule formation, and mineral-water/ice interaction during aqueous alteration. Individual chondrules of a given chondrite type show some spread in O isotopic composition, and there is overlap between chondrules from different chondrite groups and types (Clayton et al. 1991). This means that analyzing the O isotopic composition of a single chondrule does not allow one to assign it to a certain chondrite type with high confidence. This begs the question as to whether individual chondrules in a given chondrite (i) formed locally from isotopically variable precursor mixtures, or rather (ii) originated from isotopically distinct regions in the disk and were transported over large distances to the chondrite accretion region.

Support for the former interpretation comes from the observation of relict grains in chondrules. These grains are found mostly in CC chondrules and are characterized by ^{16}O-rich compositions relative to the composition of the host chondrule (Ushikubo et al. 2012). The identification of relict grains implies that CAIs and AOA-like materials were part of the precursor mix of the chondrules, and hence, that isotopic heterogeneity among individual chondrules is inherited from heterogeneities at the chondrule precursor scale, and not necessarily a result of disk-wide chondrule transport (Marrocchi et al. 2018). In contrast, the recent finding of O isotopic heterogeneity in ordinary chondrite chondrules scaling with chondrule-size may be interpreted as evidence for the transport of some small ^{16}O-rich chondrules from the CC to the NC reservoir, and hence a "leaky" NC/CC barrier (Marrocchi et al. 2024b). Isotope data for other elements will be needed to confirm an outer SS origin for these chondrules.

Chromium is a main component element, such that unlike for O, the Cr concentration and isotopic composition of chondrules is not sensitive to gas-dust interaction, but thought to trace the bulk dust component of the disk. As such, the observation of Cr isotopic variability among individual CC chondrules spanning the range of bulk NC to CC bodies was interpreted as indicating large-scale transport of chondrules from different disk regions to the accretion region of CC chondrites (e.g., Olsen et al. 2016). Likewise, the observation of pristine CC chondrules with NC-like ε^{54}Cr isotope signatures in their cores, and CI-like compositions towards their rim were interpreted as signatures of massive outward transport of NC chondrules to the outer SS, potentially coupled with inward flux of CI-like material to the NC region (Van Kooten et al. 2021). Elevated ε^{54}Cr in two chondrules from an enstatite chondrite support transport from the outer to the inner disk (Zhu et al. 2020). Moreover, the ε^{54}Cr data of the individually analyzed chondrules form a continuum filling the isotopic "gap" between NC and CC bulk bodies. Assuming that the ε^{54}Cr composition of the chondrules represents solely a source signature implies that there are chondrules sampling disk regions between the NC and CC reservoirs, where no bulk planetary body formed (or at least for which no bulk samples are available in the world's collections). Alternatively, the Cr isotope anomalies could also indicate isotopic heterogeneity at the chondrule precursor scale, i.e., heterogeneous mixing of NC and IC materials during chondrule formation in localized disk regions.

Titanium is a highly sensitive tracer of isotopic heterogeneity at the chondrule precursor scale with respect to the presence of CAIs. This is because CAIs have large Ti isotope anomalies and are strongly enriched in Ti compared to chondrules. Furthermore, in contrast to Cr (Grossman and Brearley 2005; Van Kooten et al. 2021), the Ti isotopic composition of individual chondrules is not subject to later modification by aqueous alteration on the parent body. Hence, the Ti isotopic variability observed in CC chondrules was readily assigned to variable admixing of CAIs to a NC-chondrule-like starting material at the individual chondrule level (Gerber et al. 2017; Niemeyer 1988b). This implies that NC materials and CAIs have been present in the formation region of CC chondrules, in line with the proposition that the CC reservoir represents a mixture of NC and IC materials.

Such a mixing scenario is also preferred in combined O, Ti, and Cr isotope studies of NC and CC chondrules (Fig. 12), which reveal a restricted range of isotopic compositions for NC chondrules, and a wide spread in CC chondrules, from the NC field towards the composition of CAIs and AOAs (Schneider et al. 2020; Williams et al. 2020). Furthermore, these data indicate that although isotopically NC-like material is present in NC chondrules and CC chondrules, the formation region of the NC and CC chondrules was likely distinct. This is because refractory-enriched chondrules from NC and CC chondrites have a contrasting isotopic composition, indicating a distinct origin of the refractory materials in the NC and CC reservoir (Ebert et al. 2018). This has been interpreted as supporting the existence of a strong barrier between the outer and inner SS, and a limited inward drift from the CC to the NC reservoir. However, more multi-elemental isotope anomaly data of a more diverse set of chondrules from different NC and CC chondrites are needed to address the extent of chondrule transport between the disk reservoirs, and hence the nature and permeability of the barrier more confidently.

5.3 Mixing and Transport: The Current Picture Painted by Nebular Dust Components

Although knowledge gaps remain, the isotope anomaly data from cometary and chondritic materials reveal that outer SS bodies are mixtures of isotopically diverse materials at a multitude of scales. Carbonaceous chondrites are mixtures of multiple components (e.g., CAIs, chondrules, matrix), some of which are themselves mixtures of materials of distinct nucleosynthetic heritage (e.g., Desch et al. 2018; Hellmann et al. 2023). This is particularly evident from the isotopic composition of CC chondrules, which are mixtures of NC materials and variable amounts of refractory inclusion and CI-like dust (Fig. 12). CI-chondrites (and comets) are, in turn, themselves mixtures of low and high-temperature materials of also distinct isotopic heritage (e.g., Barrat et al. 2016; Dauphas and Pourmand 2015; Frank et al. 2023; Kawasaki et al. 2022). Taken at face value, these observations call into question the well-accepted idea that CI chondrites are faithful proxies of the chemical and isotopic composition of primitive SS dust (see Sect. 4.1.2). At a higher level, they also highlight how all SS materials, even those traditionally thought of as highly primitive (comets, CI chondrites), are intimately linked and testify to the details of disk dynamics. The continued exploration of nucleosynthetic anomalies in meteorite components (which is currently in its infancy, Fig. 5) promises to bring more granularity, and hopefully clarity, to the emerging picture of disk wide transport and mixing of dust components.

6 Limitations, Open Questions, and Outlook

As discussed above, the careful investigation of nucleosynthetic anomalies in CAIs, bulk meteorites, and more recently AOAs, chondrules, and matrices have brought the most important constraints to-date on our understanding of the architecture and evolution of our early SS. This community effort over decades has now yielded a rich dataset, spanning more than 20 elements, including all geo- and cosmochemical categories, and revealing systematic behavior across them. At the same time, these momentous advances have also unveiled major questions.

First and foremost is the question of the origin of the isotopic heterogeneity observed in SS materials. The debate currently focuses on the *inheritance* and *unmixing* models (Fig. 11 & 13), which both have their respective strengths and weaknesses. The major strength of the inheritance models is that they explain with a minimum number of free parameters (3 in total: 2 for the NC reservoir and one for the IC reservoir) the global systematics seen across all elements investigated to date. Qualitatively, this is a remarkable feat, and initial quantitative testing of such models (Yap and Tissot 2023) showed that the elemental and isotopic composition of CC chondrites can be consistently reproduced within $\sim$20% for the four elements investigated (Ca, Ti, Cr, Fe). Another strength of the inheritance models is that they are physically grounded into the astronomical observations of filaments, streamers and other (sub)structures in planet-forming disks (see Sect. 2). An obvious limitation of the inheritance models is that, while the reservoirs that they invoke are physically motivated – for instance, the IC reservoir's isotopic composition is that of CAIs and AOAs while its chemical composition is chondritic – they are nonetheless hypothetical: i.e., there are no known bulk meteorite samples of the actual reservoirs. This is best encapsulated by the denomination of the "missing end-member" (Burkhardt et al. 2021; Burkhardt 2021), which describes the reservoir needed to explain the terrestrial composition, and without which the Earth ends up being an end-member itself in many isotope-isotope spaces (e.g., Fig. 8b-e). This also begs the question of whether such samples actually exist, i.e., how long the primary reservoirs persisted before their original signatures were diluted to the compositions recorded in planetary materials sampled today. Indeed, recent isotopic analyses of outer disk materials suggest that the comet-forming region has not been dominated by an IC component at the time of dust accretion, but that a streamer likely contributed primitive molecular cloud material to this region after the initial disk formation and before parent body assembly (Van Kooten et al. 2024). In the inheritance models, physico-chemical processing of SS materials is secondary to mixing and transport within the disk.

In contrast, the unmixing models do not invoke hypothetical reservoirs but hypothesize that CI chondrites, which are a known family of meteorites, represent the unprocessed primitive materials from which the SS formed. Furthermore, they build on the idea that physico-chemical gradients in the disk fundamentally impact the elemental and isotopic make-up of the evolving protoplanetary disk. Clearly, a strength of these models is that they build on the processing of a known starting material to explain the observed heterogeneity in SS materials. At the same time, and in contrast to the simplicity of the mixing processes invoked in the inheritance models, unmixing models must rely on a large number of free parameters to explain the isotope anomalies systematics in bulk meteorites and across elements with different geo- and cosmo-chemical behaviors. This is a major limitation of these models as the carriers of different types of anomalies, which are today correlated in bulk meteorites, are likely to respond differently to thermal processing. Furthermore, and as discussed in Sect. 5, new evidence is showing that CI chondrites themselves are mixtures of low- and high-T materials with distinct isotopic heritage. As such, they are unlikely to be good proxies of the unprocessed primitive SS composition.

At the heart of the debate between the inheritance and unmixing models loom some key unknowns. The first is our nearly total lack of knowledge on the nature and physico-chemical properties of the carriers of nucleosynthetic anomalies. Since the processing of these carriers is key to the actual mechanisms invoked by unmixing models, testing in detail these models remains extremely arduous, if not impossible. Efforts devoted to the identification and characterization of nucleosynthetic anomaly carriers (e.g., Dauphas et al. 2010; Qin et al. 2010) will be critical to future progress. A second issue pertains to identifying whether the trends seen in isotope-isotope space testify to a temporal and/or spatial variability of the protoplanetary disk. Current chronological constraints mainly rely on accretion ages of iron meteorites, and seem to indicate that both NC and CC irons formed coevally, pointing to a spatial separation of NC and CC materials. But accretion ages come with built-in assumptions about the size and composition of their respective parent bodies, which are not necessarily correct.

Clearly, the inheritance and unmixing models are endmember scenarios on what must in reality represent a spectrum of possibilities, with both processes contributing to different extents in shaping SS evolution. Given the current state of our knowledge on isotope anomalies and their carriers, Occam's razor points to the inheritance models as being most appealing since (i) they can explain the current data with the least number of free parameters (only 3 components needed), and (ii) astronomical observation have revealed a ubiquity of heterogeneous (sub)structures in star- and planet-forming regions, which provide a viable mechanism for delivery of compositionally heterogeneous material to the protosolar disk throughout its formation and until the latest stages of infall.

It is important to recognize that to elucidate the origin of the isotopic heterogeneity of SS materials, two complementary avenues can be pursued. The first, and most common one in the literature so far, is to focus on the physical interpretation for the observed isotopic systematics. That is, to devise models for where, why, and how heterogeneity in the early SS arose and was maintained, and how that heterogeneity was sampled by parent bodies formed across the disk across its lifetime. Another, less popular but arguably more neutral, path forward is to identify the number and nature of endmembers necessary to create the emergent systematics seen in multi-element isotope space. These endmembers would correspond to nebular dust components that mixed (Sect. 4.1) and/or unmixed (Sect. 4.2) to yield the isotopic diversity in SS materials (Yap and Tissot 2023). Although one step removed from the physical, the latter avenue serves as a logical prerequisite to the former.

At the intersection of meteoritics, astrophysics and astronomy, another important open question is that of the efficacy of disk transport, in particular in the earliest stages of disk formation and evolution. The presence of refractory crystalline solids in cometary samples and carbonaceous chondrites is in line with observations of high crystalline to amorphous dust ratios in disks of YSOs, and testifies to the outward transport of materials forming in high-T settings close to the young star. However, the nature and efficiency of this transport process remain debated, with proposed mechanisms ranging from viscous disk spreading (Ciesla 2007), large scale stochastic diffusion (Morfill and Voelk 1984), and ballistic orbits above the disk plane through X-winds (Shu et al., 2001). Further study of isotope anomalies in meteoritic components will provide additional insights to help differentiate these possibilities. For instance, the detection of excess nucleogenic isotope variations in CAIs or chondrules would be a fingerprint of irradiation during transport above the disk midplane, while the absence of any such excess would suggest embedded transport, and hence favor viscous spreading or stochastic diffusion. Exploratory work in this direction has thus far returned no positive detection in support of the latter scenarios (Ghaznavi et al. 2025; Roth et al. 2016). Likewise, the detection of materials with an isotopic composition similar to that of CAIs and AOAs, but a less-refractory chemical composition would provide additional support for the viscous disk spreading (and inheritance) model, while the absence of this hypothesized IC material would be in line with the X-Wind model. A cometary sample large enough to provide high-precision isotope anomaly data for a number of elements, in addition to the outer disk xenoliths already observed in various chondrite groups (Goodrich et al. 2021; Kebukawa et al. 2023; Van Kooten et al. 2024, 2017b,a), would help to resolve these debates (Bockelée-Morvan et al. 2022).

Another open issue with profound impact on our understanding of disk transport and mixing is whether or not the components found in chondritic meteorites represent the materials that also formed the first generation planetesimals. More specifically, it remains unclear whether achondrite parent bodies initially contained similar mixtures of refractory inclusions, chondrules, and matrix as found in chondrites. The observation of relict chondrules in primitive achondrites (Schrader et al. 2017) and similar isotope anomaly ranges in chondrites and achondrites seem to support this notion (Fig. 8), however, based on trace element patterns a recent work also argued that some iron meteorite parent bodies might have formed with up to 25 wt% CAIs (Zhang et al. 2024). If correct, this would require efficient trapping, sorting, and enrichment mechanisms in the disk, and implies the existence of CAIs with isotopic compositions different from the ones found in carbonaceous chondrites. Such CAIs with a more NC-like composition might have existed (Brennecka et al. 2020; Ebert et al. 2018), but more work is needed to pinpoint them as refractory products of the NC reservoir. CAIs aside, the most abundant component in chondrites are chondrules, and thus the question of whether chondrites and achondrites share the same starting material is directly related to the unresolved debate about the nature and timing of the chondrule forming process(es), and hence the importance of chondrules in protoplanetary disks (Marrocchi et al. 2024a; Piralla et al. 2023; Russell et al. 2018). Clearly, future work on isotope anomalies in chondrules of different chondrite types, in particular with respect to apparent trends with chondrule size (Gerber et al. 2017; Marrocchi et al. 2024b, 2022), as well as studies of matrix and metal phases of different chondrites will provide novel perspectives for linking SS dust processing and transport to astronomical observations and astrophysical concepts of disk evolution.

Our capacity to interpret the message from meteorites also depends on future breakthroughs in astronomical observations and theory, which inform our understanding of the early SS by using protoplanetary systems as analogues to the early protosolar disk. With a large sample of sources, current observations provide snapshots of disks (i) at different evolutionary stages, (ii) within a variety of star-forming environments, and (iii) undergoing stochastic events, such as accretion outbursts or infalling streamers. It remains uncertain, however, which conclusions drawn from studying these other protoplanetary systems are directly applicable to the case of the early SS. Timelines of events from astrophysical observations are difficult to pin down; the uncertainties on protostellar ages are typically on the order of several million years. Streamers themselves have only recently become an area of study, with a growing number of detections in molecular line observations of protoplanetary sources. With no unbiased samples from dedicated searches to date, streamer detections have mostly happened accidentally, in existing ALMA data, which are themselves heavily focused on low-mass star-forming regions. This should change in the near future as strategies in dedicated searches for new observational campaigns are being developed and candidate sources are identified in archival data (Gupta et al. 2023). Such future investigations will be necessary to constrain streamer occurrence rates in different star-forming environments and tackle the currently open question of the role of streamers in planet formation and their impact across the lifetime of the protoplanetary disk.

Finally, advances in the resolution of astronomical observations will be key to better understanding the process of planetary formation. To date, only the reservoirs of planetary ingredients, both solids and volatiles, have been mapped with high resolution observations ($\sim$10 AU) at (sub)millimeter wavelengths for a number of nearby systems. In contrast, the signposts of planet formation (e.g., dust substructures, embedded protoplanets) identified so far have been limited to those at much larger separations (10 s of AU) than that of the SS planets, with inferred masses of protoplanet candidates typically on the order of several Jupiter masses. Even with current facilities, knowledge of the inner 10–30 AU of the disk midplane is limited; at millimeter wavelengths, the midplane dust emission is often optically thick and not spatially resolved, preventing an accurate accounting or characterization within much of the region of interest for comparison to the protosolar nebula. Future facilities, however, such as the proposed next generation Very Large Array (ngVLA) are slated to be capable of 5 milliarcsecond angular resolution (sub-AU resolution for sources within 140 pc) and higher sensitivity at longer wavelengths than ALMA where dust emission is less likely to be optically thick (van der Marel et al. 2018). At these capabilities, the ngVLA would be able to resolve the substructures induced by $>$ 5–10 Earth mass protoplanets at separations of $\sim$ 5 AU (Ricci et al. 2018) – paving the way for unprecedented observations such as, for instance, the detection of the signposts for a proto-Jupiter mid-formation.

Supplementary Information The online version contains supplementary material available at https://doi.org/10.1007/s11214-025-01207-0.

Acknowledgements We thank Steve Desch and Yves Marrocchi for constructive reviews that helped improve the manuscript, and Alessandro Morbidelli for prompt and careful editorial handling. We are grateful to workshop organizers Herbert Palme, Jutta Zipfel, Alessandro Morbidelli, Klaus Mezger, and Dominik Hezel for their invitation and leadership, and to the International Space Science Institute (ISSI) for supporting and hosting this event. This work was further supported by a Packard Fellowship (to FLHT).

Funding Information Open Access funding enabled and organized by Projekt DEAL.

Declarations

Competing Interests The authors declare that they have no conflict of interest.

References

Adams FC (2010) Annu Rev Astron Astrophys 48:47

Akiyama E, Vorobyov EI, Baobabu Liu H, Dong R, De Leon J, Liu S-Y, Tamura M (2019) Astron J 157:165

Akram W, Schönbächler M, Sprung P, Vogel N (2013) Astrophys J 777:169

Albarède F, Telouk P, Blichert-Toft J, Boyet M, Agranier A, Nelson B (2004) Geochim Cosmochim Acta 68:2725

Alexander CMO (2019) Geochim Cosmochim Acta 254:277

ALMA Partnership, Brogan CL, Pérez LM, Hunter TR, Dent WRF, Hales AS, Hills RE, Corder S, Fomalont EB, Vlahakis C, Asaki Y, Barkats D, Hirota A, Hodge JA, Impellizzeri CMV, Kneissl R, Liuzzo E, Lucas R, Marcelino N, Matsushita S, Nakanishi K, Phillips N, Richards AMS, Toledo I, Aladro R, Broguiere D, Cortes JR, Cortes PC, Espada D, Galarza F, Appadoo DG, Ramirez LG, Humphreys EM, Jung T, Kameno S, Laing RA, Leon S, Marconi G, Mignano A, Nikolic B, Nyman L-A, Radiszcz M, Remijan A, Rodón JA, Sawada T, Takahashi S, Tilanus RPJ, Vilaro BV, Watson LC, Wiklind T, Akiyama E, Chapillon E, Monsalvo IG, Francesco JD, Gueth F, Kawamura A, Lee C-F, Luong QN, Mangum J, Pietu V, Sanhueza P, Saigo K, Takakuwa S, Ubach C, Kempen T, van Wootten A, Carrizo AC, Francke H, Gallardo J, Garcia J, Gonzalez S, Hill T, Kaminski T, Kurono Y, Liu H-Y, Lopez C, Morales F, Plarre K, Schieven G, Testi L, Videla L, Villard E, Andreani P, Hibbard JE, Tatematsu K (2015) The 2014 alma long baseline campaign: first results from high angular resolution observations toward the hl tau region*. Astrophys J Lett 808:L3

Alves J, Zucker C, Goodman AA, Speagle JS, Meingast S, Robitaille T, Finkbeiner DP, Schlafly EF, Green GM (2020) Nature 578:237

Amari S, Anders E, Virag A, Zinner E (1990) Nature 345:238

Andrews SM, Huang J, Pérez LM, Isella A, Dullemond CP, Kurtovic NT, Guzmán VV, Carpenter JM, Wilner DJ, Zhang S, Zhu Z, Birnstiel T, Bai X-N, Benisty M, Hughes AM, Öberg KI, Ricci L (2018) Astrophys J Lett 869:L41

Armitage PJ (2020) Astrophysics of planet formation, 2nd edn. Cambridge University Press, Cambridge

Arzoumanian D, André P, Könyves V, Palmeirim P, Roy A, Schneider N, Benedettini M, Didelon P, Francesco JD, Kirk J, Ladjelate B (2019) Characterizing the properties of nearby molecular filaments observed with Herschel. Astron Astrophys 621:A42

Arzoumanian D, Arakawa S, Kobayashi MIN, Iwasaki K, Fukuda K, Mori S, Hirai Y, Kunitomo M, Kumar MSN, Kokubo E (2023) Astrophys J Lett 947:L29

Asplund M, Grevesse N, Sauval AJ, Scott P (2009) Annu Rev Astron Astrophys 47:481

Bae J, Zhu Z (2018) Astrophys J 859:118

Bae J, Isella A, Zhu Z, Martin R, Okuzumi S, Suriano S (2022)

Bae J, Isella A, Zhu Z, et al (2023) Monograph 9. In: Inutsuka S, Aikawa Y, Muto T, Tomida K, Tamura M (eds) Protostars and planets VII. Astro. Soc. Pac. Conf. Ser., vol 534, pp 423–464. https://arxiv.org/pdf/2205.03935
Bally J, Langer WD (1982) Astrophys J 255:143
Barclay T, Pepper J, Quintana EV (2018) Astrophys J Suppl Ser 239:2
Barrat JA, Dauphas N, Gillet P, Bollinger C, Etoubleau J, Bischoff A, Yamaguchi A (2016) Geochim Cosmochim Acta 176:1
Bernatowicz T, Fraundorf G, Ming T, Anders E, Wopenka B, Zinner E, Fraundorf P (1987) Nature 330:728
Bigeleisen J, Goeppert-Mayer M (1947) J Chem Phys 15:261
Black DC (1972) Geochim Cosmochim Acta 36:377
Black DC, Pepin RO (1969) Earth Planet Sci Lett 6:395
Blackburn T, Alexander CMO, Carlson R, Elkins-Tanton LT (2017) Geochim Cosmochim Acta 200:201
Bland PA, Alard O, Benedix GK, Kearsley AT, Menzies ON, Watt LE, Rogers NW (2005) Proc Natl Acad Sci USA 102:13755
Blum J, Wurm G (2008) Annu Rev Astron Astrophys 46:21
Bockelée-Morvan D, Filacchione G, Altwegg K, Bianchi E, Bizzarro M, Blum J, Bonal L, Capaccioni F, Choukroun M, Codella C, Cottin H, Davidsson B, De Sanctis MC, Drozdovskaya MN, Engrand C, Galand M, Güttler C, Henri P, Herique A, Ivanovski S, Kokotanekova R, Levasseur-Regourd A-C, Miller KE, Rotundi A, Schönbächler M, Snodgrass C, Thomas N, Tubiana C, Ulamec S, Vincent J-B (2022) Exp Astron 54:1077
Bollard J, Kawasaki N, Sakamoto N, Olsen M, Itoh S, Larsen K, Wielandt D, Schiller M, Connelly JN, Yurimoto H, Bizzarro M (2019) Geochim Cosmochim Acta 260:62
Borlina CS, Weiss BP, Bryson JFJ, Armitage PJ (2022) JGR planets 127:e2021JE007139
Brennecka GA, Burkhardt C, Budde G, Kruijer TS, Nimmo F, Kleine T (2020) Science 370:837
Brownlee D (2014) Annu Rev Earth Planet Sci 42:179
Bryson JFJ, Brennecka GA (2021) Astrophys J 912:163
Budde G, Kleine T, Kruijer TS, Burkhardt C, Metzler K (2016) Proc Natl Acad Sci USA 113:2886
Budde G, Kruijer TS, Kleine T (2018) Geochim Cosmochim Acta 222:284
Budde G, Tissot FLH, Kleine T, Marquez RT (2023) Geochemistry 83:126007
Burbidge EM, Burbidge GR, Fowler WA, Hoyle F (1957) Rev Mod Phys 29:547
Burkhardt C (2021) Elements 17:395
Burkhardt C, Kleine T, Oberli F, Pack A, Bourdon B, Wieler R (2011) Earth Planet Sci Lett 312:390
Burkhardt C, Kleine T, Dauphas N, Wieler R (2012) Earth Planet Sci Lett 357–358:298
Burkhardt C, Dauphas N, Hans U, Bourdon B, Kleine T (2019) Geochim Cosmochim Acta 261:145
Burkhardt C, Spitzer F, Morbidelli A, Budde G, Render JH, Kruijer TS, Kleine T (2021) Sci Adv 7:eabj7601
Cameron AGW (1957) In: Stellar evolution, nuclear astrophysics, and nucleogenesis, 2nd edn., p 195
Cameron AGW (1962) Icarus 1:13
Carlson RW, Boyet M, Horan M (2007) Science 316:1175
Chakraborty S, Ahmed M, Jackson TL, Thiemens MH (2008) Science 321:1328
Chakraborty S, Yanchulova P, Thiemens MH (2013) Science 342:463
Charlier BLA, Parkinson IJ, Burton KW, Grady MM, Wilson CJN, Smith EGC (2017) Geochem Perspect Lett 4:35
Charlier BLA, Tissot FLH, Dauphas N, Wilson CJN (2019) Geochim Cosmochim Acta 265:413
Charlier BLA, Tissot FLH, Vollstaedt H, Dauphas N, Wilson CJN, Marquez RT (2021) Sci Adv 7:eabf6222
Chen JH, Papanastassiou DA, Wasserburg GJ (2010) Geochim Cosmochim Acta 74:3851
Choi B-G, McKeegan KD, Krot AN, Wasson JT (1998) Nature 392:577
Ciesla FJ (2007) Science 318:613
Clayton RN (1978) Annu Rev Nucl Part Sci 28:501
Clayton DD (1982) Q J R Astron Soc 23:174
Clayton DD (1983) Principles of stellar evolution and nucleosynthesis. University of Chicago Press, Chicago
Clayton RN (1993) Annu Rev Earth Planet Sci 21:115
Clayton RN (2002) Nature 415:860
Clayton RN (2003) 1.06 - oxygen isotopes in meteorites. In: Holland HD, Turekian KK (eds) Treatise on geochemistry. Pergamon, Oxford, pp 129–142
Clayton RN (2008) Rev Mineral Geochem 68:5
Clayton RN, Mayeda TK (1977) Geophys Res Lett 4:295
Clayton RN, Mayeda TK (1983) Earth Planet Sci Lett 62:1
Clayton RN, Mayeda TK (1984) Earth Planet Sci Lett 67:151
Clayton RN, Mayeda TK (1996) Geochim Cosmochim Acta 60:1999
Clayton RN, Mayeda TK (1999) Geochim Cosmochim Acta 63:2089
Clayton RN, Grossman L, Mayeda TK (1973) Science 182:485

Clayton RN, Onuma N, Mayeda TK (1976) Earth Planet Sci Lett 30:10
Clayton RN, Onuma N, Grossman L, Mayeda TK (1977) Earth Planet Sci Lett 34:209
Clayton RN, Hinton RW, Davis AM (1988) Philos Trans R Soc Lond Ser A 325:483
Clayton RN, Mayeda TK, Goswami JN, Olsen EJ (1991) Geochim Cosmochim Acta 55:2317
Colmenares MJ, Lambrechts M, Van Kooten E, Johansen A (2024) Astron Astrophys 685:A114
Dauphas N (2017) Nature 541:521
Dauphas N, Chaussidon M (2011) Annu Rev Earth Planet Sci 39:351–386
Dauphas N, Pourmand A (2015) Geochim Cosmochim Acta 163:234
Dauphas N, Schauble EA (2016) Annu Rev Earth Planet Sci 44:709
Dauphas N, Marty B, Reisberg L (2002a) Astrophys J 565:640
Dauphas N, Marty B, Reisberg L (2002b) Astrophys J 569:L139
Dauphas N, Davis AM, Marty B, Reisberg L (2004) Earth Planet Sci Lett 226:465
Dauphas N, Remusat L, Chen JH, Roskosz M, Papanastassiou DA, Stodolna J, Guan Y, Ma C, Eiler JM (2010) Astrophys J 720:1577
Dauphas N, Chen JH, Zhang JJ, Papanastassiou DA, Davis AM, Travaglio C (2014) Earth Planet Sci Lett 407:96
Dauphas N, Hopp T, Nesvorný D (2024) Icarus 408:115805
Davis AM, McKeegan KD (2014) In: Treatise on geochemistry, 2nd edn. Elsevier, Oxford, pp 361–395
Davis AM, Richter FM, Mendybaev RA, Janney PE, Wadhwa M, McKeegan KD (2015) Geochim Cosmochim Acta 158:245
Davis AM, Zhang J, Greber ND, Hu J, Tissot FLH, Dauphas N (2018) Geochim Cosmochim Acta 221:275
Desch S, Miret-Roig N (2024) Space Sci Rev 220:76
Desch SJ, Kalyaan A, Alexander CMO (2018) Astrophys J Suppl Ser 238:11
Desch SJ, Young ED, Dunham ET, Fujimoto Y, Dunlap DR (2022)
Desch SJ, Dunham ET, Herbst AK, Unterborn CT, Sharp TG, Bose M, Mane P, Williams CD (2023a) Astrophys J 953:146
Desch SJ, Dunlap DR, Dunham ET, Williams CD, Mane P (2023b) Icarus 402:115607
Desch SJ, Dunlap DR, Williams CD, Mane P, Dunham ET (2023c) Icarus 402:115611
Dong R, Fung J (2017) Astrophys J 835:146
Doyle PM, Jogo K, Nagashima K, Krot AN, Wakita S, Ciesla FJ, Hutcheon ID (2015) Nat Commun 6:7444
Dullemond CP, Penzlin ABT (2018) Dust-driven viscous ring-instability in protoplanetary disks. Astron Astrophys 609:A50
Dunham ET, Sheikh A, Opara D, Matsuda N, Liu M-C, McKeegan KD (2023) Meteorit Planet Sci 58:643
Dwarkadas VV, Dauphas N, Meyer B, Boyajian P, Bojazi M (2017) Astrophys J 851:147
Ebert S, Render J, Brennecka GA, Burkhardt C, Bischoff A, Gerber S, Kleine T (2018) Earth Planet Sci Lett 498:257
Ek M, Hunt AC, Lugaro M, Schönbächler M (2020) Nat Astron 4:273
Flores C, Ohashi N, Tobin JJ, Jørgensen JK, Takakuwa S, Li Z-Y, Lin Z-YD, Van MLR, Hoff T, Plunkett AL, Yamato Y, Sai Insa Choi J, Koch PM, Yen H-W, Aikawa Y, Aso Y, De Gregorio-Monsalvo I, Kido M, Kwon W, Lee J-E, Lee CW, Looney LW, Santamaría-Miranda A, Sharma R, Thieme TJ, Williams JP, Han I, Narayanan S, Lai S-P (2023) Astrophys J 958:98
Frank DR, Huss GR, Zolensky ME, Nagashima K, Le L (2023) Meteorit Planet Sci 58:1495
Friesen RK, Bourke TL, Francesco JD, Gutermuth R, Myers PC (2016) Astrophys J 833:204
Frossard P, Bonnand P, Boyet M, Bouvier A (2024) Geochim Cosmochim Acta 367:165
Fukai R, Yokoyama T (2019) Astrophys J 879:79
Gallino R, Arlandini C, Busso M, Lugaro M, Travaglio C, Straniero O, Chieffi A, Limongi M (1998) Astrophys J 497:388
Gammie CF (1996) Layered accretion in T Tauri disks. Astrophys J 457:355
Gárate M, Pinilla P, Haworth TJ, Facchini S (2024) Astron Astrophys 681:A84
Gerber S, Burkhardt C, Budde G, Metzler K, Kleine T (2017) Astrophys J 841:L17
Ghaznavi P, Burkhardt C, Tissot FLH, Leya I (2025) Meteorit Planet Sci 60:247
Goldreich P, Ward WR (1973) Astrophys J 183:1051
Goodrich CA, Sanborn ME, Yin Q-Z, Kohl I, Frank D, Daly RT, Walsh KJ, Zolensky ME, Young ERD, Jenniskens P, Shaddad MH (2021) Planet Sci J 2:13
Gounelle M, Meynet G (2012) Astron Astrophys 545:A4
Gounelle M, Krot AN, Nagashima K, Kearsley A (2009) Astrophys J 698:L18
Grossman L (1972) Geochim Cosmochim Acta 36:597
Grossman L (1975) Geochim Cosmochim Acta 39:433
Grossman JN, Brearley AJ (2005) Meteorit Planet Sci 40:87
Gupta A, Miotello A, Manara CF, Williams JP, Facchini S, Beccari G, Birnstiel T, Ginski C, Hacar A, Küffmeier M, Testi L, Tychoniec L, Yen H-W (2023) Astron Astrophys 670:L8

Hacar A, Tafalla M, Forbrich J, Alves J, Meingast S, Grossschedl J, Teixeira PS (2018) Astron Astrophys 610:A77
Hacar A, Hogerheijde MR, Harsono D, Portegies Zwart S, De Breuck C, Torstensson K, Boland W, Baryshev AM, Hesper R, Barkhof J, Adema J, Bekema ME, Koops A, Khudchenko A, Stark R (2020) Astron Astrophys 644:A133
Hammer M, Pinilla P, Kratter KM, Lin M-K (2019) Mon Not R Astron Soc 482:3609
Hans U, Kleine T, Bourdon B (2013) Earth Planet Sci Lett 374:204
Hashizume K, Chaussidon M (2005) Nature 434:619
Hashizume K, Takahata N, Naraoka H, Sano Y (2011) Nat Geosci 4:165
Heger A, Fröhlich C, Truran JW (2014) 2.1 - origin of the elements. In: Holland HD, Turekian KK (eds) Treatise on geochemistry, 2nd edn. Elsevier, Oxford, pp 1–14
Heidenreich JE, Thiemens MH (1983) J Chem Phys 78:892
Hellmann JL, Kruijer TS, Van Orman JA, Metzler K, Kleine T (2019) Geochim Cosmochim Acta 258:290
Hellmann JL, Hopp T, Burkhardt C, Kleine T (2020) Earth Planet Sci Lett 549:116508
Hellmann JL, Schneider JM, Wölfer E, Drążkowska J, Jansen CA, Hopp T, Burkhardt C, Kleine T (2023) Astrophys J Lett 946:L34
Henning T, Semenov D (2013) Chem Rev 113:9016
Hezel DC, Russell SS, Ross AJ, Kearsley AT (2008) Meteorit Planet Sci 43:1879
Hibiya Y, Iizuka T, Enomoto H, Hayakawa T (2023) Astrophys J Lett 942:L15
Hilton CD, Ash RD, Walker RJ (2022) Geochim Cosmochim Acta 318:112
Holland-Ashford T, Lopez LA, Auchettl K (2020) Astrophys J 889:144
Hopp T, Dauphas N, Abe Y, Aléon J, Alexander CMOD, Amari S, Amelin Y, Bajo K, Bizzarro M, Bouvier A, Carlson RW, Chaussidon M, Choi B-G, Davis AM, Di Rocco T, Fujiya W, Fukai R, Gautam I, Haba MK, Hibiya Y, Hidaka H, Homma H, Hoppe P, Huss GR, Ichida K, Iizuka T, Ireland TR, Ishikawa A, Ito M, Itoh S, Kawasaki N, Kita NT, Kitajima K, Kleine T, Komatani S, Krot AN, Liu M-C, Masuda Y, McKeegan KD, Morita M, Motomura K, Moynier F, Nakai I, Nagashima K, Nesvorný D, Nguyen A, Nittler L, Onose M, Pack A, Park C, Piani L, Qin L, Russell SS, Sakamoto N, Schönbächler M, Tafla L, Tang H, Terada K, Terada Y, Usui T, Wada S, Wadhwa M, Walker RJ, Yamashita K, Yin Q-Z, Yokoyama T, Yoneda S, Young ED, Yui H, Zhang A-C, Nakamura T, Naraoka H, Noguchi T, Okazaki R, Sakamoto K, Yabuta H, Abe M, Miyazaki A, Nakato A, Nishimura M, Okada T, Yada T, Yogata K, Nakazawa S, Saiki T, Tanaka S, Terui F, Tsuda Y, Watanabe S, Yoshikawa M, Tachibana S, Yurimoto H (2022a) Sci Adv 8:eadd8141
Hopp T, Dauphas N, Spitzer F, Burkhardt C, Kleine T (2022b) Earth Planet Sci Lett 577:117245
Huang J, Bergin EA, Öberg KI, Andrews SM, Teague R, Law CJ, Kalas P, Aikawa Y, Bae J, Bergner JB, Booth AS, Bosman AD, Calahan JK, Cataldi G, Cleeves LI, Czekala I, Ilee JD, Le Gal R, Guzmán VV, Long F, Loomis RA, Ménard F, Nomura H, Qi C, Schwarz KR, Tsukagoshi T, Van MLR, Hoff T, Walsh C, Wilner DJ, Yamato Y, Zhang K (2021) Astrophys J Suppl Ser 257:19
Hueso R, Guillot T (2005) Astron Astrophys 442:703
Ibañez-Mejia M, Tissot FLH (2021) Elements 17:379
Ireland TR (1988) Geochim Cosmochim Acta 52:2827
Ireland TR, Holden P, Norman MD, Clarke J (2006) Nature 440:776
Ireland TR, Avila J, Greenwood RC, Hicks LJ, Bridges JC (2020) Space Sci Rev 216:25
Jacobsen SB (2005) Chondrites and the protoplanetary disk. In: Krot AN, Scott ERD, Reipurth B (eds) ASP Conference Series, vol 341. ISBN: 1-58381-208-3
Jacobsen B, Yin QZ, Moynier F, Amelin Y, Krot AN, Nagashima K, Hutcheon ID, Palme H (2008) Earth Planet Sci Lett 272:353
Jacquet E, Pignatale FC, Chaussidon M, Charnoz S (2019) Astrophys J 884:32
Jansen CA, Burkhardt C, Marrocchi Y, Schneider JM, Wölfer E, Kleine T (2024) Earth Planet Sci Lett 627:118567
Jogo K, Nakamura T, Ito M, Wakita S, Zolotov MY, Messenger SR (2017) Geochim Cosmochim Acta 199:58
Johansen A, Low M-MM, Lacerda P, Bizzarro M (2015) Sci Adv 1:e1500109
Johansen A, Ronnet T, Bizzarro M, Schiller M, Lambrechts M, Nordlund Å, Lammer H (2021) Sci Adv 7:eabc0444
Johansen A, Ronnet T, Schiller M, Deng Z, Bizzarro M (2023) Astron Astrophys 671:A75
Jongejan S, Dominik C, Dullemond CP (2023) Astron Astrophys 679:A45
Joswiak DJ, Brownlee DE, Nguyen AN, Messenger S (2017) Meteorit Planet Sci 52:1612
Kawasaki N, Nagashima K, Sakamoto N, Matsumoto T, Bajo K, Wada S, Igami Y, Miyake A, Noguchi T, Yamamoto D, Russell SS, Abe Y, Aléon J, Alexander CMO, Amari S, Amelin Y, Bizzarro M, Bouvier A, Carlson RW, Chaussidon M, Choi B-G, Dauphas N, Davis AM, Di Rocco T, Fujiya W, Fukai R, Gautam I, Haba MK, Hibiya Y, Hidaka H, Homma H, Hoppe P, Huss GR, Ichida K, Iizuka T, Ireland TR, Ishikawa A, Ito M, Itoh S, Kita NT, Kitajima K, Kleine T, Komatani S, Krot AN, Liu M-C, Masuda

Y, McKeegan KD, Morita M, Motomura K, Moynier F, Nakai I, Nguyen A, Nittler L, Onose M, Pack A, Park C, Piani L, Qin L, Schönbächler M, Tafla L, Tang H, Terada K, Terada Y, Usui T, Wadhwa M, Walker RJ, Yamashita K, Yin Q-Z, Yokoyama T, Yoneda S, Young ED, Yui H, Zhang A-C, Nakamura T, Naraoka H, Okazaki R, Sakamoto K, Yabuta H, Abe M, Miyazaki A, Nakato A, Nishimura M, Okada T, Yada T, Yogata K, Nakazawa S, Saiki T, Tanaka S, Terui F, Tsuda Y, Watanabe S, Yoshikawa M, Tachibana S, Yurimoto H (2022) Sci Adv 8:eade2067

Kebukawa Y, Mathurin J, Dartois E, Dazzi A, Deniset-Besseau A, Duprat J, Remusat L, Noguchi T, Miyake A, Igami Y, Paoletti MV, Zolensky ME, Engrand C, Sandt C, Borondics F, Yamashita S, Wakabayashi D, Takeichi Y, Takahashi Y (2023) Icarus 400:115582

Keppler M, Benisty M, Müller A, Henning T, Van Boekel R, Cantalloube F, Ginski C, Van Holstein RG, Maire A-L, Pohl A, Samland M, Avenhaus H, Baudino J-L, Boccaletti A, De Boer J, Bonnefoy M, Chauvin G, Desidera S, Langlois M, Lazzoni C, Marleau G-D, Mordasini C, Pawellek N, Stolker T, Vigan A, Zurlo A, Birnstiel T, Brandner W, Feldt M, Flock M, Girard J, Gratton R, Hagelberg J, Isella A, Janson M, Juhasz A, Kemmer J, Kral Q, Lagrange A-M, Launhardt R, Matter A, Ménard F, Milli J, Mollière P, Olofsson J, Pérez L, Pinilla P, Pinte C, Quanz SP, Schmidt T, Udry S, Wahhaj Z, Williams JP, Buenzli E, Cudel M, Dominik C, Galicher R, Kasper M, Lannier J, Mesa D, Mouillet D, Peretti S, Perrot C, Salter G, Sissa E, Wildi F, Abe L, Antichi J, Augereau J-C, Baruffolo A, Baudoz P, Bazzon A, Beuzit J-L, Blanchard P, Brems SS, Buey T, De Caprio V, Carbillet M, Carle M, Cascone E, Cheetham A, Claudi R, Costille A, Delboulbé A, Dohlen K, Fantinel D, Feautrier P, Fusco T, Giro E, Gluck L, Gry C, Hubin N, Hugot E, Jaquet M, Le Mignant D, Llored M, Madec F, Magnard Y, Martinez P, Maurel D, Meyer M, Möller-Nilsson O, Moulin T, Mugnier L, Origné A, Pavlov A, Perret D, Petit C, Pragt J, Puget P, Rabou P, Ramos J, Rigal F, Rochat S, Roelfsema R, Rousset G, Roux A, Salasnich B, Sauvage J-F, Sevin A, Soenke C, Stadler E, Suarez M, Turatto M, Weber L (2018) Astron Astrophys 617:A44

Kido M, Takakuwa S, Saigo K, Ohashi N, Tobin JJ, Jørgensen JK, Aikawa Y, Aso Y, Encalada FJ, Flores C, Gavino S, De Gregorio-Monsalvo I, Han I, Hirano S, Koch PM, Kwon W, Lai S-P, Lee CW, Lee J-E, Li Z-Y, Lin Z-YD, Looney LW, Mori S, Narayanan S, Plunkett AL, Phuong NT, Insa Choi JS, Santamaría-Miranda A, Sharma R, Sheehan PD, Thieme TJ, Tomida K, Van MLR, Hoff T, Williams JP, Yamato Y, Yen H-W (2023) Astrophys J 953:190

Kita NT, Yin Q, MacPherson GJ, Ushikubo T, Jacobsen B, Nagashima K, Kurahashi E, Krot AN, Jacobsen SB (2013) Meteorit Planet Sci 48:1383

Kleine T, Budde G, Burkhardt C, Kruijer TS, Worsham EA, Morbidelli A, Nimmo F (2020) Space Sci Rev 216:55

Kleine T, Steller T, Burkhardt C, Nimmo F (2023) Icarus 397:115519

Kobayashi S, Imai H, Yurimoto H (2003) Geochem J 37:663

Kööp L, Nakashima D, Heck PR, Kita NT, Tenner TJ, Krot AN, Nagashima K, Park C, Davis AM (2018) Geochim Cosmochim Acta 221:296

Krestianinov E, Amelin Y, Yin Q-Z, Cary P, Huyskens MH, Miller A, Dey S, Hibiya Y, Tang H, Young ED, Pack A, Di Rocco T (2023) Nat Commun 14:4940

Krot AN, Nagashima K, Wasserburg GJ, Huss GR, Papanastassiou D, Davis AM, Hutcheon ID, Bizzarro M (2014) Geochim Cosmochim Acta 145:206

Krot AN, Ma C, Nagashima K, Davis AM, Beckett JR, Simon SB, Komatsu M, Fagan TJ, Brenker F, Ivanova MA, Bischoff A (2019) Geochemistry 79:125519

Kruijer TS, Burkhardt C, Budde G, Kleine T (2017) Proc Natl Acad Sci USA 201704461

Kruijer TS, Kleine T, Borg LE (2020) Nat Astron 4:32

Kuffmeier M, Mogensen TF, Haugbølle T, Bizzarro M, Nordlund Å (2016) Astrophys J 826:22

Kuffmeier M, Haugbølle T, Nordlund Å (2017) Astrophys J 846:7

Kuffmeier M, Jensen SS, Haugbølle T (2023) Eur Phys J Plus 138:272

Kuznetsova A, Hartmann L, Ballesteros-Paredes J (2015) Astrophys J 815:27

Kuznetsova A, Hartmann L, Heitsch F (2019) Astrophys J 876:33

Kuznetsova A, Bae J, Hartmann L, Low M-MM (2022) Astrophys J 928:92

Larsen KK, Trinquier A, Paton C, Schiller M, Wielandt D, Ivanova MA, Connelly JN, Nordlund A, Krot AN, Bizzarro M (2011) Astrophys J Lett 735

Larsen KK, Wielandt D, Schiller M, Krot AN, Bizzarro M (2020) Earth Planet Sci Lett 535:116088

Lauretta DS, Connolly HC, Aebersold JE, Alexander CMO, Ballouz R, Barnes JJ, Bates HC, Bennett CA, Blanche L, Blumenfeld EH, Clemett SJ, Cody GD, DellaGiustina DN, Dworkin JP, Eckley SA, Foustoukos DI, Franchi IA, Glavin DP, Greenwood RC, Haenecour P, Hamilton VE, Hill DH, Hiroi T, Ishimaru K, Jourdan F, Kaplan HH, Keller LP, King AJ, Koefoed P, Kontogiannis MK, Le L, Macke RJ, McCoy TJ, Milliken RE, Najorka J, Nguyen AN, Pajola M, Polit AT, Righter K, Roper HL, Russell SS, Ryan AJ, Sandford SA, Schofield PF, Schultz CD, Seifert LB, Tachibana S, Thomas-Keprta KL, Thompson MS, Tu V, Tusberti F, Wang K, Zega TJ, Wolner CWV, the OSIRIS-REx Sample Analysis Team (2024) Meteorit Planet Sci Maps 14227

Law CJ, Booth AS, Öberg KI (2023) Astrophys J Lett 952:L19
Lee T, Papanastassiou DA, Wasserburg GJ (1977) Astrophys J 211:L107
Lewis RS, Ming T, Wacker JF, Anders E, Steel E (1987) Nature 326:160
Leya I, Schönbächler M, Wiechert U, Krähenbühl U, Halliday AN (2008) Earth Planet Sci Lett 266:233
Lichtenberg T, Drążkowska J, Schönbächler M, Golabek GJ, Hands TO (2021) Science 371:365
Lodders K (2003) Astrophys J 591:1220
Lyons JR, Young ED (2005) Nature 435:317
Lyons JR, Gharib-Nezhad E, Ayres TR (2018) Nat Commun 9:908
Maíz Apellániz J, Pantaleoni González M, Barbá RH, Simón-Díaz S, Negueruela I, Lennon DJ, Sota A, Trigueros Páez E (2018) Astron Astrophys 616:A149
Mamajek EE, Meyer MR, Hinz PM, Hoffmann WF, Cohen M, Hora JL (2004) Astrophys J 612:496
Manara CF, Ansdell M, Rosotti GP, Hughes AM, Armitage PJ, Lodato G, Williams JP (2022)
Mandal S, Duffell PC, Polin A, Milisavljevic D (2023) Astrophys J 956:130
Marcus RA (2004) J Chem Phys 121:8201
Maréchal CN, Telouk P, Albarede F (1999) Chem Geol 156:251
Marrocchi Y, Bekaert DV, Piani L (2018) Earth Planet Sci Lett 482:23
Marrocchi Y, Piralla M, Regnault M, Batanova V, Villeneuve J, Jacquet E (2022) Earth Planet Sci Lett 593:117683
Marrocchi Y, Piralla M, Tissot FLH (2023) Astrophys J Lett 954:L27
Marrocchi Y, Jones RH, Russell SS, Hezel DC, Barosch J, Kuznetsova A (2024a) Space Sci Rev 220:69
Marrocchi Y, Longeau A, Goupil RL, Dijon V, Pinto G, Neukampf J, Villeneuve J, Jacquet E (2024b) Geochim Cosmochim Acta 371:52
Marschall R, Morbidelli A (2023) Astron Astrophys 677:A136
Martins R, Kuthning S, Coles BJ, Kreissig K, Rehkämper M (2023) Science 379:369
McDonough WF (2016) The composition of the lower mantle and core. In: Terasaki H, Fischer RA (eds) Deep Earth: Physics and Chemistry of the Lower Mantle and Core American Geophysical Union.Wiley, New York
McKeegan KD, Aléon J, Bradley J, Brownlee D, Busemann H, Butterworth A, Chaussidon M, Fallon S, Floss C, Gilmour J, Gounelle M, Graham G, Guan Y, Heck PR, Hoppe P, Hutcheon ID, Huth J, Ishii H, Ito M, Jacobsen SB, Kearsley A, Leshin LA, Liu M-C, Lyon I, Marhas K, Marty B, Matrajt G, Meibom A, Messenger S, Mostefaoui S, Mukhopadhyay S, Nakamura-Messenger K, Nittler L, Palma R, Pepin RO, Papanastassiou DA, Robert F, Schlutter D, Snead CJ, Stadermann FJ, Stroud R, Tsou P, Westphal A, Young ED, Ziegler K, Zimmermann L, Zinner E (2006) Science 314:1724
McKeegan KD, Kallio APA, Heber VS, Jarzebinski G, Mao PH, Coath CD, Kunihiro T, Wiens RC, Nordholt JE, Moses RW, Reisenfeld DB, Jurewicz AJG, Burnett DS (2011) Science 332:1528
Miller MF (2002) Geochim Cosmochim Acta 66:1881
Monga N, Desch S (2014) Astrophys J 798:9
Morbidelli A, Baillié K, Batygin K, Charnoz S, Guillot T, Rubie DC, Kleine T (2022) Nat Astron 6:72
Morbidelli A, Marrocchi Y, Ali Ahmad A, Bhandare A, Charnoz S, Commerçon B, Dullemond CP, Guillot T, Hennebelle P, Lee Y-N, Lovascio F, Marschall R, Marty B, Maury A, Tamami O (2024) Astron Astrophys 691:A147
Morbidelli A, Kleine T, Nimmo F (2025) Earth Planet Sci Lett 650:119120
Morfill GE, Voelk HJ (1984) Astrophys J 287:371
Morin GLF, Marrocchi Y, Villeneuve J, Jacquet E (2022) Geochim Cosmochim Acta 332:203
Moynier F, Day JMD, Okui W, Yokoyama T, Bouvier A, Walker RJ, Podosek FA (2012) Astrophys J 758:45
Nanne JAM, Nimmo F, Cuzzi JN, Kleine T (2019) Earth Planet Sci Lett 511:44
Neumann W, Luther R, Trieloff M, Reger PM, Bouvier A (2023) Planet Sci J 4:196
Nie NX, Chen X-Y, Hopp T, Hu JY, Zhang ZJ, Teng F-Z, Shahar A, Dauphas N (2021) Sci Adv 7:eabl3929
Nie NX, Wang D, Torrano ZA, Carlson RW, Alexander CMOD, Shahar A (2023) Science 379:372
Niemeyer S (1988a) Geochim Cosmochim Acta 52:2941
Niemeyer S (1988b) Geochim Cosmochim Acta 52:309
Niemeyer S, Lugmair GW (1984) Geochim Cosmochim Acta 48:1401
Nittler LR, Stroud RM, Alexander CMO, Howell K (2020) Meteorit Planet Sci 55:1160
Nyquist LE, Kleine T, Shih CY, Reese YD (2009) Geochim Cosmochim Acta 73:5115
Öberg KI, Guzmán VV, Walsh C, Aikawa Y, Bergin EA, Law CJ, Loomis RA, Alarcón F, Andrews SM, Bae J, Bergner JB, Boehler Y, Booth AS, Bosman AD, Calahan JK, Cataldi G, Cleeves LI, Czekala I, Furuya K, Huang J, Ilee JD, Kurtovic NT, Le Gal R, Liu Y, Long F, Ménard F, Nomura H, Pérez LM, Qi C, Schwarz KR, Sierra A, Teague R, Tsukagoshi T, Yamato Y, Van MLR, Hoff T, Waggoner AR, Wilner DJ, Zhang K (2021) Astrophys J Suppl Ser 257:1

Ohashi N, Tobin JJ, Jørgensen JK, Takakuwa S, Sheehan P, Aikawa Y, Li Z-Y, Looney LW, Williams JP, Aso Y, Sharma R, Sai Insa Choi J, Yamato Y, Lee J-E, Tomida K, Yen H-W, Encalada FJ, Flores C, Gavino S, Kido M, Han I, Lin Z-YD, Narayanan S, Phuong NT, Santamaría-Miranda A, Thieme TJ, Van MLR, Hoff T, De Gregorio-Monsalvo I, Koch PM, Kwon W, Lai S-P, Lee CW, Plunkett A, Saigo K, Hirano S, Lam KH, Mori S (2023) Astrophys J 951:8
Okuzumi S, Hirose S (2011) Modeling magnetorotational turbulence in protoplanetary disks with dead zones. Astrophys J 742(2):65
Olsen MB, Schiller M, Krot AN, Bizzarro M (2013) Astrophys J 776:L1
Olsen MB, Wielandt D, Schiller M, Van Kooten EMME, Bizzarro M (2016) Geochim Cosmochim Acta 191:118
Onyett IJ, Schiller M, Makhatadze GV, Deng Z, Johansen A, Bizzarro M (2023) Nature 619:539
Ouellette N, Desch SJ, Hester JJ (2010) Astrophys J 711:597
Pack A, Tanaka R, Hering M, Sengupta S, Peters S, Nakamura E (2016) Rapid Commun Mass Spectrom 30:1495
Palme H, Lodders K, Jones A (2014) 2.2 - Solar System abundances of the elements. In: Holland HD, Turekian KK (eds) Treatise on geochemistry, 2nd edn. Elsevier, Oxford, pp 15–36
Papanastassiou DA, Wasserburg GJ (1978) Geophys Res Lett 5:595
Parker RJ (2020) R Soc Open Sci 7:201271
Paton C, Schiller M, Bizzarro M (2013) Astrophys J 763:L40
Pineda JE, Segura-Cox D, Caselli P, Cunningham N, Zhao B, Schmiedeke A, Maureira MJ, Neri R (2020) Nat Astron 4:1158
Pineda JE, Arzoumanian D, André P, Friesen RK, Zavagno A, Clarke SD, Inoue T, Chen C-Y, Lee Y-N, Soler JD, Kuffmeier M (2022)
Pineda JE, Arzoumanian D, André P, et al (2023) Monograph 9. In: Inutsuka S, Aikawa Y, Muto T, Tomida K, Tamura M (eds) Protostars and planets VII. Astron. Soc. Pac. Conf. Ser., vol 534. ASP, San Francisco, pp 233–273. https://arxiv.org/abs/2205.03935
Pinilla P, Birnstiel T, Ricci L, Dullemond CP, Uribe AL, Testi L, Natta A (2012) Astron Astrophys 538:A114
Pinilla P, Flock M, de Juan Ovelar M, Birnstiel T (2016) Can dead zones create structures like a transition disk? Astron Astrophys 596:A81
Pinte C, Price DJ, Ménard F, Duchêne G, Christiaens V, Andrews SM, Huang J, Hill T, Van Der Plas G, Perez LM, Isella A, Boehler Y, Dent WRF, Mentiplay D, Loomis RA (2020) Astrophys J Lett 890:L9
Piralla M, Villeneuve J, Schnuriger N, Bekaert DV, Marrocchi Y (2023) Icarus 394:115427
Podio L, Ceccarelli C, Codella C, Sabatini G, Segura-Cox D, Balucani N, Rimola A, Ugliengo P, Chandler CJ, Sakai N, Svoboda B, Pineda J, De Simone M, Bianchi E, Caselli P, Isella A, Aikawa Y, Bouvier M, Caux E, Chahine L, Charnley SB, Cuello N, Dulieu F, Evans L, Fedele D, Feng S, Fontani F, Hama T, Hanawa T, Herbst E, Hirota T, Jiménez-Serra I, Johnstone D, Lefloch B, Le Gal R, Loinard L, Baobab Liu H, López-Sepulcre A, Maud LT, Maureira MJ, Menard F, Miotello A, Moellenbrock G, Nomura H, Oba Y, Ohashi S, Okoda Y, Oya Y, Sakai T, Shirley Y, Testi L, Vastel C, Viti S, Watanabe N, Watanabe Y, Zhang Y, Zhang ZE, Yamamoto S (2024) Astron Astrophys 688:L22
Poole GM, Rehkämper M, Coles BJ, Goldberg T, Smith CL (2017) Earth Planet Sci Lett 473:215
Qin LP, Alexander CMO, Carlson RW, Horan MF, Yokoyama T (2010) Geochim Cosmochim Acta 74:1122
Rauscher T, Dauphas N, Dillmann I, Frohlich C, Fulop Z, Gyurky G (2013) Rep Prog Phys 76
Regelous M, Elliott T, Coath CD (2008) Earth Planet Sci Lett 272:330
Rego ES, Dauphas N, Hopp T (2024) In: 34th annual V. M. Goldschmidt conference, Chicago, Illinois, USA
Renzo M, Zapartas E, De Mink SE, Götberg Y, Justham S, Farmer RJ, Izzard RG, Toonen S, Sana H (2019) Astron Astrophys 624:A66
Reynolds JH, Turner G (1964) Rare gases in the chondrite Renazzo. J Geophys Res 69:3263–3281
Ricci L, Isella A, Andrews SM, Birnstiel T, Cuzzi JN, D'Angelo G, Dong R, Dutrey A, Ercolano B, Estrada PR, Flock M, Li H, Liu S-F, Lyra W, Oberg K, Okuzumi S, Perez L, Turner N, van der Marel N, Wilner D, Youdin AN, Zhu Z (2018)
Rosenthal LJ, Knutson HA, Chachan Y, Dai F, Howard AW, Fulton BJ, Chontos A, Crepp JR, Dalba PA, Henry GW, Kane SR, Petigura EA, Weiss LM, Wright JT (2022) Astrophys J Suppl Ser 262:1
Rotaru M, Birck JL, Allègre CJ (1992) Nature 358:465
Roth ASG, Metzler K, Baumgartner LP, Leya I (2016) Meteorit Planet Sci 51:1256
Russell WA, Papanastassiou DA, Tombrello TA (1978) Geochim Cosmochim Acta 42:1075
Russell SS, Connolly HC Jr, Krot AN (eds) (2018) Chondrules: records of protoplanetary disk processes 1st edn. Cambridge University Press, Cambridge
Sai Insa Choi J, Yen H-W, Ohashi N, Tobin JJ, Jørgensen JK, Takakuwa S, Saigo K, Aso Y, Lin Z-YD, Koch PM, Aikawa Y, Flores C, De Gregorio-Monsalvo I, Han I, Kido M, Kwon W, Lai S-P, Lee CW, Lee J-E, Li Z-Y, Looney LW, Mori S, Phuong NT, Santamaría-Miranda A, Sharma R, Thieme TJ, Tomida K, Williams JP (2023) Astrophys J 954:67

Sakamoto N, Seto Y, Itoh S, Kuramoto K, Fujino K, Nagashima K, Krot AN, Yurimoto H (2007) Science 317:231

Savage PS, Moynier F, Boyet M (2022) Icarus 386:115172

Schiller M, Connelly JN, Glad AC, Mikouchi T, Bizzarro M (2015) Earth Planet Sci Lett 420:45

Schiller M, Bizzarro M, Fernandes VA (2018) Nature 555:507

Schiller M, Bizzarro M, Siebert J (2020) Sci Adv 6:eaay7604

Schneider JM, Burkhardt C, Marrocchi Y, Brennecka GA, Kleine T (2020) Earth Planet Sci Lett 551:116585

Schneider JM, Burkhardt C, Kleine T (2023) Astrophys J Lett 952:L25

Schrader DL, McCoy TJ, Gardner-Vandy K (2017) Geochim Cosmochim Acta 205:295

Scott ERD, Krot AN (2014) 1.2 - chondrites and their components. In: Holland HD, Turekian KK (eds) Treatise on geochemistry, 2nd edn. Elsevier, Oxford, pp 65–137

Scott ERD, Krot AN, Sanders IS (2018) Astrophys J 854:164

Segura-Cox DM, Schmiedeke A, Pineda JE, Stephens IW, Fernández-López M, Looney LW, Caselli P, Li Z-Y, Mundy LG, Kwon W, Harris RJ (2020) Nature 586:228

Sheehan PD, Tobin JJ, Federman S, Megeath ST, Looney LW (2020) Astrophys J 902:141

Sieverding A, Kresse D, Janka H-T (2023) Astrophys J Lett 957:L25

Spitzer F, Burkhardt C, Budde G, Kruijer TS, Morbidelli A, Kleine T (2020) Astrophys J Lett 898:L2

Spitzer F, Burkhardt C, Nimmo F, Kleine T (2021) Earth Planet Sci Lett 576:117211

Spitzer F, Kleine T, Burkhardt C, Hopp T, Yokoyama T, Abe Y, Aléon J, O'D Alexander CM, Amari S, Amelin Y, Bajo K, Bizzarro M, Bouvier A, Carlson RW, Chaussidon M, Choi B-G, Dauphas N, Davis AM, Di Rocco T, Fujiya W, Fukai R, Gautam I, Haba MK, Hibiya Y, Hidaka H, Homma H, Hoppe P, Huss GR, Ichida K, Iizuka T, Ireland TR, Ishikawa A, Itoh S, Kawasaki N, Kita NT, Kitajima K, Komatani S, Krot AN, Liu M-C, Masuda Y, Morita M, Moynier F, Motomura K, Nakai I, Nagashima K, Nguyen A, Nittler L, Onose M, Pack A, Park C, Piani L, Qin L, Russell SS, Sakamoto N, Schönbächler M, Tafla L, Tang H, Terada K, Terada Y, Usui T, Wada S, Wadhwa M, Walker RJ, Yamashita K, Yin Q-Z, Yoneda S, Young ED, Yui H, Zhang A-C, Nakamura T, Naraoka H, Noguchi T, Okazaki R, Sakamoto K, Yabuta H, Abe M, Miyazaki A, Nakato A, Nishimura M, Okada T, Yada T, Yogata K, Nakazawa S, Saiki T, Tanaka S, Terui F, Tsuda Y, Watanabe S, Yoshikawa M, Tachibana S, Yurimoto H (2024) Sci Adv 10:eadp2426

Squire J, Hopkins PF (2018) Mon Not R Astron Soc 477:5011

Stapper LM, Hogerheijde MR, Van Dishoeck EF, Mentel R (2022) Astron Astrophys 658:A112

Steele RCJ, Elliott T, Coath CD, Regelous M (2011) Geochim Cosmochim Acta 75:7906

Steinmeyer M-L, Woitke P, Johansen A (2023) Astron Astrophys 677:A181

Steller T, Burkhardt C, Yang C, Kleine T (2022) Icarus 386:115171

Stephan T, Trappitsch R, Hoppe P, Davis AM, Bose M, Boujibar A, Gyngard F, Hynes KM, Liu N, Nittler LR, Ogliore RC (2024) Astrophys J Suppl Ser 270:27

Sugiura N, Fujiya W (2014) Meteorit Planet Sci 49:772

Svoboda BE, Shirley YL, Traficante A, Battersby C, Fuller GA, Zhang Q, Beuther H, Peretto N, Brogan C, Hunter T (2019) Astrophys J 886:36

Tang M, Anders E (1988) Astrophys J 335:L31

Thiemens MH, Heidenreich JE (1983) Science 219:1073

Thiemens MH, Lin M (2021) Rev Mineral Geochem 86:35

Thompson SE, Coughlin JL, Hoffman K, Mullally F, Christiansen JL, Burke CJ, Bryson S, Batalha N, Haas MR, Catanzarite J, Rowe JF, Barentsen G, Caldwell DA, Clarke BD, Jenkins JM, Li J, Latham DW, Lissauer JJ, Mathur S, Morris RL, Seader SE, Smith JC, Klaus TC, Twicken JD, Van Cleve JE, Wohler B, Akeson R, Ciardi DR, Cochran WD, Henze CE, Howell SB, Huber D, Prša A, Ramírez SV, Morton TD, Barclay T, Campbell JR, Chaplin WJ, Charbonneau D, Christensen-Dalsgaard J, Dotson JL, Doyle L, Dunham EW, Dupree AK, Ford EB, Geary JC, Girouard FR, Isaacson H, Kjeldsen H, Quintana EV, Ragozzine D, Shabram M, Shporer A, Aguirre VS, Steffen JH, Still M, Tenenbaum P, Welsh WF, Wolfgang A, Zamudio KA, Koch DG, Borucki WJ (2018) Astrophys J Suppl Ser 235:38

Tissot FLH, Dauphas N, Grossman L (2016) Sci Adv 2:e1501400

Tobin JJ, Sheehan PD, Megeath ST, Díaz-Rodríguez AK, Offner SSR, Murillo NM, Van MLR, Hoff T, Van Dishoeck EF, Osorio M, Anglada G, Furlan E, Stutz AM, Reynolds N, Karnath N, Fischer WJ, Persson M, Looney LW, Li Z-Y, Stephens I, Chandler CJ, Cox E, Dunham MM, Tychoniec Ł, Kama M, Kratter K, Kounkel M, Mazur B, Maud L, Patel L, Perez L, Sadavoy SI, Segura-Cox D, Sharma R, Stephenson B, Watson DM, Wyrowski F (2020) Astrophys J 890:130

Torrano ZA, Alexander CMO, Carlson RW, Render J, Brennecka GA, Bullock ES (2024) Earth Planet Sci Lett 627:118551

Trinquier A, Birck J, Allegre CJ (2007) Astrophys J 655:1179

Trinquier A, Birck JL, Allegre CJ, Gopel C, Ulfbeck D (2008) Geochim Cosmochim Acta 72:5146

Trinquier A, Elliott T, Ulfbeck D, Coath C, Krot AN, Bizzarro M (2009) Science 324:374

Urey HC (1947) J Chem Soc 562
Ushikubo T, Kimura M, Kita NT, Valley JW (2012) Geochim Cosmochim Acta 90:242
Valdivia-Mena MT, Pineda JE, Caselli P, Segura-Cox DM, Schmiedeke A, Spezzano S, Offner S, Ivlev AV, Kuffmeier M, Cunningham N, Neri R, Maureira MJ (2024) Astron Astrophys 687:A71
van der Marel N, Matthews B, Dong R, Birnstiel T, Isella A (2018) arXiv:1809.06403
van der Marel N, Booth AS, Leemker M, Van Dishoeck EF, Ohashi S (2021) Astron Astrophys 651:L5
Van Kooten E, Moynier F (2019) Geochim Cosmochim Acta 261:248
Van Kooten EMME, Wielandt D, Schiller M, Nagashima K, Thomen A, Larsen KK, Olsen MB, Nordlund Å, Krot AN, Bizzarro M (2016) Proc Natl Acad Sci USA 113:2011
Van Kooten EMME, Nagashima K, Kasama T, Wampfler SF, Ramsey JP, Frimann S, Balogh ZI, Schiller M, Wielandt DP, Franchi IA, Jørgensen JK, Krot AN, Bizzarro M (2017a) Geochim Cosmochim Acta 205:119
Van Kooten EMME, Schiller M, Bizzarro M (2017b) Geochim Cosmochim Acta 208:1
Van Kooten EMME, Moynier F, Agranier A (2019) Proc Natl Acad Sci USA 116:18860
Van Kooten E, Cavalcante L, Wielandt D, Bizzarro M (2020) Meteorit Planet Sci 55:575
Van Kooten E, Schiller M, Moynier F, Johansen A, Haugbølle T, Bizzarro M (2021) Astrophys J 910:70
Van Kooten E, Zhao X, Franchi I, Tung P-Y, Fairclough S, Walmsley J, Onyett I, Schiller M, Bizzarro M (2024) Sci Adv 10:eadp1613
Vollstaedt H, Mezger K, Alibert Y (2020) Carbonaceous chondrites and the condensation of elements from the Solar Nebula. Astrophys J 897:82
Wang H, Weiss BP, Bai X-N, Downey BG, Wang J, Wang J, Suavet C, Fu RR, Zucolotto ME (2017) Science 355:623
Warren PH (2011) Earth Planet Sci Lett 311:93
Wasserburg GJ, Tera F, Papanastassiou DA, Huneke JC (1977) Earth Planet Sci Lett 35:294
Wasserburg GJ, Wimpenny J, Yin Q (2012) Meteorit Planet Sci 47:1980
Watson DM, Leisenring JM, Furlan E, Bohac CJ, Sargent B, Forrest WJ, Calvet N, Hartmann L, Nordhaus JT, Green JD, Kim KH, Sloan GC, Chen CH, Keller LD, d'Alessio P, Najita J, Uchida KI, Houck JR (2009) Astrophys J Suppl Ser 180:84
Wilhelm MJC, Portegies Zwart S, Cournoyer-Cloutier C, Lewis SC, Polak B, Tran A, Mac Low M-M (2023) Mon Not R Astron Soc 520:5331
Williams CD, Sanborn ME, Defouilloy C, Yin Q-Z, Kita NT, Ebel DS, Yamakawa A, Yamashita K (2020) Proc Natl Acad Sci USA 117:23426
Winkler PF, Williams BJ, Reynolds SP, Petre R, Long KS, Katsuda S, Hwang U (2014) Astrophys J 781:65
Woitke P, Drążkowska J, Lammer H, Kadam K, Marigo P (2024) Astron Astrophys 687:A65
Yang L, Ciesla FJ (2012) Meteorit Planet Sci 47:99
Yap TE, Batygin K (2024) Icarus 417:116085
Yap TE, Tissot FLH (2023) Icarus 405:115680
Yen H-W, Gu P-G, Hirano N, Koch PM, Lee C-F, Liu HB, Takakuwa S (2019) Astrophys J 880:69
Yokoyama T, Fukami Y, Okui W, Ito N, Yamazaki H (2015) Earth Planet Sci Lett 416:46
Youdin AN, Goodman J (2005) Astrophys J 620:459
Young ED (2014) Earth Planet Sci Lett 392:16
Young ED, Galy A, Nagahara H (2002) Geochim Cosmochim Acta 66:1095
Yurimoto H, Kuramoto K (2004) Science 305:1763
Zanda B, Lewin E, Humayun M (2018) In: Russell SS, Connolly HC Jr, Krot AN (eds) Chondrules: records of protoplanetary disk processes. Cambridge University Press, Cambridge, pp 122–150
Zhang JJ, Dauphas N, Davis AM, Pourmand A (2011) J Anal At Spectrom 26:2197
Zhang J, Huang S, Davis AM, Dauphas N, Hashimoto A, Jacobsen SB (2014) Geochim Cosmochim Acta 140:365
Zhang B, Chabot NL, Rubin AE (2024) Proc Natl Acad Sci USA 121:e2306995121
Zhu Z, Nelson RP, Dong R, Espaillat C, Hartmann L (2012) Astrophys J 755:6
Zhu Z, Dong R, Stone JM, Rafikov RR (2015) Astrophys J 813:88
Zhu K, Liu J, Moynier F, Qin L, Alexander CMO, He Y (2019) Astrophys J 873:82
Zhu K, Moynier F, Schiller M, Bizzarro M (2020) Astrophys J Lett 894:L26
Zinner E (2014) In: Treatise on geochemistry. Elsevier, Amsterdam, pp 181–213
Zucker C, Goodman AA, Alves J, Bialy S, Foley M, Speagle JS, Großschedl J, Finkbeiner DP, Burkert A, Khimey D, Swiggum C (2022) Nature 601:334

Authors and Affiliations

François L.H. Tissot[1] · Christoph Burkhardt[2] · Aleksandra Kuznetsova[3,4,5] · Andreas Pack[6] · Martin Schiller[7] · Fridolin Spitzer[2] · Elishevah M.M.E. Van Kooten[7] · Teng Ee Yap[1]

✉ F.L.H. Tissot
tissot@caltech.edu

✉ C. Burkhardt
burkhardtc@mps.mpg.de

A. Kuznetsova
akuznetsova@amnh.org; akuznetsova@flatironinstitute.org; aleksandra.kuznetsova@uconn.edu

A. Pack
apack@uni-goettingen.de

M. Schiller
schiller@sund.ku.dk

F. Spitzer
spitzer@mps.mpg.de

E.M.M.E. Van Kooten
elishevah.vankooten@sund.ku.dk

T.E. Yap
tyap@caltech.edu

1 The Isotoparium, Division of Geological and Planetary Sciences, California Institute of Technology, Pasadena, CA 91125, USA

2 Department for Planetary Sciences, Max Planck Institute for Solar System Research, Justus-von-Liebig-Weg 3, 37077 Göttingen, Germany

3 Department of Astrophysics, American Museum of Natural History, 200 Central Park West, New York, NY, 10024, USA

4 Center for Computational Astrophysics, Flatiron Institute, 162 5th Ave, New York, NY, 10010, USA

5 Department of Physics, University of Connecticut, 196 Auditorium Rd, Storrs, CT, 06269, USA

6 Geoscience Center, Department of Geochemistry and Isotope Geology, University of Göttingen, Goldschmidtstraße 1, 37077 Göttingen, Germany

7 Center for Star and Planet Formation, Globe Institute, University of Copenhagen, Øster Voldgade 5-7, 1350 Copenhagen, Denmark

Space Science Reviews (2025) 221:23
https://doi.org/10.1007/s11214-025-01146-w

Solar System Elemental Abundances from the Solar Photosphere and CI-Chondrites

K. Lodders[1] · **M. Bergemann**[2] · **H. Palme**[3]

Received: 16 September 2024 / Accepted: 7 February 2025 / Published online: 24 February 2025

Abstract

Solar photospheric abundances and CI-chondrite compositions are reviewed and updated to obtain representative solar system abundances of the elements and their isotopes. The new photospheric abundances obtained here lead to higher solar metallicity. Full 3D NLTE photospheric analyses are only available for 11 elements. A quality index for analyses is introduced. For several elements, uncertainties remain large. Protosolar mass fractions are H ($X = 0.7060$), He ($Y = 0.2753$), and for metals Li to U ($Z = 0.0187$). The protosolar (C+N)/H agrees within 13% with the ratio for the solar core from the Borexino experiment. Elemental abundances in CI-chondrites were screened by analytical methods, sample sizes, and evaluated using concentration frequency distributions. Aqueously mobile elements (e.g., alkalis, alkaline earths, etc.) often deviate from normal distributions indicating mobilization and/or sequestration into carbonates, phosphates, and sulfates. Revised CI-chondrite abundances of non-volatile elements are similar to earlier estimates. The moderately volatile elements F and Sb are higher than before, as are C, Br and I, whereas the CI-abundances of Hg and N are now significantly lower. The solar system nuclide distribution curves of s-process elements agree within 4% with s-process predictions of Galactic chemical evolution models. P-process nuclide distributions are assessed. No obvious correlation of CI-chondritic to solar elemental abundance ratios with condensation temperatures is observed, nor is there one for ratios of CI-chondrites/solar wind abundances.

Keywords Elemental abundances · Sun · Solar photosphere · Meteorites · Chondrites · CI-chondrites

✉ K. Lodders
lodders@wustl.edu

✉ M. Bergemann
bergemann@mpia-hd.mpg.de

1 Dept of Earth, Environmental, & Planetary Sciences and McDonnell Center for Space Sciences, Washington Univ., St. Louis, MO 63130, USA

2 Max Planck Institute for Astronomy, Heidelberg, Germany

3 Senckenberg Forschungsinstitut und Naturmuseum, Frankfurt, Germany

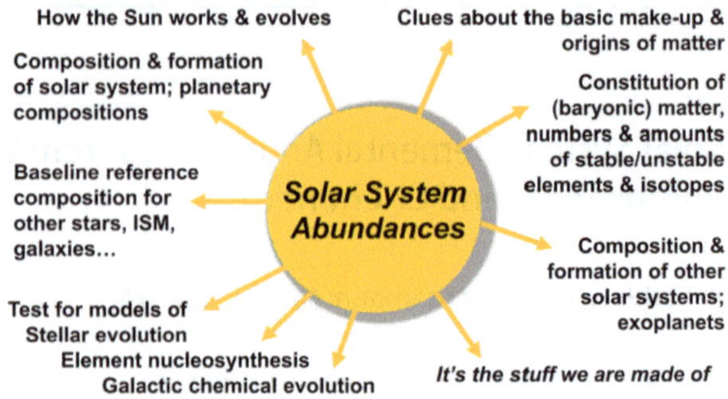

Fig. 1 Why solar system elemental abundances are important

1 Introduction

The solar system or proto-solar elemental abundances are a widely used reference set in astronomy, astrophysics, cosmochemistry, geosciences/Earth sciences and planetary sciences. Figure 1 illustrates the wide-ranging impacts of solar abundances.

The Sun's chemical composition – like that of other stars – is revealed through absorption spectra. Gustav Kirchhoff (1824 – 1887) was the first to recognize that the dark lines in the sunlight spectrum are characteristic of chemical elements in the outer cooler layers of the Sun absorbing radiation from the hotter underlying parts. First attempts to quantify the information contained in stellar and solar absorption lines were made in the early 20th century. Henry N. Russell (1929) published the first comprehensive list for photospheric abundances of 56 elements. Improvements in instrumentation and in interpretation of absorption spectra using increased knowledge of atomic properties and better modeling of the solar photosphere produced increasingly accurate compositional data. Today solar atmospheric abundances can be determined within $\pm 10\%$ to 20% for many elements.

At about the same time Victor M. Goldschmidt and others analyzed meteorites to derive the composition of average solar system matter. Many compilations followed after Goldschmidt's and Russell's lists. The Suess and Urey (1956) compilation was influential for theories of nucleosynthesis (Suess and Urey 1956; Burbidge et al. 1957; Cameron 1957). Later reviews employed improved meteoritic and photospheric analyses and emphasized the excellent agreement of photospheric abundances with abundances of carbonaceous chondrites of the Ivuna-type, the CI-chondrites (Cameron 1973; Anders and Ebihara 1982; Anders and Grevesse 1989; Palme and Beer 1993; Lodders 2003, Asplund et al. 2009, 2021; Lodders et al. 2009; Palme et al. 2014; Lodders 2020 gives a historical review). This and the fact that CI-chondrites are the most volatile element-rich chondrite group are the major arguments for using CI-chondrites as proxy for the condensable elemental abundances including those that currently cannot be determined quantitatively in the Sun. Only by combining solar and meteoritic data (plus some theoretical inputs) a complete set of solar system abundances for all naturally occurring elements and their isotopes is obtained. Sometimes "Solar System Abundances" are called cosmic abundances because of compositional similarities among some G-type stars like the Sun and other dwarf stars (e.g., Fischer and Valenti 2005; Nissen 2015; Bedell et al. 2018).

Table 1 lists sources and limitations for abundance determinations. Meteorite abundances can be determined more precisely than photospheric abundances and are therefore often preferred, however, improvements in spectroscopy and more realistic physical models for the solar atmosphere give increasingly accurate and precise photospheric abundances.

"Solar abundances" means abundances primarily obtained from spectral analyses of the solar photosphere, sunspots, and the corona. There are also in-situ measurements of the solar

Table 1 Sources for Solar System Abundances

Abundances from:	Limitations:
Sun: Photosphere & Sunspots representative of present-day solar convection envelope (CE)	68 of 83 natural occurring elements were analyzed, about 10-20 elements with nominal uncertainties <10%. Limitations are line accessibilities, atomic parameters and transition probabilities. Model atmosphere (1D vs 3D) and choices between local thermodynamic equilibrium (LTE) vs. non-LTE (NLTE) are required. The best solar system source for abundances of C, N, O are currently debated. No direct method for noble gases is available. To obtain proto-solar values from the present-day convective envelope (CE), corrections for atomic diffusion (gravitational settling and radiative acceleration) are needed.
Sun: Corona/solar wind (SW) from GENESIS representative of FIP/FIT biased photospheric values	Genesis provides direct measurements of all noble gases, but other data are limited to abundant elements. To derive photospheric values first ionization potential (FIP) & first ionization time (FIT) corrections from fast and slow SW, and solar energetic particles are required. Coronal sources corrected for FIP/FIT bias from photosphere need further settling and diffusion corrections to obtain proto-solar values (see Lodders 2020 for a review).
CI-Chondrites representative of proto-solar condensable fraction	All elements are measurable. Ultra volatile elements (H, C, N, O, noble gases) are strongly depleted in CI-chondrites. Elements are usually determined with 3-10% relative uncertainties. This requires representative sampling. Limited available material may be a problem.
Other Sources indirect and/or model-dependent	He abundance can be calculated from helioseismology, Ne from O/Ne of solar wind and B stars, Ar, Kr, Xe from nucleosynthesis systematics and abundance curve interpolations. Additional data are provided by B stars, the ISM (interstellar medium), and gas-giant planets in the solar system.

corpuscular radiation (solar wind, SW) by spacecraft (Gloeckler et al. 1998) and laboratory analyses of solar wind implanted in lunar surface materials and returned samples collected by the Genesis mission (e.g., Reisenfeld et al. 2007; Burnett 2011; Pilleri et al. 2015; Heber et al. 2021 and references therein).

"Meteoritic" or "chondritic" abundances are independent sources for bulk solar system abundances from laboratory chemical analyses of CI-chondrites. Their elemental abundances compare most closely to the composition of the solar photosphere for condensable elements, except for ultra-volatile elements H, C, N, O and noble gases which are not fully retained in meteorites and Li which is destroyed in the Sun.

The photospheric spectrum gives the composition of the present-day solar convective envelope (CE). This is not the proto-solar composition at the birth of the solar system (see Lodders 2020 for a review). Over the Sun's lifetime, elements heavier than H diffused ("settled") from the CE into the solar interior. Protosolar abundances are described below.

Currently, noble gas abundances cannot be determined from the photosphere but may be derived from the solar wind. Photospheric C, N, and O abundances are uncertain (see below). Other sources for these abundances are nearby B-type stars. Future entry probes may give atmospheric noble gas data for Saturn, Uranus, and Neptune as the Galileo probe did for Jupiter, but deviations from solar abundances are to be expected. Interpolations using the elemental (or isotopic) abundance curve as a function of atomic number (or mass number) remain useful (e.g., Goldschmidt 1937; Suess 1947a,b), and with modern nucleosynthetic systematics and Galactic chemical evolution models such interpolations give abundance estimates of Kr and Xe. The solar wind measurements from the Genesis mission yielded relative noble gas abundances and data for some other major elements (e.g., Heber et al. 2009, 2012, 2021; Vogel et al. 2011; Pepin et al. 2012; Meshik et al. 2014). The determination of H in the Genesis samples was necessary to compare solar wind to photospheric abundances relative to H (Huss et al. 2020; Heber et al. 2021). Genesis data can give the composition of the solar convection zone (with photosphere on top), but it requires full understanding of elemental fractionations between the photosphere and coronal solar wind sources. Atoms with low ionization energies (FIP < 10 eV) are more abundant in the corona and in solar energetic particles (SEP) than in the photosphere. The solar wind is ultimately derived from coronal sources, and the "FIP-bias" also applies to abundances determined from the Genesis samples.

At least two element fractionation processes must be considered to get the proto-solar or solar system elemental abundances from solar wind analyses: (1) fractionations during settling from the CE and (2) ionization potential and/or first ionization time-driven fractionations between the photosphere and the solar corona. Compositional differences exist between SEPs and the slow and fast solar winds and require model-dependent corrections for fractionations during acceleration of solar wind into different solar wind regimes in order to back-track solar wind to compositions of the CE.

2 Solar Photospheric Analyses

The analysis of the solar photosphere cannot be performed directly. No in-situ experimental probes can reach the photosphere through the extremely hot corona and chromosphere. The only sources for quantitative analysis are solar spectra. These data are accessible with ground-based facilities, such as the Kitt Peak National Observatory (KPNO) Fourier Transform Spectrometer (FTS; Kurucz et al. 1984), FTS of the Institut Astrophysik Göttingen (IAG; Reiners et al. 2016), and the Swedish Solar Telescope (SST; Pietrow et al. 2023a), and from space-based facilities, such as HINODE (Caffau et al. 2015). The positions of absorption lines can be derived from experimental or theoretical atomic and molecular data. For extracting elemental abundances, the depths and detailed shapes (so-called "profiles") of lines are measured for each chemical element. The solar optical spectrum harbors hundreds of thousands of absorption lines (the exact number is unknown) of neutral and singly ionized atoms, and lines of diverse molecules.

The lines are caused by absorption of the outgoing radiation field by the photospheric plasma. The depths and shapes of lines are defined by the complex physical structure of the photosphere, specifically by distributions of gas temperature, density, and gas velocities to a depth of about ~2000 km. The shapes of spectral lines also depend on how photons interact with gas, i.e., how much true absorption or scattering occurs. Proper calculations require detailed models of the solar photosphere (see Nordlund et al. 2009 and references therein). The synthetic spectra calculated from these models are then compared to the observed spectra to find the best model fit. The corresponding synthetic spectral model is then taken as the preferred one and the abundance of the element used to compute it as the representative solar photospheric abundance. A short summary of the currently used methods is given by Bergemann and Serenelli (2014).

Recommended photospheric abundances are listed in Table 2 with a new system flagging the solar abundances (all highly model-dependent quantities) by their accuracy and precision. Flags range from A+ (top, most reliable value, accuracy ~0.05 dex approximately 10%) to E (lowest quality, highly unreliable, accuracy worse than 0.25 dex, approximately a factor of 2). The error parameter "sigma" is partly based on this model-dependent assessment. The sigma here is the fiducial error of the value and is not related to confidence intervals. The criteria for assigning the flags to abundances are as follows:

Table 2 Solar photospheric abundances

Element	This work					Asplund et al. (2021)		
	12+log N(E/H)dex	±1σ dex	1σ%	Quality index	Notes*	12+log N(E/H)dex	±1σ dex	Difference (1) - (6)
E	(1)	(2)	(3)	(4)	(5)	(6)	(7)	(8)
H	12	0.004	0.9	NA	...	12.00	0.00	0.00
He	10.922	0.012	2.8	NA	see text	10.91	0.013	0.01
Li	1.04	0.09	23	B	see text	0.96	0.06	0.08
Be	1.21	0.14	38	C	A24, U	1.38	0.09	0.00
B	2.7	0.25	78	E	CS99, U	2.70	0.20	0.00
C	8.51	0.09	23	B	see text	8.46	0.04	0.05
N	7.98	0.11	29	B-	see text	7.83	0.07	0.15
O	8.76	0.05	12	A	see text	8.69	0.04	0.07
F	4.4	0.2	58	D	sunspot; see text	4.40	0.25	0.00
Ne	8.15	0.12	32	D	see text	8.06	0.05	0.09
Na	6.29	0.05	12	A	see text	6.22	0.03	0.07
Mg	7.58	0.05	12	A	see text	7.55	0.03	0.03
Al	6.43	0.05	12	A	see text	6.43	0.03	0.00
Si	7.56	0.05	12	A	see text	7.51	0.03	0.05
P	5.44	0.12	32	C+	see text	5.41	0.03	0.03
S	7.16	0.11	29	B-	see text	7.12	0.03	0.04
Cl	5.43	0.2	58	D	sunspot; see text	5.31	0.20	0.12
Ar	6.5	0.12	32	D	see text	6.38	0.10	0.12
K	5.09	0.09	23	B	see text	5.07	0.03	0.02
Ca	6.35	0.06	15	A-	see text	6.30	0.03	0.05
Sc	3.13	0.11	29	B-	see text	3.14	0.04	-0.01
Ti	4.97	0.11	29	B-	see text	4.97	0.05	0.00
V	3.89	0.16	45	C-	see text	3.90	0.08	-0.01
Cr	5.74	0.11	29	B-	see text	5.62	0.04	0.12
Mn	5.52	0.05	12	A	see text	5.42	0.06	0.10
Fe	7.51	0.05	12	A	see text	7.46	0.04	0.05
Co	4.95	0.11	29	B-	see text	4.94	0.05	0.01
Ni	6.24	0.06	15	A-	see text	6.20	0.04	0.04
Cu	4.24	0.11	29	B-	see text	4.18	0.05	0.06
Zn	4.55	0.11	29	B-	see text	4.56	0.05	-0.01
Ga	3.02	0.14	38	C	A21	3.02	0.05	0.00
Ge	3.62	0.14	38	C	A21	3.62	0.10	0.00
As	...	...		...	see G15	...	...	...
Se	...	...		...	...	...	...	...
Br	...	...		...	...	...	...	...
Kr	3.31	0.12	3Table 42	D	see text	3.12	0.1	0.19
Rb	2.35	0.11	29	B-	see text	2.32	0.08	0.03
Sr	2.93	0.11	29	B-	see text	2.83	0.06	0.10

Table 2 (*Continued*)

Element	This work					Asplund et al. (2021)		
	12+log N(E/H)dex	±1σdex	1σ%	Quality index	Notes*	12+log N(E/H)dex	±1σdex	Difference (1) - (6)
Y	2.3	0.06	15	A-	see text	2.21	0.05	0.09
Zr	2.68	0.11	29	B-	see text	2.59	0.04	0.09
Nb	1.47	0.14	38	C	A21, U	1.47	0.06	0.00
Mo	1.88	0.16	45	C-	A21, U	1.88	0.09	0.00
Ru	1.75	0.16	45	C-	A21, U	1.75	0.08	0.00
Rh	0.78	0.16	45	C-	A21, U	0.78	0.11	0.00
Pd	1.57	0.16	45	C-	A21, U	1.57	0.1	0.00
Ag	0.96	0.16	45	C-	A21, U	0.96	0.1	0.00
Cd	1.77	0.2	58	D	G15, U, see text	...	...	...
In	0.8	0.2	58	D	Sunspot, see text	0.80	0.2	0.00
Sn	2.02	0.16	45	C-	A21, U	2.02	0.1	0.00
Sb	...	...		...	...	...	...	...
Te	...	...		...	...	...	...	...
I	...	...		...	...	...	...	...
Xe	2.3	0.12	32	D	see text	2.22	0.05	0.08
Cs	...	...		...	...	...	...	...
Ba	2.27	0.06	15	A-	see text	2.27	0.05	0.00
La	1.1	0.16	45	C-	see text	1.11	0.04	-0.01
Ce	1.58	0.16	45	C-	G15, U	1.58	0.04	0.00
Pr	0.75	0.11	29	B-	see text	0.75	0.05	0.00
Nd	1.42	0.16	45	C-	G15, U	1.42	0.04	0.00
Sm	0.95	0.16	45	C-	G15, U	0.95	0.04	0.00
Eu	0.57	0.06	15	A-	see text	0.52	0.04	0.05
Gd	1.08	0.16	45	C-	G15, U	1.08	0.04	0.00
Tb	0.31	0.16	45	C-	G15, U	0.31	0.1	0.00
Dy	1.1	0.16	45	C-	G15, U	1.10	0.04	0.00
Ho	0.48	0.16	45	C-	G15, U	0.48	0.11	0.00
Er	0.93	0.16	45	C-	G15, U	0.93	0.05	0.00
Tm	0.11	0.16	45	C-	G15, U	0.11	0.04	0.00
Yb	0.85	0.16	45	C-	G15, U	0.85	0.11	0.00
Lu	0.1	0.16	45	C-	G15, U	0.10	0.09	0.00
Hf	0.86	0.12	32	C+	see text	0.85	0.05	0.01
Ta	...	...		...	...	...	...	...
W	0.79	0.2	58	D	A21, U	0.79	0.11	0.00
Re	...	...		...	...	...	...	...
Os	1.36	0.14	38	C	see text	1.35	0.12	0.01
Ir	1.42	0.2	58	E	G15, U, see text	...	...	...
Pt	...	...		...	...	...	...	...
Au	0.91	0.2	58	D	A21, U	0.91	0.12	0.00

Table 2 *(Continued)*

Element	This work					Asplund et al. (2021)		
	12+log N(E/H)dex	±1σdex	1σ%	Quality index	Notes*	12+log N(E/H)dex	±1σdex	Difference (1) - (6)
Hg	...	...		...	...	...	...	...
Tl	0.95	0.2	58	D	Sunspot, see text	0.92	0.17	0.03
Pb	1.95	0.2	58	D	A21, U	1.95	0.08	0.00
Bi	...	...		...	...	...	...	...
Th	[0.08]	[0.03]		**	See footnote	0.03	0.1	...
U	...	...		...	...	...	...	...

*Notes: A21: Asplund et al. (2021). A24: Amarsi et al. (2024). CS99: Cunha and Smith (1999). G15: Grevesse et al. (2015). U: uncertainty changed from value given in A21, A24, CS99, or G15. **Th from Caffau et al. (2011). This 3D value is based on a weak and blended line at 401.9 nm, and is only listed for reference here

Error from 0.04 to 0.06 dex:

A+ NLTE based on time-dependent 3D atmosphere models, comprehensive NLTE model atom (complete level system, quantum-mechanical estimates of photo-ionization cross-sections and inelastic X+H collisions), accurate log(gf) values, multiple diagnostic lines in the optical and IR solar spectrum, lack of significant blending, independent consistent estimates; sigma = 0.04 dex.

A NLTE based on time-dependent 3D atmosphere models, minor concerns about the NLTE model atom or atomic data (e.g., incomplete knowledge of collisional cross-sections); other criteria as in [A+], no independent validation; sigma = 0.05 dex.

A- NLTE based on time-dependent horizontally-averaged 3D atmosphere models; other criteria as in [1] but measurement uncertainties due to blending and/or lack of a statistically significant number of clean diagnostic lines (e.g., solar Li), and/or lack of possibility to reliably test the excitation-ionization balance (e.g., Ba, Y, Eu); sigma = 0.06 dex.

Error from 0.07 to 0.11 dex:

B+ as in group A, but more problematic atomic data (e.g., only theoretical oscillator strengths), minor differences between estimates by different groups; sigma = 0.07 dex.

B as in group A, but significant differences reported by independent groups (either in 1D or 3D calculations), despite reliable atomic and molecular data; or a mixture of 3D LTE and 3D NLTE (e.g., C, N); sigma = 0.09 dex.

B- 1D NLTE spectrum synthesis or 3D LTE + 1D NLTE modeling, no validation through direct 3D NLTE calculations; sigma = 0.11 dex.

Error from 0.12 to 0.2 dex:

C+ 1D LTE synthesis + 1D NLTE abundance correction, or direct 3D LTE spectrum synthesis, multiple diagnostic lines available, reliable atomic or molecular data, multiple estimates by different independent groups; sigma = 0.12 dex.

C 3D LTE spectrum synthesis, limited number of lines, uncertain atomic or molecular data; sigma=0.14 dex.

C- 3D LTE spectrum synthesis or 1D LTE synthesis + 3D LTE abundance correction, no independent validation; sigma = 0.16 dex.

D 1D LTE calculations, multiple diagnostic lines available, reliable atomic or molecular data; sigma=0.20 dex; or 1D LTE+3D LTE correction, but heavily blended 1 diagnostic line in the blue (e.g., W) or UV (e.g., Os, Au, Pb).

E 1D LTE calculations, very limited number of diagnostic lines and/or substantial concerns over the quality of atomic or molecular data, any other critical concern as described for individual elements below; sigma at least 0.25 dex.

3 Solar Model Atmospheres

Early compilations of the solar composition derived from absorption line spectra (Anders and Grevesse 1989; Grevesse and Sauval 1998) relied on results from simplified one-dimensional (1D) model atmospheres in hydrostatic equilibrium (HE) and local thermodynamic equilibrium (LTE). One of the main drawbacks of such models – owing to the assumption of HE – is the lack of considering turbulence and convection. Convective energy transport is usually parameterized using the mixing length theory (Böhm-Vitense 1958), whereas velocity fields are represented by an ad-hoc correction to opacity (the so-called "micro-turbulence") and an artificial broadening applied to emergent monochromatic intensities ("macro-turbulence"). These corrections are used in standard 1D HE LTE models, such as MARCS (Gustafsson et al. 2008) and Kurucz (Castelli and Kurucz 2003). In the vast majority of astrophysical research, these models serve as the basis of stellar abundance calculations, for example for the studies of other Sun-like stars.

The Sun has a convective envelope (CE) occupying roughly the outermost 30% of the Sun in radius (e.g. Serenelli et al. 2009). This envelope has a major impact on the thermodynamic structure of the atmosphere, the latter being in comparison a very thin layer of only 0.1% of the Sun. Convection manifests itself observationally through granulation (see Nordlund et al. 2009 and references therein). The scales and properties of sub-surface convection are defined by the Standard Solar Models (SSM, Basu and Antia 2004, 2008, Bahcall et al. 2004; Serenelli et al. 2009). These sophisticated models describe the evolution of the Sun from the pre-main sequence to the present. The present-day interior structure of the Sun including the depth of the convective envelope and the sound speed profile can be probed precisely by several thousands of oscillation modes measured via helioseismology methods (e.g., Christensen-Dalsgaard 2002; Basu and Antia 2008). Early tests for the depth of the solar convection zone compared abundances of Li, Be, and B of the Sun to meteorites. The isotopes of Li are destroyed at temperatures > 2.5 MK whereas Be and B (unclear whether depleted or not) require higher temperatures not attained near the bottom of the convective envelope. Far more advanced testable observables are neutrino fluxes resulting from the pp-chain and CNO cycles (Appel et al. 2022; Basilico et al. 2023). Neutrino fluxes were measured as a function of neutrino energy with the Borexino experiment, including p-p, pep, ^{7}Be, and ^{8}B, the latter two very accurately to 3.5% and 2%, respectively, providing stringent constraints on the structure of the solar interior.

Over the past decade, much work in modeling of the outer structure of the Sun concentrated on 3-dimensional (3D) Radiation-Hydro Dynamics (RHD) models (e.g., Nordlund 1982; Spruit et al. 1990; Vögler et al. 2005; Nordlund et al. 2009; Freytag et al. 2012). These simulations involve solving self-consistently the equations of radiation transfer and time-dependent (magneto)-hydrodynamics, and eliminate the need for ad-hoc user-dependent corrections, which are inherent to 1D HE models. The new generation of models are commonly referred to as "3D models", although the main physical improvement is not in multi-D geometry. Physically exact 2D or 3D replicas of a 1D model can be made. The key difference lies in the thermo-dynamic structure, including velocity fields and temperature-density-pressure inhomogeneities (caused by sub-surface convection), and the loss of radiation at the surface modeled from first principles. The 3D RHD model atmospheres vastly improve the agreement of synthetic observables with various observations of the Sun, including the observed granulation at the solar surface, the time variability and contrast of the granules, and fits to high-resolution solar spectra across the limb.

4 Non-local Thermodynamic Equilibrium (NLTE)

All stellar atmosphere and radiative transfer models, either 1D or 3D, rely heavily on atomic and molecular data. Choices have to be made on parameters such as wavelength, excitation potentials, transition probabilities, damping parameters, ionization and dissociation potentials and cross-sections, hyperfine splitting and electric dipole and magnetic quadrupole constants, partition functions and isotopic structure (in all solar analyses, fixed isotope ratios from studies of meteorites are assumed). In NLTE, further physical quantities play a role, such as rates of transitions in inelastic collisions with free electrons and charge exchange with H atoms (e.g., Barklem 2016; Belyaev et al. 2019), but also completeness of the representation of atomic and molecular systems through the energy structure and transitions between states (e.g., Mashonkina et al. 2011; Bergemann et al. 2012). Calculating lines of molecules in NLTE requires inclusion of photo-dissociation and photo-attachment (e.g., Heays et al. 2017, Hrodmarsson and Van Dishoeck 2023), as well as corresponding collisional destruction and attachment reactions. Most of these atomic and molecular datasets are theoretical and are difficult to verify experimentally, especially for collisional data (Barklem 2016). For example, for O see discussion in Bergemann et al. (2021). Atomic structure calculations rely on various assumptions about the representation of nuclear potentials, electron-electron correlations, relativistic effects, etc. (e.g., Bautista et al. 2000, 2022), and systematic effects in the solar abundance analysis are intricately tied to the quality of atomic data. For molecules the situation is worse. Currently, CH is the only molecule with detailed NLTE abundance predictions for solar atmospheric conditions (Popa et al. 2023).

For discussion of the individual elements, the following precautions apply:

- Solar photospheric abundances are not observed quantities. All methods used for the determination of solar photospheric abundances (1D LTE, 3D LTE, 3D NLTE) require theoretical modeling and depend on choices made regarding the sub-grid physics and the numerical approach. This causes differences among individual estimates of solar abundances by different groups (e.g., Asplund et al. 2009, 2021, Caffau et al. 2012; Magg et al. 2022). Only a few alternative, less model-dependent, methods to derive solar abundances exist (e.g., Socas-Navarro 2015).
- The 3D NLTE is not an objective self-consistent methodology which is uniformly and unambiguously applied by different authors, see examples and references below. The term "3D NLTE" entails a sequence of approximations assumed by different authors. In some 3D NLTE calculations (e.g., Asplund et al. 2009; Klevas et al. 2016; Bergemann et al. 2019; Amarsi et al. 2019), 3D LTE model atmospheres are used to compute 3D LTE or 3D NLTE synthetic spectra, but both with background opacities in LTE. Other calculations use time- and spatially-averaged 3D model atmospheres to compute NLTE synthetic spectra with the improvement that background opacities are handled in NLTE (Magg et al. 2022). Other studies (e.g., Caffau et al. 2011 or Asplund et al. 2021) resort to 3D LTE calculations for most elements, using 1D NLTE calculations by other authors (with 1D or <3D> models, where the brackets represent spatially-averaged models) to inform the line

selection or to correct 3D LTE values by 1D NLTE (see references below). Sometimes, 3D LTE spectra are computed and 1D NLTE corrections are applied during post-processing, with the latter computed using different means and techniques. These abundances are sometimes quoted as '3D, NLTE', '3D + NLTE', '3D - NLTE' (with a comma, a plus, or a minus signs placed in between the two shortcuts), e.g., Caffau et al. 2011 (- sign, their Table 1 for K), Grevesse et al. 2015 (+), and Scott et al. 2015a,b (+ sign, their Table 1).

- The '3D, NLTE' or '3D + NLTE', or 'NLTE corrected 3D' abundances are not necessarily physically better than 1D NLTE, or even 1D LTE. No absolute test exists for the accuracy of the resulting abundances. The precision is usually defined as the statistical scatter (1-standard deviation) between abundances obtained from various spectral lines of an element. Over the years, several diagnostic tests were developed including the excitation balance, ionization balance, line-by-line scatter, and center-to-limb variation tests (Korn et al. 2003; Bergemann et al. 2012; Lind et al. 2017; Pietrow et al. 2023b). All of these are applied to various extents in different studies, but typically a comprehensive analysis of all these tests for consistency is not available.
- All 1D hydrostatic models rely on arbitrary parameters to correct physical structures of models for the absence of gas dynamics (convection and turbulent flows) and for other physical limitations. Some models, such as the semi-empirical 1D LTE Holweger-Mueller (HM) model (Holweger and Müller 1974) used by Asplund et al. (2009, 2021), Grevesse et al. (2015), Scott et al. (2015a,b), have a much steeper temperature and pressure gradient than strictly theoretical 1D models (MARCS, MAFAGS, Kurucz). This model was constructed and tuned to achieve the best fit of solar observations in 1D LTE. NLTE effects obtained with such models can be of opposite sign and amplitude making it difficult to analyze and interpret results of different authors.

5 Beryllium

The Be value is from a 3D NLTE study by Amarsi et al. (2024) which is adopted here with an increased error. The value is based on one Be II in the far-UV. This wavelength regime is very difficult to interpret in terms of abundances, because the radiation field, and hence the UV spectrum, is formed in the chromosphere (Vernazza et al. 1981). Chromosphere is not included in standard 1D or 3D models. The diagnostic line is strongly blended by an unknown feature (Fig. 8 in Amarsi et al. 2024). Therefore, despite the use of 3D NLTE models, the error in the Be abundance could be underestimated because of strong blending, lack of chromosphere in the physical modelling, and lack of independent Be diagnostics that does not allow to verify the measurement. Hence, we recommend A(Be) = 1.21±0.14 with a larger error than that (±0.05) given by Amarsi et al. (2024).

6 Carbon, Nitrogen, and Oxygen

The elements C, N, and O make up around 60-70% by mass of all elements heavier than He and provide most of the opacity in the solar interior. Their abundances also determined the amount of condensable ices, in turn affecting oxidation states in the solar nebula materials and planet compositions (Krot et al. 2000). The amounts of C, O, and particularly the C/O ratio are also important for modeling AGB stars stellar evolution and nucleosynthesis, since the initial amounts assumed in models affect how quickly these evolve to become carbon stars.

Our recommended C abundance is the mean of C I based values from Caffau et al. (2010), Asplund et al. (2021), and Magg et al. (2022). We do not consider C abundances from molecular lines (C_2, CH, CO), because no 1D NLTE or 3D NLTE modeling is available for them and another open and poorly-understood issue is the assumption of chemical equilibrium. Magg et al. (2022) used several optical C I lines and handled self-consistently effects of blends and opacities using average 3D models. In the NLTE study by Amarsi et al. (2019), the IR and far-IR C I lines are preferred and their NLTE effects are estimated to <0.01 dex. However, Caffau et al. (2010) found the largest NLTE effects in the IR. The three optical C I lines selected by Magg et al. (2022) are neither sensitive to NLTE nor to 3D effects, and in agreement with the results of Alexeeva and Mashonkina (2015), when re-normalized to new f-values from Li et al. (2021). Caffau et al. (2010) reported significant differences between C I EWs measured by different authors. The values of Amarsi et al. (2019) and Asplund et al. (2021) are systematically lower compared to other studies, which is also discussed in Ryabchikova et al. (2022) in relation to the diagnostic C_2 and CN features. The differences between these analyses are due to unresolved systematic differences in the methodology (blends, continuum) and/or choice of the solar observational data.

For N, we adopt the values from Magg et al. (2022) corrected for 3D - <3D> difference based on Caffau et al. (2009, here -0.04 dex following their value for the 8683 Å line). This value is based on measurements of two least-blended atomic N I lines in the solar spectrum (8629, 8683 Å) using new theoretical f-values for these lines computed in the same study. Even these atomic N I lines, despite being more reliable, are blended by CN lines. As shown in M22, the result is only different by 0.011 dex, if the solar CO5BOLD or Stagger model is used. The N value by Amarsi et al. (2020) is not used because it relies on empirical rescaling of the strengths of CN features in the diagnostic N I lines. The N value adopted here is in agreement with the 3D N I value of Caffau et al. (2009) within the respective errors of both values. If we use the f-values of Amarsi et al. (2020), which are taken from Tachiev and Fischer (2002), the measurements based on the two N I lines would be in excellent agreement and give A(N) = 7.94 ± 0.017 dex. We retain the larger error, as the value is based on only two blended lines and reliable NLTE modelling of CN is needed to confirm the atomic results.

Our recommended abundance of O is based on the average of 7 values from Steffen et al. (2015; O I 777 nm lines), Caffau et al. (2013, 2015; 630, 636 nm [O I] lines), Cubas Armas et al. (2020; 630 nm line), Bergemann et al. (2021), Asplund et al. (2021), and Magg et al. (2022). We avoid multiple measurements by the same group, except when the group used different spectral indicators (e.g. 777 nm vs 630 nm, as in Caffau et al. and Steffen et al.) or different codes (Turbospectrum vs MULTI3D, as in Magg et al. vs. Bergemann et al.). No molecule-based abundances are included, as NLTE effects are unknown. Ayres et al. (2013) find A(O) = 8.78 ± 0.02 dex based on CO lines. For atomic O lines it is critical to account for 3D NLTE effects. The 1D LTE value of 8.83±0.06 dex from Grevesse and Sauval (1998) was revised to lower values once 3D NLTE calculations became possible. Asplund et al. (2009, 2021) derived the solar A(O) = 8.69 ± 0.04 dex in 3D NLTE. These low abundances received much attention, however, subsequent independent studies did not confirm them (Caffau et al. 2011; Bergemann et al. 2021; Magg et al. 2022)

The solar A(O) = 8.77 ± 0.04 dex from Magg et al. (2022) relies on the new NLTE model atom of O from Bergemann et al. (2021). In the latter paper, both the forbidden [O I] line and the permitted lines of O I were modeled in full 3D NLTE. These authors also used, for the first time, 3D NLTE formation for the critical Ni I blend in the [O I] feature, finding that the LTE assumption for Ni adopted in previous studies (Allende Prieto et al. 2001, Asplund et al. 2004, 2021) is inadequate. The solar 3D NLTE O abundance in Caffau et al. (2008)

is 8.76 ± 0.07 dex, and a further 3D NLTE estimate by the same group is 8.76 ± 0.02 dex (Steffen et al. 2015) from the analysis of center-to-limb variation of the O 777 nm lines. Caffau et al. (2013) noted that the O I forbidden line at 636 yields A(O) of 8.78 ± 0.02 dex. Magg et al. (2022) pointed out that the difference in the O abundance between CO5BOLD and Stagger 3D models does not exceed 0.015 dex. An independent estimate of the solar O abundance was proposed in Socas-Navarro et al. (2015, see also Centeno and Socas-Navarro 2008), who analyzed the polarization (Stokes V) profile of the NLTE-insensitive 630 nm [O I] line. They find A(O) = 8.86 ± 0.03 dex (Centeno and Socas-Navarro 2008), O/Ni = 210 ± 24, and the improved analysis (Cubas Armas et al. 2017, 2020) yields A(O) = 8.80 ± 0.03 dex, where estimates for the granular and intergranular regions amount to 8.83 ± 0.02 and 8.76 ± 0.02 dex, respectively. This approach only weakly depends on models and it yields the same result when different model atmospheres are used.

The recommended abundances from combining measurements by independent groups are:

A(C) = 8.51 ± 0.09 dex
A(N) = 7.94 ± 0.11 dex
A(O) = 8.76 ± 0.05 dex

Here uncertainties primarily reflect limitations of theoretical models. Problems that need to be resolved include:

- Significant difference between the cross-sections for O+H charge transfer reactions by two different groups (Barklem 2018, Belyaev et al. 2019) → effect on the O abundance at the level of 0.07 dex, which is highly significant at the level required for solar abundance diagnostics
- Lack of understanding the NLTE effects in molecules (e.g., OH, C_2, CN, CH, CO, NH lines). Calculations suggest that molecular abundances are under-estimated owing to the overlooked effect of photo-dissociation of molecules (Popa et al. 2023). The systematic bias influences all other species with low photo-dissociation potentials, such as OH, NH, CO. Lines from low-excited states of molecules like CN may be significantly affected. These are essential for reliable modelling of N I optical lines, as these are affected by CN.
- A systematic difference between the 636 and 630 nm O I line (Caffau et al. 2013) affects the O abundance at the level of 0.05 dex
- Photo-ionization cross-sections for Ni I are needed to test the size of NLTE effects on Ni, and consequently how much of an impact this has on the [O I] diagnostics.

7 Alpha-Elements: Mg, Si, Ca, and S

The elements Mg, Si, Ca, and S are mainly produced by successive He nuclei capture (hence "alpha-elements") during hydrostatic (C-, O-, and Si-burning) and explosive nucleosynthesis in massive stars (e.g. Rauscher et al. 2002). For most of these elements, detailed abundance estimates including 3D and/or NLTE effects are available (Alexeeva et al. 2018; Osorio et al. 2015; Bergemann et al. 2017; Asplund et al. 2021; Magg et al. 2022). Abundances of Mg, Si, and Ca were not reported by Caffau et al. (2011).

Solar abundances of Mg and Si are better constrained because of more accurate atomic data and a wealth of lines of different excitation potentials can be utilized. However, contrary to Fe-group elements, no reliable constraints on the abundance can be made using single-ionized species of these elements. Calcium is comparatively accurate, perhaps even more accurate than Mg and Si, because also diagnostic lines of Ca II are available, the f-values are reliable, and Ca I lines of different excitation potentials can be used.

For Mg, we use the averages from Osorio et al. (2015), Bergemann et al. (2017), Asplund et al. (2021), and Magg et al. (2022). Here we do not distinguish between 3D NLTE and <3D> NLTE because the abundances obtained with both approaches agree to better than 0.01 dex (Asplund et al. 2021, Table A1). The effects of NLTE depend on the choice of Mg I lines in the analysis (Bergemann et al. 2017), and the associated uncertainty is at least 0.05 dex. Alexeeva et al. (2018) found an imbalance between Mg I and Mg II, with Mg I lines yielding lower abundances than Mg II. The solar abundance of A(Mg) =7.66 $\pm$ 0.07 dex was reported by Osorio et al. (2015). The results of Asplund et al. (2021) disagree with other, similar studies, also their reported NLTE effects are of opposite sign compared to these studies (e.g., Bergemann et al. 2017; Osorio et al. 2015). It is not clear whether this is due to the line selection or the NLTE model atom employed by Asplund et al. (2021). Mg I lines accessible in optical and near-IR solar high-resolution solar spectra suffer from a strong sensitivity to damping and/or blends (Bergemann et al. 2017).

For Si, our recommended value is based on the mean of Asplund et al. (2021), Magg et al. (2022), and Deshmukh et al. (2022), who reported 7.51 $\pm$ 0.03 dex, 7.59 $\pm$ 0.07 dex, and 7.57 $\pm$ 0.04 dex, respectively. The choice of oscillator strengths seems to be the main source of disagreement between these three estimates. The Asplund et al. study relied on the older source of gf-values for Si I lines, whereas both latter studies used newer laboratory values from Pehlivan Rhodin (2018), and these data were later published in Pehlivan Rhodin et al. (2021). An independent, although unpublished, analysis of a benchmark metal-poor star by C. Sneden (priv. comm) supports the higher quality of the f-values determined by Pehlivan Rhodin et al. (2024). The NLTE effects in optical Si I lines are at the level of $\pm$0.01 dex or less (Bergemann et al. 2013; Amarsi and Asplund 2017). The quality of the only diagnostic Si II line at 6371.37 Å is debated, and discrepant results are obtained by different groups (e.g., Magg et al. 2022; Asplund et al. 2021). Deshmukh et al. (2022) found that the difference between 3D MHD and 3D RHD results is roughly -0.005 dex in abundance for Si I lines and max +0.015 dex for the Si II line. Hence, the magnetic field is not a major source of uncertainty in the solar Si abundance. Line selection also influences the Si abundance. For example, the 1D NLTE value quoted by Mashonkina (2020) would be A(Si) = 7.55, dex if re-normalized to Pehlivan Rhodin et al. (2024), but even higher (A(Si) = 7.60 dex) if the line list is limited to the Si I lines used by Magg et al. (2022). Asplund et al. (2021) did not provide details for their choice of Si lines, hence it is not possible to validate their finding of an ionization imbalance.

For S, only inhomogeneous and rather inconsistent estimates are available and no 3D NLTE analysis has been carried out to date. The 1D LTE estimate of 7.33 $\pm$ 0.11 dex (Grevesse and Sauval 1998) was superseded by Scott et al. (2015a), who found only 7.06 in 1D LTE, but 7.12 $\pm$ 0.03 dex in 3D LTE with a 1D NLTE abundance correction obtained from the ATLAS models using 8 S I lines in the optical and IR. However, the model atom was chosen to produce the most consistent "3D + NLTE" abundances, given 3D LTE values and ad-hoc scaled collisional data (using a so-called Sh scaling factor of 0.4). Another 3D estimate, A(S) = 7.16 $\pm$ 0.05 dex by Caffau et al. (2011). That study also used NLTE corrections based on Korotin (2009) model and pointed out that a significantly higher, A(S) = 7.30 dex can be obtained depending on the choice of diagnostic lines. None of these are self-consistent 3D NLTE analysis of S I lines. In Asplund et al. (2021), the solar S abundance was adopted from Scott et al. (2015a). The quality of the NLTE model of S is under debate, as radiative transitions and quantum-mechanical data for S+H collisions (Belyaev and Voronov 2020) have not yet been integrated into NLTE models and no rigorous analysis of the line formation of S I in the solar atmosphere in 3D NLTE has been undertaken so far.

For Ca, our value is based on averaging three recent Ca estimates (Mashonkina et al. 2017; Asplund et al. 2021; Magg et al. 2022). The 3D NLTE estimate by Asplund et al.

(2021), 6.30 ± 0.03 dex, is lower than the 3D NLTE estimate by Magg et al. 2022, A(Ca) = 6.37 ± 0.05 dex. The latter brackets other independent NLTE estimates, e.g., Mashonkina et al. (2017) with A(Ca) = 6.33 ± 0.06 dex (from Ca I lines) and 6.40 ± 0.05 dex (from Ca II lines), and Osorio et al. (2015) albeit with a larger range of line-by-line Ca abundances. Whereas the latter estimates refer to 1D MARCS models, 3D effects are strictly positive for the diagnostic lines of both ions (when compared to MARCS, Delta (3D -1D) of +0.05 dex for Ca I and +0.02 for Ca II, see Scott et al. 2015a). Thus, the solar Ca abundance is likely higher than the value by Asplund et al. (2021). Calcium results obtained with <3D> NLTE and 3D NLTE models are identical (Scott et al. 2015a, their Table 5). It is therefore expected that our value for the solar Ca abundance is reliable.

We recommend the following abundances of alpha-elements:

A(Mg) = 7.58 ± 0.05 dex
A(Si) = 7.56 ± 0.05 dex
A(S) = 7.16 ± 0.22 dex
A(Ca) = 6.35 ± 0.06 dex

The current uncertainties to be resolved include:

- Mg: Significant scatter remains between different Mg I lines in the solar spectra. The only clean feature is at 5711 Å, other features are very weak and/or blended.
- S: Full 3D NLTE calculations with a comprehensive NLTE model atom are lacking. Quantum-mechanical data for S+H collisions are available (Belyaev and Voronov 2020). These still need to be implemented into the NLTE models. Accurate photo-ionization cross-sections for S I are also missing.
- Ca: Systematic differences exist between abundances derived by different groups, even for Ca II lines that are nearly unaffected by NLTE (Scott et al. 2015a; but see Mashonkina et al. 2017). The line-by-line scatter is small, but different groups differ over 30% in the Ca abundance using the same gf-values and same models.
- Si: The disagreement of the f-values of Si I and Si II lines, specifically the values from Garz (1973) lead to systematically lower abundances of Si in Asplund et al. (2021) compared to using laboratory values from Pehlivan Rhodin et al. (2021) which are preferred in recent studies (Magg et al. 2022; Deshmukh et al. 2022).

8 Low-Charge Odd-Z Elements: Na, Al, P, K, and Sc

The solar abundances of low-charge elements, Na, Al, P, K, and Sc are still under debate. The Na abundances obtained by Zhao et al. (2016) and Asplund et al. (2021) differ by 20% for the two diagnostic Na I lines in common (at 6145, 6160 Å). The values of the latter group are 0.1 dex lower compared to Zhao et al. (2016), and the difference exceeds the line-by scatter (of 0.02) and is five times higher than the error of the atomic data for the Na I transitions. The recommended NLTE value is adopted from Zhao et al. (2016). It was derived using QM data for Na+H collisions. The total error is increased to 0.05 dex to reflect the unexplained mismatch in the estimates by the two groups. Contrary to the statement by Asplund et al. (2021), the NLTE effects in Na do not change significantly depending on the NLTE model atom. This can be seen by comparing the Zhao et al. (2016) value of the NLTE correction of -0.04 dex with the Asplund et al. (2021) value of the NLTE correction of -0.045 dex for the 6145 and 6160 Å lines in common between the two independent studies.

The solar 3D NLTE Al abundance by Nordlander and Lind (2017) used the same models and the same NLTE code as Asplund et al. (2021). This value is identical to the solar 3D NLTE value by Scott et al. (2015a) based on Al I lines, A(Al) = 6.43 ± 0.04 dex, although the uncertainty is slightly smaller. This value is adopted here. The 1D NLTE estimate of the Al abundance is only 0.03 dex lower (Scott et al. 2015a, based on MARCS). Gehren et al. (2004) computed the Al abundance in 1D NLTE, A(Al) = 6.43 dex, although they relied on a semi-empirical NLTE model atom, with a scaling factor to Al+H collisions computed using the Drawin's formula. They note the possibility of a systematic error in the van der Waals damping constants for the diagnostic Al I transitions.

The solar abundance of P remains uncertain, owing primarily to the lack of a reliable NLTE model atom and limited quality of the diagnostic lines. The earlier 1D LTE estimate by Grevesse and Sauval (1998), A(P) = 5.45 ± 0.04 dex, is higher than the 1D LTE estimate by Scott et al. (2015a). However, the solar 3D LTE P abundance of 5.46 ± 0.04 by Caffau et al. (2011) is almost identical to that of Grevesse and Sauval (1998). The value from Scott et al. (2015a) is 5.41 ± 0.03 and it relies on eight very weak P I lines in the near-IR. Similar to S I, the NLTE effects in the P I lines are likely significant and detailed NLTE modeling is needed. According to Scott et al. (2015a), 3D effects increase the P abundance by $\sim$+0.03 dex, compared to the MARCS 1D LTE result. Our value, A(P) = 5.44 ± 0.10 dex, is based on the average of C11 and S15, and the error is set to 0.10 dex due to the lack of knowledge of NLTE effects in P I lines.

For K, several estimates are available in the literature, employing different model atmospheres, atomic data, NLTE models, and sources of solar observations. The dispersion between these values is still significant, considering uncertainties tabulated by different authors. In 1D NLTE, the estimates range from 5.02 (Asplund et al. 2021) to 5.11 ± 0.01dex (Reggiani et al. 2019), although both studies used the same NLTE model atom. The latter value is corroborated by an independent analysis of Zhang et al. 2006 (5.12 ± 0.03 dex), while the value by Caffau et al. (2011) falls in the middle of these, 5.06 ± 0.04 dex. In 3D LTE, the K abundances are even more discrepant, e.g., Caffau et al. (2011) estimate A(K) of 5.26 dex, whereas the 3D LTE value by Asplund et al. (2021) is 5.12 dex. Mixed (3D LTE + <3D> NLTE) values are primarily within the range of the published 1D NLTE values. The average of 3D NLTE value from Asplund et al. (2021) and 3D NLTE corrected value from Caffau et al. (2011) is adopted here, with a conservative error due to lack of knowledge on the accuracy of atomic data, and discrepant values obtained with similar 3D and 1D model atmospheres by different groups.

The solar Sc abundance derived by different groups differs despite adopting similar models and line selections. In 1D LTE, the abundance estimates range from 2.90 ± 0.09 dex (Zhang et al. 2008) for Sc I lines to 3.21 ± 0.046 dex for Sc II (Scott et al. 2015b, the average and standard deviation are based on Table 1 in their Appendix, the Holweger-Mueller HM model in LTE). In LTE, systemically lower abundances from Sc I lines were reported (Zhang et al. 2008), however Lawler et al. (2019) obtained a perfect ionization balance for Sc I and Sc II in LTE, with A(Sc) = 3.15 ± 0.06 and 3.16 ± 0.01 dex, respectively, using the HM model. Lawler et al. (2019) comment on the unresolved systematic difference with Asplund et al. (2009) and Scott et al. (2015a) using the same HM model atmosphere. All studies consistently derive a smaller error based on Sc II lines compared to Sc I. Mashonkina and Romanovskaya (2022) analyzed 17 optical Sc II lines using experimental f-values and found A(Sc) = 3.12 ± 0.04 dex in NLTE and 3.14 = 3.12 ± 0.05 dex in LTE. These values are consistent with NLTE estimates from Zhang et al. 2008 (3.07 ± 0.04 dex for Sc II lines). The values from Scott et al. (2015a,b) were obtained by co-adding the 3D LTE abundances with 1D NLTE corrections computed using the 1D LTE MAFAGS model atmosphere (Zhang et al. 2008), but this hybrid approximation for Sc has not been validated. Renormalizing the Scott et al. (2015b) values to the new experimental log(gf) values from Lawler et al. (2019), we obtain a smaller dispersion of line-by-line abundances for Sc II.

The less NLTE sensitive Sc II lines are preferred because substantially more work on the atomic data has been performed (Pehlivan Rhodin et al. 2017; Lawler et al. 2019). No direct estimates of 3D NLTE Sc abundances are published yet.

Our recommended abundances are:
A(Na) = 6.29 ± 0.05 dex
A(Al) = 6.43 ± 0.05 dex
A(P) = 5.44 ± 0.12 dex
A(K) = 5.09 ± 0.09 dex
A(Sc) = 3.13 ± 0.11 dex

9 Fe-Peak Elements: Ti, V, Cr, Mn, Fe, Co, Ni

The iron-peak group encompasses Ti, V, Cr, Mn, Fe, Co, and Ni produced by hydrostatic Si-burning in massive stars and during explosive nucleosynthesis in SN Ia and SN II. The abundance of Fe has been extensively investigated over the past decade. Asplund et al. (2009) were among the first to determine the solar Fe abundance in 3D. Their estimate, based on 3D LTE modeling of Fe lines plus a correction for NLTE effects using a <3D> model with an unpublished model atom, is 7.50 ± 0.04 dex. Scott et al. (2015b) recommended A(Fe) = 7.47 ± 0.04 dex using a similar approach, but note a positive difference between the lines of two ionization stages, with Fe II yielding ~0.05 dex higher abundances compared to the lines of Fe I. Lind et al. (2017) used the list of Fe lines from Scott et al. (2015a,b) and a new model atom of Fe based on the QM collisional data for Fe+H collisions from Barklem 2016. They tested the center-to-limb variation (CLV) of Fe I lines across the solar disc and found the solar Fe abundance of 7.48 ± 0.04 dex.

In Caffau et al. (2011), the Fe abundance is based on the 3D LTE analysis of 15 Fe II lines with the CO5BOLD 3D model. Their estimate is A(Fe) = 7.52 ± 0.06 dex. They found the central Fe abundance is immune to the choice of transition probabilities, but the error and line-by-line scatter is sensitive to the source of f-values. The solar Fe abundance by Asplund et al. (2021), A(Fe) = 7.46 ± 0.04 rests upon the 3D NLTE calculations with the Stagger model. Sitnova et al. (2015) and Mashonkina et al. (2020) found A(Fe) = 7.54 dex from a 1D NLTE analysis. The <3D> NLTE value by Bergemann et al. (2012), based on the analysis of over 50 Fe I and Fe II lines, is 7.46 ± 0.06 dex, fully consistent with 1D NLTE. Magg et al. (2022) employed an updated NLTE model atom of Fe, as well as average CO5BOLD and Stagger models atmospheres and obtained A(Fe)= 7.50 ± 0.06 dex. 3D effects are not significant for the solar Fe abundance. Different groups arrive at different conclusions using very similar solar atmospheric models and line formation methods. The value adopted here is based on the average of Mashonkina et al. (2011), Caffau et al. (2011), Lind et al. (2017), Asplund et al. (2021), and Magg et al. (2022). The error primarily reflects the line-by-line scatter, and the residual uncertainties of the gf-values, damping, unresolved blends, and problems with continuum normalization of the data.

Among other Fe-group elements, 3D NLTE estimates are only available for Mn (Bergemann et al. 2019) and in <3D> for Ni (Magg et al. 2022). We do not consider the Mn and Co values from Asplund et al. (2021), because they used the outdated Mn and Co model atoms from Bergemann and Gehren (2007), which rely on incomplete collisional and radiative data. Updated models of Mn and Co were presented in Bergemann et al. (2019) and in Yakovleva et al. (2020). For Co and Cr, we use the 1D NLTE estimates from Bergemann et al. (2010) and Bergemann and Cescutti (2010), respectively. We increase the total uncertainty to 0.11 dex for Cr and Co to account for the absence of full 3D NLTE calculations.

Our recommended values for Fe-group elements are:
A(Ti) = 4.97 ± 0.11 dex
A(V) = 3.89 ± 0.16 dex
A(Cr) = 5.74 ± 0.11 dex
A(Mn) = 5.52 ± 0.05 dex
A(Fe) = 7.51 ± 0.05 dex
A(Co) = 4.95 ± 0.11 dex
A(Ni) = 6.24 ± 0.06 dex

10 Copper and Zinc

Neither of the two elements were previously considered in full 3D NLTE. For Cu, we recommend the 1D NLTE value from Shi et al. (2014), which was obtained using new radiative transition probabilities computed using the method by Liu et al. (2011). The NLTE value should still be taken with caution as no detailed quantum-mechanical data for Cu+H collisions are integrated into the model atom. The solar Cu abundance from Asplund et al. (2021) was adopted from Grevesse et al. (2015) and it relies on old f-values for Cu I lines from Kock and Richter (1968). As demonstrated in Shi et al. (2014), these f-values lead to a significantly lower Cu abundance, an excitation imbalance, and a larger line-to-line scatter. A comprehensive discussion of the problems of the Kock and Richter (1968) atomic data is given in Shi et al. (2014).

For Zn, the careful NLTE analysis by Sitnova et al. (2022) is preferred here over the value by Grevesse et al. (2015). The improvements in Sitnova et al. (2022) include the high quality of atomic data, NLTE models including quantum-mechanical collisional data for Zn+H and Zn+e data, and the use of f-values from measurements of Roederer and Lawler (2012). The value A(Zn) = 4.56 ± 0.05 dex in Grevesse et al. (2015) applies a NLTE correction that was computed using a model atom lacking realistic cross-sections for collisional and radiative transitions. We did not find a new estimate of the solar Zn abundance in Asplund et al. (2021). The 1D LTE estimates of both elements are systematically lower in Grevesse et al. (2015) compared to other studies mentioned above. We adopt the NLTE value from Sitnova et al. (2022) but increase the uncertainty to 0.11 dex owing to the lack of 3D NLTE calculations.
A(Cu) = 4.24 ± 0.11 dex
A(Zn) = 4.55 ± 0.11 dex

11 Neutron-Capture (Trans-Fe) Elements

Limited progress exists in improving precision and accuracy in abundance determinations for elements beyond the Fe-peak. For several elements, new f-values were determined via laboratory experiments and/or theoretical calculations. The 3D NLTE calculations are available for a few elements (Ba, Y, and Eu).

For the majority of neutron-capture elements only 3D LTE, 1D LTE and 1D NLTE calculations have been performed. The solar abundance of Rb analyzed by Korotin (2020) is 2.47 ± 0.05 dex in 1D LTE, based on both Rb I lines, which is similar to the LTE value of Grevesse et al. (2015). However, the NLTE abundance of Rb is 2.35 ± 0.05 dex, over 0.1 dex lower compared to LTE (Korotin 2020). 3D LTE estimates range from 2.47 ± 0.07 dex in Grevesse et al. (2015) to 2.44 ± 0.08 dex in Asplund et al. (2021). Mixed 3D LTE and

1D NLTE estimates are somewhat divergent, and show a large uncertainty, exceeding that of 1D NLTE values. Our recommended 1D NLTE value is from Korotin (2020), as the effect of 3D is very small (~0.005 dex). We raised the total uncertainty to 0.11 dex to account for the sparse number of diagnostic Rb I lines and the lack of full 3D NLTE estimates.

For Sr, we use the estimate from Bergemann et al. (2012), who employed NLTE calculations with a realistic model atom of Sr to provide the solar Sr abundance based on Sr I and Sr II lines. Abundances derived from lines of both ionization potentials are consistent, with the internal precision error of 0.04 dex. The NLTE effects are consistent with those obtained by Mashonkina and Gehren (2001). The value recommended by Grevesse et al. (2015) although referred to as "3D + NLTE" is based on LTE modeling with the averaged <3D> model atmosphere and co-added with a 1D NLTE correction. Due to a significant difference of 0.05 dex between their estimates for both ionization stages, their abundance is not used here. No 3D NLTE calculations for Sr have been performed yet.

For Y and Eu, we adopt the new full 3D NLTE estimates from Storm et al. (2024) who present calculations with novel quantum-mechanical atomic data for Y+H and Eu+H collisional processes. The 3D NLTE values for Y are higher than previously available 3D LTE estimates from Grevesse et al. 2015 (2.21$\pm$0.05 dex). The latter value relies on oscillator strengths from Hannaford et al. (1982), whereas the value from Storm et al. (2024) is based on new laboratory lifetime and branching fraction measurements from Palmeri et al. (2017), which is an update of Biemont et al. (2012). In LTE, the 3D - 1D differences for the diagnostic Y II lines are positive ranging from +0.03 dex to +0.07 dex, depending on the line, supporting the results from Grevesse et al. (2015) (their 3D - MARCS differences). For Eu II, our results are larger than the estimate in Grevesse et al. (2015). This is due to the full 3D NLTE radiative transfer (positive NLTE effects in Eu II, consistent with Mashonkina and Gehren 2000) as well as self-consistent 3D treatment of blending features (Si and Cr). In LTE, the 3D - 1D estimate corrections for Eu are fully consistent with the results of Mucciarelli et al. (2008).

The abundance of Zr is the NLTE-corrected average of the 3D LTE estimates from Grevesse et al. (2015) and Caffau et al. (2011). The large ionization imbalance for Zr I and Zr II in Grevesse et al. (2015) is likely due to neglect of NLTE effects. Velichko et al. (2010) found in NLTE A(Zr) = 2.63 $\pm$ 0.07 dex and consistent Zr abundances based on both ionization stages. We use their NLTE correction of +0.078 dex (Velichko et al. 2010, their Table 3, kh=0.1 recommended) for the Zr II lines representative of the selection in Grevesse et al. (2015). We find that Grevesse et al. unjustly criticize the selection of lines by Caffau et al. (2011). The recommended Zr results of Grevesse et al. (2015) are fully based on highly blended and strong Zr II lines in the UV and near-UV (see Table 1 in the Appendix of Grevesse et al. (2015) for the list of Zr II lines, and discussion in Velichko et al. 2010). We increased the error to 0.11 dex to account for the absence of self-consistent <3D> NLTE or 3D NLTE calculations for Zr lines.

For Ba, the 3D NLTE values are directly from calculations of Gallagher et al. (2020), who used the up-to-date atomic model of Ba based on quantum-mechanical Ba+H data. These values were also adopted in Asplund et al. (2021) and are higher than the 3D LTE+1D NLTE estimates from Grevesse et al. (2015).

No estimates of NLTE abundances are available for La. The La abundance (1.13 $\pm$ 0.03 dex) in Lawler et al. (2001) is based on the 1D LTE analysis of 14 near-UV and optical La II lines. We adopt the 1D value from Lawler et al. (2001) and correct it for the 3D - 1D (HM) differences (-0.03 dex) based on the estimates from Grevesse et al. (2015). We increased the error to 0.16 dex to account for the lack of 1D NLTE and 3D NLTE calculations.

A(Rb) = 2.35 $\pm$ 0.11 dex

A(Sr) = 2.93 ± 0.11 dex
A(Y) = 2.30 ± 0.06 dex
A(Zr) = 2.68 ± 0.11 dex
A(Ba) = 2.27 ± 0.06 dex
A(La) = 1.10 ± 0.16 dex
A(Eu) = 0.57 ± 0.06 dex

No new measurements are available for the solar abundances of Ga, Ge, Ce, Pr, Nd, Sm, Gd, Tb, Dy, Lu, and other n-capture elements. Of this group, the only elements with NLTE estimates of abundances in 1D available are Sr, Y, Zr, Ba, Pr, Eu, Gd, and Nd.

12 Solar Helium Abundance and Present-Day Mass Fractions of H (X), He (Y), and the Heavy Elements Li-U (Z)

The mass fractions of the elements in the photosphere or present-day solar system matter (from meteoritic and solar data) requires the knowledge of the He abundance. As described in Basu and Antia (2004, 2008) the H and He mass fractions (called X and Y, respectively) are constrained by helioseismology. The analysis depends on the ratio Z/X, which is the mass fraction Z of heavy elements from Li to U, relative to the mass fraction of H.

Basu and Antia (2004) used two model compositions for calibration with two different Z/X ratios (Z/X(mix1) = 0.0171, and Z/X(mix2) = 0.0218) and found Y is more dependent on Z/X than X. Their analyses of data from the Global Oscillation Network Group (GONG) gave X(GONG, mix1) = 0.7392 ±0.0034 and from MDI data X(MDI,mix1) = (0.7385 ±0.0034) averaging to X(mix1) = 0.7389 ±0.0048. Similarly, for mix2 the average is X(mix2) = 0.7390 ±0.0048 (assuming that for mix2, the individual uncertainties on X for GONG and MDI are the same as given for mix1, ±0.0034).

Both calibration mixtures give approximately the same value of X = 0.7389 ±0.0068. Thus, the H mass fraction from helioseismology is relatively insensitive to the mass fraction of heavy elements and is essentially constant for Z/X from 0.0171 to 0.0218 in the calibration models (Basu and Antia 2004). We adopted X = 0.7389 ±0.0068 for the present-day solar convective envelope. However, the mass fractions for He by Basu and Antia (2004) were slightly different: the averages from GONG and MDI gave Y(mix1) = 0.2485 ± 0.048 and similarly, Y(mix2) = 0.2449 ± 0.048.

The Z/X can be computed without knowledge of the He abundance from the atomic abundances by multiplying the atomic elemental abundances for elements heavier than He (atomic number z>2; here lower-case z is used to distinguish it from mass fraction of He usually written as capitalized Z) with the appropriate atomic masses:

$$\frac{\mathrm{Z}}{\mathrm{X}} = \frac{\sum_{\mathrm{z=Li-U}} \varepsilon_{\mathrm{z}} \mathrm{atwt}_{\mathrm{z}}}{\mathrm{atwt(H)}}$$

Here ε is the abundance relative to H; $\varepsilon_{z>2} = N_{z>2}/N_H$. Atomic weights (atwt) of the elements are computed using the masses of the isotopes (Wang et al. 2021; appendix Table A13) weighted by the relative isotopic composition for each element because the atomic weights can be different for solar material than for terrestrial matter usually tabulated (e.g., H, O, Ar, see Lodders 2020, 2021). We obtain $Z/X_{\mathrm{photo}} = 0.0217(\pm 8\%)$ for the photosphere, and $Z/X_{\mathrm{ss\ present}} = 0.0216(\pm 8\%)$ for the present-day "solar system". The small difference comes from the combination of meteoritic and photospheric data in the latter dataset.

The He mass fraction is from the relations Y/X = 1/X – 1 – Z/X and X + Y + Z = 1. This yields Y_{photo}= 0.2451 ±0.0069 = $Y_{\mathrm{ss\ present}}$ where the uncertainty is taken to be the same as for X as is in Basu and Antia (2004, 2008). The largest uncertainty (8%) is in Z/X from the combined uncertainties of the heavy elements in Z, of which O, Ne, C, N alone constitute 79%.

The mass fraction for the solar convection zone, taken as present-day solar system values, are X = 0.7389 ±0.0068 (±0.9%), Y = 0.2451 ±0.0069 (±2.8%), and Z = 0.0160 ±0.0013 (±8%); see below for protosolar mass fractions.

Using the mass fractions X and Y and corresponding atomic weights, the atomic He abundance is $\varepsilon_{He} = N_{He}/N_H = 0.0836 \pm 0.0025$ (2.8%) and A(He) = 12+log ε_{He} = 10.922±0.012 (2.8%).

The photospheric and present-day solar system Z/X ratios are essentially those of mix2 in the calibration models by Basu and Antia (2004), hence the mix2 calibration and results are more relevant. The approach adopting a constant X instead of Y in the mass balance equations gives the same Y (0.2450) as the average of GONG & MDI for mix2 (Y= 0.2449) and maintains proper mass balance when estimating the He abundance if Z/X is between the calibration values. A priori we did not know how the Z/X would come out. The frequently used Y = 0.2485 from the mix1 calibration cannot be used with our current Z/X, as this leads to X = 0.7355, clearly different from X in any of the calibration mixtures (mix1 and mix2) in Basu and Antia (2004, 2008).

13 Solar Noble Gases: Ne, Ar, Kr, Xe

Noble gas abundances are summarized in Table 3. Helium is discussed in the previous section. The Ne and Ar are derived from elemental abundance ratios with other elements from photospheric and meteoritic data. Krypton and Xe from isotope systematics.

The neon abundance, A(Ne) = 8.15±0.12 dex is calculated as in Lodders (2020) and Magg et al. (2022) using Ne/O = 0.244±0.05 from Young (2018) for the solar quiet transition region and the recommended photospheric O abundance. For Ar, we adopt 6.50±0.10 dex proposed in Lodders (2008). This Ar abundance is somewhat low when compared to A(Ar) = 6.56 dex from an interpolated estimate for Ar made with the semi-statistical equilibrium method described by Cameron (1973) and our recommended Si and Ca solar system abundances both from averaged scaled meteoritic and photospheric values. The A(Ar) = 6.5 dex adopted here is, however, high in comparison to other values around 6.4 dex based on solar wind values (see below). The photospheric values used here are also often higher than in Asplund et al. (2021), involving higher O and Ca values in the ratios for estimating Ne (from Ne/O) and Ar (from Ca/Ar) also yield higher Ne and Ar than theirs. The higher metallicity, also seen from different scale coupling factors (SCF, see below) also leads to higher absolute isotopic solar system abundances, and noble gas abundances estimates are higher when nuclear systematics involving other elements are used to estimate them.

Krypton and Xe are from interpolation of s-process nuclide abundances of neighboring elements using the s-process model and Galactic chemical evolution yields from Prantzos et al. (2020) which are within 3-4% of the exact match; see also section on isotopes and Figs. 15,16 below why the interpolation result is preferred. Here Kr/Xe =10.2 is consistent with the elemental Kr/Xe =10.45 on Jupiter (Mahaffy et al. 2000), which should represent the solar and proto-solar ratio, and should be compared to the lower solar wind Kr/Xe of about five.

Huss et al. (2020) derived Ne (8.060± 0.033 dex) and Ar (6.38±0.12 dex) abundances from Ne/He and Ar/He correlations with the respective He/H ratios in the four solar wind regimes as captured in the targets of the Genesis mission. Their values were essentially adopted by Asplund et al. (2021). For neon from the Ne/O ratio by Young (2018), the assumed lower O abundance from Asplund et al. (2021) also results in A(Ne) = 8.06 dex, which we believe is too low.

The Ne and Ar abundances by Huss et al. (2020) are around 25% lower than our values. For their fits, they used a solar H/He of 11.90 ± 0.17 (Basu and Antia 2008), somewhat lower than the H/He = 11.976 used here. Huss et al. note that their "approach to estimating the Ne and Ar abundances in the Sun is model independent", but it depends on the adopted photospheric He abundance (H/He ratio) which in turn is sensitive to the amounts of heavy elements, including the more abundant O and Ne. The He/H ratio used for fitting the Ne/H thus depends on the O and Ne abundances used in the He determination. Our He abundance (10.9217 dex) is for a heavy element mass fraction that includes Ne as A(Ne) = 8.15 dex; the He abundance would slightly increase to A(He) = 10.9225 dex if A(Ne) = 8.05 dex were adopted.

The fit method by Huss et al. (2020) is intriguing but seems to have some problems. Among the four regimes (bulk solar wind (their B/C), interstream wind (L), coronal hole wind (H), and coronal mass ejections (CME, their E) used for fits in their Fig. 19, the coronal hole wind (H) should be the least fractionated material from the photosphere (Huss et al. 2020). Coronal hole, interstream, and bulk solar wind plot relatively close together compared to the CME, which is the most fractionated component among the wind regimes and plots as lowest Ne/He, Ar/He, and H/He. If the coronal hole wind is the least fractionated, then photospheric values should fall near that end of the correlation, however, their derived photospheric Ne and Ar values fall at the opposite end beyond the CME values. This seems counterintuitive to expectations from solar wind fractionations of photospheric source compositions.

The Kr and Xe abundances by Meshik et al. (2020) and Asplund et al. (2021) are lower than our recommended values. Their proposed solar Kr only accounts for about 70% of the expected pure s-process nuclides ^{80}Kr and ^{82}Kr, and 84% of pure s-process ^{128}Xe and ^{130}Xe from stellar models by Prantzos et al. (2020), which is used here to obtain Kr and Xe abundances by interpolation (see isotope section below). Asplund et al. (2021) derived Xe as done in Lodders (2003) using the s-process nuclide cross-sections for Xe measured by Reifarth et al. (2002) and scaling to their ^{150}Sm abundance to obtain A(Xe) = 2.22 ± 0.10 dex. Using this procedure with the higher ^{150}Sm here gives A(Xe) = 2.297 dex, mainly because of the higher solar metallicity and larger scale-coupling factor (1.551) here than 1.51 in Asplund et al. (2021).

Meshik et al. (2020) used the "σN-curve" approach for estimating photospheric abundances of Kr and Xe (Table 3). This classical model approximation remains useful to estimate abundances in mass regions with about constant σNs between magic neutron numbers, but also has limitations (see below).

14 Meteorites and the Significance of CI-Chondrites

There are two types of meteorites, differentiated and undifferentiated ones. Differentiated meteorites are derived from once melted planetesimals, while undifferentiated meteorites, such as chondrites, never were heated to melting temperatures. They represent aggregates of primary solar system material. Their first-order uniform composition approximates the average composition of the Solar System, except for ultra-volatile elements. However, small variations in their elemental compositions divide them into different chondrite sub-groups (e.g., Krot et al. 2014; Scott and Krot 2014). These compositions reflect processing in the

Table 3 Noble Gas Abundances

	Solar A(X), dex recommended	Solar A(X), dex other authors	Reference	Bulk Solar Wind, A(X) dex	Reference
He	10.922 ± 0.012	10.914 ± 0.013	Asplund et al. (2021)	10.706 ± 0.02	Heber et al. (2021)
		10.924	Huss et al. (2020)		Huss et al. (2020)
		10.914 ± 0.05	Meshik et al. (2020)		
Ne	8.15 ± 0.12	8.06 ± 0.05	Asplund et al. (2021)	7.920 ± 0.004	Heber et al. (2021)
		8.060 ± 0.033	Huss et al. (2020)	7.92 ± 0.02	Meshik et al. (2020)
		8.06 ± 0.05	Meshik et al. (2020)		
		8.15 ± 0.11	Magg et al. (2022)		
		8.15 ± 010	Young (2018) using O from Caffau et al. (2011)		
Ar	6.50 ± 0.12	6.38 ± 0.10	Asplund et al. (2021)	6.337 ± 0.004	Heber et al. (2021)
		6.38 ± 0.12	Huss et al. (2020)	6.34 ± 0.02	Meshik et al. (2020)
		6.38 ± 0.08	Meshik et al. (2020)		
Kr	3.31 ± 0.12	3.12 ± 0.10	Asplund et al. (2021)	3.132 ± 0.009	Heber et al. (2021)
		3.24 ± 0.06	Meshik et al. (2020)	3.12 ± 0.02	Meshik et al. (2020)
Xe	2.30 ± 0.12	2.22 ± 0.05	Asplund et al. (2021)	2.474 ± 0.025	Heber et al. (2021)
		2.26 ± 0.06	Meshik et al. (2020)	2.42 ± 0.04	Meshik et al. (2020)

solar nebula prior to accretion, such as incomplete condensation, evaporation, preferred accumulation or separation of metal by magnetic forces, differential movement of fine vs. coarse grained material, etc.

15 Cosmochemical and Geochemical Classification of the Elements

The geochemical classification of the elements is based on the chemical affinity to silicate and oxides (lithophile elements), sulfides (chalcophile elements and metal alloys (siderophile elements). The cosmochemical classification is based on the relative volatility of the elements during condensation and evaporation. Condensation temperatures are calculated assuming thermodynamic equilibrium between condensed solid and nebular gas at a given total pressure. They are measures of the relative volatility of the elements. Major elements condense as minerals whereas minor and trace elements often condense in solid solution with major minerals and/or melts (but see (1) below). The temperature where 50% of an element is condensed (or evaporated) is called the 50% condensation temperature (Lodders 2003; Fegley and Schaefer 2010; Lodders et al. 2025a, this collection). Condensation temperatures only apply to H- and He-rich solar-like elemental compositions and may be very different under more oxidizing conditions than in the canonical solar nebula. Condensation temperatures calculated for solar composition should not be applied to evaporation processes in the absence of abundant H. Five components can account for the variations in the elemental abundances in primitive meteorites. In addition, their oxidation state controls abundance variations during condensation and evaporation, adding to the complexity of chondrites.

(1) *Refractory component:* The first major phases to condense from a cooling gas of solar composition are Ca, Al-oxides and minor silicates associated with a large number of refractory lithophile elements (RLE) including Al, Ti, Ca, Zr, Hf, Sc, Y and the REE (Rare Earth Elements). Trace elements condense into solid solution with each other (ultra-refractory phases) or with host phases made of more abundant elements. Refractory siderophile elements (RSE) comprise all metals with lower vapor pressures than those of Fe and Ni. They include W, Os, Re, and Ir and condense as refractory metal alloys, e.g. Palme et al. (1994).

Constant ratios among refractory elements in most chondritic meteorites allow the determination of representative abundances in CI-chondrites where parent body processes may have caused heterogenous re-distribution of some refractory mobile elements, such as for example Ca or U (see below). A notable exception of constant refractory element ratios is REE in CV-chondrites. The REE pattern of bulk Allende is fractionated relative to CI-chondrites (e.g., Stracke et al. 2012).

(2) *Mg-silicates and iron-alloy:* The major fraction of condensable matter is associated with the three most abundant elements heavier than O: Si, Mg and Fe. Iron first condenses as metal alloy (FeNi), whereas Mg and Si form forsterite (Mg_2SiO_4) which converts to enstatite ($MgSiO_3$) at lower temperatures by reaction with SiO(gas). Below 0.1 mbar, FeNi-metal condenses at lower temperatures than forsterite, at higher total pressures, FeNi-metal condenses at higher temperatures than forsterite.

(3) Moderately volatile elements have condensation temperatures between those of Mg-silicates and FeS (troilite). This includes Mn and Na. The most abundant moderately volatile element is sulfur which starts condensing at 704 K (independent of total pressure). Half of all S is condensed at 664 K (see Palme et al. 1988; Lodders 2003).

(4) Highly volatile elements have condensation temperatures below that of FeS (704 K) and above water ice (e.g. Cd, Bi, Pb).

(5) Ultra-volatile elements have condensation temperatures at and below that of water ice. This group includes H, C N, O, and the noble gases. About 20-25% of oxygen can be removed by silicate formation at higher temperature, but because the 50% condensation temperature of O (as water ice) <200 K, O is regarded as an ultra-volatile element.

16 Carbonaceous Chondrites of the Ivuna-Type (CI)

The three major types of chondritic meteorites are carbonaceous chondrites (CC), ordinary chondrites (OC), and enstatite chondrites (EC). Figure 2 illustrates compositional differences between these types. The elements in Fig. 2 represent elements in various cosmochemical groups, e.g., Al represents refractory elements such as Ca and Ti; Si the elements of intermediate volatility, and Mn and S represent moderately volatile elements. The concentration ratios in Fig. 2 are relative to solar photospheric abundances and are further normalized to Mg. Only CI-chondrites match solar abundances closely (i.e., element ratios for CI-chondrites plot at or close to unity). All other types of chondrites diverge in some way from CI-chondrite abundances and solar abundances. The close correspondence of solar abundances with CI-chondrites is the major argument why CI-chondrites are combined with solar data for the solar system composition (e.g., Anders 1971; Holweger 2001).

The major differences among chondrites are the depletions of moderately volatile elements (exemplified by S and Mn in Fig. 2) and variations in the level of refractory elemental abundances. The Al/Mg ratio (squares in Fig. 2) for CI-chondrites is similar to the Al/Mg ratio in the Sun, but Al/Mg ratios are higher in other carbonaceous chondrites and lower in

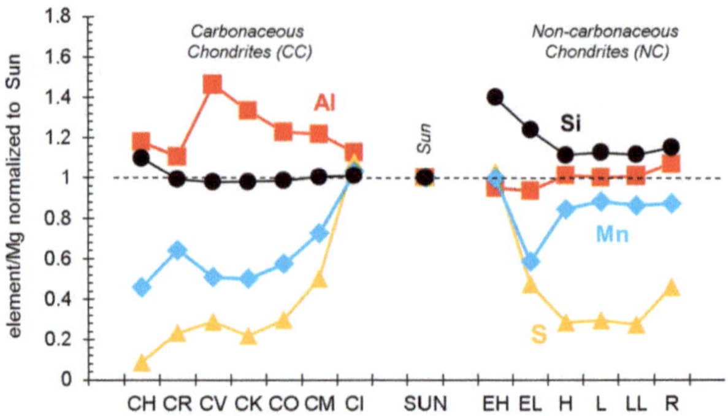

Fig. 2 *Si/Mg, Al/Mg, Mn/Mg* and *S/Mg* ratios in various chondrite groups normalized to solar abundance ratios presented in this paper. The CI-chondrites show the best overall match to solar photospheric abundances. The EH chondrites also fit solar element ratios for moderately volatile elements (represented by *S/Mg* and *Mn/Mg*), but they miss the solar *Si/Mg* ratio. Solar and CI-chondritic data are from this study, other chondrite data are from Lodders (2021)

other chondrite groups. Ratios of the moderately volatile elements Mn/Mg and S/Mg in CI-chondrites, and possibly EH-chondrites, closely match the solar ratios. The EH-chondrites come close to solar and CI-chondritic ratios for Al/Mg, S/Mg, and Mn/Mg, but their Si/Mg is much higher and their element/Mg ratios for elements more volatile than S are lower than in CI-chondrites, making them less suitable for solar system proxies.

The five CI-chondrites Orgueil, Ivuna, Alais, Tonk, and Revelstoke are observed falls of this rare meteorite group. CI-chondrites are fragile and easily break up during atmospheric entry. The most mass is preserved from Orgueil, and most chemical and isotopic analyses were done on this meteorite. Much less material is preserved from the other CI-chondrites. Problems with representative sampling may arise when only small fragments or samples (< 50 mg) can be analyzed. Over the years sensitivity and precision of analytical methods have improved significantly: 50-gram size samples were used for analyses of Orgueil in 1864 after its fall, later published analyses of Orgueil in the 1950s used 1 to 2 g samples with wet chemical gravimetric, colorimetric and spectrographic methods. Results were largely limited to major elements. Neutron activation analyses since the 1960s expanded analyses to trace elements and sample sizes dropped to the range of about 100 mg and below. Since the 1990s, methods using inductively coupled plasma (ICP), mainly with mass spectrometry (ICP-MS) considerably improved the precision of trace element analyses and allowed the analysis of very small samples (1 mg and below). The most accurate method is isotope dilution (ID) which requires no standard (see Stracke et al. 2014). The method is very labor intensive and has, so far, been applied to some lithophile and siderophile elements (see below). While the advance in analyzing small samples seems to eliminate the need for using up precious sample material, it comes at the price of non-representative sampling.

Representative sampling is required because CI-chondrites have some chemical and mineralogical heterogeneities (Alfing et al. 2019; Greshake et al. 1998; King et al. 2020; Morlok et al. 2006). CI-chondrites contain about 10-20% bound and absorbed water affecting the distribution of aqueously mobile elements, either on the parent asteroid, or on Earth when exposed to humidity. Rare variations of up to 30% may stem from accessory phases (such as carbonates, phosphates, sulfides, sulfates) which can concentrate major and trace elements. Here trace elements are elements which do not form their own minerals but occupy positions in crystal lattices normally populated by major elements, e.g., Sr can substitute for Ca in carbonates and phosphates.

The returned materials from asteroids Ryugu (Japanese Hayabusa 2 mission, Nakamura et al. 2022; Ito et al. 2022, Yokoyama et al. 2023, 2024) and Bennu (US OSIRIS-REx mission, Lauretta et al. 2024) are similar to CI-chondrites in bulk chemical and isotopic composition. These pristine samples are not discussed here due to space limitations.

17 Elemental Abundances in CI-Chondrites

A discussion of all elements is beyond the scope of this paper and will be published elsewhere (Lodders et al. 2025b). References to the data and reference codes (first letter of first author's last name and 2-digit year) are listed in the electronic appendix.

Recommended element concentrations by mass are given in μg/g (parts per million, ppm) for CI-chondrites are in Table 4. In column (2) absolute 1σ uncertainties are listed, which are converted to % uncertainties (SD%) in column (3). The quality index in column (4) gives an estimate for the reliability of the CI-concentrations with A = highest quality. It is based on 1-sigma standard deviation, element variability and mobility, and issues with analytical methods. Column (6) shows the percent deviation to data in Palme et al. (2014). Major deviations are described below. Other mass concentration units used here are weight-percent (wt%) for major elements and parts per billion, ppb, for trace elements (where 1 ppm = 1 g/ton = 1 microgram/gram (μg/g) = 0.0001 wt% = 1000 ppb).

18 Averaging Method

The chemical and mineralogical compositions of CI-chondrites are broadly similar but subtle differences exist. For obtaining average CI-compositions we used procedures similar to Lodders (2003) and Palme et al. (2014). The compiled chemical data for each CI-chondrite are screened for outliers and straight averages are calculated for each. The average CI-chondritic composition is obtained by taking the weighted average of the individual meteorite compositions where the statistical weight is given by the number of selected analyses per meteorite. This procedure makes Orgueil dominant as most of the analyses are done for it.

Orgueil analyses were used to evaluate results obtained by various analytical methods. Computing the distributions around the mean and 1-sigma standard deviation for each method provides a check for consistency among analytical methods and shows which analytical method(s) are most suitable for a given element. For some elements, outliers can be associated with particular analytical methods often involving extensive wet chemical processing or separations. Instrumental neutron activation analysis (INAA) does not require chemical processing and is superior in this regard. However, for CI-compositions INAA is limited to about 15 elements with uncertainties below 5%, including Al, Na, Mn, Cr, Sc, Fe, Co, Sm, and Ir (see Palme and Zipfel 2022).

Histograms aid to spot outliers in the distributions; and the lowest and/or highest values outside the two-sigma standard deviations (SD) range are removed when Gaussian distributions and single mode distributions are indicated. Two examples, Cr and Fe, illustrate the approach. The variations in most samples deemed representative are usually to within 10% and variations among elements are less than those between different chondrite groups (e.g., smaller than differences between CI- and CM-chondrites).

Some elements vary more than expected from the quality of analytical methods used. Most of these elements are aqueous mobile elements (see below). This includes Na, K, and

Table 4 CI-Chondrite Composition and Comparison to Solar Abundances [a]

E	Concentration by mass ± 1σ (SD)		%SD	n	SE	%SE	95%C.I.	N(E) ± 1σ Si = 10^6 atoms		Quality Index & Notes		Change from P14, %	$A(E)_{CI}$ ± 1σ [b]dex		$A(E)_{Sun}$ [c]dex	CI/Sun
H	18598	1719	9.2	16	430	2.3	916	4.86E+6	4.5E+5	B		-5.6	8.24	0.04	12.00	1.7E-4
He	9.17E-3	9.2E-4	10					0.604	0.060	B	see text	0	1.33	0.04	10.92	3E-10
Li	1.48	0.07	4.7	20	0.02	1.1	0.03	56.2	2.7	A		2.1	3.30	0.02	1.04	182
Be	0.0225	0.0013	5.8	13	0.0004	1.6	0.0008	0.658	0.038	B		2.8	1.37	0.02	1.21	1.44
B	0.744	0.095	13	5	0.042	5.7	0.118	18.1	2.3	C		-4.0	2.81	0.05	2.70	1.29
C	37813	6600	17	38	1071	2.8	2169	8.29E+5	1.45E+5	D	see text	8.7	7.47	0.07	8.51	0.09
N	1965	970	49	22	206	10.5	428	37000	18200	D		-33	6.12	0.17	7.94	0.02
O	465700	8000	1.7	4	4000	0.9	12730	7.7E+6	1.3E+3	A	see text	1.5	8.44	0.01	8.76	0.47
F	92	20.0	22	6	8.2	8.8	21.0	1280	277	D	see LF23	59	4.66	0.09	4.40	1.81
Ne	1.80E-4	1.8E-5	10					2.36E-3	2.4E-4	B	see text	0	-1.08	0.04	8.15	6E-10
Na	4960	512	10	59	67	1.3	133	56800	5900	B		-0.05	6.31	0.04	6.29	1.04
Mg	95600	3000	3.1	51	420	0.4	844	1.0E+6	32.5E+3	A		0.2	7.57	0.01	7.58	0.97
Al	8470	480	5.7	48	69	0.8	139	82700	4700	B		0.8	6.47	0.02	6.43	1.09
Si	106600	4000	4.1	26	863	0.8	1777	1.0E+6	41.3E+3	A		-0.4	7.55	0.02	7.56	0.98
P	989	89	9.1	32	16	1.6	32	8410	760	B		0.4	5.48	0.04	5.44	1.09
S	51800	4600	8.9	29	854	1.6	1750	4.26E+5	3.8E+4	B		-3.2	7.18	0.04	7.16	1.05
Cl	717	110	15	16	28	3.8	59	5330	820	C	see LF23	2.7	5.28	0.06	5.43	0.70
Ar	1.33E-3	1.3E-4	10					9.66E-3	9.7E-4	B	see text	0	-0.46	0.04	6.50	1.1E-7
K	544	41	7.5	62	5.21	1.0	10	3670	280	B		-0.33	5.12	0.03	5.09	1.06
Ca	9148	554	6.1	R				60100	3600	B	Ca/Al Tab. A5	0.41	6.33	0.03	6.35	0.96
Sc	5.76	0.37	6.4	60	0.05	0.8	0.10	33.8	2.2	B		-0.81	3.08	0.03	3.13	0.89
Ti	442	30	6.8	43	5	1.0	9	2430	170	B		-1.1	4.94	0.03	4.97	0.93
V	53.1	3.9	7.3	42	0.6	1.1	1.2	275	20	B		-2.7	3.99	0.03	3.89	1.26

Table 4 (*Continued*)

E	Concentration by mass ± 1σ (SD)		%SD	n	SE	%SE	95%C.I.	N(E) ± 1σ Si = 10^6 atoms		Quality Index & Notes		Change from P14, %	$A(E)_{CI}$ ± 1σ [b]dex		$A(E)_{Sun}$ [c]dex	CI/Sun
Cr	2616	116	4.4	54	16	0.6	32	13300	600	A	see text	-0.3	5.67	0.02	5.74	0.86
Mn	1936	153	7.9	72	18	0.9	36	9280	730	B		1.0	5.52	0.03	5.52	1.00
Fe	185000	6000	3.5	54	871	0.5	1747	872.8E+3	30.2E+3	A	see text	-0.9	7.49	0.01	7.51	0.96
Co	514	26	5.1	73	3	0.6	6	2300	120	B		0.16	4.91	0.02	4.95	0.92
Ni	11180	300	3.0	53	45	0.4	91	50200	1500	A		2.5	6.25	0.01	6.24	1.03
Cu	133	15	11	40	2	1.8	5	552	62	C		0.17	4.29	0.05	4.24	1.13
Zn	310	22	7.1	81	2	0.8	5	1250	90	B		0.5	4.65	0.03	4.55	1.25
Ga	9.54	0.68	7.1	59	0.09	0.9	0.18	36.1	2.6	B		-0.8	3.11	0.03	3.02	1.22
Ge	33.5	2.3	6.9	33	0.4	1.2	0.8	122	8	B		2.9	3.64	0.03	3.62	1.04
As	1.75	0.17	9.7	40	0.03	1.5	0.05	6.16	0.60	B		0.7	2.34	0.04	...	...
Se	21.5	0.70	3.3	32	0.1	0.6	0.3	71.7	2.3	A		5.8	3.41	0.01	...	...
Br	3.77	0.90	24	15	0.2	6.2	0.5	12.4	3.0	D	see LF23	16	2.65	0.09	...	...
Kr	5.22E-5	5.2E-6	10					1.64E-4	1.6E-5	B	see text	0	-2.23	0.04	3.31	2.9E-6
Rb	2.26	0.12	5.3	41	0.02	0.8	0.04	6.98	0.37	B		-2.4	2.39	0.02	2.35	1.11
Sr	8.04	0.27	3.4	22	0.06	0.7	0.12	24.2	0.8	A		3.3	2.93	0.01	2.93	1.01
Y	1.52	0.13	8.8	R				4.50	0.40	C	Zr/Y, Tab. A10	4.0	2.20	0.04	2.30	0.80
Zr	3.65	0.21	5.8	R				10.5	0.6	C	Zr/Hf, Tab. A7	0.4	2.57	0.02	2.68	0.78
Nb	0.271	0.022	8.2	R				0.767	0.063	C	Zr/Nb, Tab. A8	-4.4	1.44	0.03	1.47	0.92
Mo	0.947	0.100	11	23	0.021	2.2	0.043	2.60	0.27	C		-1.5	1.97	0.04	1.88	1.22
Tc	...	...	...	...	...	...	...	...	...	...		...	...	...	...	...
Ru	0.680	0.042	6.2	R				1.77	0.11	B	Os/Ru, Table 5	-1.5	1.80	0.03	1.75	1.12
Rh	0.133	0.006	4.5	11	0.002	1.4	0.004	0.341	0.015	A		1.0	1.08	0.02	0.78	2.01
Pd	0.561	0.026	4.6	36	0.004	0.8	0.009	1.39	0.06	A		0.13	1.69	0.02	1.57	1.33

Table 4 (*Continued*)

E	Concentration by mass ± 1σ (SD)		%SD	n	SE	%SE	95%C.I.	N(E) ± 1σ Si = 10^6 atoms		Quality Index & Notes		Change from P14, %	$A(E)_{CI}$ ± 1σ [b]dex		$A(E)_{Sun}$ [c]dex	CI/Sun
Ag	0.206	0.019	9.2	30	0.003	1.7	0.007	0.504	0.046	B		2.6	1.25	0.04	0.96	1.96
Cd	0.682	0.047	6.9	45	0.007	1.0	0.014	1.60	0.11	B		1.1	1.75	0.03	1.77	0.96
In	0.0781	0.0055	7.0	38	0.001	1.1	0.002	0.179	0.013	B		0.4	0.80	0.03	0.80	1.01
Sn	1.67	0.20	12	39	0.03	1.9	0.06	3.70	0.44	C		2.2	2.12	0.05	2.02	1.26
Sb	0.157	0.026	17	51	0.004	2.3	0.007	0.340	0.056	D		8.3	1.08	0.07	...	...
Te	2.30	0.09	3.9	26	0.02	0.8	0.04	4.75	0.19	A		1.0	2.23	0.02	...	...
I	0.772	0.310	40	5	0.139	18	0.385	1.60	0.64	D	see LF23	46	1.76	0.15	...	...
Xe	1.74E-4	1.7E-5	10					3.49E-4	3.49E-5	B	see text	0	-1.91	0.04	2.30	6.2E-5
Cs	0.186	0.012	6.5	58	0.002	0.8	0.003	0.368	0.024	B		-1.3	1.12	0.03	...	...
Ba	2.41	0.14	5.8	36	0.02	1.0	0.05	4.62	0.27	B		-0.5	2.22	0.02	2.27	0.88
La	0.248	0.010	4.0	28	0.002	0.8	0.004	0.470	0.019	A		2.6	1.22	0.02	1.10	1.33
Ce	0.633	0.026	4.2	28	0.005	0.8	0.010	1.19	0.05	A		2.2	1.63	0.02	1.58	1.11
Pr	0.0957	0.0042	4.4	28	0.0008	0.8	0.0016	0.179	0.008	A		1.9	0.80	0.02	0.75	1.13
Nd	0.482	0.025	5.2	29	0.005	1.0	0.010	0.880	0.046	B		1.8	1.50	0.02	1.42	1.19
Pm	...	...	...	...	...	...	...	...	...				...	...	...	...
Sm	0.156	0.008	5.2	29	0.002	1.0	0.003	0.273	0.014	B		1.6	0.99	0.02	0.95	1.09
Eu	0.0601	0.0027	4.5	28	0.0005	0.8	0.0010	0.104	0.005	A		2.2	0.57	0.02	0.57	1.00
Gd	0.211	0.010	4.9	29	0.002	0.9	0.004	0.353	0.017	A		1.8	1.10	0.02	1.08	1.04
Tb	0.0385	0.0017	4.3	24	0.0003	0.9	0.0007	0.0638	0.003	A		1.4	0.36	0.02	0.31	1.11
Dy	0.259	0.011	4.2	28	0.002	0.8	0.004	0.421	0.018	A		1.4	1.17	0.02	1.10	1.19
Ho	0.0568	0.0023	4.0	24	0.0005	0.8	0.0010	0.0908	0.004	A		0.7	0.51	0.02	0.48	1.07
Er	0.167	0.007	4.3	28	0.001	0.8	0.003	0.263	0.011	A		0.9	0.97	0.02	0.93	1.10
Tm	0.0263	0.0012	4.6	24	0.0002	0.9	0.0005	0.0410	0.002	A		0.9	0.16	0.02	0.11	1.13

Table 4 (*Continued*)

E	Concentration by mass ± 1σ (SD)		%SD	n	SE	%SE	95%C.I.	N(E) ± 1σ Si = 10^6 atoms		Quality Index & Notes		Change from P14, %	$A(E)_{CI}$ ± 1σ [b]dex		$A(E)_{Sun}$ [c]dex	CI/Sun
Yb	0.170	0.007	4.0	28	0.001	0.8	0.003	0.259	0.010	A		1.0	0.96	0.02	0.85	1.30
Lu	0.0255	0.0015	5.7	31	0.0003	1.0	0.0005	0.0384	0.002	B		1.8	0.13	0.02	0.10	1.08
Hf	0.107	0.006	5.8	R				0.158	0.009	C	Lu/Hf, Tab. A6	0.4	0.75	0.02	0.86	0.77
Ta	0.0149	0.0007	4.7	R				0.0218	0.0010	D	Nb/Ta, Tab. A9	-0.3	-0.11	0.02	...	...
W	0.101	0.012	12	R				0.144	0.017	D	Hf/W, see text	4.9	0.71	0.05	0.79	0.83
Re	0.0399	0.0017	4.1	R				0.0564	0.0023	A	Os/Re, Table 5	-0.3	0.30	0.02	...	...
Os	0.491	0.014	2.8	28	0.003	0.5	0.005	0.680	0.019	A		-0.7	1.38	0.01	1.36	1.06
Ir	0.456	0.027	5.9	70	0.003	0.7	0.006	0.625	0.037	B		-2.8	1.35	0.02	1.42	0.84
Pt	0.904	0.070	7.8	R				1.22	0.09	B	Pt/Ir, Table 5	-2.3	1.64	0.03	...	...
Au	0.150	0.013	8.8	50	0.002	1.2	0.004	0.201	0.018	B		1.6	0.85	0.04	0.91	0.88
Hg	0.293	0.106	36	3	0.061	21	0.263	0.385	0.139	D		-16	1.14	0.13	...	...
Tl	0.141	0.014	9.9	31	0.003	1.8	0.005	0.182	0.018	B		0.8	0.81	0.04	0.95	0.73
Pb	2.61	0.19	7.3	26	0.04	1.4	0.08	3.32	0.24	B		-0.2	2.07	0.03	1.95	1.33
Bi	0.113	0.007	6.2	25	0.001	1.2	0.003	0.142	0.009	B		2.5	0.70	0.03	...	...
Th	0.0304	0.0033	11	37	0.0005	1.8	0.0011	0.0345	0.0037	C		1.3	0.09	0.04	...	...
U	8.10E-3	7.70E-4	9.5	44	1.2E-4	1.4	2.3E-4	8.96E-3	8.52E-4	D		-0.1	-0.50	0.04	...	...

(a) Concentrations mainly from CI-chondrites, except when concentrations were calculated from element ratios, as noted (R). Quality index (A highest quality) is based on 1-sigma standard error (SD in %) and element variability and mobility, and issues with analytical methods. n = number of analyses included in grand mean. SD: standard deviation. SE: standard error. 95%C.I.: 95% confidence interval. LF23 = Lodders and Fegley (2023). P14 = Palme et al. (2014). (b) A(E) = 1.551+ log N(E). (c) from Table 2

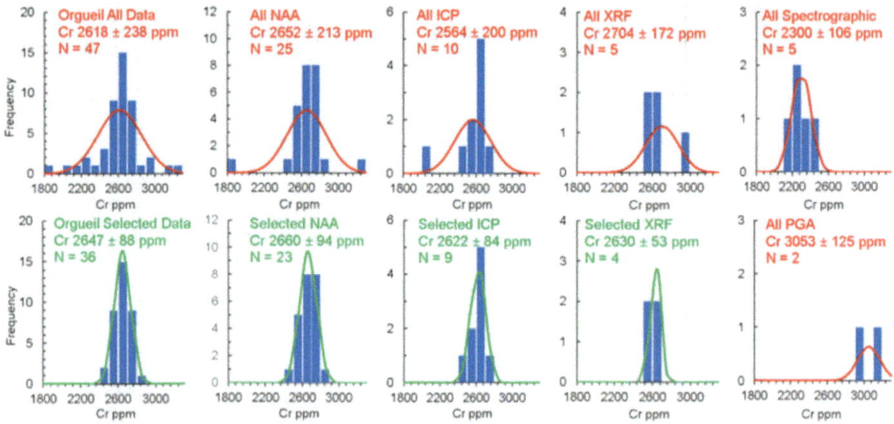

Fig. 3 Published Cr concentrations for Orgueil analyzed by five methods. Neutron activation analysis, ICP methods (mainly ICP-AES), and XRF give similar Cr concentrations

Ni analyzed by INAA, where variations due to chemical processing during analysis can be excluded. The variations of some elements are often correlated, suggesting the presence of minor phases enriched in trace elements such as CAIs, carbonates, sulfates, and phosphates. For these elements, sampling and sample sizes become relevant.

18.1 Chromium (Cr)

As an example, we discuss Cr concentrations. Chromium has similar concentrations in all CI-chondrites. In analyzing Cr in Orgueil five analytical methods were used for 47 samples as shown in Fig. 5. The first row shows histograms of literature data for Orgueil organized by analytical methods, the averages, 1-σ standard deviations and number of samples (N). The curves are calculated normal distributions functions (PDF) about the mean. Neutron activation analyses (mainly INAA) contribute around half of all data, followed by ICP methods (10; mainly ICP-AES; inductively coupled plasma-atomic emission spectrometry), and five XRF (x-ray fluorescence) analyses. These three methods give similar averages once outliers are removed (bottom row of Fig. 5). The five older spectrographic results match tightly among themselves but yield systematically lower values whereas prompt gamma ray analyses (only two) give the highest values. The latter two methods introduced bias and are excluded in the Orgueil average of 2647 ±88 ppm from 36 measurements.

Studies analyzing more than one Orgueil sample allow inter-laboratory comparisons and allow to gauge intrinsic variability. Three different sets of NAA gave: 2663 ±95 ppm (N = 6, Gooding 1979), 2645 ±85 ppm (N = 4, Kallemeyn and Wasson 1981); and 2632 ±66 ppm (N = 9, Palme and Zipfel 2022). Agreements among different groups are excellent and similar standard deviations suggest homogeneous distribution of Cr in Orgueil. Similar conclusions follow from 2618 ±43 ppm (N =6) by ICP-AES from Barrat et al. (2012) and 2630 ±53 ppm (N = 4) by XRF from Wolf and Palme (2001). Further, concentrations have no apparent dependence on sample size (see Palme and Zipfel 2022).

Figure 4 shows all literature data for Cr for the CI-chondrites Alais, Ivuna and Orgueil; the plot "All CI-chondrite data" includes Tonk (1x) and Revelstoke (2x). Only Tonk was retained for the group mean because the XRF Cr values for Revelstoke (2000 and 3200 ppm, Folinsbee et al. 1967) are at the extreme concentration limits of the overall observed

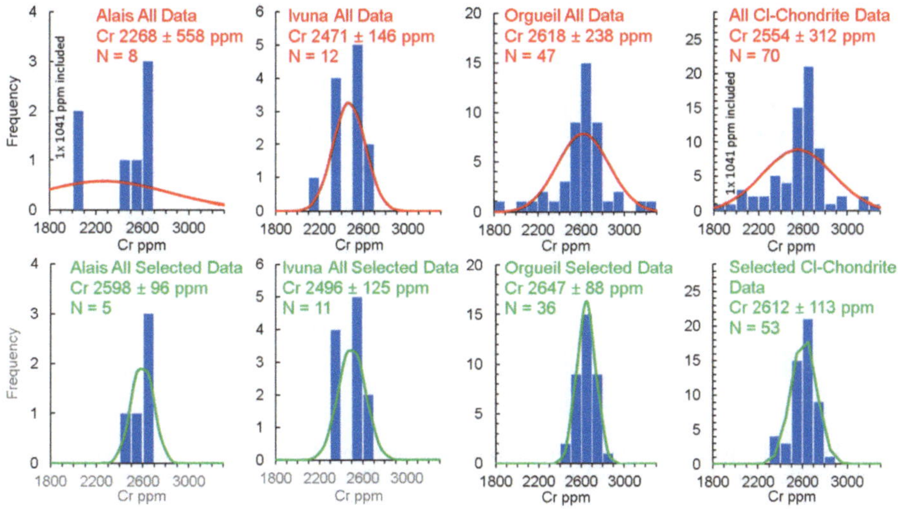

Fig. 4 Chromium analyses for all CI-chondrites from the literature. Only methods using ICP, NAA, and XRF are included (see Fig. 3). Ivuna and Orgueil agree within uncertainties

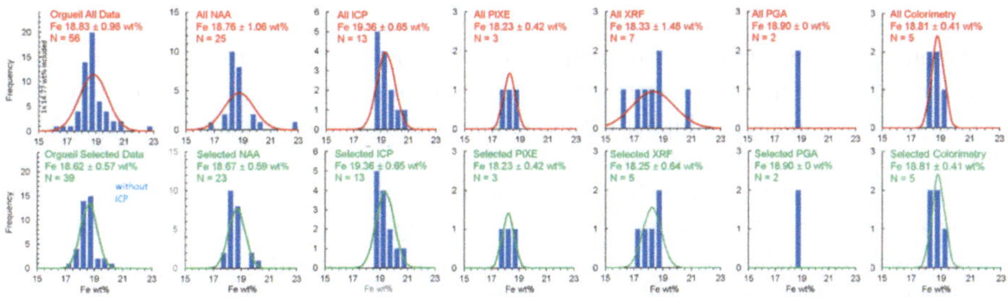

Fig. 5 Results of Fe analyses for the CI-chondrite Orgueil. Four methods give similar results. The ICP-AES/OES data systematically overestimate the Fe-contents, are right-skewed, and therefore were excluded from the average

range. As for Orgueil, no spectrographic or PGA measurements were used, which removes one measurement each for Ivuna and Alais. The Ivuna average of Cr = 2496 ±125 ppm is about 150 ppm smaller than in Orgueil but agrees within uncertainties. Alais has similar Cr = 2596 ± 95 ppm after removing two samples (1041 and 2010 ppm) which are also unusual in several other elements. The Cr content of Alais is essentially the same as Orgueil within uncertainties. Overall, three different analytical procedures give the same Cr concentrations among CI-chondrites within uncertainties and Cr is fairly homogeneously distributed in them. The combined data give a CI-chondrite Cr-content of 2612 ± 113 ppm or ± 4.3%.

18.2 Iron (Fe)

The second example of the averaging method is Fe. Figure 5 shows the raw (56) and selected (39) data of Orgueil by methods, Fig. 6 shows the raw and selected values for the individual CI-chondrites.

Most of the 56 Orgueil analyses were done by NAA, followed by ICP-AES/OES, XRF, colorimetry, and PIXE. Like Cr, Fe is an ideal element for NAA because of the long half

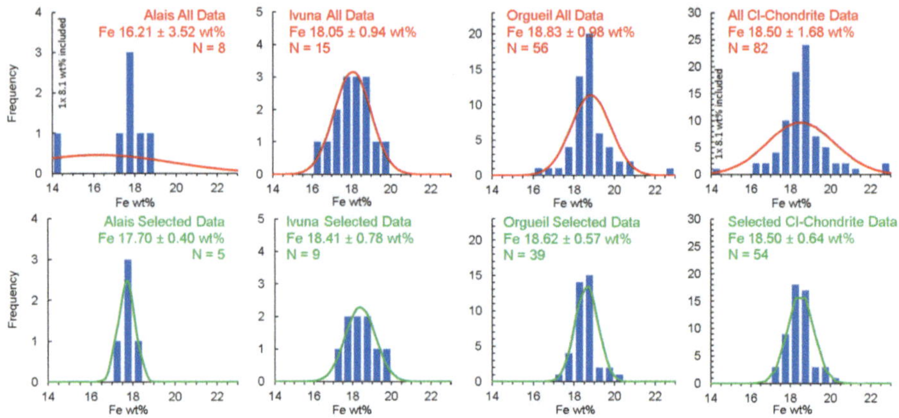

Fig. 6 Iron in CI-chondrites. Averages of Ivuna and Orgueil agree but Alais is somewhat lower. The five Alais and the few Tonk and Revelstoke analyses only have a small influence on the overall CI-chondrite average

lives of the radioactive nuclides produced during irradiation and excellent counting statistics. Averaging the Fe-contents of all 25 samples measured by NAA after removing the highest outlier and the lowest value gives 18.67 ± 0.59 wt% (3.2%SD). The good interlaboratory agreement also suggests homogeneous distribution of Fe in Orgueil.

Among the different analytical methods, results from ICP (inductively coupled plasma combined with AES/ OES (Atomic Emission spectroscopy or Optical Emission spectroscopy) are systematically higher with a right-skewed distribution than results from NAA, PIXE (particle induced x-ray emission) and XRF although they agree within combined uncertainties. Seven XRF-analyses from different labs have a comparatively large spread. The reason for the higher ICP results is unknown and might be related to matrix effects; this disagreement with other methods is not addressed in the original papers and needs to be re-visited by analysts. We excluded the ICP-data in the grand average because the unusual distribution and the higher average indicate a larger analytical issue for the ICP-AES/OES values.

Figure 6 shows iron averages for Alais, Ivuna and Orgueil. The lower Fe-content in Alais (17.7 ± 0.4 wt.%) and its differences in other major elements compared to other CI-chondrites was found by several groups, e.g., Palme and Zipfel (2022). The Fe content of Ivuna is intermediate between Alais and Orgueil. The CI-chondrite group average for iron (including Alais, but excluding ICP-data, is 18.50% ± 0.64 or ± 3.5%.

18.3 Carbon (C)

The 57 C concentrations (sources in Table A2) in Fig. 7 are bimodally distributed in CI-chondrites, this is also indicated in Orgueil samples only. The high mode is dominated by Alais samples, which include several recent analyses.

The two major C reservoirs are assorted organics, and carbonates (dolomite, breunnerite, calcite). During stepwise heating/pyrolysis under oxidizing conditions of CI-chondrites (e.g., K70, G72, G91, W85; see appendix) the low temperature (200-400 °C) C is sourced from carbonates and low-mass molecular organics, whereas C released at temperatures up to around 1350-1400 °C is associated with refractory organics. Classical chemical combustion analyses pointed to 0.9 wt% C as carbonate in Alais (B834) and for Orgueil to around 0.15 wt% (C864c), as confirmed by G88 (0.16 wt%).

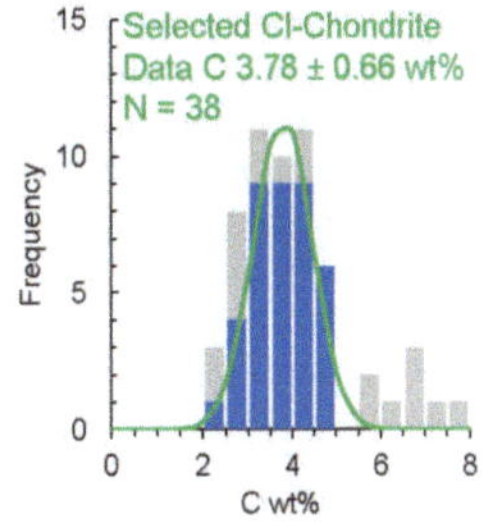

Fig. 7 Distribution of C concentrations in all CI-chondrites using various methods. The gray bars are for all data, the blue ones for selected data from combustion analyses. Stepwise-heating results are excluded, as are several high concentration results (mainly for Alais)

Several C determinations by stepwise heating were done on acid residues of CI-chondrites where carbonate C and acid-soluble organics were lost and can amount to about one percent of total C (e.g., S70, W85). This is apparent when results for Orgueil (after obvious outliers are removed) from different methods are compared: Classical combustion (C = 3.84 ± 0.47 wt%, N=8), and "element analyzer" combustion (C = 3.70 ± 0.62 wt% N=13) average to about half a percent higher than stepwise heating results (C = 3.15 ± 0.83 wt%, N=9 with a bimodal distribution). However, within the larger standard deviations, results are consistent among these methods.

The measurements by Pearson et al. (2006; P05) for Alais (5.40 wt%) yield higher C concentrations than for Orgueil (4.88 wt%), and generally somewhat higher than previous results. Alais has six out of ten C analyses with about 7 wt% (P05, W86, B834) not included in the grand mean. Pearson et al. (2006) found two data clusters for small aliquots from two Alais specimens, indicating heterogeneity. Among 33 Orgueil analyses, only three are close to 6 wt% or higher (C864a, W86, P05). Five Ivuna measurements by combustion compare well with corresponding Orgueil data, and one Tonk (C14) and one Revelstoke (F67) analysis are in the group mean.

Carbon concentrations show an inverse trend with analyzed sample mass (Fig. 8). As the mass scale in the figure is logarithmic, the use of approximate masses (as reported in the literature) seems reasonable to gauge possible dependencies. Larger Orgueil samples show lower C concentrations (Fig. 8). Alais appears similar but the old results on its two highest-mass samples (M806, T806) should not be over-interpreted. The two data clusters for Alais (P05) noted above are well resolved. Leaving the possible bias from analytical methods aside, data in Fig. 8 suggest that samples below 50 mg are prone to higher C concentrations up to twice the recommended group average of 3.78±0.66 wt%. The recommended value excludes Alais and Orgueil values >5.8 wt% based on the data distribution in Fig. 7.

18.4 Nitrogen (N)

Measured N contents of Orgueil vary from 800 to 8200 ppm in 20 analyses by different authors (Table A2). Thirteen measurements for Orgueil give 1934 ± 458 ppm, excluding the lowest value, two extreme results from P05, and all high chromatographic values from G71. Seven Alais measurements yield N = 2021 ± 521 ppm; two for Ivuna give N = 1962 ± 151 ppm. All selected values provide the grand average N = 1965 ± 447 ppm for CI-chondrites.

18.5 Oxygen (O)

Total O contents determined by fast neutron activation analysis exist for Orgueil (2x Palme and Zipfel 2022, 1x Wing 1964) and one for Ivuna (Palme and Zipfel 2022). The Orgueil

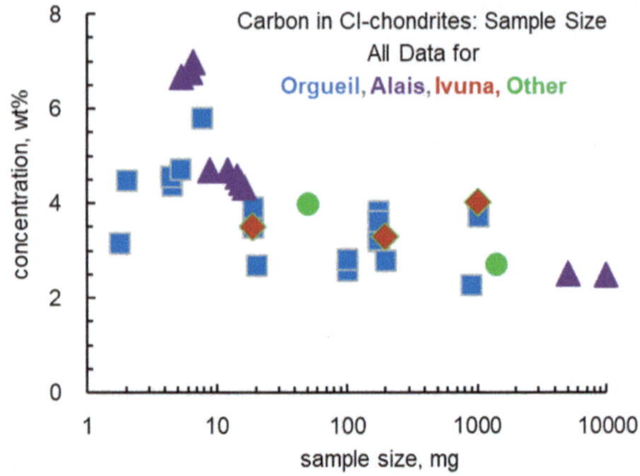

Fig. 8 Carbon concentrations as a function of logarithmic sample mass where (approximate) masses analyzed were reported. Orgueil = squares, Alais = triangles; Ivuna (*B*54, *S*70) = diamonds; Tonk (*C*14) and Revelstoke (*F*67) =circles

value by Wing (1964) is about 1 wt% lower than the average of two measurements reported by Palme and Zipfel (2022) but without further measurements, Wing's (1964) value is retained in the mean. For Alais, an estimate by difference gives 47.05% oxygen. The recommended CI-chondrite value is O = 46.57 ± 0.8 wt%.

18.6 Silicon (Si)

The average Si concentration from all CI-chondrites, Si = 10.66 ± 0.43 wt%, is from 26 analyses by fast-neutron activation analysis (FNAA) and XRF. The scatter in the distribution is caused by Alais which has lower Si than Ivuna and Orgueil. Silicon is traditionally used to anchor the atomic cosmochemical abundance scale to one million Si atoms, and element/Si ratios are frequently reported in the literature. However, Si is not routinely measured, and it is not included in several modern ICP-MS measurements. In this procedure samples are dissolved in HF and some Si may be removed. Thus, one has to rely on the adopted average concentration of Si for each CI-chondrite to derive atomic abundances for the Si-based abundance scale. Given the natural variations, one ideally should always normalize to the actual Si concentration of the sample at hand, which is not possible for most existing analyses. We recommend normalizing abundances to Mg instead of Si. Silicon is more volatile than Mg, and Si/Mg are fractionated in different meteorites and planetary objects (including bulk silicate Earth).

18.7 Rare Earth Elements (REE)

The CI-chondritic averages for the REE are based on ICP-MS and ID-MS measurements (e.g., B12, B16; B84, D15, N74, M06, P12). Outlying results, often with high concentrations, from samples analyzed by B18, E78, K73, and R93 are excluded. No NAA and PGA values (K81, P22, I12) are used because these are largely limited to La, Sm, Eu, and Lu, and would introduce heterogeneity in the weighted averages of all REE.

The REE were redistributed within CI-chondrites into carbonates, phosphates, and sulfates and sampling and sample sizes are important. In addition, analytical uncertainties remain as e.g., Nd and Lu averages from different laboratories differ. Many individual sample patterns show variations within 5-10%, depending on which samples were selected for use as normalizing values (e.g., see Barrat et al. 2012; Pourmand et al. 2012).

As quality control we used cross-correlations among the REE. Correlations of REE with their nearest neighbors can identify outlier samples with possible analytical issues (such as standards); often this is associated with the mono-isotopic REE. Another check is the consistency of the elemental data with isotope systematics of Sm-Nd for which we calculated $^{147}Sm/^{144}Nd = 0.1967$ which compares to 0.1966 (Bouvier et al. 2008) and 0.1967 (Jacobsen and Wasserburg 1984), see also Table A11.

18.8 Mobile Elements

The CI-chondrites are not a completely homogenous group in terms of chemical composition. It was long suspected (e.g., Wiik 1969; Schmitt et al. 1972; Kallemeyn and Wasson 1981; Ebihara et al. 1982) that Alais is lower in Al, Au, Br, Ca, Cr, Fe, and Si, but higher in Bi, In, Sb, and Se than Orgueil and Ivuna. We find similar differences ($\geq$5%, rarely up to 10%) for individual CI-chondrite averages, but cannot confirm lower Au and Cr or higher in Bi and Se for Alais. In addition, Alais is lower (5-10%) in Li, Rb, Re, Ru, V, Zr, and higher in Nb, Sn, W, Y, and REE than Orgueil. For some elements (e.g., Nb, Ru, V, W, Y, Zr) the differences are probably due to paucity of and difficulties in the chemical analyses, and not always real compositional differences. Ivuna has the highest concentrations of Na and Br. Variations beyond analytical uncertainties were found among individual meteorites and within a given CI-chondrite, for Na, K, and Ca (Palme and Zipfel 2022; Gooding 1979; Barrat et al. 2012).

Variable elements in CI-chondrites include the same elements accumulated abundantly in terrestrial ocean water: e.g., Na, K, Mg, Ca, Sr, B, C, S halogens, also Au and Os. Aqueous (re)distribution of mobile elements into host phases that are concentrating them account for some heterogeneity in bulk samples if carbonates, sulfates, and phosphates are sampled in varying proportions. Redistribution of sulfates during terrestrial storage at varying humidity is seen by appearance of efflorescence on CI-chondrite samples. If redistributions are within a closed system (isochemical) mobile element net loss does not occur but they influence sampling and representative sample sizes. For example, if efflorescence or veined material is avoided, the elements making such materials might be lower in bulk analyses.

The "chemical homogeneity" of CI-chondrites depends on the sample size and is different for each element because of different host phases. Most elements do not show any discernible correlation of concentration with sample sizes larger than about 20-50 mg (but see e.g., example for C above). This comparison is not completely unbiased because different methods also can affect the results.

Mobile elements have concentration histograms with either widely spread values or bi- to tri-modal distributions as shown in Fig. 9 for e.g., K, Rb, Ca, Os, Ni, and S. In Fig. 9, visual divisions were done into low (red), main (blue), and high (purple) modes for which the averages and PDF curves are shown. These divisions are non-unique but were guided by criteria to define bimodal or higher mode distributions. There is no practical analytical division into modes, which among other things depends on the chosen bin sizes. We apply Sturge's rule (Scott 2009) and Scott's rule (Scott 1979) for bin size optimization. For the mode division we use the rule that the means of the modes should be separated by more than two standard deviations and the standard deviations for each mode must be (about) the same absolute values. The recognition and division of the distributions into modes is potentially biased by analytical methods. In the ideal cases, enough reliable analyses by one or more proven methods with similar analytical uncertainties are available, but this may not be the case for each element considered. In all multi-modal cases (e.g., diagrams like in Fig. 9) the most populated mode is taken as representative average. This "main" mode (in blue) is typically the central mode in trimodal distributions.

Ca-bearing phases (carbonates, sulfates, phosphates) are important carriers of trace elements such as Sr, Ba, Sc, Zr, Hf, U, Th, and REE. Samples enriched in these elements could be explained by carbonate and phosphate accumulations. Sodium, K, and Ni enhancements often indicate deposited sulfates, depletions point to samples poor in sulfates and other salts. Element associations cannot be checked easily using trace element correlations because only a few measurements exist where major and trace elements were measured for the same sample.

18.9 Alkalis (Na, K, Rb, Cs)

Sodium, K, and Rb enhancements suggest deposited sulfates, whereas depletions point to samples poor in sulfates and other salts, as observed in matrix samples (see e.g., Morlok et al. 2006; McSween and Richardson 1977). Water soluble host phases (mainly sulfates, halides, possibly phosphates) facilitate alkali mobility, and essentially all Na and K are extractable by water from CI-chondrites (F88). Bimodality (possibly trimodality, Fig. 9) and well-defined correlations of K and Rb are known (Goles 1970). However, Cs is unimodal with a well-defined average, and no correlations of Cs with other alkalis are observed suggesting different host phases for Cs. Due to its larger ionic radius, Cs might be stronger intercalated in layer silicates than the other alkalis, minimizing redistribution during aqueous processing.

18.10 Calcium (Ca)

In CI-chondrites, Ca is a major element in carbonates (breunnerite, calcite, dolomite), phosphates, and in (rare) sulfates (e.g., gypsum). Calcium was analyzed by NAA, ICPAES, PGA, XRF and classical wet chemical analytical methods in all CI-chondrites. Following the division into modes as described above, the nominal main mode average for Orgueil is 0.893 $\pm$0.088 wt%, and for all CI-chondrites 0.897 $\pm$.080 wt%. However, the criterion requiring similar standard deviations for the modes is not fulfilled, and the division into modes is not trivial for Ca. The distribution might be better described by a continuous log-normal distribution as expected from dispersal of several different Ca-rich phases (in contrast to other elements where the variations are influenced by one main carrier phase). For a representative Ca abundance, we use element ratios as described in Palme et al. (2014) and below.

Large concentration variations become apparent when small sample volumes or areas are analyzed. Morlok et al. (2006) described different lithologies in CI-chondrites with variable chemical compositions on a scale of 50 to 100 μm, thus the estimated analyzed volume is about 10^{-6}cm^3 which is orders of magnitudes below the volume of the analyzed bulk samples (about 10-2000 mg for Ca). Their Ca contents vary by a factor of 60, from 0.02 to 1.5% Ca in 115 different fragments. Greshake et al. (1998) found large K, Ca, S, and Al variations among 35 Orgueil matrix particles of about 100 μm in size. During aqueous redistribution Ca is enriched into phosphates, sulfates, and carbonates. This causes the micrometer-scale variability. The Ca-content of bulk samples on the scale of around 10^{-1}cm^3 is about constant, but varies on the scale of 100 μm.

Matrix samples were sought out by Greshake et al. (1998) and Ca-bearing phases were rarely among their "matrix" samples. Such phases were studied by Morlok et al. (2006). In both studies, samples with Ca > 1 wt% often have high and correlating S and/or P contents. Tiny high-Ca samples can include minute phosphates, sulfates, and carbonates which can explain some trace element variations in CI-chondrite samples.

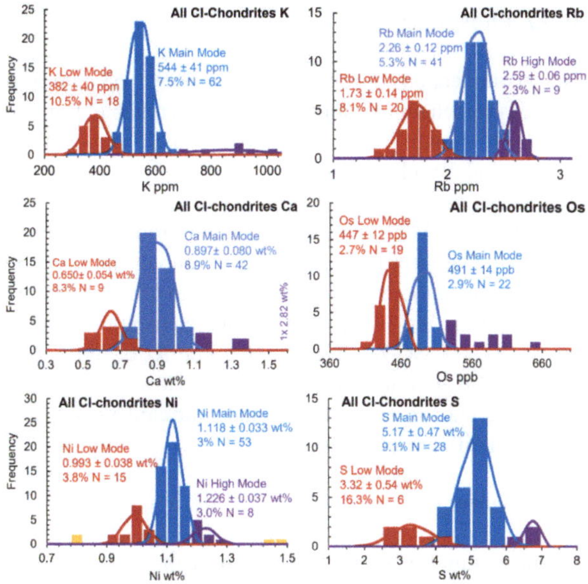

Fig. 9 Examples of mobile element concentrations in CI-chondrites

18.11 Nickel (Ni)

Nickel is typically not considered a "mobile element", however, variable Ni concentrations in CI-chondrites are not just due to potential analytical difficulties. Nickel analyses for Orgueil range from about 0.9 to 1.4 wt% and indicate bimodal distributions in NAA and ICP datasets. Similarly large variations appear in Alais and Ivuna, and are best seen in the combined dataset (Fig. 9). A critical number of analyses has to be available before different modes are clearly discernible. All data (without any outlier removal) are distributed into 2-3 modes. The criteria for the modes are fulfilled, and the main mode average gives 1.118±0.033 wt% Ni for CI-chondrites.

Tonk and Revelstoke data do not contribute to the main mode of the Ni data. The two Ni measurements (0.8, 1.3 wt%) of Revelstoke from Folinsbee et al. (1967) both fall outside the two-sigma range of the main mode mean. Their analytical method for Ni was optical emission spectrography with a stated uncertainty of about 16%. The single Tonk value by INAA (Palme and Zipfel 2022) is rather low (0.80 wt%) and compares to the lowest value for Revelstoke by Folinsbee et al. (1967).

One interpretation of the modes is soluble epsomite with isomorphous Ni substitution for Mg. Whether this sulfate is indigenous to CI-chondrites or formed in the terrestrial environment is inconclusive and related to the sulfur distribution (e.g., Gounelle and Zolensky 2001; Lodders and Fegley 2011). The sulfate is highly soluble, and Ni was found in water extracts from CI-chondrites (e.g., Thenard 1806; Berzelius 1834; Cloez 1864a,b; Fredriksson and Kerridge 1988). The low mode may represent samples that lost sulfates, and the high mode samples with sulfate deposits Samples very high in Ni (yellow bars in Fig. 9) likely stem from sulfate-veined lithologies of CI-chondrites. Nickel concentrations in sulfates reach up to 10 wt% (Fredriksson and Kerridge 1988). Addition of 1 mass% of such sulfates to samples with normal Ni concentrations of 1.12wt% increases the Ni content to 1.21 wt%, just what is found for the high mode average.

18.12 Phosphorus (P)

The recommended concentration of P = 989 ± 63 ppm from 24 analyses of all CI-chondrites excludes wet chemical values deemed less reliable. Barrat et al. (2012) measured six Orgueil samples both with ICP-AES and ICP-MS-SF and obtained reproducible results for a given sample, and found a range of 940-1080 ppm P among their six samples. Seven XRF measurements (A89, M69, mainly W01) for Orgueil range from 800-1200 ppm. The two values from A69 and M69 seem systematically too low whereas J14 is the highest value of the XRD set. The intra-laboratory spread among samples compares to inter-laboratory results by the same methods making it difficult to uncover data with underlying analytical issues.

Phosphates are likely involved for elemental distributions with "trailing" values, e.g., for Sc, Zr, Hf, Sr, Ba, Th, U and some REE. Some other elements (Nb, Ta, B, As, Sb, Na, and Mg) hint at such distributions. However, fewer trailing values for P are observed than expected for isolated phosphate enrichments but there are also fewer P analyses and analytical difficulties for P.

18.13 Sulfur (S)

The histogram for S for all CI-chondrite data shows a trimodal distribution, consistent with redistribution of S in CI-chondrites and associated changes in oxidation states of S as sulfides, elemental sulfur, hyposulfites, and sulfates. The main mode from all CI-chondrites is taken as 4.25 < S (wt%) < 6.25, which includes 29 analyses giving an average CI-chondritic S concentration of 5.18 ±0.48 wt%.

Only three S measurements for Alais are in the literature. Two older analyses gave relatively low sulfur (3.5%, Thenard 1806, 1.6% Berzelius 1834) which are nominally grouped into the low mode, and one analyses with high S (6.7%) by Wiik (reported in Mason 1962) is among the high-mode samples. The paucity of measurements and divergence of data for Alais do not provide a representative individual S value for Alais.

18.14 Other Elements with Changes Above 3% from Palme et al. (2014)

18.15 Antimony (Sb)

Compared to P14, Sb increased by 8.3%. Out of 65 Sb values for all CI-chondrites, 51 are obtained by RNAA and INAA, which vary widely (110 – 220 ppb). The Sb distribution for the NAA results is unusually right-skewed and maybe low chemical yields are to blame, but the cause is difficult to track. The RNAA by Krähenbühl et al. (1973) and Takahashi et al. (1978) are often lower than those by Ebihara et al. (1982) for aliquots of the same samples, and the Sb results remain puzzling (see Ebihara et al. 1982). The average excludes K73 and T78 and the highest (220 ppb, C73) and lowest (110 ppm, X92) NAA values. The 7 ICP-MS results give 170 ±12 ppb for Orgueil, here the low 124 ppb by W05 is an outlier. Among the 3 ID-SMSS one unusually high value was not used (250 ppb, R93). All selected values yield the recommended Sb 157 ±25 ppb (N=51) from a right-skewed PDF. If none of the NAA data are kept, the average is 161 ±22 ppb with a more gaussian-shaped histogram. Overall, Sb remains ill-determined, and we assign an overall rating of "C".

18.16 Mercury (Hg)

Mercury (Hg) is difficult to analyze and is a frequent contaminant in CI-chondrites, especially in Orgueil (e.g., Palme and Beer 1993; Palme et al. 2014). Alais and Ivuna show more plausible values. The selected value from three analyses is 0.283 ± 0.106 ppm (36%). However, this selected concentration is only 85% of the expected s-process systematics, which implies 0.35 ppm instead. Another test for the Hg abundance is the curve of the product of the neutron-cross section times the isotopic abundances of s-process elements (σN_S) as a function of mass number (shown in isotope section below). This downward stepping curve has regions of constant σN_S and pure s-process ^{198}Hg should follow that trend. However, its σN_S plots slightly above other heavy s-process only isotopes. This comparison is more complicated as nearby abundances (^{192}Pt, ^{186}Os) are also somewhat problematic.

Shirai et al. (2024) obtained Hg with NAA in four Ryugu samples averaging to 0.806 ± 0.083 ppm. Ryugu shows many compositional similarities to CI-chondrites and the carefully curated Ryugu samples are unlikely to be contaminated with Hg. However, the more than twice higher Hg concentration in Ryugu than in CI-chondrites is puzzling. In the odd-numbered nuclear abundance systematics, the ^{199}Hg abundance would then be similar to ^{197}Au, and the right-sided curvature of the prominent r-process peak would take an asymmetric form. Gold and Tl abundances are unlikely to be grossly underestimated in the existing samples, so Hg remains the exception here. Similarly, ^{198}Hg in the "pure" s-process systematics would be far too high (see Figures below). Even allowing for somewhat generally higher overall concentrations in Ryugu because of lower water contents than in CI-chondrites, the Hg in Ryugu is too high, possibly suggesting localized enrichment of Hg in the small samples analyzed. The concentrations of Hg need to be re-determined in larger (500-1000 mg), more representative samples.

Given the analytical uncertainties and high mobility of Hg, the best source for the solar system abundances would be the Sun itself but photospheric Hg determinations are not available and have their own challenges.

18.17 Halogens

The halogen (F, Cl, Br, I) abundances in CI-chondrites, the Sun, and other astronomical environs were reviewed in Lodders and Fegley (2023) and are adopted here. The problems related to halogen abundances are also discussed by Palme and Zipfel (2022). Reliable measurements on 0.5-1 gram well characterized samples are highly recommended.

18.18 Noble Gases in CI-Chondrites

The noble gas data are the same as in Lodders (2003) and Palme et al. (2014), which were computed from the principal isotope data in Anders and Grevesse (1989) and date back to Mazor et al. (1970). Although the measurements themselves are precise, we added 10% uncertainties because the actual assignment of indigenous C-chondritic noble gases is difficult amid the considerable noble wind and cosmic ray exposure contributions.

19 Element Ratio or Element Correlation Method for Refractory Elements

In some cases, we used well determined element ratios in CI-chondrites and other carbonaceous chondrites to increase the accuracy of CI-abundances. Of particular importance are element ratios used for dating, such as Os/Re, Lu/Hf, Sm/Nd, or Rb/Sr. A ratio such as the ^{176}Lu/^{177}Hf ratio can be precisely determined to within one percent, and such a ratio is readily converted to an element ratio. Absolute concentrations of Lu and Hf require additional analytical efforts and are often not given in papers focusing on age determination. Thus, one has to rely on one element usually determined with less accuracy than the ratio and calculate the other element using the precise element ratio and error propagation.

The method is particularly useful when ratios are constant in various types of chondrites. In the following we only consider refractory element ratios from carbonaceous chondrites. The isotope ratio of ^{176}Lu/^{177}Hf is a good example (see below). Both elements, Lu and Hf, are refractory elements and the calculated Lu/Hf ratio has the same uncertainty as the ^{176}Lu/^{177}Hf ratio. The CI-chondrites contain up to 15% water, more than any other chondrite group, and mobile elements may have been redistributed (see above). By using pairs of refractory elements analyzed in other carbonaceous chondrites with minimal or no aqueous alterations, and with one element being more mobile than the other it may be possible to improve the CI-abundance of the more mobile element if analyses were done on small samples, and/or sampling avoided or mistakenly incorporated phases that concentrate mobile elements.

The most accurate method for determining the concentrations of trace elements with more than one isotope is isotope dilution (ID) combined with thermal ionization mass spectrometry (TIMS) or with inductively coupled plasma mass spectrometry (ICP-MS). A calibrated tracer with an isotope ratio different from the natural isotope ratio is mixed with the dissolved sample. The isotope ratio of an element in the dissolved sample plus the ratio in the spike are then measured and concentration of the trace element can be calculated. The mass of the mixed solution is irrelevant, and no rock standards are required (e.g., Stracke et al. 2014). The method is labor intensive and is primarily used to determine the concentration for pairs of elements used for dating purposes, such as Rb/Sr, Lu/Hf etc. Most other methods require well analyzed standard rocks as reference. The ID method cannot be used for determining the abundances of monoisotopic elements, e.g., Sc, Nb, and Rh but is good for most REE and other refractory lithophile elements, such as Zr, Hf, and Ta and for siderophile refractory trace elements, like Re, Os, Ru, Ir, W, Mo. Abundances of trace elements analyzed with ID can be determined with an accuracy of one percent or less, whereas data obtained with ICP-MS methods involving dissolution of fragments or homogenized powers are less accurate, in the best cases about 3 to 5%.

Other methods are less accurate for most of the refractory elements. An exception is the monatomic Sc, which can be determined with high accuracy by neutron activation analysis. Aluminum, Ca, and Ti are typical elements for high accuracy analysis with XRF. Another problem in analyzing bulk samples of CI-chondrites is sample heterogeneity. The variability of analytical results of some major and trace elements obtained with ID is primarily the result of inhomogeneous distribution of elements within the meteorite. The contribution of analytical uncertainties is often minor. Analyzing large enough samples, e.g., a total of 1-2 grams of homogenized powder is often recommended (e.g., Morlok et al. 2006; Stracke et al. 2012). However, analyzing aliquots of homogeneous powders from larger masses is not necessarily a guarantee to avoid the nugget effect.

In the following sections we use values obtained from correlations only, if the uncertainty of an element calculated from correlations is below the uncertainty of the grand mean.

19.1 Refractory Lithophile Elements

Ca/Al Relying on published data and using the averaging procedure described above gives CI-chondritic grand average values of Ca = 0.897 ± 0.08 wt.% and Al = 0.847 ± 0.048 wt.%, with a Ca/Al ratio of 1.059. The uncertainty of 8.9% in the Ca abundance is higher than the uncertainty of Al (5.7%). Calcium is more mobile than Al and concentrates into minor phases such as carbonates (see above) which affects the Ca/Al ratio in heterogenous, small samples.

If it can be shown that other carbonaceous chondrites with fewer hydrous phases have constant Ca/Al ratios, one may assume that CI-chondrites have the same ratio. Then the Ca abundance can be calculated from the Al content in CI-chondrites and the Ca/Al ratio of the other carbonaceous chondrites. Table A5 (appendix) gives Ca/Al ratios of carbonaceous chondrites with less water than CI-chondrites. Ahrens et al. (1969, 1970) used XRF on chondritic sample masses of around one gram. Data for twelve carbonaceous chondrites reported by Ahrens (1970) give very constant Ca/Al ratios with a mean of 1.077±0.02. Earlier Ahrens et al. (1969) summarized four Orgueil datasets with highly variable Ca/Al ratios ranging from 1.0 to 2.05, confirming the high mobility of Ca in CI-chondrites.

A broader survey of the literature yields a mean Ca/Al ratio of 1.080 ± 0.023 for carbonaceous chondrites. In calculating this ratio 126 samples of carbonaceous chondrites were considered, and 17 samples were excluded because their Ca/Al ratios were outside a 2σ limit (Ahrens 1970; Kallemeyn and Wasson 1981; Jarosewich 1990; Wolf and Palme 2001; Patzer et al. 2010; Stracke et al. 2012). A summary is given in the appendix (Table A5).

Assuming that all carbonaceous chondrite groups have the same Ca/Al ratio and using an Al content of 0.847% for CI-chondrites (Table 4) a Ca content of 0.847*1.08 = 0.915 wt% is calculated for CI-chondrites. This is somewhat higher, but within the uncertainty of the grand mean Ca concentration. The combined uncertainties of the Al content and the Ca/Al ratio give an uncertainty of 6.1% for Ca, which is less than the uncertainty for Ca of the grand average (8.9%). Table 4 thus gives the Ca content from the Ca/Al correlation. The difference between the grand average and the correlation method is small (2-3%) but it serves to demonstrate the procedure.

Lu/Hf The refractory lithophile element ratio Lu/Hf connects the abundances of the refractory trivalent REE with tetravalent Zr and Hf. The Lu/Hf ratio in geochronology requires precise knowledge of the abundances of ^{176}Lu (half-life 37.1 Ga) and ^{176}Hf. From the precisely determined ^{176}Lu/^{177}Hf ratios, Lu/Hf mass ratios can be calculated. Using only isotope dilution data an average Lu/Hf of 0.2384 ± 1% is obtained (Table A6, appendix).

This Lu/Hf ratio and the absolute Lu concentration of 25.5 ± 1.5 ppb from the grand average (Tables 4 & A4) are used to calculate Hf = 107 ± 6 ppb, which is close to the grand average of Hf of 108 ± 8 ppb. We adopt the value with lower uncertainty (Table A4). The uncertainty in this Hf determination is only slightly higher than for Lu because the uncertainty in the Lu/Hf ratio is only around 1%. The grand average of Lu = 25.5 ppb is well determined. Two large, representative Orgueil samples analyzed with ID (Beer et al. 1984) gave 25.3 and 25.4 ppb Lu (Table 4), and six recent ID-analyses of Lu in Orgueil average to 25.5 ppb (Münker et al. 2025).

Zr/Hf Both elements are the first lithophile elements to condense from a hot nebular gas. Their condensation temperatures as pure oxides or in solid solution are above those of the REE and Al. Hence the Zr/Hf ratio should be constant in carbonaceous chondrites (Table A7). The grand average CI-chondrite concentration for Zr is 3.73 ± 0.30 ppm. No differences in Zr/Hf ratios between CI-chondrites and other carbonaceous chondrites are apparent (e.g., Münker et al. 2003, 2025). An average carbonaceous chondritic Zr/Hf ratio of 34.1 ± 1 is obtained from isotope dilution data (see appendix A7). This ratio is used in further calculations. With a Hf content of 107 ± 6 ppb a Zr value of 3.65 ± 0.21 ppm is calculated. The uncertainty of this value is lower than that of the grand average (± 0.30 ppm). We therefore adopt the Hf and Zr values obtained from the correlations. This example demonstrates the correlation method. The Zr/Hf of all carbonaceous chondrites has an uncertainty of 3%; using only ID data it reduces to about 1%. The combined uncertainties of the Zr/Hf ratio and the Hf content were calculated by taking the square root of the sum of the squared uncertainties in all cases. The minimum uncertainty is the uncertainty of 5.7% from Lu which enters into the Hf abundance. Thus, using the correlation method requires a single reference element with known absolute concentration.

Zr/Nb Both ratios, Lu/Hf and Zr/Hf, are identical, within uncertainties, in CI, CM, CV and CO chondrites (e.g., Münker et al. 2003, 2025; Bouvier et al. 2008). All three elements have similar high condensation temperatures: Lu (1659 K), Hf (1703 K), Zr (1764 K). But Zr and Nb (1559 K) differ by nearly 200 K in condensation temperature, encompassing the REE condensing between 1578 K and 1659 K (e.g., Lodders et al. 2009). The refractory element and REE patterns in the Allende CV chondrite are fractionated with a deficit of early condensing refractory REE (Stracke et al. 2012 and references therein), similar effects could be expected for the Zr/Nb ratio, with enhancements of the more volatile Nb. Allende (Zr/Nb = 12.56, Münker et al. 2003) has indeed a slightly lower Zr/Nb ratio than Orgueil (14.07 in Münker et al. 2003, and 14.3 in Münker et al. 2025). Stracke et al. (2012) showed that the Zr/Nb ratio is variable in 500 mg-size samples of Allende. A range of Zr/Nb ratios from 10 to 13 is observed in their data (Table A8), and these authors concluded that homogenization of 60 grams of Allende is required for determining its average Nb abundance. The variability of Nb is probably caused by the inhomogeneous distribution of group II inclusions in Allende (Mason and Taylor 1982; Stracke et al. 2012). The abundances of group II CAIs in other carbonaceous chondrite groups are unclear.

To avoid problems with the potential addition of group II inclusions we restrict element ratios involving Nb to CI- and CM-chondrites. The average Zr/Nb ratio is 13.47 ±0.78 (Table A9, appendix), leading to Nb = 0.271 ± 0.022 ppm, not far from the grand average value for Nb of 0.281±0.028 ppm. The grand average Zr/Nb ratio is 13.28 ± 1.7.

Nb/Ta Niobium is monoisotopic and cannot be analyzed with ID. This lowers the accuracy of the Nb/Ta ratio. Tantalum shows the same tendency as perhaps Nb, higher Zr/Ta ratios in CI- and CM-chondrites than in CV-chondrites. The trend is stronger for Ta than for Nb: Münker et al. (2003) have six CI and CM chondrites with a mean Nb/Ta ratio of 20.40 ± 0.77, whereas seven CV- and CK-chondrites give a mean ratio of 17.73 ± 2.15. However, data by Barrat et al. (2012) and Göpel et al. (2015) do not show this effect. Eight CI-chondrite measurements give a mean Nb/Ta of 19.6 and three CM-chondrites a mean Nb/Ta ratio of 19.3. Lu et al. (2007) did not find significant differences between CI-, CM- and CV-chondrites. More high-quality data of Nb and Ta in chondrites are needed. Adding the results of Münker et al. (2025) to the average Nb/Ta ratio of carbonaceous chondrites (Table A9) and using only CI- and CM-chondrites gives an average Nb/Ta =19.40 ±0.76, compared to the grand average Nb/Ta of 18.79. The Ta concentration of 0.0140 ppm (±9.2%) from correlations is somewhat lower than the grand average of 0.0149 ppm (± 4.7%), but with lower uncertainties. The latter value is used for CI-chondrites in Table 4.

Zr/Y The Y concentration is calculated from Zr/Y ratios (Table A10) but only a few data of this ratio are in the literature. The REE datasets often include Y, but not Zr, which is analyzed together with other HFSE elements (High Field Strength Elements). The average Zr/Y ratio from analyses by Jochum et al. (2000), Stracke et al. (2012), Barrat et al. (2012) and Göpel et al. (2015) is 2.4 ± 0.16. Insufficient data exist to address whether the Zr/Y ratio differs between CI-chondrites and other types of carbonaceous chondrites. From the CI-chondrite

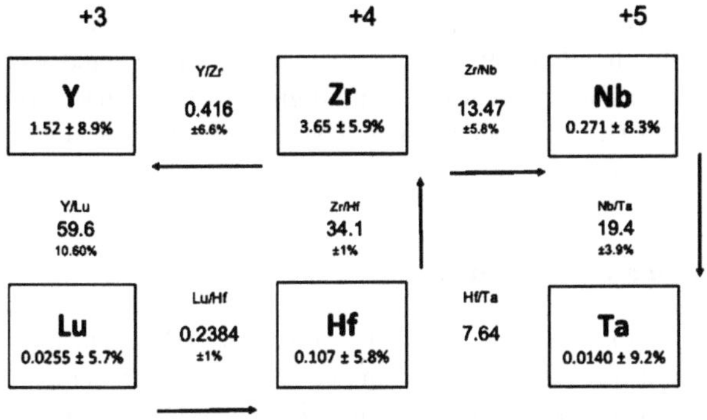

Fig. 10 Relationships among incompatible refractory lithophile elements. Starting with the absolute Lu abundance all other abundances of elements in the Figure are derived from correlations involving all types of carbonaceous chondrites, except those including Nb, where only CI and CM ratios are used. Improvement of the accuracy of Lu will improve the accuracy of all other elements in the Figure. Element ratios are listed in Tables A4 to A12

abundance of Zr 3.65 ± 0.21 ppm and the Zr/Y ratio we calculate an Y abundance of 1.52 ± 0.13 ppm, close to the grand mean of 1.51 ± 0.14 ppm. We use the value from the correlation, Y = 1.52 ± 0.13 ppm.

Sm/Nd The Sm/Nd ratio is probably the most accurately determined element ratio in chondrites because the Sm isotopes ^{146}Sm and ^{147}Sm are used for dating. The large dataset shows no significant variations among carbonaceous chondrite groups (Table A11) and averaging over all members of the CC groups is justified. Isotope dilution results and LA-ICP-MS data give a Sm/Nd elemental ratio of 0.3268 with an uncertainty <1%. The CI grand mean ratio is 0.3231 ± 0.0237 (7.3%), only about 1% below the average carbonaceous chondritic ratio from correlations.

Th/U Neither the absolute concentrations of U and Th, nor the Th/U ratio in chondrites are well known. We do not find a well-defined Th-U correlation for CI-chondrites. Uranium is sensitive to aqueous mobilization (Rocholl and Jochum 1993). Variations in Th/U ratios exist in all carbonaceous chondrites. It is unclear if CI- and CV-chondrites have the same Th/U ratio. The Th and U histograms display trailing distributions like e.g., Ba, suggesting that U and Th are in phosphates and possibly carbonates. The mean Th/U of carbonaceous chondrites (Table A12) is 3.75 ± 0.38 (10%), compared to 3.75 ± 0.54 (14%) for CI-chondrites from the grand average. The agreement of the ratios could be coincidental because wide variations in the Th/U ratio exist among chondrites.

19.2 Summary of Refractory Lithophile Elements

Figure 10 summarizes using the ratio method for elemental abundances in CI-chondrites. The main thrust is to use the information in other carbonaceous chondrites to improve the quality of CI-chondrite values.

Starting with the grand average CI-chondritic Lu content of 0.0255 ppm and using the well-defined Lu/Hf ratio leads to a Hf content of 0.107 ppm with a combined uncertainty of 5.8%. By continuing with this procedure (Fig. 10), CI-abundances for Zr, Nb, Ta, and Y can be calculated. The difference to the directly measured grand average values of these elements is in all cases within error bars. However, the uncertainties of the elements derived from correlations are often lower than those calculated from the grand averages.

The premise of constant element ratios in all groups of carbonaceous chondrites is the case for Lu/Hf and Zr/Hf ratios, but Zr/Nb and Nb/Ta ratios may be slightly different in CI- and CM- compared to CV-chondrites, as explained before. Except for ratios involving Nb we used all types of carbonaceous chondrites for averaging element ratios.

19.3 Refractory Siderophile Elements

Hf/W The Hf/W ratio connects refractory lithophile (Hf) elements with refractory siderophile (W) elements. Under the reducing conditions of the solar nebula, W condenses as metal in an alloy with other refractory metals, Re, Os, Ir etc. (Palme and Wlotzka 1976). At higher oxygen fugacities enhanced formation of gaseous WO_3 lowers condensation temperatures of W (Fegley and Palme 1985). Thus, concentrations of W in bulk chondrites and Ca, Al-rich inclusions can be variable.

The application of the Hf-W clock, using the decay of ^{182}Hf ($T_{1/2}$ = 9 Ma) into ^{182}W, led to precise measurements of ^{180}Hf/^{184}W ratios, which can be converted to Hf/W ratios. Kleine et al. (2004) analyzed 18 carbonaceous chondrites and found a mean Hf/W ratio of 1.08 ± 11%, indicating inhomogeneous distribution of W and Hf in chondrites. Analytical uncertainties are much lower (around 1%) because both ^{180}Hf and ^{184}W were analyzed by isotope dilution. Yin et al. (2002) obtained, with the same method, 1.14 for the Hf/W ratio in Allende. Braukmüller et al. (2018) reported a mean Hf/W ratio of 1.10 ± 10% for 23 carbonaceous chondrites, whereas Barrat et al. (2012) determined a mean Hf/W ratio of 0.92 ± 30% for 11 carbonaceous chondrites, significantly lower than other analysts. Using an average Hf/W ratio of 1.08 and a Hf content of 107 ppb leads to a CI-chondritic W content of 99 ppb with an estimated uncertainty of ± 12.3%, compared to 101 ± 11.9% ppb determined by the grand average. Münker et al. (2025) measured a Hf/W ratio for six Orgueil samples of 1.16 ± 0.08 and an average Hf/W CI-chondritic ratio of 1.12 ± 0.15. Kleine et al. (2004) reported data on two Orgueil samples with Hf/W ratios of 0.95 and 1.00. In view of these discrepancies, we use the grand average W-content of 101 ± 12 ppb.

The variability in W may reflect its strong sensitivity to aqueous alteration but it is also unclear if the CI-chondritic Hf/W ratio differs from those of other carbonaceous chondrites. More high-quality data for W and Hf in CI-chondrites are necessary to resolve this question (see Hellmann et al. 2024).

19.4 Refractory Metals Other than W

The other refractory metals (Ru, Re, Pt, Rh) can be scaled to Ir and Os (Table 5) using the grand means for Ir = 456 ± 27 ppb and Os = 491 ± 14 ppb. The nearly identical Os/Re ratios of Horan et al. (2003) and Fischer-Gödde et al. (2010) provide a reliable Re content (39.9 ± 1.6 ppb). The grand mean Re of 37.1 ± 3.4 ppb is lower than that from the Os/Re ratio and has significantly larger uncertainty. The Ru and Pt values are also derived from correlations. The grand mean for Rh is used for Table 4 because it is more accurate than the correlation derived Rh value.

20 Present-Day and Proto-Solar System Abundances

A scale-coupling factor is needed to combine the meteoritic and solar abundance sets to obtain a complete set of a solar system elemental abundances (combining the sets is different than renormalizing the scales to Si=10^6 atoms). The meteoritic set is relative to 10^6 Si

Table 5 Concentration Ratios and Concentrations of Refractory Metals in Carbonaceous Chondrites

Reference	N [a]	Os/Ir	Os/Re	Os/Ru	Pt/Ir	Pt/Rh
		(1)	(2)	(3)	(4)	(5)
Horan et al. (2003)	20	1.06 ±2.3%	12.37±3.1%	0.727 ±6.2%	1.91±7.2%	...
Fischer-Gödde et al. (2010)	13	1.06 ±3.6%	12.23 ±2.7%	0.716 ±3.9%	2.05±3.6%	7.24 ±7.5%
Selected samples:	33	1.06 ±2.9%	12.32 ±3.0%	0.723 ±5.3%	1.982 ±5.0%	7.24 ±7.5%

Abundances of refractory metals in CI-chondrites (ppb)

	Ir	Os	Re	Ru	Pt	Rh
Via ratios:	N/A	483 ±6.6%	**39.9 ± 4.1%**	**680 ± 6.0%**	**904 ± 7.7%**	126 ±8.0%
Via grand mean:	**456 ± 5.9%**	**491 ± 2.8%**	37.1 9.2%	648 ±10.2%	926 ±11.2%	**133 ± 4.5%**
Recommended: (Table 4)	456 ±5.9%	491 ±2.8%	39.9 ±4.1	680 ±6.0%	904 ±7.7%	133 ±4.5%

Selected values are in bold-face. Excluded from average: (1) EET 9204; (2) Kainsaz, Karoonda, one Orgueil sample; (3) Ornans; (4) Kainsaz, Ornans, Lance (5) Ninqiang. (a) N = number of samples

atoms. The solar set is normalized to $\log_{10}$ N(H) = 12. The scale-coupling factor to combine both sets is from elements that (1) have a top "A" rating for the goodness of analysis (Table 2), and (2) are more abundant non-highly volatile elements from 3D NLTE measurements in the solar photosphere: Na, Mg, Al, Si, Ca, Mn, Fe, and Ni. The neutron capture elements Y, Ba, and Eu with "A-" ratings are not included because they have only a few clean diagnostic lines and a reliable test of the excitation-ionization balance is not possible. The scale-coupling factor (a constant in log space) from the average of the eight elements is log SCF = 1.551±0.020. It is calculated from the meteoritic mass concentration C(E) for element E normalized to Si = 10^6 atoms, and converted to logarithmic values, viz.,

$\log_{10}\mathrm{SCF(E)} = \mathrm{A(E)}_{\mathrm{solar}} - \log_{10}(\mathrm{N(E)/(Si)}_{\mathrm{CI}})$ where E = Na, Mg, Al, Si, Ca, Mn, Fe, and Ni, and

$\log_{10}(\mathrm{N(E)/(Si)}_{\mathrm{CI}}) = \log_{10}\{(\mathrm{C(E)}$ [in g/g] / atwt(E) [in g/mol]) / (Si [in g/g]/ atwt(Si) [in g/mol] + 6}.

Elemental photospheric abundances are then convertible to the meteoritic abundance scale via:

$\mathrm{N(E/Si)}_{\mathrm{solar}} = 10^{(\mathrm{A(E)solar} - 1.551)}$

or CI-chondritic values to the photospheric abundance scale via:

$\mathrm{A(E)}_{\mathrm{CI}} = \log \mathrm{N(E/Si)}_{\mathrm{CI}} + 1.551$

The solar and meteoritic abundances are then easily compared (see Table 4). The meteoritic/photospheric abundance ratios are shown in Fig. 11 as a function of atomic number.

Table 6 summarizes the recommended solar system abundances; the isotope abundances are given in the appendix. Both present-day (column 3) and proto-solar (column 4) abundances in Table 6 are a combination of solar photospheric and meteoritic values as indicated in the note column. The CI-chondritic values in column (1) are from Table 4 (strictly normalized to Si = 10^6 atoms) and are used when no or uncertain photospheric values exist. Photospheric values in column (2) are obtained via $\mathrm{N(E)}=10^{\mathrm{A(E)}-1.551}$ from Table 2 and are used for H, C, N, and O; the noble gases are from Table 3. Solar and meteoritic abundances (weighted by uncertainties) were only averaged for elements (including Si) that were used to link photospheric and meteoritic abundance sets. The solar system values in column (3) were re-normalized to Si = 10^6 because the average of solar and meteoritic abundances is also used for Si (see note column) and does not exactly equal 10^6. This explains the small

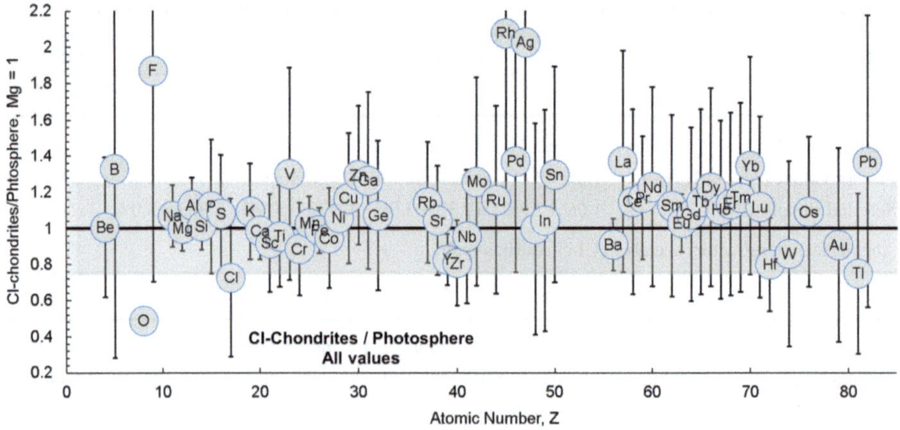

Fig. 11 Concentration ratios of CI-chondrites/photosphere for all elements where the comparison can be made as a function of atomic number. Solar F, Cl, In, and Tl are from sunspot spectra. The gray band indicates agreement within 26% (0.1 dex). The most deviant elements shown are Ag, Rh, Pd, F, B, B, V, La, and Yb. The reason (s) are long-standing analytical issues for photospheric but also CI-chondritic abundance determinations. Ratios for H, C, N and noble gases plot «0.2. These and O are low because ultra-volatile elements are not fully retained on meteorite parent bodies. Lithium is destroyed in the Sun, and is not shown; the ratio is 188 ± 50

differences in columns (1) and (3) when meteoritic values are used or differences in columns (2) and (3) when photospheric values are used.

21 Derivation of Proto-Solar System Abundances and Mass Fractions (X, Y, and Z)

Heavy element settling towards the solar interior during the Sun' lifetime depleted the convection zone and thus the photosphere in heavy elements (Li to U) relative to H. Atomic diffusion is element specific and may not be a smooth function of atomic mass, and is somewhat counteracted by radiative levitation which depends on the radiative properties of the atomic species, specifically the ionization stage (e.g., Turcotte et al. 1998; Turcotte and Wimmer-Schweingruber 2002). Within the limits of analytical precisions, depletions are the same for all elements heavier than Li. Currently there is no discernable *systematic* gradual depletion suggesting "non-uniform" gravitational settling of the elements with increasing atomic mass between meteoritic and photospheric abundances; aside of this, how would such a baseline for gravitationally undisturbed or non-levitated abundances be defined to begin with? The comparison of meteoritic and photospheric abundances in Fig. 11 might suggest some mass-dependent fractionation because several of the heaviest elements are higher in the photosphere (or lower in CI-chondrites). However, the analytical uncertainties are still too large for solar and/or meteoritic abundances involved in defining the "apparent fractionations".

The present-day photospheric abundances relative to H are lower than at the time the Sun formed. Settling corrections assume that all elements heavier than Li were reduced in the convective envelope (CE) by the same factor over time. Lithium is depleted because of pre-main sequence Li destruction and ongoing settling combined with the destruction of the fragile Li nuclei at the hot bottom of the solar convection zone. Beryllium and B

Table 6 Recommended Atomic Solar System Abundances: Present and Proto-Solar 4.567 Ga Ago

Z	E	CI-Chondrites (from Table 4)		Sun Convection Zone (mainly photosphere)		Solar System* Present		Solar System* Proto-Solar †,‡		Note	Solar System* Present		Solar System* Proto-Solar †,‡	
		N(E), 10^6 Si	$\pm\sigma$	N(E)=$10^{A(E)+1.551}$	$\pm\sigma$	N(E), 10^6 Si	$\pm\sigma$	N(E), 10^6 Si	$\pm\sigma$		log (E/H)+12	$\pm\sigma$	log (E/H)+12	$\pm\sigma$
1	H	4.86E+06	4.49E+05	2.81E+10	2.57E+08	2.81E+10	2.57E+08	2.29E+10	2.1E+08	s	12.000	0.004	12.000	0.004
2	He	0.604	0.060	2.35E+09	7E+07	2.34E+09	7E+07	2.25E+09	6E+07	s, t	10.922	0.012	10.992	0.012
3	Li	56.2	2.7	0.308	0.071	56.1	2.7	56.1	2.7	m	3.30	0.02	3.39	0.02
4	Be	0.658	0.038	0.456	0.174	0.657	0.038	0.657	0.038	m	1.37	0.02	1.46	0.02
5	B	18.1	2.3	14.1	11.0	18.1	2.3	18.1	2.3	m	2.81	0.05	2.90	0.05
6	C	8.29E+05	1.45E+05	9.10E+06	2.10E+06	9.08E+06	2.10E+06	9.08E+06	2.10E+06	s	8.51	0.09	8.60	0.09
7	N	3.70E+04	1.82E+04	2.45E+06	7.1E+05	2.44E+06	7E+05	2.44E+06	7.1E+05	s	7.94	0.11	8.03	0.11
8	O	7.67E+06	1.3E+05	1.62E+07	2.0E+06	1.62E+07	2.0E+06	1.62E+07	2.0E+06	s	8.76	0.05	8.85	0.05
9	F	1280	277	706	413	1278	277	1278	277	m	4.66	0.09	4.75	0.09
10	Ne	2.36E-03	2.4E-04	3.97E+06	1.26E+06	3.96E+06	1.26E+06	3.96E+06	1.26E+06	s, t	8.15	0.12	8.24	0.12
11	Na	56838	5868	54838	6691	55852	1414	55852	1414	a	6.30	0.01	6.39	0.01
12	Mg	1.04E+06	3E+04	1.07E+06	1.3E+05	1.04E+06	2E+04	1.04E+06	2E+04	a	7.57	0.01	7.66	0.01
13	Al	82707	4687	75698	9237	81102	4956	81102	4956	a	6.46	0.03	6.55	0.03
14	Si	1.00E+06	4E+04	1.02E+06	1.2E+05	1.00E+06	1E+04	1.00E+06	1E+04	a	7.55	0.01	7.64	0.01
15	P	8413	757	7746	2465	8395	757	8395	757	m	5.48	0.04	5.56	0.04
16	S	425635	37798	406521	117180	424747	37798	424747	37798	m	7.18	0.04	7.27	0.04
17	Cl	5326	817	7570	4428	5315	817	5315	817	m	5.28	0.06	5.37	0.06
18	Ar	0.0097	0.0010	88937	28305	88752	28305	88751	28305	s, t	6.50	0.12	6.59	0.12
19	K	3667	276	3460	797	3659	276	3664	277	m	5.12	0.03	5.20	0.03
20	Ca	60135	3641	62963	9328	60382	2000	60378	2000	a	6.33	0.01	6.42	0.01
21	Sc	33.8	2.2	38	11	33.7	2.2	33.7	2.2	m	3.08	0.03	3.17	0.03
22	Ti	2433	165	2625	757	2428	165	2428	165	m	4.94	0.03	5.03	0.03

Table 6 (*Continued*)

Z	E	CI-Chondrites (from Table 4)		Sun Convection Zone (mainly photosphere)		Solar System* Present		Solar System* Proto-Solar †,‡		Note	Solar System* Present		Solar System* Proto-Solar †,‡	
		N(E), 10^6 Si	±σ	N(E)=$10^{A(E)+1.551}$	±σ	N(E), 10^6 Si	±σ	N(E), 10^6 Si	±σ		log (E/H)+12	±σ	log (E/H)+12	±σ
23	V	275	20	218	97	274	20	274	20	m	3.99	0.03	4.08	0.03
24	Cr	13255	588	15456	4455	13227	588	13227	588	m	5.67	0.02	5.76	0.02
25	Mn	9282	734	9313	1136	9272	22	9272	22	a	5.52	0.00	5.61	0.00
26	Fe	8.73E+05	3.0E+04	9.10E+05	1.11E+05	8.74E+05	2.6E+04	8.74E+05	2.6E+04	a	7.49	0.01	7.58	0.01
27	Co	2297	116	2507	723	2292	116	2292	116	m	4.91	0.02	5.00	0.02
28	Ni	50184	1481	48875	7241	50026	926	50026	926	a	6.25	0.01	6.34	0.01
29	Cu	552	62	489	141	551	62	551	62	m	4.29	0.05	4.38	0.05
30	Zn	1251	89	998	288	1248	89	1248	89	m	4.65	0.03	4.74	0.03
31	Ga	36	3	29.4	11.2	36	3	36.0	2.6	m	3.11	0.03	3.20	0.03
32	Ge	122	8	117	45	121	8	121	8	m	3.64	0.03	3.72	0.03
33	As	6.16	0.60	...	...	6.15	0.60	6.15	0.60	m	2.34	0.04	2.43	0.04
34	Se	71.7	2.3	...	...	71.5	2.3	71.5	2.3	m	3.41	0.01	3.49	0.01
35	Br	12.4	3.0	...	...	12.4	3.0	12.4	3.0	m	2.65	0.09	2.73	0.09
36	Kr	0.000164	0.000016	57.4	18.3	57.3	18.3	57.3	18.3	t	3.31	0.12	3.40	0.12
37	Rb	6.98	0.37	6.30	1.81	6.96	0.37	7.09	0.38	m	2.39	0.02	2.49	0.02
38	Sr	24.2	0.8	23.9	6.9	24.1	0.8	24.0	0.8	m	2.93	0.01	3.02	0.01
39	Y	4.50	0.40	5.61	0.83	4.49	0.40	4.49	0.40	m	2.20	0.04	2.29	0.04
40	Zr	10.5	0.6	13.5	3.9	10.5	0.6	10.5	0.6	m	2.57	0.02	2.66	0.02
41	Nb	0.77	0.06	0.830	0.316	0.77	0.06	0.77	0.06	m	1.44	0.03	1.52	0.03
42	Mo	2.60	0.27	2.13	0.95	2.60	0.27	2.60	0.27	m	1.97	0.04	2.05	0.04
43	Tc		...	...	...			...	...	...		...	...	...
44	Ru	1.77	0.11	1.58	0.70	1.77	0.11	1.77	0.11	m	1.80	0.03	1.89	0.03

Table 6 (*Continued*)

Z	E	CI-Chondrites (from Table 4)		Sun Convection Zone (mainly photosphere)		Solar System* Present		Solar System* Proto-Solar †,‡		Note	Solar System* Present		Solar System* Proto-Solar †,‡	
		N(E), 10^6 Si	±σ	N(E)=$10^{A(E)+1.551}$	±σ	N(E), 10^6 Si	±σ	N(E), 10^6 Si	±σ		log (E/H)+12	±σ	log (E/H)+12	±σ
45	Rh	0.341	0.015	0.169	0.075	0.341	0.015	0.341	0.015	m	1.08	0.02	1.17	0.02
46	Pd	1.39	0.06	1.04	0.47	1.39	0.06	1.39	0.06	m	1.69	0.02	1.78	0.02
47	Ag	0.504	0.046	0.256	0.114	0.503	0.046	0.503	0.046	m	1.25	0.04	1.34	0.04
48	Cd	1.60	0.11	1.66	0.97	1.59	0.11	1.59	0.11	m	1.75	0.03	1.84	0.03
49	In	0.179	0.013	0.177	0.104	0.179	0.013	0.179	0.013	m	0.80	0.03	0.89	0.03
50	Sn	3.70	0.44	2.94	1.31	3.69	0.44	3.69	0.44	m	2.12	0.05	2.21	0.05
51	Sb	0.340	0.056	...	...	0.339	0.056	0.339	0.056	m	1.08	0.07	1.17	0.07
52	Te	4.75	0.19	...	...	4.74	0.19	4.74	0.19	m	2.23	0.02	2.32	0.02
53	I	1.60	0.64	...	...	1.60	0.64	1.60	0.64	m	1.76	0.15	1.84	0.15
54	Xe	3.49E-04	3.5E-05	5.61	1.79	5.60	1.79	5.60	1.79	t	2.30	0.12	2.39	0.12
55	Cs	0.368	0.024	...	...	0.367	0.024	0.367	0.024	m	1.12	0.03	1.20	0.03
56	Ba	4.62	0.27	5.24	0.78	4.61	0.27	4.61	0.27	m	2.22	0.02	2.30	0.02
57	La	0.470	0.019	0.354	0.158	0.469	0.019	0.469	0.019	m	1.22	0.02	1.31	0.02
58	Ce	1.190	0.049	1.07	0.48	1.187	0.049	1.187	0.049	m	1.63	0.02	1.71	0.02
59	Pr	0.179	0.008	0.158	0.046	0.179	0.008	0.179	0.008	m	0.80	0.02	0.89	0.02
60	Nd	0.880	0.046	0.740	0.330	0.879	0.046	0.877	0.046	m	1.50	0.02	1.58	0.02
61	Pm	0.000	...	...	...	...	...	...	...	...	...	...	...	...
62	Sm	0.273	0.014	0.251	0.112	0.273	0.014	0.274	0.014	m	0.99	0.02	1.08	0.02
63	Eu	0.104	0.005	0.104	0.015	0.104	0.005	0.104	0.005	m	0.57	0.02	0.66	0.02
64	Gd	0.353	0.017	0.338	0.151	0.352	0.017	0.352	0.017	m	1.10	0.02	1.19	0.02
65	Tb	0.0638	0.0028	0.0574	0.0256	0.0637	0.0028	0.0637	0.0028	m	0.36	0.02	0.44	0.02
66	Dy	0.421	0.018	0.354	0.158	0.420	0.018	0.420	0.018	m	1.17	0.02	1.26	0.02

Table 6 (*Continued*)

Z	E	CI-Chondrites (from Table 4)		Sun Convection Zone (mainly photosphere)		Solar System* Present		Solar System* Proto-Solar †,‡		Note	Solar System* Present		Solar System* Proto-Solar †,‡	
		N(E), 10^6 Si	$\pm\sigma$	$N(E)=10^{A(E)+1.551}$	$\pm\sigma$	N(E), 10^6 Si	$\pm\sigma$	N(E), 10^6 Si	$\pm\sigma$		log (E/H)+12	$\pm\sigma$	log (E/H)+12	$\pm\sigma$
67	Ho	0.0908	0.0037	0.0849	0.0378	0.0906	0.0037	0.0906	0.0037	m	0.51	0.02	0.60	0.02
68	Er	0.263	0.011	0.239	0.107	0.263	0.011	0.263	0.011	m	0.97	0.02	1.06	0.02
69	Tm	0.0410	0.0019	0.0362	0.0161	0.0410	0.0019	0.0410	0.0019	m	0.16	0.02	0.25	0.02
70	Yb	0.259	0.010	0.199	0.089	0.259	0.010	0.259	0.010	m	0.96	0.02	1.05	0.02
71	Lu	0.0384	0.0022	0.0354	0.0158	0.0383	0.0022	0.0384	0.0022	m	0.13	0.02	0.22	0.02
72	Hf	0.158	0.009	0.204	0.065	0.157	0.009	0.157	0.009	m	0.75	0.02	0.84	0.02
73	Ta	0.0218	0.0010	...	...	0.0217	0.0010	0.0217	0.0010	m	-0.11	0.02	-0.02	0.02
74	W	0.144	0.017	0.173	0.101	0.144	0.017	0.144	0.017	m	0.71	0.05	0.80	0.05
75	Re	0.0564	0.0023	...	...	0.0563	0.0023	0.0591	0.0024	m	0.30	0.02	0.41	0.02
76	Os	0.680	0.019	0.644	0.245	0.679	0.019	0.676	0.019	m	1.38	0.01	1.47	0.01
77	Ir	0.625	0.037	0.740	0.433	0.624	0.037	0.624	0.037	m	1.35	0.03	1.44	0.03
78	Pt	1.221	0.095	...	...	1.218	0.095	1.218	0.095	m	1.64	0.03	1.73	0.03
79	Au	0.201	0.018	0.229	0.134	0.201	0.018	0.201	0.018	m	0.85	0.04	0.94	0.04
80	Hg	0.385	0.139	...	...	0.384	0.139	0.384	0.139	m	1.14	0.13	1.22	0.13
81	Tl	0.182	0.018	0.251	0.147	0.181	0.018	0.181	0.018	m	0.81	0.04	0.90	0.04
82	Pb	3.32	0.24	2.51	1.47	3.32	0.24	3.29	0.24	m	2.07	0.03	2.16	0.03
83	Bi	0.142	0.009	...	...	0.142	0.009	0.142	0.009	m	0.70	0.03	0.79	0.03
90	Th	0.0345	0.0037	...	...	0.0344	0.0037	0.0431	0.0047	m	0.09	0.04	0.27	0.04
92	U	0.00896	0.00085	...	...	0.00894	0.00085	0.02382	0.00227	m	-0.50	0.04	0.02	0.04

*CI-chondritic values are strictly normalized to Si = 10^6 atoms. Photospheric values via $N(E)=10^{A(E)-1.551}$. **Solar system values are from the combined solar and meteoritic datasets in columns (1) and (2) and were re-normalized to Si = 10^6 because the average of solar and meteoritic values is used for Si. Note column: m = meteoritic, s = from sun, o = by other means †corrected for radioactive decay. ‡present day photospheric values corrected for element settling from the convection zone to obtain protosolar values

Table 7 Mass Fractions for Solar System Composition*

Mass Fraction	Present-Day	Protosolar
X	0.7389 ± 0.0068 (±0.9%)	0.7060 ± 0.0065 (±0.9%)
Y	0.2452 ± 0.0069 (±2.8%)	0.2753 ± 0.0077 (±2.8%)
Z	0.0160 ± 0.0013 (±8%)	0.0187 ± 0.0015 (±8%)
Z/X	0.0216 ± 0.0017 (±8%)	0.0265 ± 0.0021 (±8%)

*Composition derived from photospheric and CI-chondritic data. Mass fraction X is for H, Y for He, and Z is for the sum of Li to U. See text

could be affected by this type of mixing (e.g., Boesgaard et al. 2005, 2016), but the CI-chondritic/photospheric ratios for Be = 1.5 ± 0.6 and B = 1.3 ± 1.0 suggest no Be and B depletions in the solar photosphere.

To obtain proto-solar (= solar system) abundances, settling correction factors, SF, are applied. The protosolar mass fractions (indicated by subscript "0") and present-day ratios are related as

$Y_0/X_0 = (Y/X) \times SF(He/H)$, and $Z_0/X_0 = (Z/X) \times SF(Li/H\text{-}U/H)$.

The SF for Li-U was taken to be somewhat smaller than that for He (e.g., Lodders 2003; Piersanti et al. 2007; Asplund et al. 2009) but it might be the other way around (see Lodders 2020). Here we use SF(He/H) = 1.175 (17.5% change in He/H, $\log_{10}$SF(He/H) = 0.070 dex) and SF(Li/H-U/H) = 1.225 (22.5% change in heavier element/H ratios, $\log_{10}$SF(Li/H-U/H) = 0.0882 dex) and are based on models by Yang (2019). For more details and discussion see Lodders (2020). The SF here are similar to those found by Piersanti et al. (2007) for different solar compositions, e.g., from data in their Table 3, SF(He/H) = 1.181-1.197 and a smaller SF(Li/H-U/H) = 1.164-1.177.

Using Y_0/X_0 and Z_0/X_0, and the mass-balance relation, $X_0 = 1/(1 + Y_0/X_0 + Z_0/X_0)$, all proto-solar mass fractions in Table 7 are derived.

The atomic weights cancel in the relation of present-day and protosolar ratios and the log of settling factors can be directly applied to obtain individual protosolar abundances on the log atomic scale relative to hydrogen. The settling correction increases the present-day abundances on the scale normalized to H by a factor "SF(E/H)," meaning a constant $\log_{10}$SF(E/H) is added to the logarithmic scale with A(H) = 12:

$A(E)_{\text{proto-solar}} = 12 + \log_{10} \varepsilon_{E,\text{ proto-solar}} = A(E/H)_{\text{present}} + \log_{10}SF(E/H) = 12 + \log_{10}\varepsilon_{E,\text{ present}} + \log_{10}SF(E/H)$

This equation can be used to compute other protosolar abundances if other settling factors are preferred.

The settling corrections are applied to the element/H ratios and not to the absolute mass fraction of an element, or X, Y, and Z (this is an important difference when comparing depletions in abundance ratios and absolute mass fractions). *Absolute* mass fractions drop by $(Y_0\text{-}Y)/Y_0$ = 11% for He, and $(Z_0\text{-}Z)/Z_0$ = 14.5% for heavy elements, which are directly comparable quantities to the "depletion efficiency δ" listed by Piersanti et al. (2007). They find around 10% for each element for several different solar compositions.

The presence of long-lived (above the lifetime of the current solar system age) radioactive isotopes requires another adjustment to protosolar abundances from present-day values. The abundances of radioactive parent isotopes were adjusted for decay loss over time and the stable daughter isotopes for gain.

22 Comparison to BOREXINO C+N Abundances

A test for the adopted settling factors is the comparison of the C and N abundances from spectroscopy with those in the solar core from the BOREXINO particle physics experiment, for which Basilico et al. (2023) estimated (C+N)/H (5.81(+1.22,-0.94)10^{-4}. They found the best agreement to the photospheric (C+N)/H by Magg et al. (2022), although still only within the lower margin uncertainties. Here the photospheric (C+N)/H = 4.11(±0.79)10^{-4} agrees within error limits.

However, this comparison is flawed because the 23% loss of heavy elements (including C and N) from the CE (hence photosphere) must be considered. The solar interior essentially remains at the protosolar values plus the settling gains from the CE. The latter are very small in absolute terms because the CE is only about 2% of solar mass. The protosolar (C+N)/H = 5.03(±0.97)10^{-4} is closer to the BOREXINO value (within 13%), and indicates that the larger settling corrections applied here are plausible. To match the BOREXINO (C+N)/H, a reduction of heavy elements from the CE requires doubling the settling factor to around 42% (log SF(Li-U) = 0.151). Whether such high settling losses are consistent with standard solar models needs to be investigated. Such high settling corrections would increase the protosolar metallicity to Z_0 = 0.0216, or $[Fe/H]_{protosolar}$ = 0.13, still consistent with the metallicity dispersions of B stars (e.g., Mashonkina et al. 2020) and independent measurements in the local ISM (e.g., Ritchey et al. 2023).

23 Comparison of Solar and Meteoritic Abundances: No Trend with Condensation Temperatures

By the late 1960s, CI-chondrites were firmly identified as the best proxy for solar system abundances. Anders (1971), Holweger (2001), and others found no apparent differences in relative element abundances except for ultra-volatile elements. However, some doubts occasionally remain (e.g., Grevesse 2019). González Hernández et al. (2010), González (2014) concluded that there is a significant trend of CI-chondritic/photospheric abundance ratios with condensation temperatures, similar claims were made by Meléndez et al. (2014), Asplund et al. (2021), and Jurewicz et al. (2024).

Figure 12 shows CI-chondritic and solar abundance ratios as a function of 50% condensation temperatures for all elements where the comparison can be made. Within uncertainties solar and meteoritic abundances for most elements agree within 26% (grey bar in Fig. 12). A trend with 50% condensation temperatures could be suggested by lower ratios for refractory Y, Hf, Zr, and W, and higher volatiles (e.g., Pb, B, Ga) but considerable uncertainties remain for solar and CI-chondritic Zr, Hf, W, Pb, Ga, and B abundances (see above).

Restricting the comparison to elements with high quality solar analyses (Table 2), few elements remain as shown in Fig. 13a. A "trend" could be interpreted into these data. However, the refractory Al is out of trend, and the meteoritic Y and Ca concentrations are from element ratios here, and not strictly from CI-chondrites. Given the uncertainties, the conclusion of a volatility trend for solar photospheric and CI-chondritic data is premature and strongly depends on the elements and their uncertainties selected for comparison. However, using photospheric data from Asplund et al. (2021), Jurewicz et al. (2024) found a trend of CI-chondritic/solar abundance ratios with condensation temperatures. They also argue for a similar trend for CI-chondrite/solar wind abundance ratios but did not compare solar wind/photospheric abundance ratios as a function of condensation temperatures. All these abundances ratios as functions of condensation temperatures are compared in Fig. 13b, and

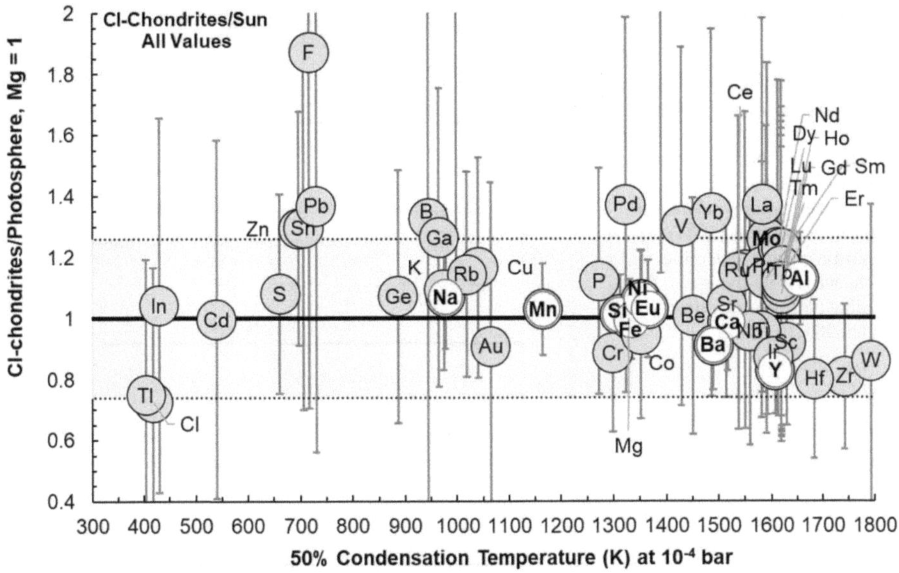

Fig. 12 CI-chondrite/solar abundance ratios as a function of 50% condensation temperatures. White symbols with bold labels are for elements with low uncertainties in the Sun. Gray shaded bar indicates agreement within 26%. Abundances are normalized to Mg

13c. Figure 13b may suggest depletions of refractory Ca and Al and enrichments of more volatile Na and K in CI-chondrites relative to solar wind, which seems to support the notion that CI-chondritic abundances are fractionated from the "solar" composition if solar is taken from either photosphere or solar wind. However, using the same arguments, the data in Fig. 13c would then suggest that the solar wind is enriched in refractories over photospheric values, or that alkalis are enriched in the photosphere over those in the solar wind. A comparison of solar wind/photospheric ratios with condensation temperature is not discussed in Jurewicz et al. (2024), and FIP corrections would not resolve this journal. The correlations in Figs. 13b and 13c appear stronger than in 13a, and assuming the arguments by Jurewicz et al. (2024) hold, the conclusion could be that solar wind abundances are volatility fractionated from the photosphere (Fig. 13c). This seems implausible in the solar plasma.

24 Isotope Systematics as Test for Abundances

One type of empirical abundance scaling and testing is to use isotopes mainly created by the slow neutron capture process (s-process) and to plot the product of their neutron cross-sections (σ) with their isotopic abundances (σN_S) versus mass number (see comprehensive summaries in Käppler et al. 1989; Reifarth et al. 2014).

In the classical s-process model approach, the "σN_S" of the s-process only nuclides give a smooth stepping curve with intervals of more or less constant σN_S which drop at mass numbers for nuclides with magic neutron numbers N = 50, 82, and 126. This is shown in Fig. 14 for our data (Table A13).

The σN_S are relatively constant along the plateau for isotopes only formed by the s-process. This approach employs Maxwellian averaged neutron capture cross sections

Fig. 13 Abundance ratios versus 50% condensation temperatures at 0.1 mbar for (a) CI-chondrites/photosphere for top-rated solar analyses (Table 2; also Fig. 12), (b) CI-chondrites/solar wind, and (c) solar wind/photospheric abundances. Solar wind data are from Jurewicz et al. (2024); photospheric and meteoritic abundances from this study. Gray shading indicates agreement within 26% (0.1 dex)

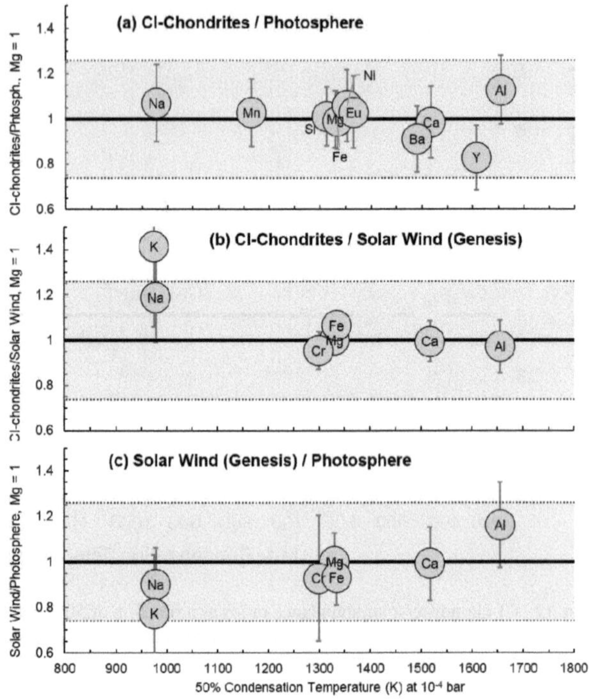

(MACS) values (e.g., from the Kadonis database) for a single thermal energy of k_BT = 30 keV for a scenario of constant temperature and neutron density. Then, elemental abundances for e.g., Kr and Xe can be estimated by fitting desired isotopic abundances (e.g., s-only isotopes for Kr or Xe) to the σN_S-curve and then scaling to elemental abundances using the isotopic composition.

The high mass range beyond the Ba region shows more scatter, and Os, and possibly Pt, seem somewhat low, however, these isotopic abundances are affected by radioactive decay or gain. The two s-process isotopes ^{176}Lu and ^{176}Hf are linked through decay of ^{176}Lu, and their sum follows the linear trend going from ^{160}Dy to ^{170}Yb to ^{204}Pb. In this region, the pure s-isotopes follow a little valley seated below the trend of isotopes with less than 70% s-process contributions (gray symbols). This and the log-scale for σN_S complicates abundance estimates and comparisons in that mass region. Furthermore, most of the "pure" s-process isotopes have small p-process contributions which are not subtracted from the overall observed isotopic abundances to obtain the "truly pure" s-process concentrations for in Fig. 14.

Reifarth et al. (2014) describe how this classical model is superseded by more sophisticated stellar models considering different conditions (temperatures, neutron fluences and densities in stars of different mass and metallicities). The classical model is useful for quickly estimating abundances in mass regions with about constant σN_s between magic neutron numbers (see Käppler et al. 1989; Reifarth et al. 2014).

Another test is the comparison of pure s-process isotopic abundances to those expected from independent models using nucleosynthetic network computations and galactic chemical evolution models (e.g., Arlandini et al. 1999; Reifarth et al. 2014). Here abundances of "pure" s-process elements are compared to results from Prantzos et al. (2020). In Fig. 15 the horizontal line at unity is for perfect agreement, the dotted lines indicate agreement within

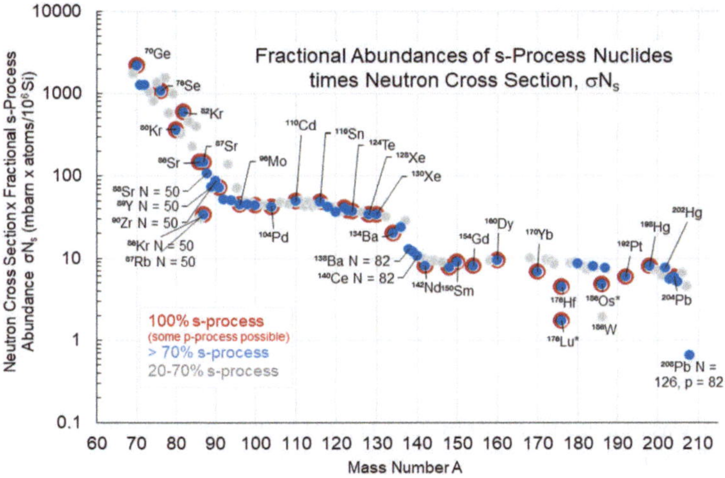

Fig. 14 The product of the neutron capture cross-sections with the isotopic abundances (σN_S) versus mass number. "Pure" s-process produced isotopes are shown with red circles. Blue symbols without rim are for isotopes with more than 70% s-process contributions, grey ones for less than 70%. Similar diagrams are discussed in Anders and Grevesse (1989) and Palme and Beer (1993). See text

Fig. 15 Comparison of observed abundances of "pure" s-process isotopes abundances with model predictions by Prantzos et al. (2020). Isotopes with purple rims can have a few percent of p-process contributions. The abundances of Kr and Xe are fitted into the trends given by neighboring isotopes

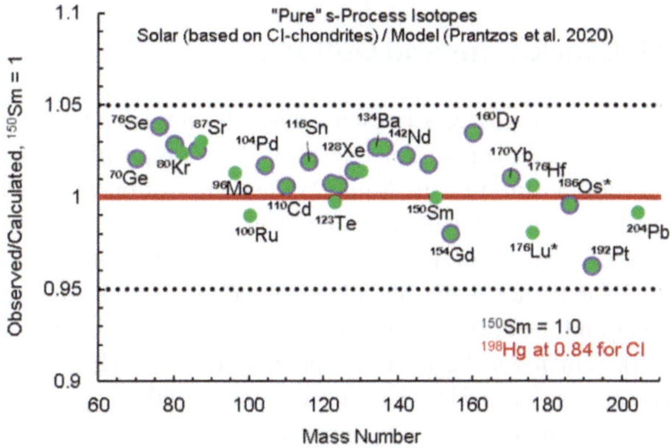

5%. The agreement for most isotopes is within 4%, except for Hg. Many of the isotopes show small excesses (1-4%) from the pure s-process abundances predicted by the model, and possibly lighter isotopes are slightly more enriched than heavy ones. These observed "excesses" are likely due to p-process contributions (Käppler et al. 1989).

Figure 15 is also used to obtain Kr and Xe elemental abundances (Table 3) by interpolation of their "pure" s-process nuclide abundance ratios between those of their neighboring s-only nuclides. The abundances derived from the interpolated values are within 4% of the exact match to the model values. They are also following the curvature in the σN_S diagram (Fig. 14). The lower Kr and Xe abundances by other authors (Table 3) only reach 70 or 84% of the model abundances.

In Fig. 15, most elements (other than Kr, Xe) show an excess (ratio > 1) which is why we interpolated the Kr and Xe abundances to neighboring "pure" s-process isotopes instead of fixing the Kr and Xe abundances to the model values. For example, about 15% p-process

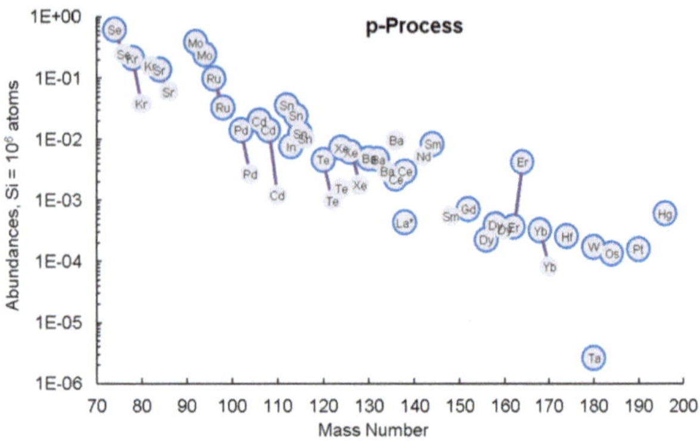

Fig. 16 Abundances of isotopes mainly made by the p-process. Symbols with solid rims indicate (*almost*) pure p-process isotopes. For other isotopes only the fractional isotopic abundances from the observed excesses to the model abundances of "pure" s-process isotopes (seen in Fig. 15) are plotted

contribution was estimated for ^{80}Kr by Käppler et al. (1989) and about 9% p-process contribution to ^{128}Xe; here 3% and 4%, resp., are needed. The declining excesses for "s-process only" nuclide abundances over the model values with increasing mass numbers (Fig. 15) mimic the declining p-process nuclide abundances in Fig. 16. Figure 16 also shows the nominal fractional contribution from the p-process to the "pure" s-isotopes as defined from the excesses to the s-process model abundances by Prantzos et al. (2020) that are apparent in Fig. 15.

25 Conclusions and Outlook

We report on progress in establishing the single chemical composition, parental to the Sun and the planets in our solar system. This composition is established from two sources: (1) The composition of the Sun, (2) the composition of CI-chondrites. The increasing agreement of solar and meteoritic abundances with nucleosynthetic calculations demonstrate the unprocessed nature (in terms of elemental composition) of this material.

Solar photospheric abundance data substantially improved over the last two decades. Solar and meteoritic abundances agree within 25% for many condensable elements. Much work remains to be done to obtain full 3D NLTE results for many elements. Eleven elements have full 3D NLTE results and excellent agreement with CI-chondrites exists for the non-volatile elements among these. Until more laboratory and theoretical work bring improvements, uncertainties in solar abundances remain relatively large for many elements and it is too early to debate a possibly abnormal composition of the Sun. The solar metallicity, especially abundances of C, N, and O, is slightly higher than previously thought with the more conservative approach chosen here. This is consistent with other independent measurements, such as helioseismology and constraints from the solar neutrino experiments.

The use of solar wind data as potential abundance standards is currently limited to the noble gases and a few major elements and depends heavily on modeling how elemental abundances become fractionated when passing through the corona to become the solar wind. Suggested elemental fractionations between photospheric, solar wind, and meteoritic abundances with condensation temperatures are just apparent at this stage and should not be over-interpreted in light of existing uncertainties with the abundances in each regime.

The abundances of the refractory lithophile elements (Y, Zr, Hf, Nb, Ta, U, Th) are much improved. However, the apparent heterogeneous distribution of some of them (e.g., U, W) in CI-meteorites needs to be better understood. All siderophile element abundances in CI-chondrites would benefit from refined analyses, especially As, Sb, Mo, W, noble metals including Au. More isotope dilution analyses of multi-isotope elements would increase the accuracy of CI-elemental abundances.

New reliable halogen analyses are needed for more than one specimen of each CI-chondrite because the halogens are among the most aqueously mobile elements. Many elements with tri-bimodal and/or trailing concentration distributions are mobile elements and the mechanism and timing of their redistribution (parent body versus museum) needs to be better understood. The dissolving and reprecipitating phases are likely carbonates and phosphates; sulfates are mainly secondary precipitates after sulfide decomposition. The trace element distributions into accessory phases require further study to assess the scale of homogeneity for mobile elements for better estimates of required sample sizes for analyses. New analytical data of samples from the asteroids Ryugu and Bennu confirm that there is much more material of CI-composition than earlier thought. The analyses of these pristine asteroidal materials will help to better understand redistribution of aqueously mobile elements in CI-chondrite parent bodies.

Supplementary Information The online version contains supplementary material available at https://doi.org/10.1007/s11214-025-01146-w.

Acknowledgements We thank ISSI for the invitations to an interesting workshop and for hosting the meeting. Work by KL is supported by NSF AST 2108172, NASA 80NSSC24K1284, and the McDonnell Center for the Space Sciences. MB is supported through the Lise Meitner grant from the Max Planck Society. This project has received funding from the European Research Council (ERC) under the European Union's Horizon 2020 research and innovation programme (Grant agreement No. 949173). We acknowledge support from the DAAD grant Project No.: 57654415. We thank the reviewers Andy Davis and Rainer Wieler and the editor Klaus Mezger for detailed comments that helped to improve the manuscript.

Funding Open Access funding enabled and organized by Projekt DEAL.

Declarations

Competing Interests The authors declare that they have no conflict of interest.

References

Ahrens LH (1970) The composition of stony meteorites (IX) abundance trends of the refractory elements in chondrites, basaltic achondrites and Apollo 11 fines. Earth Planet Sci Lett 10:1–6

Ahrens LH, Von Michaelis H, Fesq H (1969) The composition of the stony meteorites (IV) some analytical data on Orgueil, Nogoya, Ornans and Ngawi. Earth Planet Sci Lett 6:285–288

Alexeeva SA, Mashonkina LI (2015) Carbon abundances of reference late-type stars from 1D analysis of atomic C I and molecular CH lines. Mon Not R Astron Soc 453:1619–1631

Alexeeva S, Ryabchikova T, Mashonkina L, Hu S (2018) NLTE line formation for Mg I and Mg II in the atmospheres of B-A-F-G-K stars. Astrophys J 866:153

Alfing J, Patzek M, Bischoff A (2019) Modal abundances of coarse-grained (> 5 μm) components within CI-chondrites and their individual clasts–mixing of various lithologies on the CI parent body(ies). Geochemistry 79:125532

Allende Prieto C, Lambert DL, Asplund M (2001) The forbidden abundance of oxygen in the Sun. Astrophys J 556:L63–L66

Amarsi AM, Asplund M (2017) The solar silicon abundance based on 3D non-LTE calculations. Mon Not R Astron Soc 464:264–273

Amarsi AM, Barklem PS, Collet R, Grevesse N, Asplund M (2019) 3D non-LTE line formation of neutral carbon in the Sun. Astron Astrophys 624:A111

Amarsi AM, Grevesse N, Grumer J, Asplund M, Barklem PS, Collet R (2020) The 3D non-LTE solar nitrogen abundance from atomic lines. Astron Astrophys 636:A120

Amarsi AM, Ogneva D, Buldgen G, Grevesse N, Zhou Y, Barklem PS (2024) The solar beryllium abundance revisited with 3D non-LTE models. Astron Astrophys 690:A128

Anders E (1971) How well do we know "cosmic" abundances? Geochim Cosmochim Acta 35:516–522

Anders E, Ebihara M (1982) Solar System abundances of the elements. Geochim Cosmochim Acta 46:2363–2380

Anders E, Grevesse N (1989) Abundances of the elements— meteoritic and solar. Geochim Cosmochim Acta 53(197):214

Appel S, Bagdasarian Z, Basilico D, Bellini G, Benziger J, Biondi R, Caccianiga B, Calaprice F, Caminata A, Cavalcante P, Chepurnov A, Borexino Collaboration (2022) Improved measurement of solar neutrinos from the carbon-nitrogen-oxygen cycle by Borexino and its implications for the standard solar model. Phys Rev Lett 129:252701

Arlandini C, Käppeler F, Wisshak K, Gallino R, Lugaro M, Busso M, Straniero O (1999) Neutron capture in low mass asymptotic giant branch stars: cross sections and abundance signatures. Astrophys J 525:886–900

Asplund M, Grevesse N, Sauval AJ, Scott P (2009) The chemical composition of the Sun. Annu Rev Astron Astrophys 47:481–522

Asplund M, Amarsi AM, Grevesse N (2021) The chemical make-up of the Sun: a 2020 vision. Astron Astrophys 653:A141

Ayres TR, Lyons JR, Ludwig HG, Caffau E, Wedemeyer-Böhm S (2013) Is the sun lighter than the Earth? Isotopic CO in the photosphere, viewed through the lens of three-dimensional spectrum synthesis. Astrophys J 765:A46

Bahcall JN, Serenelli AM, Pinsonneault M (2004) How accurately can we calculate the depth of the solar convective zone? Astrophys J 614:464–471

Barklem PS (2016) Accurate abundance analysis of late-type stars: advances in atomic physics. Astron Astrophys Rev 24:1–54

Barrat JA, Zanda B, Moynier F, Bollinger C, Lior Zou C, Bayon G (2012) Geochemistry of CI chondrites: major and trace elements, and Cu and Zn isotopes. Geochim Cosmochim Acta 83:79–92

Basilico D, Bellini G, Benziger J, Biondi R, Caccianiga B, Calaprice F, Caminata A, Chepurnov A, D'Angelo D, Derbin A, Di Giacinto A, Borexino Collaboration (2023) Final results of Borexino on CNO solar neutrinos. Phys Rev D 108:102005

Basu S, Antia HM (2004) Constraining solar abundances using helioseismology. Astrophys J 606:L85–L88

Basu S, Antia HM (2008) Helioseismology and solar abundances. Phys Rep 457:217–283

Bautista M (2000) Theoretical Calculations of Atomic Data for Spectroscopy. Atomic Data Needs for X-ray Astronomy. Editors: Bautista, Kallmann, Pradhan. NASA CP-2000-209968, pp. 25

Bautista MA, Bergemann M, Gallego HC, Gamrath S, Palmeri P, Quinet P (2022) Atomic radiative data for oxygen and nitrogen for solar photospheric studies. Astron Astrophys 665:A18

Bedell M, Bean JL, Meléndez J, Spina L, Ramírez I, Asplund M, Alves-Brito A, dos Santos L, Dreizler S, Yong D, Monroe T, Casagrande L (2018) The chemical homogeneity of sun-like stars in the solar neighborhood. Astrophys J 865:id68

Beer H, Walter G, Macklin RL, Patchett PJ (1984) Neutron capture cross sections and Solar System abundances of 160,161Dy, 170, 171 Yb, 175,176 Lu and 176,177 Hf for the S-process analysis of the radionuclide 176Lu. Phys Rev C 30(2):464–478

Belyaev AK, Voronov YV (2020) Inelastic processes in low-energy sulfur-hydrogen collisions. Astrophys J 893:59

Belyaev AK, Voronov YV, Mitrushchenkov A, Guitou M, Feautrier N (2019) Inelastic processes in oxygen–hydrogen collisions. Mon Not R Astron Soc 487:5097–5105

Bergemann M, Cescutti G (2010) Chromium: NLTE abundances in metal-poor stars and nucleosynthesis in the Galaxy. Astron Astrophys 522:A9

Bergemann M, Gehren T (2007) Formation of Mn I lines in the solar atmosphere. Astron Astrophys 473:291–302

Bergemann M, Serenelli A (2014) Solar abundance problem. In: Niemczura E, Smalley B, Pych W (eds) Determination of atmospheric parameters of B-, A-, F-and G-type stars: lectures from the school of spectroscopic data analyses, series: GeoPlanet: Earth and planetary sciences. Springer, Cham, pp 245–258

Bergemann M, Pickering JC, Gehren T (2010) NLTE analysis of CoI/CoII lines in spectra of cool stars with new laboratory hyperfine splitting constants. Mon Not R Astron Soc 401:1334–1346

Bergemann M, Lind K, Collet R, Magic Z, Asplund M (2012) Non-LTE line formation of Fe in late-type stars—I. Standard stars with 1D and< 3D> model atmospheres. Mon Not R Astron Soc 427:27–49

Bergemann M, Kudritzki R-P, Würl M, Plez B, Davies B, Gazak Z (2013) Red supergiant stars as cosmic abundance probes. II. NLTE effects in J-band silicon lines. Astrophys J 764:id115

Bergemann M, Collet R, Amarsi AM, Kovalev M, Ruchti G, Magic Z (2017) Non-local thermodynamic equilibrium stellar spectroscopy with 1D and <3D> models. I. Methods and application to magnesium abundances in standard stars. Astrophys J 847:A15

Bergemann M, Gallagher AJ, Eitner P, Bautista M, Collet R, Yakovleva SA, Mayriedl A, Plez B, Carlsson M, Leenaarts J, Belyaev AK, Hansen C (2019) Observational constraints on the origin of the elements. I. 3D NLTE formation of Mn lines in late-type stars. Astron Astrophys 631:A80

Bergemann M, Hoppe R, Semenova E, Carlsson M, Yakovleva SA, Voronov YV, Bautista M, Nemer A, Belyaev AK, Leenaarts J, Mashonkina L (2021) Solar oxygen abundance. Mon Not R Astron Soc 508:2236–2253

Berzelius JJ (1834) Om Meteorstenar. Handl. Svenska Vetenskaps-akad. Handl. Svenska Vetenskaps-Akad: 115–183. Also in *Ann. Physik* (*Poggendorff's*) 33, 1–32, and 113–148.

Boesgaard AM, Deliyannis CP, Steinhauer A (2005) Boron depletion in F and G dwarf stars and the beryllium-boron correlation. Astrophys J 621:991–998

Boesgaard AM, Lum MG, Deliyannis CP, King JR, Pinsonneault MH, Somers G (2016) Boron abundances across the "Li-Be dip" in the Hyades cluster. Astrophys J 830:49

Böhm-Vitense E (1958) Über die Wasserstoffkonvektionszone in Sternen verschiedener Effektivtemperaturen und Leuchtkräfte. Z Astrophys 46:108–143

Bouvier A, Vervoort JD, Patchett PJ (2008) The Lu–Hf and Sm–Nd isotopic composition of CHUR: constraints from unequilibrated chondrites and implications for the bulk composition of terrestrial planets. Earth Planet Sci Lett 273:48–57

Braukmüller N, Wombacher F, Hezel DC, Escoube R, Münker C (2018) The chemical composition of carbonaceous chondrites: implications for volatile element depletion, complementarity and alteration. Geochim Cosmochim Acta 239:17–48

Burbidge EM, Burbidge GR, Fowler WA, Hoyle F (1957) Synthesis of the elements in stars. Rev Mod Phys 29:547–650

Burnett DS, Genesis Science Team (2011) Solar composition from the genesis discovery mission. Proc Natl Acad Sci 108:19147–19151

Caffau E, Ludwig HG, Steffen M, Ayres TR, Bonifacio P, Cayrel R, Freytag B, Plez B (2008) Astron Astrophys 488:1031–1046

Caffau E, Maiorca E, Bonifacio P, Faraggiana R, Steffen M, Ludwig HG, Kamp I, Busso M (2009) The solar photospheric nitrogen abundance-analysis of atomic transitions with 3D and 1D model atmospheres. Astron Astrophys 498:877–884

Caffau E, Ludwig H-G, Bonifacio P, Faraggiana R, Steffen M, Freytag B, Kamp I, Ayres TR (2010) The solar photospheric abundance of carbon. Analysis of atomic carbon lines with the CO5BOLD solar model. Astron Astrophys 514:A92

Caffau E, Ludwig HG, Steffen M, Freytag B, Bonifacio P (2011) Solar chemical abundances determined with a CO5BOLD 3D model atmosphere. Sol Phys 268:255–269

Caffau E, Bonifacio P, François P, Spite M, Spite F, Zaggia S, Ludwig HG, Steffen M, Mashonkina L, Monaco L, Sbordone L (2012) A primordial star in the heart of the Lion. Astron Astrophys 542:A51

Caffau E, Ludwig H-G, Malherbe J-M, Bonifacio P, Steffen M, Monaco L (2013) The photospheric solar oxygen project. II. Non-concordance of the oxygen abundance derived from two forbidden lines. Astron Astrophys 554:A126

Caffau E, Ludwig H-G, Steffen M, Livingston W, Bonifacio P, Malherbe J-M, Doerr HP, Schmidt W (2015) The photospheric solar oxygen project. III. Investigation of the centre-to-limb variation of the 630 nm [O I]-Ni I blend. Astron Astrophys 579:A88

Cameron AGW (1957) Nuclear reactions in stars and nucleogenesis. Publ Astron Soc Pac 69:201–222

Cameron AGW (1973) Abundances of the elements in the Solar System. Space Sci Rev 15:121–146. https://doi.org/10.1007/BF00172440

Castelli F, Kurucz RL (2003) New grids of ATLAS9 model atmospheres. In: Piskunov N, Weiss WW, Gray DF (eds) Modelling of stellar atmospheres. IAU symposium, vol 210, poster A20

Centeno R, Socas-Navarro H (2008) A new approach to the solar oxygen abundance problem. Astrophys J 682:L61

Christensen-Dalsgaard J (2002) Helioseismology. Rev Mod Phys 74:1073–1129

Cloez S (1864a) Analyse chimique de la Pierre météorique d'Orgueil. C R Math Acad Sci Paris, Sér I 59:37–40

Cloez S (1864b) Note sur la composition chimique de la Pierre météorique d'Orgueil. C R Math Acad Sci Paris, Sér I 59:986–988
Cubas Armas M, Asensio Ramos A, Socas-Navarro H (2017) Uncertainties in the solar photospheric oxygen abundance. Astron Astrophys 600:A45
Cubas Armas M, Asensio Ramos A, Socas-Navarro H (2020) Spatially resolved measurements of the solar photospheric oxygen abundance. Astron Astrophys 643:A142
Cunha K, Smith VV (1999) A determination of the solar photospheric boron abundance. Astrophys J 512:1006–1013
Deshmukh SA, Ludwig HG, Kučinskas A, Steffen M, Barklem PS, Caffau E, Dobrovolskas V, Bonifacio P (2022) The solar photospheric silicon abundance according to CO5BOLD-investigating line broadening, magnetic fields, and model effects. Astron Astrophys 668:A48
Ebihara M, Wolf R, Anders E (1982) Are C1 chondrites chemically fractionated? A trace element study. Geochim Cosmochim Acta 46:1849–1861
Fegley B, Palme H (1985) Evidence for oxidizing conditions in the solar nebula from Mo and W depletions in refractory inclusions in carbonaceous chondrites. Earth Planet Sci Lett 72:311–326
Fegley B, Schaefer L (2010) Principles and perspectives in cosmochemistry. In: Goswami A, Reddy B (eds) Astrophysics and Space Science Proceedings. Springer, Berlin, pp 347–377. https://doi.org/10.1007/978-3-642-10352-0_7
Fischer DA, Valenti J (2005) The planet-metallicity correlation. Astrophys J 622:1102–1117
Fischer-Gödde M, Becker H, Wombacher F (2010) Rhodium, gold and other highly siderophile element abundances in chondritic meteorites. Geochim Cosmochim Acta 74:356–379
Folinsbee RE, Douglas JAV, Maxwell JA (1967) Revelstoke, a new type I carbonaceous chondrite. Geochim Cosmochim Acta 31:1625–1635
Fredriksson K, Kerridge JF (1988) Carbonates and sulfates in CI chondrites: formation by aqueous activity on the parent body. Meteoritics 23:35–44
Freytag B, Steffen M, Ludwig HG, Wedemeyer-Böhm S, Schaffenberger W, Steiner O (2012) Simulations of stellar convection with CO5BOLD. J Comput Phys 231:919–959
Gallagher AJ, Bergemann M, Collet R, Plez B, Leenaarts J, Carlsson M, Yakovleva SA, Belyaev AK (2020) Observational constraints on the origin of the elements-II. 3D non-LTE formation of Ba II lines in the solar atmosphere. Astron Astrophys 634:A55
Garz T (1973) Absolute oscillator strengths of SI I lines between 2500 A and 9000 A. Astron Astrophys 26:47–477
Gehren T, Liang YC, Shi JR, Zhang HW, Zhao G (2004) Abundances of Na, Mg and Al in nearby metal-poor stars. Astron Astrophys 413:1045–1063
Gloeckler G, Cain J, Ipavich FM, Tums EO, Bedini P, Fisk LA, Zurbuchen TH, Bochsler P, Fischer J, Wimmer-Schweingruber RF, Geiss J, Kallenbach R (1998) Investigation of the composition of solar and interstellar matter using solar wind and pickup ion measurements with SWICS and SWIMS on the ACE spacecraft. Space Sci Rev 86:497–539. https://doi.org/10.1023/A:1005036131689
Goldschmidt VM (1937) Geochemische Verteilungsgesetze der Elemente, IX Die Mengenverhältnisse der Elemente und der Atom-Arten. Skr Nor Vidensk Akad Oslo I Math Mathurvid Kl 1937(4):99–101
Goles G (1970) Potassium. In: Mason B (ed) Handbook of elemental abundances in meteorites. Gordon and Breach Science Publ., NY
González Hernández JI, Israelian G, Santos NC, Sousa S, Delgado-Mena E, Neves V, Udry S (2010) Searching for the signatures of terrestrial planets in solar analogs. Astrophys J 720:1592–1602
González G (2014) Solar System chemical abundances corrected for systematics. Mon Not R Astron Soc Lett 443(1):L99–L103
Gooding JL (1979) Petrogenetic properties of chondrules in unequilibrated H-, L-, and LL-group chondritic meteorites. Ph. D. Thesis, Univ. of New, Mexico, Albuquerque
Göpel C, Birck JL, Galy A, Barrat JA, Zanda B (2015) Mn–Cr systematics in primitive meteorites: Insights from mineral separation and partial dissolution. Geochim Cosmochim Acta 156:1–24
Gounelle M, Zolensky ME (2001) A terrestrial origin for sulfate veins in CI1 chondrites. Meteorit Planet Sci 36:1321–1329
Greshake A, Kloeck W, Arndt P, Maetz M, Flynn GJ, Bajt S, Bischoff A (1998) Heating experiments simulating atmospheric entry heating of micrometeorites: clues to their parent body sources. Meteorit Planet Sci 33:267–290
Grevesse N (2019) The solar chemical composition: past and present. Bull Soc R Sci Liège 88:5–14
Grevesse N, Sauval AJ (1998) Standard solar composition. Space Sci Rev 85:161–174. https://doi.org/10.1023/A:1005161325181
Grevesse N, Scott P, Asplund M, Sauval AJ (2015) The elemental composition of the Sun. III. The heavy elements Cu to Th. Astron Astrophys 573:A27

Gustafsson B, Edvardsson B, Eriksson K, Jørgensen UG, Nordlund Å, Plez B (2008) A grid of MARCS model atmospheres for late-type stars. I. Methods and general properties. Astron Astrophys 486:A951
Hannaford P, Lowe RM, Grevesse N, Biemont E, Whaling W (1982) Oscillator strengths for YI and Y II and the solar abundance of yttrium. Astrophys J 261:736–746
Heays AN, Bosman AV, Van Dishoeck EF (2017) Photodissociation and photoionisation of atoms and molecules of astrophysical interest. Astron Astrophys 602:A105
Heber VS, Wieler R, Baur H, Olinger C, Friedmann TA, Burnett DS (2009) Noble gas composition of the solar wind as collected by the Genesis mission. Geochim Cosmochim Acta 73:7414–7432
Heber VS, Baur H, Bochsler P, McKeegan KD, Neugebauer M, Reisenfeld DB, Wieler R, Wiens RC (2012) Isotopic mass fractionation of solar wind: evidence from fast and slow solar wind collected by the Genesis mission. Astrophys J 759:id121
Heber VS, McKeegan KD, Steele RC, Jurewicz AJ, Rieck KD, Guan Y, Wieler R, Burnett DS (2021) Elemental abundances of major elements in the solar wind as measured in Genesis targets and implications on solar wind fractionation. Astrophys J 907:id15
Hellmann JL, Van Orman JA, Kleine T (2024) Hf-W isotope systematics of enstatite chondrites: parent body chronology and origin of Hf-W fractionations among chondritic meteorites. Earth Planet Sci Lett 626:118518
Holweger H (1977) The solar Na/Ca and S/Ca ratios: a close comparison with carbonaceous chondrites. Earth Planet Sci Lett 34:152–154
Holweger H (2001) Photospheric abundances, problems, updates, implications. In: In: Wimmer-Schweingruber RF (ed) Solar and galactic composition. AIP conference proceedings, vol 598. American Institute of Physics, College Park, pp 23–30
Holweger H, Müller EA (1974) The photospheric barium spectrum: solar abundance and collision broadening of Ba II lines by hydrogen. Sol Phys 39:19–30
Horan MF, Walker RJ, Morgan JW, Grossman JN, Rubin AE (2003) Highly siderophile elements in chondrites. Chem Geol 196:5–20
Hrodmarsson HR, Van Dishoeck EF (2023) Photodissociation and photoionization of molecules of astronomical interest-updates to the Leiden photodissociation and photoionization cross section database. Astron Astrophys 675:A25
Huss GR, Koeman-Shields E, Jurewicz AJ, Burnett DS, Nagashima K, Ogliore R, Olinger CT (2020) Hydrogen fluence in Genesis collectors: implications for acceleration of solar wind and for solar metallicity. Meteorit Planet Sci 55:326–351
Ito M, Tomioka N, Uesugi M, Yamaguchi A, Shirai N, Ohigashi T, Liu MC, Greenwood RC, Kimura M, Imae N, Uesugi K (2022) A pristine record of outer Solar System materials from asteroid Ryugu's returned sample. Nat Astron 6:1163–1171
Jacobsen SB, Wasserburg GJ (1984) Sm-Nd isotopic evolution of chondrites and achondrites, II. Earth Planet Sci Lett 67:137–150
Jarosewich E (1990) Chemical analyses of meteorites: a compilation of stony and iron meteorite analyses. Meteoritics 25:323–337
Jurewicz AGJ, Amarsi AM, Burnett DS, Grevesse N (2024) Differences in elemental abundances between CI chondrites and the solar photosphere. Meteorit Planet Sci 59:3193–3214
Kallemeyn GW, Wasson JT (1981) The compositional classification of chondrites: I. The carbonaceous chondrite groups. Geochim Cosmochim Acta 45:1217–1230
Käppler F, Beer H, Wisshak K (1989) S-process nucleosynthesis-nuclear physics and the classical model. Rep Prog Phys 52:945
King AJ, Phillips KJH, Strekopytov S, Vita-Finzi C, Russell SS (2020) Terrestrial modification of the Ivuna meteorite and a reassessment of the chemical composition of the CI type specimen. Geochim Cosmochim Acta 268:73–89
Kleine T, Mezger K, Palme H, Münker C (2004) The W isotope evolution of the bulk silicate Earth: constraints on the timing and mechanisms of core formation and accretion. Earth Planet Sci Lett 228:109–123
Klevas J, Kučinskas A, Steffen M, Caffau E, Ludwig HG (2016) Lithium spectral line formation in stellar atmospheres - the impact of convection and NLTE effects. Astron Astrophys 586:A156
Kock M, Richter J (1968) Experimentelle Übergangswahrscheinlichkeiten und die solare Häufigkeit des Kupfers. Z Astrophys 69:180
Korn AJ, Shi J, Gehren T (2003) Kinetic equilibrium of iron in the atmospheres of cool stars. III. The ionization equilibrium of selected reference stars. Astron Astrophys 407:691–703
Korotin SA (2009) The effects of deviations from LTE in sulphur lines for late-type stars. Astron Rep 53:651–659
Korotin SA (2020) Non-LTE effects in rubidium lines in cool stars. Astron Lett 46(8):541–549

Krähenbühl U, Morgan JM, Ganapathy R, Anders E (1973) Abundance of 17 trace elements in carbonaceous chondrites. Geochim Cosmochim Acta 37:1353–1370
Krot AN, Fegley B, Palme H, Lodders K (2000) Meteoritical and astrophysical constraints on the oxidation state of the solar nebula. In: Mannings V, Boss AP, Russell SS (eds) Protostars and planets IV. University of Arizona Press, Tucson, pp 1019–1054
Krot AN, Keil K, Scott ERD, Goodrich CA, Weisberg MK (2014) Classification of meteorites and their genetic relationships. Treatise on geochemistry. Meteor Cosmochem Proc 1:1–63
Kurucz RL, Furenlid I, Brault J, Testerman L (1984) National Solar Observatory atlas no. 1.: solar flux atlas from 296 to 1300 nm. NOAO, Tucson
Lauretta DS, Connolly HC Jr, Aebersold JE, Alexander CMOD, Ballouz RL, Barnes JJ, Bates HC, Bennett CA, Blanche L, Blumenfeld EH, Clemett SJ (2024) Asteroid (101955) Bennu in the laboratory: properties of the sample collected by OSIRIS-REx. Meteorit Planet Sci 59:2453–2486
Lawler JE, Bonvallet G, Sneden C (2001) Experimental radiative lifetimes, branching fractions, and oscillator strengths for La II and a new determination of the solar lanthanum abundance. Astrophys J 556:452–460
Lawler JE, Sneden C, Nave G, Wood MP, Cowan JJ (2019) Transition probabilities of Sc I and Sc II and Scandium abundances in the Sun. Arcturus, and HD 84937. Astrophys J Suppl Ser 241:id21
Lind K, Amarsi AM, Asplund M, Barklem PS, Bautista M, Bergemann M, Collet R, Kiselman D, Leenaarts J, Pereira TMD (2017) Non-LTE line formation of Fe in late-type stars - IV. Modelling of the solar centre-to-limb variation in 3D. Mon Not R Astron Soc 468:4311–4322
Liu YP, Gao C, Zeng JL, Shi JR (2011) Atomic data of Zn I for the investigation of element abundances. Astron Astrophys 536:A51
Lodders K (2003) Solar System abundances and condensation temperatures of the elements. Astrophys J 591:1220–1247
Lodders K (2008) The solar argon abundance. Astrophys J 674:607–611
Lodders K (2020) Solar elemental abundances. Oxford research encyclopedia of planetary science. Oxford University Press, London. https://doi.org/10.1093/acrefore/9780190647926.013.145
Lodders K (2021) Relative atomic Solar System abundances, mass fractions, and atomic masses of the elements and their isotopes, composition of the solar photosphere, and compositions of the major chondritic meteorite groups. Space Sci Rev 217:A44. https://doi.org/10.1007/s11214-021-00825-8
Lodders K, Fegley B (2011) Chemistry of the Solar System. Royal Society of Chemistry, Cambridge
Lodders K, Fegley B (2023) Solar System abundances and condensation temperatures of the halogens fluorine, chlorine, bromine, and iodine. Geochemistry 83(1):125957
Lodders K, Palme H, Gail HP (2009) Abundances of the elements in the Solar System. In: Trümper JE (ed) Landolt-Börnstein. New series, vol VI/4B, Chap. 4.4. Springer, Berlin, pp 560–630
Lodders K, et al (2025b) The composition of CI-chondrites. (in preparation)
Lodders K, Fegley B, Mezger K, Ebel DS (2025a) Condensation and the volatility trend of the Earth. Space Sci Rev 221
Lu Y, Makishima A, Nakamura E (2007) Coprecipitation of Ti, Mo, Sn and Sb with fluorides and application to determination of B, Ti, Zr, Nb, Mo, Sn, Sb, Hf and Ta by ICP-MS. Chem Geol 236(1–2):13–26
Maas ZG, Pilachowski CA, Hinkle K (2016) Chlorine abundances in cool stars. Astrophys J 152:id196
Magg E, Bergemann M, Serenelli A, Bautista M, Plez B, Heiter U, Gerber JM, Ludwig HG, Basu S, Ferguson JW, Gallego HC (2022) Observational constraints on the origin of the elements-IV. Standard composition of the Sun. Astron Astrophys 661:A140
Mahaffy PR, Niemann HB, Alpert A, Atreya SK, Demick J, Donahue TM, Harpold DN, Owen TC (2000) Noble gas abundance and isotope ratios in the atmosphere of Jupiter from the Galileo Probe Mass Spectrometer. J Geophys Res, Planets 105(E6):15061–15071
Mashonkina L (2020) Non-local thermodynamic equilibrium line formation for Si i-ii-iii in A-B stars and the origin of Si ii emission lines in ι Her. Mon Not R Astron Soc 493:6095–6108
Mashonkina L, Gehren T (2000) Barium and europium abundances in cool dwarf stars and nucleosynthesis of heavy elements. Astron Astrophys 364:249–264
Mashonkina L, Gehren T (2001) Heavy element abundances in cool dwarf stars: an implication for the evolution of the Galaxy. Astron Astrophys 376:232–247
Mashonkina LI, Romanovskaya AM (2022) Scandium abundance in F–G–K stars in a wide metallicity range. Astron Lett 48:455–468
Mashonkina L, Gehren T, Shi JR, Korn AJ, Grupp F (2011) A non-LTE study of neutral and singly-ionized iron line spectra in 1D models of the Sun and selected late-type stars. Astron Astrophys 528:A87
Mashonkina L, Jablonka P, Sitnova T, Pakhomov Y, North P (2017) The formation of the Milky Way halo and its dwarf satellites; a NLTE-1D abundance analysis-II. Early chemical enrichment. Astron Astrophys 608:A89
Mashonkina L, Ryabchikova T, Alexeeva S, Sitnova T, Zatsarinny O (2020) Chemical diversity among A-B stars with low rotational velocities: non-LTE abundance analysis. Mon Not R Astron Soc 499:3706–3719

Mason B (1962) The carbonaceous chondrites. Space Sci Rev 1:621–646. https://doi.org/10.1007/BF00212446

Mason B, Taylor SR (1982) Inclusions in the Allende meteorite. Smithson Contrib Earth Sci 25:1–30

Mazor E, Heymann D, Anders E (1970) Noble gases in carbonaceous chondrites. Geochim Cosmochim Acta 34:781–824

McSween HY Jr, Richardson SM (1977) The composition of carbonaceous chondrite matrix. Geochim Cosmochim Acta 41:1145–1161

Meléndez J, Schirbel L, Monroe TR, Yong D, Ramírez I, Asplund M (2014) HIP 114328: a new refractory-poor and Li-poor solar twin. Astron Astrophys 567:L3

Meshik A, Hohenberg C, Pravdivtseva O, Burnett D (2014) Heavy noble gases in solar wind delivered by Genesis mission. Geochim Cosmochim Acta 127:326–347

Meshik A, Pravdivtseva O, Burnett D (2020) Refined composition of Solar Wind xenon delivered by Genesis NASA mission: comparison with xenon captured by extraterrestrial regolith soils. Geochim Cosmochim Acta 276:289–298

Morlok A, Bischoff A, Stephan T, Floss C, Zinner E, Jessberger EK (2006) Brecciation and chemical heterogeneities of CI chondrites. Geochim Cosmochim Acta 70:5371–5394

Mucciarelli A, Carretta E, Origlia L, Ferraro F.R. (2008) The chemical composition of red giant stars in four intermediate-age clusters of the Large Magellanic Cloud. Astron J 136:375–388

Münker C, Pfänder JA, Weyer S, Buchl A, Kleine T, Mezger K (2003) Evolution of planetary cores and the Earth-Moon system from Nb/Ta systematics. Science 301:84–87

Münker C, et al (2025) High field strength elements in chondrites and their components. Geochem Perspect Lett. (in press)

Nakamura E, Kobayashi K, Tanaka R, Kunihiro T, Kitagawa H, Potiszil C, Ota T, et al (2022) On the origin and evolution of the asteroid Ryugu: a comprehensive geochemical perspective. Proc Jpn Acad Ser B: Phys Biol Sci 98:227–282

Nissen PE (2015) High-precision abundances of elements in solar twin stars-trends with stellar age and elemental condensation temperature. Astron Astrophys 579:52

Nordlander T, Lind K (2017) Non-LTE aluminium abundances in late-type stars. Astron Astrophys 607:A75

Nordlund A (1982) Numerical simulations of the solar granulation. I-basic equations and methods. Astron Astrophys 107:1–10

Nordlund Å, Stein RF, Asplund M (2009) Solar surface convection. Living Rev Sol Phys 6:1–117

Osorio Y, Barklem PS, Lind K, Belyaev AK, Spielfiedel A, Guitou M, Feautrier N (2015) Mg line formation in late-type stellar atmospheres-I. The model atom. Astron Astrophys 579:A53

Palme H, Beer H (1993) Abundances of the elements in the Solar System. In: Voigt HH (ed) Landolt-Bornstein, Group VI. Astronomy and Astrophysics Vol. 3, Extension and Supplement to Vol. 2, Subvolume A. Springer, Berlin, pp 196–221. https://doi.org/10.1007/978-3-540-88055-4_34

Palme H, Wlotzka F (1976) A metal particle from a Ca, Al-rich inclusion from the meteorite Allende, and the condensation of refractory siderophile elements. Earth Planet Sci Lett 33:45–60

Palme H, Zipfel J (2022) The composition of CI chondrites and their contents of chlorine and bromine; results from instrumental neutron activation analysis. Meteorit Planet Sci 57:317–333

Palme H, Larimer JW, Lipschutz ME (1988) Moderately volatile elements. In: Kerridge JF, Matthews MS (eds) Meteorites and the early Solar System. The University of Arizona Press, Tucson, pp 436–461

Palme H, Hutcheon ID, Spettel B (1994) Composition and origin of refractory-metal-rich assemblages in a Ca–Al-rich Allende inclusion. Geochim Cosmochim Acta 58:495–513

Palme H, Lodders K, Jones A (2014) Solar System abundances of the elements. In: Davis AM (ed) Planets, asteroids, comets and the Solar System. Treatise on geochemistry, vol 2, pp 15–36

Patzer A, Pack A, Gerdes A (2010) Zirconium and hafnium in meteorites. Meteorit Planet Sci 45:1136–1151

Pearson VK, Sephton MA, Franchi IA, Gibson JM, Gilmour I (2006) Carbon and nitrogen in carbonaceous chondrites: elemental abundances and stable isotopic compositions. Meteorit Planet Sci 41:1899–1918

Pehlivan Rhodin A (2018) Experimental and computational atomic spectroscopy for astrophysics: Oscillator strengths and lifetimes for Mg I, Si I, Si II, Sc I, and Sc II. Ph.D. Thesis, Lund University, Sweden

Pehlivan Rhodin A, Hartman H, Nilsson H, Jönsson P (2017) Experimental and theoretical oscillator strengths of Mg I for accurate abundance analysis. Astron Astrophys 598:A102

Pehlivan Rhodin A, Hartman H, Nilsson H, Jönsson P (2021) Accurate and experimentally validated transition data for Si I and Si II. Astron Astrophys 682:A184

Pehlivan Rhodin A, Hartman H, Nilsson H, Jönsson P (2021) Accurate and experimentally validated transition data for Si I and Si II. Astron Astrophys 682:A184

Pehlivan Rhodin A, Hartman H, Nilsson H, Jönsson P (2024) Accurate and experimentally validated transition data for Si I and Si II. Astron Astrophys 682:A184

Pepin RO, Schlutter DJ, Becker RH, Reisenfeld DB (2012) Helium, neon, and argon composition of the solar wind as recorded in gold and other Genesis collector materials. Geochim Cosmochim Acta 89:62–80

Piersanti L, Straniero O, Cristallo S (2007) A method to derive the absolute composition of the Sun, the Solar System, and the stars. Astron Astrophys 462:1051–1062

Pietrow AG, Hoppe R, Bergemann M, Calvo F (2023b) Solar oxygen abundance using SST/CRISP center-to-limb observations of the O I 7772 Å line. Astron Astrophys 672:L6

Pietrow AGM, Kiselman D, Andriienko O, Petit dit de la Roche DJM, Díaz Baso CJ, Calvo F (2023a) Center-to-limb variation of spectral lines and continua observed with SST/CRISP and SST/CHROMIS. Astron Astrophys 671:A130

Pilleri P, Reisenfeld DB, Zurbuchen TH, Lepri ST, Shearer P, Gilbert JA, von Steiger R, Wiens RC (2015) Variations in solar wind fractionation as seen by ACE/SWICS and the implications for GENESIS mission results. Astrophys J 812:id.1

Popa SA, Hoppe R, Bergemann M, Hansen CJ, Plez B, Beers TC (2023) Non-local thermodynamic equilibrium analysis of the methylidyne radical molecular lines in metal-poor stellar atmospheres. Astron Astrophys 670:A25

Pourmand A, Dauphas N, Ireland TJ (2012) A novel extraction chromatography and MC-ICP-MS technique for rapid analysis of REE, Sc and Y: Revising CI-chondrite and Post-Archean Australian Shale (PAAS) abundances. Chem Geol 291:38–54

Prantzos N, Abia C, Cristallo S, Limongi M, Chieffi A (2020) Chemical evolution with rotating massive star yields II. A new assessment of the solar s-and R-process components. Mon Not R Astron Soc 491:1832–1850

Rauscher T, Heger A, Hoffman RD, Woosley SE (2002) Nucleosynthesis in massive stars with improved nuclear and stellar physics. Astrophys J 576(1):323–348

Reggiani H, Amarsi AM, Lind K, Barklem PS, Zatsarinny O, Bartschat K, Fursa DV, Bray I, Spina L, Meléndez J (2019) Non-LTE analysis of KI in late-type stars. Astron Astrophys 627:A177

Reifarth R, Heil M, Käppeler F, Voss F, Wisshak K, Becvář F, Krticka M, Gallino R, Nagai Y (2002) Stellar neutron capture cross sections of 128, 129, 130 Xe. Phys Rev C 66:064603

Reifarth R, Lederer C, Käppeler F (2014) Neutron reactions in astrophysics. J Phys G, Nucl Part Phys 41:053101

Reiners A, Mrotzek N, Lemke U, Hinrichs J, Reinsch K (2016) The IAG solar flux atlas: accurate wavelengths and absolute convective blueshift in standard solar spectra. Astron Astrophys 587:A65

Reisenfeld DB, Burnett DS, Becker RH, Grimberg AG, Heber VS, Hohenberg CM, Jurewicz AJG, Meshik A, Pepin RO, Raines JM, Schlutter DJ, Wieler R, Wiens RC, Zurbuchen TH (2007) Elemental abundances of the bulk solar wind: analyses from genesis and ACE. Space Sci Rev 130:79–86. https://doi.org/10.1007/s11214-007-9215-1

Ritchey AM, Jenkins EB, Shull JM, Savage BD, Federman SR, Lambert DL (2023) The distribution of metallicities in the local galactic interstellar medium. Astrophys J 952:id57

Rocholl A, Jochum KP (1993) Th, U and other trace elements in carbonaceous chondrites: implications for the terrestrial and solar-system Th/U ratios. Earth Planet Sci Lett 117:265–278

Roederer IU, Lawler JE (2012) Detection of elements at all three R-process peaks in the metal-poor star HD 160617. Astrophys J 750(1):id76

Russell HN (1929) On the composition of the sun's atmosphere. Astrophys J 70:11–28

Ryabchikova T, Piskunov N, Pakhomov Y (2022) Using molecular lines to determine carbon and nitrogen abundances in the atmospheres of cool stars. Atoms 10:103

Schmitt RA, Goles GG, Smith RH, Osborn TW (1972) Elemental abundances in stone meteorites. Meteoritics 7:131–214

Scott DW (1979) On optimal and data-based histograms. Biometrika 66:605–610

Scott DW (2009) Sturges' rule. WIREs Wiley's Interdiscip Rev Comp Stat 1:303–306. https://doi.org/10.1002/wics.35

Scott ERD, Krot AN (2014) Chondrites and their components. Treatise on geochemistry. Meteor Cosmochem Proc 1:65–137

Scott P, Asplund M, Grevesse N, Bergemann M, Sauval AJ (2015a) The elemental composition of the Sun. II. The iron group elements Sc to Ni. Astron Astrophys 573:A26

Scott P, Grevesse N, Asplund M, Sauval AJ, Lind K, Takeda Y, Collet R, Trampedach R, Hayek W (2015b) The elemental composition of the Sun. I. The intermediate mass elements Na to Ca. Astron Astrophys 573:A25

Serenelli AM, Basu S, Ferguson JW, Asplund M (2009) New solar composition: the problem with solar models revisited. Astrophys J 705:L123

Shi JR, Gehren T, Zeng JL, Mashonkina L, Zhao G (2014) Statistical equilibrium of copper in the solar atmosphere. Astrophys J 782:id80

Shirai N, Ito M, Yamaguchi A, Tomioka N, Uesugi M, Imae N, Kimura M, Greenwood R, Liu MC, Ohigashi T, Sekimoto S (2024) Mercury (Hg) in Ryugu particles and implications for the origin of volatile elements in early Earth. Research Square preprint. https://doi.org/10.21203/rs.3.rs-4002901/v1

Sitnova T, Zhao G, Mashonkina L, Chen Y, Liu F, Pakhomov Y, Tan K, Bolte M, Alexeeva S, Grupp F, Shi JR (2015) Systematic non-LTE study of the $-2.6 \leq$[Fe/H]≤ 0.2 F and G dwarfs in the solar neighborhood. I. Stellar atmosphere parameters. Astrophys J 808(2):148

Sitnova TM, Yakovleva SA, Belyaev AK, Mashonkina LI (2022) Non-LTE abundances of zinc in different spectral type stars and the Galactic [Zn/Fe] trend based on quantum-mechanical data on inelastic processes in zinc-hydrogen collisions. Mon Not R Astron Soc 515:1510–1523

Socas-Navarro H (2015) The solar oxygen abundance from an empirical three-dimensional model. Astron Astrophys 577:id25

Spruit HC, Nordlund A, Title AM (1990) Solar convection. Annu Rev Astron Astrophys 28:263–301. https://doi.org/10.1146/annurev.aa.28.090190.001403

Steffen M, Prakapavičius D, Caffau E, Ludwig HG, Bonifacio P, Cayrel R, Kučinskas A, Livingston WC (2015) The photospheric solar oxygen project-IV. 3D-NLTE investigation of the 777 nm triplet lines. Astron Astrophys 583:A57

Storm N, Barklem PS, Yakovleva SA, Belyaev AK, Palmeri P, Quinet P, Lodders K, Bergemann M, Hoppe R (2024) 3D NLTE modelling of Y and Eu-Centre-to-limb variation and solar abundances. Astron Astrophys 683:A200

Stracke A, Palme H, Gellissen M, Münker C, Kleine T, Birbaum K, Günther D, Bourdon B, Zipfel J (2012) Refractory element fractionation in the Allende meteorite: implications for solar nebula condensation and the chondritic composition of planetary bodies. Geochim Cosmochim Acta 85:114–141

Stracke A, Scherer EE, Reynolds BC (2014) Application of isotope dilution in geochemistry. Treatise Geochem 15:71–86. https://doi.org/10.1016/B978-0-08-095975-7.01404-2

Suess HE (1947a) Über kosmische Kernhäufigkeiten. I. Mitteilung, Einige Häufigkeitsregeln und ihre Anwendung bei der Abschätzung der Häufigkeitswerte für die mittelschweren und schweren Elemente. Z Naturforsch 2a:311–321

Suess HE (1947b) Über kosmische Kernhäufigkeiten II. Mitteilung, Einzelheiten in der Häufigkeitsverteilung der mittelschweren und schweren Kerne. Z Naturforsch 2a:604–608

Suess HE, Urey HC (1956) Abundances of the elements. Rev Mod Phys 28:53–74

Tachiev GI, Fischer CF (2002) Breit-Pauli energy levels and transition rates for nitrogen-like and oxygen-like sequences. Astron Astrophys 385(2):716–723

Takahashi H, Janssens M-J, Morgan JW, Anders E (1978) Further studies of trace elements in C3 chondrites. Geochim Cosmochim Acta 42:97–106

Thenard LJ (1806) Analyse d'un aérolithe tombée dans l'arrondissement d'Alais, le 15 Mars 1806. Ann Chim Phys 59:103–110

Turcotte S, Wimmer-Schweingruber RF (2002) Possible in situ tests of the evolution of elemental and isotopic abundances in the solar convection zone. J Geophys Res 107(A12):1442

Turcotte S, Richter J, Michaud G, Iglesia CA, Rogers FJ (1998) Consistent solar evolution model including diffusion and radiative acceleration effects. Astrophys J 504:539–558

Velichko AB, Mashonkina LI, Nilsson H (2010) Formation of Zr I and II lines under non-LTE conditions of stellar atmospheres. Astron Lett 36:664–679

Vernazza JE, Avrett EH, Loeser R (1981) Structure of the solar chromosphere. III. Models of the EUV brightness components of the quiet sun. Astrophys J Suppl Ser 45:635

Vogel N, Heber Veronika VS, Baur H, Burnett DS, Wieler R (2011) Argon, krypton, and xenon in the bulk solar wind as collected by the Genesis mission. Geochim Cosmochim Acta 75:3057–3071

Vögler A, Shelyag S, Schüssler M, Cattaneo F, Emonet T, Linde T (2005) Simulations of magneto-convection in the solar photosphere-equations, methods, and results of the MURaM code. Astron Astrophys 429:335–351

Wang M, Huang WJ, Kondev FG, Audi G, Naimi S (2021) The AME 2020 atomic mass evaluation (II). Tables, graphs and references. Chin Phys C 45(3):030003

Wiik HB (1969) On the regular discontinuities in the composition of meteorites. Comment Phys-Math Soc Sci Fenn 34:135–145

Wing J (1964) Simultaneous determination of oxygen and silicon in meteorites and rocks by nondestructive activation analysis with fast neutrons. Anal Chem 36:559–564

Wolf D, Palme H (2001) The Solar System abundances of phosphorus and titanium and the nebular volatility of phosphorus. Meteorit Planet Sci 36:559–571

Yakovleva SA, Belyaev AK, Bergemann M (2020) Cobalt-hydrogen atomic and ionic collisional data 8(3):34

Yang W (2019) Rotating solar models with low metal abundances as good as those with high metal abundances. Astrophys J 873:id18

Yin Q, Jacobsen SB, Yamashita K, Blichert-Toft J, Télouk P, Albarede F (2002) A short timescale for terrestrial planet formation from Hf–W chronometry of meteorites. Nature 418:949–952

Yokoyama T, Nagashima K, Hakai I, Young ED, Abe Y, Aleon J, Alexander CMO'D, et al (2023) Samples returned from the asteroid Ryugu are similar to Ivuna-type carbonaceous chondrites. Science 379:eabn7850

Yokoyama T, Dauphas N, Fukai R, Usui T, Tachibana S, Schönbächler M, Busemann H, Abe M, Yada T (2024) The elemental abundances of Ryugu: assessment of chemical heterogeneities and the nugget effect. arXiv preprint. arXiv:2405.04500

Young PR (2018) Element abundance ratios in the quiet Sun transition region. Astrophys J 855:id15

Zhang HW, Gehren T, Butler K, Shi JR, Zhao G (2006) Potassium abundances in nearby metal-poor stars. Astron Astrophys 457:645–650

Zhang HW, Gehren T, Zhao G (2008) A non-local thermodynamic equilibrium study of scandium in the Sun. Astron Astrophys 481:489–497

Zhao G, Mashonkina L, Yan HL, Alexeeva S, Kobayashi C, Pakhomov Y, Shi JR, Sitnova T, Tan KF, Zhang HW, Zhang JB (2016) Systematic non-LTE study of the $-2.6\leq$[Fe/H]≤ 0.2 F and G dwarfs in the solar neighborhood. II. Abundance patterns from Li to Eu. Astrophys J 833:225

Space Science Reviews (2024) 220:69
https://doi.org/10.1007/s11214-024-01102-0

Chondrule Properties and Formation Conditions

Yves Marrocchi[1] · Rhian H. Jones[2] · Sara S. Russell[3] · Dominik C. Hezel[4] ·
Jens Barosch[5,6] · Aleksandra Kuznetsova[7,8]

Received: 20 December 2023 / Accepted: 19 August 2024 / Published online: 5 September 2024

Abstract
Chondrules are iconic sub-millimeter spheroids representing the most abundant high-temperature dust formed during the evolution of the circumsolar disk. Chondrules have been the subject of a great deal of research, but no consensus has yet emerged as to their formation conditions. In particular, the question of whether chondrules are of nebular or planetary origin remains largely debated. Building upon decades of chondrule investigation and recent headways in combining petrographic observations and O−Ti−Cr isotopic compositions, we here propose a comprehensive vision of chondrule formation. This holistic approach points toward a nebular origin of both NC and CC chondrules, with repetitive high-temperature recycling processes controlling the petrographic and isotopic diversities shown by chondrules. Chondrule precursors correspond to mixing between (i) early-formed refractory inclusions $\pm$ NC-like dust and (ii) previous generation of chondrules $\pm$ CI-like material. Chondrule formation took place under open conditions with gas-melt interactions with multi-species gas (H_2O, Mg, SiO) playing a key role for establishing their characteristics. Petrographic and isotopic systematics do not support disk-wide transport of chondrules but point toward local formation of chondrules within their respective accretion reservoirs. Altogether, this shows that several generations of genetically-related chondrules (i.e., deriving from each other) co-exist in chondrites. In addition to supporting the nebular brand of chondrule-forming scenarios, this argues for repetitive and extremely localized heating events for producing chondrules.

Keywords Chondrules · Chondrites · Protoplanetary disk · Isotopes · Recycling processes

1 Introduction

Chondrules are submillimeter to millimeter-scale spheroids composed predominantly of silicate minerals, Fe-Ni metal beads, sulfides and glassy mesostases. They represent the most abundant high-temperature component in non-carbonaceous chondrites (NC, corresponding to ordinary chondrites, enstatite chondrites, Rumuruti chondrites and Kakangari chondrites) and carbonaceous chondrites (CC, composed of CM, CO, CV, CK, CL, CR, CH, CB chondrites), thus implying that both the inner (sampled by NC) and outer (sampled by CC) solar system were affected by the chondrule formation process. Although the timing of the onset of chondrule formation is the subject of an intense debate (Bollard et al. 2017; Fukuda et al. 2022; Piralla et al. 2023; Siron et al. 2022), all chronometers point to formation processes

Extended author information available on the last page of the article

spanning several million years. Chondrules thus represent invaluable witnesses of dust production and processing during the evolution of the solar protoplanetary disk.

Despite the relatively simple major mineralogy of chondrules, countless and disparate models have been proposed to describe their formation. This is due to the difficulty of (i) integrating all the chemical, petrographic, and isotopic signatures of chondrules and (ii) deciphering the complex processes at play during chondrule formation: precursor heterogeneity, open-system gas-melt interactions, potential large-scale transport and secondary alteration modifications. Another difficulty stands in the lack of astrophysical model to generate chondrule-forming shock (Desch et al. 2012). This has led to development of a plethora of models, either considering chondrule production in nebular environments (e.g., shock waves, lightning discharges; Desch and Connolly 2002; Hubbard and Ebel 2018; Kaneko et al. 2023; Morris and Boley 2018) or planetary settings (e.g., solid debris of collisions, impact splashes; Johnson et al. 2015; Libourel and Krot 2007; Lichtenberg et al. 2018; Sanders and Scott 2012). Models also exist where chondrules formed through the melting of nebular dust by eccentric large planetary embryos (Morris et al. 2012). This situation could appear unsolvable but recent years have seen important progress through revealing new, overlooked petrographic characteristics and isotopic constraints. Here, we highlight recently reported constraints and discuss implications regarding chondrule formation conditions, building upon decades of chondrule investigation.

2 Petrographic Characteristics of Chondrules

Chondrules are described according to their mineralogy and petrographic characteristics (e.g., Connolly and Jones 2016; Hewins et al. 2005; Zanda 2004). NC and CC chondrules have common properties (except for CH and CB chondrites), and they are described together in this section. They are broadly divided into two varieties, porphyritic and non-porphyritic. Porphyritic textures are more abundant in all chondrite groups. The term porphyritic is used to describe the texture of an igneous rock, in which larger crystals occur in a fine-grained matrix. The terminology of igneous rocks can be applied to chondrules which have textures produced during cooling from a molten droplet. A chondrule with a porphyritic texture typically has larger grains of olivine and/or pyroxene, both of which are Fe- and Mg-bearing silicate minerals. The remaining interstitial material, which represents the final liquid present during cooling, is either glassy or fine-grained, and is referred to as mesostasis. Mesostasis has higher abundances of elements such as Ca, Al, Na, K than olivine and pyroxene: these are elements that constitute the mineral feldspar. Feldspar may or may not be present, so the general description of mesostasis is that it is typically feldspathic in composition. Non-porphyritic chondrules are described further below.

There are two main types of porphyritic chondrules, designated type I and type II (Scott and Taylor 1983). Type I chondrules are more reduced than type II chondrules. This is reflected in the Fe/Mg ratio of olivine and pyroxene, as well as the abundance of Fe-Ni metal and chromite (a Cr-oxide). A second part of the description of a porphyritic chondrule defines its olivine/pyroxene ratio. Chondrules with high olivine content (>90%) are designated as A, chondrules with high pyroxene content (>90%) are designated as B, and more equal ratios are designated as AB. For example, a type IAB chondrule is reduced and contains similar amounts of olivine and pyroxene grains. Porphyritic chondrules are also often described using O and P for olivine and pyroxene, i.e., PO = porphyritic olivine, POP = porphyritic olivine and pyroxene, PP = porphyritic pyroxene, analogous to A, AB and B respectively. However, chondrule textures are mostly studied in 2D meteorite sections and determined proportions of olivine and pyroxene are not always accurate, leading to misclassification of textural types (Barosch et al. 2020).

Pristine, unaltered type I porphyritic chondrules have low FeO content: the fayalite content of olivine (Fa) and the ferrosilite content of pyroxene (Fs) (where Fa = atomic Fe/(Mg + Fe) and Fs = atomic Fe/(Mg + Fe + Ca)) are typically < 5 mole%, although the formal definition is Fa < 10 mole%. Type I porphyritic chondrules typically contain numerous small, subhedral, olivine grains, low-Ca pyroxene (clinoenstatite), interstitial glassy or microcrystalline mesostasis, and variable abundances of rounded grains of Fe, Ni metal (kamacite) (Fig. 1a-c). Low-Ca pyroxene commonly has narrow Ca-pyroxene overgrowths on individual grains and often poikilitically encloses olivine grains. Pyroxene is mostly concentrated in the outer part of type I chondrules, forming an outer shell (Fig. 1a; e.g., Barosch et al. 2019, 2020; Friend et al. 2016; Tissandier et al. 2002). This observation, along with evidence such as zoning in olivine identified using cathodoluminescence and X-ray maps (Fig. 2, Nagahara et al. 2008; Libourel and Portail 2018; Marrocchi et al. 2019a, 2018), has been used to argue for open-system behavior of type I chondrule melts, and continuous interactions between chondrule melt and surrounding gas during chondrule crystallization.

Unaltered type II porphyritic chondrules contain olivine and/or pyroxene with a range of grain sizes, including larger, well-formed (euhedral) olivine and/or pyroxene grains (Fig. 1d-f). Olivine and pyroxene are more FeO-rich than in type I chondrules and they are usually strongly zoned, for example olivine with Fa = 10-15 mole% in the center of a grain to Fa > 30 mole% at the edge. This zoning is a consequence of disequilibrium growth as the chondrule cools, and it is expected to be more pronounced in more FeO-rich systems than in type I chondrules. Type II chondrules typically have a microcrystalline mesostasis, and low abundances of chromite as well as metal/sulfide grains, predominantly sulfides. Olivine in type II chondrules can be skeletal in form, a texture that represents disequilibrium growth (Fig. 1e, f). In type II POP and PP (type IIAB, IIB) chondrules, pyroxene is not typically concentrated in the outer part, and olivine and pyroxene are intergrown (Jones 1996). For type II chondrules, there are currently no convincing arguments for open-system behavior and gas-melt interactions during chondrule melting and cooling.

Non-porphyritic chondrules have textures in which olivine and pyroxene grains are elongate, a sign of rapid crystal growth. Rapid growth occurs when nucleation is inhibited, as a combined result of two factors: the extent of heating above the liquidus temperature (i.e., the temperature at which the entire chondrule is melted) which destroys potential nucleation sites, and a rapid cooling rate which results in supersaturation of the melt. In barred olivine (BO) chondrules, single crystals of olivine grow in parallel plates which appear as a set of bars in cross section (Weisberg 1987; Fig. 1g). Several such crystals can occur in one chondrule. Mesostasis between the olivine bars is often glassy. Radiating pyroxene (RP) textures consist of very thin needles of pyroxene that radiate from a single nucleation point in a fan-like texture (Fig. 1h). Cryptocrystalline (CC) chondrules are extremely fine-grained, and individual needles are randomly oriented (Fig. 1i).

Porphyritic chondrules commonly contain relict grains (e.g., Nagahara 1981; Rambaldi 1981; Jones 2012; Ruzicka et al. 2008). These are identifiable grains that were present in the precursor assemblage and survived the melting event. Most relict grains are olivine, although pyroxene relict grains can also occur. In type II chondrules, relict forsterite grains are easily identifiable in back-scattered electron images, as core regions of more FeO-rich olivine grains that have grown around the relicts during cooling (Fig. 1d-f). Type I chondrules contain "dusty" relict olivine grains which are also easily identified (Fig. 1c). These have a dark, dusty appearance in thin sections because they contain numerous sub-micrometer blebs of iron metal that exsolved from an originally FeO-rich relict olivine grain as a result of reduction, presumably during formation of the host type I chondrule. Relict grains of similar composition to those grown from the host chondrule melt are more difficult to detect, but they can be observed using techniques such as cathodoluminescence and high-resolution chemical mapping (Jones and Carey 2006; Libourel et al. 2022; Marrocchi et al. 2018). Relict grains appear to be derived from previous generations of chondrules, or from amoeboid olivine aggregates (AOAs: a type of refractory inclusion, Krot et al. 2004; Grossman and Steele 1976; Weisberg et al. 2004). They therefore attest to multiple heating events in chondrule-forming regions. Another feature ascribed to multiple heating events is igneous rims on chondrules. These are layers that overlay the host chondrule, interpreted as a second melting event during which dust that was adhering to the chondrule was melted and cooled under similar conditions to the main chondrule-forming event.

The chondrules described above are dominated by olivine and pyroxene, and can be described as ferromagnesian chondrules. A further category of chondrules is aluminium-rich chondrules (ARC), which includes plagioclase-rich chondrules (PRC) named after their dominant mineral, plagioclase feldspar (Krot and Keil 2002). Aluminium-rich chondrules have compositions intermediate between refractory inclusions and ferromagnesian chondrules. Bischoff and Keil (1984) defined ARCs as having bulk compositions with >10 wt% Al_2O_3, but MacPherson and Huss (2005) suggested relaxing that definition in favor of including chondrules which contain primary igneous phases with high Al/Mg ratios such as plagioclase, Al-rich glass, spinel, and Al-diopside. ARCs include a wide variety of textures and compositions, including porphyritic, non-porphyritic and glass-rich chondrules (Fig. 1j, k). Non-porphyritic ARCs include barred olivine chondrules with Na-Al-rich glass between MgO-rich olivine bars (e.g., MacPherson and Huss 2005).

Chondrules in the metal-rich CB and CH chondrites have unusual non-porphyritic textures and chemistry. They are thought to have formed by condensation of liquid in an impact plume (e.g., Campbell et al. 2002; Krot et al. 2012, 2010, 2005a; Rubin et al. 2003). These chondrules have cryptocrystalline (CC) and skeletal olivine-pyroxene (SOP) textures, which do not contain metal or relict grains, indicating that they formed from completely molten droplets (Fig. 1i is from a CH chondrite; Fig. 1l is from a CBa chondrite). CH chondrites also contain a population of porphyritic ferromagnesian chondrules, but CB chondrites only contain CC and SOP chondrules. Chondrules in CBa chondrites are cm-sized whereas in CBb and CH chondrites they are significantly smaller, sub-millimeter.

Several properties of chondrules can be used to infer their thermal histories, including peak temperatures and cooling rates (Jones et al. 2018). Observed properties such as the presence of clinoenstatite (a specific form of enstatite that is only formed during rapid cooling), chemical and isotopic zoning in mineral grains, presence of plagioclase, and survival of relict grains, can all place quantitative constraints on thermal histories. In addition, laboratory experiments designed to reproduce chondrule textures, as well as chemical compositions of mineral grains, play an essential role in defining formation conditions. Porphyritic textures result when a chondrule is heated to a maximum temperature close to, but below, its liquidus temperature (1400–1700 °C, depending on composition), and cooled at initial rates between about 10 and 1000 °C/h. Non-porphyritic chondrules are produced when peak temperatures exceed the liquidus slightly (for barred/dendritic textures) and significantly (radiating textures) and chondrules cool at rates around 500 – 3000 °C/h. Aluminium-rich chondrules formed under similar conditions to ferromagnesian chondrules, with peak temperatures of 1400–1500 °C, and cooling rates of 10–500 °C/h for porphyritic chondrules and possibly higher (1000 °C/h) for barred chondrules (Tronche et al. 2007).

As noted previously, the above description of the petrography of chondrules includes chondrules from NC and CC isotopic groups. However, in detail there are differences among the chondrule populations in different chondrite groups, some of which can be viewed on the

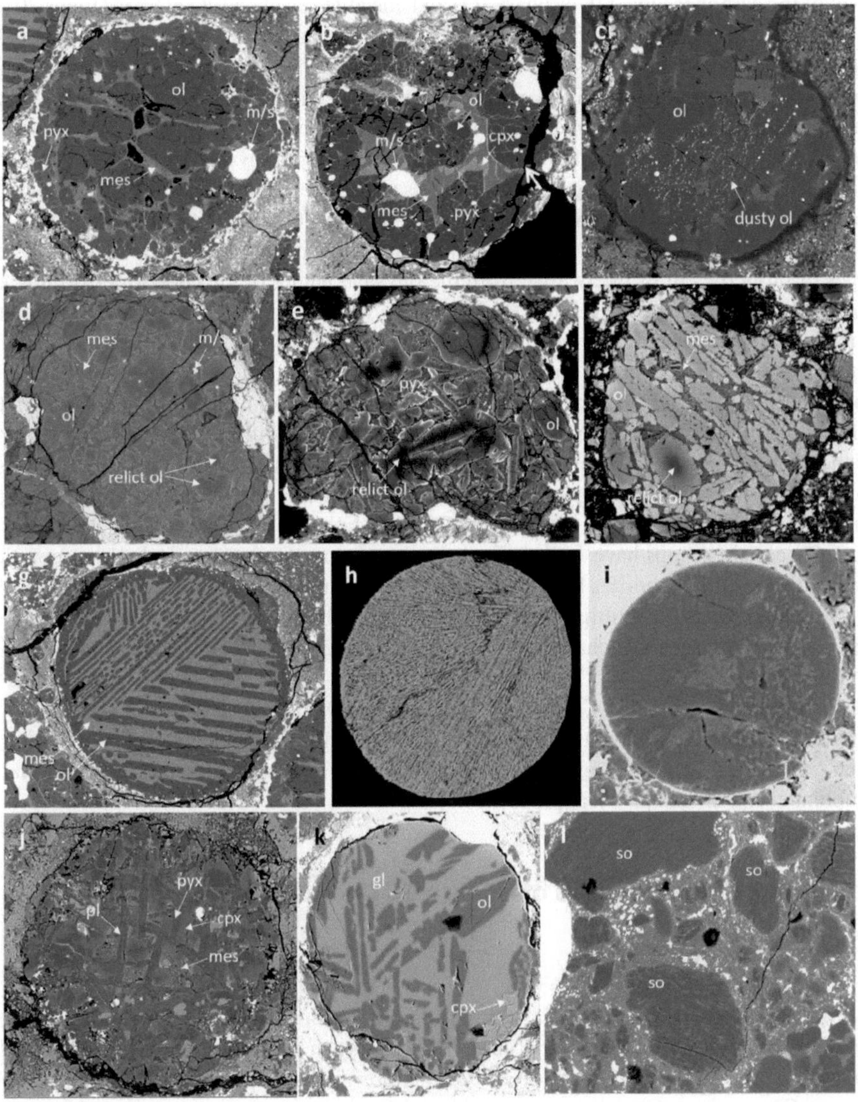

Fig. 1 Diversity of chondrule textures. All images are back-scattered electron (BSE) images apart from (h) which is an elemental Mg X-ray map. In a BSE image, grey level is a function of average atomic number, for example Fe, Ni metal and iron sulfide are white, forsterite (Mg-rich olivine) is dark grey and FeO-rich olivine is medium grey. a) Type IA porphyritic olivine (PO) chondrule; b) Type IB porphyritic pyroxene (PP) chondrule; c) type IA porphyritic olivine (PO) chondrule with a dusty relict olivine grain; d) type IIA porphyritic olivine (PO) chondrule with relict forsteritic cores in some grains; e) type IIAB porphyritic olivine / pyroxene (POP) chondrule, with skeletal relict olivine; f) type IIA porphyritic olivine (PO) chondrule with elongate, skeletal olivine grains; g) barred olivine (BO) chondrule with two small adhering chondrules; h) radial pyroxene (RP) chondrule with elongate grains of pyroxene; i) cryptocrystalline (CC) chondrule, mineral phases too small to identify; j) plagioclase-rich chondrule (PRC); k) glassy aluminium-rich chondrule (ARC); l) skeletal olivine (SO) chondrules in CBa chondrite, Weatherford. All chondrules are several hundred micrometres in diameter, except (i) which is about 150 mm in diameter. Image (l) is about 1 mm across. Abbreviations: ol = olivine, pyx = low-Ca pyroxene, cpx = Ca-rich pyroxene, mes = mesostasis, m/s = metal and/or sulfide, pl = plagioclase, gl = glass, so = skeletal olivine

level of the carbonaceous (C), ordinary (O) and enstatite (E) chondrite classes (Jones 2012). While the CC isotopic group consists of only C chondrites, the NC isotopic group includes the OC, EC, and R chondrites and the chondrule populations in these classes often differ, most notably because of the highly reduced nature of E chondrites (Jacquet et al. 2018). As an example, it is possible to make a general statement that NC chondrites, including OC and EC, have a higher proportion of non-porphyritic vs. porphyritic chondrules compared with CC. However, other properties can only sensibly be compared between CC and OC chondrites, for example CCs have a higher ratio of type I/type II chondrules than OCs, and relict forsterite grains in type II chondrules are more abundant in CCs than in OCs. These types of properties most likely arise from the physical nature of the chondrule-forming environment, which could be related to differences in inner vs. outer solar system regions. There are also chemical differences among chondrite classes, for example olivine grains in type II PO chondrules in OCs have FeO contents and Mn/Fe ratios distinct from type II PO chondrules in CCs (Berlin et al. 2011; Jones 2012). Such chemical differences may reflect differences in the composition of condensed solids, and / or differences in oxygen fugacity of the ambient gas. In addition, within each class of chondrites, some chondrule properties vary among groups. For example, mean chondrule diameters vary among chondrite groups, but the range in CCs overlaps the range in OCs and ECs (Rubin 2010; Jones 2012). Within the CCs, certain groups have well-defined characteristics such as chondrule rim properties: chondrules in CM chondrites have wide fine-grained rims (Metzler et al. 1992), chondrules in CV chondrites have multiple igneous rims (Rubin 1984), and chondrules in CR chondrites have igneous rims including silica-rich rims (Hezel et al. 2003; Krot and Keil 2002; Smith and Jones 2024). The Mn/Fe ratio in olivine in type II PO chondrules is higher in CR chondrites than in other CC groups (Berlin et al. 2011; Jones 2012). These characteristics that are unique to individual chondrite groups indicate that the chondrule population in each group was localized, and each chondrule-forming region had slightly but identifiably different chemical and physical properties.

3 Isotopic Characteristics of NC and CC Chondrules

3.1 Type I Chondrules

Since the pioneering works on Calcium-Aluminium-rich inclusions (CAIs) in the seventies by the Chicago group (Clayton et al. 1973), oxygen isotopes are commonly used for quantifying the physico-chemical processes that prevailed during the evolution of the solar system. Oxygen isotope ratios are reported in delta notation, or parts-per-thousand fractionation relative to Vienna Standard Mean Ocean Water (VSMOW; Baertschi 1976): $\delta^{17,18}O$ (‰) = $[(R_{sample}/R_{VSMOW}) - 1] \times 1{,}000$, where R = $^{17,18}O/^{16}O$. Data are reported using oxygen three-isotope plot ($\delta^{17}O$ *vs.* $\delta^{18}O$). Data are also reported as $\Delta^{17}O$ (= $\delta^{17}O - 0.52 \times \delta^{18}O$), approximating the difference in $\delta^{17}O$ relative to the terrestrial fractionation line.

In early times of geochemistry, no *in-situ* technique was available and determining the oxygen isotopic composition of chondrules required their mechanical separation. Although bulk measurements of individual chondrules are likely affected by secondary alteration processes that could have slightly modified their oxygen isotopic compositions (especially that of the glassy mesostasis, Kita et al. 2010; Ushikubo et al. 2012), they are nevertheless interesting and reveal several important features:

1- Type I CC chondrules show large mass-independent variations (Fig. 3a, Clayton and Mayeda 1999, 1984; Jabeen et al. 2019; Jones et al. 2004; Rowe et al. 1994; Rubin et al.

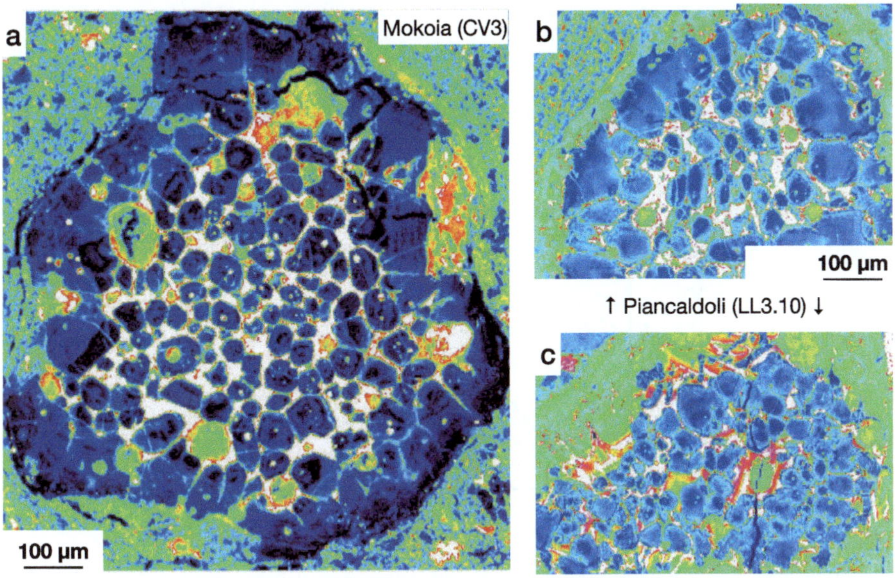

Fig. 2 X-ray map of Ti distribution in PO chondrules of the carbonaceous chondrite CV3 Mokoia (**a**) and the ordinary chondrite LL3 Piancaldoli (**b & c**). These maps reveal internal structures and emphasizing that chondrules are radially zoned with outer Ti-rich olivine grains and interiors dominated by smaller Ti-poor olivine grains surrounded by mesostasis. Ti-rich overgrowths commonly armor Ti-poor olivine cores. Data from (Piralla et al. 2021) for OC chondrules and unpublished data for Mokoia

1990; Weisberg et al. 1993). In the oxygen three-isotope diagram, these variations define a general single array along a line of $\delta^{17}O = [0.985 \pm 0.029] \times \delta^{18}O - [3.52 \pm 0.08]$ that plot in-between the primitive chondrule minerals line (PCM; Ushikubo et al. 2012; Zhang et al. 2022) and the carbonaceous chondrite anhydrous mineral line (CCAM;, Fig. 3a, N = 119, MSWD = 4.5). Of note, all CC chondrules have $\Delta^{17}O < 0‰$ (Fig. 3a). We note that they plot slightly to the right of the PCM, this position probably resulting from secondary alteration processes.

2- Such mass-independent variations are also observed within a given CC. For example, type I chondrules from the CV3 Allende, the most studied CC (N = 64), also show large mass-independent variations (Clayton and Mayeda 1984; Jabeen et al. 2019; Rubin et al. 1990).

3- Type I NC chondrules show narrower isotopic variations (Fig. 3a, Bridges et al. 1999; Clayton et al. 1991). With few exceptions (6 out of 106), all NC chondrules show $\Delta^{17}O > 0‰$ and cluster to the left of the PCM (Fig. 3a). They however define, within error, mass-independent variations with $\delta^{17}O = [0.837 \pm 0.213] \times \delta^{18}O - [1.11 \pm 1.09]$.

4- A key observation that has largely remained below the radar screen stands in the size-dependent oxygen isotopic compositions of chondrules in ordinary chondrites (Fig. 3b, Clayton et al. 1991). It appears that OC chondrules with size $\leq 300\ \mu m$ have statistically distinct, ^{16}O-rich isotopic composition compared to their larger counterparts (Fig. 3b, Clayton et al. 1991).

Secondary ion mass spectrometry (SIMS) enables the determination of *in-situ* oxygen isotopic compositions of chemically-characterized minerals within chondrules, as well as isolated olivine grains in chondrite matrices (hereafter IOGs, Kita et al. 2010; Jacquet et al. 2021; Jones et al. 2004; Marrocchi et al. 2016; Ushikubo et al. 2012). With SIMS, high

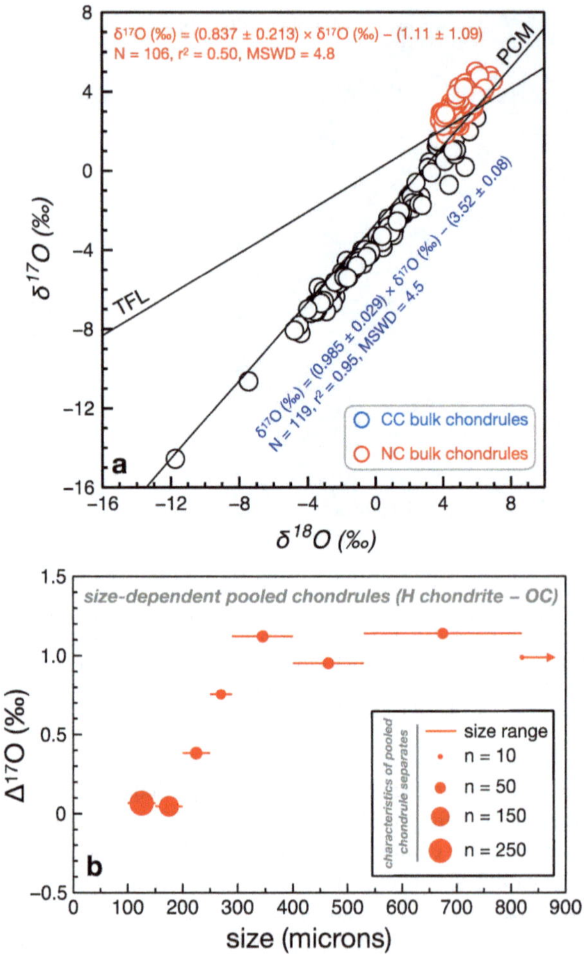

Fig. 3 (a) Oxygen three-isotope plot for bulk mechanically-separated chondrules in various carbonaceous (CM, CV, CK, and CR) and non-carbonaceous (L, LL, H, EH and EL) chondrites. Bulk CC chondrules show large O-isotopic variations that define a well-resolved with slope of 0.985 ± 0.029 and intercept of -3.52 ± 0.08 (2 s) plotting in-between the PCM and CCAM lines. Bulk NC chondrules show narrower O-isotopic variations with $\Delta^{17}O > 0$‰ (for 100 out of 106 measurements). They define a trend with slope of 0.837 ± 0.213 and intercept of -1.11 ± 1.09 (2 s). Data from Clayton et al. 1991; Mayeda et al. 1988; Rubin et al. 1990; Rowe et al. 1994, Weisberg et al. 2003; Jones et al. 2004. **(b)** $\Delta^{17}O$ of sized NC chondrules separated from the H chondrite Dhajala (data from Clayton et al. 1991). This reveals that chondrules with size < 300 μm have significantly lower $\Delta^{17}O$ that point in the direction of CC chondrules. Of note, the $\Delta^{17}O$ value for a given size range corresponds to pooled chondrule measurements (ranging from n = 8 for chondrules with size > 820 μm to n = 260 for chondrules with sizes spanning from 100 to 150 μm; Clayton et al. 1991)

precision isotopic analyses of chemically-characterized minerals are possible with minimal destruction, and they allow the detection of small-scale heterogeneities that would be concealed in bulk analyses. The most targeted minerals are olivine and low-Ca pyroxene, whereas high-Ca pyroxene, plagioclase and glassy mesostasis have been less studied due to small sizes and/or incipient aqueous alteration. The large body of data available today makes it possible to establish isotopic systematics.

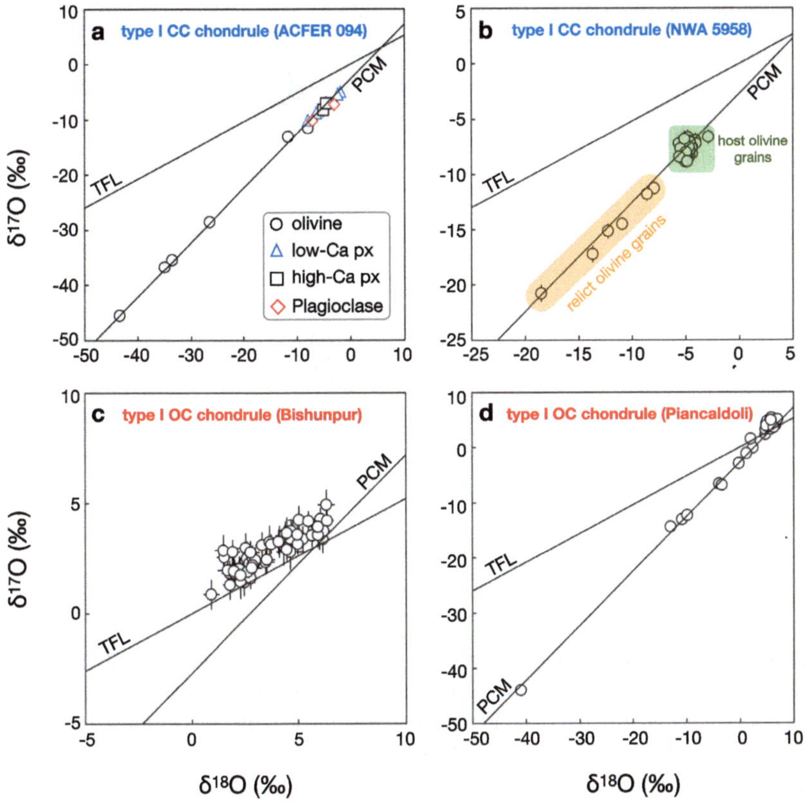

Fig. 4 (a) Example of SIMS interphase O-isotope data within a chondrule from the C-ung. Acfer 094 (data from Ushikubo et al. 2012). Heterogeneous O-isotopic compositions are observed for olivine grains whereas other mineral phases are isotopically homogeneous. **(b)** Representative example from the C-ung. NWA 5958 of CC chondrule olivine grains showing large, mass-independent, O-isotopic variations. Relict olivine grains correspond to minerals with $\Delta^{17}O$ values outside 3SD of averaged homogeneous chondrule phase $\Delta^{17}O$ (here only olivine, data from Marrocchi et al. 2018). Of note, this chondrule is characterized by olivine crystals showing 120° triple junctions. **(c)** Oxygen isotopic composition of olivine grains in chondrule Ch-6 from LL3.15 Bishunpur. All olivine grains in this chondrule correspond to host olivines, with relatively constant $\Delta^{17}O = 0.9 \pm 0.8$‰ but variable $\delta^{18}O$ values from 0.9 to 6.4‰. This correlation reveals mass-dependent fractionation within this chondrule. Data from (Piralla et al. 2021). **(d)** Oxygen isotopic composition of olivine grains in chondrule Ch-6 from LL3.10 Piancaldoli (Marrocchi et al. 2020). This chondrule contains seven relicts with variable ^{16}O enrichments. Host grains define a good correlation with $\delta^{17}O = (1.002 \pm 0.111) \times \delta^{18}O - (1.72 \pm 0.61)$ (MSWD = 2.52, $r^2 = 0.73$, n = 50, data from Piralla et al. 2021)

1- The different minerals within a single type I CC chondrule commonly show uniform isotopic compositions (Tenner et al. 2018). However, they occasionally define mass-independent variations plotting along the PCM with $\Delta^{17}O < 0$‰ (Fig. 4a, b, Marrocchi et al. 2018; Ushikubo et al. 2012). These mass-independent variations are mostly carried by olivine grains that frequently show variable compositions within individual chondrules (Fig 4a, b, Chaumard et al. 2021, 2018; Hertwig et al. 2018; Jones et al. 2004; Marrocchi et al. 2022, 2021, 2019a, 2018; Marrocchi and Chaussidon 2015; Rudraswami et al. 2011; Schneider et al. 2020; Tenner et al. 2019, 2015, 2013; Ushikubo et al. 2012; Ushikubo and Kimura 2021; Williams et al. 2020). Conversely, low-Ca pyroxene, high-Ca pyroxene and plagioclase grains are overwhelmingly isotopically uniform (Tenner et al. 2018). Olivine isotopic

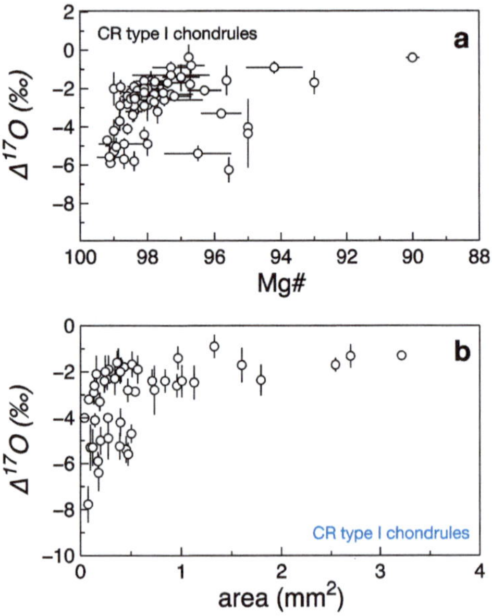

Fig. 5 **(a)** Positive trend between chondrule Mg#'s and host $\Delta^{17}O$ for CR chondrules. Data from Connolly and Huss 2010, Schrader et al. 2013, 2014, 2017, Tenner et al. 2015. **(b)** Oxygen isotopic compositions of CR type I chondrules as a function of their sizes. Small CR chondrules have $\Delta^{17}O$ values similar to those reported for CM-CV- CO chondrules, whereas large CR chondrules are enriched in $^{17,18}O$. Data are from Marrocchi et al. 2022, Schrader et al. 2013, 2017, 2014, Tenner et al. 2015

variations led to the isotopic definitions of *host* and *relict* olivine grains with (i) host grains corresponding to average O-isotope composition of minerals with similar $\Delta^{17}O$ within a single chondrule (Fig. 4b) and (ii) relict grains being those with $\Delta^{17}O$ values beyond analytical uncertainties (Fig. 4b, Ushikubo et al. 2012). With few exceptions (e.g., Chaumard et al. 2021; Kita et al. 2010; Marrocchi et al. 2018), relict olivine grains are systematically ^{16}O-rich compared to host olivine crystals (Fig. 4). These two populations of olivine grains are generally interpreted as reflecting (i) inheritance of mineral grains present in the chondrule precursors (relicts) and (ii) crystallization from the final melt whose composition is buffered by gas-melt interaction processes (hosts: Jones et al. 2004; Ushikubo et al. 2012). Of note, type I CV chondrule silicates also show large, mass-dependent silicon isotopic variations similar to those observed in AOAs (Marrocchi et al. 2019b; Villeneuve et al. 2020).

2- Type I chondrules in CR have a high degree of O-isotope homogeneity per chondrule (Tenner et al. 2018). Compared to other CCs, CR chondrules show greater variations in $\Delta^{17}O$ values ranging from -6 to - 2‰ (Schrader et al. 2013, 2014, 2017; Tenner et al. 2015, 2018), which define a positive trend with the Mg# of chondrules (Fig. 5a; Tenner et al. 2015, 2018). It has also been noticed that two chondrule populations characterized by different sizes and oxygen isotopic compositions co-exist in CR chondrites with the typically larger having $\Delta^{17}O > -3.5$‰ whereas the smaller chondrules (i.e., with areas < 1 mm^2) have $\Delta^{17}O < -3.5$‰, similar to chondrules in other CCs (Fig. 5b; Marrocchi et al. 2022).

3- Olivine grains in type I OC chondrules are predominantly isotopically uniform in $\Delta^{17}O$ (Fig. 4c), although a few rare chondrules show relict olivine grains plotting along the mass-independent PCM line (Fig. 4d, Kita et al. 2010; Piralla et al. 2021; Schneider et al. 2020; Williams et al. 2020). Similar to CC chondrules, other phases than olivine appear more isotopically homogeneous with relatively constant $\Delta^{17}O$ (Kita et al. 2010; Tenner et al. 2018). In agreement with bulk oxygen isotopic data, *in-situ* measurements reveal size-dependent oxygen isotopic variations in chondrules of ordinary chondrites with chondrules smaller than ~ 300 microns having CC-like sub-TFL $\Delta^{17}O$ values (Fig. 6, Marrocchi et al. 2024). Other NC chondrites (i.e., R and EC chondrites) also have olivine showing more

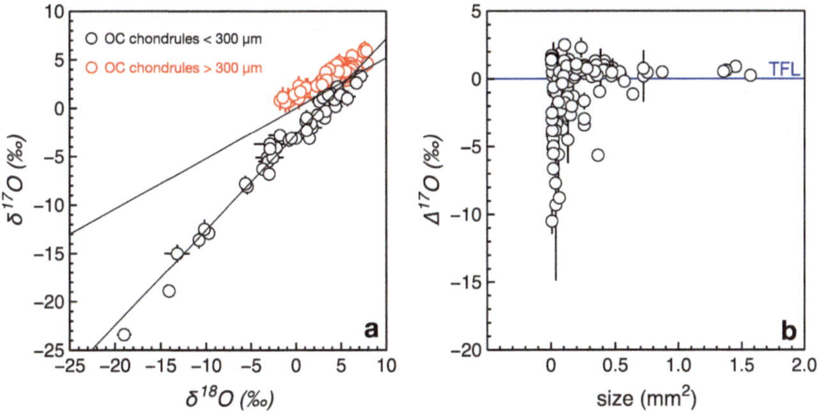

Fig. 6 (a) Oxygen isotopic compositions of olivine grains in OC chondrules with average $\Delta^{17}O < 0‰$ (red circles) and those with $\Delta^{17}O > 0‰$ (black circles). The former show mass-independent isotopic variations whereas the latter display mass-dependent isotopic variations (Marrocchi et al. 2024). (b) Average $\Delta^{17}O$ values of chondrules as a function of their surface areas in chondrites NWA 5206, Piancaldoli, Semarkona, Bishunpur, Chainpur, Dar al Gani 369, Dar al Gani 378 and Dar al Gani 327. Data from Marrocchi et al. 2024 and references therein

variable $\Delta^{17}O$ compared to low-Ca pyroxene (Fig. S1, Greenwood et al. 2000; Miller et al. 2017; Pack et al. 2004; Regnault et al. 2022; Weisberg et al. 2021, 2011). In R chondrites, IOGs appear more ^{16}O-rich than chondrules (Regnault et al. 2022), thus also suggesting size-controlled $\Delta^{17}O$ variations (Marrocchi et al. 2024). Most of olivine and low-Ca pyroxene in Kakangari (K-group) chondrules are ^{16}O-poor and plot on or close to the TFL (Nagashima et al. 2015). Limited mass-dependent variations of a few ‰ plotting to the left of the PCM appear to be a recurring characteristic of silicates in OC chondrules (Kita et al. 2010; Piralla et al. 2021) but this is not as pronounced in EC, R and K chondrites (Greenwood et al. 2000; Miller et al. 2017; Nagashima et al. 2015; Pack et al. 2004; Regnault et al. 2022; Weisberg et al. 2021, 2011).

4- Considering only olivine grains, *in-situ* isotopic data of type I CC chondrules/IOGs and OC chondrules show large mass-independent variations with $\Delta^{17}O$ ranging by ∼27‰ (from -23.0 to +3.6‰; Fig. 7) whereas EC chondrule and R chondrules/IOGs show more limited variations of ∼20 and 10‰, respectively (Fig. S1). Compared to other CCs, CR chondrites show distinct signatures with olivine grains showing smaller $\Delta^{17}O$ variations that span only ∼10‰ (Marrocchi et al. 2022; Schrader et al. 2014; Tenner et al. 2019, 2015). In addition to extreme negative values, another major difference between chondrite types lies in the abundance of relict olivine grains relative to their respective host counterparts. Based on the threshold defined by Ushikubo et al. (2012), ∼50% of analyzed olivine grains correspond to isotopic relicts in COs and the C-ung. chondrites Acfer 094 whereas 5 to 24% were reported for other CCs (Fig. 7, Chaumard et al. 2018; Hertwig et al. 2018; Marrocchi et al. 2019a, 2018; Tenner et al. 2018). Lower abundances of 3 and 7% were estimated for OCs (Kita et al. 2010; Piralla et al. 2021) whereas isotopic relict olivine grains appear either anecdotal or absent in the notably less studied EC and R chondrites (Fig. S1, Regnault et al. 2022; Weisberg et al. 2021, 2011).

5- Despite showing peculiar isotopic characteristics, relict olivine grains can be identified that have no clear major- and minor-element signatures and/or textural features (Jones et al. 2004; Ushikubo et al. 2012). High-resolution X-ray mapping and high-current quantitative electron microprobe analyses revealed that relict olivine grains in both CC and OC

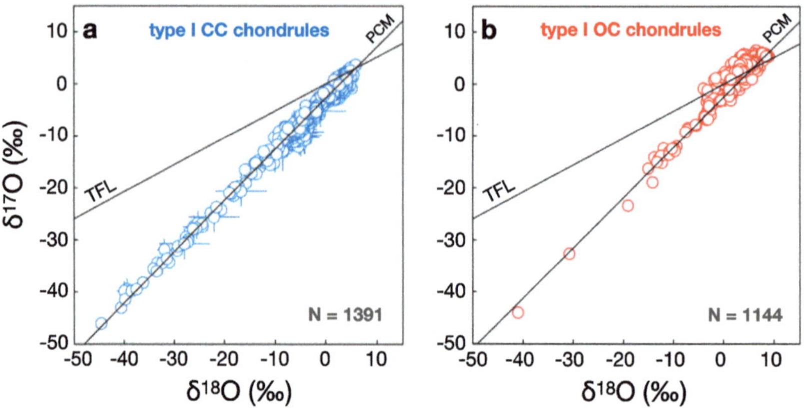

Fig. 7 (a) Oxygen isotopic compositions of olivine grains measured in a large set of CC chondrules. They show large and continuous O-isotopic variations spanning ~27‰ in $\Delta^{17}O$. Altogether, they define a well-resolved, mass-independent correlation with $\delta^{17}O = (0.989 \pm 0.005) \times \delta^{18}O - (3.08 \pm 0.03)$ (MSWD = 1.5, $r^2 = 0.98$, n = 1391). Data from (Chaumard et al. 2021, 2018; Hertwig et al. 2018; Jacquet et al. 2021; Jones et al. 2004; Marrocchi et al. 2024, 2022, 2021, 2019a, 2018; Rudraswami et al. 2011; Schneider et al. 2020; Tenner et al. 2019, 2015, 2013; Ushikubo et al. 2012; Ushikubo and Kimura 2021; Williams et al. 2020). **(b)** Oxygen isotopic compositions of olivine grains measured in a large set of OC chondrules. Although some OC chondrules display mass-dependent variations, the whole set of data a correlation close to mass-independent variations with $\delta^{17}O = (0.962 \pm 0.016) \times \delta^{18}O - (1.22 \pm 0.07)$ (MSWD = 1.6, $r^2 = 0.92$, n = 889). Data from Kita et al. 2010; Piralla et al. 2021

chondrules are systematically Al-Ca-Ti-poor, with concentrations lower than most (but not all) of those measured in host olivine grains (Fig. 8, Marrocchi et al. 2022, 2019a, 2018; Piralla et al. 2021). They also tend to have higher Mg# compared to host olivine grains, although variable Mg# was reported for relict olivine grains in different chondrules of the same chondrite (Marrocchi et al. 2019a).

6- When considering only O-isotopic compositions of host olivine grains, a clear dichotomy emerges with both CC and NC chondrules showing statistically different average $\Delta^{17}O$ of -5.8 ± 1.8 and +0.7 ± 0.9‰, respectively (Fig. 9a). In detail, small variations of the average $\Delta^{17}O$ are observed among both CC and NC. CO-CM-CV type I chondrules show a $\Delta^{17}O$ distribution largely dominated by a mode at -6‰ whereas CR chondrites display a bimodal distribution with $\Delta^{17}O$ modes of -6 and -2‰ and dominated by the latter (Fig. 9b). R, EC and OC chondrites show slightly different unimodal $\Delta^{17}O$ distributions (Fig. 9b).

7- Nucleosynthetic isotope anomalies have recently arisen as a powerful tool for understanding chondrule formation (Gerber et al. 2017; Schneider et al. 2020; Williams et al. 2020). Contrary to oxygen isotopes, nucleosynthetic isotope anomalies are not the result of isotope fractionation, but instead arise from the heterogeneous distribution of presolar matter derived from multiple nucleosynthetic sources. They are commonly reported in the ε-unit notation as parts-per-ten-thousand deviations from terrestrial standard values. The $\varepsilon^{54}Cr$ nucleosynthetic anomalies of type I chondrules from most CC and NC groups have been extensively studied in the last decade (N = 119, Connelly et al. 2012; van Kooten et al. 2020, 2016; Olsen et al. 2016; Schneider et al. 2020; Williams et al. 2020; Yamashita et al. 2010; Zhu et al. 2020, 2019). Type I chondrules from all CC types show variable ^{54}Cr excesses that differ significantly from their respective host chondrite compositions (Fig. 10, Connelly et al. 2012; van Kooten et al. 2020, 2016; Olsen et al. 2016; Schneider et al. 2020; Williams et al. 2020; Yamashita et al. 2010; Zhu et al. 2019). CO-CM-CV-CK chondrules show variations spanning almost 3 ε units, despite having similar $\varepsilon^{54}Cr$ average 0.41 ± 0.61 (Connelly

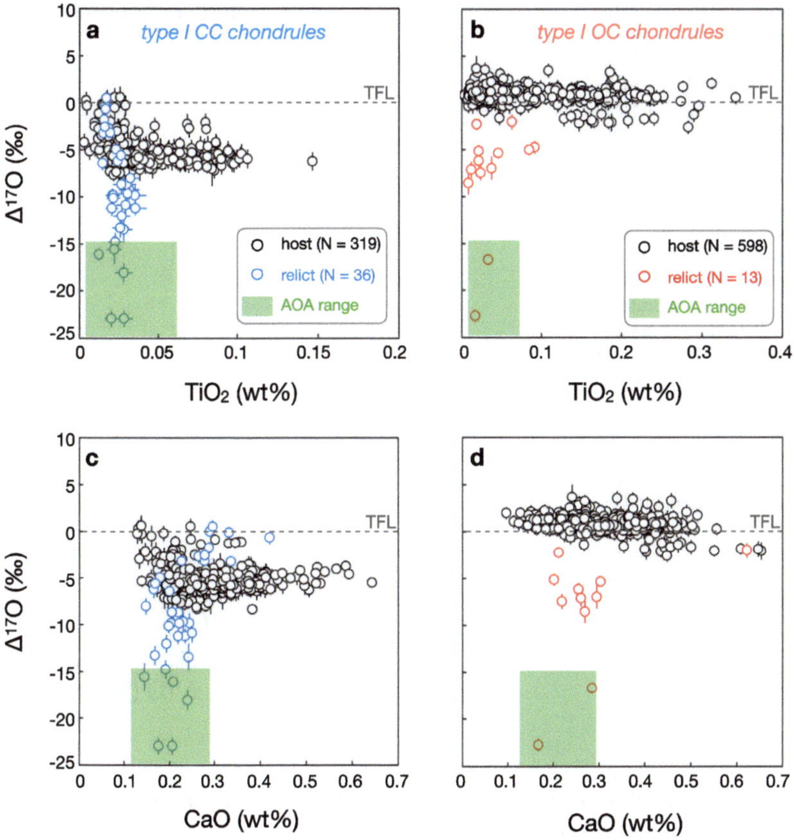

Fig. 8 TiO_2 and CaO concentrations of relict and host olivine grains plotted against their $\Delta^{17}O$ for both CC (**a&c**) and OC (**b&d**) chondrules. This reveals that relict olivine grains are commonly depleted in refractory elements relative to host olivine crystals, which are characterized by variable concentrations of these elements. Data from Marrocchi et al. 2019a, 2018; Piralla et al. 2021

et al. 2012; van Kooten et al. 2020; Olsen et al. 2016; Schneider et al. 2020; Williams et al. 2020; Zhu et al. 2019). Conversely, CB and CR chondrites show more uniform ε^{54}Cr with higher average values of 1.32 ± 0.16 and 1.42 ± 0.15, respectively (van Kooten et al. 2016; Olsen et al. 2016; Schneider et al. 2020; Yamashita et al. 2010). Although less studied than CC counterparts (N = 16), EC and OC type chondrules show different characteristics with more uniform ε^{54}Cr that are similar to the compositions of their host chondrites (with the notable exception of two EC chondrules, Fig. 10, Schneider et al. 2020; Zhu et al. 2020). Most studies only report chondrule ε^{54}Cr but Δ^{17}O-ε^{50}Ti-ε^{54}Cr compositions were also reported for the same chondrules in OC, EC and CV-CK-CR chondrules (N = 37, Schneider et al. 2020; Williams et al. 2020). The ε^{50}Ti- ε^{54}Cr diagram reveals a significant difference between NC and CC chondrules, with (i) none of the NC chondrules plotting in the field of bulk CC chondrites and (ii) CC chondrules showing a much wider spread than the field of bulk CC chondrites, including some that overlap with the bulk NC values (Fig. 11). Of note, potential ^{54}Cr zoning between chondrule cores and edges was also reported in the CV3 Leoville (van Kooten et al. 2021). ^{54}Fe nucleosynthetic excesses have been also reported in

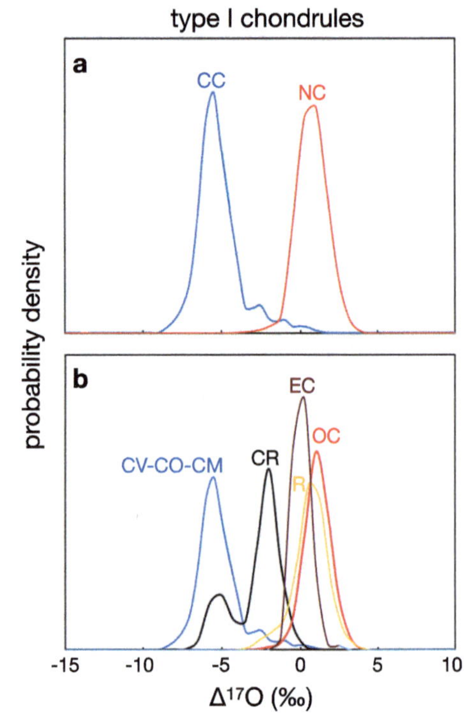

Fig. 9 (a) Curves showing the probability density functions of $\Delta^{17}O$ of host olivine grains in chondrules from CO-CM-CV, CR, R, EC and OC chondrites. **(b)** Similar plot but combining all data for both carbonaceous and non-carbonaceous chondrites. This reveals an isotopic dichotomy with NC chondrules showing ^{16}O-poor isotopic compositions (with average compared to CC chondrules (+0.75 ± 0.94‰ *vs*. 5.83 ± 1.81, respectively). Data from (Chaumard et al. 2021, 2018; Hertwig et al. 2018; Jacquet et al. 2021; Jones et al. 2004; Marrocchi et al. 2022, 2021, 2019a, 2018; Rudraswami et al. 2011; Schneider et al. 2020; Tenner et al. 2019, 2015, 2013; Ushikubo et al. 2012; Ushikubo and Kimura 2021; Williams et al. 2020)

only two CR chondrules (Schiller et al. 2020) but likely represent a systematic signatures of CC chondrules (Marrocchi et al. 2023a).

3.2 Type II Chondrules

FeO-rich olivine grains in OC type II chondrules show homogenous O-isotopic compositions with similar $\Delta^{17}O$ to those in their type I counterparts (Kita et al. 2010). On the other hand, FeO-rich olivine grains in CC type II chondrules have ^{16}O-poor compositions compared to Mg-rich host olivine grains in type I chondrules of the same chondrite type (Fig. 12a, b, Chaumard et al. 2021, 2018; Hertwig et al. 2018; Rudraswami et al. 2011; Tenner et al. 2018, 2015, 2013; Ushikubo et al. 2012). FeO-rich olivine grains in CR type II chondrules tend to have more ^{16}O-poor isotopic compositions than olivine grains in type II chondrules of other CCs (Fig. 12a, b, c). Isotopic differences between FeO-rich olivine grains and Mg-rich relicts were also reported for silicon isotopes with CV type II chondrules silicates varying by less than 2‰ in $\delta^{30}Si$ whereas type I chondrule silicates show larger variations of ~10‰ (Villeneuve et al. 2020). Of note, no data are available for nucleosynthetic anomalies in type II chondrules.

In both CO and CR chondrites, Mg-rich relict olivine grains embedded within FeO-rich olivine phenocrysts in type II chondrules show similar $\Delta^{17}O$ to host olivine grains in type I chondrules of the same respective chondrites (Figs. 12d, e, Pinto et al. 2024). In particular, Mg-rich relict grains in CR type II chondrules display the same bi-modal distribution as that defined by Mg-rich host olivine grains of type I chondrules (Fig. 12e, Pinto et al. 2024).

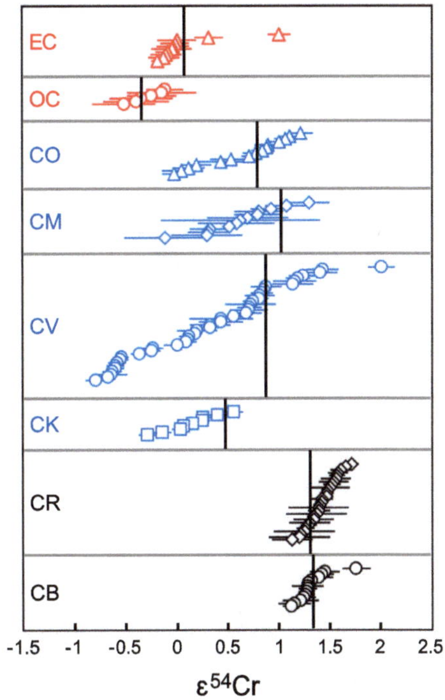

Fig. 10 Nucleosynthetic ε^{54}Cr anomalies in individual type I chondrules from CCs (CO, CM, CV, CK, CR, and CB chondrites) and NCs (EC and OC chondrites). Compositions of respective bulk chondrites are shown as solid black lines; shaded areas around these lines indicate 2 s.d. uncertainties of the bulk values. Note that the average ε^{54}Cr values of CV, CO, and CM chondrules are indistinguishable from each other whereas the average ε^{54}Cr of CR chondrules is significantly higher. Data from Bollard et al. 2019; Connelly et al. 2012; van Kooten et al. 2020, 2016; Schneider et al. 2020; Williams et al. 2020; Yamashita et al. 2010; Zhu et al. 2019

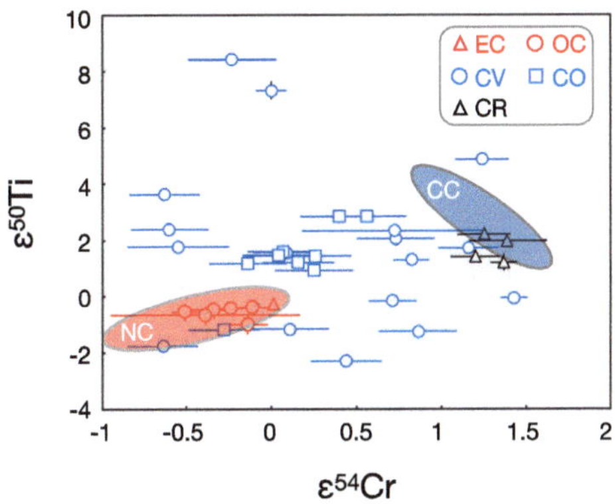

Fig. 11 ε^{50}Ti, and ε^{54}Cr of individual chondrules in comparison to bulk NC and CC meteorites. The NC (red) and CC (blue) fields represent the range of anomalies measured for bulk meteorites from the NC and CC reservoir, respectively. NC chondrules display relatively homogeneous ε^{50}Ti and ε^{54}Cr, which are indistinguishable from the compositions of their respective host chondrites, while CC chondrules have more variable ε^{50}Ti and ε^{54}Cr. Note that although the compositions of CC and NC chondrules may overlap for either ε50Ti or ε54Cr, in multi-isotope space no CC chondrule (but two) plots within the NC field, and no NC chondrule plots within the CC field. Data from Schneider et al. 2020; Williams et al. 2020

3.3 Aluminium-Rich Chondrules

CC and OC ARCs have different petrographic characteristics with the former commonly containing CAI-AOA-like relict grains whereas as the latter are often glassy chondrules (Nagahara et al. 2008; Ebert et al. 2022). They also show different isotopic characteristics.

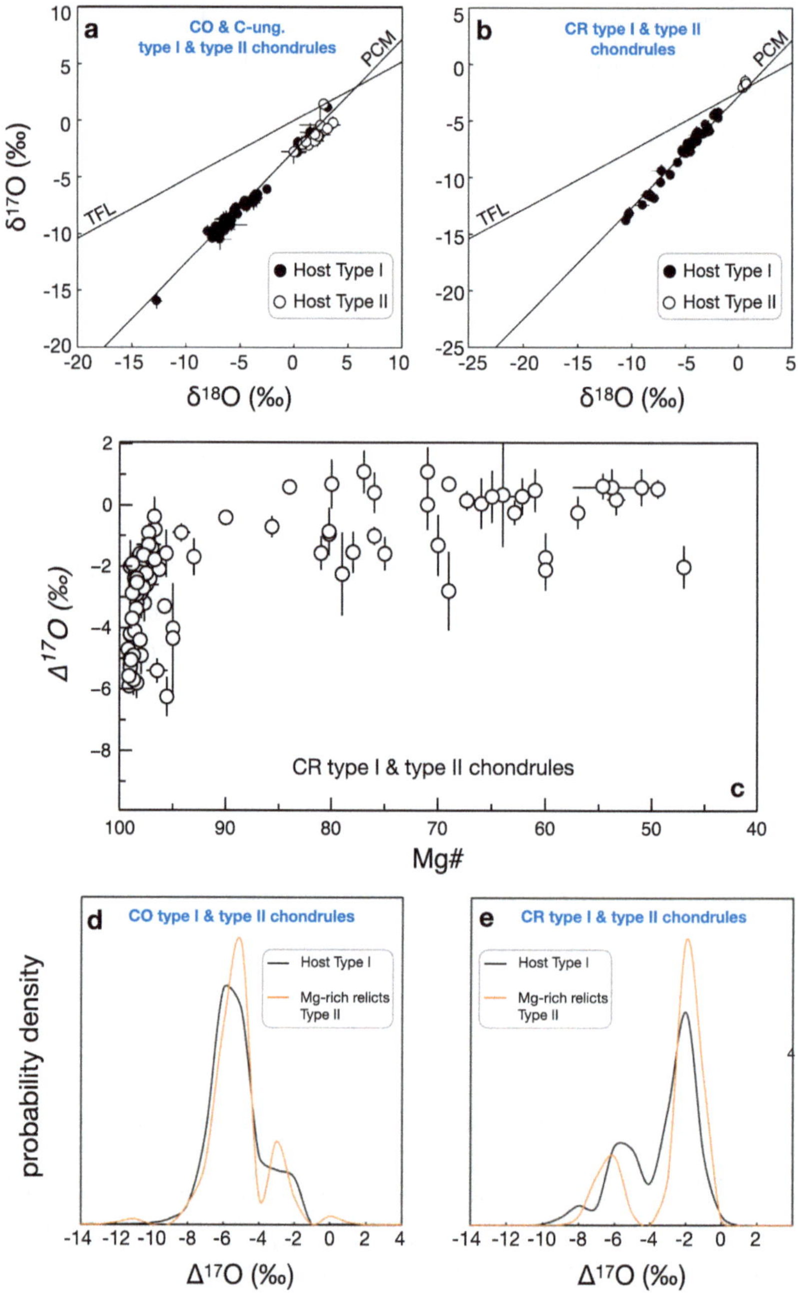

Fig. 12 Oxygen isotopic compositions of host olivine grains in both type I and type II chondrules from the CO and C-ung. (**a**) and CR chondrites (**b**). (**c**) Chondrule Mg#'s and host $\Delta^{17}O$ for type I and type II CR chondrules. Data from Connolly and Huss 2010, Schrader et al. 2013, 2014, 2017, Tenner et al. 2015. Curves showing the probability density function of $\Delta^{17}O$ of type I host olivines and type II Mg-relict grains in both CO (**d**) and CR chondrites (**e**). Data from (Fukuda et al. 2022; Kunihiro et al. 2004; Libourel et al. 2023, 2022; Pinto et al. 2024; Schrader et al. 2013; Tenner et al. 2013)

ARC chondrule minerals in the Dar al Gani 083 CO3.1 chondrite have heterogeneous oxygen isotopic compositions with $\Delta^{17}O$ ranging from -24.6 to -0.3‰ whereas those in OC ARCs are more homogeneous with $\Delta^{17}O$ spanning from -4.5 to + 1.9‰ (Ebert et al. 2022; Russell et al. 2000). ARCs also display differences in $\varepsilon^{50}Ti$ with (i) OC ARC having no ^{50}Ti excess and (ii) CC ARC displaying large ^{50}Ti excesses with $\varepsilon^{50}Ti$ up to +14.5 (Ebert et al. 2018)

4 Key Questions Regarding Chondrule Formation

4.1 What Were the Local Conditions of Chondrule Formation Processes?

Chondrules contain chemical information about the local conditions in the protoplanetary disk where chondrule heating events took place. Petrologic evidence allows us to infer parameters such as the ambient temperature and the partial pressures of various gas species, which have a direct bearing on interpretation of local dust densities.

The ambient temperature in chondrule-forming regions can be inferred from examining the abundances of moderately volatile elements present in chondrules. Moderately volatile elements are those that have 50% condensation temperatures (T_c) in the range 1353–664 K, corresponding to the range between condensation of Fe-Ni metal and FeS (Lodders 2003; Palme et al. 2014). Of relevance to chondrules, moderately volatile elements Mn, K, Na, P, and S are all observed at measurable levels (e.g., Jacquet et al. 2015; Jones 2012; Jones et al. 2005; Nagahara et al. 2008). If we make the assumptions that these elements were present in chondrule precursor material at about the same concentration as in the present chondrules, and that chondrule bulk compositions reflect precursor compositions, we can infer that the ambient temperature was lower than the condensation temperature of the element observed. However, chondrules were very likely open systems during the time that they were melted (e.g., Ebel et al. 2018; Hewins and Zanda 2012), meaning that they were susceptible to (i) evaporative loss of volatile elements and (ii) gain of volatile elements through condensation from the gas into chondrule melts. For example, evaporation of Na occurs rapidly from chondrule melts (Wick and Jones 2012; Yu and Hewins 1998) and condensation of silica into chondrule melts has been proposed to explain pyroxene-rich zones around the edges of type I chondrules (Barosch et al. 2019; Friend et al. 2016; Libourel et al. 2006; Tissandier et al. 2002). To complicate matters further, observed abundances of moderately volatile elements may have been modified from their primary compositions: there is clear evidence that Na and K were mobilized and added to chondrules as a result of secondary alkali metasomatism in meteorite parent bodies, and S is highly mobile through solid state diffusion at low metamorphic temperatures (Grossman and Brearley 2005; Rubin 2013).

Since sulfides are common primary constituents of chondrules, the ambient temperature must have been below 664 K, i.e., around 400 °C. For instance, $\sim$13% of Semarkona chondrules could contain primary troilite embedded in mafic silicate grains (Rubin et al. 1999). However, many type I chondrules have low S contents (e.g., Jacquet et al. 2013) so this constraint may not apply to all chondrules. In general, type I chondrules have lower moderately volatile element abundances than type II chondrules (e.g., Jacquet et al. 2015). Moderately volatile lithophile elements (i.e., those that prefer to enter silicate and oxide minerals), such as Mn, Na, K, and P, are also demonstrably present in primary chondrule melts because they are present in minerals including olivine, pyroxene and plagioclase that grew from the melt (e.g., Alexander et al. 2008; Jacquet et al. 2015; Lewis and Jones 2019; Wick and

Jones 2012). The presence of primary K and Na (50% Tc values of 1006, and 958 K respectively, Lodders 2003) define ambient temperatures slightly higher than the constraint from S. The presence of primary Cl may lower the ambient temperature constraint significantly. In particular, chondrules in enstatite chondrites have notably high Cl contents, up to several wt% (Grossman et al. 1985; Lin 2022; Piani et al. 2016). The 50% T_c for chlorine has been discussed at length, because of uncertainty about the host condensate phase. It has recently been revised from 948 K (condensation into sodalite, Lodders 2003) to 427 K (Lodders and Fegley 2023). In detail, Cl begins to condense as chlorapatite at 648 K, but only a limited amount of Cl can condense into chlorapatite because of limited availability of P. Further condensation of Cl only occurs when halite (NaCl) becomes stable at 430 K, and Cl is 50% condensed at 427 K. Thus, significant amounts of primary Cl in chondrules in enstatite chondrites might indicate ambient temperature below 430 K (i.e., around 150 °C). Lower amounts of Cl in chondrules in other chondrite groups could be consistent with ambient temperatures similar to the constraint inferred from sulfur.

Moderately volatile elements can also be used to infer local gas densities during chondrule formation. Alexander et al. (2008) demonstrated that Na was present in chondrule melts throughout crystallization of individual olivine grains. From this, they calculated equilibrium vapor pressures of Na over chondrule melts (values $>10^{-4}$ bar, Desch et al. 2012), and inferred solids densities $>10^4$–10^5 times nebula densities. In turn, this requires chondrule-forming regions >4000 km in radius, which Alexander et al. (2008) argue would be self-gravitating.

Stable isotope measurements provide additional support for a model in which chondrules were molten within high surrounding dust densities. Alexander and Wang (2001) showed that olivine phenocrysts in chondrules have iron isotope compositions that are indistinguishable from bulk measurements. Later, more precise whole-chondrule measurements showed that there is some variation in iron isotope compositions, but that this variation is much less than expected for free evaporation, suggesting that the iron may have back-reacted into the chondrule under high pressure conditions (Hezel et al. 2010). Sulfur could also experience back-reaction and reach a concentration limit, the sulfur content at sulfide saturation, at which an immiscible iron sulfide liquid separates from the silicate melt (Marrocchi et al. 2016; Marrocchi and Libourel 2013; Piani et al. 2016). Similarly, Mg isotope compositions of chondrules show only very small fractionation effects ruling out widespread open system evaporation during chondrule formation (Young and Galy 2004), as do K and Si isotope compositions (Hezel et al. 2010; Yu et al. 2003). Ebel et al. (2018) showed that the suppression of isotopic fractionation in several elements is likely due to equilibration with the surrounding hot gas, requiring either a very high partial pressure ($\sim$1 bar) of hydrogen in the surrounding gas, or greatly enhanced dust concentrations. This enables the evaporated chondrule material to back-react and re-equilibrate with the chondrule, suppressing isotopic fractionation. Models of chondrule heating and gas-solid exchange suggest that to reduce isotopic fractionation to the levels observed would require dust enrichments 1000x canonical at pressures of 10^{-3} bar (Ebel et al. 2018; Fedkin and Grossman 2013) and chondrule-forming regions 150 to 6000 km in radius (Cuzzi and Alexander 2006).

Several other properties of chondrules point to high dust/gas ratios in chondrule-forming regions, where "dust" refers to solids of CI composition (Ebel et al. 2018). High FeO contents in type II chondrules require high total pressure (P_{tot}) and high dust enrichments at temperatures where dust and gas could be expected to equilibrate, $\geq$ 1200 K (Ebel and Grossman 2000). Solids in the chondrule formation regions must have been enriched in dust by factors of 15–100 (by numbers of atoms), relative to solar, at a total pressure of P_{tot} = 10^{-3} bars, and even higher enrichments are required for lower P_{tot} (Ebel 2006; Ebel and

Grossman 2000). For type II chondrules to retain their bulk FeO contents, and for metal melts to be stable in type I chondrules, enrichments >1,000 times solar even at 10^{-3} bars are needed (Alexander 2004; Alexander and Ebel 2012; Ebel and Grossman 2000; Tenner et al. 2015).

Chondrule cooling rates also require elevated dust/gas ratios. An individual chondrule heated to 2000 K in free space will radiate away its internal heat energy in less than a second, and for a solar solids/gas ratio, it will radiate its heat in $\sim 10^3$ seconds (Desch et al. 2012). Chondrule-formation in regions of higher opacity is therefore required, although the control on cooling rates is complicated. For example, in the shock wave model the main control on slowing chondrule cooling rates is the fact that the gas is also hot, and chondrule cooling rate is inversely correlated with gas density (Desch et al. 2012; Desch and Connolly 2002; Morris and Desch 2010).

4.2 Were Conditions the Same for Forming NC and CC Type I Chondrules?

At first order this could appear as a weird question since type I chondrules from different chondrite groups show specific properties, including varying abundances of chondrule textural types, chondrule sizes and bulk compositions, oxygen isotopic compositions and nucleosynthetic anomaly signatures (Fig. 13, Jones 2012; Schneider et al. 2020; Tenner et al. 2018). In addition, the fundamental isotopic dichotomy of the solar system revealed by nucleosynthetic isotopic anomalies (e.g., ^{48}Ca, ^{50}Ti, ^{54}Cr, and Mo and Ni isotopes, Burkhardt et al. 2021) imply that the protoplanetary disk was already separated in two different reservoirs at the start of chondrule formation (Fukuda et al. 2022; Piralla et al. 2023). It would therefore appear possible that NC and CC type I chondrules formed under different conditions and/or from different precursors. However, the essential petrographic properties of NC and CC porphyritic olivine-rich type I chondrules are very similar. In particular, the recent applications of high-resolution cathodoluminescence (CL) and X-ray mapping techniques have revealed that most of NC and CC porphyritic type I chondrules show similar textures characterized by chemically-zoned trace and minor element contents and asymmetric growth of outer chondrule crystals (Fig. 2; Libourel et al. 2023, 2022; Libourel and Portail 2018; Marrocchi et al. 2022, 2019a, 2018; Piralla et al. 2021; Regnault et al. 2022; Schnuriger et al. 2022). These features are interpreted as evidence for gas-melt interactions, hence open-system conditions, during chondrule formation (Libourel and Portail 2018; Marrocchi et al. 2018), and the similarity between NC and CC reveals that type I chondrules formed by a near similar process throughout the disk over several millions of years.

However, different scenarios have been proposed to explain the nature and origin of porphyritic type I chondrule precursors, which are proposed to correspond to:

(i) solid debris generated by collisional impacts between differentiated planetesimals (Faure et al. 2012; Libourel and Krot 2007).

(ii) fully molten droplets consisting of silicate melts and liquid Fe-Ni metal blebs. Such droplets would correspond to impact splashes produced by energetic collisions between early-formed planetesimals (Libourel et al. 2023, 2022; Lichtenberg et al. 2018; Sanders and Scott 2012).

(iii) partial melting of calcium-aluminum-rich inclusions (CAIs) and ameboid olivine aggregates (AOAs) previously condensed early in disk history (Ebert et al. 2018; Marrocchi et al. 2019a; Russell et al. 2005; Ruzicka et al. 2012, Whattam et al. 2008, 2022, Whattam and Hewins 2009).

(iv) fine-grained dust of similar bulk composition to the final chondrule composition, with a limited number of large precursor grains that are preserved as relicts in the final chondrule.

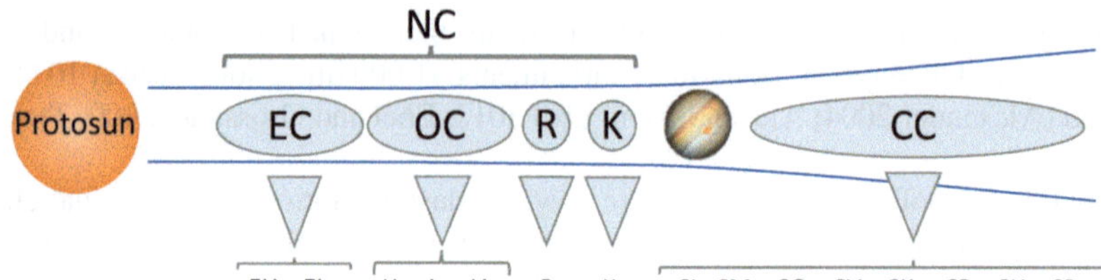

Group	EH	EL	H	L	LL	R	K	CI	CM	CO	CV	CK	CR	CH	CBa	CBb
Chondrule abundance (%)[a]	90			80		>40	20-30	0	20	40	45	15	55	70	40	30
Mean chondrule size (mm)	0.2	0.6	0.3	0.5	0.6	0.4	0.6	-	0.3	0.15	1.0	0.8	0.7	<0.1	5	0.5
% Porphyritic chondrules	85			85						> 95				20	0	1
% Type II chondrules[b]	0			70-80			<1			< 10[c]					< 5	
Mean $\Delta^{17}O$, type I chondrules	0‰			+1‰		+0.6‰	0‰			-5.83‰			-2‰		-2.2‰[d]	
Mean $\Delta^{17}O$, type II chondrules				+1‰		+0.6‰	0‰			-2 ‰			-1 ‰			
Mean $\varepsilon^{54}Cr$ (per 10^4)	0.07			-0.28			-0.44			0.41			1.42		1.32	

Fig. 13 Interpretation of chondrule formation regions in the protoplanetary disk, and a summary of chondrule properties in individual chondrite groups. Two isotopic reservoirs are defined based on nucleosynthetic isotope anomalies: the carbonaceous chondrite (CC) reservoir, thought to represent the outer disk, beyond Jupiter, and the non-carbonaceous (NC) reservoir, thought to represent the inner disk. The NC reservoir includes the enstatite chondrites (EC), ordinary chondrites (OC), Rumuruti-like (R) chondrites, and Kakangari-like (K) chondrites. There are multiple groups of EC, OC and CC. Chondrule abundance, size and frequency are from Jones (2012) and Scott and Krot (2014), and references therein. Mean isotope compositions are based on references given in this paper, plus Weisberg et al. (1996), and Zhu et al. (2023) for K chondrites. Footnotes: [a)] CI and CM are highly altered by aqueous fluids, destroying chondrules. [b)] % of porphyritic chondrules. [c)] A range of values exists for CM chondrites, 10-40%; CO and CV are < 10%. [d)] This corresponds to the $\Delta^{17}O$ value of homogeneous impact-generated chondrules. Porphyritic chondrules show more scattered values (ranging from -6 to + 4‰)

Recent works have made some progress in deciding which of the above is the most likely, by combining petrographic observations and O, Ti and/or Cr isotopic compositions of type I chondrules. It is generally stated that the majority of type I NC and CC chondrules are characterized by olivine grains having homogeneous O-isotope compositions (Tenner et al. 2018). However, this (at least partially) results from an analytical bias as only few *in-situ* measurements are generally performed within a single chondrule (i.e., usually n < 10, Chaumard et al. 2018; Kunihiro et al. 2005, 2004; Rudraswami et al. 2011; Schrader et al. 2014), leading to a possible underestimation of isotopically heterogeneous chondrules. Recent multiple measurements (n > 30 in individual chondrules) revealed that CC type I chondrules (and to a lesser extent NC type I chondrules) are characterized by a more common presence of (i) intra-chondrule mass-independent oxygen isotopic variations (Fig. 4b, d) and (ii) ^{16}O-rich relict olivine grains with chemical characteristics different from those of host olivine grains (Fig. 8, Kita et al. 2010; Tenner et al. 2018; Marrocchi et al. 2019a, 2018; Piralla et al. 2021). Such isotopic features are inconsistent with derivation of type I NC and CC chondrule precursors from differentiated planetesimals as previously argued from 120° triple junctions between olivine crystals (Faure et al. 2012; Libourel and Krot 2007), as it would result in very limited and mass-dependent O-isotopic variations (Eiler 2001; Richet et al. 1977). Rather, this evidence points toward ^{16}O-rich precursors that subsequently experienced interactions with a gas enriched in $^{17,18}O$ during the high-temperature chondrule-forming event(s). Relict olivine grains are however more abundant in type I CC chondrules than their NC counterparts, the latter commonly showing homogeneous $\Delta^{17}O$ and small mass-dependent variations (Fig. 4c,7, Kita et al. 2010; Piralla et al. 2021). This attests that type I NC chondrules were affected by more complex evaporation/recondensation processes during gas-melt interactions and likely formed in more dynamic regions characterized by higher temperatures and repetitive melting events (Alexander 1994; Piralla et al. 2021).

Although their existence has long been attested (Jones et al. 2004; Nagahara 1981; Rambaldi 1981), relict olivine grains in type I chondrules have long appeared to have no clear major-element signatures and/or textural features that could reveal their origin (Jones et al. 2004; Ushikubo et al. 2012). On the basis of trace element data, it was however proposed that type I chondrule olivine could have formed by melting AOAs and subsequent olivine-melt fractionation during chondrule crystallization (Ruzicka et al. 2012). More recently, high-current electron microprobe quantitative measurements showed that ^{16}O-rich relict grains are systematically Ca-Al-Ti-poor relative to host olivine crystals in both NC and CC type I porphyritic chondrules (Fig. 8, Marrocchi et al. 2019a, 2018; Piralla et al. 2021). These characteristics are similar to those reported in AOAs (Jacquet and Marrocchi 2017; Komatsu et al. 2015; Krot et al. 2004; Ruzicka et al. 2012; Weisberg et al. 2004), thus making these early condensates plausible candidates as chondrule precursors. In theory, such features can also be accommodated in the framework of the impact splash model with olivine crystallization occurring from chemically and isotopically evolving magma droplets via gas-melt interactions. However, this would require producing initial magma droplets with $\Delta^{17}O < -20‰$ whereas no known chondrites or achondrites with bulk $\Delta^{17}O < -6‰$ have been reported (Greenwood et al. 2020).

Furthermore, bulk chondrule compositions, rare earth elemental patterns and nucleosynthetic anomalies also suggest that Al-rich chondrules in CC likely derive from CAIs (Ebert and Bischoff 2016; Huss et al. 2001; Rubin 2004; Russell et al. 2005; Williams et al. 2020). Altogether, this suggests that most refractory inclusions likely experienced recycling through type I chondrules in both NC and CC reservoirs, thus making CAIs and AOAs currently observed in chondrites lucky survivors of the disk evolution (Dunham et al. 2023; Krot 2019; Itoh et al. 2007). Notably, some of the observed relict olivine grains in both NC and CC type I chondrules have more ^{16}O-poor compositions than AOA olivine grains (Fig. 8). They could result from an analytical bias (i.e., relict grain might have a small domain compared to SIMS pit size) or correspond to earlier generations of chondrules (Marrocchi et al. 2019a; Piralla et al. 2021; Ruzicka et al. 2007; Schrader et al. 2018). Chondrule production and recycling in later-formed chondrules appears to have been more enhanced in the inner disk as attested by the (i) lower abundance of AOA-like relict olivine grains in NC chondrules (Kita et al. 2010; Piralla et al. 2021) and (ii) the higher modal abundances of matrix and refractory inclusions in CCs (Fig. 13, Scott and Krot 2014).

Although CAIs and AOAs appear as appealing precursors, their distinct chemical and isotopic compositions compared to chondrules requires additional precursors and/or processes for generating the final NC and CC chondrules. Hence, the distinct O-isotope compositions of refractory inclusions and chondrules require addition of ferromagnesian dust and open-system gas-melt interactions (Kita et al. 2010; Marrocchi et al. 2019a, 2018; Piralla et al. 2021; Ruzicka et al. 2012; Tenner et al. 2018). Along similar lines, the Si-poor compositions of refractory inclusions compared to chondrules means addition of dust and interactions with a SiO-rich gas are required to produce pyroxene-dominated chondrule compositions (Krot et al. 2004; Ruzicka et al. 2012; Tissandier et al. 2002). Furthermore, the fact that mechanically-separated type I porphyritic CC chondrules have varying $\varepsilon^{50}Ti$ and $\varepsilon^{54}Cr$ plotting in-between CAIs/AOAs and NC chondrites implies that they cannot only derive from AOAs without adding NC material (Fig. 11, Schneider et al. 2020). The presence of NC-like dust in the CC reservoir is generally interpreted as either resulting from (i) outward migration of NC chondrule fragments (Hertwig et al. 2019; Schrader et al. 2020; Williams et al. 2020; Zhang et al. 2022) or (ii) change in the isotopic composition of latter infalling material from the solar system's parental molecular cloud (Nanne et al. 2019; Schneider et al. 2020). Importantly, these isotopic constraints are also extremely difficult to accommodate in the impact splash model as it would require (i) small-scale heterogeneity in the colliding planetesimals or (ii) the ingestion of unmelted and unvaporized refractory inclusions by each melt droplet. Compared to CC chondrules, separated NC chondrules have homogeneous ε^{50}Ti and ε^{54}Cr similar to the compositions of their respective host chondrites (Fig. 11, Schneider et al. 2020). If NC chondrules derive from AOA-like precursors, their particular ε^{50}Ti and ε^{54}Cr imply that AOAs condensed after the isotopic composition of the inner disk had changed toward NC values (Nanne et al. 2019). Interestingly, this change is recorded in CAIs whose Mo isotopic compositions range from CC to NC values (Brennecka et al. 2020). Furthermore, a few rare NC CAIs show no enrichment in ^{50}Ti relative to their host chondrites (Ebert et al. 2018). NC refractory inclusions can thus represent a fraction of NC chondrule precursors in addition to the NC-like dust inherited from late infalling material (Nanne et al. 2019).

Taken together, this suggests that both NC and CC porphyritic type I chondrules were formed through the melting of a mix of early condensates and NC-like dust followed by subsequent gas-melt interactions. Recent mass balance calculations using tellurium isotopes and different nucleosynthetic anomalies (i.e., ^{54}Cr and ^{54}Fe) suggest that the CC chondrule-forming regions were characterized by a homogeneous refractory inclusions/NC-dust ratio (Hellmann et al. 2020; Marrocchi et al. 2023a). The nature of the gas chondrules interacted with likely corresponds to a mixture of different oxygen-bearing molecules, already present in the gas phase (i.e., CO) and/or originating from solid evaporation (i.e., H_2O, SiO, Mg, S) induced by the formation of chondrule in dust-enriched regions of the disk (Friend et al. 2016; Hezel et al. 2006; Libourel et al. 2006; Marrocchi and Chaussidon 2015; Tenner et al. 2018; Thomassin et al. 2023). Oxygen isotopic data point toward an enhanced chondrule-forming rate in the inner disk after its composition was dominantly modified by late infall-material from the molecular cloud. This is consistent with the expected thermal stratification of the disk (e.g., Yang et al. 2013) and implies that both NC and CC porphyritic type I chondrules result from the melting of previously-formed nebular dust. This requires a protracted heating mechanism, efficient in both the inner and outer disk. This could be achieved under different conditions such as bow shocks (Morris et al. 2012), lightning (Kaneko et al. 2023), local (potentially impact-generated) shock waves (Hood et al. 2009) or current sheets (Lebreuilly et al. 2023).

4.3 Did Chondrules Produce Chondrules?

Absolute lead-lead (Pb–Pb) ages and aluminum-magnesium (^{26}Al–^{26}Mg) ages relative to CAIs point toward protracted chondrule formation, even within a single chondrite group (Bollard et al. 2019, 2017; Fukuda et al. 2022; Hertwig et al. 2019, Nagashima et al. 2014, 2018, Pape et al. 2019; Piralla et al. 2023; Schrader et al. 2017; Siron et al. 2022, 2021; Tenner et al. 2019; Ushikubo et al. 2013; Villeneuve et al. 2009). This implies that early-formed chondrules resided > 100 kyr to Myr embedded in the disk before they accreted to larger bodies. During this time, early-formed chondrules could have suffered (i) collisions with other chondrules resulting in compound objects (Jacquet 2021; Krot et al. 2017) and/or (ii) repeated high-temperature thermal event(s). This implies that some of the current chondrules could have been formed from previous generations of chondrules that were recycled during disk evolution.

On the basis of chemical evidence, relict grains in type II chondrules are likely derived from type I chondrules, and dusty relict olivine in type I chondrules is plausibly derived from type II chondrules (Jones 2012, 1996). Oxygen isotopic compositions support this view, at least for relict forsterite in type II chondrules (Fig. 12). A few ^{16}O-rich relict olivine grains, mostly in NC chondrules, show $\Delta^{17}O$ values higher than those commonly measured in AOAs (Jones et al. 2004; Krot 2019; Marrocchi et al. 2019a, 2018; Piralla et al. 2021; Tenner et al. 2018). This could result from isotopic re-equilibration between AOA-like relict grains and chondrule melts that are evolving via gas-melt interactions. An alternative interpretation lies in some early chondrules having been recycled to form new generations of chondrules (Alexander 1994). This last assumption is strongly support by the recent finding of size-dependent oxygen isotopic variations in chondrules of ordinary chondrites (Fig. 6), with chondrules smaller than $\sim$ 300 microns having chemically-similar olivine grains than relict grains measured in larger OC chondrules (Marrocchi et al. 2024).

This recycling process can be quantitatively estimated by considering that (i) the different chondrule-forming events were independent and rare and (ii) the chondrule/matrix ratio of a given chondrite is representative of the chondrule/dust ratio of the reservoir from which it formed. Then, the number n of chondrule-forming events obeys a Poisson distribution (Jacquet et al. 2015) and the fraction of chondrule-hosted particles that have undergone at least two chondrule-forming events can be estimated (see Marrocchi et al. 2019a for further details). Different estimates of the fraction of recycled chondrules (f_{recycled}) were obtained with f_{recycled} of (i) 12 to 22‰ for CC chondrules (Marrocchi et al. 2019a) and (ii) ~82% for OC chondrules (Piralla et al. 2021). Consequently, debris of earlier chondrules appear to represent a minority among CC relict grains, and other precursors such as CAIs and AOAs must be considered. This is in good agreement with (i) independent mass balance calculations based on oxygen isotopic compositions of CC chondrules (Hezel and Palme 2007), (ii) trace and minor element systematics of CC chondrule silicates (Ruzicka et al. 2007) and (iii) nucleosynthetic anomaly signatures of CC chondrules (see previous section, Schneider et al. 2020; Williams et al. 2020). Notably, the OC chondrule-forming region was more dynamic, resulting in more recycling processes of early formed chondrules (Alexander 1994). This does not in itself constitute a surprise considering the dominant modal abundance of chondrules in this specific type of chondrite (Jones 2012, Fig. 13). Also, type II chondrules in OCs have lower FeO contents than those in CCs, and lower abundances of clearly identifiable relict forsterite grains: both of these observations are consistent with more extensive recycling. Altogether, this implies that chondrules can form from chondrules, thus blurring the identification of the initial chondrule precursors.

Compared to other CC chondrules, CR chondrules are characterized by abundant Fe-Ni metal beads, multiple igneous rims including metal-rich rims and silica-rich rims, a Mn/Fe ratio in olivine in type II PO chondrules intermediate between OC chondrules and chondrules in other CC groups (Berlin et al. 2011; Jones 2012), and rarer ^{16}O-rich olivine grains in type I chondrules (Jones 2012; Charles et al. 2018; Marrocchi et al. 2022; Smith and Jones 2024; Tenner et al. 2015; Weisberg et al. 1995, 1993). In addition, CR chondrules (i) become depleted in ^{16}O as their Mg#'s increase (Fig. 5a; Tenner et al. 2015, 2018), (ii) have host olivine grains with more ^{16}O-poor isotopic compositions ($\Delta^{17}O \sim$ -2‰ vs. -6‰ for other CCs, Fig. 9b) and (iii) show higher and less variable $\varepsilon^{54}Cr$ than in other CC groups (van Kooten et al. 2016; Olsen et al. 2016; Schneider et al. 2020, Fig. 10). This is generally ascribed to a formation of CR chondrules from different precursor material than other CC chondrules, either because they formed at larger heliocentric distance (van Kooten et al. 2020, 2016) and/or at later times (Budde et al. 2018; Schrader et al. 2017; Tenner et al. 2019). The monotonic increase of host chondrule $\Delta^{17}O$ as chondrule Mg#'s decrease (for Mg# ~99.2 to ~94, Fig. 5a, Tenner et al. 2018) was modeled assuming it could be produced by variable H_2O ice proportions and dust-to-gas ratio in chondrule precursors (Tenner et al. 2015, 2018). The mass balance predicts that most type I CR chondrules could form (i) at dust-to-gas ratios between 100 and 200 and (ii) H_2O ice abundances within chondrule precursors ranging from anhydrous to ~75% of the atomic abundance within CI dust (Tenner et al. 2018). Although this approach successfully reproduces the $\Delta^{17}O$-Mg# trend, it does not address the peculiar ^{54}Cr isotopic compositions of CR chondrules compared to other CCs (Fig. 10, Schneider et al. 2020).

Recent CL and isotopic characterization of type I CR porphyritic chondrules show that small chondrules have textural features and $\Delta^{17}O$ values similar to those of other CCs (i.e., $\Delta^{17}O \sim$ -6‰) whereas larger chondrules display more complex textures and higher $\Delta^{17}O$ values around -2‰ (Marrocchi et al. 2022). Furthermore, large CR chondrules are characterized by relict olivine grains with O-isotopic signatures characteristic of small CR and other CC chondrules (Marrocchi et al. 2023b, 2022; Schrader et al. 2017; Tenner et al. 2015). This indicates that large CR chondrules formed late out of a CR reservoir already populated by smaller CC chondrules (Marrocchi et al. 2022). This is borne out by Al-Mg dating as large CR chondrules with $\Delta^{17}O \sim$-2% formed ∼1–2 Myr after smaller CR chondrules whose formation timing was estimated to be 2 to 3 Myr after CAIs (Tenner et al. 2019). Taken together, these observations point toward multiple chondrule-forming events in the CR reservoir with small chondrules acting as precursors for the formation of large metal-rich CR chondrules (Marrocchi et al. 2022). Mass balance calculations based on O, Te and ^{54}Cr isotopic compositions indicates that ∼70-80 wt% of CI-like matrix was added to small CC-like chondrules for producing their late larger counterparts (Bryson and Brennecka 2021; Hellmann et al. 2023; Marrocchi et al. 2022). This thus attests that the precursors of the majority of CR chondrules included pre-existing chondrules, illustrating the complex history of chondrule formation during disk evolution.

Compared to type I, type II chondrules are characterized by FeO-rich silicates and more chondritic abundances of moderately volatile elements (Jones 1990). This attests to their formation under oxidizing conditions that could be produced (i) in regions with (enhanced dust/gas ratios, Schrader et al. 2013) or (ii) from precursors with low carbon contents (Connolly et al. 1994). Similar modelling than for type I CR chondrules (Tenner et al. 2015) predicts formation of type II chondrules at significantly higher dust enrichments (∼2,500), from dusts with CI chondritic H_2O abundances (Tenner et al. 2015). Of note, such high dust enrichments are difficult to achieve by dynamic models of the protoplanetary disk.

Alternatively, it has also been proposed that type II chondrules are derived from their type I counterparts through nebular oxidation shocks (Villeneuve et al. 2020, 2015) or within impact-generated vapor plumes (Libourel et al. 2023). As detailed above, the latter process appears difficult to reconcile with the presence of olivine grains with $\Delta^{17}O < -15$‰ and variable ^{50}Ti and ^{54}Cr nucleosynthetic anomalies in chondrule from the same chondrites (Schneider et al. 2020). Furthermore, Mg-rich relict olivine grains in type II chondrules show similar oxygen and silicon isotopic compositions to forsteritic grains in type I chondrules (Fig. 9, Tenner et al. 2018; Villeneuve et al. 2020). This indicates that at least some type II chondrules derived from their type I counterparts. Overall, this shows that (i) recycling was common during the evolution of the protoplanetary disk and (ii) chondrules were in some cases their own precursors.

4.4 Did Chondrules Accrete Locally or Experience Large-Scale Transport?

Dust transport is an important process during disk history as attested by the distribution of refractory inclusions, which are thought to have formed close to the Sun. Two key observations relevant to considering transport of refractory inclusions are: (i) refractory inclusions are predominantly found in CCs and have low abundances in OCs and ECs (Hezel et al. 2008; Dunham et al. 2023) and (ii) there are occasional observations of CAI-AOA-chondrule-like objects in CI chondrites and cometary samples (Defouilloy et al. 2017; Frank et al. 2023;

Joswiak et al. 2012; Leshin et al. 1997; Morin et al. 2022; Nakamura et al. 2008; Nakashima et al. 2012; Piralla et al. 2020; Zolensky et al. 2006). Considering constraints on chondrule formation discussed above, several petrologic and isotopic observations weaken models implying disk-wide transport, in favour of local formation and agglomeration processes.

Each chondrite group is characterized by chondrules showing specific petrographic characteristics (Jones 2012) and the different chondrite groups define positive correlation between different properties such as chondrule size, the percentage chondrule with igneous rims, the thickness of these rims or the percentage of enveloping compound chondrules (Rubin 2010). These observations suggest that each chondrite group sampled separated chondrule reservoirs. These observations represent strong constraints suggesting multiple separate reservoirs in the disk. A similar conclusion was reached based on (i) the positive correlation observed among CCs between the evaluated percentages of rimmed chondrules and the modal abundance of matrix material (Pinto et al. 2022) and (ii) statistical analyses of bulk chondrule compositions (Hezel and Parteli 2018). The range of chondrule ages in individual chondrite groups suggests that separate reservoirs were maintained over millions of years (Bollard et al. 2017; Fukuda et al. 2022; Piralla et al. 2023; Siron et al. 2021). Another major constraint on localised chondrule formation relates to deciphering whether chondrules formed (i) together with matrix in a single reservoir or (ii) separately from matrix, with later admixing of a matrix unrelated to the chondrule formation processes. This "complementarity" question has been debated for a long time, with no agreement so far (e.g., Hezel et al. 2018; Zanda et al. 2018). Average compositions of chondrules and matrix in individual chondrite groups have different chemical compositions, but if considered together, a number of elements have overall bulk chondrite average compositions close to the CI bulk composition. This has been used as an argument for the formation of chondrules and matrix in the same location (e.g., Hezel and Palme 2010; Palme et al. 2015). A similar argument has been made for complementary isotope ratios between chondrules and matrix (e.g., Budde et al. 2016). However, alternative models propose that bulk chondrite compositions are better reproduced by considering that a unique CI-composition matrix was added to various amounts of chondrules (e.g., Anders 1964; Zanda et al. 2018) The latter model implies that chondrules and chondrite matrices are not genetically related, but formed in different reservoirs (Zanda et al. 2018). Recent observations of Te-^{54}Cr and Te-^{54}Fe isotopic correlations suggests that all carbonaceous chondrites contain a portion of CI-like matrix, and that chondrules and this CI-like matrix formed from isotopically distinct material originating from different regions of the disk (Hellmann et al. 2020; Marrocchi et al. 2023a). Similar arguments have been made from chemical evidence (Braukmüller et al. 2018; Jacquet et al. 2016) Solving the question of complementarity has the potential to strongly support astrophysical models that require either of these two scenarios, thereby providing constraints for the formation conditions, and ultimately mechanism of chondrules.

The question of whether chondrule experienced transport during disk evolution or accreted locally can also be addressed using bulk and *in-situ* isotopic compositions of NC and CC chondrules. Based on the presence in CC chondrules of olivine grains with typical OC Δ^{17}O values, it has been proposed that some OC chondrule fragments migrated from the inner disk outwards to the CC chondrule-forming regions (Schrader et al. 2020). Similarly, the presence in the CV and CK carbonaceous chondrites of chondrules with negative and positive ε^{54}Cr, with oxygen isotopic compositions respectively plotting above and below the PMC line, has been interpreted as evidence of the presence of NC chondrules (Williams et al. 2020; Zhang et al. 2022). Furthermore, the fact that variable ε^{54}Cr values observed among CO-CM-CV chondrules (Fig. 10) cover the range measured among bulk NC and CC meteorites has been argued that chondrules' ε^{54}Cr reflect their specific formation location within the disk, followed by subsequent disk-wide transport to the accretion region of their respective host chondrites (Olsen et al. 2016). It should also be noted that the ^{54}Cr systematics of CV chondrules has also recently been interpreted as the result of chemical equilibration between chondrules and fine-grained matrix during secondary alteration (van Kooten et al. 2021). In such a model, CV chondrule precursors correspond to ^{54}Cr-poor OC-like chondrules formed in the inner disk and that subsequently experienced massive outward chondrule transport.

However, other considerations of isotopic compositions are more consistent with limited transport. If we put the question of the formation of chondrules into the global context of the isotopic dichotomy of the solar system, the presence of NC-like dust in the outer CC reservoir does not represent an irrefutable proof of outward chondrule migration. Importantly, the isotopic dichotomy exists for lithophile (Ti, Cr) and siderophile (Ni, Mo) as well as refractory (Ti, Mo) and non-refractory (Ni, Cr, Zn) elements (Burkhardt et al. 2021). In addition, the CC/NC isotopic offset is seen for all elements showing anomalies, including those not enriched in CAIs. Another key constraint is that for all elements displaying the NC-CC dichotomy, the CC meteorites always plot between the isotopic compositions of CAIs/AOAs and NC meteorites (Nanne et al. 2019; Burkhardt et al. 2021; Jansen et al. 2024). Altogether, these observations indicate that the isotopic variations neither reflects thermal processing of presolar carriers in the disk, nor the heterogeneous distribution of isotopically anomalous CAIs/AOAs (Jansen et al. 2024). These isotopic variations rather reflect the distinct isotopic composition of later infalling material from the solar system's parental molecular cloud, which affected the inner and outer regions of the disk differently. In such scenario, the isotopic composition of the CC reservoir is the result of mixing between early-formed CAIs/AOAs and late-infalling material with NC-like isotopic compositions (Nanne et al. 2019). As the isotopic composition changed early in disk's history (< 0.5 Myr, Nanne et al. 2019), the ^{50}Ti-^{54}Cr isotopic variations among CC chondrules likely reflect local precursor heterogeneities, resulting from mixing between AOAs and NC-like dust, the latter corresponding to late infall material (Schneider et al. 2020) rather than material transported from the internal disk (Williams et al. 2020).

The formation of CV chondrules (from the CV3 Leoville) from OC precursors transported (formed in the inner disk, van Kooten et al. 2021) followed by subsequent chondrule/matrix equilibration is also not by Cr mass-balance calculations. Taking into account the modal abundance, Cr concentration and ε^{54}Cr values of refractory inclusions, chondrules, fine-grained rims and matrix in Leoville, the calculated bulk ε^{54}Cr is at odds with that measured for the bulk sample of Leoville (0 vs. +0.8, van Kooten et al. 2021; Qin et al. 2010). Contrary to the statement of van Kooten et al. (2021), invoking Cr transfer from fine-grained rims to chondrules cannot help as it would require that all the Leoville matrix would be composed of fine-grained rims. Conversely, the measured bulk ε^{54}Cr value can only be reproduced considering chondrules with an average ε^{54}Cr of $\sim$0.6, which is similar to those measured in other CM, CO and more altered CV chondrites (van Kooten et al. 2020; Olsen et al. 2016; Schneider et al. 2020; Williams et al. 2020; Zhu et al. 2019).

Altogether, this shows that the question of disk-wide transport cannot be tied to isotopic data for a single element but requires multi-isotopic approaches combining O-Ti-Cr systems. For example, although it would be tempting to attribute an outer solar system origin to the small OCs chondrules showing CC-like Δ^{17}O values (Fig. 6b, Marrocchi et al. 2024), the measurement of their nucleosynthetic signatures is absolutely fundamental to determine in which reservoir they formed. Indeed, nucleosynthetic anomalies have revealed drastic differences between NC and CC chondrules: NC chondrules have homogeneous ε^{50}Ti and ε^{54}Cr values similar to those of their host chondrites, whereas CC chondrules have more variable values that sometimes greatly differ from those of their host chondrites (Schneider et al.

2020). In addition, considering multi-isotopic space (ε^{50}Ti, ε^{54}Cr and Δ^{17}O), none (but two) of the CC chondrules plot in the NC chondrite isotopic field and vice versa (Schneider et al. 2020). Although this does not totally rule out occasional transport of NC chondrules into the CC reservoir (Williams et al. 2020; Zhang et al. 2022), the multi-isotope constraints on the NC/CC isotopic dichotomy do not suggest large-scale transport of chondrules. Instead, this support local formation of CC chondrules from isotopically heterogenous precursors composed of early-formed refractory inclusions and NC-like dust corresponding to late infall material incorporated before the CC chondrule formation processes (Nanne et al. 2019; Schneider et al. 2020). To reconcile this with the idea that refractory inclusions have undergone significant radial transport, it is possible that refractory inclusions are transported by a different mechanism, such as disk winds originating close to the Sun, rather than throughout the disk midplane which might be the only transport mechanism available for chondrules.

4.5 How to Deal with the Different Models?

Due to the relatively short heating and cooling timescales inferred, chondrule formation is likely a non-steady state process in which a large amount of energy is dissipated thermally before cooling back down to an ambient equilibrium temperature. Proposed mechanisms for chondrule formation have been focused on investigating physical mechanisms by which large amounts of kinetic energy (collisions, shocks) or magnetic energy (e.g., lightning discharge) can be dissipated locally. The viability of proposed models has been discussed extensively in the literature (see Desch et al. 2012 for a review), in which proofs of concept of specific mechanisms are evaluated by their ability to produce the requisite amount and approximate duration of heating and cooling under some assumed ambient "nebular conditions".

However, the ability to satisfy the inferred thermochemical constraints is dependent not only on the heating mechanism but on the local nebular conditions: the gas density, the solid density (for both larger grain chondrule precursors and the smaller grain constituents of the matrix), and the equilibrium temperature. This dependence is in large part due to the fact that protoplanetary disk environments cool radiatively and are thus dependent on the local opacity, a function of the density, temperature, and composition of the nebular material. For example, Desch et al. (2005) found that the local density of chondrule precursors directly affects the efficiency with which a thin shocked zone of nebular material can cool. From an astrophysical perspective, physically self-consistent models for protoplanetary disks are radially and temporally varying in gas density, solid density, and equilibrium temperature. Given the diversity of chondrule compositions and the range of chondrule ages, chondrule formation mechanisms likely operate across a broad extent of the disk, representing a variety of ambient equilibrium conditions.

So far, no proposed model of chondrule formation satisfies the numerous physical, chemical, isotopic, and chronological constraints. However, the entire set of constraints does not universally apply to all chondrule populations. Although different populations share many key characteristics and must have formed by a similar process, there is no clear evidence that they were formed by only one heating mechanism. While such a "simpler" assumption may have been preferred in the past (e.g., Rubin 2000), the possibility that there was more than one chondrule forming mechanism is increasingly being explored (e.g., Krot et al. 2022). Important support for this hypothesis comes from the study of CB-CH chondrules (Krot et al. 2005a). CB-CH chondrites contain a population of magnesian, non-porphyritic chondrules with unique chemical and mineralogical characteristics. They formed contemporaneously during a brief energetic event postdating CAIs by ~5 Ma (Bollard et al. 2015; Krot et al. 2005b; Pravdivtseva et al. 2017; Wölfer et al. 2023). The origin of these chondrites and their chondrules is best explained in an impact scenario: They probably formed from a vapor-melt impact plume generated by the collision of planetesimals (Krot et al. 2005a; Wasson and Kallemeyn 1990).

While this single-stage heating mechanism best produces the unique characteristics of CB-CH chondrules, it is unable to explain chondrule formation in other chondrite groups. For example, the mechanism cannot account for chondrules that were heated more than once such as those with relict grains, igneous rims, and enveloping compound chondrules often found in CV, CK and CR chondrites (Rubin 2010; Jacquet 2021). Formation of these chondrules requires a repeatable heating mechanism like shock waves or lightning in a nebular setting. As it becomes increasingly evident that chondrules formed in very different locations, astrophysical environments, and over large timescales in a dynamic, evolving protoplanetary disk, it will be important to define which heating mechanism best explains a (sub)set of constraints.

For nebular mechanisms, considering that chondrule formation occurs in both the inner and outer disk, we can broadly identify the disk regimes in which viable mechanisms operate by examining the range and variation of physical conditions spanning the inner and outer disk. For a steady-state viscous disk, the expected range of physical conditions across the spatial extent of the protosolar nebula can be largely understood in terms of the expected radial variation of the temperature and surface density profile of the dust and that of the gas. The equilibrium temperature of passively irradiated disks goes as $T \propto r^{-1/2}$, which means that ambient temperatures for chondrules forming in the inner disk at ($r \sim 0.5$ au) vs. those in the outer disk ($r \sim 10$ au) will vary by a factor of 5. For sufficiently high accretion rates, relevant early on in the disk evolution, T(r=0.5)/T(r=10) could be as high as 10 due to viscous heat dissipation preferentially heating the midplane of the inner disk (Harsono et al. 2015).

Surface density variation for power-law self-similar profiles of viscously evolving disks (Lynden-Bell and Pringle 1974) of the form $\Sigma(r) = \Sigma_c(r/r_c)^{-\gamma} exp(-r/r_c)^{2-\gamma}$ where standard assumptions for the power-law exponent range between 1 and 2, similarly implies the ambient gas density should vary by factors of 20 - 400 between the inner and outer disk. Taken together, the ambient nebular pressure between the outer and inner disk should vary by at least a factor of 100 - 4000.

Additionally, the dynamics of chondrule precursors (assuming $\sim$ millimeter-sized particles) will also strongly vary as a function of radius, related to the aerodynamic coupling of solids, parameterized by the Stokes number St. St is inversely proportional to the local gas surface density and the expected radial drift velocity $u_d = \eta v_k/(St + 1/St)$ or $\eta \propto (H/R)^2 dlnP/dlnr$ are strong functions of the density and pressure gradients and thus depend on the radial location in the disk. The result of such a drift dominated regime is strong differential transport between the solids and the gas, an effect observed in the systematic differences of the dust radii vs. molecular gas radii (where Rgas/Rdust $\sim$ 2-3) in millimeter surveys of protoplanetary disks in nearby star-forming regions (Ansdell et al. 2018). Drift-dominated evolution in some cases can be expected to enhance the dust-to-gas ratio in the inner disk relative to the outer disk by nearly an order of magnitude (Birnstiel et al. 2012). In addition, millimeter-sized grains can be expected to vertically settle to 20-30% of the gas scale height, depending on the amount of turbulent stirring (Dullemond and Dominik 2004), contributing to enhancements of the midplane dust-to-gas ratio by a factor of 3-5. Further enhancement of the dust-to-gas ratio above this background during chondrule formation in the nebular scenario would require local perturbations beyond the steady-state.

Given the range of ambient conditions across the disk, in addition to providing a source for rapid heating, an additional constraint on chondrule formation mechanisms may be that such mechanisms must operate efficiently in the physical regimes of a steady-state disk or themselves accompany (or even cause) perturbations to the steady-state that meet the required thermochemical constraints. Proposed heating mechanisms operate in distinct physical regimes, directly or indirectly tied to location in the disk by virtue of the assumed disk model and physical origin. Mechanisms that rely on the dissipation of magnetic energy, for instance, are generally restricted to regions in the disk in which the magnetic resonance instability (MRI) can operate, generally the hot inner disk where material can be thermally ionized (T> 800 K) as in the short circuit instability (Hubbard et al. 2012) or in the hot diffuse regions in the upper layers of the outer disk atmosphere where heating by current sheets or ohmic dissipation could provide a source of heating (McNally et al. 2014; Muranushi et al. 2012).

The efficiency of heating and cooling in nebular shocks on the other hand depends on a more complicated relationship of the proposed driving mechanism and the disk properties and local conditions. Models of chondrule formation investigating the heating of chondrule precursors by their passage through nebular shocks at ambient conditions corresponding roughly to heliocentric distances of about $r \sim 1$ au find that fairly strong shocks (Mach numbers of 7-9) can reach chondrule melting temperatures and require enhanced chondrule densities (enhancements of about an order of magnitude above solar metallicities) to cool efficiently (Morris and Desch 2010; Boley et al. 2013). There is a broad range of shock driving mechanisms that can operate in protoplanetary disk environments, either from waves driven by perturbers in the disk or as a result of waves from disk mechanisms such as gravitational instability (originally suggested as the physical mechanism behind the nebular shock scenario; Ciesla and Hood 2002). There is a wide diversity of candidate perturbers, from smaller bodies such as planetesimals or planetary embryos on eccentric orbits which could drive bow shocks (Hood and Weidenschilling 2012; Morris et al. 2012; Nagasawa et al. 2019) to giant planets driving spiral waves that steepen into shocks near their Lindblad resonances (Lyra et al. 2016, Bodenan et al. 2020). The viability of nebular chondrule formation scenarios, especially that of the nebular shock scenario, is strongly dependent on where and when in the disk's lifetime the proposed physical driving mechanisms can operate.

Untangling the local conditions of chondrule formation is further complicated by results from recent astrophysical studies which suggest that even smooth steady-state assumptions may not be representative of observed planet-forming disks. High-resolution observations of the millimeter emission from nearby protoplanetary sources with ALMA have revealed millimeter grains concentrated into annular and spiral structures (Andrews et al. 2018) and often differentially distributed when compared to the molecular gas (Law et al. 2021). Models have shown that such structure formation is a natural consequence of many disk dynamical processes −planetary perturbers, giant planet gap opening, envelope infall, vertical shear instability or MHD zonal flows (see Bae et al. 2023 for a review)− by virtue of perturbing the local disk pressure gradient creating local regions of rarefaction and concentration due to the differential aerodynamic drift of solids relative to the gas (Pinilla et al. 2012).

Substructures and their driving mechanisms may play an important role in the transport and distribution of the reservoir of larger grains such as chondrule precursors. Dust trapping by pressure bumps to produce rings can serve to separate and store solids in reservoirs resistant to inward drift, as well as locally increasing the dust-to-gas ratio and creating preferential conditions for rapid dust growth and planetesimal formation through the streaming instability or direct gravitational collapse (Drążkowska and Dullemond 2014; Carrera and Simon 2022). Robust pressure bumps at the edges of gaps opened by massive planets have been shown to lead to the rapid formation of planetary cores (Lau et al. 2022), though gap-opening generally requires the presence of Neptune to Jupiter mass perturbers (Dong and

Fung 2017) and is a strong function of the assumed disk viscosity. However, planetary mass perturbers can drive disk structure formation through spiral waves at even lower masses; where a thermal mass, $M_{th} = (H/R)^3 M_*$, is defined as the point at which perturbers drive spiral waves that develop into shocks close to the planet. Linear analyses and numerical simulations have shown that multiple rings and gaps can form from spiral wakes even down to subthermal (0.1 Mth) masses (Bae et al. 2017 Miranda and Rafikov 2019), though the exact mass threshold is likely dependent on the details of the disk model, particularly the thermodynamics (Miranda and Rafikov 2020). Taking 0.1 M_{th} as a nominal threshold at which to consider the effect of planets on the structure and dynamics of the disk material, for a typical protosolar disk temperature profile, the range of relevant perturber masses would extend down to 1 - 10 Earth masses from 0.1 AU to 50 AU, respectively.

Meteorites cannot be assumed to uniformly sample the original reservoir of solids in the early protosolar disk. Thus, we cannot assume that chondrites reflect the entire range of conditions in the disk nor can we assume that chondrules record representative thermochemical conditions of solid protosolar material. If we consider the protosolar disk as a dynamic system with differential transport and trapping of solids, chondrites may be rapidly assembled from material stored in reservoirs, subject to highly local thermochemical processing events. Many of the heating mechanisms explored require the presence of larger bodies, implying that chondrule formation is contemporaneous with ongoing planet formation in the disk. Chondrules may serve as a record of dynamic disk processes at work during planet formation or even could be considered a direct by-product of planet formation.

5 Conclusions

Thanks to decades of investigation and recent breakthroughs in petrographic and isotopic constraints, the conditions of chondrule formation appear at last emerging from the fog. All these data cannot be used individually (e.g., the cooling rate) but require a holistic approach, which is the only way to provide a comprehensive model of chondrule formation. In this review, we combined mineralogical, petrographic, chemical and multi-element isotopic constraints from which we propose the following inferences:

1- Recycling processes appear to be a key mechanism at the origin of type I porphyritic chondrules. Al-rich and porphyritic chondrules commonly show ^{16}O-rich relict olivine grains whose chemical characteristics suggest CAIs and AOAs as likely precursors. This inference is consistent with the varying ^{50}Ti-^{54}Cr excesses reported in individual CC chondrules. CC chondrules cannot only derive from refractory precursors but also require the addition of NC-like dust followed by gas-melt interactions with multi-species gas (H_2O, Mg, SiO) during the chondrule-forming events.

2- Chondrules do not only derive from refractory precursors but represent in some cases their own precursors. CR and NC type I chondrules have chemical and isotopic characteristics pointing toward multiple recycling processes involving previous generation of chondrules and punctual addition of CI-like matrix. Similar evidence suggests that type II chondrules could also derive from previously-formed type I chondrules.

3- Altogether, the current data strongly support that both NC and CC chondrules derive from nebular dust (excluding some CH-CB chondrules, which formed in an impact environment). They also do not support important disk-wide transport of chondrules across the NC and CC reservoirs, and indicate that no significant transport occurred during disk evolution. Within the NC and the CC reservoirs, chondrules in different chondrite groups have different characteristics. Chondrules thus experienced local formation from isotopically heterogeneous precursors within their respective accretion reservoirs.

4- The presence of several generations of genetically-related populations of chondrules in chondrites requires repetitive and extremely localized formation processes that only affect a fraction of the available dust. These high-temperature episodes were more efficient in the inner disk although they also occurred late in the outer disk history.

5- The chondrule heating mechanism has long been the subject of debate. Although there has been much emphasis on satisfying thermochemical constraints on chondrule formation, it is equally important for a viable heating mechanism to satisfy constraints on local nebular conditions including gas density, solid density, and the equilibrium temperature, as well as the extent of dust transport possible in different disk environments. Recent advances in understanding the temporal and spatial evolution of disk structures help to establish disk locations where conditions are appropriate for different chondrule-forming mechanisms to operate.

Supplementary Information The online version contains supplementary material available at https://doi.org/10.1007/s11214-024-01102-0.

Acknowledgements We thank Alan Rubin and anonymous reviewer for constructive reviews and Jutta Zipfel for efficient editorial handling. YM deeply thanks Emmanuel Jacquet, Maxime Piralla, Thorsten Kleine, Christophe Burkhardt and Alessandro Morbidelli for helpful discussion. This is CRPG contribution #2888.

Declarations

Competing Interests The authors declare that they have no conflict of interest.

References

Alexander CMOD (1994) Geochim Cosmochim Acta 58:3451
Alexander CMOD (2004) Geochim Cosmochim Acta 68:3943
Alexander CMOD, Ebel DS (2012) Meteorit Planet Sci 47:1157
Alexander CMOD, Wang J (2001) Meteorit Planet Sci 36:419
Alexander CMOD, Grossman JN, Ebel DS, Ciesla FJ (2008) Science 320:1617
Anders E (1964) Space Sci Rev 3:583
Andrews SM, Huang J, Pérez LM, Isella A, Dullemond CP, Kurtovic NT, Guzmán VV, Carpeter JM, Wilner DJ, Zhang S, Zhu Z, Birnstiel T, Bai X-N, Benisty M, Hughes AM, Öberg KI, Ricci L (2018) Astrophys J Lett 869:L41
Ansdell M, Williams JP, Trapman L, Terwisga SEV, Facchini S, Manara CF, Marel NVD, Miotello A, Tazzari M, Hogerheijde M, Guidi G, Testi L, Dishoeck EFV (2018) Astrophys J Lett 859:21
Bae J, Zhu Z, Hartmann L (2017) Astrophys J 850:201
Bae J, Isella A, Zhu Z, Martin R, Okuzumi S, Suriano S (2023) In: Protostars and planets VII
Baertschi P (1976) Earth Planet Sci Lett 31:341
Barosch J, Hezel DC, Ebel DS, Friend P (2019) Geochim Cosmochim Acta 249:1
Barosch J, Hezel DC, Sawatzki L, Halbauer L, Marrocchi Y (2020) Meteorit Planet Sci 55:993
Berlin J, Jones RH, Brearley AJ (2011) Meteorit Planet Sci 46:513

Birnstiel T, Klahr H, Ercolano B (2012) Astron Astrophys 539:A148
Bischoff A, Keil K (1984) Geochim Cosmochim Acta 48:693
Bodénan J-D, Surville C, Szulágyi J, Mayer L, Schönbächler M (2020) Astrophys J 901:60
Boley AC, Morris MA, Desch SJ (2013) Astrophys J 776:101
Bollard J, Connelly JN, Bizzarro M (2015) Meteorit Planet Sci 50:1197
Bollard J, Connelly JN, Whitehouse MJ, Pringle EA, Bonal L, Jørgensen JK, Nordlund A, Moynier F, Bizzarro M (2017) Sci Adv 3:e1700407
Bollard J, Kawasaki N, Sakamoto N, Olsen M, Itoh S, Larsen K, Wielandt D, Schiller M, Connelly JN, Yurimoto H, Bizzarro M (2019) Geochim Cosmochim Acta 260:62
Braukmüller N, Wombacher F, Hezel DC, Escoube R, Münker C (2018) Geochim Cosmochim Acta 239:17
Brennecka GA, Burkhardt C, Budde G, Kruijer TS, Nimmo F, Kleine T (2020) Science 370:837
Bridges JC, Franchi IA, Sexton AS, Pillinger CT (1999) Geochim Cosmochim Acta 63:945
Bryson JFJ, Brennecka GA (2021) Astrophys J 912:163
Budde G, Burkhardt C, Brennecka GA, Fischer-Gödde M, Kruijer TS, Kleine T (2016) Earth Planet Sci Lett 454:293
Budde G, Kruijer TS, Kleine T (2018) Geochim Cosmochim Acta 222:284
Burkhardt C (2021) Elements 17:395
Burkhardt C, Spitzer F, Morbidelli A, Budde G, Render JH, Kruijer TS, Kleine T (2021) Sci Adv 7:eabj7601
Campbell AJ, Humayun M, Weisberg MK (2002) Geochim Cosmochim Acta 66:647
Carrera D, Simon JB (2022) Astrophys J Lett 933:L10
Charles CRJ, Robin P-YF, Davis DW, McCausland PJA (2018) Meteorit Planet Sci 53:935
Chaumard N, Defouilloy C, Kita NT (2018) Geochim Cosmochim Acta 228:220
Chaumard N, Defouilloy C, Hertwig AT, Kita NT (2021) Geochim Cosmochim Acta 299:199
Ciesla F, Hood LL (2002) Icarus 158:281
Clayton RN, Mayeda TK (1984) Earth Planet Sci Lett 67:151
Clayton RN, Mayeda TK (1999) Geochim Cosmochim Acta 63:2089
Clayton RN, Grossman L, Mayeda TK (1973) Science 182:485
Clayton RN, Mayeda TK, Goswami JN (1991) Geochim Cosmochim Acta 55:2317
Connelly JN, Bizzarro M, Krot AN, Nordlund A (2012) Science 338:651
Connolly HC, Huss GR (2010) Geochim Cosmochim Acta 74:2473
Connolly HC, Jones RH (2016) J Geophys Res, Planets 121:1885
Connolly HC, Hewins RH, Ash RD, Zanda B, Lofgren GE, Bourot-Denise M (1994) Nature 371:136
Cuzzi JN, Alexander CMOD (2006) Nature 441:483
Defouilloy C, Nakashima D, Joswiak DJ, Brownlee DE, Tenner TJ, Kita NT (2017) Earth Planet Sci Lett 465:145
Desch SJ, Connolly HC (2002) Meteorit Planet Sci 37:183
Desch SJ, Ciesla FJ, Hood LL, Nakamoto T (2005) In: Chondrites and the protoplanetary disk, p 849
Desch SJ, Morris MA, Connolly HC, Boss AP (2012) Meteorit Planet Sci 47:1139
Dong R (董若冰), Fung J (馮澤之) (2017) Astrophys J Lett 835:146
Drążkowska J, Dullemond CP (2014) Astron Astrophys 572:A78
Dullemond CP, Dominik C (2004) Astron Astrophys 421:1075
Dunham ET, Sheikh A, Opara D, Matsuda N, Liu M-C, McKeegan KD (2023) Meteorit Planet Sci 58:643
Ebel DS (2006) In: Meteorites and the early Solar System II, p 253
Ebel DS, Grossman L (2000) Geochim Cosmochim Acta 64:339
Ebel DS, Alexander CMOD, Libourel G (2018) In: Chondrules, p 151
Ebert S, Bischoff A (2016) Geochim Cosmochim Acta 177:182
Ebert S, Render J, Brennecka GA, Burkhardt C, Bischoff A, Gerber S, Kleine T (2018) Earth Planet Sci Lett 498:257
Ebert S, Nagashima K, Bischoff A, Berndt J, Krot AN (2022) Geochim Cosmochim Acta 336:448
Eiler JM (2001) Rev Mineral Geochem 43:319
Faure F, Tissandier L, Libourel G, Mathieu R, Welsch B (2012) Earth Planet Sci Lett 1:319
Fedkin AV, Grossman L (2013) Geochim Cosmochim Acta 112:226
Frank DR, Huss GR, Zolensky ME, Nagashima K, Le L (2023) Meteorit Planet Sci 58:1495
Friend P, Hezel DC, Mucerschi D (2016) Geochim Cosmochim Acta 173:198
Fukuda K, Tenner TJ, Kimura M, Tomioka N, Siron G, Ushikubo T, Chaumard N, Hertwig AT, Kita NT (2022) Geochim Cosmochim Acta 322:194
Gerber S, Burkhardt C, Budde G, Metzler K, Kleine T (2017) Astrophys J Lett 841:L17
Greenwood JP, Rubin AE, Wasson JT (2000) Geochim Cosmochim Acta 64:3897
Greenwood RC, Burbine TH, Franchi IA (2020) Geochim Cosmochim Acta 277:377
Grossman JN, Brearley AJ (2005) Meteorit Planet Sci 40:87
Grossman L, Steele IM (1976) Geochim Cosmochim Acta 40:149

Grossman JN, Rubin AE, Rambaldi ER, Rajan RS, Wasson JT (1985) Geochim Cosmochim Acta 49:1781
Harsono D, Van Dishoeck EF, Bruderer S, Li Z-Y, Jørgensen JK (2015) Astron Astrophys 577:A22
Hellmann JL, Hopp T, Burkhardt C, Kleine T (2020) Earth Planet Sci Lett 549:116508
Hellmann JL, Schneider JM, Wölfer E, Drążkowska J, Jansen CA, Hopp T, Burkhardt C, Kleine T (2023) Astrophys J Lett 946:L34
Hertwig AT, Defouilloy C, Kita NT (2018) Geochim Cosmochim Acta 224:116
Hertwig AT, Kimura M, Ushikubo T, Defouilloy C, Kita NT (2019) Geochim Cosmochim Acta 253:111
Hewins RH, Zanda B (2012) Meteorit Planet Sci 47:1120
Hewins RH, Connolly HC, Lofgren GE, Libourel G (2005) In: Chondrites and the protoplanetary disk, p 286
Hezel DC, Palme H (2007) Geochim Cosmochim Acta 71:4092
Hezel DC, Palme H (2010) Earth Planet Sci Lett 294:85
Hezel DC, Parteli EJR (2018) Astrophys J 863:54
Hezel DC, Palme H, Brenker FE, Nasdala L (2003) Meteorit Planet Sci 38:1199
Hezel DC, Palme H, Nasdala L, Brenker FE (2006) Geochim Cosmochim Acta 70:1548
Hezel DC, Russell SS, Ross AJ, Kearsley AT (2008) Meteorit Planet Sci 43:1879
Hezel DC, Needham AW, Armytage R, Georg B, Abel RL, Kurahashi E, Coles BJ, Rehkämper M, Russell SS (2010) Earth Planet Sci Lett 296:423
Hezel DC, Bland PA, Palme H, Jacquet E, Bigolski J (2018) In: Russell SS, Connolly HC Jr, Krot AN (eds) Chondrules, p 91
Hood LL, Weidenschilling SJ (2012) Meteorit Planet Sci 47:1715
Hood LL, Ciesla FJ, Artemieva NA, Marzari F, Weidenschilling SJ (2009) Meteorit Planet Sci 44:327
Hubbard A, Ebel DS (2018) In: Russell SS, Connolly HC Jr, Krot AN, Chondrules I (eds) Chondrules, p 400
Hubbard A, McNally CP, Mac Low M-M (2012) Astrophys J 761:58
Huss GR, MacPherson GJ, Wasserburg GJ, Russell SS, Srinivasan G (2001) Meteorit Planet Sci 36:975
Itoh S, Russell SS, Yurimoto H (2007) Meteorit Planet Sci 42:1241
Jabeen I, Kusakabe M, Nagao K, Ali A (2019) Meteorit Planet Sci 54:431
Jacquet E (2021) Geochim Cosmochim Acta 296:18
Jacquet E, Marrocchi Y (2017) Meteorit Planet Sci 52:2672
Jacquet E, Paulhiac-Pison M, Alard O, Kearsley AT, Gounelle M (2013) Meteorit Planet Sci 48:1981
Jacquet E, Alard O, Gounelle M (2015) Geochim Cosmochim Acta 155:47
Jacquet E, Barrat J-A, Beck P, Caste F, Gattacceca J, Sonzogni C, Gounelle M (2016) Meteorit Planet Sci 51:851
Jacquet E, Piani L, Weisberg MK (2018) In: Chondrules, p 175
Jacquet E, Piralla M, Kersaho P, Marrocchi Y (2021) Meteorit Planet Sci 56:13
Jansen C, Burkhardt C, Marrocchi Y, Schneider JM, Wölfer E, Kleine T (2024) Earth Planet Sci Lett 627:118567
Johnson BC, Minton DA, Melosh HJ, Zuber MT (2015) Nature 517:339
Jones RH (1990) Geochim Cosmochim Acta 54:1785
Jones RH (1996) Geochim Cosmochim Acta 60:3115
Jones RH (2012) Meteorit Planet Sci 47:1176
Jones RH, Carey ER (2006) Am Mineral 91:1664
Jones RH, Leshin LA, Guan Y, Sharp ZD (2004) Geochim Cosmochim Acta 68:3423
Jones RH, Grossman JN, Rubin AE (2005) In: Chondrites and the protoplanetary disk, vol 341, p 241
Jones RH, Villeneuve J, Libourel G (2018) In: Chondrules, vol 57
Joswiak DJ, Brownlee DE, Matrajt G, Westphal AJ, Snead CJ, Gainsforth Z (2012) Meteorit Planet Sci 47:471
Kaneko H, Sato K, Ikeda C, Nakamoto T (2023) Astrophys J 947:15
Kita NT, Nagahara H, Tachibana S, Tomomura S, Spicuzza MJ, Fournelle JH, Valley JW (2010) Geochim Cosmochim Acta 74:6610
Komatsu M, Fagan TJ, Mikouchi T, Petaev MI, Zolensky ME (2015) Meteorit Planet Sci 50:1271
Krot AN (2019) Meteorit Planet Sci 54:1647
Krot AN, Keil K (2002) Meteorit Planet Sci 37:91
Krot AN, Petaev MI, Russell SS, Itoh S, Fagan TJ, Yurimoto H, Chizmadia L, Weisberg MK, Komatsu M, Ulyanov AA, Keil K (2004) Chem Erde Geochem 64:185
Krot AN, Amelin Y, Cassen P, Meibom A (2005a) Nature 436:989
Krot AN, Fagan TJ, Nagashima K, Petaev MI, Yurimoto H (2005b) Geochim Cosmochim Acta 69:1873
Krot AN, Nagashima K, Yoshitake M, Yurimoto H (2010) Geochim Cosmochim Acta 74:2190
Krot AN, Nagashima K, Petaev MI (2012) Geochim Cosmochim Acta 83:159
Krot AN, Nagashima K, van Kooten EMME, Bizzarro M (2017) Geochim Cosmochim Acta 201:185
Krot AN, Petaev MI, Nagashima K, Dobrică E, Johnson BC, Cashion MD, Rubin A (2022) Meteorit Planet Sci 57:352

Kruijer TS, Kleine T, Borg LE (2020) Nat Astron 4:32
Kunihiro T, Rubin AE, McKeegan KD, Wasson JT (2004) Geochim Cosmochim Acta 68:3599
Kunihiro T, Rubin AE, Wasson JT (2005) Geochim Cosmochim Acta 69:3831
Lau TCH, Drążkowska J, Stammler SM, Birnstiel T, Dullemond CP (2022) Astron Astrophys 668:A170
Law CJ, Loomis RA, Teague R, Öberg KI, Czekala I, Andrews SM, Huang J, Aikawa Y, Alarcón F, Bae J, Bergin EA, Bergner JB, Boehler Y, Booth AS, Bosman AD, Calahan JK, Cataldi G, Cleeves LI, Furuya K, Guzmán VV, Ilee JD, Le Gal R, Liu Y, Long F, Ménard F, Nomura H, Qi C, Schwarz KR, Sierra A, Tsukagoshi T, Yamato Y, Van 'T Hoff MLR, Hoff T, Walsh C, Wilner DJ, Zhang K (2021) Astrophys J Lett 257:3
Lebreuilly U, Mac Low M-M, Commerçon B, Ebel DS (2023) Astron Astrophys 675:A38
Leshin LA, Rubin AE, McKeegan KD (1997) Geochim Cosmochim Acta 61:835
Lewis JA, Jones RH (2019) Meteorit Planet Sci 54:72
Libourel G, Krot AN (2007) Earth Planet Sci Lett 254:1
Libourel G, Portail M (2018) Sci Adv 4:eaar3321
Libourel G, Krot AN, Tissandier L (2006) Earth Planet Sci Lett 251:232
Libourel G, Nagashima K, Portail M, Krot AN (2022) Geochim Cosmochim Acta 319:73
Libourel G, Nagashima K, Portail M, Krot AN (2023) Geochim Cosmochim Acta 346:102
Lichtenberg T, Golabek GJ, Dullemond CP, Schönbächler M, Gerya TV, Meyer MR (2018) Icarus 302:27
Lin Y (2022) Prog Earth Planet Sci 9:28
Lodders K (2003) Astrophys J 591:1220
Lodders K, Fegley B (2023) Geochemistry 83:125957
Lynden-Bell D, Pringle JE (1974) Mon Not R Astron Soc 168:603
Lyra W, Richert AJW, Boley A, Turner N, Mac Low M-M, Okuzumi S, Flock M (2016) Astrophys J 817:102
MacPherson GJ, Huss GR (2005) Geochim Cosmochim Acta 69:3099
Marrocchi Y, Chaussidon M (2015) Earth Planet Sci Lett 430:308
Marrocchi Y, Libourel G (2013) Geochim Cosmochim Acta 119:117
Marrocchi Y, Chaussidon M, Piani L, Libourel G (2016) Sci Adv 7:e1601001
Marrocchi Y, Villeneuve J, Batanova V, Piani L, Jacquet E (2018) Earth Planet Sci Lett 496:132
Marrocchi Y, Euverte R, Villeneuve J, Batanova V, Welsch B, Ferrière L, Jacquet E (2019a) Geochim Cosmochim Acta 247:121
Marrocchi Y, Villeneuve J, Jacquet E, Piralla M, Chaussidon M (2019b) Proc Natl Acad Sci USA 116:23461
Marrocchi Y, Bonal L, Gattacceca J, Piani L, Beck P, Greenwood R, Eschrig J, Basque A, Nuccio PM, Martin FF (2020) Meteorit Planet Sci 55:1924
Marrocchi Y, Avice G, Barrat J-A (2021) Astrophys J Lett L9
Marrocchi Y, Piralla M, Regnault M, Batanova V, Villeneuve J, Jacquet E (2022) Earth Planet Sci Lett 593:117683
Marrocchi Y, Jacquet E, Neukampf J, Villeneuve J, Zolensky ME (2023b) Meteorit Planet Sci 58:195
Marrocchi Y, Piralla M, Tissot FLH (2023a) Astrophys J Lett 954:L27
Marrocchi Y, Longeau A, Goupil RL, Dijon V, Pinto G, Neukampf J, Villeneuve J, Jacquet E (2024) Geochim Cosmochim Acta 371:52
Mayeda TK, Clayton RN, Kring DA, Davis AM (1988) Meteoritics 23:288
McNally CP, Hubbard A, Yang C-C, Low M-MM (2014) Astrophys J 791:62
Metzler K, Bischoff A, Stöffler D (1992) Geochim Cosmochim Acta 56:2873
Miller KE, Lauretta DS, Connolly HC, Berger EL, Nagashima K, Domanik K (2017) Geochim Cosmochim Acta 209:24
Miranda R, Rafikov RR (2019) Astrophys J 875:37
Miranda R, Rafikov RR (2020) Astrophys J 892:65
Morin GLF, Marrocchi Y, Villeneuve J, Jacquet E (2022) Geochim Cosmochim Acta 332:203
Morris MA, Boley AC (2018) In: Chondrules, p 375
Morris MA, Desch SJ (2010) Astrophys J 722:1474
Morris MA, Boley AC, Desch SJ, Athanassiadou T (2012) Astrophys J 752:27
Muranushi T, Okuzumi S, Inutsuka S (2012) Astrophys J 760:56
Nagahara H (1981) Nature 292:135
Nagahara H, Kita NT, Ozawa K, Morishita Y (2008) Geochim Cosmochim Acta 72:1442
Nagasawa M, Tanaka KK, Tanaka H, Nomura H, Nakamoto T, Miura H (2019) Astrophys J 871:110
Nagashima K, Krot AN, Huss GR (2014) Geochem J 48:561
Nagashima K, Krot AN, Huss GR (2015) Geochim Cosmochim Acta 151:49
Nagashima K, Kita NT, Luu T-H (2018) In: Chondrules, p 247
Nakamura T, Noguchi T, Tsuchiyama A, Ushikubo T, Kita NT, Valley JW, Zolensky ME, Kakazu Y, Sakamoto K, Mashio E, Uesugi K, Nakano T (2008) Science 321:1664

Nakashima D, Ushikubo T, Joswiak DJ, Brownlee DE, Matrajt G, Weisberg MK, Zolensky ME, Kita NT (2012) Earth Planet Sci Lett 355:357
Nanne JAM, Nimmo F, Cuzzi JN, Kleine T (2019) Earth Planet Sci Lett 511:44
Ogliore RC, Huss GR, Nagashima K, Butterworth AL, Gainsforth Z, Stodolna J, Westphal AJ, Joswiak D, Tyliszczak T (2012) Astrophys J Lett 745:L19
Olsen MB, Wielandt D, Schiller M, Van Kooten EMME, Bizzarro M (2016) Geochim Cosmochim Acta 191:118
Pack A, Yurimoto H, Palme H (2004) Geochim Cosmochim Acta 68:1135
Palme H, Spettel B, Hezel DC (2014) Geochemistry 74:507
Palme H, Hezel DC, Ebel DS (2015) Earth Planet Sci Lett 411:11
Pape J, Mezger K, Bouvier A-S, Baumgartner LP (2019) Geochim Cosmochim Acta 244:416
Piani L, Marrocchi Y, Libourel G, Tissandier L (2016) Geochim Cosmochim Acta 195:84
Pinilla P, Birnstiel T, Ricci L, Dullemond CP, Uribe AL, Testi L, Natta A (2012) Astron Astrophys 538:A114
Pinto GA, Marrocchi Y, Jacquet E, Olivares F (2022) Meteorit Planet Sci 57:1004
Pinto GA, Jacquet E, Corgne A, Olivares F, Villeneuve J, Marrocchi Y (2024) Geochim Cosmochim Acta 364:65
Piralla M, Marrocchi Y, Verdier-Paoletti MJ, Vacher LG, Vilelneuve J, Piani L, Bekaert DV, Gounelle M (2020) Geochim Cosmochim Acta 269:451
Piralla M, Villeneuve J, Batanova V, Jacquet E, Marrocchi Y (2021) Geochim Cosmochim Acta 313:295
Piralla M, Villeneuve J, Schnuriger N, Bekaert DV, Marrocchi Y (2023) Icarus 394:115427
Pravdivtseva O, Meshik A, Hohenberg CM, Krot AN (2017) Geochim Cosmochim Acta 201:320
Qin L, Alexander CMOD, Carlson RW, Horan MF, Yokoyama T (2010) Geochim Cosmochim Acta 74:1122
Rambaldi ER (1981) Nature 293:558
Regnault M, Marrocchi Y, Piralla M, Villeneuve J, Batanova V, Schnuriger N, Jacquet E (2022) Meteorit Planet Sci 57:122
Richet P, Bottinga Y, Javoy M (1977) Annu Rev Earth Planet Sci 5:65
Rowe MW, Clayton RN, Mayeda TK (1994) Geochim Cosmochim Acta 58:5341
Rubin AE (1984) Geochim Cosmochim Acta 48:1779
Rubin AE (2000) Earth-Sci Rev 50:3
Rubin AE (2004) Am Mineral 89:867
Rubin AE (2010) Geochim Cosmochim Acta 74:4807
Rubin AE (2010) Geochim Cosmochim Acta 74:4807
Rubin AE (2013) Meteorit Planet Sci 48:445
Rubin AE, Wasson JT, Clayton RN, Mayeda TK (1990) Earth Planet Sci Lett 96:247
Rubin AE, Sailer AL, Wasson JT (1999) Geochim Cosmochim Acta 63:2281
Rubin AE, Kallemeyn GW, Wasson JT, Clayton RN, Mayeda TK, Grady M, Verchovsky AB, Eugster O, Lorenzetti S (2003) Geochim Cosmochim Acta 67:3283
Rudraswami NG, Ushikubo T, Nakashima D, Kita NT (2011) Geochim Cosmochim Acta 75:7596
Russell SS, MacPherson GJ, Leshin LA, McKeegan KD (2000) Earth Planet Sci Lett 184:57
Russell SS, Krot AN, Huss GR, Keil K (2005) In: Chondrites and the protoplanetary disk, p 317
Ruzicka A, Hiyagon H, Hutson M, Floss C (2007) Earth Planet Sci Lett 257:274
Ruzicka A, Floss C, Hutson M (2008) Geochim Cosmochim Acta 72:5530
Ruzicka A, Floss C, Hutson M (2012) Geochim Cosmochim Acta 79:79
Sanders IS, Scott ERD (2012) Meteorit Planet Sci 47:2170
Schiller M, Bizzarro M, Siebert J (2020) Sci Adv 6:eaay7604
Schneider JM, Burkhardt C, Marrocchi Y, Brennecka GA, Kleine T (2020) Earth Planet Sci Lett 551:116585
Schnuriger N, Cartier C, Villeneuve J, Batanova V, Regnault M, Marrocchi Y (2022) Meteorit Planet Sci 57:1018
Schrader DL, Connolly HC Jr, Lauretta DS, Nagashima K, Huss GR, Davidson J, Domanik KJ (2013) Geochim Cosmochim Acta 101:302
Schrader DL, Nagashima K, Krot AN, Ogliore RC, Hellebrand E (2014) Geochim Cosmochim Acta 132:50
Schrader DL, Nagashima K, Krot AN, Ogliore RC, Yin Q-Z, Amelin Y, Stirling CH, Kaltenbach A (2017) Geochim Cosmochim Acta 201:275
Schrader DL, Nagashima K, Waitukaitis SR, Davidson J, McCoy TJ, Connolly HC, Lauretta DS (2018) Geochim Cosmochim Acta 223:405
Schrader DL, Nagashima K, Davidson J, McCoy TJ, Ogliore RC, Fu RR (2020) Geochim Cosmochim Acta 282:133
Scott ERD, Krot AN (2014) In: Treatise on geochemistry, vol 1, p 65
Scott ERD, Taylor GJ (1983) J Geophys Res 88:B275
Siron G, Fukuda K, Kimura M, Kita NT (2021) Geochim Cosmochim Acta 293:103
Siron G, Fukuda K, Kimura M, Kita NT (2022) Geochim Cosmochim Acta 324:312

Smith A, Jones RH (2023) Meteorit Planet Sci 59:685
Smith A, Jones RH (2024) Geochim Cosmochim Acta 374:106
Tenner TJ, Ushikubo T, Kurahashi E, Kita NT, Nagahara H (2013) Geochim Cosmochim Acta 102:226
Tenner TJ, Nakashima D, Ushikubo T, Kita NT, Weisberg MK (2015) Geochim Cosmochim Acta 148:228
Tenner TJ, Ushikubo T, Nakashima D, Schrader DL, Weisberg MK, Kimura M, Kita NT (2018) In: Oxygen isotope characteristics of chondrules from recent studies by secondary ion mass spectrometry
Tenner TJ, Nakashima D, Ushikubo T, Tomioka N, Kimura M, Weisberg MK, Kita NT (2019) Geochim Cosmochim Acta 260:133
Thomassin D, Piani L, Villeneuve J, Caumon M-C, Bouden N, Marrocchi Y (2023) Earth Planet Sci Lett 616:118225
Tissandier L, Libourel G, Robert F (2002) Meteorit Planet Sci 37:1377
Trinquier A, Elliott T, Ulfbeck D, Coath C, Krot AN, Bizzarro M (2009) Science 324:374
Tronche EJ, Hewins RH, MacPherson GJ (2007) Geochim Cosmochim Acta 71:3361
Ushikubo T, Kimura M (2021) Geochim Cosmochim Acta 293:328
Ushikubo T, Kimura M, Kita NT, Valley JW (2012) Geochim Cosmochim Acta 90:242
Ushikubo T, Nakashima D, Kimura M, Tenner TJ, Kita NT (2013) Geochim Cosmochim Acta 109:280
van Kooten EMME, Wielandt D, Schiller M, Nagashima K, Thomen A, Larsen KK, Olsen MB, Nordlund A, Krot AN, Bizzarro M (2016) Proc Natl Acad Sci USA 113:2011
van Kooten EMME, Cavalcante L, Wielandt D, Bizzarro M (2020) Meteorit Planet Sci 55:575
van Kooten EMME, Schiller M, Moynier F, Johansen A, Haugbølle T, Bizzarro M (2021) Astrophys J 910:70
Villeneuve J, Chaussidon M, Libourel G (2009) Science 325:985
Villeneuve J, Libourel G, Soulié C (2015) Geochim Cosmochim Acta 160:277
Villeneuve J, Marrocchi Y, Jacquet E (2020) Earth Planet Sci Lett 9
Warren PH (2011) Earth Planet Sci Lett 311:93
Wasson JT, Kallemeyn GW (1990) Earth Planet Sci Lett 101:148
Weisberg MK (1987) J Geophys Res 92:E663
Weisberg MK, Prinz M, Clayton RN, Mayeda TK (1993) Geochim Cosmochim Acta 57:1567
Weisberg MK, Prinz M, Clayton RN (1995) Antarct Meteor 8:11
Weisberg MK, Prinz M, Clayton RN, Mayeda TK, Grady MM, Franchi I, Pillinger CT, Kallemeyn GW (1996) Geochim Cosmochim Acta 60:4253
Weisberg MK, Prinz M, Clayton RN, Mayeda TK (2003) Geochim Cosmochim 57:1567
Weisberg MK, Connolly HC, Ebel DS (2004) Meteorit Planet Sci 39:1741
Weisberg MK, Ebel DS, Connolly HC, Kita NT, Ushikubo T (2011) Geochim Cosmochim Acta 75:6556
Weisberg MK, Kita NT, Fukuda K, Siron G, Ebel DS (2021) Geochim Cosmochim Acta 300:279
Whattam SA, Hewins RH (2009) Geochim Cosmochim Acta 73:5460
Whattam SA, Hewins RH, Cohen BA, Seaton NC, Prior DJ (2008) Earth Planet Sci Lett 269:200
Whattam SA, Hewins RH, Seo J, Devouard B (2022) Geochim Cosmochim Acta 319:30
Wick MJ, Jones RH (2012) Geochim Cosmochim Acta 98:140
Williams CD, Sanborn ME, Defouilloy C, Yin Q-Z, Kita NT, Ebel DS, Yamakawa A, Yamashita K (2020) Proc Natl Acad Sci USA 117:23426
Wölfer E, Budde G, Kleine T (2023) Geochim Cosmochim Acta 361:288
Yamashita K, Maruyama S, Yamakawa A, Nakamura E (2010) Astrophys J 723:20
Yang L, Ciesla FJ, Alexander CMOD (2013) Icarus 226:256
Young ED, Galy A (2004) In: Reviews in mineralogy and geochemistry, pp 197–230
Yu Y, Hewins RH (1998) Geochim Cosmochim Acta 62:159
Yu Y, Hewins RH, Alexander CMOD, Wang J (2003) Geochim Cosmochim Acta 67:773
Zanda B (2004) Earth Planet Sci Lett 224:1
Zanda B, Lewin E, Humayun M (2018) In: Chondrules: records of protoplanetary disk processes. Cambridge University Press, Cambridge, pp 122–150
Zhang M, Fukuda K, Spicuzza MJ, Siron G, Heimann A, Hammerstrom AJ, Kita NT, Ushikubo T, Valley JW (2022) Chem Geol 608:121016
Zhu K, Liu J, Moynier F, Qin L, Alexander CMOD, He Y (2019) Astrophys J 873:82
Zhu K, Moynier F, Schiller M, Bizzarro M (2020) Astrophys J 894:L26
Zhu K, Schiller M, Moynier F, Groen M, Alexander CMOD, Davidson J, Schrader DL, Bischoff A, Bizzarro M (2023) Geochim Cosmochim Acta 342:156
Zolensky ME, Zega TJ, Yano H, Wirick S, Westphal AJ, Weisberg MK, Weber I, Warren JL, Velbel MA, Tsuchiyama A, Tsou P, Toppani A, Tomioka N, Tomeoka K, Teslich N, Taheri M, Susini J, Stroud R, Stephan T, Stadermann FJ, Snead CJ, Simon SB, Simionovici A, See TH, Robert F, Rietmeijer FJM, Rao W, Perronnet MC, Papanastassiou DA, Okudaira K, Ohsumi K, Ohnishi I, Nakamura-Messenger K, Nakamura T, Mostefaoui S, Mikouchi T, Meibom A, Matrajt G, Marcus MA, Leroux H, Lemelle L, Le L, Lanzirotti A, Langenhorst F, Krot AN, Keller LP, Kearsley AT, Joswiak D, Jacob D, Ishii H,

Harvey R, Hagiya K, Grossman L, Grossman JN, Graham GA, Gounelle M, Gillet P, Genge MJ, Flynn G, Ferroir T, Fallon S, Ebel DS, Dai ZR, Cordier P, Clark B, Chi M, Butterworth AL, Brownlee DE, Bridges JC, Brennan S, Brearley A, Bradley JP, Bleuet P, Bland PA, Bastien R (2006) Science 314:1735

Authors and Affiliations

Yves Marrocchi[1] · Rhian H. Jones[2] · Sara S. Russell[3] · Dominik C. Hezel[4] · Jens Barosch[5,6] · Aleksandra Kuznetsova[7,8]

✉ Y. Marrocchi
yves.marrocchi@univ-lorraine.fr

✉ R.H. Jones
rhian.jones-2@manchester.ac.uk

1 Centre de recherches pétrographiques et géochimiques (CRPG), CNRS, UMR 7358, Nancy, France

2 Department of Earth and Environmental Sciences, The University of Manchester, Manchester, M13 9PL, UK

3 Planetary Materials Group, Department of Earth Sciences, Natural History Museum, Cromwell Road, London SW7 5BD, UK

4 Institut für Geowissenschaften, Goethe-Universität Frankfurt, Altenhöferallee 1, Frankfurt, 60437, Germany

5 Department of Geoscience, University of Wisconsin-Madison, Madison, WI 53706, USA

6 School of Geosciences, The University of Edinburgh, Edinburgh EH9 3FE, UK

7 American Museum of Natural History, 200 Central Park West, New York, NY 10024, USA

8 Center for Computational Astrophysics, Flatiron Institute, New York, NY 10010, USA

Space Science Reviews (2024) 220:74
https://doi.org/10.1007/s11214-024-01107-9

Is There a Genetic Relationship Between Chondrules and Matrix?

Elishevah M.M.E. van Kooten[1] · Adrian Brearley[2] · Denton S. Ebel[3,4,5] · Conel M.O'D. Alexander[6] · Marina E. Gemma[7,3] · Dominik C. Hezel[8]

Received: 22 December 2023 / Accepted: 4 September 2024 / Published online: 24 September 2024

Abstract
Chondritic components such as chondrules and matrix are the key time capsules that can help us understand the evolution and dynamics of the protoplanetary disk from which the Solar System originated. Knowledge of where and how these components formed and to what extent they were transported in the gaseous disk provides major constraints to astrophysical models that investigate planet formation. Here, we explore whether chondrules and matrix are genetically related to each other and formed from single reservoirs per chondrite group or if every chondrite represents a unique proportion of components transported from a small number of formation reservoirs in the disk. These 'static versus dynamic disk' interpretations of cosmochemical data have profound implications for the accretion history of the planets in the Solar System. To fully understand the relationship between chondrules and matrix and their potential "complementarity", we dive into the petrological nature and origin of matrix, the chemical and isotopic compositions of chondrules and matrix and evaluate these data considering the effect of secondary alteration observed in chondrites and the potential complexity of chondrule formation. Even though we, the authors, have used different datasets and arrived at differing interpretations of chondrule-matrix relationships in the past, this review provides clarity on the existing data and has given us new directions towards future research that can resolve the complementarity debate.

✉ E.M.M.E. van Kooten
elishevah.vankooten@sund.ku.dk

1 Centre for Star and Planet Formation, Globe Institute, University of Copenhagen, Øster Voldgade 5-7, 1350 Copenhagen, Denmark

2 Department of Earth & Planetary Sciences, University of New Mexico, Albuquerque, NM 87131, USA

3 Department of Earth and Planetary Sciences, American Museum of Natural History, New York, NY 10024, USA

4 Department of Earth and Environmental Sciences, Columbia University, New York, USA

5 Graduate Center of City University of New York, New York, NY, USA

6 Carnegie Institution of Washington, 5241 Broad Branch Road, Washington DC, 20015, USA

7 Department of Geosciences, Stony Brook University, Stony Brook, NY, 11974, USA

8 Goethe-Universität Frankfurt, Altenhöferallee 1, 60438 Frankfurt am Main, Germany

Keywords Complementarity · Matrix · Chondrule

1 Introduction: What Is Complementarity and (Why) Is There a Problem?

Chondrites are aggregates of components that formed in the protoplanetary disk surrounding the young Sun. The major chondrite components are chondrules, matrix, and to a lesser extent, refractory inclusions (CAIs: calcium-, aluminum-rich inclusions, and AOAs: amoeboid-olivine-aggregates), and opaque phases such as metal, magnetite or sulfide (in rare cases up to 70 vol.% metal). A key cosmochemical constraint on the formation, evolution and dynamics of the protoplanetary disk is whether these individual components formed together in local disk regions, or whether they formed in separate regions of the protoplanetary disk and were later transported, mixed, and accreted into chondrite parent bodies. For example, it is crucial to understand if terrestrial planets accreted only from local feeding zones or after significant radial mass transport of dust and pebbles (Fig. 1). Moreover, the debate over whether there is a genetic relationship between chondrules and matrix (i.e., complementarity) has far-reaching astrophysical implications for how disks function and planetesimals form. Chondrules and matrix generally represent the main mass of chondrites, so we focus on understanding the potential genetic relationships between these two components, although we do not disregard the potential contributions of refractory inclusions to the chondrule (and matrix) forming regions.

Chemical complementarity is defined as opposing deviations of element ratios in chondrules and matrix of a chondrite with a CI bulk chondrite value of the considered element. We assume a close to CI chondrite chemical composition of the initial dust inventory of the protoplanetary disk, since CI chondrites most closely match the solar photosphere for the non-volatile elements (Lodders 2021; Barrat et al. 2012; Frank et al. 2023). Thermal processing of CI-like dust may have altered its volatile composition, but as we restrict ourselves to the moderately volatile and more refractory elements, this has no bearing on our discussion. As the reader will notice throughout the article, this definition for elemental ratios is not applicable to *isotopic* complementarity since bulk chondrites are generally isotopically not CI-like due to inherited heterogeneities from the molecular cloud and/or processing of dust in the disk. Moreover, since non-carbonaceous chondrites typically have minor fractions of matrix and non-solar bulk ratios for many elements, we only discuss carbonaceous chondrites in this article. The complementarity model assumes that chondrules and matrix initially formed from a dust reservoir with a solar (CI-like) composition and, respectively, became sub- and supersolar during chondrule formation events (Palme et al. 2015). Complementarity requires that chondrules formed in close proximity to the matrix that envelopes them and that despite very large differences in grain size the two remained coupled until accretion into their parent bodies.

The interpretation of observed complementary chondrule-matrix compositions largely hinges on whether the deviations in the average compositions of chondrules and matrix from the solar composition can be explained by exchange of materials prior to, during or after chondrule formation. We approach the debate by first defining the nature and origin of the matrix and chondrules, as well as discussing the existing and controversial elemental and isotopic datasets. One of the aims of this study is trying to understand whether these datasets are consistent with one another and to assess the pros and cons of the various approaches adopted to date. Following this, we will test these datasets in the framework of secondary alteration in the chondrite parent body and/or from terrestrial weathering. We then place

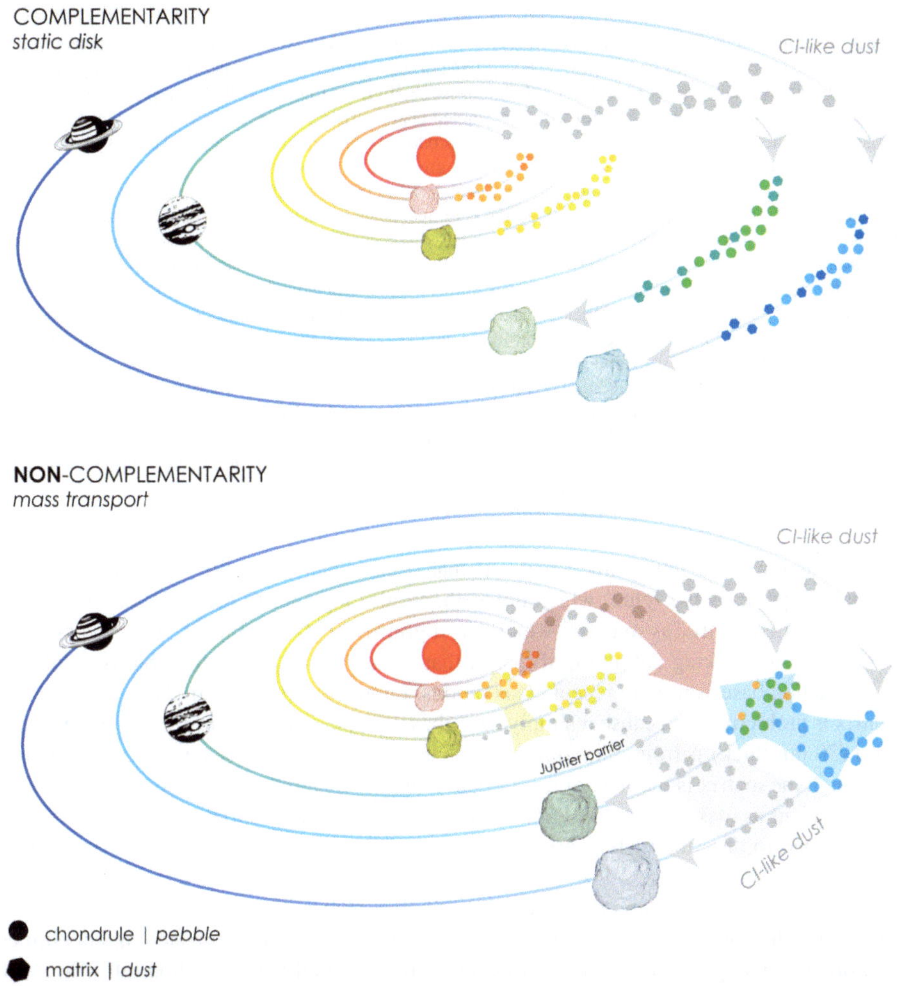

Fig. 1 A schematic representation of the early Solar System, with two opposing models of chondrule (pebble) and matrix (dust) mass transport. *A*) Matrix and chondrules form co-genetically from CI-like dust and accrete together in local reservoirs. Mixing between reservoirs is limited and matrix and chondrules are complementary to each other. *B*) Transport of chondrules by mass outflows and inward flux of CI-like dust combined with 'pebble traps' (Whipple 1972) result in non-complementary relationships between chondrules and matrix

these data in the context of astrophysical models of protoplanetary disk dynamics. Finally, we aim to construct several lines of future research that should be followed to ultimately create a paradigm for the relationship between chondrules and matrix.

2 What Is the Petrological Nature of Matrix and Chondrules?

2.1 Definition of Matrix

Any consideration of the question of compositional complementarity between chondrules and matrix needs to consider the exact definition of matrix. Matrix has traditionally been

Fig. 2 Backscattered electron SEM, STEM and TEM images showing the characteristics of matrix in different pristine carbonaceous chondrites. *A*) BSE image of well-defined fine-grained rim around a porphyritic type IAB chondrule in the ALH 77307 CO3 chondrite. *B*) High magnification BSE image of a fine-grained rim in ALH 77307 showing the extremely fine-grained characteristics of the matrix. The bright grains, the largest component in the matrix, are magnetite grains produced by secondary alteration. The rest of the matrix consists of an extremely fine-grained, heterogeneous groundmass. Coarser-grained silicate minerals even a few microns in size such as olivine and pyroxene are extremely rare. *C*) BSE image of a typical region of matrix in the QUE 99177 CR chondrite showing the heterogeneous characteristics of the matrix at a fine scale. Chondrule fragments or isolated olivines and pyroxenes are present locally (lower Z phases within the matrix) but are typical <3 microns in size. *D*) Dark-field STEM image of region of matrix in the QUE 99177 CR chondrite that contains a high abundance of nanosulfides (sulf - grains with high Z contrast) embedded in a groundmass of amorphous Fe-rich silicate material (Am). Crystalline silicate minerals are absent in this field of view. *E*) Bright-field TEM image of matrix in the unique carbonaceous chondrite, Acfer 094 showing the much higher abundance of very fine-grained crystalline silicates, mainly olivine (OL) in this case, although pyroxenes are also commonly present. The olivines are embedded in a matrix of amorphous Fe-rich silicate material (am). *F*) Dark-field STEM image of matrix in the CR chondrite MET 00426 showing the presence of common organic matter grains (OM and white arrows, low Z contrast) embedded within amorphous Fe-rich silicate. In this region of the matrix the abundance of nanosulfides (high Z grains) is much lower than in the region of matrix of QUE 99177 shown in panel D

defined as the dark, fine-grained material that occurs interstitially to chondrules and essentially represents the 'glue' which holds chondrules together to form a rock. Scott et al. (1988) defined matrix as material which has a grain size of <5 micron. This has become a widely used working definition of matrix that has stood the test of time but is worthy of re-evaluation based on more recent observations. Firstly, using the term 'dark' to define the matrix is problematic, because it simply describes its appearance in transmitted light in standard (30 μm thick) petrographic thin sections, not an inherent characteristic of the matrix. Although the matrix is the host of most organic matter in chondrites, the concentrations are not sufficient to make the matrix dark. Therefore, it is most appropriate to just use the term fine-grained to describe matrix, especially since many modern studies do not involve thin section microscopy.

The issue of grain size is also important. Studies of the most pristine chondrites show that matrix has two major types of occurrences: fine-grained rims surrounding chondrules and other coarse-grained objects, and as interstitial material between chondrites. In some chondrites a significant proportion of the matrix occurs as fine-grained rims, whereas in other chondrites, fine-grained rims are relatively rare. For example, Metzler et al. (1992) showed that in some rare CM chondrites that had escaped brecciation, essentially all the fine-grained matrix occurred as rims around chondrules, refractory inclusions, etc., a feature that they termed a primary accretionary texture. With increasing brecciation, the abundance of fine-grained rims decreased due to regolith comminution and the fine-grained component of the chondrites became dominated by a clastic matrix that consisted of chondrule fragments, mineral grains derived from chondrules, and material derived from fine-grained rims. Most CM chondrites are breccias (Metzler et al. 1992; Lentfort et al. 2021; Bischoff et al. 2017) and therefore matrix in these meteorites is a complex mixture of fine-grained matrix and material derived from chondrules and other coarse-grained objects. On the other hand, in CO chondrites such as ALH 77307, many chondrules have fine-grained rims (**Fig. 2A**), but there is also a primary interstitial matrix component that does not occur as fine-grained rims (e.g., Brearley 1993). In general, CO3 chondrites have escaped the extensive brecciation that has affected the CM chondrites. In the CR chondrites, fine-grained rims are rare and most of the matrix is interstitial to chondrules and other coarse-grained objects (e.g., Weisberg et al. 1993; Abreu and Brearley 2010; Schrader et al. 2011). Although CR chondrites are

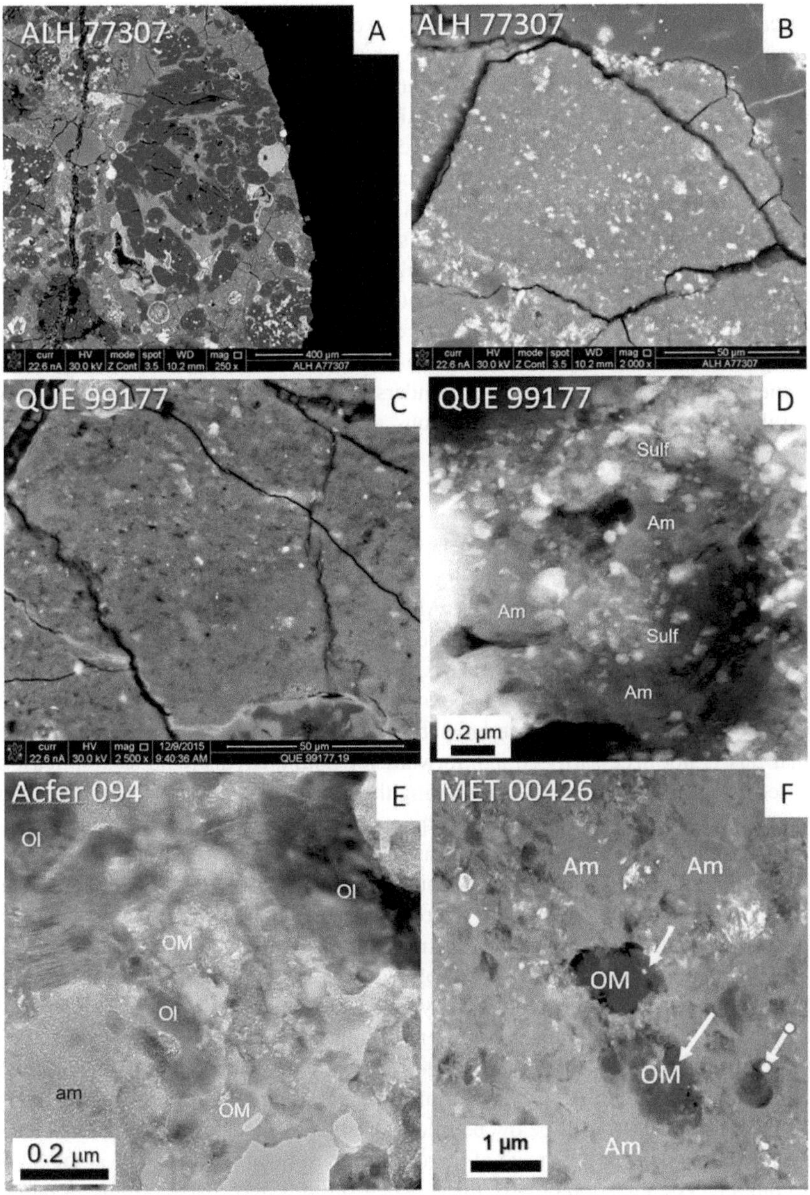

Fig. 2 (*Continued*)

clearly breccias (Bischoff et al. 1993; Weisberg et al. 1993), the extensive comminution that affected most CM chondrites has not occurred in the CRs and their matrices consist largely of extremely fine-grained material with minor amounts of chondrule fragments.

This work provides a framework for reevaluating what we define as matrix and hence the grain size (diameter) of what constitutes matrix material. It has become increasingly apparent that in the most pristine chondrites that have escaped significant brecciation, mineral grains with sizes >1 micron are rare (**Fig. 2B and C**). In addition, electron probe microanalytical (EPMA) studies of olivines and pyroxenes from the matrices of a range of type 3 ordinary and carbonaceous chondrites (i.e., grain sizes >3 microns) (Matsunami 1984; Alexander et al. 1989b; Frank et al. 2014; Patzek et al. 2021) show that essentially all these grains have major and minor element compositions that were consistent with derivation from either type I or type II chondrules. This suggests then that the coarser-grained material (i.e., greater than ~2 microns in size) that may occur embedded within matrix should not be considered to be a part of matrix but originates from chondrules that presumably underwent fragmentation in the solar nebula and were subsequently accreted with chondrules. Additional fragmentation likely occurred as a result of regolith processes during accretion. These observations lead to the conclusion that a better definition of matrix in terms of grain size, especially when considering the question of complementarity, is that it consists almost exclusively of material that is submicron in character. This material most likely represents the component of nebular dust that must be understood fully to constrain whether matrix and chondrules have complementary compositions, a task which is by no means straightforward.

2.2 The Nature of Matrix

In considering the question of whether chondrules and matrix are complementary, it is important to have a complete understanding of the petrologic, chemical, and isotopic characteristics of both types of components. Although chondrules have been studied in extensive detail and their characteristics are generally well understood in the different chondrite groups, the same cannot be said of matrix, although there have been important advances over the last 20 years. There are a number of reasons why this is the case. First, although it seems to be relatively straightforward to obtain the elemental composition of matrix, it turns out that this is not necessarily straightforward for both analytical and sample reasons, as discussed later (Sect. 3). Second, determining the mineralogical characteristics of matrix is hampered by its fine-grained nature that requires highly specialized instrumentation such as transmission electron microscopy (TEM) that is not widely available. Third, there have been significant misconceptions about matrix that arise from a lack of understanding of the effects of thermal metamorphism and aqueous alteration on the characteristics of matrix (Sect. 5). Due to its fine-grained characteristics, matrix is highly sensitive to both heating and interaction with aqueous fluids. Although petrologic type 3 chondrites are considered to be the least altered of all chondrites, within the type 3 s, it is only the very lowest type 3 s, with petrologic type <3.1, that come close to preserving the pre-accretionary characteristics of matrix. For example, Grossman and Brearley (2005) demonstrated that there are important and detectable changes in the Cr content of ferroan olivine in type II chondrules in UOCs and CO chondrites that occur below petrological grade 3.1. Matrix shows significant compositional variations over this range of petrologic types, notably a decrease in the S content due to rapid recrystallization of very fine-grained sulfides into coarser grains. A number of studies (e.g., Alexander et al. 1989a,b; Brearley 1993; Brearley and Jones 1998; Dobrică and Brearley 2020a; Krot et al. 1995, 1997) demonstrate that by petrologic type 3.1, the matrix is essentially completely altered and has lost all of its primary textural and mineralogic characteristics. The compositional changes that may accompany these mineralogical changes are incompletely understood at present (see Sect. 5).

Despite these caveats, there is a significant body of evidence that at least some chondrites do retain important information about the characteristics of matrix which likely do reflect the primary characteristics of nebular dust. Over the past three decades, a small number of chondrites in several of the carbonaceous chondrite groups and the ordinary chondrites have been recognized to show minimal degrees of thermal metamorphism. Recognizing these specific chondrites has become possible through a combination of different kinds of analyses. For instance, the Cr in olivine technique (Grossman and Brearley 2005) has provided a straightforward method for identifying very low petrologic type chondrites. In many cases, the pristine nature of these chondrites has been confirmed by subsequent studies of isotopically anomalous presolar silicate grain abundances (e.g., Nguyen and Zinner 2004; Floss and Stadermann 2009; Bose et al. 2012; Nittler et al. 2019). High abundances of presolar silicates have become a definitive signature of relatively pristine chondrites, because of these silicates' high susceptibility to destruction by thermal metamorphism and aqueous alteration (e.g., Floss and Haenecour 2016).

There are now a handful of very low petrologic type 3 chondrites that provide important insights into matrix that has not been extensively modified by thermal metamorphism. These include the anomalous, but heavily weathered, C chondrite Acfer 094 (3.00), the CO3 chondrites DOM 08006 (3.00), and ALHA 77307 (3.01), and regions of matrix in the LL3.00 chondrite Semarkona (Greshake 1997; Davidson et al. 2019a; Brearley 1993; Dobrică and Brearley 2020a). Additionally, there are now a number of meteorites that are considered to be petrologic type 2 chondrites that exhibit minimal evidence of aqueous alteration and also preserve matrices that are comparatively unaltered, such as the CM chondrites Paris (Hewins et al. 2014; Leroux et al. 2015), Asuka 12169, and Asuka 12236 (Kimura et al. 2020; Noguchi et al. 2021), as well as the CR chondrites QUE 99177, MET 00426, and GRV 021710 (Abreu and Brearley 2010; Le Guillou and Brearley 2014; Davidson et al. 2019b). It is, however, important to recognize that almost all CM chondrites are breccias and consist of clasts that have experienced variable degrees of alteration, as documented in Paris (Hewins et al. 2014). Moreover, the Antarctic finds as well as Acfer 094 have undergone weathering to varying degrees (Jull et al. 1988; Velbel 1988; Bland et al. 2005; Tyra et al. 2007).

The sample suite of minimally altered chondritic meteorites is, therefore, still very limited and few studies have used these samples to address the question of chemical complementarity (Ebel et al. 2016; Patzer et al. 2021, 2022, 2023). However, these meteorites have provided a much clearer picture of the constituents of matrix that provide a basis for evaluating if and how matrix is related to chondrules. Observations of these samples also provide a much more coherent view of matrix across different chondrite groups that was previously obscured by the fact that studies of matrix were carried out on type 3 chondrites with varying petrologic types. As a result, the pervasive view of matrix until the early 1990s was that it consisted predominantly of ferroan olivine that was the product of nebular condensation (e.g., Nagahara 1984; Palme and Fegley 1990), although a wide range of other mechanism had also been proposed (e.g., Housley and Cirlin 1983; Kojima and Tomeoka 1995; Krot et al. 1997). This conclusion has now been demonstrated to be incorrect. The ferroan olivine in matrices of type 3 carbonaceous and ordinary chondrites is principally the product of metamorphic recrystallization of amorphous material (e.g., Alexander et al. 1989a; Dobrică and Brearley 2020b).

2.3 Characteristics of Matrix in the Most Pristine Chondritic Meteorites

Although there is wide disagreement on the exact formation mechanism of chondrules, there is general agreement that chondrules were all likely formed by similar heating mechanisms in most chondrite groups (with the exception of the CB and CH chondrites; Krot et al. 2005). Until recently, this type of relatively coherent view was lacking for matrix. Indeed, a different model existed for the formation of matrix in every chondrite group (e.g., Scott et al. 1988). The lack of any consistent formation mechanism that might be relevant to matrix in all chondrite groups arose from the study of matrix in chondrites that had experienced variable degrees of thermal metamorphism, even if in some cases this was relatively low based on apparent petrologic type. But as noted above, even type 3.1 chondrites do not contain pristine matrix (Brearley and Jones 1998).

This situation has changed significantly as a result of studies of the low petrologic type 3 chondrites and minimally altered type 2 chondrites. It is now apparent that matrix in all the chondrite groups shares a remarkable number of similarities, suggesting that like chondrules, matrix is likely the product of similar processes in all the chondrite groups. This is a very important step forward in understanding matrix and the relationship of matrix to chondrules. These characteristics will be discussed below.

Our concept of matrix is slowly becoming more sophisticated. This is also extremely important for understanding the question of complementarity. Most studies have viewed matrix as some kind of homogeneous component that consists of just one kind of material. However, the recognition that matrix consists of different kinds of materials that are the result of different kinds of processes that may have occurred at different times and locations within the protoplanetary disk is slowly emerging. Alexander (2005) was among the first to point out that matrix should be considered as consisting of a mixture of different components with potentially different origins, consistent with the observation that matrix consists of a highly unequilibrated mixture of different materials (e.g., Brearley 1993). Indeed, it is reasonable to propose that matrix should actually be viewed almost as we view chondrites on a macroscopic scale, i.e., consisting of different components, such as CAIs, chondrules, Fe, Ni metal etc., but on the submicron scale. The distinct components that represent constituents of matrix are discussed below and are based on analyses carried out on both macroscopic samples of matrix and by microbeam and nanobeam techniques such as nanoSIMS and TEM.

2.4 Major Components of Matrix

Amorphous silicates Alexander et al. (1989b) first reported the presence of amorphous silicates in the matrices of UOCs. The first occurrence in a type 3 carbonaceous chondrite was described by Brearley (1993) in the CO3 chondrite ALHA 77307. Subsequent studies have demonstrated that the widespread occurrence of amorphous silicates is a characteristic of the matrices of Acfer 094 (**Fig. 2E**; Greshake 1997; Ohtaki et al. 2021), as well as the CM and CR chondrites noted above and pristine regions of the matrix of Semarkona (Dobrică and Brearley 2020a). This material is widespread and is consistently iron-rich. In some chondrites, the amorphous silicates appear to occur as distinct domains that might represent discrete, primary nebular dust grains (e.g., Yamato 791198 CM2; Chizmadia and Brearley 2008). These distinct domains are defined by the presence or absence of sulfide grains. However, in other chondrites, such as the CO and CR chondrites, these domains are not so readily apparent. Nevertheless, at the submicron scale, there are clearly significant heterogeneities (**Fig. 2F**). These amorphous silicates have been suggested to have affinities to GEMS (glass with embedded metal sulfide) that occur in IDPs (Bradley 1994; Leroux et al. 2015), although in detail, there are important differences between the two types of materials. As discussed in detail by Le Guillou and Brearley (2014), Le Guillou et al. (2014) and Ohtaki et al. (2021), the amorphous silicates in even the least altered chondrites have undergone hydration and oxidation, and consequently have been extensively modified since their formation. The absence of metal grains in these GEMS-like materials within chondrite matrices and the high Fe contents of the amorphous silicates further distinguishes them from GEMS in IDPs (Ohtaki et al. 2021).

Nanophase Fe-sulfides Another ubiquitous feature of the most pristine chondrites is the occurrence of nanophase Fe and Fe-, Ni-sulfides, principally pyrrhotite and pentlandite (**Figs. 2D and F**). These phases occur embedded with the Fe-rich amorphous silicates and can have variable distributions and abundance at the submicron scale. In some chondrites, distinct boundaries exist between sulfides embedded in GEMS-like materials and sulfide-free domains containing, for example, amorphous silicates and organics. In other chondrites, this distinction between sulfide and sulfide-rich domains is not as well defined.

Crystalline anhydrous silicates A common misconception from early studies of matrix was that it was dominated by Fe-rich olivine. This is now known to be incorrect. The abundance of crystalline anhydrous silicates is variable but generally low in pristine chondrite matrices. The highest abundance is in Acfer 094 with $\sim$30 vol.%. In comparison the CR, CO and CM chondrites contain typically $<$15 vol.%. However, more crystal-rich regions of matrix have been observed, such as in fine-grained rims in some CM chondrites (e.g., QUE 97990; Brearley 2016). The distribution and characteristics of anhydrous silicates is variable, but they are dominated by Mg-rich olivines and pyroxenes, the latter occurring in both ortho and clinoenstatite forms, and sometimes as disordered intergrowths of the two. These anhydrous silicates occur as isolated grains or aggregates, some of which have textures suggesting that they have been annealed to some degree but not to the extent that they have undergone significant grain growth. Grain sizes of individual grains and aggregates are rarely larger than a micron in size and are more typically 0.3 to 0.7 microns. Although compositional data are still relatively sparse, all chondrite groups contain matrix olivine and pyroxenes which are predominantly Mg-rich. There is only rare evidence of Fe-bearing olivines and pyroxenes. Remarkably, evidence of anhydrous minerals such as those that are constituents of CAIs is extremely rare although some evidence for microCAI-like objects has been reported in Acfer 094 (Bland et al. 2007). The apparent lack of refractory minerals such as hibonite, melilite, perovskite and spinel, for example, may be a combination of the fact that the volume of matrix that has been studied at the TEM scale is extremely low and that these minerals may be heterogeneously distributed.

Presolar grains It is now fully recognized that the fine-grained matrices of chondrites are the hosts of the presolar grain populations in chondrites, including diamonds, silicon carbide, graphite, TiN, etc., as well as presolar silicate grains (Zinner 2014; Hoppe et al. 2022; Nittler and Ciesla 2016). NanoSIMS studies of matrices now routinely identify these grains in situ in matrix in carbonaceous and ordinary chondrites using isotopic mapping techniques (e.g., Nguyen and Zinner 2004; Nguyen et al. 2007; Floss and Stadermann 2009; Bose et al. 2012; Nittler et al. 2021) The highest abundances are present in the most pristine carbonaceous chondrites (as noted above) and also in unequilibrated ordinary chondrites such as Semarkona (Barosch et al. 2022). These grains represent a distinct component of matrix, which, although present at low abundances (10 s to 100 s of ppm in the most pristine chondrites, e.g., Floss and Haenecour 2016) signifies the presence of a presolar component that escaped thermal processing within the protoplanetary disk. The abundances of the major types of presolar grains are fairly uniform amongst the most primitive chondrite matrices (e.g., Huss and Lewis 1995; Davidson et al. 2014; Nittler et al. 2021), except presolar silicates that have been largely destroyed in CIs.

Organic matter Numerous studies have shown that a wide diversity of organic matter is present in chondritic meteorites, especially the carbonaceous chondrites (e.g., Cronin et al. 1988; Alexander et al. 2017a). Two distinct types of organic matter have been identified, insoluble and soluble in acids, the former being the most abundant and representing $\sim$70% of the total organic material (Alexander et al. 2017a). Recent studies have shown convincingly that the insoluble organic matter is predominantly located within the fine-grained matrices of chondrites (e.g., Garvie and Buseck 2007; Zega et al. 2010; Le Guillou et al. 2014; Le Guillou and Brearley 2014; Changela et al. 2018) and occurs with a variety of morphologies and grain sizes ranging from the micron down to the nanometer scale. The finest grained organic matter occurs intimately intergrown with inorganic components of matrix, including amorphous material, phyllosilicates, sulfides, oxides, carbides, etc. In situ nanoSIMS location and analysis of insoluble organic matter grains in pristine chondritic meteorites has shown that a significant number of grains show isotopic anomalies for H (D/H), N ($^{15}N/^{14}N$), and C ($^{13}C/^{12}C$) (e.g., Busemann et al. 2006; De Gregorio et al. 2010; Floss et al. 2014; Alexander et al. 2017a). These isotopic anomalies for different elements show no specific correlations but suggest different formation locations and origins for different organic grains that are still poorly understood. However, the elevated D/H ratios found in organic grains in all chondrites are suggestive of presolar (e.g., Robert and Epstein 1982; Yang and Epstein 1983, 1984; Kerridge 1983; Busemann et al. 2006) or outer solar system sources (Remusat et al. 2006, 2010). The widespread survival of isotopically anomalous organic matter grains in pristine chondrites that are highly sensitive to thermal processing demonstrates that this material, like presolar grains and amorphous silicates, is an additional distinct component that has escaped the relatively high temperature required to form chondrules. The total abundances of organic material normalized to matrix abundance in all primitive chondrite matrices are roughly CI-like (Alexander et al. 2007).

Implications These mineralogical and isotopic data support arguments of Alexander (2019), who noted that matrix is not a single component but should be considered as a complex mixture of highly unequilibrated materials that formed by distinct processes in different astrophysical environments, including presolar and inner and outer protoplanetary disk environments. These materials were mixed in different proportions prior to accretion. The bulk of the silicate material in pristine chondrite matrices, namely the amorphous FeO-rich silicates (with and without embedded sulfides), conceivably has a genetic relationship to chondrules. They potentially represent a condensed evaporative residue produced by volatilization during chondrule formation (e.g., Wasson 2008). This is plausibly supported by, for example, the CR chondrites, which are dominated by extremely sulfide-poor chondrules embedded within a very sulfide-rich matrix (Abreu and Brearley 2010). Alternatively, an origin unrelated to chondrules has been proposed for these amorphous silicates with embedded sulfides through links to GEMS from IDPs (Bradley et al. 2022) or reflecting mixtures of interstellar matter and material made by heating in the disk such as by FU Orionis outbursts (Alexander et al. 2017b). Regardless of their origin, the FeO-rich silicate component was clearly mixed with an organic and presolar grain bearing component that is unrelated to chondrule formation. An important observation here, is that the bulk chondritic matrix has roughly CI-like organic and presolar grain abundances, implying that the amorphous material was included in this component and was not added later to significantly dilute the CI-like matrix.

2.5 The Petrological Complexity of Chondrules

Complementary compositions between bulk matrix and bulk chondrules have been observed that together amount to a solar bulk chondrite composition (Sect. 3). Within the complementarity model, this implies that all chondrules within a given chondrite formed co-genetically with the matrix. If this model is correct, the observed complementarity should be reconcilable with the petrological complexity and the underlying formation mechanism of chondrules. This includes the presence of mineralogically and compositionally different chondrule types (i.e., Mg-rich, FeO-rich, Al-rich) within the same chondrite and the fact that many chondrules underwent multiple accretion and melting events (Fox and Hewins 2005; Rubin 2010; Rubin and Krot 1996; Wasson 1993) as evidenced by, for example, their igneous rims (Krot and Wasson 1995; Rubin 2010, 1984; Rubin and Wasson 1987) and relict chondrule grains (e.g., Rambaldi 1981; Nagahara 1981; Jones 2012; Marrocchi et al. 2018). Here, we explore how multiple melting events change the chemical composition of chondrules and whether the petrological range of chondrules can be explained by formation in a single reservoir.

2.5.1 Multiple Melting Events

Relatively small chondrules from CO and CM chondrites typically record a seemingly singular melting event, in which the chondrule shows a homogeneous texture. Even so, FeO-poor relict grains in type II chondrules are suggested fragments from a previous generation of type I chondrules (Schrader and Davidson 2017). The presence of dusty olivine in type I chondrules shows that they contained fragments of type II chondrules in their precursors, and, hence, the chondrule represents multiple chondrule melting events. CM and CO chondrites contain relict grains in 12% and 48% of the type II chondrules, respectively (Schrader et al. 2015; Schrader and Davidson 2017; Tenner et al. 2013), suggesting that a significant portion of the chondrules is derived from a non-solar starting composition. This means that either a previous chondrule melting would have produced a chondrule with a non-solar composition (i.e., by volatile or physical loss of a component) or that material with solar composition accreted to non-solar relict grains and melted to produce a secondary chondrule. Even for completely molten chondrules without relict grains, it is possible that these chondrules also experienced multiple melting events of which only evidence for the last is preserved in the chondrule record. As such, in the complementarity model, chondrules and matrix are required to experience repeated melting events together to maintain their observed complementary compositions. To keep chondrules and fine-grained dust physically together in the turbulent protoplanetary disk, as required by complementarity, means that these melting events occurred at rapid intervals. In this case, the matrix would have to have experienced multiple thermal processing events as well.

The matter becomes more complicated for larger chondrules observed in CV and CR chondrites (**Fig. 3**). These chondrules can exhibit multiple metal and silicate rim structures, with different textures (i.e., barred/radial versus porphyritic) and temperature profiles. One of the most complex examples of these onion shell structures is shown in **Fig. 3B**, where three metal and four silicate rims can be observed in a single CR chondrule. This chondrule is suggested to have formed from the accretion of different metal and silicate dust populations that were flash heated in highly local events (Hobart et al. 2015). Alternatively, similar multilayered chondrules may instead represent enveloping compound chondrules produced by collisions and sectioned near the contact (Jacquet et al. 2021). Although not all chondrules in CV and CR chondrites show the same degree of complexity, a significant percentage of them reflect at least two melting events and, thus, two accretion events. Hewins et al. (2011) noted that although chondrules clearly experienced multiple accretion events, it has so far not been elucidated at what time intervals these events occurred. Hence, constraining the relative timing of these events (preferably in combination with nucleosynthetic isotope systematics, see Sect. 4) would significantly increase our understanding of the locality and frequency of chondrule formation and, therefore, on the likelihood of single reservoir formation of chondrules and matrix.

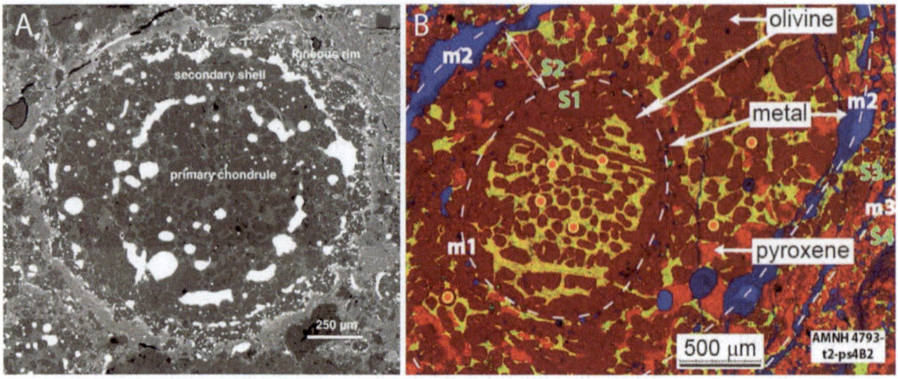

Fig. 3 Chondrules exhibiting complex structures with multiple zoning and rims, which is indicative of multiple melting events. *A*) Chondrule from the LAP 02342 CR2 chondrite with a primary and secondary shell and a surrounding igneous rim, all separated by metal rims (from Rubin 2010). *B*) Complex chondrule from CR2 chondrite Acfer 139 with four shells (s) and metal (m) rims (from Hobart et al. 2015), with Si = red, Ca = green, Fe = blue. Olivine in pyroxene-bearing layer S2 grew epitaxialy on olivine in pyroxene-free layer S1, the two separated by a thin layer M1 of numerous metal grains

2.5.2 FeO-Poor, FeO-Rich and Al-Rich Chondrules

Chondrules contain many different textures, petrologies and compositions and it is not well understood whether these variations are attributable to local or large-scale differences in the chondrule-forming environments. The observed complementarity includes a non-solar bulk matrix composition that is offset by a bulk chondrule composition, which includes FeO-poor type I chondrules, the less abundant FeO-rich type II and Al-rich (>10 wt.% Al) chondrules (Ebel et al. 2008). Within the complementarity model, this requires all chondrules to form together from the same local reservoir. Alternatively, it is, for example, possible that the subsolar matrix is complementary to a supersolar chondrule population dominated by FeO-poor chondrules (Hezel et al. 2010). If these are mixed with hypothetical solar FeO-rich chondrules, this will produce 'hybrid complementarity' (Jacquet et al. 2016; Goldberg et al. 2015). If correct, a significant portion of a chondrite ($<10\%$ of carbonaceous and $<50\%$ of ordinary) is formed independently of the complementary chondrule and matrix components. It has not been well established whether FeO-poor and FeO-rich chondrules formed co-genetically. They have been proposed to have formed together in heterogeneous formation regions with sizes that limited diffusive equilibration on chondrule formation timescales (i.e, redox conditions; Cuzzi and Alexander 2006; Libourel et al. 2023) or sequentially (i.e., relict Mg-rich olivine found in FeO-rich chondrules, Connolly and Huss 2010; Villeneuve et al. 2015; Wasson and Rubin 2003; Jacquet et al. 2015). For Al-rich chondrules, it is generally accepted that the compositions of these chondrules reflect the addition of refractory inclusions to the precursor inventory (Krot and Keil 2002). The inference of similar thermodynamic formation conditions between ferromagnesian and Al-rich chondrules has been used to conclude that these objects formed during the same or at least similar heating events as typical chondrules (Tronche et al. 2007; Zhang et al. 2020). If Al-rich chondrules formed co-genetically with ferromagnesian chondrules, this requires a nugget effect in the precursor dust from which the chondrules formed (i.e., a random distribution of Al- and Ca-rich refractory grains). In contrast, co-genetic relationships between FeO-rich and FeO-poor chondrules could be established by O and nucleosynthetic isotope studies of these objects since their compositional variations are thought to result from environmental conditions rather than nugget effects in the precursor inventory. In sum, the evidence presented so far suggests that the petrological variation of chondrule populations poses no restriction in the complementarity model.

3 Evidence from the Chemical Compositions of Chondrules and Matrix

To apply Baldy's Law, the idea that "some of it plus the rest of it equals all of it", to the chemistry of chondrites and their components requires knowledge of the bulk compositions of the chondrite groups, the relative fractions of "clasts" (mostly chondrules) and matrix among the different groups, and knowledge of the average compositions of the clasts and matrix in each group. There is a large literature on bulk chondrite compositions that is reviewed elsewhere (Lodders 2021). Recent work has refined the measurement of component abundances as fractions of total area of polished sections. Analysis of individual chondrule and bulk matrix is challenging and the source of debate. The compositions of bulk chondrules and matrix are different for most elements studied so far. Differences in element compositions are often compared as element ratios, as these are independent from absolute element concentrations that can be modified by variations in the abundances of major elements.

3.1 Bulk Major Element Compositions of Chondrites

Chondrites have bulk compositions that for some element ratios are similar to CI chondritic compositions. This is primarily the case for the more refractory elements, such as Mg/Si, Ca/Al, or Ti/Al. Other majorn elements, such as Fe, are readily disturbed by terrestrial weathering. A number of carbonaceous chondrite groups have nearly CI-like Mg/Si ratios, whereas their bulk chondrule and matrix compositions do not have the CI-like Mg/Si ratios. This constellation – chondrules and matrix with different element ratios, while the bulk chondrite has CI-like ratios – seems an unlikely outcome of mixing two reservoirs – i.e., a chondrule and a matrix reservoir – with different compositions (Fig. 1). This special case of different chondrule-matrix compositions with at the same time about bulk chondrite CI chondritic compositions has been termed 'complementarity' throughout the literature. This complementarity has so far been reported primarily for carbonaceous and R chondrites. In this section, we focus on the Mg/Si ratios of bulk chondrites, chondrules and matrix, since other elements such as Ca, Ti and Al can be affected by contributions from CAIs or by redistribution during secondary alteration (i.e., Fe, S and Ca; see Sect. 5).

3.2 Bulk Chondrite Compositions from Volatility Patterns

While the bulk chemical compositions of chondrites, by group, are quite similar, they differ from the Solar bulk composition as approximated by CI chondrites (Lodders and Fegley 1998; Braukmüller et al. 2018; Lodders 2020, 2021). Grossman (1996) argued that each chondrite group accreted from a distinct reservoir established prior to chondrule formation and prior to chondrite accretion, a statement of the single-reservoir hypothesis. As noted by multiple authors (most recently by Braukmüller et al. 2018), bulk carbonaceous chondrites show roughly monotonic decreases in elemental abundances, relative to the CI composition, with increasing volatility (i.e., their 50% condensation temperatures under canonical conditions, Lodders 2003) until at condensation temperatures of $\leq$800 K their abundances plateau. These depletion patterns had previously been interpreted as the product of a partial condensation process in the inner Solar System with a radial thermal gradient, where chondrites formed in distinct reservoirs reflected by the degree of their (moderately) volatile element depletions (Wasson and Chou 1974; Wai and Wasson 1977). Some of these models have postulated these depletion patterns to be the result of mixing of chondrules with volatile depleted compositions and matrices with CI-like compositions (e.g., Anders 1977; Alexander 2005, 2019). Seemingly ruling out this model, the bulk matrix compositions of many carbonaceous chondrites were found to be volatile-depleted, relative to CI chondrites (Bland et al. 2005). However, matrix analyses of relatively unaltered carbonaceous chondrites are much more CI-like (Hewins et al. 2014; Zanda et al. 2018; van Kooten et al. 2019), unlike their more altered counterparts (Bland et al. 2005; van Kooten et al. 2019). This suggests that the matrices of carbonaceous chondrites at the time of accretion were roughly CI-like, which would be consistent with presolar grain and organic abundances, and that they were not subject to heating and volatile loss during chondrule formation. Alteration may not be the only cause of deviations from CI. The addition of chondrule and refractory inclusion materials to the primordial matrix, as defined here, would also cause apparent fractionations in analyzed areas (Alexander 2005; Bland et al. 2007; Jacquet et al. 2016; Braukmüller et al. 2018). If the matrices of chondrites were CI-like at the time of accretion, it requires that the matrix come from some distant (not quantified) reservoir. Nevertheless, even if the bulk matrix did not experience (moderately) volatile element loss relative to CI chondrites, as we will discuss below this component undoubtedly has a sub-chondritic Mg/Si ratio that has been connected to chondrule-matrix complementarity.

3.3 Chondrule/Matrix Component Abundances

To assess whether the chemical compositions of chondrite components are complementary in each group, it is important to know the correct abundance ratios of chondrules, refractory inclusions, and matrix in each of the chondrite classes. There is relatively little original data on this topic, most based on optical point counting. In Fig. 4, we summarize the recently published results for CM (Fendrich and Ebel 2021; Patzer et al. 2023), CO and CV (Ebel et al. 2016; Patzer et al. 2021), CR (Ebel et al. 2024; Patzer et al. 2022), K (Barosch et al. 2020), and LL chondrites (Ebel et al. 2024)These data are based on either image analysis of stacked (registered) X-ray maps of large areas of thin and thick sections, or point-counting of thin sections with an electron microprobe. Results for the abundances of chondrules, matrix and refractory inclusions agree, within the uncertainties and variability within meteorites and groups, with earlier studies (e.g., reviews by Weisberg et al. 2006, based on Grossman 1988, citing McSween 1977a,b, 1979 and Huss et al. 1981). The broad literature remains in agreement on the central point, that the different chondrite classes have highly variable chondrule/matrix ratios.

3.4 The Mg/Si Ratios

In this section, we will explore the Mg and Si elemental abundances of chondrules and matrix relative to the bulk chondrite and CI chondrites. For individual data points and methods used, the reader is referred to the supplementary data Table **S1**.

3.4.1 The Matrix

Hezel and Palme (2010) and references therein observed subsolar Mg/Si ratios in the matrix of CV, CR, CO and CM chondrites. They noted that "absolute concentrations from broad

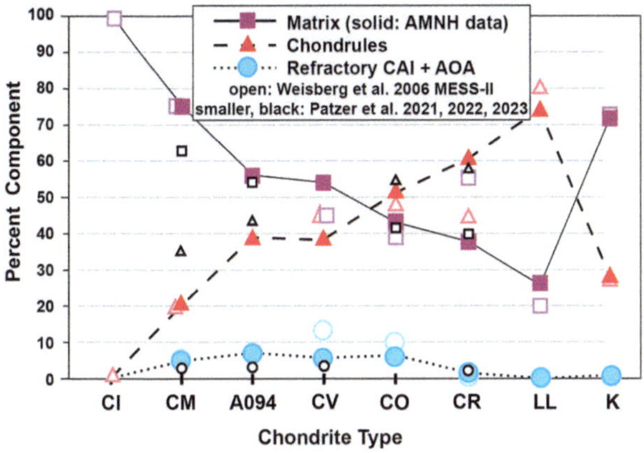

Fig. 4 The percent component of chondrules, refractory inclusions and matrix by surface area in CI, CM, Acfer 094, CV, CO, CR, LL, and K chondrites (references in text). Lines start at CI (99% matrix). Data is compared to Table 1 of Weisberg et al. (2006, large open symbols) and Patzer et al. (2021, 2022, 2023; smaller, black open symbols, with Renazzo data only, for CR). Note that the percent component of chondrules and matrix in CM chondrites from Patzer et al. (2023) are for the least altered CMs, removing much of the uncertainty associated with estimating matrix abundance the CMs. Adapted from Ebel et al. (2024, their Fig. 9)

beam matrix analyses are unreliable as matrix can be porous and contains sulfide and metal" which are calculated as FeO, but that "the Mg/Si ratios are unaffected by these analytical limitations". Similarly, later research on chondrule and matrix compositions from relatively unaltered CO, CM, CV and CR chondrites have found chondritic to subsolar Mg/Si ratios of the matrix (Hewins et al. 2014; Ebel et al. 2016, 2024; van Kooten et al. 2019; Patzer et al. 2021, 2022, 2023). The averaged matrix compositions of carbonaceous chondrite groups are very similar and all appear to be richer in Si relative to CI chondrites (**Fig. 5B**). It has been questioned whether this increase reflects a systematic analytical error or if it is a real feature of the matrix. Zanda et al. (2018) compared results of different analytical methods for measuring compositions of matrix and chondrules. They pointed out that "caution needs to be exercised when data obtained from different analytical methods are used as a basis for understanding chondritic compositions", based on a comparison of broad beam EPMA on Orgueil with earlier wet chemical results. One reason for caution is that the most commonly used technique, EPMA, assumes that the sample is physically and chemically homogeneous, which is not the case for the porous and polyphase matrix. Other techniques may also suffer from artifacts associated with the analysis of the polyphase matrix. Zanda et al. (2018) found that their Mg/Si wt.% ratio for Orgueil measured by EPMA was 10-20% lower than that measured by bulk chemical methods. Comparison with CI chondrites is an extreme case, as these are the most hydrous, volatile-rich, and most porous of all chondrites (Macke et al. 2011). Huss et al. (1981), Huss (1979) attempted to test the accuracy of broad-beam EPMA of matrix by analyzing a pressed pellet of powdered peridotite as a proxy for ordinary chondrite matrix. The normalized analysis of the pellet agreed with the published data, except for the MgO (7% too high) and the S concentration (25% too low). Another comparison was made between broad-beam EPMA (Kimura and Ikeda 1998; McSween and Richardson 1977) and wet chemical analysis (Clarke et al. 1971) for Allende matrix (**Table 1**). They noted that "weight per cent Mg/Si in matrices is constant (0.82 ± 0.05) but less than ratios derived from bulk analyses". McSween and Richardson (1977) compared the EPMA Allende matrix analyses to wet chemical analysis of a <100 μm sieved fraction by Clarke et

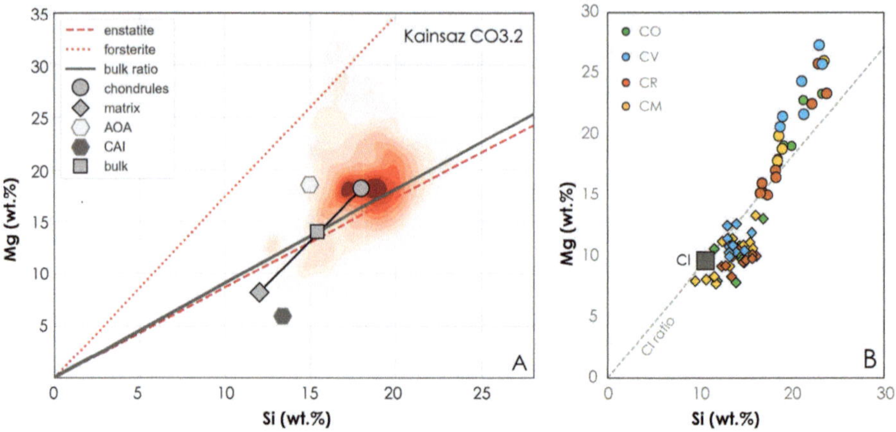

Fig. 5 Si and Mg contents of chondrules, matrix and bulk for CO, CM, CR and CV chondrites. *A*) Weighted density distribution plot for CO3.2 chondrite Kainsaz, using 3367 chondrules from Ebel et al. (2016). Isolated olivine grains in the matrix are counted as chondrules and metal grains in the matrix are counted as matrix. *B*) Compiled Si and Mg contents for averaged chondrules and matrix from individual CO, CV, CR and CM chondrites with data from Hezel and Palme (2010 and references therein), Hewins et al. (2014) and Patzer et al. (2021, 2022, 2023). See supplementary Table S1 for data and methods used. Diamonds are matrix, spheres are chondrules and squares are bulk analyses

al. (1971), reporting that "agreement between the two methods of analysis is generally very good". However, a <100 μm fraction does not meet the definition of matrix of <5 μm given by Scott et al. (1988). Moreover, Allende is a hydrothermally altered CV chondrite that has probably lost porosity relative to unaltered chondrite matrix (Krot et al. 1995). Nevertheless, at face value, the matrix Mg/Si ratio obtained by broad-beam EPMA is higher than that obtained by wet analysis (Table 1), where we would expect an opposite result if we were to explain the subsolar matrix as an analytical artifact. Hence, we conclude that carbonaceous chondrite matrix is in fact subsolar for its Mg/Si ratio. Although the data at hand does not favor an analytical artifact as a cause for subsolar Mg/Si ratios, a comprehensive comparison between broad-beam and focused spot EPMA, LA-ICP and wet chemical analyses of the various chondritic matrices (altered and unaltered) is needed to understand the potential caveats for using these techniques.

3.4.2 Chondrules

The diversity of compositions among chondrules in the carbonaceous chondrites is much larger than seen in ordinary chondrite chondrules (Lobo et al. 2014, Fig. 5A). Chondrules generally contain higher proportions of Mg and Si than matrix in all groups, but their diversity makes it difficult to obtain the statistical power to constrain their aggregate or mean composition in any chondrite.

A wide variety of EPMA techniques have been used to analyze chondrule compositions (see supplementary data Table S1). Hezel and Palme (2010) calculated bulk chondrule compositions (10 in CO, 13 in CR, 8 in CV and CM) by modal reconstruction from broad beam EPMA and modes derived from element maps. An "absolute error" correction was applied to 2D data based on Hezel and Kießwetter (2010). Such an approach may not be representative of the Mg/Si ratio in chondrules, since surficial sections may be more pyroxene-rich with Mg/Si ratios closer to unity, than equatorial sections. Ebel et al. (2016) analyzed many

Table 1 Comparison of Si, Mg and Al abundances in bulk and matrix of Allende (CV3) for three techniques. Data are presented as oxides and recalculated to atom. wt.%. Broad beam EPMA of Allende matrix yielded a higher Mg/Si ratio than the only wet chemical analysis available (Clarke et al. 1971, analysis by Jarosewich) but a lower Al/Si ratio. Data are from Clarke et al. (1971, C+ 70), Jarosewich et al. (1987, J+ 87), Stracke et al. (2012, S+ 12), McSween and Richardson (1977, MR 77), and Kimura and Ikeda (1998, KI 98)

	wet chemistry			XRF	broad beam EMPA	
	C+70 bulk	J+87 bulk	C+70 matrix	S+12 bulk	MR77 matrix	KI98 matrix
SiO_2	34.23	34.27	33.11	34.24	28.00	27.90
MgO	24.62	24.51	21.42	24.77	20.20	19.00
Al_2O_3	3.27	3.35	3.07	3.00	2.30	2.10
Si	16.00	16.02	15.48	16.00	13.09	13.04
Mg	14.85	14.78	12.92	14.94	12.18	11.46
Al	1.73	1.77	1.62	1.59	1.22	1.11
Mg/Si	0.928	0.923	**0.835**	0.934	**0.930**	**0.879**
Al/Si	0.108	0.110	0.105	0.099	0.093	0.085

more chondrules using spot EPMA analyses to measure bulk compositions across thousands of chondrules in CO and CV chondrites, and millions of matrix spots, corrected in Fig. 5A for Kainsaz (CO3.2) using calibration curves on standards (Crapster-Pregont and Ebel 2020). For CO chondrites, the Mg/Si ratios of the matrix are consistently more CI-like with increasing petrological grade, whereas the chondrules show the opposite effect (Ebel et al. 2016, their Fig. 10). Similar to the Mg/Si ratios of the matrix, the overall supersolar Mg/Si ratios of chondrules in carbonaceous chondrites are observed by various authors and through different methods (see supplementary data Table S1). It is unclear to what extent the density differences between different chondrule phases (i.e., olivine, pyroxene, mesostasis, sulfide and metal) influence the Mg and Si concentrations determined by EPMA broad-beam analyses. For example, Hezel and Palme (2010) commented that an increasing amount of metal/sulfide in the chondrule results in higher concentrations of Mg and Si. This may be reflected in the relatively higher Mg and Si concentrations of CV chondrules (Fig. 5B), which typically contain more metal/sulfide within the chondrules. However, although metal and sulfide in the chondrules affect the absolute Mg and Si concentrations, the Mg/Si ratio should remain constant. Therefore, the offset in Mg and Si from the CI-line as observed in Fig. 5, is not an analytical artifact. As commented above, the huge spread of chondrule compositions makes it that much harder to average and compare the composition of the chondrules with that of the matrix and the bulk with the aim to make a significant statement about chondrule-matrix complementarity. Where statistical power is possible, the "gold standard" of this kind of work is the reproducibility of chondrite bulk compositions from determination of the abundances (area fraction), and compositions of many refractory inclusions, chondrules, and matrix. In the next section, we will discuss these results and whether they point to a genetic relationship between chondrules and matrix.

3.4.3 Interpreting the Mg/Si Ratio of Chondrites and Their Components

As discussed above, the Mg/Si ratios of CV, CR, CM and CO chondrite matrices appear to be subsolar and this is a primary feature. However, the overall chemical compositions of these matrices are CI-like and do not exhibit significant volatile loss or gain (see Sect. 3.3). At the same time, in contrast to the matrices, there is a large spread in the Mg/Si ratios of chondrules. Chondrule formation is generally accepted to be an open system process, where interaction with the gas determines a large fraction of the chondrule chemistry (Friend et al. 2016; Barosch et al. 2019). For example, the mineralogical zonation in chondrules can be readily explained by the progressive condensation of SiO gas into the chondrule melt since SiO remains longer in the gas phase compared to other major elements (Ebel and Grossman 2000). As such, the chondrule composition reflects the shift from a nominally olivine to enstatite composition and a corresponding decrease in the Mg/Si ratio of a chondrule (Friend et al. 2016). Hence, the range of Mg/Si ratio in chondrules is likely to reflect this melt-gas interaction and appears to be a generic feature of chondrule formation. This results in bulk chondrule populations from different chondrite groups (including CR, CO, CM and CV) with supersolar to solar Mg/Si ratios between the forsterite and enstatite correlations (Fig. 5A). If the SiO gas condenses onto the matrix that is more coupled to the gas, the initially CI-like matrix becomes subsolar for Mg/Si.

Another interpretation that has been put forward to explain a subsolar Mg/Si in matrix is the loss of a component that stoichiometrically resembles forsterite (Hezel et al. 2018a; Patzer et al. 2022). If this forsteritic component were then incorporated into chondrules or their precursors prior to or during their formation, this could explain supersolar Mg/Si ratios of CV chondrules (Hezel et al. 2018a). It is unclear what kind of physical process can be invoked to produce this complementarity (see Sect. 6 and 7), especially since the depletion in the matrix is restricted to forsterite and does not involve loss of volatiles relative to the CI starting composition. Patzer et al. (2022) proposed that the matrix could have lost a forsterite component by aerodynamic sorting, given that large cluster IDPs contain large forsterite grains (adding up to roughly solar compositions) that are absent in smaller IDPs with supersolar volatile compositions. Alternatively, apparent forsterite-enrichments in chondrules (and depletions in matrix) have been interpreted as cases of 'mistaken identity', where forsterite grains in the matrix have been counted as chondrule fragments, even though they are related to the matrix fraction or represent small AOA-like fragments (Russell et al. 2010). Distinguishing chondrule-related forsterite (i.e., type I chondrule phenocrysts) from olivine grains in the matrix that were not part of a chondrule is no straightforward process, since both types can have compositional relationships to AOAs (Jacquet and Marrocchi 2017). Recent X-ray imaging of forsterite grains in the Murchison matrix has shown co-genetic relationships of these grains with chondrules (Perotti et al. 2021), in agreement with O isotope and trace element analyses of these fragments (Jacquet et al. 2021 and references therein). If we can extrapolate these results to most $\sim$5 µm forsterite grains, the 'mistaken identity' hypothesis does not hold, since removing these forsterite grains from the matrix counts will only result in a more subsolar Mg/Si ratio of the matrix. This implies that a better understanding of the origin of micron- to submicron-sized forsterite (and Mg-rich pyroxene) grains located in the matrix can lead to tighter constraints on genetic relationships between chondrules and matrix.

Alternative solutions to subsolar Mg/Si ratios for the matrix do not require a complementary relationship with chondrules. CI-like dark clasts from polymict ureilites, HED meteorites, ordinary and CR chondrites also contain solar to subsolar Mg/Si ratios (Fig. 6; Patzek et al. 2018). These clasts do not contain chondrules and some of these clasts have proposed outer disk origins (Nittler et al. 2019; Kebukawa et al. 2019; van Kooten et al. 2024), showing a continuum in chemically CI-like dust with different (nucleosynthetic) isotope compositions (van Kooten et al. 2024). As such, the changes from solar to subsolar Mg/Si ratios could reflect a change in the dust composition of the outer disk (Fig. 6).

In summary, a few key areas emerge on which we should focus our attention to correctly interpret the Mg/Si ratios of chondrules and matrix. First, although it appears that preliminary tests comparing EPMA with wet chemical analyses of matrix cannot explain its

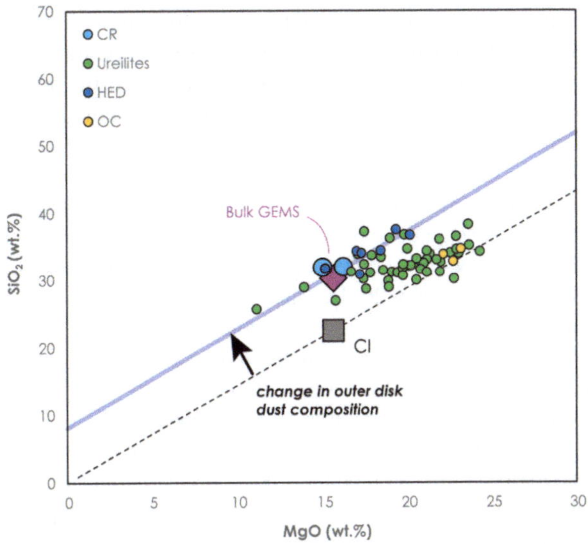

Fig. 6 The MgO versus SiO_2 contents of CI-like dark clasts from CR chondrites, polymict ureilites, HED meteorites and ordinary chondrites (data from Patzek et al. 2018). These clasts do not contain chondrules and show solar to subsolar Mg/Si ratios. Bulk GEMS compositions (Keller and Messenger 2011; Ishii 2019) show the same subsolar Mg/Si ratios (plotted using a CI chondrite MgO content) as CI-like clasts from CR chondrites, indicating a change in outer disk material (van Kooten et al. 2024)

subsolar Mg/Si ratio, further research is needed to systematically understand the effects of porosity and polyphase mineralogy that can offset EPMA from wet analyses data. Secondly, the broad definition of matrix used throughout literature (grains < 5–100 μm) can include or exclude forsterite (but also pyroxene, metal, and sulfide) grains that have a poorly defined origin. A deeper understanding of the relation of these grains to chondrules or matrix is critical to define the final Mg/Si ratios of these components. Finally, a detailed micron- to nanoscale investigation of these grains in pristine chondrites and their potential relationship to other primary matrix components can help constrain their abundance and origin and their importance in establishing subsolar Mg/Si ratios in the matrix of carbonaceous chondrites.

4 Evidence from Mass-Dependent and Mass-Independent Isotope Systematics

Mass-dependent isotope variations can help us understand the physical processes of evaporation and condensation in the chondrule forming reservoir, with a side note that these processes with respect to chondrule formation are not yet fully understood (Krot et al. 2018; Jones et al. 2018). For example, for free evaporation of moderately volatile elements (MVEs) from chondrules during high temperature melting events, an isotopically heavy chondrule residue is predicted (Ebel et al. 2018). However, for multiple isotope systems (i.e., Zn, K, Rb, Te, Ga, Cu) we observe (or infer from bulk compositions) isotopically light chondrules, relative to CI (Hellmann et al. 2020; Nie et al. 2021; Pringle et al. 2017; Pringle and Moynier 2017; van Kooten and Moynier 2019; Kato and Moynier 2017; Luck et al. 2003, 2005; Hu et al. 2023), which is usually interpreted as either physical loss of chondrule melt components (i.e., sulfide), interaction with an isotopically light gas (i.e., SiO) or kinetic isotope fractionation during condensation from an isotopically 'normal' gas (Alexander et al. 2022). Moreover, the presence of Na, K, S, and Fe in chondrules in some chondrites all point to their formation in relatively dense environments that enabled chondrule melts to equilibrate with their vapor after relatively little evaporation (Alexander et al. 2022, 2008).

In addition to mass-dependent variations, mass independent isotope variations are also observed in chondritic components. These isotope signatures represent nucleosynthetic variations that are independent of any naturally occurring mass-dependent isotope fractionation processes (i.e., evaporation, condensation, fractional crystallization, etc.). As such, variations in mass independent isotope signatures are generally thought to reflect spatio-temporal changes in the dust isotope composition of the protoplanetary disk that are the result of incomplete mixing or unmixing of dust. For example, the distinct μ^{54}Cr values of bulk non-carbonaceous and carbonaceous chondrites are generally accepted to represent accretion of their parent bodies at different times and orbital distances from the protoSun (Trinquier et al. 2009, 2007; Warren 2011). Ungrouped achondrites have been linked to chondritic counterparts by their similar Cr isotope signatures (Sanborn et al. 2019). Although having the same nucleosynthetic signatures in all elements does not require an origin from the same parent body, it does imply formation within the same isotopic reservoir (even though there are no clear spatial constraints on these reservoirs).

Extrapolating to chondritic components, we can infer a genetic link between chondrules and matrix if these components possess *identical* nucleosynthetic isotope signatures, assuming the matrix and chondrules formed from the same precursor material. This is because if chondrules formed from the same dust inventory as the matrix, processes such as evaporation and/or condensation will not erase the mass independent signature. The only way for chondrules to have changed their nucleosynthetic compositions prior to or during chondrule formation would have been by physically gaining or losing (without equilibrating with the remaining melt) a mineral phase(s) that was a carrier of a nucleosynthetic anomaly.

4.1 Mass-Dependent Isotope Signatures

Provided that there was relatively unrestricted evaporation (e.g., Rayleigh-fractionation-like behavior), mass-dependent heavy isotope enrichments of chondrules are predicted (Ebel et al. 2018). This is particularly true for (moderately) volatile elements (i.e., those with 50% condensation temperatures below ~1350 K), based on their elemental depletions, relative to CI, in chondrules from CV, CO, CM and CR chondrites (Kong and Palme 1999; Rubin and Wasson 1988, 1987; van Kooten et al. 2019). Significant elemental depletions, relative to CI, are also observed for Fe and other siderophile elements such as Ni and Co, which largely reflects physical metal/sulfide loss from the chondrules (which may have been partially retained as metal rims around chondrules). Major elements Mg, Fe and Si, which have relatively high condensation temperatures (~1300 K, Ebel 2006), have been analyzed for their mass-dependent isotope compositions in bulk aggregates of separated chondrules and individual chondrules from CV, CM and CR chondrites. For these elements, there are modest ranges of mass-dependent isotopic variations that are both positive and negative relative to the terrestrial value and CI chondrites (Fig. 7; Bouvier et al. 2013; Hezel et al. 2010, 2018b; Kadlag et al. 2021; Olsen et al. 2016; van Kooten et al. 2020). For some Al-rich chondrules, there is a clear contribution of refractory inclusions shifting the δ^{25}Mg of the chondrules to more extreme values (Bouvier et al. 2013; Olsen et al. 2016). For Mg isotopes, no correlation is observed between the elemental compositions and textures of the chondrules (barred versus porphyritic), and the Mg isotope variability (Olsen et al. 2016). The small variations in δ^{25}Mg values, but lack of correlation with Al/Mg ratios, in CV and CR chondrules have been suggested to either reflect 1) complex histories of evaporation and condensation for the chondrules (Olsen et al. 2016) and/or subsequent secondary alteration (Bouvier et al. 2013), or 2) the accretion of precursor materials with variable δ^{25}Mg values (Olsen et al. 2016). The former scenario (i.e., evaporation/condensation) is also favored to explain the

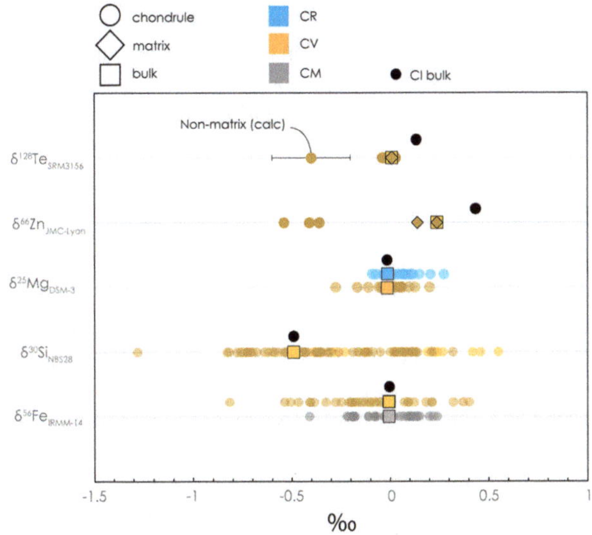

Fig. 7 Mass-dependent isotope compositions of chondrules (spheres), matrix (diamond) and bulk chondrites (square) for CR (blue), CV (yellow) and CM (gray) chondrites. Data is compiled for Fe (Hezel et al. 2010, 2018b), Mg (Olsen et al. 2016), Si (Kadlag et al. 2021), Zn (van Kooten and Moynier 2019) and Te (Hellmann et al. 2020). We also show the non-matrix fraction (i.e., chondrules and refractory inclusions) calculated by Hellmann et al. (2020). CI chondrite isotope compositions are δ^{30} $Si_{CI} = -0.48 \pm 0.01$‰ (Onyett et al. 2023), $\delta^{25}Mg_{CI} = 0.01 \pm 0.04$ (Larsen et al. 2011; Bizzarro et al. 2011), $\delta^{56}Fe_{CI} = 0.06 \pm 0.03$‰ (Schiller et al. 2020), and plot at the same locations as CM, CV and CR bulk chondrites. $\delta^{128}Te_{CI} = 0.15 \pm 0.04$‰ (Hellmann et al. 2020) and $\delta^{66}Zn_{CI} = 0.48 \pm 0.06$‰ (Luck et al. 2005; Pringle et al. 2017) are typically enriched relative to bulk chondrites

Fe isotope data of bulk CV and CM chondrules (Hezel et al. 2010, 2018b). Here, the added complexity is that besides possibly experiencing some evaporative Fe loss, chondrules also experienced metal/sulfide loss, which can potentially create opposing directions in isotopic fractionation. Silicon isotope variations in CV chondrules have also been interpreted as the result of an evaporation process, complicated by the interaction of chondrules with an isotopically light gas (Kadlag et al. 2021). Collectively, these results point towards a generic chondrule formation process, where the bulk reservoir of all chondrites or the individual reservoirs of each chondrite group, had a similar isotope composition of these elements to CI chondrites, which was then subjected to the chondrule forming process described above, creating negative and positive isotope fractionations relative to the bulk value.

Given the depletion of MVE abundances in CC chondrules (and bulk chondrites), relative to CI, mass-dependent isotope fractionation of MVEs such as S, Zn, Cu, Rb, K and Te are expected to be large if they were lost via free evaporation (up to 10 s of permil, depending on the degree of depletion). The potential levels of mass fractionation are large enough that, if present, they should be seen in the isotopic compositions of bulk chondrites. This is the case even if the more volatile-rich matrix dilutes the signal, unless the matrix is the complement to the chondrules. In fact, no systematic heavy isotope enrichments that correlate with MVE elemental depletions have been reported in bulk chondrites. On the contrary, for many, but not all MVEs, their bulk isotopic compositions are light isotope enriched, relative to CI. The lack of such large heavy isotope deviations in the bulk chondrite compositions, as well as in some cases the observed or inferred light MVE isotope signatures of chondrules, have been proposed to reflect 1) equilibrium between chondrules and surrounding gas near peak formation temperatures to suppress heavy isotope fractionation, and 2) kinetic light isotope fractionation due to incomplete recondensation during cooling, and/or loss of MVE-bearing metals and sulfides (Pringle et al. 2017; van Kooten and Moynier 2019).

Many of the MVEs are potentially mobile during hydrothermal alteration and thermal metamorphism. Thus, measurements in situ or of physically separated components must be interpreted with caution (see Sect. 5). For instance, the K isotope systematics of chondrules, matrix and CAIs in Allende (CV3.6) can be mainly attributed to secondary processes (Jiang et al. 2021). No systematic heavy K isotope enrichments that correlated with K depletions were found in Semarkona (LL3.00) chondrules after selection of samples with the least petrologic evidence for K exchange between chondrules and matrix (Alexander and Grossman 2005). However, a separate study of Semarkona chondrules found that all the measured chondrule glasses had exchanged O isotopes with the matrix (Kita et al. 2010). That chondrules in a type 3.00 have extensively exchanged O isotopes with the matrix should give one pause when studying samples of higher petrologic type. Sulfur appears to have been relatively immobile at least in the matrix of Semarkona (Grossman and Brearley 2005). Tachibana and Huss (2005) measured S isotopes in Semarkona chondrules and found no evidence for significant isotopic mass fractionation, although alteration of metal/sulfide grains in many chondrules suggests that they may have been open to S exchange with the matrix too. Similar observations were made for Zn isotopes in chondrules from Allende (CV3.6) and the less altered but shock heated Leoville (CV3.1). Zinc isotope systematics from Allende are more variable and show higher δ^{66}Zn values that are closer to the bulk CV (Pringle et al. 2017) than in Leoville, which contains chondrules with consistently light Zn isotope signatures (van Kooten et al. 2021; van Kooten and Moynier 2019). The Zn isotope compositions from Leoville chondrules are interpreted as primary signatures (or at least as upper limits) and agree with Te isotope signatures calculated for the non-matrix fraction of CV chondrites (Fig. 7, Hellmann et al. 2020). The Zn isotope composition of the Leoville matrix fraction is indistinguishable from the bulk value (van Kooten and Moynier 2019).

A different approach for inferring chondrule and matrix compositions that is less sensitive to internal parent body redistribution of elements is to correlate bulk chondrite mass-dependent isotope compositions with estimates of their chondrule and matrix fractions. For Zn, Cu, K, Rb, Te, Sn and Ga, the bulk carbonaceous chondrite isotopic compositions are best explained by two-component mixing of a chondrule endmember consisting of isotopically light Zn, Cu, K, Rb, Te, Sn and Ga, and a CI-like matrix endmember (Luck et al. 2003, 2005; Nie et al. 2021; Hellmann et al. 2020; Hu et al. 2023; Koefoed et al. 2023) (Fig. 8). This is consistent with combined chemical, Zn and Cr isotope analyses of CV chondrite components (Fig. 9, see Sect. 4.3; van Kooten et al. 2021; van Kooten and Moynier 2019).

In summary, the MVE isotope systematics of carbonaceous chondrites are best explained by chondrule formation and subsequent combination with unmodified CI-like dust. The term 'unmodified' applies to the very limited amount of thermal processing this dust had to experience to maintain a CI-like MVE isotope composition. From there the question arises: is complementarity then just a matter of scale? How far did the matrix component in a chondrite have to have been from the chondrule reservoir to have preserved its MVE inventory?

4.2 Nucleosynthetic Isotope Variations

Mass independent variations are thought to reflect spatial and/or temporal variations in the isotopic composition of dust in the disk. Other than the gross distinction between the inner and outer Solar System, the scales at which these variations occurred have been the subject of much speculation but are still largely unknown. Nevertheless, mass independent

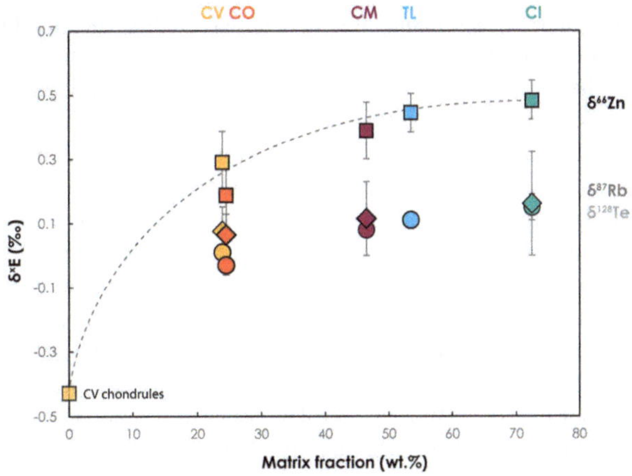

Fig. 8 The mass-dependent isotope ratios in delta notation of Zn (Luck et al. 2005; Pringle et al. 2017), Rb (Nie et al. 2021) and Te (Hellmann et al. 2020) from chondrites with different matrix fractions in weight percent (from Alexander 2019). Errors represent 2SD of the mean value. We also plot the average CV chondrule composition (van Kooten and Moynier 2019). Note that a mixing line between the matrix and chondrule endmember is hyperbolic, since chondrules have a much lower Zn concentration (<5 ppm) than the matrix (∼300 ppm)

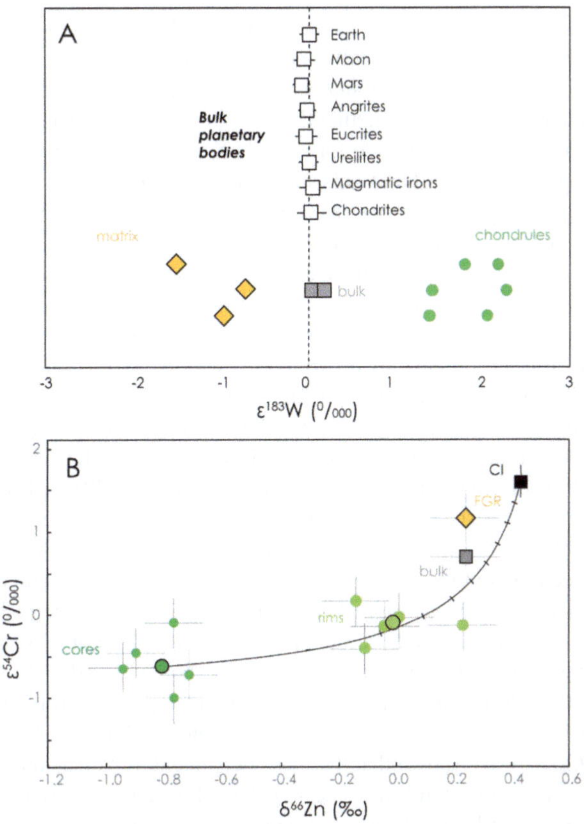

Fig. 9 *A)* The ε^{183}W values of bulk planetary bodies (ε^{183}W = 0) and the bulk, chondrule and matrix isotope compositions of Allende (from Budde et al. 2016), *B)* The ε^{54}Cr versus δ^{66}Zn isotope composition of CV chondrule cores ($n = 5$ chondrules, with large dark green sphere as average), their corresponding igneous rims, the bulk CV value and a fine-grained dust rim (FGR) surrounding one of the chondrules (modified after van Kooten et al. 2021). The bulk CV chondrite composition can be explained by mixing between inner Solar System derived chondrules that underwent progressive accretion of CI-like dust. The black mixing line shows 10% increments between mixing the chondrule cores and CI-like dust. The bulk CV plots at approximately 40% matrix, in agreement with estimates from Alexander (2019)

isotopic variations can be used to infer mass transport between reservoirs. Mass independent isotope variations of bulk Solar System materials are typically small and require high precision analyses that necessitate the use of relatively large samples. The individual analysis of chondrules and matrix is size-limited and, as a result, has been mainly carried out on CV chondrites because their chondrules are large compared to other chondrite groups and CVs are relatively abundant. Hence, in understanding the direction and magnitude of mass transport of chondritic components in the disk through their nucleosynthetic isotope signatures, we have so far been restricted to a very small sampling of the Solar System that may well be overprinted by secondary processes (see Sect. 5).

The nucleosynthetic W isotope systematics of Allende CV chondrules and matrix have so far been the greatest isotopic evidence for complementarity (Fig. 9A). While bulk Solar System materials, including CV chondrites, exhibit very limited variation in μ^{183}W, the chondrule and matrix fractions of Allende are enriched and depleted relative to the bulk, respectively (Budde et al. 2016). As discussed above, a requirement of nucleosynthetic complementarity should be that matrix and chondrules have *identical* μ^{183}W signatures, since they form from the same presolar dust reservoir. However, Budde et al. (2016) have argued that 1) within the same reservoir, size sorting can occur that can preferentially enrich the chondrules in ^{183}W-rich presolar grains (i.e., metal, sulfide or SiC), or 2) chondrules do not completely melt and then expel the presolar grains somehow. It is unclear how these models of chondrule formation would work in nature since the size sorting would require a larger scale over which complementarity operated to the extent that we must wonder if it can still be considered as such. For model 2, it begs the question why and in which relative abundance (to isotopically 'normal' grains) these relict presolar grains were expelled. Notably, CR chondrites contain significantly larger bulk μ^{183}W values (Budde et al. 2018). This raises the question whether this isotope system emulates complementarity, or if it advocates for a general chondrule formation process, where the μ^{183}W signature of all chondrules is supersolar and all matrix is subsolar and together forms a random isotopic mixture. An alternative explanation that has been put forward to explain the mirroring W isotope signatures is through secondary alteration (see Sect. 5).

As the key tracers of (nucleosynthetic) isotope variability in the Solar System, Cr, Ti, and O isotopes are frequently used in combination to pinpoint mainly spatial differences in the accretion of (a)chondrite parent bodies (Trinquier et al. 2009; Warren 2011). Since O and Ti isotopes each have their own complications with regards to the interpretation of the nature of chondrule and matrix precursor materials (i.e., secondary alteration, O precursors and gas interaction, refractory inclusions), we will continue discussing primarily the nucleosynthetic Cr isotope variations. Chromium isotope analyses have been carried out on individual chondrules from CV and CR chondrites (Olsen et al. 2016). The observed range of μ^{54}Cr values is very restricted and positive for CR chondrules, whereas CV chondrules are highly variable and essentially cover the entire observed range of the Solar System materials. This has been interpreted as formation of CR chondrules in the outer Solar System (van Kooten et al. 2016) and formation of CV chondrules throughout the disk and subsequent mass transport to their final accretion region (Olsen et al. 2016) or as within a highly heterogeneous dust reservoir at their accretion location (Gerber et al. 2017; Williams et al. 2020; Schneider et al. 2020). Some of the nucleosynthetic Cr isotope variability in CV chondrules has also been attributed to AOA addition to the chondrules (Schneider et al. 2020). However, Hellmann et al. (2020) have shown that the mass-dependent Te versus nucleosynthetic Cr isotope composition of bulk chondrites is best explained by a two-component mixing model between CI-like matrix and ^{54}Cr-depleted chondrules. Recent analyses of individual CV chondrules from Leoville and their surrounding igneous and fine-grained accretion rims have shown the importance of Cr mobility during thermal metamorphism (see Sect. 5; van Kooten et al. 2021). These analyses show that relatively unaltered CV chondrules display negative μ^{54}Cr values, like ordinary chondrite chondrules, but become progressively enriched in ^{54}Cr towards their outer rims (Fig. 9B). Note that Leoville chondrules, although significantly less altered than, for example, Allende, could still have experienced some Cr redistribution and these values should be considered as upper limits. As has been verified by Nie et al. (2021) through MVE isotope analyses (Fig. 8), this points to a two-component mixing between ^{54}Cr-rich CI-like dust and ^{54}Cr-poor chondrules. This has been interpreted as requiring significant mass transport of either chondrules to the outer disk or CI-like dust to the inner disk (Schiller et al. 2020, 2018; van Kooten et al. 2021) and does not support the complementarity model. Note that the Cr isotope matrix analyses only represent the fine-grained matrix surrounding the chondrules and not the coarser interchondrule matrix. The latter has a composition similar to the bulk CV chondrite and it is suggested to reflect matrix plus chondrule and refractory fragments that were processed on the parent body (van Kooten et al. 2021).

Collectively, with chondrule formation being a complex and poorly understood process, it is not straightforward to interpret the mass-dependent and nucleosynthetic isotope signatures of chondrules in the framework of complementarity. Working towards a solid understanding of isotopic complementarity requires 1) using unaltered samples, 2) chemical data of chondrules that allows us to assess the degree of elemental/phase loss from the chondrule relative to CI chondrites, 3) correlated mass-dependent and mass-bias corrected isotope data of chondrules and matrix, 4) isotopic analyses of chondrite groups other than CVs.

4.3 Radiometric Isotope Anomalies

The decay of long- and short-lived radionuclides has been used in cosmochemistry to obtain absolute (i.e., U-corrected Pb-Pb ages) and relative (i.e., ^{26}Al-^{26}Mg, ^{53}Mn-^{53}Cr, ^{182}Hf-^{182}W etc.) ages of chondritic components. For a review about the caveats using each of these dating systems, the reader is referred to other works (Connelly et al. 2017; McSween and Huss 2022). Here, we briefly discuss the Pb-Pb and Al-Mg ages obtained for chondrules so far, assuming that these ages are accurate, and their implications for the chondrule-matrix complementarity debate.

From the perspective of U-corrected Pb-Pb ages obtained for individual chondrules from CV and CR chondrites, chondrules formed in a timeframe encompassing the entire disk lifetime of approximately 4 Myr (Bollard et al. 2019, 2017; Connelly et al. 2012). For CR chondrites, an offset between the U-corrected Pb-Pb ages of chondrules and the modeled accretion age of the CR parent body is observed. Neglecting the poor statistics of individual chondrule Pb-Pb ages for a moment, the mean CR chondrule formation age lies around 2.5 Myr after CAI formation (Bollard et al. 2019), whereas the accretion age is proposed to be around 4 Myr (Doyle et al. 2015; Schrader et al. 2015; Budde et al. 2018). If we assume that both ages are correct, this implies that CR chondrules were stored in the protoplanetary disk for >1 Myr, a timeframe that would have allowed a significant mass flux of fresh, 'anomalous' dust and pebbles to the CR accretion region. As such, the Pb-Pb ages of CR chondrules do not favor a complementarity model. For the CV chondrites, very few individually Pb-Pb dated chondrules exist (Bollard et al. 2019), and all have ages that are within 1 Myr after CAI formation. This agrees with pooled ages of Allende chondrules (4566.6 ± 1.0 Ma; Amelin and Krot 2007). The accretion age of CV chondrites is estimated to have been between 2-3 Myr after CAI formation (Doyle et al. 2015), suggesting again a gap between chondrule formation and accretion that allows for significant mass transport to the CV chondrite accretion region during that time.

In contrast to their Pb-Pb ages, ^{26}Al-^{26}Mg ages of chondrules from the most primitive chondrites exhibit peak distributions close or coinciding with their accretion ages. For example, the observed peak of CC chondrules lies between 2-2.5 Myr after CAI formation (Pape et al. 2019; Villeneuve et al. 2009) From this perspective, storage of chondrules before accretion is not required and complementarity is more likely to be achieved in these circumstances. The reported Al-Mg ages of CR chondrules (3.5-4.0 Myr after CAI formation; Devin et al. 2017) also show overlap with their accretion ages. However, Tenner et al. (2019) have reported that FeO-rich and FeO-poor CR chondrules formed at different time intervals, with a >1 Myr gap, again allowing for the possibility of mass transport during that time. Hence, resolving the age debate regarding the validity of either Pb-Pb or Al-Mg ages would go a long way in tightening the constraints around the complementarity debate.

5 Parent Body Alteration: The Possibility of Apparent Complementarity

A major issue in understanding early disk evolution including dust processing, chondrule formation and mass transport, is separating these primary processes from subsequent secondary alteration in chondrite parent bodies. Establishing potential genetic links between chondrules and matrix requires the analyses of relatively unaltered samples and insights into the earliest stages of alteration, including the mobility of elements under various alteration conditions as well as their sources and sinks within chondrites. Studies investigating complementarity have so far focused dominantly on relatively altered chondrites, such as Allende (CV3.6 or higher), Murchison (CM2.5), Jbilet Winselwan (a shock heated CM2.1-2.4) and Renazzo (CR2.5). These chondrites have already experienced mild or more extensive thermal metamorphism and/or aqueous alteration, but it is not well understood how this affects each element that is the subject of a complementarity study. For example, some elements such as Cr, S, Fe, Ca, K, and Na are mobilized very early on, whereas other more refractory elements such as Al and Ti are typically less mobile. An assumption usually made about secondary alteration is that while the matrix is typically affected first (McSween 1979; Bunch and Chang 1980; Browning et al. 1996), it behaves as a closed system until relatively late in the alteration process when the matrix starts to interact with chondrules and other chondritic components at a local scale (Stracke et al. 2012). Hence, up to a certain stage (i.e., type 3 and 2 chondrites), even though the mineralogy of the matrix is completely altered (Fig. 10), it is assumed we can use these chondrites to infer genetic relationships between *bulk* chondrules and matrix. Likewise, nucleosynthetic isotope compositions of chondrules and matrix are often assumed to reflect primary precursor dust signatures irrespective of the degree of secondary alteration. While this may be true for bulk compositions of chondrites, secondary alteration can result in mass-dependent isotope fractionation according to equilibrium laws or diffusion-related non-equilibrium isotope fractionation, processes that are not typically corrected for when reporting nucleosynthetic isotope data. In the following section, we discuss secondary alteration according to different element groups, the scale of alteration and the effect on the observed complementarity.

5.1 Siderophile Elements (Fe, W, Mo)

Metal and sulfides are the main carriers of siderophile elements in chondrites. Therefore, the assessment of chemical complementarity between chondrules and matrix within this group of elements must consider the redistribution of these mineral phases during secondary alteration (and potentially during weathering). In the case of nucleosynthetic isotope complementarity, this assessment becomes more complex since small non-metal/sulfide phases (i.e., SiC, spinel and other presolar grains) may exist that together carry the anomalous components and behave differently from the metal and sulfide during secondary alteration.

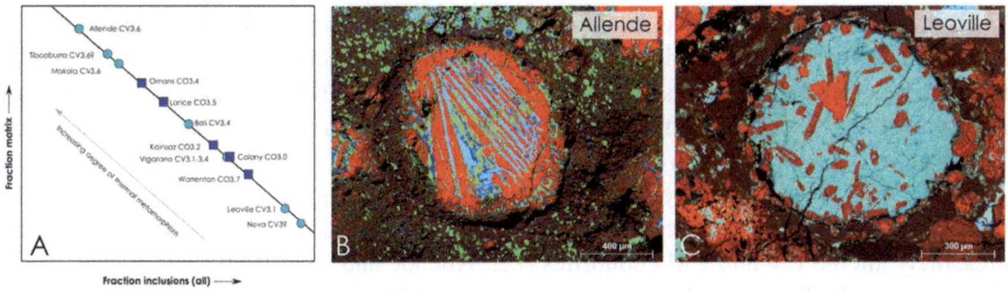

Fig. 10 Textural and compositional changes during progressive thermal metamorphism. *A*) after Fig. 2 from (Ebel et al. 2016). In general, the fraction of the CV matrix increases with increasing petrological degree. Chondrules are eaten up and reprocessed into coarse-grained intra-chondrule matrices. CO chondrite fractions may be subject to a size sorting effect of increasing chondrule diameter with depth on the parent body and, hence, the petrological degree (Pinto et al. 2021). Similarly, the trend for CV chondrites has been suggested to reflect the pre-accretionary history of oxidized and reduced subgroups (Gattacceca et al. 2020). *B*) MgCaAl (RGB) elemental map of Allende CV3.6 showing coarsening of the matrix, including Ca-rich pyroxene (green). The chondrule rims are being replaced with matrix-like material. *C*) MgCaAl map of Leoville CV3.1 showing an Al-rich chondrule in an apparently pristine state, including an outer rim of fine-grained matrix and coarser intra-chondrule matrix containing chondrule and CAI-like fragments

As discussed above, siderophile elements and their isotopes are potential tools for testing complementarity (i.e., Fe contents, Hf-W contents, W and Mo isotope systematics). Since metal is one of the first phases to be altered during progressive oxidation by either fluid-assisted metamorphism or aqueous alteration, it is reasonable to assume that most, if not all, chondrites investigated for complementarity are affected by this process. If metal oxidation were to occur within a closed system represented by the matrix or the chondrules, this effect would be negligible. However, if siderophile elements were to transfer from chondrules to matrix and/or vice versa, this would pose a considerable problem to assess the degree of complementarity.

A complication in using siderophile elements is that metal grains in chondrules and their rims, as well as isolated larger grains in the matrix, are generally thought to be genetically related (Jacquet et al. 2013; van Kooten et al. 2022), where the larger matrix metal grains are suggested to be expelled from chondrules during their formation and reaccreted with the matrix. Hence, complementarity of siderophile elements is proposed to be achieved by physical redistribution of metal from chondrules to matrix when both components are present in the same reservoir (Hezel et al. 2018a). Alternatively, isolated metal grains that were roughly half the size of co-genetic chondrules (Skinner and Leenhouts 1993) were transported along with chondrules to their final accretion region by aerodynamic sorting (Kuebler et al. 1999; Liffman 2005), which does not require a genetic link between chondrules and matrix (van Kooten et al. 2022). In that sense, these larger metal grains represent metal-rich chondrules, in agreement with observations of igneous rims around the 'isolated' metal grains in the CM chondrite Paris (van Kooten et al. 2022). In contrast, the elemental composition of small metal grains (<50 μm) observed in the CM chondrite Maribo are significantly different from the larger metal grains, requiring an alternate origin (van Kooten et al. 2022). Furthermore, TEM images of Maribo matrix show abundant submicron-sized metal grains are present in the matrix, which all could be isotopically and chemically different from the larger chondrule-like metal grains (van Kooten et al. 2018). Hence, the composition of the unaltered fine-grained matrix may be considerably changed when chondrule metal grains start leaching siderophile elements into the matrix. These elements can be sourced from the isolated metal or from so-called armored chondrules that have metal rims connecting the chondrule with the matrix (van Kooten et al. 2019).

From a chemical perspective, oxidation and leaching of components in chondrule-derived metal and transport to the matrix would directly result in an increase of siderophile elements to the matrix, producing supersolar Fe/Mg and W/Hf ratios. In Mg versus Fe space, for relatively unaltered CV and CM matrices (i.e., Leoville and Maribo; Fig. 1 in van Kooten et al. 2019), an increase in Fe relative to CI chondrites is observed, but no change in Mg content. This increase in Fe is suggested to be related to the first stage alteration of the matrix by oxidation of chondrule rim metal grains (i.e., armored chondrules) to the matrix. With increasingly altered CV and CM chondrites (i.e., Allende and Cold Bokkeveld, Fig. 1 in van Kooten et al. 2019), a correlated decrease in Fe and Mg in matrix is observed that is the result of exchange between chondrules and matrix (Hanowski and Brearley 2001; Krot et al. 1995; Patzer et al. 2021, 2022, 2023). Hence, the unaltered matrix of CV and CM chondrites is suggested to have a CI-like Fe/Mg ratio. We note that this process has not been verified for other siderophiles, but caution should be taken when investigating complementarity with these elements. From an isotope perspective, it is unclear what the carrier phases of various W and Mo isotopes are and how they would behave with respect to each other during secondary alteration. It has been suggested that secondary chondrule-matrix exchange either involving selective dissolution of presolar carriers of s-process-depleted Mo and W in the matrix (Sanders and Scott 2022), or oxidation of chondrule metal (Alexander 2019) has resulted in the apparent isotope complementarity between chondrules and matrix in Allende. In CR chondrites (which notably have a distinct bulk μ^{183}W from other Solar System materials), metal is found to be the most ^{183}W-depleted phase (Budde et al. 2018). As such, the chondrite matrix would become increasingly ^{183}W-depleted, relative to its starting composition, during secondary alteration. Finally, as evidenced by μ^{182}W isochrons of Allende and Vigarano (a less altered CV chondrite), the matrix from Vigarano shows the same μ^{182}W value as the bulk chondrite (Becker et al. 2015), whereas Allende has a subsolar μ^{182}W for the matrix, suggesting that secondary alteration has caused the shift in μ^{182}W.

5.2 Mg and Si

The subersolar Mg/Si ratio of carbonaceous chondrite matrices relative to the supersolar Mg/Si ratios of (mainly CV) chondrules is often used as a pillar of complementarity (see Sect. 3). This assumes that the alteration of the matrix, which is observed to at least some degree even in the least altered chondrites, occurred in a closed system. However, loss of Si and Mg during progressive alteration of the matrix (and at later stage also from the chondrules; McSween 1979; Bunch and Chang 1980; Browning et al. 1996) is predicted for reactions that behave isovolumetrically, meaning that the volumes of the reaction products (e.g., serpentine) are equal to that of the reactants (olivine or pyroxene). Observations of pseudomorphic crystals of phyllosilicates replacing olivine and pyroxene in the matrix of CM chondrites are evidence of isovolumetric reactions. Therefore, the alteration from an anhydrous matrix to a hydrated serpentine-rich one is thought to proceed as follows (Velbel 2014):

$$5Mg_2SiO_4 + 8H^+ + 2H_2O \rightarrow 2Mg_3Si_2O_5(OH)_4 + 4Mg^{2+} + H_4SiO_4(aq)$$

In this reaction, the resulting serpentine has a Mg:Si ratio of 3:2, whereas olivine has a Mg:Si ratio of 2:1. Hence, during isovolumetric serpentinization in an open system, the matrix is predicted to become enriched in Si over Mg. At later stages of aqueous alteration, the phyllosilicates become increasingly enriched in Mg over Fe, a process that is suggested to be facilitated by the alteration of olivine phenocrysts in chondrules. Hence, for a relatively unaltered chondrite matrix, we would expect to see subsolar Mg/Si ratios, whereas more altered matrices would exhibit Mg/Si ratios that approximate the bulk chondrite. This agrees with observations from CV, CM (Fig. S7 in van Kooten et al. 2019) and CO chondrites (Ebel et al. 2016). In addition, recent data from relatively unaltered CO, CM and CR chondrite matrices show subsolar Mg/Si ratios (Patzer et al. 2023, 2022, 2021), although this has been attributed to ∼10% forsterite loss prior to accretion by the authors. Nevertheless, such serpentinization reactions represent endmember approximations in complex systems of water-rock interactions of which the scale is unknown. Hence, serpentinization of the matrix has the potential to change its Mg/Si ratio but the magnitude and direction of this change is dependent on the environmental parameters of alteration.

5.3 Refractory Lithophiles (Ca, Sr, Ti and Al)

The abundances of refractory lithophile elements, such as Ca, Sr, Al, and Ti, are high in CAIs. Refractory inclusions in carbonaceous chondrites are not rare (∼0.1–3 vol.%; Krot 2019) and, thus, can be important sources of these elements to supply the matrix during aqueous alteration and fluid-assisted thermal metamorphism. For example, observations from Allende CAIs show that Ca, Sr and, to a lesser extent, Al were removed from the inclusions and presumably gained by the surrounding matrix (Krot et al. 2021). Calcium, for instance, was redistributed into carbonates and Ca-pyroxenes in the CV matrix (Murakami and Ikeda 1994). Importantly, this textural alteration of the matrix also poses problems in identifying and isolating the primary fine-grained matrix component. In addition to the mobilization of refractory lithophiles in CAIs, significant addition of more volatile elements such as Fe, Si, Na, K and Cl is reported. Titanium, like Al, is relatively immobile during progressive thermal metamorphism. While detailed petrological descriptions have been made of secondary alteration and mineral replacement in CAIs from carbonaceous (Hashimoto and Grossman 1987; Hutcheon and Newton 1981; Keller and Buseck 1991; Krot et al. 2021, 1995) and even non-carbonaceous chondrite groups (Dunham et al. 2023), little is known about the bulk CAI elemental flux. In the framework of complementarity, the latter is important to constrain the role of CAIs as source and sink during secondary alteration. Finally, chondrule mesostasis can also be an important carrier of some refractory lithophiles (e.g., Al-rich chondrules) and is readily altered. Leaching of Al and Ca from chondrule mesostasis has been observed in even relatively unaltered chondrules (Hanowski and Brearley 2001; Brearley 2006; Brearley and Krot 2013; Patzer et al. 2023). Collectively, this implies that a significant fraction of the variability observed in the Ca and Al contents of altered chondrules and matrix (i.e., Allende and Y-86751; Hezel and Palme 2007, 2008) can be ascribed to redistribution (i.e., Ca precipitation in the matrix) and coarsening (i.e., analyses including Al-rich refractory fragments in the matrix). This agrees with Ca and Al chondrule and matrix analyses from the Efremovka CV3.1 chondrite, showing distributions around the solar Ca/Al ratio for both components (Hezel and Palme 2008).

5.4 Moderately Volatile Elements (Cr, Zn)

Nucleosynthetic Cr isotope systematics are a reliable tracer of genetic heritage in bulk Solar System materials (Trinquier et al. 2009, 2007; Warren 2011). This tracer has also been measured inbulk chondrule and matrix fractions of Allende (Kadlag et al. 2019), in individual chondrules of CV, CK, CM, CR and CH/CB chondrites (Olsen et al. 2016; van Kooten et al. 2021, 2020, 2016; Zhu et al. 2019; Williams et al. 2020; Schneider et al. 2020) and on individual matrix areas in CV chondrites (van Kooten et al. 2021). It has been known for some time, however, that Cr is easily mobilized at the onset of thermal metamorphism (Grossman and Brearley 2005) and the Cr_2O_3 contents of fayalitic olivine in type II chondrules can be used to compare the relative degree of alteration between chondrites of a given parent body (Davidson et al. 2019b). Hence, only the most unaltered chondrite samples (petrological type <3.1) should be used for the analyses of individual chondritic components. Indeed, van Kooten et al. (2021) have demonstrated that nucleosynthetic Cr isotope signatures of chondrules homogenize with the matrix towards the bulk CV composition during progressive thermal metamorphism. Hence, while bulk matrix and chondrule fractions separated from Allende now have identical Cr isotope signatures (Kadlag et al. 2019), this need not imply a genetic relationship. In contrast, Leoville chondrules exhibit distinct Cr isotope compositions from their surrounding fine-grained rims (van Kooten et al. 2021), suggesting that the chondrules and matrix, at least in reduced CV chondrites, are not genetically related.

Similar observations have been made for mass-dependent Zn isotope signatures of Allende and Leoville chondrule and matrix fractions. Allende chondrules display a relatively large variability in the Zn isotope signatures of individual chondrules ($\sim$1‰; Pringle et al. 2017), whereas individual Leoville chondrules have more restricted Zn isotope values (van Kooten and Moynier 2019). This can be attributed to increasing sulfurization during thermal metamorphism, where the Zn content of chondrules increases by precipitation of sulfides within the chondrules (Zn being a chalcophile element). Other chalcophiles may exhibit similar behavior (e.g., Te, Sn, Cd) and caution in selecting meteorites should be applied when studying these elements.

In sum, we have demonstrated that secondary alteration can have a non-negligible effect on creating apparent complementary relationships between chondrules and matrix for some elements. However, as of now, we have a very limited understanding of what happens for each element (chemically and isotopically) during progressive thermal and aqueous alteration. To objectively determine the chondrule-matrix relationships, it is key to know the source and sink, the mobility during various P-T-fO_2 conditions and open/closed system behavior of each element.

6 The Chemical and Physical Implications of Complementarity

6.1 A Single Reservoir Process and Chondrule Formation

An important requirement of complementarity is that chondrules and matrix were processed co-genetically in the same dust reservoir. However, the final chemical outcome, the chondrule and matrix compositions observed, depends very much on the mechanism(s) of chondrule formation, the feedback from the surrounding gas and the spatial scales at which the heating event(s) occurred. For example, if we consider chondrules to reflect 'CI-like dust minus a (volatile) component', matrix should consist of a complementary 'CI-like dust plus a (volatile) component' (Fig. 11). In this context, if matrix is thermally processed alongside chondrules, why would matrix remain CI-like and not gain this volatile component from chondrules, and how does matrix maintain its primitive nature (e.g., presolar grain abundance, organic composition, volatile-rich inventory)? If we consider the nature of unaltered chondritic matrix, it generally consists of amorphous (Fe-rich) silicates with embedded nanosulfides, complex organic matter, rare anomalous presolar grains and anhydrous silicates (see Sect. 2). As noted earlier, although this amorphous silicate material has been suggested to have affinities to GEMS in IDPs (GEMS-like), there are several significant caveats in relating GEMS (from IDPs) to GEMS-like materials (from chondrites) through secondary alteration (Leroux et al. 2015; Ohtaki et al. 2021). An important question is then

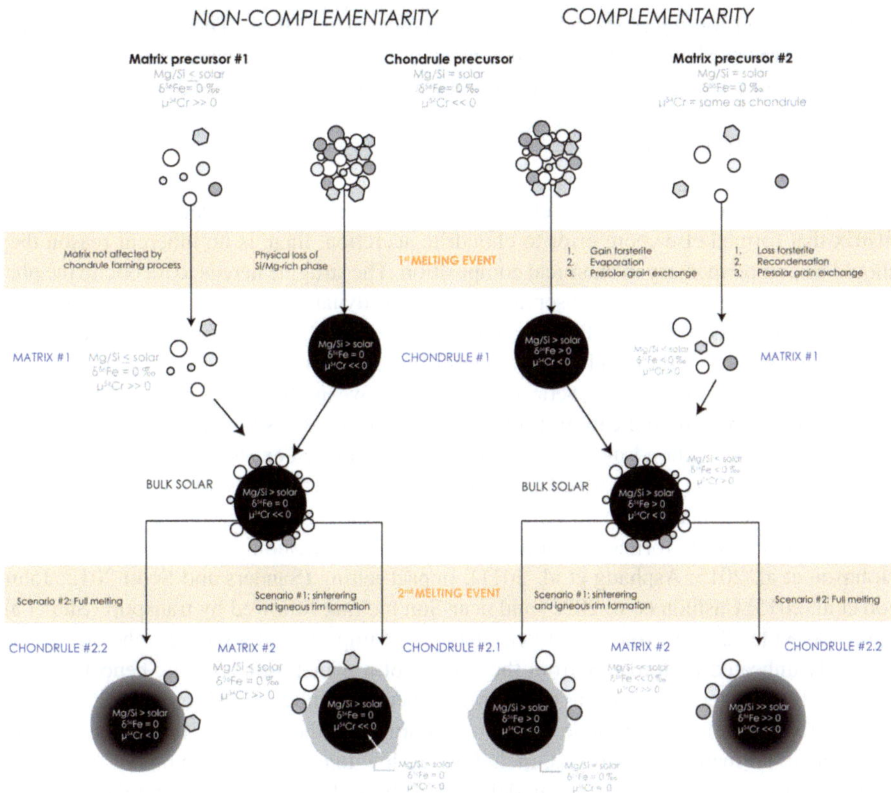

Fig. 11 A schematic representation of chondrule formation and accretion in a complementarity and non-complementarity scenario and implications for the Mg/Si ratios, mass-dependent and mass-independent isotope signatures

whether the amorphous silicate fraction of primitive matrix represents a disequilibrium condensation product from the solar nebula (Brearley 1993; Greshake 1997; Wasson 2008) or if they are altered GEMS-like material that were inherited from the interstellar medium (Ishii et al. 2018), or a combination of both (i.e., thermal processing of interstellar GEMS by FU Orionis outbursts; Alexander et al. 2017a). If we consider the first, these amorphous silicate materials were formed alongside chondrules in the same heating event, during what has been called 'chaotic' disequilibrium condensation (Rietmeijer and Nuth 2013) where the condensation sequence and the resulting composition are difficult to predict but is generally thought to be consistent with GEMS-like materials in chondritic matrix (Leroux et al. 2015; Ohtaki et al. 2021). The complex organic matter and presolar grains are then suggested to be a late addition to this thermally processed matrix, but only represent a small fraction of the matrix (5-10%; Alexander et al. 2017a). As such, GEMS-like amorphous silicates make up the bulk of the matrix, which has been overall characterized as CI-like (Alexander 2019; Zanda et al. 2018; van Kooten et al. 2019), possibly with the loss of 10% forsterite to explain the sub-solar Mg/Si ratios (Alexander 2019; Patzer et al. 2021, 2022, 2023). An important question is then whether these proposed condensation products would have solar compositions, or if they should exhibit some volatile loss. One requirement would be that the heating occurs in an environment where the solid condensates remained fine-grained enough to stay coupled to the gas during cooling (Alexander 2005; Alexander et al. 2017a). This is quite different to what seems to have happened during chondrule formation where densities were high enough to stabilize quite large liquid droplets and not condense very fine-grained solids.

6.2 Astrophysical Implications

If chondrules formed in specific "chondrule-forming regions" and were then mixed with matrix that formed elsewhere prior to chondrite accretion, there is no inherent reason they should be complementary in chemical composition. The struggle here is to reconcile the phenomenology of chondrites, our observations, with the dynamical models that might describe protoplanetary disk processes. Plate tectonic theory took 60 years to reconcile observed geological phenomena with geophysical models of mantle convection.

Whether or not there is a genetic relationship between chondrules and matrix has far-reaching astrophysical implications for how protoplanetary disks function and planetesimals accreted material. If chondrule-matrix complementary indeed exists, this places several important constraints on the theorized and modeled aspects of Solar System formation.

First, chondrules must be melted locally andin the same reservoir as matrix, ruling out various chondrule-formation models (Wood 1996b), including planetesimal collisions (Johnson et al. 2015; Asphaug et al. 2011), impact jetting (Sanders and Scott 2012; Johnson et al. 2015; Cashion et al. 2022) and near-Sun melting followed by transport (Shu et al. 1996; Brownlee 2014). Note that the mechanism of impact jetting includes the ejection of relatively unheated protomatrix from the surface of a planetesimal and is, henceforth, not necessarily considered as a caveat to the complementarity model (Cashion et al. 2022). Secondly, other models may be able to melt chondrules in the presence of matrix, but to a point where all matrix is destroyed, including the low-temperature components commonly found in pristine meteorites. These models include bow-shocks generated by relatively large planet(esimal)s (Hood and Horanyi 1993; Hood 1998; Desch and Connolly 2002; Morris and Desch 2010; Morris and Boley 2018), shock waves from clumpy accretion (Boss and Graham 1993), gravitational instabilities (Wood 1996a; Boss and Durisen 2005) and current sheets produced by magnetorotational instability, which have not been shown to sufficiently localize heating (McNally et al. 2013; Hubbard and Ebel 2018; Lebreuilly et al. 2023). Hence, the complementarity model either requires an extremely localized heating mechanism or the later addition of a small amount of unheated non-complementary matrix component consistent with the conclusion of Braukmüller et al. (2018). Finally, chondrites of each class must accrete from their individual reservoirs on timescales faster than mixing of components between different reservoirs (Alexander et al. 2008; Alexander and Ebel 2012). Jones (2012) found this timescale problem "one of the biggest conundrums" in our understanding, if we consider the formation of CAIs 2-3 Myrs prior to chondrule formation based on Al-Mg dating. This "storage problem" is a challenge for disk dynamical models.

Formation of chondrules coexisting with matrix in a single reservoir appears to require non-shock, non-collisional astrophysical mechanisms for chondrule formation that have not yet been conceived or fully investigated (Hubbard and Ebel 2018). To date, large scale disk models rarely address radiative transfer, local ionization, and/or vertical disk structure. Recent ALMA observations indicate strong segregation of protoplanetary disks into radial zones separated by "pressure bumps" and/or planets (Dullemond et al. 2018). Such regions, if persistent (Jiang and Ormel 2021), could support a vertical structure that would engender strongly different physical-chemical regimes between a low-density region above one scale height and the dense, cold midplane. Heating at high vertical disk heights, in low-density regions ionized by external photons could provide a sufficiently local mechanism

Table 2 An overview of various constraints used in cosmochemistry to validate or disprove complementarity between chondrules and matrix. H = hybrid complementarity

	Complementarity	Non-complementarity	Validation
The nature of matrix			
GEMS-like material	IF: *high T condensation origin*	IF: *cold condensation origin + alteration*	Investigating primitive matrix on a nanoscale
Crystalline anhydrous silicates	H	x	
Presolar grains	H	x	
Organic matter	H	x	
Ices		x	
The nature of chondrules			
Multiple melting events	?	?	
Igneous rims		x	
Chondrule populations	x	x	
Chemical composition			
Mg/Si ratio	IF: *exchange of forsterite*	IF: *change in disk dust composition*	Understanding the relationship of >5 μm
Volatility patterns	CR chondrites	x	matrix-embedded grains to matrix or
Isotopes			
Mass-dependent (Si, Mg, Fe)	?	?	Interconnection between chemistry, mass-
Mass-dependent (MVE)	IF: *exchange of MVE phases*	x	dependent and nucleosynthetic isotope
Mass-independent	IF: *exchange of presolar grains*	x	signatures of chondrules
			High-prescion timeline of chondrule melting
Radiogenic	IF: Al-Mg ages are correct	IF: Pb-Pb ages are correct	and accretion events

for chondrule formation. However, some chondrules retain volatile elements that would be isotopically fractionated in a low solid density, high-temperature regime (Alexander et al. 2008; Ebel et al. 2018), a feature that is not observed for numerous isotope systems (see Sect. 4).

7 How Can We Resolve the Complementarity Debate?

Our detailed discussion about the issues we face in resolving the complementarity debate, has left us with a few key follow-up points that we can address with future research (Table 2):

- Understanding the formation pathways of GEMS-like matrix from either cold condensation the interstellar medium or 'hot' disequilibrium condensation in the protoplanetary disk will allow us to constrain whether chondrules and matrix formed independently or co-genetically, respectively. This requires expanding on the restricted amount of nanoscale petrological descriptions we have of highly primitive, unaltered matrices.
- In that respect, it is essential to restrict complementarity research to the most unaltered chondrites available to us, especially since we have a very limited understanding of elemental and isotopic redistribution and potential open system behavior during secondary alteration on the chondrite parent body and during terrestrial weathering. We should also consider that sample preparation can modify the composition of a sample. For instance, during polishing of samples, polishing compounds can be pressed into matrix, some phases may be soluble in lubricants/solvents that are used, others may be preferentially plucked because they are hard or removed because they are soft.
- A few key areas emerge on which we should focus our attention to correctly interpret the Mg/Si ratios of chondrules and matrix. First, although it appears that preliminary tests comparing EPMA with wet analyses of matrix cannot explain its subsolar Mg/Si ratio, further research is needed to systematically understand the effects of porosity and polyphase mineralogy that can offset EPMA from wet analyses data. Secondly, the broad definition of matrix used throughout the literature (grains < 5–100 μm) can include or exclude forsterite (but also pyroxene, metal and sulfide) grains that have a poorly defined origin. A deeper understanding of the relation of these grains to chondrules or matrix is critical to define the final Mg/Si ratios of these components. Finally, a detailed micron- to nano-scale investigation of these grains in pristine chondrites and their potential relationship to other primary matrix components can help constrain their abundance and origin

and their importance in establishing subsolar Mg/Si ratios in the matrix of carbonaceous chondrites.

- Chondrule formation is a complex and poorly understood process, making it difficult to interpret the mass-dependent and nucleosynthetic isotope signatures of chondrules in the framework of complementarity. Improving on this journal requires 1) chemical data of chondrules that allows us to assess the degree of elemental/phase loss from the chondrule relative to CI chondrites, 2) correlated mass-dependent and mass-bias corrected isotope data of chondrules and matrix, 3) analyses of other chondrites groups than CV chondrites for isotope analyses.
- Constraining the relative timing and precursor material of multiple accretion events reflected by the onion-shell petrology of chondrules (i.e., igneous rims, primary and secondary chondrule cores) using consistent dating in combination with nucleosynthetic isotope systematics would significantly increase our understanding of the locality and frequency of chondrule formation and, therefore, on how we interpret the observed complementarity.
- Comparing cosmochemical datasets (i.e., chemical and isotope composition of chondrules and matrix versus chondrule/matrix abundance ratio and chondrule size distributions in chondrites) to dust and pebble growth and mass transport models that may apply in protoplanetary disks with real constraints from astronomical observations in the era of ALMA and JWST telescopes.

Supplementary Information The online version contains supplementary material available at https://doi.org/10.1007/s11214-024-01107-9.

Acknowledgements We thank two anonymous referees for their helpful comments that have been an improvement to this manuscript. The organizers of the International Space Science Institute workshop "Evolution of the Solar System: Constraints from Meteorites" (June 5-9, 2023, Bern) are thanked for their dedication to facilitate the writing of the review chapters.

Funding Villum Young Investigator Grant (project no. 53024, EvK).

Declarations

Competing Interests The authors report no conflict of interests.

References

Abreu NM, Brearley AJ (2010) Early Solar System processes recorded in the matrices of two highly pristine CR3 carbonaceous chondrites, MET 00426 and QUE 99177. Geochim Cosmochim Acta 74:1146–1171. https://doi.org/10.1016/j.gca.2009.11.009

Alexander CMO'D (1994) Trace element distributions within ordinary chondrite chondrules: implications for chondrule formation conditions and precursors. Geochim Cosmochim Acta 58:3451–3467. https://doi.org/10.1016/0016-7037(94)90098-1

Alexander CMO'D (1995) Trace element contents of chondrule rims and interchondrule matrix in ordinary chondrites. Geochim Cosmochim Acta 59:3247–3266. https://doi.org/10.1016/0016-7037(95)00208-H

Alexander CMO'D (2005) Re-examining the role of chondrules in producing the elemental fractionations in chondrites. Meteorit Planet Sci 40:943–965. https://doi.org/10.1111/j.1945-5100.2005.tb00166.x

Alexander CMO'D (2019) Quantitative models for the elemental and isotopic fractionations in chondrites: the carbonaceous chondrites. Geochim Cosmochim Acta. https://doi.org/10.1016/j.gca.2019.02.008

Alexander CMO'D, Ebel DS (2012) Questions, questions: can the contradictions between the petrologic, isotopic, thermodynamic, and astrophysical constraints on chondrule formation be resolved? Meteorit Planet Sci 47:1157–1175. https://doi.org/10.1111/j.1945-5100.2011.01308.x

Alexander CMO'D, Grossman JN (2005) Alkali elemental and potassium isotopic compositions of Semarkona chondrules. Meteorit Planet Sci 40:541. https://doi.org/10.1111/j.1945-5100.2005.tb00962.x

Alexander CMO'D, Barber DJ, Hutchison R (1989a) The microstructure of Semarkona and Bishunpur. Geochim Cosmochim Acta 53:3045–3057. https://doi.org/10.1016/0016-7037(89)90180-4

Alexander CMO'D, Hutchison R, Barber DJ (1989b) Origin of chondrule rims and interchondrule matrices in unequilibrated ordinary chondrites. Earth Planet Sci Lett 95:187–207. https://doi.org/10.1016/0012-821X(89)90096-4

Alexander CMOD, Fogel M, Yabuta H, Cody GD (2007) The origin and evolution of chondrites recorded in the elemental and isotopic compositions of their macromolecular organic matter. Geochim Cosmochim Acta 71(17):4380–4403. ISSN 0016-7037. https://doi.org/10.1016/j.gca.2007.06.052.

Alexander CMO'D, Grossman JN, Ebel DS, Ciesla FJ (2008) The formation conditions of chondrules and chondrites. Science 320:1617–1619. https://doi.org/10.1126/science.1156561

Alexander CMO'D, Cody GD, De Gregorio BT, Nittler LR, Stroud RM (2017a) The nature, origin and modification of insoluble organic matter in chondrites, the major source of Earth's C and N. Geochemistry 77:227–256. https://doi.org/10.1016/j.chemer.2017.01.007

Alexander CMO'D, Nittler LR, Davidson J, Ciesla F (2017b) Measuring the level of interstellar inheritance in the solar protoplanetary disk. Meteorit Planet Sci 52:1797–1821. https://doi.org/10.1111/maps.12891

Alexander CMO'D, Wynn JG, Bowden R (2022) Sulfur abundances and isotopic compositions in bulk carbonaceous chondrites and insoluble organic material: clues to elemental and isotopic fractionations of volatile chalcophiles. Meteorit Planet Sci 57:334–351. https://doi.org/10.1111/maps.13746

Amelin Y, Krot AN (2007) Pb isotopic age of the Allende chondrules. Meteorit Planet Sci 42:1321–1335

Anders E (1977) Critique of "nebular condensation of moderately volatile elements and their abundances in ordinary chondrites" by Chien M. Wai and John T. Wasson. Earth Planet Sci Lett 36:14–20. https://doi.org/10.1016/0012-821X(77)90183-2

Asphaug E, Jutzi M, Movshovitz N (2011) Chondrule formation during planetesimal accretion. Earth Planet Sci Lett 308:369–379. https://doi.org/10.1016/j.epsl.2011.06.007

Barosch J, Hezel DC, Ebel DS, Friend P (2019) Mineralogically zoned chondrules in ordinary chondrites as evidence for open system chondrule behaviour. Geochim Cosmochim Acta 249:1–16. https://doi.org/10.1016/j.gca.2019.01.018

Barosch J, Ebel DS, Hezel DC, Alpert S, Palme H (2020) Formation of chondrules and matrix in Kakangari chondrites. Earth Planet Sci Lett 542:116286. https://doi.org/10.1016/j.epsl.2020.116286

Barosch J, Nittler LR, Wang J, Dobrică E, Brearley AJ, Hezel DC, Alexander CMO'D (2022) Presolar O- and C-anomalous grains in unequilibrated ordinary chondrite matrices. Geochim Cosmochim Acta 335:169–182. https://doi.org/10.1016/j.gca.2022.08.027

Barrat JA, Zanda B, Moynier F, Bollinger C, Liorzou C, Bayon G (2012) Geochemistry of CI chondrites: major and trace elements, and Cu and Zn isotopes. Geochim Cosmochim Acta 83:79–92. https://doi.org/10.1016/j.gca.2011.12.011

Becker M, Hezel DC, Schulz T, Elfers B-M, Münker C (2015) Formation timescales of CV chondrites from component specific Hf–W systematics. Earth Planet Sci Lett 432:472–482. https://doi.org/10.1016/j.epsl.2015.09.049

Bischoff A, Palme H, Ash RD, Clayton RN, Schultz L, Herpers U, Stoffler D, Grady MM, Pillinger CT, Spettel B, Weber H, Grund T, Endress M, Weber D (1993) Paired Renazzo-type (CR) carbonaceous chondrites from the Sahara. Geochim Cosmochim Acta 57:1587–1603. https://doi.org/10.1016/0016-7037(93)90014-N

Bischoff A, Ebert S, Metzler K, Lentfort S (2017) Breccia classification of CM chondrites. In: 80th ann. meet. Met. soc., July 23-28, 2017, Santa Fe, New Mexico. LPI contribution, vol 1987. id.6089

Bizzarro M, Paton C, Larsen K, Schiller M, Trinquier A, Ulfbeck D (2011) High-precision Mg-isotope measurements of terrestrial and extraterrestrial material by HR-MC-ICPMS-implications for the relative and absolute Mg-isotope composition of the bulk silicate Earth. J Anal At Spectrom 26:565–577. https://doi.org/10.1039/C0JA00190B

Bland PA, Alard O, Benedix GK, Kearsley AT, Menzies ON, Watt LE, Rogers NW (2005) Volatile fractionation in the early Solar System and chondrule/matrix complementarity. Proc Natl Acad Sci USA 102:13755–13760. https://doi.org/10.1073/pnas.0501885102

Bland PA, Stadermann FJ, Floss C, Rost D, Vicenzi EP, Kearsley AT, Benedix GK (2007) A cornucopia of presolar and early Solar System materials at the micrometer size range in primitive chondrite matrix. Meteorit Planet Sci 42:1417–1427

Bollard J, Connelly JN, Whitehouse MJ, Pringle EA, Bonal L, Jørgensen JK, Nordlund Å, Moynier F, Bizzarro M (2017) Early formation of planetary building blocks inferred from Pb isotopic ages of chondrules. Sci Adv 3:e1700407. https://doi.org/10.1126/sciadv.1700407

Bollard J, Kawasaki N, Sakamoto N, Olsen M, Itoh S, Larsen K, Wielandt D, Schiller M, Connelly JN, Yurimoto H, Bizzarro M (2019) Combined U-corrected Pb-Pb dating and ^{26}Al-^{26}Mg systematics of individual chondrules – evidence for a reduced initial abundance of ^{26}Al amongst inner Solar System chondrules. Geochim Cosmochim Acta 260:62–83. https://doi.org/10.1016/j.gca.2019.06.025

Bonal L, Quirico E, Bourot-Denise M, Montagnac G (2006) Determination of the petrologic type of CV3 chondrites by Raman spectroscopy of included organic matter. Geochim Cosmochim Acta 70:1849–1863. https://doi.org/10.1016/j.gca.2005.12.004

Bose M, Floss C, Stadermann FJ, Stroud RM, Speck AK (2012) Circumstellar and interstellar material in the CO3 chondrite ALHA77307: an isotopic and elemental investigation. Geochim Cosmochim Acta 93:77–101. https://doi.org/10.1016/j.gca.2012.06.027

Boss AP, Durisen RH (2005) Sources of shock waves in the protoplanetary disk. In: Krot AN, Scott ERD, Reipurth B (eds) Chondrites and the protoplanetary disk. ASP Conference Series, vol 341, pp 821–838

Boss AP, Graham JA (1993) Clumpy disk accretion and chondrule formation. Icarus 106:168–178. https://doi.org/10.1006/icar.1993.1164

Bouvier A, Wadhwa M, Simon SB, Grossman L (2013) Magnesium isotopic fractionation in chondrules from the Murchison and Murray CM2 carbonaceous chondrites. Meteorit Planet Sci 48:339–353. https://doi.org/10.1111/maps.12059

Bradley JP (1994) Nanometer-scale mineralogy and petrography of fine-grained aggregates in anhydrous interplanetary dust particles. Geochim Cosmochim Acta 58:2123–2134. https://doi.org/10.1016/0016-7037(94)90290-9

Bradley JP, Ishii HA, Bustillo K, Ciston J, Ogliore R, Stephan T, Brownlee DE, Joswiak DJ (2022) On the provenance of GEMS, a quarter century post discovery. Geochim Cosmochim Acta 334:323–338. https://doi.org/10.1016/j.gca.2022.06.036

Braukmüller N, Wombacher F, Hezel DC, Escoube R, Münker C (2018) The chemical composition of carbonaceous chondrites: implications for volatile element depletion, complementarity and alteration. Geochim Cosmochim Acta 239:17–48. https://doi.org/10.1016/j.gca.2018.07.023

Brearley AJ (1993) Matrix and fine-grained rims in the unequilibrated CO3 chondrite, ALHA77307: origins and evidence for diverse, primitive nebular dust components. Geochim Cosmochim Acta 57:1521–1550. https://doi.org/10.1016/0016-7037(93)90011-K

Brearley A (2006) The action of water. In: Lauretta DS, McSween HY (eds) Meteorites and the early Solar System II, pp 587–624. https://doi.org/10.2307/j.ctv1v7zdmm.35

Brearley AJ (2016) Diverse anhydrous silicates in a fine-grained rim in the weakly altered CM2 chondrite queen elizabeth range 97990: evidence for the localized preservation of pristine nebular dust in CM chondrites. In: 79th annual meeting of the meteoritical society, held 7-12 August, 2016, Berlin, Germany. LPI contribution vol 1921, id.6190

Brearley AJ, Jones RH (1998) Chondritic meteorites. In: Papike JJ (ed) Planetary materials, reviews in mineralogy, vol 36. Mineralogical Society of America, Washington, p 39

Brearley AJ, Krot AN (2013) Metasomatism in chondritic meteorites. In: Harlov DE, Austrheim H (eds) Metasomatism and the chemical transformation of rock: the role of fluids in terrestrial and extraterrestrial processes. Springer, Berlin, pp 653–782

Browning LB, McSween HY Jr, Zolensky ME (1996) Correlated alteration effects in CM carbonaceous chondrites. Geochim Cosmochim Acta 60:26212633. https://doi.org/10.1016/0016-7037(96)00121-4

Brownlee D (2014) The stardust mission: analyzing samples from the edge of the Solar System. Annu Rev Earth Planet Sci 42:179–205. https://doi.org/10.1146/annurev-earth-050212-124203

Budde G, Kleine T, Kruijer TS, Burkhardt C, Metzler K (2016) Tungsten isotopic constraints on the age and origin of chondrules. Proc Natl Acad Sci 113:2886–2891. https://doi.org/10.1073/pnas.1524980113

Budde G, Kruijer TS, Kleine T (2018) Hf-W chronology of CR chondrites: implications for the timescales of chondrule formation and the distribution of ^{26}Al in the solar nebula. Geochim Cosmochim Acta 222:284–304. https://doi.org/10.1016/j.gca.2017.10.014

Bunch TE, Chang S (1980) Carbonaceous chondrites II. Carbonaceous chondrite phyllosilicates and light element geochemistry as indicators of parent body processes and surface conditions. Geochim Cosmochim Acta 44:1543–1577. https://doi.org/10.1016/0016-7037(80)90118-0

Busemann H, Young AF, Alexander CMO'D, Hoppe P, Mukhopadhyay S, Nittler LR (2006) Interstellar chemistry recorded in organic matter from primitive meteorites. Science 312:727. https://doi.org/10.1126/science.1123878

Cashion MD, Johnson BC, Krot AN, Kretke KA, Wakita S, Davidson TM (2022) Chondrule formation via impact jetting in the icy outer Solar System. Icarus 384:115110. https://doi.org/10.1016/j.icarus.2022.115110

Changela HG, Le Guillou C, Bernard S, Brearley AJ (2018) Hydrothermal evolution of the morphology, molecular composition, and distribution of organic matter in CR (Renazzo-type) chondrites. Meteorit Planet Sci 53:1006–1029. https://doi.org/10.1111/maps.13045

Chizmadia LJ, Brearley AJ (2008) Mineralogy, aqueous alteration, and primitive textural characteristics of fine-grained rims in the Y-791198CM2 carbonaceous chondrite: TEM observations and comparison to ALHA81002. Geochim Cosmochim Acta 72:602–625. https://doi.org/10.1016/j.gca.2007.10.019

Clarke RS, Jarosewich E, Mason B, Nelen J, Gomez M, Hyde JR (1971) Allende, Mexico, Meteorite Shower. https://doi.org/10.5479/si.00810274.5.1

Connelly JN, Bizzarro M, Krot AN, Nordlund Å, Wielandt D, Ivanova MA (2012) The absolute chronology and thermal processing of solids in the solar protoplanetary disk. Science 338:651–655. https://doi.org/10.1126/science.1226919

Connelly JN, Bollard J, Bizzarro M (2017) Pb–Pb chronometry and the early Solar System. Geochim Cosmochim Acta 201:345–363. https://doi.org/10.1016/j.gca.2016.10.044

Connolly HC, Huss GR (2010) Compositional evolution of the protoplanetary disk: oxygen isotopes of type-II chondrules from CR2 chondrites. Geochim Cosmochim Acta 74:2473–2483. https://doi.org/10.1016/j.gca.2010.01.005

Crapster-Pregont EJ, Ebel DS (2020) Reducing supervision of quantitative image analysis of meteorite samples. Microsc Microanal 26:63–75. https://doi.org/10.1017/S1431927619015216

Cronin JR, Pizzarello S, Cruikshank DP (1988) Organic matter in carbonaceous chondrites, planetary satellites, asteroids and comets. In: Kerridge JF, Matthews MS (eds) Meteorites and the early Solar System. University of Arizona Press, Tucson, pp 819–857

Cuzzi J, Alexander C (2006) Chondrule formation in particle-rich nebular regions at least hundreds of kilometres across. Nature 441:483–485. https://doi.org/10.1038/nature04834

Davidson J, Schrader DL, Alexander CMO'D, Lauretta DS, Busemann H, Franchi IA, Greenwood RC, Connolly HC, Domanik KJ, Verchovsky A (2014) Petrography, stable isotope compositions, microRaman spectroscopy, and presolar components of Roberts Massif 04133: a reduced CV3 carbonaceous chondrite. Meteorit Planet Sci 49:2133–2151. https://doi.org/10.1111/maps.12377

Davidson J, Alexander CMO'D, Stroud RM, Busemann H, Nittler LR (2019a) Mineralogy and petrology of dominion range 08006: a very primitive CO3 carbonaceous chondrite. Geochim Cosmochim Acta 265:259–278. https://doi.org/10.1016/j.gca.2019.08.032

Davidson J, Schrader DL, Alexander CMO, Nittler LR, Bowden R (2019b) Re-examining thermal metamorphism of the Renazzo-like (CR) carbonaceous chondrites: insights from pristine Miller range 090657 and shock-heated graves Nunataks 06100. Geochim Cosmochim Acta 267:240–256. https://doi.org/10.1016/j.gca.2019.09.033

De Gregorio BT, Stroud RM, Nittler LR, Alexander CMO'D, Kilcoyne ALD, Zega TJ (2010) Isotopic anomalies in organic nanoglobules from comet 81P/Wild 2: comparison to Murchison nanoglobules and isotopic anomalies induced in terrestrial organics by electron irradiation. Geochim Cosmochim Acta 74:4454–4470. https://doi.org/10.1016/j.gca.2010.05.010

Desch SJ, Connolly HC Jr (2002) A model of the thermal processing of particles in solar nebula shocks: application to the cooling rates of chondrules. Meteorit Planet Sci 37:183–207. https://doi.org/10.1111/j.1945-5100.2002.tb01104.x

Dobrică E, Brearley AJ (2020a) Amorphous silicates in the matrix of Semarkona: the first evidence for the localized preservation of pristine matrix materials in the most unequilibrated ordinary chondrites. Meteorit Planet Sci 55:649–668. https://doi.org/10.1111/maps.13458

Dobrică E, Brearley AJ (2020b) Iron-rich olivine in the unequilibrated ordinary chondrite, MET 00526: earliest stages of formation. Meteorit Planet Sci 55:2652–2669. https://doi.org/10.1111/maps.13610

Doyle PM, Jogo K, Nagashima K, Krot AN, Wakita S, Ciesla FJ, Hutcheon ID (2015) Early aqueous activity on the ordinary and carbonaceous chondrite parent bodies recorded by fayalite. Nat Commun 6:7444. https://doi.org/10.1038/ncomms8444

Dullemond CP, et al (2018) The Disk Substructures at High Angular Resolution Project (DSHARP). VI. Dust trapping in thin-ringed protoplanetary disks. Astrophys J Lett 869:L46. https://doi.org/10.3847/2041-8213/aaf742

Dunham ET, Sheikh A, Opara D, Matsuda N, Liu M-C, McKeegan KD (2023) Calcium–aluminum-rich inclusions in non-carbonaceous chondrites: abundances, sizes, and mineralogy. Meteorit Planet Sci 58:643–671. https://doi.org/10.1111/maps.13975

Ebel DS (2006) Condensaton of rocky material in astrophysical environments. In: Lauretta D, McSween HY Jr (eds) Meteorites and the early Solar System II. University Arizona, Tucson, pp 253–277

Ebel DS, Grossman L (2000) Condensation in dust-enriched systems. Geochim Cosmochim Acta 64:339–366. https://doi.org/10.1016/S0016-7037(99)00284-7

Ebel DS, Weisberg MK, Hertz J, Campbell AJ (2008) Shape, metal abundance, chemistry and origin of chondrules in the Renazzo (CR) chondrite. Meteorit Planet Sci 43:1725–1740. https://doi.org/10.1111/j.1945-5100.2008.tb00639.x

Ebel DS, Brunner C, Konrad K, Leftwich K, Erb I, Lu M, Rodriguez H, Crapster-Pregont EJ, Friedrich JM, Weisberg MK (2016) Abundance, major element composition and size of components and matrix in CV, CO and Acfer 094 chondrites. Geochim Cosmochim Acta 172:322–356. https://doi.org/10.1016/j.gca.2015.10.007

Ebel DS, Alexander CMO'D, Libourel G (2018) Vapor-melt exchange - constraints on chondrule formation conditions and processes. In: Russell S, Krot AN, Connolly HC Jr (eds) Chondrules: records of protoplanetary disk processes. Cambridge University Press, Cambridge, pp 151–174

Ebel DS, Gemma ME, Alpert SP, Bayron J, Lobo AH, Weisberg MK (2024) Abundance, sizes, and major element compositions of components in CR and LL chondrites: formation of CC and NC bodies from single reservoirs, Meteorit Planet Sci 59:2276–2295. https://doi.org/10.1111/maps.14191

Fendrich KV, Ebel DS (2021) Comparison of the Murchison CM2 and Allende CV3 chondrites. Meteorit Planet Sci 56:77–95. https://doi.org/10.1111/maps.13623

Floss C, Haenecour P (2016) Presolar Silicate Abundances in the Unequilibrated Ordinary Chondrites Meteorite Hills 00526 and Queen Alexandra Range 97008. 79th Annual Meeting of the Meteoritical Society, held 7-12 August, 2016 in Berlin, Germany. LPI Contribution No. 1921, id.6015

Floss C, Stadermann F (2009) Auger nanoprobe analysis of presolar ferromagnesian silicate grains from primitive CR chondrites QUE 99177 and MET 00426. Geochim Cosmochim Acta 73:2415–2440. https://doi.org/10.1016/j.gca.2009.01.033

Floss C, Le Guillou C, Brearley A (2014) Coordinated NanoSIMS and FIB-TEM analyses of organic matter and associated matrix materials in CR3 chondrites. Geochim Cosmochim Acta 139:1–25. https://doi.org/10.1016/j.gca.2014.04.023

Fox GE, Hewins RH (2005) Chondrule reheating experiments and relict olivine. Geochim Cosmochim Acta 69:2441–2449. https://doi.org/10.1016/j.gca.2004.11.014

Frank DR, Zolensky ME, Le L (2014) Olivine in terminal particles of stardust aerogel tracks and analogous grains in chondrite matrix. Geochim Cosmochim Acta 142:240–259. https://doi.org/10.1016/j.gca.2014.05.037

Frank DR, Huss GR, Zolensky ME, Nagashima K, Le L (2023) Calcium-aluminum-rich inclusion found in the Ivuna CI chondrite: are CI chondrites a good proxy for the bulk composition of the Solar System? Meteorit Planet Sci 58:1495–1511. https://doi.org/10.1111/maps.14083

Friend P, Hezel DC, Mucerschi D (2016) The conditions of chondrule formation, part II: open system. Geochim Cosmochim Acta 173:198–209. https://doi.org/10.1016/j.gca.2015.10.026

Garvie LAJ, Buseck PR (2007) Prebiotic carbon in clays from Orgueil and Ivuna (CI), and Tagish Lake (C2 ungrouped) meteorites. Meteorit Planet Sci 42:2111–2117. https://doi.org/10.1111/j.1945-5100.2007.tb01011.x

Gattacceca J, Bonal L, Sonzogni C, Longerey J (2020) CV chondrites: more than one parent body. Earth Planet Sci Lett 547:1164667. https://doi.org/10.1016/j.epsl.2020.116467. Monthly Notices of the Royal Astronomical Society, 452, 01

Gerber S, Burkhardt C, Budde G, Metzler K, Kleine T (2017) Mixing and transport of dust in the early solar nebula as inferred from titanium isotope variations among chondrules. Astrophys J Lett 841:L17. https://doi.org/10.3847/2041-8213/aa72a2

Goldberg AZ, Owen JE, Jacquet E (2015). Chondrule transport in protoplanetary discs. https://doi.org/10.1093/mnras/stv1610

Greshake A (1997) The primitive matrix components of the unique carbonaceous chondrite Acfer 094: a TEM study. Geochim Cosmochim Acta 61:437–452. https://doi.org/10.1016/S0016-7037(96)00332-8

Grossman JN (1988) Formation of chondrules. In: Kerridge JF, Matthews MS (eds) Meteorites and the early Solar System, ch. 9.3. University of Arizona Press, Tucson, pp 680–696

Grossman JN (1996) Chemical fractionations of chondrites: signatures of events before chondrule formation. In: Hewins R, Jones RH, Scott ERD (eds) Chondrules and the protoplanetary disk, ch. 26, pp 243–253. University of Cambridge, Cambridge

Grossman JN, Brearley AJ (2005) The onset of metamorphism in ordinary and carbonaceous chondrites. Meteorit Planet Sci 40:87–122. https://doi.org/10.1111/j.1945-5100.2005.tb00366.x

Hanowski NP, Brearley AJ (2001) Aqueous alteration of chondrules in the CM carbonaceous chondrite, Allan Hills 81002: implications for parent body alteration. Geochim Cosmochim Acta 65:495–518. https://doi.org/10.1016/S0016-7037(00)00552-4

Hashimoto A, Grossman L (1987) Alteration of Al-rich inclusions inside amoeboid olivine aggregates in the Allende meteorite. Geochim Cosmochim Acta 51:1685–1704. https://doi.org/10.1016/0016-7037(87)90348-6

Hellmann JL, Hopp T, Burkhardt C, Kleine T (2020) Origin of volatile element depletion among carbonaceous chondrites. Earth Planet Sci Lett 549:116508. https://doi.org/10.1016/j.epsl.2020.116508

Hewins RH, Jones R, Scott E (2011) Chondrules and the protoplanetary disk. Cambridge University Press, Cambridge, p 372

Hewins RH, Bourot-Denise M, Zanda B, Leroux H, Barrat J-A, Humayun M, Göpel C, Greenwood RC, Franchi IA, Pont S, Lorand J-P, Cournède C, Gattacceca J, Rochette P, Kuga M, Marrocchi Y, Marty B (2014) The Paris meteorite, the least altered CM chondrite so far. Geochim Cosmochim Acta 124:190–222. https://doi.org/10.1016/j.gca.2013.09.014

Hezel DC, Kießwetter R (2010) Quantifying the error of 2D bulk chondrule analyses using a computer model to simulate chondrules (SIMCHON). Meteorit Planet Sci 45:555–571. https://doi.org/10.1111/j.1945-5100.2010.01040.x

Hezel DC, Palme H (2007) Pre-accretionary distribution of Ca and Al between matrix and chondrules in CV chondrites. In: 38th Annual Lunar and Planetary Science Conference

Hezel DC, Palme H (2008) Constraints for chondrule formation from Ca-Al distribution in carbonaceous chondrites. Earth Planet Sci Lett 265:716–725. https://doi.org/10.1016/j.epsl.2007.11.003

Hezel DC, Palme H (2008) Constraints for chondrule formation from Ca–Al distribution in carbonaceous chondrites. Earth Planet Sci Lett 265:716–725. https://doi.org/10.1016/j.epsl.2007.11.003

Hezel DC, Palme H (2010) The chemical relationship between chondrules and matrix and the chondrule matrix complementarity. Earth Planet Sci Lett 294:85–93. https://doi.org/10.1016/j.epsl.2010.03.008

Hezel DC, Needham AW, Armytage R, Georg B, Abel RL, Kurahashi E, Coles BJ, Rehkämper M, Russell SS (2010) A nebula setting as the origin for bulk chondrule Fe isotope variations in CV chondrites. Earth Planet Sci Lett 296:423–433. https://doi.org/10.1016/j.epsl.2010.05.029

Hezel DC, Bland PA, Palme H, Jacquet E, Bigolski J (2018a) Composition of Chondrules and Matrix and Their Complementary Relationship in Chondrites. Chondrules: Records of Protoplanetary Disk Processes. https://doi.org/10.1017/9781108284073.004

Hezel DC, Wilden JS, Becker D, Steinbach S, Wombacher F, Harak M (2018b) Fe isotope composition of bulk chondrules from Murchison (CM2): constraints for parent body alteration, nebula processes and chondrule-matrix complementarity. Earth Planet Sci Lett 490:31–39. https://doi.org/10.1016/j.epsl.2018.03.013

Hobart KK, Crapster-Pregont EJ, Ebel DS (2015) Decoding the history of a layered chondrule through olivine grain orientation measurements using EBSD. In: Lunar and Planetary Science Conference XLVI, #1978

Hood LL (1998) Thermal processing of chondrule precursors in planetesimal bow shocks. Meteorit Planet Sci 33:97–107. https://doi.org/10.1111/j.1945-5100.1998.tb01611.x

Hood LL, Horanyi M (1993) The nebular shock wave model for chondrule formation: one-dimensional calculations. Icarus 106:179–189. https://doi.org/10.1006/icar.1993.1165

Hoppe P, Leitner J, Kodolányi J, Borrmann S, Jones AP (2022) Dust from supernovae and their progenitors in the solar nebula. Nat Astron 6:1027–1034. https://doi.org/10.1038/s41550-022-01737-5

Housley RM, Cirlin EH (1983) On the alteration of Allende chondrules and the formation of matrix. In: Chondrules and their origins (A85-26528 11-91). Lunar and Planetary Institute, Houston, pp 145–161.

Hu Y, Moynier F, Yang X (2023) Volatile-depletion processing of the building blocks of Earth and Mars as recorded by potassium isotopes. Earth Planet Sci Lett 620:118319. https://doi.org/10.1016/j.epsl.2023.118319

Hubbard A, Ebel DS (2018) Evaluating non-shock, non-collisional models for chondrule formation. In: Russell S, Krot AN, Connolly HC Jr (eds) Chondrules: records of protoplanetary disk processes. Cambridge University Press, Cambridge, pp 400–427

Huss GR (1979) The matrix of unequilibrated ordinary chondrites: Implications for the origin and history of chondrites. Master's thesis, Univ. of New Mexico, pp. 1-139

Huss GR, Lewis RS (1995) Presolar diamond, SiC, and graphite in primitive chondrites: abundances as a function of meteorite class and petrologic type. Geochim Cosmochim Acta 59:115–160. https://doi.org/10.1016/0016-7037(94)00376-W

Huss GR, Keil K, Taylor GJ (1981) The matrices of unequilibrated ordinary chondrites: implications for the origin and history of chondrites. Geochim Cosmochim Acta 45:33–51. https://doi.org/10.1016/0016-7037(81)90262-3

Hutcheon ID, Newton RC (1981) MG isotopes, mineralogy and mode of formation of secondary phases in c3 refractory inclusions. In: Lunar and Planetary Science Conference XII, pp 491–493

Ishii HA (2019) Comparison of GEMS in interplanetary dust particles and GEMS-like objects in a stardust impact track in aerogel. Meteorit Planet Sci 54:202–219. https://doi.org/10.1111/maps.13182

Ishii HA, Bradley JP, Bechtel HA, Brownlee DE, Bustillo KC, Ciston J, Cuzzi JN, Floss C, Joswiak DJ (2018) Multiple generations of grain aggregation in different environments preceded Solar System body formation. Proc Natl Acad Sci 115:6608–6613. https://doi.org/10.1073/pnas.1720167115

Jacquet E, Marrocchi Y (2017) Chondrule heritage and thermal histories from trace element and oxygen isotope analyses of chondrules and amoeboid olivine aggregates. Meteorit Planet Sci 52:2672–2694. https://doi.org/10.1111/maps.12985

Jacquet E, Alard O, Gounelle M (2012) Chondrule trace element geochemistry at the mineral scale. Meteorit Planet Sci 47:1695–1714. https://doi.org/10.1111/maps.12005

Jacquet E, Paulhiac-Pison M, Alard O, Kearsley AT, Gounelle M (2013) Trace element geochemistry of CR chondrite metal. Meteorit Planet Sci 48:1981–1999. https://doi.org/10.1111/maps.12212

Jacquet E, Alard O, Gounelle M (2015) Trace element geochemistry of ordinary chondrite chondrules: the type I/type II chondrule dichotomy. Geochim Cosmochim Acta 155:47–67. https://doi.org/10.1016/j.gca.2015.02.005

Jacquet E, Barrat J-A, Beck P, Caste F, Gattacceca J, Sonzogni C, Gounelle M (2016) Northwest Africa 5958: a weakly altered CM-related ungrouped chondrite, not a CI3. Meteorit Planet Sci 51:851–869. https://doi.org/10.1111/maps.12628

Jacquet E, Piralla M, Kersaho P, Marrocchi Y (2021) Origin of isolated olivine grains in carbonaceous chondrites. Meteorit Planet Sci 56:13–33. https://doi.org/10.1111/maps.13583

Jarosewich E, Clarke RS, Barrows JN (1987) Allende Meteorite Reference Sample. https://doi.org/10.5479/si.00810274.27.1

Jiang H, Ormel CW (2021) Survival of ALMA rings in the absence of pressure maxima. Mon Not R Astron Soc 505:1162–1179. https://doi.org/10.1093/mnras/stab1278

Jiang Y, Koefoed P, Pravdivtseva O, Chen H, Li C-H, Huang F, Qin L-P, Liu J, Wang K (2021) Early Solar System aqueous activity: K isotope evidence from Allende. Meteorit Planet Sci 56:61–76. https://doi.org/10.1111/maps.13588

Johnson BC, Minton DA, Melosh HJ, Zuber MT (2015) Impact jetting as the origin of chondrules. Nature 517:339–341. https://doi.org/10.1038/nature14105

Jones RH (2012) Petrographic constraints on the diversity of chondrule reservoirs in the protoplanetary disk. Meteorit Planet Sci 47:1176–1190. https://doi.org/10.1111/j.1945-5100.2011.01327.x

Jones RH, Villeneuve J, Libourel G (2018) Thermal histories of chondrules: petrologic observations and experimental constraints. In: Russell SS, Connolly HC Jr, Krot AN (eds) Chondrules: records of protoplanetary disk processes. Cambridge University Press, Cambridge, pp 57–90

Jull AJT, Cheng S, Gooding JL, Velbel MA (1988) Rapid growth of magnesium-carbonate weathering products in a stony meteorite from Antarctica. Science 242:417–419. https://doi.org/10.1126/science.242.4877.417

Kadlag Y, Becker H, Harbott A (2019) Cr isotopes in physically separated components of the Allende CV3 and Murchison CM2 chondrites: implications for isotopic heterogeneity in the solar nebula and parent body processes. Meteorit Planet Sci 54:2116–2131. https://doi.org/10.1111/maps.13375

Kadlag Y, Tatzel M, Frick DA, Becker H, Kühne P (2021) In situ Si isotope and chemical constraints on formation and processing of chondrules in the Allende meteorite. Geochim Cosmochim Acta 304:234–257. https://doi.org/10.1016/j.gca.2021.04.022

Kato C, Moynier F (2017) Gallium isotopic evidence for extensive volatile loss from the Moon during its formation. Sci Adv 3:e1700571. https://doi.org/10.1126/sciadv.1700571

Kebukawa Y, Ito M, Zolensky ME, et al (2019) A novel organic-rich meteoritic clast from the outer Solar System. Sci Rep 9:3169. https://doi.org/10.1038/s41598-019-39357-1

Keller LP, Buseck PR (1991) Calcic micas in the Allende meteorite: evidence for hydration reactions in the early solar nebula. Science 252:946–949. https://doi.org/10.1126/science.252.5008.946

Keller LP, Messenger S (2011) On the origins of GEMS grains. Geochim Cosmochim Acta 75:5336–5365. https://doi.org/10.1016/j.gca.2011.06.040

Kerridge JF (1983) Isotopic composition of carbonaceous-chondrite kerogen: evidence for an interstellar origin of organic matter in meteorites. Earth Planet Sci Lett 64:186–200. https://doi.org/10.1016/0012-821X(83)90203-0

Kimura M, Ikeda Y (1998) Hydrous and anhydrous alterations of chondrules in Kaba and Mokoia CV chondrites. Meteorit Planet Sci 33:1139–1146. https://doi.org/10.1111/j.1945-5100.1998.tb01718.x

Kimura M, Imae N, Komatsu M, Barrat JA, Greenwood RC, Yamaguchi A, Noguchi T (2020) The most primitive CM chondrites, Asuka 12085, 12169, and 12236, of subtypes 3.0–2.8: Their characteristic features and classification. Polar Science 100565. https://doi.org/10.1016/j.polar.2020.100565

Kita NT, Nagahara H, Tachibana S, Tomomura S, Spicuzza MJ, Fournelle JH, Valley JW (2010) High precision SIMS oxygen three isotope study of chondrules in LL3 chondrites: role of ambient gas during chondrule formation. Geochim Cosmochim Acta 74:6610–6635. https://doi.org/10.1016/j.gca.2010.08.011

Koefoed P, Barrat J-A, Pravdivtseva O, Alexander CMO, Lodders K, Ogliore R, Wang K (2023) The potassium isotopic composition of CI chondrites and the origin of isotopic variations among primitive planetary bodies. Geochim Cosmochim Acta 358:49–60. https://doi.org/10.1016/j.gca.2023.07.025

Kojima T, Tomeoka K (1995) Allende dark inclusions affected to different degrees by aqueous alteration. Meteoritics 30:529

Kong P (1999) Compositional and genetic relationships between chondrules, chondrule rims, metal and matrix in the Renazzo chondrite. Geochim Cosmochim Acta 63:3673–3682. https://doi.org/10.1016/S0016-7037(99)00110-6

Kong P, Palme H (1999) Compositional and genetic relationship between chondrules, chondrule rims, metal, and matrix in the Renazzo chondrite. Geochim Cosmochim Acta 63:3673–3682. https://doi.org/10.1016/S0016-7037(99)00110-6

Krot AN (2019) Refractory inclusions in carbonaceous chondrites: records of early Solar System processes. Meteorit Planet Sci 54:1647–1691. https://doi.org/10.1111/maps.13350

Krot AN, Keil K (2002) Anorthite-rich chondrules in CR and CH carbonaceous chondrites: genetic link between calcium-aluminum-rich inclusions and ferromagnesian chondrules. Meteorit Planet Sci 37:91–111. https://doi.org/10.1111/j.1945-5100.2002.tb00797.x

Krot AN, Wasson JT (1995) Igneous rims on low-FeO and high-FeO chondrules in ordinary chondrites. Geochim Cosmochim Acta 59:4951–4966. https://doi.org/10.1016/0016-7037(95)00337-1

Krot AN, Scott ERD, Zolensky ME (1995) Mineralogical and chemical modification of components in CV3 chondrites: nebular or asteroidal processing? Meteoritics 30:748–775. https://doi.org/10.1111/j.1945-5100.1995.tb01173.x

Krot AN, Scott ERD, Zolensky ME (1997) Origin of fayalitic olivine rims and lath-shaped matrix olivine in the CV3 chondrite Allende and its dark inclusions. Meteoritics 32:31–49

Krot AN, Amelin Y, Cassen P, Meibom A (2005) Young chondrules in CB chondrites from a giant impact in the early Solar System. Nature 436:989–992. https://doi.org/10.1038/nature03830

Krot AN, Nagashima K, Libourel G, Miller KE (2018) Multiple mechanisms of transient heating events in the protoplanetary disk: evidence from precursors of chondrules and igneous Ca, al-rich inclusions. In: Russell SS, Connolly HC Jr, Krot AN (eds) Chondrules: records of protoplanetary disk processes. Cambridge University Press, Cambridge, pp 11–56

Krot AN, Petaev MI, Nagashima K (2021) Infiltration metasomatism of the Allende coarse-grained calcium-aluminum-rich inclusions. Prog Earth Planet Sci 8:61. https://doi.org/10.1186/s40645-021-00437-4

Kuebler KE, McSween HY, Carlson WD, Hirsch D (1999) Sizes and masses of chondrules and metal–troilite grains in ordinary chondrites: possible implications for nebular sorting. Icarus 141:96–106. https://doi.org/10.1006/icar.1999.6161

Larsen KK, Trinquier A, Paton C, Schiller M, Wielandt D, Ivanova MA, Connelly JN, Nordlund Å, Krot AN, Bizzarro M (2011) Evidence for magnesium isotope heterogeneity in the solar protoplanetary disk. Astrophys J Lett 735:L37. https://doi.org/10.1088/2041-8205/735/2/L37

Le Guillou C, Brearley A (2014) Relationships between organics, water and early stages of aqueous alteration in the pristine CR3.0 chondrite MET 00426. Geochim Cosmochim Acta 131:344–367. https://doi.org/10.1016/j.gca.2013.10.024

Le Guillou C, Bernard S, Brearley AJ, Remusat L (2014) Evolution of organic matter in Orgueil, Murchison and Renazzo during parent body aqueous alteration: in situ investigations. Geochim Cosmochim Acta 131:368–392. https://doi.org/10.1016/j.gca.2013.11.020

Lebreuilly U, Mac Low M-M, Commerçon B, Ebel DS (2023) Dust dynamics in current sheets within protoplanetary disks I. isothermal models including ambipolar diffusion and Ohmic resistivity. Astron Astrophys 675:A38. https://doi.org/10.1051/0004-6361/202141840

Lentfort S, Bischoff A, Ebert S, Patzek M (2021) Classification of CM chondrite breccias—implications for the evaluation of samples from the OSIRIS-REx and Hayabusa 2 missions. Meteorit Planet Sci 56:127–147. https://doi.org/10.1111/maps.13486

Leroux H, Cuvillier P, Zanda B, Hewins RH (2015) GEMS-like material in the matrix of the Paris meteorite and the early stages of alteration of CM chondrites. Geochim Cosmochim Acta 170:247–265. https://doi.org/10.1016/j.gca.2015.09.019

Libourel G, Nagashima K, Portail M, Krot AN (2023) On the significance of oxygen-isotope variations in chondrules from carbonaceous chondrites. Geochim Cosmochim Acta 346:102–120. https://doi.org/10.1016/j.gca.2023.01.026

Liffman K (2005) Chondrule and metal grain size sorting from jet flows. Meteorit Planet Sci 40:123–138. https://doi.org/10.1111/j.1945-5100.2005.tb00367.x

Lobo A, Wallace SW, Ebel DS (2014) Modal abundances, chemistry, and sizes of clasts in the Semarkona (LL3.0) chondrite by X-ray map analysis. Lunar and Planetary Science Conference 45. Abstract #1423

Lodders K (2003) Solar System abundances and condensation temperatures of the elements. Astrophys J 591:1220–1247

Lodders K (2020) Solar elemental abundances. In: The Oxford research encyclopedia of planetary science. Oxford University Press, London. https://doi.org/10.1093/acrefore/9780190647926.013.145

Lodders K (2021) Relative atomic Solar System abundances, mass fractions, and atomic masses of the elements and their isotopes, composition of the solar photosphere, and compositions of the major chondritic meteorite groups. In: Lammer H, Marty B, Zerkle AL, Blanc M, O'Neill H, Kleine T (eds) Reading Terrestrial Planet Evolution in Isotopes and Element Measurements. Space Sci Rev 217:44. https://doi.org/10.1007/s11214-021-00825-8.

Lodders K, Fegley B (1998) The planetary scientist's companion. Oxford University Press, London

Luck JM, Othman DB, Barrat JA, Albarède F (2003) Coupled 63Cu and 16O excesses in chondrites. Geochim Cosmochim Acta 67:143–151. https://doi.org/10.1016/S0016-7037(02)01038-4

Luck J-M, Othman DB, Albarède F (2005) Zn and Cu isotopic variations in chondrites and iron meteorites: early solar nebula reservoirs and parent-body processes. Geochim Cosmochim Acta 69:5351–5363. https://doi.org/10.1016/j.gca.2005.06.018

Macke RJ, Consolmagno GJ, Britt DT (2011) Density, porosity, and magnetic susceptibility of carbonaceous chondrites. Meteorit Planet Sci 46:1842–1864. https://doi.org/10.1111/j.1945-5100.2011.01298.x

Marrocchi Y, Villeneuve J, Batanova V, Piani L, Jacquet E (2018) Oxygen isotopic diversity of chondrule precursors and the nebular origin of chondrules. Earth Planet Sci Lett 496:132–141. https://doi.org/10.1016/j.epsl.2018.05.042

Matsunami S (1984) The chemical compositions and textures of matrices and chondrule rims of eight unequilibrated ordinary chondrites. Mem Natl Inst Polar Res, Spec Issue 35:126–148

McNally CP, Hubbard A, Mac Low M-M, Ebel DS, D'Alessio P (2013) Mineral processing by short circuits in protoplanetary disks. Astrophys J Lett 767:L2. https://doi.org/10.1088/2041-8205/767/1/L2

McSween HY Jr (1977a) Carbonaceous chondrites of the ornans type: a metamorphic sequence. Geochim Cosmochim Acta 41:477–491. https://doi.org/10.1016/0016-7037(77)90286-1

McSween HY Jr (1977b) Petrographic variations among the carbonaceous chondrites of the Vigarano type. Geochim Cosmochim Acta 41:1777–1790. https://doi.org/10.1016/0016-7037(77)90210-1

McSween HY Jr (1979) Alteration in CM carbonaceous chondrites inferred from modal and chemical variations in matrix. Geochim Cosmochim Acta 43:1761–1770. https://doi.org/10.1016/0016-7037(79)90024-3

McSween HY Jr, Huss G (2022) Chronology of the Solar System from radioactive isotopes. In: Cosmochemistry, 2nd edn. Cambridge University Press, Cambridge, pp 238–270. https://doi.org/10.1017/9781108885263.011

McSween HY Jr, Richardson SM (1977) The composition of carbonaceous chondrite matrix. Geochim Cosmochim Acta 41:1145–1161. https://doi.org/10.1016/0016-7037(77)90110-7

Metzler K, Bischoff A, Stöffler D (1992) Accretionary dust mantles in CM chondrites: evidence for solar nebula processes. Geochim Cosmochim Acta 56:2873–2897. https://doi.org/10.1016/0016-7037(92)90365-P

Morris MA, Boley AC (2018) Formation of chondrules by shock waves. In: Russell S, Krot AN, Connolly HC Jr (eds) Chondrules. Cambridge University Press, Cambridge, pp 375–399

Morris MA, Desch SJ (2010) Thermal histories of chondrules in solar nebula shocks. Astrophys J 722:1474–1494. https://doi.org/10.1088/0004-637X/722/2/1474

Murakami T, Ikeda Y (1994) Petrology and mineralogy of the Yamato-86751 CV3 chondrite. Meteoritics 29:397–409. https://doi.org/10.1111/j.1945-5100.1994.tb00604.x

Nagahara H (1981) Evidence for secondary origin of chondrules. Nature 292:135–136. https://doi.org/10.1038/292135a0

Nagahara H (1984) Matrices of type 3 ordinary chondrites-primitive nebular records. Geochim Cosmochim Acta 48:2581–2595. https://doi.org/10.1016/0016-7037(84)90307-7

Nguyen AN, Zinner E (2004) Discovery of ancient silicate stardust in a meteorite. Science 303:1496–1499. https://doi.org/10.1126/science.1094389

Nguyen AN, Stadermann FJ, Zinner E, Stroud RM, Alexander CMO, Nittler LR (2007) Characterization of presolar silicate and oxide grains in primitive carbonaceous chondrites. Astrophys J 656:1223–1240. https://doi.org/10.1086/510612

Nie NX, Chen X-Y, Hopp T, Hu JY, Zhang ZJ, Teng F-Z, Shahar A, Dauphas N (2021) Imprint of chondrule formation on the K and Rb isotopic compositions of carbonaceous meteorites. Sci Adv 7:eabl3929. https://doi.org/10.1126/sciadv.abl3929

Nittler LR, Ciesla F (2016) Astrophysics with extraterrestrial materials. Annu Rev Astron Astrophys 54:53–93. https://doi.org/10.1146/annurev-astro-082214-122505

Nittler LR, Stroud RM, Trigo-Rodríguez JM, De Gregorio BT, Alexander CMO, Davidson J, Moyano-Cambero CE, Tanbakouei S (2019) A cometary building block in a primitive asteroidal meteorite. Nat Astron 3:659–666. https://doi.org/10.1038/s41550-019-0737-8

Nittler LR, Alexander CMO, Patzer A, Verdier-Paoletti MJ (2021) Presolar stardust in highly pristine CM chondrites Asuka 12169 and Asuka 12236. Meteorit Planet Sci 56:260–276. https://doi.org/10.1111/maps.13618

Noguchi T, Yasutake M, Tsuchiyama A, Miyake A, Kimura M, Yamaguchi A, Imae N, Uesugi K, Takeuchi A (2021) Mineralogy of fine-grained matrix, fine-grained rim, chondrule rim, and altered mesostasis of a chondrule in Asuka 12169, one of the least altered CM chondrites. Polar Sci 29:100727. Special Issue on "The Sixth International Symposium on Arctic Research (ISAR-6)". https://doi.org/10.1016/j.polar.2021.100727

Ohtaki KK, Ishii HA, Bradley JP, Villalon KL, Davis AM, Stephan T, Bustillo KC, Ciston J (2021) Search for meteoritic GEMS I: comparison of amorphous silicates in Paris and Acfer 094 chondrite matrices and in anhydrous chondritic interplanetary dust particles. Geochim Cosmochim Acta 310:320–345. https://doi.org/10.1016/j.gca.2021.05.042

Olsen MB, Wielandt D, Schiller M, Van Kooten EMME, Bizzarro M (2016) Magnesium and ^{54}Cr isotope compositions of carbonaceous chondrite chondrules – insights into early disk processes. Geochim Cosmochim Acta 191:118–138. https://doi.org/10.1016/j.gca.2016.07.011

Onyett IJ, Schiller M, Makhatadze GV, Deng Z, Johansen A, Bizzarro M (2023) Silicon isotope constraints on terrestrial planet accretion. Nature 619:539–544. https://doi.org/10.1038/s41586-023-06135-z

Palme H, Fegley B Jr (1990) High-temperature condensation of iron-rich olivine in the solar nebula. Earth Planet Sci Lett 101:180–195. https://doi.org/10.1016/0012-821X(90)90152-N

Palme H, Hezel DC, Ebel DS (2015) The origin of chondrules: constraints from matrix composition and. matrix-chondrule complementarity. Earth Planet Sci Lett 411:11–19. https://doi.org/10.1016/j.epsl.2014.11.033

Pape J, Mezger K, Bouvier A-S, Baumgartner LP (2019) Time and duration of chondrule formation: constraints from 26Al-26Mg ages of individual chondrules. Geochim Cosmochim Acta 244:416–436. https://doi.org/10.1016/j.gca.2018.10.017

Patzek M, Bischoff A, Visser R, John T (2018) Mineralogy of volatile-rich clasts in brecciated meteorites. Meteorit Planet Sci 53:2519–2540. https://doi.org/10.1111/maps.13175

Patzek M, Bischoff A, Ludwig T, Whitehouse M, Trieloff M, Visser R, John T (2021) O-Isotope Signatures of Olivine and Pyroxene Grains in C1 and CM-Like Clasts. 84th Annual Meeting of the Meteoritical Society, 6096

Patzer A, Bullock ES, Alexander CMO (2021) Testing models for the compositions of chondrites and their components: I. CO chondrites. Geochim Cosmochim Acta 304:119–140. https://doi.org/10.1016/j.gca.2021.04.004

Patzer A, Bullock ES, Alexander CMO (2022) Testing models for the compositions of chondrites and their components: II. CR chondrites. Geochim Cosmochim Acta 319:1–29. https://doi.org/10.1016/j.gca.2021.12.021

Patzer A, Bullock ES, Alexander CMO (2023) Testing models for the compositions of chondrites and their components: III. CM chondrites. Geochim Cosmochim Acta 359:30–45. https://doi.org/10.1016/j.gca.2023.08.021

Perotti G, Sørensen HO, Haack H, Andersen AC, Sanchez DF, van Kooten EMME, Tsai EHR, Dalby KN, Holler M, Grolimund D, Hassenkam T (2021) Thermal history of matrix forsterite grains from Murchison based on high-resolution tomography. Astrophys J 922:256. https://doi.org/10.3847/1538-4357/ac26bc

Pinto GA, Marrocchi Y, Morbidelli A, Charnoz S, Varela MA, Soto K, Martinez R, Olivares F (2021) Constraints on planetesimal accretion inferred from particle-size distribution in CO chondrites. Astrophys J Lett 917:L25. https://doi.org/10.3847/2041-8213/ac17f2

Pringle EA, Moynier F (2017) Rubidium isotopic composition of the Earth, meteorites, and the Moon: evidence for the origin of volatile loss during planetary accretion. Earth Planet Sci Lett 473:62–70. https://doi.org/10.1016/j.epsl.2017.05.033

Pringle EA, Moynier F, Beck P, Paniello R, Hezel DC (2017) The origin of volatile element depletion in early Solar System material: clues from Zn isotopes in chondrules. Earth Planet Sci Lett 468:62–71. https://doi.org/10.1016/j.epsl.2017.04.002

Rambaldi ER (1981) Relict grains in chondrules. Nature 293:558–561. https://doi.org/10.1038/293558a0

Remusat L, Palhol F, Robert F, Derenne S, France-Lanord C (2006) Enrichment of deuterium in insoluble organic matter from primitive meteorites: a Solar System origin? Earth Planet Sci Lett 243:15–25. https://doi.org/10.1016/j.epsl.2005.12.010

Remusat L, Guan Y, Wang Y, Eiler JM (2010) Accretion and preservation of D-rich organic particles in carbonaceous chondrites: evidence for important transport in the early Solar System nebula. Astrophys J 713:1048–1058. https://doi.org/10.1088/0004-637X/713/2/1048

Rietmeijer FJM, Nuth JA (2013) Petrologic constraints on amorphous and crystalline magnesium silicates: dust formation and evolution in selected Herbig Ae/Be systems. Astrophys J 771:34. https://doi.org/10.1088/0004-637X/771/1/34

Robert F, Epstein S (1982) The concentration and isotopic composition of hydrogen, carbon and nitrogen in carbonaceous meteorites. Geochim Cosmochim Acta 46:81–95. https://doi.org/10.1016/0016-7037(82)90293-9

Rubin AE (1984) Coarse-grained chondrule rims in type 3 chondrites. Geochim Cosmochim Acta 48:1779–1789. https://doi.org/10.1016/0016-7037(84)90032-2

Rubin AE (2010) Physical properties of chondrules in different chondrite groups: implications for multiple melting events in dusty environments. Geochim Cosmochim Acta 74:4807–4828. https://doi.org/10.1016/j.gca.2010.05.018

Rubin AE, Krot AN (1996) Multiple heating of chondrules. International conference: Chondrules and the protoplanetary disk, p. 173–180

Rubin AE, Wasson JT (1987) Chondrules, matrix and coarse-grained chondrule rims in the Allende meteorite: origin, interrelationships and possible precursor components. Geochim Cosmochim Acta 51:1923–1937. https://doi.org/10.1016/0016-7037(87)90182-7

Rubin AE, Wasson JT (1988) Chondrules and matrix in the Ornans CO3 meteorite: possible precursor components. Geochim Cosmochim Acta 52:425–432. https://doi.org/10.1016/0016-7037(88)90098-1

Russell SDJ, Longstaffe FJ, King PL, Larson TE (2010) The oxygen-isotope composition of chondrules and isolated forsterite and olivine grains from the Tagish Lake carbonaceous chondrite. Geochim Cosmochim Acta 74:2484–2499. https://doi.org/10.1016/j.gca.2010.01.018

Sanborn ME, Wimpenny J, Williams CD, Yamakawa A, Amelin Y, Irving AJ, Yin Q-Z (2019) Carbonaceous achondrites Northwest Africa 6704/6693: milestones for early Solar System chronology and genealogy. Geochim Cosmochim Acta 245:577–596. https://doi.org/10.1016/j.gca.2018.10.004

Sanders IS, Scott ERD (2012) The origin of chondrules and chondrites: debris from low-velocity impacts between Molten planetesimals? Meteorit Planet Sci 47:2170–2192. https://doi.org/10.1111/maps.12002

Sanders IS, Scott ERD (2022) Complementary nucleosynthetic isotope anomalies of Mo and W in chondrules and matrix in the Allende carbonaceous chondrite: The case for hydrothermal metamorphism and its implications. Meteorit Planet Sci 57:450–471. https://doi.org/10.1111/maps.13742

Schiller M, Bizzarro M, Fernandes VA (2018) Isotopic evolution of the protoplanetary disk and the building blocks of Earth and the Moon. Nature 555:507–510. https://doi.org/10.1038/nature25990

Schiller M, Bizzarro M, Siebert J (2020) Iron isotope evidence for very rapid accretion and differentiation of the proto-Earth. Sci Adv 6:eaay7604. https://doi.org/10.1126/sciadv.aay7604

Schneider JM, Burkhardt C, Marrocchi Y, Brennecka GA (2020) Early evolution of the solar accretion disk inferred from Cr-Ti-O isotopes in individual chondrules. Earth Planet Sci Lett 551:116585. https://doi.org/10.1016/j.epsl.2020.116585

Schrader DL, Davidson J (2017) CM and CO chondrites: a common parent body or asteroidal neighbors? Insights from chondrule silicates. Geochim Cosmochim Acta 214:157–171. https://doi.org/10.1016/j.gca.2017.07.031

Schrader DL, Franchi IA, Connolly HC Jr, Greenwood RC, Lauretta DS, Gibson JM (2011) The formation and alteration of the Renazzo-like carbonaceous chondrites I: implications of bulk-oxygen isotopic composition. Geochim Cosmochim Acta 75:308–325. https://doi.org/10.1111/maps.12402

Schrader DL, Connolly HC Jr, Lauretta DS, Zega TJ, Davidson J, Domanik KJ (2015) The formation and alteration of the Renazzo-like carbonaceous chondrites III: toward understanding the genesis of ferromagnesian chondrules. Meteorit Planet Sci 50:15–50. https://doi.org/10.1111/maps.12402

Schrader DL, Nagashima K, Krot AN, Ogliore RC, Yin Q-Z, Amelin Y, Stirling CH, Kaltenbach A (2017) Distribution of 26Al in the CR chondrite chondrule-forming region of the protoplanetary disk. Geochim Cosmochim Acta 201:275–302. https://doi.org/10.1016/j.gca.2016.06.023

Scott ERD, Barber DJ, Alexander CM, Hutchinson R, Peck JA (1988) Primitive material surviving in chondrites: matrix. In: Kerridge JF, Matthews MS (eds) Meteorites and the early Solar System. University of Arizona Press, Tucson, pp 718–745

Shu F, Shang H, Lee T (1996) Toward an astrophysical theory of chondrites. Science 271:1545–1552. https://doi.org/10.1126/science.271.5255.1545

Skinner WR, Leenhouts JM (1993) Size distributions and aerodynamic equivalence of metal chondrules and silicate chondrules in Acfer 059. In: Twenty-fourth Lunar and Planetary Science Conference. Part 3: N-Z, Lunar and Planetary Inst., p 1315-1316

Stracke A, Palme H, Gellissen M, Münker C, Kleine T, Birbaum K, Günther D, Bourdon B, Zipfel J (2012) Refractory element fractionation in the Allende meteorite: implications for solar nebula condensation and the chondritic composition of planetary bodies. Geochim Cosmochim Acta 85:114–141. https://doi.org/10.1016/j.gca.2012.02.006

Tachibana S, Huss GR (2005) Sulfur isotope composition of putative primary troilite in chondrules from Bishunpur and Semarkona. Geochim Cosmochim Acta 69:3075–3097. https://doi.org/10.1016/j.gca.2004.06.025

Tenner TJ, Ushikubo T, Kurahashi E, Kita NT, Nagahara H (2013) Oxygen isotope systematics of chondrule phenocrysts from the CO3.0 chondrite Yamato 81020: evidence for two distinct oxygen isotope reservoirs. Geochim Cosmochim Acta 102:226–245. https://doi.org/10.1016/j.gca.2012.10.034

Tenner TJ, Nakashima D, Ushikubo T, Tomioka N, Kimura M, Weisberg MK, Kita NT (2019) Extended chondrule formation intervals in distinct physicochemical environments: evidence from Al-Mg isotope systematics of CR chondrite chondrules with unaltered plagioclase. Geochim Cosmochim Acta 260:133–160. https://doi.org/10.1016/j.gca.2019.06.023

Trinquier A, Birck J-L, Allègre CJ (2007) Widespread ^{54}Cr heterogeneity in the inner Solar System. Astrophys J 655:1179. https://doi.org/10.1086/510360

Trinquier A, Elliott T, Ulfbeck D, Coath C, Krot AN, Bizzarro M (2009) Origin of nucleosynthetic isotope heterogeneity in the solar protoplanetary disk. Science 324:374–376. https://doi.org/10.1126/science.1168221

Tronche EJ, Hewins RH, MacPherson GJ (2007) Formation conditions of aluminum-rich chondrules. Geochim Cosmochim Acta 71:3361–3381. https://doi.org/10.1016/j.gca.2007.03.037

Tyra MA, Farquhar J, Wing BA, Benedix GK, Jull AJT, Jackson T, Thiemens MH (2007) Terrestrial alteration of carbonate in a suite of Antarctic CM chondrites: evidence from oxygen and carbon isotopes. Geochim Cosmochim Acta 71:782–795. https://doi.org/10.1016/j.gca.2006.10.023

van Kooten E, Moynier F (2019) Zinc isotope analyses of singularly small samples (<5 ng Zn): investigating chondrule-matrix complementarity in Leoville. Geochim Cosmochim Acta 261:248–268. https://doi.org/10.1016/j.gca.2019.07.022

van Kooten E, Wielandt D, Schiller M, Nagashima K, Thomen A, Larsen KK, Olsen MB, Nordlund Å, Krot AN, Bizzarro M (2016) Isotopic evidence for primordial molecular cloud material in metal-rich carbonaceous chondrites. Proc Natl Acad Sci 113:2011–2016. https://doi.org/10.1073/pnas.1518183113

van Kooten EMME, Cavalcante LL, Nagashima K, Kasama T, Balogh ZI, Peeters Z, Hsiao SS-Y, Shang H, Lee D-C, Lee T, Krot AN, Bizzarro M (2018) Isotope record of mineralogical changes in a spectrum of aqueously altered CM chondrites. Geochim Cosmochim Acta 237:79–102. https://doi.org/10.1016/j.gca.2018.06.021

van Kooten E, Moynier F, Agranier A (2019) A unifying model for the accretion of chondrules and matrix. Proc Natl Acad Sci 116:18860–18866. https://doi.org/10.1073/pnas.1907592116

van Kooten E, Cavalcante L, Wielandt D, Bizzarro M (2020) The role of bells in the continuous accretion between the CM and CR chondrite reservoirs. Meteorit Planet Sci 55:1–16. https://doi.org/10.1111/maps.13459

van Kooten E, Schiller M, Moynier F, Johansen A, Haugbølle T, Bizzarro M (2021) Hybrid accretion of carbonaceous chondrites by radial transport across the Jupiter barrier. Astrophys J 910:70. https://doi.org/10.3847/1538-4357/abd9c8

van Kooten EMME, Kubik E, Siebert J, Heredia BD, Thomsen TB, Moynier F (2022) Metal compositions of carbonaceous chondrites. Geochim Cosmochim Acta 321:52–77. https://doi.org/10.1016/j.gca.2022.01.008

van Kooten EMME, Zhao X, Franchi I, Tung PY, Fairclough S, Walmsley J, Onyett I, Schiller M, Bizzarro M (2024) The nucleosynthetic fingerprint of the outermost protoplanetary disk and early Solar System dynamics. Sci Adv 10:eadp1613. https://doi.org/10.1126/sciadv.adp1613

Velbel MA (1988) The distribution and significance of evaporitic weathering products on Antarctic meteorites. Meteoritics 23:151–159. https://doi.org/10.1111/j.1945-5100.1988.tb00910.x

Velbel MA (2014) Stoichiometric reactions describing serpentinization of anhydrous primary silicates: a critical appraisal, with application to aqueous alteration of chondrule silicates in CM carbonaceous chondrites. Clays Clay Miner 62:126–136. https://doi.org/10.1346/CCMN.2014.0620205

Villeneuve J, Chaussidon M, Libourel G (2009) Homogeneous distribution of ^{26}Al in the Solar System from the Mg isotopic composition of chondrules. Science 325:985–988. https://doi.org/10.1126/science.1173907

Villeneuve J, Libourel G, Soulié C (2015) Relationships between type I and type II chondrules: implications on chondrule formation processes. Geochim Cosmochim Acta 160:277–305. https://doi.org/10.1016/j.gca.2015.03.033

Wai CM, Wasson JT (1977) Nebular condensation of moderately volatile elements and their abundances in ordinary chondrites. Earth Planet Sci Lett 36:1–13. https://doi.org/10.1016/0012-821X(77)90182-0

Warren PH (2011) Stable isotopes and the noncarbonaceous derivation of ureilites, in common with nearly all differentiated planetary materials. Geochim Cosmochim Acta 75:6912–6926. https://doi.org/10.1016/j.gca.2011.09.011

Wasson JT (1993) Constraints on chondrule origins. Meteoritics 28:14–28. https://doi.org/10.1111/j.1945-5100.1993.tb00244.x

Wasson JT (2008) Evaporation of nebular fines during chondrule formation. Icarus 195:895–907. https://doi.org/10.1016/j.icarus.2008.01.017

Wasson JT, Chou CL (1974) Fractionation of moderately volatile elements in ordinary chondrites. Meteoritics 9:69–84. https://doi.org/10.1111/j.1945-5100.1974.tb00063.x

Wasson JT, Rubin AE (2003) Ubiquitous low-FeO relict grains in type II chondrules and limited overgrowths on phenocrysts following the final melting event. Geochim Cosmochim Acta 67:2239–2250. https://doi.org/10.1016/S0016-7037(03)00023-1

Weisberg MK, Prinz M, Clayton RN, Mayeda TK (1993) The CR (Renazzo-type) carbonaceous chondrite group and its implications. Geochim Cosmochim Acta 57:1567–1586. https://doi.org/10.1016/0016-7037(93)90013-M

Weisberg MK, McCoy TJ, Krot AN (2006) Systematics and evaluation of meteorite classification. In: Lauretta D, McSween HY Jr (eds) Meteorites and the early Solar System II, pp 19–52. University of Arizona, Tucson

Whipple FL (1972) On certain aerodynamic processes for asteroids and comets. In: Elvius A (ed) From plasma to planet. Proceedings of the twenty-first Nobel symposium, p 211

Williams CD, Sanborn ME, Defouilloy C, Yin Q-Z, Kita NT, Ebel DS, Yamakawa A, Yamashita K (2020) Chondrules reveal large-scale outward transport of inner Solar System materials in the protoplanetary disk. Proc Natl Acad Sci 117:23426–23435. https://doi.org/10.1073/pnas.2005235117

Wood JA (1996a) Processing of chondritic and planetary material in spiral density waves in the nebula. Meteoritics 31:641–645. https://doi.org/10.1111/j.1945-5100.1996.tb02037.x

Wood JA (1996b) Unresolved issues in the formation of chondrules and chondrites. In: Hewins RH, Jones RH, Scott ERD (eds) Chondrules and the protoplanetary disk. Cambridge University Press, Cambridge, pp 55–70

Yang J, Epstein S (1983) Interstellar organic matter in meteorites. Geochim Cosmochim Acta 47:2199–2216. https://doi.org/10.1016/0016-7037(83)90043-1

Yang J, Epstein S (1984) Relic interstellar grains in Murchison meteorite. Nature 311:544–547. https://doi.org/10.1038/311544a0

Zanda B, Lewin É, Humayun M (2018) the Chondritic Assemblage. Chondrules: Records of Protoplanetary Disk Processes. https://doi.org/10.1017/9781108284073.005

Zega TJ, Alexander CMOD, Busemann H, Nittler LR, Hoppe P, Stroud RM, Young AF (2010) Mineral associations and character of isotopically anomalous organic material in the Tagish Lake carbonaceous chondrite. Geochim Cosmochim Acta 74:5966–5983. https://doi.org/10.1016/j.gca.2010.07.018

Zhang M, Lin Y, Tang G, Liu Y, Leya I (2020) Origin of Al-rich chondrules in CV chondrites: incorporation of diverse refractory components into the ferromagnesian chondrule-forming region. Geochim Cosmochim Acta 272:198–217. https://doi.org/10.1016/j.gca.2019.12.011

Zhu K, Liu J, Moynier F, Qin L, Alexander CMO'D, He Y (2019) Chromium isotopic evidence for an early formation of chondrules from the Ornans CO chondrite. Astrophys J 873:82. https://doi.org/10.3847/1538-4357/aafe79

Zinner E (2014) Presolar grains. In: Davis AM (ed) Meteorites and cosmochemical processes. Elsevier, Oxford, pp 181–213

Space Science Reviews (2025) 221:97
https://doi.org/10.1007/s11214-025-01216-z

Initial Conditions of Planet Formation: Time Constraints from Small Bodies and the Lifetime of Reservoirs in the Solar Protoplanetary Disk

Maria Schönbächler[1] · Audrey Bouvier[2] · Noriko T. Kita[3] · Thomas S. Kruijer[4]

Received: 16 April 2025 / Accepted: 5 September 2025 / Published online: 17 October 2025

Abstract
This review explores the timescales of the initial phase of planet formation, from nebular dust (CAIs and chondrules) to planetesimal accretion and differentiation, using evidence from meteorite research. Aluminium-Mg systematics of CAIs indicate either an extended period of CAI formation ($\sim$0.3 Ma) or an initial ^{26}Al heterogeneity, with evidence supporting a homogeneous ^{26}Al abundance in the protoplanetary disk. Thermal and aqueous alteration on the parent body can disturb the U-Pb and Al-Mg chronometers in chondrules. Focusing on relatively robust isochron data from plagioclase of pristine (types $\leq$3.05) chondrites indicates a shift in chondrule formation locations, moving from the inner to the outer disk over time. Ages of basaltic achondrites show that silicate differentiation on small bodies was well underway within the first few million years (Ma) of our solar system. Their age record, however, reveals inconsistencies between different chronometers, partly caused by (i) secondary disturbances due to thermal metamorphism, aqueous alteration, or impacts, (ii) the presence of xenolithic minerals, and (iii) potentially variable initial ^{26}Al abundances due to disturbances at the mineral scale. Nucleosynthetic isotope data indicate that parent bodies of iron and stony meteorites formed in two distinct regions within the protoplanetary disk: the inner, non-carbonaceous (NC) and the outer, carbonaceous (CC) region. Based on Hf-W chronometry it has been demonstrated that NC and CC parent bodies of magmatic iron meteorites segregated their cores within $\sim$1–3 Ma after CAI formation, implying that parent body accretion occurred within $<$1 Ma in both reservoirs. Combining accretion ages with nucleosynthetic data further reveals that, at first order, NC and CC reservoirs in the solar protoplanetary disk were established within 1 Ma and existed over several Ma with limited exchange between them. In the CR chondrite accretion region of the disk, planetary bodies formed over at least 3 Ma, while in most other regions, formation spanned at least 1 Ma, with minimal changes in nucleosynthetic isotope compositions. Aerodynamical size sorting of dust likely introduced or amplified some of these variations.

Keywords Meteorite chronology · Planetesimal formation · Nucleosynthetic data · ^{26}Al distribution · Protoplanetary disk · Lifetime of isotopic reservoirs

Extended author information available on the last page of the article

1 Introduction

The accretion and differentiation of 10-100 km sized planetesimals and larger planetary embryos during the first few million years (Ma) of solar system history represents an essential step towards the growth of larger planetary bodies like the Earth. However, the formation of planetesimals and embryos from small nebular dust constitutes a step in the planet formation process that is notoriously hard to constrain, and the so-called meter-size barrier has been a challenge in models of terrestrial planet formation (Chambers 2004; Nimmo et al. 2018; Raymond and Morbidelli 2022). Vital insights into this problem and, more generally, into planetesimal accretion and differentiation can be obtained from meteorites.

Moreover, the final compositions of planets are influenced at different stages during planet formation: (i) by the composition of dust and gas collected during the infall stage of the disk, (ii) through dust transport and thermal processing (e.g., moving snow line) during the infall stage and within the protoplanetary disk, (iii) through planetesimal formation whose compositions depend on the time and location in the evolving disk, and (iii) finally, by mixing of material during protoplanet growth by collisions of the planetesimals and pebble accretion.

In the past decade, the exploration of early solar system processes has made significant advances in various disciplines. Astronomical observations of protoplanetary disks by the Atacama Large Millimeter Array (ALMA) telescope have provided detailed insights into the structures around new-born stars (e.g., Andrews et al. 2016). Astrophysical models have succeeded in defining new scenarios on how planetesimals and planets are formed (e.g., streaming instabilities and pebble accretion). In cosmochemistry, novel data from meteorites using various isotopic systems have further constrained the timing of dust, planetesimal and planet formation as well as their composition at the different stages.

In this contribution, we will address the initial stages of the protoplanetary disk by reviewing ages of (i) the oldest known material of the solar system preserved in primitive meteorites, i.e. Calcium-Aluminium-rich Inclusions (CAI) and chondrules, (ii) basaltic achondrites, and (iii) magmatic iron meteorites. In contrast to primitive meteorites, basaltic achondrites and iron meteorites originate from differentiated bodies that underwent melting and have preserved crucial evidence about their accretion and differentiation early in solar system history. These bodies experienced partial melting and differentiation to form core, mantle, and crust layers, which can be dated using radioactive decay systems.

CAIs and chondrules The elemental abundances and mineralogy of CAIs match those expected for the first ∼5% material that condenses out of a nebular gas of solar composition based on thermodynamic calculations (e.g., Davis and Richter 2014). CAIs are generally assumed to have formed near the Sun, because regions near the young Sun were heated by the energy released from material accreting onto it, leading to the production of solar gas and a transitional region where the gas can cool and condense.

While CAIs have relatively low abundances (<3–10%; Hezel et al. 2008) in chondrites, chondrules are one of the main components (up to 80%; Jones 2012) of primitive chondrites. They are droplets of crystallized silicate melt formed by rapid heating of dust aggregates to high temperatures (>1800 K) followed by hours or days of cooling (e.g., Jones 2012). The formation of CAIs and chondrules occurred in the first few Ma of solar system history, corresponding to the estimated lifetime of the protoplanetary disk (e.g., Evans et al. 2009; Hunt et al. 2022). By definition, the age of CAIs marks the time zero (t_0) of our solar system, and as such the age of our solar system (e.g., Amelin et al. 2010; Bouvier and Wadhwa 2010; Connelly et al. 2012; Desch et al. 2023b; Piralla et al. 2023). Here, we review recent progress on CAI and chondrule chronology using the U-Pb and Al-Mg decay systems that can resolve small time differences (≤0.1 Ma).

Basaltic achondrites These differentiated meteorites are igneous rocks (melts or cumulates) or breccias of igneous rock fragments from the silicate-dominated crust and mantle of large asteroids (i.e., > 20 km in radius) or other planetary bodies (Mars and Moon). The elemental compositions of the basaltic achondrites and their parental melts are fractionated (i.e., enriched in incompatible elements) compared to chondritic compositions (Mittlefehldt 2014). Additionally, impact processes can have modified their compositions by crystal assimilation and brecciation processes (e.g., Rider-Stokes et al. 2023). This complex history needs to be considered when attributing a geological meaning to the ages or for the comparison of ages determined with different chronometers (e.g., K-Ar, Rb-Sr, Sm-Nd, U-Pb) on the same sample. For example, the chronological records of magmatic processes that took place on the meteorite parent bodies can vary across radiometric systems due to differences in elemental diffusion rates. These differences affect the temperature at which diffusion becomes negligible and the system closes, marking the time recorded by each specific decay system. Diffusion rates also contribute to the distinct responses of each decay system to impact-induced changes such as phase transformation, re-heating, and possible re-melting depending on the energy associated with the event and the size of the impacted object (Bogard 1995). As a result, different chronometers may date different events related to the formation and evolution of meteorite parent bodies and ages must be interpreted carefully to accurately link them to the corresponding geological events.

Using a combination of long-lived radiometric (LLR) systems (U-Pb) and short-lived radionuclides (SLR) (Al-Mg, Mn-Cr, Hf-W), the formation of early-formed silicate-enriched reservoirs can be constrained with a precision better than one million years. These decay systems provide complementary information on planetary processes due to the contrasting geochemical behaviour of the parent and daughter isotopes. The Al-Mg system (half-life ~0.705 Ma) is particularly useful to date silicate differentiation as well as CAIs and chondrules. The Hf-W system (half-life ~8.9 Ma) is utilised to date metal-silicate differentiation due to W being a siderophile (preserving the W isotopic composition of the core at the time of its formation) and Hf a lithophile element. In addition, the Hf-W system can also date silicate differentiation. The Mn-Cr system (half-life ~3.7 Ma) provides constraints on nebular processes associated with the volatile nature of Mn ($T_c = 1158$ K) compared to Cr ($T_c = 1296$ K) (Lodders 2003) and Mn/Cr fractionation associated with igneous, metamorphic or aqueous alteration processes.

Here we will focus on the chronological record of basaltic achondrites that are used as time anchors to deduce a consistent chronology of early solar system events. Within this framework, we will address the age of CAIs and the magmatic activity on small bodies, which can in turn be used to infer the time of parent body accretion. These age comparisons also prompt the question whether the short-lived nuclide ^{26}Al was distributed homogeneously or heterogeneously within our solar system. This has implications for using the Al-Mg decay system for dating, in addition to parent body differentiation (metal-silicate and silicate-silicate), because the radioactive decay of ^{26}Al is the major heat source for internal heating in the early solar system (Hevey and Sanders 2006).

Iron meteorites Additional vital insights into planetesimal accretion and differentiation come from isotopic studies of iron meteorites. Magmatic iron meteorites consist mainly of metallic iron or Fe-Ni alloys and are interpreted to be fragments of metal cores of protoplanets (Scott 1972; Scott and Wasson 1975). Consequently, magmatic iron meteorites are the only hand specimens of planetary core material in our collection of extraterrestrial samples and provide an opportunity to directly study the differentiation, cooling, and crystallization history of the metallic cores of small differentiated solar system bodies. Trace element signatures of iron meteorites indicate that the metal cores of magmatic iron meteorite parent bodies formed by segregation and crystallization of metallic melts (Scott and Wasson 1975). The classification of iron meteorites, as well as their formation, crystallization, and cooling history, have been discussed in several comprehensive reviews (Chabot and Haack 2006; Goldstein et al. 2009; Benedix et al. 2014; Scott 2020; Bouvier et al. 2025). Here, we focus on accretion and core formation of the iron meteorite parent bodies and their bearing on planetesimal formation and the early evolution of the solar protoplanetary disk. While the accretion time of chondrite parent bodies is constrained by the minimum age of individual chondrules measured in each group, the time of accretion of differentiated meteorite parent bodies has been deduced from geochemical and thermal modelling for bodies internally heated by ^{26}Al decay (Hevey and Sanders 2006; Sugiura and Fujiya 2014; Neumann et al. 2018).

1.1 Nucleosynthetic Isotope Variations

The nucleosynthetic isotope composition of each planetary body is unique, acting as a distinctive identifier for asteroids and planets, with rare exceptions, such as the Earth-Moon system (e.g., Fig. 1). These signatures are powerful tracers to determine genetic relationships between meteorite groups, samples returned by space missions, and planets like Earth and Mars (e.g., Yokoyama et al. 2023; Barnes et al. 2025). They also offer insights into the source regions of meteorites in the solar protoplanetary disk as well as into transport and mixing processes within the disk. Combining nucleosynthetic isotope data with radiometric ages of meteorites and other cosmochemical evidence has proven to be a powerful approach to constrain planet formation. For example, this combination was used to constrain the lifetime of major reservoirs within the solar protoplanetary disk (Warren 2011; Kruijer et al. 2017, 2020), which has important implications for the large scale evolution of the early solar system as well as for the formation timescales and accretion mode of Jupiter (e.g., Morbidelli et al. 2016; Kruijer et al. 2017; Alibert et al. 2018; Brasser and Mojzsis 2020). Additionally, integration of cosmochemical constraints with astrophysical models helped constrain how the evolution of the protoplanetary disk affected the timing and mechanisms of planet formation (e.g., Brasser and Mojzsis 2020; Lichtenberg et al. 2021; Morbidelli et al. 2022; Johansen et al. 2023).

The variations in nucleosynthetic isotope compositions are inherited from stardust (i.e., presolar grains) formed by earlier star generations. These grains, found in primitive chondrites, have extreme and unique isotopic compositions that reflect their stellar formation site. This dust survived in the interstellar medium (ISM) and was mixed into the solar protoplanetary disk together with dust grown in the ISM, the latter with approximately solar system composition (Ek et al. 2020). The incorporation of different amounts of stardust, depending on the formation region in the disk, led to fundamental isotopic differences between all known planetary materials. Depending on the geochemical properties of an element and its carrier phase, parent body processes (e.g., aqueous alteration) can redistribute the nucleosynthetic signal and lead to small scale variations in chondrites (e.g., Cr, Os; Yokoyama et al. 2023, 2011). Additionally, incomplete digestion of presolar grains during bulk sample analysis can produce similar effects (e.g., Schönbächler et al. 2025).

Earlier studies discovered that carbonaceous chondrites exhibit distinctly higher ^{54}Cr and ^{50}Ti abundances compared to other meteorite groups (Trinquier et al. 2007, 2009; Leya

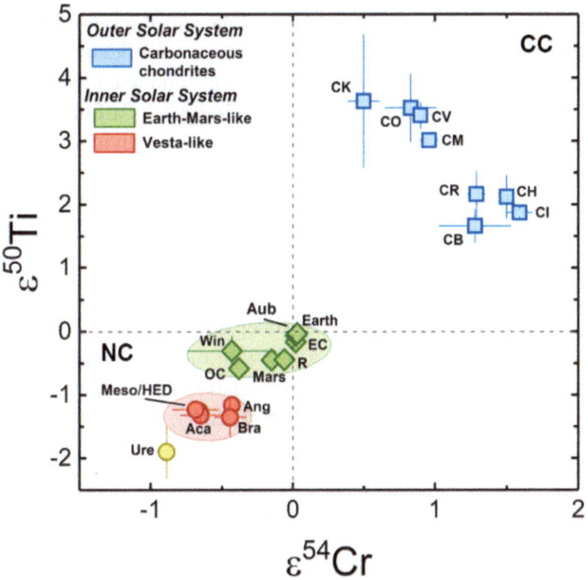

Fig. 1 Averages for different meteorite groups for ε^{50}Ti versus ε^{54}Cr. In addition to the NC-CC dichotomy (see main text), the NC reservoir displays two distinct clusters (red and green ellipses), which are also distinct from ureilites (yellow). The red and green ellipses comprise the Vesta-like meteorites and the Earth-Mars like meteorites, respectively. Uncertainties are student-t 95% confidence intervals. Epsilon is defined as the deviations of the ^{50}Ti/^{47}Ti or ^{54}Cr/^{52}Cr ratio of the sample from the corresponding terrestrial standards expressed in parts per 10,000. Abbreviations: Aub - aubrites, EC - enstatite chondrites, Win - winonaites, OC - ordinary chondrites, R - R-chondrites, Meso - mesosiderites, Aca - acapulcoites, Ang - angrites, Bra - brachinites, Ure - ureilites. From Rüfenacht et al. (2023)

et al. 2008), which is particularly evident when ^{54}Cr is plotted versus ^{50}Ti (Trinquier et al. 2009). These early studies highlighted two main reservoirs from which early solar system bodies formed: the volatile-rich carbonaceous chondrites that formed outside the snowline and more volatile-depleted material from the inner solar system. These two reservoirs were termed "carbonaceous (CC)" and "non-carbonaceous (NC)" by Warren (2011), who also speculated that carbonaceous chondrites may have formed outside Jupiter's orbit. These early studies already indicated that these two reservoirs contain both differentiated and undifferentiated bodies based on Ti, Cr and Ni isotope data (Trinquier et al. 2007, 2009; Leya et al. 2008; Steele et al. 2011, 2012). Subsequent work extended the dichotomy to Mo isotopes and further highlighted that the NC-CC dichotomy is not limited to primitive meteorites but also includes differentiated meteorites, including iron meteorites, which are found in both the NC and CC reservoirs (Budde et al. 2016; Kruijer et al. 2017). Among other elements, the dichotomy has been extended to Fe, Ca and Zn (Schiller et al. 2015b; Hopp et al. 2022a,b; Steller et al. 2022; Savage et al. 2022; Martins et al. 2023). How much this dichotomy is reflected in other heavier elements, which show nucleosynthetic isotope variations, can be debated. For iron meteorites, distinct isotope compositions have been reported as evidence for the dichotomy, including siderophile elements such as Ni, Ru, and W (e.g., Kruijer et al. 2017; Nanne et al. 2019; Worsham et al. 2019; Spitzer et al. 2025). Many heavier elements display a continuous trend of nucleosynthetic isotope compositions and have some overlaps of NC and CC materials (e.g., Pd, Nd and W; Worsham et al. 2019; Ek et al. 2020; Frossard et al. 2021; Spitzer et al. 2025). Despite overlaps, the NC and CC data each consistently define their specific regions in the isotope space, highlighting the close genetic relationships within the individual CC and NC reservoirs (Burkhardt et al. 2021; Rüfenacht et al. 2023). In summary, a substantial body of evidence underscores the NC-CC dichotomy as a key feature of solar system formation and evolution. As a first order observation, the solar system material available for laboratory studies can be divided into two types: (i) carbonaceous chondrites and related achondrites, which sample the outer solar system (CC reservoir) and are enriched in supernova-derived isotopes, and (ii) non-carbonaceous chondrites and related achondrite from the inner solar system (NC reservoir), which are generally enriched in s-process isotopes.

The preservation of the NC-CC dichotomy indicates an early spatial separation within the solar protoplanetary disk separating the inner and the outer solar system for several Ma (Kruijer et al. 2017; Morbidelli et al. 2022) with minor exchange (e.g., Williams et al. 2020). This in turn requires a dynamical barrier against mixing and homogenization, such as the early formation of Jupiter (Morbidelli et al. 2016; Budde et al. 2016; Kruijer et al. 2017; Alibert et al. 2018), a pressure maximum within the disk (Brasser and Mojzsis 2020) possibly related to the water ice line (Charnoz et al. 2021; Lichtenberg et al. 2021; Izidoro et al. 2022; Morbidelli et al. 2022) or a combination of these processes. The finding that this dichotomy is observed for not only for chondrites, but also iron meteorites and basaltic achondrites provides strong evidence that differentiated meteorite parent bodies formed in the inner (NC) and the outer (CC) protoplanetary disk (e.g., Kruijer et al. 2017).

A closer examination of the nucleosynthetic data reveals isotopic differences within each of the two major reservoirs (NC-CC dichotomy) (e.g., Trinquier et al. 2009; Steele et al. 2012; Akram et al. 2015; Fischer-Gödde et al. 2015; Kruijer et al. 2017; Ek et al. 2020; Frossard et al. 2021; Render et al. 2022). Based on new high-precision Ti and Cr isotope measurements, Rüfenacht et al. (2023) proposed three sub-reservoirs within the inner solar system (NC) reservoir (Fig. 1). They proposed that these sub-reservoirs may mirror disk substructures such as secondary gaps and rings, potentially created by spiral arms emitted from the growing Jupiter and/or Saturn. The distinct isotopic compositions of these reservoirs may result from thermal processing of dust in the disk (Trinquier et al. 2009; Akram et al. 2015; Ek et al. 2020) and/or temporal isotopic variations caused by infalling material from an isotopically heterogeneous molecular cloud (Nanne et al. 2019; Burkhardt et al. 2019) and/or thermal effects during infall (Van Kooten et al. 2016). The variations were likely amplified by dust sorting due pressure bumps and gaps in the protoplanetary disk (Hutchison et al. 2022; Rüfenacht et al. 2023).

In this review, we combine age constraints from CAIs, chondrules and achondrites, with nucleosynthetic isotope data to reassess, whether nucleosynthetic variations reflect (i) the isotopic make-up of their formation location in the solar protoplanetary disk, which remained stable over time or (ii) a signal of evolving isotopic compositions over time in the disk such that each planetary body capture a snapshot of the evolving isotopic composition of the disk at the time of their formation. The goal is to evaluate how long isolated rings can exist in the disk, which is crucial for understanding planet formation. Astronomical observations (ALMA) show rings reflecting dust of >100 μm to mm-size, but we lack information on the ring lifetime, making this an important question to address.

2 Chronology of Refractory Inclusions and Chondrules

Formation of CAIs and chondrules in primitive chondrites occurred in the first few Ma corresponding to the lifetime of the protoplanetary disk (e.g., Evans et al. 2009). Here, we focus on recent progress on CAI and chondrule chronology using the U-Pb and Al-Mg systems that can resolve small time differences relevant to the timing and extent of their formation ($\leq$0.1 Ma). The reader may refer to previous review articles, Kita and Ushikubo (2012), Kita et al. (2013) and Nagashima et al. (2018), for comprehensive reviews of earlier studies.

2.1 U-Pb Chronometry

The U-Pb chronometer uses the radioactive decay chains of two U isotopes to two Pb isotopes: ^{238}U-^{206}Pb, ^{235}U-^{207}Pb with decay constants λ_{238} of 1.55125×10^{-10} and λ_{235} of 9.8485×10^{-10} y^{-1}, respectively (Steiger and Jäger 1977). The formation ages (t) of CAIs and chondrules provide absolute timescales for their formation events (often referred as "absolute age") if highly accurate and precise radiogenic (^{207}Pb*/^{206}Pb*) ratios and U isotope ratios (^{235}U/^{238}U) of the samples are determined using the Eq. (2.1).

$$\frac{^{207}\mathrm{Pb}^*}{^{206}\mathrm{Pb}^*} = \frac{^{235}\mathrm{U}}{^{238}\mathrm{U}} \times \frac{e^{\lambda_{235}t} - 1}{e^{\lambda_{238}t} - 1} \tag{2.1}$$

The symbol * refers to daughter isotope produced solely by radiogenic ingrowth. Highly accurate estimates of radiogenic ^{207}Pb*/^{206}Pb* ratios require samples with extremely low ^{204}Pb/^{206}Pb ratios of $<10^{-3}$ after the repeated acid leaching procedures, which aim at removing common or unradiogenic (initial or exogenous) Pb (e.g., Amelin 2006). A spread in Pb isotopic compositions is needed to deduce the ^{207}Pb*/^{206}Pb* radiogenic ratio from the linear Pb-Pb isochron regression or for two-point model ages using the Pb isotopic composition of troilite from the Canyon Diablo iron meteorite (used as the initial Pb isotopic composition of the solar system because of the absence of U in troilite; Amelin 2006). Variations in ^{235}U/^{238}U caused by ^{247}Cm short-lived radioactive isotope decay in CAIs can bias the Pb-Pb ages up to a few Ma unless accurate U isotope measurements are carried out (Brennecka et al. 2010; Tissot et al. 2016). Published data of CAIs from CV3 chondrites with corresponding U isotope analyses show a small range of U-Pb ages with a mean value of 4567.30 ± 0.16 Ma (n = 5, Amelin et al. 2010; Connelly et al. 2012). A report of an internal U-corrected Pb-Pb isochron on mineral separates of a compact type A CAI from NWA 6991 is, however, notably older at 4567.94 ± 0.31 Ma (Bouvier et al. 2011). Hence, CAI ages show that they are the oldest dated materials in the solar system. In contrast, chondrules from both ordinary chondrites (OC) and CR chondrites show a larger range of Pb-Pb formation ages from 4567.5 Ma to 4563 Ma (Fig. 2, Bollard et al. 2017), suggesting that some chondrules may have formed within $\leq$0.5 Ma from CAI formation and continued to form for 3-4 Ma, while most chondrules in OC are older than those in CR chondrites.

2.2 Al-Mg Chronology of CAIs

2.2.1 Bulk CAI Al-Mg Isochron

The short-lived nuclide ^{26}Al decays to ^{26}Mg (half-life $t_{1/2} = 0.705$ Ma; Nishiizumi 2004) and the decay system is used as a chronometer under the assumption of a homogeneous initial ^{26}Al abundance in the early solar system (e.g., Kita et al. 2013). Many studies use a half-life of 0.717 Ma which is the weighted mean of literature estimates (Samworth et al. 1972; Norris et al. 1983; Middleton et al. 1983; Thomas et al. 1984). We adopt 0.705 Ma according to Nishiizumi (2004) who estimated the ^{26}Al half-life to 0.708 ± 0.017 Ma by

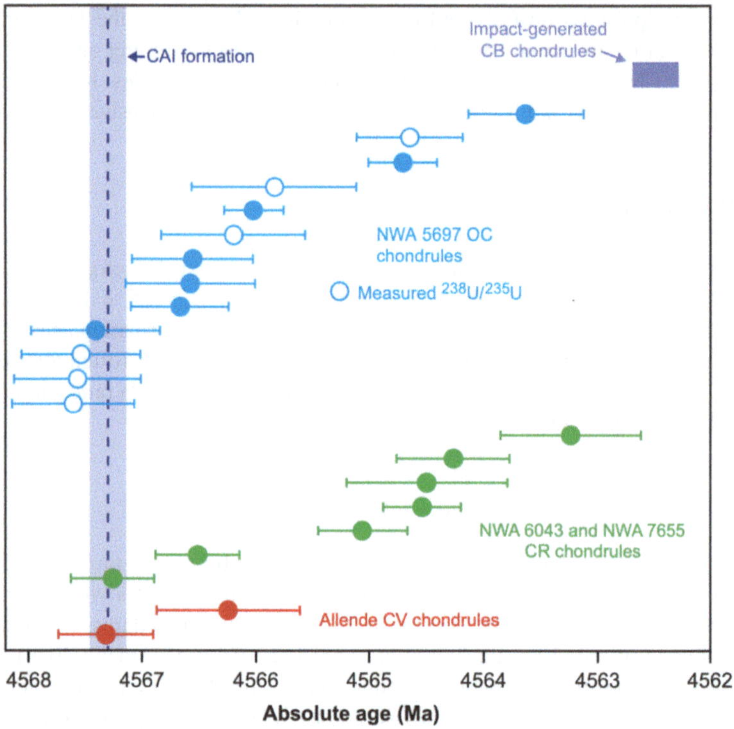

Fig. 2 Pb-Pb ages of individual chondrules from ordinary chondrite (NWA 5697, L3.10), CR chondrites and a CV chondrite (Bollard et al. 2017). The shaded area indicates the timing of CAI formation (4567.30 ±0.16 Ma; Connelly et al. 2012). CB chondrule data are shown for comparison (4562.49 ±0.21 Ma; Bollard et al. 2015). The CB chondrules are considered to have formed in a planetesimal impact plume (Krot et al. 2005; Bollard et al. 2015)

using three direct ^{26}Al measurements by mass spectrometry (Samworth et al. 1972; Norris et al. 1983; Middleton et al. 1983). This is within uncertainty of the value 0.705 Ma widely used in the cosmogenic nuclide community (Nishiizumi 2004). Earlier studies by Jacobsen et al. (2008) and Larsen et al. (2011) show that multiple bulk Al-Mg analyses of CAIs and amoeboid olivine aggregates (AOA) plot on a single isochron regression line with the slopes corresponding to a $(^{26}\mathrm{Al}/^{27}\mathrm{Al})_0$ ratio of $\sim 5.2 \times 10^{-5}$, which is often referred as the *canonical ratio* of CAIs. The subscript 0 refers to the initial ratio at the time of object formation. More recently, Larsen et al. (2020) reported bulk CAI data from CR chondrites, which also exhibit an indistinguishable $(^{26}\mathrm{Al}/^{27}\mathrm{Al})_0$ ratio of 5.1 ($\pm$ 0.2) $\times 10^{-5}$ but with significantly lower initial $(\delta^{26}\mathrm{Mg}^*)_0$ as the intercept of the Al-Mg isochron regression line. Three data sets are summarized in Fig. 3 and suggest that the ^{26}Al abundance in the formation location of CAIs seems to be reasonably homogeneous, but Mg isotopes were heterogeneously distributed. Similarly low $(\delta^{26}\mathrm{Mg}^*)_0$ values have been reported from internal isochron studies of individual CAIs, which are likely due to nucleosynthetic anomalies of Mg isotopes (Wasserburg et al. 2012; MacPherson et al. 2017; Larsen et al. 2020). Widespread Mg isotope heterogeneity among CAIs may hinder us from determining the slope of the bulk CAI isochrons with high precision.

Conventionally, we use the $(^{26}\mathrm{Al}/^{27}\mathrm{Al})_0$ of 5.252 ($\pm$ 0.019) $\times 10^{-5}$ from CV CAIs determined by Larsen et al. (2011) as that of the initial solar system at time of 0 and estimate

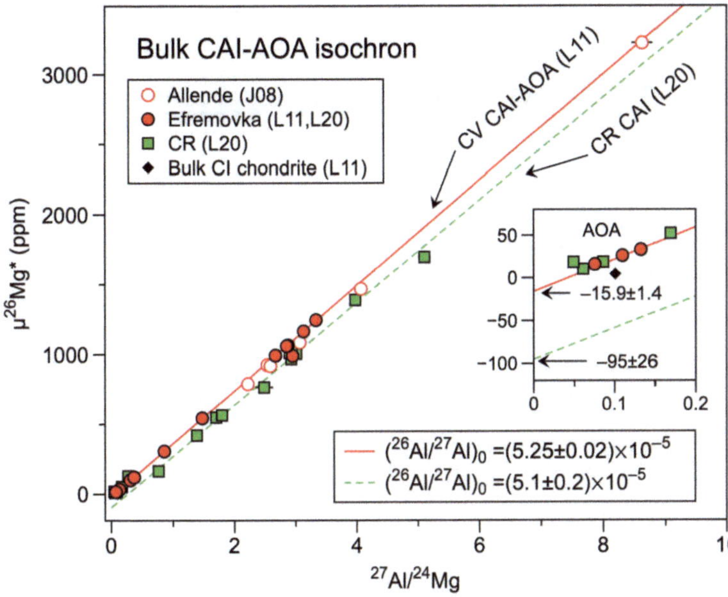

Fig. 3 Bulk Al-Mg analyses of CAIs and AOA from CV chondrites (Allende and Efremovka) and CR chondrites (J08 = Jacobsen et al. 2008; L11, L20 = Larsen et al. 2011, 2020). Regression lines are for CV CAI-AOAs (solid line) and CR CAIs (dashed line). The AOAs from CR chondrites (inlet) plot on the same regression line as the CV CAI-AOA isochron

a ^{26}Al-^{26}Mg age relative to CAI formation using the Equation (2.2) below.

$$\Delta t_{\mathrm{CAI}}(Ma) = \ln\left[\frac{\left(\frac{^{26}\mathrm{Al}}{^{27}\mathrm{Al}}\right)_{0,\mathrm{CAIs}}}{\left(\frac{^{26}\mathrm{Al}}{^{27}\mathrm{Al}}\right)_{0,\mathrm{Samples}}}\right] \times \frac{0.705}{\ln(2)}. \tag{2.2}$$

2.2.2 Internal Al-Mg Isochrons of CAIs

MacPherson et al. (2012) determined internal Al-Mg isochrons of multiple types of CAIs in CV chondrites and reported uniform $(^{26}\mathrm{Al}/^{27}\mathrm{Al})_0$ ratios of $\sim 5.2 \times 10^{-5}$ for condensate CAIs and a range from 5.2×10^{-5} to 4.2×10^{-5} for melted CAIs. These results suggested that CAIs experienced high temperature reheating events typically for 0.2 Ma after their initial formation. A summary of $(^{26}\mathrm{Al}/^{27}\mathrm{Al})_0$ ratios determined from internal isochrons of CAIs are displayed in Fig. 4 including data from other carbonaceous chondrites. Data from CV chondrites show $(^{26}\mathrm{Al}/^{27}\mathrm{Al})_0$ ratios of (4-5) $\times 10^{-5}$ regardless of CAI types, such as types A (melilite-rich) and B (melilite-fassaite). Many condensate CAIs, such as Fluffy Type A (FTA) and spinel-rich fine-grained CAIs, exhibit $(^{26}\mathrm{Al}/^{27}\mathrm{Al})_0$ consistent with canonical ratios, while others have lower values in the same range as igneous CAIs (e.g., Kawasaki et al. 2017, 2019, 2020). CAIs in Acfer 094, CR, CO and CM chondrites are mostly non-igneous fine-grained CAIs that are likely nebular condensates and reveal a range of $(^{26}\mathrm{Al}/^{27}\mathrm{Al})_0$ ratios similar to those of igneous CAIs in CV. These data collectively indicate that condensate CAIs formed contemporaneously with igneous CAIs (e.g., Kawasaki et al. 2021) over $\sim$0.3 Ma. Several CAIs in Fig. 4 display significantly lower $(^{26}\mathrm{Al}/^{27}\mathrm{Al})_0$ ratios suggesting an extended formation and/or reprocessing period of CAIs up to $\sim$0.7 Ma

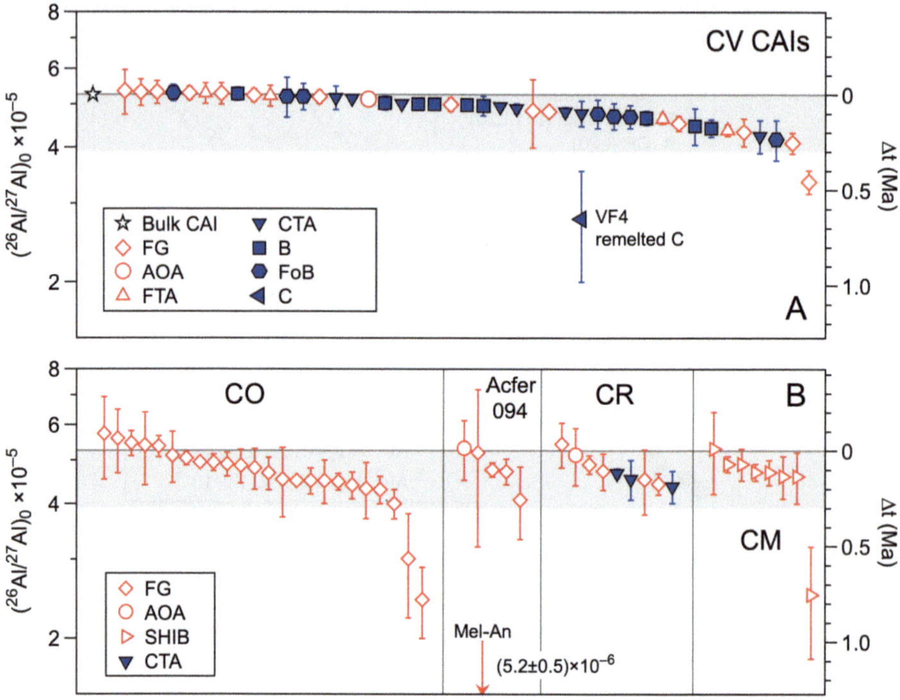

Fig. 4 Compilation of initial $(^{26}Al/^{27}Al)_0$ ratios determined from Al-Mg internal isochrons of CAIs and AOAs. The majority of data are indistinguishable from, or within 20%, of the canonical ratios defined by bulk CV CAIs. Two CAIs each from CV chondrites and Acfer 094 show primary and secondary ages. Data source: Jacobsen et al. 2008; Makide et al. 2009; MacPherson et al. 2010, 2012, 2017, 2018, 2020, 2022; Kita et al. 2012; Wasserburg et al. 2012; Kööp et al. 2016b; Kawasaki et al. 2017, 2018, 2019, 2020, 2021; Ushikubo et al. 2017; Liu et al. 2019; Simon et al. 2019; Hertwig et al. 2021; Fukuda et al. 2021; Dunham et al. 2022. Abbreviation of CAI types: AOA: amoeboid olivine aggregate, B: type B, C: type C, CTA: compact type A, FG: fine-grained, FoB: forsterite-bearing type B, FTA: fluffy type A, SHIB: spinel-hibonite. Condensates (AOA, FG, FTA, SHIB) and igneous CAIs (B, FoB, CTA, C) are shown as open and filled symbols, respectively

(MacPherson et al. 2012). In rare cases, CAIs were partially reset ~2.5 Ma later (Ushikubo et al. 2017).

A small fraction of CAIs exhibits little or no excess $\delta^{26}Mg^*$, suggesting an $(^{26}Al/^{27}Al)_0$ ratio of $<5 \times 10^{-5}$ at the time of their formation. These CAIs, often referred to as FUN CAIs, usually have large nucleosynthetic isotope anomalies e.g., in ^{48}Ca and ^{50}Ti, and mass-dependent fractionation of O, Mg, and Si isotopes (e.g., Kööp et al. 2016a; Park et al. 2017; Williams et al. 2017; Marin-Carbonne et al. 2023). Holst et al. (2013) obtained a well-defined Al-Mg isochron with an $(^{26}Al/^{27}Al)_0$ of 2.9 ($\pm$ 0.2) $\times 10^{-6}$ from a FUN CAI STP-1 in Allende (CV3) that has high $(^{182}Hf/^{180}Hf)_0$ of 0.96 ($\pm$ 0.11) $\times 10^{-4}$, indistinguishable from the initial $(^{182}Hf/^{180}Hf)_0$ ratio of 1.018 ($\pm$ 0.043) $\times 10^{-4}$ of a bulk (non-FUN) CAI isochron, reported by Kruijer et al. (2014a,b). Based on these observations, sub-canonical $(^{26}Al/^{27}Al)_0$ ratios of FUN CAIs are not considered to reflect late formation/reprocessing but instead likely represent early formed CAIs in the protoplanetary disk with a heterogeneous ^{26}Al distribution, possibly due to late injection of ^{26}Al in the protoplanetary disk or due to selective preservation of presolar isotope signatures (Desch et al. 2023a).

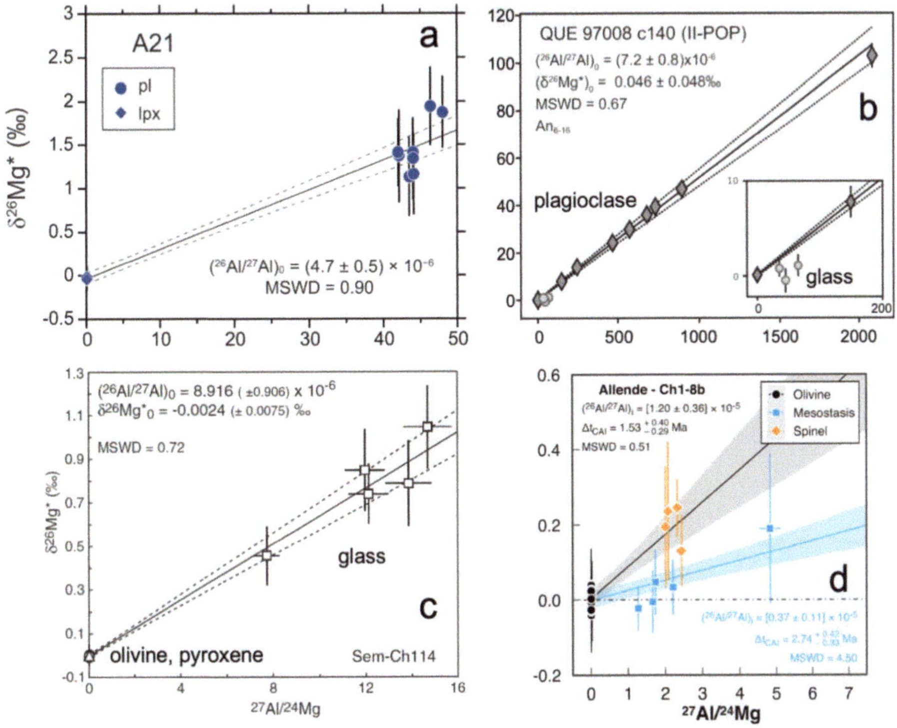

Fig. 5 Internal Al-Mg isochrons of chondrules. (a) A 12236 (CM2.9) using anorthite (Fukuda et al. 2022). (b) QUE 97008 (L3.05) using albitic plagioclase and glass (Siron et al. 2022). (c) Semarkona (LL3.00) using glassy mesostasis (Villeneuve et al. 2009). (d) Allende (CV3.6) using spinel and mesostasis (Piralla et al. 2023)

2.3 Al-Mg Chronology of Chondrules

Chondrules predominantly consist of mafic minerals, olivine and pyroxene, and Al-rich phases occur as glassy mesostasis and plagioclase. Spinel is relatively rare but present in chondrules of CV and other carbonaceous chondrites. Most Al-Mg analyses of chondrules were obtained by targeting plagioclase or high Al/Mg glass that show resolvable excess δ^{26}Mg* ($\geq$1‰) using $\leq$10 µm small SIMS spot beams (e.g., Siron et al. 2021, 2022). A few studies utilized high precision ($\leq$0.1‰) Mg isotope analyses of Mg-rich mesostasis glass in ordinary chondrite chondrules (Villeneuve et al. 2009; Pape et al. 2019) and spinel crystals in carbonaceous chondrites (Piralla et al. 2023) with smaller excess δ^{26}Mg* ($\leq$1‰). Examples of Al-Mg isochron diagrams using three different approaches are displayed in Fig. 5. Ideally, such Al-Mg analyses are completed solely on chondrules from the least metamorphosed chondrites (subtypes $\leq$ 3.10) because Mg diffusion and/or alteration of Al-rich phases even occur at mild metamorphism in the parent body (e.g., Van Orman et al. 2014). A summary of recent results is compiled in Fig. 6.

2.3.1 Chondrule Al-Mg Ages Using Plagioclase and High Al/Mg Glass

Plagioclase commonly occurs in chondrules of carbonaceous chondrites, and less commonly in ordinary chondrites. The composition of plagioclase is typically near the pure anorthite

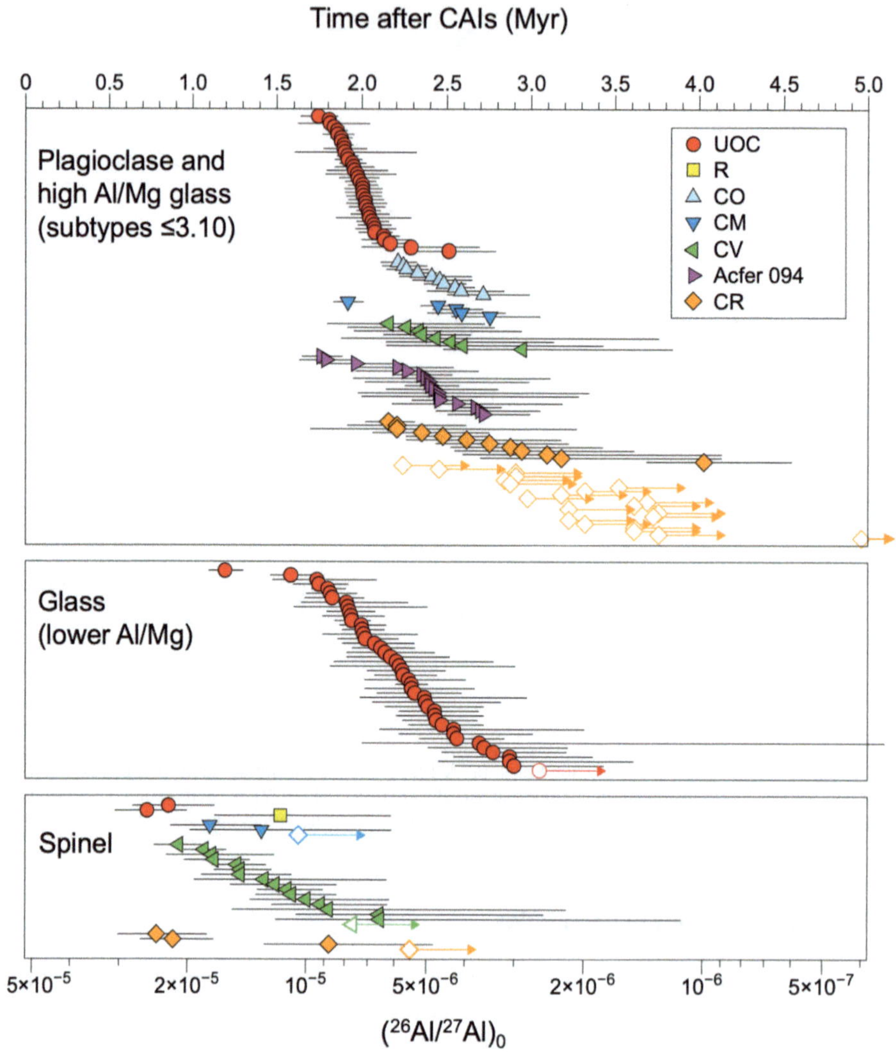

Fig. 6 Compilation of Al-Mg data of chondrules obtained from internal isochrons. Open symbols reflect data with upper limit only. Data sources are Ushikubo et al. (2013), Nagashima et al. (2014, 2017), Schrader et al. (2017), Tenner et al. (2019), Bollard et al. (2019), Hertwig et al. (2019), Siron et al. (2021, 2022), Fukuda et al. (2022) for plagioclase and high Al/Mg glass analyses, Villeneuve et al. (2009) and Pape et al. (2019) for low Al/Mg glass analyses, and Piralla et al. (2023) for spinel analyses. UOC - unequilibrated ordinary chondrites

endmember for FeO-poor (type I) chondrules in carbonaceous chondrites, while Na-rich plagioclase is found in FeO-rich (type II) chondrules. In OCs, coarse-grained anorthite-bearing chondrules occur with a wide range of FeO compositions in mafic minerals (Siron et al. 2021). They were used in the first Al-Mg study of chondrules and were initially considered to be igneous clasts from a differentiated body (Hutcheon and Hutchison 1989). Self-diffusion of Mg varies significantly with plagioclase composition, with no Mg diffusion at temperatures $\leq$ 450 °C for anorthite, while the Al-Mg system is significantly affected at 400 °C

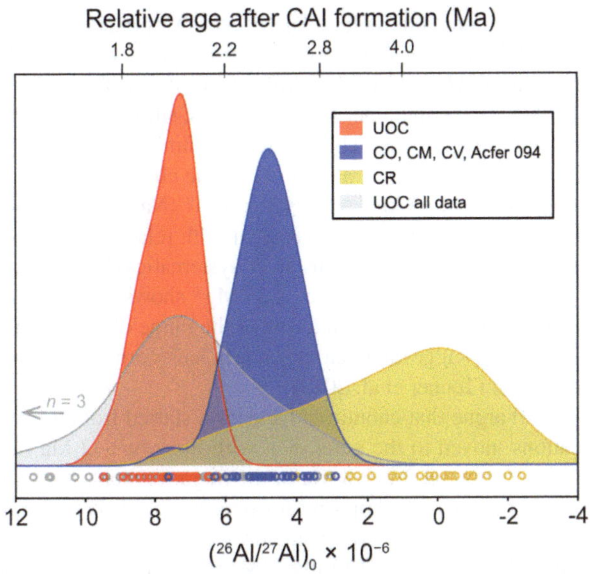

Fig. 7 Kernel density estimates of $(^{26}Al/^{27}Al)_0$ ratios for chondrules from UOC, major CC (CO, CM, CV, Acfer 094) and CR chondrites (Fukuda et al. 2022) from SIMS analyses of plagioclase and high Al/Mg glass. Data from all chondrules including low Al/Mg glass analyses are shown as grey area for comparison. Data sources are Ushikubo et al. (2013), Nagashima et al. (2014, 2017), Schrader et al. (2017), Tenner et al. (2019), Hertwig et al. (2019), Siron et al. (2021, 2022), Fukuda et al. (2022) for plagioclase and high Al/Mg glass analyses. For UOC, data include those from Hutcheon and Hutchison (1989), Kita et al. (2000), Rudraswami and Goswami (2007), Rudraswami et al. (2008), Villeneuve et al. (2009), Mishra et al. (2010), Pape et al. (2019), Bollard et al. (2019)

in albite (LaTourrette and Wasserburg 1998; Van Orman et al. 2014). Kimura et al. (2008) estimated the maximum metamorphic temperatures to be ≤500 °C for metals in chondrites with subtypes ≤3.05. Therefore, the Al-Mg system in anorthite-bearing chondrules from the lowest subtypes (≤3.05) likely remained closed since their formation. In addition, Al-Mg studies of chondrules with albitic plagioclase from subtypes ≤3.05 show well-defined isochrons (Fig. 5b, Siron et al. 2022), suggesting that the Al-Mg system in such chondrules was not disturbed. Siron et al. (2022) also reported a well-defined isochron for 3 chondrules with high Al/Mg glassy mesostasis from type 3.05 chondrites. However, analyses of glass in 2 chondrules with albitic plagioclase plot below the isochron regression lines defined from plagioclase analyses (Fig. 5b). Low degrees of aqueous alterations and thermal metamorphism can modify a part of chondrule mesostasis (e.g., Lewis and Jones 2019), such that careful sample documentation and data evaluation are required for such studies.

The $(^{26}Al/^{27}Al)_0$ ratios of chondrules determined based on plagioclase and high Al/Mg glass analyses differ systematically among different chondrite groups (Fig. 7). The $(^{26}Al/^{27}Al)_0$ ratio of chondrules using plagioclase and high Al/Mg glass in OC range from (6-9) $\times 10^{-6}$ and many of them agree within analytical uncertainties, suggesting that chondrules formed between ~1.8 and ~2.1 Ma after CAIs. In particular, Na-rich chondrules studied by Siron et al. (2022) show indistinguishable ages at 2.03 ± 0.10 Ma after CAIs, which is similar to the estimated accretion age of parent bodies of ordinary chondrite using thermal models with ^{26}Al as a heat source (~2.15 ± 0.10 Ma after CAIs; Sugiura and Fujiya 2014; Blackburn et al. 2017). Chondrules in OC forming regions accreted immediately (≤0.1 Ma) to the parent bodi(es), consistent with the formation of OC chondrules in highly dense disk regions (Alexander et al. 2008). The $(^{26}Al/^{27}Al)_0$ ratios of chondrules from CO, CM, CV, and Acfer 094 chondrites are mostly (3-6) $\times$ 10^{-6}, corresponding to formation ages of 2.2–2.8 Ma after CAIs. A few chondrules from Acfer 094 and CM chondrites exhibit higher $(^{26}Al/^{27}Al)_0$ ratios and older formation ages, similar to those in OC and their O isotope signatures, and mineral compositions are also similar to OC (Hertwig et al. 2019; Fukuda et al. 2022). In CR chondrites, a majority of the chondrules lack resolvable excess $\delta^{26}Mg^*$ with an upper limit of $(^{26}Al/^{27}Al)_0$ $\leq 3 \times 10^{-6}$ (Nagashima et al. 2014; Schrader et al. 2017; Tenner et al. 2019), which is consistent with relatively late formation of chondrules in CR chondrites as indicated by their Hf-W systematics (Budde et al. 2018). A small subset of CR chondrules with resolvable excess $\delta^{26}Mg^*$ shows similarities to chondrules in CO, CM, CV, and Acfer 094 chondrites in terms of the range of $(^{26}Al/^{27}Al)_0$, O isotope ratios, and mineral chemistry, suggesting that they might have formed in a similar environment or share a common origin (Tenner et al. 2019).

Fukuda et al. (2022) argue that chondrule formation started from the inner disk and that the formation locations moved to the outer disk with time; at $\sim$ 2 Ma (after CAIs) in the OC chondrite forming region, at 2.2–2.8 Ma in the major carbonaceous chondrite forming region, and at $\geq$2.8 Ma in the CR chondrite forming region. It is interesting to find older chondrules with NC-like oxygen isotope signatures in CM and Acfer 094 (Fukuda et al. 2022; Hertwig et al. 2019) as well as "CO-like" (Marrocchi et al. 2022) older chondrules in CR chondrites (Tenner et al. 2019). Their presence may indicate the transport of chondrules from the inner to the outer disk, where older chondrules became a part of chondrule precursors (e.g., Hertwig et al. 2019; Schrader et al. 2020; Williams et al. 2020). However, the early formation of proto-Jupiter ($\leq$1 Ma after CAIs; Kruijer et al. 2017) might have hampered the outward transport of chondrule-sized objects. Alternatively, the isotope signatures and environments of later forming chondrules had changed locally within individual disk regions (e.g., Tenner et al. 2019; Marrocchi et al. 2022).

2.3.2 Chondrule Al-Mg Ages Using Low Al/Mg Glass

Villeneuve et al. (2009) and Pape et al. (2019) conducted high precision Mg isotope analyses (precisions $\leq$0.1‰) of chondrules in OC (subtypes 3.01-3.10) by targeting glassy mesostasis with low Al/Mg ratios using large SIMS spots ($\sim$30 μm). Their reported range of $(^{26}Al/^{27}Al)_0$ overlaps with those determined for a similar suite of meteorites by Siron et al. (2021, 2022), who analysed plagioclase and high Al/Mg glass. However, analyses of low Al/Mg mesostasis define a wider range of $(^{26}Al/^{27}Al)_0$ ratios from $\sim 1.6 \times 10^{-5}$ to $\sim 3 \times 10^{-6}$, corresponding to 1.2 Ma to 3 Ma after CAIs. In addition, two chondrules do not show correlated $\delta^{26}Mg^*$ excesses and a slope of $\sim$0. The majority of analyses were obtained on the fine-grained mixture of glassy mesostasis and Mg-rich micro-crystallites, in which radiogenic ^{26}Mg might have been redistributed. However, a deviation from the isochron regression lines cannot be identified by statistical evaluation due to the generally small $\delta^{26}Mg^*$ excess (typically $\leq$0.3‰). Alternatively, Pape et al. (2021) argue that the younger ages of $\sim$3 Ma after CAIs represent the timing of remelting based on detailed petrological examination of multi-layered chondrules.

2.3.3 Chondrule Al-Mg Ages Using Spinel

Spinel ($MgAl_2O_4$) is rare in chondrules. The majority of the Al-Mg data obtained on spinel are from chondrules of CV chondrites because these meteorites contain spinel more frequently than other chondrite groups. However, spinel in CV chondrites could be metamorphosed (loss of radiogenic ^{26}Mg or Fe-Mg exchange by diffusion) and/or be a relict phase from CAI and AOA precursors that already accumulated radiogenic ^{26}Mg prior to chondrule formation. Therefore, Piralla et al. (2023) restricted their Al-Mg analyses of spinel to FeO-poor (type I) chondrules without metamorphic zoning in Fe-Mg, and with O isotope ratios in agreement with olivine in the same chondrule. The resulting Al-Mg ages (Fig. 6) are systematically older than those using plagioclase and glass by $\sim$1 Ma. Piralla et al. (2023) argue that the self-diffusion in spinel is slower than in plagioclase and glass, such that the Al-Mg age determination using spinel is the more robust approach. They concluded that chondrules started to form $\sim$0.8 Ma after CAIs and continued to form for 2-3 Ma during the entire lifetime of the protoplanetary disk. Such a long duration of chondrule formation is in principle consistent with the range in U-Pb ages reported for individual chondrules (Bollard et al. 2017). However, whereas the Al-Mg ages reveal a clear time gap between CAIs and chondrules, this is not evident from the U-Pb ages of individual chondrules.

As discussed earlier, the self-diffusion rate of Mg in plagioclase varies with composition, and in anorthite, which commonly occurs in type I chondrules, is actually similar to that of spinel (Sheng et al. 1992; LaTourrette and Wasserburg 1998; Van Orman et al. 2014). Major element compositions of anorthite-rich plagioclase in types $\leq$3.05 chondrites show silica-excesses and high abundances of MgO (0.5-1%), likely preserving Mg since their formation at high temperatures. Therefore, the systematic difference in Al-Mg ages between spinel and anorthite-rich plagioclase in type I chondrules is not expected for the least metamorphosed $\leq$ 3.05 chondrites. Spinel Al-Mg ages obtained by Piralla et al. (2023) are mainly from CV chondrites (subtypes $\geq$3.1), which display significant variations in MgO and FeO (Schnuriger et al. 2022), likely due to Fe-Mg diffusional exchange. MacPherson et al. (2012) reported compelling evidence for the re-distribution of ^{26}Mg excesses between anorthite and spinel in Vigarano (CV3) type B CAI "F1" due to parent body thermal metamorphism. In this CAI, the anorthite likely lost radiogenic ^{26}Mg due to thermal metamorphism (δ^{26}Mg* $\leq$10‰), which was redistributed into spinel enclosed in anorthite with the δ^{26}Mg* values 1-2‰ higher than the isochron regression line defined by melilite, fassaite, and spinel enclosed in melilite and fassaite. Self-diffusion of Mg in spinel is comparable to that in anorthite and the Fe-Mg inter-diffusion rate is even faster (Van Orman and Crispin 2010; Van Orman et al. 2014). In chondrules, spinel is often in contact with mesostasis (Schnuriger et al. 2022) that could have higher ^{27}Al/^{24}Mg ratios than spinel (> 2.5). During parent body metamorphism, radiogenic ^{26}Mg is likely re-distributed in a similar manner as in type B CAI F1 resulting in δ^{26}Mg* enhancements. Among the chondrules studied by Piralla et al. (2023), only two were from subtypes 3.00-3.01 chondrites and both show very high $(^{26}Al/^{27}Al)_0$ of $\sim 2 \times 10^{-5}$; Ch1 in Acfer 094 (ungroup C3.00) and Ch10 in Semarkona (LL3.01). Analyses of glass in Ch10 (Semarkona) do not show resolvable ^{26}Mg excesses, possibly the excesses were re-distributed by very low degrees of metamorphism, although Mg isotope exchange between spinel and glass is unlikely. Oxygen isotope ratios of Ch1 of Acfer 094 exhibit large variations in δ^{18}O and δ^{17}O along a slope 1 line with spinel being more ^{16}O-rich than olivine. This suggests that spinel is a relict phase from CAIs. For Ch10, no O isotope data are available, such that it is unclear if spinel in Ch10 also represents a relict from CAIs. To clarify this issue, further studies are required for spinel in chondrules combining Al-Mg chronology with detailed petrography, mineral chemistry, O isotope ratios, and Mg isotope analysis of mesostasis.

2.4 Disagreements of U-Pb and Al-Mg Ages Among CAIs and Chondrules

The initial $(^{26}Al/^{27}Al)_0$ values of most CAIs vary as much as 20-30% from the canonical value, suggesting either the extended period of CAI formation ($\sim$0.3 Ma) or heterogeneity of initial ^{26}Al abundance among CAIs. Such short formation intervals among CAIs are consistent with a narrow range of U-Pb ages of CAIs (4567.30 ± 0.16 Ma; n = 5, Amelin et al. 2010; Connelly et al. 2012). Initial $(^{26}Al/^{27}Al)_0$ ratios of chondrules determined from plagioclase isochrons in pristine chondrites (types ≤3.05) suggest systematic changes in the chondrule formation location with time, possibly related to growth of protoplanets. However, U-Pb ages of chondrules show a significantly larger range compared to those of Al-Mg ages. Both the U-Pb and Al-Mg systematics of CAIs and chondrules might be easily disturbed by a low degree thermal and aqueous alteration. Earlier formation of chondrules (<1.7 Ma after CAIs) cannot be fully ruled out, although the chronometers were likely disturbed due to heat generated in the parent bodies.

3 The Importance of a Detailed Chronology of Basaltic Achondrites

Basaltic achondrites (excluding martian and lunar meteorites) are mafic igneous rocks or breccias composed of igneous rock fragments from the crust and mantle of large asteroids. The parent bodies of basaltic achondrites experienced partial melting and differentiation into a core, mantle, and crust. The basaltic achondrites and their parental melts are enriched in incompatible elements compared to chondritic compositions (Mittlefehldt 2014).

The chronological records of crust-mantle differentiation and secondary processes such as impact events can be obtained from LLR systems (e.g., K-Ar, Rb-Sr, Sm-Nd, U-Pb). Short-lived radionuclide systems (e.g., Al-Mg, Mn-Cr, Hf-W), with their shorter half-lives, generally provide better age precision. The chronological records of meteorites can be selectively modified across different radiometric systems due to the varying diffusion behaviours of elements during thermal events. Despite limited constraints on diffusion parameters associated with chemical species of mineral hosts of interest (e.g., Cr diffusion kinetics in spinel, Posner et al. 2016; Mg diffusion in anorthite, LaTourrette and Wasserburg 1998; Van Orman et al. 2014), individual age dates from LLR or SLR systems provide important constraints on the thermal history of rock samples from crystallisation to impact events. The individual dates can serve as input for numerical models to constrain the accretion time and size of achondrite parent bodies (e.g., Golabek et al. 2014; Neumann et al. 2018). An assessment of the significance of the individual dates needs, however, to be carried out because some chronometers may be affected by secondary processes, hence, indicating a secondary event or possibly a partial reset; the latter implies that the age does not hold geological significance. If the dates from several radiometric systems correspond to rapid cooling of the magmatic system and have not been altered by secondary processes, they can be compared across different meteorites formed in distinct regions of the protoplanetary disk. This comparison helps constrain the initial abundances and distribution of SLR such ^{26}Al (e.g., Schiller et al. 2015a; Desch et al. 2023b).

3.1 Timescales of Silicate Differentiation

The timescales of the formation of silicate-enriched reservoirs can be constrained with a precision below 1 Ma by using the two long radioactive decay chains of $^{238}U(\rightarrow^{206}Pb)$ and $^{235}U(\rightarrow^{207}Pb)$ isotopes to deduce ages based on the radiogenic $^{207}Pb^{*}$-$^{206}Pb^{*}$ compositions. The radiogenic $^{207}Pb^{*}/^{206}Pb^{*}$ ratio provides the highest precision (at best ±100,000 years, Amelin and Yin 2025) for LLR systems, but it requires the measurement and use of the corresponding $^{238}U/^{235}U$ ratio of planetary materials. The $^{238}U/^{235}U$ ratio can vary due to radioactive decay of ^{247}Cm (half-life ∼15.6 Ma) or due to mass-dependent isotopic fractionation during chemical reactions and secondary processes (which can be significant if associated with a change in U redox state) (Brennecka et al. 2010; Goldmann et al. 2015; Tissot et al. 2016; Huyskens et al. 2025). The ages can be combined with high-resolution chronology using SLR. Two SLR systems are particularly useful to date silicate differentiation. These are (i) the ^{26}Al-^{26}Mg system that provides constraints on condensation, melting and crust-mantle formation due to contrasting condensation temperatures ($T_{50} = 1653$ K for Al and $T_{50} = 1336$ K for Mg) and compatibility in felsic or mafic minerals; and (ii) the ^{53}Mn-^{53}Cr system that provides insights into nebular processes owing to the volatile nature of Mn ($T_{50} = 1158$ K) compared to Cr ($T_{50} = 1296$ K) (Lodders 2003) as well as Mn/Cr fractionation linked to igneous, metamorphic, or aqueous alteration processes. While the short-lived ^{182}Hf-^{182}W system has also provided key constraints on silicate differentiation history of planetesimals, the focus of this section is explicitly on constraints provided by the ^{53}Mn-^{53}Cr and ^{26}Al-^{26}Mg systems.

Primitive (e.g., brachinites, acapulcoites) and basaltic achondrites (e.g., basaltic eucrites, quenched angrites, NWA 6704 grouplet), have early crystallisation ages within the first 10 Ma of solar system history (e.g., see Bouvier et al. 2025 for a review). Some achondrites even pre-date chondrule formation in the NC region such as Erg Chech 002 (EC 002) ungrouped achondrite, with reported Al-Mg ages ranging from 1.8 to 2.1 Ma after CAIs (Barrat et al. 2021; Reger et al. 2023; Connelly et al. 2023) (see further discussion on EC 002 chronological records in Sect. 3.3). Later crystallisation ages, up to 50 Ma after CAIs, are found for the cumulate eucrites and indicate an extended period of plutonic magmatism on Vesta (e.g., Bouvier et al. 2015; Roszjar et al. 2016). Such a prolonged period of magmatism was also recently reported from younger Pb-Pb ages of Ca-phosphates at 4514 ± 30 Ma that were found in the dunitic cumulative angrite NWA 8535 (Rider-Stokes et al. 2024). Later major impact events were also identified, e.g., responsible for the mesosiderite formation at 4525.39 ± 0.85 Ma (Haba et al. 2019).

3.2 Chronological Anchors for Short-Lived Radionuclides

The U-isotope-corrected Pb-Pb ages of rapidly cooled and pristine (unshocked) basaltic achondrites are the only radiometric ages in absolute time scale that can be obtained with high enough precision to be compared to the relative SLR ages (e.g., Brennecka and Wadhwa 2012). They are, however, dependent on knowing precisely the U isotopic composition of the analysed mineral (e.g., Brennecka et al. 2010; Reger et al. 2023). There should also be an absence of secondary disturbance by planetary or weathering processes that are commonly observed in meteorite finds (Tissot et al. 2017; Bouvier et al. 2025). If undisturbed, the U-corrected Pb-Pb ages of individual meteorites may then be compared against relative SLR age differences between these relative chronometers and to model the initial abundance of the respective radioactive isotopes at a given time; traditionally taken as t_0, the time of condensation of refractory inclusions (i.e., CAIs) and formation of the solar system (e.g., Desch et al. 2023b).

A particular petrological suite of achondrites has been used as a potential reference point for relative ages determined by various SLR systems, including the Al-Mg, Mn-Cr, and Hf-W decay systems. The angrites, in particular the volcanic types (e.g., D'Orbigny and Sahara 99555), formed around 4 to 5 Ma after CAIs (Brennecka and Wadhwa 2012; Tissot et al. 2017). They cooled rapidly after crystallization, facilitating synchronized records across different chronometers, because the time for diffusion was very limited. The volcanic angrites generally lack reports of significant major shock effects, which could have altered their primary chemical and isotopic compositions. Recently, however, moderate to high shock effects were reported for NWA 1670 (containing olivine xenocrysts with fractures, mosaicism and undulatory extinction) and NWA 7203 (melt shock veins associated with amorphous anorthite and younger Pb-Pb ages of silicophosphates at 4543 ± 19 Ma) (Hayashi et al. 2022). Distinct isotopic compositions in $\Delta^{17}O$ in olivine xenocrysts of several angrites support isotopic mixing (Rider-Stokes et al. 2023). The higher initial ^{26}Al abundance in spinel compared to plagioclase in D'Orbigny indicates a disturbance of the Al-Mg chronological records in this quenched angrite, and questions its use as an anchor for SLR systems (Deligny et al. 2024).

Plutonic angrites formed approximately 7 Ma after their volcanic counterparts, after the extinction of ^{26}Al (hence no Al-Mg ages are reported), and experienced slower cooling, as evidenced by their coarse-grained textures (Brennecka and Wadhwa 2012; Datta et al. 2024). The diabasic angrite NWA 12320 has pyroxene with a Pb-Pb age of ∼4564 Ma that is similar to quenched angrites, and 1.3 ± 0.6 Ma younger U-Pb phosphate ages suggesting rather slow cooling or later reheating (Datta et al. 2024). The U-corrected Pb-Pb ages of both quenched and plutonic angrites align with their Hf-W and Mn-Cr systematics and this indicates a homogeneous distribution of ^{182}Hf and ^{53}Mn in the formation regions of CAIs and the angrite parent body (Kleine et al. 2012; Kruijer et al. 2014a,b; Zhu et al. 2019). Falling between the quenched and plutonic angrites, a group of brecciated angrites, including NWA 2999 and NWA 6291, shows a mix of coarse- and fine-grained textures and metal-rich lithologies. The ^{147}Sm-^{143}Nd age of NWA 2999 angrite is 3925 ± 125 Ma (Sanborn et al. 2015), which is significantly younger than its U-isotope corrected Pb–Pb age of ∼4560 Ma (Brennecka and Wadhwa 2012) and the ages provided by the short-lived Hf–W and Mn–Cr chronometers (Shukolyukov and Lugmair 2008; Kleine et al. 2012).

The initial $^{182}Hf/^{180}Hf$ ratio of the solar system was determined at $(1.018 \pm 0.043) \times 10^{-4}$ based on a Hf-W isochron of bulk CAIs (Kruijer et al. 2014a,b). The solar system's $^{53}Mn/^{55}Mn_0$ ratio cannot be directly measured in CAIs, and was estimated at $(7 \pm 1) \times 10^{-6}$ using U-corrected Pb-Pb ages and SLR systematics of quenched and plutonic angrites, with a relatively large uncertainty due to the variability in CAI U-corrected Pb-Pb ages (Tissot et al. 2017; Sanborn et al. 2019). Early igneous activity on planetesimals suggests that ^{26}Al was the major radioactive heat source for the first few Ma of solar system history. The initial ^{26}Al abundance is a critical parameter to constrain numerical models of planetary accretion and differentiation. The initial and canonical abundance of ^{26}Al was unambiguously determined from tens of individual and mineral isochrons of "normal" CAIs to be $\sim 5.2 \times 10^{-5}$ (Jacobsen et al. 2008; MacPherson et al. 2010; Larsen et al. 2011, 2020) (see Sect. 2.2 on CAIs). Discrepancies in estimating the initial ^{26}Al abundance have been found when comparing the ^{26}Al-^{26}Mg and U-corrected Pb-Pb ages for individual CAIs against single (CR or LL) chondrules or achondrites (NC angrites, Asuka 881314, Erg Chech 002, and CC NWA 2976 and NWA 6704 grouplets) (Schiller et al. 2015a; Bollard et al. 2019; Sanborn et al. 2019; Wimpenny et al. 2019; Connelly et al. 2023; Krestianinov et al. 2023; Reger et al. 2023). From chronological studies of the SAH 99555 and D'Orbigny angrites as well as Erg Chech 002, it was suggested that ^{26}Al could be heterogeneously distributed with variable degrees of depletion, up to a factor of 3 to 4 compared to the canonical value, and within the NC formation region (Connelly et al. 2023; Krestianinov et al. 2023). It was proposed that this heterogeneous distribution within the protoplanetary disk was subsequently modified over time by inward-moving ^{26}Al-rich dust of solar CI composition from the outer to the inner disk (Schiller et al. 2015a,b; Bollard et al. 2019). If ^{26}Al was present in abundance lower than the canonical value ($\sim 5.2 \times 10^{-5}$), such as the abundances that have been reported in NC achondrite forming regions (i.e., $(1.5\text{-}2.5) \times 10^{-5}$) and ^{26}Al decay was the primary energy source for planetesimal heating, it would restrict the period of accretion of differentiated planetesimal to within the first 100,000 yr after CAIs to reach melting temperature and a magma ocean stage (Schiller et al. 2015a; Bollard et al. 2019). If so, an ^{26}Al difference between unrelated objects would not have any chronological meaning (Bollard et al. 2019).

Taking a closer look at angrite isotopic records, two independent ^{26}Al-^{26}Mg studies of the D'Orbigny angrite suggested a possible age adjustment of 0.3 ± 0.1 Ma towards younger ages, which corresponds to a ~20% difference in the calculated initial abundance (Spivak-Birndorf et al. 2009; Schiller et al. 2015a,b). Uranium isotopic compositions vary between quenched and plutonic angrite groups and evidence of non-secular equilibrium values in ^{234}U associated with U remobilization during terrestrial alteration is also reported in angrites (most angrites being desert finds) (Tissot et al. 2017). Differences in U isotopic compositions were identified between acid-washed minerals (e.g., pyroxenes) and phosphates in plutonic angrites, but not in all angrites (Datta et al. 2024; Huyskens et al. 2025). The U-corrected Pb-Pb age of the quenched angrite D'Orbigny does not seem to be affected by U isotopic effects considering mineral or whole-rock fractions (Huyskens et al. 2025). Angra dos Reis and NWA 4590 have, however, lower ^{238}U/^{235}U ratios of ~137.78 for leached or unleached whole-rock samples compared to ~137.80 and 137.83 for pyroxene residues and phosphates, respectively, that should be used to correct the Pb-Pb ages of each corresponding phase (Huyskens et al. 2025; Tissot et al. 2017). In contrast, the diabasic angrite NWA12320 does not show such an effect between whole-rock and acid washes containing soluble phosphates (Datta et al. 2024). This inter-mineral variation for U isotopes was also measured between leached pyroxene fractions (^{238}U/^{235}U $= 137.77 \pm 0.03$) used for the Pb-Pb isochron regression and whole-rock compositions (^{238}U/^{235}U $= 137.82 \pm 0.01$) of Erg Chech 002, leading to a 0.56 ± 0.28 Ma younger Pb-Pb age adjustment (Reger et al. 2023). The reported leached bulk-rock compositions of Erg Chech 002 (^{238}U/^{235}U from 137.81 to 137.83) are higher than the mean proposed for chondrules, chondrites, and achondrites of ~137.79 (Connelly et al. 2012). Regardless of the origin of U isotopic variation between minerals in achondrites (i.e., ^{247}Cm decay or isotopic fractionation), the U isotopic composition of acid washes of mineral separates or whole-rock should be systematically measured to calculate corresponding Pb-Pb isochron ages, because it may lead to an age adjustment up to ~0.6 Ma in achondrites (Reger et al. 2023; Huyskens et al. 2025).

With the increased number of SLR and LLR chronometry studies of CAIs, individual chondrules and basaltic achondrites, Desch et al. (2023b) used a statistical approach (based on three angrites, two ungrouped NC achondrites NWA 7325 and Asuka 881394, and two ungrouped CC achondrites NWA 6704 and 2976) to constrain the age of the solar system (t_0) to a value of 4568.4 ± 0.2 Ma. Two key assumptions were made in this study: (i) homogeneous distribution of ^{26}Al and other extinct radionuclides, and (2) no inclusion of the U-corrected Pb-Pb ages of CAIs. For all NC achondrites, a systematic lower Pb-Pb age of 0.2 Ma was used based on correlations of estimated pyroxene modal abundances and the measured whole-rock U isotopic compositions following the recommendation of Tissot et al. (2017). Other achondrites such as NWA 7325 (paired with NWA 8486) also indicate a correlation, but this is hampered by large uncertainties on the Pb-Pb ages due to the low U contents of these meteorites, resulting in a less-radiogenic Pb isotopic composition. Consequently, the U isotopic compositions of the samples could not be measured and are assumed to correspond to the mean of achondrites ~137.79 (Connelly et al. 2012; Goldmann et al. 2015). Using a similar approach based on D'Orbigny Al-Mg and U-corrected Pb-Pb systematics, Piralla et al. (2023) also proposed an older age of ~ 4568.7 +0.8/−0.6 Ma for solar system formation. Such older t_0 age than the reported ages for CAIs implies that CAI Pb-Pb ages may have been reset by later thermal events or that ^{26}Al was heterogeneously distributed between CAIs and achondrites' respective formation regions (Schiller et al. 2015a).

3.3 The Quest for a Robust Short-Lived Radionuclide Anchor

Due to remaining ambiguities to reconcile CAI and angrite age records, other possible anchors for ^{26}Al-^{26}Mg systematics have been investigated among NC achondrites. Three ungrouped achondrites (GRA 06128, NWA 11119, Erg Chech 002) with evolved and alkaline-rich composition (e.g., Srinivasan et al. 2018; Barrat et al. 2021) sample NC asteroids and indicate even earlier crust formation (within 2 to 3 Ma after CAIs) compared to NC basaltic angrites and eucrites ($\sim$3 to 10 Ma after CAIs). They can therefore potentially provide constraints on some of the earliest reservoirs of planetesimal formation and their corresponding initial ^{26}Al abundance. Additionally, two ungrouped CC achondrite grouplets (NWA 011/NWA 2976 and NWA 6704) provide a comparison between the NC and CC reservoirs and their ^{26}Al abundance. The U-corrected Pb-Pb age of the paired NWA 7325 and NWA 8486 ungrouped NC achondrites is, however, not precise enough and was based on an average U composition for achondrites that may affect its accuracy (4563.9 $\pm$ 1.7 Ma; Desch et al. 2023b).

One critical meteorite found in 2020 is the ungrouped NC achondrite Erg Chech 002 (EC 002). It is particularly suitable as a potential anchor and for interlaboratory studies. Its diverse mineralogy (Na-rich and K-rich feldspars, pyroxenes, olivine, chromite, phosphate) and enriched elemental composition are advantageous to use multiple isotopic systems and small sample masses. Moreover, it is fairly accessible for investigations due to a large total mass initially reported at 31.8 kg. Several independent studies of the EC 002 have been published indicating the oldest crystallization ages for an achondrite using the Al-Mg, Mn-Cr, and U-corrected Pb-Pb chronometers. The age estimates, however, vary outside their reported uncertainties for the different radionuclide systems. Four reported internal ^{53}Mn-^{53}Cr isochrons for EC 002 fragments provide initial ^{53}Mn/^{55}Mn corresponding to formation ages that are either contemporaneous with CAIs to 0.7 $\pm$ 0.6, 1.3 $\pm$ 0.7, or 1.88 $\pm$ 0.42 Ma after CAIs, some of which do not overlap within uncertainties (Anand et al. 2022; Zhu et al. 2022; Anand and Mezger 2023; Yang et al. 2025). All reported bulk fragments have also distinct nucleosynthetic ε^{54}Cr values ranging from -0.35 ± 0.06 to -1.02 ± 0.16 that indicate a heterogeneous composition of the EC 002 meteorite (Anand et al. 2022; Zhu et al. 2022; Krestianinov et al. 2023; Yang et al. 2025). There is no concurrent report of Mn-Cr and Al-Mg systematics on the same fragment of EC 002. Finally, due to the absence of direct measurement of ^{53}Mn/^{55}Mn in CAIs and discrepancy between the measured U-corrected Pb-Pb ages of CAIs and modelled t_0 for the formation of the solar system based on achondrites (Fig. 8), an additional $\pm$1.1 Ma uncertainty must be considered for all Mn-Cr modelled ages. The ^{26}Al/^{27}Al initial abundances obtained by in-situ SIMS and solution analysis MC-ICP-MS correspond to crystallization ages of $\sim$2.1 and $\sim$1.8 Ma after CAIs, respectively. The Al-Mg and Mn-Cr isochron age differences could be associated with the different employed analytical methods or from xenotlithic materials as ^{54}Cr heterogeneities were found between individual sample fragments (Yang et al. 2025). For instance, SIMS analyses target individual feldspar grains with high Al/Mg ratios using a small spot size, which depends on accurate standardization for Al/Mg ratios. By comparison, solution MC-ICP-MS analysis has low matrix effects due to the separation of Mg prior to isotopic analysis. This method, however, has to pool and dissolve several milligrams of mechanically separated feldspar grains with a minor proportion of mixed mineral grains such as pyroxenes, which dramatically lowers the measured Al/Mg ratios and corresponding ^{26}Mg* excesses. Alternatively, the $\sim$0.3 Ma age gap may be significant and resolvable. It may correspond to two generations of feldspars such as oligoclase (Gattacceca et al. 2021) (possibly anorthoclase; Barrat et al. 2021) and albititic plagioclase (containing lamellae of K-feldspar), because the latter were potentially

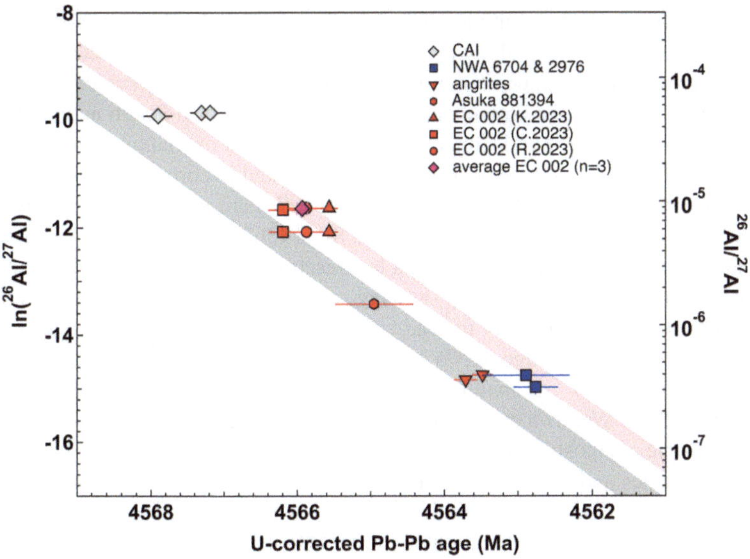

Fig. 8 The $^{26}Al/^{27}Al$ ratio (in natural log scale) at the time of isotopic closure vs. $^{238}U/^{235}U$-corrected ^{207}Pb-^{206}Pb ages of individual CAIs (Amelin et al. 2010; Bouvier et al. 2011; Connelly et al. 2012), and NC achondrites (Sahara 99555 and D'Orbigny angrites, Asuka 881397, EC 002, and CC achondrites NWA 6704 and NWA 2976). For EC 002, the U-corrected Pb-Pb ages from three independent studies (Reger et al. 2023; Connelly et al. 2023; Krestianinov et al. 2023) are shown both against the $^{26}Al/^{27}Al$ ratio obtained by MC-ICP-MS measurements ($\sim 8.65 \times 10^{-6}$ by Connelly et al. 2023 and $\sim 8.89 \times 10^{-6}$ by Fang et al. 2022 and Reger et al. 2023) and by in-situ SIMS ($\sim 5.72 \times 10^{-6}$) measurements, respectively. NC- and CC-type achondrites are shown in red and blue, respectively, except for the weighted average Pb-Pb age of EC 002, which is indicated in pink. The shaded areas represent the decay of ^{26}Al (slope of -0.967) over time using EC 002 weighted average (pink) or angrite (grey) Pb-Pb ages as anchors, respectively. Modified after Krestianinov et al. (2023)

preferentially analysed by SIMS. The Pb-Pb ages also vary from 4565.56 ± 0.12 Ma (including whole-rock and mineral separate leachates and residues, using $^{238}U/^{235}U = 137.829 \pm 0.005$; Krestianinov et al. 2023), to 4565.87 ± 0.30 Ma (pyroxene leachates and residue, using $^{238}U/^{235}U = 137.766 \pm 0.027$; Reger et al. 2023), and 4566.19 ± 0.20 Ma (whole-rock leachates and residue, using $^{238}U/^{235}U = 137.813 \pm 0.010$; Connelly et al. 2023). The difference in U-corrected Pb-Pb ages of EC 002 may be explained (similarly to what has been described above for angrites) by the variability in U isotopic compositions measured in unleached and leached whole-rock and leached pyroxene samples (Reger et al. 2023). The difference in measured $^{238}U/^{235}U$ ratios of EC 002 whole-rocks and pyroxenes causes a $\sim$0.6 Ma offset on Pb-Pb age calculations. Additionally, variations in the contribution of the three Pb isotopic components (initial, radiogenic, and common Pb contamination) and their corresponding $^{207}Pb^*/^{206}Pb^*$ compositions may explain the 0.63 ± 0.23 Ma gap for two reported Pb-Pb ages (4566.19 ± 0.20 Ma versus 4565.56 ± 0.12 Ma) (Connelly et al. 2023; Krestianinov et al. 2023) with two similar U isotopic whole-rock compositions (contributing to a difference of only $\sim$0.17 Ma in the Pb-Pb age correction). Out of the three Pb isotope datasets, two are consistent (Reger et al. 2023; Connelly et al. 2023) and one is distinct (Krestianinov et al. 2023) indicating an additional common Pb-like component. The age differences between the two consistent Pb isotope datasets are related to the selection of the U isotopic composition for Pb-Pb age calculation. Alternatively, the EC 002 parent body may be heterogeneous as shown by nucleosynthetic ^{54}Cr variations (Yang et al.

2025) obtained from different samples. The ^{54}Cr data of the individual samples analysed for their U-corrected Pb-Pb chronology are, however, not available to evaluate, if they derive or do not derive from the same source material. Combining all the evidence, model ages for canonical CAIs using the corresponding solution U-corrected Pb-Pb and Al-Mg analyses from Reger et al. (2023) and Connelly et al. (2023) yield overlapping ages of 4567.7 ± 0.3 Ma, and 4568.0 ± 0.2 Ma, respectively. The Al-Mg and U-corrected ages of EC 002 and ungrouped CC achondrites (NWA 6704 and NWA 2976) are moreover consistent (Fig. 8). The younger U-corrected Pb-Pb CAI ages (Fig. 8) of 4567.30 ± 0.16 Ma (n = 5) obtained on Allende and Efremovka may represent a late resetting event (possibly related to thermal metamorphism on the CV parent body) as suggested by Desch et al. (2023b). Alternatively, both CC and NC achondrites may instead be consistent with a younger t_0 that is closer to the unpublished U-corrected age of NWA 6991 compact type A CAI mineral separates with a 4.9×10^{-5} initial ^{26}Al/^{27}Al abundance and U-corrected Pb-Pb age of 4567.9 ± 0.30 Ma (Bouvier et al. 2011; Wadhwa et al. 2014).

The U isotopic composition of EC 002 leached whole-rock was measured by three independent laboratories and yielded a mean of 137.82 for the ^{238}U/^{235}U ratio (Connelly et al. 2023; Reger et al. 2023; Krestianinov et al. 2023). This value is higher than the average composition of ∼137.79 proposed for achondrites (Connelly et al. 2012). This value is also higher than the U isotopic compositions reported for angrites from both petrological groups with 137.80 ± 0.01 and 137.77 ± 0.01 for two quenched angrites and four plutonic angrites, respectively (Tissot et al. 2017). These measurements therefore argue against using a mean value for whole-rock achondrites formed in the NC formation reservoir. Additionally, such resolved variations in ^{238}U/^{235}U ratios of bulk achondrites add complexities in using bulk U isotopic compositions for pyroxene Pb-Pb dating. We therefore recommend that the U isotopic composition of each bulk or mineral leached sample used for Pb-Pb dating should be measured to avoid possible discrepancy due to U isotopic variation between mineral phases (i.e. as shown in Reger et al. 2023 and discussed in Huyskens et al. 2025).

Alternatively, the systematics of angrites and CAIs can be interpreted in relation to the initial ^{26}Al distribution and heterogeneities between planetary formation reservoirs. By comparing the difference of Pb-Pb ages between angrites and CAIs, a lower initial ^{26}Al/^{27}Al$_0$ abundance, as low as $(1.3 \pm 0.2) \times 10^{-5}$ (∼25% of the accepted canonical value obtained from CAIs), was proposed for the parent reservoir where the angrite parent body formed (Schiller et al. 2015a). For EC 002, Connelly et al. (2023) suggested a value of $(2.5 \pm 0.6) \times 10^{-5}$ for the initial ^{26}Al abundance (∼50% of the canonical value) in its formation reservoir calculated at the time of formation of Efremovka CAIs (Connelly et al. 2012). Using a comparison between EC 002 and angrites, Krestianinov et al. (2023) argued that volcanic angrites indicated a 3 to 4 times lower ^{26}Al abundance than EC 002 (Fig. 8). These two independent studies of EC 002 are therefore at odds. Additionally, both EC 002 and angrite parent bodies were formed in the NC region based on Tm and nucleosynthetic ^{54}Cr data (Barrat et al. 2021; Zhu et al. 2022). There is therefore a need for additional isotopic and chronological data from pristine achondrites to clarify the initial abundance and distribution of ^{26}Al in the inner and outer regions of the protoplanetary disk. Mechanisms must also be proposed to explain how isotopic reservoirs with initial ^{26}Al abundances differing by factors of 2–3 could coexist in the NC reservoir (e.g., Iizuka et al. 2025).

The Al-Mg and Hf-W age intervals between CAIs and angrites appear concordant (Kruijer et al. 2014a,b). In addition, the initial ^{26}Al/^{27}Mg and initial ^{182}Hf/^{180}Hf ratios correlate for CAIs, CV chondrules, CR chondrules, and angrites, i.e., across different formation regions from both the NC and CC reservoirs (Budde et al. 2018). These important observations argue against a high degree of ^{26}Al heterogeneity within the protoplanetary disk in contrast to proposals from several chondrules or achondrite studies (e.g., Larsen et al. 2011; Schiller et al. 2015a; Bollard et al. 2019). The initial $^{182}Hf/^{180}Hf$ ratio of EC 002 has not been determined yet for comparison against angrites and CAIs.

The inconsistencies between Al-Mg and U-Pb age differences between CAIs and angrites raise therefore doubts regarding (i) the significance of either U-corrected Pb-Pb ages measured in CAIs or (ii) of both the ^{26}Al-^{26}Mg systematics and the corresponding Pb-Pb ages of angrites used together as anchors for constraining the initial age (t_0) of the solar system. Reger et al. (2023) found a good agreement between their results from EC 002 with both the CC achondrite grouplets (NWA 6704 and 2976) and the NWA 6991 B4 CAI, suggesting therefore ^{26}Al homogeneity, considering all uncertainties related to the measurements. This interpretation removes the need to call for a large initial ^{26}Al heterogeneities at the 30-50% level as previously proposed by angrite and chondrule studies (Schiller et al. 2015a,b; Bollard et al. 2019) or by a EC 002 study (Krestianinov et al. 2023). Although the more limited resolution of the Hf-W system hampers ruling out all ^{26}Al heterogeneity, the concordance of these systems indicates that ^{26}Al heterogeneity was limited to $<20\%$ (Budde et al. 2018), i.e., much less than the $\sim 80\%$ heterogeneity proposed by e.g., Schiller et al. (2015a).

Radioactive decay systems possess different closure temperatures due to element specific diffusion rates in minerals (Ito and Ganguly 2006; Amelin 2006; Posner et al. 2016). Using a combination of both LLR and SLR systems, planetary events from igneous crystallization to thermal metamorphic processes (cooling or impact-related) can be identified in achondrites. The presence of xenolithic olivine with distinct O isotopic compositions (Rider-Stokes et al. 2023) or higher initial ^{26}Al abundances in spinel compared to plagioclase (Deligny et al. 2024) may also explain some of the age inconsistencies observed between mineral and whole-rock Pb-Pb, Al-Mg or Mn-Cr systematics (Fig. 8). Additionally, the assumption that basaltic achondrites cooled rapidly enough to align all LLR and SLR chronometers, and that they evolved without secondary disturbances like thermal metamorphism, aqueous alteration, or impact events, may be inaccurate. A prolonged cooling history could produce some of the age disparities observed with high-precision chronology in achondrites. Thus, it is imperative to further investigate the LLR and SLR systematics in various mineral hosts of EC 002 and other achondrites that crystallized within 5 Ma after CAIs. Such detailed mineralogical records will enable us to detect and document age disturbances and as such to refine the understanding of the formation age of the solar system and the initial ^{26}Al abundances within the solar nebula. Such detailed chronological records also provide a way to improve numerical models for thermal evolution, accretion time, and parent body size of meteorites (e.g., Neumann et al. 2023).

4 Hf-W Chronometry of Iron Meteorites – Timescales of Accretion and Core Formation on Differentiated Planetesimals

4.1 Iron Meteorites: Remnants of Planetesimal Cores

Vital insights into planetesimal accretion and differentiation come from isotopic studies of iron meteorites. Magmatic iron meteorites consist mainly of metallic iron or Fe-Ni alloys and are interpreted to be fragments of metal cores of protoplanets (Palme and O'Neill 2014; Scott 1972; Scott and Wasson 1975). Consequently, magmatic iron meteorites are the only hand specimens of planetary core material in our collection of extraterrestrial samples and provide an opportunity to directly study the differentiation, cooling, and the crystallization history of the metallic cores of small differentiated solar system bodies. Trace element signatures of iron meteorites indicate that the metal cores of magmatic iron meteorite parent bodies formed by segregation and subsequent cooling and crystallization of metallic melts (Scott and Wasson 1975). The classification of iron meteorites, as well as their formation, crystallization, and cooling history, have been discussed in several comprehensive reviews (Benedix et al. 2014; Chabot and Haack 2006; Goldstein et al. 2009; Scott 2020). Here, we restrict our focus on the accretion and core formation history of iron meteorite parent bodies and their bearing on planetesimal formation and the early evolution of the solar protoplanetary disk.

4.2 Chemical Classification and Solar Nebula Heritage of Iron Meteorites

4.2.1 Classification of Iron Meteorites

Magmatic iron meteorites are classified into several different groups (e.g., IC, IIAB, IIC, IID, IIG, IIIAB, IIIE, IIIF, IVA, IVB) based on their chemical composition, and each group is thought to sample a distinct parent body (Scott and Wasson 1975). Specifically, the groups are primarily distinguished by different contents of moderately volatile elements, such as Ga and Ge (relative to Ni) (Fig. 9a). The order-of-magnitude variations in volatile element depletion may arise from chemical fractionations induced by processes within the solar nebula, suggesting that the conditions of parent body accretion varied in time and/or space, but may alternatively be the result of energetic collisions among iron meteorite parent bodies causing heating and losses of volatile elements. Strong evidence that magmatic iron meteorites formed following crystallization of metallic melt comes from their trace element signatures (e.g., Scott and Wasson 1975). For instance, in a diagram of Ir vs. Ni, the magmatic iron meteorite groups display very well-defined distinct arrays consistent with the trends expected during fractional crystallization of a large volume of metallic melt in planetesimal cores. By contrast, non-magmatic iron meteorites (groups IAB, IIICD, IIE) have trace element abundances that are incompatible with fractional crystallization and, thus, likely do not derive from fully formed planetary cores (e.g., Scott and Wasson 1975; Benedix et al. 2000; Wasson and Kallemeyn 2002; Worsham et al. 2016; Hunt et al. 2018). Instead, the evolution of non-magmatic iron meteorites appears to be more complex, involving metal-silicate separation driven by a combination of internal (endogenic) processes and impact-induced events (Hunt et al. 2018; Worsham et al. 2017). Because of the scientific importance of planetesimal core–derived iron meteorites, this section will mostly focus on magmatic rather than non-magmatic iron meteorite parent bodies. Finally, while most iron meteorites belong to one of the aforementioned groups, there are numerous individual iron meteorites that are chemically distinct and do not fit into any established group - they are referred to as 'ungrouped' iron meteorites.

4.2.2 Nucleosynthetic Isotope Variations Measured in Iron Meteorites

As discussed in Sect. 1.1, nucleosynthetic isotope signatures allow genetic links between planetary materials to be determined and provide insights into the source regions of meteorites in the solar protoplanetary disk. In particular, the NC-CC dichotomy observed in nucleosynthetic isotope signatures demonstrates that meteorites are derived from two genetically distinct reservoirs. For iron meteorites, the NC-CC dichotomy is most clearly observed in their Mo isotope compositions, where NC iron meteorites (IAB, IIE, IC, IIAB, IIIAB, IIIE, IVA) and CC iron meteorites (IIC, IID, IIF, IIIF, IVB, SBT) plot on two distinct

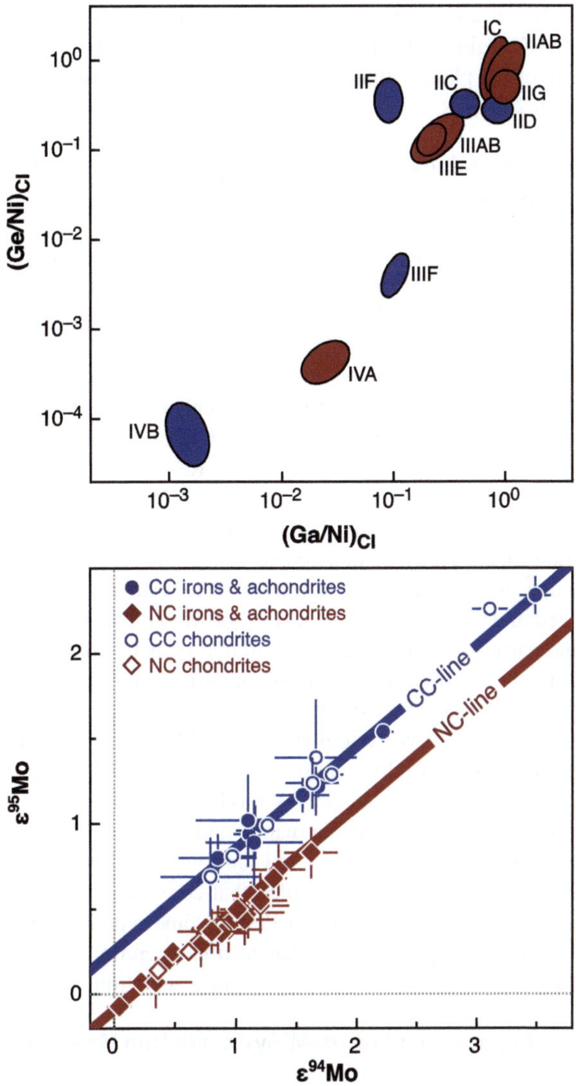

Fig. 9 Chemical and isotopic characteristics of iron meteorite groups (a) Volatile element depletions in iron meteorite groups expressed as CI chondrite–normalized Ge/Ni vs. Ga/Ni. (b) Molybdenum isotope dichotomy of iron meteorites and other meteorites expressed as ε^{95}Mo vs. ε^{94}Mo where CC (blue) and NC (red) iron meteorite groups define two distinct s-process mixing lines with identical slopes but distinct intercept values. Data sources: Goldstein et al. (2009), Scott (1972), Palme and O'Neill (2014), Borg et al. (2022), Budde et al. (2016, 2019), Kruijer and Kleine (2017), Kruijer et al. (2014b), Spitzer et al. (2020). Figure panel b from Kruijer et al. (2020)

parallel s-process mixing lines (Fig. 9b). Other studies have also reported Mo isotopic data for iron meteorites and are generally consistent with the data shown in Fig. 9b. The same Mo isotope dichotomy is also observed for NC chondrites (e.g., ordinary, enstatite) and CC (i.e., carbonaceous) chondrites, as well as other NC and CC achondrites (e.g., Budde et al. 2019; Yokoyama et al. 2019). Additionally, the NC-CC dichotomy for iron meteorites has been observed for multiple elements (e.g., Mo, Ni, Zn, Fe) (Steele et al. 2012; Budde et al. 2016; Kruijer et al. 2017; Poole et al. 2017; Nanne et al. 2019; Yokoyama et al. 2019; Worsham et al. 2019; Spitzer et al. 2020; Hopp et al. 2022a,b; Steller et al. 2022; Martins et al. 2023). A recent study reported additional nucleosynthetic Mo isotope compositions for an extensive suite of ungrouped iron meteorites (Spitzer et al. 2025). These results reveal that Mo isotopic signatures of the ungrouped irons cover the same range of compositions as established iron meteorite groups, indicating that the ungrouped irons can be assigned to either an NC or a CC heritage. Notably, non-magmatic iron meteorites appear to be exclusive to the NC region of the protoplanetary disk. This may be due to the more fragile and porous nature of CC planetesimals, which likely inhibited the formation of impact melt pools (Spitzer et al. 2025).

4.3 Chronology of Metal-Silicate Separation on Magmatic Iron Meteorite Parent Bodies

Precise determination of the timing of core formation for magmatic iron meteorite parent bodies is essential for determining their accretion rates and for identifying the heat sources of differentiation (Kleine et al. 2005; Kleine and Rudge 2011; Neumann et al. 2012; Kruijer et al. 2014b; Lichtenberg et al. 2021), and also has pivotal implications for reconstructing the large-scale evolution of the solar protoplanetary disk (Kruijer et al. 2017; Kleine et al. 2020; Kruijer et al. 2020; Lichtenberg et al. 2021). The core crystallization and cooling history of iron meteorite parent bodies have been studied using long-lived and short-lived chronometers, including the ^{187}Re-^{187}Os, ^{207}Pb-^{206}Pb, ^{107}Pb-^{107}Ag, ^{53}Mn-^{53}Cr, and ^{182}Hf-^{182}W chronometers (e.g., Chen and Wasserburg 1990; Shen et al. 1996; Smoliar et al. 1996; Horan et al. 1998; Blichert-Toft et al. 2010; Theis et al. 2013; Matthes et al. 2015; Hunt et al. 2022). A detailed review of the chronology of iron meteorites is beyond the scope of this article but collectively the results imply that their parent body cores crystallized and cooled within a few tens of Ma after CAI formation. To determine the formation and earliest evolution of differentiated planetesimals, it is useful to precisely constrain the time of core formation on iron meteorite parent bodies. The timescales of core formation on differentiated meteorite parent bodies can be constrained using the short-lived and now-extinct ^{182}Hf-^{182}W system ($t_{1/2} = 8.9$ Ma). This is because both Hf and W are refractory elements but have different geochemical affinities during metal-silicate separation. Because Hf is strongly lithophile and W moderately siderophile, core–mantle differentiation results in high Hf/W in the silicate mantle and Hf/W of essentially zero in the core. Therefore, the Hf-W system is uniquely suited for constraining timescales of core formation on differentiated planetary bodies, provided that core formation occurred within the lifetime of ^{182}Hf (see e.g., Harper and Jacobsen 1996; Kleine et al. 2009). In addition, because core formation is relatively closely tied to planetesimal accretion, the Hf-W system also provides indirect insights into the timescales of planetary accretion.

4.4 Application of the Hf-W System to Iron Meteorites

The application of the Hf-W system to iron meteorites necessitates the measurement of the ^{182}W/^{184}W composition of metal samples (commonly expressed in ε^{182}W as the parts-per-10,000 deviation relative to terrestrial mantle standard values). This is because the metal core, because of its low Hf/W, retains the ε^{182}W composition at the time of core formation. A two-stage model age of core segregation in the parent bodies of iron meteorites relative to CAI formation can thus be calculated by comparing the measured ε^{182}W representative of the iron meteorite parent body to the evolution of a reservoir with chondritic ^{180}Hf/^{184}W (Fig. 10). This is conveniently expressed using the following equation:

$$t = \frac{1}{\lambda} ln\left(\frac{\varepsilon^{182}W_{chondrites} - \varepsilon^{182}W_{SSI}}{\varepsilon^{182}W_{iron} - \varepsilon^{182}W_{SSI}}\right) \tag{4.1}$$

where $\varepsilon^{182}W_{iron}$ is the measured (radiogenic) value of an iron meteorite, $\varepsilon^{182}W_{SSI}$ is the initial value of CAIs (-3.49 ± 0.07; Kruijer et al. 2014a), $\varepsilon^{182}W_{chondrites}$ is the value of

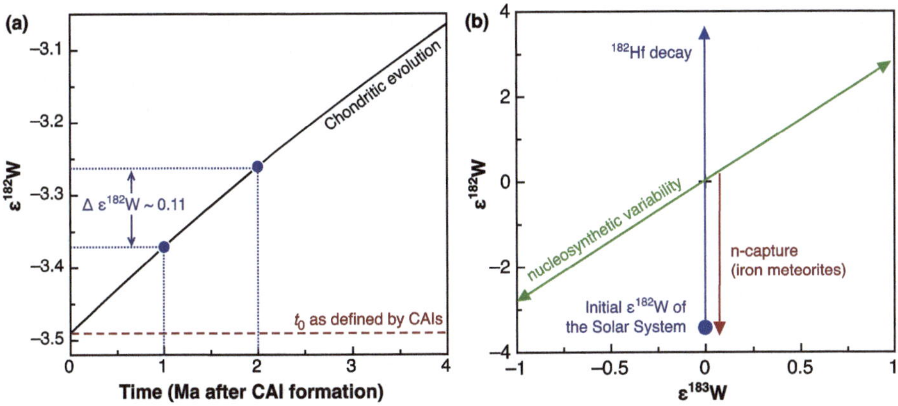

Fig. 10 (a) Illustration of two-stage Hf-W model ages for iron meteorites. The measured ε^{182}W of iron meteorites (blue circles) can be directly compared to the evolution line for chondrites (solid line) to obtain model ages of metal segregation relative to the formation of CAIs (t_0). (b) Possible processes causing W isotope variations in iron meteorites shown in ε^{182}W vs. ε^{183}W space (i.e., ^{182}W/^{184}W vs. ^{183}W/^{184}W, internally normalized to ^{186}W/^{184}W, in parts-per-10^4 relative to terrestrial standard values). Distinguished are radiogenic contributions from ^{182}Hf decay, neutron capture effects during cosmic ray exposure, and nucleosynthetic W isotope variability which results in correlated ε^{182}W vs. ε^{183}W variations (Burkhardt et al. 2012; Kruijer et al. 2014a). n-capture refers to neutron capture

chondritic meteorites, and λ is the ^{182}Hf decay constant of 0.078 ± 0.002 Ma^{-1} (Vockenhuber et al. 2004). The average measured value of carbonaceous chondrites of −1.91 ± 0.08 is commonly assumed to universally represent the ε^{182}W of chondrites (Kleine et al. 2004). However, more recently, it has been argued that using a bulk enstatite chondrite-like ε^{182}W of −2.23 ± 0.13 (corresponding to a ^{180}Hf/^{184}W of 1.06 ± 0.16) is more realistic for calculating Hf-W model ages of NC iron meteorite groups (Hellmann et al. 2019; Hellmann et al. 2024) instead of the ε^{182}W value of −1.91 ± 0.08 (and corresponding ^{180}Hf/^{184}W of ~1.35) commonly used for CC iron meteorites (Kleine et al. 2004; Kruijer et al. 2014b).

The above approach for calculating two-stage Hf-W model ages assumes that ^{182}W variations in iron meteorites are solely radiogenic in origin. However, this assumption is not necessarily accurate because two additional processes, beyond the radiogenic contributions from ^{182}Hf decay, can produce ^{182}W anomalies in iron meteorites (Fig. 10b). These processes are (i) nucleosynthetic W isotope heterogeneity and (ii) secondary neutron capture effects induced during exposure of the meteoroids to galactic cosmic rays (Leya and Masarik 2013; Leya et al. 2003; Masarik 1997). Nucleosynthetic isotope heterogeneity leads to coupled anomalies in ε^{182}W and ε^{183}W, and thus, the magnitude of nucleosynthetic ε^{182}W variability can be quantified by measuring ε^{183}W in the same samples (Qin et al. 2008; Burkhardt et al. 2012; Kruijer et al. 2014a). By contrast, quantifying the superimposed effects of secondary neutron capture on W isotopes is more complex as these effects solely appear in (mass bias–corrected) ε^{182}W. However, as discussed below, accurately quantifying cosmogenic effects is vital for an accurate interpretation of radiogenic ε^{182}W signatures in iron meteorites.

4.5 Tungsten Isotope Studies of Iron Meteorites and Quantification of Cosmogenic Effects

The earliest comprehensive W isotope study of iron meteorites revealed that they have substantial deficits in ^{182}W, indicating that their parent bodies segregated their cores within ca.

5 Ma of each other (Horan et al. 1998). Subsequent, studies conducted at higher precision revealed that iron meteorites exhibit variable ^{182}W deficits, which in some cases appeared to be as low as, or even lower than, the initial ε^{182}W of CAIs (Kleine et al. 2005; Markowski et al. 2006; Scherstén et al. 2006). At face value, these results imply that iron meteorite parent bodies segregated their cores at the beginning of solar system history, i.e., predating the formation of most chondrules and other meteorite parent bodies. However, unaccounted secondary neutron capture effects induced during cosmic ray exposure pose a major issue and cause net downward shift in ε^{182}W by up to 1 ε, leading to spuriously old Hf-W ages. Initially, several studies employed indirect proxies to quantify these superimposed cosmogenic effects, including published cosmic ray exposure ages and cosmogenic noble gases (Kleine et al. 2005; Markowski et al. 2006; Scherstén et al. 2006; Qin et al. 2008). However, cosmogenic noble gases are mainly produced at higher energies near the surface of a meteoroid, while W isotopes are primarily affected by neutron capture effects at (epi)thermal energies occurring at larger depths (Masarik 1997; Leya and Masarik 2013). Therefore, noble gas proxies only provide an indirect means of quantifying cosmogenic effects on W isotopes. Nevertheless, in some cases, cosmogenic noble gases are helpful in identifying iron meteorite samples with minor to absent neutron capture effects, but such samples are rare (Kruijer et al. 2012). The challenges caused by cosmogenic W isotope effects were overcome with the development of direct neutron capture proxies, including Os, and in particular, Pt isotopes (Kruijer et al. 2013; Wittig et al. 2013). Variations in ε^{196}Pt (^{196}Pt/^{195}Pt, internally normalized to ^{198}Pt/^{195}Pt) are caused solely by cosmic ray–induced neutron capture effects (predominantly on ^{195}Pt) and can, therefore, be used to assess the magnitude of neutron capture–induced shifts in ε^{182}W. The development of direct neutron dosimeters coincided with significant analytical improvements in the precision of W isotope data obtained by MC-ICP-MS (Willbold et al. 2011; Kruijer et al. 2012) and N-TIMS (Touboul et al. 2012) so that the ε^{182}W of metal samples could now routinely be determined with a precision of ca. 5 ppm (95% conf.).

Kruijer et al. (2013) reported combined high-precision Pt and W isotope data determined by MC-ICP-MS for the IID and IVB iron meteorite groups that displayed empirical, neutron capture-induced ε^{182}W vs. ε^{196}Pt correlations with precise regression-derived pre-exposure ε^{182}W values for a given group (Fig. 11). Subsequent studies reported combined Pt and W isotope compositions determined by MC-ICP-MS for the major (IIAB, IIIAB, IVA) (Kruijer et al. 2014b) and the minor (IC, IIC, IIF, IIIE, IIIF) iron meteorite groups (Kruijer et al. 2017). Spitzer et al. (2021) identified small nucleosynthetic Pt isotope anomalies in some ungrouped iron meteorites and proposed that the most appropriate pre-exposure ε^{196}Pt of meteorites is -0.06 ± 0.01 instead of zero as was previously assumed. This results in a ca. 0.1 ε upward correction of the pre-exposure ε^{182}W values of iron meteorites initially reported in Kruijer et al. (2014b). Most recently, Spitzer et al. (2025) reported W isotope data for an extensive suite of NC and CC ungrouped iron meteorites, significantly increasing the number of iron meteorite parent bodies studied for their W isotope systematics.

While most of the combined Pt-W isotope studies were performed by MC-ICP-MS, others have used W isotopes by N-TIMS in combination with Os isotopes to quantify cosmogenic effects on ^{182}W signatures of iron meteorites. This approach was applied to IC iron meteorites (Tornabene et al. 2023), IIC iron meteorites (Tornabene et al. 2020), and the South Byron Trio (SBT) (Hilton et al. 2019). Chiappe et al. (2023) reported combined Pt and W isotopic data for a suite of IIIE irons measured by N-TIMS. Overall, different approaches to correct for cosmogenic effects (Pt vs. Os isotopes) and different mass spectrometry techniques (MC-ICP-MS vs. N-TIMS) have yielded W isotope results and inferred pre-exposure ε^{182}W for each group that are in excellent agreement.

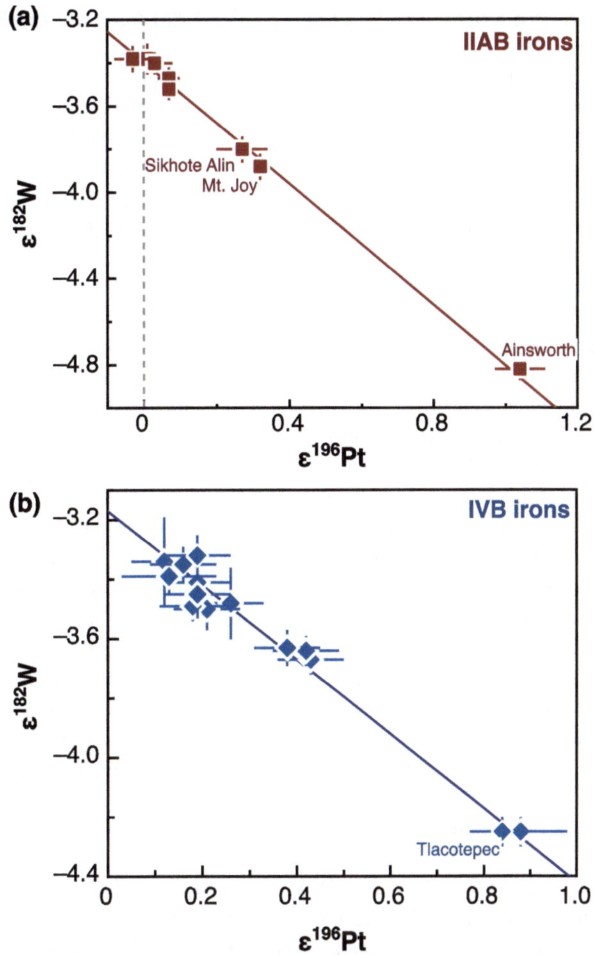

Fig. 11 ε^{182}W vs. ε^{196}Pt for the (a) IIAB and (b) IVB iron meteorite groups. Solid lines are best-fit regressions through the data. The pre-exposure ε^{182}W for each group is defined by the intercept of the Pt-W isotope correlation at $\varepsilon^{196}\text{Pt} = 0$. Similar well-defined correlations and precisely determined pre-exposure ε^{182}W were obtained for other magmatic iron meteorite groups and main-group pallasites (Kruijer et al. 2014b, 2017, 2022). Error bars represent external uncertainties (2 SD for Pt and 95% confidence for W). Figure redrawn after Kruijer et al. (2014b)

Pre-exposure ε^{182}W have also been determined for pallasites (Kruijer et al. 2022), a group of stony-iron meteorites, as well as for the non-magmatic iron meteorite groups, including IAB (Worsham et al. 2017; Hunt et al. 2018) and IIE iron meteorites (Kruijer and Kleine 2019). Kruijer et al. (2022) found that main-group pallasites exhibit pre-exposure ε^{182}W and Mo isotopic compositions that are indistinguishable from IIIAB iron meteorites. Thus, there is a strong chemical, isotopic, and chronological link between IIIAB irons and main-group pallasites, implying that main-group pallasite metal likely originated in the IIIAB core. Subsequent catastrophic impact disruption of this core would then have caused impact–induced mixing of metal and silicates, resulting in the formation of main-group pallasites (Kruijer et al. 2022; Tarduno et al. 2012). Collectively, the non-magmatic iron meteorites exhibit within-group variations in pre-exposure ε^{182}W, reflecting a more complex evolution involving endogenic heating and differentiation, followed by impact-induced metal-silicate separation. Consequently, the Hf-W systematics of non-magmatic irons only provide indirect constraints on the accretion and core formation history of their parent bodies and, therefore, will not be further discussed here.

4.6 Pre-Exposure ε^{182}W and Hf-W Model Ages of Core Formation

A summary of the pre-exposure ε^{182}W of the magmatic iron meteorite groups alongside the corresponding two-stage Hf-W model ages of core formation is given in Fig. 12. Several key observations can be made. First, small but resolvable differences in pre-exposure ε^{182}W exist between different magmatic iron meteorite groups with a total range of about 0.25 ε (Kruijer et al. 2014b, 2017; Hilton et al. 2019), where the IC irons (NC) have the lowest ε^{182}W (-3.34 ± 0.03) and the IVB irons (CC) have the highest ε^{182}W (-3.10 ± 0.05) (Fig. 12a). After taking into account cosmogenic effects, there are no iron meteorite groups with ε^{182}W lower than the CAI initial as was the case before cosmogenic effects could be adequately quantified. Second, while both the NC and CC reservoirs exhibit variable pre-exposure ε^{182}W, the CC iron meteorite groups have, on average, higher and more uniform ε^{182}W than the NC iron meteorites (Kruijer et al. 2017; Hilton et al. 2019). This translates to a range in two-stage Hf-W model ages of metal segregation ranging of ~1.5 to 3 Ma after CAI formation for NC iron meteorite groups (assuming an enstatite chondrite like ^{180}Hf/^{184}W) and a somewhat higher but a more uniform range of ~3-4 Ma for CC iron meteorite groups (assuming a carbonaceous chondrite-like ^{180}Hf/^{184}W) (Fig. 12). In addition, with a few exceptions, the ε^{182}W and corresponding Hf-W model ages of both NC and CC ungrouped iron meteorites generally overlap with those of the iron meteorite groups in each reservoir (Spitzer et al. 2025). Third, pre-exposure ε^{182}W are inversely correlated with the degree of moderately volatile depletion and the S content inferred for each core (Kruijer et al. 2014b; Hilton et al. 2019; Spitzer et al. 2021). The relation between ε^{182}W and S contents is most evident for the NC iron meteorite groups, while CC iron meteorite groups define a shallower and somewhat less well-defined correlation (Fig. 12c). Given that the melting temperature of Fe-Ni cores is strongly influenced by their bulk S contents (Wasson and Huber 2006), the ε^{182}W vs. S correlation suggests that the time of core formation is related to the melting temperature of the parent body (Kruijer et al. 2014b). Specifically, in S-rich bodies (e.g., IIAB, IID), most of the metal melted relatively early at the Fe-FeS eutectic (~1250 K), resulting in (on average) earlier core formation times. By contrast, in volatile-poor bodies (i.e., IVA, IVB), most of the Fe metal would only have melted after the extraction of silicate melt (at ~1800 K), resulting in (on average) later times of core formation. This conceptual model can successfully explain the extent of ε^{182}W variability seen among NC iron meteorite groups and perhaps also among individual CC iron meteorite groups (Fig. 12c). A corollary of this finding is that the Hf-W timescales of core formation on iron meteorite parent bodies, and by inference modelled accretion times, are most reliably assessed on the basis of ε^{182}W compositions measured in volatile-rich groups (e.g., NC groups IC and IIAB; CC groups IID, IIF) (Kruijer et al. 2017). In addition to the relationship between pre-exposure ε^{182}W and bulk S content, CC iron meteorite parent bodies have, on average, higher pre-exposure ε^{182}W than NC iron meteorite bodies, translating into calculated Hf-W ages that are ~1 Ma younger (Fig. 12a, 12c). This becomes evident when comparing the ε^{182}W of parent bodies with similar S contents (and volatile contents) (Fig. 12c). Specifically, volatile-rich CC bodies (e.g., IIC, IID, IIF) exhibit pre-exposure ε^{182}W that are on average ~0.2 ε^{182}W higher than volatile-rich NC bodies (e.g., IC, IIAB) (Kruijer et al. 2017). Thus, different melting temperatures of the NC and CC parent bodies cannot (solely) explain the apparent small difference in Hf-W ages between the NC and CC reservoirs. This observation also extends to NC and CC ungrouped iron meteorites (Spitzer et al. 2025). Varying origins for the ~1 Ma difference in Hf-W ages between NC and CC bodies have been proposed, including (i) earlier accretion times in the inner (NC) compared to the outer (CC) protoplanetary disk (Kruijer et al. 2017; Hilton et al. 2019), (ii) more oxidizing conditions and higher water ice contents in CC compared to NC bodies resulting in

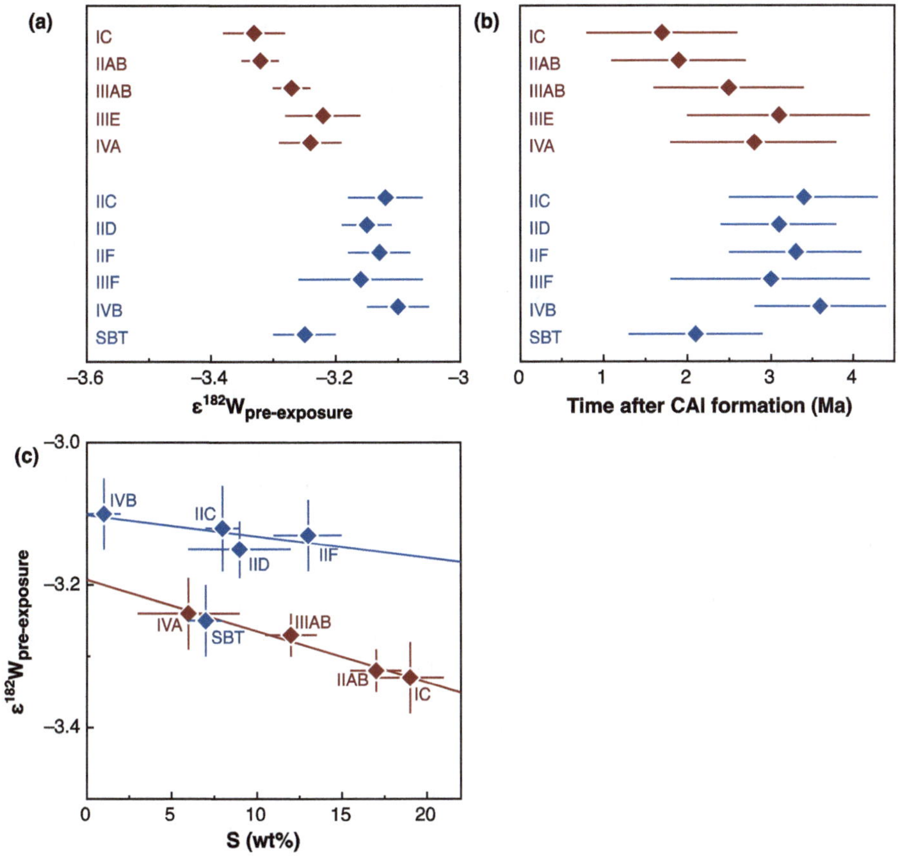

Fig. 12 (a) Pre-exposure ε^{182}W for magmatic iron meteorite groups derived from the NC reservoir (red) and CC reservoir (blue). (b) Two-stage Hf-W model ages of NC and CC iron meteorite groups, where model ages of NC bodies were calculated assuming an enstatite chondrite ε^{182}W of -2.23 ± 0.13 (Hellmann et al. 2024) but those of CC bodies using a carbonaceous chondrite ε^{182}W of -1.9 $\pm$ 0.1 (Kleine et al. 2004). (c) Pre-exposure ε^{182}W vs. inferred S concentrations (wt%) for the core of each parent body (see Table S7 in Spitzer et al. 2021 and references therein). Pre-exposure ε^{182}W are from Kruijer et al. (2014b, 2017), Hilton et al. (2019), Kruijer and Kleine (2019), Chiappe et al. (2023), Tornabene et al. (2023, 2020) and were recalculated assuming a pre-exposure ε^{196}Pt of -0.06 ± 0.01 (Spitzer et al. 2021)

delayed times of core formation in CC over NC bodies (Spitzer et al. 2021), and (iii), higher bulk Hf/W in CC compared to NC bodies resulting in distinct ε^{182}W for identical times of core formation (Kruijer et al. 2014b; Hellmann et al. 2019; Hellmann et al. 2024).

Through a comprehensive Hf-W isochron study of enstatite chondrites, Hellmann et al. (2024) demonstrated that (bulk) enstatite chondrites, and by inference, other NC chondrite precursors, have lower Hf/W than carbonaceous chondrites, and illustrated that enstatite chondrites are a good estimate for the average Hf/W of primitive chondritic material from the inner (NC) disk. Consequently, the lower Hf/W of enstatite chondrites may also be more representative of NC iron meteorite bodies. Using a (time-integrated) ε^{182}W of -2.23 ± 0.13 determined for enstatite chondrites yields Hf-W model ages that are about 0.4-0.7 Ma younger than those obtained using a carbonaceous chondrite-like Hf/W (Fig. 12b), at least in part, explaining the ε^{182}W differences between NC and CC iron meteorite groups, and closing the gap in Hf-W model ages for the parent bodies. However, although not fully resolved in all cases, (volatile-rich) CC bodies still exhibit slightly younger Hf-W ages than (volatile-poor) NC bodies, which could either reflect a later accretion time or a delayed onset of melting and core formation in CC bodies compared to NC bodies.

Estimates of the accretion timescales of iron meteorites parent bodies are obtained by combining their Hf-W model ages of core formation with thermal models for bodies heated by ^{26}Al decay ($t_{1/2} = 0.7$ Ma) (Hevey and Sanders 2006; Qin et al. 2008; Kruijer et al. 2014b; Hilton et al. 2019; Kaminski et al. 2020). Using a relatively simple thermal model involving conductive heating (Kruijer et al. 2017), this approach yields unequivocally early but slightly distinct accretion ages of <0.5 Ma for NC iron meteorite bodies and 0.5-1 Ma for CC bodies.

A later accretion time for CC compared to NC iron meteorite parent bodies would be consistent with the idea that accretion rates in the outer protoplanetary disk are expected to be nominally slower than in the inner disk. However, Spitzer et al. (2021) noted that CC iron meteorite parent bodies have systematically lower Fe/Ni than their NC counterparts, suggesting that the conditions of core formation were slightly more oxidizing in CC compared to NC iron parent bodies. This is an important observation given that CC bodies likely formed at or beyond the snowline (e.g., Kruijer et al. 2017; Lichtenberg et al. 2021; Izidoro et al. 2022; Morbidelli et al. 2022), allowing them to incorporate water ice. The presence of water ice may have delayed melting on CC parent bodies mainly by reducing the concentration of heat-producing ^{26}Al. Additionally, it would have promoted oxidation, resulting in core formation under more oxidizing conditions in CC bodies compared to NC bodies, explaining the difference in observed Fe/Ni. Subsequent work revealed that NC bodies also have significant oxidized Fe in their mantles (Grewal et al. 2024), resulting in overlapping ranges of fO_2 among NC and CC iron meteorite parent bodies. Nevertheless, using a slightly different approach for estimating oxidation states, Spitzer et al. (2025) found that, although the fO_2 values of CC and NC iron meteorite parent bodies still partially overlap, the average fO_2 of CC bodies is slightly but significantly higher than that of NC bodies, consistent with higher water ice fractions in CC over NC bodies.

Using a more advanced thermal model that considers variable water ice fractions, Spitzer et al. (2021) found that a higher water ice fraction indeed results in a delayed onset of melting and metal segregation because it lowers the concentration of heat-producing ^{26}Al and enhances heat transfer through water convection. This thermal modelling suggests that, despite slightly different core formation ages, the NC and CC bodies accreted contemporaneously, within <1 Ma after CAI formation (Kruijer et al. 2017; Spitzer et al. 2021). The uncertainties associated with the Hf-W data, and especially with the thermal models used for determining accretion ages, currently do not allow any putative differences in accretion time for these objects to be resolved.

Finally, the accretion time inferred for iron meteorite parent bodies is earlier than the time that most chondrite parent bodies accreted, implying that differentiated planetesimals preceded that of un-melted planetesimals (Fig. 13). This finding is consistent with the notion that ^{26}Al decay constituted the primary heat source driving planetesimal differentiation in the early solar system. Specifically, differentiated parent bodies (e.g., like those of iron meteorites) accreted sufficiently early for ^{26}Al decay to be an efficient heat source for melting and differentiation, whereas chondrite parent bodies accreted too late for ^{26}Al-induced melting to occur (Kleine et al. 2005; Hevey and Sanders 2006; Kleine and Rudge 2011).

The Hf-W chronometry of iron meteorites has far-reaching implications for our understanding of the large-scale evolution of the solar protoplanetary disk. The finding that both NC and CC iron meteorite parent bodies accreted within <1 Ma implies that planetesimal

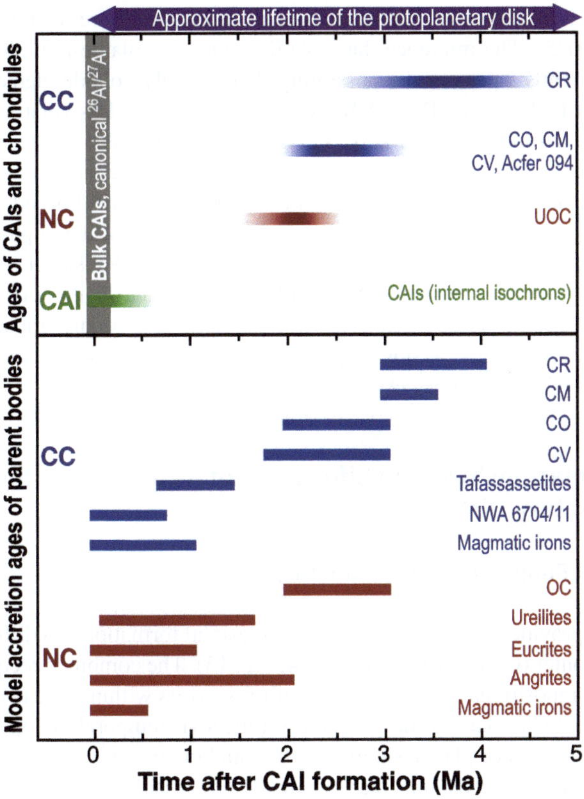

Fig. 13 Summary of chronological constraints on early Solar System evolution, distinguishing materials derived from NC and CC reservoirs. Upper panel displays ages of CAIs and chondrules derived from unequilibrated ordinary chondrites (UOC), and various carbonaceous chondrite groups (CV, CO, CM, CR, Acfer 094), determined using Al-Mg chronometry. All relative ages are displayed relative to the U-Pb ages of CAIs (4567.30 ± 0.16 Ma or 4567.94 ± 0.31 Ma; see main text). The estimated lifetime of the solar protoplanetary disk is indicated for context. Lower panel depicts accretion timescales of meteorite parent bodies, inferred from radiogenic ages of meteorites. Accretion ages of iron meteorite and achondrite parent bodies are inferred from model ages for differentiation combined with thermal modelling for internal heating of the parent bodies by ^{26}Al decay (see main text). Accretion timescales for chondrite parent bodies are based on Al–Mg, Hf–W and Pb–Pb ages obtained for chondrules (see panel above and corresponding references in main text), and on the chronology of alteration products combined with thermal modelling. Horizontal bars reflect uncertainties in accretion age estimates, not the duration of accretion. Lower panel was modified after Kruijer et al. (2020)

accretion occurred very early (within <1 Ma) at distinct radial locations and in both the inner (NC) and outer (CC) solar protoplanetary disk (Fig. 13). Therefore, the early accretion ages of iron meteorites imply that the spatial separation between the inner (NC) and the outer (CC) solar system was already present within less than 1 Ma after solar system formation (Kruijer et al. 2017) (also see Sect. 1.1 and 5). In addition, core formation in NC and CC bodies likely occurred over slightly distinct timescales (Fig. 12b) because of different oxidation states and water ice fractions in the NC and CC reservoirs (Spitzer et al. 2021). Nevertheless, the moderately volatile element abundances in iron meteorites (e.g., Scott and Wasson 1975) indicate that both the NC and CC reservoirs contained planetesimals with highly variable degrees of volatile element depletion (Kruijer et al. 2014a,b, 2017). Somewhat unexpectedly, the most extreme volatile element depletions appear to be more common in the CC reservoir than in the NC reservoir when ungrouped iron meteorites are included in the analysis (Spitzer et al. 2025). The inference that volatile-poor and volatile-rich NC and CC bodies accreted concurrently rules out the possibility that strongly volatile element depleted iron meteorite parent bodies (e.g., IVA, IVB) accreted earlier than less volatile depleted bodies (e.g., IIAB, IID) (Kruijer et al. 2014a,b). This implies that variable depletions in moderately volatile elements among iron meteorites do not reflect possible increasing condensation of moderately volatile elements in the protoplanetary disk over time. Instead, they must reflect more local processes within the protoplanetary disk, such as localized physical removal of the gas phase during condensation (e.g., Sengupta et al. 2022; Vollstaedt et al. 2020), parent bodies accreting variable contributions of volatile-rich and volatile-poor precursors (e.g., Hellmann et al. 2020; Nie et al. 2021), or secondary, catastrophic impact disruptions of iron meteorite parent bodies causing magma degassing and/or vaporization (e.g., Yang et al. 2007; Horan et al. 2012; Matthes et al. 2018).

5 Genetic Relationships and Life Spans of Reservoirs in the Solar Protoplanetary Disk

5.1 Lifetime of Rings in the Protoplanetary Disk

Iron meteorites provide evidence for coeval planetesimal formation and differentiation in the inner (NC) and outer (CC) disk region (Sect. 4, Fig. 13). The combined record of nucleosynthetic isotope compositions indicates further sub-reservoirs within these two regions (Fig. 1; Rüfenacht et al. 2023). This could potentially reflect (i) temporal isotopic changes in the disk over the time span of planetesimal accretion and/or (ii) different formation location in the disk with distinct compositions that remained largely isolated over time. In the scenario of temporal variations, correlations of the accretion ages with the nucleosynthetic isotope compositions are expected (e.g., Sugiura and Fujiya 2014). The nucleosynthetic Ti and Cr isotope data are among those, which exhibit the largest variations relative to their analytical uncertainties, thus providing the highest resolution for silicate materials. Considering the Ti isotope compositions versus accretion time of various meteorite groups (Fig. 14), bodies with nucleosynthetic compositions similar to CR chondrites formed over a period of $\sim$1 Ma (Tafassasset parent body; Ma et al. 2022) to $\sim$4 Ma (CR chondrite parent body; Sugiura and Fujiya 2014; Neumann et al. 2024). The CR-associated CB chondrites are of impact origin (Krot et al. 2005; Bollard et al. 2015). If the Ti isotope compositions indeed represent distinct rings in the protoplanetary disk where planetesimals formed, this implies that the compositions of the rings stayed approximately constant from $<$1 to $\sim$4 Ma in the CR region with an increasing amount of material locked in planetesimals/asteroids. This further argues for limited exchange between the ring reservoirs to preserve these distinct compositions.

Within these reservoirs, even smaller heterogeneities exist (Fig. 1, Fig. 14). For example, it was argued that Tafassasset and CR chondrites, despite their overall strong isotopic similarities, might have accreted material with slightly different isotopic compositions (Ma et al. 2022). Moreover, CR chondrites fall in the same area as CI chondrites in the ^{50}Ti-^{54}Cr isotope graph (Fig. 1), but these chondrite groups are clearly distinguishable and well resolved from each other based on ^{54}Fe and Ni isotope data (Steele et al. 2012; Schiller et al. 2020; Hopp et al. 2022a,b). Thus, there are nucleosynthetic heterogeneities within the broader reservoirs defined by Ti-Cr isotope data (Fig. 1, Fig. 14). Despite these small-scale heterogeneities within the reservoirs, the nucleosynthetic data are consistent with long-lasting

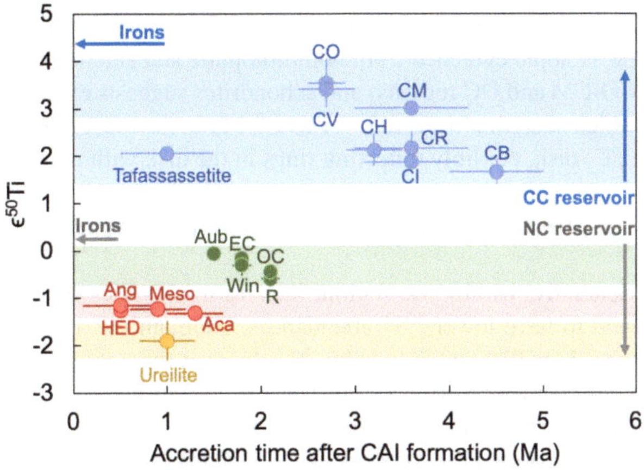

Fig. 14 Titanium isotope compositions versus accretion times of various meteorite groups. Accretion ages have relatively large uncertainties and depend on model assumptions. Data from Rüfenacht et al. 2023; accretion ages from Sugiura and Fujiya 2014; Ma et al. 2022; Neumann et al. 2024). The coloured bars show the reservoirs/rings defined by Rüfenacht et al. (2023). Black horizontal arrows indicate the accretion intervals determined for magmatic iron meteorites in the CC and NC regions (see Sect. 4). For abbreviations see Fig. 1

reservoirs in the disk with limited dust exchange between each other. Some minor exchange is indicated, e.g., by NC chondrules found in later formed CC chondrites (Marrocchi et al. 2022; Hertwig et al. 2019; Schrader et al. 2020; Williams et al. 2020). Moreover, fine dust (<10-100 μm) is generally well coupled to the gas and is able to move through the gaps (e.g., Hutchison et al. 2022). This fine dust fraction originated from the outer disk regions and moved inwards with the gas during the later stages of the nebula (Weidenschilling 1997; Desch et al. 2018; Hutchison et al. 2022). This material could be represented in our collections of extraterrestrial materials by CI chondrites (e.g., Schiller et al. 2020) and CI-related samples returned from the rubble-pile asteroids Ryugu and Bennu, because they mostly contain fine-grained material (Yokoyama et al. 2023; Lauretta et al. 2024; Barnes et al. 2025), originally coming from regions beyond where carbonaceous chondrites formed. However, the proportion of small, freely moving dust in the disk, which is enriched in presolar grains, must have remained approximately constant in each reservoir over time to preserve the, at first order, approximately constant isotopic composition of the bodies that formed within each reservoir.

This reasoning for static rings over time can be extended to evidence from Fe, Ni and Mo isotope data (Steele et al. 2011, 2012; Nanne et al. 2019; Worsham et al. 2019; Schiller et al. 2020; Hopp et al. 2022a,b). Compared to lithophile elements like Ti, these elements offer the advantage of exhibiting both lithophile and siderophile tendencies, and thus these data can be obtained for stony, stony-iron and iron meteorite groups. This expands the range of meteorite groups for direct comparison. The resolution offered by these systems (e.g., Fe, Ni) is generally more limited because the isotopic variations are often only marginally larger than the analytical uncertainties. Nevertheless, Ni and Fe isotope data (Steele et al. 2012; Nanne et al. 2019; Schiller et al. 2020; Hopp et al. 2022a,b) reveal distinct similarities between (i) CV-CO-CM chondrites and the magmatic irons IID, IIIF, IVB and between (ii) OCs and the iron meteorite groups IIAB, IIIAB and IVA. The Mo isotope data generally supports these links, but allow for additional heterogeneity, although CM chondrites are difficult to assess because the CM Mo data point seems to be an outlier based on correlation relative to Zr and Ru isotope data (Akram et al. 2015; Fischer-Gödde et al. 2015). In summary, Fe, Ni and Mo isotope data are consistent with the reservoirs proposed based on lithophile elements (Rüfenacht et al. 2023) and also suggest small additional heterogeneities within the reservoirs. The indicated relationships between chondrites and magmatic iron meteorite groups combined with accretion ages (Fig. 14) provide evidence for prolonged accretion in the CV-CO-CM region of <1–3 Ma and in the OC region of <1–2.2 Ma. Considering the overall dichotomy, the isotopic evidence from both lithophile and siderophile elements in chondrites (CR, CV-CO-CM and OC regions) and achondrites suggests extended and coeval time periods during which accretion occurred: <1 to 2.2 Ma in the inner (NC) disk, and <1 to 4 Ma in the outer (CC) disk, possibly reflecting rings in the disk with relatively little dust transfer.

5.2 Size Sorting of Dust as Source of Nucleosynthetic Variations?

Generally, within one region, later formed bodies (chondrites and primitive achondrites) tend to have lower ^{50}Ti abundances, although this is at the limit of what can be resolved (Fig. 14). Taken at face value, this hints at the possibility of a change in isotopic compositions over time within a ring. This can be considered as evidence of dust sorting in the rings (Hutchison et al. 2022; Hellmann et al. 2023; Rüfenacht et al. 2023), although the influence of parent body processing (Yokoyama et al. 2011; Goderis et al. 2015), late infall (Van Kooten et al. 2016; Nanne et al. 2019) and selective destruction of presolar grains (Trinquier et al. 2009; Akram et al. 2015) or material with solar system compositions (Ek et al. 2020) in hotter regions closer to the Sun cannot be entirely excluded. Nevertheless, rings formed due to pressure bumps in the protoplanetary disk (e.g., Dullemond et al. 2018), which preferentially trap larger particles, thereby naturally leading to dust sorting (Hutchison et al. 2022). It has been recognized for some time that aerodynamic size sorting of chondrules led to the similar size ranges of chondrules and refractory inclusions observed in each chondrite group (Cuzzi et al. 2001; Jones 2012). Moreover, carbonaceous chondrites are a mixture of matrix, chondrules and CAIs with each component having distinct isotopic compositions (Zanda et al. 2006; Hellmann et al. 2020; Morton et al. 2024). Hellmann et al. (2023) proposed that the variable mixtures of CAIs, chondrules and fine-grained matrix in CV, CO and CM chondrites can be explained by the aerodynamic properties of these components, which affect their preferential incorporation into chondrite parent bodies during planetesimal formation by streaming instability. This results in the formation of matrix-poor bodies first, and later bodies incorporate more fine-grained material from the same region. This scenario can explain, for example, the higher ε^{50}Ti values and the older accretion age of the CAI and chondrule-rich CO and CV relative to CM chondrites (Fig. 1). This is due to the fact that the ε^{50}Ti enrichments in carbonaceous chondrites are strongly influenced by the presence of CAIs and chondrules (Leya et al. 2008, 2009; Trinquier et al. 2009; Williams et al. 2020). Hutchison et al. (2022) used a different approach to model nucleosynthetic isotope variations. Presolar grains with their extreme isotope compositions are typically small ($<$ 10 μm) in chondrites, while larger dust particles exhibit more solar-like isotope compositions. Neglecting grain growth, 3-D simulations of disks suggest that trapping larger dust in pressure bumps can cause nucleosynthetic isotope variations within the rings, because they affect the mixing ratios of small presolar and larger solar system-like grains and aggregates (Hutchison et al. 2022). These different dust mixtures can translate into the distinct isotope compositions of planetesimals formed in these regions.

6 Summary

Small solar system bodies are key to understanding planet formation because they set the stage for collisional planetary accretion after the gas of the protoplanetary disk has dissipated. Pebble accretion is only possible for larger bodies while the nebular gas is still present. The accretion and differentiation of planetesimals during the initial few Ma of our solar system represent a crucial phase that strongly influenced the final compositions of the terrestrial planets. Here we reviewed the initial stages of planet formation starting from small nebular particles to planetesimal accretion and differentiation using key evidence derived from the study of meteorites.

Valuable insights into the dust of the protoplanetary disk can be obtained from CAI and chondrules using the U-Pb and Al-Mg chronometers. CAIs exhibit a notable range of $^{235}U/^{238}U$ ratios. These variations can skew the Pb-Pb ages by up to a few Ma unless precise U isotope measurements are performed on the same samples (Brennecka et al. 2010). Published data on CAIs from CV3 chondrites, along with corresponding U isotope analyses, show a narrow range of U-Pb ages, with an average value of 4567.30 ± 0.16 Ma (Amelin et al. 2010; Connelly et al. 2012). However, a report of an internally U-corrected Pb-Pb isochron on mineral separates indicates a slightly older age of 4567.94 ± 0.31 Ma (Bouvier et al. 2011).

The initial $(^{26}Al/^{27}Al)_0$ ratio of CAIs varies by 20-30% from the canonical value. This implies either an extended period of CAI formation (~0.3 Ma) or initial ^{26}Al heterogeneity among CAIs. Nevertheless, the evidence is consistent with the idea that the ^{26}Al abundance in the early solar system was likely homogeneous across both the inner and outer disk regions. It is important to note that the U-Pb and Al-Mg decay systems in chondrules might be easily disturbed by low degree thermal and aqueous alteration because of the ease of self-diffusion in the most relevant host phases (i.e., plagioclase, glass, spinel). Moreover, this diffusion highly depends on the specific chemical compositions of these phases. The ages derived from the U-Pb and Al-Mg systems can also be affected by recycling and reheating in the protoplanetary disk. The initial $(^{26}Al/^{27}Al)_0$ ratio of chondrules, determined from plagioclase isochrons of pristine (types ≤3.05) chondrites seems to be relatively robust and indicates a systematic shift in the chondrule formation location over time, moving from the inner to the outer disk, assuming a homogeneous ^{26}Al distribution. The ages suggest chondrule formation occurred at ~2 Ma after CAIs in the OC region, at 2.2-2.8 Ma in the major carbonaceous chondrite-forming region, and at ≥2.8 Ma in the CR region. Chondrule formation may be linked to the growth of protoplanets at different locations. Earlier formation of chondrules (<1.7 Ma after CAIs) is indicated e.g., by Pb-Pb ages and cannot be fully ruled out, although the U-Pb and Al-Mg chronometers of these chondrules were potentially disturbed due to internal heating of their parent bodies or thermal recycling in the disk.

The detailed chronology of differentiated basaltic achondrites serves as a powerful approach to date planetary processes from igneous crystallization to thermal metamorphism. The age record shows that these processes were well underway within the first few Ma of our solar system, while the solar nebula was still present. Such detailed chronological records help constrain the early planetary thermal evolution, accretion times, and meteorite parent body size.

The age record of basaltic achondrites, however, reveals substantial inconsistencies e.g., between mineral and whole-rock Pb-Pb, Al-Mg or Mn-Cr systematics. This can have several reasons. First, radioactive decay systems possess different closure temperatures due to element-specific diffusion rates in minerals. Hence, slow cooling will inevitably lead to different ages depending on the chronometers used. The assumption that basaltic achondrites cooled quickly enough to align the ages of different chronometers, and that they evolved without secondary disturbances such as thermal metamorphism, aqueous alteration, or impact events, may be inaccurate and thus could account for some of the observed age disparities. Second, the presence of xenolithic olivine (Rider-Stokes et al. 2023) or higher initial ^{26}Al abundances in spinel (Deligny et al. 2024) may also contribute to the observed age difference. Third, due to the sensitivity of Pb-Pb ages to the corresponding U isotopic composition, it is recommended to analyse the U isotopic composition of each leached sample to prevent differences caused by U isotope variations between mineral phases. In summary, further chronological investigations on the various host minerals of basaltic achondrites that crystallized within 5 Ma after CAI formation are crucial. This will help identify age disturbances and refine our understanding of the formation age of our solar system and the initial abundance and distribution of ^{26}Al in the solar nebula.

Isotopic studies of iron meteorites provide vital insights into planetesimal accretion and differentiation. Based on nucleosynthetic isotope data it has become apparent that iron meteorites derive from two distinct regions within the solar protoplanetary disk, with some parent bodies originating from the inner (NC) disk and others from the outer (CC) disk. The timescales of core formation on iron meteorite parent bodies have been determined using the short-lived ^{182}Hf-^{182}W decay system. Accurate application of Hf-W chronometry requires highly precise W isotope measurements, complemented by Pt isotope analyses to account for cosmic-ray-induced effects on W isotopes. After correcting for cosmogenic effects, the pre-exposure ^{182}W values exhibit slight variability: NC iron meteorite groups exhibit lower ^{182}W values and earlier core formation ages (1.5–3 Ma after CAI formation) than CC groups with higher ^{182}W values and later core formation ages (3–4 Ma after CAI). For NC iron meteorite groups, an inverse correlation is observed between pre-exposure ^{182}W and volatile element depletion, including sulfur content. This observation has been interpreted to reflect that S-rich parent bodies underwent earlier core formation due to lower melting temperatures, whereas core formation in volatile-poor bodies was delayed due to their higher average melting temperatures. Moreover, on average, CC groups exhibit higher ^{182}W values and younger Hf-W model ages ($>\sim 1$ Ma younger than NC groups), which may reflect earlier accretion in the inner (NC) than outer (CC) disk, more oxidizing conditions and higher water ice content in CC over NC bodies, or differences in bulk Hf/W ratios. Regardless of the exact interpretation, thermal modelling based on ^{26}Al decay reveals that both NC and CC parent bodies accreted within <1 Ma after CAI formation. Thus, parent body accretion in both, the NC and CC reservoirs, started early, indicating that a spatial separation between the inner NC and outer CC regions was established within <1 Ma. Moreover, differentiated planetesimals—such as those that produced iron meteorites—accreted earlier than the parent bodies of chondrites. Chondrites formed too late for ^{26}Al-induced melting and differentiation to occur. These findings highlight the critical role of ^{26}Al as the primary heat source driving early planetary differentiation.

The combination of accretion ages with nucleosynthetic data indicates that, at first order, reservoirs in protoplanetary disks are a long-lived feature with limited exchange between the different reservoirs. Evidence suggests that in the CR chondrite accretion region, new planetary bodies accreted over at least 3 Ma, whereas in most other regions formation occurred over at least 1 Ma with only minor variations in nucleosynthetic isotope compositions over time. Accretion took place during coeval time periods of <1 to 2.2 Ma in the inner (NC) disk, and <1 to 4 Ma in the outer (CC) disk. Accepting a disk where large grains (<100 μm) are trapped in pressure bumps forming ring structures and planetesimal formation by streaming instabilities in these rings, implies that size sorting of dust becomes inevitable. Size sorting likely introduced or amplified nucleosynthetic isotope variations now observed between different planetary bodies. These isotope variations may additionally be affected by late infall (Van Kooten et al. 2016; Nanne et al. 2019), selective thermal destruction of presolar grains (Trinquier et al. 2009) or destruction of material with solar system compositions (Ek et al. 2020) in hotter regions closer to the Sun. The latter scenario is particularly attractive for explaining the tight correlation that is observed for many isotope systems (e.g., Cr, Ti; Fig. 1) for inner (NC) disk materials. These correlations are indicative of two-component mixing (Ek et al. 2020; Palme and Mezger 2024). As discussed in

Ek et al. (2020), the parental molecular cloud may have contained abundant refractory dust grains with organic-rich icy mantles (Allamandola et al. 1999). Supernova shockwaves in the ISM can implant refractory elements into these mantles (King et al. 2012). Unlike silicates, these volatile mantles are more susceptible to photoevaporation, allowing a broader temperature range for thermal processing in the disk while preserving refractory presolar grains. Removing these icy ISM-grown mantles will affect the isotope mixing ratio of the two-component mixture, while the elemental ratios stay nearly constant. This process naturally accounts for the observed trends. Moreover, removing any material with solar system composition - from icy mantles grown in the ISM to sulfides, metal or silicates - will lead to the observed correlation between nucleosynthetic data of different elements.

Acknowledgements We thank ISSI and the conveners for organising the workshop on "Evolution of the Solar System: Constraints from Meteorites". We are also grateful to editor Klaus Mezger, Vinciane Debaille and an anonymous reviewer for providing constructive suggestions to improve this work. MS acknowledges support by the Swiss National Science Foundation (Projects 200020_179129 and 200021_208079). TK acknowledges support from the Laboratory Directed Research and Development Program at Lawrence Livermore National Laboratory (Grants: 23-ERD-003 and 26-ERD-015).

Funding Information Open access funding provided by Swiss Federal Institute of Technology Zurich.

Declarations

Competing Interests The authors declare that they have no competing interests.

References

Akram W, Schönbächler M, Bisterzo S, Gallino R (2015) Zirconium isotope evidence for the heterogeneous distribution of s-process materials in the Solar System. Geochim Cosmochim Acta 165:484–500

Alexander CMO, Grossman JN, Ebel DS, Ciesla FJ (2008) The formation conditions of chondrules and chondrites. Science 320:1617–1619

Alibert Y, Venturini J, Helled R, Ataiee S, Burn R, Senecal L, Benz W, et al (2018) The formation of Jupiter by hybrid pebble–planetesimal accretion. Nat Astron 2:873–877

Allamandola LJ, Bernstein MP, Sandford SA, Walker RL (1999) Evolution of interstellar ices. Space Sci Rev 90:219–232

Amelin Y (2006) The prospect of high-precision Pb isotopic dating of meteorites. Meteorit Planet Sci 41:7–17

Amelin Y, Yin QZ (2025) Recent progress and future prospects of the early solar system chronology. Natl Sci Rev 12(9):nwaf281

Amelin Y, Kaltenbach A, Iizuka T, Stirling CH, Ireland TR, Petaev M, Jacobsen SB (2010) U–Pb chronology of the Solar System's oldest solids with variable $^{238}U/^{235}U$. Earth Planet Sci Lett 300:343–350

Anand A, Mezger K (2023) Early Solar System chronology from short-lived chronometers. Geochemistry 83:126004

Anand A, Kruttasch PM, Mezger K (2022) ^{53}Mn-^{53}Cr chronology and $\varepsilon^{54}Cr$-$\Delta^{17}O$ genealogy of Erg Chech 002: the oldest andesite in the Solar System. Meteorit Planet Sci 57:2003–2016

Andrews SM, Wilner DJ, Zhu Z, Birnstiel T, Carpenter JM, Pérez LM, Bai X-N, et al (2016) Ringed substructure and a gap at 1 au in the nearest protoplanetary disk. Astrophys J Lett 820:L40

Barnes JJ, Nguyen AN, Abernethy FAJ, et al. (2025) The variety and origin of materials accreted by Bennu's parent asteroid. Nat Astron. https://doi.org/10.1038/s41550-025-02631-6

Barrat JA, Chaussidon M, Yamaguchi A, Beck P, Villeneuve J, Byrne DJ, Broadley MW, Marty B (2021) A 4,565-My-old andesite from an extinct chondritic protoplanet. Proc Natl Acad Sci USA 118(11):e2026129118.

Benedix GK, Haack H, McCoy TJ (2014) Iron and stony-iron meteorites. In: Davis AM (ed) Treatise on geochemistry, 2nd edn. vol 1. Elsevier, Oxford, pp 267–285

Blackburn T, Alexander CMO, Carlson R, Elkins-Tanton LT (2017) The accretion and impact history of the ordinary chondrite parent bodies. Geochim Cosmochim Acta 200:201–217

Blichert-Toft J, Moynier F, Lee C-TA, Telouk P, Albarède F (2010) The early formation of the IVA iron meteorite parent body. Earth Planet Sci Lett 296:469–480

Bogard DD (1995) Impact ages of meteorites: a synthesis. Meteoritics 30:244–268

Bollard J, Connelly JN, Bizzarro M (2015) Pb-Pb dating of individual chondrules from the CB_a chondrite Gujba: assessment of the impact plume formation model. Meteorit Planet Sci 50:1197–1216

Bollard J, Connelly JN, Whitehouse MJ, Pringle EA, Bonal L, Jørgensen JK, Nordlund Å, Moynier F, Bizzarro M (2017) Early formation of planetary building blocks inferred from Pb isotopic ages of chondrules. Sci Adv 3:e1700407

Bollard J, Kawasaki N, Sakamoto N, Olsen M, Itoh S, Larsen K, Wielandt D, Schiller M, Connelly JN, Yurimoto H, Bizzarro M (2019) Combined U-corrected Pb-Pb dating and ^{26}Al-^{26}Mg systematics of individual chondrules – evidence for a reduced initial abundance of ^{26}Al amongst inner Solar System chondrules. Geochim Cosmochim Acta 260:62–83

Borg LE, Brennecka GA, Kruijer TS (2022) The origin of volatile elements in the Earth-Moon system. Proc Natl Acad Sci USA 119:e2115726119

Bouvier A, Wadhwa M (2010) The age of the Solar System redefined by the oldest Pb–Pb age of a meteoritic inclusion. Nat Geosci 3:637–641

Bouvier A, Brennecka GA, Wadhwa M (2011) Absolute chronology of the first solids in the Solar System Workshop on the Formation of the First Solids in the Solar System, Abs. #9054

Bouvier A, Blichert-Toft J, Boyet M, Albarède F (2015) ^{147}Sm-^{143}Nd and ^{176}Lu-^{176}Hf systematics of eucrite and angrite meteorites. Meteorit Planet Sci 50:1896–1911

Bouvier A, Bermingham KR, Füri E (2025) Planetary materials: a record of early Solar System events to planetary processes. In: Anbar A, Weis D (eds) Treatise on geochemistry, 3rd edn, vol 7. Elsevier, Oxford, pp 203–256

Brasser R, Mojzsis SJ (2020) The partitioning of the inner and outer Solar System by a structured protoplanetary disk. Nat Astron 4:492–499

Brennecka GA, Wadhwa M (2012) Uranium isotope compositions of the basaltic angrite meteorites and the chronological implications for the early Solar System. Proc Natl Acad Sci USA 109:9299–9303

Brennecka GA, Weyer S, Wadhwa M, Janney PE, Zipfel J, Anbar AD (2010) ^{238}U/^{235}U variations in meteorites: extant ^{247}Cm and implications for Pb-Pb dating. Science 327:449–451

Budde G, Burkhardt C, Brennecka GA, Fischer-Gödde M, Kruijer TS, Kleine T (2016) Molybdenum isotopic evidence for the origin of chondrules and a distinct genetic heritage of carbonaceous and non-carbonaceous meteorites. Earth Planet Sci Lett 454:293–303

Budde G, Kruijer TS, Kleine T (2018) Hf-W chronology of CR chondrites: implications for the timescales of chondrule formation and the distribution of ^{26}Al in the solar nebula. Geochim Cosmochim Acta 222:284–304

Budde G, Burkhardt C, Kleine T (2019) Molybdenum isotopic evidence for the late accretion of outer Solar System material to Earth. Nat Astron 3:736–741

Burkhardt C, Kleine T, Dauphas N, Wieler R (2012) Nucleosynthetic tungsten isotope anomalies in acid leachates of the Murchison chondrite: implications for hafnium-tungsten chronometry. Astrophys J 753:L6

Burkhardt C, Dauphas N, Hans U, Bourdon B, Kleine T (2019) Elemental and isotopic variability in Solar System materials by mixing and processing of primordial disk reservoirs. Geochim Cosmochim Acta 261:145–170

Burkhardt C, Spitzer F, Morbidelli A, Budde G, Render JH, Kruijer TS, Kleine T (2021) Terrestrial planet formation from lost inner Solar System material. Sci Adv 7:eabj7601

Chabot NL, Haack H (2006) Evolution of asteroidal cores. In: Lauretta DS, McSween HY (eds) Meteorites and the early Solar System II, p 747

Chambers JE (2004) Planetary accretion in the inner Solar System. Earth Planet Sci Lett 223:241–252

Charnoz S, Avice G, Hyodo R, Pignatale FC, Chaussidon M (2021) Forming pressure traps at the snow line to isolate isotopic reservoirs in the absence of a planet. Astron Astrophys 652:A35

Chen JH, Wasserburg GJ (1990) The isotopic composition of Ag in meteorites and the presence of ^{107}Pd in protoplanets. Geochim Cosmochim Acta 54:1729–1743

Chiappe EM, Ash RD, Walker RJ (2023) Age, genetics, and crystallization sequence of the group IIIE iron meteorites. Geochim Cosmochim Acta 354:51–61

Connelly JN, Bizzarro M, Krot AN, Nordlund Å, Wielandt D, Ivanova MA (2012) The absolute chronology and thermal processing of solids in the solar protoplanetary disk. Science 338:651–655

Connelly J, Bollard J, Amsellem E, Schiller M, Larsen K, Bizzarro M (2023) Evidence for very early planetesimal formation and $^{26}Al/^{27}Al$ heterogeneity in the protoplanetary disk. Astrophys J Lett 952:L33

Cuzzi JN, Hogan RC, Paque JM, Dobrovolskis AR (2001) Size-selective concentration of chondrules and other small particles in protoplanetary nebula turbulence. Astrophys J 546:496

Datta C, Amelin Y, Krestianinov E, Irving AJ, Williams IS (2024) Disparate Pb-isotopic ages of silicate and phosphate minerals in the diabasic angrite northwest Africa 12320. Icarus 412:115979

Davis AM, Richter FM (2014) Condensation and evaporation of Solar System materials. In: Turekian KK, Holland HD (eds) Treatise on geochemistry, 2nd edn. Elsevier, Oxford, pp 335–360

Deligny C, Piralla M, Villeneuve J, Füri E, Marrocchi Y (2024) Potential chronological disturbance of the D'Orbigny angrite inferred from discordant ^{26}Al ages. Astrophys J Lett 975:L16

Desch SJ, Kalyaan A, Alexander CMD (2018) The effect of Jupiter's formation on the distribution of refractory elements and inclusions in meteorites. Astrophys J Suppl Ser 238:11

Desch SJ, Dunham ET, Herbst AK, Unterborn CT, Sharp TG, Bose M, Mane P, Williams CD (2023a). Origin of low-$^{26}Al/^{27}Al$ corundum/hibonite inclusions in meteorites. Astrophys J 953:146

Desch SJ, Dunlap DR, Dunham ET, Williams CD, Mane P (2023b) Statistical chronometry of meteorites. I. A test of ^{26}Al homogeneity and the Pb-Pb age of the Solar System's t = 0. Icarus 402:115607

Dullemond CP, Birnstiel T, Huang J, Kurtovic NT, Andrews SM, Guzmán VV, Pérez LM, et al (2018) The disk substructures at high angular resolution project (DSHARP). VI. Dust trapping in thin-ringed protoplanetary disks. Astrophys J Lett 869:L46

Dunham ET, Wadhwa M, Desch SJ, Liu MC, Fukuda K, Kita N, Hertwig AT, Hervig RL, Defouilloy C, Simon SB, Davidson J, Schrader DL, Fujimoto Y (2022) Uniform initial $^{10}Be/^{9}Be$ inferred from refractory inclusions in CV3, CO3, CR2, and CH/CB chondrites. Geochim Cosmochim Acta 324:194–220

Ek M, Hunt AC, Lugaro M, Schönbächler M (2020) The origin of s-process isotope heterogeneity in the solar protoplanetary disk. Nat Astron 4:273–281

Evans NJ, Dunham MM, Jørgensen JK, Enoch ML, Merín B, Van Dishoeck EF, Alcalá JM, Myers PC, Stapelfeldt KR, Huard TL, Allen LE, Harvey PM, Van Kempen T, Blake GA, Koerner DW, Mundy LG, Padgett DL, Sargent AI (2009) The spitzer c2d legacy results: star-formation rates and efficiencies; evolution and lifetimes. Astrophys J 181:321–350

Fang L, Frossard P, Boyet M, Bouvier A, Barrat J-A, Chaussidon M, Moynier F (2022) Half-life and initial Solar System abundance of 146Sm determined from the oldest andesitic meteorite. Proc Natl Acad Sci USA 119:e2120933119

Fischer-Gödde M, Burkhardt C, Kruijer TS, Kleine T (2015) Ru isotope heterogeneity in the solar protoplanetary disk. Geochim Cosmochim Acta 168:151–171

Frossard P, Guo Z, Spencer M, Boyet M, Bouvier A (2021) Evidence from achondrites for a temporal change in Nd nucleosynthetic anomalies within the first 1.5 million years of the inner Solar System formation. Earth Planet Sci Lett 566:116968

Fukuda K, Hiyagon H, Fujiya W, Kagoshima T, Itano K, Iizuka T, Kita NT, Sano Y (2021) Irradiation origin of ^{10}Be in the solar nebula: evidence from Li-Be-B and Al-Mg isotope systematics, and REE abundances of CAIs from Yamato-81020 CO3.05 chondrite. Geochim Cosmochim Acta 293:187–204

Fukuda K, Tenner TJ, Kimura M, Tomioka N, Siron G, Ushikubo T, Chaumard N, Hertwig AT, Kita NT (2022) A temporal shift of chondrule generation from the inner to outer Solar System inferred from oxygen isotopes and Al-Mg chronology of chondrules from primitive CM and CO chondrites. Geochim Cosmochim Acta 322:194–226

Gattacceca J, McCubbin FM, Grossman J, Bouvier A, Bullock E, Chennaoui Aoudjehane H, Debaille V, D'Orazio M, Komatsu M, Miao B, Schrader DL (2021) The meteoritical bulletin, no. 109. Meteorit Planet Sci 56:1626–1630

Goderis S, Brandon AD, Mayer B, Humayun M (2015) s-process Os isotope enrichment in ureilites by planetary processing. Earth Planet Sci Lett 431:110–118

Golabek GJ, Bourdon B, Gerya TV (2014) Numerical models of the thermomechanical evolution of planetesimals: application to the acapulcoite-lodranite parent body. Meteorit Planet Sci 49(6):1083–1099

Goldmann A, Brennecka G, Noordmann J, Weyer S, Wadhwa M (2015) The uranium isotopic composition of the Earth and the Solar System. Geochim Cosmochim Acta 148:145–158

Goldstein JI, Scott ERD, Chabot NL (2009) Iron meteorites: crystallization, thermal history, parent bodies, and origin. Geochemistry 69:293–325

Grewal DS, Nie NX, Zhang B, et al (2024) Accretion of the earliest inner Solar System planetesimals beyond the water snowline. Nat Astron 8:290–297

Haba MK, Wotzlaw J-F, Lai Y-J, Yamaguchi A, Schönbächler M (2019) Mesosiderite formation on asteroid 4 Vesta by a hit-and-run collision. Nat Geosci 12:510–515

Harper CL, Jacobsen SB (1996) Evidence for ^{182}Hf in the early Solar System and constraints on the timescale of terrestrial accretion and core formation. Geochim Cosmochim Acta 60:1131–1153

Hayashi H, Mikouchi T, Kim NK, Park C, Sano Y, Takenouchi A, Yamaguchi A, Kagi H, Bizzarro M (2022) Unique igneous textures and shock metamorphism of the northwest Africa 7203 angrite: implications for crystallization processes and the evolutionary history of the angrite parent body. Meteorit Planet Sci 57:105–121

Hellmann JL, Kruijer TS, Van Orman JA, Metzler K, Kleine T (2019) Hf-W chronology of ordinary chondrites. Geochim Cosmochim Acta 258:290–309

Hellmann JL, Hopp T, Burkhardt C, Kleine T (2020) Origin of volatile element depletion among carbonaceous chondrites. Earth Planet Sci Lett 549:116508

Hellmann JL, Schneider JM, Wölfer E, Drążkowska J, Jansen CA, Hopp T, Burkhardt C, et al (2023) Origin of isotopic diversity among carbonaceous chondrites. Astrophys J Lett 946:L34

Hellmann JL, Van Orman JA, Kleine T (2024) Hf-W isotope systematics of enstatite chondrites: parent body chronology and origin of Hf-W fractionations among chondritic meteorites. Earth Planet Sci Lett 626:118518

Hertwig AT, Kimura M, Ushikubo T, Defouilloy C, Kita NT (2019) The ^{26}Al-^{26}Mg systematics of FeO-rich chondrules from Acfer 094: two chondrule generations distinct in age and oxygen isotope ratios. Geochim Cosmochim Acta 253:111–126

Hertwig AT, Liu M-C, Brearley AJ, Simon SB (2021) Petrographic and isotopic investigations of two unusual Ca-Al-rich inclusions from primitive CO3 chondrites. Geochim Cosmochim Acta 296:75–96

Hevey PJ, Sanders IS (2006) A model for planetesimal meltdown by ^{26}Al and its implications for meteorite parent bodies. Meteorit Planet Sci 41:95–106

Hezel DC, Russell SS, Ross AJ, Kearsley AT (2008) Modal abundances of CAIs: implications for bulk chondrite element abundances and fractionations. Meteorit Planet Sci 43:1879–1894

Hilton CD, Bermingham KR, Walker RJ, McCoy TJ (2019) Genetics, crystallization sequence, and age of the south Byron Trio iron meteorites: new insights to carbonaceous chondrite (CC) type parent bodies. Geochim Cosmochim Acta 251:217–228

Holst JC, Olsen MB, Paton C, Nagashima K, Schiller M, Wielandt D, Larsen KK, Connelly JN, Jørgensen JK, Krot AN, Nordlund Å, Bizzarro M (2013) ^{182}Hf–^{182}W age dating of a ^{26}Al-poor inclusion and implications for the origin of short-lived radioisotopes in the early Solar System. Proc Natl Acad Sci USA 110:8819–8823

Hopp T, Dauphas N, Abe Y, Aléon J, Alexander CMOD, Amari S, Amelin Y, et al (2022a) Ryugu's nucleosynthetic heritage from the outskirts of the Solar System. Sci Adv 8:eadd8141

Hopp T, Dauphas N, Spitzer F, Burkhardt C, Kleine T (2022b) Earth's accretion inferred from iron isotopic anomalies of supernova nuclear statistical equilibrium origin. Earth Planet Sci Lett 577:117245

Horan MF, Smoliar MI, Walker RJ (1998) ^{182}W and ^{187}Re-^{187}Os systematics of iron meteorites: chronology for melting, differentiation, and crystallization in asteroids. Geochim Cosmochim Acta 62:545–554

Horan MF, Carlson RW, Blichert-Toft J (2012) Pd–Ag chronology of volatile depletion, crystallization and shock in the Muonionalusta IVA iron meteorite and implications for its parent body. Earth Planet Sci Lett 351–352:215–222

Hunt AC, Cook DL, Lichtenberg T, Reger PM, Ek M, Golabek GJ, Schönbächler M (2018) Late metal-silicate separation on the IAB parent asteroid: constraints from combined W and Pt isotopes and thermal modelling. Earth Planet Sci Lett 482:490–500

Hunt AC, Theis KJ, Rehkämper M, Benedix GK, Andreasen R, Schönbächler M (2022) The dissipation of the solar nebula constrained by impacts and core cooling in planetesimals. Nat Astron 6:812–818

Hutcheon ID, Hutchison R (1989) Evidence from the Semarkona ordinary chondrite for ^{26}Al heating of small planets. Nature 337:238–241

Hutchison MA, Bodénan J-D, Mayer L, Schönbächler M (2022) Presolar grain dynamics: creating nucleosynthetic variations through a combination of drag and viscous evolution. Mon Not R Astron Soc 512:5874–5894

Huyskens MH, Amelin Y, Yin Q-Z, Iizuka T (2025) ^{238}U/^{235}U isotopic variations in angrites and their constituent minerals. Geochim Cosmochim Acta 399:205–220

Iizuka T, Hibiya Y, Yoshihara S, Hayakawa T (2025) Timescales of Solar System formation based on Al–Ti isotope correlation by supernova ejecta. Astrophys J Lett 979:L29

Ito M, Ganguly J (2006) Diffusion kinetics of Cr in olivine and ^{53}Mn–^{53}Cr thermochronology of early Solar System objects. Geochim Cosmochim Acta 70:799–809

Izidoro A, Dasgupta R, Raymond SN, Deienno R, Bitsch B, Isella A (2022) Planetesimal rings as the cause of the Solar System's planetary architecture. Nat Astron 6:357–366

Jacobsen B, Yin Q, Moynier F, Amelin Y, Krot AN, Nagashima K, Hutcheon ID, Palme H (2008) ^{26}Al–^{26}Mg and ^{207}Pb–^{206}Pb systematics of Allende CAIs: canonical solar initial ^{26}Al/^{27}Al ratio reinstated. Earth Planet Sci Lett 272:353–364

Johansen A, Ronnet T, Schiller M, Deng Z, Bizzarro M (2023) Anatomy of rocky planets formed by rapid pebble accretion-ii. Differentiation by accretion energy and thermal blanketing. Astron Astrophys 671:A75

Jones RH (2012) Petrographic constraints on the diversity of chondrule reservoirs in the protoplanetary disk. Meteorit Planet Sci 47:1176–1190

Kaminski E, Limare A, Kenda B, Chaussidon M (2020) Early accretion of planetesimals unraveled by the thermal evolution of the parent bodies of magmatic iron meteorites. Earth Planet Sci Lett 548:116469

Kawasaki N, Itoh S, Sakamoto N, Yurimoto H (2017) Chronological study of oxygen isotope composition for the solar protoplanetary disk recorded in a fluffy Type A CAI from Vigarano. Geochim Cosmochim Acta 201:83–102

Kawasaki N, Simon SB, Grossman L, Sakamoto N, Yurimoto H (2018) Crystal growth and disequilibrium distribution of oxygen isotopes in an igneous Ca-Al-rich inclusion from the Allende carbonaceous chondrite. Geochim Cosmochim Acta 221:318–341

Kawasaki N, Park C, Sakamoto N, Park SY, Kim HN, Kuroda M, Yurimoto H (2019) Variations in initial ^{26}Al/^{27}Al ratios among fluffy Type A Ca–Al-rich inclusions from reduced CV chondrites. Earth Planet Sci Lett 511:25–35

Kawasaki N, Wada S, Park C, Sakamoto N, Yurimoto H (2020) Variations in initial ^{26}Al/^{27}Al ratios among fine-grained Ca-Al-rich inclusions from reduced CV chondrites. Geochim Cosmochim Acta 279:1–15

Kawasaki N, Itoh S, Sakamoto N, Simon SB, Yamamoto D, Yurimoto H (2021) Oxygen and Al-Mg isotopic constraints on cooling rate and age of partial melting of an Allende Type B CAI, Golfball. Meteorit Planet Sci 56:1224–1239

Kimura M, Grossman JN, Weisberg MK (2008) Fe-Ni metal in primitive chondrites: indicators of classification and metamorphic conditions for ordinary and CO chondrites. Meteorit Planet Sci 43:1161–1177

King AJ, Henkel T, Rost D, Lyon IC (2012) Trace element depth profiles in presolar silicon carbide grains. Meteorit Planet Sci 47:1624–1643

Kita NT, Ushikubo T (2012) Evolution of protoplanetary disk inferred from ^{26}Al chronology of individual chondrules. Meteorit Planet Sci 47:1108–1119

Kita NT, Nagahara H, Togashi S, Morishita Y (2000) A short duration of chondrule formation in the solar nebula: evidence from 26Al in Semarkona ferromagnesian chondrules. Geochim Cosmochim Acta 64:3913–3922

Kita NT, Ushikubo T, Knight KB, Mendybaev RA, Davis AM, Richter FM, Fournelle JH (2012) Internal 26Al-26Mg isotope systematics of a type B CAI: remelting of refractory precursor solids. Geochim Cosmochim Acta 86:37–51

Kita NT, Yin Q-Z, Macpherson GJ, Ushikubo T, Jacobsen B, Nagashima K, Kurahashi E, Krot AN, Jacobsen SB (2013) ^{26}Al-^{26}Mg isotope systematics of the first solids in the early Solar System. Meteorit Planet Sci 48:1383–1400

Kleine T, Rudge J (2011) Chronometry of meteorites and the formation of the Earth and Moon. Elements 7:41–46

Kleine T, Mezger K, Munker C, Palme H, Bischoff A (2004) Hf-182-W-182 isotope systematics of chondrites, eucrites, and Martian meteorites: chronology of core formation and early mantle differentiation in Vesta and Mars. Geochim Cosmochim Acta 68:2935–2946

Kleine T, Mezger K, Palme H, Scherer E, Münker C (2005) Early core formation in asteroids and late accretion of chondrite parent bodies: evidence from ^{182}Hf-^{182}W in CAIs, metal-rich chondrites, and iron meteorites. Geochim Cosmochim Acta 69:5805–5818

Kleine T, Touboul M, Bourdon B, Nimmo F, Mezger K, Palme H, Jacobsen SB, Yin Q-Z, Halliday AN (2009) Hf–W chronology of the accretion and early evolution of asteroids and terrestrial planets. Geochim Cosmochim Acta 73:5150–5188

Kleine T, Hans U, Irving AJ, Bourdon B (2012) Chronology of the angrite parent body and implications for core formation in protoplanets. Geochim Cosmochim Acta 84:186–203

Kleine T, Budde G, Burkhardt C, Kruijer T, Worsham E, Morbidell A, Nimmo F (2020) The non-carbonaceous–carbonaceous meteorite dichotomy. Space Sci Rev 216:1–27

Kööp L, Davis AM, Nakashima D, Park C, Krot AN, Nagashima K, Tenner TJ, Heck PR, Kita NT (2016a) A link between oxygen, calcium and titanium isotopes in ^{26}Al-poor hibonite-rich CAIs from Murchison and implications for the heterogeneity of dust reservoirs in the solar nebula. Geochim Cosmochim Acta 189:70–95

Kööp L, Nakashima D, Heck PR, Kita NT, Tenner TJ, Krot AN, Nagashima K, Park C, Davis AM (2016b) New constraints on the relationship between ^{26}Al and oxygen, calcium, and titanium isotopic variation in the early Solar System from a multielement isotopic study of spinel-hibonite inclusions. Geochim Cosmochim Acta 184:151–172

Krestianinov E, Amelin Y, Yin Q-Z, Cary P, Huyskens MH, Miller A, Dey S, Hibiya Y, Tang H, Young ED (2023) Igneous meteorites suggest Aluminium-26 heterogeneity in the early Solar Nebula. Nat Commun 14:4940

Krot AN, Amelin Y, Cassen P, Meibom A (2005) Young chondrules in CB chondrites from a giant impact in the early Solar System. Nature 436:989–992

Kruijer TS, Kleine T (2017) Tungsten isotopes and the origin of the Moon. Earth Planet Sci Lett 475:15–24

Kruijer TS, Kleine T (2019) Age and origin of IIE iron meteorites inferred from Hf-W chronology. Geochim Cosmochim Acta 262:92–103

Kruijer TS, Sprung P, Kleine T, Ley I, Burkhardt C, Wieler R (2012) Hf–W chronometry of core formation in planetesimals inferred from weakly irradiated iron meteorites. Geochim Cosmochim Acta 99:287–304

Kruijer TS, Fischer-Godde M, Kleine T, Sprung P, Leya I, Wieler R (2013) Neutron capture on Pt isotopes in iron meteorites and the Hf-W chronology of core formation in planetesimals. Earth Planet Sci Lett 361:162–172

Kruijer TS, Kleine T, Fischer-Godde M, Burkhardt C, Wieler R (2014a) Nucleosynthetic W isotope anomalies and the Hf-W chronometry of Ca-Al-rich inclusions. Earth Planet Sci Lett 403:317–327

Kruijer TS, Touboul M, Fischer-Godde M, Bermingham KR, Walker RJ, Kleine T (2014b) Protracted core formation and rapid accretion of protoplanets. Science 344:1150–1154

Kruijer TS, Burkhardt C, Budde G, Kleine T (2017) Age of Jupiter inferred from the distinct genetics and formation times of meteorites. Proc Natl Acad Sci USA 114:6712–6716

Kruijer TS, Kleine T, Borg LE (2020) The great isotopic dichotomy of the early Solar System. Nat Astron 4:32–40

Kruijer TS, Burkhardt C, Borg LE, Kleine T (2022) Tungsten and molybdenum isotopic evidence for an impact origin of pallasites. Earth Planet Sci Lett 584:117440

Larsen KK, Trinquier A, Paton C, Schiller M, Wielandt D, Ivanova MA, Connelly JN, Nordlund Å, Krot AN, Bizzarro M (2011) Evidence for magnesium isotope heterogeneity in the solar protoplanetary disk. Astrophys J Lett 735:L37

Larsen KK, Wielandt D, Schiller M, Krot AN, Bizzarro M (2020) Episodic formation of refractory inclusions in the Solar System and their presolar heritage. Earth Planet Sci Lett 535:116088

LaTourrette T, Wasserburg GJ (1998) Mg diffusion in anorthite: implications for the formation of early Solar System planetesimals. Earth Planet Sci Lett 158:91–108

Lauretta DS, Connolly HC Jr, Aebersold JE, Alexander CMOD, Ballouz RL, Barnes JJ, Bates HC, et al (2024) Asteroid (101955) Bennu in the laboratory: properties of the sample collected by OSIRIS-REx. Meteorit Planet Sci 59:2453–2486

Lewis JA, Jones RH (2019) Primary feldspar in the Semarkona LL 3.00 chondrite: constraints on chondrule formation and secondary alteration. Meteorit Planet Sci 54:72–89

Leya I, Masarik J (2013) Thermal neutron capture effects in radioactive and stable nuclide systems. Meteorit Planet Sci 48:665–685

Leya I, Wieler R, Halliday AN (2003) The influence of cosmic-ray production on extinct nuclide systems. Geochim Cosmochim Acta 67:529–541

Leya I, Schönbächler M, Wiechert U, Krähenbühl U, Halliday AN (2008) Titanium isotopes and the radial heterogeneity of the Solar System. Earth Planet Sci Lett 266:233–244

Leya I, Schönbächler M, Krähenbühl U, Halliday AN (2009) New titanium isotope data for Allende and Efremovka CAIs. Astrophys J 702:1118–1126

Lichtenberg T, Drażkowska J, Schönbächler M, Golabek GJ, Hands TO (2021) Bifurcation of planetary building blocks during Solar System formation. Science 371:365–370

Liu M-C, Han J, Brearley AJ, Hertwig AT (2019) Aluminum-26 chronology of dust coagulation and early Solar System evolution. Sci Adv 5:eaaw3350

Lodders K (2003) Solar System abundances and condensation temperatures of the elements. Astrophys J 591:1220

Ma N, Neumann W, Néri A, Schwarz WH, Ludwig T, Trieloff M, Klahr H, et al (2022) Early formation of primitive achondrites in an outer region of the protoplanetary disc. Geochem Perspect Lett 23:33–37

MacPherson GJ, Bullock ES, Janney PE, Kita NT, Ushikubo T, Davis AM, Wadhwa M, Krot AN (2010) Early solar nebula condensates with canonical, not supracanonical, initial $^{26}Al/^{27}Al$ ratios. Astrophys J 711:L117–L121

MacPherson GJ, Kita NT, Ushikubo T, Bullock ES, Davis AM (2012) Well-resolved variations in the formation ages for Ca-Al-rich inclusions in the early Solar System. Earth Planet Sci Lett 331–332:43–54

MacPherson GJ, Bullock ES, Tenner TJ, Nakashima D, Kita NT, Ivanova MA, Krot AN, Petaev MI, Jacobsen SB (2017) High precision Al–Mg systematics of forsterite-bearing Type B CAIs from CV3 chondrites. Geochim Cosmochim Acta 201:65–82

MacPherson GJ, Defouilloy C, Kita NT (2018) High-precision Al–Mg isotopic systematics in USNM 3898 – the benchmark “ALL” for initial 87Sr/86Sr in the earliest Solar System. Earth Planet Sci Lett 491:238–243

MacPherson GJ, Krot AN, Nagashima K (2020) Al-Mg isotopic study of spinel-rich fine-grained CAIs. Meteorit Planet Sci 55:2519–2538

MacPherson GJ, Krot AN, Kita NT, Bullock ES, Nagashima K, Ushikubo T, Ivanova MA (2022) The formation of Type B CAIs: Evolution from Type A CAIs. Geochim Cosmochim Acta 321:343–374

Makide K, Nagashima K, Krot AN, Huss GR, Hutcheon ID, Bischoff A (2009) Oxygen- and magnesium-isotope compositions of calcium–aluminum-rich inclusions from CR2 carbonaceous chondrites. Geochim Cosmochim Acta 73:5018–5050

Marin-Carbonne J, McKeegan KD, Davis AM, MacPherson GJ, Mendybaev RA, Richter FM (2023) In situ oxygen, magnesium, and silicon isotopic compositions of the FUN inclusion Vigarano 1623-5. Meteorit Planet Sci 58:546–571

Markowski A, Quitté G, Halliday AN, Kleine T (2006) Tungsten isotopic compositions of iron meteorites: chronological constraints vs. cosmogenic effects. Earth Planet Sci Lett 242:1–15

Marrocchi Y, Piralla M, Regnault M, Batanova V, Villeneuve J, Jacquet E (2022) Isotopic evidence for two chondrule generations in CR chondrites and their relationships to other carbonaceous chondrites. Earth Planet Sci Lett 593:117683

Martins R, Kuthning S, Coles BJ, Kreissig K, Rehkämper M (2023) Nucleosynthetic isotope anomalies of zinc in meteorites constrain the origin of Earth's volatiles. Science 379:369–372

Masarik J (1997) Contribution of neutron-capture reactions to observed tungsten isotopic ratios. Earth Planet Sci Lett 152:181–185

Matthes M, Fischer-Gödde M, Kruijer TS, Leya I, Kleine T (2015) Pd–Ag chronometry of iron meteorites: correction of neutron capture-effects and application to the cooling history of differentiated protoplanets. Geochim Cosmochim Acta 169:45–62

Matthes M, Fischer-Gödde M, Kruijer TS, Kleine T (2018) Pd-Ag chronometry of IVA iron meteorites and the crystallization and cooling of a protoplanetary core. Geochim Cosmochim Acta 220:82–95

Middleton R, Klein J, Raisbeck GM, Yiou F (1983) Accelerator mass spectrometry with ^{26}Al. Nucl Instrum Methods 218:430

Mishra RK, Goswami JN, Tachibana S, Huss GR, Rudraswami NG (2010) ^{60}Fe and ^{26}Al in chondrules from unequilibrated chondrites: implications for early Solar System processes. Astrophys J 714:L217–L221

Mittlefehldt D (2014) Achondrites. In: Davis AM (ed) Treatise on geochemistry, vol 1. 2nd edn. Elsevier, Oxford, pp 235–265

Morbidelli A, Bitsch B, Crida A, Gounelle M, Guillot T, Jacobson S, Johansen A, Lambrechts M, Lega E (2016) Fossilized condensation lines in the Solar System protoplanetary disk. Icarus 15:368–376

Morbidelli A, Baillie K, Batygin K, Charnoz S, Guillot T, Rubie DC, Kleine T (2022) Contemporary formation of early Solar System planetesimals at two distinct radial locations. Nat Astron 6:72–79

Morton EM, Pickard H, Wombacher F, Huang Y, Palk E, Martins R, Kuthning S, et al (2024) Volatile element depletion of carbonaceous chondrites—insights from mass-dependent zinc, cadmium, and tellurium isotope variations. Astrophys J 977:53

Nagashima K, Krot AN, Huss GR (2014) ^{26}Al in chondrules from CR2 chondrites. Geochem J 48:561–570

Nagashima K, Krot AN, Komatsu M (2017) ^{26}Al–^{26}Mg systematics in chondrules from Kaba and Yamato 980145 CV3 carbonaceous chondrites. Geochim Cosmochim Acta 201:303–319

Nagashima K, Kita NT, Luu T-H (2018) ^{26}Al–^{26}Mg systematics of chondrules. In: Russell AA, Connolly Jr HC, Kort AN (eds) Chondrules: Records of protoplanetary disk processes, Cambridge University Press, Cambridge, pp 247–275

Nanne JAM, Nimmo F, Cuzzi JN, Kleine T (2019) Origin of the non-carbonaceous–carbonaceous meteorite dichotomy. Earth Planet Sci Lett 511:44–54

Neumann W, Breuer D, Spohn T (2012) Differentiation and core formation in accreting planetesimals. Astron Astrophys 543:A141

Neumann W, Henke S, Breuer D, Gail H-P, Schwarz WH, Trieloff M, Hopp J, Spohn T (2018) Modeling the evolution of the parent body of acapulcoites and lodranites: a case study for partially differentiated asteroids. Icarus 311:146–169

Neumann W, Luther R, Trieloff M, Reger PM, Bouvier A (2023) Fitting thermal evolution models to the chronological record of Erg Chech 002 and modeling the ejection conditions of the meteorite. Planet Sci J. 4:196

Neumann W, Ma N, Bouvier A, Trieloff M (2024) Recurrent planetesimal formation in an outer part of the early Solar System. Sci Rep 14:14017

Nie NX, Chen XY, Hopp T, Hu JY, Zhang ZJ, Teng FZ, et al (2021) Imprint of chondrule formation on the K and Rb isotopic compositions of carbonaceous meteorites. Sci Adv 7(49):eabl3929

Nimmo F, Kretke K, Ida S, Matsumura S, Kleine T (2018) Transforming dust to planets. Space Sci Rev 214:101

Nishiizumi K (2004) Preparation of ^{26}Al AMS standards. Nucl Instrum Methods Phys Res, Sect B, Beam Interact Mater Atoms 223–224:388–392

Norris TL, Gancarz AJ, Rokop DJ, Thomas KW (1983) Paper presented at the Proc. 14th Lunar Planet. Sci. Conf. J Geophys Res 88:B331

Palme H, Mezger K (2024) Nucleosynthetic isotope variations in chondritic meteorites and their relationship to bulk chemistry. Meteorit Planet Sci 59:382–394

Palme H, O'Neill HSC (2014) Cosmochemical estimates of Mantle composition. In: Holland HD, Turekian KK (eds) Treatise on geochemistry, 2nd edn. Elsevier, Oxford, pp 1–39

Pape J, Mezger K, Bouvier AS, Baumgartner LP (2019) Time and duration of chondrule formation: constraints from ^{26}Al-^{26}Mg ages of individual chondrules. Geochim Cosmochim Acta 244:416–436

Pape J, Rosén ÅV, Mezger K, Guillong M (2021) Primary crystallization and partial remelting of chondrules in the protoplanetary disk: petrographic, mineralogical and chemical constraints recorded in zoned type-I chondrules. Geochim Cosmochim Acta 292:499–517

Park C, Nagashima K, Krot AN, Huss GR, Davis AM, Bizzarro M (2017) Calcium-aluminum-rich inclusions with fractionation and unidentified nuclear effects (FUN CAIs): II. Heterogeneities of magnesium isotopes and ^{26}Al in the early Solar System inferred from in situ high-precision magnesium-isotope measurements. Geochim Cosmochim Acta 201:6–24

Piralla M, Villeneuve J, Schnuriger N, Bekaert DV, Marrocchi Y (2023) A unified chronology of dust formation in the early Solar System. Icarus 394:115427

Poole GM, Rehkämper M, Coles BJ, Goldberg T, Smith CL (2017) Nucleosynthetic molybdenum isotope anomalies in iron meteorites–new evidence for thermal processing of solar nebula material. Earth Planet Sci Lett 473:215–226

Posner ES, Ganguly J, Hervig R (2016) Diffusion kinetics of Cr in spinel: experimental studies and implications for ^{53}Mn–^{53}Cr cosmochronology. Geochim Cosmochim Acta 175:20–35

Qin L, Dauphas N, Wadhwa M, Masarik J, Janney PE (2008) Rapid accretion and differentiation of iron meteorite parent bodies inferred from 182Hf–182W chronometry and thermal modeling. Earth Planet Sci Lett 273:94–104

Raymond SN, Morbidelli A (2022) Planet formation: key mechanisms and global models. In: Biazzo K, Bozza V, Mancini L, Sozzetti A (eds) Demographics of exoplanetary systems: lecture notes of the 3rd advanced school on exoplanetary science. Springer, Cham, pp 3–82

Reger PM, Roebbert Y, Neumann W, Gannoun A, Regelous M, Schwarz WH, Ludwig T, Trieloff M, Weyer S, Bouvier A (2023) Al-Mg and U-Pb chronological records of Erg Chech 002 ungrouped achondrite meteorite. Geochim Cosmochim Acta 343:33–48

Render J, Brennecka GA, Burkhardt C, Kleine T (2022) Solar System evolution and terrestrial planet accretion determined by zr isotopic signatures of meteorites. Earth Planet Sci Lett 595:117748

Rider-Stokes B, Greenwood RC, Anand M, White L, Franchi IA, Debaille V, Goderis S, Pittarello L, Yamaguchi A, Mikouchi T (2023) Impact mixing among rocky planetesimals in the early Solar System from angrite oxygen isotopes. Nat Astron 7:836–842

Rider-Stokes BG, Anand M, White LF, Darling JR, Tartèse R, Whitehouse MJ, Franchi I, Greenwood RC, Degli-Alessandrini G (2024) The impact history and prolonged magmatism of the angrite parent body. Meteorit Planet Sci 59:23–39

Roszjar J, Whitehouse MJ, Srinivasan G, Mezger K, Scherer EE, Van Orman JA, Bischoff A (2016) Prolonged magmatism on 4 Vesta inferred from Hf–W analyses of eucrite zircon. Earth Planet Sci Lett 452:216–226

Rudraswami NG, Goswami JN (2007) 26Al in chondrules from unequilibrated L chondrites: onset and duration of chondrule formation in the early Solar System. Earth Planet Sci Lett 257:231–244

Rudraswami NG, Goswami JN, Chattopadhyay B, Sengupta SK, Thapliyal AP (2008) 26Al records in chondrules from unequilibrated ordinary chondrites: II. Duration of chondrule formation and parent body thermal metamorphism. Earth Planet Sci Lett 274:93–102

Rüfenacht M, Morino P, Lai Y-J, Fehr MA, Haba MK, Schönbächler M (2023) Genetic relationships of Solar System bodies based on their nucleosynthetic Ti isotope compositions and sub-structures of the solar protoplanetary disk. Geochim Cosmochim Acta 355:110–125

Samworth EA, Warburton EK, Engelbertink GAP (1972) Phys Rev 5:138

Sanborn ME, Carlson RW, Wadhwa M (2015) ^{147,146}Sm–^{143,142}Nd, ^{176}Lu–^{176}Hf, and ^{87}Rb–^{87}Sr systematics in the angrites: implications for chronology and processes on the angrite parent body. Geochim Cosmochim Acta 171:80–99

Sanborn ME, Wimpenny J, Williams CD, Yamakawa A, Amelin Y, Irving AJ, Yin Q-Z (2019) Carbonaceous achondrites northwest Africa 6704/6693: milestones for early Solar System chronology and genealogy. Geochim Cosmochim Acta 245:577–596

Savage PS, Moynier F, Boyet M (2022) Zinc isotope anomalies in primitive meteorites identify the outer Solar System as an important source of Earth's volatile inventory. Icarus 386:115–172

Scherstén A, Elliott T, Hawkesworth C, Russell S, Masarik J (2006) Hf–W evidence for rapid differentiation of iron meteorite parent bodies. Earth Planet Sci Lett 241:530–542

Schiller M, Connelly JN, Glad AC, Mikouchi T, Bizzarro M (2015a) Early accretion of protoplanets inferred from a reduced inner Solar System ^{26}Al inventory. Earth Planet Sci Lett 420:45–54

Schiller M, Paton C, Bizzarro M (2015b) Evidence for nucleosynthetic enrichment of the protosolar molecular cloud core by multiple supernova events. Geochim Cosmochim Acta 149:88–102

Schiller M, Bizzarro M, Siebert J (2020) Iron isotope evidence for very rapid accretion and differentiation of the proto-Earth. Sci Adv 6:eaay7604

Schnuriger N, Cartier C, Villeneuve J, Batanova V, Regnault M, Marrocchi Y (2022) Spinel in CV chondrules: investigating precursor legacy and chondrule thermal histories. Meteorit Planet Sci 57:1018–1037

Schönbächler M, Fehr MA, Yokoyama T, Gautam I, Nakanishi N, Abe Y, Aléon J, et al (2025) Zirconium isotope composition indicates s-process depletion in samples returned from asteroid Ryugu. Meteorit Planet Sci 60:3–16

Schrader DL, Nagashima K, Krot AN, Ogliore RC, Yin Q, Amelin Y, Stirling CH, Kaltenbach A (2017) Distribution of ^{26}Al in the CR chondrite chondrule-forming region of the protoplanetary disk. Geochim Cosmochim Acta 201:275–302

Schrader DL, Nagashima K, Davidson J, McCoy TJ, Ogliore RC, Fu RR (2020) Outward migration of chondrule fragments in the early Solar System: O-isotopic evidence for rocky material crossing the Jupiter Gap? Geochim Cosmochim Acta 282:133–155

Scott ERD (1972) Chemical fractionation in iron meteorites and its interpretation. Geochim Cosmochim Acta 36:1205–1236

Scott ERD (2020) Iron meteorites: composition, age, and origin. Oxford University Press, London

Scott ERD, Wasson JT (1975) Classification and properties of iron meteorites. Rev Geophys 13:527–546

Sengupta D, Estrada PR, Cuzzi JN, Humayun M (2022) Depletion of moderately volatile elements by open system loss in the early solar nebula. Astrophys J 932:82

Shen JJ, Papanastassiou DA, Wasserburg GJ (1996) Precise Re-Os determinations and systematics of iron meteorites. Geochim Cosmochim Acta 60:2887–2900

Sheng YJ, Wasserburg GJ, Hutcheon ID (1992) Self-diffusion of magnesium in spinel and in equilibrium melts: constraints on flash heating of silicates. Geochim Cosmochim Acta 56:2535–2546

Shukolyukov A, Lugmair G (2008) Mn-Cr chronology of eucrite CMS 04049 and angrite NWA 2999. In: 39th Annual Lunar and Planetary Science Conference

Simon SB, Krot AN, Nagashima K (2019) Oxygen and Al-Mg isotopic compositions of grossite-bearing refractory inclusions from CO3 chondrites. Meteorit Planet Sci 54:1362–1378

Siron G, Fukuda K, Kimura M, Kita NT (2021) New constraints from ^{26}Al-^{26}Mg chronology of anorthite bearing chondrules in unequilibrated ordinary chondrites. Geochim Cosmochim Acta 293:103–126

Siron G, Fukuda K, Kimura M, Kita NT (2022) High precision ^{26}Al-^{26}Mg chronology of chondrules in unequilibrated ordinary chondrites: evidence for restricted formation ages. Geochim Cosmochim Acta 324:312–345

Smoliar MI, Walker RJ, Morgan JW (1996) Re-Os ages of group IIA, IIIA, IVA, and IVB iron meteorites. Science 271:1099–1102

Spitzer F, Burkhardt C, Budde G, Kruijer TS, Morbidelli A, Kleine T (2020) Isotopic evolution of the inner Solar System inferred from Molybdenum isotopes in meteorites. Astrophys J 898:L2

Spitzer F, Burkhardt C, Nimmo F, Kleine T (2021) Nucleosynthetic Pt isotope anomalies and the Hf-W chronology of core formation in inner and outer Solar System planetesimals. Earth Planet Sci Lett 576:117211

Spitzer F, Burkhardt C, Kruijer TS, Kleine T (2025) Comparison of the earliest NC and CC planetesimals: evidence from ungrouped iron meteorites. Geochim Cosmochim Acta 397:134–148

Spivak-Birndorf LJ, Wadhwa M, Janney PE (2009) ^{26}Al–^{26}Mg systematics in D'Orbigny and Sahara 99555 angrites: implications for high-resolution chronology using extinct chronometers. Geochim Cosmochim Acta 73:5202–5211

Srinivasan P, Dunlap DR, Agee CB, Wadhwa M, Coleff D, Ziegler K, Zeigler R, McCubbin FM (2018) Silica-rich volcanism in the early Solar System dated at 4.565 Ga. Nat Commun 9(1):3036

Steele RC, Elliott T, Coath CD, Regelous M (2011) Confirmation of mass-independent Ni isotopic variability in iron meteorites. Geochim Cosmochim Acta 75:7906–7925

Steele RC, Coath CD, Regelous M, Russell S, Elliott T (2012) Neutron-poor nickel isotope anomalies in meteorites. Astrophys J 758:59

Steiger RH, Jäger E (1977) Subcommission on geochronology: convention on the use of decay constants in geo-and cosmochronology. Earth Planet Sci Lett 36:359–362

Steller T, Burkhardt C, Yang C, Kleine T (2022) Nucleosynthetic zinc isotope anomalies reveal a dual origin of terrestrial volatiles. Icarus 386:115171

Sugiura N, Fujiya W (2014) Correlated accretion ages and ε^{54}Cr of meteorite parent bodies and the evolution of the solar nebula. Meteorit Planet Sci 49:772–787

Tarduno JA, Cottrell RD, Nimmo F, Hopkins J, Voronov J, Erickson A, Blackman E, Scott ERD, McKinley R (2012) Evidence for a dynamo in the main group pallasite parent body. Science 338:939
Tenner TJ, Nakashima D, Ushikubo T, Tomioka N, Kimura M, Weisberg MK, Kita NT (2019) Extended chondrule formation intervals in distinct physicochemical environments: evidence from Al-Mg isotope systematics of CR chondrite chondrules with unaltered plagioclase. Geochim Cosmochim Acta 260:133–160
Theis KJ, Schönbächler M, Benedix GK, Rehkämper M, Andreasen R, Davies C (2013) Palladium–silver chronology of IAB iron meteorites. Earth Planet Sci Lett 361:402–411
Thomas JH, Rau RL, Skelton RT, Kavanagh RW (1984) Phys Rev C 30:385
Tissot FLH, Dauphas N, Grossman L (2016) Origin of uranium isotope variations in early solar nebula condensates. Sci Adv 2
Tissot FL, Dauphas N, Grove TL (2017) Distinct $^{238}U/^{235}U$ ratios and REE patterns in plutonic and volcanic angrites: geochronologic implications and evidence for U isotope fractionation during magmatic processes. Geochim Cosmochim Acta 213:593–617
Tornabene HA, Hilton CD, Bermingham KR, Ash RD, Walker RJ (2020) Genetics, age and crystallization history of group IIC iron meteorites. Geochim Cosmochim Acta 288:36–50
Tornabene HA, Ash RD, Walker RJ, Bermingham KR (2023) Genetics, age, and crystallization history of group IC iron meteorites. Geochim Cosmochim Acta 340:108–119
Touboul M, Puchtel IS, Walker RJ (2012) ^{182}W evidence for long-term preservation of early mantle differentiation products. Science 335:1065–1069
Trinquier A, Birck J-L, Allegre CJ (2007) Widespread ^{54}Cr heterogeneity in the inner Solar System. Astrophys J 655:1179
Trinquier A, Elliott T, Ulfbeck D, Coath C, Krot AN, Bizzarro M (2009) Origin of nucleosynthetic isotope heterogeneity in the solar protoplanetary disk. Science 324:374–376
Ushikubo T, Nakashima D, Kimura M, Tenner TJ, Kita NT (2013) Contemporaneous formation of chondrules in distinct oxygen isotope reservoirs. Geochim Cosmochim Acta 109:280–295
Ushikubo T, Tenner TJ, Hiyagon H, Kita NT (2017) A long duration of the ^{16}O-rich reservoir in the solar nebula, as recorded in fine-grained refractory inclusions from the least metamorphosed carbonaceous chondrites. Geochim Cosmochim Acta 201:103–122
Van Kooten EM, Wielandt D, Schiller M, Nagashima K, Thomen A, Larsen KK, Olsen MB, et al (2016) Isotopic evidence for primordial molecular cloud material in metal-rich carbonaceous chondrites. Proc Natl Acad Sci USA 113:2011–2016
Van Orman JA, Crispin KL (2010) Diffusion in oxides. Rev Mineral Geochem 72:757–825
Van Orman JA, Cherniak DJ, Kita NT (2014) Magnesium diffusion in plagioclase: dependence on composition, and implications for thermal resetting of the ^{26}Al-^{26}Mg early Solar System chronometer. Earth Planet Sci Lett 385:79–88
Villeneuve J, Chaussidon M, Libourel G (2009) Homogeneous distribution of ^{26}Al in the Solar System from the Mg isotopic composition of chondrules. Science 325:985–988
Vockenhuber C, Oberli F, Bichler M, Ahmad I, Quitté G, Meier M, Halliday AN, Lee DC, Kutschera W, Steier P, Gehrke RJ, Helmer RG (2004) New half-life measurement of ^{182}W: improved chronometer for the early Solar System. Phys Rev Lett 93:172501
Vollstaedt H, Mezger K, Alibert Y (2020) Carbonaceous chondrites and the condensation of elements from the solar nebula. Astrophys J 897:82
Wadhwa M, Kita N, Nakashima D, Bullock E, MacPherson G, Bouvier A (2014) High precision ^{26}Al-^{26}Mg systematics for an almost pristine refractory inclusion: implications for the absolute age of the Solar System. In: 45th Lunar and Planetary Science Conference
Warren PH (2011) Stable-isotopic anomalies and the accretionary assemblage of the Earth and Mars: a subordinate role for carbonaceous chondrites. Earth Planet Sci Lett 311:93–100
Wasserburg GJ, Wimpenny J, Yin Q (2012) Mg isotopic heterogeneity, Al-Mg isochrons, and canonical $^{26}Al/^{27}Al$ in the early Solar System. Meteorit Planet Sci 47:1980–1997
Wasson JT, Huber H (2006) Compositional trends among IID irons; their possible formation from the P-rich lower magma in a two-layer core. Geochim Cosmochim Acta 70:6153–6167
Wasson JT, Kallemeyn GW (2002) The IAB iron-meteorite complex: a group, five subgroups, numerous grouplets, closely related, mainly formed by crystal segregation in rapidly cooling melts. Geochim Cosmochim Acta 66:2445–2473
Weidenschilling S (1997) The origin of comets in the solar nebula: a unified model. Icarus 127:290–306
Willbold M, Elliott T, Moorbath S (2011) The tungsten isotopic composition of the Earth's mantle before the terminal bombardment. Nature 477:195–198
Williams CD, Ushikubo T, Bullock ES, Janneyevans PE, Hines RR, Kita NT, Hervig RL, MacPherson GJ, Mendybaev RA, Richter FM, Wadhwa M (2017) Thermal and chemical evolution in the early Solar System as recorded by FUN CAIs: Part I – Petrology, mineral chemistry, and isotopic composition of Allende FUN CAI CMS-1. Geochim Cosmochim Acta 201:25–48

Williams CD, Sanborn ME, Defouilloy C, Yin Q-Z, Kita NT, Ebel DS, Yamakawa A, Yamashita K (2020) Chondrules reveal large-scale outward transport of inner Solar System materials in the protoplanetary disk. Proc Natl Acad Sci USA 117:23426–23435

Wimpenny J, Sanborn ME, Koefoed P, Cooke IR, Stirling C, Amelin Y, Yin Q-Z (2019) Reassessing the origin and chronology of the unique achondrite Asuka 881394: implications for distribution of ^{26}Al in the early Solar System. Geochim Cosmochim Acta 244:478–501

Wittig N, Humayun M, Brandon AD, Huang S, Leya I (2013) Coupled W-Os-Pt isotope systematics in IVB iron meteorites: in situ neutron dosimetry for W isotope chronology. Earth Planet Sci Lett 361:152–161

Worsham EA, Bermingham KR, Walker RJ (2016) Siderophile element systematics of IAB complex iron meteorites: new insights into the formation of an enigmatic group. Geochim Cosmochim Acta 188:261–283

Worsham EA, Bermingham KR, Walker RJ (2017) Characterizing cosmochemical materials with genetic affinities to the Earth: genetic and chronological diversity within the IAB iron meteorite complex. Earth Planet Sci Lett 467:157–166

Worsham EA, Burkhardt C, Budde G, Fischer-Gödde M, Kruijer TS, Kleine T (2019) Distinct evolution of the carbonaceous and non-carbonaceous reservoirs: Insights from Ru, Mo, and W isotopes. Earth Planet Sci Lett 521:103–112

Yang J, Goldstein JI, Scott ERD (2007) Iron meteorite evidence for early formation and catastrophic disruption of protoplanets. Nature 446:888–891

Yang B, Pang R, Wang Q, Zhang A, Du W, Qin L (2025) Highly heterogeneous parent body of the rare andesitic Erg Chech 002 meteorite revealed by the revisited Mn–Cr isotopic systematics. Planet Sci J 6:73

Yokoyama T, Alexander CMD, Walker RJ (2011) Assessment of nebular versus parent body processes on presolar components present in chondrites: evidence from osmium isotopes. Earth Planet Sci Lett 305:115–123

Yokoyama T, Nagai Y, Fukai R, Hirata T (2019) Origin and evolution of distinct Molybdenum isotopic variabilities within carbonaceous and non-carbonaceous reservoirs. Astrophys J 883:62

Yokoyama T, Nagashima K, Nakai I, Young ED, Abe Y, Aléon J, Alexander CMOD, et al (2023) Samples returned from the asteroid Ryugu are similar to Ivuna-type carbonaceous meteorites. Science 379:eabn7850

Zanda B, Hewins RH, Bourot-Denise M, Bland PA, Albarède F (2006) Formation of solar nebula reservoirs by mixing chondritic components. Earth Planet Sci Lett 248:650–660

Zhu K, Moynier F, Wielandt D, Larsen KK, Barrat J-A, Bizzarro M (2019) Timing and origin of the angrite parent body inferred from Cr isotopes. Astrophys J Lett 877:L13

Zhu K, Becker H, Li S-J, Fan Y, Liu X-N, Elliott T (2022) Radiogenic chromium isotope evidence for the earliest planetary volcanism and crust formation in the Solar System. Mon Not R Astron Soc Lett 515:L39

Authors and Affiliations

Maria Schönbächler[1] · Audrey Bouvier[2] · Noriko T. Kita[3] · Thomas S. Kruijer[4]

✉ M. Schönbächler
mariasc@ethz.ch

[1] Institute for Geochemistry and Petrology, Department of Earth and Planetary Sciences, ETH Zurich, Clausiusstrasse 25, Zurich, Switzerland

[2] Bayerisches Geoinstitut, Universität Bayreuth, Bayreuth 95447, Germany

[3] Department of Geoscience, University of Wisconsin-Madison, Madison, WI 53706, USA

[4] Cosmochemical and Isotopic Signatures Group, Lawrence Livermore National Laboratory, Livermore, CA 94550, USA

Space Science Reviews (2025) 221:7
https://doi.org/10.1007/s11214-025-01135-z

Metasomatic Alteration of Type 3 Ordinary and Carbonaceous Chondrites

A.N. Krot[1] · M.I. Petaev[2] · L. Piani[3] · Y. Marrocchi[3] · W. Fujiya[4] · O.V. Pravdivtseva[5] · E. Dobrică[1] · L.G. Vacher[6] · A.J. King[7] · M. Lee[8] · E. Van Kooten[9] · B. Jacobsen[10] · C.M.O'D. Alexander[11] · A. Bischoff[12] · A.J. Brearley[13] · C. Le Guillou[14] · L. Remusat[15] · J. Leitner[16,17] · G.R. Huss[1]

Received: 2 April 2024 / Accepted: 28 December 2024 / Published online: 20 January 2025

Abstract
Metasomatism refers to the process during which a pre-existing rock undergoes compositional and mineralogical transformations associated with chemical reactions triggered by the reaction of fluids which invade the protolith. It changes chemical compositions of minerals, promotes their dissolution and precipitation of new minerals. In this paper, we review metasomatic alteration of type 3 ordinary (H, L, LL) and carbonaceous (CV, CO, CK) chondrites, including (*i*) secondary mineralization, (*ii*) physicochemical conditions, (*iii*) chronology (^{53}Mn-^{53}Cr, ^{26}Al-^{26}Mg, ^{129}I-^{129}Xe) of metasomatic alteration, (*iv*) records of metasomatic alteration in H, O, N, C, S, and Cl isotopic systematics, (*v*) effects of metasomatic alteration on O- and Al-Mg-isotope systematics of primary minerals in chondrules and refractory inclusions, and (*vi*) sources of water ices in metasomatically altered CV, CO, and ordinary chondrites, and outline future studies.

Keywords Carbonaceous chondrites · Ordinary chondrites · Metasomatic alteration · Water · Oxygen · Hydrogen · Nitrogen and sulfur isotopes · ^{53}Mn-^{53}Cr chronology · ^{129}I-^{129}Xe chronology

1 Introduction

There are several classes of chondritic meteorites comprising a number of groups: carbonaceous (CI, CM, CR, CO, CV, CK, CR, CH, CB, CY, CL), ordinary (H, L, LL), enstatite (EH, EL), Kakangari (K), and Rumuruti (R) (Krot et al. 2014a). Some chondrites cannot be classified into the existing groups and are named ungrouped or unique, e.g., Acfer 094 (Newton et al. 1995). CV chondrites are subdivided into three subgroups, oxidized Allende-like (CV_{oxA}), oxidized Bali-like (CV_{oxB}), and reduced (CV_{red}), which may or may not represent fragments of the same parent asteroid (Weisberg et al. 2006; Gattacceca et al. 2020). Virtually all chondrites experienced aqueous alteration and/or thermal metamorphism in their parent asteroids. Based on the degree of aqueous alteration and thermal metamorphism, chondrites are classified into petrologic types from 1 to <3 and from 3 to 6, respectively. Based on the distribution of chromium in chondrule ferroan olivines (Grossman and Brearley 2005), Raman parameters of matrix organic matter (Bonal et al. 2006, 2007), sulfur

Extended author information available on the last page of the article

abundance in matrix (Grossman and Brearley 2005), composition of FeNi-metal (Kimura et al. 2007), matrix modal mineralogy (Brearley 1993; Greshake 1997), and abundance of presolar grains and noble gas components (Huss and Lewis 1995; Nguyen et al. 2010; Leitner et al. 2016; Nittler et al. 2018), type 3 chondrites are further subdivided into subtypes from 3.0 to 3.9; even more detailed classification is used for type 3.0–3.1 chondrites. We note that this classification scheme is rather simplistic, because aqueous alteration and thermal metamorphism are closely related processes and virtually all thermally metamorphosed type 3 chondrites experienced aqueous alteration (Bonal et al. 2020). The least altered and the least thermally metamorphosed chondrites are type 3.00. There are only a few type 3.00 chondrites known, including Semarkona (LL), Northwest Africa (NWA) 8276 (L), Dominion Range (DOM) 08006 (CO), and Acfer 094 (C-ungrouped) (Fig. SM1; Newton et al. 1995; Grossman and Brearley 2005; Alexander et al. 2018a; Davidson et al. 2019; Zanetta et al. 2022; Krämer Ruggiu et al. 2022; Barosch et al. 2022).

Here we review metasomatic alteration of type 3 ordinary (H, L, LL) and carbonaceous (CV, CO, CK) chondrites, including (*i*) secondary mineralization, (*ii*) physicochemical conditions and (*iii*) chronology (^{53}Mn-^{53}Cr, ^{26}Al-^{26}Mg, ^{129}I-^{129}Xe) of metasomatic alteration; (*iv*) records of metasomatic alteration in H, O, N, C, S, and Cl isotope systematics; (*v*) effects of metasomatic alteration on O- and Al-Mg-isotope systematics of primary minerals in chondrules and refractory inclusions; (*vi*) sources of water ices in metasomatically altered CV, CO, and ordinary chondrites, and outline future studies.

2 Primary and Secondary Mineralization in Type 3 Chondrites

2.1 Primary Mineralogy of Chondritic Components in Primitive Chondrites

Chondrites are composed of several major components, including refractory inclusions [Ca,Al-rich inclusions (CAIs) and amoeboid olivine aggregates (AOAs) – aggregates of CAIs and fortsterite+FeNi-metal condensates], chondrules, matrix, FeNi-metal, Fe±Ni-sulfides, and organic matter. Refractory inclusions consist of minerals that formed by gas-solid condensation above $\sim$ 1300 K or by melting of gas-solid condensates in a gas of approximately solar chemical composition, i.e., highly reduced: H_2O/H_2 ratio of $\sim 6.6\times10^{-4}$ corresponding to $\sim$ 7 log units below the iron-wüstite (IW) buffer (e.g., Krot 2019). As a result, primary oxide and silicate minerals in CAIs and AOAs from type 3.00 chondrites are Fe- and Na-free (Table SM1). Pristine refractory inclusions contain no oxidized and low-temperature secondary minerals, such as magnetite, nepheline, sodalite, etc.

Pristine chondrules consist of minerals (Table SM2) formed by melting of solid precursors during transient heating events at $\sim$ 1600–2000 K in a gas much more oxidized than solar, IW−3 to IW−1, and contain no secondary minerals such as magnetite, phyllosilicates, nepheline, etc. Chondrules experienced gas-melt interaction before and during crystallization resulting in re-condensation and zoned distribution of moderately volatile elements, such as Si, Mn, Cr, Na, and K (Libourel et al. 2006). Because these elements are also mobile during aqueous/metasomatic alteration (Grossman and Brearley 2005), this may complicate distinguishing nebular and asteroidal records in chondrules.

Pristine matrices are composed predominantly of amorphous ferromagnesian silicates, sub-micron forsterite and enstatite grains, FeNi-metal, Fe-sulfides, organics, presolar grains, and have abundant porosity (Table SM2; Greshake 1997; Grossman and Brearley 2005; Davidson et al. 2019). Because crystalline silicates are common in pristine matrices but virtually absent in interstellar medium (ISM), some fraction of matrix materials must have experienced thermal processing in the protoplanetary disk (PPD), most likely during formation of refractory inclusions and chondrules. We note that thermal processing of amorphous silicates in the PPD (*i*) below 1100 K that resulted in formation of crystalline silicates and/or (*ii*) below 900 K that may have modified their primordial oxygen-isotope composition without formation of crystalline silicates are also possible (Ishizaki et al. 2023; Sakurai et al. 2023; Yamamoto et al. 2024). The abundance of primordial, molecular cloud material in pristine chondrite matrices that avoided thermal processing in the PPD remains controversial (Alexander et al. 2017). Due to its extremely fine-grained nature and high porosity, matrix can be easily altered; e.g., amorphous material even in the most primitive carbonaceous chondrite Acfer 094 experienced hydration (Matsumoto et al. 2019).

Refractory inclusions, chondrules, and matrices having different mineralogies, textures, chemistry, and grain-sizes responded differently to metasomatic alteration.

2.2 Initial Stages of Alteration of Chondritic Components in Type 3.0–3.1 Chondrites

The initial stages of metasomatic alteration are largely recorded by matrices and chondrules in ordinary, CV, and CO chondrites of petrologic types 3.0–3.1. The petrographic and chemical effects of metasomatic alteration include (*i*) dissolution of chondrule mesostasis and plagioclase, (*ii*) zoning of Na and Ca in chondrule mesostasis and plagioclase, (*iii*) replacement of chondrule mesostasis by phyllosilicates, and (*iv*) redistribution of matrix nanosulfides (Grossman et al. 2000; Grossman and Brearley 2005; Lewis and Jones 2019). Typical secondary minerals formed during the initial stages of metasomatic alteration of type 3 chondrites are listed in Table SM3; their textural occurrences are shown in Figures SM2–SM7.

FeNi-metal experienced carburization to form FeNi-carbides (Figs. SM2a,b; Hutchison et al. 1987; Alexander et al. 1989; Taylor et al. 1981; Shibita 1996; Krot et al. 1997; Keller 1998; Krot and Todd 1998; Davidson et al. 2019):

(R1) $CO_{(g)} + H_2O_{(g)} \rightarrow CO_{2(g)} + H_{2(g)}$
(R2) $3Fe_{(s)} + CO_{2(g)} + 2H_{2(g)} \rightarrow Fe_3C_{(s)} + 2H_2O_{(g)}$

FeNi-metal and carbides were subsequently oxidized to form magnetite (Fig. SM2):

(R3) $Fe_3C_{(s)} + 6H_2O_{(g)} \rightarrow Fe_3O_{4(s)} + 6H_{2(g)} + CO_{2(g)}$
(R4) $3Fe_{(s)} + 4H_2O_{(g)} \rightarrow Fe_3O_{4(s)} + 4H_{2(g)}$

Note that reactions R1–4 do not require aqueous fluid and may have occurred either in the solar nebula or the chondrite parent bodies. For example, FeNi-carbides described in anhydrous interplanetary dust particles (IDPs) could be by-products of Fischer-Tropsch synthesis in the solar nebula (Christoffersen and Buseck 1983; Bradley et al. 1984; Bradley 1994). Oxidation of metal to magnetite is thermodynamically possible in the canonical solar nebula below < 400 K (Fegley and Prinn 1989) and was experimentally studied by Hong and Fegley (1998) (see Sect. 5.1). There are, however, mineralogical evidence suggesting formation of FeNi-carbides and magnetite in the chondrite parent asteroids (Krot et al. 1997). (*i*) FeNi-carbides replace metal nodules in chondrules and form veins crosscutting their fine-grained rims (Figs. SM2c–f). (*ii*) Two types of magnetite can be distinguished: (*a*) "dirty", Cr-bearing magnetite that formed by *in situ* oxidation of Cr-bearing FeNi-metal nodules (Figs. SM2c, 4b):

(R5) $3Fe_{(s)} + 4H_2O_{(aq,\,g)} \rightarrow Fe_3O_{4(s)} + 4H_{2(g)}$

and (*b*) "clean", Cr-free magnetite overgrowing oxidized nodules (Figs. SM2c, 4b) and crosscutting fine-grained rims around chondrules (Figs. SM2c–f, 5b,e). This magnetite formed by precipitation of Fe dissolved in aqueous solution:

(R6) $3Fe^{2+}{}_{(aq)} + 4H_2O_{(aq)} \rightarrow Fe_3O_{4(s)} + 6H^{+}{}_{(aq)} + H_{2(g)}$,

Marrocchi et al. (2016) interpreted clean Cr-free magnetite overgrowing opaque nodules in chondrules from the oxidized Bali-like CV3.1 chondrite Kaba as igneous, i.e., as a result of crystallization FeSO melts. Formation of FeSO melt requires drastic oxidizing conditions, too oxidizing even for chondrule formation, IW+1 to IW+2 (see Sect. 2.1). The lack of magnetite-bearing chondrules in the reduced CV3 chondrites and the differences in O-isotope compositions between Cr-free magnetite and chondrule phenocrysts in Kaba chondrules (Marrocchi et al. 2016; Krot et al. 2022a) contradict this interpretation (see Sect. 7.1.2).

In addition, framboidal magnetite occasionally observed in matrix of the oxidized CV3.1 chondrites Kaba and Mokoia also formed by direct precipitation from aqueous fluid (Keller and Buseck 1990; Tomeoka and Buseck 1990). The existence of aqueous solution during alteration of type 3.0–3.1 chondrites is supported by hydration and replacement of amorphous silicates by phyllosilicates and ferroan olivine (Fig. SM3).

Subsequently, the Cr-free magnetite is preferentially replaced by Mn-bearing fayalite (Fig. SM4):

(R7) $2Fe_3O_{4(s)} + 3SiO_{2(aq)} + 3xMn^{2+}{}_{(aq)} + 2H_{2(g)} \rightarrow 3(Fe_{2-x}Mn_x)_2SiO_{4(s)} + 3xFe^{+}{}_{(aq)} + 2H_2O_{(aq)}$

Isolated fayalite grains in matrices and pore space of AOAs and fine-grained CAIs, as well as fayalite grains overgrowing chondrule olivines and fine-grained matrix-like rims (Figs. SM5–6) apparently formed by direct precipitation from aqueous solutions:

(R8) $xFe^{2+}{}_{(aq)} + (2-x)Mn^{2+}{}_{(aq)} + SiO_{2(aq)} + 2H_2O_{(aq)} \rightarrow (Fe,Mn)_2SiO_{4(s)} + 4H^{+}{}_{(aq)}$

In Kaba and MacAlpine Hills (MAC) 88107 (CO3.1-like), fayalite often associates with hedenbergite (Fig. SM5) that may have formed by direct precipitation from aqueous fluid and by replacement of fayalite:

(R9) $Fe^{2+}{}_{(aq)} + Ca^{2+}{}_{(aq)} + 2SiO_{2(aq)} + 2H_2O_{(aq)} \rightarrow CaFeSi_2O_{6(s)} + 4H^{+}{}_{(aq)}$
(R10) $Fe_2SiO_{4(s)} + SiO_{2(aq)} + Ca^{2+}{}_{(aq)} \rightarrow CaFeSi_2O_{6(s)} + Fe^{2+}{}_{(aq)}$

Manganese-bearing fayalite is critical for ^{53}Mn-^{53}Cr dating of metasomatic alteration of UOCs, COs, and CVs (see Sect. 6.1). It is also important for constraining O-isotope composition of aqueous fluid, and, potentially, temperature of metasomatic alteration (see Sect. 7.1). We note, however, that reactional relationships between fayalite and magnetite (Fig. SM4) suggest these minerals may not have been in isotopic equilibrium with the same fluid. Therefore, temperature estimates based on O-isotope compositions of fayalite and magnetite must be taken with caution.

In the reduced CV3s (Efremovka, Leoville, Vigarano) and MAC 88107, magnetite, fayalite, and hedenbergite associate with kirschsteinite (Fig. SM7) (Krot et al. 2000; Ganino and Libourel 2017; MacPherson et al. 2017) that may have formed by replacement of magnetite and by direct precipitation from aqueous solution:

(R11) $Fe_3O_{4(s)} + Ca_{(aq)}{}^{2+} + SiO_{2(aq)} + H_{2(g)} + 2H^{+}{}_{(aq)} \rightarrow CaFeSiO_{4(s)} + 2Fe^{2+}{}_{(aq)} + 2H_2O_{(aq)}$
(R12) $Ca^{2+}{}_{(aq)} + Fe^{2+}{}_{(aq)} + SiO_{2(aq)} + 2H_2O_{(aq)} \rightarrow CaFeSiO_{4(s)} + 4H^{+}{}_{(aq)}$

Kirschsteinite (kirschsteinite-monticellite solid solution) grains in CVs are often chemically-zoned (Fig. SM7b) suggesting evolution of chemical composition of aqueous fluid:

(R13) $CaFeSiO_{4(s)} + xMg^{2+}{}_{(aq)} \rightarrow Ca(Fe_{1-x}Mg_x)SiO_{4(s)} + xFe^{2+}{}_{(aq)}$

Kirschsteinite is a critical mineral for ^{53}Mn-^{53}Cr chronology of metasomatic alteration of the reduced CVs (see Sect. 6.2), and potentially important for constraining O-isotope composition of aqueous fluid in their parent body. However, no SIMS (secondary ion mass-spectrometry) standards for kirschsteinite are currently available.

In contrast to matrices and chondrules, most primary minerals in refractory inclusions from 3.0–3.1 chondrites show little evidence for metasomatic alteration. The only exception are Ca-aluminates, grossite and krotite, which can be replaced by hercynite that is commonly Zn-bearing (Fig. SM8; Krot et al. 2019b; Ebert et al. 2024; Han et al. 2024):

(R14) $CaAl_2O_{4(s)} + xFe^{2+}{}_{(aq)} + (1-x)Zn^{2+}{}_{(aq)} \rightarrow (Fe,Zn)Al_2O_{4(s)} + Ca^{2+}{}_{(aq)}$
(R15) $CaAl_4O_{7(s)} + 2xFe^{2+}{}_{(aq)} + (2-2x)Zn^{2+}{}_{(aq)} + H_2O_{(aq)} \rightarrow 2(Fe_xZn_{1-x})Al_2O_{4(s)} + Ca^{2+}{}_{(aq)} + 2H^{+}{}_{(aq)}$

2.3 Advanced Stages of Alteration in Type 3 Chondrites of Higher Petrologic Subtypes

Studies of secondary mineralization in type 3 carbonaceous chondrites as a function of their petrologic subtypes are largely focused on the formation of nepheline in refractory inclusions and chondrules in CO3.0–3.8 s, where it formed by replacement of melilite, plagioclase, and chondrule mesostasis (Kojima et al. 1995; Tomeoka and Itoh 2004; Itoh et al. 2004; Brearley and Krot 2012; Wick and Jones 2012). There is a correlation between the amounts of nepheline in the CO chondritic components and the degree of thermal metamorphism of their host meteorites (Kojima et al. 1995; Itoh and Tomeoka 1998; Russell et al. 1998; Tomeoka and Itoh 2004), suggesting the nephelinization resulted from fluid-assisted thermal metamorphism. Mineralogical changes during metasomatic alteration of CV3 and CK3 chondrites are poorly known, because (*i*) well established metamorphic sequence among CV3s is currently absent (Bonal et al. 2006) and (*ii*) CKs of petrologic type <3.7 are not known.

Bridges et al. (1997) and Lewis et al. (2022) showed that evidence of metasomatism is present throughout the metamorphic sequence in OCs, and thus was a ubiquitous process in the OC parent bodies. In types 3.0–3.8, chondrule glassy mesostasis experienced dissolution and replacement by phyllosilicates, which survived only in type 3.0–3.1. Oxidation of phosphorous exsolved from FeNi-metal resulted in formation of merrillite and apatite; apatite typically replaces merrillite. Primary calcic plagioclase in chondrules was altered to various degrees by sodalite, scapolite, and nepheline in petrologic types 3.2–3.9, and to albite in types 3.6–5 (see also Bischoff and Keil 1983, 1984, 2012). Plagioclase also developed alteration features such as Ca-Na zoning (Grossman et al. 2000), micropores, and alteration lamellae in types 3–4. Sodic plagioclase is present in minor amounts as a primary phase, but also formed from the crystallization of chondrule mesostasis glass (types 3.2–3.9), and predominantly through albitization reactions in calcic plagioclase (types 3.6–5). K-feldspar occurs in albite in types 3.6–6 as fine-scale exsolution lamellae and as larger patches. Not all K-feldspar in OCs is secondary: e.g., coarse-grained K-feldspar (Or_{95}) occurs in achondritic clasts of unknown origin in Adzhi-Bogdo (LL3–6) breccia (Bischoff et al. 1993). In type 4–6 Ls and LLs, mesostasis glass crystallized into secondary plagioclase of diverse composition ($An_{<85}$). Equilibration of plagioclase is observed through type 5 and by type 6, plagioclase is chemically equilibrated at $\sim An_{11}$. Apatite evolves to Cl-apatite with a minor F and no OH. There are, however, exceptions: e.g., apatite in Renchen (L5–6) contains 3.4 wt% F and only 0.47 wt% Cl (Bischoff et al. 2019). Based on these observations, Lewis et al. (2022) concluded that aqueous alteration during prograde metamorphism of OCs produced most of the alteration and equilibration features in their plagioclase. During retrograde metamorphism, high temperature, short duration infiltration of anhydrous, alkali- and halogen-bearing fluids caused incorporation of potassium into plagioclase that subsequently exsolved.

2.4 Metasomatic Alteration of Igneous CAIs in Allende ($CV_{oxA} > 3.6$)

Coarse-grained, cm-sized igneous CAIs are present exclusively in CV and CK chondrites. Several types of igneous CAIs are recognized: Compact Type A (CTA), Type B (B), Forsterite-bearing Type B (FoB), and Type C (C) (Fig. SM9). These types consist of similar primary minerals, but have different modal mineralogy and bulk chemical compositions (MacPherson 2014). Because of their large sizes, the common presence in a widely available CV3 chondrite fall Allende (CV_{oxA}>3.6), and the importance for studying the early solar system chronology, the secondary mineralogy mainly of Allende igneous CAIs is characterized in detail (e.g., Krot et al. 2021).

Most of the secondary minerals in Allende igneous CAIs are Mg-rich and anhydrous (Table SM4). As a result, both nebular and asteroidal models have been proposed to explain their formation. According to the nebular models (Hashimoto and Grossman 1987; McGuire and Hashimoto 1989; Keller and Buseck 1991; Ushikubo et al. 2007a,b), secondary minerals formed during gas-solid reactions in the solar nebula. According to the asteroidal models (Brearley and Krot 2012; Krot et al. 2021), these minerals formed during aqueous fluid – rock interaction in the CV parent asteroid. Both nebular and asteroidal processes have been invoked to explain alteration of CV CAIs by Fagan et al. (2007).

Secondary minerals providing clear evidence for involvement of the CO_2- and H_2O-bearing fluid in their formation include clintonite, montmorillonite, margarite, biopyriboles, calcite, and tilleyite (Table SM4; Fig. SM10). Involvement of aqueous solutions is also indicated by the presence of veins of secondary minerals crosscutting primary anorthite and melilite, and by dissolution textures of anorthite and melilite with open space filled by secondary minerals (Fig. SM11).

Allende igneous CAIs are surrounded by Wark-Lovering (WL) rims that originated in the CAI-forming region (Wark and Lovering 1976), which are in turn surrounded by fine-grained matrix-like rims composed of secondary ferroan olivine and nepheline (Fig. SM9e), and by a layer of secondary CaFe-rich silicates (salite-hedenbergite pyroxenes, andradite, and wollastonite (Fig. SM9f). The altered regions of primary melilite are depleted in Ca and enriched in Si. There is a zoned distribution of secondary Na- and Cl-bearing minerals in the Allende CAIs: wadalite/adrianite is found almost exclusively in åkermanite-rich cores of Type B CAIs, whereas sodalite and nepheline are concentrated in their gehlenitic peripheries (Figs. SM9d,e). These observations provide clear evidence for an open-system behavior of several elements (Si, Na, Cl, Ca, and Al) during *in situ* metasomatic alteration of these CAIs: Si, Na, and Cl were added, whereas Ca and some Al were removed from the inclusions. The mobility of Al during the alteration is consistent with the presence of grossular, kushiroite, and Na-rich melilite in veins crosscutting primary anorthite grains inside the inclusions (Fig. SM11c). The outward transport of dissolved Al from CAIs and the influx of Na, K, and Si with an aqueous fluid equilibrated with the Allende matrix could explain the abundant nepheline grains in fine-grained rims around altered Allende CAIs (Fig. SM9e):

(R16) $Na^{+}_{(aq)} + Al^{3+}_{(aq)} + SiO_{2(aq)} + 2H_2O_{(aq)} \rightarrow NaAlSiO_{4(s)} + 4H^{+}_{(aq)}$

The zoned distribution of nepheline and sodalite in Allende CAIs and the lack of sodalite in fine-grained rims around these CAIs may indicate an earlier formation of nepheline than sodalite, consistent with observations of Che and Brearley (2021). Calcium released during alteration of melilite interacted with Fe-, Mg- and Si-enriched fluid that most likely was in equilibrium with the Allende fine-grained matrix (Krot et al. 2001; Ganino and Libourel 2017). This process resulted in precipitation of CaFe-rich silicates (hedenbergite, andradite, wollastonite, and salite-hedenbergite pyroxene) at the boundary between fine-grained rims and the Allende matrix:

(R17) $Ca^{2+}{}_{(aq)} + Fe^{2+}{}_{(aq)} + 2SiO_{2(aq)} + 2H_2O_{(aq)} \rightarrow CaFeSi_2O_{6(s)} + 4H^+{}_{(aq)}$
(R18) $3Ca^{2+}{}_{(aq)} + 2Fe^{3+}{}_{(aq)} + 3SiO_{2(aq)} + 6H_2O_{(aq)} \rightarrow Ca_3Fe_2Si_3O_{12(s)} + 12H^+{}_{(aq)}$
(R19) $Ca^{2+}{}_{(aq)} + SiO_{2(aq)} + H_2O_{(aq)} \rightarrow CaSiO_{3(s)} + 2H^+{}_{(aq)}$
(R20) $Ca^{2+}{}_{(aq)} + 0.5Fe^{2+}{}_{(aq)} + 0.5Mg^{2+}{}_{(aq)} + 2SiO_{2(aq)} + 2H_2O_{(aq)} \rightarrow Ca(Fe,Mg)Si_2O_{6(s)} + 4H^+{}_{(aq)}$

Secondary mineral assemblages in Allende igneous CAIs are largely controlled by the elements mobilized by the fluid and chemical composition of melilite replaced, that ranges from nearly pure gehlenite ($Åk_{<2}$) to $Åk_{\sim 80}$. For example, gehlenitic melilite in CTA CAIs and in mantles of Type B1 CAIs is often crosscut by veins of grossular and anorthite with minor spinel and corundum (Figs. SM11a, 12). Mass-balance equations describing these reactions are:

(R21) $\underset{ghl}{2Ca_2Al_2SiO_{7(s)}} + 3SiO_2 \rightarrow \underset{grl}{Ca_3Al_2Si_3O_{12(s)}} + \underset{an}{CaAl_2Si_2O_{8(s)}}$
(R22) $\underset{ghl\text{–}åk_{ss}}{3Ca_2Al_2SiO_{7(s)} + Ca_2MgSi_2O_{7(s)}} + SiO_2 \rightarrow \underset{grl}{Ca_3Al_2Si_3O_{12(s)}} + \underset{sp}{MgAl_2O_{4(s)}} + 2CaO$
(R23) $\underset{ghl}{3Ca_2Al_2SiO_{7(s)}} + 3SiO_2 \rightarrow \underset{grl}{2Ca_3Al_2Si_3O_{12(s)}} + \underset{cor}{Al_2O_3}$
(R24) $\underset{ghl\text{–}åk_{ss}}{3Ca_2Al_2SiO_{7(s)} + 2Ca_2MgSi_2O_{7(s)}} + H_2O \rightarrow \underset{grl}{Ca_3Al_2Si_3O_{12}} + \underset{cln}{CaMg_2Al_4SiO_{10}(OH)_2} + 6CaO + 3SiO_2$

Åkermanite-rich melilite in cores of Type B1 CAIs and in Type C CAIs is replaced by coarse-grained grossular, monticellite, wollastonite, Al-diopside, and minor forsterite (Figs. SM13–14):

(R25) $\underset{ghl\text{–}åk_{ss}}{Ca_2MgSi_2O_{7(s)} + Ca_2Al_2SiO_{7(s)}} + SiO_2 \rightarrow \underset{grl}{Ca_3Al_2Si_3O_{12(s)}} + \underset{mnl}{CaMgSiO_{4(s)}}$
(R26) $\underset{ghl\text{–}åk_{ss}}{Ca_2MgSi_2O_{7(s)} + 2Ca_2Al_2SiO_{7(s)}} + SiO_2 \rightarrow \underset{grl}{Ca_3Al_2Si_3O_{12(s)}} + \underset{di\text{–}CTs_{ss}}{CaMgSiO_{4(s)} + CaAl_2SiO_{6(s)}} + CaO$
(R27) $\underset{ghl\text{–}åk_{ss}}{3Ca_2MgSi_2O_{7(s)} + Ca_2Al_2SiO_{7(s)}} + \underset{an}{2CaAl_2Si_2O_{8(s)}} \rightarrow \underset{grl}{3Ca_3Al_2Si_3O_{12(s)}} + \underset{mnl}{CaMgSiO_{4(s)}} + \underset{fo}{Mg_2SiO_{4(s)}}$
(R28) $\underset{ghl\text{-}åk_{ss}}{Ca_2Al_2SiO_{7(s)} + 3Ca_2MgSi_2O_{7(s)}} + SiO_2 \rightarrow \underset{grl}{Ca_3Al_2Si_3O_{12(s)}} + \underset{mnl}{3CaMgSiO_{4(s)}} + \underset{wol}{2CaSiO_{3(s)}}$

Because of large compositional variations of primary melilite in the Allende igneous CAIs, and, possibly, of local variations of fluid composition, different assemblages of secondary minerals can be observed even within a single inclusion (Fig. SM13).

Secondary minerals in Allende igneous CAIs recorded multistage metasomatic alteration. For example, (*i*) melilite in a FoB CAI was replaced by grossular, monticellite, and wollastonite (Fig. SM14a). In the peripheral part of the same CAI, grossular and monticellite were subsequently replaced by Al-diopside (Fig. SM14b):

(R29) $\underset{\text{grl}}{Ca_3Al_2Si_3O_{12(s)}} + \underset{\text{mnl}}{CaMgSiO_{4(s)}} \rightarrow \underset{\text{di–CTs}_{ss}}{CaMgSiO_{4(s)} + CaAl_2SiO_{6(s)}} + 2CaO + 2SiO_2$

(*ii*) Secondary mineral assemblage of monticellite, Al-diopside, and wollastonite replacing åkermanitic melilite in a Type B2 CAI is crosscut by a monomineralic vein of wollastonite (Fig. SM14c):

(R30) $CaO + SiO_2 \rightarrow \underset{\text{wol}}{CaSiO_{3(s)}}$

(*iii*) Secondary anorthite in the peripheral portions of CAIs is often replaced by sodalite, nepheline, and ferroan olivine (Figs. 6c,d in Krot et al. 2021):

(R31) $\underset{\text{an}}{CaAl_2Si_2O_{8(s)}} + 3Na_2O + 2NaCl \rightarrow \underset{\text{sod}}{Na_8(Al_6Si_6O_{24})Cl_{2(s)}} + CaO$

(R32) $\underset{an}{CaAl_2Si_2O_{8(s)}} + Na_2O \rightarrow \underset{\text{nph}}{2NaAlSiO_{4(s)}} + CaO$

(R33) $MgO + FeO + SiO_2 \rightarrow \underset{\text{fa–fo}_{ss}}{(Fe, Mg)_2SiO_{4(s)}}$

Secondary minerals also provide evidence for local variations in chemical composition of metasomatic fluid. These include (*i*) ferroan grossular overgrowing Fe-poor/free grossular (Fig. SM15a); (*ii*) ferroan monticellite overgrowing Fe-poor/free monticellite (Fig. SM15b), and (*iii*) hedenbergite and andradite overgrowing Fe-poor/free Al-diopside (Fig. SM15c). These late-formed Fe-bearing/rich secondary minerals are typically Ti-free. Only rare occurrences of Ti-rich garnet, hutcheonite, overgrowing grossular have been described in the Allende Type B CAIs (Fig. SM15d). These observations suggest that mobility of titanium in the Allende aqueous fluid was very limited.

2.5 Metasomatic Alteration of Igneous CAIs in CK3.7–3.8 Chondrites

Coarse-grained igneous CAIs in CK3.7–3.8s have similar textures, primary mineralogies, and isotopic compositions to those from Allende (Chaumard et al. 2014; Shollenberger et al. 2018; Krot et al. 2024, 2023), but experienced higher degree of metasomatic alteration under much more oxidizing conditions, FMQ+2 to FMQ+4.5 *vs*. IW to ~FMQ, respectively (Righter and Neff 2007).

Only three igneous CAIs (CTA, B, FoB) have been studied in detail in CK3s so far (Fig. SM16; Krot et al. 2024). Secondary minerals identified in these CAIs are listed in Table SM5. In contrast to the commonly observed secondary monticellite, Na-melilite, nepheline, sodalite, and wollastonite in Allende igneous CAIs, none of these minerals have been identified in the CK3 CAIs studied. In addition, secondary grossular, Al-diopside, spinel, and olivine in CK3 CAIs are enriched in FeO, and secondary plagioclase contains significant Na_2O content (up to 85 mol% Ab).

Melilite is almost completely replaced by Fe-bearing grossular, Fe-bearing Al-diopside, clintonite, CaNa-plagioclase, FeMg-spinel, FeMg-olivine, and wadalite; anorthite is less altered (Figs. SM17–19). Primary Ti-rich pyroxenes (grossmanite and fassaite) are corroded by FeAl-diopside, ilmenite, titanite, FeMg-spinel, Ti-bearing grossular, and clintonite (Figs. SM17f, 18c–e). Low-Ti Al-diopside in a FoB CAI shows little evidence for alteration (Fig. SM19). Primary spinel and forsterite experienced Fe-Mg interexchange to various degrees. Coarse secondary grossular and Al-diopside in the CAIs cores are often chemically-zoned: the grain edges are enriched in TiO_2 (up to 3 and 7 wt%, respectively) suggesting local mobilization of Ti during metasomatic alteration (Figs. SM17c,d). Some grossular grains are overgrown by ferroan grossular with a sharp compositional boundary between them suggesting multistage formation of this mineral (Figs. SM17e, 18f).

The CK igneous CAIs are crosscut by veins of FeMg-olivine, CaNa-plagioclase, FeAl-diopside, Ti-bearing grossular, and occasionally contain geodes composed of these minerals (Figs. SM18b,f). These CAIs are also surrounded by a Ca-rich layer composed of chemically-zoned FeAl-diopside, Fe-grossular, and andradite; edges of the grossular and Al-diopside grains contain significant TiO_2 contents (up to 0.7 wt%) suggesting removal of Al and Ti from the host inclusions.

2.6 Metasomatic Alteration of Dark Inclusions in Allende

Chondritic lithic fragments, commonly observed in CV3s and named dark inclusions (DIs), appear to have experienced much higher degree of the metasomatic alteration than the host meteorites suggesting their alteration predated final accretion of the CV parent asteroid. Although originally the origin of DIs was attributed to nebular processes – condensation, aggregation, and alteration – under highly-oxidized, non-solar conditions (Kurat et al. 1989; Johnson et al. 1990), it was subsequently shown that alteration of DIs resulted in extensive mobilization of Ca, Na, K, and S and could have occurred in the CV parent asteroid in zones of high fluid permeability (Krot et al. 2001; Brearley and Krot 2012) prior to local impact brecciation and mixing. In this section, we will touch on only one aspect of Allende DIs – redistribution of Ca and formation of Ca-rich rims around them.

Figure SM20 shows three completely altered Allende DIs, $4302-1, IV-1$, and $3b-1$, which consist entirely of secondary minerals – ferroan olivine, nepheline, and CaFe-pyroxenes, and contain chondrule pseudomorphs (Fig. SM20d). In addition, the inclusion $4301-2$ is crosscut by numerous CaFe-pyroxene veins. The DIs are surrounded by a thick layer of hedenbergite+wollastonite+andradite. High abundance of CaFe-silicate objects is also observed in the Allende host outside the CaFe-rich rim of $4302-1$. The thickness of the CaFe-silicate layer around DIs is inversely proportional to the abundance of CaFe-pyroxenes inside their host inclusions suggesting loss of Ca from the previously altered DIs and its precipitation along the boundary with the Allende meteorite. Therefore, the Allende rimmed DIs, like CaFe-silicate layers around Allende CAIs (Fig. SM9), provide an important evidence for *in situ* alteration of Allende.

3 Modification of Organic Matter During Metasomatic Alteration

Organic material occurs in ordinary and carbonaceous chondrites (Yang and Epstein 1983; Robert et al. 1987; Alexander et al. 1998, 2007, 2012; Aponte et al. 2017, 2019), and mainly consists of insoluble organic matter (IOM); the soluble fraction (SOM) is very low in abundance. The IOM is located within the matrix, and is finely distributed through the mineral phases. The elemental and isotopic composition of IOM evolves with increasing petrologic type (see also Sects. 7.2 & 7.3), as well as its structural organization.

3.1 Abundance, Elemental and Isotopic Compositions of Insoluble Organic Matter

In comparison to lower grade (type 2 and 1) chondrites (CIs, CMs, and CRs), the IOM contents of CO3s, CV3s and UOCs are lower, which is mainly due to lower proportions of matrix in them. When normalized to the matrix contents, the IOM carbon content of the least metamorphosed type 3 chondrites is roughly comparable to those of CMs and CIs at $\sim$2 wt% (Grady et al. 1989; Alexander et al. 2007). However, the H/C, N/C and O/C ratios of IOM decrease with increasing petrologic type, indicating progressive loss of these heteroatoms as the structure becomes more aromatic in response to thermal stress (Remusat et al. 2019). For terrestrial samples, such as kerogens in sedimentary rocks, this process is called carbonization (Vandenbroucke and Largeau 2007).

The carbon-isotope compositions of IOM become progressively heavier with heating intensity in the parent body (Alexander et al. 2007). This is consistent with the usual process of "cracking" that organic macromolecules experience under thermal stress (Sackett 1978) in which the residual IOM gets heavier with preferential loss of ^{12}C-bearing functional groups (the ^{13}C bonds are stronger than ^{12}C bonds).

The nitrogen-isotope compositions of IOM are more variable, but scatter around $\delta^{15}N$ of $\sim 0‰$ and do not clearly correlate with the petrologic type. Heating experiments performed on the IOM from CIs and CMs also do not show evolution of N-isotope composition (Remusat et al. 2019).

The hydrogen-isotope compositions of IOM in UOCs exhibit a dramatic increase in δD (up to $\sim$ 12,000‰) with increasing of petrologic type (Alexander et al. 2007). Flash pyrolysis to 600 °C of the IOM from the L3.3 chondrite Grosvenor Mountains (GRO) 95502 resulted in a sharp increase in the δD of the residue (Remusat et al. 2016). This could be due to (*i*) the presence of minute amounts of extremely D-rich and thermally resistant organic moieties in the IOM of UOCs (Aléon 2010; Remusat et al. 2016) or (*ii*) the loss of thermally labile functional groups that are hydroscopic and/or easily exchange hydrogen isotopes (e.g., OH or acid groups) during IOM isolation (Alexander et al. 2017). Hydrogen-isotope fractionation associated with the progressive oxidation of Fe by H_2O and loss of H_2 are also proposed to explain the progressive D-enrichment of OC IOM (Alexander et al. 2010). Relative to IOM in UOCs, the δD values are much lower in the IOM of COs and CVs. In the reduced CVs, δD only slightly increases with petrologic type (Alexander et al. 2007). This would imply that the putative D-rich, thermally resistant component in the UOC IOM is not present in the IOM of the COs and CVs. It is also problematic for the Fe oxidation explanation for the D-enrichments in the UOC IOM since there is clear petrologic evidence for oxidation of Fe in the COs and CVs (see Sect. 2.2).

3.2 Structural Organization of IOM: A Sensitive Tool of the Peak Metamorphic Temperature

Raman spectroscopy has turned out to be the instrument of choice to evaluate the extent of thermal metamorphism in type 3 chondrites (Michel-Levy and Lautie 1981; Quirico et al. 2003, 2009, 2014; Bonal et al. 2006, 2007, 2016, 2020, 2007). As for terrestrial sample analogues, the chondritic IOM is described as a disordered macromolecule and the so-called Raman "defect" bands are sensitive to the structural organization degree and to the size of the polyaromatic units. As largely documented for terrestrial samples (see Buseck and Beyssac 2014 for a review), with increasing peak metamorphic temperature, the defects progressively diminish and the size of the aromatic domains increases. Raman spectroscopy thus allows for classification of the petrologic grade of chondrites independently of the mineralogy. However, a precise cross comparison among chondrite groups is possible only if the accreted precursors were comparable, and if the conditions during metamorphism (duration, T, P, fO_2, etc.) were similar (Beyssac et al. 2003; Quirico et al. 2009). Unfortunately, Raman spectroscopy cannot be used as a thermometer, since we have a limited knowledge on the nature of the accreted organic precursors, and that simulating long term evolution over geological timescales in the laboratory is highly challenging. An attempt has been made to use Raman spectra to estimate peak metamorphic temperatures in chondrites by correlating them with independent temperature estimates for specific meteorites (Busemann et al. 2007), but Bonal et al. (2016) questioned this approach.

The structural evolution of IOM under temperature increase can also be documented by high resolution TEM (transmission electron microscopy), which reveals the size of the aromatic units and the spacing between aromatic layers in them as the IOM becomes increasingly carbonized (Rouzaud and Oberlin 1990). It has been shown that the typical size of the aromatic units increases in COs, CVs, and UOCs between type 3 and 3.6, from dimensions of $\sim$1 nm to $\sim$5–10 nm, and the mean interlayer spacing decreases to values that are close to those of graphite (Smith and Buseck 1981; Rietmeijer and MacKinnon 1985; Remusat et al. 2008; Le Guillou et al. 2012). Nuclear magnetic resonance (NMR) spectroscopy shows that it is the aliphatic functional groups that are lost first during metamorphism, while in carbon XANES (X-ray absorption near edge structure) spectra the increasing absorption intensity of an exciton is directly linked to the growth of aromatic units (Cody et al. 2008).

3.3 Soluble Organic Matter

In type 3 chondrites, the abundance of soluble organic compounds is much lower than in meteorites of lower petrologic types and much more restricted in their diversity (Schmitt-Kopplin et al. 2012). As a result, the SOM in type 3 chondrites has received much less attention compared to low petrologic type counterparts. A few amino acids were reported in COs and CVs, with concentrations around ppm level ($\sim$10$\times$ lower than in CMs, $\sim$300$\times$ lower than in CRs). The predominance of amine terminal amino acids in COs and CVs is interpreted as the result of metasomatism in their parent bodies (Burton et al. 2012). A few monocarboxylic acids ranging from C1 to C6 have been reported in Allende, whereas acetic acid only has been detected in investigated COs and CVs (Aponte et al. 2019). However, the C-isotope compositions of the monocarboxylic acids in Allende indicate that they are dominated by terrestrial contamination. Aliphatic amines are also much less abundant in CVs and COs compared to CIs, CMs and CRs, with concentrations being one to three orders of magnitude lower (Aponte et al. 2017). In both cases, a lower abundance is likely the result of metasomatism in the parent body. Free polycyclic aromatic hydrocarbons (PAHs) show a similar pattern of decreasing abundance with increasing petrologic type (Elsila et al. 2005). In both COs and CVs, the lower abundances of SOM are almost certainly the result of metasomatic alteration in their parent bodies.

An extensive series of Mg-metalorganic compounds has been identified in thermally altered type 3 chondrites, but are virtually absent in type 1 and 2 chondrites (Ruf et al. 2017). This class of compounds is believed to be thermally stable, and hence can be used as a fingerprint for metasomatism in the parent bodies.

3.4 Open Questions

There is a clear effect of thermal metamorphism on organics in type 3 chondrites, which evolve chemically, isotopically and structurally. This transformation is interpreted as the result of long-term peak metamorphic temperature, likely related to ^{26}Al radioactive decay. It serves as a signature of the petrologic type and is a precious tool to classify chondrites.

Some questions remain open:

(*i*) The evolution mechanism (loss of hetero-atoms and structural reorganization) is assessed from measurements obtained on the bulk IOM but has not so far been studied *in situ*. This could reveal the organic-inorganic association and the role of the former on the maturation using scanning transmission X-ray microscopy (STXM) and TEM. This is, however, made difficult by the increasing scarcity of organic particles within the matrix of higher petrologic types.

(*ii*) The origin of the precursors of OM in different chondrite groups and whether CVs, COs, and UOCs accreted the same mix of organic precursors remain unclear.

(*iii*) Whereas the role of fluids during metamorphism has been documented in UOCs, CVs and COs for inorganic phases, it is difficult to tell whether or not such fluids played a role on the organic evolution, or if temperature alone is responsible for the observed transformation. Experimentally identifying specific reaction pathways related to the presence of oxidizing fluids at temperature above 300 °C could potentially offer a way to resolve this question.

4 Modification of Presolar Grains Abundances and Abundances of Isotopically Anomalous Noble Gas Components Associated with Them During Aqueous/Metasomatic Alteration

Presolar grains are stellar condensates that form in the outflows of evolved stars (red giants, asymptotic red giants) and in the ejecta of stellar explosions (novae, supernovae), were incorporated in the protosolar molecular cloud (PMC), and can be found in chondrites and IDPs (e.g., Lewis et al. 1987; Tang and Anders 1988; Choi et al. 1998a; Zinner et al. 2005; Zinner 2014; Floss and Haenecour 2016). Presolar grains are distinguished from solar system material by their highly anomalous isotopic compositions and have been identified *in situ* with ion imaging techniques using either a NanoSIMS or SIMS+SCAPS (stacked CMOS-type active pixel sensor) detector. They are typically ~100−600 nm in size and include SiC, graphite, spinel, corundum, hibonite, and silicates (e.g., Nagashima et al. 2004; Zinner 2014; Nittler and Ciesla 2016).

In chondrites, presolar grains occur exclusively in interchondrule matrix and fine-grained chondrule rims, with apparently no systematic differences in the abundances among these locations (Leitner et al. 2016; Floss and Haenecour 2016). There are, however, indications of different alteration paths for interchondrule matrix and fine-grained rims in CRs and COs, resulting in differing presolar silicate abundances in matrix and rims (Leitner et al. 2016; Floss and Haenecour 2016).

Investigation of the abundance and distribution of presolar grains in different chondrite groups and petrologic types can shed light on possible heterogeneities in the PMC and/or PPD, and effects of metasomatic alteration and thermal metamorphism in the chondrite parent asteroids. To compare the distribution of presolar grains identified by ion imaging in matrices of different chondrites, they are commonly reported as matrix-normalized abundances. Presolar nanodiamond, SiC, and graphite contain relatively high abundance of isotopically anomalous noble gas components. As a result, the abundance of these presolar grains can be also inferred from whole-rock noble gas measurements (Huss 1990; Huss and Lewis 1995; Huss et al. 2003). Alexander (2005) argued that the relative abundances of presolar grains and their matrix-normalized absolute abundances are roughly CI-like in all chondrites. However, presolar SiC grain abundance estimates based on noble gas contents are inconsistent with this: they are typically lower (Huss and Lewis 1995; Huss et al. 2003). Alexander et al. (1990) and Russell et al. (1997) hypothesized that the presence of oxidants and a fluid to enhance diffusion and reaction may make it possible to degas/destroy the SiC grains at relatively low temperatures. Later Riebe et al. (2016) found that SiC grains in samples of Tagish Lake (C2ung) that experienced aqueous alteration to different degrees have different releases of noble gases when subjected to HNO_3 during Closed System Step

Etching (CSSE), a commonly used technique for extracting noble gases. This finding is in agreement with studies by Wieler et al. (1992), Floss and Stadermann (2009a), and Davidson et al. (2014) who showed that in CR2s the concentration of SiC grains determined by NanoSIMS raster ion imaging is higher than SiC concentrations derived from noble gas concentrations. Davidson et al. (2014) suggested that aqueous alteration in the CR parent body might have affected and degassed the outermost gas-rich layers of SiC grains, resulting in the lower estimate of SiC abundance based on noble gas concentrations.

Presolar silicate grains (typically amorphous), the most abundant type of presolar grains, make up to percent levels (on average $\sim$ 400 ppm) of anhydrous IDPs (Floss et al. 2006; Busemann et al. 2009; Davidson et al. 2012; Alexander et al. 2017) and have a factor of a few lower abundance in type 3 carbonaceous chondrites (Nittler et al. 2018). The highest abundance of presolar silicates reported are in Acfer 094 and Chwichiya 002 (C3.00; 150–200 ppm), CO3.00 chondrites Allan Hills (ALHA)77307 (190 ppm) and DOM 08006 (240 ppm), minimally altered CRs (160–220 ppm), and the CM chondrite A-12169 ($\sim$275 ppm) (Nguyen et al. 2007, 2010, 2009b, 2009, 2015, 2018; Nittler et al. 2018, 2021; Davidson et al. 2019; Smith et al. 2023). Recent studies of UOCs showed similarly high presolar silicate abundances, although their distribution appears to be more heterogeneous. Presolar silicate abundances in relatively pristine matrices Semarkona (LL3.00), MET 00526 (L/LL3.05) and Queen Alexandra Range (QUE) 97008 (L3.05) are 150, 55–275, and 145 ppm, (see Fig. SM21 and references therein).

A systematic study of the inferred abundances of presolar SiC, diamond, and graphite in meteorites of various class and petrologic type showed correlations between the abundance patterns within a meteorite group and the metamorphic history of the host meteorite (Huss and Lewis 1995). Alexander et al. (1990) confirmed for a set of OCs the presence of isotopically anomalous phases only for specimens of petrologic type 3.6 or below, with destruction during metamorphism being the most likely and straightforward explanation. Besides SiC, presolar refractory oxide grains have been identified in grain separates from Tieschitz (H/L3.6) (Nittler et al. 1997, 2008). When compared the fraction of presolar Al-oxides among the total number of analyzed Al-oxide grains for the Tieschitz separates, $0.27\pm0.04\%$ (Nittler et al. 1997) and $1.59^{+0.44}_{-0.35}\%$ (Nittler et al. 2008), respectively, with those for ordinary chondrites of lower petrologic types, e.g., Krymka (LL3.2) with $0.85\pm0.11\%$, no significant difference is evident: i.e., the occurrence of refractory presolar grains appears to be, on a first order estimate, unchanged up to a petrologic type of $\sim$3.6.

Presolar silicates are sensitive indicators of aqueous alteration and thermal metamorphism (e.g., Zinner 2014) leading to rapid decreases in their abundance, decreases in the ratio of presolar silicates to oxides, and increases in the FeO contents in presolar silicates. However, abundances of presolar grains, especially of silicates, in chondrites of petrologic types >3.0 are not as thoroughly studied as in carbonaceous chondrites of lower petrologic types. Most of the available data are from OCs with petrologic types ranging from 3.00 to 3.2 (Fig. SM21).

Generally, the study of presolar grain abundances has the potential to support the identification of more complex, and potentially small, alteration processes affecting the fine-grained fractions of chondritic meteorites. Statistically significant variations of presolar silicate abundances could, in combination with other micro- and nanoanalytical techniques, provide additional information on the alteration state of a given area of a meteorite. The presolar silicate-to-oxide ratio can also be of limited value. Typically, presolar oxide abundances are very low (a few ppm), resulting in low grain numbers, which in turn are responsible for large errors for the respective ratios. Thus. statistically significant silicate-to-oxide ratios are only available for very thoroughly studied meteorites, or if the specimen contains unusually high abundances of oxides (e.g., Leitner et al. 2012). A better approach might therefore be to monitor the presolar SiC abundances simultaneously, which are typically in the range of a few tens of ppm (e.g., Davidson et al. 2014); however, this requires even more effort, since both O-anomalous and C-anomalous grain abundances have to be measured. Such investigations are rather time-intensive, even more so, if presolar grain abundances are low. In addition, investigation of presolar silicates and oxides by AES (Auger Electron Spectroscopy, e.g., Floss and Stadermann 2009b; Bose et al. 2010) and TEM-EELS (Transmission Electron Microscopy-based Electron Energy Loss Spectroscopy) can yield information on their Fe contents and the oxidation state of iron in these grains, and thus, allow inferences about the alteration state of their host materials.

5 Experimental Studies and Physicochemical Analysis of Metasomatic Alteration Reactions

There are experimental and theoretical approaches in understanding physicochemical conditions of metasomatic alteration that affected type 3 ordinary, CO, and CV chondrites. Both are discussed below.

5.1 Experiments

Experimental studies are largely focused on the origin of FeNi-carbides (Llorca and Casanova 1998), magnetite (Hong and Fegley 1998), fayalite (Dobrică et al. 2022), and phyllosilicates (Tomeoka and Kojima 1995; Kojima and Tomeoka 1999, 2000) commonly observed in chondrules and matrices of type 3.0–3.2 chondrites, and alteration of melilite, anorthite, and enstatite described in refractory inclusions and chondrules in type 3 chondrites of higher petrologic subtypes (Nomura and Miyamoto 1998; Ohnishi and Tomeoka 2007; Ichimura et al. 2017). There are only a few experimental studies focused on the effects of alteration on noble gases concentrations and isotopic compositions (Nakasyo et al. 2000). Various types of starting materials were used in the experiments: synthetic and natural (meteorite fragments) samples, powder and block reactants, single crystals or polymineralic samples.

Llorca and Casanova (1998) studied reaction between kamacite grains and H_2+CO gas mixture in a nebular-type environment (H_2/CO = 250:1; 5×10^{-4} atm total pressure (P), and temperature (T) of 473K). Carbon deposition, hydrocarbon production in the C_1–C_4 range, and the formation of an ϵ-carbide phase occurs when FeNi-metal particles are exposed to the gas mixture for 10^3 hours, suggesting that gas-solid reactions in the solar nebula during CO hydrogenation represent a plausible mechanism for the formation of FeNi-carbides and carbonaceous material in IDPs.

The calculated magnetite formation temperatures in the canonical solar nebula are 357–384K, depending on the distribution of C between CO and CH_4, but formation of magnetite in the solar nebula may be kinetically inhibited (Fegley and Prinn 1989). Hong and Fegley (1998) studied oxidation of FeNi-metal from the Gibeon meteorite to magnetite *via* the net reaction $3Fe_{(s)} + 4H_2O_{(g)} \rightarrow Fe_3O_{4(s)} + 4H_{2(g)}$ at ambient atmospheric P, at T = 364–715K, and a gas mixture with H_2/H_2O molar ratios of $\sim$4–41. The obtained data were used to model the reaction time of FeNi-metal grains in the solar nebula as a function of grain size and T. Although the calculated reaction times for 0.1–1 μm radius metal grains are within theoretical lifetimes of the solar nebula (0.1–10 Ma), these estimates may provide only the lower limits (Hong and Fegley 1998) suggesting that formation of magnetite under nebular conditions (lower T, P, and higher H_2/H_2O ratio) seems unlikely.

Dobrică et al. (2022) performed a hydrothermal experiment using a mixture of FeO-rich amorphous silicates, Fe-metal powder, and water as starting components that was heated at $T = 493$K for 6 days. Multiple mineral phases were observed in the experimental products, including elongated porous crystals of pure fayalite, greenalite $[(Fe^{2+},Fe_{3+})_{2-3}Si_2O_5(OH)_4]$, Fe-rich amorphous silicate, and iron oxide. The experiment supports a hydrothermal growth of fayalite in metasomatically altered type 3 chondrites (Zolotov et al. 2006). The lack of Mg in starting amorphous silicates is unfortunate, because matrices of primitive chondrites are dominated by FeMg-amorphous silicates which are often replaced by fayalite (Figs. SM3c,d).

Ohnishi and Tomeoka (2007) performed hydrothermal alteration reactions of enstatite at variable pH (0–14), T (373–573K), and reaction times (24–336 h). Under acidic conditions, no significant alteration occurred, whereas under neutral and alkaline conditions serpentine and saponite formed in different proportions. The amounts of phyllosilicates tend to increase with increasing pH, T, and run duration. There is also a tendency for saponite to form at higher pH and T and under longer run durations compared to serpentine. Ohnishi and Tomeoka (2007) concluded that alteration of CVs occurred at higher T and more alkaline conditions than that of CIs.

Tomeoka and Kojima (1995) and Kojima and Tomeoka (1999, 2000) performed long-time (7 weeks) hydrothermal alteration experiments of Allende at 723K, 800 bars pressure, under acidic and neutral conditions. It was found that pH and precursor mineralogy are important factors controlling phyllosilicate mineralogy. (*i*) *Under acidic conditions*, phyllosilicates replacing low-Ca pyroxene are dominated by a chlorite-group phyllosilicate containing 2–10 wt% Al_2O_3; serpentine and talc are minor. Phyllosilicates replacing matrix olivine consist of chlorite and serpentine and contain 8–10 wt% Al_2O_3. Phyllosilicates replacing chondrule mesostasis are dominated by a high-Al (12–20 wt% Al_2O_3) phyllosilicate. CaFe-rich pyroxenes in the matrix are extensively replaced by saponite. Matrix altered under acidic conditions is depleted in Ca, Na, K, and S compared to the Allende matrix. (*ii*) *Under neutral conditions*, phyllosilicates produced are mainly low-Al chlorite containing up to 2.5 wt% Na_2O; high-Al phyllosilicates were not observed. Diopside in CAIs and CaFe-rich pyroxenes in the matrix remain largely unaltered. Melilite and anorthite in CAIs are replaced by Al-saponite. Most common phyllosilicates in naturally altered CV3 are Fe-rich saponite and intergrowths of Na-mica with serpentine or saponite; chlorite is minor. Matrix altered under neutral conditions shows no significant differences in bulk major element compositions compared to the Allende matrix. By comparing the mineralogy and chemical compositions of naturally and artificially altered CVs, Kojima and Tomeoka (1999, 2000) concluded that alteration of CVs took place under variable pH.

In the experiments by Nomura and Miyamoto (1998) and Ichimura et al. (2017), pure gehlenite, diopside, spinel, and CaNa-plagioclase mixed with SiO_2, $\pm Al_2O_3$ were heated in alkaline and acid solutions at T of ~473–673K. Spinel and diopside showed no evidence for alteration. Gehlenite was replaced by hydrogrossular $[Ca_3Al_2(SiO_4)_{3-x}(OH)_{4x}]$, hydrosodalite $[Na_8Al_6Si_6O_{24}(OH)_2 \times 2H_2O]$, analcime $(NaAlSi_2O_6 \times H_2O)$, tobermorite $(Ca_5Si_6O_{18}H_2 \times 4H_2O)$ and nepheline hydrate $(NaAlSiO_4 \times 0.5H_2O)$; CaNa-plagioclase was replaced by Na-zeolites and analcime. Upon heating above 673 K, these minerals were transformed into grossular, dmisteinbergite, nepheline, sodalite, and wollastonite. Although none of the intermediate hydrated secondary phases have been observed in Allende CAIs, their formation cannot be excluded (e.g., Enokido et al. 2023). Studies of CAIs from CVs less metamorphosed than Allende are required to test this hypothesis.

Hydrothermal experiments performed on Allende at T of 473 K and water/rock of 1/50 showed that the concentration of noble gases decreased (Nakasyo et al. 2000). The loss of noble gases also affected the isotopic composition of xenon, apparently due to the preferential alteration of nanodiamond containing mineral phase/phases relative to the trapped xenon component carrier ones.

5.2 Physicochemical Analysis of Metasomatic Alteration Reactions in CV Chondrites

5.2.1 Transport of Elements in Aqueous Solution and Steam During Metasomatic Alteration

It is inferred that CVs experienced fluid-rock interaction during progressive metamorphism (Krot et al. 1995, 1998). The common presence of phyllosilicates in Kaba ($CV_{OxB}3.1$) and nearly complete absence of hydrous minerals in Allende ($CV_{OxA}>3.6$) suggest the aqueous alteration in the CV parent asteroid occurred at progressively increasing temperatures and was followed by thermal metamorphism resulting in loss of aqueous solutions and dehydration of phyllosilicates in the CV_{OxA} chondrites.

Thermodynamic modeling of metasomatic alteration in a multisystem compositionally similar to the CVs (Petaev and Mironenko 1997) shows that there are two distinct stages of the alteration process. At low water/rock ratio, <0.1−0.3, no aqueous solution is present in the system, because all water is consumed by the oxidation and hydration reactions. Hydrogen released by these reactions forms large volumes of gaseous phase that contains trace amounts of volatile anionic elements (H, O, S, Cl, and C). At this stage, an aqueous alteration could proceed as an isochemical process without appreciable redistribution of cationic elements. At higher water/rock ratio, an aqueous solution coexists with the altered rock and is capable of redistributing cationic elements such as Na, Ca, Mg, Fe, Si, and Al among CV chondrite components.

The source of water in chondrite parent bodies was probably water ice. Upon heating, ice sublimates, melts and evaporates to form liquid water and steam, if the temperature exceeds boiling point (Fig. SM22). Liquid water and steam react with minerals and amorphous phases and leach some elements. Initially pure water is transformed into aqueous solution with SiO_2, Fe^{2+}, Mn^{2+}, Mg^{2+}, Ca^{2+}, Cl^-, Na^+, HCO_3^- etc. dissolved in it. Aqueous solution can transport dissolved elements at all temperatures. The ability of steam to transport elements depends on P and T. Supercritical fluid can efficiently transport elements, but measured Allende tensile strength values implies that at $P_{gas} > 100$ bars the rock would break and gas would be released (Grimm and McSween 1989). Therefore, the presence of supercritical fluid in the CV asteroid seems unlikely. Low density subcritical "wet" steam ($P_{H2O} = P_{H2O,equil}$, $T < 647$ K) or superheated "dry" steam ($P_{H2O} < P_{H2O,equil}$, $T > 647$ K) can transport some elements dissolved in the solution.

The concentration of a solute in steam depends on its T, P, and density. The distribution coefficients of solutes between the "parental" aqueous solution and the steam are low (Fig. SM23). Steam condensation would produce a "secondary" solution nearly free of solutes which cannot be efficiently transported by the steam. Solubility of iron in water under oxidizing conditions is low; therefore, equilibrium concentration of iron in the steam is close to 0. Calcium concentration in the steam is expected to be also low because of low distribution coefficients of $CaCl_2$ and $CaSO_4$ between steam and aqueous solution. Therefore, neither Fe nor Ca can be transported by steam. Note that rapid mobilization of Ca in aqueous solution in Allende at temperature < 473 K was experimentally demonstrated (Jones and Brearley 2006).

Metasomatic alteration was a non-equilibrium, multistage process with many reactions observed not going to a completion (see Sects. 2.2−2.6). Therefore, we used equilibrium approach in evaluating intensive parameters of the observed reactions assuming they reached steady state. The calculated values of fugacities and activities of gaseous and aqueous species for each reaction are compared with each other and with independent estimates of these parameters when available. Gaseous and solid solutions were treated as ideal. For aqueous species, at any T, total pressure (P_{tot}) $\approx$ pressure of saturated water vapor ($P_{\text{sat H2O vapor}}$); i.e., gaseous phase involved in alteration reactions is nearly pure H_2O steam equilibrated with an aqueous solution.

5.2.2 Fe-C-O-H System: Oxidation and Carburization of FeNi-Metal and Formation of FeNi-Carbides and Magnetite

Oxidation and carburization of FeNi-metal could have taken place by a gaseous phase only, without aqueous solutions (see Sect. 2.2). The gas-phase in the Fe-C-O-H system may contain up to three major C-bearing species (CO, CO_2, and CH_4), depending on its chemical composition and P-T conditions. Carbon monoxide, CO, is dominant at high T, while methane, CH_4, is the most thermodynamically stable C-bearing species at low T and moderate P characteristic of asteroidal alteration of type 3 chondrites as well as at low T and low P characteristic of the solar nebula. Therefore, if chemical equilibrium among gaseous species is reached in H-dominated nebular or asteroidal gases, then essentially all C will form CH_4. However, in the cooling solar nebula the conversion of CO to CH_4 is kinetically inhibited (Fegley and Prinn 1989), thus CO remains the most abundant C-bearing species even at low T. Studies of hydrothermal gases in the H_2O-H_2-CO_2-CO-CH_4 system suggest that in hydrothermal natural systems formation of CH_4 is also kinetically inhibited (Chiodini and Marini 1998). Therefore, in the analysis of solid-gas equilibria, CH_4 can be neglected.

The partitioning of C between gaseous CO and CO_2 is governed by the redox reaction:

(R34) $CO_{(g)} + H_2O_{(g)} = CO_{2(g)} + H_{2(g)}$.

Concentration of either gaseous species can be used as an intensive parameter along with the T and P_{tot} of the system. Under oxidizing conditions necessary to stabilize magnetite, the CO_2/CO ratio in the gaseous phase is $\sim$ 1 at high T; below $\sim$900K, it is rapidly increasing with falling T. Here we choose the H_2O/H_2 ratio and the concentration of CO_2 (X_{CO2}) as intensive parameters. Then the CO_2/CO ratio in the gaseous phase can be calculated using the equation:

(R35) $log\ (CO_2/CO) = 2041.5/T - 1.90 + log\ (H_2O/H_2)$.

There are four Fe-carbides – orthorhombic Fe_2C (η-carbide), hexagonal Fe_7C_3 (ϵ-carbide), monoclinic Fe_5C_2 (χ-carbide or Hägg-carbide), and orthorhombic Fe_3C (θ-carbide or cementite, Ni-free analogue of meteoritic cohenite); their thermodynamic properties are poorly known (see *Gibbs free energies of Fe carbides in Supplementary Material to Sect. 5.2.2*).

The observed replacement of metal by magnetite and carbides in the Fe-C-O-H system can be described by the following chemical reactions:

(R36) $Fe_{(s)} + 4\ H_2O_{(g)} = Fe_3O_{4(s)} + 4H_{2(g)}$
(R37) $3Fe_{(s)} + CO_{2(g)} + 2H_{2(g)} = Fe_3C_{(s)} + 2H_2O_{(g)}$
(R38) $5Fe_{(s)} + 2CO_{2(g)} + 4H_{2(g)} = Fe_5C_{2(s)} + 4H_2O_{(g)}$

Similar reactions can be written for mutual replacements of metal, magnetite, cohenite, Hägg–carbide, wüstite ($Fe_{0.947}O$), and siderite ($FeCO_3$) (see *Fe-C-O-H system: Oxidation and carburization of FeNi-metal and formation of FeNi-carbides and magnetite in Supplementary Material to Sect. 5.2.2*).

Phase relations in the Fe-C-O-H system at different partial pressure of CO_2 (P_{CO2}) as a function of T and H_2O/H_2 ratio are shown in Figure SM24. Low- and mid-T portions of the diagrams have the same topology at all P_{CO2}: large stability fields of magnetite and Hägg-carbide are terminated by siderite at low T. As P_{CO2} increases, the stability fields of siderite and Hägg-carbide increase at the expense of metal and magnetite; the Hägg-carbide—siderite—magnetite invariant point moves toward higher T and more oxidizing conditions (higher H_2O/H_2 ratios).

The topology of the high-T portions of the diagrams is P_{CO2}-dependent. At p_{CO2} = 0.05, the Hägg-carbide, unaltered (low-Ni) metal, magnetite, and wüstite are stable at T = 812K, H_2O/H_2 = 0.30, and CO_2/CO = 1.21. At lower P_{CO2} values (Figs. SM24a,b), the invariant assemblage consists of metal, Hägg-carbide, and magnetite; the T and H_2O/H_2 ratio of the invariant point are decreasing as the P_{CO2} decreases. If P_{CO2} exceeds 0.05 bar, the invariant assemblage of metal, Hägg-carbide, and magnetite becomes unstable; metal and Hägg-carbide coexist with wüstite instead (Fig. SM24c). A decrease in X_{Fe} has the opposite effect on the phase boundaries due to enlargement of the metal stability field; e.g., altered metal is in equilibrium with magnetite and metastable Hägg-carbide even at P_{CO2} = 0.05 (Fig. SM24d). A decrease in X_{mgt} would increase the magnetite stability field, but the amount of impurities in magnetite is so small, that the X_{mgt} never drops below ~0.98; this would result in a negligible shift (0.008 log unit) of the magnetite phase boundaries.

High-T portions of all diagrams include metastable fields of cohenite and/or Hägg-carbide (Fig. SM24). Cohenite is metastable at all T. Hägg carbide becomes metastable relative to unaltered metal (X_{Fe} = 0.95) and carbon above ~700K according to reaction:

(R39) $Fe_5C_{2(s)} = 5Fe_{(s)} + 2C_{(s)}$

The assemblage of altered metal (X_{Fe} = 0.5) and Hägg-carbide is metastable above ~ 650K.

The lack of wüstite in FeNi-metal-carbide-magnetite assemblages places upper limits on the T (812K), P_{CO2} (0.05 bar), and H_2O/H_2 ratio (0.30) of the initial stage of alteration of type 3 chondrites. However, at such conditions the Hägg-carbide would be metastable. To form a stable mineral assemblage of unaltered metal, magnetite, and Hägg-carbide, the T should be below 700K and P_{CO2} in the gaseous phase less than ~ 0.002 bar (Fig. SM24a). No cohenite will be formed at this stage of alteration. Because both the T and H_2O/H_2 ratio of the invariant assemblage of metal, magnetite, and Hägg-carbide are decreasing with decreasing P_{CO2} values, this assemblage can be formed either in a nebular (low P_{CO2} values) or asteroidal (higher P_{CO2} values) environment.

5.2.3 Ca-Fe-Mg-Si-O-H System: Formation of Pure Magnetite, Fayalite, Hedenbergite, Andradite, and Kirschsteinite

The formation conditions of these minerals have been first evaluated by Krot et al. (1998). Due to space limitation, an updated discussion of the reactions involved and their physicochemical parameters is presented in the *Supplementary Material to Sect. 5.2.3*. Overall, fayalite is stable in equilibrium with relatively reducing gaseous phase and Fe^{2+}-rich aqueous solution. An increase of Ca^{2+} concentration in aqueous solution will result in formation of either hedenbergite or kirschsteinite at high or low SiO_2 concentrations in aqueous solution, respectively. Magnetite requires oxidizing gaseous phase and Fe^{2+}-rich aqueous solution. Andradite would be the only mineral equilibrated with oxidizing gaseous phase and Ca^{2+}-rich aqueous solution.

Thermochemical equilibria in closed gas-solution-rock system (Zolotov et al. 2006; Jogo et al. 2009) suggest that fayalite content in olivine is a function of T, P, water/rock ratio, and ratios of activities of Mg and Fe in aqueous solution (Fig. SM26). Pure fayalite is stable at lower T and Fe^{2+}/Mg^{2+} ratio in aqueous solution, and higher water/rock ratio and P than FeMg-olivine. In equilibrium, fayalite, magnetite, CaFe-pyroxenes, FeNi-carbides, and phyllosilicates coexist.

5.2.4 Ca-Fe-Mg-Si-O-H System: Formation of Ca-Rich Rims Around Allende Dark Inclusions

Petrographic observations (see Sect. 2.6) indicate that Ca-rich rims around Allende DIs grew *in situ*, at the interface between the DIs and the host Allende after the breccia formation of Allende. Because the growth of the Ca-rich rims requires significant transport (re-mobilization) of Ca, Fe, Si, and Mg, aqueous solutions must have played an important role in this process.

Physicochemical conditions of Ca-rich rim formation have been first evaluated by Krot et al. (2001) and are reviewed in some detail in the *Supplementary Material to Sect. 5.2.4*. The analysis of observed mineral-solution reactions revealed that the aqueous solutions locally equilibrated either with DIs or Allende matrix should have had Fe^{2+}/Ca^{2+} ratios different by a factor of $\sim$1.2. This difference will result in diffusional fluxes of Ca^{2+} ions from the DIs towards the host Allende and Fe^{2+} ions in opposite direction, which will drive the reactive system toward equilibrium. Such compositional gradients and diffusional exchange of Ca and Fe between the solutions could exist only if the solutions are stagnant in pores of the host Allende and Allende DIs. Mixing of such solutions at the Allende–DI interface will increase concentrations of both Ca^{2+} and Fe^{2+} ions above the solubility limit of a CaFe-rich phase, which will precipitate in the cavities along the interface as a rim.

The rim formation will continue until matrix olivines in the DI and Allende become equilibrated, or the Ca-rich objects in the DI are exhausted, or when an aqueous solution dries out. Petrographic observations indicate that neither complete exhaustion of the DIs in Ca, nor complete equilibrium between the DIs and Allende have been achieved. As a result, the growth of the Ca-rich rims was probably terminated by disappearance of the aqueous solution.

5.2.5 Ca-Fe-Mg-Zn-H-O System: Formation of Zn-Bearing Hercynite and Ilmenite

Substitution of Ca by Fe$\pm$Zn in krotite, grossite, and perovskite in CAIs indicates arrival Fe- and Zn-bearing aqueous solution from the matrix, perhaps, before the formation of CaFe-rich silicate rims around CAIs. As the Ca^{2+}/Fe^{2+} ratio in the solution diminishes due to gradual removal of Ca^{2+}, krotite and grossite are altered first, followed by perovskite, well before the formation of Fe-bearing grossular (Figs. SM29a,b). An accompanying change of Mg^{2+}/Fe^{2+} ratio should increase Fe content in spinel and olivine first, followed by formation monticellite-kirschsteinite solid solution (Fig. SM29c).

5.2.6 Alteration of Igneous CAIs in Allende

The altered patches and veins in mantles and cores of the Allende igneous CAIs are dominated by Ca,Al,Mg-rich phases which are enriched in SiO_2 and depleted in CaO compared to the primary phases (Tables SM1, 4). Such mineralogy requires an efficient addition of Si to and removal of Ca from CAIs. Most secondary FeMg-silicates in CAI interiors – forsterite, monticellite, grossular, and Al-diopside – are essentially Fe-free compared to the ferroan olivine, kirschsteinite, andradite, and salite–hedenbergite pyroxene in fine-grained rims and neighboring matrix. It appears that for some reason the Fe, abundant in an aqueous solution circulating through the Allende matrix (Krot et al. 1998; Ganino and Libourel 2017), has not been efficiently delivered to the interiors of the Allende igneous CAIs, which prior to the alteration were very compact. The same may be true for Na and Cl which are predicted to have high concentrations in an aqueous solution equilibrated with matrix (Petaev and Mironenko 1997), but in wadalite replacing åkermanitic melilite in Type B CAIs, these elements are decoupled from each other.

High concentrations of SiO_2, Na, and Cl are expected in both steam and aqueous solution. Therefore, these elements can be supplied to CAIs by steam equilibrated with aqueous solution without addition of Fe and Ca insoluble in steam. Reactions involving SiO_2, CO_2, Na, and Cl do not require aqueous solution. Reactions involving Ca, Fe, and Mg need aqueous solution for addition or removal of these elements. This may explain Fe-rich secondary minerals in matrix and FeO-poor secondary minerals in CAIs. Sodium and chlorine are not expected to be decoupled, unless steam formed at early stage of alteration contains only Cl-bearing species such as HCl and NH_4Cl which may be present in accreted ices (Zolotov and Mironenko 2007). This may explain formation of Na-free wadalite in CAI cores.

The inferences about the involvement of an aqueous solution or H_2O steam or both in the alteration process can be made based on a physicochemical analysis of the observed and/or possible alteration reactions and the textural relationships and mineral crystallization sequences in altered refractory inclusions. The textures and mineral chemistry of altered areas leave no doubt that an aqueous alteration was a non-equilibrium, multistage process, with many observed reactions not going to completion (see Sect. 2.4). Given the multitude of such non-equilibrium effects, considering them would make the problem intractable. Instead, an equilibrium approach in evaluating intensive parameters of the observed alteration reactions is employed here, assuming that they have reached a steady state. Then, in order to evaluate the plausibility of these reactions, the calculated values of fugacities and activities of different gaseous and aqueous species for each reaction are compared (for details see *Supplementary Material to Sect. 5.2.6.*). Since large number of cations and anions were involved in alteration reactions, the observed reactions can be split into several subsystems based on chemical elements added to and/or removed from primary mineral assemblages: **(1)** SiO_2, **(2)** CO_2, **(3)** SiO_2, CO_2, and Ca, and **(4)** Na$\pm$Cl and Ca$\pm SiO_2$. The results of modeling are shown in Figures SM30−33.

(1) The reactions involving addition of either gaseous H_4SiO_4 or aqueous SiO_2 (Fig. SM30), can proceed forward at any T and P because the required H_4SiO_4 fugacities in steam and activities of SiO_2 in aqueous solution are either much lower than or comparable with the solubility of SiO_2 in the steam or in the aqueous solution formed by its condensation.

Of special value here are the reactions describing possible alteration of nepheline ($NaAlSiO_4$) to analcime ($NaAlSi_2O_{6(s)} \times H_2O$):

$$(R55_{aq})\quad NaAlSiO_{4(s)} + SiO_{2(aq)} + H_2O_{(liq)} = NaAlSi_2O_{6(s)} \times H_2O_{(s)}$$
$$(R55_{g})\quad NaAlSiO_{4(s)} + H_4SiO_{4(g)} = NaAlSi_2O_{6(s)} \times H_2O_{(s)} + H_2O_{(g)}$$

Experimental studies at $T - P_{H2O}$ conditions identical to those considered here (Wilkin and Barnes 1998) showed that analcime is a stable phase, which can be dissolved or replaced by other phases if activities of SiO_2 and cations in the aqueous solution change. Therefore, the lack of analcime in the CV chondrites effectively places an upper limit on the H_4SiO_4 fugacity in the steam (Fig. SM30a) and $aSiO_2$ in aqueous solution (Fig. SM30b) implying that during CAIs' alteration, neither steam nor aqueous solution was SiO_2-saturated. Because the upper limit of aqueous $aSiO_2$ is 3−6 orders of magnitude higher than that in the aqueous solution made by *in situ* condensation of steam (Fig. SM30b), a solution with a higher concentration of SiO_2, tentatively labeled as "external" aqueous solution, can be involved in CAI's alteration. Such a solution must come to a CAI from outside (matrix?) as a liquid (Krot et al. 2001; Ganino and Libourel 2017). The lower limits on the H_4SiO_4 fugacity and $a SiO_2$ is set by reactions of melilite replacement by anorthite+grossular and by grossular+monticellite+wollastionite, respectively, which are widespread in the Allende igneous CAIs (Figs. SM12a, 14a).

(2) The phase equilibria involving loss or gain of CO_2:

(R56) $\underset{\text{calcite}}{3CaCO_{3(s)}} + \underset{\text{wollastonite}}{2CaSiO_{3(s)}} = \underset{\text{tilleyite}}{Ca_5Si_2O_7(CO_3)_{2(s)}} + CO_{2(g)}$

(R57) $\underset{\text{tilleyite}}{Ca_5Si_2O_7(CO_3)_{2(s)}} = \underset{\text{spurrite}}{Ca_5Si_2O_8(CO_3)_{(s)}} + CO_{2(g)}$

($R58_{aq}$) $CaCO_{3(s)} + SiO_{2(aq)} = CaSiO_{3(s)} + CO_{2(g)}$

($R58_g$) $CaCO_{3(s)} + H_4SiO_{4(g)} = CaSiO_{3(s)} + CO_{2(g)} + 2H_2O_{(g)}$

and the occurrence of tilleyite, wollastonite, and calcite within the same CAI (Fig. SM10a,c) allow to estimate of X_{CO2} (P_{CO2}) in steam as well as the activities of Ca^{2+} and HCO_3^- ions in aqueous solution (Fig. SM31). In Figure SM31, reactions involving melilite are shown for comparison; dashed lines denote metastable reactions. Relative stabilities of calcite and wollastonite are greatly dependent on the presence or lack of aqueous solution. If steam does not condense, stabilization of calcite requires high amounts of CO_2 (Fig. SM31a). If steam condenses, even very low (ppm) levels of CO_2 would result in precipitation of calcite. The general lack of calcite and common presence of wollastonite among melilite alteration products suggests that at the initial stages of melilite alteration silica was delivered as H_4SiO_4. Precipitation of calcite requires an aqueous solution with concentrations of $[HCO_3^-] \approx 2\times[Ca^{2+}]$ of $\sim 10^{-5}-10^{-4}$ mol/kg$_{H2O}$ (Fig. SM31b) and rather low mole fraction of CO_2 in the steam (Fig. SM31), which seems to be reasonable for a C-bearing system such as carbonaceous chondrites. At such conditions, the observed replacement of melilite by grossular+monticellite+forsterite can readily proceed, providing the released CaO reacts with CO_2 to form calcite, regardless of whether an aqueous solution saturated in calcite is present or not.

(3) Many alteration reactions require addition of $CO_2 \pm SiO_2$ accompanied by removal of $Ca \pm SiO_2$ (Fig. SM32). While both CO_2 and SiO_2 can be added or removed in either steam (Figs. SM32a,b) or aqueous solution (Figs. SM32c,d), the very low solubility of Ca in steam requires the presence of an aqueous solution, which would absorb the released Ca as either dissolved Ca bicarbonate, Ca^{2+} + $2HCO_3^-$ or calcite. If SiO_2 is delivered in the steam, the latter must condense to form an aqueous solution, initially Ca-free, before alteration reactions can proceed. In such a case, the amount of aqueous solution is expected to be rather small, with its Ca absorption capacity being limited. The calcium concentration in the aqueous solution will affect the sequence of alteration reactions, with the alteration of gehlenitic melilite to anorthite being the most sensitive to the Ca build up. If such aqueous solution was not effectively drained before becoming calcite saturated, one would expect to see widespread calcite associated with altered silicates. While calcite is present at least in some altered Allende CAIs, it tends to occur in nodules hinting for an effective removal of Ca^{2+} and HCO_3^- with drained solution, which can accumulate in cavities and fill them with calcite and other CO_2-bearing minerals upon evaporation of water to form nodules.

(4) The formation of nepheline and sodalite (Fig. SM33) requires an addition of Na and Cl and removal of Ca (reactions R31 and R32) and addition or removal SiO_2 (reactions $R59_{aq}$ and $R60_{aq}$):

($R59_{aq}$) $2Ca_2Al_2SiO_{7(s)} + 3SiO_{2(aq)} + 2Na^+{}_{(aq)} = Ca_3Al_2Si_3O_{12(s)} + 2NaAlSiO_{4(s)} + Ca^{2+}{}_{(aq)}$

$$(R60_{aq})\ 2CaAl_2Si_2O_{8(s)} + 2Na^+_{(aq)} + 2CO_{2(g)} + H_2O_{(aq)} = 2NaAlSiO_{4(s)} + Al_2O_{3(s)} + 2Ca^{2+}_{(aq)} + 2HCO_3^-{}_{(aq)} + 2SiO_{2(aq)}$$

Sodium, chlorine, and SiO_2 can be transported by either steam or aqueous solution, but the necessity to remove Ca requires the presence of aqueous solution. As discussed above, the latter can be formed either by steam condensation or arrive from the matrix. The substitution of anorthite by sodalite at higher T requires a solution with rather low Na (and Cl) concentrations that can be formed by the steam condensation. At lower T, an 'external' aqueous solution with higher Na and Cl concentration is required. Other reactions can proceed only in the presence of 'external' aqueous solution with high concentrations of Na and Cl. It is interesting to note that the presence of widespread Cl-bearing and Na-free wadalite in many altered Type B CAIs (Fig. SM9d) points to the decoupling of Na and Cl during at least early stages of aqueous alteration. In this case, the Cl could be supplied by the steam containing HCl and NH_4Cl but neither NaCl nor NaOH.

6 Chronology of Metasomatic Alteration

The chronology of metasomatic alteration provides important constraints on the setting of alteration (nebular *vs.* asteroidal), the accretion time, and the thermal histories of chondrite parent asteroids.

6.1 ^{53}Mn-^{53}Cr Chronology of Fayalite

Doyle et al. (2015) reported high precision SIMS analyses of nearly pure fayalite in EET 90161 (L3.05), A-881317 ($CV_{oxB}3$) and MAC 88107 (CO3.1-like). To ensure proper correction for relative sensitivities (RSF) between $^{55}Mn^+$ and $^{52}Cr^+$ ions, synthetic fayalite (Fa_{99}) doped with Mn and Cr was used as a standard. The excess of ^{53}Cr in fayalite from these meteorites correlates with the $^{55}Mn/^{52}Cr$ ratio, indicating *in situ* decay of ^{53}Mn. The inferred initial $^{53}Mn/^{55}Mn$ ratios [$(^{53}Mn/^{55}Mn)_0$] in fayalite from EET 90161, A-881317 and MAC 88107 are $(4.35\pm1.23)\times10^{-6}$, $(3.07\pm0.44)\times10^{-6}$, and $(2.58\pm0.21)\times10^{-6}$, respectively (Fig. SM34). Note that since RSF strongly depends on fayalite content in olivine (Fig. SM35), the reported $(^{53}Mn/^{55}Mn)_0$ in fayalite using San Carlos olivine (Fa_{10}) as a SIMS standard are incorrect (Table SM6).

^{53}Mn is a short-lived radionuclide, therefore a time anchor with a known $^{53}Mn/^{55}Mn$ ratio and U-corrected Pb-Pb absolute age is required to translate the inferred $(^{53}Mn/^{55}Mn)_0$ to absolute ages. The D'Orbigny angrite with the measured U-corrected Pb-Pb absolute age of 4563.37±0.25 Ma (Brennecka and Wadhwa 2012) is often used as the time anchor. We note, however, that two different $(^{53}Mn/^{55}Mn)_0$ ratios have been reported for this meteorite, $(3.24\pm0.04)\times10^{-6}$ (Glavin et al. 2004) and $(3.54\pm0.18)\times10^{-6}$ (McKibbin et al. 2015). The former is based on whole-rock and mineral separates measurements using Thermal Ionization Mass Spectrometry (TIMS); the latter is based on *in situ* measurements of olivine and kirschsteinite using SIMS. When anchored to $(^{53}Mn/^{55}Mn)_0$ of D'Orbigny reported by Glavin et al. (2004), these ratios correspond to fayalite formation at $2.4^{+1.8}_{-1.3}$, $4.2^{+0.8}_{-0.7}$, and $5.1^{+0.5}_{-0.4}$ Ma after CV CAIs, respectively. If we use $(^{53}Mn/^{55}Mn)_0$ reported by McKibbin et al. (2015), these ages will be younger by ~0.5 Ma.

6.2 ^{53}Mn-^{53}Cr Chronology of Kirschsteinite

MacPherson et al. (2017) reported high precision SIMS analyses of kirschsteinite ($CaFeSiO_4$) in the reduced CV3s Vigarano and Efremovka. To ensure proper correction for relative sensitivities between $^{55}Mn^+$ and $^{52}Cr^+$ ions, synthetic kirschsteinite doped with Mn and Cr was used as a standard. Well-resolved ^{53}Cr excesses correlate with ^{55}Mn/^{52}Cr indicative of *in situ* decay of ^{53}Mn; the inferred initial ratio (^{53}Mn/^{55}Mn)$_0$ is $(3.71\pm0.50)\times10^{--6}$ (Fig. SM36). When anchored to ^{53}Mn-^{53}Cr relative and U-corrected ^{207}Pb-^{206}Pb absolute ages of the D'Orbigny angrite (Glavin et al. 2004; Brennecka and Wadhwa 2012), this ratio corresponds to kirschsteinite formation $3.2^{+08}_{-0.7}$ Ma after CV CAIs. If we use (^{53}Mn/^{55}Mn)$_0$ reported by McKibbin et al. (2015) as an anchor, this age will be younger by ~0.5 Ma. The kirschsteinite data are consistent within error with the data for aqueously-formed fayalite from the CV_{oxB}3 chondrite A-881317 as reported by Doyle et al. (2015), supporting the idea that CaFe-silicates in CV3s are cogenetic with fayalite and formed during metasomatic alteration in the CV parent asteroid(s).

6.3 ^{26}Al-^{26}Mg Chronology of Secondary Minerals in CV3 Igneous CAIs

There are a number of secondary minerals in the Allende igneous CAIs with high ^{27}Al/^{24}Mg ratio, >20, including anorthite, corundum, grossular, nepheline, and sodalite, and therefore suitable for search for a resolvable excess of radiogenic ^{26}Mg (^{26}Mg*) using *in situ* SIMS measurements. Most analyses of these minerals showed no detectable ^{26}Mg* [(^{26}Al/^{27}Al)$_0$ $<5\times10^{-6}$] suggesting formation > 3 Ma after primary CAI minerals (e.g., Hsu et al. 2006). This formation age is consistent with ^{53}Mn-^{53}Cr ages of secondary fayalite and kirschsteinite in CV3s, and, thus suggesting metasomatic alteration of Allende CAIs occurred in an asteroidal setting. However, Fagan et al. (2007) reported resolvable ^{26}Mg* in several grossular and plagioclase analyses of Type B and CTA CAIs in Allende which correspond to the inferred (^{26}Al/^{27}Al)$_0$ of $\sim4.5\times10^{-5}$ (see their Fig. 8). Fagan et al. (2007) concluded that Allende CAIs experienced a protrackted, episodic alteration: some alteration occurred during the formation of CAIs; some continued in the CV parent asteroid. This interpretation was subsequently challenged by Krot et al. (2010) who demonstrated that ^{26}Mg* in secondary grossular could have been inherited from primary gehlenitic melilite replaced by this grossular, and, thus has no chronological significance (Fig. SM37).

Ushikubo et al. (2007a,b) reported ^{26}Mg* in anorthite of an Allende Type B CAI corresponding to ^{26}Al/^{27}Al)$_0$ of $(1.2\pm0.5)\times10^{-5}$ and interpreted this result as an evidence for a late-stage nebular episode of alteration. Krot et al. (2021) showed subsequently that anorthite measured by Ushikubo et al. (2007a,b) is in fact primary, igneous and its Al-Mg isotope systematics was probably disturbed during thermal metamorphism (see Sect. 6.4) and, therefore has no chronological significance.

To summarize, ^{26}Al-^{26}Mg systematics of secondary minerals in Allende CAIs suggest that metasomatic alteration was a late-stage process, > 3 Ma after crystallization of primary minerals.

6.4 Modification of ^{26}Al-^{26}Mg Systematics in Primary Minerals of CAIs and Chondrules from Metasomatically Altered Type 3 Chondrites

Metasomatic alteration could have disturbed Al-Mg isotope systematics in some primary phases in chondrules and refractory inclusions of type 3 chondrites. In type 3.0–3.1 UOCs, COs and CVs, most CAIs and AOAs contain ^{26}Mg* due to *in situ* decay of ^{26}Al and define internal isochrons with the canonical ($^{26}Al/^{27}Al)_0$ of $\sim 5\times 10^{-5}$ (e.g., MacPherson et al. 1995; Ito and Messenger 2010; Ushikubo et al. 2017). These also include CAIs which experienced postcrystallization O-isotope exchange in melilite, anorthite, grossite, and krotite (e.g., Simon et al. 2019), suggesting that Mg self-diffusion in these minerals under $P - T - f_{O2}$ conditions of metasomatic alteration of type 3.0−3.1 chondrites is much slower than O self-diffusion.

In CAIs from more metamorphosed chondrites such as Allende (CV>3.6), the $(^{26}Al/^{27}Al)_0$ of anorthite show a wide spread, ranging from the canonical to ~ 0, whereas other minerals in the same CAIs plot along the canonical isochron (e.g., Jacobsen et al. 2008). The observed spread in $(^{26}Al/^{27}Al)_0$ in anorthite is most likely due to redistribution of Mg isotopes through Mg-self diffusion (Fig. SM38) during thermal metamorphism that resulted in homogenization of Mg isotopes and disturbance of its ^{26}Al-^{26}Mg systematics (LaTourette and Wasserburg 1998; Ito and Messenger 2010). Although both melilite and anorthite in Allende igneous CAIs experienced postcrystallization O-isotope exchange (see Sect. 7.1.4.3), the ^{26}Al-^{26}Mg systematics in melilite are undisturbed, suggesting its O self-diffusion in melilite is much faster than Mg self-diffusion under $P - T - f_{O2}$ conditions of metasomatic alteration experienced by Allende.

In coarse-grained igneous CAIs from NWA 4964 (CK3.8) and NWA (CK3.7), fluid-assisted thermal metamorphism resulted in the disturbance of ^{26}Al-^{26}Mg systematics to various degree in melilite, grossmanite, AlTi-diopside, and spinel (Fig. SM39; Krot et al. 2023).

In chondrules from type 3.0−3.1, ^{26}Al-^{26}Mg systematics of chemically pristine plagioclase (having excess of Si and high Mg content; Tenner et al. 2019) appears to be undisturbed, even when plagioclase experienced extensive O-isotope exchange (Nagashima et al. 2017, 2018, 2018, 2021, 2022). In contrast, ^{26}Al-^{26}Mg systematics of chondrule glassy mesostasis and plagioclase replaced by nepheline in weakly metamorphosed chondrites (3.05−3.1) can be reset resulting in younger crystallization ages (Van Orman et al. 2014; Nagashima et al. 2017; Piralla et al. 2023).

6.5 ^{36}Cl-^{36}S Systematics of Sodalite and Wadalite in Allende and Ningqiang (C3-Ungrouped)

^{36}Cl is a short-lived radionuclide that decays to ^{36}Ar (98.1%, β^-) and ^{36}S (1.9%, ϵ and β^+) with a half-life of 0.3 Ma. Early ^{36}Cl studies of early solar system materials primarily focused on the search for ^{36}Ar excesses (Villa et al. 1981; Göbel et al. 1982; Murty et al. 1997; Rai et al. 2003) and provided no conclusive evidence for the presence of ^{36}Cl in the early solar system. Turner et al. (2013) first reported ^{36}Ar excesses produced by the decay of ^{36}Cl in sodalite from the Allende CAI *Pink Angel*. The initial $^{36}Cl/^{35}Cl$ ratio ($(^{36}Cl/^{35}Cl)_0$) in sodalite that formed $\sim$ 7 Ma after CAI formation (based on ^{129}I-^{129}Xe chronology) is $(1.9\pm0.5)\times10^{-8}$. No evidence for ^{26}Al, $(^{26}Al/^{27}Al)_0 < 1.7\times10^{-6}$, was found in the *Pink Angel* sodalite (Hsu et al. 2006). Conversely, studies of the ^{36}Cl-^{36}S systematics in sodalite and wadalite in CAIs and chondrules from Allende and Ningqiang showed clear excesses of ^{36}S correlated with $^{35}Cl/^{34}S$ ratios, indicative for *in situ* decay of ^{36}Cl (Lin et al. 2005; Hsu et al. 2006; Ushikubo et al. 2007a,b; Jacobsen et al. 2011). The inferred $(^{36}Cl/^{35}Cl)_0$ in these minerals ranges from $(1.81\pm0.13)\times10^{-5}$ to $(3.7\pm0.8)\times10^{-6}$. The ^{26}Al-^{26}Mg systematics in the same sodalite and grossular-wadalite mineral assemblage showed no resolvable excesses of ^{26}Al, with upper limits on $(^{26}Al/^{27}Al)_0$ ranging between 0.4×10^{-6} and 7.0×10^{-6}. This corresponds to formation ages between 3 to 4 Ma after CAI formation, suggesting that sodalite and wadalite in the CV3 parent body formed after or close to the time ^{26}Al had gone extinct.

Large excesses of ^{36}S were reported in sodalite of a heavily metasomatically altered Allende CAI *Curious Marie*. These excesses do not correlate with the $^{35}Cl/^{34}S$ ratios, suggesting late closed-system homogenization of ^{36}S excesses after ^{36}Cl became extinct (Tang et al. 2017). The ^{26}Al-^{26}Mg systematics of *Curious Marie* shows nearly flat, elevated $^{26}Mg^*$ (~40‰) and extremely high bulk $^{27}Al/^{24}Mg$ ratio (~95). Assuming $\delta^{26}Mg_0 = 0$, the inferred model $(^{26}Al/^{27}Al)_0$ is $(6.2\pm0.9)\times10^{-5}$. These observations suggest that either the secondary phases in *Curious Marie* formed early (<0.15 Ma after CAI formation) when ^{26}Al was still at the canonical level, as favored by Tang et al. (2017), or the alteration took place after ^{26}Al had completely decayed away and $^{26}Mg^*$ had been homogenized.

No ^{36}S excesses were found in several CAIs and chondrules from CVs (Plagge et al. 2006; Nakashima et al. 2008) and in a halite from the Zag (H3–6) meteorite (Nakashima et al. 2010).

There is no agreement on the origin of the ^{36}Cl excesses. The proposed models include stellar nucleosynthesis (Lin et al. 2005; Tang et al. 2017) and irradiation by solar energetic particles (SEP) (Hsu et al. 2006; Jacobsen et al. 2011; Wasserburg et al. 2011; Leya et al. 2018). The initial abundance of ^{36}Cl and decoupling from ^{26}Al in same or co-existing minerals, as well as the "young" I-Xe ages in *Pink Angle* sodalite, indicate (*i*) the production of ^{36}Cl by late-stage SEP irradiation in the PPD, and (*ii*) the production of ^{36}Cl is unrelated to the origin of ^{26}Al (Jacobsen et al. 2011). Jacobsen et al. (2011) concluded that ^{36}Cl was largely produced by irradiation of a volatile-rich reservoir [$^{35}Cl(n,\gamma)^{36}Cl$)], possibly HCl-bearing water ices (Zolotov and Mironenko 2007), in an optically thin region of the PPD adjacent to the region in which the CV chondrite parent asteroid accreted while the Sun was a weak T Tauri star. Subsequently, ^{36}Cl accreted into the Allende together with condensed water ices. Note that Leya et al. (2018) concluded that the correlation lines in the diagram $^{36}S/^{34}S$ and $^{35}Cl/^{34}S$ are not isochrons but mixing between water that contains chlorine (likely as ice) that has been irradiated either by galactic cosmic rays or SEP and sulfur (without any chlorine) with solar isotopic composition.

More work is needed to understand the origin of ^{36}Cl in the secondary metasomatic minerals in chondrites. For example, (*i*) wadalite is present in a CTA CAI from NWA 4964 (CK3.8) (Fig. SM17a) but has not been analyzed for its ^{36}Cl-^{36}S systematics yet. (*ii*) Large excesses of ^{36}S correlating with $^{35}Cl/^{34}S$ ratios, with $(^{36}Cl/^{35}Cl)_0$ as high as $(9.4\pm0.8)\times10^{-5}$, have been observed in lawrencite ($FeCl_2$) in an EL3 clast from the polymict ureilite Almahata Sitta (Feng et al. 2012). These preliminary results have not been published and need to be reinvestigated; they could have implication for the origin and distribution of ^{36}Cl in the early solar system as well as for gas dissipation in the PPD (Jacobsen and Desch 2023).

6.6 ^{129}I-^{129}Xe Chronology of the CV, CO, and CK Chondritic Components

The ^{129}I-^{129}Xe chronometer is based on the decay of now-extinct ^{129}I to ^{129}Xe with half-life of 16.14 Ma (García-Toraño et al. 2018), where the ratio of the accumulated daughter ^{129}Xe to stable ^{127}I in the host mineral reflects the iodine isotopic ratio $^{129}I/^{127}I$ at closure to Xe loss. This ratio is measured as the ratio of radiogenic ^{129}Xe ($^{129}Xe^*$) present in the sample to $^{128}Xe^*$ produced by artificial neutron irradiation of ^{127}I. This relative chronometer can be tied to the absolute age through the Shallowater aubrite. The absolute age of Shallowater has not been measured directly due to the low U content; instead it is derived from the observed correlation between ^{129}I-^{129}Xe and ^{207}Pb-^{206}Pb ages in a number of samples. Currently, the re-evaluated absolute I–Xe age of Shallowater is 4562.4±0.2 Ma (Pravdivtseva et al. 2017a).

I-Xe dating cannot be done *in situ*: separation of individual minerals or chondritic components is required. The separated polymineralic chondritic components could potentially contain multiple I-carrier phases with different closure times. The I-Xe ages have been reported for single plagioclase and pyroxene grains (Pravdivtseva et al. 2009, 2013), phyllosilicate separates (Krot et al. 2006), and magnetite separates (Pravdivtseva et al. 2017b). All these minerals occur as secondary phases in metasomatically altered type 3 chondrites (Tables SM3–5). Such phases, when they are small, cannot be cleanly separated, which could make interpretation of I–Xe ages difficult. Nevertheless, the age information in such complex samples can be successfully resolved if, during step-wise heating, radiogenic Xe from different host phases is released at significantly different temperatures, resulting in multiple I-Xe isochrons.

I-Xe ages of magnetites separated from several CV3s, CO3s, and CK3–5 s and of Allende CAIs, chondrules, matrix, and DIs from several CVs relative to Shallowater are shown in Figure SM40 and listed in Tables SM7–8. It appears that formation of magnetite in CVs, COs, and CKs started nearly contemporaneously with the formation of magnetite in Orgueil (CI1; Pravdivtseva et al. 2017b) and Murchison (CM2; Pravdivtseva et al. 2013), ~0–2 Ma prior to Shallowater.

I-Xe ages of the Allende matrix and most Allende DIs are within the range of the CV magnetite. These samples yielded high-precision I-Xe ages suggesting similar I-carrier phase. Two DIs, one from Allende and another from Efremovka have older I-Xe ages (5.7±0.4 Ma and 4.92±1.8 Ma prior to Shallowater, respectively), suggesting a different, more refractory, I-carrier phase that was apparently not affected by later metasomatic alteration.

The heavily-altered fine-grained CAIs from Allende, with the exception of the CAI *Curious Marie*, have nearly identical release profiles for radiogenic ^{129}Xe* and ^{128}Xe* suggesting the same I-carrier, most likely sodalite (Kirschbaum 1988; Pravdivtseva et al. 2003). The major release peak of ^{129}Xe* in *Curious Marie* is also consistent with sodalite being the I-carrier phase. However, high-resolution step-wise pyrolysis of this CAI demonstrated that the I-Xe system survived intact in two different mineral phases, suggesting that metasomatic alteration affected them to different degrees. The fission composition in low-T extractions is consistent with Xe from ^{244}Pu while Xe from neutron-induced fission of ^{235}U resided in the high-T release peak, supporting two different I-carrier mineral phases.

I-Xe ages of the Allende fine-grained CAIs are ~ 3 Ma after Shallowater suggesting the I-carrier phase (sodalite) recorded later stage of metasomatic alteration compared to the Allende matrix and DIs. I-Xe ages of an altered chondrule, an altered coarse-grained CAI, and one of the I-carrier secondary phases in *Curious Marie* have intermediate ages, ~0–1 Ma after Shallowater (Fig. SM40). Two DIs from the reduced CV3s Vigarano and Leoville have the youngest I-Xe ages, ~ 9 Ma after Shallowater. Overall, CV3 components recorded a prolonged duration of metasomatic alteration of the CV parent asteroid, ~ 11 Ma.

6.7 ^{129}I-^{129}Xe Chronology of Chondrules from Type 3 Ordinary Chondrites

Among the 43 LL3 chondrules studied (Gilmour et al. 2006; Fig. SM41; Table SM9 and references therein), 10 have yielded two distinct isochrons, with higher temperature releases corresponding to the older apparent I–Xe ages, probably related to chondrule formation in the lower metamorphic grade chondrules. The I–Xe ages of the oldest OC chondrules cluster tightly at 2.6±1.8 Ma after CV CAIs. Lower temperature I–Xe ages generally reflect post-formational alteration. Not only do LL chondrules tend to be younger with increasing metamorphic grade, but the spread between high- and low-temperature I–Xe ages increases also, suggesting longer or multiple alteration events for higher metamorphic grade LL meteorites. Within each OC group, the I–Xe ages of chondrules from the higher metamorphic grades tend to be younger, indicating longer post-formational processing or multiple secondary alteration events (Caffee et al. 1982; Pravdivtseva et al. 2002, 2004).

6.8 Accretion Time and Thermal History of CV, CO, CK, and OC Parent Asteroids

The ^{129}I-^{129}Xe ages of magnetites from CV3s, CO3s, and CK3−5 s indicate the accretion of their parent bodies no later than ∼3 Ma after CV CAIs. Furthermore, the ^{129}I-^{129}Xe ages of altered CAIs and DIs in CV3s suggest a prolonged duration of metasomatic alteration, longer than 11 Ma, although the carrier phases and corresponding closure temperatures of I-Xe isotopic systematics are not well-constrained. The latter requires a relatively large parent asteroid for CV chondrites.

Thermal evolution models of asteroids heated by ^{26}Al decay by Hevey and Sanders (2006) showed that asteroids as large as 50 km in radius can sustain high T close to peak T for longer than 10 Ma. Doyle et al. (2015) modelled the thermal evolution of the CV, CO, and L chondrite parent bodies heated by ^{26}Al decay. Peak T in asteroids with radius larger than 30 km are predominantly determined by their accretion time (the amount of ^{26}Al) but not by the parent body size. Thus, Doyle et al. (2015) determined the accretion time of these bodies, which are assumed to be >30 km in radius, to be 2.4−2.7 Ma, 2.1−2.4 Ma, and 1.6−1.8 Ma after CV CAIs for the CV, CO, and L chondrite parent bodies, respectively, so that the central T of these bodies reach the estimated peak metamorphic T (∼600 °C, ∼600 °C, and ∼950 °C, respectively). They showed that fayalite can form ∼0.6 Ma later than the accretion of the L chondrite parent body, or ∼1.5−2.5 Ma later than the accretion of the CV and CO chondrite parent bodies in their outer regions where the peak T are no higher than 300 °C and suitable for fayalite formation without later Fe-Mg interdiffusion. These ages are consistent with the ^{53}Mn-^{53}Cr ages of fayalite and kirschsteinite.

6.9 Future Studies

I-Xe ages of the Allende CV_{oxA}>3.6 components dominate the data set for CV3s (Fig. SM40), demonstrating a need for a broader I-Xe study of chondritic components from the reduced and oxidized CVs of different petrologic types. Although small single chondrules could yield high-precision I-Xe ages (Caffee et al. 1982), the I-Xe ages of CV, CO, and CK chondrules are virtually absent. Resonance ionization techniques, such as the RELAX instrument developed by the University of Manchester group (Gilmour et al. 2006; Crowther et al. 2008), yield greater sensitivity than conventional mass spectrometry and provide unique opportunities for small samples and single-mineral studies, which would be critical for identifying I-carrier phases in metasomatically altered chondritic components.

7 Records of Metasomatic Alteration in Oxygen, Hydrogen, Nitrogen, Carbon, Sulfur, and Chlorine Isotopic Systematics of Type 3 Chondrites and Their Components

7.1 Oxygen Isotopes

7.1.1 Introduction

Oxygen has three stable isotopes: ^{16}O, ^{17}O, and ^{18}O. Oxygen-isotope compositions of the solar system materials are typically expressed as permil deviations from the Vienna Standard

Mean Ocean Water (VSMOW) value, $\delta^{17}O$ and $\delta^{18}O$: $\delta^{i}O = ((^{i}O/^{16}O)_{sample}/(^{i}O/^{16}O)_{VSMOW} - 1)\times1000)$, where i = 17, 18. On a three-isotope oxygen diagram, $\delta^{17}O$ *vs.* $\delta^{18}O$, all terrestrial samples plot along a line with slope of ~ 0.52, called the terrestrial fractionation (TF) line, whereas extraterrestrial materials deviate from this line; the deviation is expressed as $\Delta^{17}O = \delta^{17}O - 0.52\times\delta^{18}O$. Bulk oxygen-isotope composition of chondrites is one of the most important parameters used for their classification (Clayton 2008; Yurimoto et al. 2008; Ireland et al. 2020).

Ultraviolet CO self-shielding either in the PMC and/or in the outer regions of the PPD (outside ~ 30 AU) resulting in formation of ^{16}O-rich CO and $^{17,18}O$-enriched water followed by silicate dust – H_2O_{ice} – CO_{gas} fractionation and evaporation is the currently favored mechanism invoked to explain the observed range of $\Delta^{17}O$ in the solar system (Clayton 2002; Yurimoto and Kuramoto 2004; Lyons and Young 2005; Krot et al. 2020).

Chondrules (phenocrysts and mesostasis) and fine-grained refractory inclusions in unmetamorphosed chondrites (petrologic type ≤3.0) have internally uniform $\Delta^{17}O$ and on a three-isotope oxygen diagram plot along ~ slope-1 line, named the primitive chondrule mineral (PCM) line (Fig. SM42; Ushikubo et al. 2012, 2017, 2018). Secondary minerals formed during metasomatic alteration of type 3 chondrites deviate from the slope-1 and are out of isotope equilibrium with the primary minerals (see Sects. 7.1.2 & 7.1.3). In addition, primary minerals chondrules and refractory inclusions in chondrites of petrologic type > 3.0 are often isotopically heterogeneous (see Sect. 7.1.4). The nature of this heterogeneity remains controversial. The proposed mechanisms include condensation from or gas-melt exchange with nebular gas of variable O-isotope composition, and isotope exchange with metasomatic fluids in the chondrites parent asteroids (e.g., Yurimoto et al. 2008; Kawasaki et al. 2018; Krot et al. 2022b).

In this section, we summarize O-isotope compositions of the aqueously/metasomatically-formed minerals in type 3 chondrites and discuss possible effects of metasomatic alteration on O-isotope compositions of the primary minerals in their chondrules and refractory inclusions.

7.1.2 Oxygen-Isotope Compositions of Magnetite and Fayalite in Type 3.0–3.1 Chondrites

Although Acfer 094 avoided metasomatic alteration in its parent asteroid, it contains intergrowths, named cosmic symplectites (COS) composed of magnetite, pentlandite, troilite, Na_2SO_4, and elemental sulfur (Sakamoto et al. 2007; Matsumoto et al. 2022). Magnetite in COS is characterized by the anomalous O-isotope composition ($\delta^{17,18}O$ ~ +180‰) and is interpreted to have recorded O-isotope composition of the primordial water in the PPD, and, hence, providing a strong argument supporting the CO self-shielding model (Yurimoto and Kuramoto 2004; Lyons and Young 2005; Sakamoto et al. 2007). The origin of COS, however, remains unclear: they may have formed by (*i*) oxidation and sulfidization of FeNi-metal in the PPD or in a pre-existing planetesimal (Sakamoto et al. 2007), (*ii*) high temperature interaction between FeNi-metal and O-Na-S-rich material (hot corrosion) (Matsumoto et al. 2022), and (*iii*) decomposition of amorphous ferric-hydroxysulfate-hydrate formed by interaction of sulfuric acid and ferric iron in a cryofluid environment on the Acfer 094 parent asteroid (Hashimoto and Nakano 2023).

Oxygen-isotope compositions of magnetite and fayalite formed during metasomatic alteration of type 3 UOCs, COs, CVs, and CKs are shown in Figures SM43–45. On the diagrams $\delta^{17}O$ *vs.* $\delta^{18}O$, magnetite in fayalite in type 3.0–3.2 UOCs plot along mass-dependent fractionation lines with $\Delta^{17}O$ of ~ +4.3‰ which correspond to $\Delta^{17}O$ of aqueous fluids in their parent asteroids. There are no resolvable differences in the inferred $\Delta^{17}O$

of aqueous fluids in H, L, and LL chondrites or among chondrites of different petrologic subtypes (Fig. SM43). Magnetite and fayalite in CO3.05–3.1 s and MAC 88107 (CO3.1-like) chondrites plot along mass-dependent fractionation lines with $\Delta^{17}O$ of ~ −0.2‰ and ~ −1.8‰, respectively (Figs. SM44). In contrast, magnetite and fayalite in the CV_{oxB} chondrites, Kaba (CV3.1), Mokoia (CV~3.6), and Asuka (A)-881317 (CV3), show significant ranges in $\Delta^{17}O$, from ~ 0 to ~ −4‰ from ~ +3.5 to ~ −2‰, respectively (Figs. SM45a,b), suggesting either O-isotope evolution of aqueous fluid or heterogeneous O-isotope compositions of water ices that accreted into CV parent asteroid(s).

Oxygen-isotope compositions of magnetite in metamorphosed (petrologic types 3.6–3.8) CVs and CKs are shown in Figures SM45c,d. High precision data for magnetite in CKs have very limited ranges of $\delta^{18}O$ (from −2.5 to +2.5‰) and $\Delta^{17}O$ (~ −3‰), possibly reflecting O-isotope equilibration with matrix silicates during thermal metamorphism. Data for magnetite from Allende and Ningqiang (C3-ungrouped) are less precise; within analytical uncertainty, their $\Delta^{17}O$ values are similar to those of CK magnetites; $\delta^{18}O$ values are more variable.

To summarize, there are differences in $\Delta^{17}O$ of fayalite and magnetite in UOCs, COs, CVs, and CKs suggesting different $\Delta^{17}O$ of aqueous fluids, and, possibly water ices, in their parent asteroids. In the CV_{oxB} chondrites, fayalite and magnetite show a range of $\Delta^{17}O$, possibly reflecting evolution of O-isotope composition of aqueous fluids. The inferred $\Delta^{17}O$ of aqueous fluids are very different from $\Delta^{17}O$ of COS that is hypothesized to represent O-isotope compositions of primordial water ices in the PPD.

7.1.3 Oxygen-Isotope Compositions of Secondary Minerals in Igneous CAIs from Allende and CK3.7–3.8 Chondrites

Oxygen-isotope compositions of the *in situ* SIMS measured secondary minerals in coarse-grained igneous CAIs from Allende (CV_{oxA}>3.6) and CK3.7–3.8 s are shown in Figures SM46a and SM46b, respectively. The data plot along mass-dependent fractionation lines with indistinguishable $\Delta^{17}O$ of −3±2‰ (average ± 2SD) and −3.7±1.9‰, respectively. Although all minerals were measured with matrix-matched standards, there are some differences in chemical compositions of secondary minerals in CV and CK CAIs (see above) and the standards, which may explain some differences in their $\delta^{18}O$ values.

7.1.4 Oxygen-Isotope Heterogeneity of Refractory Inclusions and Chondrules in Metasomatically Altered Type 3 Chondrites

Fine-Grained CAIs and AOAs Amoeboid olivine agggregates and fine-grained CAIs in unmetamorphosed ordinary and carbonaceous chondrites (petrologic type ≤3.0) have internally uniform $\Delta^{17}O$, typically solar-like, ~ −24±2‰ (Figs. SM42, 47–49) consistent with their formation in a gas of ~ solar oxygen-isotope composition. In contrast, AOAs and fine-grained CAIs in metamorphosed chondrites are isotopically heterogeneous: corundum, hibonite, spinel, forsterite, AlTi-diopside, and enstatite have ^{16}O-rich solar-like compositions, whereas Zr- and Sc-rich oxides and silicates, grossite, krotite, perovskite, melilite, anorthite, and perovskite are ^{16}O-depleted to various degrees (Figs. SM47–49). The $\Delta^{17}O$ values of the most ^{16}O-depleted primary minerals overlap with $\Delta^{17}O$ values of metasomatic fluids inferred from O-isotope compositions of secondary minerals (see Sects. 7.1.2 & 7.1.3). These observations and the fact that the O-isotope heterogeneity is inconsistent with a condensation sequence of minerals in refractory inclusions suggest that O-isotope heterogeneity in fine-grained CAIs and AOAs resulted from isotope exchange during aqueous fluid–rock interaction in the chondrite parent asteroids (Simon et al. 2019; Krot et al. 2019a,b,c; Ebert et al. 2020; Han et al. 2024).

Igneous Inclusions Coarse-grained igneous CAIs are found exclusively in CVs and CKs. Both groups experienced metasomatic alteration and thermal metamorphism and lack type 3.0 members. Like the fine-grained CAIs in metamorphosed ordinary and carbonaceous chondrites, the coarse-grained igneous CAIs in CV3.1−4 s and CK3.7−3.8 s have internally heterogeneous O-isotope compositions: hibonite, spinel, forsterite, and low-Ti Al-diopside have solar-like ^{16}O-rich ($\Delta^{17}O \sim -24\pm2$‰) compositions, whereas grossite, krotite, perovskite, melilite, anorthite, and Ti-rich pyroxenes (fassaite, grossmanite) are ^{16}O-depleted to various degrees (Fig. SM50). There is no agreement on the nature of this heterogeneity. It could have resulted from **(1)** O-isotope exchange during multistage, incomplete melting of CAIs in the nebular gas of variable $\Delta^{17}O$ (Yurimoto et al. 1998; Ito et al. 2004; Kawasaki et al. 2015, 2018; Aléon 2016, 2018) and/or **(2)** during aqueous fluid−rock interaction in the CV and CK parent asteroids (Yurimoto et al. 2008; Krot et al. 2022b, 2023).

(1) Igneous CAIs experienced melting in the solar nebula and thus could have experienced gas-melt O-isotope exchange. The early generation of gaseous reservoirs with variable $\Delta^{17}O$ has been advocated in several papers (Yurimoto et al. 1998; Itoh and Yurimoto 2003; Aléon et al. 2007; J. Simon et al. 2005, 2011, 2016; Kawasaki et al. 2012, 2018; Aléon 2016, 2018, 2018, 2020, 2020). If igneous CAIs experienced O-isotope exchange with the nebular gas of variable $\Delta^{17}O$, the distribution of oxygen isotopes in these CAIs must be consistent with igneous crystallization sequence studied experimentally (Stolper 1982). However, this is generally not the case. For example, in Type B CAIs eutectic melt regions composed of co-crystallized highly-åkermanitic melilite ($Åk_{65-71}$), anorthite, low-Ti ($\sim 2-5$ wt% TiO_2) fassaite, and spinel, the spinel and fassaite are similarly ^{16}O-rich ($\Delta^{17}O \sim -24$‰), whereas melilite and anorthite are ^{16}O-depleted to various degrees ($\Delta^{17}O$ up to ~ -1‰; see (see Figs. 11a,b in Krot et al. 2022b), suggesting post-crystallization O-isotope exchange.

The only mineral phase in igneous CAIs where igneous chemical zoning correlates with O-isotope zoning is AlTi-pyroxenes: Ti-rich pyroxene cores are ^{16}O-depleted to various degrees relative to Ti-poor pyroxene edges having solar-like $\Delta^{17}O$ (Kawasaki et al. 2018; MacPherson et al. 2022; Krot et al. 2022b). However, it remains unclear whether this O-isotope zoning was established during igneous crystallization or resulted from postcrystallization O-isotope exchange. Several observations may support the latter mechanism. (*i*) High-Ti pyroxene inclusions inside spinel grains were isolated from interaction with external O-isotope reservoir and preserved ^{16}O-rich compositions (Figs. SM50a,d, 51). (*ii*) O-isotope zoning is observed in igneously zoned Ti-rich pyroxenes in FUN (fractionation and unidentified nuclear effects) CAIs which experienced melting, evaporation and crystallization in nearly vacuum (Krot et al. 2014b; Mendybaev et al. 2013, 2017) and therefore must have avoided gas-melt isotope exchange. (*iii*) There is an apparent correlation of $\Delta^{17}O$ in pyroxenes and thermal history of the host meteorites: pyroxenes with similar chemical compositions are more ^{16}O-depleted in CAIs from more metamorphosed chondrites, CK3.7−3.8 and NWA 8418 (CV4) (Fig. SM51).

(2) Oxygen-isotope composition of aqueous fluid in Allende (CV_{oxA}>3.6) and CK3.7−3.8 s is constrained by O-isotope compositions of their secondary minerals (see Sect. 7.1.3 and Fig. SM46). Postcrystallization O-isotope exchange of igneous CAIs with the fluid appears to be consistent with several observations: (*i*) $\Delta^{17}O$ of ^{16}O-depleted primary minerals (melilite, anorthite, high-Ti pyroxenes, grossite, krotite) approach $\Delta^{17}O$ of the fluid (Figs. SM47−49). (*ii*) On $\delta^{17}O$ *vs.* $\delta^{18}O$ plots (Fig. SM50), the ^{16}O-depleted primary minerals deviate to the right from a nebular slope-1 PCM line, i.e., towards the inferred compositions of aqueous fluids (Krot et al. 2022b). (*iii*) Oxygen self-diffusion in anorthite under wet conditions is several magnitudes higher than under dry conditions (Fig. SM52) and may explain its ^{16}O-poor composition.

To summarize, the coarse-grained igneous CAIs in CVs and CKs experienced post-crystallization O-isotope exchange to various degrees with an external ^{16}O-poor reservoir ($\Delta^{17}O \sim -3\pm2‰$), most likely aqueous fluid in the CV and CK chondrite parent asteroids. The isotope exchange appears to be mineralogically not petrologically controlled: i.e. variations in O-isotope compositions of minerals within individual CAIs are inconsistent with their inferred crystallization sequence. Oxygen-isotope compositions of melilite, anorthite, Ti-rich fassaite, perovskite, grossite, krotite, and davisite were affected by this process to various degrees. Hibonite, spinel, low-Ti fassaite, Al-diopside, and forsterite retained their original O-isotope compositions established during igneous crystallization of coarse-grained CAIs. This does not preclude O-isotope exchange in some igneous CAIs during melting in the solar nebula (e.g., some FoB and FUN CAIs having relatively ^{16}O-depleted compositions, $\Delta^{17}O > -20‰$, and CAIs melted during chondrule formation), but such inclusions appear to be rare (Kawasaki et al. 2012; Krot et al. 2005, 2014b).

Chondrules Chondrule phenocrysts and mesostasis in the least altered and the least metamorphosed carbonaceous chondrite Acfer 094 (C3.00) have internally uniform O-isotope compositions (Fig. SM42). Glassy mesostasis and plagioclase in chondrules from even weakly altered and metamorphosed ordinary and carbonaceous chondrites (Semarkona (LL3.00), Yamato (Y)-81020 (CO3.05), and Kaba (CV_{oxB}3.1) are ^{16}O-depleted relative to chondrule phenocrysts (Fig. SM53; see also Ebert et al. 2022). The Δ^{17}O values of mesostasis and plagioclase approach ^{17}O of aqueous fluid inferred from secondary magnetite and fayalite (see Sect. 7.1.2) suggesting they experienced O-isotope exchange with the fluid (Kita et al. 2010; Ebert et al. 2022; Fukuda et al. 2022; Krot et al. 2022a,b).

7.1.5 Future Studies

Oxygen-isotope compositions of aqueous fluid in the reduced CV3s is not known yet, but can be constrained by measuring secondary kirschsteinite. Development of the matrix-matched mineral standards, including kirschsteinite, for SIMS measurements is required to constrain the δ^{18}O values of aqueous fluids, and, potentially, crystallization temperature of secondary mineral parageneses.

To test the hypotheses of O-isotope exchange in igneous CAIs from CVs and CKs – gas-melt and fluid-rock – experimental data on oxygen self-diffusion in CAI minerals under proper physicochemical conditions ($P - T - fO_2$–water/rock ratio) of metasomatic alteration of CVs and CKs are required.

7.2 Hydrogen Isotopes

7.2.1 Bulk Chondrite Hydrogen-Isotope Compositions

The first bulk H-isotope compositions measured in UOCs were obtained about 40 years ago (Robert et al. 1979; McNaughton et al. 1982; Yang and Epstein 1983) and shown to present high variabilities with δD relative to SMOW, from $\sim$ –140 up to +3000‰. UOCs of petrologic type $\leq$ 3.2 generally contain the highest amounts of hydrogen (up to 0.1 wt% H) and are enriched in deuterium compared to UOCs of higher petrologic types (Alexander et al. 2012; Vacher et al. 2020; Grant et al. 2023; Table SM10, Fig. SM54a). The least metamorphosed LL3.00 Semarkona has the highest amount of hydrogen among non-weathered UOCs (average H content of 0.07$\pm$0.04 wt%) and the highest and more variable δD value of 2000$\pm$745‰ (Table SM10).

Metasomatized CVs, COs, and CKs are depleted in hydrogen compared to CIs, CMs, and CRs that experienced relatively low temperature aqueous alteration and present moderate δD variations with values ranging from –115 to +240‰, with the exception of one aliquot of Ornans (CO3.4) having a δD of +2150‰ (Kerridge 1985) (Table SM10 and references therein, Fig. SM54b).

7.2.2 Hydrogen-Isotope Compositions of Organics

As described in Sect. 3, the IOM is the major carrier of C, N and noble gases, and one of the two major carriers of hydrogen in type 3 chondrites. In the most primitive chondrites, the IOM is strongly enriched in δD compared to bulk rocks (Alexander et al. 2007, 2012, 2021) and shows heterogeneities with μm-sized areas both enriched and depleted in D compared to the average IOM value (Busemann et al. 2006; Remusat et al. 2006). With increasing temperatures during parent body alteration, the H/C ratio of the IOM decreases due to the re-organization of the organic matter structure, the loss of heteroatoms, and the development of large aromatic domains (see Sect. 3.1; Alexander et al. 2007, 2014, 2017, 2009). In carbonaceous chondrites, the hydrogen loss is associated with the destruction of μm-scale heterogeneities (Remusat et al. 2008) and likely to H-isotope fractionation (Alexander et al. 2010, 2014).

In CVs and COs, the IOM shows low-to-moderate D-enrichments relative to the bulk rocks compared to the more water-rich CIs, CMs, and CRs with δD ranging from +60 to +1400‰ and from +130 to +700‰, respectively (Alexander et al. 2007). Although there is only moderate D-increase for the CV IOM and no clear trend exists for the CO IOM δD compositions with petrological type or H contents (Alexander et al. 2007), the relatively low δD and H contents of these meteorites compared to those of CMs, CIs, and CRs could reflect the effects of parent body processes under increasing temperatures.

In UOCs, the IOM is enriched in δD (from +3000 to 12,000‰) (Alexander et al. 2010) compared to those of other chondrites and to the UOC bulk compositions (Yang and Epstein 1983; Alexander et al. 2007, 2010, 2021). However, the IOM appears to be more homogenous than in CCs showing only rare μm-sized D-rich anomalies (Remusat et al. 2016). The presence of a positive correlation between the IOM δD and its C/H ratio (Fig. SM55a) was initially proposed to result from isotopic exchanges between IOM and aqueous fluids following D enrichments of water through Rayleigh-type fractionations after metal oxidation during alteration processes in the parent asteroid (Alexander et al. 2010). However, this hypothesis is at odds with the absence of general D-enrichments in bulk UOCs with increasing the metamorphism degree (Fig. SM55a; Vacher et al. 2020; Marrocchi et al. 2020; Piani et al. 2021; Grant et al. 2023). Moreover, the IOM of ordinary chondrites seems to contain a D-rich thermally resistant component, whose proportions would be enhanced during heating due to the loss of more D-poor organic material (Figs. SM55b,c; Remusat et al. 2016). The presence of this D-rich component in UOCs that is unsampled by carbonaceous chondrites argues for a difference in the organic reservoirs of OC and CC IOMs.

7.2.3 Hydrogen-Isotope Compositions of Hydrated Silicates

Like for CMs, CIs and CRs (for details see Lee et al. this journal), the δD composition of water-bearing minerals in the oxidized Bali-like CVs (Kaba, Bali, and Grosnaja) was estimated by SIMS with an analytical protocol using Cs^+ primary ions and found to share a common δD composition of $\sim -80\pm100$‰ (Piani and Marrocchi 2018). The D-enrichment of the CV water-bearing minerals compared to those of CMs can be interpreted as the evidence that CVs accreted ices with a distinct hydrogen-isotope composition than those in

CMs or CRs. However, the additional presence of generally δD-depleted organic matter in CVs compared to that in CMs could also be seen as the result from isotopic exchanges between water and organics during parent body metasomatism (Piani and Marrocchi 2018). The exchange reactions could be favored in CVs compared to CMs by the higher T and/or different alteration mechanisms, possibly implying steam instead of, or in addition to, liquid fluids (see Sect. 5).

For COs, the δD of water-bearing minerals was measured only in DOM 08006 (CO3.00) and found to be +306±135‰ (Piani et al. 2021), a value slightly higher than for CVs. More measurements in this group of chondrites would be needed to determine if this unique value is representative for the water in all COs; only meteorites with low terrestrial weathering degrees should be selected for analyses to avoid terrestrial contamination (Alexander et al. 2018a).

Highly heterogeneous δD with extreme values up to 10,000‰ have been measured in the H-bearing matrix of Semarkona (LL3.00) (Fig. SM56; Deloule and Robert 1995; Piani et al. 2015). The highest δD values in the matrix exceed by far the isotopic composition observed in primitive OC IOMs (Alexander et al. 2007; Remusat et al. 2016) and are not associated with C-rich areas indicating that their carrier is inorganic, likely hydrous silicates (Fig. SM56; Piani et al. 2015). The identification of the extreme δD variations associated with inorganic phases in Semarkona indicate that the parent body of LL-type ordinary chondrites have likely accreted some D-rich material derived from the molecular cloud and having escaped complete homogenization with hydrogen of the PPD (Aléon 2010; Piani et al. 2021). The nature of the D-rich inorganic carrier, possibly phyllosilicates or hydrated amorphous silicates (Dobrică and Brearley 2021), has yet to be characterized.

7.3 Carbon- and Nitrogen-Isotope Compositions of the Main C- and N-Bearing Phases

The carbon-isotope compositions of carbonates in UOCs, CVs, and COs (δ^{13}C range from –23 to +5‰) were determined from CO_2 released after phosphoric acid treatments of the whole rocks and found to be ^{13}C-depleted compared to carbonates in CIs, CMs, and CRs (Grady et al. 1988; Alexander et al. 2015). The limited quantities of carbon measured (e.g., 0.02 wt% C in Semarkona; Alexander et al. 2015) are consistent with the absence or scarcity of carbonates observed in these chondrites (e.g., Hutchison et al. 1987). The low δ^{13}C values of carbonates in CVs, COs and UOCs could be explained either by their formation at higher temperature and/or by more oxidized fluids (higher CO_2 fractions) than for the CIs, CMs, and CRs (Alexander et al. 2015).

In UOCs, the C-isotope compositions of organic matter were estimated using stepped heating combustion (Grady et al. 1989). Removing the terrestrial weathering contaminants that can constitute an important part of the low temperature released gases of chondrite finds, the δ^{13}C of the indigenous organic compounds were estimated to be of ∼ –28‰ with no clear variation trend observed according to the chondrite petrographic type (Grady et al. 1989).

The carbon- and nitrogen-isotope compositions of CV and CO IOM range from –18 to –5‰ for δ^{13}C and from –65 to +17‰ for δ^{15}N (Alexander et al. 2007). Due to the high abundance of IOM among the known C-bearing components in CVs and COs, the bulk δ^{13}C and δ^{15}N values of CVs and COs appear to be dominated by the IOM (Greenwood and Franchi 2004; Alexander et al. 2015).

Although FeNi-carbides are commonly observed in the least metamorphosed COs, CVs, and UOCs (see Sect. 2.2), their C-isotope compositions are not known yet.

7.3.1 Future Studies

Trace amounts of carbonates appear to be present in CVs, COs and UOCs when measured in whole rocks (Grady et al. 1988), but no *in situ* measurement were ever reported for their carbon-isotope compositions. Such measurements should be possible with SIMS at least in Semarkona (LL3.00), in which 15 μm-sized calcites were observed (Hutchison et al. 1987) and would allow a better understanding of the conditions of secondary alteration. In addition, SIMS measurements of C-isotope compositions of FeNi-carbides in CVs, COs, and UOCs would be also important.

Although the organic matter (including both SOM and IOM) and carbonates are major carbonaceous constituents in chondrites, their reported abundances do not account for more than ~60% of the total carbon content of the rock (Smith and Kaplan 1970; Alexander et al. 2015). Investigations should aim to identify the exact nature and origin of the missing carbonaceous component, which possibly consists of some kind of organic compounds lost during the IOM isolation process (Alexander et al. 2015).

Finally, the nature of the inorganic H-bearing components at the origin of the D-rich signatures observed in UOCs and the amplitude of their modifications during parent body processes require thorough investigations that could be achieved by coupling high resolution mineralogical observations using TEM and isotope imaging using NanoSIMS.

7.4 Sulfur Isotopes

With four stable isotopes (^{32}S, ^{33}S, ^{34}S and ^{36}S), sulfur can be used as a probe of the astrophysical environment of mass-independent photochemistry (Thiemens 2006). The sulfur isotope ratios are usually expressed as δ^{33}S, δ^{34}S, and δ^{36}S ($\delta^{n}S = [(^{n}S/^{32}S)/(^{n}S/^{32}S_{V\text{-}CDT}) - 1] \times 1000$; with n being the mass of the considered isotope and V–CDT: Vienna Canyon Diablo Troilite used as reference. Briefly, the extents of mass-independent anomalies for the different isotopes are thought to depend on the energy of the photons responsible for the photolysis of the major S-bearing phase in the gas of the solar nebula, H_2S (Chakraborty et al. 2013).

The highest mass-independent variations of sulfur reported so far in chondrites have been found to be associated with the ^{16}O-poor COS in Acfer 094 (weighted means of $\Delta^{33}S = +3.84 \pm 0.72‰$ and $\Delta^{36}S = -6.05 \pm 2.25‰$ with $\Delta^{33}S = \delta^{33}S + 1000 \times [(\delta^{33}S/1000+1)^{0.515} - 1]$ and $\Delta^{36}S = \delta^{36}S + 1000 \times [(\delta^{36}S/1000+1)^{1.9} - 1]$; Vacher et al. 2021). These values are inconsistent with the energy of ultraviolet-photons coming from the protosun and could imply the presence of massive O and B stars close to the solar system's parent molecular cloud (Vacher et al. 2021). Except in COS of Acfer 094, no significant mass-dependent or mass-independent isotopic variations were reported for type 3 chondrites so far (Tachibana and Huss 2005; Marrocchi et al. 2016; Alexander et al. 2022).

Large mass-dependent isotope fractionation effects (δ^{34}S from −8 to +6‰) have been described in aqueously formed sulfides from CIs, CMs, and volatile-rich clasts in polymict ureilites and HED meteorites (Visser et al. 2019). Aqueously formed magnetite and fayalite in chondrules and matrices of UOCs, CVs, and COs commonly associate with elongated sulfide (troilite or pyrrhotite?) grains, which may have formed by direct precipitation from aqueous fluid (Fig. SM57). Sulfur-isotope compositions of sulfides associated with Cr-free magnetite in Kaba chondrules and interpreted as igneous in origin (Marrocchi et al. 2016; but see Sect. 2.2 for an alternative explanation) show a small range of δ^{34}S, from $-1 \pm 0.4‰$ to $+1.4 \pm 0.4‰$. Elongated sulfides associated with fayalite have yet to be analyzed.

7.5 Sources of Water in UOCs, COs, and CVs

The main process controlling the O-isotope composition of the CV and CO parent asteroids is related to isotopic exchange between ^{16}O-enriched anhydrous silicates and ^{17,18}O-enriched aqueous fluid (Fig. SM58; Marrocchi et al. 2018). The initial O-isotope composition of aqueous fluid and its abundance (water/rock ratio) in the CV and CO asteroids, however, are poorly constrained. It is commonly assumed that COS in Acfer 094 having anomalously ^{17,18}O-enriched compositions ($\delta^{17,18}$O $\sim$ +180‰; Δ^{17}O $\sim$ +85‰; Sakamoto et al. 2007) and sulfur mass-independent isotope anomalies (Δ^{33}S $\sim$ +4‰ and Δ^{36}S $\sim$ −6‰; Vacher et al. 2021) recorded the primordial O-isotope composition of water ice in the PPD providing evidence for the CO self-shielding in the PMC and/or outer disk (Yurimoto and Kuramoto 2004; Lyons and Young 2005; Krot et al. 2020). The ^{17,18}O-depleted compositions of aqueously formed minerals in UOCs and CCs (see Sects. 7.1.2 & 7.1.3, and Lee et al. 2025, this collection) provide no evidence that such primordial water ice directly accreted into chondrite parent bodies, and suggest instead that prior to accretion into chondrite parent asteroids O-isotope composition of water ice evolved towards the Earth-like value (Δ^{17}O $\sim$ 0‰) through either partial evaporation and isotope exchange with ^{16}O-enriched silicates and CO gas or/and isotopic evolution of water with time during CO self-shielding in the outer disk (Yurimoto and Kuramoto 2004; Lyons and Young 2005; Lyons et al. 2009; Krot et al. 2015; Alexander et al. 2017; Marrocchi et al. 2018).

Once accreted and melted, the chondritic water evolved towards ^{16}O-rich composition by exchange with anhydrous minerals and amorphous phases (Marrocchi et al. 2018; see also Lee et al. 2025, this collection). The initial water ice that accreted into chondrite asteroids is thus expected to have had a Δ^{17}O value equivalent to or higher than any aqueously formed minerals present in chondrites (Krot et al. 2015; Fujiya 2018; Marrocchi et al. 2018; Alexander 2019a,b). In the case of weakly metamorphosed CO3.0−3.1 s and CV_{oxB}3.1 s, the highest Δ^{17}O values of aqueously formed minerals ranging from $\sim$ −2 to $\sim$ +3.5‰ (Figs. SM44−45) put the lower limits on the initial Δ^{17}O of water ice in COs and CVs. Alexander (2019b) suggested that (*i*) different carbonaceous chondrites, including CVs and COs, could have originated from a mixture of the same four components: (1) chondrule/chondrule precursor component that partially lost FeNi-metal and volatiles, (2) refractory inclusion component that has a refractory inclusion-like bulk composition, (3) anhydrous and reduced but otherwise CI-like matrix, and (4) water ice with relatively high Δ^{17}O (+3.5‰) and δ^{18}O (18–21‰). (*ii*) Variations in the amount of water ice accreted by different chondrite groups could be responsible for differences in their bulk O-isotope compositions and O-isotope compositions of aqueously formed minerals. For example, the ^{16}O-enriched bulk compositions of COs compared to CVs would result from lower amount of the water ice component accreted by COs (Alexander 2019b). Nonetheless, the observed differences in Δ^{17}O values of fayalite in Kaba (CV_{oxB}3.1) and Elephant Moraine (EET) 90043 (CO3.1), $\sim$ +3.5‰ and −0.3‰, respectively (Figs. SM45a, 44c), may suggest that CVs accreted a more ^{17,18}O-rich water ice than COs (Marrocchi et al. 2018). Using the estimated O-isotope composition of the CV_{oxB} aqueous fluids during precipitation of pure fayalite and the inferred water/rock ratio modeled for its formation (0.1−0.2; Zolotov et al. 2006), Marrocchi et al. (2018) calculated the initial O-isotope composition of the CV_{oxB} aqueous fluid (δ^{18}O $\geq$ +60‰, Δ^{17}O > 15‰) suggesting a potential contribution of water ice from the outer solar system.

Hydrogen-isotope compositions estimated for hydrated minerals in CVs could also be the results of the parent body evolution from an initially D-poor CM-like water toward D-richer values through isotope exchanges with D-rich organics that would be favored by the high alteration temperatures of CV parent body (Piani and Marrocchi 2018). Alternatively, the moderate δD values measured for both water and IOM in CVs and COs could have resulted from their early accretion in the disk at temperatures allowing a more efficient re-equilibration with the D-poor solar gas than for CIs, CMs, and CRs that accreted later or at higher heliocentric distances (Piani et al. 2021). Further isotopic measurements on natural samples and experimental constraints on isotope exchange reactions are needed to better differentiate primary isotopic signatures from secondary isotope modifications taking place in the parent bodies during metasomatic alteration.

As for weakly metamorphosed CVs and COs, the Δ^{17}O of ices accreted in the UOC parent bodies should be equal or higher than the highest value inferred from the alteration fluids at the origin of the secondary minerals, fayalite and magnetite (i.e., $\Delta^{17}\mathrm{O} \sim +5‰$; Fig. SM43). However, the ranges of δ^{18}O and Δ^{17}O measured in bulk UOCs cannot be reconciled with a similar water ice O-isotope composition as those suggested for CCs, and Δ^{17}O values as high as $\sim$16‰ were proposed for the OC initial water ice (Alexander 2019a). Additionally, the presence of D-rich IOM, including a D-rich thermally resistant component (Sect. 7.2.2), and the D-rich and isotopically heterogeneous hydrated silicates measured in Semarkona (Sect. 7.2.3) support the interpretation that OCs inherited a higher fraction of the primordial (i.e., molecular cloud) water ice and/or organics compared to CCs (Deloule and Robert 1995; Aléon 2010; Piani et al. 2015; Remusat et al. 2016). The inferred Δ^{17}O and δD values of water ice and δD composition of IOM seem to indicate that OCs have recorded lower degrees of re-equilibration between the primordial molecular cloud molecules and protosolar H_2 than most CCs, which might be at odds with their supposed formation locations in the inner and outer solar system, respectively (Warren 2011; Kruijer et al. 2017). However, the temperatures regulating the efficiency of re-equilibration between the primordial molecules and protosolar H_2 are not only dependent on the formation distance of the chondritic parent body but also on the thermal evolution of the disk with time and the nature of interactions with the parent molecular cloud. Strong differences exist in models trying to infer the H-isotope evolution of the disk water (see the discussion in Krot et al. 2015 and references therein); with no systematic dependences on the heliocentric distance (Yang et al. 2013; Jacquet and Robert 2013). Nonetheless, although the re-equilibration timescale could be different for oxygen and hydrogen isotopes (Lyons and Young 2005; Yang et al. 2013), no model allows a comprehensive description of the O- and H-isotope distribution for water throughout the disk evolution, nor does it explain the isotopic differences between OCs and CCs. Further attempts should be made in the future (*i*) to bring reliable constraints on isotope distribution in planetary materials, notably by deconvoluting the isotopic contribution of parent body processes from the hydrogen- and oxygen-isotope signatures of the primary ice in chondrites, and (*ii*) to estimate the spatial/temporal evolution of water hydrogen and oxygen-isotope compositions in the context of the recent disk dynamical evolution models.

8 Conclusions

1. Type 3 ordinary and CV, CO, and CK carbonaceous chondrites experienced aqueous fluid – rock interaction at $\sim$100–300 °C during progressive thermal metamorphism followed by loss of the fluid and nearly dry metamorphism. Both aqueous solution and steam played an important role in mobilization of rock-forming elements and resulted in metasomatic alteration of these meteorites. The alteration was multistage and produced different minerals and mineral assemblages in chondrite matrices, chondrules, and refractory inclusions reflecting differences in primary mineralogy, textures, and chemical compositions of these chondritic

components as well as temporal and local variations in fluid chemical composition. Detailed mineralogical studies of the secondary minerals and mineral assemblages in type 3 chondrites of different petrologic subtypes using FE SEM and focused ion beam (FIB) + TEM combined with physicochemical analysis and thermodynamic calculations could potentially constrain conditions during metasomatic alteration ($P - T - fO_2$-*water/rock ratio*-$aSiO_2 - aCa - aFe\ldots$).

2. Experimental syntheses of secondary minerals (grossular, wollastonite, nepheline, sodalite, dmisteinbergite) observed in metasomatically altered CAIs in Allende (CV_{oxA}>3.6) resulted in formation of intermediate hydrated phases (e.g., hydrogrossular, hydrosodalite, analcime, nepheline hydrate) not observed in Allende. Studies of CAIs from less metamorphosed CVs are required to understand how relevant these experiments are. Although a hydrothermal experiment with Fe-metal, amorphous ferroan silicates and water did produce pure fayalite commonly observed in UOCs, CVs and COs of petrologic type 3.0–3.2, an experimental synthesis of FeMg-olivine from amorphous FeMg-silicates commonly observed in matrix of primitive chondrites is needed to understand a broad range of matrix olivine compositions ($Fa_{\sim 40-100}$) in these meteorites.

3. The origin of the precursors of organic matter in different chondrite groups and whether CVs, COs, and UOCs accreted the same mix of organic precursors remain unclear. Although IOM, SOM, and carbonates are the major carbon-bearing constituents in chondrites, their reported abundances do not account for more than ~60% of the chondrites total carbon content. Identification of the exact nature and the origin of the missing carbonaceous component is urgently needed. The role of metasomatic fluids in the evolution of organic matter in UOCs, COs, and CVs is not known. *In situ* studies of IOM using STXM and TEM and experimentally identifying specific reaction pathways related to the presence of oxidizing fluids at temperature above 300 °C could potentially offer a way to shed some light on this problem.

4. The ^{53}Mn-^{53}Cr chronology of fayalite and kirschsteinite in UOCs, CVs, and COs measured with SIMS using matrix-matched standards suggests metasomatic alteration in their parent asteroids occurred ~2–5 Ma after formation of CAIs. However, the obtained data are very limited and do not allow the duration of metasomatic alteration in the UOC, CO, and CV parent asteroids to be constrained. The ^{129}I-^{129}Xe ages of magnetites separated from CV3s, CO3s, and CK3–5 s are similar to those from CMs and CIs, ~3–5 Ma after CAIs, suggesting nearly contemporaneous aqueous activity in the carbonaceous chondrites parent asteroids. The ^{129}I-^{129}Xe ages of polymineralic samples (matrix, chondrules, CAIs, dark inclusions) from CV chondrites (mostly from Allende) recorded prolonged duration of their metasomatic alteration, > 11 Ma. However, the ^{129}I-carrier phases and corresponding closure temperatures of I-Xe isotope systematics of these samples are not well-constrained yet. The ^{53}Mn-^{53}Cr and ^{129}I-^{129}Xe chronologies of metasomatic alteration and thermal evolution models of asteroids heated by ^{26}Al decay suggest that the CV, CO, and ordinary chondrite parent bodies accreted 2.4–2.7 Ma, 2.1–2.4 Ma, and 1.6–1.8 Ma after CV CAIs, respectively.

5. The ^{26}Al-^{26}Mg systematics of secondary minerals having high Al/Mg ratio (anorthite, grossular, nepheline, sodalite) places only lower limits on their formation ages (> 3 Ma after formation of CAIs). The reported large excesses of radiogenic ^{26}Mg in some secondary minerals were probably inherited from the primary minerals after nearly complete decay of ^{26}Al and, thus, provide no chronological information. High precision Al-Mg isotope measurements of cogenetic secondary minerals (mineral parageneses) showing a large range of Al/Mg ratio (e.g., grossular, anorthite, monticellite, forsterite, corundum) can be used to construct internal Al-Mg isochrons and potentially date their formation.

6. The ^{26}Al-^{26}Mg systematics of chondrule glasses can be disturbed in metasomatically altered chondrites even of low petrologic type (3.0–3.2) diminishing their chronological significance. Although chemically primitive chondrule plagioclase in some type 3.0–3.1 chondrites experienced aqueous/metasomatic alteration resulting in formation of phyllosilicates, there is no clear evidence that this process affected its ^{26}Al-^{26}Mg systematics. Therefore, the observed ranges of (^{26}Al/^{27}Al)$_0$ in plagioclase-bearing chondrules from type 3.0–3.1 chondrites are probably chronologically meaningful (but see Piralla et al. 2023). In chondrites of higher petrologic type, the ^{26}Al-^{26}Mg systematics of plagioclase in chondrules and refractory inclusions, and even of melilite in CAIs can be disturbed and should be considered with caution in chronological interpretations.

7. The Δ^{17}O values of metasomatically formed minerals in UOCs, COs, CVs and CKs range from $\sim$ +5 to $\sim$ −4‰ suggesting differences in Δ^{17}O of aqueous fluids and, possibly, water ices that accreted into their parent asteroids. Development of the matrix-matched mineral standards for SIMS measurements is required to constrain the δ^{18}O values of aqueous fluids, and, potentially, crystallization temperature of secondary mineral parageneses. The fact that the inferred Δ^{17}O values of aqueous fluids in the UOCs, COs, CVs, and CKs parent asteroids are distinctly different from the hypothesized Δ^{17}O of primordial water ices recorded by cosmic symplectites in Acfer 094 ($\sim$ +80‰) suggests either extensive thermal processing of water ices in the PPD accompanying by isotopic exchange with ^{16}O-enriched CO and anhydrous silicates or/and isotopic evolution of water with time during CO self-shielding in the outer disk.

8. Chondrule phenocrysts and plagioclase or glassy mesostasis as well as refractory inclusions in the least metamorphosed (type 3.00) chondrites have internally uniform Δ^{17}O values. Chondrules and refractory inclusions in UOCs, COs, CVs, and CKs of higher petrologic type (> 3.0) have internally heterogeneous Δ^{17}O: glassy mesostasis, plagioclase, melilite, grossite, krotite, Ti-rich pyroxenes, perovskite and Sc- and Zr-rich oxides and silicates are ^{16}O-depleted relative to olivine, low-Ca pyroxene, hibonite, spinel, and low-Ti pyroxenes which retained their initial O-isotope compositions. The Δ^{17}O values of the ^{16}O-depleted phases approach the inferred Δ^{17}O values of aqueous fluid suggesting that the observed O-isotope heterogeneity most likely have resulted from isotopic exchange with aqueous fluids during metasomatic alteration. Most of the ^{16}O-depleted minerals show no evidence for disturbance of their Al-Mg isotope systematics, suggesting that Mg self-diffusion coefficients in these minerals under $P-T-f_{O2}$ conditions of metasomatic alteration is much slower than oxygen self-diffusion. Therefore, CAIs with the canonical (^{26}Al/^{27}Al)$_0$ and internally heterogeneous O-isotope compositions do not necessarily provide evidence for significant fluctuations in O-isotope composition of nebular gas in the CAI-forming region. At the same time, isotopic exchange with nebular gas of variable O-isotope composition cannot be excluded. Experimentally defined oxygen self-diffusion in CAI- and chondrule-like minerals under the inferred $P-T-fO_2$-*water/rock* ratio conditions during metasomatic alteration are required to test both hypotheses.

9. Carbon-isotope compositions of carbonates and FeNi-carbides in UOCs, CVs, and COs can potentially provide constraints on the origin of water ices in their parent asteroids, but are not known yet. These phases can be analyzed *in situ* by SIMS.

10. The main process controlling the O-isotope composition of the CV and CO parent asteroids is related to isotopic exchange between ^{16}O-enriched anhydrous silicates and ^{17,18}O-enriched aqueous fluid. The initial O-isotope of water ice and its abundance in the CV and CO chondrite parent asteroids are poorly constrained: its Δ^{17}O value must have been equal or higher that the highest Δ^{17}O of aqueously formed minerals in CV and CO chondrites, i.e., > +3.2‰ and > −0.3‰, respectively. The calculated water/rock ratios in

CV and CO asteroids are 0.1–0.2 and 0.01–0.1 (Marrocchi et al. 2018); these calculations, however, are highly model-dependent.

Parent asteroids of ordinary chondrites appear to have accreted low abundance of water ice, possibly, because of Jupiter blocking the inward flow of ice (Morbidelli et al. 2016), but with higher $\Delta^{17}O$ value than that in CVs and COs, $\geq$ +5‰. These observations and the presence of D-rich IOM, including a D-rich thermally resistant component, and the D-rich and isotopically heterogeneous hydrated silicates measured in Semarkona (LL3.00) may indicate that (*i*) UOCs inherited a higher fraction of the primordial (molecular cloud) water ice and/or organics compared to CCs, and (*ii*) UOCs recorded lower degrees of re-equilibration between the primordial molecular cloud molecules and protosolar H_2 than most CCs. This interpretation appears to contradict the assumed formation locations of OCs and CCs in the inner and outer solar system (Warren 2011; Kruijer et al. 2017). Further attempts should be made to deconvolute the isotopic contribution of parent body processes from the hydrogen and oxygen isotopic signatures of the primary ice in chondrites, and to estimate the spatial/temporal evolution of water hydrogen- and oxygen-isotope compositions in the context of the recent disk dynamical evolution models (Desch et al. 2018; Burkhardt et al. 2019; Jacquet et al. 2019; Kleine et al. 2020).

Supplementary Information The online version contains supplementary material available at https://doi.org/10.1007/s11214-025-01135-z.

Funding Open Access funding enabled and organized by Projekt DEAL.

Declarations

Competing Interests The authors declare that they have no conflict of interest.

References

Aléon J (2010) Multiple origins of nitrogen isotopic anomalies in meteorites and comets. Astrophys J 722:1342–1351

Aléon J (2016) Oxygen isotopes in the early protoplanetary disk inferred from pyroxene in a classical type B CAI. Earth Planet Sci Lett 440:62–70

Aléon J (2018) Closed system oxygen isotope redistribution in igneous CAIs upon spinel dissolution. Earth Planet Sci Lett 482:324–333

Aléon J, El Goresy A, Zinner E (2007) Oxygen isotope heterogeneities in the earliest protosolar gas recorded in a meteoritic calcium–aluminium-rich inclusion. Earth Planet Sci Lett 263:114–127

Alexander CMO (1995) Trace element contents of chondrule rims and interchondrule matrix in ordinary chondrites. Geochim Cosmochim Acta 59:3247–3266

Alexander CMO (2005) Re-examining the role of chondrules in producing the elemental fractionations in chondrites. Meteorit Planet Sci 40:943–965

Alexander CMO (2019a) Quantitative models for the elemental and isotopic fractionations in the chondrites: the non-carbonaceous chondrites. Geochim Cosmochim Acta 254:246–276

Alexander CMO (2019b) Quantitative models for the elemental and isotopic fractionations in chondrites: the carbonaceous chondrites. Geochim Cosmochim Acta 254:277–309

Alexander CMO, Barber DJ, Hutchison R (1989) The microstructure of Semarkona and Bishunpur. Geochim Cosmochim Acta 53:3045–3057

Alexander CMO, Arden JW, Ash RD, Pillinger CT (1990) Presolar components in the ordinary chondrites. Earth Planet Sci Lett 99:220–229

Alexander CMO, Russell SS, Arden JW, Ash RD, Grady MM, Pillinger CT (1998) The origin of chondritic macromolecular organic matter: a carbon and nitrogen isotope study. Meteorit Planet Sci 33:603–622

Alexander CMO, Fogel ML, Yabuta H, Cody GD (2007) The origin and evolution of chondrites recorded in the elemental and isotopic compositions of their macromolecular organic matter. Geochim Cosmochim Acta 71:4380–4403

Alexander CMO, Newsome SD, Fogel ML, Nittler LR, Busemann H, Cody GD (2010) Deuterium enrichments in chondritic macromolecular material-implications for the origin and evolution of organics, water and asteroids. Geochim Cosmochim Acta 74:4417–4437

Alexander CMO, Fogel ML, Howard KT, Herd CDK, Nittler LR (2012) The provenances of asteroids, and their contributions to the volatile inventories of the terrestrial planets. Science 337:721–723

Alexander CMO, Cody GD, Kebukawa Y, Bowden R, Fogel ML, Kilcoyne ALD, Nittler LR, Herd CDK (2014) Elemental, isotopic, and structural changes in Tagish Lake insoluble organic matter produced by parent body processes. Meteorit Planet Sci 49:503–525

Alexander CMO, Bowden R, Fogel ML, Howard KT (2015) Carbonate abundances and isotopic compositions in chondrites. Meteorit Planet Sci 50:810–833

Alexander CMO, Nittler LR, Davidson J, Ciesla FJ (2017) Measuring the level of interstellar inheritance in the solar protoplanetary disk. Meteorit Planet Sci 52:1797–1821

Alexander CMO, Greenwood RC, Bowden R, Gibson JM, Howard KT, Franchi IA (2018a) A mutli-technique search for the most primitive CO chondrites. Geochim Cosmochim Acta 221:406–420

Alexander CMO, McKeegan KD, Altwegg K (2018b) Water reservoirs in small planetary bodies: meteorites, asteroids, and comets. Space Sci Rev 214:36

Alexander CMO, Wynn JC, Bowden R, Scott E (2022) Sulfur abundances and isotopic compositions in bulk carbonaceous chondrites and insoluble organic material: Clues to elemental and isotopic fractionations of volatil chalcophiles. Meteorit Planet Sci 57:334–351

Aponte JC, Abreu NM, Glavin DP, Dworkin JP, Elsila JE (2017) Distribution of aliphatic amines in CO, CV, and CK carbonaceous chondrites and relation to mineralogy and processing history. Meteorit Planet Sci 52:2632–2646

Aponte JC, Woodward HK, Abreu NM, Elsila JE, Dworkin JP (2019) Molecular distribution, ^{13}C-isotope, and enantiomeric compositions of carbonaceous chondrite monocarboxylic acids. Meteorit Planet Sci 54:415–430

Barosch J, Nittler LR, Wang J, Dobrică E, Brearley AJ, Hezel DC, Alexander CMO (2022) Presolar O- and C-anomalous grains in unequilibrated ordinary chondrite matrices. Geochim Cosmochim Acta 335:169–182

Beyssac O, Goffé B, Petitet JP, Froigneux E, Moreau M, Rouzaud JN (2003) On the characterization of disordered and heterogeneous carbonaceous materials by Raman spectroscopy. Spectrochim Acta, Part A, Mol Biomol Spectrosc 59:2267–2276

Bischoff A, Keil K (1983) Catalog of Al-rich chondrules, inclusions and fragments in ordinary chondrites. Special Publication No. 22, UNM, Institute of Meteoritics, Albuquerque, 1–33

Bischoff A, Keil K (1984) Al-rich objects in ordinary chondrites: related origin of carbonaceous and ordinary chondrites and their constituents. Geochim Cosmochim Acta 48:693–709

Bischoff A, Geiger T, Palme H, Spettel B, Schultz L, Scherer P, Schlüter J, Lkhamsuren J (1993) Mineralogy, chemistry, and noble gas contents of Adzhi-Bogdo – an LL3-6 chondritic breccia with L-chondritic and granitoidal clasts. Meteoritics 28:570–578

Bischoff A, Barrat J-A, Berndt J, Borovicka J, Burkhardt C, Busemann H, Hakenmüller J, Heinlein D, Hertzog J, Kaiser J, Maden C, Meier MMM, Morino P, Pack A, Patzek M, Reitze MP, Rüfenacht M, Schmitt-Kopplin P, Schönbächler M, Spurny P, Weber I, Wimmer K, Zikmund T (2019) The Renchen L5-6 chondrite breccia – the first confirmed meteorite fall from Baden-Württemberg (Germany). Geochem Chem Erde 79:125525

Boata G (1954) The isotopic composition of hydrogen and carbon in the carbonaceous chondrites. Geochim Cosmochim Acta 6:209–220

Bodénan J-D, Starkey NA, Russell SS, Wright IP, Franchi IA (2020) One of the earliest refractory inclusions and its implications for Solar System history. Geochim Cosmochim Acta 286:214–226

Bonal L, Quirico E, Bourot-Denise M, Montagnac G (2006) Determination of the petrologic type of CV3 chondrites by Raman spectroscopy of included organic matter. Geochim Cosmochim Acta 70:1849–1863

Bonal L, Bourot-Denise M, Quirico E, Montagnac G, Lewin E (2007) Organic matter and metamorphic history of CO chondrites. Geochim Cosmochim Acta 71:1605–1623

Bonal L, Quirico E, Flandinet L, Montagnac G (2016) Thermal history of type 3 chondrites from the Antarctic meteorite collection determined by Raman spectroscopy of their polyaromatic carbonaceous matter. Geochim Cosmochim Acta 189:312–337

Bonal L, Gattacceca J, Gerenne A, Eschrig J, Rochette P, Rugiu L (2020) Water and heat: new constraints on the evolution of the CV chondrite parent body. Geochim Cosmochim Acta 276:363–383

Bose M, Floss C, Stadermann FJ (2010) An investigation into the origin of Fe-rich presolar silicates in Acfer 094. Astrophys J 714:1624–1636

Bose M, Zega TJ, Williams P (2014) Assessment of alteration processes on circumstellar and interstellar grains in Queen Alexandra Range 97416. Earth Planet Sci Lett 399:128–138

Bradley JP (1994) Nanometer-scale mineralogy and petrography of fine-grained aggregates in anhydrous interplanetary dust particles. Geochim Cosmochim Acta 58:2123–2134

Bradley JP, Brownlee DE, Fraundorf P (1984) Carbon components in interplanetary dust: evidence for formation by heterogeneous catalysis. Science 223:56–58

Brearley AJ (1993) Matrix and fine-grained rims in the unequilibrated CO3 chondrite, ALHA77307: origins and evidence for diverse, primitive nebular dust components. Geochim Cosmochim Acta 57:1521–1550

Brearley AJ, Krot AN (2012) Metasomatism in the early Solar System: the record from chondritic meteorites. In: Metasomatism and the chemical transformation of rock. Lecture notes in Earth system sciences, pp 659–789

Brennecka GA, Wadhwa M (2012) Uranium isotope compositions of the basaltic angrite meteorites and the chronological implications for the early Solar System. Proc Natl Acad Sci USA 109:9299–9303

Bridges JC, Alexander CMO, Hutchison R, Franchi IA, Pillinger CT (1997) Sodium-, chlorine-rich mesostases in Chainpur (LL3) and Parnallee (LL3) chondrules. Meteorit Planet Sci 32:555–565

Burkhardt C, Dauphas N, Ulrik H, Bourdon B, Thorsten K (2019) Elemental and isotopic variability in Solar System materials by mixing and processing of primordial disk reservoirs. Geochim Cosmochim Acta 261:145–170

Burton AS, Elsila JE, Callahan MP, Martin MG, Glavin DP, Johnson NM, Dworkin JP (2012) A propensity for n-u-amino acids in thermally altered Antarctic meteorites. Meteorit Planet Sci 47:374–386

Buseck PR, Beyssac O (2014) From organic matter to graphite: graphitization. Elements 10:421–426

Busemann H, Young AF, Alexander CMO, Hoppe P, Mukhopadhyay S, Nittler LR (2006) Interstellar chemistry recorded in organic matter from primitive meteorites. Science 312:727–730

Busemann H, Alexander CMO, Nittler LR (2007) Characterization of insoluble organic matter in primitive meteorites by microRaman spectroscopy. Meteorit Planet Sci 42:1387–1416

Busemann H, Nguyen AN, Cody GD, Hoppe P, Kilcoyne ALD, Stroud RM, Zega TJ, Nittler LR (2009) Ultra-primitive interplanetary dust particles from the comet 26P/Grigg-Skjellerup dust stream collection. Earth Planet Sci Lett 288:44–57

Caffee MW, Hohenberg CM, Swindle TD, Hudson B (1982) I-Xe ages of individual Bjurböle chondrules. J Geophys Res, Solid Earth 87:A307–A317

Chakraborty S, Jackson TL, Ahmed M, Thiemens MH (2013) Sulfur isotopic fractionation in vacuum UV photodissociation of hydrogen sulfide and its potential relevance to meteorite analysis. Proc Natl Acad Sci USA 110:17650–17655

Chaumard N, Devouard B, Bouvier A, Wadhwa M (2014) Metamorphosed calcium-aluminum-rich inclusions in CK carbonaceous chondrites. Meteorit Planet Sci 49:419–452

Che S, Brearley AJ (2021) The formation and alteration history of a forsterite-bearing Type C CAI from Allende: evidence for a Type B CAI precursor, and implications for fluid-assisted metasomatism on the CV chondrite parent body. Geochim Cosmochim Acta 293:277–307

Chiodini G, Marini L (1998) Hydrothermal gas equilibria: the H_2O-H_2-CO_2-CO-CH_4 system. Geochim Cosmochim Acta 62:2673–2687

Choi B-G, Huss GR, Wasserburg GJ, Gallino R (1998a) Presolar corundum and spinel in ordinary chondrites: origins from AGB stars and a supernova. Science 282:1284

Choi B-G, McKeegan KD, Krot AN, Wasson JT (1998b) Extreme oxygen-isetope compositions in magnetite from unequilibrated ordinary chondrites. Nature 392:577–579

Christoffersen R, Buseck PR (1983) Epsilon carbide: a low-temperature component of interplanetary dust particles. Science 222:1327–1329

Clayton RN (2002) Solar System: self-shielding in the solar nebula. Nature 415:860–861

Clayton RN (2008) Oxygen isotopes in the early Solar System – a historical perspective. In: MacPherson GJ (ed) Oxygen in the Solar System. Rev. Mineral. Geochem., vol 68, pp 5–14

Clayton RN, Onuma N, Grossman L, Mayeda TK (1977) Distribution of the presolar component in Allende and other carbonaceous chondrites. Earth Planet Sci Lett 34:209–224

Cody GD, Alexander CMO, Yabuta H, Kilcoyne ALD, Araki T, Ade H, Dera P, Fogel M, Mysen BO (2008) Organic thermometry for chondritic parent bodies. Earth Planet Sci Lett 272:446–455

Costa F, Chakraborty S (2008) The effect of water on Si and O diffusion rates in olivine and implications for transport properties and processes in the upper mantle. Phys Earth Planet Inter 166:11–29
Crowther SA, Mohapatra RK, Turner G, Blagburn DJ, Kehm K, Gilmour JD (2008) Characteristics and applications of RELAX, an ultrasensitive resonance ionization mass spectrometer for xenon. J Anal At Spectrom 23:938–947
Davidson J, Busemann H, Franchi IA (2012) A NanoSIMS and Raman spectroscopic comparison of interplanetary dust particles from comet Grigg-Skjellerup and non-Grigg Skjellerup collections. Meteorit Planet Sci 47:1748–1771
Davidson J, Busemann H, Nittler LR, Alexander CMO, Orthous-Daunay F-R, Franchi IA, Hoppe P (2014) Abundances of presolar silicon carbide grains in primitive meteorites determined by NanoSIMS. Geochim Cosmochim Acta 139:248–266
Davidson J, Alexander CMO, Stroud RM, Busemann H, Nittler LR (2019) Mineralogy and petrology of dominion range 08006: a very primitive CO3 carbonaceous chondrite. Geochim Cosmochim Acta 265:259–278
Deloule E, Robert F (1995) Interstellar water in meteorites? Geochim Cosmochim Acta 59:4695–4706
Desch SJ, Kalyaan A, Alexander CMO (2018) The effect of Jupiter's formation on the distribution of refractory elements and inclusions in meteorites. Astrophys J 238:11
Dobrică E, Brearley AJ (2011) Earliest stages of metamorphism and aqueous alteration observed in the fine-grained materials of two unequilibrated ordinary chondrites. Lunar Planet Sci 42:2092
Dobrică E, Brearley AJ (2020) Amorphous silicates in the matrix of Semarkona: the first evidence for the localized preservation of pristine matrix materials in the most unequilibrated ordinary chondrites. Meteorit Planet Sci 54:1973–1989
Dobrică E, Brearley AJ (2021) Iron-rich olivine in the unequilibrated ordinary chondrite, MET 00526: earliest stages of formation. Meteorit Planet Sci 55:2652–2669
Dobrică E, Nuth JA, Brearley AJ (2022) Fayalite formation through hydrothermal experiments: insights into early fluid-assisted aqueous alteration processes on asteroids. Meteorit Planet Sci 57:381–391
Doyle PM, Jogo K, Nagashima K, Krot AN, Wakita S, Ciesla FJ, Hutcheon ID (2015) Early aqueous activity on the carbonaceous and ordinary chondrite parent asteroids recorded by secondary fayalite. Nat Commun 6:1–10
Doyle PM, Jogo K, Nagashima K, Huss GR, Krot AN (2016) Mn-Cr relative sensitivity factor in ferromagnesian olivines defined for Sims measurements with a Cameca ims-1280 ion microprobe: implications for dating secondary fayalite. Geochim Cosmochim Acta 174:102–121
Dyl KA, Bischoff A, Ziegler K, Young ED, Wimmer K, Bland PA (2012) Early Solar System hydrothermal activity in chondritic asteroids on 1-10-year timescales. Proc Natl Acad Sci 109:18306–18311
Ebert S, Nagashim K, Krot AN, Bischoff A (2020) Oxygen-isotope heterogeneity in the northwest Africa 3358 (H3.1) refractory inclusions – fluid-assisted isotopic exchange on the H-chondrite parent body. Geochim Cosmochim Acta 282:98–112
Ebert S, Nagashima K, Bischoff A, Berndt J, Krot AN (2022) Mineralogy, petrology, and oxygen isotopic compositions of aluminum-rich chondrules from unequilibrated ordinary and the Dar al Gani 083 (CO3.1) chondrite. Geochim Cosmochim Acta 336:448–468
Ebert S, Nagashima K, Krot AN, Patzek M, Bischoff A (2024) Oxygen isotopic variation in the CAI-forming region recorded by a single refractory inclusion from the CO3.1 carbonaceous chondrite Dar Al Gani 083. Astrophys J 966:10
Enokido Y, Nakamura T, Matsumoto M, Miyake A, Shibuya T, Park C, Zolensky M (2023) Mineralogical alteration of a type A CAI from Allende CV3 chondrite: formation of secondary dmisteinbergite and its phase transition to anorthite. Meteorit Planet Sci 58:405–420
Fagan TJ, Guan Y, MacPherson GJ (2007) Al-Mg isotopic evidence for episodic alteration of Ca-Al-rich inclusions from Allende. Meteorit Planet Sci 42:1221–1240
Farver JR (1989) Oxygen self-diffusion in diopside with applications to cooling rate determination. Earth Planet Sci Lett 92:322–330
Fayolle EC, Öberg KI, Cuppen HM, Visser R, Linnartz H (2012) Laboratory H_2O:CO_2 ice desorption: entrapment and its parameterization with an extended three-phase model. EAS Publ Ser 58:327–331
Fegley B Jr, Prinn RG (1989) Solar nebula chemistry: implications for volatiles in the Solar System. In: Weaver H, Danly L (eds) The formation and evolution of planetary systems. Cambridge University Press, Cambridge, pp 171–211
Feng L, El Goresy A, Zhang J, Hao J, Boyet M, Lin Y (2012) Excess ^{36}S in lawrencite and nitrogen isotopic compositions of sinoite from Almahata Sitta MS-17 EL3 chondrite fragment. Lunar Planet Sci 43:1766
Floss C, Haenecour P (2016) Presolar silicate grains: abundances, isotopic and elemental compositions, and the effects of secondary processing. Geochem J 50:3–15
Floss C, Stadermann FJ (2009a) High abundances of circumstellar and interstellar C-anomalous phases in the primitive CR3 chondrites QUE 99177 and MET 00426. Astrophys J 697:1242–1255

Floss C, Stadermann FJ (2009b) Auger Nanoprobe analysis of presolar ferromagnesian silicate grains from primitive CR chondrites QUE 99177 and MET 00426. Geochim Cosmochim Acta 73:2415–2440

Floss C, Stadermann FJ, Bradley JP, Dai ZR, Bajt S, Graham G, Lea AS (2006) Identification of isotopically primitive interplanetary dust particles: a NanoSIMS isotopic imaging study. Geochim Cosmochim Acta 70:2371–2399

Foustoukos DI, Alexander CMO, Cody GD (2021) H and N systematics in thermally altered chondritic insoluble organic matter: an experimental study. Geochim Cosmochim Acta 300:44–64

Fujiya W (2018) Oxygen isotopic ratios of primordial water in carbonaceous chondrites. Earth Planet Sci Lett 481:264–272

Fukuda K, Tenner TJ, Kimura M, Tomioka N, Siron G, Ushikubo T, Chaumard N, Hertwig AT, Kita NT (2022) A temporal shift of chondrule generation from the inner to outer Solar System inferred from oxygen isotopes and Al-Mg chronology of chondrules from primitive CM and CO chondrites. Geochim Cosmochim Acta 322:194–226

Ganino C, Libourel G (2017) Reduced and unstratified crust in CV chondrites parent body. Nat Commun 8:1–10

García-Toraño E, Altzitzoglou T, Auerbach P, Bé M-M, Bobin C, Cassette P, Chartier F, Dersch R, Fernández M, Isnard H, Kossert K, Lourenço V, Nähle O, Nonell A, Peyrés V, Pommé S, Rozkov A, Sánchez-Cabezudo A, Sochorová J (2018) The half-life of ^{129}I. Appl Radiat Isot 140:157–162

Gattacceca J, Bonal L, Sonzogni C, Longerey J (2020) CV chondrites: more than one parent body. Earth Planet Sci Lett 547:116467

Giletti BJ, Semet MP, Yund RA (1978) Studies in diffusion – III. Oxygen in feldspars: an ion microprobe determination. Geochim Cosmochim Acta 42:45–57

Gilmour JD, Ash RD, Hutchison R, Bridges RC, Lyon IC, Turner G (1995) Iodine-xenon studies of Bjurböle and Parnallee using RELAX. Meteorit Planet Sci 30:405–411

Gilmour JD, Crowther SA, Blagburn DJ, Turner G (2006) Recent and future developments of the RELAX mass spectrometer. Geochim Cosmochim Acta 70:A204

Gilmour JD, Whitby JA, Turner G, Bridges JC, Hutchison R (2010) The iodine-xenon system in clasts and chondrules from ordinary chondrites: implications for early Solar System chronology. Meteorit Planet Sci 35:445–455

Glavin DP, Kubny A, Jagoutz E, Lugmair GW (2004) Mn-Cr isotope systematics of the D'Orbigny angrite. Meteorit Planet Sci 39:693–700

Glavin DP, Alexander CMO, Aponte ACM, Dworkin JP, Elsila JE, Yabuta H (2018) The origin and evolution of organic matter in carbonaceous chondrites and links to their parent bodies. In: Primitive meteorites and asteroids. Elsevier, Amsterdam, pp 205–271

Göbel R, Begemann F, Ott U (1982) On neutron-induced and other noble gases in Allende inclusions. Geochim Cosmochim Acta 46:1777–1792

Grady MM, Wright IP, Carr RH, Poths J, Pillinger CT (1988) Differences in isotopic composition of carbonaceous components in enstatite chondrites. Earth Planet Sci Lett 87:293–302

Grady MM, Wright IP, Pillinger CT (1989) A preliminary investigation into the nature of carbonaceous material in ordinary chondrites. Meteoritics 24:147–154

Grant H, Tartèse R, Jones R, Piani L, Marrocchi Y, King A, Rigaudier T (2023) Bulk mineralogy, water abundance, and hydrogen isotope composition of unequilibrated ordinary chondrites. Meteorit Planet Sci 58:1365–1381

Greenwood RC, Franchi IA (2004) Alteration and metamorphism of CO3 chondrites: evidence from oxygen and carbon isotopes. Meteorit Planet Sci 39:1823–1838

Greenwood RC, Franchi IA, Kearsley AT, Alard O (2010) The relationship between CK and CV chondrites. Geochim Cosmochim Acta 74:1684–1705

Greshake A (1997) The primitive matrix components of the unique carbonaceous chondrite Acfer 094: a TEM study. Geochim Cosmochim Acta 61:437–452

Grimm RE, McSween HY Jr (1989) Water and the thermal evolution of carbonaceous chondrite parent bodies. Icarus 82:244–280

Grossman JN, Brearley AJ (2005) The onset of metamorphism in ordinary and carbonaceous chondrites. Meteorit Planet Sci 61:437–452

Grossman JN, Alexander CMO, Wang J, Brearley AJ (2000) Bleached chondrules: evidence for widespread aqueous processes on the parent asteroids of ordinary chondrites. Meteorit Planet Sci 35:467–486

Haenecour P, Floss C, Zega TJ, Croat TK, Wang A, Jolliff BL, Carpenter P (2018) Presolar silicates in the matrix and fine-grained rims around chondrules in primitive CO3.0 chondrites: evidence for pre-accretionary aqueous alteration of the rims in the solar nebula. Geochim Cosmochim Acta 221:379–405

Hahn JH, Zenobi R, Bada JL, Zare RN (1988) Application of two-step laser mass spectrometry to cosmogeochemistry: direct analysis of meteorites. Science 239:1523–1525

Han J, Nagashima K, Park C, Krot AN, Keller LP (2024) Grossite-bearing refractory inclusions from reduced CV chondrites: mineralogical and oxygen isotopic constraints on the parent body alteration history. Geochim Cosmochim Acta 385:100–117. https://doi.org/10.1016/j.gca.2024.09.001

Hashimoto A, Grossman L (1987) Alteration of Al-rich inclusions inside amoeboid olivine aggregates in the Allende meteorite. Geochim Cosmochim Acta 51:1685–1704

Hashimoto A, Nakano Y (2023) Sulfuric acid as a cryofluid and oxygen isotope reservoir of planetesimals. Icarus 398:115535

Hevey PJ, Sanders IS (2006) A model for planetesimal meltdown by 26Al and its implication for meteorite parent body. Meteorit Planet Sci 41:95–106

Hohenberg CM, Pravdivtseva OV (2004) I–Xe dating: from adolescence to maturity. Chem Erde 68:339–351

Hohenberg CM, Pravdivtseva OV (2008) I–Xe dating: from adolescence to maturity. Chem Erde 68:339–351

Hohenberg CM, Pravdivtseva OV, Meshik AP (2004) Trapped Xe and I-Xe ages in aqueously altered CV3 meteorites. Geochim Cosmochim Acta 68:4745–4763

Hohenberg CM, Pravdivtseva OV, Meshik AP (2004) Trapped Xe and I-Xe ages in aqueously altered CV3 meteorites. Geochim Cosmochim Acta 68:4745–4763

Hong Y, Fegley B Jr (1998) Experimental studies of magnetite formation in the solar nebula. Meteorit Planet Sci 33:1101–1112

Hoppe P, Leitner J, Kodolányi J (2015) New constraints on the abundances of silicate and oxide stardust from supernovae in the Acfer 094 meteorite. Astrophys J Lett 808:L9

Hsu W, Guan Y, Leshin L, Ushikubo T, Wasserburg GJ (2006) A let episode of irradiation in theearly Solar System: evidence from extinct ^{36}Cl and ^{26}Al in meteorites. Astrophys J 640:525–529

Hua X, Huss GR, Tachibana S, Sharp TG (2005) Oxygen, silicon, and Mn-Cr isotopes of fayalite in the Kaba oxidized CV3 chondrite: constraints for its formation history. Geochim Cosmochim Acta 69:1333–1348

Huss GR (1990) Ubiquitous interstellar diamond and SiC in primitive chondrites – abundances reflect metamorphism. Nature 347:159–162

Huss GR, Lewis RS (1995) Presolar diamond, SiC, and graphite in primitive chondrites: abundances as a function of meteorite class and petrologic type. Geochim Cosmochim Acta 59:115–160

Huss GR, Meshik AP, Smith JB, Hohenberg CM (2003) Presolar diamond, silicon carbide, and graphite in carbonaceous chondrites: implications for thermal processing in the solar nebula. Geochim Cosmochim Acta 67:4823–4848

Hutcheon ID, Krot AN, Keil K, Phinney DL, Scott ERD (1998) ^{53}Mn-^{53}Cr dating of fayalite formation in the CV3 chondrite Mokoia: evidence for asteroidal alteration. Science 282:1865–1867

Hutchison R, Alexander CMO, Barber DJ (1987) The Semarkona meteorite: first recorded occurrence of smectite in an ordinary chondrite, and its implications. Geochim Cosmochim Acta 51:1875–1882

Ichimura S, Seto Y, Tomeoka K (2017) Nepheline formation in chondrite parent bodies: verification through experiments. Geochim Cosmochim Acta 210:114–131

Ireland TR, Avila J, Greenwood RC, Hicks LJ, Bridges JC (2020) Oxygen isotopes and sampling of the Solar System. Space Sci Rev 216:25

Ishizaki L, Tachibana S, Okamoto T, Yamamoto D, Ida S (2023) Effective reaction temperatures of irreversible dust chemical reactions in a protoplanetary disk. Astrophys J 957:47

Ito M, Messenger S (2010) Thermal metamorphic history of a Ca, Al-rich inclusion constrained by high spatial resolution Mg isotopic measurements with NanoSIMS 50L. Meteorit Planet Sci 45:583–595

Ito M, Nagasawa H, Yurimoto H (2004) Oxygen isotopic Sims analysis in Allende CAI: details of the very early thermal history of the Solar System. Geochim Cosmochim Acta 68:2905–2923

Itoh D, Tomeoka K (1998) Na-bearing Ca-Al-rich inclusions in four CO3 chondrites, Kainsaz, Ornans, Lance, and Warrenton. In: 23rd Symposium on Antarctic Meteorites. NIPR, Tokyo, pp 42–44

Itoh S, Yurimoto H (2003) Contemporaneous formation of chondrules and refractory inclusions in the early Solar System. Nature 423:728–731

Itoh S, Kojima H, Yurimoto H (2004) Petrography and oxygen isotopic compositions in refractory inclusions from CO chondrites. Geochim Cosmochim Acta 68:183–194

Itoh S, Russel SS, Yurimoto H (2007) Oxygen and magnesium compositions of amoeboid olivine aggregates from Semarkona LL3.0 chondrite. Meteorit Planet Sci 42:1241–1247

Jacobsen B, Desch SJ (2023) The timing of ^{36}Cl production in the early Solar System and implications for gas dissipation in the protoplanetary disk. Ann Meteorit Soc Meet 86:6312

Jacobsen B, Yin Q-Z, Moyniera F, Amelin Y, Krot AN, Nagashima K, Hutcheon ID, Palme H (2008) ^{26}Al-^{26}Mg and ^{207}Pb-^{206}Pb systematics of Allende CAIs: canonical solar initial ^{26}Al/^{27}Al ratio reinstated. Earth Planet Sci Lett 272:353–364

Jacobsen B, Matzel J, Hutcheon ID, Krot AN, Nagashima K, Ishii HA, Ramon E, Meyer BS, Davis AM, Ciesla FJ, Yin Q-Z (2011) Formation of short-lived radionuclides during epoch of late-stage irradiation in the protoplanetary disk. Astrophys J 731:L28–L31

Jacquet E, Robert F (2013) Water transport in protoplanetary disks and the hydrogen isotopic composition of chondrites. Icarus 223:722–732

Jacquet E, Pignatale FC, Chaussidon M, Charnoz S (2019) Fingerprints of the protosolar cloud collapse in the Solar System. II. Nucleosynthetic anomalies in meteorites. Astrophys J 884:32

Jogo K, Nakamura T, Noguchi T, Zolotov MY (2009) Fayalite in the Vigarano CV3 carbonaceous chondrite: occurrences, formation age and conditions. Earth Planet Sci Lett 287:320–328

Jogo K, Nakamura T, Ito M, Wakita S, Zolotov MY, Messenger SR (2017) Mn–Cr ages and formation conditions of fayalite in CV3 carbonaceous chondrites: constraints on the accretion ages of chondritic asteroids. Geochim Cosmochim Acta 199:58–74

Johnson CA, Prinz M, Weisberg MK, Clayton RN, Mayeda TK (1990) Dark inclusions in Allende, Leoville, and Vigarano: evidence for nebular oxidation of CV3 constituents. Geochim Cosmochim Acta 54:819–830

Jones CL, Brearley AJ (2006) Experimental aqueous alteration of the Allende meteorite under oxidizing conditions: constraints on asteroidal alteration. Geochim Cosmochim Acta 70:1040–1058

Kanawati RB, Hertkorn N, Yin Q-Z, Moritz F, Harir M, Lucio M, Michalke B, Wimpenny J, Shilobreeva S, Bronsky B, Saraykin V, Gabelica Z, Gougeon RD, Quirico E, Ralew S, Jakubowski T, Haack H, Gonsior M, Jenniskens P, Hinman NW, Schmitt-Kopplin P (2017) Previously unknown class of metalorganic compounds revealed in meteorites. Proc Natl Acad Sci USA 114:2819–2824

Kawasaki N, Sakamoto N, Yurimoto H (2012) Oxygen isotopic and chemical zoning of melilite crystals in a type A Ca–Al-rich inclusion of Efremovka CV3 chondrite. Earth Planet Sci Lett 47:2084–2093

Kawasaki N, Kato C, Itoh S, Wakaki S, Ito M, Yurimoto H (2015) ^{26}Al–^{26}Mg chronology and oxygen isotope distributions of multiple melting for a type C CAI from Allende. Geochim Cosmochim Acta 169:99–114

Kawasaki N, Simon SB, Grossman L, Sakamoto N, Yurimoto H (2018) Crystal growth and disequilibrium distribution of oxygen isotopes in an igneous Ca-Al-rich inclusion from the Allende carbonaceous chondrite. Geochim Cosmochim Acta 221:318–341

Keil K, Zucolotto ME, Krot AN, et al (2015) The Vicência meteorite fall: a new unshocked (S1) weakly metamorphosed (3.2) LL chondrite. Meteorit Planet Sci 50:1089–1111

Keller LP (1998) A transmission electron microscope study of iron-nickel carbides in matrix of the Semarkona unequilibrated ordinary chondrite. Meteorit Planet Sci 33:913–919

Keller LP, Buseck PR (1990) Aqueous alteration in the Kaba CV3 carbonaceous chondrite. Geochim Cosmochim Acta 54:2113–2120

Keller LP, Buseck PR (1991) Calcic micas in the Allende meteorite: evidence for hydration reactions in the early solar nebula. Science 252:946–949

Kerridge JF (1985) Carbon, hydrogen and nitrogen in carbonaceous chondrites: abundances and isotopic compositions in bulk samples. Geochim Cosmochim Acta 49:1707–1714

Kimura M, Grossman JN, Weisberg MK (2007) Fe-Ni metal in primitive chondrites: indicators of classification and metamorphic conditions for ordinary and CO chondrites. Meteorit Planet Sci 43:1161–1177

Kirschbaum C (1988) Carrier phases for iodine in the Allende meteorite and their associated ^{129}Xe/^{129}I ratios; a laser microprobe study. Geochim Cosmochim Acta 52:679–699

Kita NT, Nagahara H, Tachibana S, Tomomura S, Spicuzza MJ, Fournelle JH, Valley JW (2010) High precision Sims oxygen three isotope study of chondrules in LL3 chondrites: role of ambient gas during chondrule formation. Geochim Cosmochim Acta 74:6610–6635

Kleine T, Budde G, Burkhardt C, Kruijer TS, Worsham EA, Morbidelli A, Nimmo F (2020) The non-carbonaceous – carbonaceous meteorite dichotomy. Space Sci Rev 216:55

Kojima T, Tomeoka K (1999) Experimental hydrothermal alteration of the Allende CV3 chondrite under acidic and neutral conditions. Meteorit Planet Sci 34:A67

Kojima T, Tomeoka K (2000) Transmission electron micrscopy of phyllosilicates in experimentally altered Allende: comparison to naturally altered CV3 chondrites. Meteorit Planet Sci 35:A90

Kojima T, Yada S, Tomeoka K (1995) Ca-Al-rich inclusions in three Antarctic CO3 chondrites, Yamato-81020 Yamato-82050 and Yamato-790992: record of low temperature alteration. Antarct Meteor Res 8:79

Kolodny Y, Kerridge JF, Kaplan IR (1980) Deuterium in carbonaceous chondrites. Earth Planet Sci Lett 46:149–158

Kööp L, Nakashima D, Heck PR, Kita NT, Tenner TJ, Krot AN, Nagashima K, Park C, Davis AM (2018) A multielement isotopic study of refractory FUN and F CAIs: mass-dependent and mass-independent isotope effects. Geochim Cosmochim Acta 221:296–317

Krämer Ruggiu L, Devouard B, Gattacceca J, Bonal L, Leroux H, Eschrig J, Borschneck D, King AJ, Beck P, Marrocchi Y, Debaille V, Hanna RD, Grauby O (2022) Detection of incipient aqueous alteration in carbonaceous chondrites. Geochim Cosmochim Acta 336:308–331

Krot AN (2019) Refractory inclusions in carbonaceous chondrites: records of early Solar System processes. Meteorit Planet Sci 54:1647–1692

Krot AN, Todd CS (1998) Metal-carbide-magnetite-fayalite association in a Bali-like clast in the reduced CV3 chondrite breccia Vigarano. Meteorit Planet Sci 34:A88–A89

Krot AN, Scott ERD, Zolensky ME (1995) Mineralogic and chemical variations among CV3 chondrites and their components: nebular and asteroidal processing. Meteoritics 30:748–775

Krot AN, Zolensky ME, Wasson JT, Scott ERD, Keil K, Ohsumi K (1997) Carbide-magnetite-bearing type 3 ordinary chondrites. Geochim Cosmochim Acta 61:219–237

Krot AN, Zolensky ME, Keil K, Scott ERD, Nakamura K (1998) Secondary Ca-Fe-rich minerals in the Bali-like and Allende-like oxidized CV3 chondrites and Allende dark inclusions. Meteorit Planet Sci 33:623–645

Krot AN, Brearley AJ, Ulyanov AA, Biryukov VV, Swindle TD, Keil K, Mittlefehldt DW, Scott ERD, Clayton RN, Mayeda TK (1999) Mineralogy, petrography, bulk chemical, iodine-xenon, and oxygen-isotopic composition of dark inclusions in the reduced CV3 chondrite Efremovka. Meteorit Planet Sci 34:67–89

Krot AN, Brearley AJ, Petaev MI, Kallemeyn GW, Sears DWG, Benoit PH, Hutcheon ID, Zolensky ME, Keil K (2000) Evidence for in situ growth of fayalite and hedenbergite in MacAlpine Hills 88107, ungrouped carbonaceous chondrite related to CM-CO clan. Meteorit Planet Sci 35:1365–1387

Krot AN, Petaev MI, Meibom A, Keil K (2001) In situ growth of Ca-rich rims around Allende dark inclusions. Cosmochem Iss Geochem Int 36:351–368

Krot AN, Yurimoto H, Hutcheon ID, MacPherson GJ (2005) Relative chronology of CAI and chondrule formation: evidence from chondrule-bearing igneous CAIs. Nature 434:998–1001

Krot AN, Hutcheon ID, Brearley AJ, Pravdivtseva OV, Petaev MI, Hohenberg CM (2006) Timescales for secondary alteration of chondritic meteorites. In: Lauretta D, McSween H (eds) Meteorites and the early Solar System II. University of Arizona Press, Tucson, pp 525–555

Krot AN, Nagashima K, Hutcheon ID, Ishii HA, Yin Q-Z, Davis AM, Simon SB (2010) Mineralogy, petrography, oxygen and magnesium isotopic compositions and formation age of grossular-bearing assemblages in the Allende CAIs. Lunar Planet Sci 41:1406

Krot AN, Keil K, Goodrich C, Weisberg MK, Scott ERD (2014a) Classification of meteorites. In: Davis AM (ed) Meteorites and cosmochemical processes. Treatise on geochemistry (eds. H. D. Holland and K. K. Turekian), vol 1. Elsevier, Oxford, pp 1–63

Krot AN, Nagashima K, Wasserburg GJ, Huss GR, Papanastassiou D, Davis AM, Hutcheon ID, Bizzarro M (2014b) Calcium-aluminum-rich inclusions with fractionation and unknown nuclear effects (FUN CAIs): I. Mineralogy, petrology, and oxygen-isotope compositions. Geochim Cosmochim Acta 145:206–247

Krot AN, Nagashima K, Alexander CMO, Ciesla FJ, Fujiya W, Bonal L (2015) Sources of water and aqueous activity on the chondrite parent asteroids. In: Asteroids IV. University of Arizona Press, Tucson, pp 21–32

Krot AN, Nagashima K, Fintor K, Pál-Molnár E (2018) Evidence for oxygen isotopic exchange in chondrules from Kaba (CV3.1) carbonaceous chondrite during aqueous fluid – rock interaction on the CV parent asteroid. Acta Geogr Geol Meteorol Debr: Geol Geom Phys Geogr Ser 13:137–149

Krot AN, Ma C, Nagashima K, Davis AM, Beckett JR, Simon SB, Komatsu M, Fagan TJ, Brenker F, Ivanova MA, Bischoff A (2019b) Mineralogy, petrography, and oxygen isotopic compositions of ultrarefractory inclusions from carbonaceous chondrites. Geochemistry 79:125519

Krot AN, Nagashima K, Fintor K, Pál-Molnár E (2019c) Evidence for oxygen-isotope exchange in refractory inclusions from Kaba (CV3.1) carbonaceous chondrite during fluid-rock interaction on the CV parent asteroid. Geochim Cosmochim Acta 246:419–435

Krot AN, Nagashima K, Simon SB, Ma C, Connolly HC Jr, Huss GR, Davis AM, Bizzarro M (2019a) Mineralogy, petrography, and oxygen and aluminum-magnesium isotope systematics of grossite-bearing refractory inclusions. Geochemistry 79:125529

Krot AN, Nagashima K, Lyons JR, Lee J-E, Bizzarro M (2020) Oxygen isotopic heterogeneity in the early Solar System inherited from the protosolar molecular cloud. Sci Adv 6:eaay2724

Krot AN, Petaev MI, Nagashima K (2021) Infiltration metasomatism of the Allende coarse-grained calcium-aluminum-rich inclusions. Prog Earth Planet Sci 8:61

Krot AN, Doyle P, Nagashima K, Dobrică E (2022a) Mineralogy, petrology and oxygen-isotope compositions of magnetite±fayalite assemblages in CO3, CV3 and LL3 chondrites. Meteorit Planet Sci 57:392–428

Krot AN, Nagashima K, MacPherson GJ, Ulyanov AA (2022b) On the nature of oxygen-isotope heterogeneity of igneous calcium-aluminum-rich inclusions in CV carbonaceous chondrites. Geochim Cosmochim Acta 332:327–354

Krot AN, Nagashima K, Dunn TL, Petaev MI, Ma C (2023) Mineralogy, petrography, oxygen- and aluminum-magnesium isotope systematics of igneous CAIs from CK3 chondrites. In: 86th Meteoritic Soc Meet, vol 6116

Krot AN, Dunn TL, Petaev MI, Ma C, Nagashima K, Zipfel J (2024) Metasomatic alteration of coarse-grained igneous calcium-aluminum-rich inclusions from CK3 carbonaceous chondrites. Meteorit Planet Sci 59:809–835

Kruijer TS, Burkhardt C, Budde G, Kleine T (2017) Age of Jupiter inferred from the distinct genetics and formation times of meteorites. Proc Natl Acad Sci USA 114:6712–6716

Kurat G, Palme H, Brandstätter F, Huth J (1989) Allende xenolith AF: undisturbed record of condensation and aggregation of matter in the solar nebula. Z Naturforsch 44a:988–1004

LaTourette T, Wasserburg GJ (1998) Mg diffusion in anorthite: implications for the formation of early Solar System planetesimals. Earth Planet Sci Lett 158:91–108

Le Guillou C, Rouzaud JN, Bonal L, Quirico E, Derenne S, Remusat L (2012) High resolution TEM of chondritic carbonaceous matter: metamorphic evolution and heterogeneity. Meteorit Planet Sci 47:345–362

Lee MR, Alexander CMO, Bischoff A et al (2025) Low-temperature aqueous alteration of chondrites. Space Sci Rev 221. https://doi.org/10.1007/s11214-024-01132-8

Leitner J, Vollmer C, Hoppe P, Zipfel J (2012) Characterization of Presolar Material in the CR Chondrite Northwest Africa 852. Astrophys J 745:38–53

Leitner J, Vollmer C, Floss C, Zipfel J, Hoppe P (2016) Ancient stardust in fine-grained chondrule dust rims from carbonaceous chondrites. Earth Planet Sci Lett 434:117–128

Lewis AL, Jones RH (2019) Primary feldspar in the Semarkona LL3.00 chondrite: constraints on chondrule formation and secondary alteration. Meteorit Planet Sci 54:72–89

Lewis RS, Ming T, Wacker JF, Anders E, Steel E (1987) Interstellar diamonds in meteorites. Nature 326:160–162

Lewis AL, Jones RH, Brearley AJ (2022) Plagioclase alteration and equilibration in ordinary chondrites: metasomatism during thermal metamorphism. Geochim Cosmochim Acta 316:201–229

Leya I, Masarik J, Lin Y (2018) Alteration of CAIs as recorded by $^{36}S/^{34}S$ as a function of $^{35}Cl/^{34}S$. Meteorit Planet Sci 53:1252–1266

Libourel G, Krot AN, Tissandier L (2006) Role of gas-melt interaction during chondrule formation. Earth Planet Sci Lett 251:232–240

Lin Y, Guan Y, Leshin LA, Ouyang Z, Wang D (2005) Short-lived chlorine-36 in a Ca- and Al-rich inclusion from the Ningqiang carbonaceous chondrite. Proc Natl Acad Sci 102:1306–1311

Llorca J, Casanova I (1998) Formation of carbides and hydrocarbons in chondritic interplanetary dust particles: a laboratory study. Meteorit Planet Sci 33:243–251

Lyons JR, Young ED (2005) O self-shielding as the origin of oxygen isotope anomalies in the early solar nebula. Nature 435:317–320

Lyons JR, Bergin EA, Ciesla FJ, Davis AM, Desch SJ, Hashizume K, Lee J-E (2009) Timescales for the evolution of oxygen isotope compositions in the solar nebula. Geochim Cosmochim Acta 73:4998–5017

MacPherson GJ (2014) Calcium-aluminum-rich inclusions in chondritic meteorites. In: Davis AM (ed) Meteorites, comets and planets. Treatise on geochemistry (eds. H. D. Holland and K. K. Turekian), vol 1. Elsevier, Oxford, pp 139–179

MacPherson GJ, Davis AM, Zinner EK (1995) The distribution of ^{26}Al in the early Solar System – a reappraisal. Meteoritics 30:365–386

MacPherson GJ, Kita NT, Ushikubo T, Bullock ES, Davis AM (2012) Well-resolved variations in the formation ages for Ca-Al-rich inclusions in the early Solar System. Earth Planet Sci Lett 331:43–54

MacPherson GJ, Nagashima K, Krot AN, Doyle PM, Ivanova MA (2017) ^{53}Mn-^{53}Cr chronology of Ca-Fe silicates in CV3 chondrites. Geochim Cosmochim Acta 201:260–274

MacPherson GJ, Krot AN, Kita NT, Bullock ES, Nagashima K, Ushikubo T, Ivanova MA (2022) The formation of type B CAIs: evolution from type A CAIs. Geochim Cosmochim Acta 321:343–374

MacPherson GJ, Nagashima K, Krot AN, Kuehner SM, Irving AJ, Ziegler K, Mallozzi L, Corrigan C, Pitt D (2023) Northwest Africa 8418: the first CV4 chondrite. Meteorit Planet Sci 58:135–157

Marrocchi Y, Chaussidon M, Piani L, Libourel G (2016) Early scattering of the solar protoplanetary disk recorded in meteoritic chondrules. Sci Adv 2:11601001

Marrocchi Y, Bekaert DV, Piani L (2018) Origin and abundance of water in carbonaceous asteroids. Earth Planet Sci Lett 482:23–32

Marrocchi Y, Bonal L, Gattacceca J, Piani L, Beck P, Greenwood R, Eschrig J, Basque A, Nuccio PM, Martin FF (2020) The Piancaldoli meteorite: a forgotten primitive LL3.10 ordinary chondrite. Meteorit Planet Sci 55:1924–1935

Matsumoto M, Tsuchiyama A, Nakato A, Matsuno J, Miyake A, Kataoka A, Ito M, Tomioka N, Kodama Y, Uesugi K, Takeuchi A, Nakano T, Vaccaro E (2019) Discovery of fossil asteroidal ice in primitive meteorite Acfer 094. Sci Adv 5:eaax5078

Matsumoto M, Tsuchiyama A, Miyake A, Ito M, Matsuno J, Uesugi K, Takeuchi A, Kodama Y, Yasutake M, Vaccaro E (2022) Three-dimensional microstructure and mineralogy of a cosmic symplectite in the Acfer 094 carbonaceous chondrite: implication for its origin. Geochim Cosmochim Acta 323:220–241

McGuire AV, Hashimoto A (1989) Origin of zoned fine-grained inclusions in the Allende meteorite. Geochim Cosmochim Acta 53:1123–1133

McKeegan KD, Leshin LA, Russel SS, MacPherson GJ (1998) Oxygen isotopic abundances in calcium-aluminum-rich inclusion from ordinary chondrites: implications for nebular heterogeneity. Science 280:414–418

McKeegan KD, Kallio APA, Heber VS, Jarzebinski G, Mao PH, Coath CD, Kunihiro T, Wiens RC, Nordholt JE, Moses RW Jr, Reisenfeld DB, Jurewicz AJG, Burnett DS (2011) The oxygen isotopic composition of the Sun inferred from captured solar wind. Science 332:1528–1532

McKibbin SJ, Ireland TR, Amelin Y, Holden P (2015) Mn–Cr dating of Fe- and Ca-rich olivine from 'quenched' and 'plutonic' angrite meteorites using secondary ion mass spectrometry. Geochim Cosmochim Acta 157:13–27

McNaughton NJ, Borthwick J, Fallick AE, Pillinger CT (1981) Deuterium/hydrogen ratios in unequilibrated ordinary chondrites. Nature 294:639–641

McNaughton NJ, Hinton RW, Pillinger CT, Fallick AE (1982) D/H ratios of some ordinary and carbonaceous chondrites. Meteoritics 17:252

Mendybaev RA, Richter FM, Georg RB, Janney PE, Spicuzza MJ, Davis AM, Valley JW (2013) Experimental evaporation of Mg- and Si-rich melts: implications for the origin and evolution of FUN CAIs. Geochim Cosmochim Acta 123:368–384

Mendybaev RA, Williams CD, Spicuzza MJ, Richter FM, Valley JW, Fedkin AY, Wadhwa M (2017) Thermal and chemical evolution of early Solar System solids as seen by Allende FUN CAI CMS-1: part II – laboratory evaporation of potential CMS-1 precursor. Geochim Cosmochim Acta 201:49–64

Michel-Levy MC, Lautie A (1981) Microanalysis by Raman spectroscopy of carbon in the Tieschitz chondrite. Nature 292:321–322

Morbidelli A, Bitsch B, Crida A, Gounelle M, Guillot T, Jacobson S, Johansen A, Lambrechts N, Lega E (2016) Fossilized condensation lines in the Solar System protoplanetary disk. Icaris 267:368–376

Murty SVS, Goswami JN, Shukolyukov YA (1997) Excess ^{36}Ar in the Efremovka meteorite: a strong hint for the presence of ^{36}Cl in the early Solar System. Astrophys J 475:L65–L68

Nagashima K, Krot AN, Yurimoto H (2004) Stardust from primitive meteorites. Nature 428:921–924

Nagashima K, Huss GR, Krot AN, Yurimoto H (2011) Disturbance of magnesium isotopes in anorthite from an Allende CAI inferred from magnesium isotope mapping with isotope microscope. Lunar Planet Sci 42:2447

Nagashima K, Krot AN, Komatsu M (2017) ^{26}Al-^{26}Mg systematics in chondrules from Kaba and Yamato 980145 CV3 chondrites. Geochim Cosmochim Acta 201:303–319

Nagashima K, Kita NT, Luu T-H (2018) ^{26}Al-^{26}Mg systematics of chondrules. In: Russell SS, Connolly HC, Krot AN (eds) Chondrules: records of protoplanetary disk processes. Cambridge University Press, Cambridge, pp 247–275

Nakashima D, Ott U, Hoppe P, El Goresy A (2008) Search for extinct ^{36}Cl: vigarano CAIs, the Pink Angel from Allende, and a Ningqiang chondrule. Geochim Cosmochim Acta 72:6141–6153

Nakashima D, Ott U, Hoppe P (2010) Search for extinct chlorine-36: a halite grain from the Zag meteorite. Ann Meteorit Soc Meet 73:5289

Nakasyo E, Maruoka T, Matsumoto T, Matsuda J-I (2000) A laboratory experiment on the influence of aqueous alteration on noble gas compositions in the Allende meteorite. Antarct Meteor Res 13:135

Newton J, Bischoff A, Arden JW, Franchi IA, Geiger T, Pillinger CT (1995) Acfer 094, a uniquely primitive carbonaceous chondrite from the Sahara. Meteoritics 30:47–56

Nguyen AN, Stadermann FJ, Zinner E, Stroud RM, Alexander CMO, Nittler LR (2007) Characterization of presolar silicate and oxide grains in primitive carbonaceous chondrites. Astrophys J 656:1223–1240

Nguyen AN, Nittler LR, Stadermann FJ, Stroud RM, Alexander CMO (2010) Coordinated analyses of presolar grains in the Allan Hills 77307 and Queen Elizabeth Range 99177 meteorites. Astrophys J 719:166–189

Nittler LR, Ciesla FJ (2016) Astrophysics and extraterrestrial materials. Annu Rev Astron Astrophys 54:53–93

Nittler LR, Alexander CMO, Gao X, Walker RM, Zinner E (1997) Stellar sapphires: the properties and origins of presolar Al_2O_3 in meteorites. Astrophys J 483:479–495

Nittler LR, Alexander CMO, Gallino R, Hoppe P, Nguyen AN, Stadermann FJ, Zinner EK (2008) Aluminum-, calcium- and titanium-rich oxide stardust in ordinary chondrite meteorites. Astrophys J 682:1450–1478

Nittler LR, Alexander CMO, Davidson J, Riebe MEI, Stroud RM, Wang J (2018) High abundances of presolar grains and ^{15}N-rich organic matter in CO3.0 chondrite dominion range 08006. Geochim Cosmochim Acta 226:107–131

Nittler LR, Alexander CMO, Patzer A, Verdier-Paoletti MJ (2021) Presolar stardust in highly pristine CM chondrites Asuka 12169 and Asuka 12236. Meteorit Planet Sci 56:260–276

Nomura K, Miyamoto M (1998) Hydrothermal experiments on alteration of Ca–Al-rich inclusions (CAIs) in carbonaceous chondrites: implication for aqueous alteration in parent asteroids. Geochim Cosmochim Acta 62:3575–3588
Oba Y, Naraoka H (2009) Elemental and isotope behavior of macromolecular organic matter from CM chondrites during hydrous pyrolysis. Meteorit Planet Sci 44:943–953
Ohnishi I, Tomeoka K (2007) Hydrothermal alteration experiments of enstatite: implications for aqueous alteration of carbonaceous chondrites. Meteorit Planet Sci 42:49–61
Okumura F, Mimura K (2011) Gradual and stepwise pyrolyses of insoluble organic matter from the Murchison meteorite revealing chemical structure and isotopic distribution. Geochim Cosmochim Acta 75:7063–7080
Petaev MI, Mironenko MV (1997) Thermodynamic modeling of aqueous alteration in CV chondrites. In: Zolensky ME, Krot AN, Scott ERD (eds) Workshop on parent-body and nebular modification of chondritic materials. LPI Technical Report 97-02, 49–50
Piani L, Marrocchi Y (2018) Hydrogen isotopic composition of water in CV-type carbonaceous chondrites. Earth Planet Sci Lett 504:64–71
Piani L, Robert F, Remusat L (2015) Micron-scale D/H heterogeneity in chondrite matrices: a signature of the pristine Solar System water? Earth Planet Sci Lett 415:154–164
Piani L, Marrocchi Y, Vacher LG, Yurimoto H, Bizzarro M (2021) Origin of hydrogen isotopic variations in chondritic water and organics. Earth Planet Sci Lett 567:117008
Piralla M, Villneuve J, Schnuriger N, Bekaert DV, Marrocchi Y (2023) A unified chronology of dust formation in the early Solar System. Icarus 394:115427
Plagge M, Ott U, Hoppe P (2006) Search for extinct ^{36}Cl in an Allende CAI. Lunar Planet Sci 37:1287
Pravdivtseva OV, Amelin Y, Hohenberg CM, Meshik AP (2002) I–Xe and Pb–Pb ages of Richardton chondrules. Geochim Cosmochim Acta 66:A614
Pravdivtseva OV, Krot AN, Hohenberg CM, Meshik AP, Weisberg MK, Keil K (2003) The I-Xe record of alteration in the Allende CV chondrite. Geochim Cosmochim Acta 67:5011–5026
Pravdivtseva OV, Amelin Y, Meshik AP, Hohenberg CM (2004) I–Xe and Pb–Pb ages of individulal Elenovka (L5) chondrules. Geochim Cosmochim Acta 68:A760
Pravdivtseva O, Meshik A, Hohenberg CM, Petaev MI (2009) Interpreting the I-Xe system in individual silicate grains from Toluca IAB. Meteorit Planet Sci 44:1787–1796
Pravdivtseva O, Meshik A, Hohenberg CM, Kurat G (2013) I-Xe ages of Campo del Cielo silicates as a record of the complex early history of the IAB parent body. Meteorit Planet Sci 48:2480–2490
Pravdivtseva OV, Krot AN, Hohenberg CM (2017b) I–Xe dating of aqueous alteration in the CI chondrite Orgueil: I. Magnetite and ferromagnetic separates. Geochim Cosmochim Acta 227:38–47
Pravdivtseva OV, Meshik AP, Hohenberg CM, Krot AN (2017a) I–Xe systematics of the impact plume produced chondrules from the CB carbonaceous chondrites: implications for the half-life value of ^{129}I and absolute age normalization of ^{129}I–^{129}Xe chronometer. Geochim Cosmochim Acta 201:320–330
Quirico E, Raynal PI, Bourot-Denise M (2003) Metamorphic grade of organic matter in six unequilibrated ordinary chondrites. Meteorit Planet Sci 38:795–811
Quirico E, Montagnac G, Rouzaud JN, Bonal L, Bourot-Denise M, Duber S, Reynard B (2009) Precursor and metamorphic condition effects on Raman spectra of poorly ordered carbonaceous matter in chondrites and coals. Earth Planet Sci Lett 287:185–193
Quirico E, Orthous-Daunay FR, Beck P, Bonal L, Brunetto R, Dartois E, Pino T, Montagnac G, Rouzard J-N, Engrand C, Duprat J (2014) Origin of insoluble organic matter in type 1 and 2 chondrites: new clues, new questions. Geochim Cosmochim Acta 136:80–99
Rai VK, Murty SVS, Ott U (2003) Noble gases in ureilites: cosmogenic, radiogenic, and trapped components. Geochim Cosmochim Acta 67:4435–4456
Remusat L, Palhol F, Robert F, Derenne S, France-Lanord C (2006) Enrichment of deuterium in insoluble organic matter from primitive meteorites: a Solar System origin? Earth Planet Sci Lett 243:15–25
Remusat L, Le Guillou C, Rouzaud JN, Binet L, Derenne S, Robert F (2008) Molecular study of insoluble organic matter in Kainsaz CO3 carbonaceous chondrite: comparison with CI and CM IOM. Meteorit Planet Sci 43:1099–1111
Remusat L, Piani L, Bernard S (2016) Thermal recalcitrance of the organic D-rich component of ordinary chondrites. Earth Planet Sci Lett 435:36–44
Remusat L, Bonnet J-Y, Bernard S, Buch A, Quirico E (2019) Molecular and isotopic behavior of insoluble organic matter of the Orgueil meteorite upon heating. Geochim Cosmochim Acta 263:235–249
Riebe MEI, Foustoukos DI, Alexander CMO, Steele A, Cody GD, Mysen BO, Nittler LR (2020) The effects of atmospheric entry heating on organic matter in interplanetary dust particles and micrometeorites. Earth Planet Sci Lett 540:116266
Riebe MEI, Busemann H, Alexander CMO, Herd CDK, Maden C, Wieler R (2016) Effects of aqueous alteration on primordial noble gas in Tagish Lake (C2-ungr.). Meteorit Planet Sci 51:6211

Rietmeijer FJM, MacKinnon IDR (1985) Poorly graphitized carbon as a new cosmothermometer for primitive extraterrestrial materials. Nature 315:733–736

Righter K, Neff KE (2007) Temperature and oxygen fugacity constraints on CK and R chondrites and implications for water and oxidation in the early Solar System. Polar Sci 1:25–44

Robert F, Epstein S (1982) The concentration and isotopic composition of hydrogen, carbon and nitrogen in carbonaceous meteorites. Geochim Cosmochim Acta 46:81–95

Robert F, Merlivat L, Javoy M (1979) Deuterium concentration in the early Solar System: hydrogen and oxygen isotope study. Nature 282:785–789

Robert F, Javoy M, Halbout J, Dimon B, Merlivat L (1987) Hydrogen isotope abundances in the Solar System. Part 1: unequilibrated chondrites. Geochim Cosmochim Acta 51:1787–1805

Rouzaud J-N, Oberlin A (1990) The characterization of coals and cokes by transmission electron microscopy. In: Charcosset H, Nickel-Pepin-Donat B (eds) Advanced methodologies in coal characterization. Elsevier, Amsterdam, pp 311–355

Rubin AE (1998) Correlated petrologic and geochemical characteristics of CO3 chondrites. Meteorit Planet Sci 33:385–391

Ruf A, Kanawati B, Hertkorn N, Yin Q-Z, Moritz F, Harir M, Lucio M, Michalke B, Wimpenny J, Shilobreeva S, Bronsky B, Saraykin V, Gabelica Z, Gougeon RD, Quirico E, Ralew S, Jakubowski T, Haack H, Gonsior M, Jenniskens P, Hinman NW, Schmitt-Kopplin P (2017) Previously unknown class of metalorganic compounds revealed in meteorites. Proc Natl Acad Sci USA 114:2819–2824

Russell SS, Ott U, Alexander CMO, Zinner EK, Arden JW, Pillinger CT (1997) Presolar silicon carbide from the Indarch (EH4) meteorite: comparison with silicon carbide populations from other meteorite classes. Meteorit Planet Sci 32:719–732

Russell SS, Huss GR, Fahey AJ, Greenwood RC, Hutchison R, Wasserburg GJ (1998) An isotopic and petrologic study of calcium–aluminum-rich inclusions from CO3 meteorites. Geochim Cosmochim Acta 62:689–714

Ryerson FJ, McKeegan KD (1994) Determination of oxygen self-diffusion in åkermanite, anorthite, diopside, and spinel: implications for oxygen isotopic anomalies and the thermal histories of Ca-Al-rich inclusions. Geochim Cosmochim Acta 58:3713–3734

Sackett WM (1978) Carbon and hydrogen isotope effects during the thermocatalytic production of hydrocarbons in laboratory simulation experiments. Geochim Cosmochim Acta 42:571–580

Sakamoto N, Seto Y, Itoh S, Kuramoto K, Fujino K, Nagashima K, Krot AN, Yurimoto H (2007) Oxygen isotope evidence for remnants of the early Solar System primordial water. Science 317:231–233

Sakurai R, Yamamoto D, Takigawa A, Tachibana S (2023) Crystallization kinetics of amorphous ferromagnesian silicate with olivine-like composition. Lunar Planet Sci 54:2011

Schmitt-Kopplin P, Harir M, Kanawati B, Tziozis D, Hertkorn N, Gabelica Z (2012) Chemical footprint of the solvent soluble extraterrestrial organic matter occluded in Sołtmany ordinary chondrite. Meteorites 2:79–92

Shibita Y (1996) Opaque minerals in Antarctic CO3 carbonaceous chondrites Yamato-74153, -790992, -791717, -81020, -81025, and Allan Hills-77307. Proc NIPR Symp Antartc Meteorit 9:79–96

Shollenberger QR, Borg LE, Render J, Ebert S, Bischoff A, Russell SS, Brennecka GA (2018) Isotopic coherence of refractory inclusions from CV and CK meteorites: evidence from multiple isotope systems. Geochim Cosmochim Acta 228:62–80

Simon JI, Young ED, Russell SS, Tonui EK, Dyl KA, Manning CE (2005) A short timescale for changing oxygen fugacity in the solar nebula revealed by high-resolution ^{26}Al–^{26}Mg dating of CAI rims. Earth Planet Sci Lett 238:272–283

Simon JI, Hutcheon ID, Simon SB, et al (2011) Oxygen isotope variations at the margin of a CAI records circulation within the solar nebula. Science 331:1175–1178

Simon JI, Matzel JEP, Simon SB, Hutcheon ID, Ross DK, Weber PK, Grossman L (2016) Oxygen isotopic variations in the outer margins and Wark-Lovering rims of refractory inclusions. Geochim Cosmochim Acta 186:242–276

Simon SB, Krot AN, Nagashima K (2019) Oxygen and Al-Mg isotopic compositions of grossite-bearing refractory inclusions from CO3 chondrites. Meteorit Planet Sci 54:1362–1378

Siron G, Fukuda K, Kimura M, Kita NT (2021) New constraints from ^{26}Al–^{26}Mg chronology of anorthite bearing chondrules in unequilibrated ordinary chondrites. Geochim Cosmochim Acta 293:103–126

Smith PPK, Buseck PR (1981) Graphitic carbon in the Allende meteorite: a microstructural study. Science 212:322–324

Smith JW, Kaplan IR (1970) Endogenous carbon in carbonaceous meteorites. Science 167:1367–1370

Smith LR, Haenecour P, Barnes J, Dominik K, Neuman M, Wang K, Ogliore R, Koefoed P (2023) Abundance of presolar grains in the C3.00-ungrouped chondrite Chwichiya 002. Lunar Planet Sci 54:2425

Stolper E (1982) Crystallization sequences of Ca–Al-rich inclusions from Allende: an experimental study. Geochim Cosmochim Acta 46:2159–2180

Sutton S, Alexander CMO, Bryant A, Lanzirotti A, Newville M, Cloutis EA (2017) The bulk valence state of Fe and the origin of water in chondrites. Geochim Cosmochim Acta 211:115–132

Swindle TD, Caffee MW, Hohenberg CM, Lindstrom MM (1983) I-Xe studies of individual Allende chondrules. Geochim Cosmochim Acta 47:2157–2177

Swindle TD, Caffee MW, Hohenberg CM (1988) I-Xe studies of Allende inclucions: EGGs and the Pink Angel. Geochim Cosmochim Acta 52:2215–2229

Tachibana S, Huss GR (2005) Sulfur isotope composition of putative primary troilite in chondrules from Bishunpur and Semarkona. Geochim Cosmochim Acta 69:3075–3097

Tang M, Anders E (1988) Isotopic anomalies of Ne, Xe, and C in meteorites. II. Interstellar diamond and SiC: carriers of exotic noble gases. Geochim Cosmochim Acta 52:1235–1244

Tang H, Liu M-C, McKeegan KD, Tissot FLH, Dauphas N (2017) In situ isotopic studies of the U-depleted Allende CAI Curious Marie: pre-accretionary alteration and the co-existence of ^{26}Al and ^{36}Cl in the early solar nebula. Geochim Cosmochim Acta 207:1–18

Taylor GJ, Okada A, Scott ERD, Rubin AE, Huss GR, Keil K (1981) The occurrence and implications of carbide-magnetite assemblages in unequilibrated ordinary chondrites. In: Proc. Lunar Planet. Sci. Conf., vol 12, pp 1076–1078

Tenner TJ, Ushikubo T, Nakashima D, Schrader DL, Weisberg MK, Kimura M, Kita N (2018) Oxygen isotope characteristics of chondrules from recent studies by secondary ion mass spectrometry. In: Russel SS, Connolly HC Jr, Krot AN (eds) Chondrules: records of the protoplanetary disk processes. Cambridge University Press, Cambridge, pp 196–247

Tenner TJ, Nakashima D, Ushikubo T, Tomioka N, Kimura M, Weisberg MK, Kita NT (2019) Extended chondrule formation intervals in distinct physicochemical environments: evidence from Al-Mg isotope systematics of CR chondrites chondrules with unaltered plagioclase. Geochim Cosmochim Acta 260:133–160

Thiemens MH (2006) History and applications of mass-independent isotope effects. Annu Rev Earth Planet Sci 34:217–262

Tomeoka K, Buseck PR (1990) Phyllosilicates in the Mokoia CV carbonaceous chondrite: evidence for aqueous alteration in an oxidizing environment. Geochim Cosmochim Acta 54:1745–1754

Tomeoka K, Itoh D (2004) Sodium-metasomatism in chondrules in CO3 chondrites: relationship to parent body thermal metamorphism. Meteorit Planet Sci 39:1359–1373

Tomeoka K, Nomura K, Takeda H (1992) Na-bearing Ca-Al-rich inclusions in the Yamato-791717 CO carbonaceous chondrite. Meteoritics 27:136–143

Turner G, Crowther SA, Burgess R, Gilmour JD, Kelley SP, Wasserburg GJ (2013) Short lived ^{36}Cl and its decay products ^{36}Ar and ^{36}S in the early solar system. Geochim Cosmochim Acta 123:358–367

Ushikubo T, Guan Y, Hiyagon H, Sugiura N, Leshin LA (2007a) ^{36}Cl, ^{26}Al, and O isotopes in an Allende type B2 CAI: implications for multiple secondary alteration events in the early Solar System. Meteorit Planet Sci 42:1267–1279

Ushikubo T, Guan Y, Hiyagon H, Sugiura N, Leshin LA (2007b) ^{36}Cl, ^{26}Al, and O isotopes in an Allende type B2 CAI: implications for multiple secondary alteration events in the early Solar System. Meteorit Planet Sci 42:1267–1279

Ushikubo T, Kimura M, Kita NT, Valley JW (2012) Primordial oxygen isotope reservoirs of the solar nebula recorded in chondrules in Acfer 094 carbonaceous chondrite. Geochim Cosmochim Acta 90:242–264

Ushikubo T, Tenner TJ, Hiyagon H, Kita NT (2017) A long duration of the ^{16}O-rich reservoir in the solar nebula, as recorded in fine-grained refractory inclusions from the least metamorphosed carbonaceous chondrites. Geochem Cosmochim Acta 201:103–122

Vacher LG, Piani L, Rigaudier T, Thomassin D, Florin G, Piralla M, Marrocchi Y (2020) Hydrogen in chondrites: influence of parent body alteration and atmospheric contamination on primordial components. Geochim Cosmochim Acta 281:53–66

Vacher LG, Ogliore RC, Jones C, Liu N, Fike DA (2021) Cosmic symplectite recorded irradiation by nearby massive stars in the Solar System's parent molecular cloud. Geochim Cosmochim Acta 309:135–150

Van Orman JA, Cherniak DJ, Kita NT (2014) Magnesium diffusion in plagioclase: dependence on composition, and implications for thermal resetting of the ^{26}Al–^{26}Mg early Solar System chronometer. Earth Planet Sci Lett 385:79–88

Vandenbroucke M, Largeau C (2007) Kerogen origin, evolution and structure. Org Geochem 38:719–833

Villa IM, Huneke JC, Papanastassiou DA, Wasserburg GJ (1981) The Allende pink angel: chronological constraints from Xe, Ar, and Mg. In: Proc. Lunar sci. conf., vol 12, pp 1115–1117

Visser R, John T, Patzek M, Bischoff A, Whitehouse MJ (2019) Sulfur isotope study of sulfides in CI, CM, C2ung chondrites and volatile-rich clasts–evidence for different generations and reservoirs of sulfide formation. Geochim Cosmochim Acta 261:210–223

Vollmer C, Hoppe P, Stadermann FJ, Floss C, Brenker FE (2009) NanoSIMS analysis and Auger electron spectroscopy of silicate and oxide stardust from the carbonaceous chondrite Acfer 094. Geochim Cosmochim Acta 73:7127–7149

Wark DA, Lovering JF (1976) Marker events in the early evolution of the Solar System: evidence from rims on Ca-Al-rich inclusions in carbonaceous chondrites meteorites. In: Proc. Lunar sci. conf., vol 8, pp 95–112

Warren PH (2011) Stable isotopic anomalies and the accretionary assemblage of the Earth and Mars: a subordinate role for carbonaceous chondrites. Earth Planet Sci Lett 311:93–100

Wasserburg GJ, Hutcheon ID, Aléon J, Ramon EC, Krot AN, Nagashima K, Brearley AJ (2011) Extremely Na- and Cl-rich chondrule from the CV3 carbonaceous chondrite Allende. Geochim Cosmochim Acta 75:4752–4770

Weisberg MK, McCoy TJ, Krot AN (2006) Systematics and evaluation of meteorite classification. In: Lauretta DS, McSween HY Jr (eds) Meteorites and the early Solar System II, pp 19–52

Wick MJ, Jones RH (2012) Formation conditions of plagioclase-bearing type I chondrules in CO chondrites: a study of natural samples and experimental analogs. Geochim Cosmochim Acta 98:140–159

Wieler R, Anders E, Baur H, Lewis RS, Signer P (1992) Characterisation of Q-gases and other noble gas components in the Murchison meteorite. Geochim Cosmochim Acta 56:2907–2921

Yamamoto D, Kawasaki N, Tachibana S, Ishizaki L, Sakurai R, Yurimoto H (2024) An experimental simulation of oxygen isotope exchange reaction between amorphous silicate dust and carbon monoxide gas in the early Solar System. Geochim Cosmochim Acta 374:93–105

Yang J, Epstein S (1983) Interstellar organic matter in meteorites. Geochim Cosmochim Acta 47:2199–2216

Yang L, Ciesla FJ, Alexander CMO (2013) The D/H ratio of water in the solar nebula during its formation and evolution. Icarus 226:256–267

Yesiltas M, Young J, Glotch TD (2021) Thermal metamorphic history of Antarctic CV3 and CO3 chondrites inferred from the first-and second-order Raman peaks of polyaromatic organic carbon. Amer Mineralogist: J Earth Planet Mater 106:506–517

Yurimoto H, Kuramoto K (2004) Molecular cloud origin for the oxygen−isotope heterogeneity in the Solar System. Science 305:1763–1766

Yurimoto H, Morioka M, Nagasawa H (1989) Diffusion in single crystals of melilite. I. Oxygen Geochim Cosmochim Acta 53:2387–2394

Yurimoto H, Ito M, Nagasawa H (1998) Oxygen isotope exchange between refractory inclusion in Allende and solar nebula gas. Science 282:1874–1877

Yurimoto H, Krot AN, Choi B-G, Aléon J, Kunihiro T, Brearley AJ (2008) Oxygen isotopes of chondritic components. In: MacPherson GJ (ed) Oxygen in the Solar System. Rev. Mineral. Geochem., vol 68, pp 141–187

Zanetta P-M, Le Guillou C, Leroux H, Zanda B, Hewins R, Bellino G (2022) Processes and temperatures of FGR formation in chondrites. Geochim Cosmochim Acta 319:94–117

Zheng Y-F (1991) Calculation of oxygen isotope fractionation in metal oxides. Geochim Cosmochim Acta 55:2299–2307

Zheng Y-F (1993) Calculation of oxygen isotope fractionation in anhydrous silicate minerals. Geochem Cosmochim Acta 57:1079–1091

Zheng Y-F (2011) On the theoretical calculations of O isotope fractionation factors for carbonate-water systems. Geochem J 45:341–354

Zinner E (2014) Presolar grains. In: Davis AM (ed) Meteorites and cosmochemical processes, 2nd edn. Treatise on geochemistry, vol 1. Elsevier, Amsterdam, pp 181–213

Zinner E, Nittler LR, Hoppe P, Gallino R, Straniero O, Alexander CMO (2005) Oxygen, magnesium and chromium isotopic ratios of presolar spinel grains. Geochim Cosmochim Acta 69:4149–4165

Zolotov MY, Mironenko MV (2007) Hydrogen chloride as a source of acid fluids in parent bodies of chondrites. Lunar Planet Sci 38:2340

Zolotov MY, Mironenko MV, Shock EL (2006) Thermodynamic constraints on fayalite formation on parent bodies of chondrites. Meteorit Planet Sci 41:1775–1796

Authors and Affiliations

A.N. Krot[1] · M.I. Petaev[2] · L. Piani[3] · Y. Marrocchi[3] · W. Fujiya[4] · O.V. Pravdivtseva[5] · E. Dobrică[1] · L.G. Vacher[6] · A.J. King[7] · M. Lee[8] · E. Van Kooten[9] · B. Jacobsen[10] · C.M.O'D. Alexander[11] · A. Bischoff[12] · A.J. Brearley[13] · C. Le Guillou[14] · L. Remusat[15] · J. Leitner[16,17] · G.R. Huss[1]

✉ J. Leitner
jan.leitner@mpic.de

A.N. Krot
sasha@higp.hawaii.edu

L. Piani
laurette.piani@univ-lorraine.fr

[1] Hawai'i Institute of Geophysics & Planetology, School of Ocean & Earth Science & Technology, University of Hawai'i at Mānoa, Honolulu, HI, 96822, USA

[2] Department of Earth & Planetary Sciences, Harvard University, Cambridge, MA, 02138, USA

[3] CRPG, UMR 7358 CNRS, Université de Lorraine, 54500, Vandoeuvre-lès-Nancy, France

[4] Faculty of Science, Ibaraki University, 2-1-1 Bunkyo, Mito, Ibaraki, 310-8512, Japan

[5] Physics Department, Washington University, St. Louis, MO, 63130, USA

[6] University of Grenoble Alpes, CNRS, IPAG, 38000, Grenoble, France

[7] Planetary Materials Group, Department of Earth Sciences, Natural History Museum, Cromwell Road, London, SW7 5BD, UK

[8] School of Geographical & Earth Sciences, University of Glasgow, Gregory Building, Lilybank Gardens, Glasgow, G12 8QQ, UK

[9] Centre for Star & Planet Formation & Natural History Museum of Denmark, University of Copenhagen, DK-1350, Copenhagen, Denmark

[10] Lawrence Livermore National Laboratory, 7000 East Ave, Livermore, CA, 94550, USA

[11] Earth & Planets Laboratory, Carnegie Institution of Washington, 5241 Broad Branch Road, NW, Washington, DC, 20015, USA

[12] Institut für Planetologie, Wilhelm-Klemm-Str. 10, Münster, 48149, Germany

[13] University of New Mexico, Albuquerque, NM, 87131-0001, USA

[14] CNRS, INRAE, Centrale Lille, UMR 8207 - UMET - Unité Matériaux et Transformations, University of Lille, F-59000, Lille, France

[15] IMPMC, UMR CNRS 7590, Sorbonne Université, MNHN, CP52 57 rue Cuvier, 75231, Paris Cedex 05, France

[16] Geo- & Cosmochemistry Group, Institute of Earth Sciences, Heidelberg University, Im Neuenheimer Feld 234-236, D-69120, Heidelberg, Germany

[17] Max Planck Institute for Chemistry, Particle Chemistry Department, Hahn-Meitner-Weg 1, 55128, Mainz, Germany

Space Science Reviews (2025) 221:11
https://doi.org/10.1007/s11214-024-01132-8

Low-Temperature Aqueous Alteration of Chondrites

Martin R. Lee[1] · Conel M. O'D. Alexander[2] · Addi Bischoff[3] · Adrian J. Brearley[4] · Elena Dobrică[5] · Wataru Fujiya[6] · Corentin Le Guillou[7] · Ashley J. King[8] · Elishevah van Kooten[9] · Alexander N. Krot[5] · Jan Leitner[10,11] · Yves Marrocchi[12] · Markus Patzek[3] · Michail I. Petaev[13] · Laurette Piani[12] · Olga Pravdivtseva[14] · Laurent Remusat[15] · Myriam Telus[16] · Akira Tsuchiyama[17,18,19] · Lionel G. Vacher[20]

Received: 31 January 2024 / Accepted: 18 December 2024 / Published online: 4 February 2025

Abstract
Chondritic meteorites (chondrites) contain evidence for the interaction of liquid water with the interiors of small bodies early in Solar System history. Here we review the processes, products and timings of the low-temperature aqueous alteration reactions in CR, CM, CI and ungrouped carbonaceous chondrites, the asteroids Ryugu and Bennu, and hydrated dark clasts in different types of meteorites. We first consider the nature of chondritic lithologies and the insights that they provide into alteration conditions, subdivided by the mineralogy and petrology of hydrated chondrites, the mineralogy of hydrated dark clasts, the effects of alteration on presolar grains, and the evolution of organic matter. We then describe the properties of the aqueous fluids and how they reacted with accreted material as revealed by physicochemical modelling and hydrothermal experiments, the analysis of fluid inclusions in aqueously formed minerals, and isotope tracers. Lastly, we outline the chronology of aqueous alteration reactions as determined using the ^{53}Mn-^{53}Cr and ^{129}I-^{129}Xe systems.

Keywords Carbonaceous chondrites · Aqueous alteration · Clasts · Presolar grains · Organic matter · Physicochemical modelling · Hydrothermal experiments · Fluid inclusions · Stable isotopes · ^{53}Mn-^{53}Cr chronology · ^{129}I-^{129}Xe chronology

1 Introduction

The radiogenic heating of small bodies early in Solar System history produced liquid water from the melting of accreted water-rich ices. The principal products of the reaction of water with other accreted materials were phyllosilicate minerals whose abundance helps to classify such altered carbonaceous chondrites into petrologic types 1 and 2. Samples of these hydrated lithologies are available as the CR (Renazzo-like), CM (Mighei-like), and CI (Ivuna-like) meteorites, ungrouped carbonaceous chondrites, clasts that occur in a wide variety of meteorite groups, and grains returned from the Cb-type asteroid Ryugu and B-type asteroid Bennu. Hydrated lithologies include interplanetary dust particles and Antarctic micrometeorites (e.g., Bradley 2014), but they are not covered here.

Extended author information available on the last page of the article

This article starts by describing the petrological and mineralogical characteristics of meteorites and clasts that have undergone low-temperature aqueous alteration, and the insights that their properties can provide into the physicochemical conditions and environments of aqueous alteration. The effects of liquid water on two of the originally accreted constituents of carbonaceous chondrite parent bodies, presolar grains and organic matter, is then described. Next, this article explores the properties of the aqueous fluids, and how they reacted with the chondritic parent bodies (e.g., chemistry, temperature, water-rock (W/R) ratio, duration of water-rock interaction) as investigated by physicochemical modelling and hydrothermal experiments, by analyzing samples of water preserved as fluid inclusions, and through measurement of the oxygen, hydrogen, carbon, nitrogen and sulfur isotope compositions of bulk rocks and constituent minerals. This article finishes by describing the chronology of aqueous alteration as revealed by the ^{53}Mn-^{53}Cr and ^{129}I-^{129}Xe systems.

2 Mineralogy and Petrology of Hydrated Carbonaceous Chondrites

This section includes the three groups of hydrated chondrites (CR, CM, CI), several of the more well-studied ungrouped carbonaceous chondrites (Tagish Lake, Tarda, Flensburg), and samples that have been recently returned from asteroids Ryugu and Bennu. It finishes with a description of thermally altered carbonaceous chondrites.

2.1 CR (Renazzo-Like) Chondrites

The CR chondrites are characterised by high abundances of presolar silicate grains (except heavily hydrated CRs), very elevated D/H ratios of their organic matter, and a solar Co/Ni ratio of Fe-Ni metal (Kallemeyn and Wasson 1981; Krot et al. 2002; Floss and Haenecour 2016). This record is preserved despite the CR chondrites exhibiting significant variations in their degree of alteration, but show no evidence of thermal metamorphism, with the exception of the shock-heated CR Graves Nunataks (GRA) 06100 (Abreu and Bullock 2013). The vast majority of CRs are Saharan and Antarctic finds. Only two falls are known: Al Rais and Renazzo (potentially three if Bells is considered a CR; Marrocchi et al. 2023a). Although most CRs are predominantly composed of chondrules and have a low modal abundance of matrix ($\sim$70 *vs* $\sim$30 vol. %), rare matrix-rich CR chondrites like Al Rais also exist ($\sim$70 vol. % matrix) (Schrader et al. 2011). Like CM chondrites, CRs are commonly brecciated and contain so-called dark inclusions that have been widely documented, as well as clasts of more highly altered lithologies (Endress et al. 1994).

With the exception of the almost completely altered CR1 chondrites Al Rais and Grosvenor Mountains (GRO) 95577 (Weisberg and Huber 2007; Schrader et al. 2011), CRs typically belong to petrologic type 2 (Schrader et al. 2011; Harju et al. 2014). However, this classification is more nuanced because some CR chondrites, such as Queen Alexandra Range (QUE) 99177, Meteorite Hills (MET) 00426 and Miller Range (MIL) 090657, exhibit characteristics more typical of type 3.00 chondrites (e.g., Abreu and Brearley 2010). CRs can be classified as type 2 (e.g., Harju et al. 2014; Howard et al. 2015), but have experienced significantly less alteration than almost all CM2 chondrites. Many CR chondrites have minimally altered glassy chondrule mesostasis and Fe-Ni metal beads in chondrules (Weisberg et al. 1993; Abreu and Brearley 2010) as well as high abundances of presolar silicate grains that are known to be highly sensitive to aqueous alteration (e.g., Floss and Stadermann 2009; Floss and Haenecour 2016) (Sect. 4). The generally lower degrees of alteration of the CRs are reflected in their low bulk H contents (0.14–0.68 wt.% H) compared to CM2 chondrites (0.9–1.5 wt.% H) (Alexander et al. 2012). The low degree of aqueous alteration makes the CR chondrites of special importance for studying the earliest interaction of aqueous fluids with chondritic materials. Some CRs, such as Renazzo, GRO 95577 and Al Rais, show more advanced alteration with glassy chondrule mesostases being completely replaced by phyllosilicates (Weisberg et al. 1993).

For most CR chondrites, aqueous alteration effects are seen mainly in the fine-grained matrix that occurs interstitially to chondrules; fine-grained matrix-like chondrule rims are comparatively rare in CR chondrites. Scanning electron microscope (SEM) studies show that the dominant secondary alteration minerals that form in the matrix are framboidal magnetites [$Fe^{2+}Fe^{3+}{}_2O_4$] and Ca carbonates (Fig. 1) with grain sizes of 10s of microns. The abundance of these minerals increases with increasing degree of aqueous alteration (Harju et al. 2014). Transmission electron microscope (TEM) studies of several different CR chondrite matrices (e.g., Abreu and Brearley 2010; Le Guillou and Brearley 2014; Le Guillou et al. 2015a; Changela et al. 2018; Davidson et al. 2019; Vollmer et al. 2020a) show that the least altered CRs contain matrix that is dominated by Fe-rich amorphous silicate material that contains abundant nanophase Fe,Ni sulfides and is complexly intergrown with insoluble organic matter (Le Guillou and Brearley 2014; Floss et al. 2014). The amorphous silicate material is both hydrated (Bonal et al. 2013; Le Guillou and Brearley 2014) and highly oxidized (Le Guillou et al. 2015a). As aqueous alteration of matrix advances, either due to an increase in temperature or increased availability of aqueous fluid, the amorphous silicate material progressively transforms to fine-grained hydrous phyllosilicates (an intergrowth of serpentine [$(Mg,Fe)_3Si_2O_5(OH)_4$] and smectite (saponite) [$(Ca,Na)_{0.3}(Mg,Fe^{2+})_3(Si,Al)_4O_{10}(OH)_2 \cdot 4H_2O$]). The increase in phyllosilicate abundance with increasing aqueous alteration has been tracked by both X-ray diffraction (XRD) (Howard et al. 2015) and TEM studies (Abreu and Brearley 2010; Le Guillou and Brearley 2014; Changela et al. 2018; Vollmer et al. 2020a). Generally, alteration mostly affected the matrices of CR chondrites forming mineral assemblages that are similar to those of CI chondrites, although more limited in extent. The effects of aqueous alteration on matrix are also tracked by the abundance of presolar silicate grains (Sect. 4).

Petrographic evidence for alteration of chondrules in minimally altered CR chondrites is limited to the replacement of chondrule mesostasis glass in narrow zones a few 10s to a 100 μm wide in contact with matrix. A unique feature of the alteration in CR chondrites, which has not been observed in any other chondrite group, is the presence of so-called smooth rims (Harju et al. 2014; Martínez and Brearley 2022), 5–50 μm in thickness, surrounding some chondrules (Fig. 1B). The smooth rims are present in even the least altered CR chondrites such as QUE 99177 and MET 00426. Martínez and Brearley (2022) showed that smooth rims in QUE 99177 formed by the replacement of silica (cristobalite, SiO_2) within Silica-Rich Igneous Rims (SIRs) that occur surrounding some type I chondrules (Fig. 1B) (Noguchi 1995; Krot et al. 2004). Smooth rims consist of a very Fe-rich amorphous silicate phase (Brearley and Jones 2016; Martínez and Brearley 2022) and show no evidence for phyllosilicates. Secondary phases in CR2 chondrites include framboidal magnetite, Ca carbonates, fine-grained sulfides and phyllosilicates corresponding to intergrowths between dominant serpentine and saponite (Weisberg et al. 1993; Zolensky et al. 1993) (Fig. 1C, D). During more advanced alteration, dolomite [$CaMg(CO_3)_2$] also forms, and in the CR1 chondrites siderite [$FeCO_3$] can be present.

Isotopic data are very informative about the evolution of CC lithologies. Bulk O-isotope studies of a large number of CR chondrites (mostly Antarctic finds) (Schrader et al. 2011) show a progressive increase in $\delta^{17}O$ and $\delta^{18}O$ that appears to correlate well with petrographic evidence of aqueous alteration, such as the abundance of phyllosilicates

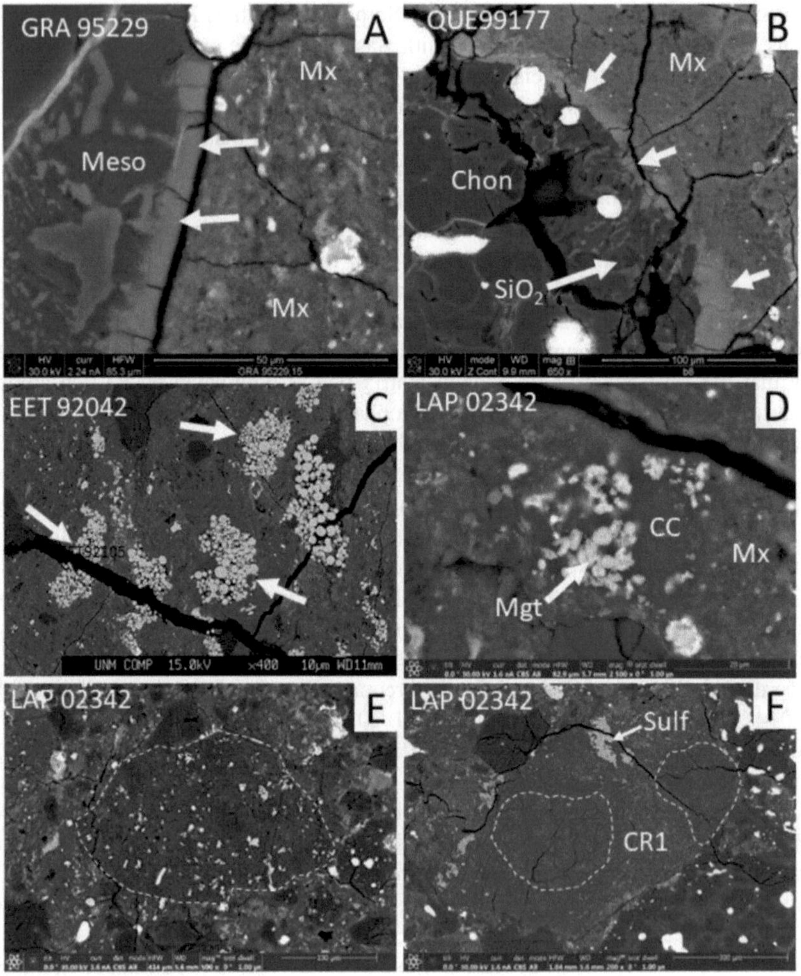

Fig. 1 Backscattered electron (BSE) images of alteration features and clasts in CR chondrites. (A) Thin alteration rind (indicated by white arrows) on the edge of a type IIA chondrule in GRA 95229. The alteration rind has replaced mesostasis glass where it is in direct contact with the matrix (Mx). In this example, the alteration rind is <5 μm in thickness and is enriched in FeO compared with the adjacent unaltered mesostasis (Meso). (B) Smooth rim associated with a silica-rich igneous rim (SIR) in a type IA porphyritic olivine chondrule in QUE 99177. The smooth rim (indicated by white arrows and with slightly higher BSE contrast) has replaced abundant silica crystals (SiO_2) in the SIR which is also in contact with the adjacent fine-grained matrix (Mx). (C) Abundant, large magnetite framboids (indicated by white arrows) in a region of matrix in the Elephant Moraine (EET) 92042 CR chondrite. (D) Calcite grain (CC) intergrown with extremely fine-grained framboidal magnetite (Mgt) within the matrix of LaPaz Icefield (LAP) 02342. (E) Example of a dark clast (optically dark, but a distinct lithic clast with very few or no anhydrous silicates; outlined by white dashed line) of a C1 lithology consisting of a very fine-grained matrix, with coarse-grained magnetite and sulfides in LAP 02342 (compare Sect. 3). (F) A CR1 clast within LAP 02342 (CR2.8 – Harju et al. 2014). The clast consists of parts of completely pseudomorphed chondrules (outlined by white dashed line) that have been replaced by serpentine with a uniformly low contrast and large masses of Fe,Ni sulfide (Sulf)

$\delta^{18}O = [(^{18}O/^{16}O)_{sample}/(^{18}O/^{16}O)_{standard} - 1] \times 1000$, $\delta^{17}O = [(^{17}O/^{16}O)_{sample}/(^{17}O/^{16}O)_{standard} - 1] \times 1000$. The bulk O-isotope compositions lie on a well-defined linear array on a three O-isotope plot with a slope of 0.70, with QUE 99177 defining the lightest O-isotope composition of $\sim -4‰$ and $\delta^{18}O = \sim -2.3‰$ and Al Rais the heaviest ($\delta^{17}O = \sim 4.3‰$ and $\delta^{18}O = \sim 10.8‰$). $\Delta^{17}O$ ($= \delta^{17}O - 0.52 \times \delta^{18}O$), a measure of deviation from the terrestrial fractionation line (TFL), ranges from ~ -3 to $-0.4‰$ (Sect. 8). These data demonstrate that most CR chondrites have relatively low degrees of aqueous alteration, consistent with their petrographic characteristics, and that the two known CR chondrite falls, Renazzo and Al Rais, are among the most heavily altered CR chondrites. It is estimated that the CRs had W/R mass ratios of 0.10–0.40 (Sect. 8.1.4).

The O-isotope composition of secondary minerals (Jilly-Rehak et al. 2018) provides constraints on the conditions of aqueous alteration in CR chondrites. Carbonates and magnetite show a range of $\delta^{18}O \approx 9$ to 35‰ in calcite [$CaCO_3$, trigonal], from $\delta^{18}O \approx 23$ to 27‰ in dolomite, and from $\delta^{18}O \approx -18$ to 5‰ in magnetite. The widest variation in O-isotope compositions occurs in the minimally altered CR chondrites such as QUE 99177 and MIL 090292, which Jilly-Rehak et al. (2018) proposed is due to low W/R ratios and progressive evolution in the isotopic compositions of the fluid. In this scenario, carbonate grains grew at different times and reflect this isotopic evolution that may also be caused by local variations in fluid composition. In more heavily altered CRs, e.g., Al Rais and the CR1 GRO 95577, the carbonate O-isotope compositions have a more restricted range, suggesting that they all formed from the same isotopically homogeneous fluid. Assuming closed-system aqueous alteration on the CR chondrite parent body, and that isotopic equilibrium was attained between carbonate and magnetite, Jilly-Rehak et al. (2018) calculated a global temperature of aqueous alteration from ~55 to 88 °C, based on the parallel arrays of calcite and magnetite O-isotope compositions, and Marrocchi et al. (2018) estimated temperatures of between 35 and 110 °C (Sect. 8.1.4). The chronology of aqueous alteration of the CRs is discussed along with other CC groups in Sect. 9.

2.2 CM (Mighei-Like) Chondrites

Among the carbonaceous chondrites, the CM chondrites are the most abundant group, representing around 30% of falls and finds. CM-like matter also constitutes an important fraction of xenoliths discovered in meteorites (Gounelle et al. 2003; Patzek et al. 2018; Marrocchi et al. 2021a; King et al. 2022) (Sect. 3), attesting to the importance of these types of meteorites in the formation and the evolution of the Solar System. CM chondrites are dark, highly brecciated meteorites, in which lithic clasts show various degrees of aqueous alteration (Fig. 2, e.g., Metzler et al. 1992; Bischoff et al. 2006, 2017; Lindgren et al. 2013; Verdier-Paoletti et al. 2019; Lentfort et al. 2021; Kerraouch et al. 2022; Suttle et al. 2022a). The main components of CMs are abundant fine-grained matrix (~70 vol. %), small chondrules (0.194 mm; Floyd et al. 2024) and refractory inclusions, and abundant secondary minerals. Most chondrules are surrounded by fine-grained rims (FGRs; Metzler et al. 1992).

The CM chondrites represent the largest group of hydrated meteorites with water contents ranging from 3 to almost 10 wt.% (Vacher et al. 2020; Marrocchi et al. 2023b). They span a wide range of alteration degrees, from less altered (i.e., CM2) down to heavily altered (i.e., CM1; Zolensky et al. 1997), with several minimally altered samples being recognized over the last decade (Fig. 2, Hewins et al. 2014; Marrocchi et al. 2014; Kimura et al. 2020). These mildly altered CMs are characterized by well-preserved Fe-Ni metal beads and glassy chondrule mesostases as well as porous aggregates of amorphous silicates dispersed throughout the fine-grained matrix (Leroux et al. 2015; Kimura et al. 2020). More altered CMs contain a variety of secondary phases produced through aqueous alteration processes, including carbonates, sulfides, oxides, and hydroxides (Fig. 2, Brearley 2006). The most characteristic secondary phases in CM chondrites are tochilinite-cronstedtite intergrowths (TCIs, Fig. 2E), occurring as complex assemblages dispersed throughout the chondrules, FGRs, and matrix (Mackinnon and Zolensky 1984; Pignatelli et al. 2017). Tochilinite is a mineral with an incommensurate structure based on two layers: (i) a mackinawite-type $Fe_{1-x}S$ sulfide layer and (ii) a brucite- or amakinite-type hydroxide layer, depending on the composition. Its ideal formula is $2(Fe_{1-x}S)n(Fe,Mg,Al,Ca)(OH)_2$ with $1.58 < n < 1.85$ (Zolensky and MacKinnon 1986) and $0.08 < x < 0.28$ (Gubaidulina et al. 2007), where n and x correspond to the hydroxide layer content and the vacancies of the sulfide layer, respectively. Cronstedtite is a T-O phyllosilicate with a general formula $(Fe^{2+}_{3-x}Fe^{3+}_{x})(Si_{2-x}Fe^{3+}_{x})O_5(OH)_4$ with $0 < x < 0.8$ in terrestrial crystals (Pignatelli et al. 2016, 2017).

The majority of CM2s are dominated by phyllosilicates in the range 73–79 vol. % (average = 75 vol. %) with Fe-cronstedtite (43–50 vol. %) being more abundant than Mg-serpentine (25–33 vol. %) (Howard et al. 2015). The only exception is Cold Bokkeveld where the abundance of Mg-serpentine (49–59 vol. %) exceeds that of Fe-cronstedtite (19–27 vol. %). For fall samples there is a limited range in olivine (6.9 vol. %) and pyroxene (5 vol. %) whereas find samples show larger anhydrous silicate abundances that can reach up to 30 vol %. Modal abundances of the remaining identified phases also show a limited range: calcite (0–1.3 vol. %); gypsum [$CaSO_4.2H_2O$] (0–1.6 vol. %, but likely terrestrial contamination); magnetite (1.1–2.4 vol. %); pentlandite [$(Fe,Ni)_9S_8$] (0–2.1 vol. %) and pyrrhotite [$Fe_{1-x}S$] (1–3.8 vol. %). More altered samples (CM1) contain 6–10 vol. % olivine + pyroxene and 86–88 vol. % total phyllosilicate. Magnetite (0.6–5.2 vol. %), sulfide (0.6–3.9 vol. %), calcite (0–1.9 vol. %) and gypsum (0–0.8 vol. %) are minor phases across all samples.

Calcite is present in all CMs and is often the most abundant carbonate mineral. Aragonite [$CaCO_3$, orthorhombic] also occurs in the less altered meteorites (Barber 1981; Lee and Ellen 2008; Lee et al. 2014) and dolomite in the more highly altered ones (Rubin et al. 2007; de Leuw et al. 2010; Lee et al. 2014). Breunnerite [$(Fe,Mg)CO_3$], and ankerite [$Ca(Mg,Fe,Mn)(CO_3)_2$] have been described only from the highly altered QUE 93005 (Lee et al. 2012). Several generations of calcite have been recognised from their petrographic characteristics and O-isotope compositions. Tyra et al. (2012) defined Types 1 and 2, with Vacher et al. (2018) adding Type 0, and subdividing Types 1 and 2 into two subtypes. The type 1 calcite grains often display cathodoluminescence (CL) zoning that reflects changes in the concentrations of Mn and Fe of the precipitating fluids during crystal growth. Fujiya et al. (2020) showed that the O-isotope composition of calcite can also differ significantly between the zones of individual crystals in Yamato-791198 such that later zones have a lower $\delta^{18}O$ and $\Delta^{17}O$ than earlier ones. These isotopic differences reflect the evolution of the O-isotope composition of aqueous solutions during their progressive reaction with anhydrous silicates and show that the growth rate of calcite crystals was slower than the rate of isotopic changes in parent body fluids.

The liquid water that was responsible for aqueous alteration of the CM chondrites came from the melting of accreted ices. The main source of heat was the decay of ^{26}Al, although collisions may also have contributed (Rubin 2012; Vacher et al. 2018). The temperatures of the fluids responsible for alteration have been investigated extensively using a range of geothermometers including clumped O-isotopes (~20–71 °C, Guo and Eiler 2007; ~5–101 °C, Clog et al. 2024), phyllosilicate-carbonate O-isotope fractionation (< ~20 °C, Clayton and Mayeda 1984), magnetite-carbonate O-isotope fractionation (125 ± 60 °C, Telus et al. 2019), and modeling of carbonate O-isotope compositions (0–130 °C, Alexander et al. 2015; 113 ± 54 °C, Verdier-Paoletti et al. 2017, 2019, Vacher et al. 2019a). These measurements

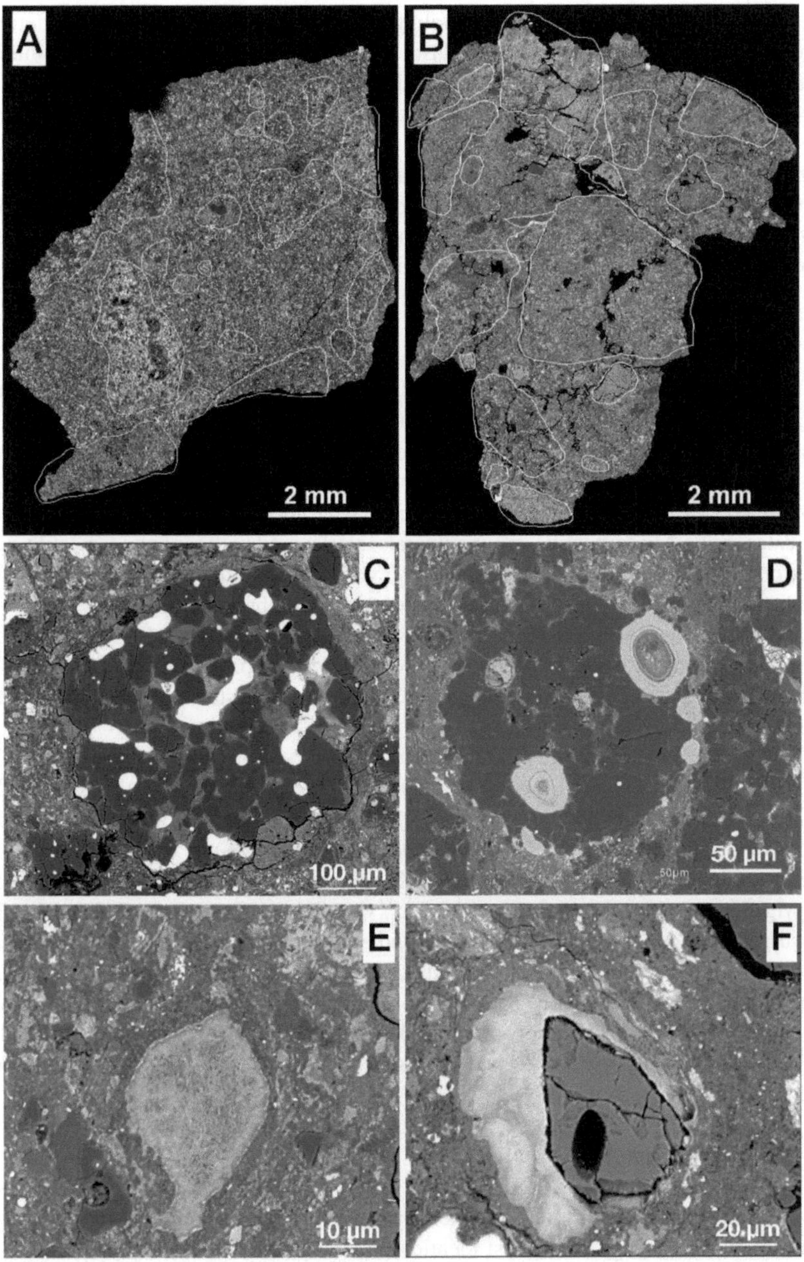

Fig. 2 BSE images of CM meteorites. (A) and (B) are the CM breccias ALH 85013 and LON 94101 (sample: co. 36), respectively. Fragments with different degrees of alteration are embedded in a fine-grained clastic matrix. From Lentfort et al. (2021). (C) A type I chondrule in the Paris that contains well-preserved metal beads but altered mesostasis. (D) Altered Fe-Ni metal beads within a type I chondrule of Murchison. (E) A tochilinite-cronstedtite intergrowth (TCI) in Paris. (F) BSE image of Ca-carbonate ($CaCO_3$) surrounded by TCIs within the matrix of Paris. Black spot corresponds to a SIMS pit

typically provide only one temperature per meteorite, whereas the CM parent bodies will have undergone at least one heating-cooling cycle, and possibly several considering the occurrence of carbonates in different clasts with various degrees of alteration within the same CM breccia and if collisions generated thermal pulses. The aqueous alteration products could have formed during heating and/or cooling, and Vacher et al. (2019b) have proposed that carbonates crystallized early during heating whereas tochilinite formed shortly before or after the parent body reached its peak temperature. The duration of aqueous alteration of any one CM is difficult to quantify, although Velbel et al. (2012) estimates that serpentinization of chondrule olivine would need timescales of hundreds of days to more than a decade. The chronology of aqueous alteration is discussed in Sect. 9.

2.3 CI (Ivuna-Like) Chondrites

The rare CI carbonaceous chondrites are all classified as petrologic type 1 having experienced pervasive aqueous alteration. They have primitive bulk chemical compositions that are almost identical to the solar photosphere, making them of significant cosmochemical importance (e.g., Anders and Grevesse 1989). The main component is an abundant (>90%) fine-grained (<1 µm) matrix of phyllosilicates (e.g., Tomeoka and Buseck 1988). The phyllosilicates are Fe-Mg serpentine-group minerals interlayered with saponite (smectite-group). Other phases include oxides (magnetite, ~3–6 vol. %), Fe-sulfides (pyrrhotite, pentlandite, cubanite [$CuFe_2S_3$], ~1–7 vol. %), carbonates (dolomite, breunnerite, calcite, ~0.1–2 vol. %), and phosphates (Alfing et al. 2019) (Fig. 3). The scarcity of intact CAIs and chondrules implies the parent body either accreted in a region of the protoplanetary disk nearly lacking them, or that the primary mineralogy and textures were destroyed by aqueous alteration (Frank et al. 2023).

The CI chondrites were altered at temperatures of ~50 to >150 °C, W/R mass ratios of 0.38–0.43, and a pH and Eh of ~7–10 and ~ −0.2 to −0.8, respectively (e.g., Zolensky et al. 1989). These conditions are largely supported by isotopic and petrographic constraints (e.g., Clayton and Mayeda 1999). Other minerals commonly reported in the CI chondrites are sulphates (e.g., gypsum [$CaSO_4 \cdot 2H_2O$], epsomite [$MgSO_4 \cdot 7H_2O$], and bloedite [$Na_2Mg(SO_4)_2 \cdot 4H_2O$]) and ferrihydrite [$Fe^{3+}_{10}O_{14}(OH)_2$] (Gounelle and Zolensky 2001). The sulphates occur as white flecks, flakes, and efflorescence on sample surfaces and infilling veins within the interior of the meteorites, while the ferrihydrite is intimately intergrown with phyllosilicates. The porous and volatile-rich nature of CI chondrites makes them susceptible to terrestrial modification, and the sulphates and ferrihydrite probably formed from reactions between the meteorites and atmospheric O_2 and H_2O (Gounelle and Zolensky 2001).

2.4 (162173) Ryugu

The mineralogy, petrography, and chemical properties of samples returned from the Cb-type asteroid (162173) Ryugu are almost identical to the CI chondrites (Yokoyama et al. 2023; Nakamura et al. 2023). In hand specimen, these samples are dark, highly friable, and have bulk densities up to ~1.8 g cm^{-3}. Most of the examined particles are breccias, but typically they consist of an abundant (~90%) fine-grained (<1 µm) matrix of phyllosilicates (interlayered serpentine and saponite) in which are embedded coarser (~10's–100's µm in size) grains, fragments, and clusters of oxides (~4 vol. %), Fe-sulfides (~3 vol. %), carbonates (~3 vol. %), and phosphates (<1 vol. %). Sulphates and ferrihydrite are absent from pristine particles not exposed to Earth's atmosphere, seemingly confirming the terrestrial origin of

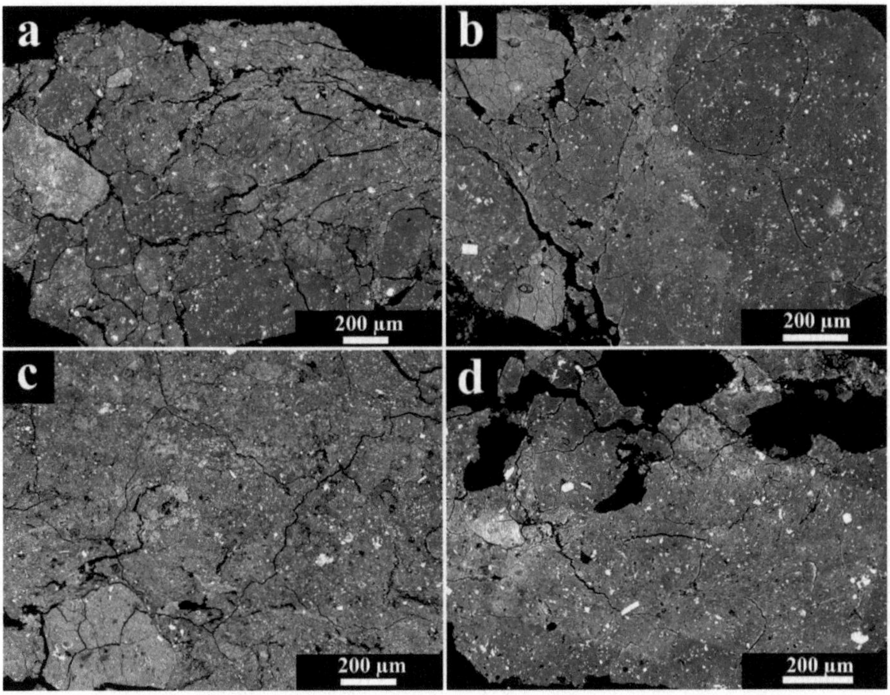

Fig. 3 Representative BSE images of typical textures of CI chondrites, demonstrating brecciation. (a) Orgueil, (b) Ivuna, (c) Alais, and (d) Tonk. The white spots are magnetites and pyrrhotite typically embedded within a fine-grained intergrowth of phyllosilicates. The degree of brecciation decreases in the sequence: Orgueil (a) > Ivuna (b) > Alais (c) ∼ Tonk (d). From Alfing et al. (2019)

these phases in the CI chondrites. Small (<30 μm) fragments of CAI- and chondrule-like materials have been identified in some Ryugu particles (e.g., Nakamura et al. 2023).

Alteration temperatures of < ∼50 °C for Ryugu particles have been estimated from the O-isotope composition of dolomite-magnetite assemblages (Yokoyama et al. 2023) and the composition and structure of Fe-sulfide grains (Nakamura et al. 2023). Thermodynamic modelling suggests a fluid Eh < −0.45 V, pH > ∼8.5, and W/R ratios of 0.2–0.9 for the main highly altered lithology (Nakamura et al. 2023). Fluid inclusions trapped within pyrrhotite grains retain H_2O and CO_2, consistent with Ryugu's parent body accreting beyond the water and carbon dioxide snowlines at heliocentric distances of > ∼3–4 au (Nakamura et al. 2023, Sect. 7). The aqueous alteration chronology of Ryugu is discussed in Sect. 9.

2.5 (101955) Bennu

Initial characterization of samples returned from the B-type asteroid (101955) Bennu has shown that they are dark, with morphologies that range from angular to sub-rounded or hummocky (Lauretta et al. 2024). The mineralogy is dominated by phyllosilicates (serpentine and saponite, ∼80 vol.%), with minor amounts of Fe-sulfides (∼10 vol.%), oxides (∼5 vol.%), carbonates (∼3 vol.%), and olivine (∼2 vol.%). Mg,Na-phosphates are also present, often occurring as higher reflectance coatings, veins, or individual grains within the matrix. Overall, the mineralogy and petrography of Bennu particles are consistent with extensive aqueous alteration and closely resemble samples of Ryugu and low petrologic type carbonaceous chondrites.

2.6 Ungrouped Hydrated Carbonaceous Chondrites

2.6.1 Tagish Lake

Tagish Lake is an ungrouped C2 carbonaceous chondrite whose chemical and isotopic composition, mineralogy, and petrographic characteristics suggest affinities to both the CI and CM groups (Brown et al. 2000; Zolensky et al. 2002). The meteorite is a breccia containing multiple lithologies including carbonate-poor, carbonate-rich, magnetite- and sulfide-rich, and carbonate-rich siderite-dominated (Zolensky et al. 2002; Izawa et al. 2010). These lithologies have a fine-grained matrix composed of serpentine and saponite together with magnetite and sulfides within which are rare altered chondrules and CAIs, anhydrous silicate grains, carbonate nodules that range widely in composition (calcite-dolomite-siderite), magnetite, and Fe-Ni sulfides (Brown et al. 2000; Zolensky et al. 2002). The mineralogy and petrography of Tagish Lake shows that it is petrologic type 2 with different lithologies showing contrasting degrees of aqueous alteration under largely isochemical conditions (Greshake et al. 2005; Herd et al. 2011; Alexander et al. 2014; Blinova et al. 2014a). Despite the intensity of alteration, Blinova et al. (2014b) showed that one sample (TL5b) contains minimally aqueously altered amorphous material of a nebular origin that is analogous to amorphous material in some CR and CM chondrites (Chizmadia and Brearley 2008; Abreu and Brearley 2010). Tagish Lake has also incorporated exogeneous CM1 materials (Zolensky et al. 2002).

2.6.2 Tarda

Tarda fell in Morocco in 2020 and is classified as a C2 ungrouped chondrite. It has a bulk O-isotope composition, and enrichments in deuterium (D), ^{15}N, and ^{13}C that suggest a close genetic relationship to Tagish Lake (Marrocchi et al. 2021b). Tarda contains highly altered chondrules and isolated olivine grains, rare CAIs, magnetite clusters, Fe-sulfides (pyrrhotite, pentlandite, $\sim$8 vol. %), and carbonates (dolomite, calcite, siderite). However, the most abundant component is a fine-grained matrix of phyllosilicates ($\sim$72 vol. %), which comprise serpentines, smectites, and an interlayered serpentine/smectite phase. The composition of Fe-sulfides in Tarda constrains its alteration temperature to $\sim$100–135 °C.

2.6.3 Flensburg

Flensburg is a C1 ungrouped chondrite breccia that was observed to fall in Germany in 2019 (Bischoff et al. 2021). It shares some mineralogical and chemical properties with highly altered members of the CM, CR, and CI groups, but has textures and bulk H, C, and N compositions that differ from other carbonaceous chondrites. Flensburg consists of chondrule pseudomorphs, clusters of Fe-sulfides (pyrrhotite, troilite [FeS], pentlandite, $\sim$4 vol. %), magnetite ($\sim$2 vol. %), and carbonate grains (calcite, dolomite, $\sim$4 vol. %) within a matrix of serpentine-rich phyllosilicates ($\sim$90 vol. %). Fe-sulfide compositions are consistent with formation at temperatures of $\sim$150–200 °C, and bulk O-isotope indicate that Flensburg was altered at a relatively low W/R ratio. The ^{53}Mn-^{53}Cr ages of Flensburg carbonates are discussed in Sect. 9.

2.7 Aqueously and Thermally Altered Chondrites

A number of aqueously altered chondrites also experienced thermal metamorphism at temperatures of ~300 °C to >750 °C. This heating resulted in the dehydration of phyllosilicates and crystallization of olivine and pyroxene, and the modification and depletion of volatile species (e.g., water, noble gases, and organics) (Nakamura 2005). Thermal metamorphism must have occurred after the main period of aqueous alteration since phyllosilicates are dehydrated in these meteorites, but it remains unclear whether the process was a single event or episodic. Mineral textures and the structure of organics indicate that the metamorphism was short-lived, on the order of hours to several years, implying that the heat source was impacts and/or solar radiation rather than radiogenic decay.

3 Mineralogy of Hydrated Dark Clasts

3.1 Introduction

3.1.1 What Are Dark Clasts?

Most chondritic meteorites, but also a significant number of achondrites, are breccias (i.e., mixtures of different rock types within a single specimen). Most of these breccias likely represent rock units from a single parent body (genomict breccias), but many chondrites and achondrites also include xenolithic clasts that were derived from a different Solar System objects. The most obvious and easily visible clasts in meteorite breccias are the so-called dark inclusions or clasts. In general, most of these clasts are not chondritic in origin and the term "dark clast" does not provide any mineralogic or genetic information (Bischoff et al. 2006). Except for the carbonaceous chondrite groups and polymict ureilites, dark clasts in most brecciated meteorite groups are dominated by (1) shock-darkened objects, (2) fragments of fine-grained brecciated materials, (3) metal-troilite-rich clasts, (4) fine-grained, matrix-like cognate inclusions (related to the host rock but with a different alteration degree), and (5) fragments of shock melts with abundant tiny metal/sulfide grains (Bischoff et al. 2006). Dark clasts in this work exclusively refer to hydrated clasts in all kinds of chondritic and achondritic breccias and will be subdivided into C1- and C2-type clasts based on their degree of aqueous alteration (Fig. 4).

3.1.2 Classification of Hydrated Dark Clasts

C1 clasts represent the highest degree of hydrous modifications and include fragments of the brecciated CI-chondrite type, homogeneous fragments that could occur as clasts in CI-breccias, and C1-like fragments (not related to CI chondrites). All C1-like clasts either contain no or only very minor (<2 vol. %) anhydrous phases (e.g., olivine [$(Mg,Fe)_2SiO_4$], pyroxene, metal; Endress 1994; Morlok et al. 2006; Patzek et al. 2018; Alfing et al. 2019; Morin et al. 2022) (Fig. 4a-c). C2 clasts include typical fragments of brecciated CM-related clasts since most (if not all) CM chondrites are breccias (Fig. 4d). However, not all anhydrous mineral-bearing (or chondrule-containing) clasts are CM-related and thus, the neutral description 'C2-type' is more appropriate.

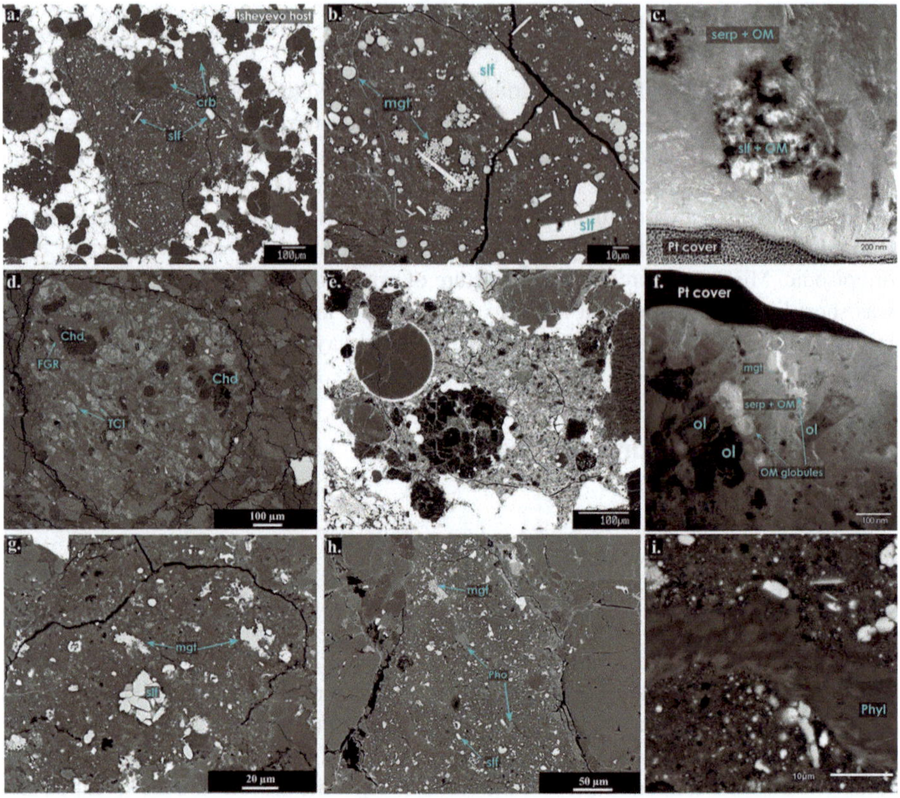

Fig. 4 Hydrated dark clasts from various Solar System objects, including (a) C1 clast from Isheyevo (van Kooten et al. 2017a) including abundant sulfide (slf) and complex carbonates (crb) (b) zoom-in of panel A: framboidal and single magnetite (mgt) grains. (c) TEM cross-section of Isheyevo C1 clast showing a fine-grained serpentine-rich matrix (serp) with diffuse organic matter (OM) and pockets of sulfide with more granular OM (from van Kooten et al. 2017a). (d) C2 clast from polymict eucrite NWA 7542 that contains Mg-rich chondrules (Chd) and mineral fragments surrounded by fine-grained rims (FGR). Fe-rich lumps of tochilinite (TCI), sulfide and carbonate are spread throughout the clast and are embedded in a fine-grained matrix (Patzek et al. 2018). (e) C2 clast from Isheyevo with a large cryptocrystalline and another porphyritic chondrule with sulfide and metal rim, respectively. Anhydrous silicates dominate the matrix (van Kooten et al. 2017a). (f) TEM cross-section of C2 clast from Isheyevo showing globular OM, olivine (ol) grains and small magnetite grains (mgt). (g) C1 clast from ordinary H chondrite SAH 98645 with abundant magnetite (mgt) and (platy) sulfide (slf) in a phyllosilicate matrix. (h) C1 clast from polymict ureilite Dar al Gani (DaG) (Patzek et al. 2018) with a similar mineralogy as panel g, but also containing phosphates (Pho) and large phyllosilicates (phyl, panel i)

3.1.3 Occurrence of Hydrated Dark Clasts

Only very few reports describe the occurrence of hydrated clasts in non-carbonaceous chondrites (e.g., Zolensky et al. 2003, 2016; Funk et al. 2011; Briani et al. 2012; Greshake 2014; Bischoff et al. 2018; Patzek et al. 2018) (Fig. 4d). Phyllosilicate-rich clasts have not been found in enstatite chondrites and members of the Kakangari grouplet. The main reason that C1 or C2 clasts have not yet been found in the most reduced non-carbonaceous (NC, Warren 2011) chondrite groups might be the small number of samples among the recovered meteorites. Ordinary chondrites contain rare C1 clasts, which are most abundant in H-chondrites (e.g., Zolensky et al. 2003; Funk et al. 2011; Briani et al. 2012; Bischoff et al. 2018) and possibly represent very primitive outer Solar System materials (Kebukawa et al. 2019).

Hydrated clasts are absent from many achondrites, including brachinites, acapulcoites, lodranites, winonaites, and angrites. Polymict ureilites are known to contain numerous clasts (e.g., Prinz et al. 1987; Brearley and Prinz 1992; Ikeda et al. 2000; Ikeda and Prinz 2001; Goodrich and Keil 2002; Ikeda et al. 2003; Goodrich et al. 2004; van Kooten et al. 2017b; Patzek et al. 2018) including genomict ones related to ureilite petrogenesis, but also xenolithic clasts such as chondrules (or fragments thereof), C2 clasts (van Kooten et al. 2017b) and C1 clasts with a high abundance of phyllosilicates (Zolensky and Ivanov 2003; Zolensky et al. 1996b; van Kooten et al. 2017b; Patzek et al. 2018; Visser et al. 2018; Goodrich et al. 2019; Bischoff et al. 2021). Based on their H- and Cr-isotope systematics, the ureilite C1 clasts may represent some of the most primitive Solar System materials (van Kooten et al. 2017b; Patzek et al. 2020). C1 material has been described within several HED meteorites (e.g., Brearley 1993; Zolensky et al. 1996a). These clasts resemble CR chondrite matrix and are suggested to represent fossil micrometeorites (Gounelle et al. 2003, 2005).

To our knowledge, phyllosilicate-bearing C1 or C2 clasts have not been reported from CO, CV and CK chondrites. However, dark clasts within these chondrite groups have frequently been described (Fruland et al. 1978; Bischoff et al. 1988; Palme et al. 1989; Kojima et al. 1993; Krot et al. 1997, 1999; Itoh and Tomeoka 2003) and could represent C1 or C2 clasts that experienced metasomatism. During this process they may have lost water, and phyllosilicates were transformed into Fe-rich olivine (Krot et al. 1997, 1999). CM chondrites are well known to be extensively brecciated (e.g., Metzler et al. 1992; Lindgren et al. 2013; Lentfort et al. 2021; Kerraouch et al. 2021, 2022), and it is unclear if some of the clasts represent C2-type xenoliths. Metal-rich carbonaceous chondrites (i.e., CR, CB, and CH) contain the most abundant hydrated clasts of all carbonaceous and non-carbonaceous chondrite groups (up to 25 vol. % in CR2 chondrite Al Rais; Weisberg et al. 1993; Bischoff et al. 1993; Endress et al. 1994), but also contain rarer C2 clasts (Bonal et al. 2010; van Kooten et al. 2017a). The highly anomalous ^{15}N-enrichments in Isheyevo (CH/CB) C2 clasts have been interpreted as reflecting an accretion location in the outermost protoplanetary disk (Briani et al. 2009; Bonal et al. 2010; van Kooten et al. 2016, 2017a). Likewise, C1 clasts with high ^{15}N-enrichments have been discovered in samples returned from the Cb-asteroid Ryugu (Nguyen et al. 2023). Recently, dark clasts from the CR chondrite NWA 14250 have been attributed an outer disk origin, representing a nucleosynthetic isotope endmember for the dust composition of the protoplanetary disk (van Kooten et al. 2024).

3.2 Petrology and Mineralogy of Dark, Phyllosilicate-Bearing Clasts

3.2.1 Anhydrous Components in C2 and C1 Clasts

Ordinary chondrites: Among the 24 hydrous clasts found in a study of H-group regolith breccias (Fig. 4g), Briani et al. (2012) reported on the occurrence of three larger C2 clasts. The smaller C1 microxenoliths (<1 mm) include olivine and pyroxene grains, but chondrules were not observed. The larger C2-like fragments (>1 mm) contain serpentinized relict chondrules (Briani et al. 2012). These chondrules are typically <500 μm in size, but due to scarcity and heavily altered nature the size range is likely biased. Compositionally, the matrix of these C2 clasts is CM-like.

Ureilites: Rare porphyritic, Mg-rich chondrules (<200 μm) have been identified in C1-like clasts in ureilites (van Kooten et al. 2017b; Patzek et al. 2018). Some of these chondrules are rounded and well-preserved (Patzek et al. 2018), but the bulk consists of heavily altered, irregularly shaped chondrules. Ureilites contain C2 clasts with chondrules and CAIs (>200 μm) in a heavily hydrated sulfide-rich matrix. In contrast to C1-like clasts, these chondritic materials have Mg- and Cr-isotope signatures related to NC chondrites and that are most similar to Rumuruti chondrites (van Kooten et al. 2017b) (Fig. 4h).

HED: C2 clasts in HED meteorites are sub-mm to mm in size and contain chondrules and olivine/pyroxene grains and fragments thereof (Wilkening 1973; Chou et al. 1976; Mittlefehldt 1994, 2015; Zolensky et al. 1996a; Gounelle et al. 2003). In general, there is a variability in the mineralogy within the clasts reaching sizes of up to a few mm with varying abundances of chondrules (and fragments thereof), olivine and pyroxene grains, as well as TCIs, sulfides, and carbonates (e.g., Buchanan et al. 1993; Gounelle et al. 2003, 2005; Patzek et al. 2018).

Metal-rich carbonaceous chondrites: For metal-rich carbonaceous chondrites, C2 clasts have only been observed in Isheyevo (Fig. 4e) and the CR chondrite Northwest Africa (NWA) 14250 (van Kooten et al. 2024). Few microchondrules are present in clasts from the CR chondrite El Djouf 001 (Weisberg and Prinz 1991; Endress et al. 1994; Patzek et al. 2018, 2020) and CH chondrite Allan Hills (ALH) 85085 (Greshake et al. 2002) but are not comparable to the 100-200 μm sized chondrules in Isheyevo and 100-500 μm sized chondrules in NWA 14250 (van Kooten et al. 2017a, 2024; Gattacceca et al. 2022). Hence, we consider dark clasts with rare microchondrules as C1-type objects, or possibly, as an intermediate petrology between C1 and C2 clasts. The C2 clasts in the Isheyevo meteorite contain both microchondrules (<50 μm) and larger chondrules (100–200 μm) (Fig. 4e). For the latter, both porphyritic olivine (PO) chondrules and non-porphyritic chondrules with a glassy mesostasis have been observed, which are typically perfectly round but can also occur as fragments. The glassy chondrules are usually surrounded by metal or sulfide rims. The matrix of C2 clasts contains abundant olivine and pyroxene grains.

3.2.2 Matrix Composition and Mineralogy of C1 and C2 Clasts

The mineralogy of C1 clasts generally consists of a phyllosilicate-rich matrix, and includes magnetite framboids as well as isolated grains of magnetite, sulfides, and occasionally accessory phases such as carbonates, phosphate, and chromite [$FeCr_2O_4$]. Chondrules and fragments of anhydrous silicates (mostly forsterite-rich olivine [Mg_2SiO_4]) have only rarely been identified (Patzek et al. 2018). The matrix of C2 clasts is more variable and is typically linked to either CM-like compositions (including tochilinite) and CR-like compositions (including magnetite). We highlight below some variations between different clasts, which illuminate the impact of aqueous alteration and/or precursor materials.

Ordinary chondrites: Submillimeter-size C1 clasts in ordinary chondrites are divided into (1) hydrous magnetite-rich, (2) semi-hydrous magnetite-rich, and (3) semi-hydrous magnetite-poor (Briani et al. 2012). While all are dominated by a fine-grained phyllosilicate matrix with pyrrhotite laths, spheroidal, hexagonal and framboidal magnetite and rare dolomite and breunnerite, the second and third groups have abundant olivine and pyroxene. In the third group, Fe-Ni metal is replaced by magnetite, anhydrous silicates are less abundant and pyrrhotite dominates over troilite.

Ureilites: C1 clasts are dominated by fine-grained phyllosilicate-rich matrix and varying abundances of magnetite, sulfide, carbonate, and isolated grains of olivine and pyroxene (rarely chondrules) indicating intense aqueous alteration under slightly different alteration conditions (Patzek et al. 2018).

HED: Zolensky et al. (1996a) state that these clasts possess important similarities to matrices of CR2 chondrites with a phyllosilicate-rich matrix in which olivine grains and aggregates are embedded along with framboidal magnetite and carbonates. In general, the C1 clasts in HEDs are mineralogically similar to C1 clasts in ureilites including the variable occurrence of individual minerals such as olivine, pyroxene, magnetite, lath-shaped sulfides, and carbonates (Patzek et al. 2018).

Metal-rich carbonaceous chondrites (*CR, CH, CB*): Electron microprobe imaging of hydrated C1 clasts in metal-rich carbonaceous chondrites shows that these clasts are dominated by phyllosilicates (typically serpentine) and contain abundant accessory phases such as magnetite (framboidal and micron-sized individual grains), sulfides (mainly laths of pyrrhotite and pentlandite), carbonates (typically dolomite) and to lesser extent phosphates and chromite (Endress et al. 1994; Greshake et al. 2002; Bonal et al. 2010; van Kooten et al. 2017a; Patzek et al. 2018). Detailed analyses of the phyllosilicate matrix mineralogy by TEM show areas of serpentine, interwoven with organic matter, as well as regions covered by sulfides and larger organics (van Kooten et al. 2017a). TEM sections of C1-like dark clasts in NWA 14250 show very limited aqueous alteration and consist predominantly of organic-sulfide-rich regions (resembling GEMS-like materials and amorphous silicates (van Kooten et al. 2024)).

Detailed accounts of the matrix mineralogy in C2 clasts from metal-rich carbonaceous chondrites stem solely from the CH/CB chondrite Isheyevo (Bonal et al. 2010; van Kooten et al. 2017a). Secondary alteration products such as sulfides, magnetite and carbonates are absent in C2 clasts, but coarse phyllosilicates, although rare, are present with compositions indistinguishable from C1 clasts. C2 clasts contain abundant anhydrous silicates and Fe-Ni metal beads that are embedded in the matrix (<100 μm). Most of these silicates are Mg-rich low-Ca pyroxenes (En_{97-99}), whereas olivines are less abundant and compositionally more variable (Fo_{83-98}) (van Kooten et al. 2017a). TEM analyses of the matrix show very fine-grained phyllosilicates that are embedded together with organic matter (diffuse and globular) and magnetite grains between anhydrous silicates, some of which show alteration at the edges towards phyllosilicates.

3.2.3 Changes in Organic Matter with Aqueous Alteration

Due to size limitations, detailed studies of the organic inventory of hydrated clasts are so far limited to those observed within the Isheyevo chondrite (van Kooten et al. 2017a) and to the H chondrite Zag clast (Kebukawa et al. 2019). TEM imaging of Isheyevo C2-like clasts shows the presence of small organic nanoglobules ($\sim$100 nm diameter) with increased C=O group functionality relative to globules in CR, CM, and CI chondrites (Le Guillou et al. 2014) (Fig. 4f). De Gregorio et al. (2013) show correlations between nanoglobule size, functional group chemistry and degree of meteorite hydration that identify the importance of increasing aqueous alteration on the growth of nanoglobules in CR and CM chondrites. Hence, the small nanoglobules in C2 clasts from Isheyevo have been interpreted to be primitive organic components formed in the cold regions of the interstellar medium. The more hydrated C1 clasts in Isheyevo are depleted in C=O bonding and enriched in N-H functional groups, whereas the C2 clasts contain relatively more C-N groups. It is uncertain, from the structural information alone, if the decrease in C=O and C-N bonding environments in C1 relative to C2 clasts is the result of increasing aqueous alteration, but correlated N- and H-isotope data of these clasts suggest that they accreted isotopically different organic precursors and ices (i.e., NH_3 and HCN; van Kooten et al. 2017a). The pristine dark clasts in NWA 14250 invoke the presence of ISM-derived D- and ^{15}N-rich organic matter that is coated with D-poor and extremely ^{15}N-rich ices. As such, the ^{15}N-enrichments observed in metal-rich carbonaceous chondrite dark clasts are likely outer disk signatures from vertical mixing of small icy grains (van Kooten et al. 2024).

3.3 Hydration of Clasts

3.3.1 What Is the Protolith of These Clasts?

Variations in the overall mineralogy of hydrated clasts involve different abundances of accessory phases and the presence of anhydrous minerals such as olivine, pyroxene, and metal grains. If we compare this general description of C1 clasts to CI chondrites, we observe many similarities to CI chondrites and samples returned from asteroid Ryugu (e.g., Yokoyama et al. 2023). The CI chondrites are chemically heterogeneous objects at small scales ($<$ 200 mg; Barrat et al. 2012; see also Morlok et al. 2006), which suggests variations in mineralogy. This small size could potentially be a reason for the observed variations in mineralogy of the hydrated clasts. *In situ* S-isotope studies of sulfides in C1 clasts and CI chondrites indicate a difference in the isotopic reservoir and therefore it appears unlikely that CI chondrites and C1-like clasts from various groups share a common origin (Visser et al. 2019). This view is further strengthened by the bulk isotopic composition of hydrated clasts in ureilites (van Kooten et al. 2017b; Patzek et al. 2019) and Isheyevo (van Kooten et al. 2017a). Thus, the protoliths for different groups of C1 clasts possess a similar initial mineralogy, although the accreted material formed in unique isotopic environments.

In some C1 clasts, we find evidence for pseudomorphs of chondrules that have been heavily altered. From bulk CI chondrites, it is clear from the absence of chondrule pseudomorphs that a relationship to C2 chondrites by progressive aqueous alteration is ruled out. Very few refractory inclusions have been reported from CI chondrites (Frank et al. 2023). Moreover, nucleosynthetic isotope signatures of bulk chondrites have established different accretion regions for their respective parent bodies. Even for the unique C1 chondrite Flensburg, which has many overlapping characteristics with CM and CI chondrites, a genetic relationship to CI chondrites is unlikely (Bischoff et al. 2021). Unless we can observe a progressive mineralogical sequence between C1 and C2 clasts, as well as identical nucleosynthetic isotope fingerprinting, we cannot confirm the link between these clasts. For example, Isheyevo C1 and C2 clasts share the same nucleosynthetic Cr- and Mg-isotope systematics, but their organic structure and isotope composition excludes a shared origin on the same parent body (van Kooten et al. 2017a). In polymict ureilites, C1 and C2 clasts have very different Cr- and Mg-isotope compositions, as well as a different matrix composition and mineralogy (van Kooten et al. 2017b).

3.3.2 Timing and Location of Clast Hydration

It is important to constrain whether hydrated clasts were altered on a larger body with fluid circulation or if the alteration occurred in a closed system similar to IDPs, including possible interaction with a gaseous environment in the disk or by cosmic ray interaction. It is also possible for these clasts to have avoided melting of accreted ices (i.e., in the absence of live ^{26}Al), but having experienced alteration within the host parent body. Furthermore, clasts could have also undergone multiple alteration events on their own precursor parent bodies, and subsequently in the host parent body rendering it important to constrain their alteration histories.

Hydrated clasts from the CH/CB chondrite Isheyevo record an aqueous alteration age (i.e., 4 Myr after CAI formation) that is about 1 Myr older than the upper limit accretion age of CB chondrite Gujba (i.e., with chondrule ages of 5 Myr after CAI formation). Hence, these clasts were altered before incorporation into the host parent body, and presumably, given the continuous nature of H-isotope systematics, on the same parent body (van Kooten et al. 2017a).

For hydrated clasts in other meteorites, the timing of incorporation can, so far, only be determined from textural observations. For example, hydrated clasts in type >5 ordinary chondrites were clearly incorporated after thermal metamorphism (>50 Myr after CAI formation; Blackburn et al. 2017), but experienced hydration before this.

Some clasts appear to be overprinted by hydrothermal alteration of the host parent body: dark clasts within CO3 and CV3 chondrites have frequently been described (e.g., Fruland et al. 1978; Bischoff et al. 1988; Palme et al. 1989; Kojima et al. 1993; Krot et al. 1997, 1999; Itoh and Tomeoka 2003) and could represent C1- or C2-like clasts that experienced metasomatism during which phyllosilicates were transformed into Fe-rich olivine (e.g., Krot et al. 1997, 1999). For such clasts, it may be difficult, if not impossible, to constrain earlier alteration events.

In general, it seems likely that there were many similar small, water-rich parent bodies. Many were probably destroyed by impact and fragments (clasts) were included in the regolith of other parent bodies (ureilites, HEDs, OCs) or re-accreted with other components to form new bodies (e.g., CR, CH, asteroid 2008TC3).

There are a number of open questions that need to be addressed in future studies, and are described in the Supplementary Text.

4 Low-Temperature Alteration Effects on Presolar Grains

Primitive meteorite matrices, micrometeorites, and IDPs, as well as cometary and asteroid dust returned by NASA and JAXA missions, contain low concentrations of refractory dust grains which condensed in the ejecta of evolved stars or stellar explosions (novae and supernovae) before the formation of the Sun and the Solar System. This "presolar" dust is characterized by highly anomalous isotopic compositions representing nucleosynthetic fingerprints of their stellar parents. After injection into the interstellar medium (ISM), these grains were exposed to highly energetic irradiation, shockwaves from nearby supernova explosions, and grain-grain-collisions. Presolar grains found today in primitive Solar System materials largely escaped alteration and homogenization processes in the early Solar System (e.g., Zinner 2014; Nittler and Ciesla 2016). In contrast to the more refractory presolar dust species (silicon carbide [SiC], graphite, silicon nitride [Si_3N_4], oxides), silicates cannot be separated chemically from their meteoritic hosts (Nguyen and Zinner 2004). Instead, they have to be analyzed *in situ* or among physically separated matrix grains with high spatial resolution techniques like NanoSIMS ion imaging (e.g., Hoppe et al. 2013), or SIMS+SCAPS (stacked CMOS-type active pixel sensor) detector systems (e.g., Nagashima et al. 2004). It is estimated that the dust fraction of the presolar molecular cloud from which our Solar System formed consisted of $\sim$3% circumstellar grains; while 97% of the population formed in the ISM. In the following, the term "presolar" refers to *circumstellar* presolar grains (Zhukovska et al. 2016; Hoppe et al. 2017; Alexander et al. 2017a). Silicates are the most abundant presolar dust species ($\leq$250 ppm in meteorite matrix, up to the percent level in individual IDPs) and are large enough for the isotopic analysis of individual grains (e.g., Floss and Haenecour 2016; Nittler et al. 2018, 2021), followed by SiC (several tens ppm), refractory oxides (a few ppm to a few ten ppm), graphite (a few ppm), and Si_3N_4 (at the ppb-level) (e.g., Davidson et al. 2014a; Zinner 2014).

4.1 Presolar Grain Abundances and Petrologic Type of the Host Material

The various presolar mineral phases are affected differently by low-temperature parent body aqueous alteration or thermal metamorphism. Therefore, comparing the abundances of the different presolar grain species should yield information on the degree of alteration of the grains and their host material, although several conditions have to be met: (i) the presolar grain inventories of all fine-grained primitive Solar System materials were largely the same prior to any alteration processes; (ii) the sample size (i.e., the scanned area per meteorite) has to be sufficiently large to monitor potential heterogeneous distributions of presolar dust within the matrix of a given meteorite ($> 10,000\ \mu m^2$ per meteorite).

Silicates are more susceptible to aqueous alteration than more refractory presolar grain species like Al-rich oxides and SiC. Thus, with progressive alteration, presolar silicate abundances will decrease (e.g., Floss and Haenecour 2016) (Fig. 5). Although we observe a general trend in the presolar silicate abundances that supports this concept, this is not always the case (Fig. 5). This is evident for the CRs and CM chondrites. Among the CRs, the degree of aqueous alteration varies widely (Sect. 2.1). There is an ongoing discussion whether the CR chondrites with high presolar grain abundances (MET 00426, QUE 99177, GRV 021710) are petrologic type 2 or 3. Matrix in MET 00426 and QUE 99177 shares many similarities with the fine-grained material in ALH 77307 (CO3.00) and Acfer 094 (C-ung3.00) (Abreu and Brearley 2010), and the O-rich presolar grain abundances are also comparable (Fig. 5). However, there is evidence for aqueous alteration (Le Guillou and Brearley 2014), which would argue against a classification as type 3. Nevertheless, from a presolar grain perspective, the silicate stardust abundances in MET 00426, QUE 99177, and GRV 021710 are clearly higher than those in "moderately" altered CR2 chondrites, and also in the CM chondrites, with the exception of Asuka (A-) 12169, which is also considered to be "very primitive" (Fig. 5).

Thus, a more sensitive scale for the degree of aqueous alteration experienced by CR chondrites is desirable and could ultimately allow differences to be discerned in the presolar grain abundances due to contrasting degrees of alteration and inherent heterogeneities in the distribution of presolar dust. Several studies have attempted to establish more refined classification schemes (Alexander et al. 2013; Harju et al. 2014; Howard et al. 2015). These approaches do not provide petrological classifications that reflect the vast differences in presolar grain abundances and distributions in CR chondrites. For more extensively altered CRs, the scale devised by Harju et al. (2014) is consistent with the significantly lower presolar silicate and oxide abundances observed for Renazzo (CR2.4) (Fig. 5). In contrast, e.g., Alexander et al. (2013) assigned a petrologic type to Renazzo which puts it in line with the most pristine CR chondrites (CR2.5), indicating less aqueous alteration than in QUE 99177 (CR2.4). Several possible explanations for this apparent discrepancy exist: (i) the CRs might have been heterogeneously altered, and in addition, variations in alteration temperatures could have affected the fraction of surviving presolar silicates and oxides (Alexander et al. 2013); (ii) TEM investigations showed that amorphous silicates in the minimally altered CR chondrite MET 00426 contain water (Le Guillou and Brearley 2014), which has been ascribed to the simultaneous accretion of dust and ice. The ice could have produced localized centers of aqueous alteration on sublimation, which would explain the survival of larger amounts of presolar grains in MET 00426 and the other minimally altered CRs than in the more extensively altered CR2 chondrites, which might have experienced higher alteration temperatures. This model could also be applicable to the CM chondrites. Here, the presolar silicate and oxide abundances for petrologic types $\leq$2.6 are generally low and suffer from comparably large statistical errors (due to small grain numbers), while A-12169, the less altered regions of Paris (estimated $\geq$2.8), and A-12236, show higher abundances (Verdier-Paoletti et al. 2020; Nittler et al. 2021). The bulk O-isotope compositions of fine-grained matrix of CR chondrites (Schrader et al. 2014) appear to correlate quite well with both the degree of alteration of the respective meteorite and the abundances of presolar O-anomalous

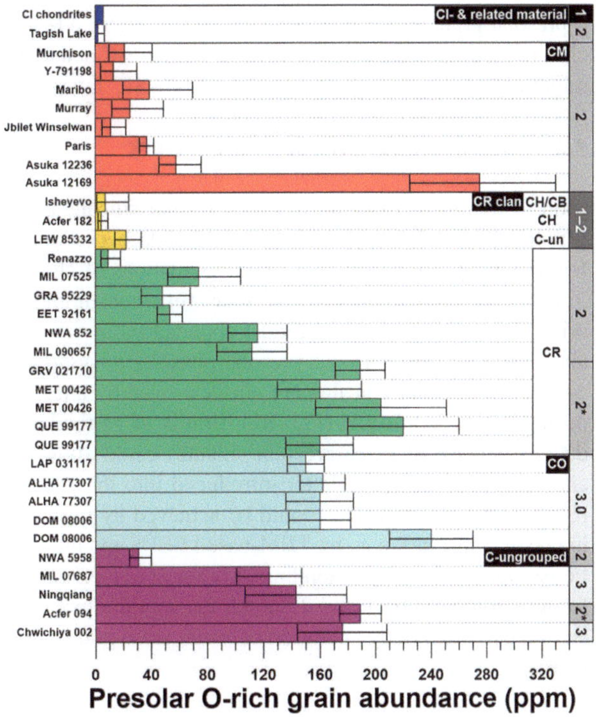

Fig. 5 Matrix-normalized presolar O-rich grain abundances together with the petrologic types of the host meteorites. All errors are 1 sigma. Data sources: Barosch et al. (2022); Marhas et al. (2006); Leitner et al. (2012, 2016, 2018, 2020); Verdier-Paoletti et al. (2020); Nittler et al. (2018, 2020, 2021); Davidson et al. (2014b, 2015); Zhao et al. (2011, 2013); Floss and Stadermann (2009); Nguyen et al. (2010); Haenecour et al. (2018); Vollmer et al. (2009); Smith et al. (2023). *These meteorites are officially classified as type 2, but suggestions that they be considered as type 3/3.0 are supported by their high presolar silicate abundances

grains. So far, there is only little overlap between the sample sets, and differing results for individual samples from the same meteorite seem to indicate heterogeneous alteration, and more work is needed.

A combination of low- and high-temperature alteration could potentially "reset" petrologic type and be responsible for comparably low presolar silicate abundances in several chondrites that appear to be pristine at a first glance. LEW 85332, an ungrouped member of the CR clan, has been classified as C3.0–3.1 (Rubin and Kallemeyn 1990). However, its matrix was aqueously altered, and subsequently heated to ~500–600 °C (Brearley 1997; Tonui et al. 2002), and despite its apparent low petrologic type, it has a low presolar silicate abundance (~22 ppm; Leitner et al. 2018).

The abundances of presolar oxides are typically low (a few ppm) and the small grain numbers are responsible for large errors for the respective ratios. Thus, statistically significant silicate/oxide ratios are only available for very thoroughly studied meteorites, or specimens that contain unusually high abundances of presolar oxides (e.g., Leitner et al. 2012, 2020 and references therein). Haenecour et al. (2016) investigated the silicate/oxide ratio of sub-micrometer-sized matrix silicates and oxides in LAP 031117 (CO3.0). The elemental compositions and compositional range of both presolar and non-presolar silicates showed no significant differences. This indicates that the sub-micrometer-sized non-presolar silicate and oxide populations could also be used to assess the degree of alteration of a given matrix area, but with better statistics.

A more promising approach might be to use the presolar SiC abundances instead, which are typically in the range of a few tens of ppm (e.g., Davidson et al. 2014a) (Fig. S1). However, this requires even more effort, since both O-anomalous and C-anomalous grain abundances have to be measured. Such investigations are relatively time-consuming, especially if presolar grain abundances are low. Moreover, for a few meteorites, as well as for a matrix clast in material from asteroid Ryugu, higher SiC abundances have been observed; this fact might complicate the use of presolar silicate/SiC ratios as indicators of matrix alteration.

4.2 Remaining Questions and Future Work

More comprehensive studies of the presolar silicate and oxide inventories of moderate to heavily altered chondritic matrix would yield better statistics and allow monitoring of differences between the presolar grain populations of chondrites of petrologic types 1, 2 and 3. However, such studies will be more time-consuming than those of more pristine meteorites, owing to the lower stardust abundances in the target material, especially when the abundances of carbonaceous presolar phases are investigated in addition.

Even if presolar silicates are not destroyed by aqueous alteration, their chemical compositions can be affected (Fig. S2). Iron is of particular interest in this context, and more detailed studies of Fe in presolar silicates could allow discrimination between "primary" and "secondary" Fe, the latter introduced into the grains by aqueous alteration processes (Supplementary text). This could be achieved by measurements of the Fe oxidation states in presolar silicates, e.g., by TEM-based EELS and high-spatial resolution measurements of Fe-isotopes. Generally, extending the dataset of elemental compositions of presolar silicates from petrologic type 1 and 2 chondrites would allow searching for systematic differences caused by parent body alteration between these grains and those from highly primitive meteorites.

5 Evolution of Organic Matter During Low-Temperature Aqueous Alteration

5.1 Introduction

Organic matter likely underwent modification during aqueous alteration. It is divided into soluble organic material (SOM) and insoluble organic matter (IOM). The SOM is extracted using water or organic solvents, while the IOM is the residue recovered after demineralization. IOM shares similarities with type III terrestrial kerogen (Quirico and Bonal 2019). Unlike kerogens, IOM and SOM exhibit enrichments in D and ^{15}N (Robert and Epstein 1982; Kerridge et al. 1987; Alexander et al. 2007). Chondritic organic matter was likely synthesized at very low temperatures (<40 K) in the local interstellar medium or in the outer solar nebula (Robert and Epstein 1982; Kerridge 1983; Yang and Epstein 1983, 1984; Alexander et al. 1998, 2007; Remusat et al. 2006). IOM displays elemental, structural and isotopic variations within and between chondritic groups (e.g., Alexander et al. 2017b). However, the causes of these variations are still debated. Do they originate from different mixtures of heterogeneous precursor materials, or are they the products of aqueous alteration in the chondrite parent bodies? (Alexander et al. 2007, 2010, 2017b; Remusat et al. 2010; Orthous-Daunay et al. 2013; Le Guillou et al. 2014; Vollmer et al. 2014, 2020b; Vinogradoff et al. 2017; Changela et al. 2018).

5.2 The Impact of Aqueous Alteration on the Organic Matter

The CR, CM, and CI chondrites, along with the ungrouped C2 Tagish Lake chondrite, exhibit varying degrees of aqueous alteration, offering valuable insights into the evolution of organic matter through aqueous alteration. Pearson et al. (2002) first noticed the spatial association between organic matter and clay minerals using SEM and osmium labelling, that was later confirmed by scanning transmission X-ray microscopy (STXM) and TEM studies (Garvie and Buseck 2007; Le Guillou et al. 2014; Vinogradoff et al. 2017). Le Guillou et al. (2014) reported diffuse organic components intimately distributed among phyllosilicates and connected to fractures, likely indicating fluid transport at the micrometer scale. Changela et al. (2018) compared CR2 with CR1 chondrites (Al Rais and GRO 95577) and interpreted the observed molecular changes as cracking occurring in the presence of fluid at temperature above 150 °C.

Other indications of the influence of aqueous alteration on organic matter are evident in the diverse lithologies of the Tagish Lake chondrite, which have experienced varying alteration conditions (Zega et al. 2010; Blinova et al. 2014b). Alexander et al. (2014) found a correlation between the degree of alteration of the host lithology and decreases in both the δD values and bulk H/C, accompanied by increases in aromaticity of their IOM. These authors suggested that the IOM in Tagish Lake lost its D-rich material and/or underwent isotopic exchange with water possibly as a result of short-term heating. Vinogradoff et al. (2017) also observed similar trends between the IOM extracted from the weakly altered CM Paris and the more altered CM Murchison. Laser desorption ionization coupled to ultra-high-resolution mass spectrometry on various IOM from CM chondrites reveals the gradual loss of heteroatoms and increase in aromaticity with increasing alteration degree (Laurent et al. 2022a). It also shows that the IOM of CMs and CIs preserve records of slightly different precursors.

The diverse chondritic groups might have accreted IOM of different compositions, suggesting that aqueous alteration only played a minor role. Remusat et al. (2010) reported D-rich IOM hotspots in the matrices of CR, CM, and CI chondrites, suggesting minimal or no isotopic diffusion has occurred between the organic matter and hydrous minerals. Conversely, Alexander et al. (2017b) argued that water-organic exchange is slower than diffusion into the grains. Other evidence from the alkyl abundance (CH_2 and CH_3) in aqueously altered chondritic IOM shows no correlation with alteration degree (Orthous-Daunay et al. 2013), but instead reflects thermal processing or precursor heterogeneity (Quirico et al. 2014).

5.3 Soluble Organic Matter

The SOM in carbonaceous chondrites is composed of huge diversity of compounds, with tens of thousands of molecules containing C, H, N, O and S (Schmitt-Kopplin et al. 2010), a diversity also found in Ryugu samples (Schmitt-Kopplin et al. 2023). There have been many studies focusing on amino acids, nucleobases, carboxylic acids and amines (see Glavin et al. 2018 for a review), while recent papers have reported the distribution of polycyclic aromatic hydrocarbons (PAHs) in CM chondrites (Sephton et al. 2000; Kalpana et al. 2021; Graham et al. 2022; Lecasble et al. 2022). If aqueous alteration seems to have a subtle impact on the molecular distribution of these chemically recalcitrant molecules, their H-isotope composition has likely been disturbed by exchange with the fluid during aqueous alteration (Lecasble et al. 2023). Their C-isotope signatures, however, are inherited from pre-accretion synthesis (Naraoka et al. 2000).

Amino acids, which have been studied extensively due to their biological interest, are found in over 90 different forms, including isomers (Glavin et al. 2018). Their abundances significantly vary within and between chondrite groups, with the highest contents reported in some CR chondrites (up to 300 ppm vs. 80 ppm in the Murchison meteorite). In the

CM and CR groups, the amino acid content decreases with increasing aqueous alteration (Martins et al. 2007; Glavin et al. 2020), suggesting that aqueous alteration might cause amino acid decomposition or introduce analytical bias, as more altered chondrites contain more phyllosilicates that could trap amino acids. Amino acid yields increase when the water extract is subjected to acid hydrolysis, likely indicating the presence of small peptides or amino acids precursors in carbonaceous chondrites (Serra et al. 2022). The amino acids form through various synthetic routes, including the Strecker synthesis, Michael addition or formose reaction (Elsila et al. 2012; Vinogradoff et al. 2020b). Some amino acids can exhibit enantiomeric excesses (Engel and Nagy 1982), with excesses of the L-form, notably of the non-proteaginous isovaline, reported in various carbonaceous chondrites (Glavin and Dworkin 2009). These excesses tend to be larger in more altered meteorites, suggesting enantiomeric excess amplification during parent body alteration (Glavin and Dworkin 2009), though irradiation by circular polarized light could also explain the enantiomeric excess (Modica et al. 2014).

5.4 Laboratory Experiments

Experimental studies can help to better understand the behaviour of organics during hydrothermal alteration. Two approaches have been followed: (i) study of the evolution of molecules akin to those detected in chondrites, including SOM and IOM, (ii) assumptions about the nature of accreted molecules.

Hydrous pyrolysis experiments on IOM from Murray and Murchison at $\sim$300 °C revealed a decrease of the H/C, δD and δ^{15}N values, and a lower aliphatic content (Yabuta et al. 2007; Oba and Naraoka 2009). The release of PAHs and carboxylic acids in the aqueous solutions showed that IOM can be a source of SOM. Laurent et al. (2022b) subjected the IOM of the CM Paris to aqueous alteration at 150 °C for $\sim$50 days. They observed that the fraction of aromatic C increases with alteration, while D-rich hotspots are almost unaffected. The temperature and duration of the experiments affect the molecular and isotope evolution of the IOM, with rapid modifications of the isotopic, elemental and structural properties occurring at $\geq$250 °C (Foustoukos et al. 2021). Kebukawa et al. (2021) suggested that D/H isotopic exchange between D-rich IOM and D-poor water must occur slowly at low temperatures (probably close to 0 °C) to prevent equilibration of the D/H ratio between organic matter and water on timescales of millions of years.

Formaldehyde (H_2CO) or hexamethylenetetramine (HMT, $(CH_2)_6N_4$) could have been accreted directly, as they are found in the interstellar medium, comets or ice irradiation experiments. Upon heating in the presence of water, both go through the formose reaction and polymerize. This reaction has also been proposed to account for the formation of sugars (Furukawa et al. 2019). Cody et al. (2011) showed that formaldehyde in aqueous solution at 250 °C could form IOM and globule-like morphologies similar to chondritic IOM. Follow up studies investigated the effect of NH_3 in the original mixtures, as well as the kinetic evolution of the products (Kebukawa et al. 2013; Kebukawa and Cody 2015). HMT follows a similar behaviour (Vinogradoff et al. 2017). The presence of phyllosilicates modifies the chemical pathways, lowering IOM yield and trapping organic matter within their interlayer (Vinogradoff et al. 2020a; Viennet et al. 2022). Amino acids are also produced in all these experiments (Kebukawa et al. 2017; Vinogradoff et al. 2020b).

5.5 Remaining Questions and Future Work

Critical questions relating to the effects of low-temperature alteration on organic matter are described in the Supplementary text.

6 Physicochemical Modelling and Hydrothermal Alteration Experiments

Important questions remain regarding the physicochemical conditions (e.g., pH, fO_2, fluid composition, etc.), mechanisms, and rates of alteration in chondrite parent bodies. Two main approaches have been used to address these questions: (1) thermodynamic and kinetic modelling of alteration processes in systems of different compositions and/or (2) alteration experiments carried out under controlled settings. Both approaches have some limitations. In thermodynamic modelling, the most serious limitations are: (*i*) the lack of appropriate thermodynamic data for the minerals/phases of interest and (*ii*) the lack of complete chemical equilibrium in the system even though some coexisting phases may have been in local equilibrium. The major limitation of laboratory experiments are: (*i*) slow rates of alteration reactions and (*ii*) a large number of physicochemical parameters controlling such reactions. As a result, a comprehensive experimental study of these processes becomes economically prohibitive. However, a proper combination of both approaches, with thermodynamic and kinetic modelling followed by intelligently designed experiments, provides insights into relative sequences of complicated alteration reactions and physicochemical conditions under which the observed reactions most likely occurred.

6.1 Physicochemical Modelling

Two major types of thermodynamic modelling used in simulation of alteration processes are (1) evaluation of observed alteration reactions at a range of pressure (P), temperature (T), and activities of involved phases and ions (e.g., Alexander et al. 1989; Krot et al. 1998) or (2) calculation of full chemical equilibrium in a system of assumed composition at specified T and P (e.g., Zolensky et al. 1989; Rosenberg et al. 2001; Schulte and Shock 2004; Zolotov et al. 2006). The latter type of modelling is briefly reviewed here; details can be found in the original publications referenced below.

Zolensky et al. (1989) were the first to use an equilibrium code (Eq. 3/6) to model interaction between liquid water containing dissolved CO_2 with concentrations of total C ranging from 10^{-8} to 10^{-2} moles and varying amounts of an anhydrous mineral assemblage (olivine + orthopyroxene + glass + spinel + chromite + Fe-Ni metal + Fe sulfides) similar to either CM or CV chondrites at temperatures of 1, 25, 50, 99, and 150 °C. While the pressures are not explicitly reported, these should have been at least equal to the saturated pressures of H_2O vapor at these temperatures, *i.e.* 0.007, 0.03, 0.12, 0.98, and 4.75 bars, respectively. It was found that the alteration mineralogy of the CM chondrites is best reproduced by interaction of the anhydrous precursor mineralogically similar to the CM chondrites with water containing C in a wide range of concentrations at temperatures of 1 to ~25 °C. The alteration mineralogy is essentially the same for a wide range of W/R ratios. Under these conditions solution pH changes from 7 to >12, with most alteration taking place at a pH greater than ~10. Solution Eh changes from −0.5 to −7.5 V, with increasing alteration. The mineralogy of the CI chondrites can be well-reproduced by alteration of either CM or CV anhydrous precursor at temperatures of 50–150 °C and high C concentrations in aqueous solution of 10^{-3}–10^{-2} moles for a wide range of W/R ratios. Under these conditions solution pH and Eh change from 7 to ~10 and from −0.3 to −0.8 V, respectively, as alteration progresses

A follow up study by Rosenberg et al. (2001) focussed on more detailed modelling of CM chondrite alteration using the REACT code, which is very similar to Eq. 3/6. The calculations were performed at 25 °C using two different precursors, 'reduced' and 'oxidized', with most Fe being present as metal in the former and fayalite [Fe_2SiO_4] in the latter, and about the same amount of Fe sulfide in both. The mineralogy of the precursors is qualitatively similar to that of Zolensky et al. (1989), but their chemical compositions differ significantly. The results of modelling are generally consistent with those of Zolensky et al. (1989). The alteration mineralogy of the 'reduced' and 'oxidized' precursors is mostly similar. The main difference is a release of large volumes of H_2 gas in the former that has to be lost (open system behaviour) to allow alteration reactions to proceed further and produce Fe-rich silicates. Rosenberg et al. (2001) noted that the aqueous solutions produced by such alteration are 'dominated by Ca and Si and are strongly basic and reducing'.

Schulte and Shock (2004) used the Eq. 3/6 code to model reactions of a rock of the average CM chondrite composition with a variety of different aqueous solutions of unspecified compositions at temperatures of 2–200 °C and pressures of saturated H_2O vapor, focussing on the possible formation of soluble organic matter during aqueous alteration. Their results are generally consistent with the studies mentioned above. They concluded that "alteration of the CM parent body occurred at temperatures perhaps as high as ~150 °C at a W/R ratio greater than 1 and involved a fluid with concentrations of C and N greater than 0.1 molal. The fluid was initially more oxidizing than the rock and the pH of the fluid was much greater than neutral."

Since 2006, Zolotov and colleagues (Zolotov 2012, 2014, 2017; Zolotov and Mironenko 2007, 2008a, 2008b; Zolotov et al. 2006, 2015) investigated water-rock interaction in several systems which emulate aqueous alteration in CR, CM, CI, CV and ordinary chondrites. They used the equilibrium GEOCHEQ code equipped with a module accounting for mineral dissolution kinetics in addition to phase equilibria. The open and closed chemical systems explored include major rock-forming and volatile elements (O, H, C, Cl, S, Mg, Fe, Ca, Na, K, Mn, Si, Al, Ni, Co, Cr, P) partitioned among solid, gaseous, and aqueous solutions at temperatures and pressures up to ~300 °C and 300 bars, respectively, W/R ratios up to 1000, and a wide ranges of assumed porosities of unaltered precursors. So far it is the most comprehensive physicochemical study of aqueous alteration of primitive meteorites and their putative or sampled (asteroid Ryugu; Nakamura et al. 2023) parent bodies. In addition to mineralogical changes, which are generally consistent with the previous results described above, they report a great deal of information on chemical composition and speciation of gaseous and aqueous solutions and their relative amounts that is very useful for planning laboratory experiments on hydrothermal alteration. The major finding of these studies are (1) depending on local W/R ratio and T during interaction of primary anhydrous chondritic minerals with H_2O-rich ices, H_2O can be present as mostly vapor or vapor + aqueous solution or mostly aqueous solution; (2) the mineralogies of alteration products are relatively independent of starting 'chondritic' composition and are mainly controlled by the local W/R ratios, T, amounts of anion-forming elements such as Cl, C, N in a water source/ice, and the dissolution kinetics of primary phases; (3) the dissolution rates of primary minerals increase in order forsterite → enstatite/diopside [$Mg_2Si_2O_6$/$CaMgSi_2O_6$] → feldspar → Fe-Ni metal resulting, as alteration progresses, in formation of magnetite + Fe-rich serpentine first, followed by Mg,Fe-serpentine + magnetite + Ca-Na-K-Fe-bearing saponite ± pyrrhotite, and finally Fe-rich phyllosilicates + fayalite when solutions is consumed; (4) the pH of the aqueous solution changes from neutral or slightly acidic (if Cl is present in the ices) at the beginning to progressively more alkaline; (5) the dominant solutes are Na^+, Cl^-, HCO_3^- with subordinate amounts of Ca^{2+}, Mg^{2+}, Fe^{2+}, etc.; (6) oxidation of sulfides is kinetically inhibited, so formation of sulfates requires photochemical and radiolytic processes that produce O-rich oxidants either derived from or accreted with ices.

6.2 Hydrothermal Alteration Experiments

These experiments aim to replicate alteration conditions inferred from meteorites, tracking mineralogical and chemical evolution at every stage. Comparing experimental products with meteorites can test models of alteration and constrain key factors such as the initial mineralogy, timescales, fluid compositions, and W/R mass ratios.

The original mineralogies and initial stages of alteration are difficult to constrain in highly altered CM and CI chondrites where little of the precursor material remains. Nonetheless, mineralogical observations of the least altered type 3 carbonaceous chondrites suggest that Fe-Ni metal grains, anhydrous silicates and/or amorphous silicates were precursors to secondary minerals (Abreu and Brearley 2010; Davidson et al. 2019; Dobrică and Brearley 2020). That the matrices of the least altered chondrites often contain amorphous silicate grains that are susceptible to aqueous alteration (Leroux et al. 2015) has led several studies to use amorphous materials as reactants in alteration experiments (e.g., Nakamura-Messenger et al. 2011). Le Guillou et al. (2015b) produced micrometer-thick layers of synthetic amorphous or crystalline silicates of fayalitic composition and exposed them under anoxic conditions to deionized water at a W/R mass ratio of 150, and T of 90 °C for two weeks or 190 °C for two hours. At both temperatures, the surfaces of the amorphous films were altered into an amorphous assemblage with a Fe-rich serpentine and a Mg-rich phyllosilicate composition (similar in composition to saponite), replicating some important aspects of the amorphous Fe-rich material and oxidized phyllosilicates reported in CR chondrites.

To understand the formation mechanisms of TCIs in CM chondrites (see Sect. 2.2), Vacher et al. (2019b) conducted hydrothermal experiments on synthetic glass, iron metal (Fe^0) particles and forsterite at 80 °C for 30–60 days, at a W/R mass ratio of 10 under anoxic conditions. The experiments produced crystals of tochilinite and cronstedtite from the reaction between synthetic glass, Fe^0 and S-bearing or S-free water, respectively, and demonstrate that TCIs are likely formed from the dissolution of amorphous silicates and Fe-Ni metal in CM chondrite matrices.

Forsterite and enstatite are the main anhydrous silicates in carbonaceous chondrites, often occurring as chondrule phenocrysts. Ohnishi and Tomeoka (2007) explored hydration of enstatite, altering synthesized Mg-enstatite grains at temperatures of 100–300 °C and pH of 0–14 for between 1–14 days. Alteration of enstatite was only observed under neutral to alkaline conditions, with the abundance of phyllosilicates (serpentine and saponite, similar to those found in CI chondrites) increasing with temperature and/or run duration. This suggests that a variety of Mg-Fe-rich minerals, in particular enstatite and Mg-Fe olivine, act as precursors during extensive aqueous alteration to promote the formation of Mg-bearing serpentine and/or saponite.

6.2.1 The Timescales of Aqueous Alteration

The Mn-Cr and I-Xe chronometers indicate that alteration in carbonaceous chondrites lasted for extended periods of time ($\sim$2–15 My after CV CAIs; Fujiya et al. 2012; Jilly-Rehak et al. 2017; Pravdivtseva et al. 2018) (Sect. 9). However, experiments applicable to carbonaceous chondrites suggest that alteration may have occurred on short timescales, even at low temperatures. For example, Le Guillou et al. (2015b) reported that the reaction rate of a crystalline film is ten times slower than an amorphous film, suggesting that under low temperature conditions (90 °C) chondritic amorphous silicates are probably altered rapidly relative to the timescales on which asteroid interiors were heated by radiogenic decay (i.e., within a few million years). Jones and Brearley (2006) hydrated cubes of a CV chondrite under highly oxidising conditions at W/R mass ratios of 1:1 to 9:1 and temperatures of 100–200 °C for 7–180 days. Altered cubes had Ca- and Mg-sulphates, Ca-carbonates, and Fe-(oxy)hydroxides on exterior surfaces, while interiors contained serpentine interlayered with saponite and an amorphous SiO_2-rich phase. This assemblage resembles the CI chondrites (except the sulphates, which form due to terrestrial weathering, e.g., Imae et al. 2024), and demonstrates that phyllosilicate formation within asteroids was likely rapid, on the order of days to weeks at temperatures of ~150 °C, and 100s to 1000s of years at temperatures of <100 °C.

6.2.2 What Was the Composition of the Alteration Fluid and Was the Alteration Under a Closed or Open System?

Determining fluid compositions using hydrothermal experiments has received little attention. Vacher et al. (2019b) used S-rich (pH = 11.5) and S-free (pH = 6.5) fluids to form tochilinite and cronstedtite, respectively. The pH has an important influence on dissolution of silicate glass; in S-free experiments the Mg/Si ratio of the solutions was higher (i.e., less Si was dissolved), permitting the precipitation of Mg-rich minerals, whereas cronstedtite was only formed with S-free fluids and reactants with high Fe contents. Tochilinite was only identified in experiments using S-rich fluids. Vacher et al. (2019b) proposed a model for TCI formation whereby the precipitation of tochilinite caused the S activity of the fluid to decrease to the point at which Fe-Si-rich phyllosilicates could precipitate.

Models generally assume alteration in a closed system due to the apparent isochemical composition of altered carbonaceous chondrites and their low permeability (Bland et al. 2009). Such properties indicate limited fluid flow and chemical transport inside asteroids. Alternative models predict large-scale transport over km scales in an open system through exhalation flow from the interior to the surface of the asteroid (Young et al. 1999, 2003; Palguta et al. 2010). These models are supported by meter-sized carbonate veins at the surface of B-type asteroid Bennu (Kaplan et al. 2020). Suttle et al. (2022b) altered CO chondrite chips in Teflon reaction vessels, with gases able to escape and produce a partially open system. Chondrules in the chips remained largely unaltered, whereas the matrix displayed localized alteration features including Fe-enrichments (leached from Fe-Ni metal), decreased porosity, and the opening of large ~100 µm wide channels. Alteration products included Fe-(oxyhydroxides), magnetite, and fayalite, plus minor Fe-sulfides, carbonates, and sulphates, with the alteration similar in stye to the Fe-alkali metasomatism reported in CV chondrites.

Kikuchi et al. (2022) altered synthetic chondrite mixtures at reducing conditions within Teflon bottles. To approximate the loss of H gas from an asteroid, the bottles were deliberately opened every 5–7 days to remove the headspace air. Saponite encrusted primary minerals in all experiments, with its formation taking 1 day at 80 °C and 200 days at 25 °C. It was proposed that phyllosilicate formed via direct transformation (i.e., replacement of the initial phase without an intermediate) or a two-step process involving an SiO_2-rich intermediate phase following the dissolution and recrystallization of silicates. Serpentine was not detected, with the experimental products most closely resembling the mildly altered CO, CR, and CM chondrites.

6.2.3 Remaining Questions and Future Work

Despite extensive data regarding alteration processes having been accumulated through laboratory experiments, numerous questions persist. Conducting further experiments could provide new insights, including: (1) was water completely consumed during aqueous alteration, or did the alteration occur in an open system; (2) can mineral textures provide information concerning kinetics and fluid compositions; (3) how does terrestrial weathering affect the formation of secondary phases; (4) do isotopic exchanges occur during low-temperature aqueous processes; (5) what are the triggers for the formation of secondary phases, considering the complexity and variability observed due to multiple episodes of alterations; (6) how would the addition of anionic elements (C, N, S) to the experimental solutions affect the secondary mineralogy?

7 Fluid Inclusions in Aqueously Formed Minerals

Fluid inclusions in a mineral retain the aqueous fluid that was present when the mineral was formed and are fossils of the aqueous fluid that mediated aqueous alteration. They provide information on the composition of the ices that caused the aqueous alteration, and alteration conditions. Two major types of fluid inclusions that are not contaminated with terrestrial water or other liquids are currently found in extraterrestrial samples: those in halite (NaCl) clasts in H chondrites (Zolensky et al. 1999c; Whitby et al. 2000; Rubin et al. 2002), and those in minerals in an aqueously altered carbonaceous chondrite (Tsuchiyama et al. 2021) and related material from the asteroid Ryugu (Nakamura et al. 2023).

7.1 Fluid Inclusions in Halite Clasts in H Chondrites

Two H chondrites, Monahans (1998) (hereafter Monahans) (H5) and Zag (H3-6) are regolith breccias and have clasts consisting of blue to purple halite (NaCl) crystals up to 1 cm in size (mostly a few mm in Monahans and a few hundred μm in Zag). Fluid inclusions were found in these halite crystals for the first time in extraterrestrial materials (Fig. S3a) (reviewed by Zolensky et al. 2017). The Monahans halite contains inclusions of sylvite (KCl) (Rubin et al. 2002). The color is thought to be due to solar/galactic cosmic ray irradiation or exposure to ^{40}K beta decay of the sylvite (Zolensky et al. 2017). They have early Solar System formation ages (4.66 ± 0.08 to 4.03 ± 0.05 Ga, Zolensky et al. 1999c; Bogard et al. 2001; Whitby et al. 2000).

Fluid inclusions in the halite grains are mostly <10 μm in size (Zolensky et al. 1999c; Rubin et al. 2002) and sometimes have bubbles inside ($\sim 25\%$ of Monahans; Zolensky et al. 1999c). Faceted cubic negative crystals are also observed (Fig. S3a). Raman analysis revealed a significant peak at 3400 cm^{-1}, corresponding to an aqueous salt solution (not pure water), whereas CO_2, N_2, and CH_4 were not detected (Zolensky et al. 2017). The freezing and melting temperatures of the fluid inclusions suggest the absence of CO_2 and likely indicate the presence of (Fe^{2+}, Ca^{2+}, or Mg^{2+}) in addition to Na^+ and K^+, respectively (Zolensky et al. 1999c). Their formation temperature estimated from the bubble volume was <100 °C (possibly <50 °C). The O- and H-isotope compositions of the fluids measured using cryo-SIMS varied widely among the individual fluid inclusions, $-400‰ < \delta D < +1300‰$ and $-20‰ < \Delta^{17}O < +30‰$ (Yurimoto et al. 2014), indicating isotopic disequilibrium before incorporation into the halite. This variation is explained by the mixing of carbonaceous chondritic water (δD-poor, low $\Delta^{17}O$) and outer Solar System (cometary) water (δD-rich, high $\Delta^{17}O$) during rock-water interaction. The presence of some of these cations in the Monahans fluid inclusions, as well as a large assemblage of organic molecules, was subsequently demonstrated by ToF-SIMS (Bodnar et al. 2019).

A variety of mineral inclusions have also been found in halite: olivine, high- and low-Ca pyroxenes, feldspars, magnetite, Fe-Ni sulfides, Fe-Ni metals, lepidocrocite (FeO(OH)), carbonates, diamond, apatite [$Ca_5(PO_4)_3(F,OH,Cl)$], phyllosilicates, and zeolites (Fries et al. 2011; Zolensky et al. 2017). Halite also contains a variety of carbonaceous materials and organic matter: macromolecular C, aliphatic materials, and amino acid precursors (Fries et al. 2011; Kebukawa et al. 2016; Chan et al. 2018). These are thought to be organic matter precipitated from brines along with halite in the parent body. Similar organic matter has been found in dark clasts in Monahans and Zag and has been linked to fluid inclusions (Kebukawa et al. 2016, 2019, 2023).

The formation temperature of fluid inclusions shows that the halites with fluid inclusions were incorporated into the H chondrites after thermal metamorphism (approximately 700 °C for H5). This is thought to be related to cryovolcanism of ice-bearing small bodies, where the eruption of brine containing organic matter and rocky mantle components occurred (Fries et al. 2013; Zolensky et al. 2017). C-type asteroid Ceres is a possible source because of the similarity of mineralogy between inclusions in the halite and the Ceres regolith observed by Dawn and of the intersecting orbits of Ceres and 6 Hebe proposed as the origin of the H chondrites.

7.2 Potential Fluid Inclusions in Carbonaceous Chondrites

Carbonaceous chondrites that have undergone aqueous alteration contain minerals that grew from aqueous fluids. Thus, fluid inclusions in these minerals have been sought in CIs, CMs, and the Tagish Lake meteorite after careful sample preparation to avoid water contamination; potential fluid inclusions have been found in carbonates, sulfides, olivine, and enstatite (Zolensky et al. 2017). Some of these contain bubbles, but because of their small size (<a few μm), they have not been confirmed as fluid inclusions by chemical analysis.

X-ray tomography (XCT), a nondestructive analytical technique, is highly effective in the search for fluid inclusions. In particular, microsampling of a region of interest (ROI) using a focused ion beam (FIB) and imaging of the ROI with X-ray nanotomography (XnCT) make it possible to search for even small fluid inclusions. By combining two different XnCT techniques using synchrotron radiation, namely dual-energy tomography (DET) for absorption images (Tsuchiyama et al. 2013) and scanning-imaging X-ray microscopy (SIXM) for phase-shift images (Takeuchi et al. 2013) (DET-SIXM method), many mineral phases, plus organic matter and aqueous fluid, can be identified with an effective spatial resolution of ∼200 nm (Matsumoto et al. 2019). A fluid inclusion (∼7 μm) with a bubble in terrestrial quartz microsampled by FIB from a thin section was clearly imaged using DET-SIXM (Yoshida et al. 2016).

By using this method, a number of inclusions < ∼5 μm were observed in calcite from the Sutter's Mill (CM) chondrite (Tsuchiyama et al. 2014, 2021; Zolensky et al. 2014, 2017). However, no fluid was observed in micrometer-sized inclusions, although some had facets (negative crystals) which may indicate that they once contained aqueous fluid (Fig. S3b). IOM inclusions with bubbles (Fig. S3c) were confirmed using microRaman, TEM, STIM-XANES, and NanoSIMS analyses performed after XnCT (Tsuchiyama et al. 2017). IOM inclusions also occur in enstatite (Tsuchiyama et al. 2017). This is consistent with the observation that IOM inclusions are universally present in CM carbonates (calcite and dolomite) (Chan et al. 2017). Many inclusions found in Fe sulfides and Fe-Ni sulfides are irregular in shape and connected outward in three dimensions (Tsuchiyama et al. 2014).

7.3 Fluid Nanometer-Sized Inclusion in Sutter's Mill (CM) Calcite

Numerous nanometer-size inclusions (<1 μm) were identified by XnCT (Fig. S3b) (Tsuchiyama et al. 2021) but their contents could not be examined owing to spatial resolution limitations. Well-developed {001} facets of these inclusions observed by TEM (Fig.

S3d) indicate the possibility that aqueous fluid remained in some inclusions. Aqueous fluid was identified in one of them (Fig. S3e) by selected area electron diffraction (SAED) patterns of the inclusion in cryo-TEM, showing additional reflection spots identified as CO_2 ice or CO_2 hydrate ($CO_2 \cdot 5.75H_2O$) at −100 °C. According to the phase diagram of the CO_2-H_2O system (Longhi 2005), this fluid is enriched in CO_2 ($CO_2/H_2O > \sim 0.15$). Some nanometer-sized particles of Na-Mn and Fe sulfates, and FeOOH (or hematite [Fe_2O_3]), that occur as trapped or daughter crystals in some inclusions, suggest that the fluid was brine. The estimated pressure at which the CO_2-rich fluid was incorporated during aqueous alteration within the Sutter's Mill parent body was $P > \sim 50$ bars (possible >100 bars), suggesting that fluid uptake likely occurred in the interiors of relatively large bodies of >100 km (possibly >200 km).

CO_2-rich ice has been proposed as the origin of C in the CO_2-rich fluid, in which case the Sutter's Mill parent body may have formed outside the CO_2 snow line and inside the CO snow line (~75 K and ~ 22 K, respectively; Okuzumi et al. 2016). This is consistent with the Grand Tack model (e.g., Walsh et al. 2012), which suggests that the parent body formed outside the orbit of Jupiter in the early Solar System and then moved inward, as well as with the isotope dichotomy of Solar System materials (e.g., Warren 2011).

7.4 Fluid Inclusions in Ryugu Samples and CI Chondrite?

The Hayabusa2 spacecraft collected samples from the surface/subsurface of the Cb-type asteroid Ryugu and returned them to Earth in 2020 (Tachibana et al. 2022). They were identified as CI meteorites or related materials in the initial analysis (e.g., Nakamura et al. 2023). Potential fluid inclusions of a few µm were found in pyrrhotite grains by XnCT (Fig. S3f) and confirmed by cryo-TOF-SIMS (Nakamura et al. 2023). Secondary ion signals detected inside the inclusion indicate that it was filled with a fluid containing H_2O, CO_2, S species, and N- and Cl-bearing organic compounds. The presence of CO_2 indicates that Ryugu's parent body also formed outside the CO_2 snowline. There are many possible IOM inclusions in pyrrhotite, dolomite, breunnerite, and apatite grains; however, no potential fluid inclusions have been identified thus far (Tsuchiyama et al. 2024). Fluid inclusions in dolomite have also been explored in Ivuna (CI) using DET-SIXM and cryo-TEM (Tsuchiyama et al. 2019). Ivuna dolomite also had numerous micrometer- and nanometer-sized inclusions, similar to calcite in Sutter's Mill. Some micrometer-sized inclusions had facets, but they were empty. Nanometer-sized inclusions also had facets, but no fluid inclusions were detected using cryo-TEM. Aqueous fluid would have once been present in these inclusions but is now missing not only by mechanical disturbance but also due to terrestrial weathering after falling to Earth (e.g., Nakamura et al. 2023).

7.5 Remaining Questions and Future Work

See the Supplementary text.

8 Isotope Tracers

8.1 Oxygen Isotopes

Oxygen has three stable isotopes: ^{16}O (99.757%), ^{17}O (0.038%) and ^{18}O (0.205%). Oxygen-isotope compositions are expressed in delta units relative to the Vienna Standard Mean

Ocean Water (VSMOW, $^{17}O/^{16}O = 0.0003829$ and $^{18}O/^{16}O = 0.0020052$). The deviation from the TFL, a line of slope ~0.52 defined by common terrestrial solids and waters (Robert et al. 1992), is expressed as $\Delta^{17}O = \delta^{17}O - 0.52 \times \delta^{18}O$. Hence, mass-dependent variations are characterized by constant $\Delta^{17}O$ while mass-independent fractionation presents variable $\Delta^{17}O$. In the three-oxygen isotope diagram, $\delta^{17}O$ *vs.* $\delta^{18}O$, anhydrous chondritic constituents (e.g., CAIs, amoeboid olivine aggregates, chondrules) plot along a slope-1 line, referred to as the Primitive Chondrule Minerals line (PCM; Ushikubo et al. 2012). This line is thought to represent mixing between ^{16}O-depleted (H_2O) and ^{16}O-enriched (CO) reservoirs in the protoplanetary disk, which may have been produced by CO self-shielding by UV light in the protosolar molecular cloud and/or in the outer part (>5 AU) of the protoplanetary disk (Yurimoto and Kuramoto 2004; Lyons and Young 2005). A striking feature of most Solar System materials, excluding refractory inclusions and so-called cosmic symplectite, is that their O-isotope compositions cluster closer to the TFL ($\Delta^{17}O \sim 0‰$) than to the solar composition ($\Delta^{17}O \sim -29‰$; McKeegan et al. 2011). Potential explanations include: (i) the initial silicates inherited from the molecular cloud had a bulk $\Delta^{17}O \sim 0‰$ (Krot et al. 2010), or (ii) the bulk composition of the initial silicates and water ices had $\Delta^{17}O \sim 0‰$ and this composition has imprinted on solar materials by thermal processing in regions with elevated (dust + ice)/gas ratios (Alexander et al. 2017b) and/or because H_2O isotopically exchanged with amorphous interstellar silicates much faster than CO (Yamamoto et al. 2018).

8.1.1 Bulk Compositions

In a plot of $\delta^{17}O$ *vs* $\delta^{18}O$ (Fig. 6A), whole rock and matrix samples of hydrated chondrites (i.e., CR, CM, CI, C1-ungrouped like Flensburg, and C2-ungrouped like Adelaide or MAC 87300) show significant departure from the line of slope 1 defined by unaltered refractory inclusions and chondrules. CI and C2-ungrouped chondrites show narrow O-isotope variations falling on or near the TFL and are higher in both $\delta^{17}O$ and $\delta^{18}O$ than anhydrous silicates and oxides in these meteorites (Fig. 6A; Clayton and Mayeda 1984, 1999; Ushikubo and Kimura 2021; Marrocchi et al. 2021b; Kawasaki et al. 2022; Morin et al. 2022). Furthermore, CM and CR chondrites with varying degrees of alteration have O-isotope compositions defining lines of intermediate slopes between the PCM and the TFL (Fig. 6A; Clayton and Mayeda 1984, 1999; Schrader et al. 2011; Marrocchi et al. 2018).

These continuous isotopic trends with extent of alteration do not correspond to the primary characteristics established during disk evolution but relate to secondary alteration processes that took place in asteroidal settings (Clayton and Mayeda 1984). They reveal that the main process controlling the O-isotope compositions of CM and CR chondrites is dictated by different W/R ratios involving ^{16}O-rich anhydrous silicates and $^{17,18}O$-rich fluids (Clayton and Mayeda 1984, 1999; Brearley 2006; Zolensky et al. 2008; Krot et al. 2015; Marrocchi et al. 2018; Alexander 2019). This generates bulk O-isotope compositions that are progressively enriched in $^{17,18}O$ with increasing degree of alteration. Assuming anhydrous chondritic protoliths plot along the PCM line, the CM and CR secondary isotopic trends offer the possibility to estimate the initial O-isotope composition of their respective anhydrous protoliths. Such a method implied initial bulk $\Delta^{17}O$ values in the CM and CR protoliths of $-4.5 \pm 0.6‰$ and $-2.1 \pm 0.6‰$, respectively (Piralla et al. 2020) probably representing different chondrule populations sampled by these chondrites (Tenner et al. 2015; Marrocchi et al. 2022, 2024).

8.1.2 *In situ* Analysis of Aqueously Formed Minerals

The proportions of the original components of chondritic parent bodies (refractory inclusions, chondrules, Fe-Ni metal beads, sulfides, fine-grained matrix, and water ices) are

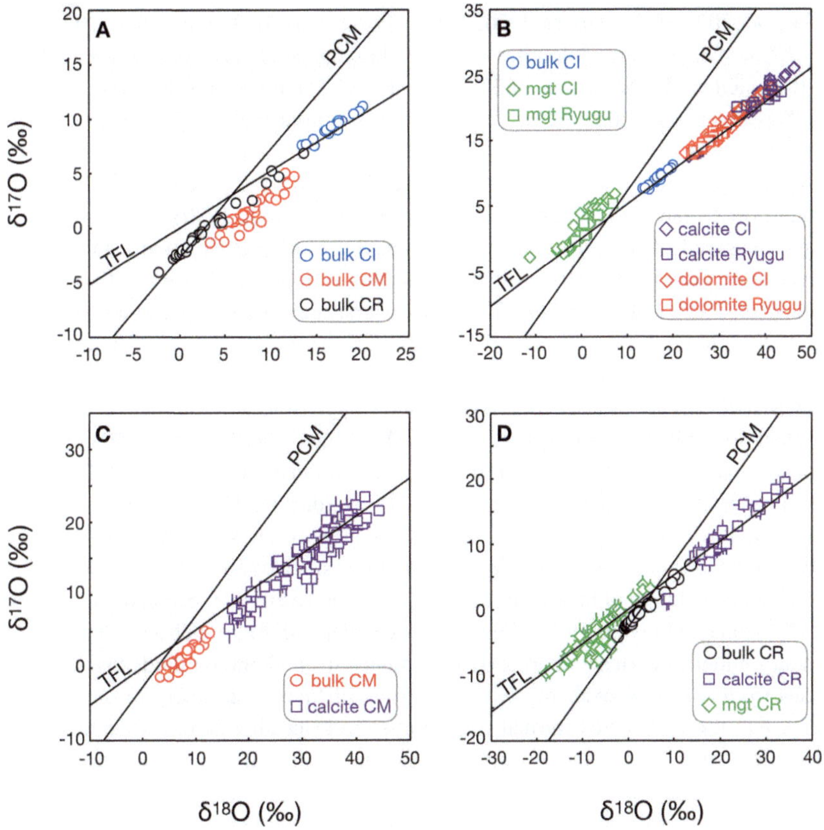

Fig. 6 A) Oxygen three-isotope plot for bulk CI (including Ryugu), CM and CR chondrites. Data from Clayton and Mayeda (1999), Schrader et al. (2011), Hewins et al. (2014), Yokoyama et al. (2023). B) Oxygen-isotope composition of bulk CI chondrites and Ryugu samples and their secondary minerals (calcite, dolomite, magnetite (mgt)). Data from Clayton and Mayeda (1999), Piralla et al. (2020), Nakamura et al. (2022), Yokoyama et al. (2023), McCain et al. (2023), Fujiya et al. (2023). C) Oxygen-isotope compositions of bulk CM chondrites and representative isotopic compositions of CM calcite grains. Data from Clayton and Mayeda (1999), Lindgren et al. (2017), and Vacher et al. (2019a). D) Oxygen-isotope compositions of bulk CR chondrites and CR calcite and magnetite grains. Data from Clayton and Mayeda (1999) and Jilly-Rehak et al. (2018). Published carbonate data are listed in Table S1

highly variable between different chondrite groups. Heat released from the decay of short-lived ^{26}Al led to the melting of the water ice, resulting in the establishment of prolonged fluid alteration processes within asteroids commencing shortly after their accretion (Sect. 9). Aqueous fluids affected the anhydrous primary chondritic components, leading to the formation of a large spectrum of secondary minerals whose nature is dependent on the physicochemical conditions and the duration of alteration (Sects. 2, 3, 6). Although the nature and abundances of secondary phases vary among CR, CM, CI, C1- and C2-ungrouped chondrites, they share two minerals whose O-isotope compositions can be analyzed using bulk techniques (Grady et al. 1988; Alexander et al. 2015) and/or *in situ* measurements by secondary ion mass spectrometry (SIMS): carbonate (Table S1) and magnetite.

In CI chondrites and Ryugu samples, Ca-carbonate grains (calcite and dolomite) plot on or slightly above the TFL and have higher $\delta^{18}O$ compared to the bulk values (Fig. 6B;

Piralla et al. 2020; Nakamura et al. 2022; Yokoyama et al. 2023; Fujiya et al. 2023; McCain et al. 2023). On average, dolomites are more homogeneous and have lower $\delta^{18}O$ values than the calcites ($\delta^{18}O_{average}$ of 30.2 ± 4.4‰ *vs.* 37.2 ± 6.6‰, respectively; Piralla et al. 2020; Nakamura et al. 2022; Yokoyama et al. 2023; Fujiya et al. 2023; McCain et al. 2023). Magnetite grains are ^{16}O-enriched compared to bulk values (i.e., $\Delta^{17}O = 1.70 \pm 0.05$‰; Clayton and Mayeda 1999) and were historically considered as out of equilibrium with carbonates (i.e., having different $\Delta^{17}O$; Clayton and Mayeda 1999). However, the recent isotopic appraisal of Ryugu samples has revealed a less clear situation: calcite, dolomite and magnetite grains have $\Delta^{17}O$ possibly indicating isotopic equilibrium within uncertainty (Piralla et al. 2020; Nakamura et al. 2022; Yokoyama et al. 2023; Fujiya et al. 2023; McCain et al. 2023), although microstructural features of some magnetite grains indicate that they formed under disequilibrium conditions (Dobrică et al. 2023).

In CM and CR chondrites, Ca-carbonates show large isotopic variations ($\delta^{18}O$ varying by more than 30‰ plotting on either side of the TFL). Interestingly, in both chondrite groups, carbonates extend the trends defined by bulk rocks and matrices (Fig. 6C, D; Benedix et al. 2003; Tyra et al. 2012, 2016; Fujiya et al. 2015; Verdier-Paoletti et al. 2017, 2019; Jilly-Rehak et al. 2017; Vacher et al. 2016, 2017, 2018, 2019a). These continuous trends confirm that O-isotope exchange between ^{16}O-rich anhydrous silicates and $^{17,18}O$-rich water controls the O-isotope evolution of bulk CMs and CRs, and their respective secondary alteration minerals (Clayton and Mayeda 1984, 1999; Marrocchi et al. 2018). In such a scheme, $^{17,18}O$-rich secondary minerals correspond to early precipitates from alteration fluids that had suffered the least from O-isotope exchange with anhydrous silicates. Conversely, lower $\Delta^{17}O$ values indicate that the carbonates formed following significant alteration, and as the O-isotope compositions of fluids evolved toward the O-isotope composition of anhydrous silicates and fluid (Verdier-Paoletti et al. 2017).

Magnetite grains are ubiquitous alteration products in CR matrices (Weisberg et al. 1993; Fig. 6D). Their O-isotope compositions plot along the TFL (with the notable exception of those in the CR MIL 09292 that have $\Delta^{17}O \sim -3$‰) with $\delta^{18}O$ ranging from −18‰ to +5‰ (Jilly-Rehak et al. 2017). CM chondrite magnetite grains have similar $\delta^{18}O$ values ranging from −14‰ to +4‰ (Telus et al. 2019). A few CR and CM chondrites have magnetite and Ca-carbonates plotting on a single mass-fractionation line possibly allowing the equilibrium temperature to be estimated (Jilly-Rehak et al. 2017; Marrocchi et al. 2018; Telus et al. 2019). The bulk O-isotope measurements of magnetite grains separated from CIs show similar $\Delta^{17}O \sim +1.5$‰ (Rowe et al. 1994) as those determined by SIMS (Nakamura et al. 2022; Yokoyama et al. 2023; McCain et al. 2023).

Of note, *in situ* SIMS measurements of the O-isotope compositions of carbonates are affected by complex matrix effects that require specific sets of standards, especially for quantifying the effects of minor elements (i.e., up to −0.3‰ per wt.% MgO for carbonates poor in Fe, Mn, Rollion-Bard and Marin-Carbonne 2011; Śliwiński et al. 2017). Assessing this potential instrumental bias is not straightforward as: (1) carbonates are rarely chemically and isotopically homogeneous, and (2) the instrumental bias is non-linear as a function of the concentrations of minor elements. These specificities are sources of uncertainties and call for the development of well-characterized carbonate standards for precisely determining the O-isotope compositions. This is particularly true for breunnerite, a rare Mg-Fe-carbonate commonly observed in CI chondrites. In addition, SIMS measurements of magnetite grains are affected by crystal orientation effects that are not always taken into account when measuring their O-isotope compositions (Huberty et al. 2010).

8.1.3 Oxygen Isotope Compositions of Carbonaceous Chondrite Water Ices

Determining the initial O-isotope compositions of water-ice grains accreted by chondritic asteroids is not an easy task as water-rock interactions continuously modified the O-isotope compositions of altering fluids. Although several approaches have been attempted for estimating the initial water ice compositions, most have focused on CM chondrites as bulk and Ca-carbonate compositions define a trend with varying $\Delta^{17}O$ values.

Based on bulk O-isotope compositions and a two-stage model of gas-solid isotopic exchange, Clayton and Mayeda (1984, 1999) estimated the $\delta^{18}O$ and $\Delta^{17}O$ of the primordial CM water to be ~30.3‰ and ~4.4‰, respectively. A different estimate for the initial CM water composition was based on mass-balance calculations ($\delta^{18}O = 55$‰ and $\Delta^{17}O = 6.4$‰; Fujiya 2018). Estimates of the O-isotope composition of fluid at the time of phyllosilicate formation and carbonate precipitation were also reported using (1) clumped-isotope thermometry of CO_2 ($\delta^{18}O$ ranging from 2‰ to 8.1‰ and $\Delta^{17}O$ ranging from −1.4‰ to −0.5‰; Guo and Eiler 2007) and (2) water extracted under stepped pyrolysis ($\delta^{18}O = 11.5$‰ and $\Delta^{17}O = 0.6$‰; Baker et al. 2002). Alexander (2019) proposed that all CCs accreted ices with a similar O-isotope composition of about $\delta^{18}O = 30.9$‰ and $\Delta^{17}O = 3.5$‰. Interestingly, such estimates define a line ($R^2 = 0.987$; Fig. S4) with a slope that is indistinguishable, within error, from the bulk-matrix-carbonate trend, but with a different intercept (Verdier-Paoletti et al. 2017). This implies that the initial O-isotope composition of CM water was likely $^{17,18}O$-rich and does not plot along the slope-1 line but is shifted to the right from it. To the first order, this appears inconsistent with a self-shielding scenario, but the meaning is uncertain and needs to be investigated further.

8.1.4 Thermal Conditions of Aqueous Alteration and Water/Rock Ratio

Secondary minerals, such as carbonate and magnetite, represent direct proxies of the asteroidal fluids from which they formed and can, in theory, be used to decipher their formation temperatures (Clayton and Mayeda 1984). However, determining these temperatures requires knowledge of the O-isotope compositions of their parental fluids, which itself requires knowledge of the carbonate/magnetite precipitation temperatures, leading to a seemingly circular problem. This issue can be overcome when secondary phases with similar $\Delta^{17}O$ are present in a given chondrite, which allow alteration temperatures to be estimated (assuming equilibrium that is not so straightforward). This approach led to (1) contrasting estimates of alteration temperatures of CM chondrites ranging from 0-20 °C (Benedix et al. 2003) to 80–120 °C (Baker et al. 2002) and (2) more coherent results for CR chondrites with temperatures of 55–90 °C (Jilly-Rehak et al. 2018) and 35–110 °C (Marrocchi et al. 2018). A similar approach for the CI-related Ryugu samples gave estimates of ~0-80 °C (Yokoyama et al. 2023; McCain et al. 2023; Nakamura et al. 2022; Kita et al. 2024). Other estimates for CM chondrites are based on (1) clumped-isotope thermometry of CO_2 (Δ^{47}, which corresponds to anomalous enrichments of mass 47 (i.e., $^{13}C^{18}O^{16}O$ in CO_2 derived from H_3PO_4 digestion of carbonates, Ghosh et al. 2006): 20–70 °C (Guo and Eiler 2007), ~5-51 °C and ~75–101 °C for calcite and dolomite, respectively (Clog et al. 2024), (2) bulk $\delta^{13}C$ and $\delta^{18}O$ trends in carbonates: 0–130 °C (Alexander et al. 2015), and (3) the distance between the respective trends defined by CM water and bulk + carbonates: 50–300 °C (Verdier-Paoletti et al. 2017; Vacher et al. 2019b). Discrepancies could in part be due to the lack of standard materials with compositions perfectly matching the chondritic minerals to correct the SIMS instrumental mass fractionation or relate to the possibly incorrect estimate of equilibrium between components used for the temperature calculation.

Water/rock ratios of chondritic parent bodies can be estimated by combining bulk and mineral O-isotope data (Alexander 2019 and references therein). These estimates indicate variable W/R mass ratios for hydrated chondrite groups with CR = 0.10–0.40, CM = 0.10–0.40, and CI = 0.38–0.43 (Alexander 2019 and references therein).

8.2 Hydrogen Isotopes

8.2.1 Bulk Compositions

Hydrogen has two stable isotopes ^{1}H (99.9885%) and ^{2}H (deuterium: D) (0.0115%). The H-isotope composition is reported in the delta notation relative to the SMOW (D/H = 0.00015576). Bulk H-isotope compositions have been measured in a large range of phyllosilicate-rich chondrites including CR, CM, CI, and C1- and C2-ungrouped CCs either by step heating (e.g., Robert and Epstein 1982; Eiler and Kitchen 2004) or using an elemental analyzer coupled with an isotope ratio mass spectrometer (EA-IRMS) allowing permil-level measurements on aliquots of a few milligrams (e.g., Alexander et al. 2012, 2018; Vacher et al. 2020; Marrocchi et al. 2023b). In water-rich chondrites, bulk δD values range from −300‰ in the ungrouped C1 chondrite Flensburg (Bischoff et al. 2021) to > +600‰ in CRs and some C-ungrouped chondrites (see McCubbin and Barnes 2019; Piani et al. 2021 for compilations).

In CM and CR chondrites, bulk H-isotope measurements reveal the presence of positive and distinct correlations between the δD values and C/H ratios (Fig. S5A; Alexander et al. 2012). These correlations were interpreted to reflect mixing lines between a D-poor and C-free component (presumably the water from which the hydrated minerals formed) with D-rich organic matter. The H-isotope intercepts were used to estimate the δD of the initial water components accreted by the CM and CR parent bodies, which are several hundreds of permil lower than the bulk compositions (Alexander et al. 2012; Fig. 7 and Table S2). The bulk H-isotope analyses performed after degassing to remove atmospheric water (Vacher et al. 2020; Lee et al. 2021) produce a CM mixing line in δD versus C/H space that is consistent with the existing IOM being the C-rich and D-rich endmember of the mixture (Fig. S5B; Marrocchi et al. 2023b). This result suggests that only limited isotopic exchange occurred during aqueous alteration in CM chondrites (a few tens of permil at most) and that the primary H-isotope signatures in IOM and water were preserved during alteration (Remusat et al. 2010; Marrocchi et al. 2023b). On the other hand, the H-isotope variations observed among CM IOM (ranging from ∼800‰ to ∼1200‰; Alexander et al. 2010) cannot be explained by this model, and either require that there was H_2O-IOM isotopic exchange or heterogeneity in the IOM accreted by CMs.

8.2.2 Hydrogen Isotopes in Hydrated Minerals

Hydrogen isotopes are powerful tracers of the origin of chondritic water-ice and organic molecules, as those inherited from the molecular cloud should be enriched in D by ∼2–3 orders of magnitude relative to those formed or re-equilibrated during the evolution of the protoplanetary disk (Ceccarelli et al. 2014). However, direct measurements of δD signatures in water or organics are challenging because chondrites reflect complex mixtures of organics and hydrated minerals that cannot be physically separated. *In situ* analyses at the micrometer-scale by SIMS allows for determination of the approximate H-isotope compositions from hydrated minerals to be obtained using two different analytical protocols.

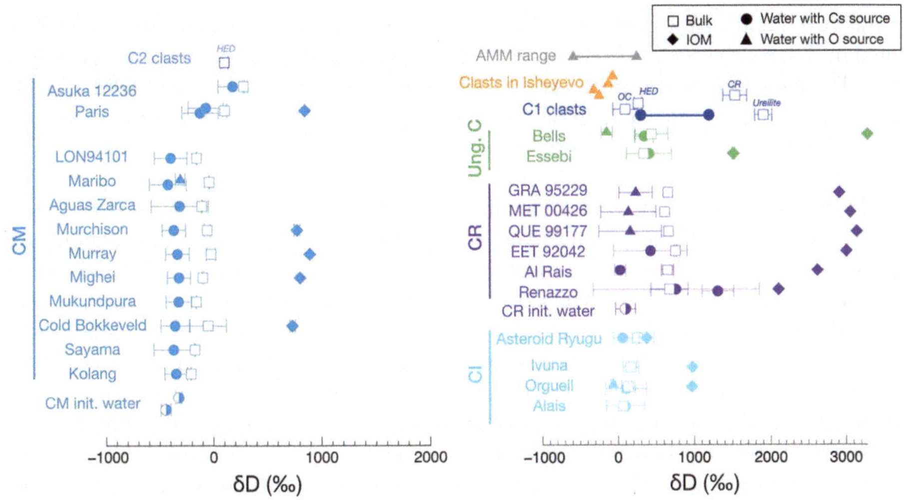

Fig. 7 Comparison of the H-isotope composition (expressed in delta notation against standard mean ocean water, SMOW) measured in bulk, IOM and hydrous silicates in hydrated CR, CM, CI, and ungrouped CCs, hydrated clasts, Ryugu samples and Antarctic micrometeorites (AMMs). Hydrous silicate data were obtained by SIMS using two different measurement protocols, with an O (triangle symbols) and a cesium (round symbols) primary beam, respectively. CM chondrites are listed from the more altered at the bottom to the least altered to the top. References are in Table S2 in the supplementary materials. For AMMs and water in C1 clasts, the symbols and horizontal line indicates the range of δD values. The host meteorites of the C1 and C2 clasts are indicated above the corresponding symbol (for more detailed see Table S2)

A first protocol proposed by Deloule and Robert (1995) takes advantage of the preferential emission of H^+ and D^+ from hydrated silicates relative to organics when using a O^- source for sputtering the sample surface (at least one order of magnitude lower; Bonal et al. 2010). Due to the large proportions of H-rich minerals in the matrix of aqueously altered chondrites, the δD values measured in this way should approximate the bulk compositions of hydrated minerals (Fig. 7 and references therein). However, the C content should be monitored to avoid the risk of local contributions from D-rich organics to the measured δD values.

With a Cs^+ primary beam, the difference of H^- and D^- emissions between hydrated silicates and organics is less pronounced (Bonal et al. 2010; Piani et al. 2012) and thus both phases contribute to the measured δD. The second protocol is based on the simultaneous measurement of the δD and C/H ratios. Depending on the location of the measurement in the chondrite matrix, the δD and C/H ratios vary with the relative amounts of hydrated minerals and organic matter defining a mixing line between a D-poor, C-free endmember (hydrated minerals) and a D-rich and C-rich organic endmember (Piani et al. 2018). The extrapolation to the C-free endmember (C/H = 0) allows the δD of hydrated minerals to be estimated without H contributions from adjacent organics (Fig. 7 and references therein). The limitation of this protocol is the large error ($\pm$100‰ for a given chondrite δD) coming from extrapolations to the zero intercepts. Additionally, the H-isotope compositions of individual coarse-grained hydrated silicates and chondrule glasses were measured in CR chondrites using a Cs^+ ion beam (Bonal et al. 2013; Shimizu et al. 2021).

Chondrites measured by the two protocols are in good agreement implying that the two techniques produce reliable results (Fig. 7). Nonetheless, the H-isotope measurements by SIMS in chondrites would gain in accuracy with the development of hydrated silicate standards covering the large range of D/H variations present in chondritic minerals (several hundred of ‰). Except for CI chondrites, for which hydrated minerals dominate the H budget, the δD values of hydrated minerals are systematically lower than bulk and IOM (Fig. 7). Like for IOM (e.g., Alexander et al. 2017b), distinct δD values are observed for water in the different chondrite groups with $\delta D{\sim}{-}350‰$ in CM, $\sim+100‰$ in CI and CR, and $\sim+400‰$ for the ungrouped Bells and Essebi (Alexander et al. 2012; Piani et al. 2021).

Hydrated clasts have been identified in the majority of chondrite and achondrite groups as C1-like and C2 objects that underwent aqueous alteration in their own parent bodies and possibly also in their host bodies (Sect. 3). Hydrogen-isotope studies of C1 dark clasts in ordinary chondrites (OC), HED meteorites, ureilites, CR and CH/CB chondrites have been carried out by NanoSIMS using primary O^- and Cs^+ sources. Patzek et al. (2020) found high bulk δD values for the aforementioned clasts (300–2000‰), relative to most of the other CCs (Fig. 7). The ureilite and CR clasts are among the most D-rich bulk Solar System materials, and their anomalous H-isotope signatures have been interpreted as reflecting the accretion of abundant D-rich ices and organics relative to CCs. Like bulk values, the ureilite and CR clast water ($\delta D{\sim}300$–$1200‰$) derived from the zero intercepts of correlations between D/H and Si/H or C/H ratios are significantly increased relative to CCs. In contrast, Isheyevo lithic clasts that contain the highest ^{15}N-enrichments found in Solar System materials and indicative of outer Solar System signatures (Briani et al. 2009; Bonal et al. 2010; van Kooten et al. 2017a), have phyllosilicate compositions that are comparable to other CCs ($\delta D{\sim}{-}100/-300‰$; van Kooten et al. 2017a; Fig. 7).

In CM chondrites, for which samples with different degrees of aqueous alteration are available, the δD values measured *in situ* for hydrated minerals are independent to the extent of aqueous alteration, indicating the CM chondrites accreted D-poor water ice with $\delta D{\sim}{-}350‰$ (Fig. 7). Certain regions of Paris and Asuka 12236 are the least altered CM lithologies (Hewins et al. 2014; Kimura et al. 2020) and are amongst the exceptions in displaying unusually D-rich H-isotope compositions possibly reflecting the presence of another H-bearing component in addition to organics and water (Piani et al. 2018; Marrocchi et al. 2023b). Hydrated amorphous silicates, that are ubiquitous in these two chondrites, were proposed as a H-bearing component that could retain H acquired before the onset of CM parent-body alteration (Marrocchi et al. 2023b).

8.2.3 Origin of H-Isotope Variations of Organic Matter and Water Among Chondrite Groups

The highly variable H-isotope signatures of hydrated minerals and organics among chondritic materials (Fig. 7) can be interpreted in two ways. They can either be the result of secondary processes occurring in asteroids and disturbing the initial isotope signatures, or they reflect the heterogeneous distributions and compositions of water ice and organics in the protoplanetary disk at the time of asteroid accretion. In asteroids, the accreted ice, once melted, percolates among the solid phases. The main process thought to have affected the H-isotope composition in H-rich chondrites is H-isotope exchange between D-rich organic components and D-poor aqueous fluids (e.g., Alexander et al. 2010). Such exchanges are rapid and almost complete in laboratory experiments conducted at $\geq$250 °C, resulting in a decrease of the IOM H/C and δD in the presence of D-poor water (Yabuta et al. 2007; Oba and Naraoka 2009; Kebukawa et al. 2021; Foustoukos et al. 2021) (Sect. 5.4). A progressive decrease of IOM H/C and δD has been observed in lithologies of the C2 Tagish Lake that experienced varying degrees of aqueous alteration and elevated temperatures for relatively short periods of time (Alexander et al. 2014). However, no evidence for H-isotope exchange between organic particles and water retained in the surrounding hydrated minerals was observed at the micrometer scale in the matrix (Remusat et al. 2010) nor H-isotope evolution in CM chondrites of increasing alteration degrees (Fig. 7). Temperatures might explain the differences in exchange efficiency for water and organics in laboratory experiments performed above 150 °C (Oba and Naraoka 2009; Kebukawa et al. 2021; Foustoukos et al. 2021). The experimentally determined D–H exchange between organic matter and water being relatively fast, alteration temperatures close to 0 °C would likely be necessary to maintain H-isotope disequilibrium between organic matter and water for millions of years (Kebukawa et al. 2021). These low-temperatures are at odd with the peak temperatures commonly estimated for aqueous alteration (50–300 °C; Sect. 8.1.4); the discrepancy between experiments and natural samples constitute a major issue for the interpretation of the isotopic data.

If not resulting from asteroidal processes, the variable H-isotope signatures of organics and water in the different chondrites could reflect an isotopically heterogeneous distribution in the protoplanetary disk, possibly inherited from multiple episodes of infall from the surrounding molecular cloud to the disk (Piani et al. 2021; Kuznetsova et al. 2022). Because the preservation of D-rich interstellar water and/or organics in the disk is only possible below their respective sublimation temperatures (160 K and 350–450 K, respectively), the H-isotope signatures of chondritic materials could depend on both the timing and location at which their parent asteroids formed, both varying with the thermal evolution of the disk (Yang et al. 2013; Jacquet and Robert 2013).

8.3 Carbon, Nitrogen, and Sulfur Isotopes

8.3.1 Carbon Isotopes in Carbonates and Conditions of Parent Body Alteration

The bulk C- and O-isotope compositions of carbonates in CI chondrites have been measured for Orgueil and Ivuna (Grady et al. 1988; Alexander et al. 2015). The data are consistent with, on average, the carbonates having formed from a single fluid at temperatures of ~0–40 °C (Alexander et al. 2015). From *in situ* measurements obtained in Ivuna and Ryugu samples, calcites show large variations for C ($\delta^{13}C = 65–108‰$ relative to the Vienna Peedee Belemnite; $^{13}C/^{12}C = 0.01123720$) and O-isotopes (Sect. 8.1.2), while dolomite is more homogeneous ($\delta^{13}C = 65–75‰$; Fujiya et al. 2023). The isotope variations in calcite are thought to result from their early precipitation over a large range of prograde temperatures and of fO_2 associated with the evolution of the proportions of CH_4, CO, and CO_2 in the fluid. In contrast, the isotopically homogeneous dolomite grains seem to have formed later, possibly during retrograde cooling, from a fluid that was close to equilibrium with a CO_2-dominated gas (Fujiya et al. 2023).

In CM chondrites, C-isotopes in carbonates measured either by bulk techniques (Grady et al. 1988; Alexander et al. 2015; Tyra et al. 2016; Vacher et al. 2020), SIMS (Lee et al. 2013; Fujiya et al. 2015; Vacher et al. 2017; Telus et al. 2019) or using clumped isotopes (Sect. 8.1.4). Guo and Eiler (2007) show a large range of $\delta^{13}C$, from ~25‰ to ~80‰. Positive correlations for $\delta^{13}C$ and $\delta^{18}O$ observed in bulk carbonates have been used to infer that on average they precipitated from a single fluid under variable temperatures of 0–130 °C that roughly correlate with extent of alteration for many but not all meteorites (Alexander et al. 2015). On the other hand, negative correlations for $\delta^{13}C$ and $\delta^{18}O$ observed using clumped isotopes of a smaller sample of meteorites have been interpreted as the result of the progressive ^{13}C-enrichment in the fluid due to the loss of ^{12}C-enriched CH_4 gas (Guo and Eiler 2007). *In situ* measurements of individual carbonate grains performed by SIMS revealed a more complex picture with (1) calcite being more heterogeneous than dolomite, and (2) no systematic link between O- and C-isotope variations (Fig. S6). The latter possibly indicates that carbonates in CM chondrites probably result from several episodes of carbonate dissolution and reprecipitation (e.g., Tyra et al. 2012).

8.3.2 Carbon, Nitrogen, and Sulfur Isotopes: Implication for Understanding the Nature of Chondritic Water Ices

The range of $\delta^{13}C$ values in chondrite carbonates can be related to the nature of the primary C-bearing precursors and could provide clues on the formation location of their parent bodies. The highest $\delta^{13}C$ values measured in CI carbonates call for the presence of ^{13}C-rich ices (Fujiya et al. 2023) that might have also been sampled by comets (Hässig et al. 2017). The ^{13}C-rich compositions of ices, preserved in the colder regions of the Solar System, could originate from self-shielding processes during CO photodissociation in the solar nebula or its parent molecular cloud (e.g., Lyons et al. 2018). Similarly, the C2 Tagish Lake was found to contain both calcite and dolomite with $\delta^{13}C$ values that were as high as in the CI chondrite carbonates (~70‰; Fujiya et al. 2019), but lower than the higher values measured recently in Ryugu and Ivuna calcites (up to 108‰; Fujiya et al. 2023; Fig. S7). Associated with the unusually high abundance of carbonates (Grady et al. 2002), the isotopic composition of Tagish Lake could indicate its parent body accreted a relatively high proportion of ^{13}C-rich CO_2 ices and thus possibly formed beyond the CO_2 snowline (Fujiya et al. 2019). However, precisely deducing the chondrite formation location based on the inferred amounts of CO_2 in water ice is difficult because some amounts of CO_2 can be trapped in H_2O ice even at temperatures higher than the CO_2 condensation temperature (e.g., Fayolle et al. 2012).

Soluble organic compounds have been proposed to be candidates for the C source of CM chondrite carbonates due to their comparable isotopic ranges (0–65‰; Vacher et al. 2017; Fig. S7). However, the high abundance of carbonates would require much more abundant soluble organics than measured in any chondrite (Aponte et al. 2016). ^{13}C-rich ices could also be the source of CM carbonate C (Telus et al. 2019), but the $\delta^{13}C$ composition of ice accreted by CMs would have to be less enriched in ^{13}C than in CIs or Tagish Lake chondrites suggesting fractionation processes operating in the parent bodies and/or the sampling of different C-bearing ice reservoirs during accretion.

Differences in the S-isotope compositions of sulfides ($\delta^{34}S \sim -2$‰ and $+1$‰ in CMs and CIs, respectively, relative to the Vienna-Canyon Diablo Troilite) were suggested to derive from the sampling of two distinct S reservoirs in CI and CM chondrites (Visser et al. 2019). The presence of mass-independent variations observed in the inorganic phases of CM chondrites are likely derived from photochemistry occurring in the nebula (Labidi et al. 2017). The heterogeneity of mass-independent S-isotope signatures among different CM chondrites and between fragments of the CM chondrite Murchison argues for heterogeneous accretion of at least two S-bearing components with distinct isotopic signatures (Labidi et al. 2017). Nonetheless, analysis of $\delta^{34}S$ in a large suite of bulk chondrites (mainly CR and CM) and their IOM have revealed the absence of large, systematic variations in $\delta^{34}S$ among chondrite groups or petrological types (Alexander et al. 2022).

In CM and CO chondrites, bulk $\delta^{15}N$ value decreases with increasing degree of alteration, likely due the progressive removal of an organic ^{15}N-rich component during alteration (Pearson et al. 2006). The presence of extreme $\delta^{15}N$ variations in the hydrated clasts of the CB/CH chondrite Isheyevo with D-depletion compared to organic matter were proposed to derive from the heterogeneous accretion of NH_3 and HCN-bearing ices during the cooling of the protoplanetary disk (van Kooten et al. 2017a). Such ice reservoirs do not appear to have been sampled by other water-rich CCs.

Overall, the isotopic compositions of H, C, O and, possibly, S could be consistent with the fact that the ices sampled by the parent bodies of the different chondrites originate from distinct reservoirs in the protoplanetary disk, but efforts must continue to quantify the influence of parent body processes on the isotopic signatures.

8.4 Remaining Questions and Future Work

Studies coupling precise mineralogical observations and O-isotope analyses with proper standards (i.e., effects of trace elements) in alteration phases are needed to validate the use of the O-isotopes in estimating the temperatures of aqueous alteration in chondrites. *In situ* C-isotope analyses of CR carbonates would also be of help to characterize the alteration conditions of their parent body.

Efforts should be undertaken to reconcile the temperature estimates obtained from different geochemical constraints and relate them to the complex (prograde/retrograde) thermal history of the chondrite parent asteroids.

9 Chronology of Aqueous Alteration

In this section, we provide an overview of the existing chronological data regarding low-temperature aqueous alteration of CR, CM, CI, and ungrouped carbonaceous chondrites, hydrated clasts in other meteorites, and surface materials of asteroid Ryugu. The chronology of aqueous alteration in these materials is crucial for constructing thermal evolution models of their parent bodies. Since the 1980s, it has been inferred from ^{87}Rb-^{87}Sr isotopic systematics that aqueous alteration is one of the earliest geological processes in the Solar System and took place within 100 Myr after its birth (Macdougall et al. 1984). Furthermore, the evidence for *in situ* decay of ^{53}Mn, a short-lived radionuclide with a half-life of 3.7 Myr, in carbonate minerals in CI chondrites confirmed that there was water activity in their parent body within a few tens of Myr after the birth of the Solar System (Endress et al. 1996). Therefore, the chronology of low-temperature aqueous alteration can be best explored by using short-lived radionuclides, which normally provide higher temporal resolution than long-lived radionuclides. Thus, we here focus on the chronology using ^{53}Mn-^{53}Cr ($t_{1/2} = 3.7$ Myr) and ^{129}I-^{129}Xe ($t_{1/2} = 16.1$ Myr) decay systems.

9.1 ^{53}Mn-^{53}Cr Chronology of Carbonates in CR, CM, CI, and Ungrouped Carbonaceous Chondrites, Hydrated Clasts in Other Meteorites, and Ryugu Samples

In general, radiometric dating of secondary minerals works well only if these minerals have high abundance ratios of parent/daughter nuclides resulting from the chemical fractionation during aqueous alteration. Therefore, the application of ^{53}Mn-^{53}Cr dating of low-temperature aqueous alteration is predominantly limited to carbonate minerals, which incorporate Mn^{2+} into their crystal structure but hardly contain any Cr^{3+}.

Carbonate minerals can commonly be found in CR, CM, CI, and ungrouped carbonaceous chondrites, hydrated clasts in other meteorites, and Ryugu samples (e.g., Johnson and Prinz 1993; Riciputi et al. 1994; Endress and Bischoff 1996; Endress et al. 1996; Ivanova et al. 2008; Lee et al. 2013; Jilly-Rehak et al. 2017; Bischoff et al. 2021; Nakamura et al. 2023). Calcium carbonate (calcite or aragonite), Ca- and Mg-carbonate (dolomite), and Mg-, Fe-, and Mn-carbonate (breunnerite) are the most common carbonate minerals (Sect. 2).

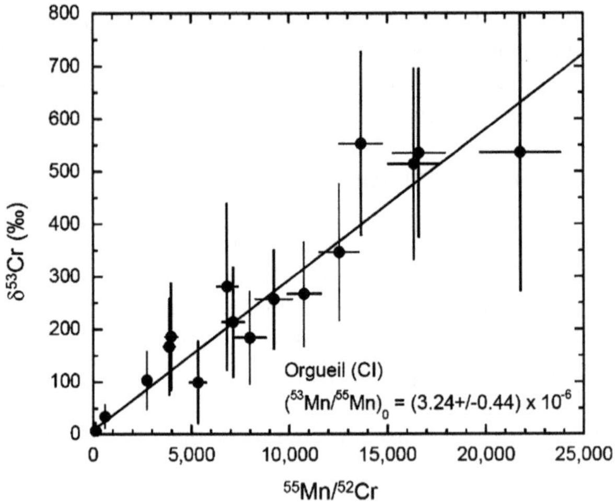

Fig. 8 ^{53}Mn-^{53}Cr isochron diagram of dolomite in the Orgueil CI chondrite. Note that the ^{55}Mn/^{52}Cr ratios were estimated using the RSF of a synthetic calcite standard (Sugiura et al. 2010), which may lead to systematic bias. Data from Fujiya et al. (2013)

Their ^{53}Mn-^{53}Cr systematics record the timing of carbonate formation during aqueous alteration; later disturbance of ^{53}Mn-^{53}Cr systematics in carbonates by Mn and Cr diffusion is unlikely because the peak temperatures experienced by these meteorites (metamorphosed CI and CM chondrites are not discussed here) are low (<220–240 °C) (Busemann et al. 2007; Sect. 8). Notably, carbonates in CM chondrites often preserve zoning in Mn (and other minor elements) within single grains, indicating that the Mn was not homogenized after the carbonate formation (e.g., Lee et al. 2014; Fujiya et al. 2020).

Since the pioneering work by Endress et al. (1996), the ^{53}Mn-^{53}Cr ages of carbonates have been measured *in situ* using SIMS. For ^{53}Mn-^{53}Cr dating, ^{53}Cr/^{52}Cr and ^{55}Mn/^{52}Cr ratios must be measured, and the slope of the regression line of the data in ^{53}Cr/^{52}Cr *vs.* ^{55}Mn/^{52}Cr space (^{53}Mn-^{53}Cr isochron diagram) corresponds to the initial ^{53}Mn/^{55}Mn ratio [(^{53}Mn/^{55}Mn)$_0$] when the Mn-Cr isotope system achieved closure (Fig. 8). The SIMS measurements of ^{55}Mn/^{52}Cr ratios are challenging because the so-called matrix effect of SIMS influences the observed $^{55}Mn^+/^{52}Cr^+$ ion intensity ratios relative to the true ^{55}Mn/^{52}Cr ratios, the extent of which depends on the target material (matrix). Therefore, a matrix-matched standard material with a known ^{55}Mn/^{52}Cr ratio must be used to determine the relative sensitivity factor (RSF: the $^{55}Mn^+/^{52}Cr^+$ ion intensity ratio divided by the true ^{55}Mn/^{52}Cr ratio) and to calculate the ^{55}Mn/^{52}Cr ratios of unknown samples from the observed $^{55}Mn^+/^{52}Cr^+$ ion intensity ratio.

The importance of using a proper carbonate standard was not recognized by the earliest studies. Natural carbonates including those in meteorites contain little Cr, which makes the preparation of a proper carbonate standard difficult. Therefore, most work until the 2010s used silicate (e.g., San Carlos olivine: SCol) standards, leading to a systematic uncertainty in ^{55}Mn/^{52}Cr ratios, and therefore, in the (^{53}Mn/^{55}Mn)$_0$ of carbonates. Sugiura et al. (2010) first showed that the RSF of synthetic Mn- and Cr-bearing calcite is ~30% lower than that of SCol, implying that the ^{53}Mn-^{53}Cr ages obtained using SCol appear ~2.2 Myr older than those obtained using the synthetic calcite standard.

In Fig. S8 and Table S3, we summarize the (^{53}Mn/^{55}Mn)$_0$ of carbonates in CR, CM, CI, and ungrouped carbonaceous chondrites, hydrated clasts in the Isheyevo CH/CB-like meteorite, and Ryugu samples and their relative formation ages after CV CAIs (Δt_{CAI}). Table S3 also includes the corresponding absolute ages and standard materials used. In Figure S8, the ages obtained using proper (i.e., matrix-matched) standards are shown in filled symbols. It should be noted that we calculated these absolute ages by using the U-corrected Pb-Pb age of 4563.37 ± 0.25 Ma (Brennecka and Wadhwa 2012) and the $(^{53}Mn/^{55}Mn)_0$ of $(3.54 \pm 0.18) \times 10^{-6}$ (McKibbin et al. 2015) of the D'Orbigny angrite. Because ^{53}Mn is a short-lived radionuclide, a time anchor with a known $^{53}Mn/^{55}Mn$ ratio and Pb-Pb absolute age is required to translate the inferred $(^{53}Mn/^{55}Mn)_0$ to absolute ages. Therefore, the absolute ages of carbonates depend on the selection of a time anchor, which potentially introduces uncertainty. The above Pb-Pb age of D'Orbigny was obtained using measured $^{238}U/^{235}U$ ratios and seems to be the most reliable age among those reported so far. Another $(^{53}Mn/^{55}Mn)_0$ of D'Orbigny, $(3.24 \pm 0.04) \times 10^{-6}$ was reported by Glavin et al. (2004). If we use this number as an anchor, the absolute ages of carbonates will be older by ~0.5 Myr. The calculation of the time after CAI formation includes further complication; we need the absolute age of CAIs, which is still under significant debate (e.g., Desch et al. 2023). Here, we use the absolute age of 4567.30 Ma for CAI formation reported by Connelly et al. (2012) to calculate the Δt_{CAI} values.

An overview of the previously reported ^{53}Mn-^{53}Cr ages of carbonates in CR, CM, CI, ungrouped carbonaceous chondrites, hydrated clasts in other meteorites, and Ryugu samples, is described here. Calcite and dolomite in CM chondrites have been dated to ~4–6 Myr after CV CAI formation (Fujiya et al. 2012), which is consistent with more recent data (4 ± 2 Myr after CV CAIs) for carbonates in CMs and CIs, and CM and CI clasts in other meteorite types (Visser et al. 2020). The ^{53}Mn-^{53}Cr ages for dolomite and breunnerite grains in CI chondrites range from ~3–7 Myr after CV CAIs, with the petrography and intricate chemical zoning in carbonates recording how the composition of the fluid evolved as alteration progressed (Endress et al. 1996; Hoppe et al. 2007; Fujiya et al. 2013). The ^{53}Mn-^{53}Cr ages of carbonates in CR chondrites have been obtained for a small number of samples (Jilly-Rehak et al. 2017). This is a challenging task, because carbonates in CR chondrites are much rarer than in CM chondrites, tend to be significantly smaller, and are all calcite with low concentrations of Mn. However, relatively large polycrystalline calcite grains, up to 100 μm in size, occur in Renazzo and grains of Mn-bearing dolomite occur in dark inclusions and in the CR1 chondrite GRO 95577. Jilly-Rehak et al. (2017) showed that calcite in Renazzo, and a dolomite in a dark inclusion from Renazzo, show ^{53}Cr excesses, consistent with the presence of live ^{53}Mn during crystallization of both carbonates. These excesses indicate crystallization ages of ~4–5 Myr after CV CAIs, very similar to ages measured for CI and CM chondrite carbonates. In contrast, dolomite in GRO 95577 is significantly younger with formation ages of ~12 Myr after CV CAIs, indicating that hydrothermal systems could be maintained for extended durations on the CR chondrite parent body. Calcite and dolomite in Flensburg formed simultaneously at 2.7 ± 1.0 after CV CAIs, making them, together with calcite in the Jbilet Winselwan heated CM chondrite (~2.6 Myr after CV CAIs), the oldest ^{53}Mn-^{53}Cr ages yet measured for chondrites. The ^{53}Mn-^{53}Cr ages of carbonate grains in Ryugu samples obtained in different laboratories are not consistent: Yokoyama et al. (2023) and Nakamura et al. (2022) obtained formation ages of ~3–6 Myr after CV CAIs for dolomite, which is largely consistent with ages for carbonates in CR, CM, CI chondrites, whereas McCain et al. (2023) reported ages as old as ~0.4 Myr after CV CAIs.

There are few ^{53}Mn-^{53}Cr ages that have been obtained using proper standards (Fig. S8). In particular, most Mn-Cr ages of dolomite have been estimated without using a dolomite standard, because the laboratory synthesis of Mn- and Cr-bearing dolomite is more difficult than that of calcite (Dunohue et al. 2019). Another approach to produce carbonate standards is Cr-ion implantation into natural carbonates by a particle accelerator (Steele et al. 2017; McCain et al. 2020). A few dolomite ages have been obtained using carbonate standards with implanted Cr, and they appear somewhat older than those obtained with synthetic calcite standards. Notably, two ages of Ryugu carbonates obtained using carbonate standards with implanted Cr are very old (4566.9 Ma or 0.4 Myr after CAIs; McCain et al. 2023). The validity of these ages will be discussed in the supplementary materials and should be examined in future work.

The ^{53}Mn-^{53}Cr ages of carbonates obtained with proper standards in carbonaceous chondrites discussed here range from 3 to 5 Myr after CV CAIs (Fig. S8). These observations indicate that carbonate formation (and possibly, the onset of aqueous alteration) was almost simultaneous among water-rich planetesimals. This result, in turn, implies that these bodies accreted almost simultaneously. Fujiya et al. (2012, 2013) argued that CM and CI chondrite parent bodies formed 3–4 Myr after CV CAIs based on a thermal evolution model where the energy of ^{26}Al decay was taken into account as the heat source. On the other hand, the old ages of Ryugu carbonates (McCain et al. 2023) suggest that the Ryugu parent body may have accreted very early, within 0.4 Myr of CV CAIs. Early formation of the Ryugu parent body must have resulted in high peak temperatures. However, Ryugu samples preserve phyllosilicates without apparent signatures of dehydration and have not been heated above $\sim$100 °C (Yokoyama et al. 2023; Nakamura et al. 2023). This discrepancy may be resolved if the Ryugu parent body was smaller than $\sim$20 km in diameter, or if it was larger than $\sim$20 km in diameter but disrupted before the internal temperatures rose to the dehydration temperature (McCain et al. 2023). Another scenario would be fluid convection which may have occurred within the Ryugu parent body and efficiently transferred internal heat to its outer regions (e.g., Palguta et al. 2010). In any case, these arguments demonstrate that the Mn-Cr ages of carbonates provide important insights into the formation age, size, and thermal history of water-rich planetesimals in the early Solar System.

9.2 ^{129}I-^{129}Xe Chronology of Magnetite in CI Chondrites

The ^{129}I-^{129}Xe chronometer is based upon the decay of now-extinct ^{129}I, where the ratio of the accumulated daughter ^{129}Xe to stable ^{127}I in the host mineral reflects the I-isotope ratio ^{129}I/^{127}I at closure to Xe loss. It is measured as the ratio of radiogenic ^{129}Xe to ^{128}Xe produced by artificial neutron irradiation. Since none of the parent nuclei remains, ^{129}I-^{129}Xe is by nature a relative chronometer but, when referenced to a standard of a known age, it becomes an absolute chronometer reflecting a true closure time. It is a logical technique to apply to early Solar System processes like aqueous alteration; the short half-life of ^{129}I (16.14 ± 0.12 Myr) gives the I-Xe chronometer the precision needed for high temporal resolution (Hohenberg and Pravdivtseva 2008). I-Xe dating cannot be done *in situ*, and samples to be analyzed have to be separated from their host meteorites.

Samples are typically irradiated to the fluence of about 2×10^{19} thermal neutrons/cm^2. This ensures that the ratio of I-derived radiogenic ^{129}Xe to ^{128}Xe is close to 1 and thus could be measured with high precision. Each irradiation has its unique ratio of ^{128*}Xe (production by neutron irradiation) to ^{127}I, which would be nominally established by estimation of the neutron capture probability, thus yielding the chronometrically significant ^{129}I/^{127}I ratio at closure of the iodine host mineral. However, direct monitoring of the neutron capture probability is impractical. An alternative calibration procedure by means of a reference meteorite standard, the Shallowater aubrite, is the usual method used in I-Xe dating. This method yields the highest precision of both relative (to the irradiation standard) and absolute closure times.

The absolute age of Shallowater has not been measured directly due to the low U content; instead, it is derived from the observed correlation between I–Xe and Pb–Pb ages in several samples. This approach allows fine tuning of the absolute age normalization when the new data becomes available. Currently, the re-evaluated absolute I–Xe age of Shallowater is 4562.4 ± 0.2 Ma (Pravdivtseva et al. 2017). The only way for chronometric data to have a meaningful interpretation is to know what mineral phase is being analyzed. Identification of iodine host phases is not easy when these components are small, and the separation of pure mineral phases is not possible. For CI chondrites, magnetite is known to be an iodine carrier, and another carrier(s) appears to exist in ferromagnetic separates (Hohenberg et al. 2000; Pravdivtseva et al. 2018).

The ^{129}I-^{129}Xe ages of pure magnetite chemically separated from the Orgueil (CI) chondrite have been reported by Lewis and Anders (1975) and revisited by Pravdivtseva et al. (2018). The refined I-Xe ages are consistent with each other within uncertainties and correspond to 4564.3 $\pm$ 0.3 Ma (2.9 $\pm$ 0.3 Myr after CAIs) (Pravdivtseva et al. 2018). On the other hand, the I-Xe ages of ferromagnetic separates prepared with a hand magnet from a finely-ground powder are younger than the magnetite ages by 3.4 Myr (Pravdivtseva et al. 2018) or by 4.9 Myr (Hohenberg et al. 2000). Based on these ages, it has been argued that the I-Xe ages of the ferromagnetic separates can be explained by the mixture of magnetite and a hypothetical, not yet identified second carrier phase. These ages also suggest that aqueous alteration of Orgueil started at $\sim$2.9 Myr after CAIs and lasted for at least 5 Ma. The former value is in reasonably good agreement with ^{53}Mn-^{53}Cr ages of carbonates in carbonaceous chondrites (Fig. S8).

9.3 Remaining Questions and Future Work

See the Supplementary text.

Supplementary Information The online version contains supplementary material available at https://doi.org/10.1007/s11214-024-01132-8.

Acknowledgements AJK is funded by UK Research and Innovation (UKRI) grant MR/T020261/1. ANK acknowledges Emerging Worlds NASA grant number 80NSSC23K0253. AJB acknowledges funding from NASA Emerging Worlds grant 80NSSC21K1757. LP thanks the support of the Agence Nationale de la Recherche (ANR) through grant ANR-19-CE31-0027-01 HYDRaTE (PI LP) and of the Centre national d'études spatiales (CNES). MRL acknowledges support from the Science and Technology Facilities Council (STFC) through grant ST/W001128/1. JL acknowledges support from the German Research Foundation (DFG) through grant LE 3279/3-1. LGV acknowledges the support from CNES through a post-doctoral grant.

Author Contribution The article was conceived and structured by ANK. The authors of the various sections are as follows. Mineralogy and petrology of hydrated carbonaceous chondrites: AJB, AJK, YM, MRL. Mineralogy of hydrated dark clasts: AB, EvK, MP. Presolar grains: JL. Evolution of organic matter during low-temperature alteration: LV, CLG, LR, CMO'DA. Physicochemical modelling: MIP. Hydrothermal alteration experiments: AJK, ED, LV. Fluid inclusions in aqueously formed minerals: AT. Isotope tracers: LP, YM, EvK, ANK, CMO'DA, WF, MT, LV. Chronology of aqueous alteration: WF, OP. The sections were compiled by MRL, and all authors contributed to final editing.

Declarations

Competing Interests The authors declare that they have no conflict of interest.

References

Abreu NM, Brearley AJ (2010) Early solar system processes recorded in the matrices of two highly pristine CR3 carbonaceous chondrites, MET 00426 and QUE 99177. Geochim Cosmochim Acta 74:1146–1171

Abreu NM, Bullock E (2013) Opaque assemblages in CR2 Graves Nunataks (GRA) 06100 as indicators of shock-driven hydrothermal alteration in the CR chondrite parent body. Meteorit Planet Sci 48:2406–2429

Alexander CMO'D (2019) Quantitative models for the elemental and isotopic fractionations in chondrites: the carbonaceous chondrites. Geochim Cosmochim Acta 254:277–309

Alexander CMO'D, Barber DJ, Hutchison R (1989) The microstructure of Semarkona and Bishunpur. Geochim Cosmochim Acta 53:3045–3057

Alexander CMO'D, Russell SS, Arden JW, et al (1998) The origin of chondritic macromolecular organic matter: a carbon and nitrogen isotope study. Meteorit Planet Sci 33:603–622

Alexander CMO'D, Fogel M, Yabuta H, et al (2007) The origin and evolution of chondrites recorded in the elemental and isotopic compositions of their macromolecular organic matter. Geochim Cosmochim Acta 71:4380–4403

Alexander CMO'D, Newsome SD, Fogel ML, et al (2010) Deuterium enrichments in chondritic macromolecular material—implications for the origin and evolution of organics, water and asteroids. Geochim Cosmochim Acta 74:4417–4437

Alexander CMO'D, Bowden R, Fogel M, et al (2012) The provenances of asteroids, and their contributions to the volatile inventories of the terrestrial planets. Science 337:721–723

Alexander CMO'D, Howard KT, Bowden R, et al (2013) The classification of CM and CR chondrites using bulk H, C and N abundances and isotopic compositions. Geochim Cosmochim Acta 123:244–260

Alexander CMO'D, Cody GD, Kebukawa Y, et al (2014) Elemental, isotopic, and structural changes in Tagish Lake insoluble organic matter produced by parent body processes. Meteorit Planet Sci 49:503–525

Alexander CMO'D, Bowden R, Fogel ML, et al (2015) Carbonate abundances and isotopic compositions in chondrites. Meteorit Planet Sci 50:810–833

Alexander CMO'D, Cody GD, De Gregorio BT, et al (2017b) The nature, origin and modification of insoluble organic matter in chondrites, the major source of Earth's C and N. Geochemistry 77:227–256

Alexander CMO'D, Nittler LR, Davidson J, et al (2017a) Measuring the level of interstellar inheritance in the solar protoplanetary disk. Meteorit Planet Sci 52:1797–1821

Alexander CMO'D, Greenwood RC, Bowden R, et al (2018) A mutli-technique search for the most primitive CO chondrites. Geochim Cosmochim Acta 221:406–420

Alexander CMO'D, Wynn JG, Bowden R (2022) Sulfur abundances and isotopic compositions in bulk carbonaceous chondrites and insoluble organic material: clues to elemental and isotopic fractionations of volatile chalcophiles. Meteorit Planet Sci 57:334–351

Alfing J, Patzek M, Bischoff A (2019) Modal abundances of coarse-grained (>5 μm) components within CI-chondrites and their individual clasts – mixing of various lithologies on the CI parent body(ies). Geochem 79:125532

Anders E, Grevesse N (1989) Abundances of the elements: meteoritic and solar. Geochim Cosmochim Acta 53:197–214

Aponte JC, McLain HL, Dworkin JP, et al (2016) Aliphatic amines in Antarctic CR2, CM2, and CM1/2 carbonaceous chondrites. Geochim Cosmochim Acta 189:296–311

Baker L, Franchi IA, Wright IP, et al (2002) The oxygen isotopic composition of water from Tagish Lake: its relationship to low-temperature phases and to other carbonaceous chondrites. Meteorit Planet Sci 37:977–985

Barber DJ (1981) Matrix phyllosilicates and associated minerals in C2M carbonaceous chondrites. Geochim Cosmochim Acta 45:945–970

Barosch J, Nittler LR, Wang J, et al (2022) Presolar O- and C-anomalous grains in unequilibrated ordinary chondrite matrices. Geochim Cosmochim Acta 335:169–182

Barrat J-A, Zanda B, Moynier F, et al (2012) Geochemistry of CI chondrites: major and trace elements, and Cu and Zn isotopes. Geochim Cosmochim Acta 83:79–92

Benedix GK, Leshin LA, Farquhar J, et al (2003) Carbonates in CM2 chondrites: constraints on alteration conditions from oxygen isotopic compositions and petrographic observations. Geochim Cosmochim Acta 67:1577–1588

Bischoff A, Palme H, Spettel B, et al (1988) The chemical composition of dark inclusions from the Allende meteorite. Lunar Planet Sci Conf 19:88–89

Bischoff A, Palme H, Ash RD, et al (1993) Paired Renazzo-type (CR) carbonaceous chondrites from the Sahara. Geochim Cosmochim Acta 57:1587–1603

Bischoff A, Scott ERD, Metzler K, et al (2006) Nature and origins of meteoritic breccias. In: Lauretta DS, McSween HY Jr (eds) Meteorites and the early Solar System II, University of Arizona Press, Tucson, pp 679–712

Bischoff A, Ebert S, Metzler K, et al (2017) Breccia classification of CM chondrites. Meteorit Planet Sci 52:A26, #6089
Bischoff A, Schleiting M, Wieler R, et al (2018) Brecciation among 2280 ordinary chondrites – constraints on the evolution of their parent bodies. Geochim Cosmochim Acta 238:516–541
Bischoff A, Alexander CMO'D, Barrat J-A, et al (2021) The old, unique C1 chondrite Flensburg – insight into the first processes of aqueous alteration, brecciation, and the diversity of water-bearing parent bodies and lithologies. Geochim Cosmochim Acta 293:142–186
Blackburn T, Alexander CMO'D, Carlson R, et al (2017) The accretion and impact history of the ordinary chondrite parent bodies. Geochim Cosmochim Acta 200:201–217
Bland PA, Jackson M, Coker RF, et al (2009) Why aqueous alteration in asteroids was isochemical: high porosity $\neq$ high permeability. Earth Planet Sci Lett 287:559
Blinova AI, Herd CDK, Duke MJM (2014b) Testing variations within the Tagish Lake meteorite—II: whole-rock geochemistry of pristine samples. Meteorit Planet Sci 49:1100–1118
Blinova AI, Zega TJ, Herd CDK, et al (2014a) Testing variations within the Tagish Lake meteorite—I: mineralogy and petrology of pristine samples. Meteorit Planet Sci 49:473–502
Bodnar RJ, Dolocan A, Zolensky ME, et al (2019) Organic compounds in early Solar System aqueous fluid. Meteorit Planet Sci 54, supplement 4:abstract 6064
Bogard DD, Garrison DH, Masarik J (2001) The Monahans chondrite and halite: argon-39/argon-40 age, solar gases, cosmic-ray exposure ages, and parent body regolith neutron flux and thickness. Meteorit Planet Sci 36:107–122
Bonal L, Huss GR, Krot AN, et al (2010) Chondritic lithic clasts in the CB/CH-like meteorite Isheyevo: fragments of previously unsampled parent bodies. Geochim Cosmochim Acta 74:2500–2522
Bonal L, Alexander CMO'D, Huss GR, et al (2013) Hydrogen isotopic composition of the water in CR chondrites. Geochim Cosmochim Acta 106:111–133
Bradley JP (2014) Early solar nebula grains—interplanetary dust particles. In: Davis AM (ed) Treatise of geochemistry, 2nd edn. vol 1. Elsevier, Amsterdam, pp 287–308
Brearley AJ (1993) Carbonaceous chondrites clasts in the Kapoeta howardite. Lunar Planet Sci Conf 24:183–184
Brearley AJ (1997) Phyllosilicates in the matrix of the unique carbonaceous chondrite Lewis Cliff 85332 and possible implications for the aqueous alteration of CI chondrites. Meteorit Planet Sci 32:377–388
Brearley AJ (2006) The action of water. In: Lauretta DS, McSween HY Jr (eds) Meteorites and the early Solar System II. University of Arizona Press, Tucson, pp 587–624
Brearley AJ, Jones RH (2016) Dissolution and replacement of cristobalite by amorphous Fe-rich silicate gel in silica bearing Type I chondrules in the Elephant Moraine (EET) 92042 CR2 carbonaceous chondrite. Lunar Planet Sci 47:abstract #1897
Brearley A, Prinz M (1992) CI chondrite-like clasts in the Nilpena polymict ureilite: implications for aqueous alteration processes in CI chondrites. Geochim Cosmochim Acta 56:1373–1386
Brennecka GA, Wadhwa M (2012) Uranium isotope compositions of the basaltic angrite meteorites and the chronological implications for the early Solar System. Proc Natl Acad Sci 109:9299–9303
Briani G, Gounelle M, Marrocchi Y, et al (2009) Pristine extraterrestrial material with unprecedented nitrogen isotopic variation. Proc Natl Acad Sci USA 106(26):10522–10527
Briani G, Gounelle M, Bourot-Denise M, et al (2012) Xenoliths and microxenoliths in H chondrites: sampling the zodiacal cloud in the asteroid Main Belt. Meteorit Planet Sci 47:880–902
Brown PG, et al (2000) The fall, recovery, orbit, and composition of the Tagish Lake meteorite: a new type of carbonaceous chondrite. Science 290:320–325
Buchanan PC, Zolensky ME, Reid AM (1993) Carbonaceous chondrite clasts in the howardites Bholghati and EET87513. Meteoritics 28:659–669
Busemann H, Alexander CMO'D, Nitter LR (2007) Characterization of insoluble organic matter in primitive meteorites by microRaman spectroscopy. Meteorit Planet Sci 42:1387–1416
Ceccarelli C, Caselli P, Bockelée-Morvan D, et al (2014) Deuterium fractionation: the Ariadne's Thread from the precollapse phase to meteorites and comets today. In: Protostars and planets VI. University of Arizona Press, Tucson
Chan QHS, Zolensky ME, Bodnar RJ, et al (2017) Investigation of organo-carbonate associations in carbonaceous chondrites by Raman spectroscopy. Geochim Cosmochim Acta 201:392–409
Chan QHS, Zolensky ME, Kebukawa Y, et al (2018) Organic matter in extraterrestrial water-bearing salt crystals. Sci Adv 4:eaao3521
Changela HG, Le Guillou C, Bernard S, et al (2018) Hydrothermal evolution of the morphology, molecular composition, and distribution of organic matter in CR (Renazzo-type) chondrites. Meteorit Planet Sci 53:1006–1029
Chizmadia LJ, Brearley AJ (2008) Mineralogy, aqueous alteration, and primitive textural characteristics of fine-grained rims in the Y-791198 CM2 carbonaceous chondrite: TEM observations and comparison to ALHA81002. Geochim Cosmochim Acta 72:602–625

Chou C-L, Boynton WV, Bild RW, et al (1976) Trace element evidence regarding a chondritic component in howardite meteorite. Lunar Planet Sci Conf 7:3501–3518

Clayton RN, Mayeda TK (1984) The oxygen isotope record in Murchison and other carbonaceous chondrites. Earth Planet Sci Lett 67:151–166

Clayton RN, Mayeda TK (1999) Oxygen isotope studies of carbonaceous chondrites. Geochim Cosmochim Acta 63:2089–2104

Clog M, Lindgren P, Modestou S, et al (2024) Clumped isotope and $\Delta_{17}O$ measurements of carbonates in CM carbonaceous chondrites: new insights into parent body thermal and fluid evolution. Geochim Cosmochim Acta 369:1–16

Cody GD, Heying E, Alexander CM, et al (2011) Establishing a molecular relationship between chondritic and cometary organic solids. Proc Natl Acad Sci USA 108:19171–19176

Connelly JN, Bizzarro M, Krot AN, et al (2012) The absolute chronology and thermal processing of solids in the solar protoplanetary disk. Science 338:651–655

Davidson J, Busemann H, Nittler LR, et al (2014a) Abundances of presolar silicon carbide grains in primitive meteorites determined by NanoSIMS. Geochim Cosmochim Acta 139:248–266

Davidson J, Nittler LR, Alexander CMO'D, et al (2014b) Presolar materials and nitrogen isotope anomalies in the unique carbonaceous chondrite Miller Range 07687. Lunar Planet Sci 45:abstract #1603

Davidson J, Alexander CMO'D, Schrader DL, et al (2015) Miller Range 090657: a very pristine Renazzo-like (CR) Carbonaceous Chondrite. Lunar Planet Sci 46:abstract #1376

Davidson J, Schrader DL, Alexander CMO'D, et al (2019) Re-examining thermal metamorphism of the Renazzo-like (CR) carbonaceous chondrites: insights from pristine Miller Range 090657 and shock-heated Graves Nunataks 06100. Geochim Cosmochim Acta 267:240

De Gregorio BT, Stroud RM, Nittler LR, et al (2013) Isotopic and chemical variation of organic nanoglobules in primitive meteorites. Meteorit Planet Sci 48(5):904–928

de Leuw S, Rubin AE, Schmidt AK, et al (2010) Carbonates in CM chondrites: complex formational histories and comparison to carbonates in CI chondrites. Meteorit Planet Sci 45:513–530

Deloule E, Robert F (1995) Interstellar water in meteorites? Geochim Cosmochim Acta 59:4695–4706

Desch SJ, Dunlap DR, Dunham ET, et al (2023) Statistical chronometry of meteorites. I. A test of $_{26}$Al homogeneity and the Pb-Pb age of the solar system's $t = 0$. Icarus 402:115607

Dobrică E, Brearley AJ (2020) Amorphous silicates in the matrix of Semarkona: the first evidence for the localized preservation of 'pristine' matrix materials in the most unequilibrated ordinary chondrites. Meteorit Planet Sci 54:1973–1989

Dobrică E, Ishii HA, Bradley JP, et al (2023) Nonequilibrium spherulitic magnetite in the Ryugu samples. Geochim Cosmochim Acta 346:65–75

Dunohue PH, Huss GR, Nagashima K (2019) New synthetic carbonates for investigation of manganese-chromium chronology by secondary ion mass spectrometry. Lunar Planet Sci 50: abstract #1959

Eiler JM, Kitchen N (2004) Hydrogen isotope evidence for the origin and evolution of the carbonaceous chondrites. Geochim Cosmochim Acta 68:1395–1411

Elsila JE, Charnley SB, Burton AS, et al (2012) Compound-specific carbon, nitrogen, and hydrogen isotopic ratios for amino acids in CM and CR chondrites and their use in evaluating potential formation pathways. Meteorit Planet Sci 47:1517–1536

Endress M (1994) Mineralogische und chemische Untersuchungen an CI Chondriten – Ein Modell zur mineralogisch-kosmochemischen Evolution der(s) CI-Mutterkörper(s). PhD Thesis, University of Muenster

Endress M, Bischoff A (1996) Carbonates in CI chondrites: clues to parent body evolution. Geochim Cosmochim Acta 60:489–507

Endress M, Keil K, Bischoff A, et al (1994) Origin of dark clasts in the Acfer 059/El Djouf 001 CR2 chondrite. Meteoritics 29:26–40

Endress M, Zinner E, Bischoff A (1996) Early aqueous activity on primitive meteorite parent bodies. Nature 379:701–703

Engel MH, Nagy B (1982) Distribution and enantiomeric composition of amino acids in the Murchison meteorite. Nature 296:837–840

Fayolle EC, Öberg KI, Cuppen HM (2012) Laboratory H_2O:CO_2 ice desorption: entrapment and its parameterization with an extended three-phase model. EAS Publ Ser 58:327–331

Floss C, Haenecour P (2016) Presolar silicate grains: abundances, isotopic and elemental compositions, and the effects of secondary processing. Geochem J 50:3–15

Floss C, Stadermann FJ (2009) Auger nanoprobe analysis of presolar ferromagnesian silicate grains from primitive CR chondrites QUE 99177 and MET 00426. Geochim Cosmochim Acta 73:2415–2440

Floss C, Le Guillou C, Brearley A (2014) Coordinated NanoSIMS and FIB-TEM analyses of organic matter and associated matrix materials in CR3 chondrites. Geochim Cosmochim Acta 139:1–25

Floyd CJ, Benito S, Martin P-E, et al (2024) Chondrule sizes within the CM carbonaceous chondrites and measurement methodologies. Meteorit Planet Sci 59:2769–2788

Foustoukos DI, Alexander CMO'D, Cody GD (2021) H and N systematics in thermally altered chondritic insoluble organic matter: an experimental study. Geochim Cosmochim Acta 300:44–64

Frank DR, Huss GR, Zolensky ME, et al (2023) Calcium-aluminium-rich inclusion found in the Ivuna CI chondrite: are CI chondrites a good proxy for the bulk composition of the solar system? Meteorit Planet Sci 58:1495–1511

Fries M, Zolensky M, Steele A (2011) Mineral inclusion in Monahans and Zag halites: evidence of the originating body. Meteorit Planet Sci 46:A70

Fries M, Messenger S, Steele A, et al (2013) Do we already have samples of Ceres? H chondrite halites and the Ceres–Hebe link. Meteorit Planet Sci 48:A133

Fruland RM, King EA, McKay DS (1978) Allende dark inclusions. Lunar Planet Sci Conf Proc, 9:1305–1329

Fujiya W (2018) Oxygen isotopic ratios of primordial water in carbonaceous chondrites. Earth Planet Sci Lett 481:264–272

Fujiya W, Sugiura N, Hotta H, et al (2012) Evidence for the late formation of hydrous asteroids from young meteoritic carbonates. Nat Commun 3:627

Fujiya W, Sugiura N, Sano Y, et al (2013) Mn-Cr ages of dolomites in CI chondrites and the Tagish Lake ungrouped carbonaceous chondrite. Earth Planet Sci Lett 362:130–142

Fujiya W, Sugiura N, Marrocchi Y, et al (2015) Comprehensive study of carbon and oxygen isotopic compositions, trace element abundances, and cathodoluminescence intensities of calcite in the Murchison CM chondrite. Geochim Cosmochim Acta 161:101–117

Fujiya W, Hoppe P, Ushikubo T, et al (2019) Migration of D-type asteroids from the outer Solar System inferred from carbonate in meteorites. Nat Astron 3:910–915

Fujiya W, Aoki Y, Ushikubo T, et al (2020) Carbon isotopic evolution of aqueous fluids in CM chondrites: clues from in-situ isotope analyses within calcite grains in Yamato-791198. Geochim Cosmochim Acta 274:246–260

Fujiya W, Kawasaki N, Nagashima K, et al (2023) Carbonate record of temporal change in oxygen fugacity and gaseous species in asteroid Ryugu. Nat Geosci 16:675–682

Funk C, Bischoff A, Schlüter J (2011) Xenoliths in carbonaceous and ordinary chondrites. Meteorit Planet Sci 46:A71

Furukawa Y, Chikaraishi Y, Ohkouchi N, et al (2019) Extraterrestrial ribose and other sugars in primitive meteorites. Proc Natl Acad Sci USA 116:24440–24445

Garvie LA, Buseck PR (2007) Prebiotic carbon in clays from Orgueil and Ivuna (CI), and Tagish Lake (C2 ungrouped) meteorites. Meteorit Planet Sci 42:2111–2117

Gattacceca J, McCubbin FM, Grossman J, et al (2022) The Meteoritical Bulletin, No. 110. Meteorit Planet Sci 57:2102–2105

Ghosh P, Adkins J, Affek H (2006) $_{13}$C-$_{18}$O bonds in carbonate minerals: a new kind of paleothermometer. Geochim Cosmochim Acta 70:1439–1456

Glavin DP, Dworkin JP (2009) Enrichment of the amino acid L-isovaline by aqueous alteration on CI and CM meteorite parent bodies. Proc Natl Acad Sci USA 106:5487–5492

Glavin DP, Kubny A, Jagoutz E, et al (2004) Mn-Cr isotope systematics of the D'Orbigny angrite. Meteorit Planet Sci 39:693–700

Glavin DP, Alexander CMD, Aponte JC, et al (2018) The origin and evolution of organic matter in carbonaceous chondrites and links to their parent bodies. In: Primitive meteorites and asteroids. Elsevier, Amsterdam, pp 205–271

Glavin DP, McLain HL, Dworkin JP, et al (2020) Abundant extraterrestrial amino acids in the primitive CM carbonaceous chondrite Asuka 12236. Meteorit Planet Sci 55:1979–2006

Goodrich CA, Keil K (2002) Feldspathic and other unusual clasts in polymict ureilite DaG 165. Lunar Planet Sci 33:abstract #1777

Goodrich CA, Scott ERD, Fioretti AM (2004) Ureilitic breccias: clues to the petrologic structure and impact disruption of the ureilite parent asteroid. Chem Erde 64:283–327

Goodrich CA, Zolensky M, Fioretti AM, et al (2019) The first samples from Almahata Sitta showing contacts between ureilitic and chondritic lithologies: implications for the structure and composition of asteroid 2008 TC3. Meteorit Planet Sci 54:2769–2813

Gounelle M, Zolensky ME (2001) A terrestrial origin for sulphate veins in CI1 chondrites. Meteorit Planet Sci 36:1321–1329

Gounelle M, Zolensky ME, Liou J-C, et al (2003) Mineralogy of carbonaceous chondritic microclasts in howardites: identification of C2 fossil micrometeorites. Geochim Cosmochim Acta 67:507–527

Gounelle M, Engrand C, Alard O, et al (2005) Hydrogen isotopic composition of water from fossil micrometeorites in howardites. Geochim Cosmochim Acta 69:3431–3443

Grady MM, Wright IP, Carr RH (1988) Differences in isotopic composition of carbonaceous components in enstatite chondrites. Earth Planet Sci Lett 87:293–302
Grady MM, Verchovsky AB, Franchi IA (2002) Light element geochemistry of the Tagish Lake CI2 chondrite: comparison with CI1 and CM2 meteorites. Meteorit Planet Sci 37:713–735
Graham HV, Elsila J, Dworkin JP, et al (2022) Deuterium isotope fractionation of polycyclic aromatic hydrocarbons in meteorites as an indicator of interstellar/protosolar processing history. Life 12:1368
Greshake A (2014) A strongly hydrated microclast in the Rumuruti chondrite NWA 6828: implications for the distribution of hydrous material in the solar system. Meteorit Planet Sci 49:824–841
Greshake A, Krot AN, Meibom A, et al (2002) Heavily-hydrated lithic clasts in CH chondrites and related, metal-rich chondrites Queen Alexandra Range 94411 and Hammadah al Hamra 237. Meteorit Planet Sci 37:281–293
Greshake A, Krot AN, Flynn GJ, et al (2005) Fine-grained dust rims in the Tagish Lake carbonaceous chondrite: evidence for parent body alteration. Meteorit Planet Sci 40:1413–1431
Gubaidulina TV, Chistyakova NI, Rusakov VS (2007) Mössbauer study of layered iron hydroxysulfides: tochilinite and valleriite. Bull Russ Acad Sci, Phys 71:1269–1272
Guo W, Eiler JM (2007) Temperatures of aqueous alteration and evidence for methane generation on the parent bodies of the CM chondrites. Geochim Cosmochim Acta 71:5565–5575
Haenecour P, Floss C, Zega TJ, et al (2016) Auger spectroscopy analysis of submicron-sized silicate grains in chondrites: insight into their aqueous and thermal alteration history. Meteorit Planet Sci 51:A306
Haenecour P, Floss C, Zega TJ, et al (2018) Presolar silicates in the matrix and fine-grained rims around chondrules in primitive CO3.0 chondrites: evidence for pre-accretionary aqueous alteration of the rims in the solar nebula. Geochim Cosmochim Acta 221:379–405
Harju ER, Rubin AE, Ahn I, et al (2014) Progressive aqueous alteration of CR carbonaceous chondrites. Geochim Cosmochim Acta 139:267–292
Hässig M, Altwegg K, Balsiger H, et al (2017) Isotopic composition of CO_2 in the coma of 67P/Churyumov-Gerasimenko measured with ROSINA/DFMS. Astron Astrophys 605:A50
Herd CDK, et al (2011) Origin and evolution of prebiotic organic matter as inferred from the Tagish Lake meteorite. Science 332:1304–1307
Hewins RH, Bourot-Denise M, Zanda B, et al (2014) The Paris meteorite, the least altered CM chondrite so far. Geochim Cosmochim Acta 124:190–222
Hohenberg CM, Pravdivtseva OV (2008) I–Xe dating: from adolescence to maturity. Chem Erde 68:339–351
Hohenberg CM, Pravdivtseva O, Meshik A (2000) Reexamination of anomalous I-Xe ages: Orgueil and Murchison magnetites and Allegan feldspar. Geochim Cosmochim Acta 64:4257–4261
Hoppe P, Macdougall JD, Lugmair GW (2007) High spatial resolution ion microprobe measurements refine chronology of carbonate formation in Orgueil. Meteorit Planet Sci 42:1309–1320
Hoppe P, Cohen S, Meibom A (2013) NanoSIMS: technical aspects and applications in cosmochemistry and biological geochemistry. Geostand Geoanal Res 37:111–154
Hoppe P, Leitner J, Kodolányi J (2017) The stardust abundance in the local interstellar cloud at the birth of the Solar System. Nat Astron 1:617–620
Howard KT, Alexander CMO'D, Schrader DL, et al (2015) Classification of hydrous meteorites (CR, CM and C2 ungrouped) by phyllosilicate fraction: PSD-XRD modal mineralogy and planetesimal environments. Geochim Cosmochim Acta 149:206–222
Huberty JM, Kita NT, Kozdon R, et al (2010) Crystal orientation effects in $\delta_{18}O$ for magnetite and hematite by SIMS. Chem Geol 276:269–283
Ikeda Y, Prinz M (2001) Magmatic inclusions and felsic clasts in the Dar al Gani 319 polymict ureilite. Meteorit Planet Sci 36:481–499
Ikeda Y, Prinz M, Nehru CE (2000) Lithic and mineral clasts in the Dar al Gani (DAG) 319 polymict ureilite. Antarct Meteor Res 13:177–221
Ikeda Y, Kita NT, Morishita Y, et al (2003) Primitive clasts in the Dar al Gani 319 polymict ureilite: precursors of the ureilites. Antarct Meteor Res 16:105–127
Imae N, et al (2024) Mineralogical approach on laboratory weathering of uncontaminated Ryugu particles: comparison with Orgueil and perspective for storage and analysis. Meteorit Planet Sci 59:1705–1722
Itoh D, Tomeoka K (2003) Dark inclusions in CO3 chondrites: new indicators of parent body processes. Geochim Cosmochim Acta 67:153–169
Ivanova MA, et al (2008) The Isheyevo meteorite: mineralogy, petrology, bulk chemistry, oxygen, nitrogen, carbon isotopic compositions, and $_{40}Ar$-$_{39}Ar$ ages. Meteorit Planet Sci 43:915–940
Izawa MRM, Flemming RL, King PL, et al (2010) Mineralogical and spectroscopic investigation of the Tagish Lake carbonaceous chondrite by X-ray diffraction and infrared reflectance spectroscopy. Meteorit Planet Sci 45:675–698
Jacquet E, Robert F (2013) Water transport in protoplanetary disks and the hydrogen isotopic composition of chondrites. Icarus 223(2):722–732

Jilly-Rehak CE, Huss GR, Nagashima K (2017) $_{53}$Mn–$_{53}$Cr radiometric dating of secondary carbonates in CR chondrites: timescales for parent body aqueous alteration. Geochim Cosmochim Acta 201:224–244

Jilly-Rehak CE, Huss GR, Nagashima K, et al (2018) Low-temperature aqueous alteration on the CR chondrite parent body: implications from in situ oxygen-isotope analyses. Geochim Cosmochim Acta 222:230–252

Johnson CA, Prinz M (1993) Carbonate compositions in CM and CI chondrites, and implications for aqueous alteration. Geochim Cosmochim Acta 57:2843–2852

Jones CL, Brearley AJ (2006) Experimental aqueous alteration of the Allende meteorite under oxidizing conditions: constraints on asteroidal alteration. Geochim Cosmochim Acta 70:1040–1058

Kallemeyn GW, Wasson JT (1981) The compositional classification of chondrites—I. The carbonaceous chondrite groups. Geochim Cosmochim Acta 45:1217–1230

Kalpana MS, Babu EVSSK, Mani D, et al (2021) Polycyclic aromatic hydrocarbons in the Mukundpura (CM2) chondrite. Planet Space Sci 198:105177

Kaplan HH, et al (2020) Bright carbonate veins on asteroid (101955) Bennu: implications for aqueous alteration history. Science 370:eabc3557

Kawasaki N, Nagashima K, Sakamoto N, et al (2022) Oxygen isotopes of anhydrous primary minerals show kinship between asteroid Ryugu and comet 81P/Wild2. Sci Adv 8:eade2067

Kebukawa Y, Cody GD (2015) A kinetic study of the formation of organic solids from formaldehyde: implications for the origin of extraterrestrial organic solids in primitive Solar System objects. Icarus 248:412–423

Kebukawa Y, Kilcoyne AD, Cody GD (2013) Exploring the potential formation of organic solids in chondrites and comets through polymerization of interstellar formaldehyde. Astrophys J 771:19

Kebukawa Y, Zolensky ME, Fries M, et al (2016) STXM-XANES analysis of organic matter in dark clasts and halite crystals in Zag and Monahans meteorite. Lunar Planet Sci 47:abstract #1802

Kebukawa Y, Chan QH, Tachibana S, et al (2017) One-pot synthesis of amino acid precursors with insoluble organic matter in planetesimals with aqueous activity. Sci Adv 3:e1602093

Kebukawa Y, Ito M, Zolensky ME, et al (2019) A novel organic-rich meteoritic clast from the outer Solar System. Sci Rep 9:3169

Kebukawa Y, Kobayashi S, Kawasaki N, et al (2021) Hydrogen isotopic exchange kinetics between organic matter and water: implications for chemical evolution during meteorite parent body processing. Meteorit Planet Sci 56:440–454

Kebukawa Y, Mathurin J, Dartoiset E, et al (2023) Complex mixture of organic matter in a xenolithic clast from the Zag meteorite revealed by coordinated analyses using AFM-IR, NanoSIMS and STXM/XANES. Icarus 400:115582

Kerraouch I, Bischoff A, Zolensky ME, et al (2021) The polymict carbonaceous breccia Aguas Zarcas: a potential analogue to samples being returned by the OSIRIS-REx and Hayabusa2 missions. Meteorit Planet Sci 56:277–310

Kerraouch I, Kebukawa Y, Bischoff A, et al (2022) Heterogeneous nature of the carbonaceous chondrite breccia Aguas Zarcas – cosmochemical characterization and origin of new carbonaceous chondrite lithologies. Geochim Cosmochim Acta 334:155–186

Kerridge JF (1983) Isotopic composition of carbonaceous-chondrite kerogen: evidence for an interstellar origin of organic matter in meteorites. Earth Planet Sci Lett 64:186–200

Kerridge JF, Chang S, Shipp R (1987) Isotopic characterization of kerogen-like material in the Murchison carbonaceous chondrite. Geochim Cosmochim Acta 51:2527–2540

Kikuchi S, Shibuya T, Abe M, Uematsu K (2022) Experimental chondrite–water reactions under reducing and low-temperature hydrothermal conditions: implications for incipient aqueous alteration in planetesimals. Geochim Cosmochim Acta 319:151–167

Kimura M, Imae N, Komatsu M, et al (2020) The most primitive CM chondrites, Asuka 12085, 12169, and 12236, of subtypes 3.0–2.8: their characteristic features and classification. Polar Sci 26:100565

King AJ, Daly L, Rowe J, et al (2022) The Winchcombe meteorite, a unique and pristine witness from the outer solar system. Sci Adv 8:eabq3925

Kita NT, et al (2024) Disequilibrium oxygen isotope distribution among aqueously altered minerals in Ryugu asteroid returned samples. Meteorit Planet Sci 59:2097–2116

Kojima T, Tomeoka K, Takeda H (1993) Unusual dark clasts in the Vigarano CV3 carbonaceous chondrite: record of parent body process. Meteoritics 28:649–658

Krot AN, Scott ERD, Zolensky ME (1997) Origin of fayalitic olivine rims and lath-shaped matrix olivine in the CV3 chondrite Allende and its dark inclusions. Meteorit Planet Sci 32:31–49

Krot AN, Petaev MI, Zolensky M, et al (1998) Secondary Ca-Fe-rich minerals in the Bali-like and Allende-like oxidized CV3 chondrites and Allende dark inclusions. Meteorit Planet Sci 33:623–645

Krot AN, Brearley AJ, Ulyanov AA, et al (1999) Mineralogy, petrography, bulk chemical, iodine-xenon, and oxygen-isotopic compositions of dark inclusions in the reduced CV3 chondrite Efremovka. Meteorit Planet Sci 34:67–89

Krot AN, Meibom A, Weisberg MK, et al (2002) The CR chondrite clan: implications for early solar system processes. Meteorit Planet Sci 37:1451–1490
Krot AN, Libourel G, Goodrich CA, et al (2004) Silica-rich igneous rims around magnesian chondrules in CR carbonaceous chondrites: evidence for condensation origin from fractionated nebular gas. Meteorit Planet Sci 39:1931–1955
Krot AN, Nagashima K, Ciesla FJ, et al (2010) Oxygen isotopic composition of the sun and mean oxygen isotopic composition of the protosolar silicate dust: evidence from refractory inclusions. Astrophys J 713:1159–1166
Krot AN, Nagashima K, Alexander CMO'D, et al (2015) Sources of water and aqueous activity on the chondrite parent asteroids. In: Asteroids IV (vol. 1, issue January 2009) University of Arizona Press, Tucson, pp 21–32
Kuznetsova A, Bae J, Hartmann L, et al (2022) Anisotropic infall and substructure formation in embedded disks. Astrophys J 928:92
Labidi J, Farquhar J, Alexander CMO'D, et al (2017) Mass independent sulfur isotope signatures in CMs: implications for sulfur chemistry in the early solar system. Geochim Cosmochim Acta 196:326–350
Laurent B, Maillard J, Afonso C, et al (2022a) Diversity of chondritic organic matter probed by ultra-high resolution mass spectrometry. Geochem Perspect Lett 22:31–35
Laurent B, Remusat L, Viennet JC, et al (2022b) Preservation of the isotope signatures in chondritic IOM during aqueous alteration. Geochem Perspect Lett 23:28–32
Lauretta DS, Connolly HC Jr, et al (2024) Asteroid (101955) Bennu in the laboratory: properties of the sample collected by OSIRIS-REx. Meteorit Planet Sci 59:2453–2486
Le Guillou C, Brearley A (2014) Relationships between organics, water and early stages of aqueous alteration in the pristine CR3.0 chondrite MET 00426. Geochim Cosmochim Acta 131:344–367
Le Guillou C, Bernard S, Brearley AJ, et al (2014) Evolution of organic matter in Orgueil, Murchison and Renazzo during parent body aqueous alteration: in situ investigations. Geochim Cosmochim Acta 131:368–392
Le Guillou C, Changela HG, Brearley AJ (2015a) Widespread oxidized and hydrated amorphous silicates in CR chondrites matrices: implications for alteration conditions and H_2 degassing of asteroids. Earth Planet Sci Lett 420:162–173
Le Guillou C, Dohmen R, Rogalla D, et al (2015b) New experimental approach to study aqueous alteration of amorphous silicates at low reaction rates. Chem Geol 412:179–192
Lecasble M, Remusat L, Viennet JC, et al (2022) Polycyclic aromatic hydrocarbons in carbonaceous chondrites can be used as tracers of both pre-accretion and secondary processes. Geochim Cosmochim Acta 335:243–255
Lecasble M, Bernard S, Viennet JC, et al (2023) Influence of hydrothermal asteroidal conditions on the molecular structure and isotopic compositions of polycyclic aromatic hydrocarbons. Icarus 401:115603
Lee MR, Ellen R (2008) Aragonite in the Murray (CM2) carbonaceous chondrite: implications for parent body compaction and aqueous alteration. Meteorit Planet Sci 43:1219–1231
Lee MR, Lindgren P, Sofe MR, et al (2012) Extended chronologies of aqueous alteration in the CM2 carbonaceous chondrites: evidence from carbonates in Queen Alexandra Range 93005. Geochim Cosmochim Acta 92:148–169
Lee MR, Sofe MR, Lindgren P, et al (2013) The oxygen isotope evolution of parent body aqueous solutions as recorded by multiple carbonate generations in the Lonewolf Nunataks 94101 CM2 carbonaceous chondrite. Geochim Cosmochim Acta 121:452–466
Lee MR, Lindgren P, Sofe MR (2014) Aragonite, breunnerite, calcite and dolomite in the CM carbonaceous chondrites: high fidelity recorders of progressive parent body aqueous alteration. Geochim Cosmochim Acta 144:126–156
Lee MR, Daly L, Floyd C, et al (2021) CM carbonaceous chondrite falls and their terrestrial alteration. Meteorit Planet Sci 56:34–48
Leitner J, Vollmer C, Hoppe P, et al (2012) Characterization of presolar material in the CR Chondrite Northwest Africa 852. Astrophys J 745:38–53
Leitner J, Vollmer C, Floss C, et al (2016) Ancient stardust in fine-grained chondrule dust rims from carbonaceous chondrites. Earth Planet Sci Lett 434:117–128
Leitner J, Hoppe P, Zipfel J (2018) A study of presolar material in hydrated lithic clasts from metal-rich carbonaceous chondrites. Meteorit Planet Sci 53:204–231
Leitner J, Metzler K, Vollmer C, et al (2020) The presolar grain inventory of fine-grained chondrule rims in the Mighei-type (CM) chondrites. Meteorit Planet Sci 55:1176–1206
Lentfort S, Bischoff A, Ebert S, et al (2021) Classification of CM chondrite breccias—implications for the evaluation of samples from the OSIRIS-REx and Hayabusa 2 missions. Meteorit Planet Sci 56:127–147
Leroux H, Cuvillier P, Zanda B, et al (2015) GEMS-like material in the matrix of the Paris meteorite and the early stages of alteration of CM chondrites. Geochim Cosmochim Acta 170:247–265

Lewis RS, Anders E (1975) Condensation time of the solar nebula from the extinct 129I in primitive meteorites. Proc Natl Acad Sci USA 72:268–273
Lindgren P, Lee MR, Sofe MR, et al (2013) Clasts in the CM2 carbonaceous chondrite Lonewolf Nunataks 94101: evidence for aqueous alteration prior to complex mixing. Meteorit Planet Sci 48:1074–1090
Lindgren P, Lee MR, Starkey NA, et al (2017) Fluid evolution in CM carbonaceous chondrites tracked through the oxygen isotopic compositions of carbonates. Geochim Cosmochim Acta 204:240–251
Longhi J (2005) Phase equilibria in the system CO_2-H_2O I: new equilibrium relations at low temperatures. Geochim Cosmochim Acta 69:529–539
Lyons JR, Young ED (2005) CO self-shielding as the origin of oxygen isotope anomalies in the early solar nebula. Nature 435:317–320
Lyons JR, Gharib-Nezhad E, Ayres TR (2018) A light carbon isotope composition for the sun. Nat Commun 9:908
Macdougall JD, Lugmair GW, Kerridge JF (1984) Early solar system aqueous activity: Sr isotope evidence from the Orgueil CI meteorite. Nature 307:249–251
Mackinnon IDR, Zolensky ME (1984) Proposed structures for poorly characterized phases in C2M carbonaceous chondrite meteorites. Nature 309:240–242
Marhas KK, Hoppe P, Stadermann FJ, et al (2006) The distribution of presolar grains in CI and CO meteorites. Lunar Planet Sci 37:abstract #1959
Marrocchi Y, Gounelle M, Blanchard I, et al (2014) The Paris CM chondrite: secondary minerals and asteroidal processing. Meteorit Planet Sci 49:1232–1249
Marrocchi Y, Bekaert DV, Piani L (2018) Origin and abundance of water in carbonaceous asteroids. Earth Planet Sci Lett 482:23–32
Marrocchi Y, Avice G, Barrat J-A (2021b) The Tarda meteorite: a window into the formation of D-type asteroids. Astrophys J Lett 913:L9
Marrocchi Y, Delbo M, Gounelle M (2021a) The astrophysical context of collision processes in meteorites. Meteorit Planet Sci 56:1406–1421
Marrocchi Y, Piralla M, Regnault M, et al (2022) Isotopic evidence for two chondrule generations in CR chondrites and their relationships to other carbonaceous chondrites. Earth Planet Sci Lett 593:117683
Marrocchi Y, Jacquet E, Neukampf J, et al (2023a) From whom bells tolls: reclassifying bells among CR chondrites and implications for the formation conditions of CR parent bodies. Meteorit Planet Sci 58:195–206
Marrocchi Y, Rigaudier T, Piralla M, et al (2023b) Hydrogen isotopic evidence for nebular pre-hydration and the limited role of parent-body processes in CM chondrites. Earth Planet Sci Lett 611:118151
Marrocchi Y, Jones RH, Russell SS, et al (2024) Chondrule properties and formation conditions. Space Sci Rev 220:69
Martínez M, Brearley AJ (2022) Smooth rims in Queen Alexandra Range (QUE) 99177: fluid–chondrule interactions and clues on the geochemical conditions of the primordial fluid that altered CR carbonaceous chondrites. Geochim Cosmochim Acta 325:39–64
Martins Z, Alexander CMO'D, Orzechowska GE, et al (2007) Indigenous amino acids in primitive CR meteorites. Meteorit Planet Sci 42:2125–2136
Matsumoto M, Tsuchiyama A, Nakato A, et al (2019) Discovery of fossil asteroidal ice in primitive meteorite Acfer 094. Sci Adv 5:eaax5078
McCain KA, Liu M-C, McKeegan KD (2020) Calibration of matrix-dependent biases in isotope and trace element analyses of carbonate minerals. J Vac Sci Technol B 38:044005
McCain KA, Matsuda N, Liu MC, et al (2023) Early fluid activity on Ryugu inferred by isotopic analyses of carbonates and magnetite. Nat Astron 7:309–317
McCubbin FM, Barnes JJ (2019) Origin and abundances of H_2O in the terrestrial planets, Moon, and asteroids. Earth Planet Sci Lett 526:115771
McKeegan KD, Kallio APA, Heber VS, et al (2011) The oxygen isotopic composition of the Sun inferred from captured solar wind. Science 332:1528–1532
McKibbin SJ, Ireland TR, Amelin Y, et al (2015) Mn–Cr dating of Fe- and Ca-rich olivine from 'quenched' and 'plutonic' angrite meteorites using Secondary Ion Mass Spectrometry. Geochim Cosmochim Acta 157:13–27
Metzler K, Bischoff A, Stöffler D (1992) Accretionary dust mantles in CM-chondrites: evidence for nebula processes. Geochim Cosmochim Acta 56:2873–2897
Mittlefehldt DW (1994) The genesis of diogenites and HED parent body petrogenesis. Geochim Cosmochim Acta 58:1537–1552
Mittlefehldt DW (2015) Asteroid (4) Vesta: I. The howardite-eucrite-diogenite (HED) clan of meteorites. Geochem 75:155–183
Modica P, Meinert C, de Marcellus P, et al (2014) Enantiomeric excesses induced in amino acids by ultraviolet circularly polarized light irradiation of extraterrestrial ice analogs: a possible source of asymmetry for prebiotic chemistry. Astrophys J 788:79

Morin GLF, Marrocchi Y, Villeneuve J, et al (2022) $_{16}$O-rich anhydrous silicates in CI chondrites: implications for the nature and dynamics of dust in the solar accretion disk. Geochim Cosmochim Acta 332:203–219

Morlok A, Bischoff A, Stephan T, et al (2006) Brecciation and chemical heterogeneities of CI chondrites. Geochim Cosmochim Acta 70:5371–5394

Nagashima K, Krot AN, Yurimoto H (2004) Stardust silicates from primitive meteorites. Nature 428:921–924

Nakamura T (2005) Post-hydration thermal metamorphism of carbonaceous chondrites. J Mineral Petrol Sci 100:260–272

Nakamura E, Kobayashi K, Tanaka R, et al (2022) On the origin and evolution of the asteroid Ryugu: a comprehensive geochemical perspective. Proc Jpn Acad Ser B 98:227–282

Nakamura T, Matsumoto M, Amano K, et al (2023) Formation and evolution of carbonaceous asteroid Ryugu: direct evidence from returned samples. Science 379:eabn8671

Nakamura-Messenger K, Clemett SJ, Messenger S, et al (2011) Experimental aqueous alteration of cometary dust. Meteorit Planet Sci 46:843–856

Naraoka H, Shimoyama A, Harada K (2000) Isotopic evidence from an Antarctic carbonaceous chondrite for two reaction pathways of extraterrestrial PAH formation. Earth Planet Sci Lett 184:1–7

Nguyen AN, Zinner E (2004) Discovery of ancient silicate stardust in a meteorite. Science 303:1496–1499

Nguyen AN, Nittler LR, Stadermann FJ, et al (2010) Coordinated analyses of presolar grains in the Allan Hills 77307 and Queen Elizabeth Range 99177 meteorites. Astrophys J 719:166–189

Nguyen N, et al (2023) Abundant presolar grains and primordial organics preserved in carbon-rich exogenous clasts in asteroid Ryugu. Sci Adv 9:eadh1003

Nittler LR, Ciesla FJ (2016) Astrophysics and extraterrestrial materials. Annu Rev Astron Astrophys 54:53–93

Nittler LR, Alexander CMO'D, Davidson J, et al (2018) High abundances of presolar grains and 15N-rich organic matter in CO3.0 chondrite Dominion Range 08006. Geochim Cosmochim Acta 226:107–131

Nittler LR, Stroud RM, Alexander CMO'D, et al (2020) Presolar grains in primitive ungrouped carbonaceous chondrite northwest Africa 5958. Meteorit Planet Sci 55:1160–1175

Nittler LR, Alexander CMO'D, Patzer A, et al (2021) Presolar stardust in highly pristine CM chondrites Asuka 12169 and Asuka 12236. Meteorit Planet Sci 56:260–276

Noguchi T (1995) Petrology and mineralogy of the PCA 91082 chondrite and its comparison with the Yamato-793495 (CR) chondrite. Proc NIPR Symp Antarct Mets 8:33–62

Oba Y, Naraoka H (2009) Elemental and isotope behavior of macromolecular organic matter from CM chondrites during hydrous pyrolysis. Meteorit Planet Sci 44:943–953

Ohnishi I, Tomeoka K (2007) Hydrothermal alteration experiments of enstatite: implications for aqueous alteration of carbonaceous chondrites. Meteorit Planet Sci 42:49–61

Okuzumi S, Momose M, Sirono S, et al (2016) Sintering-induced dust ring formation in protoplanetary disks: application to the HL Tau disk. Astrophys J 821:82

Orthous-Daunay FR, Quirico E, Beck P, et al (2013) Mid-infrared study of the molecular structure variability of insoluble organic matter from primitive chondrites. Icarus 223:534–543

Palguta J, Schubert G, Travis BJ (2010) Fluid flow and chemical alteration in carbonaceous chondrite parent bodies. Earth Planet Sci Lett 296:235–243

Palme H, Kurat G, Spettel B, et al (1989) Chemical composition of an unusual xenolith of the Allende meteorite. Z Naturforsch 44a:1005–1014

Patzek M, Bischoff A, Visser R, et al (2018) Mineralogy of volatile-rich clasts in brecciated meteorites. Meteorit Planet Sci 53:2519–2540

Patzek M, Bischoff A, Hoppe P, et al (2019) Oxygen and hydrogen isotopic evidence for the existence of several C1 parent bodies in the early solar system. Lunar Planet Sci 50:abstract #1779

Patzek M, Hoppe P, Bischoff A, et al (2020) Hydrogen isotopic composition of CI- and CM-like clasts from meteorite breccias – sampling unknown sources of carbonaceous chondrite materials. Geochim Cosmochim Acta 272:177–197

Pearson VK, Sephton MA, Kearsley AT, et al (2002) Clay mineral-organic matter relationships in the early solar system. Meteorit Planet Sci 37:1829–1833

Pearson VK, Sephton MA, Franchi IA, Gibson JM, Gilmour I (2006) Carbon and nitrogen in carbonaceous chondrites: elemental abundances and stable isotopic compositions. Meteorit Planet Sci 41:1899–1918

Piani L, Remusat L, Robert F (2012) Determination of the H isotopic composition of individual components in fine-scale mixtures of organic matter and phyllosilicates with the nanoscale secondary ion mass spectrometry. Anal Chem 84:10199–10206

Piani L, Yurimoto H, Remusat L (2018) A dual origin for water in carbonaceous asteroids revealed by CM chondrites. Nat Astron 2:317–323

Piani L, Marrocchi Y, Vacher LG, et al (2021) Origin of hydrogen isotopic variations in chondritic water and organics. Earth Planet Sci Lett 567:117008

Pignatelli I, Marrocchi Y, Vacher LG, et al (2016) Multiple precursors of secondary mineralogical assemblages in CM chondrites. Meteorit Planet Sci 51:785–805

Pignatelli I, Marrocchi Y, Mugnaioli E, et al (2017) Mineralogical, crystallographic and redox features of the earliest stages of fluid alteration in CM chondrites. Geochim Cosmochim Acta 209:106–122

Piralla M, Marrocchi Y, Verdier-Paoletti MJ, et al (2020) Primordial water and dust of the Solar System: insights from in situ oxygen measurements of CI chondrites. Geochim Cosmochim Acta 269:451–464

Pravdivtseva O, Meshik A, Hohenberg CM, et al (2017) I–Xe systematics of the impact plume produced chondrules from the CB carbonaceous chondrites: implications for the half-life value of 129I and absolute age normalization of 129I–129Xe chronometer. Geochim Cosmochim Acta 201:320–330

Pravdivtseva O, Krot AN, Hohenberg CM (2018) I-Xe dating of aqueous alteration in the CI chondrite Orgueil: I. Magnetite and ferromagnetic separates. Geochim Cosmochim Acta 227:38–47

Prinz M, Weisberg MK, Nehru CE, et al (1987) EET 83309, a polymict ureilite: recognition of a new group. Lunar Planet Sci Conf 18:802

Quirico E, Bonal L (2019) Organic matter in interplanetary dusts and meteorites. In: Cavalazzi B, Westall F (eds) Biosignatures for astrobiology. Advances in astrobiology and biogeophysics. Springer, Cham

Quirico E, Orthous-Daunay F, Beck P, et al (2014) Origin of insoluble organic matter in type 1 and 2 chondrites: new clues, new questions. Geochim Cosmochim Acta 136:80–99

Remusat L, Palhol F, Robert F, et al (2006) Enrichment of deuterium in insoluble organic matter from primitive meteorites: a solar system origin? Earth Planet Sci Lett 243:15–25

Remusat L, Guan Y, Wang Y, et al (2010) Accretion and preservation of D-rich organic particles in carbonaceous chondrites: evidence for important transport in the early solar system nebula. Astrophys J 713:1048–1058

Riciputi LR, McSween HY Jr, Johnson CA, et al (1994) Minor and trace element concentrations in carbonates of carbonaceous chondrites, and implications for the compositions of coexisting fluids. Geochim Cosmochim Acta 58:1343–1351

Robert F, Epstein S (1982) The concentration and isotopic composition of hydrogen, carbon and nitrogen in carbonaceous meteorites. Geochim Cosmochim Acta 46:81–95

Robert F, Rejou-Michel A, Javoy M, et al (1992) Oxygen isotopic homogeneity of the Earth: new evidence. Earth Planet Sci Lett 108:1–9

Rollion-Bard C, Marin-Carbonne J (2011) Determination of SIMS matrix effects on oxygen isotopic compositions in carbonates. J Anal At Spectrom 26:1285–1289

Rosenberg ND, Browning L, Bourcier WL (2001) Modeling of aqueous alteration of CM carbonaceous chondrites. Meteorit Planet Sci 36:239–244

Rowe MW, Clayton RN, Mayeda TK (1994) Oxygen isotopes in separated components of CI and CM meteorites. Geochim Cosmochim Acta 58:5341–5347

Rubin AE (2012) Collisional facilitation of aqueous alteration of CM and CV carbonaceous chondrites. Geochim Cosmochim Acta 90:181–194

Rubin AE, Kallemeyn GW (1990) Lewis Cliff 85332: a unique carbonaceous chondrite. Meteoritics 25:215–225

Rubin AE, Zolensky ME, Bodnar RJ (2002) The halite-bearing Zag and Monahans (1998) meteorite breccias: shock metamorphism, thermal metamorphism and aqueous alteration on the H-chondrite parent body. Meteorit Planet Sci 37:125–141

Rubin AE, Trigo-Rodriguez JM, Huber H, et al (2007) Progressive aqueous alteration of CM carbonaceous chondrites. Geochim Cosmochim Acta 71:2361–2382

Schmitt-Kopplin P, Gabelica Z, Gougeonet RD, et al (2010) High molecular diversity of extraterrestrial organic matter in Murchison meteorite revealed 40 years after its fall. Proc Natl Acad Sci USA 107:2763–2768

Schmitt-Kopplin P, Hertkorn N, Harir M, et al (2023) Soluble organic matter molecular atlas of Ryugu reveals cold hydrothermalism on C-type asteroid parent body. Nat Commun 14:6525

Schrader DL, Franchi IA, Connolly HC Jr, et al (2011) The formation and alteration of the Renazzo-like carbonaceous chondrites I: implications of bulk-oxygen isotopic composition. Geochim Cosmochim Acta 75:308–325

Schrader DL, Davidson J, Greenwood RC, et al (2014) A water–ice rich minor body from the early Solar System: the CR chondrite parent asteroid. Earth Planet Sci Lett 407:48–60

Schulte M, Shock E (2004) Coupled organic synthesis and mineral alteration on meteorite parent bodies. Meteorit Planet Sci 39:1577–1590

Sephton MA, Pillinger CT, Gilmour I (2000) Aromatic moieties in meteoritic macromolecular materials: analyses by hydrous pyrolysis and δ13C of individual compounds. Geochim Cosmochim Acta 64:321–328

Serra C, Lange J, Remaury QB, et al (2022) Integrative analytical workflow to enhance comprehensive analysis of organic molecules in extraterrestrial objects. Talanta 243:123324. 324

Shimizu K, Alexander CMO'D, Hauri EH, et al (2021) Highly volatile element (H, C, F, Cl, S) abundances and H isotopic compositions in chondrules from carbonaceous and ordinary chondrites. Geochim Cosmochim Acta 301:230–258

Śliwiński MG, Kitajima K, Kozdon R, et al (2017) In situ δ_{13}C and δ_{18}O microanalysis by SIMS: a method for characterizing the carbonate components of natural and engineered CO_2-reservoirs. Int J Greenh Gas Control 57:116–133

Smith LR, Haenecour P, Barnes J, et al (2023) Abundance of presolar grains in the C3.00-ungrouped chondrite Chwichiya 002. Lunar Planet Sci 54:abstract #2425

Steele RCJ, Heber VS, McKeegan KD (2017) Matrix effects on the relative sensitivity factors for manganese and chromium during ion microprobe analysis of carbonate: implications for early Solar System chronology. Geochim Cosmochim Acta 201:245–259

Sugiura N, Ichimura K, Fujiya W, et al (2010) Mn/Cr relative sensitivity factors for synthetic calcium carbonate measured with a NanoSIMS ion microprobe. Geochem J 44:e11–e16

Suttle MD, Daly L, Jones RH, et al (2022a) The Winchcombe meteorite—a regolith breccia from a rubble pile CM chondrite asteroid. Meteorit Planet Sci 59:1043–1067

Suttle MD, King AJ, Ramkissoon NK, et al (2022b) Alteration conditions on the CM and CV parent bodies – insights from hydrothermal experiments with the CO chondrite Kainsaz. Geochim Cosmochim Acta 318:83–111

Tachibana S, et al (2022) Pebbles and sand on asteroid (162173) Ryugu: in situ observation and particles returned to Earth. Science 375:1011–1016

Takeuchi A, Uesugi K, Suzuki Y (2013) Three-dimensional phase-contrast X-ray microtomography with scanning-imaging X-ray microscope optics. J Synchrotron Radiat 20:793–800

Telus M, Alexander CMO'D, Hauri EH, et al (2019) Calcite and dolomite formation in the CM parent body: insight from in situ C and O isotope analyses. Geochim Cosmochim Acta 260:275–291

Tenner TJ, Nakashima D, Ushikubo T (2015) Oxygen isotope ratios of FeO-poor chondrules in CR3 chondrites: influence of dust enrichment and H_2O during chondrule formation. Geochim Cosmochim Acta 148:228–250

Tomeoka K, Buseck PR (1988) Matrix mineralogy of the Orgueil CI carbonaceous chondrite. Geochim Cosmochim Acta 52:1627–1640

Tonui E, Zolensky M, Lipschutz M (2002) Petrography, mineralogy and trace element chemistry of Yamato-86029, Yamato-793321 and Lewis Cliff 85332: Aqueous alteration and heating events. Antarct Meteor Res 15:38–58

Tsuchiyama A, Nakano T, Uesugi K, et al (2013) Analytical dual-energy microtomography: a new method for obtaining three-dimensional mineral phase images and its application to Hayabusa samples. Geochim Cosmochim Acta 116:5–16

Tsuchiyama A, Miyake A, Zolensky M, et al (2014) Search for fluid inclusions in a carbonaceous chondrite using a new X-ray microtomography technique combined with FIB sampling. Meteorit Planet Sci 49:A404

Tsuchiyama A, Kitayama A, Miyake A, et al (2017) Organic material inclusions in mineral grains of carbonaceous chondrites. In: Annual meeting of Japan association of mineralogical sciences. Abstract in Japanese

Tsuchiyama A, Hashino M, Kitayama A, et al (2019) Dolomite in the Ivuna meteorite (CI) and search for its fluid inclusions. In: Annual meeting of Japan association of mineralogical sciences. Abstract in Japanese

Tsuchiyama A, Miyake A, Okuzumi S, et al (2021) Discovery of primitive CO_2-bearing fluid in an aqueously altered carbonaceous chondrite. Sci Adv 7:eabg9707

Tsuchiyama A, Matsumoto M, Matsuno J, et al (2024) Three-dimensional textures of Ryugu samples and their implications for the evolution of aqueous alteration in the Ryugu parent body. Geochim Cosmochim Acta 375:146–172

Tyra MA, Farquhar J, Guan Y, Leshin LA (2012) An oxygen isotope dichotomy in CM2 chondritic carbonates – a SIMS approach. Geochim Cosmochim Acta 77:383–395

Tyra M, Brearley A, Guan Y (2016) Episodic carbonate precipitation in the CM chondrite ALH 84049: an ion microprobe analysis of O and C isotopes. Geochim Cosmochim Acta 175:195–207

Ushikubo T, Kimura M (2021) Oxygen-isotope systematics of chondrules and olivine fragments from Tagish Lake C2 chondrite: implications of chondrule-forming regions in protoplanetary disk. Geochim Cosmochim Acta 293:328–343

Ushikubo T, Kimura M, Kita NT, et al (2012) Primordial oxygen isotope reservoirs of the solar nebula recorded in chondrules in Acfer 094 carbonaceous chondrite. Geochim Cosmochim Acta 90:242–264

Vacher LG, Marrocchi Y, Verdier-Paoletti MJ (2016) Inward radial mixing of interstellar water ices in the solar protoplanetary disk. Astrophys J 827:L1

Vacher LG, Marrocchi Y, Villeneuve J, et al (2017) Petrographic and C & O isotopic characteristics of the earliest stages of aqueous alteration of CM chondrites. Geochim Cosmochim Acta 213:271–290

Vacher L, Marrocchi Y, Villeneuve J, et al (2018) Collisional and alteration history of the CM parent body. Geochim Cosmochim Acta 239:213–234

Vacher LG, Piralla M, Gounelle M, et al (2019a) Thermal evolution of hydrated asteroids inferred from oxygen isotopes. Astrophys J Lett 882:L20

Vacher LG, Truche L, Faure F, et al (2019b) Deciphering the conditions of tochilinite and cronstedtite formation in CM chondrites from low temperature hydrothermal experiments. Meteorit Planet Sci 54:1870–1889

Vacher LG, Piani L, Rigaudier Y, et al (2020) Hydrogen in chondrites: influence of parent body alteration and atmospheric contamination on primordial components. Geochim Cosmochim Acta 281:53–66

van Kooten EMME, Wielandt D, Schiller M, et al (2016) Isotopic evidence for primordial molecular cloud material in metal-rich carbonaceous chondrites. Proc Natl Acad Sci USA 113:2011–2016

van Kooten EMME, Nagashima K, Kasama T, et al (2017a) A divergent heritage for complex organics in Isheyevo lithic clasts. Geochim Cosmochim Acta 205:119–148

van Kooten EMME, Schiller M, Bizzarro M (2017b) Magnesium and chromium isotope evidence for initial melting by radioactive decay of $_{26}$Al and late stage impact-melting on the ureilite parent body. Geochim Cosmochim Acta 208:1–23

van Kooten EMME, et al (2024) The nucleosynthetic fingerprint of the outermost protoplanetary disk and early Solar System dynamics. Sci Adv 10:eadp1613

Velbel MA, Tonui EK, Zolensky ME (2012) Replacement of olivine by serpentine in the carbonaceous chondrite Nogoya (CM2). Geochim Cosmochim Acta 87:117–135

Verdier-Paoletti MJ, Marrocchi Y, Avice G, et al (2017) Oxygen isotope constraints on the alteration temperatures of CM chondrites. Earth Planet Sci Lett 458:273–281

Verdier-Paoletti MJ, Marrocchi Y, Vacher LG, et al (2019) Testing the genetic relationship between fluid alteration and brecciation in CM chondrites. Meteorit Planet Sci 54:1692–1709

Verdier-Paoletti MJ, Nittler LR, Wang J (2020) New estimation of presolar grain abundances in the Paris meteorite. Lunar Planet Sci 51:abstract #2523

Viennet JC, Le Guillou C, Remusat L, et al (2022) Experimental investigation of Fe-clay/organic interactions under asteroidal conditions. Geochim Cosmochim Acta 318:352–365

Vinogradoff V, Le Guillou C, Bernard S, et al (2017) Paris vs. Murchison: impact of hydrothermal alteration on organic matter in CM chondrites. Geochim Cosmochim Acta 212:234–252

Vinogradoff V, Le Guillou C, Bernard S, et al (2020a) Influence of phyllosilicates on the hydrothermal alteration of organic matter in asteroids: experimental perspectives. Geochim Cosmochim Acta 269:150–166

Vinogradoff V, Remusat L, McLain HL, et al (2020b) Impact of phyllosilicates on amino acid formation under asteroidal conditions. ACS Earth Space Chem 4:1398–1407

Visser R, John T, Menneken M, et al (2018) Temperature constraints by Raman spectroscopy of organic matter in volatile-rich clasts and carbonaceous chondrites. Geochim Cosmochim Acta 241:38–55

Visser R, John T, Patzek M, et al (2019) Sulfur isotope study of sulfides in CI, CM, C2ung chondrites and volatile-rich clasts–evidence for different generations and reservoirs of sulfide formation. Geochim Cosmochim Acta 261:210–223

Visser R, John T, Whitehouse MJ, et al (2020) A short-lived $_{26}$Al induced hydrothermal alteration event in the outer solar system: constraints from Mn/Cr ages of carbonates. Earth Planet Sci Lett 547:116440

Vollmer C, Hoppe P, Stadermann FJ, et al (2009) NanoSIMS analysis and Auger electron spectroscopy of silicate and oxide stardust from the carbonaceous chondrite Acfer 094. Geochim Cosmochim Acta 73:7127–7149

Vollmer C, Kepaptsoglou D, Leitner J, et al (2014) Fluid-induced organic synthesis in the solar nebula recorded in extraterrestrial dust from meteorites. Proc Natl Acad Sci USA 111:15338–15343

Vollmer C, Leitner J, Kepaptsoglou D, et al (2020b) Isotopic compositions, nitrogen functional chemistry, and low-loss electron spectroscopy of complex organic aggregates at the nanometer scale in the carbonaceous chondrite Renazzo. Meteorit Planet Sci 55:1293–1319

Vollmer C, Pelka M, Leitner J, et al (2020a) Amorphous silicates as a record of solar nebular and parent body processes—a transmission electron microscope study of fine-grained rims and matrix in three Antarctic CR chondrites. Meteorit Planet Sci 55:1491–1508

Walsh KJ, Morbidelli A, Raymond SN, et al (2012) Populating the asteroid belt from two parent source regions due to the migration of giant planets—"The Grand Tack". Meteorit Planet Sci 47:1941–1947

Warren PH (2011) Stable-isotopic anomalies and the accretionary assemblage of the Earth and Mars: a subordinate role for carbonaceous chondrites. Earth Planet Sci Lett 311:93–100

Weisberg MK, Huber H (2007) The GRO 95577 CR1 chondrite and hydration of the CR parent body. Meteorit Planet Sci 42:1495–1503

Weisberg MK, Prinz M (1991) El Djouf 001: a new CR2 chondrite. Meteoritics 26:406
Weisberg MK, Prinz M, Clayton RN, et al (1993) The CR (Renazzo-type) carbonaceous chondrite group and its implications. Geochim Cosmochim Acta 57:1567–1586
Whitby J, Burgess R, Turner G, et al (2000) Extinct $_{129}$I in halite from a primitive meteorite: evidence for evaporite formation in the early Solar System. Science 288:1819–1821
Wilkening LL (1973) Foreign inclusions in stony meteorites – I. Carbonaceous chondritic xenoliths in the Kapoeta howardite. Geochim Cosmochim Acta 37:1985–1989
Yabuta H, Williams LB, Cody GD, et al (2007) The insoluble carbonaceous material of CM chondrites: a possible source of discrete organic compounds under hydrothermal conditions. Meteorit Planet Sci 42:37–48
Yamamoto D, Kuroda M, Tachibana S, et al (2018) Oxygen isotopic exchange between amorphous silicate and water vapor and its implications for oxygen isotopic evolution in the early Solar System. Astrophys J 865:98
Yang J, Epstein S (1983) Interstellar organic matter in meteorites. Geochim Cosmochim Acta 47:2199–2216
Yang J, Epstein S (1984) Relic interstellar grains in Murchison meteorite. Nature 311:544–547
Yang L, Ciesla FJ, Alexander CMO'D (2013) The D/H ratio of water in the solar nebula during its formation and evolution. Icarus 226:256–267
Yokoyama T, Nagashima K, Nakai I, et al (2023) Samples returned from the asteroid Ryugu are similar to Ivuna-type carbonaceous meteorites. Science 379:eabn7850
Yoshida K, Hirajima T, Miyake A, et al (2016) Combined FIB microsampling and X-ray microtomography: a powerful tool for the study of tiny fluid inclusions. Eur J Mineral 28:245–256
Young ED, Ash RD, England P, et al (1999) Fluid flow in chondritic parent bodies: deciphering the compositions of planetesimals. Science 286:1331–1335
Young ED, Zhang K, Schubert G (2003) Conditions for pore water convection within carbonaceous chondrite parent bodies — implications for planetesimal size and heat production. Earth Planet Sci Lett 213:249–259
Yurimoto H, Kuramoto K (2004) Molecular cloud origin for the oxygen isotope heterogeneity in the solar system. Science 1979(305):1763–1767
Yurimoto H, Itoh S, Zolensky M, et al (2014) Isotopic compositions of asteroidal liquid water trapped in fluid inclusions of chondrites. Geochem J 48:549–560
Zega TJ, Alexander CMO'D, Busemann H, et al (2010) Mineral associations and character of isotopically anomalous organic material in the Tagish Lake carbonaceous chondrite. Geochim Cosmochim Acta 74:5966–5983
Zhao X, Floss C, Stadermann FJ, et al (2011) Continued investigation of presolar silicate grains in the carbonaceous chondrite Ningqiang. Lunar Planet Sci 42:abstract #1982
Zhao X, Floss C, Lin Y, et al (2013) Stardust investigation into the CR chondrite Grove Mountain 021710. Astrophys J 769:49–64
Zhukovska S, Dobbs C, Jenkins EB, et al (2016) Modeling dust evolution in galaxies with a multiphase, inhomogeneous ISM. Astrophys J 831:147–161
Zinner E (2014) Presolar grains. In: Davis AM (ed) Meteorites and cosmochemical processes. Elsevier, Amsterdam, pp 181–213
Zolensky M, Ivanov A (2003) The Kaidun microbreccia meteorite: a harvest from the inner and outer Asteroid belt. Chem Erde 63:185–246
Zolensky ME, MacKinnon IDR (1986) Microstructures of cylindrical tochilinites. Am Mineral 71:1201–1209
Zolensky ME, Bourcier WL, Gooding JL (1989) Aqueous alteration on the hydrous asteroids: results of Eq. 3/6 computer simulations. Icarus 78:411–425
Zolensky ME, Barrett R, Browning L (1993) Mineralogy and composition of matrix and chondrule rims in carbonaceous chondrites. Geochim Cosmochim Acta 57:3123–3148
Zolensky ME, Ivanov AV, Yang SV, et al (1996b) The Kaidun meteorite: mineralogy of an unusual CM1 lithology. Meteorit Planet Sci 31:484–493
Zolensky ME, Weisberg MK, Buchanan PC, et al (1996a) Mineralogy of carbonaceous chondrite clasts in HED achondrites and the Moon. Meteorit Planet Sci 31:518–537
Zolensky ME, Mittlefehldt DW, Lipschutz ME, et al (1997) CM chondrites exhibit the complete petrologic range from type 2 to 1. Geochim Cosmochim Acta 61:5099–5115
Zolensky ME, Bodnar RJ, Gibson EK, et al (1999c) Asteroidal water within fluid inclusion-bearing halite in an H5 chondrite, Monahans (1998). Science 285:1377–1379
Zolensky ME, Nakamura K, Gounelle M, et al (2002) Mineralogy of Tagish Lake: an ungrouped type 2 carbonaceous chondrite. Meteorit Planet Sci 37:737–761
Zolensky ME, Clayton RN, Mayeda T, et al (2003) Carbonaceous chondrite clasts in the halite-bearing H5 chondrite Zag. Meteorit Planet Sci 38:A114

Zolensky ME, Krot AN, Benedix G (2008) Record of low-temperature alteration in asteroids. Rev Mineral Geochem 68:429–462
Zolensky M, Mikouchi T, Fries M, et al (2014) Mineralogy and petrography of C asteroid regolith: the Sutter's Mill CM meteorite. Meteorit Planet Sci 49:1997–2016
Zolensky ME, Fries M, Utas J, et al (2016) C chondrite clasts in H chondrite regolith breccias: something different. Meteorit Planet Sci 51:A691
Zolensky ME, Bodnar J, Yurimoto H, et al (2017) The search for and analysis of direct samples of early Solar System aqueous fluids. Philos Trans R Soc Lond A 375:20150386
Zolotov MYu (2012) Aqueous fluid composition in CI chondritic materials: chemical equilibrium assessments in closed systems. Icarus 220:713–729
Zolotov MYu (2014) Formation of brucite and cronstedtite-bearing mineral assemblages on Ceres. Icarus 228:13–26
Zolotov MYu (2017) Aqueous origins of bright salt deposits on Ceres. Icarus 296:289–304
Zolotov MYu, Mironenko MV (2007) Hydrogen chloride as a source of acid fluids in parent bodies of chondrites. Lunar Planet Sci 38:abstract #2340
Zolotov MYu, Mironenko MV (2008a) Early alteration of matrices in parent bodies of CI/CM carbonaceous chondrites: kinetic-thermodynamic modeling. Lunar Planet Sci 39:abstract #1998
Zolotov MYu, Mironenko MV (2008b) Aqueous alteration oof CM2 chondrites evaluated with kinetic models. Meteorit Planet Sci 43:A5261
Zolotov MYu, Mironenko MV, Shock EL (2006) Thermodynamic constraints on fayalite formation on parent bodies of chondrites. Meteorit Planet Sci 41:1775–1796
Zolotov MYu, Morlok A, Libourel G (2015) Microchemical environments of aqueous alteration in CR chondrites: chemical equilibrium models. Lunar Planet Sci 46:abstract #1470

Authors and Affiliations

Martin R. Lee[1] · Conel M. O'D. Alexander[2] · Addi Bischoff[3] · Adrian J. Brearley[4] · Elena Dobrică[5] · Wataru Fujiya[6] · Corentin Le Guillou[7] · Ashley J. King[8] · Elishevah van Kooten[9] · Alexander N. Krot[5] · Jan Leitner[10,11] · Yves Marrocchi[12] · Markus Patzek[3] · Michail I. Petaev[13] · Laurette Piani[12] · Olga Pravdivtseva[14] · Laurent Remusat[15] · Myriam Telus[16] · Akira Tsuchiyama[17,18,19] · Lionel G. Vacher[20]

✉ M.R. Lee
Martin.Lee@Glasgow.ac.uk

C.M.O'D. Alexander
calexander@carnegiescience.edu

A. Bischoff
bischoa@uni-muenster.de

A.J. Brearley
brearley@unm.edu

E. Dobrică
dobrica@hawaii.edu

W. Fujiya
wataru.fujiya.sci@vc.ibaraki.ac.jp

C. Le Guillou
corentin.san@gmail.com

A.J. King
a.king@nhm.ac.uk

E. van Kooten
elishevah.vankooten@sund.ku.dk

A.N. Krot
sasha@higp.hawaii.edu

J. Leitner
jan.leitner@mpic.de

Y. Marrocchi
yves.marrocchi@univ-lorraine.fr

M. Patzek
markus.patzek@uni-muenster.de

M.I. Petaev
mpetaev@fas.harvard.edu

L. Piani
laurette.piani@univ-lorraine.fr

O. Pravdivtseva
olga@physics.wustl.edu

L. Remusat
laurent.remusat@mnhn.fr

M. Telus
mtelus@ucsc.edu

A. Tsuchiyama
hime305@icloud.com

L.G. Vacher
lionel.vacher@univ-grenoble-alpes.fr

[1] School of Geographical & Earth Sciences, University of Glasgow, Glasgow G12 8QQ, UK

[2] Earth & Planets Laboratory, Carnegie Institution of Washington, 5241 Broad Branch Road NW, Washington, DC 20015, USA

[3] Institut für Planetologie, University of Münster, Wilhelm-Klemm-Str. 10, D-48149 Münster, Germany

[4] Department of Earth & Planetary Sciences, University of New Mexico, Albuquerque, NM, USA

[5] Hawai'i Institute of Geophysics & Planetology, The University of Hawai'i at Mānoa, Honolulu, HI 96822, USA

[6] Faculty of Science, Ibaraki University, 2-1-1 Bunkyo, Mito, Ibaraki 310-8512, Japan

[7] Université de Lille, CNRS, INRAE, Centrale Lille, UMR 8207-UMET-Unité Matériaux et Transformations, F-59000 Lille, France

[8] Planetary Materials Group, Department of Earth Sciences, Natural History Museum, Cromwell Road, London SW7 5BD, UK

[9] Centre for Star and Planet Formation, Globe Institute, University of Copenhagen, DK-1350 Copenhagen, Denmark

[10] Institute of Earth Sciences, Heidelberg University, Im Neuenheimer Feld 234-236, D-69120, Heidelberg, Germany

[11] Max Planck Institute for Chemistry, Particle Chemistry Department, Hahn-Meitner-Weg 1, 55128 Mainz, Germany

[12] Université de Lorraine, CNRS, CRPG, UMR 7358, Nancy, France

[13] Department of Earth and Planetary Sciences, Harvard University, Cambridge 02138, USA

[14] Physics Department, Washington University, St. Louis, MO 63130, USA

[15] CNRS–Museum National d'Histoire Naturelle, Laboratoire de Minéralogie et Cosmochimie du Museum–UMR 7202, Case 52, 57 rue Cuvier, 75231 Paris Cedex 05, France

[16] Earth and Planetary Sciences, University of California Santa Cruz, Santa Cruz CA 95064, USA

[17] Research Organization of Science & Technology, Ritsumeikan University, Shiga 525-8577, Japan

[18] Chinese Academy of Sciences (CAS) Key Laboratory of Mineralogy & Metallogeny/Guangdong Provincial Key Laboratory of Mineral Physics and Materials, Guangzhou Institute of Geochemistry, CAS, Guangzhou 510640, China

[19] CAS Center for Excellence in Deep Earth Science, Guangzhou 510640, China

[20] University Grenoble Alpes, CNRS, IPAG, 38000 Grenoble, France

Space Science Reviews (2025) 221:44
https://doi.org/10.1007/s11214-025-01168-4

Sample Return Missions: Rosetta Stones Returned from the First Small Bodies in the Solar System

Tomoki Nakamura[1] · Cecile Engrand[2] · Michael Zolensky[3] · Victoria E. Hamilton[4] · Abigail A. Fraeman[5]

Received: 5 September 2024 / Accepted: 28 April 2025 / Published online: 12 May 2025

Abstract

It is now possible to bring back samples from planetary bodies of the Solar System other than the moon. This research method enables a direct link between astronomical observations and meteorite analyses, which were previously disconnected. The Hayabusa, Hayabusa2, and OSIRIS REx sample return missions have provided detailed information on the composition of S-, C-, and B-type asteroids, respectively, and the processes by which they were formed. This paper reviews the results of these three asteroid sample return missions, and also introduces the next Martian moon sample return mission MMX. Although sample returns are currently achieved only from near-Earth objects in the Solar System, it is hoped that in the future it will be possible to collect samples from outer Solar System objects and even from small objects flying from outside the Solar System.

1 Introduction

Asteroids and comets are survivors of the first small bodies (planetesimals) formed in the protoplanetary disk about 4.6 billion years ago. Diverse small solid particles were present in the disk. Asteroids and comets retain the characteristics of these solid particles in the disk region where they formed. Therefore, meteorites from asteroids and interplanetary dust from comets reveal details of the chemical evolution of the protoplanetary disk and formation and evolution of small bodies in the early Solar System.

In order to understand the distribution of elements and materials in the protoplanetary disk, it is necessary to determine the distribution of asteroids and comets and their respective elemental compositions and constituent materials. In this respect, the most important

✉ T. Nakamura
tomoki.nakamura.a8@tohoku.ac.jp

1 Department of Earth Science, Graduate School of Science, Tohoku University Aoba, Sendai, Miyagi 980-8578, Japan

2 IJCLab, University Paris-Saclay-CNRS, 91405 Orsay Cedex, France

3 Astromaterials Research and Exploration Science, NASA Johnson Space Center, Houston, TX 77058, USA

4 Solar System Science and Exploration Division, Southwest Research Institute, Boulder, CO 80302, USA

5 Jet Propulsion Laboratory, California Institute of Technology, Pasadena, CA 91109, USA

point is how accurately we can estimate the constituent materials of asteroids and comets from their reflectance spectra. Comparison of the spectra of meteorites and asteroids has led to investigation of the relationship between the meteorite's chemical type and the asteroid's spectral type (e.g., Britt et al. 1992; DeMeo et al. 2022). However, before the asteroid sample return missions, it was not possible to identify the specific type of asteroids from which the meteorite derived except in a few exceptional cases, such as eucrites from V-type asteroid that is rich in pyroxene and CM carbonaceous chondrites from Cgh and Ch asteroids with 0.7 micron absorption, although the latter relationship needs further confirmation (e.g., Vilas and Gaffey 1989). In order to confirm the asteroid-meteorite relationship with confidence, it was necessary to perform sample returns from asteroids for which spectra had been characterized.

One may think that the composition of asteroids can be estimated by comparing the reflectance spectra of asteroids and meteorites, as mentioned above, without conducting sample return missions. However, there are three problems to be overcome in estimating the composition of asteroids based solely on their reflectance spectra: 1. the spectra of many asteroids are limited to the visible wavelength range, and thus the composition of asteroids, carbonaceous asteroids in particular, is difficult to identify because of a limited number of absorption lines (Bus and Binzel 2002a,b; DeMeo et al. 2009). 2. The spectra of asteroids are affected by space weathering (e.g., Lantz et al. 2017), and the spectra of meteorites are affected by terrestrial weathering (e.g., Amano et al. 2023). These two processes make significant changes in meteorite and asteroid spectra. 3. Spectral properties in the visible to near-infrared are strongly affected by physical properties such as grain size and porosity (e.g., Cloutis et al. 2018), which make both asteroid and meteorite spectral slopes different.

A sample return mission is the only way to directly compare the spectrum of an asteroid taken from a spacecraft with the analysis results of samples recovered from the asteroid. From this comparison, we can know the "correct" answer as to what kind of molecular structures and mineralogical properties are actually responsible for the various absorptions and features, respectively, in the asteroid's spectra. In this way, sample return missions provide important insights into the above three issues that cannot be obtained by any other methods. Therefore, sample return missions have a critical impact not only on the field of meteorite science but also on the field of astronomical observations.

The importance of sample returns extends beyond confirming asteroid-meteorite relationships and deserves a more comprehensive exploration, particularly regarding the preservation of geological context and avoidance of terrestrial contamination. Unlike meteorites, which lose their original geological setting and are subject to rapid contamination on Earth, returned samples retain their pristine state. This needs then very careful curation conditions when the samples are back on Earth. This is especially critical for organic-rich materials. Organic compounds in meteorites are highly vulnerable to degradation and contamination from terrestrial exposure, which can obscure interpretations of their origin and pre-solar chemistry. However, by returning material directly from asteroids like Bennu and Ryugu, sample return missions provide access to unaltered extraterrestrial samples within their native geological context. This enables a more accurate understanding of the distribution and nature of organic molecules, crucial for studying early organic chemistry and prebiotic processes in the Solar System.

This paper describes the scientific results obtained from the now completed asteroid sample return missions. Finally, the sample return mission from Martian moon Phobos will be briefly introduced. The sample return mission from the short-period comet 81P/Wild2 is described in detail in another paper (Zolensky et al. 2024) published in this same special issue.

2 Four Sample Return Missions and General Properties of the Target Small Bodies

Four sample return missions from small bodies in the Solar System have been conducted in Japan and the United States of America (Table 1). There have been three asteroid missions (Hayabusa, Hayabusa2, and OSIRIS Rex) and one comet mission (Stardust). In addition, two sample return missions are planned. Japan Aerospace Exploration Agency (JAXA) developed the Martian Moon eXploration (MMX) mission to take samples from one of the two Martian moons, Phobos, whose origin and formation process are unknown so far (Kuramoto et al. 2022). China National Space Administration plans a sample return mission to a near Earth S-type asteroid (469219) Kamo'oalewa. The spacecraft Tianwen-2 will launch in 2025 and return to the Earth 2.5 years later. Kamo'oalewa is a small asteroid (69.45 m × 58.49 m × 51.78 m) with grain size < 2 cm regolith, and rapidly-rotating (period is 27.37 minutes) (Zhang et al. 2024). The visible to near-infrared reflectance spectrum shows a 0.984 μm absorption, similar to LL chondrites (Zhang et al. 2024). Table 1 summarizes the characteristics of the target objects and the outline of each mission except the Chinese Tianwen-2 mission. Figure 1 shows images of targeted objects at approximately the same scale. Figure 2A shows the visible to near-infrared reflectance spectra of target bodies.

The target objects of all missions are near-Earth asteroids and short-period comets that have visited near-Earth orbit. The spacecraft did not recover samples from Main Belt asteroids or the Kuiper Belt. While one-way missions have reached distant Kuiper Belt objects (New Horizons mission: Guo and Farquhar 2008), sample return missions that obviously require a round trip between the Earth and the target object have only brought back samples from objects inside the orbit of Mars. If the sample return mission to the Mars satellite Phobos (Fig. 1) by the MMX mission planned by Japan (Table 1) (Kuramoto et al. 2022) and that to Mars planned by the United States of America (Muirhead et al. 2020) succeed, they will break the previous round-trip distance record.

3 Hayabusa Mission

The Hayabusa mission (Table 1: Kawaguchi et al. 2008) was primarily an engineering mission started in 1995 with the aim of making a round-trip between Earth and a small near-Earth object (the target body was not specified at the beginning) in the Solar System. An S-type asteroid was subsequently selected as the target object because S-type asteroids are abundant in the asteroid belt and considered to be the parent bodies of the most Earth-bound ordinary chondrites, due to the similarity of the reflectance spectra of the meteorites and the asteroids (e.g., Binzel et al. 1996). However, this relationship had not been confirmed at that time.

3.1 General Properties of the Asteroid Itokawa

S-type asteroid 25143 Itokawa (hereafter referred to as Itokawa), which was selected as the mission target in 2000, is a small near-Earth asteroid with a maximum dimension of about 500 m. Itokawa was named after Dr. Hideo Itokawa, the first Japanese rocket developer. Ground-based radar observations of Itokawa in 2001 on its approach to the Earth revealed a rugby ball shape and 12-hour rotation period (Ostro et al. 2004). The so-called YORP effect, a radiation recoil torque affecting the rotation state of small asteroids (Rubincam 2000), is expected to halve Itokawa's rotation period in 50-90 thousand years (Scheeres et al. 2007).

Table 1 Summary of sample return missions from small sorlar system bodies

Mission	Stardust	Hayabusa	Hayabusa2	OSIRIS-REx	MMX
Type of small body	Short-period comet	S-type asteroid	C-type asteroid	B-type asteroid	Martian moon (D-type asteroid?)
Name	81P/Wild2	Itokawa	Ryugu	Bennu	Phobos
Size (km)	$3.3 \times 4.0 \times 5.5$	$0.54 \times 0.29 \times 0.21$	$1.04 \times 1.02 \times 0.88$	$0.56 \times 0.54 \times 0.50$	$25.90 \times 22.60 \times 18.32$
Density (g/cm^3)	0.6-0.8	1.9	1.19	1.19	1.86
Rotation period (h)	-	12.1	7.6	4.3	7.6
Orbital period (days)	2340	557	475	437	7.6 hours around Mars
Semi-major axis (au)	3.45	1.32	1.19	1.13	1.52
Reflectance at 2 μm wavelength (%)	-	17	2	2	5
Launch date (yyyy/mm/dd)	1999/2/7	2003/5/9	2014/12/3	2016/9/8	2026 summer (TBD*)
Arrival (yyyy/mm/dd)	2004/1/2	2005/9/12	2018/6/27	2018/12/3	2027 August (TBD)
Length of stay around the target (month)	Flyby	3	18	30	40 (TBD)
Earth return date (yyyy/mm/dd)	2006/1/15	2010/6/13	2020/12/5	2023/9/24	2031 winter (TBD)
First sample collection (yyyy/mm/dd)	2004/1/2	2005/11/20	2019/2/21	2020/10/20	TBD
Second sample collection (yyyy/mm/dd)	-	2005/11/25	2019/7/11	-	TBD
Quantity of recovered sample	many small particles	many small particles	5.4 g	121.6 g	
Major constituent minerals	Olivine Pyroxene Sulfide	Olivine Pyroxene Plagioclase	Phyllosilicates Carbonates Magnetite	Phyllosilicates Carbonates Magnetite	
Sampling methods	aerogel capture	projectile shoot	projectile shoot	gas blow	core burial gas blow
Extended mission target objects	9P/Tempel1	-	Asteroid 2001 CC21 Asteroid 1998 KY26	Apophis (Asteroid 99942)	TBD
References	[1], [2], [3], [4], [5], [6]	[3], [7], [8], [9]	[3], [10], [11], [12], [13], [14]	[3], [15], [16], [17], [18], [19], [20], [21], [22]	[23], [24], [25], [26], [27]

[1] Duxbury et al. (2004) [2] Davidsson and Gutiérrez (2006) [3] https://ssd.jpl.nasa.gov/tools/sbdb_lookup.html#/?sstr=81P&view=OPDCA [4] Brownlee et al. (2003) [5] Veverka et al. (2013) [6] Brownlee et al. (2004) [7] Fujiwara et al. (2006) [8] Yoshikawa et al. (2021) [9] Abe M. et al. (2006a), Abe S. et al. (2006b) [10] Watanabe et al. (2017) [11] Tsuda et al. (2020) [12] Kikuchi et al. (2020) [13] Hirabayashi et al. (2021) [14] Kitazato et al. (2019) [15] Lauretta et al. (2019a, 2019b) [16] DellaGiustina et al. 2023 [17] Simon et al. (2020) [18] Lauretta et al. (2024) [19] Lauretta et al. (2015) [20] Williams et al. (2018) [21] Wibben et al. (2024) [22] Hamilton et al. (2024) [23] Ernst et al. 2023 [24] Allen and Cox (2000) [25] Fraeman et al. (2012) [26] Kuramoto et al. (2022) [27] Nakamura T. et al. (2021). *TBD = to be determined

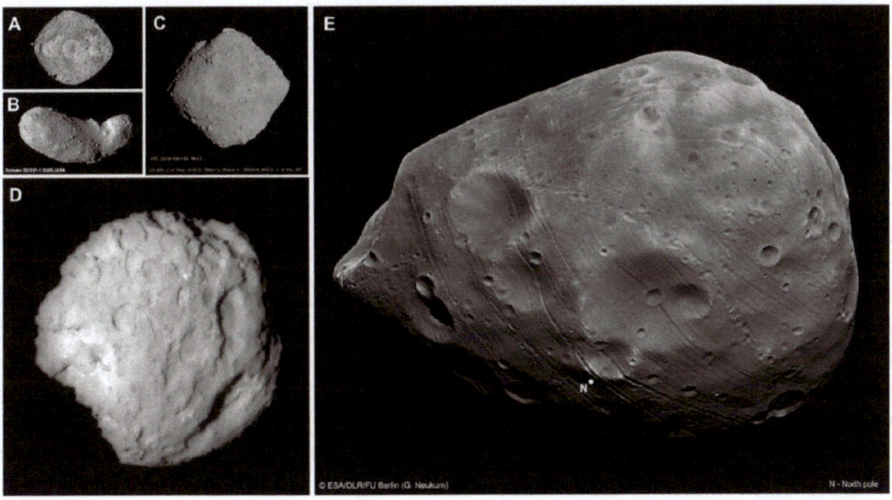

Fig. 1 Targets of the past and future sample return missions in a scale roughly proportional to real size. (A) Bennu, (B) Itokawa (C) Ryugu (D) 81P/Wild2, and (E) Phobos (maximum dimension of ~27 km). Images from Bennu (https://www.nasa.gov/news-release/nasa-spacecraft-provides-insight-into-asteroid-bennus-future-orbit/), Wild 2 (https://solarsystem.nasa.gov/stardust/photo/cometwild2.html), Itokawa (JAXA digital archives P-043-12077), Ryugu (JAXA digital archives P100011856), and Phobos from North pole (https://mars.nasa.gov/resources/6148/crater-chain-on-phobos/)

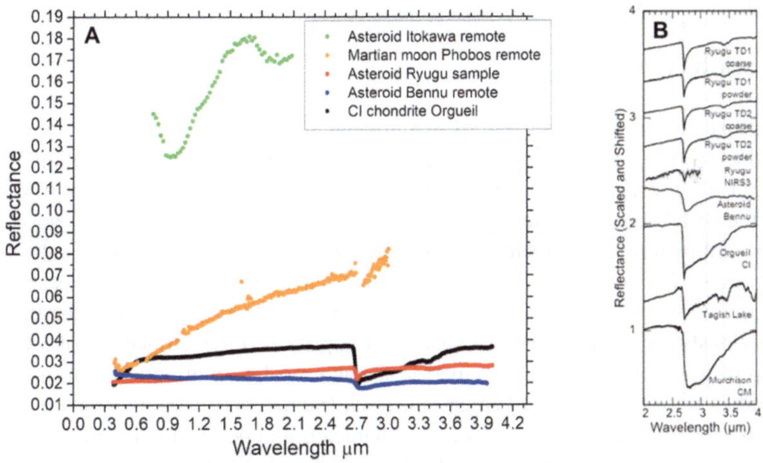

Fig. 2 (A) Reflectance spectra of Solar System small bodies for sample return missions and a CI chondrite meteorite spectrum for comparison. Data of Itokawa from Abe M. et al. (2006a), Phobos from Fraeman et al. (2012), Ryugu and CI chondrite from Nakamura T. et al. (2022), and Bennu from Simon et al. (2020). (B) Variation of 2.7-μm absorption among Ryugu samples, Ryugu remote sensing, Bennu remote sensing, and carbonaceous chondrite meteorites. Data from Nakamura T. et al. (2022)

Hayabusa was launched in 2003, followed by acceleration by an Earth flyby, and arrived at Itokawa in 2005 (Table 1). Hayabusa was the first spacecraft to make an interplanetary flight using ion engines (Kuninaka et al. 2007). Itokawa's surface topography and reflectance spectra were observed from a home position at an altitude of 7 km and a proximity observation at an altitude of 3 to 4 km (e.g., Kawaguchi et al. 2008). Macroscopically, Itokawa has an elongated shape, with a neck at about one-third of its long axis, and a small head and thick body that resemble a sea-otter (Fig. 1B). The size is 535 × 294 × 209 m (Demura et al. 2006), and the bulk density of the object is 1.95 ± 0.14 g/cm^3 (Abe S. et al. 2006b), which means that about 40% of the total volume in the object is void space when compared to the bulk density of ordinary chondrite meteorites (Fujiwara et al. 2006).

3.2 Impact History

Eighty percent of the surface of Itokawa is covered by large and small boulders (from a few meters to ~50 m Saito et al. 2006). The shapes and structures of the boulders are similar to those of rock fragments produced by impact experiments (Nakamura A. et al. 2008). This suggests that Itokawa is a rubble-pile asteroid formed by the reaccumulation of fragments of the original larger body that had been destroyed by a catastrophic impact (Fujiwara et al. 2006), and precise simulations of the impact destruction and reaccumulation of Itokawa's parent body have been performed using N-body numerical codes (Michel and Richardson 2013).

The age of the catastrophic giant impact has not been determined. Analyses of the recovered samples have provided both relatively young ages (1.3 ± 0.3 Ga by $^{39}Ar/^{40}Ar$ age (Park et al. 2015) and 1.51 ± 0.85 Ga by U-Pb age (Terada et al. 2018)) and relatively old ages (>2.3 Ga billion years by $^{39}Ar/^{40}Ar$ age (Jourdan et al. 2017) and >4.2 Ga by $^{39}Ar/^{40}Ar$ age (Jourdan et al. 2023)). It has been suggested that the reason for such long lifetimes for Itokawa-sized small asteroids is due to the shock-absorbing nature of the rubble pile bodies (Jourdan et al. 2023).

3.3 Regolith Evolution

Detailed analysis of 3D images reveals that Itokawa's surface has undergone extensive landslides, due probably to impact-generated vibration, resulting in the fluidization of the regolith and the formation of the Muses Sea region (Fig. 3A), a smooth terrain extending around the body's center of gravity (Miyamoto et al. 2007). The Muses Sea region is composed of millimeter- to centimeter-sized pebbles (Fig. 3B: Yano et al. 2006), indicating that size sorting of the constituent particles occurred during fluidization. This means that the surface of an asteroid with low gravity is much more actively modified by the fluidization than previously expected. When the spacecraft approached the surface to collect samples, close-up imaging of the surface was performed at a resolution of up to 6 mm/px. Many of the boulders were found to be aggregates of rock fragments of varying strength or composition (thus breccias), each about 10 cm in diameter, and detailed comparisons have been made with the surface of larger meteorites (Noguchi et al. 2010).

Visible camera observations from the spacecraft identified bright and dark regions on Itokawa's surface. The dark regions were slightly redder than the bright regions, and the slope of the reflectance spectra was steeper (Saito et al. 2006). The dark regions were distributed over the entire surface of the asteroid, while the bright areas were mostly distributed on inclined slopes and inside craters. This is thought to be due to the fact that vibrations caused by meteorite impacts induced landslides and craters to move and excavated surface materials, exposing bright regions in the deeper layers (Saito et al. 2006). Reflectance spectra in the near-infrared wavelength region show that the spectra of the dark regions have shallower infrared absorption of silicates than those of the bright regions (Abe M. et al. 2006a). From these results, it was inferred that the dark regions are regions that have been

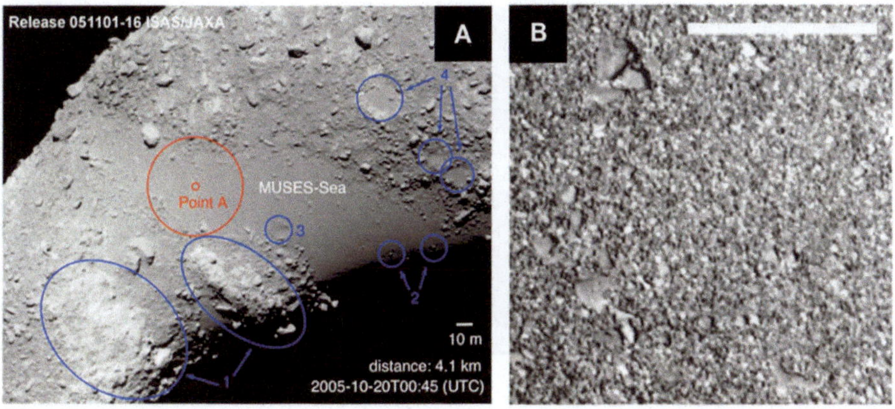

Fig. 3 (A) A whole view of smooth terrain Muses sea area. Touch down site 1 (TD1) is shown as "Point A". Overall, the Muses sea area has a dark surface, while some areas enclosed by blue circles 1 are bright areas. (B) Close-up image of TD1 area in Muses sea taken from 63 m altitude. The Muses Sea is composed of relatively homogeneous, size-sorted small pebbles ranging from several centimeters to sub-centimeter scales. Scale bar is 1 m. Images from Yano et al. (2006)

altered by space weathering (Saito et al. 2006; Hiroi et al. 2006). In particular, the Muses Sea region was a dark region affected by space weathering (Fig. 3A). The 1-μm feature of the near-infrared spectra indicated that Itokawa is an olivine- and pyroxene-dominated surface (olivine/(olivine + pyroxene) $= 70 \sim 80\%$), consistent with ordinary chondrite minerals of LL 5-6 (Abe M. et al. 2006a).

3.4 Sample Collection

Sampling operations were twice performed on the Muses Sea area in November 2005 (Yano et al. 2006; Kawaguchi et al. 2008). The recovery method involved a metal projectile being fired during the final descent of the spacecraft, and the mobilized crushed rock fragments ascended within a sample horn 1 meter tall and were stored in a sample catcher installed inside the spacecraft (Yano et al. 2006). In this process, the spacecraft only touched the surface of the asteroid for a few seconds, so it was defined as a touch-down sample recovery rather than an actual landing. The touch-down operation of the spacecraft was fraught with problems, and sample recovery was extremely difficult. Touch down was successful on both occasions, but no metal projectiles were fired, and it was not known whether the samples were actually recovered until the capsule was opened back on Earth (Kawaguchi et al. 2008).

After the two touch-down operations, the spacecraft experienced many problems and was unable to return to Earth in 2007 as scheduled; instead, it returned to Earth in 2010, three years later than scheduled. A large, state-of-the-art facility to receive samples and perform curation activities was established at ISAS in Japan (Yada et al. 2014). The sample catcher, which was where captured samples should have been, was opened at the Curation Facility without exposing samples to atmosphere (Nakamura T. et al. 2011), and minute amounts of very small particles (mostly <0.1 mm) were found inside. Approximately 1500 asteroid particles were identified by separating about 3000 particles from the catcher over a period of about 5 months and analyzing individual particles with an electron microscope (Nakamura T. et al. 2011). Thus, approximately half of the separated particles proved to be Al metal grains from the walls of the catcher itself. Many more particles have since

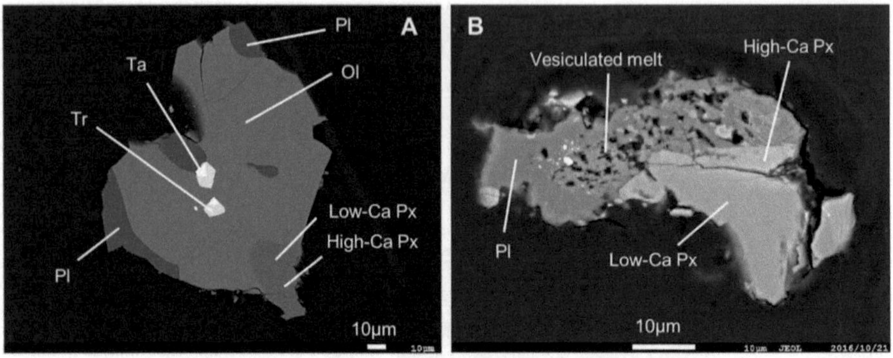

Fig. 4 (A) Typical Itokawa particle consisting of well-crystalline olivine, low- and high-Ca pyroxene, plagioclase, and inclusions of taenite and troilite. Silicates compositions are homogeneous and co-existing two pyroxenes indicates equilibrated temperature of 800 °C (Nakamura T. et al. 2011). (B) Itokawa particle consisting of low- and high-Ca pyroxene, plagioclase, and vesiculated melt. The melt was probably formed by an impact. Abbreviation: Ol = olivine, Low-Ca px = low-Ca pyroxene, High-Ca px = high-Ca pyroxene, Pl = plagioclase, Ta = taenite, and Tr = troilite

been identified, and catalogs are available from JAXA curation (https://curation.isas.jaxa.jp/curation/hayabusa/index.html).

3.5 General Properties of Return Sample

Typical Itokawa particles separated from the catcher are shown in Fig. 4. Initial analyses were performed using ~50 particles (Nakamura T. et al. 2011; Yurimoto et al. 2011; Noguchi et al. 2011; Ebihara et al. 2011; Nagao et al. 2011; Tsuchiyama et al. 2011). The results indicate that the Itokawa particles are mainly composed of silicate minerals (olivine, low-Ca pyroxene, high-Ca pyroxene, plagioclase), phosphates (principally merrillite), troilite (FeS), metallic iron (kamacite, taenite), etc. (Nakamura T. et al. 2011). Water-bearing minerals and organic matter are absent. Recent studies have detected 698 ppm and 988 ppm water in two Itokawa pyroxene grains (Jin and Bose 2019), suggesting that water was somehow supplied to the asteroid before the formation of pyroxene. Sylvite and halite were also found on the surface of two Itokawa particles (Noguchi et al. 2014b). Che and Zega (2023) found nanometer-sized NaCl crystals in association with secondary albitic plagioclase, suggesting a coupled formation process through aqueous fluid activity. This evidence implies that asteroid Itokawa may have experienced hydrothermal processes and challenges the assumption that S-type asteroids, such as Itokawa, were entirely dry compared to C-type asteroids. Instead, they suggest that S-type asteroids may have once supported active hydrothermal systems and could have contributed water to terrestrial planets.

The chemical compositions of the silicate minerals are almost homogeneous, which means that Itokawa underwent long-term thermal metamorphism in the interior of the parent asteroid. The average compositional range of olivine is $Fa_{28.6\pm1.1}$, Low-Ca pyroxene is $Fs_{23.1\pm2.2}Wo_{1.8\pm1.7}$, High-Ca pyroxene is $Fs_{8.9\pm1.6}Wo_{43.5\pm4.5}$, and plagioclase is $Ab_{83.9\pm1.3}Or_{5.5\pm1.2}$ (Nakamura T. et al. 2011). Based on the mean composition and low compositional variability of these silicate minerals, many particles were identified as being LL5~6 chondrite material. The chemical composition of metallic iron (Nakamura T. et al. 2011), oxygen isotope ratios (Yurimoto et al. 2011), bulk elemental abundance (Ebihara et al. 2011), and sample bulk density 3.4 g/cm^3 (determined by high-resolution synchrotron

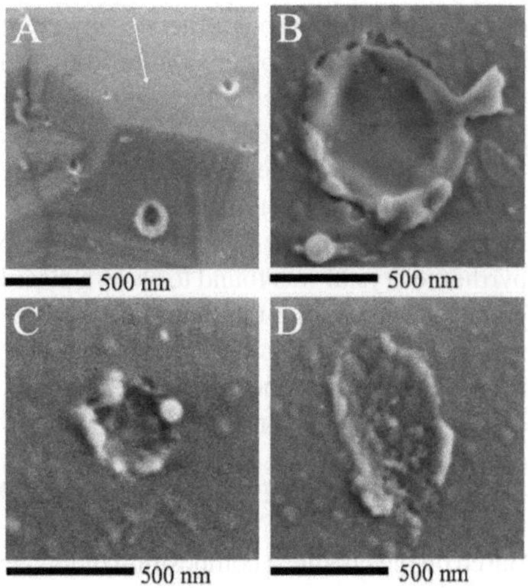

Fig. 5 Microcraters found on the surface of Itokawa particles indicating high-velocity impact of interplanetary dust particles (from Matsumoto et al. 2018)

CT images and microbalance; Tsuchiyama et al. 2011) are also consistent with this interpretation. It was also shown that the constituent minerals and the chemical compositions of the particles recovered from the first and second touchdown sites were nearly identical (Nakamura T. et al. 2014). This result is consistent with Itokawa's global homogeneous reflectance spectra obtained by the spacecraft (Abe M. et al. 2006a), confirming that asteroid Itokawa is composed almost entirely of LL5∼6 chondrite material. Based on the results of remote sensing observations and analyses of the samples returned by the Hayabusa mission, we conclude that the parent bodies of ordinary chondrites are S-type asteroids, as long suspected.

3.6 Space Weathering

Small asteroids such as Itokawa cannot hold an atmosphere. Therefore, various interactions with materials and radiation in interplanetary space are directly recorded on an asteroid's surface. One of these is the impact events. The recovered Itokawa samples record shock metamorphism of various intensities (Fig. 4B: also see Zolensky et al. 2022), and the impact stage ranges S2∼3 (Stöffler et al. 1991). In addition, interplanetary dust has impacted surface materials and formed micro-craters (Nakamura E. et al. 2012). Matsumoto et al. (2018) observed 900 micro-craters (Fig. 5) and concluded that the majority of these were probably formed by secondary-ejecta impacts and the actual crater impact flux seems to be similar between Itokawa and the Moon. The ^{21}Ne cosmic-ray exposure ages of the recovered samples indicate that Itokawa developed its current surface ($< \sim 1$ m depth) relatively recently (<8 Ma: Nagao et al. 2011).

Observations from the spacecraft showed that Muses Sea was reddish-black, and the recovered sample was expected to record space weathering, which was indeed observed (Noguchi et al. 2011). The surfaces of the recovered grains are covered with two altered zones: zone I is an amorphous silicate layer containing nano-phase Fe (npFe). Zone II is mostly crystalline silicate but partially amorphized area. The very surface (5 to 15 nm thick) of zone I is covered with sulfur-bearing Fe-rich nanoparticles (npFe) (Noguchi et al. 2011). The npFe in Zones I and II are thought to have been formed by the reduction of FeO in ferromagnesian silicates by hydrogen supplied by solar wind implantation. On the other hand, sulfur-bearing Fe-rich nanoparticles at zone I may have been formed by the deposition of FeS-containing vapor generated by collision events on the Itokawa surface vaporizing indigenous troilite or other sulfides from the impactors (Noguchi et al. 2011). The formation of tiny opaque particles at the surface of olivine, pyroxene, and plagioclase has been shown to cause darker and redder reflectance spectra of the asteroid.

In addition to zones I and II, vesicular layers were identified in the space weathered surfaces (Thompson et al. 2014; Noguchi et al. 2014a). The vesicles were probably formed by separation of solar-wind derived He implanted in this zone (Noguchi et al. 2014a). A pyrrhotite crystal was found to show a disordered rim that is sulfur-depleted and npFe metal grains are decorating the outermost surface (Keller and Berger 2014). In addition, on the surface of some Itokawa particles, small ($\sim$1 μm) metallic iron whiskers are found. It has been proposed that they are a decomposition product formed through irradiation of iron sulfide by energetic ions of the solar wind (Matsumoto et al. 2020).

3.7 Formation and Evolution Process of Asteroid Itokawa

Based on the chemical composition of low- and high-Ca pyroxene coexisting in Itokawa particles, the thermal metamorphic temperature was estimated to be about 800 °C (Nakamura T. et al. 2011). If the heating source was ^{26}Al, the parent body of Itokawa must have been larger than 20 km in radius and accreted at a period between 1.9 and 2.2 Myr after CAI formation, to satisfy mineralogic and isotopic evidence derived from Itokawa particles (Wakita et al. 2014). In summary, the Itokawa parent body formed in the early Solar System, was heated to a temperature of about 800 °C and then cooled slowly. Itokawa's parent body was then disrupted by a large impact, and its debris re-accumulated to form the second-generation body called Itokawa. Since then, the surface of the asteroid has been continuously modified by collisions and radiation, and it is only recently that the current surface conditions have been achieved. It remains unclear how and when Itokawa's parent asteroid was disrupted and when Itokawa accreted from some of the remnants.

4 Hayabusa2 Mission

The Hayabusa2 mission (Table 1) aimed to bring to Earth surface samples from the C-type asteroid Ryugu, a near-Earth asteroid (Fig. 1C: Watanabe et al. 2017). Ryugu was classified as Cb-type based on the SMASSII taxonomy and C/F-type following the Tholen taxonomy (e.g., Tatsumi et al. 2020). Pinilla-Alonso et al. (2016) showed that the near-infrared spectrum of Ryugu is similar to the Eulalia and Polana families in the asteroid belt. The collisional lifetime of Ryugu (300-500 Myr) is similar to or less than the breakup time of these families 830 $\pm$ 150 Myr and 1400 $\pm$ 150 Myr for Eulalia and Polana, respectively, and therefore Ryugu might be a member of these families (Sugita et al. 2019).

The spacecraft had almost the same design and functionality as that of the first Hayabusa mission, and the instruments on board were also similar except for the absence of the X-ray fluorescence instrument (Tsuda et al. 2013). The sample recovery method was also the same (Table 1: Sawada et al. 2017). The Hayabusa2 was launched in 2014, made an Earth swing-by in 2015, arrived at the asteroid in 2018, and started observations from an altitude of about 20 km (Tsuda et al. 2020).

4.1 General Properties of the Asteroid Ryugu

Ryugu exhibits a characteristic flattened, spinning top shape (Fig. 1C). Ryugu used to rotate with a shorter period ($\sim$3.5 h) than today, which is thought to have resulted in a spinning top shape from a spheroidal body (Watanabe et al. 2019). It was suggested that the YORP effect is responsible for the speed up in rotation (Walsh 2018). The density of the asteroid is 1.2 g/cm^3 (Table 1: Watanabe et al. 2019), which is very low compared to the density of the recovered samples of 1.79 ± 0.08 g/cm^3 (Nakamura T. et al. 2022), indicating that the asteroid interior is more than 30% void space. Thermal inertia ranges from 200 to 500 (av. 300) $J.m^{-2}.s^{-0.5}.K^{-1}$ (Okada et al. 2020). This value is much lower than that obtained from the returned samples (<1 cm in size) (890 $J.m^{-2}.s^{-0.5}.K^{-1}$: Nakamura T. et al. 2022). This difference is likely due to the presence of large cracks in the surface rocks of the asteroid, which would not be detectable in the < 1 cm sample, and which reduce the heat transfer to the entire body.

Ryugu's surface is composed of many large and small boulders (from a few meters to ~160 m; Fig. 1C and 6A), and there are no smooth terrains such as Itokawa's Muses Sea region. Most of the boulders show similar reflectance spectra, indicating that they formed by the destruction and partial reassembly of a relatively homogeneous original parent asteroid, and therefore Ryugu, like Itokawa, is a rubble pile asteroid (Watanabe et al. 2019). The boulder size distribution and shapes on the entire surface of Ryugu were determined: the shapes of Ryugu's boulders resemble laboratory impact fragments and thus support the interpretation that the boulders were produced by the disruption of Ryugu's parent asteroid (Michikami et al. 2019).

4.2 Reflectance Spectra

The reflectance of Ryugu is $1.60 \pm 0.15\%$ at 0.55 μm wavelength in standard geometry (Sugita et al. 2019), which is among the lowest in examined Solar System objects (Fig. 2). The visible and near-infrared spectra are very flat, with the only absorption identified at 2.7 microns (Kitazato et al. 2019). The visible and near-infrared spectral slopes (from 0.5 to 0.85 μm wavelengths) do not differ much depending on the location, but there are red regions with positive slopes (Fig. 6AB: slope > 0.10/μm) and blue regions with near-zero or negative slopes (slope < 0.10/μm), with the blue regions selectively located near the equator and in the polar regions (Sugita et al. 2019). This result indicates that Ryugu's surface was originally made up of blue material, but it has become reddish overall due, probably, to space weathering. The equatorial region, which used to have extended outward more in the past due to then higher rotation velocity, was also a red region. However, as Ryugu's rotation slowed, the red surface near the equator appears to have undergone a landslide and drifted down toward the mid-latitudes, exposing the blue material below the surface (Sugita et al. 2019).

The near-infrared spectra of Ryugu are even more uniform than the visible spectra (Fig. 2AB). The 2.7 μm absorption from hydrous minerals was detected globally, indicating that hydrous silicates are present all across Ryugu (Kitazato et al. 2019). Riu et al. (2021) analyzed NIR spectra of many craters and found that crater floors show a lower reflectance and a deeper 2.7 μm band than their periphery, suggesting that crater floors are more hydrated.

Compared to the spectra of CI and CM hydrous carbonaceous chondrites, Ryugu is darker and has a weaker 2.7 μm absorption feature (Kitazato et al. 2019). Heating experiments of CI and CM chondrites have shown that the reflectance and the 2.7-μm absorption decreases with increasing heating temperature (e.g. Hiroi et al. 1996). Therefore, it was pointed out that Ryugu may be hydrous carbonaceous chondrite material that had been heated to about 500 °C and that the hydrous minerals were incompletely dehydrated (Kitazato et al. 2019; Sugita et al. 2019). Other possibilities pointed to hydrated chondritic interplanetary dust particles (IDPs) that experienced lesser aqueous alteration (and were only partially hydrated)

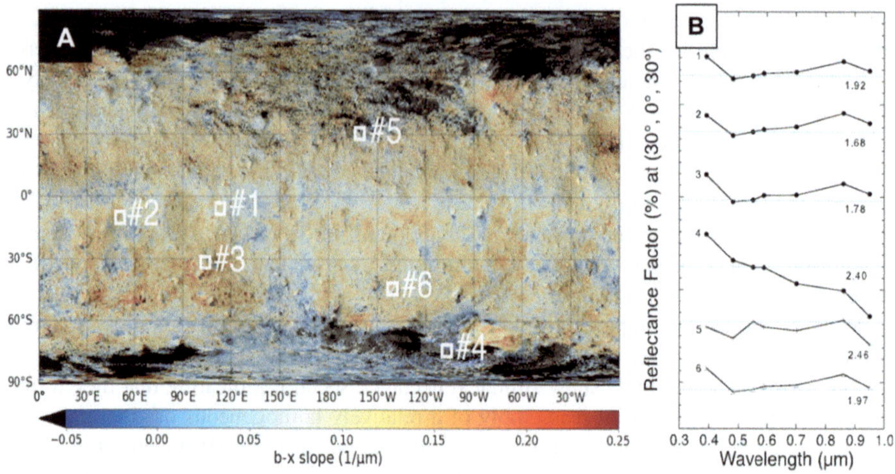

Fig. 6 Multi-band colors of Ryugu surfaces from Sugita et al. 2019. (A) Spectral slope map (μm^{-1}) superposed on a v-band image map, showing redder and bluer regions. The bluer regions locate preferentially equatorial ridge. #1to #6 regions correspond to typical regolith, dark regolith, dark/rugged boulder, Otohime boulder (the largest boulder measures ~130 m), bright/mottled boulder, and bright/smooth boulder, respectively. (B) The reflectance spectra of typical morphologic/color features on Ryugu #1 to #6 in panels (A)

of probable asteroidal origin (Kitazato et al. 2019; Sugita et al. 2019; Okada et al. 2020). However, the Ryugu samples recovered on Earth proved to be rich in hydrous silicates with no evidence of heating above ~100 °C, nor were they mineralogically similar to chondritic IDPs, hydrated or otherwise (Yokoyama et al. 2022; Nakamura T. et al. 2022). The reasons for the discrepancy between the interpretation of the reflectance spectrum of asteroid Ryugu and the characteristics of the actual samples are discussed later.

4.3 Regolith Evolution and Impact Experiment

An impact experiment on Ryugu's surface was performed using a small (~30 cm in size) impactor with a Cu plate (Arakawa et al. 2020), released from the Hayabusa2 spacecraft between the two sampling efforts. The impact velocity was ~2 km/s and produced an artificial crater with a diameter > 10 m. Subsurface material was ejected from the crater and deposited over a rather wide area due to low gravity (Arakawa et al. 2020). The surface age of Ryugu has been estimated based on the crater size distribution: $(1.58 \pm 0.47) \times 10^8$ years in case for a surface with cohesion or $(8.9 \pm 2.5) \times 10^6$ years for a cohesionless surface (Sugita et al. 2019). The fact that the artificial crater was formed in the gravity-dominated regime on a cohesionless surface supports the latter young age estimation (Arakawa et al. 2020). The young age is also consistent with the cosmic-ray exposure age (on the order of 10^6 years: Okazaki et al. 2022b) determined based on cosmogenic noble gas concentrations measured in the returned samples.

It was expected that the vibration induced by the impact experiment would fluidize the surface of the regolith around the artificial crater and cause resurfacing, but in fact the surface layer was hardly mobilized by the impact. This fact contrasts with the fluid movement of fine-grained material toward the gravity center on Itokawa's surface, which in the case of Ryugu may be due to the rapid attenuation of seismic waves due to the porous nature of the constituent rocks (Nishiyama et al. 2021).

4.4 Sample Collection

Selection of touch-down sites of the spacecraft was performed using images and spectral data taken at altitudes of 20 km, 6 km, and 5 km (Kikuchi et al. 2020). Hayabusa2 performed multiple low-altitude (~40 m) descending observations, during which spectral and morphological observations were conducted with the highest possible spatial resolutions down to 1 cm/pixel (Fig. 7). The Ryugu surface is well mixed with red and blue areas, but some areas are dominated by blue material and others by red material, depending on the location. The first touchdown sampling site was bluer than the global average (Fig. 7C from Morota et al. 2020). In the region around this first touchdown site, the reddish areas are limited to some boulders, while most boulder surfaces are blue (Fig. 7B and C). These observations are compatible with the interpretation that Ryugu's boulders were originally bluer, with redder materials produced by surface modification processes such as space weathering (Morota et al. 2020).

The second landing site was selected near the artificial crater produced by the impact experiment (Kikuchi et al. 2021), where it was confirmed that material ejected from the crater had accumulated (Tachibana et al. 2022). Unlike the troubled Hayabusa's touchdown operations, in which no bullets were fired due to a series of problems, the sample collection of Hayabusa2 was performed twice without problems, and the camera attached to the base of the sampling horn observed that the surface rocks were crushed with many fragments being scattered by the projectile firing (Tachibana et al. 2022).

The return capsule landed in the Woomera Desert in Australia, as did the Hayabusa mission capsule (Table 1). The day after the capsule returned to Earth, a trace amount of gas was collected from the sample container (Okazaki et al. 2022a). The isotopic ratios of helium and neon in the gas indicated that the gas was a mixture of solar wind and Earth's atmosphere (Okazaki et al. 2022c). The capsule was then quickly transported to Japan, where the container was opened and the recovered samples were initially examined (sample weighing, spectral analysis, etc.) for about six months at the ISAS curation facility developed specifically for Ryugu samples (Yada et al. 2022). A total of 5.4 g of samples were collected by the Hayabusa2 spacecraft, with the largest individual fragment being about 1 cm in diameter. Most of the collected particles were smaller than a few millimeters in size. Spectral measurements in the visible and near-infrared wavelength regions showed that the samples exhibit 2.7 μm absorption of hydrous silicates (Fig. 2AB), consistent with asteroid spectra obtained from the spacecraft (Yada et al. 2022; Pilorget et al. 2022).

4.5 General Properties of Return Sample

The morphology and polished surface of a large Ryugu sample is shown in Fig. 8A. Following the initial description work at ISAS, initial analyses were conducted by eight groups led by Japanese leaders, and the bulk chemical composition was found to be very primitive, close to solar and CI chondrites (e.g., Yokoyama et al. 2022; Nakamura E. et al. 2022; Ito et al. 2022). Muon beams were applied for the first time to the analysis of samples recovered from asteroids (Osawa et al. 2023), permitting an elemental composition, including the light elements C, N, O, which matched CI chondrites for most elements. However, oxygen was found to be about 1/3 less abundant than in CI meteorites (Nakamura T. et al. 2022; Ninomiya et al. 2024). Petrographic analysis indicated that the Ryugu sample is similar to CI chondrites, with the samples being dominated by hydrous silicates and carbonates that indicate pervasive aqueous alteration and lacking millimeter-sized chondrules (Fig. 8A and B from Nakamura T. et al. 2022). The samples contain a high concentration of carbon (4.6

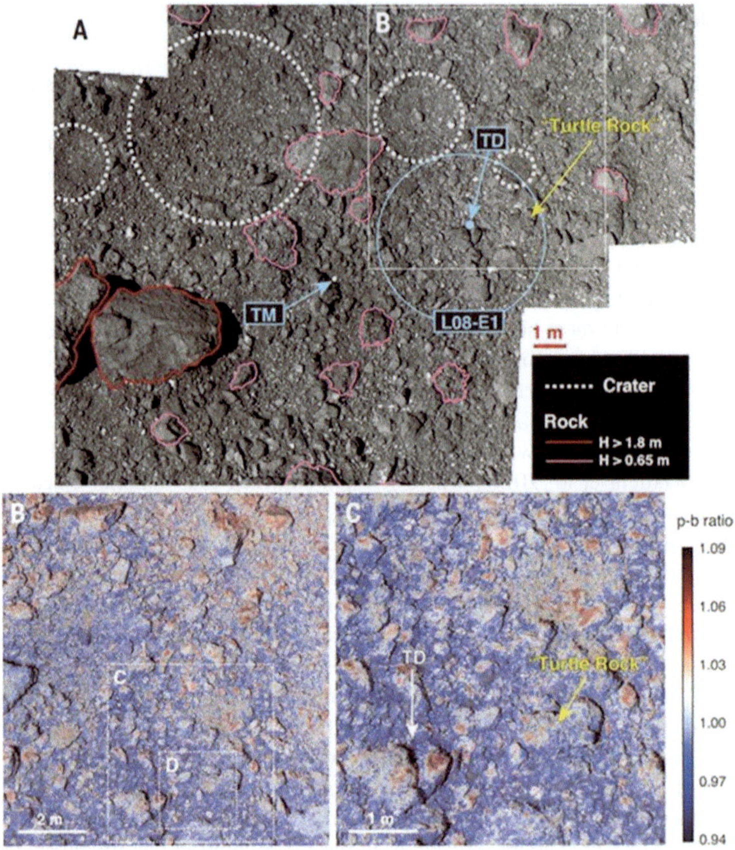

Fig. 7 Morphology and colors around the 1st touch down site L08-E1 from Morota et al. 2020. (A) Boulder and crater distribution. The light blue arrows indicate the location of the target marker (TM) and the touch-down point of the sampler horn (TD). The light blue circle indicates the L08-E1 area. The white dashed circles indicate craters. Boulder heights (H) were estimated from their shadow lengths; those with $H > 1.8$ m are outlined in red and those with $H > 0.65$ m are outlined in pink. The boulder nicknamed Turtle Rock is indicated by the yellow arrow. The white box indicates the region shown in later panels. (B and C) Blue and red area distribution obtained during the touchdown rehearsal operation, from two different altitudes. The dashed box C in (B) indicates regions shown in (C)

wt%), mostly in the form of insoluble organic matter, and bio-related molecule soluble organic matter such as amino acids (Yabuta et al. 2023; Naraoka et al. 2023). In addition, the samples contain high concentrations of primordial noble gases such as Ar, Kr, and Xe (Okazaki et al. 2022b). Concentrations are 1.5×10^{-6}, 2.0×10^{-8}, and 1.9×10^{-8} cm^3 STP g^{-1} for ^{36}Ar, ^{84}Kr, and ^{132}Xe, respectively, and these concentrations are close to the highest value found in a CI chondrite (Okazaki et al. 2022b).

4.6 Aqueous Alteration

Spectral observations from the spacecraft revealed a global abundance of hydrated silicates (Kitazato et al. 2019). The recovered samples were also found to be rich in hydrated silicates, and the almost complete absence of anhydrous silicates indicates that the Ryugu parent body was rich in liquid water at some earlier time, resulting in extensive aqueous alteration over

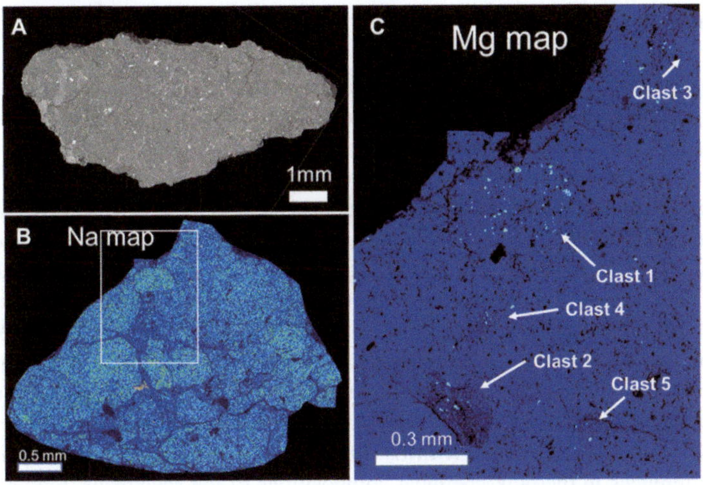

Fig. 8 Internal texture of the largest 2nd touch down sample C0002 from Nakamura T. et al. 2022. (A) CT image of the largest cross section indicating the absence of normal-size (~1 mm) chondrules and CAIs observed in other chondrites. (B) Na map of the real cross section taken by electron microscope, showing that C0002 is a breccia of many small clasts with variable Na concentrations. (C) Mg map of a portion enclosed by white box in (B). There are many clasts containing bright spots of Mg which are olivine and pyroxene. These clasts are less altered compared with surrounding major lithology

a wide area of the entire asteroid (Yokoyama et al. 2022; Nakamura T. et al. 2022). The hydrous silicates are mainly Mg-rich saponite and serpentine with Mg/(Mg + Fe) ~ 0.8-0.9, and locally small amounts of Mg-chlorite. Carbonate minerals are dominated by dolomite and breunnerite. Phosphorus phase is dominated by apatite, but small amounts of phosphides such as $(Fe, Ni)_3P$ also occur. Iron sulfides are mostly pyrrhotite with some pentlandite (Nakamura T. et al. 2022; Noguchi et al. 2023). Aliphatic organics and phyllosilicates occur together indicating a genetic relationship (Ito et al. 2022; Dionnet et al. 2023; Aléon-Toppani et al. 2024). All of these major constituent minerals were formed by aqueous alteration, and their elemental compositions, isotopic compositions, and crystal structures should record the physico-chemical conditions during the alteration.

Water temperatures during the alteration were estimated from the oxygen isotope ratios of co-existing carbonate minerals and magnetite, which were low between 0 and 20 °C (McCain et al. 2023) at the onset of the alteration, but later increased to between 9 and 20 °C (Nakamura E. et al. 2022) and 37 ± 10 °C (Yokoyama et al. 2022). In addition, alteration temperatures have been inferred from mineralogical evidence. Based on the composition of coexisting pentlandite and pyrrhotite, the temperature was estimated to be about 25 °C and, based on the site occupancy of Fe and Ni in the pentlandite crystal structure, it was about 20 °C (Nakamura T. et al. 2022). The change in carbonate mineral type (from Ca carbonate to CaMgFe carbonate) with the progression of alteration is also inferred from the oxygen and carbon isotope ratios of various carbonate minerals (Fujiya et al. 2023). The Ryugu samples preserve rock fragments with different degrees of the alteration, which apparently recorded the multistage geological evolution of the Ryugu parent body (Yamaguchi et al. 2023).

4.7 Carbonate Formation Age

Aqueous alteration was a fundamental process on water-rich planetesimals such as Ryugu, Bennu, and other carbonaceous chondrite parent bodies. This process led to the formation of secondary carbonates wherein Mn was strongly enriched relative to Cr. Such highly elevated Mn/Cr ratios in carbonates manifest themselves in large excesses of radiogenic ^{53}Cr today, making carbonates ideal targets for dating aqueous alteration using the Mn-Cr chronometer. In combination with thermal evolution models, these carbonate ages can be used to infer the accretion ages of the planetesimals. By applying ^{53}Mn-^{53}Cr chronometry to the carbonate minerals, the ages of carbonate formation were estimated to be 1.8 Myr (McCain et al. 2023), 2.6 Myr (Nakamura E. et al. 2022), and 5.2 Myr (Yokoyama et al. 2022) after CAI formation. The discrepancies in ages are likely related to complications that are inherent to the measurement technique employed by these studies: Mn and Cr have distinct ionization sensitivities and thus fractionate systematically during simultaneous measurements by secondary ion mass spectrometer (SIMS). The relative sensitivities are furthermore amplified by differences in the bulk chemical compositions of the specimens, requiring matrix-matched reference standards to avoid obtaining inaccurate Mn-Cr ages. The three studies, however, have used distinct reference standard materials, which introduced systematic shifts in the measured Mn/Cr ratios of Ryugu's carbonates.

The reference problems were solved by Sunagawa et al. (2024). They synthesized Mn-, Cr-, and Fe-bearing crystalline dolomite as reference materials and evaluated the relative sensitivity factor (RSF) of Mn/Cr for SIMS analysis. Using a new RSF, they corrected dolomite age determined by Yokoyama et al. 2022 from 5.2 Myr to 4.5 Myr after CAI formation. They also pointed out that 1.8 Myr (McCain et al. 2023) is too early and gets older if corrected by proper RSF values. It should also be noted that this debate has been resolved also by measuring the Cr isotopic composition by thermal ionization mass spectrometer and the Mn/Cr ratio by isotope dilution inductively coupled plasma mass spectrometer using bulk samples (Tanaka et al. 2024). They find an aqueous alteration age of 4.1 Myr after CAIs, which is almost identical with the age estimated by Sunagawa et al. (2024).

The carbonate formation ages (~4 Myr after CAIs) indicate that Ryugu's parent body was formed in the early Solar System, followed by hydrothermal alteration episode(s) that happened on the parent body a few million years after CAI formation. Thermodynamic equilibrium calculations were performed using the bulk chemical composition of Ryugu samples for the case of the water temperature range of 0, 20 and 40 °C, and the coexisting mineral species and mineral abundances were estimated as a function of water rock ratios of the Ryugu parent asteroid (Nakamura T. et al. 2022). By comparing the mineral species and mineral abundances in the actual samples, it was estimated that the highly-altered major. lithology that makes up the majority of Ryugu samples was formed by water-rock-weight ratios of 0.2-0.9 (Nakamura T. et al. 2022), corresponding to 17 ~ 47 wt% of water content in Ryugu's parent asteroid.

4.8 Soluble Organic Matter

Ryugu contains a carbon abundance of 4.63 ± 0.23 wt%, consisting of ~65% organic carbon and ~35% of the carbon in carbonates. The organic carbon abundance amounts to 3.08 ± 0.3 wt%, which is similar to the values observed in the Ivuna chondrite (Yokoyama et al. 2022). The bulk nitrogen abundance of Ryugu samples is ~0.15 wt%, which leads to a N/C atomic ratio of 0.01 to 0.035, a range of value that is comparable to that in primitive carbonaceous chondrites.

The organic signature of Ryugu samples was analyzed both for the soluble organic matter (SOM) and insoluble organic matter (IOM). The characterization of the SOM of Ryugu samples by FT-ICR (Fourier Transform - Ion Cyclotron Resonance) mass spectrometry and by GC×GC - HRMS (multidimensional gas chromatography coupled with high resolution mass spectrometry) showed a very high molecular diversity, and similarities with that of CI chondrites (Naraoka et al. 2023; Schmitt-Kopplin et al. 2023; Aponte et al. 2023). Low temperature and water-rich conditions on the parent asteroid are suggested by the absence of organo-magnesium species and high contents of sulfidic and nitrogen-rich compounds, as well as ammonium ions. PAHs are present in a structural continuum of carbon saturations and oxidations, suggesting several episodes of low temperature aqueous alteration, or a potential origin of these PAHs from the interstellar medium (Naraoka et al. 2023; Schmitt-Kopplin et al. 2023; Aponte et al. 2023; Zeichner et al. 2023). Hashiguchi et al. (2023) found that the spatial distributions in Ryugu samples of CHO, CHN, CHON, CHO-Na molecular ions are not strictly similar, suggesting different timing of precipitation of different transportation mechanisms and/or possible interactions between organics and minerals.

Amino acids were identified in chambers A and C Ryugu samples, with glycine and β-alanine being detected at the ppm level (Naraoka et al. 2023; Potiszil et al. 2023; Parker et al. 2022, 2023). There is no enantiomeric excess detected in the amino acid distributions, strongly suggesting the absence of terrestrial contamination (Parker et al. 2022, 2023). Aliphatic amines (e.g. C1-C5 chains) were also identified in Ryugu samples, with a decreasing abundance when increasing the chains' length (Naraoka et al. 2023; Parker et al. 2022, 2023). The concentrations and relative distributions of amino acids and aliphatic amines in Ryugu samples are somewhat different from that in Orgueil (Parker et al. 2023). This might be due to different initial compositions, or to different alteration processes on the asteroid or on Earth. The β-alanine/glycine ratio for samples from chamber A and chamber C samples are 0.57 and 1.03, respectively (Potiszil et al. 2023). As this ratio decreases with increasing aqueous alteration, this result suggests that the chamber A sample underwent a larger extent of aqueous alteration than the chamber C sample. Uracil was also found in Ryugu samples by Oba et al. (2023). Uracil being one of the four nucleobases in ribonucleic acid, this highlights the importance of such a molecule for a prebiotic delivery on the early Earth.

As a whole, the SOM components are less abundant than in type 2 chondrites like CM meteorites, and are more similar to that of CI chondrites, which is in agreement with the chemical and mineralogical composition of Ryugu samples being close to that of CI chondrites. Various long duration aqueous alteration processes on the Ryugu parent body are probably needed to explain the different types of SOMs identified in Ryugu samples (e.g. Naraoka et al. 2023).

4.9 Insoluble Organic Matter

Organic matter was also studied in whole-rock samples by in situ spectro-imaging techniques, and insoluble organic matter (IOM) extracted from Ryugu samples by acid etching was analyzed by the same techniques. The characterization methods included Fourier-transform infrared microscopy (μ-FTIR), infrared nano-spectroscopy (AFM-IR), scanning transmission X-Ray spectroscopy (STXM-XANES), and Raman spectroscopy (Yabuta et al. 2023; Dartois et al. 2023; Kebukawa et al. 2024; Mathurin et al. 2024; Phan et al. 2024; Bonal et al. 2024; Quirico et al. 2024; Ito et al. 2022; De Gregorio et al. 2024).

The μ-FTIR spectra of the Ryugu organic matter shows similarity with that of carbonaceous chondrites, both in whole-rock and IOM samples (Fig. 9, Yabuta et al. 2023; Kebukawa et al. 2024; Dartois et al. 2023; Quirico et al. 2024). The AFM-IR and STXM-XANES analyses that could study the distribution of organic matter at the sub-μm level

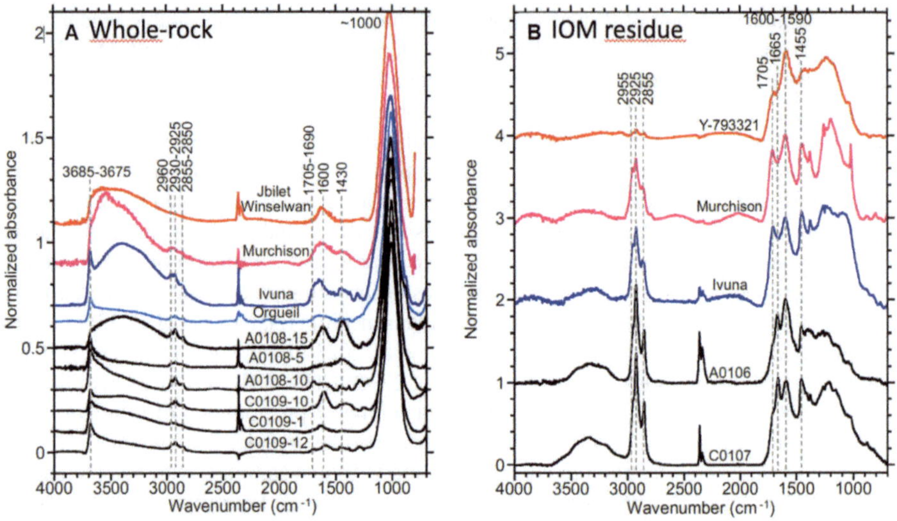

Fig. 9 (a) Infrared transmission spectra of six Ryugu intact grains from A0108 and C0109 (black), CM chondrites Jbilet Winselwan (red) and Murchison (pink), and CI chondrites Orgueil (light blue) and Ivuna (dark blue). All the spectra were baseline corrected using spline curves and normalized by the peak height of the band at $\sim$1000 cm^{-1}. (b) Same as (a) but for insoluble organic matter (IOM) residues from Ryugu samples A0106 and C0107 compared with IOM from CM chondrites Y-793321 and Murchison, and CI chondrite Ivuna. The spectra were baseline corrected using spline curves and normalization to the peak height of the aromatic C=C band at $\sim$1600 cm^{-1}. (Adapted from Yabuta et al. 2023)

revealed that the organic matter in whole-rock samples is present in two forms, as a diffuse phase intermixed with the phyllosilicate matrix, and in the form of organic patches, which often resemble nanoglobules found in primitive carbonaceous chondrites (Fig. 10). The diffuse phase shows a signature dominated by aromatic C=C, with the presence of carbonyl C=O and aliphatic CHx. The organic patches have a composition which is enriched in C=O and aliphatic CHx compared to the diffuse phase which could be explained by a formation of these organic patches (and nanoglobules) by irradiation of ices in the protoplanetary disk (Mathurin et al. 2024).

The composition of IOM extracted from Ryugu samples in chamber A and chamber C show differences in their N content, with average formulas being $C_{100}N_1O_{10}S_{0.8}$ and $C_{100}N_2O_{10}S_{0.8}$, respectively (Yabuta et al. 2023).

Raman and μ-FTIR analyses of Ryugu samples did not show evidence for long duration thermal metamorphism, but variabilities between samples may have recorded low amount of heating possibly due to impacts on the asteroidal parent body (Bonal et al. 2024; Quirico et al. 2024).

4.10 Nitrogen Bearing Matter

An unexpected N-H rich component was identified early in Ryugu samples by reflectance spectroscopy. This phase was first described as a NH_4-rich phase, maybe in the form of ammoniated phyllosilicates, NH_4 hydrated salts, or N-rich organics (Pilorget et al. 2022). Based on an in-depth analysis, Pilorget et al. (2024) then reported the discovery of a unique class of phosphorus-rich grains with major biochemical potential. These grains are up to a few hundreds of μm in size, and have a hydrated-ammonium-magnesium-phosphorus (HAMP)-rich composition. Their compositions suggest an outer Solar System origin. On the prebiotic

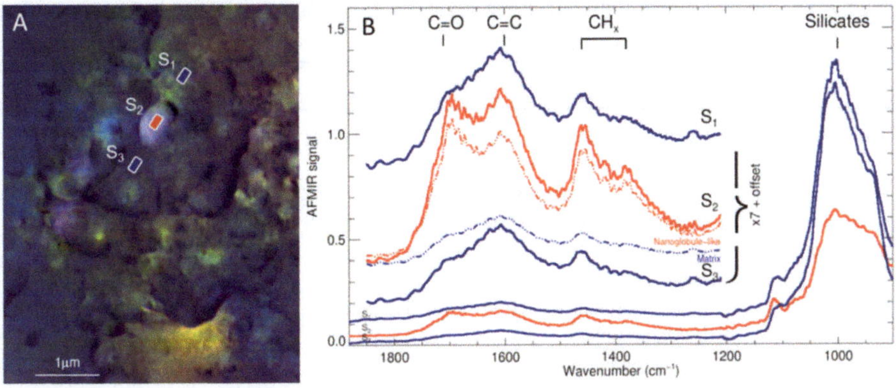

Fig. 10 Whole-rock Ryugu sample A0108-19 crushed on a diamond window. (a) Infrared nanospectroscopy (AFM-IR) RGB image of showing the contribution of C=O (red, 1720 cm^{-1}); C=C (green, 1600 cm^{-1}); CH_X (blue, peaked at 1460 cm^{-1}). (b) Spectra corresponding to the areas S1 to S3 shown in (a), showing the presence of a diffuse phase (blue spectra), and of a nanoglobule-like feature visible in the S2 location in (a) (red spectra). Adapted from Mathurin et al. (2024)

Earth, an input of such phosphorus and nitrogen-rich phases could have played an important role in the biochemical evolution of organic matter. Based on X-ray diffraction studies of Ryugu samples, Viennet et al. (2023) suggested that N-rich organics could be trapped in the interlayer space of smectite in Ryugu samples. N-rich outlier phases were also observed by μ-FTIR in IOM samples by Kebukawa et al. (2024) and Quirico et al. (2024). The compositions of these N-rich phases were not well constrained, but Kebukawa et al. (2024) propose a N-H functionality (likely amide) for one of the N-rich phases.

Although being exposed to space at the surface of asteroid Ryugu, organic components, including soluble organic matter, are preserved in Ryugu samples. This demonstrates the robustness of these compounds with regard to space weathering, and the possible delivery of prebiotic molecules from interplanetary materials to the early Earth.

4.11 Less-Altered Lithology

Most of the Ryugu samples are breccias, consisting of small rock fragments ranging from a few tens of microns to a few millimeters in size (Fig. 8B and C). The majority of the rock fragments show highly-altered lithology as described above, while some fragments are less altered (Fig. 8C). The main characteristics of these less-altered fragments are that they contain fine particles (mostly less than 10 µm) of anhydrous silicates such as olivine and low- and high-Ca pyroxene (Fig. 8C) and that their mineral combination is different from that of the altered fragments (Table 2: Nakamura T. et al. 2022). The least altered fragments contain amorphous silicate similar to GEMS, which is not contained in the less altered fragments (Nakamura T. et al. 2022). The mineral assemblages and their relative abundances of the less-altered fragments are similar to those expected under alteration conditions with low water-rock ratios (water-rock weight ratio 0.1-0.2) based on thermodynamic calculations (Nakamura T. et al. 2022). Therefore, the less altered fragments were located in relatively water-poor areas in the Ryugu parent body, and thus retain more information about Ryugu's precursor materials, i.e., information at the time of accretion (Nakamura T. et al. 2022). In fact, a high concentration of presolar grain SiC was found in one Ryugu fragment with a low degree of alteration (Nguyen et al. 2023).

Table 2 Constituent mienrals of the less altered and altered fragments in Ryugu samples (data from Nakamura T. et al. 2022)

	Less-altered fragments	Altered fragments
Anhydrous silicates	small (<10 micron) olivine, low-Ca and high Ca pyroxene, spinel, and perovskite	none
Hydarated silicates	amorphous silicates (in the least-altered fragments) to poorly crystalline phyllosilicates	saponite and serpentine
Carbonates	minor amounts of Ca carbonate (calcite and aragonite)	dolomite and breunnerite
P phases	minor phosphide $(Fe, Ni)_2P$ or $(Fe, Ni)_3P$ and NaMg phosphate	apatite
FeS phases	submicron pyrrhotite	sumicron pyrrotite and pentlandite and large hexagonal pyrrhotite

NaMg phosphate is found as small particles (∼ a few microns or less) in the least altered clasts and as relatively large veins in the less altered clasts (Fig. 11A). The texture and occurrence indicate that the phosphate formed during an early stage of aqueous alteration. The chemical composition in wt% of the NaMg phosphate is: MgO 26.34, P_2O_5 47.48, Na_2O 1.74, FeO 0.44, Cl 0.20, total 76.2, showing an atomic ratio of Mg/P = ∼1, which indicates a possible formula of Mg_2P_2 O_7-4(H_2O) or Mg(PO_3OH)-1∼2(H_2O) (Ma et al. 2022). TEM observation revealed that the NaMg-phosphate is amorphous and vesiculated (Fig. 11B: Nakamura T. et al. 2022). EBSD analysis confirms that the present Mg-phosphate is amorphous (Ma et al. 2022). These results suggest that the NaMg phosphate was originally enriched in volatiles such as water and or organics which later evaporated, resulting in the vesiculated texture. The daytime temperature of Ryugu's surface at its current orbit is ∼80 °C (Shimaki et al. 2020) and the NaMg phosphate might have been heated and vesiculated at this temperature. NaMg phosphate is also widely found in the Bennu sample and has been found to be similarly amorphous (Barnes et al. 2024). Possible similar amorphous Mg phosphates have recently been reported in the CI chondrite meteorites (Mikouchi et al. 2024).

4.12 Chondrules, AOAs, and CAIs

High-resolution (<1 μm) synchrotron radiation CT analysis revealed that chondrules and CAIs of sizes that are abundant in chondrites are not present (Fig. 8A and also see Ito et al. 2022). On the other hand, a small number of very small (<30 μm) chondrules and CAIs were found in the aforementioned less altered Ryugu fragments (Nakamura T. et al. 2022). The chondrules are type-I POP and BO, and the CAIs are composed of spinel, hibonite, and perovskite (Nakamura T. et al. 2022). Oxygen isotopic analysis by SIMS showed that all CAIs have a ^{16}O-rich composition, which is the same isotopic composition as that of CAIs in carbonaceous chondrites (Nakashima et al. 2023). On the other hand, the chondrules showed ^{16}O-rich and ^{16}O-poor compositions, with the ^{16}O-rich compositions similar to AOA and the ^{16}O-poor compositions similar to chondrules (Nakashima et al. 2023). This result is consistent with the presence of the two types (^{16}O-rich and ^{16}O-poor) of small olivine particles scattered in many Ryugu samples (Kawasaki et al. 2022; Liu et al. 2022; Nakamura E. et al. 2022). The formation process of high-temperature objects in Ryugu samples such as CAIs, chondrules, and AOAs, as well as their accretion process to the Ryugu parent body, are not well understood and therefore require further study.

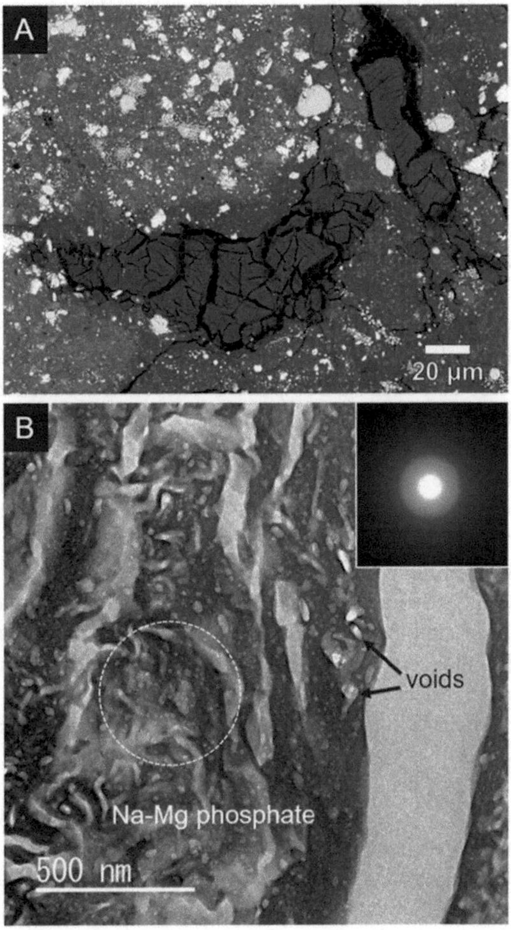

Fig. 11 NaMg phosphate in C0002 from Nakamura T. et al. 2022. (A) Back-scattered electron image of NaMg phosphate with wrinkled texture. (B) Bright field TEM image of the NaMg phosphate. The NaMg phosphate exhibits a vesicular texture suggesting vaporization of volatile molecules. The inset is a selected area electron diffraction pattern obtained from the area indicated by a white circle suggesting that the N-Mg phosphate is an amorphous phase

4.13 Formation Region in the Protoplanetary Disk

There is much evidence that Ryugu's parent body formed in a low-temperature region in the protoplanetary disk, including the presence of ^{15}N-rich organic matter (Ito et al. 2022), the presence of liquid water containing CO_2 (Nakamura T. et al. 2022), and the abundance of carbonate minerals in the entire recovered samples (Loizeau et al. 2023). Based on the above observations, Ryugu's parent body is considered to have formed at >10 au below -200 °C outside the CO_2 snow line. The fact that Ryugu contains few chondrules and CAIs, and that they are as small as most chondrules and CAIs in comet samples from 81P Wild2 (Zolensky et al. 2006; Nakamura T. et al. 2008) is also consistent with their formation at the outer edge of the Solar System.

Ryugu samples show nucleosynthetic Fe isotope anomalies similar to Ivuna-type CI chondrites, which are distinct from all other carbonaceous chondrites, suggesting that Ryugu and CIs constitute an isotopically distinct third reservoir located at the outermost protoplanetary disk (Hopp et al. 2022). Cu and Zn isotope signatures also indicate that Ryugu and CI belong to the distinct third reservoir (Paquet et al. 2023). This suggests that there has been an isotopic trichotomy between NC, CC, and Ryugu/CI in the Solar System. On the other hand, an interpretation that Ryugu and CI are part of the CC based on the Te-Fe isotopic correlation has been proposed (Marrocchi et al. 2023). In addition, Spitzer et al. (2025) reported nucleosynthetic Ni isotope anomalies, similar to Fe, but suggested that CI and Ryugu might have formed in the same reservoirs as the other CC. and the Fe and Ni isotopic composition in CI and Ryugu reflects higher accretion efficiencies of small FeNi metal grains than other CC. Future isotopic analyses of the less-altered fragments will help us to understand whether Ryugu is material from a distinct third reservoir in the protoplanetary disk.

4.14 Space Weathering and Terrestrial Weathering

Observations of asteroids from the ground and space have progressed, and asteroids have been classified based on their spectral characteristics (Bus and Binzel 2002a, 2002b; DeMeo et al. 2009). Comparison of spectra between carbonaceous chondrites and C- and D-type asteroids suggests that these asteroids are made of carbonaceous chondrites or similar materials. For the first time in the Hayabusa2 mission provided the first opportunity to obtain both the spectra of asteroids and the spectra of samples recovered from the same asteroids, and compare them in detail. The samples from the first Hayabusa mission were too minute for this to be usefully done. As a result, unexpected results were obtained: (1) Visible and near-infrared spectra of the asteroid obtained by the spacecraft (Sugita et al. 2019; Kitazato et al. 2019) and the spectra of the recovered sample are roughly consistent (Pilorget et al. 2022), but the 2.7-μm absorption depth from hydrous minerals is about twice as deep in the samples (Fig. 2B). (2) The spectra of the CI chondrite Orgueil and Ryugu samples, which are nearly identical in mineralogy and petrology, differ significantly (Fig. 2A), with the Ryugu samples being darker, monotonous, and having a less 2.7-μm absorption (Nakamura T. et al. 2022).

Since these two problems have a significant impact on future observations of asteroids and the interpretation of the obtained spectra, they were examined in detail. The conclusion is that the cause of these spectral discrepancies is space weathering of asteroids and terrestrial weathering of meteorites. The main cause of the discrepancy in (1) is space weathering. The topmost surfaces of asteroids are constantly exposed to solar wind and cosmic dust bombardment, causing space weathering and various surface transformations (Noguchi et al. 2023; Matsumoto et al. 2024). Therefore, asteroid spectra obtained from spacecraft are characterized by heating and incomplete dehydration due to space weathering. On the other hand, the recovered samples were crushed by a metal sphere at the time of sample collection, exposing a fresh surface. Therefore, the spectra of recovered samples are dominated by spectra that have not been subjected to space weathering (Matsuoka et al. 2023). This is why the spectra of the recovered samples are different from those of the remote observation.

Meanwhile, the spectral discrepancy in (2) is due to terrestrial weathering: after falling to the Earth, CI chondrites rapidly react with the Earth's oxygen and water, resulting in significant weathering. As a result, Fe^{2+} in the hydrous mineral saponite is oxidized (Roskosz et al. 2024) and becomes brighter (Nakamura T. et al. 2022). In addition, the oxidation of small pyrrhotite crystals forms bright hydrous Fe sulfate, which brightens the overall sample and deepens the absorption at 2.7-3.0-μm (Amano et al. 2023). Thus, the spectra of CI chondrites change as they are weathered on the ground, and their spectra differ significantly from those of CI materials that have not been weathered at all, such as Ryugu. That terrestrial weathering is the cause of the spectral discrepancy is proved by the fact that when CI chondrites are heated in a reducing environment to a temperature at which hydrous minerals do not decompose (about 300 °C), the spectra of the heated CI and the Ryugu sample are almost identical (Amano et al. 2023). The degree of spectral change of CI materials is

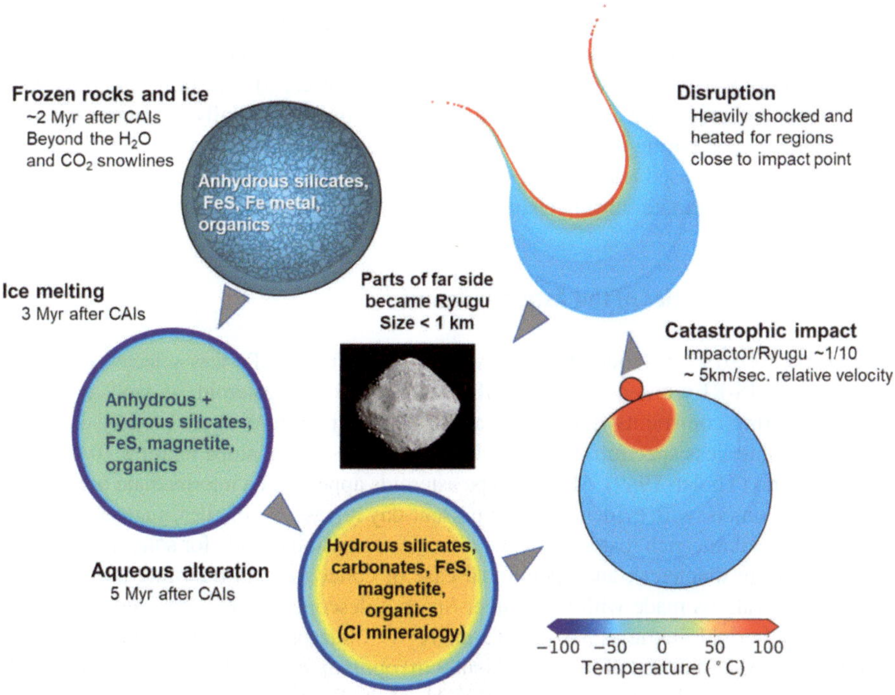

Fig. 12 Formation and alteration history of asteroid Ryugu (Nakamura T. et al. 2022). The temperature distribution inside the asteroid and the temporal variation of temperature were estimated by numerical simulation using the density and specific heat of the actual Ryugu samples. They are also consistent with the formation ages and formation temperatures of the carbonate minerals in the samples

greater by terrestrial weathering than by space weathering. Thus, the spectra of asteroids composed of these reactive materials should be interpreted with caution when we compare them to corresponding meteorite spectra.

Asteroid Ryugu is the first asteroid for which the physical properties of recovered samples were measured in detail: Density, mechanical properties, and thermal properties were determined (Nakamura T. et al. 2022; Ishizaki et al. 2023). A thermal evolution model of Ryugu's parent body with ^{26}Al as a heat source was constructed using the specific heat, thermal diffusivity, and water-rock ratio estimated from the mineral abundances of the Ryugu samples (Nakamura T. et al. 2022). A simulation of the collisional destruction of the Ryugu parent body was also performed using the measured sample strength and elastic velocities (Nakamura T. et al. 2022). A model of the formation, evolution, and collisional destruction of the Ryugu parent body was constructed using these sample measurements (Fig. 12).

5 OSIRIS-REx Mission

The primary scientific objective of the Origins, Spectral Interpretation, Resource Identification, and Security-Regolith Explorer (OSIRIS-REx) mission (Table 1) is to return and analyze a sample of pristine regolith from a primitive, carbonaceous asteroid to understand the initial stages of planet formation and primordial sources of organics delivered to Earth that may have contributed to life on our planet (Lauretta et al. 2015). Additional science objectives include: 1) mapping the global chemistry, mineralogy, physical properties, and "geology" of the asteroid, 2) documenting the sampling site down to sub-cm scales for texture, mineralogy, volatile chemistry, and spectral properties, 3) measuring the Yarkovsky effect (which changes the asteroid's orbit over time) on a potentially hazardous asteroid, and 4) directly comparing the integrated global characteristics of a primitive, carbonaceous asteroid to ground-based telescopic data of known asteroids. Here we provide an overview of pre-arrival and remote sensing characteristics of Bennu as well as early sample analysis results.

5.1 General Properties of the Asteroid Bennu

The asteroid (101955) Bennu (Fig. 1A, see spectra in Fig. 2AB) was selected as the target for OSIRIS-REx based on its classification as a B-type asteroid, as well as mission and engineering constraints (Lauretta et al. 2015). B-type asteroids are broadly characterized by a featureless spectrum with a neutral to bluish spectral slope and low (0.08 – 1.0) visible albedo (Tholen 1989). Many B-type asteroids appear to be intermediate between asteroids and comets, with evidence for the present-day release of volatiles, suggesting Bennu may contain volatile-rich material. Bennu also meets mission criteria for solar distance, size (~490 m mean diameter), and operational ease of collecting and returning the sample to Earth. Observations made with the Spitzer Space Telescope in 2007 provided an estimate of Bennu's thermal inertia as 310 ± 70 J m^2 K^1 $s^{0.5}$, interpreted as suggesting that the regolith would be comprised of fine gravel (sub-cm) optimal for sampling (Emery et al. 2014). Because Bennu's orbit brings it within 0.002 AU of the Earth, it has one of the highest probabilities of impacting the Earth of any known asteroid; as such, Bennu offers an important opportunity to study a potentially hazardous asteroid (PHA).

5.2 Scientific Instruments Onboard the Spacecraft

The OSIRIS-REx spacecraft carries a complement of scientific instruments designed to characterize Bennu as fully as possible to enable sampling site selection, provide context for the returned sample, and better understand carbonaceous Solar System objects. The OSIRIS-REx Camera Suite (OCAMS) consists of three cameras for observing the asteroid at distances ranging from 500,000 km to 2 m (Rizk et al. 2018). One of the cameras, MapCam, allows for mapping of the asteroid at four wavelengths (470, 550, 770, and 860 nm) simulating the Eight-Color Asteroid Survey (ECAS) filter system (b', v, w, and x bands, respectively), allowing for spatially resolved color characterization and direct comparison with Earth-based, multispectral observations. OSIRIS-REx carries two hyperspectral point spectrometers for the identification of minerals, organic chemistry, and measurement of surface temperature. The OSIRIS-REx Visible and Infrared Spectrometer (OVIRS) measures the reflected light spectrum from 0.4 to 4.0 μm (Reuter et al. 2018). The OSIRIS-REx Thermal Emission Spectrometer (OTES) measures the emitted radiance of the asteroid from 5.5 – 50 μm (Christensen et al. 2018). The OSIRIS-REx Laser Altimeter (OLA) provides laser ranging data for the mapping of surface topography and allowing construction of a 3D model of Bennu at the highest resolution recorded on any small body to date (Daly et al. 2017). The Regolith X-ray imaging spectrometer (REXIS) is a student experiment from the Massachusetts Institute of Technology and Harvard-Smithsonian Center for Astrophysics, designed to map elemental abundances on Bennu (Masterson et al. 2018). OSIRIS-REx also conducts radio science for probing the asteroid's gravity, internal structure, and mass distribution (Scheeres et al. 2016). To collect the crucial regolith sample, OSIRIS-REx employed the Touch and Go Sampling Mechanism (TAGSAM). The TAGSAM is designed to collect a bulk sample consisting of particles up to roughly 2 cm in diameter with a total mass substantially greater than the mission requirement of 60 g. Small, presumably space weathered, surface particles at the uppermost surface are collected by 24 stainless steel contact pads, each ∼1.75 cm in diameter (Bierhaus et al. 2018).

5.3 Small Particle Ejection from the Asteroid

Upon arrival at Bennu in 2018, OSIRIS-REx immediately made unexpected discoveries. In one case, the asteroid periodically ejects small particles from a few mm up to 10 cm in size, making it an "active asteroid" (Lauretta et al. 2019b; Chesley et al. 2020). These particles most commonly follow suborbital trajectories (65%), although some complete more than one revolution around the asteroid (20%) and only a small fraction (15%) manage to escape Bennu's gravitational field (Chesley et al. 2020). The ejection events can occur at any local time of day, although most occur in the afternoon and at night (between 12:00 – 24:00 local solar time). Cause(s) of these events that have been considered include meteoroid impacts (Bottke et al. 2020), temperature change-driven sublimation of water ice, or thermal fracturing of rock (Rozitis et al. 2020). Of these, water ice sublimation appears to be the least plausible based on surface temperatures at ejection sites (Rozitis et al. 2020). In another case, resolved imaging of the surface quickly ascertained that the nature of the asteroid's regolith was not as expected; rather than being dominated by cm-scale particles, it is covered in meter-scale rocks and larger boulders with very few areas of regolith dominated by cm-scale particles (Lauretta et al. 2019a, 2019b). However, a few smooth-looking areas up to ∼20 m in size are visible, suggesting that smaller particles are present.

5.4 Morphology and Reflectance Spectroscopy

In terms of global physical properties, Bennu is extremely dark, with a global geometric albedo of ∼4.4% (at 0.55 μm) although there are spatial variations in albedo that result in a range of albedos from ∼3.3% to more than 15% (DellaGiustina et al. 2019). Globally averaged spectral properties include a generally blue slope over 98% of the surface with the other 2% having a flatter, less blue slope (Barucci et al. 2020). Temperature measurements refined the globally averaged thermal inertia as $\sim 350 \pm 20$ J m^{-2} K^{-1} s$^{-0.5}$ (DellaGiustina et al. 2019), also with spatial variations ranging from ∼180 to 700 J m^{-2} K^{-1} s$^{-0.5}$. These variations in albedo, spectral, and thermophysical properties collectively distinguish two major lithologies (DellaGiustina et al. 2020; Rozitis et al. 2020). Some boulders appear to be breccias that are composed of clasts bound by a dark matrix material and are consistent with impact origins (Walsh et al. 2019). Mass movement affecting the upper ∼10 meters of the surface, generally migrating towards the equator, is attributable to gravitational and rotational forces (Jawin et al. 2020). The asteroid exhibits the predicted spinning-top shape with a rotation period of ∼4.3 hours and a density of ∼1.19 g/cm^3 (Scheeres et al. 2019). This low density, in combination with the asteroid's shape and topography, is consistent with Bennu having a "rubble pile" structure with substantial (∼50%) macroporosity, if it is assumed to be comprised of primarily CI- or CM-like carbonaceous chondrite material (Barnouin et al. 2019); therefore, Bennu is probably reaccumulated material produced during a major asteroid collision in the Main Belt, from either the Eulalia or the "new" Polana asteroid family, between 0.7 and 2 Gyr ago (Walsh et al. 2019).

The globally averaged visible to near infrared (VNIR) spectral properties of Bennu (Fig. 2A) provide evidence that Bennu's parent asteroid underwent significant aqueous alteration to produce materials that are like the most aqueously altered carbonaceous chondrites and that the present surface is space weathered. The asteroid's blue slope has been attributed to the presence of fine-particulate magnetite and/or insoluble organic material; it is also commonly associated with larger-particle-size samples and possibly space weathering (Cloutis et al. 2011a, 2011b; Thompson et al. 2019). The dominantly blue spectral slope (0.4 - 0.7 μm) with relatively redder young craters on the surface are consistent with space weathering of a hydrated substrate that becomes increasingly blue with age (e.g., Clark et al. 2023). A spectral feature at 0.55 μm is suggestive of the presence of magnetite (Lauretta et al. 2019a, 2019b). The globally averaged spectrum also exhibits a so-called 3-μm hydration band that is geographically widespread (Hamilton et al. 2019) and relatively invariant in shape and depth (Barucci et al. 2020). The exact position of this feature is $\sim 2.74 \pm 0.01$ μm (Hamilton et al. 2019), consistent with low petrologic type CM chondrites containing abundant Mg-serpentine and "clay" minerals (Takir et al. 2013). The mean hydrogen content of the water and hydroxyl groups in the hydrated silicates is estimated as 0.71 ± 0.16 wt% (Praet et al. 2021). A complex of features around 3.4 μm are indicative of the presence of C-bearing materials, specifically carbonate minerals and organic compounds (Simon et al. 2020; Kaplan et al. 2020). These investigators found that organic signatures do not correlate in location with any other spectral or physical characteristics and variations may be due to differences in abundance, recent exposure, or space weathering. Identified carbonate phases include calcite, Mg-, Fe-bearing carbonate, and dolomite, and in many cases, these phases are associated with cm-wide, m-long relatively bright lineaments in boulders and are interpreted as veins that formed by extensive fluid circulation on spatial scales not evident in carbonaceous chondrites (Kaplan et al. 2020).

Thermal infrared (TIR) spectra exhibit evidence of substantial aqueous alteration on Bennu's parent body as well as evidence for variability in the presence of dust across the surface. The TIR spectra are subdivided into two spectral types, T1 and T2, based on a distinctly different spectral slope between 987 and 814 cm^{-1} ($\sim$10.1 and 12.3 μm). This difference is attributed to the variable presence of a minor amount of fine particulate dust rather than a compositional difference, where a $\sim$10 μm thick deposit of particles less than about 50 μm in diameter can produce such a difference in laboratory data (Hamilton et al. 2021; Graff 2003). Although T1 appears to be the least dusty, neither appears to be entirely free of dust. The spectral types share a common Christiansen feature location at $\sim$1090 cm^{-1} ($\sim$9.2 μm), and T1 is dominated by silicate stretching and bending modes with minima near 1000 cm^{-1} (10.0 μm) and 440 cm^{-1} ($\sim$22.7 μm), respectively, which are diagnostic of phyllosilicate minerals like the serpentine and saponite found in the most aqueously altered carbonaceous chondrites (Hamilton et al. 2021). Fundamental features of magnetite at 555 and 340 cm^{-1} ($\sim$18.0 and 29.4 μm) are observed (Hamilton et al. 2019) and lend credence to the interpretation that the relatively blue spectrum of Bennu at global scales is attributable, in part, to magnetite, analogous to predictions for other B-type asteroids (Yang and Jewitt 2010). A peak near 528 cm^{-1} (18.9 μm) is diagnostic of the proportions of phyllosilicate and anhydrous minerals (e.g., olivine and pyroxene); based on the presence of this feature, the bulk mineralogy of the surface of Bennu is dominated by Mg-phyllosilicates (>80 vol% of silicates) with less than 10 vol% anhydrous silicates (Hamilton et al. 2021). These fractions and the presence of magnetite are consistent with petrologic types 1 and 2 on the Van Schmus and Wood (1967) scale or $\leq$2.4 on the Rubin et al. (2007) scale. The absence of a distinct Mg-OH feature that is observed in carbonaceous chondrites near 625 cm^{-1} (16.0 μm) may be indicative of a non-Mg endmember (Fe-bearing) phyllosilicate composition, modest heating, disorder, a particle size effect, or obscuration by surface dust (Hamilton et al. 2021).

Spectral data also reveal the presence of exogenous materials on Bennu's surface. Relatively bright (13- to 40-sigma above the mean) boulders more than 1.5 m in diameter were observed in OCAMS images in the equatorial to southern latitudes and exhibit a deflection at 1 μm (DellaGiustina et al. 2021). Hyperspectral data in the areas around these boulders exhibit 1- and 2 μm pyroxene features that are inconsistent with chondritic material and instead compare favorably to Howardite-Eucrite-Diogenite (HED) meteorites that are linked to the asteroid 1 Vesta (e.g., Binzel and Xu 1993). These boulders, and others found subsequently, likely represent an infall of fragments of Vesta to Bennu or its parent body and imply that if Bennu is a uniform mixture of parent body and S-type materials, the mass of the exogenous materials is equivalent to a sphere 36-40 meters in diameter (DellaGiustina et al. 2021; LeCorre et al. 2021).

5.5 Sample Collection

Data collected during the asteroid operations phase of the mission were used to select a site on Bennu for sampling. Because of the unexpectedly rugged nature of the asteroid surface, a limited number of locations met requirements for spacecraft navigation and safety as well as having small enough particles to be collected by the TAGSAM. The selected sample site, "Nightingale", is located inside a crater named Hokioi in the northern hemisphere. Nightingale exhibits spectral compositional characteristics consistent with the majority of Bennu, appears to contain both lithologic types, and has a slightly redder normal albedo that is attributed to the crater being young and less space weathered than the majority of the asteroid's surface. The first sampling attempt, on 20 October 2020, was successful and revealed a number of interesting phenomena, which are summarized here and described in detail by Lauretta et al. (2022).

The TAGSAM is mounted on an arm with a telescoping spring; when TAGSAM contacted the surface, nitrogen gas was released, stirring the surface materials and guiding them into the sampling container. Spacecraft thrusters then maneuvered the spacecraft backwards away from the surface. Data collected during the TAG event show that TAGSAM penetrated 48.8 cm into the subsurface, affirming low-gravity experiments and modeling of regolith materials having little cohesion. Imaging before, during, and after the sample collection showed significant disturbance of surface materials and evidence for differing rock strengths. After sampling, the Hokioi crater region was revisited by the spacecraft instruments to enable analysis of changes in color, composition, topography, and thermophysical properties. The area where sampling occurred was observed to be darker and redder than before the sampling event, with a decrease in reflectance of about 5%, and is interpreted as having exposed even fresher materials enriched in organic compounds. Although several instruments reported the deposition of particulates on their optical surfaces, the surface could still be observed and no significant changes in spectral character or thermal properties at the sampling site were identified.

In the days after sample collection, imaging of the TAGSAM confirmed the presence of a substantial amount of material in the collection chamber, as well as particles captured in the stainless steel, hook-and-loop contact pads (designed to sample the uppermost, space-weathered surface). Imaging also revealed that small stones lodged between the mylar flap closure and the sample chamber were allowing the escape of sample material. The last estimate of sample mass prior to stowage of TAGSAM, eight days after sampling, was 250 ± 101 g, a loss of approximately 67 g since a mass estimate made two days after sampling (Lauretta et al. 2022).

5.6 Sample Curation Procedure

The Sample Return Capsule (SRC) successfully landed at the Utah Test and Training Range (UTTR) on 24 September 2023 and, within hours, was transported to a temporary clean

Fig. 13 Sample container of OSIRIS-REx Touch-and-Go Sample Acquisition Mechanism. Contact pads around ring (between screws) are ∼17 mm in diameter. In the inner ring, small pebbles can be seen at upper left holding open the mylar flap. Pebbles and fine particulates are visible on top of the mylar flap at right. The presence of coarse and fine particles confirms remote sensing interpretations. Image source: https://www.asteroidmission.org/?latest-news=nasas-osiris-rex-achieves-sample-mass-milestone

room facility at UTTR before being flown to Houston, TX and the clean curation facility at NASA Johnson Space Center (JSC). The opening of the SRC, disassembly of the TAGSAM, mass measurement and containerization of sample material, and processing and allocations of sample material for scientific analysis all occur at JSC. The TAGSAM head after removal from the SRC is shown in Fig. 13. The total mass returned was determined to be 121.6 g, more than double the mission's requirement to return 60 g (Lauretta et al. 2024), but less than the minimum mass (including uncertainty) of 149 g determined from the in-situ measurement. Because of the observed escape of material, the TAGSAM was stowed earlier than planned and the method used for the in-situ estimation differed from the planned sample mass measurement approach (Ma et al. 2024). The difference between the in situ and final (JSC) mass measurements may be attributable to sample loss during stowage (where the estimate of 2.2 g min^{-1} represented a lower bound; Lauretta et al. 2022), unforeseen factors affecting the spacecraft's measurement accuracy, or both.

5.7 General Characteristics of Return Sample

The OSIRIS-REx science team developed a hypothesis-driven Sample Analysis Plan (Lauretta et al. 2023) There are 12 top-level hypotheses that are fundamental to the mission science and relate to remote observations of Bennu prior to and during the operations phase of the mission and to our understanding of carbonaceous chondrites and the formation and

Fig. 14 Examples of stone morphologies observed in Bennu sample. (A) hummocky, (B) angular, and (C) mottled. After Lauretta et al. (2024)

evolution of the Solar System. The 12 hypotheses (and their sub-hypotheses) fall into five broad categories: 1) the fundamental nature of Bennu, 2) the pre-accretionary environment and formation of Bennu's parent asteroid, 3) the scattering of Bennu's parent body into the inner Solar System, geological activity, surface modification, and catastrophic disruption, 4) Bennu's formation, dynamical, and geological evolution, and 5) sample collection operations and Earth return. Published results to date from the earliest Bennu sample analyses focus on testing the first hypothesis (and its five sub-hypotheses) in the Sample Analysis Plan, which posits that remote sensing observations of Bennu accurately characterized its mineral, chemical and physical properties. In fact, the sample has both validated observations made at the asteroid and revealed new information; the summary below is derived from Lauretta et al. (2024).

Particles in the returned sample (Fig. 14) exhibit morphologies (angular and hummocky) that resemble and may be related to the morphologies of boulders observed on the asteroid's surface and likely represent the two lithologies identified there. Particles with a mottled appearance may be related to bright veins observed in some boulders. These three particle types exhibit different densities, where the hummocky stones are the least dense (1.55 ± 0.07 g/cm^3), followed by the angular stones (1.69 ± 0.04 g/cm^3), and the mottled stones being the densest (1.77 ± 0.04 g/cm^3). Preliminary assessment of how these density values translate to a macroporosity for Bennu are presented by Ryan et al. (2024). Those investigators determined Bennu's macroporosity is somewhat lower (20–33%) than prior estimates (25–50%, above); but do not change the interpretation that Bennu is a rubble pile asteroid. Although the albedo of Bennu materials at $\sim$1 μm is like that observed at Bennu, the spectrum of aggregated particles from 0.4 to 2.5 μm is considerably redder than global average Bennu; this may represent a lesser degree of space weathering, particle size effects, compositional variation, or a textural effect, but more analyses are needed to definitively establish the origin of this difference.

The material returned from Bennu is comprised dominantly of hydrous minerals (phyllosilicates) with lesser amounts of magnetite and carbonate as predicted. Indirectly inferred from remote sensing was the presence of Fe,Ni-sulfides, which are also present. Measurements of hydrogen content range from 0.86–0.98 wt% (including uncertainties), which just overlaps (at the low end) the remote sensing-derived value of 0.71 ± 0.16 wt%); the variance may be attributable to the differences in shape and position between the 3-μm hydration band in asteroid and sample spectra (Lauretta et al. 2024; Praet et al. 2021). Anhydrous phases such as olivine and pyroxene were not observed and were expected to be present at less than 10 vol%; forsteritic olivine and low-Ca pyroxene are found in the sample at generally minor abundances along with spinel, chromite, and ilmenite. The identification in the sample of Mg,Na- and Ca-phosphates at trace abundances and on particle surfaces was not anticipated and its origin as part of an evaporite sequence has since been described in detail (McCoy et al. 2025); whether phosphates are identifiable in Bennu spectra is a subject of further study.

No exogenic materials resembling HED meteorites have been identified in the sample yet, but there are strong similarities with aqueously altered carbonaceous chondrites, specifically the CI group and Tarda (C2-ung), in terms of modal mineralogy. Bulk elemental analysis of material from inside and outside the TAGSAM exhibit slightly different values for 10 elements that are known to be mobile and could support the open hydrothermal system proposed based on carbonate-bearing veins in Bennu boulders. Organic compounds are present as expected, with a high (relative to typical carbonaceous chondrites) total C content of 4.5 – 4.7 wt%. The samples are volatile rich, containing more carbon, nitrogen, and ammonia than most meteorites or samples of Ryugu (Glavin et al. 2025). Analysis of organic nanoglobules and PAHs does not show any indication of significant thermal alteration. Oxygen isotope analysis places Bennu near the chemically primitive CI and CY chondrites and Ryugu, but with a slightly higher $\Delta^{17}O$ value. Presolar materials (SiC and graphite) have been identified at abundances consistent with other unheated chondrite samples.

Bennu materials exhibit physical properties that suggest they may differ from carbonaceous meteorites, such as higher porosity and lower density. The bulk density of hummocky particles is relatively low, like that of Orgueil (CI1) and supports the idea that materials like these are unable to withstand impact with the Earth's atmosphere.

Overall, the materials returned from asteroid Bennu confirm predictions from ground-based telescopic and orbital remote sensing of a highly aqueously altered, C-enriched asteroid. The identification of hydrated phosphates indicates a more complex fluid chemistry than would otherwise be inferred. Although preliminary analysis of particles suggests they can be related to boulder types on Bennu that are thought to have different physical properties, more data are required to be certain of these links and produce updated values of properties such as asteroid macroporosity. Although the sampled surface of Bennu is extensively aqueously altered, it still retains some components from the protoplanetary disk. Even with the overall success of remote sensing observations, laboratory observations of the returned sample reveal detailed characteristics that would not have been known otherwise and demonstrates the extraordinary value of having samples of planetary bodies to help understand their complex histories.

6 MMX Mission

Visible to near-infrared reflectance spectra of Phobos and Deimos (Fig. 2A) are similar in reflectance and shape to those of D-type and T-type asteroids, which are probably made up of primitive carbonaceous chondrite-like material (Rivkin et al. 2002; Fraeman et al. 2012). The size of the two moons is also in the range of asteroids (Table 1E), and therefore they are thought to be captured asteroids. If the moons are captured asteroids, then they might have formed beyond the snowline near the young Jupiter and transferred to the inner region across the snowline. If this is valid, then Phobos and Deimos may be representatives of carriers that delivered water and organics to the inner planets. There are many hypotheses for the origin of the two moons. They are divided into two major models, each of which predicts an expected material composition of the moons; one is chondritic, and the other is non-chondritic. The major model of non-chondritic composition is giant impact model (e.g., Craddock 2011; Rosenblatt et al. 2016; Hyodo et al. 2017) in which a small body collided with Mars, making a large crater on the Mars and a dust disk around Mars that consisted of impact ejecta. The two moons are expected to have formed by accretion of the disk material and thus they are small and have near-circular equatorial orbits (Rosenblatt et al. 2016; Hyodo et al. 2017). Temperature, cooling rates, and chemical composition of the material in the disk are estimated based on thermodynamic condensation models (Pignatale et al. 2018).

The MMX mission aims to determine the origin of the Martian moons based on the remote sensing observation of Phobos and Deimos for about three years and the laboratory analysis of Phobos returned samples (Kuramoto et al. 2022). It is important to confirm from MMX's remote sensing spectroscopic observations of the whole surface and the sample collection sites that the samples to be brought back to Earth are representative material of Phobos (or Phobos-indigenous materials). The key observations to constrain the origin of Phobos and Deimos are: 1. to estimate the elemental abundances of Phobos globally (Lawrence et al. 2019) and determine whether it is chondritic or non-chondritic composition (including Martian components or not); 2. to perform global high-spatial resolution spectroscopy to obtain visible and near-infrared reflectance spectra to determine if hydrous minerals are ubiquitously distributed on Phobos and Deimos (Barucci et al. 2021; Kameda et al. 2021), and 3. to perform a variety of analyses on returned Phobos samples mirroring those done for the Hayabusa2 returned samples in order to determine the source material of Phobos and formation and alteration conditions based on mineralogy and isotopic composition (Fujiya et al. 2021). If these observations and analyses are successful, we should be able to determine the origin and formation process of the Martian moons within a few years from the sample return expected in 2031.

Funding Information This work was supported by JSPS KAKENHI Grant Number 24H00259 to TN. CE is supported by CNRS and the Centre National d'Etudes Spatiales (CNES) in the frame of the MIAMI-H2 project. MZ is supported by NASA's Laboratory Analysis of Returned Samples Program. VEH is supported by NASA under contract NNM10AA11C issued through the New Frontiers Program. AAF is supported by NASA's MMX Participating Scientist Program, and carried out research at the Jet Propulsion Laboratory, California Institute of Technology under a contract with NASA (80NM0018D0004).

Declarations

Competing Interests The authors declare that they have no conflict of interest.

References

Abe S, Mukai T, Hirata N, et al (2006b) Mass and local topography measurements of Itokawa by Hayabusa. Science 312:1344–1347. https://doi.org/10.1126/science.1126272

Abe M, Takagi Y, Kitazato K, et al (2006a) Near-infrared spectral results of asteroid Itokawa from the Hayabusa spacecraft. Science 312:1334–1338. https://doi.org/10.1126/science.1125718

Aléon-Toppani A, Brunetto R, Dionnet Z, et al (2024) Correlated IR-SEM-TEM studies of three different grains from Ryugu: from the initial material to post-accretional processes. Geochim Cosmochim Acta. https://doi.org/10.1016/j.gca.2024.02.006

Allen CW, Cox AN (2000) Allen's astrophysical quantities. Springer, Berlin, p 294

Amano K, Matsuoka M, Nakamura T, et al (2023) Reassigning CI chondrite parent bodies based on reflectance spectroscopy of samples from carbonaceous asteroid Ryugu and meteorites. Sci Adv 9:eadi3789. https://doi.org/10.1126/sciadv.adi3789

Aponte JC, Dworkin JP, Glavin DP, et al (2023) PAHs, hydrocarbons, and dimethylsulfides in asteroid Ryugu samples A0106 and C0107 and the Orgueil (CI1) meteorite. Earth Planets Space 75:28. https://doi.org/10.1186/s40623-022-01758-4

Arakawa M, Saiki T, Wada K, et al (2020) An artificial impact on the asteroid (162173) Ryugu formed a crater in the gravity-dominated regime. Science 368:67–71. https://doi.org/10.1126/science.aaz1701

Barnes JJ, Haenecour P, Smith LR, et al (2024) Coordinated analysis of phosphates in samples from asteroid (101955) Bennu. In: 55th Lunar and Planetary Science Conference (LPSC). Lunar and Planetary Institute. https://ntrs.nasa.gov/citations/20240000339

Barnouin OS, Daly MG, Palmer EE, et al (2019) Shape of (101955) Bennu indicative of a rubble pile with internal stiffness. Nat Geosci 12(4):247–252. https://doi.org/10.1038/s41561-019-0330-x

Barucci MA, Hasselmann PH, Praet A, et al (2020) OSIRIS-REx spectral analysis of (101955) Bennu by multivariate statistics. Astron Astrophys 637:L4. https://doi.org/10.1051/0004-6361/202038144

Barucci MA, Reess JM, Bernardi P, et al (2021) MIRS: an imaging spectrometer for the MMX mission. Earth Planets Space 73:1–28. https://doi.org/10.1186/s40623-021-01423-2

Bierhaus EB, Clark BC, Harris JW, et al (2018) The OSIRIS-REx spacecraft and the touch-and-go sample acquisition mechanism (TAGSAM). Space Sci Rev 2014:107. https://doi.org/10.1007/s11214-018-0521-6

Binzel RP, Xu S (1993) Chips off of asteroid 4 vesta: evidence for the parent body of basaltic achondrites meteorites. Science 260:186–191. https://doi.org/10.1126/science.260.5105.186

Binzel RP, Bus SJ, Burbine TH, et al (1996) Spectral properties of near-Earth asteroids: evidence for sources of ordinary chondrite meteorites. Science 273:946–948. https://doi.org/10.1126/science.273.5277.946

Bonal L, Quirico E, Montagnac G, et al (2024) The thermal history of Ryugu based on Raman characterization of Hayabusa2 samples. Icarus 408:115826. https://doi.org/10.1016/j.icarus.2023.115826

Bottke WF, Moorhead AV, Connolly HC Jr, et al (2020) Meteoroid impacts as a source of Bennu's particle ejection events. J Geophys Res Planets 125:e2019JE006282. https://doi.org/10.1029/2019JE006282

Britt DT, Tholen DJ, Bell JF, et al (1992) Comparison of asteroid and meteorite spectra: classification by principal component analysis. Icarus 99:153–166. https://doi.org/10.1016/0019-1035(92)90179-B

Brownlee DE, Tsou P, Anderson JD, et al (2003) Stardust: comet and interstellar dust sample return mission. J Geophys Res Planets 108:8111. https://doi.org/10.1029/2003JE002087

Brownlee DE, Horz F, Newburn RL (2004) Surface of young Jupiter family comet 81P/Wild 2: view from the Stardust spacecraft. Science 304:1764–1769. https://doi.org/10.1126/science.1097899

Bus SJ, Binzel RP (2002a) Phase II of the small main-belt asteroid spectroscopic survey: the observations. Icarus 158:106–145. https://doi.org/10.1006/icar.2002.6857

Bus SJ, Binzel RP (2002b) Phase II of the small main-belt asteroid spectroscopic survey: a feature-based taxonomy. Icarus 158:146–177. https://doi.org/10.1006/icar.2002.6856

Che S, Zega TJ (2023) Hydrothermal fluid activity on asteroid Itokawa. Nat Astron 7(9):1063–1069. https://doi.org/10.1038/s41550-023-02012-x

Chesley SR, French AS, Davis AB, et al (2020) Trajectory estimation for particles observed in the vicinity of (101955) Bennu. J Geophys Res Planets 125:e2019JE006363. https://doi.org/10.1029/2019JE006363

Christensen PR, Hamilton VE, Mehall GL, et al (2018) The OSIRIS-REx Thermal Emission Spectrometer (OTES) instrument. Space Sci Rev 214:87. https://doi.org/10.1007/s11214-018-0513-6

Clark BE, Sen A, Zou XD, et al (2023) Overview of the search for signs of space weathering on the low-albedo asteroid (101955) Bennu. Icarus 400:115563. https://doi.org/10.1016/j.icarus.2023.115563

Cloutis EA, Hiroi T, Gaffey MJ, Alexander CMO'D, Mann P (2011a) Spectral reflecctance properties of carbonacous chondrites: 1. CI chondrites. Icarus 212:180–209. https://doi.org/10.1016/j.icarus.2010.12.009

Cloutis EA, Hudon P, Hiroi T, Gaffey MJ, Mann P (2011b) Spectral reflectance properties of carbonaceous chondrites: 2. CM chondrites. Icarus 216:309–346. https://doi.org/10.1016/j.icarus.2011.09.009

Cloutis EA, Pietrasz VB, Kiddell C (2018) Spectral reflectance "deconstruction" of the Murchison CM2 carbonaceous chondrite and implications for spectroscopic investigations of dark asteroids. Icarus 305:203–224. https://doi.org/10.1016/j.icarus.2018.01.015

Craddock RA (2011) Are Phobos and Deimos the result of a giant impact? Icarus 211:1150–1161. https://doi.org/10.1016/j.icarus.2010.10.023

Daly MG, Barnouin OS, Dickinson C, et al (2017) The OSIRIS-REx laser altimeter (OLA) investigation and instrument. Space Sci Rev 212:899–924. https://doi.org/10.1007/s11214-017-0375-3

Dartois E, Kebukawa Y, Yabuta H, et al (2023) Chemical composition of carbonaceous asteroid Ryugu from synchrotron spectroscopy in the mid- to far-infrared of Hayabusa2-returned samples. Astron Astrophys 671:A2. https://doi.org/10.1051/0004-6361/202244702

Davidsson BJ, Gutiérrez PJ (2006) Non-gravitational force modeling of comet 81P/Wild 2: I. A nucleus bulk density estimate. Icarus 180:224–242. https://doi.org/10.1016/j.icarus.2005.07.023

De Gregorio BT, Cody GD, Stroud RM, et al (2024) Variations of organic functional chemistry in carbonaceous matter from the asteroid 162173 Ryugu. Nat Commun 15(1):7488. https://doi.org/10.1038/s41467-024-51731-w

DellaGiustina DN, Emery JP, et al (2019) Properties of rubble pile asteroid (101955) Bennu from OSIRIS-REx imaging and thermal analysis. Nat Astron 4:341–351. https://doi.org/10.1038/s41550-019-0731-1

DellaGiustina DN, et al (2020) Variations in color and reflectance on the surface of asteroid (101955) Bennu. Science 370:eabc3660. https://doi.org/10.1126/science.abc3660

DellaGiustina DN, et al (2021) Exogenic basalt on asteroid (101955) Bennu. Nat Astron 5:31–38. https://doi.org/10.1038/s41550-020-1195-z

DellaGiustina DN, Nolan MC, Polit AT, et al (2023) OSIRIS-APEX: an OSIRIS-REx extended mission to asteroid apophis. Planet Sci J 4:198. https://doi.org/10.3847/PSJ/acf75e

DeMeo FE, Binzel RP, Slivan SM (2009) An extension of the Bus asteroid taxonomy into the near-infrared. Icarus 202:160–180. https://doi.org/10.1016/j.icarus.2009.02.005

DeMeo FE, Burt BJ, Marsset M, et al (2022) Connecting asteroids and meteorites with visible and near-infrared spectroscopy. Icarus 380:114971. https://doi.org/10.1016/j.icarus.2018.01.015

Demura H, Kobayashi S, Nemoto E, et al (2006) Pole and global shape of 25143 Itokawa. Science 312:1347–1349. https://doi.org/10.1126/science.1126574

Dionnet Z, Rubino S, Aléon-Toppani A (2023) Three-dimensional multiscale assembly of phyllosilicates, organics, and carbonates in small Ryugu fragments. Meteorit Planet Sci. https://doi.org/10.1111/maps.14068

Duxbury TC, Newburn RL, Brownlee DE (2004) Comet 81P/Wild 2 size, shape, and orientation. J Geophys Res Planets 109:E12S02. https://doi.org/10.1029/2004JE002316

Ebihara M, Sekimoto S, Shira N, et al (2011) Neutron activation analysis of a particle returned from asteroid Itokawa. Science 333:1119–1121. https://doi.org/10.1126/science.1207865

Emery JP, Fernández YR, Kelley MSP, et al (2014) Thermal infrared observations and thermophysical characterization of OSIRIS-REx target asteroid (101955) Bennu. Icarus 234:17–35. https://doi.org/10.1016/j.icarus.2014.02.005

Ernst CM, Daly RT, Gaskell RW (2023) High-resolution shape models of Phobos and Deimos from stereophotoclinometry. Earth Planets Space 75:103. https://doi.org/10.1186/s40623-023-01814-7

Fraeman AA, Arvidson RE, Murchie SL, et al (2012) Analysis of disk-resolved OMEGA and CRISM spectral observations of Phobos and Deimos. J Geophys Res Planets 117:E00J15. https://doi.org/10.1029/2012JE004137

Fujiwara A, Kawaguchi J, Yeomans DK, et al (2006) The rubble-pile asteroid Itokawa as observed by Hayabusa. Science 312:1330–1334. https://doi.org/10.1126/science.1125841

Fujiya W, Furukawa Y, Sugahara H, et al (2021) Analytical protocols for Phobos regolith samples returned by the Martian Moons eXploration (MMX) mission. Earth Planets Space 73:1–24. https://doi.org/10.1186/s40623-021-01438-9

Fujiya W, Kawasaki N, Nagashima K, et al (2023) Carbonate record of temporal change in oxygen fugacity and gaseous species in asteroid Ryugu. Nat Geosci 16:675–682. https://doi.org/10.1038/s41561-023-01226-y

Glavin DP, Dworkin JP, Alexander CMO'D, et al (2025) Abundant ammonia, nitrogen-rich soluble organic matter in samples from asteroid (101955) Bennu. Nat Astron 9:199–210. https://doi.org/10.1038/s41550-024-02472-9

Graff TG (2003) Effects of dust coatings on visible, near-infrared, thermal emission and Mössbauer spectra: Implications for mineralogical remote sensing of Mars. MS Thesis, Arizona State University

Guo Y, Farquhar RW (2008) New Horizons Mission design. Space Sci Rev 140:49–74. https://doi.org/10.1007/s11214-007-9242-y

Hamilton VE, Simon AA, Christensen PR, et al (2019) Evidence for widespread hydrated minerals on asteroid (101955) Bennu. Nat Astron 3:332–340. https://doi.org/10.1038/s41550-019-0722-2

Hamilton VE, Christensen PR, Kaplan HH, et al (2021) Evidence for limited compositional and particle size variation on asteroid (101955) Bennu from thermal infrared spectroscopy. Astron Astrophys 650:A120. https://doi.org/10.1051/0004-6361/202039728

Hamilton VE, Keller LP, Haenecour P (2024) Mineralogy of Bennu from spectral analysis of the sample returned by OSIRIS-REx. In: 55th Lunar and Planetary Science Conference

Hashiguchi M, Aoki D, Fukushima K, et al (2023) The spatial distribution of soluble organic matter and their relationship to minerals in the asteroid (162173) Ryugu. Earth Planets Space 75:73. https://doi.org/10.1186/s40623-023-01792-w

Hirabayashi M, Hirata N, Kamata S, et al (2021) Hayabusa2 extended mission: rendezvous with 1998 KY26, one of the most common but unexplored near-Earth asteroids. In: Copernicus meetings. https://doi.org/10.5194/epsc2021-442

Hiroi T, Zolensky ME, Pieters CM, et al (1996) Thermal metamorphism of the C, G, B, and F asteroids seen from the 0.7 μm, 3 μm, and UV absorption strengths in comparison with carbonaceous chondrites. Meteorit Planet Sci 31:321–327. https://doi.org/10.1111/j.1945-5100.1996.tb02068.x

Hiroi T, Abe M, Kitazato K, et al (2006) Developing space weathering on the asteroid 25143 Itokawa. Nature 443:56–58. https://doi.org/10.1038/nature05073

Hopp T, Dauphas N, Abe Y, et al (2022) Ryugu's nucleosynthetic heritage from the outskirts of the Solar System. Sci Adv 8:eadd8141. https://doi.org/10.1126/sciadv.add8141

Hyodo R, Genda H, Charnoz S, Rosenblatt P (2017) On the impact origin of Phobos and Deimos. I. Thermodynamic and physical aspects. Astrophys J 845:125. https://doi.org/10.3847/1538-4357/aa81c4

Ishizaki T, Nagano H, Tanaka S (2023) Measurement of microscopic thermal diffusivity distribution for Ryugu sample by infrared lock-in periodic heating method. Int J Thermophys 44:51. https://doi.org/10.1007/s10765-023-03158-6

Ito M, Tomioka N, Uesugi M, et al (2022) A pristine record of outer Solar System materials from asteroid Ryugu's returned sample. Nat Astron 6:1163–1171. https://doi.org/10.1038/s41550-022-01745-5

Jawin ER, Walsh KJ, Barnouin OS, McCoy TJ, et al (2020) Global patterns of recent mass movement on asteroid (101955) Bennu. J Geophys Res Planets 125:e2020JE006475. https://doi.org/10.1029/2020JE006475

Jin Z, Bose M (2019) New clues to ancient water on Itokawa. Sci Adv 5(5):eaav8106. https://doi.org/10.1126/sciadv.aav8106

Jourdan F, Timms NE, Eroglu E, et al (2017) Collisional history of asteroid Itokawa. Geology 45:819–822. https://doi.org/10.1130/G39138.1

Jourdan F, Timms NE, Nakamura T, et al (2023) Rubble pile asteroids are forever. Proc Natl Acad Sci USA 120:e2214353120. https://doi.org/10.1073/pnas.2214353120

Kameda S, Ozaki M, Enya K, et al (2021) Design of telescopic nadir imager for geomorphology (TENGOO) and observation of surface reflectance by optical chromatic imager (OROCHI) for the Martian Moons Exploration (MMX). Earth Planets Space 73:1–14. https://doi.org/10.1186/s40623-021-01462-9

Kaplan HH, Lauretta DS, Simon AA, et al (2020) Bright carbonate veins on asteroid (101955) Bennu: implications for aqueous alteration history. Science 370:abc3557. https://doi.org/10.1126/science.abc3557

Kawaguchi JI, Fujiwara A, Uesugi T (2008) Hayabusa—its technology and science accomplishment summary and Hayabusa-2. Acta Astronaut 62:639–647. https://doi.org/10.1016/j.actaastro.2008.01.028

Kawasaki N, Nagashima K, Sakamoto N, et al (2022) Oxygen isotopes of anhydrous primary minerals show kinship between asteroid Ryugu and comet 81P/Wild2. Sci Adv 8:eade2067. https://doi.org/10.1126/sciadv.ade2067

Kebukawa Y, Quirico E, Dartois E, et al (2024) Infrared absorption spectra from organic matter in the asteroid Ryugu samples: some unique properties compared to unheated carbonaceous chondrites. Meteorit Planet Sci 59:1845–1858. https://doi.org/10.1111/maps.14064

Keller LP, Berger EL (2014) A transmission electron microscope study of Itokawa regolith grains. Earth Planets Space 66:1–7. https://doi.org/10.1186/1880-5981-66-71

Kikuchi S, Watanabe SI, Saiki T, et al (2020) Hayabusa2 landing site selection: surface topography of Ryugu and touchdown safety. Space Sci Rev 216:1–67. https://doi.org/10.1007/s11214-020-00737-z

Kikuchi S, Saiki T, Takei Y, et al (2021) Hayabusa2 pinpoint touchdown near the artificial crater on Ryugu: trajectory design and guidance performance. Adv Space Res 68:3093–3140. https://doi.org/10.1016/j.asr.2021.07.031

Kitazato K, Milliken RE, Iwata T, et al (2019) The surface composition of asteroid 162173 Ryugu from Hayabusa2 near-infrared spectroscopy. Science 364:272–275. https://doi.org/10.1126/science.aav7432

Kuninaka H, Nishiyama K, Funaki I, et al (2007) Powered flight of electron cyclotron resonance ion engines on Hayabusa explorer. J Propuls Power 23:544–551. https://doi.org/10.2514/1.25434

Kuramoto K, Kawakatsu Y, Fujimoto M, et al (2022) Martian Moons Exploration MMX: sample return mission to Phobos elucidating formation processes of habitable planets. Earth Planets Space 74:12. https://doi.org/10.1186/s40623-021-01545-7

Lantz C, Brunetto R, Barucci MA, et al (2017) Ion irradiation of carbonaceous chondrites: a new view of space weathering on primitive asteroids. Icarus 285:43–57. https://doi.org/10.1016/j.icarus.2016.12.019

Lauretta DS, Bartels AE, Barucci MA (2015) The OSIRIS-REx target asteroid (101955) Bennu: constraints on its physical, geological, and dynamical nature from astronomical observations. Meteorit Planet Sci 50:834–849. https://doi.org/10.1111/maps.12353

Lauretta DS, DellaGiustina DN, Bennett CA, et al (2019a) The unexpected surface of asteroid (101955) Bennu. Nature 568:55–60. https://doi.org/10.1038/s41586-019-1033-6
Lauretta DS, Hergenrother CW, et al (2019b) Episodes of particle ejection from the surface of the active asteroid (101955) Bennu. Science 366:eaay3544. https://doi.org/10.1126/science.aay3544
Lauretta DS, Adam CD, Allen AJ, et al (2022) Spacecraft sample collection and subsurface excavation of asteroid (101955) Bennu. Science 377:285–291. https://doi.org/10.1126/science.abm1018
Lauretta DS, Connolly HCC Jr, Grossman JN, Polit AT, the OSIRIS-REx Sample analysis team (2023) OSIRIS-REx Sample Analysis Plan – Revision 3.0. https://doi.org/10.48550/arXiv.2308.11794. arXiv: 2308.11794
Lauretta DS, Connolly HC, Aebersold JE, et al (2024) Asteroid (101955) Bennu in the laboratory: properties of the sample collected by OSIRIS-REx. Meteorit Planet Sci 59:2453–2486. https://doi.org/10.1111/maps.14227
Lawrence DJ, Peplowski PN, Beck AW, et al (2019) Measuring the elemental composition of Phobos: the Mars-moon exploration with GAmma rays and NEutrons (MEGANE) investigation for the Martian Moons eXploration (MMX) mission. Earth Space Sci 6:2605–2623. https://doi.org/10.1029/2019EA000811
LeCorre L, Reddy V, Bottke WF, et al (2021) Characterization of exogenic boulders on near-Earth asteroid (101955) Bennu from OSIRIS-REx color images. Planet Sci J 2:114. https://doi.org/10.3847/PSJ/abfbe2
Liu MC, McCain KA, Matsuda N, et al (2022) Incorporation of ^{16}O-rich anhydrous silicates in the protolith of highly hydrated asteroid Ryugu. Nat Astron 6:1172–1177. https://doi.org/10.1038/s41550-022-01762-4
Loizeau D, Pilorget C, Riu L, et al (2023) Constraints on Solar System early evolution by MicrOmega analysis of Ryugu carbonates. Nat Astron 7:391–397. https://doi.org/10.1038/s41550-022-01870-1
Ma C, Nakamura T, Mikouchi T, et al (2022) Mg-phosphate from asteroid Ryugu: an original H_2O-rich phase. In: 85th annual meeting of the meteoritical society, p 6134. https://www.hou.usra.edu/meetings/metsoc2022/pdf/6134.pdf
Ma H, Skeen M, Olds R, et al (2024) Alternative sample mass measurement technique for OSIRIS-REx sample collection phase. In: Sandnas M, Spencer DB (eds) Proceedings of the 44th annual American astronautical society guidance, navigation, and control conference, 2022. Advances in the astronautical sciences, vol 179. Springer, Cham, pp 1537–1549. https://doi.org/10.1007/978-3-031-51928-4_85
Marrocchi Y, Piralla M, Tissot FL (2023) Iron isotope constraints on the structure of the early Solar System. Astrophys J Lett 954:L27. https://doi.org/10.3847/2041-8213/acefd1
Masterson RA, Chodas M, Bayley L, et al (2018) Regolith X-ray imaging spectrometer (REXIS) aboard the OSIRIS-REx Asteroid Sample Return Mission. Space Sci Rev 214:48. https://doi.org/10.1007/s11214-018-0483-8
Mathurin J, Bejach L, Dartois E, et al (2024) AFM-IR nanospectroscopy of nanoglobule-like particles in Ryugu samples returned by the Hayabusa2 mission. Astron Astrophys 684:A198. https://doi.org/10.1051/0004-6361/202347435
Matsumoto T, Hasegawa S, Nakao S, et al (2018) Population characteristics of submicrometer-sized craters on regolith particles from asteroid Itokawa. Icarus 303:22–33. https://doi.org/10.1016/j.icarus.2017.12.017
Matsumoto T, Harries D, Langenhorst F, et al (2020) Iron whiskers on asteroid Itokawa indicate sulfide destruction by space weathering. Nat Commun 11:1117. https://doi.org/10.1038/s41467-020-14758-3
Matsumoto M, Matsuno J, Tsuchiyama A, et al (2024) Microstructural and chemical features of impact melts on Ryugu particle surfaces: records of interplanetary dust hit on asteroid Ryugu. Sci Adv 10:eadi7203. https://doi.org/10.1126/sciadv.adi7203
Matsuoka M, Kagawa EI, Amano K, et al (2023) Space weathering acts strongly on the uppermost surface of Ryugu. Commun Earth Environ 4:335. https://doi.org/10.1038/s43247-023-00991-3
McCain KA, Matsuda N, Liu MC, et al (2023) Early fluid activity on Ryugu inferred by isotopic analyses of carbonates and magnetite. Nat Astron 7:309–317. https://doi.org/10.1038/s41550-022-01863-0
McCoy TJ, Russell SS, Zega TJ, et al (2025) An evaporite sequence from ancient brine recorded in Bennu samples. Nature 637:1072–1077. https://doi.org/10.1038/s41586-024-08495-6
Michel P, Richardson DC (2013) Collision and gravitational reaccumulation: possible formation mechanism of the asteroid Itokawa. Astron Astrophys 554:L1. https://doi.org/10.1051/0004-6361/201321657
Michikami T, Honda C, Miyamoto H, et al (2019) Boulder size and shape distributions on asteroid Ryugu. Icarus 331:179–191. https://doi.org/10.1016/j.icarus.2019.05.019
Mikouchi T, Yoshida H, Hayashi S, et al (2024) Amorphous magnesium phosphate in asteroid Ryugu samples and CI chondrites. In: 86th annual meeting of the meteoritical society, p 6319. https://www.hou.usra.edu/meetings/metsoc2024/pdf/6319.pdf
Miyamoto H, Yano H, Scheeres DJ, et al (2007) Regolith migration and sorting on asteroid itokawa. Science 316:1011–1014. https://doi.org/10.1126/science.1134390

Morota T, Sugita S, Cho Y, et al (2020) Sample collection from asteroid (162173) Ryugu by Hayabusa2: implications for surface evolution. Science 368:654–659. https://doi.org/10.1126/science.aaz6306

Muirhead BK, Nicholas A, Umland J (2020) Mars sample return mission concept status. In: 2020 IEEE Aerospace Conference. IEEE, pp 1–8. https://doi.org/10.1109/AERO47225.2020.9172609

Nagao K, Okazaki R, Nakamura T, et al (2011) Irradiation history of Itokawa regolith material deduced from noble gases in the Hayabusa samples. Science 333:1128–1131. https://doi.org/10.1126/science.1207785

Nakamura AM, Michikami T, Hirata N, et al (2008) Impact process of boulders on the surface of asteroid 25143 Itokawa—fragments from collisional disruption. Earth Planets Space 60:7–12. https://doi.org/10.1186/BF03352756

Nakamura T, Noguchi T, Tsuchiyama A, et al (2008) Chondrulelike objects in short-period comet 81P/Wild 2. Science 321:1664–1667. https://doi.org/10.1126/science.1160995

Nakamura T, Noguchi T, Tanaka M, et al (2011) Itokawa dust particles: a direct link between S-type asteroids and ordinary chondrites. Science 333:1113–1116. https://doi.org/10.1126/science.1207758

Nakamura E, Makishima A, Moriguti T, et al (2012) Space environment of an asteroid preserved on micrograins returned by the Hayabusa spacecraft. Proc Natl Acad Sci USA 109:E624–E629. https://doi.org/10.1073/pnas.111623610

Nakamura T, Nakato A, Ishida H, et al (2014) Mineral chemistry of MUSES-C regio inferred from analysis of dust particles collected from the first-and second-touchdown sites on asteroid Itokawa. Meteorit Planet Sci 49:215–227. https://doi.org/10.1111/maps.12247

Nakamura T, Ikeda H, Kouyama T, et al (2021) Science operation plan of Phobos and Deimos from the MMX spacecraft. Earth Planets Space 73:227. https://doi.org/10.1186/s40623-021-01546-6

Nakamura E, Kobayashi K, Tanaka R, et al (2022) On the origin and evolution of the asteroid Ryugu: a comprehensive geochemical perspective. Proc Jpn Acad Ser B 98:227–282. https://doi.org/10.2183/pjab.98.015

Nakamura T, Matsumoto M, Amano K, et al (2022) Formation and evolution of carbonaceous asteroid Ryugu: direct evidence from returned samples. Science 379:eabn8671. https://doi.org/10.1126/science.abn8671

Nakashima D, Nakamura T, Zhang M, et al (2023) Chondrule-like objects and Ca-Al-rich inclusions in Ryugu may potentially be the oldest Solar System materials. Nat Commun 14:532. https://www.nature.com/articles/s41467-023-36268-8

Naraoka H, Takano Y, Dworkin JP, et al (2023) Soluble organic molecules in samples of the carbonaceous asteroid (162173) Ryugu. Science 379:eabn9033. https://doi.org/10.1126/science.abn9033

Nguyen AN, Mane P, Keller LP, et al (2023) Abundant presolar grains and primordial organics preserved in carbon-rich exogenous clasts in asteroid Ryugu. Sci Adv 9:eadh1003. https://doi.org/10.1126/sciadv.adh1003

Ninomiya K, Osawa T, Terada K, et al (2024) Quantification of bulk elemental composition for C-type asteroid Ryugu samples with nondestructive elemental analysis using muon beam. Meteorit Planet Sci 59:2044–2057. https://doi.org/10.1111/maps.14135

Nishiyama G, Kawamura T, Namiki N, et al (2021) Simulation of seismic wave propagation on asteroid Ryugu induced by the impact experiment of the Hayabusa2 mission: limited mass transport by low yield strength of porous regolith. J Geophys Res Planets 126:e2020JE006594. https://doi.org/10.1029/2020JE006594

Noguchi T, Tsuchiyama A, Hirata N, et al (2010) Surface morphological features of boulders on asteroid 25143 Itokawa. Icarus 206:319–326. https://doi.org/10.1016/j.icarus.2009.09.002

Noguchi T, Nakamura T, Kimura M, et al (2011) Incipient space weathering observed on the surface of Itokawa dust particles. Science 333:1121–1125. https://doi.org/10.1126/science.1207794

Noguchi T, Kimura M, Hashimoto T, et al (2014a) Space weathered rims found on the surfaces of the Itokawa dust particles. Meteorit Planet Sci 49:188–214. https://doi.org/10.1111/maps.12111

Noguchi T, Kimura M, Hashimoto T, et al (2014b) Sylvite and halite on particles recovered from 25143 Itokawa: a preliminary report. Meteorit Planet Sci 49:1305–1314. https://doi.org/10.1111/maps.12333

Noguchi T, Matsumoto T, Miyake A, et al (2023) A dehydrated space-weathered skin cloaking the hydrated interior of Ryugu. Nat Astron 7:170–181. https://doi.org/10.1038/s41550-022-01841-6

Oba Y, Koga T, Takano Y, et al (2023) Uracil in the carbonaceous asteroid (162173) Ryugu. Nat Commun 14:1292. https://doi.org/10.1038/s41467-023-36904-3

Okada T, Fukuhara T, Tanaka S (2020) Highly porous nature of a primitive asteroid revealed by thermal imaging. Nature 579:518–522. https://doi.org/10.1038/s41586-020-2102-6

Okazaki R, Marty B, Busemann H, et al (2022b) Noble gases and nitrogen in samples of asteroid Ryugu record its volatile sources and recent surface evolution. Science 379:eabo0431. https://doi.org/10.1126/science.abo0431

Okazaki R, Miura YN, Takano Y, et al (2022c). First asteroid gas sample delivered by the Hayabusa2 mission: a treasure box from Ryugu. Sci Adv 8:eabo7239. https://doi.org/10.1126/sciadv.abo7239

Okazaki R, Yamanouchi S, Shimada K, et al (2022a) Methods and tools for handling, transportation, weighing, and pelletization applied to the initial analysis of volatile components in the Hayabusa2 samples. Earth Planets Space 74:190. https://doi.org/10.1186/s40623-022-01747-7

Osawa T, Nagasawa S, Ninomiya K, et al (2023) Development of nondestructive elemental analysis system for Hayabusa2 samples using muonic X-rays. ACS Earth Space Chem 7:699–711. https://doi.org/10.1021/acsearthspacechem.2c00303

Ostro SJ, Benner LA, Nolan MC, et al (2004) Radar observations of asteroid 25143 Itokawa (1998). Meteorit Planet Sci 39:407–424. https://doi.org/10.1111/j.1945-5100.2004.tb00102.x

Paquet M, Moynier F, Yokoyama T, et al (2023) Contribution of Ryugu-like material to Earth's volatile inventory by Cu and Zn isotopic analysis. Nat Astron 7:182–189. https://doi.org/10.1038/s41550-022-01846-1

Park J, Turrin BD, Herzog GF, et al (2015) 40Ar/39Ar age of material returned from asteroid 25143 Itokawa. Meteorit Planet Sci 50:2087–2098. https://doi.org/10.1111/maps.12564

Parker ET, Chan QHS, Glavin DP, Dworkin JP (2022) Non-protein amino acids identified in carbon-rich Hayabusa particles. Meteorit Planet Sci 57:776–793. https://doi.org/10.1111/maps.13794

Parker ET, McLain HL, Glavin DP, et al (2023) Extraterrestrial amino acids and amines identified in asteroid Ryugu samples returned by the Hayabusa2 mission. Geochim Cosmochim Acta 347:42–57. https://doi.org/10.1016/j.gca.2023.02.017

Phan VTH, Beck P, Rebois R, et al (2024) In situ investigation of an organic micro-globule and its mineralogical context within a Ryugu "sand" grain. Meteorit Planet Sci 59:1983–2001. https://doi.org/10.1111/maps.14122

Pignatale FC, Charnoz S, Rosenblatt P, et al (2018) On the impact origin of Phobos and Deimos. III. Resulting composition from different impactors. Astrophys J 853:118. https://doi.org/10.3847/1538-4357/aaa23e

Pilorget C, Okada T, Hamm V, et al (2022) First compositional analysis of Ryugu samples by the MicrOmega hyperspectral microscope. Nat Astron 6:221–225. https://doi.org/10.1038/s41550-021-01549-z

Pilorget C, Baklouti D, Bibring JP, et al (2024) Phosphorus-rich grains in Ryugu samples with major biochemical potential. Nat Astron 8:1529–1535. https://doi.org/10.1038/s41550-024-02366-w

Pinilla-Alonso N, de León J, Walsh KJ, et al (2016) Portrait of the Polana-Eulalia family complex: surface homogeneity revealed from near-infrared spectroscopy. Icarus 274:231–248. https://doi.org/10.1016/j.icarus.2016.03.022

Potiszil C, Ota T, Yamanaka M (2023) Insights into the formation and evolution of extraterrestrial amino acids from the asteroid Ryugu. Nat Commun 14:1482. https://doi.org/10.1038/s41467-023-37107-6

Praet A, Barucci MA, Clark BE, et al (2021) Hydrogen abundance estimation and distribution on (101955) Bennu. Icarus 363:114427. https://doi.org/10.1016/j.icarus.2021.114427

Quirico E, Bonal L, Kebukawa Y, et al (2024) Compositional heterogeneity of insoluble organic matter extracted from asteroid Ryugu samples. Meteorit Planet Sci 59:1907–1924. https://doi.org/10.1111/maps.14097

Reuter DC, Simon AA, Hair J, et al (2018) The OSIRIS-REx Visible and InfraRed Spectrometer (OVIRS): spectral maps of the asteroid Bennu. Space Sci Rev 214:1–22. https://doi.org/10.1007/s11214-018-0482-9

Riu L, Pilorget C, Milliken R, et al (2021) Spectral characterization of the craters of Ryugu as observed by the NIRS3 instrument on-board Hayabusa2. Icarus 357:114253. https://doi.org/10.1016/j.icarus.2020.114253

Rivkin AS, Brown RH, Trilling DE, et al (2002) Near-infrared spectrophotometry of Phobos and Deimos. Icarus 156:64–75. https://doi.org/10.1006/icar.2001.6767

Rizk B, d'Aubigny CD, Golish D, et al (2018) OCAMS: the OSIRIS-REx camera suite. Space Sci Rev 214:26. https://doi.org/10.1007/s11214-017-0460-7

Rosenblatt P, Charnoz S, Dunseath KM, et al (2016) Accretion of Phobos and Deimos in an extended debris disc stirred by transient moons. Nat Geosci 9:581–583. https://doi.org/10.1038/ngeo2742

Roskosz M, Beck P, Viennet JC, et al (2024) The iron oxidation state of Ryugu samples. Meteorit Planet Sci 59:1925–1946. https://doi.org/10.1111/maps.14098

Rozitis B, Emery JP, Siegler MA, et al (2020) Implications for ice stability and particle ejection from high-resolution temperature modeling of asteroid (101955) Bennu. J Geophys Res Planets 125:e2019JE006323. https://doi.org/10.1029/2019JE006323

Rubin AE, Trigo-Rodriguez JM, Huber H, Wasson JT (2007) Progressive aqueous alteration of CM carbonaceous chondrites. Geochim Cosmochim Acta 71:2361–2382. https://doi.org/10.1016/j.gca.2007.02.008

Rubincam DP (2000) Radiative spin-up and spin-down of small asteroids. Icarus 148:2–11. https://doi.org/10.1006/icar.2000.6485

Ryan AJ, Ballouz RJ, Macke HC, et al (2024) Physical and thermal properties of OSIRIS-REx samples: insight into the evolution of Bennu and its regolith. Lunar Planet Sci LV:1594.

Saito J, Miyamoto H, Nakamura R, et al (2006) Detailed images of asteroid 25143 Itokawa from Hayabusa. Science 312:1341–1344. https://doi.org/10.1126/science.1125722

Sawada H, Okazaki R, Tachibana S, et al (2017) Hayabusa2 sampler: collection of asteroidal surface material. Space Sci Rev 208:81–106. https://doi.org/10.1007/s11214-017-0338-8

Scheeres DJ, Abe M, Yoshikawa M, et al (2007) The effect of YORP on Itokawa. Icarus 188:425–429. https://doi.org/10.1016/j.icarus.2006.12.014

Scheeres DJ, Hesar SG, Tardivel S, et al (2016) The geophysical environment of Bennu. Icarus 276:116–140. https://doi.org/10.1016/j.icarus.2016.04.013

Scheeres DJ, McMahon JW, French AS, et al (2019) The dynamic geophysical environment of (101955) Bennu based on OSIRIS-REx measurements. Nat Astron 3:352–361. https://doi.org/10.1038/s41550-019-0721-3

Schmitt-Kopplin P, Hertkorn N, Harir M, et al (2023) Soluble organic matter molecular atlas of Ryugu reveals cold hydrothermalism on C-type asteroid parent body. Nat Commun 14:6525. https://doi.org/10.1038/s41467-023-42075-y

Shimaki Y, Senshu H, Sakatani N, et al (2020) Thermophysical properties of the surface of asteroid 162173 Ryugu: infrared observations and thermal inertia mapping. Icarus 348:113835. https://doi.org/10.1016/j.icarus.2020.113835

Simon AA, Kaplan HH, Hamilton VE, et al (2020) Widespread carbon-bearing materials on near-Earth asteroid (101955) Bennu. Science 370:eabc3522. https://doi.org/10.1126/science.abc3522

Spitzer F, Kleine T, Burkhardt C, et al (2025) The Ni isotopic composition of Ryugu reveals a common accretion region for carbonaceous chondrites. Sci Adv 10:eadp2426. https://doi.org/10.1126/sciadv.adp24

Stöffler D, Keil K, Scott ERD (1991) Shock metamorphism of ordinary chondrites. Geochim Cosmochim Acta 55:3845–3867. https://doi.org/10.1016/0016-7037(91)90078-J

Sugita S, Honda R, Morota T, et al (2019) The geomorphology, color, and thermal properties of Ryugu: implications for parent-body processes. Science 364:eaaw0422. https://doi.org/10.1126/science.aaw0422

Sunagawa S, Fujiya W, Kawasaki N, et al (2024) Update on the ^{53}Mn-^{53}Cr ages of dolomite in the Ivuna CI chondrite and asteroid Ryugu sample. Geochim Cosmochim Acta 382:40–50. https://doi.org/10.1016/j.gca.2024.08.013

Tachibana S, Sawada H, Okazaki R, et al (2022) Pebbles and sand on asteroid (162173) Ryugu: in situ observation and particles returned to Earth. Science 375:1011–1016. https://doi.org/10.1126/science.abj8624

Takir D, Emery JP, McSween HY Jr, et al (2013) Nature and degree of aqueous alteration in CM and CI carbonaceous chondrites. Meteorit Planet Sci 48:1618–1637. https://doi.org/10.1111/maps.12171

Tanaka R, Ratnayake DM, Ota T, et al (2024) Unraveling the Cr isotopes of Ryugu: an accurate aqueous alteration age and the least thermally processed solar system material. Astrophys J 965:52. https://doi.org/10.3847/1538-4357/ad276a

Tatsumi E, Domingue D, Schröder S, et al (2020) Global photometric properties of (162173) Ryugu. Astron Astrophys 639:A83. https://doi.org/10.1051/0004-6361/201937096

Terada K, Sano Y, Takahata N, et al (2018) Thermal and impact histories of 25143 Itokawa recorded in Hayabusa particles. Sci Rep 8:11806. https://doi.org/10.1038/s41598-018-30192-4

Tholen DJ (1989) Asteroid taxonomic classifications. In: Binzel RP, et al (eds) Asteroids II. University of Arizona Press, Tucson, pp 1139–1150

Thompson MS, Christoffersen R, Zega TJ, et al (2014) Microchemical and structural evidence for space weathering in soils from asteroid Itokawa. Earth Planets Space 66:89. https://doi.org/10.1186/1880-5981-66-89

Thompson MS, Loeffler MG, Morris RV, et al (2019) Spectral and chemical effects of simulated space weathering of the Murchison CM2 carbonaceous chondrite. Icarus 319:499–511. https://doi.org/10.1016/j.icarus.2018.09.022

Tsuchiyama A, Uesugi M, Matsushima T, et al (2011) Three-dimensional structure of Hayabusa samples: origin and evolution of Itokawa regolith. Science 333:1125–1128. https://doi.org/10.1126/science.1207807

Tsuda Y, Yoshikawa M, Abe M, et al (2013) System design of the Hayabusa 2—asteroid sample return mission to 1999 JU3. Acta Astronaut 91:356–362. https://doi.org/10.1016/j.actaastro.2013.06.028

Tsuda Y, Saiki T, Terui F, et al (2020) Hayabusa2 mission status: landing, roving and cratering on asteroid Ryugu. Acta Astronaut 171:42–54. https://doi.org/10.1016/j.actaastro.2020.02.035

Van Schmus WR, Wood JA (1967) A chemical-petrologic classification for the chondritic meteorites. Geochim Cosmochim Acta 31:747–765

Veverka J, Klaasen K, A'Hearn M, et al (2013) Return to comet Tempel 1: overview of Stardust-NExT results. Icarus 222:424–435. https://doi.org/10.1016/j.icarus.2012.03.034

Viennet JC, Roskosz M, Nakamura T, et al (2023) Interaction between clay minerals and organics in asteroid Ryugu. Geochem Perspect Lett 25:8–12. https://doi.org/10.7185/geochemlet.2307

Vilas F, Gaffey MJ (1989) Phyllosilicate absorption features in main-belt and outer-belt asteroid reflectance spectra. Science 246:790–792. https://doi.org/10.1126/science.246.4931.790
Wakita S, Nakamura T, Ikeda T, et al (2014) Thermal modeling for a parent body of Itokawa. Meteorit Planet Sci 49:228–236. https://doi.org/10.1111/maps.12174
Walsh KJ (2018) Rubble pile asteroids. Annu Rev Astron Astrophys 56:593–624. https://doi.org/10.1146/annurev-astro-081817-052013
Walsh KJ, Jawin ER, Ballouz R-L, et al (2019) Craters, boulders and regolith of (101955) Bennu indicative of an old and dyn amic surface. Nat Geosci 4:242–246. https://doi.org/10.1038/s41561-019-0326-6
Watanabe S, Tsuda Y, Yoshikawa M, et al (2017) Hayabusa2 mission overview. Space Sci Rev 208:3–16. https://doi.org/10.1007/s11214-017-0377-1
Watanabe S, Hirabayashi M, Hirata N, et al (2019) Hayabusa2 arrives at the carbonaceous asteroid 162173 Ryugu—a spinning top–shaped rubble pile. Science 364:268–272. https://doi.org/10.1126/science.aav8032
Wibben DR, Levine A, Russell JR, et al (2024) OSIRIS-REx Earth return & entry: targeting strategy and maneuver performance. In: 46th annual AAS guidance, navigation and control (GN&C) conference
Williams B, Antreasian P, Carranza E, et al (2018) OSIRIS-REx flight dynamics and navigation design. Space Sci Rev 214:1–43. https://doi.org/10.1007/s11214-018-0501-x
Yabuta H, Cody GD, Engrand C, et al (2023) Macromolecular organic matter in samples of the asteroid (162173) Ryugu. Science 379:eabn9057. https://doi.org/10.1126/science.abn9057
Yada T, Fujimura A, Abe M, et al (2014) Hayabusa-returned sample curation in the planetary material sample curation facility of JAXA. Meteorit Planet Sci 49:135–153. https://doi.org/10.1111/maps.12027
Yada T, Abe M, Okada T, et al (2022) Preliminary analysis of the Hayabusa2 samples returned from C-type asteroid Ryugu. Nat Astron 6:214–220. https://doi.org/10.1038/s41550-021-01550-6
Yamaguchi A, Tomioka N, Ito M, et al (2023) Insight into multi-step geological evolution of C-type asteroids from Ryugu particles. Nat Astron 7:398–405. https://doi.org/10.1038/s41550-023-01925-x
Yang B, Jewitt D (2010) Identification of magnetite in B-type asteroids. Astron J 140:692–698. https://doi.org/10.1088/0004-6256/140/3/692
Yano H, Kubota T, Miyamoto H, et al (2006) Touchdown of the Hayabusa spacecraft at the muses sea on Itokawa. Science 312:1350–1353. https://doi.org/10.1126/science.1126164
Yokoyama T, Nagashima K, Nakai I, et al (2022) Samples returned from the asteroid Ryugu are similar to Ivuna-type carbonaceous meteorites. Science 379:eabn7850. https://doi.org/10.1126/science.abn7850
Yoshikawa M, Kawaguchi J, Fujiwara A, et al (2021) The Hayabusa mission. In: Longobardo A (ed) Sample return missions. Elsevier, Amsterdam, pp 123–146. https://doi.org/10.1016/B978-0-12-818330-4.00006-9
Yurimoto H, Abe KI, Abe M, et al (2011) Oxygen isotopic compositions of asteroidal materials returned from Itokawa by the Hayabusa mission. Science 333:1116–1119. https://doi.org/10.1126/science.1207776
Zeichner SS, Aponte JC, Bhattacharjee S, et al (2023) Polycyclic aromatic hydrocarbons in samples of Ryugu formed in the interstellar medium. Science 382:1411–1416. https://doi.org/10.1126/science.adg6304
Zhang P, Zhang G, Yan X, et al (2024) (469219) Kamo'oalewa, a Space Weathering Matured Small NEA: Target of the Tianwen-2 Sample-Return Mission. Under Review, https://doi.org/10.21203/rs.3.rs-4281601/v1
Zolensky ME, Zega TJ, Yano H, et al (2006) Mineralogy and petrology of comet 81P/Wild 2 nucleus samples. Science 314:1735–1739. https://doi.org/10.1126/science.1135842
Zolensky M, Martinez J, Sitzman S, et al (2022) Measuring the shock stage of Itokawa and asteroid regolith grains by electron backscattered diffraction, optical petrography, and synchrotron X-ray diffraction. Meteorit Planet Sci 57:1060–1078. https://doi.org/10.1111/maps.13808
Zolensky M, Engrand C, Nakamura T, et al (2024) Composition and mineralogy of nuclei material of short period comets revealed by recent spacecraft missions. Space Sci Rev 220:79. https://doi.org/10.1007/s11214-024-01111-z

Space Science Reviews (2024) 220:79
https://doi.org/10.1007/s11214-024-01111-z

Composition and Mineralogy of Nuclei Material of Short Period Comets Revealed by Recent Spacecraft Missions

Michael Zolensky[1] · Cecile Engrand[2] · Tomoki Nakamura[3] · Denton Ebel[4]

Received: 3 April 2024 / Accepted: 10 September 2024 / Published online: 2 October 2024

Abstract

We summarize mineralogical and compositional results from lab and in-situ spacecraft investigations of samples of comets 81P/Wild 2 and 67P/Churyumov-Gerasimenko which provide constraints to early solar system evolution. Wild 2 cometary solids were mineralogically similar to anhydrous chondritic Interplanetary Dust Particles with the exception that GEMS have not between reliably reported in the former. Important similarities include the very wide range of olivine and pyroxene bulk compositions (including minor elements), which probably reflects a wide range of sampled grain formation environments. The abundance of presolar materials in the Wild 2 samples is also approximately equal to that in anhydrous chondritic Interplanetary Dust Particles rather than being dramatically higher. High-temperature aggregates including chondrules (mainly type II), CAI and AOAs are also present in Wild 2, but probably at less than 10% overall total abundance. There was a pre-mission expectation that comet solids would be dominated by either non-crystalline materials, nanoscale crystalline materials, or presolar grains. None of these expectations proved to be correct. Organic matter was only well preserved in a few examined samples. These exhibit complex organic bonds, including mainly alkenes, aromatic C = C and carboxyl C = O, in a variety of textures including nanoglobules. The organic matter in Wild 2 samples show spectral similarities with that of insoluble organic matter (IOM) from carbonaceous chondrites. The chronology of the collected solids has not been well established, but there is no evidence for captured live ^{26}Al, which could indicate accretion after this nuclide had effectively completely decayed. The implications of these major results are that Wild 2 sampled widely across the entire solar nebula, much wider than any other known astromaterials, potentially at a relatively late time compared to the oldest available astromaterials. A final caution is that the captured Wild 2 materials described here possibly derived from a small number of interior regions of a single Jupiter Family comet. There is also limited mineralogical information from comet 67P provided by the Rosetta Mission. The investigated 67P dust particles also contained about 45 weight% of solid organic matter, and is similar to the IOM extracted from carbonaceous chondrites. The H/C ratio of 67P IOM is about twice that of most carbonaceous chondrite IOMs and about 5 times that measured in asteroid Ryugu whole-rock samples, but lower than the H/C ratio of volatile species measured in the 67P coma. This high H/C ratio in the coma species is in turn compatible with that of the soluble organic matter in investigated carbonaceous chondrites such as Murchison. The average N/C ratio of 67P dust particles is about one order of magnitude lower than the

Extended author information available on the last page of the article

solar photosphere value, but it is compatible with that of primitive carbonaceous chondrite meteorites and of Ryugu IOM.

1 Introduction

Thanks to the Stardust mission, we can now perform mineralogical and petrographic analyses of samples derived directly from the Jupiter-family comet 81P/Wild 2. The Stardust spacecraft captured comet coma dust particles close to Wild 2, relative velocity 6.2 km/sec, at 1.86 AU from the sun. A few definitions should be given here. A 'particle' is usually defined as a small object that behaves as a single unit in terms of its transport and physical properties. A 'grain', or 'crystallite' is a smaller domain of solid matter such as a single crystal. Thus, a particle can contain one or it can contain numerous grains. An 'aggregate' always contains multiple grains. The inner dust coma encountered the spacecraft at 235 km from the nucleus, so particles were captured within hours of their release from the nucleus. Dust collection was accomplished using $2 \times 4 \times 3$ cm tiles of low-density silica aerogel, with a graded density from <0.01 g/cm3 at the surface to 0.05 g/cm^3. The collector was returned to Earth on January 15, 2006. Wild 2 coma particles have been characterized with regards to chemical composition, mineralogy, crystallinity and crystal structure, and grain size. To varying degree all these characteristics have been significantly affected by the capture process at 6.2 km/sec into silica aerogel. Based on imaging and data from the spacecraft Whipple shield, the coma particles sampled by the Stardust Mission may have derived from just a few active jets, which would have sampled only a small volume of the interior of the comet nucleus (Brownlee et al. 2006). However, the available data are insufficient to definitely establish this possibility - we will just never know. Excellent earlier reviews of Wild 2 results are available (Hanner and Zolensky 2010; Brownlee et al. 2012; Brownlee 2014; Westphal et al. 2017).

The Rosetta mission studied comet 67P/Churyumov-Gerasimenko between August 2014 and September 2016. Three dust analyzers were onboard Rosetta that could characterize dust in the size range from ~50 nm to mm sizes. MIDAS (Micro-Imaging Dust Analysis System) was an atomic force microscope that could image the structure of the cometary dust with a resolution down to ~8 nm). GIADA (Grain Impact Analyser and Dust Accumulator) was a dust analyzer that detected several tens of grains with masses between ~10 μg to ~0.1 mg. COSIMA (COmetary Secondary Ion Mass Analyser) was a time-of-flight mass spectrometer, which performed analyses of dust collected on 1 cm^2 targets. The resolution of COSIMA was ~25 μm, providing the composition of cometary dust particles at the sub-mm scale.

2 Chemical Properties

2.1 Elemental Composition

The only direct measurement of the elemental composition of comet coma dust prior to the Stardust Mission came from impact-ionization mass-spectrometers (Kissel 1986) that encountered comet Halley on the Giotto and the two VEGA spacecraft in 1986. These measurements were within a factor of two of the CI meteorite abundances (Jessberger 1999), although the accuracy of these results is limited by being based on a total sample mass of less than a nanogram, and the fact that the spectrometers could never be calibrated because

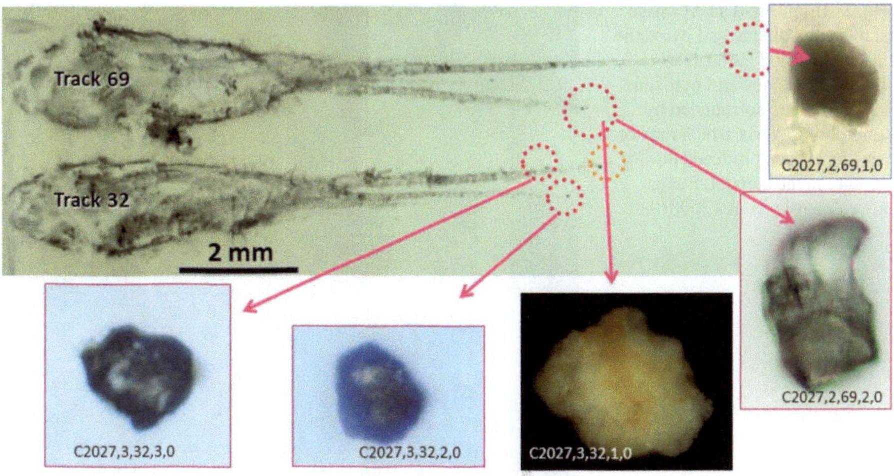

Fig. 1 Transmitted light images of tracks 32 and 69 and enlargements of 5 mineralogically diverse terminal particles. NASA JSC images

of the particle sizes, their terrific impact speed and the complexities of impact ionization of actual comet particles.

The Wild 2 coma particles were disaggregated and partially melted and partially vaporized during capture in silica aerogel, producing tracks up to about 5 mm in length (Brownlee et al. 2006) (Fig. 1). The aerogel was produced by the Jet Propulsion Lab, and carried some carbon, Fe and other trace elements residual from the production and bakeout processing. This silica aerogel was the cleanest, and most transparent that was known at the time, and was this well suited to the role of capture medium. Many but by no means all of the captured particles appear to have all been aggregates, where the densest and strongest components of the particles were deposited as aggregates down to single grains along and at the ends of the tracks (Figs. 1 & 2). Weaker and more volatile components were deposited as melt and vapor deposits along the entire lengths of the tracks. Some tracks were apparently made by nearly single grains (Nakamura-Messenger et al. 2011). A good reference on the use of track shape to infer the makeup of the impacting particle is Kearsley et al. 2012. The bulk elemental compositions of entire particle "tracks" were measured by Time-of-Flight Secondary Ion Mass Spectrometry (ToF-SIMS) and synchrotron X-ray Fluorescence (SXRF) for elements having atomic number $\geq$16 (Flynn et al. 2006). One of the investigated tracks was split open exposing the interior to analysis by ToF-SIMS, permitting measurement of lower-Z elements, particularly Mg and Al. Since Si and O are the major elements in silica aerogel collection media, neither element could be determined in the tracks. Major elements in comet particle impact residues in craters in aluminum foil (which surrounded the aerogel tray frame) were determined using Energy Dispersive X-ray analyses (SEM-EDX) and ToF-SIMS. The crater residues provided information on light elements, especially Mg and Si, that could not be determined in the SXRF analysis of tracks, with the exception of Al, which was the main component of the foil. Combining analytical results from the cometary material in the tracks and in the craters provided the most complete bulk chemical analysis available of the Wild 2 coma particles (Flynn et al. 2006). The bulk composition of 23 particle tracks were measured by SXRF, with a calculated total $\sim$115 ng of coma material. The crater residue analyzed by SEM-EDX resulted from the impact of $\sim$215 nanograms of

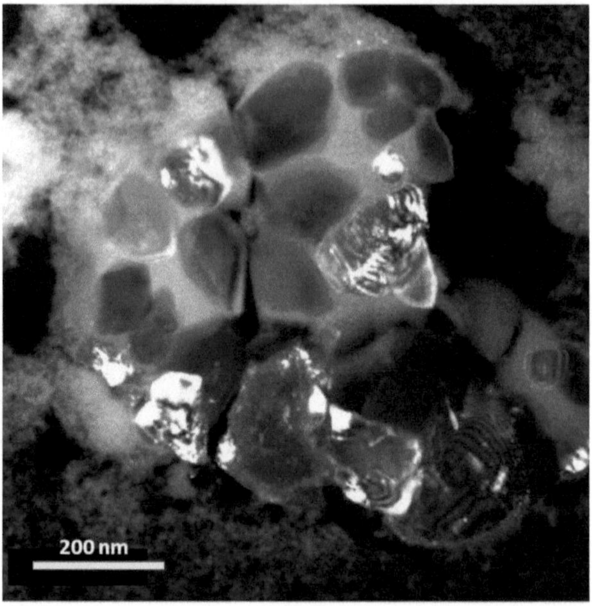

Fig. 2 Bright-field TEM image showing an olivine and pyroxene in sample C2027,2,69,1,1. Individual olivine and pyroxene crystallites are surrounded by amorphous matrix which may or may have been produced during capture in the aerogel. Image after Zolensky et al. (2008)

material. Taken together, ~300 nanograms of coma material were analyzed in the investigation. However, this was still two orders-of-magnitude more than analyzed at comet Halley. In addition, these Wild 2 samples are largely still available for additional (including repeat) measurements.

This task was complicated by the element heterogeneity since the impacting particles were aggregates of material having diverse compositions. Elements were normalized to Fe to permit comparison to meteorite compositions. For example, most Wild 2 particle element/Fe ratios for Track 19 (C2115,25,19,0,0) were observed to vary by two orders-of-magnitude along the track. Flynn et al. (2006) calculated whole track compositions by adding individual element abundances at specific points along the track to that of the single terminal particle. The calculated Fe-normalized whole track composition for Track 19 was consistent with CI bulk composition for S, Cr, Mn, and Ni, but enriched in the moderately-volatile elements Cu, Zn, Ga and Se, and depleted in Ca and Ge. Another problem was correctly applying X-ray absorption corrections. Captured Wild 2 particles tend to be "cocooned" by melted or compressed silica aerogel (see Nakamura-Messenger et al. 2011 for good examples of this phenomenon), complicating the correction for X-ray absorption, especially for lighter elements.

Flynn et al. (2006) ultimately calculated the Wild 2 coma grain mean composition by summing the measured abundance of each element over all 23 tracks. Most element/Fe ratios were observed to be consistent with the CI ratios at the 1-sigma confidence level. The exceptions are S, which was significantly depleted, and Cu, Zn, and Ga, which were enriched. These results for the Wild 2 particles are consistent with but significantly extend the Halley measurements. For example, the trace elements Cu, Zn, Ga, Ge and Se were measured for the first time.

The analyzed Wild 2 coma particles were enriched in the moderately-volatile elements Cu, Zn, and Ga compared to CI, which was also reported for anhydrous chondritic interplanetary dust particles (IDPs) (Flynn et al. 1996), some (possibly most) of which have a cometary origin (Brownlee et al. 1993). The CI meteorite element abundances are taken to represent the Solar Nebula composition for non-volatile elements because of the good agreement between CI abundances and the composition of the Solar photosphere (Anders and Grevesse 1989; Lodders 2020, 2021), although not all elements in the solar photosphere are as precisely determined as they are in IDPs, CI chondrites and Wild 2 coma particles.

The dust particles of comet 67P/Churyumov-Gerasimenko (hereafter 67P) were mostly mm-sized and moving at rather slow speed (a few m/s) (Rotundi et al. 2015). The 67P particles were aggregates of grains with sizes down to $\sim$30 nm (Mannel et al. 2019), and fragmented upon impact onto the collecting COSIMA targets (Langevin et al. 2016; Lasue et al. 2019).

COSIMA collected more than 35,000 particles and could analyze about 250 of them. The ROSINA gas phase mass spectrometer also measured the composition for elements like Na, Si, K, Ca and Fe of the surface of 67P cometary dust particles sputtered by the solar wind and of particles in the coma accidentally entering the ROSINA instrument. The compositions of these particles were chondritic for Na, enriched in K and Si compared to the chondritic value, and depleted in Ca (Wurz et al. 2015; Rubin et al. 2022). However, these results are complicated somewhat by the fact that while the refractories given in the Wurz et al. (2015) paper were compatible with sputtering by solar wind, the Rubin et al. (2022) data (except for Si) were collected during perihelion, when solar wind could not reach the surface of the comet. Accordingly, care must be taken when combining results from these two investigations.

The composition of $\sim$30 particles for a larger set of elements could be deduced from the analyses of the COSIMA instrument. The elemental compositions of these 67P dust particles measured by COSIMA showed some variability, with the average composition found to be broadly chondritic with enrichments in C, N, O and Si compared to the chondritic value, and depletions in Mg and Ca (Fig. 1) (Bardyn et al. 2017; Sansberro et al. 2022). 67P dust particles contained a large proportion (about 45 weight% of high molecular weight solid organic matter, mixed with a mineral matrix, yielding a C/Si atomic ratio of 5.5 + 1.4/-1.2 (Bardyn et al. 2017; Fray et al. 2016). This atomic ratio is compatible with the photosphere value of 7.19 ± 0.83, and is about one order of magnitude larger than the CI value of 0.76 ± 0.10 (Lodders 2021). Regarding comet Wild 2, the carbon concentration could not be quantified in 81P/Wild 2 samples because of the high-speed collection of the samples in aerogel. A low C concentration was however observed, with very few preserved C-rich phases. This could have resulted from the high speed (and somewhat destructive) collection of the dust, or it could be that the dust collected by Stardust was not representative of the comet as a whole (Westphal et al. 2017).

The H/C and N/C ratios could be measured by COSIMA in 67P dust. The H/C ratio is 1.04 ± 0.16 (Isnard et al. 2019), and N/C $= 0.035 \pm 0.011$ (Fray et al. 2017) (see also paragraph below "cometary organic matter"). Ammonium salts were recently discovered in the ROSINA data of 67P, and their presence at the surface of the comet is compatible with spectra taken with VIRTIS-M (Visible and Infrared Thermal Imaging Spectrometer, Mapping channel) (Altwegg et al. 2020; Poch et al. 2020; Altwegg et al. 2022). These ammonium salts (in particular ammonium hydrosulfides and fluorides), that were not measured in the 67P dust particles by COSIMA could account for the apparent missing nitrogen reservoir in the COSIMA data. In that case, these salts would have sublimated before analysis in COSIMA. Phosphorus and fluorine were also detected in the dust particles by COSIMA (Gardner et al. 2020).

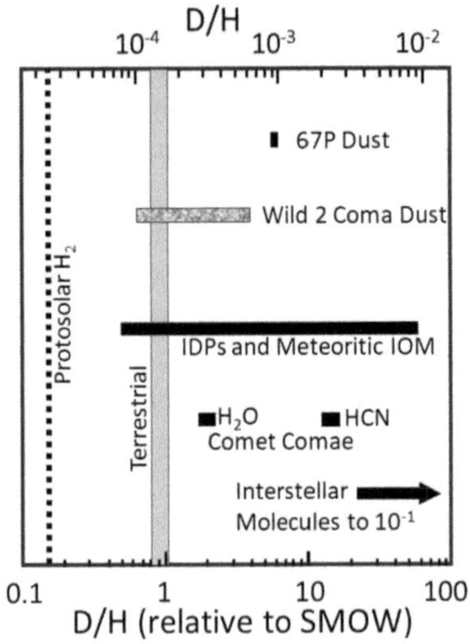

Fig. 3 Comparison of the D/H ratios of 5 Wild 2 coma dust particles (McKeegan et al. (2006) and 25 67P dust particles (Paquette et al. 2021), along with measured values for comet comae water and HCN, anhydrous chondritic IDPs, meteoritic insoluble organic material (IOM), and interstellar molecules. Figure is after McKeegan et al. (2006), which has the references to the other values

2.2 Isotopic Compositions

The Stardust and Rosetta missions enabled the measurement of isotopic ratios on dust particles from comets Wild 2 and 67P, respectively. The hydrogen isotopic composition (D/H ratio) of Wild 2 samples varies from the terrestrial D/H value (1.5576×10^{-4}) up to about three times this value in submicrometric zones (McKeegan et al. 2006; De Gregorio et al. 2010, 2011; Matrajt et al. 2008; Stadermann et al. 2008) (Fig. 3). The D/H ratio was measured in 25 dust particles from 67P by COSIMA, yielding a value of $D/H = (1.57 \pm 0.54) \times 10^{-3}$ (Paquette et al. 2021). This value is about ten times larger than the terrestrial value. Such high D/H values have also been observed in organic molecules in cometary atmospheres (Meier et al. 1998; Müller et al. 2022).

The nitrogen isotopic composition of Wild 2 dust particles is not very anomalous, with values reaching about 1.5 times the terrestrial value in submicrometric regions (McKeegan et al. 2006). In most cases, there is no correlation between the nitrogen and hydrogen isotopic compositions of cometary dust particles.

The sulfur isotopic composition of comet Wild 2 samples is compatible with the reference value (Heck et al. 2012; Ogliore et al. 2012a,b), but a cosmic symplectite showed enrichment in ^{33}S (Nguyen et al. 2017). This composition could result from photochemical irradiation of solar nebular gas. The sulfur isotopic composition (^{34}S and ^{32}S) of 67P dust was measured by COSIMA, and is compatible with the reference value (Paquette et al. 2017). The ^{33}S isotope could not be quantified by COSIMA, due to mass interferences.

The oxygen isotopic composition of Wild 2 samples is compatible with that of carbonaceous chondrites and Ryugu samples for the ferromagnesian minerals, and as well as for the refractory minerals, which show ^{16}O enrichments (Nakamura et al. 2008; Nakashima et al. 2012, 2023; Joswiak et al. 2017; Ogliore et al. 2015; Defouilloy et al. 2017; Zhang et al. 2023, 2024). The $\Delta^{17}O$ composition of Wild 2 minerals is negatively correlated with their Mg content with O isotopes on the PCM line (see Zhang et al. 2024, for the most recent tabulation). This result supports formation of these particular grains in a very oxidized environment followed by outward migration to the comet accretion region. The COSIMA instrument could only measure the $^{18}O/^{16}O$ isotopic ratio in 67P dust particles (the ^{17}O was not determinable due to mass interference with $O^{16}H$) (Paquette et al. 2018). Given the large error bars associated to the measured ratio (which is inherent to the analytical method) the measured values for 67P are compatible with the terrestrial value as well as the range of values measured in meteorites.

3 Mineralogy

3.1 Introduction

Most results to date have been obtained from only the coarser-grained portions of captured particles, which have been estimated to represent approximately 40% of the impacting comet coma grains on average (Flynn et al. 2006). However, this estimate was formed 20 years ago and should be reevaluated. In only a few cases thus far have we seen the true mineralogy of the finest-grained materials. Fortunately, in at least two recorded instances accumulations of the finest-grained, nano-scale Wild 2 materials firmly attached to coarser grains survived to the ends of tracks and are available for analysis to provide a window onto the nature of the finest-grained coma material. Also, fine-grained materials located in the upper portions of some tracks survived intact. Thus, our understanding of the finest-grained materials from Wild 2 is very incomplete and will only be revealed in full by future work. Readers should understand that the conclusions reached in this review largely pertain to the fraction of Wild 2 minerals coarser grained than 1 μm. Researchers have by now investigated approximately 225 separate particle capture tracks to varying degrees and have obtained a basic understanding of the these. However, practically every Wild 2 grain has possessed unique characteristics, so undoubtedly future work will reveal different materials and broaden the understanding of Wild 2 mineralogy. In addition, it appears that all captured grains derived from only two or three highly collimated jets originating from sites inside of the nucleus, which suggests that the mineralogical results could be unrepresentative of the nucleus as a whole. The mineralogy of examined Wild 2 coma grains is summarized in Table 1, and there compared to three other unequilibrated astromaterials that have not suffered significant aqueous alteration - namely CV3 and unequilibrated ordinary chondrites (UOCs), and anhydrous chondritic IDPs.

The captured particles are, in general, aggregates of submicron grains with a bulk chondritic composition, with most aggregates also containing one or more much larger individual crystals of (most commonly) olivine, pyroxene and Fe-Ni sulfides up to $\sim$10 μm in size (Zolensky et al. 2006; Frank et al. 2014; Joswiak et al. 2017). Summarizing the results from the studied tracks, $\sim$33% are dominated by olivine grains, $\sim$24% by low-calcium pyroxene, $\sim$10% by a fairly equal amount of olivine and pyroxene, and $\sim$33% are dominated by other minerals, mainly Fe-Ni sulfides. A major result of analyses to date is that crystalline materials are abundant in Wild 2 coma grains, dominating the coarse-grained materials. Brief summaries of the critical minerals follow.

3.2 Olivine

Olivine and/or pyroxene are present in the majority of Wild 2 particles, with observed grain sizes ranging from submicrometer to over 10 μm. The olivine exhibits an unusually wide compositional range of Fo_{3-100} (Fig. 4), with minor element abundances similar to the matrix olivine found in some unequilibrated carbonaceous chondrites and unequilibrated ordinary chondrites (Frank et al. 2014; Joswiak et al. 2012, 2017) (Fig. 5). The wide range of Fo content (Mg#) in Wild 2 is very similar to that observed in anhydrous chondritic IDPs

Table 1 Mineralogy of Wild 2 Compared to Unequilibrated Meteorites and Anhydrous Chondritic IDPs

Major phases	Wild 2	CV3[1]	UOCs	Anhydrous IDPs
Olivine	Fo_{5-100}	Fo_{4-100}[2]	Fo_{45-100}[2]	Fo_{45-100}[3]
Low Ca Pyroxene	En_{52-98}	$En_{?\text{-}100}$[4]	$En_{?\text{-}100}$[4]	En_{50-100}[5]
High Ca Pyroxene	Aug, Pig, Diop, Kos	Aug, Pig, Diop[4]	Aug, Pig, Diop[4]	Aug, Pig, Diop[5] Kos[3]
Plagioclase	yes	yes[4]	yes[4]	yes[5]
Phosphates	Apat	Apat[4]	Apat, Merr[4]	Apat[5]
Sulfides	Troi, Pyrr, Pent, Spha Cubanite (rare)	Troi, Pent[4]	Troi, Pent[4]	Troi,Pyrr, Pent, Spha[5]
Oxides	Mag (rare), Ilm, Chr, Spinel	Mag, Ilm, Chr, Spinel[4]	Mag, Ilm, Chr, Spinel, Rut[4]	Mag, Ilm, Chr, Spinel[5]
Glass	?	yes[4]	yes[4]	yes[5]
Fe-Ni Metal	?	yes[4]	yes[4]	yes[5]
Carbonates	rare	rare[4]	none noted	none noted
Crystalline Silica	Tri, Cri	none noted	none noted	none noted
Roedderite	rare	none noted	none noted	none noted
Schreibersite	rare	yes	rare	rare
Multiphase Components				
Chondrules	yes	yes[4]	yes[4]	none noted[5]
CAI	yes	yes[4]	yes[4]	Yes[5]
AOAs	possibly	yes[4]	yes[4]	none noted
GEMS	?	none noted	none noted	very abundant[5]

Abbreviations: IDPs: Interplanetary Dust Particles, UOCs: Unequilibrated Ordinary Chondrites, Aug: augite, Pig: pigeonite, Diop: diopside, Kos: kosmochlore, Apat: apatite, Merr: merrilite, Troi: troilite, Pyrr: pyrrhotite, Pent: pentlandite, Spa: sphalerite (or wurtzite), Mag: magnetite, Ilm: ilmenite, Chr; chromite, Tri: tridymite, Cri: cristobalite, CAI: Calcium-Aluminum-rich Inclusions, AOAs: amoeboid olivine aggregates, GEMS: glass with embedded metal and sulfides (called Granular- and Fundamental-Units by [5]).

[1] Includes both reduced and oxidized CV3s

[2] Frank et al. (2014)

[3] This paper, and included references

[4] Brearley and Jones (1998)

[5] Rietmeijer (1998)

[6] Rubin and Ma (2021)

(Zolensky and Barrett 1994; Zolensky 2022), although minor elements of IDP olivines have not been well measured. However, we note that the similarity of anhydrous chondritic IDP olivine major element compositions to that for Wild 2 suggests that a lack of a composition distrubution peak for forsterite (Fo_{97-100}) is a general property of comets, not limited to Wild 2.

Wild 2 olivines include some with very elevated MnO, Al_2O_3 and Cr_2O_3 contents, up to 5.1, 0.85 and 0.95 wt%, respectively (Frank et al. 2014). About 25% of these Mn- and Cr-rich olivines contain $\ll$1% FeO. Olivines with enrichments in these elements have been reported in carbonaceous chondrites, micrometeorites, Ryugu and CI chondrite meteorites, and chondritic IDPs, though they are rare (Klöck et al. 1989; Rietmeijer 1998; Gounelle et al.

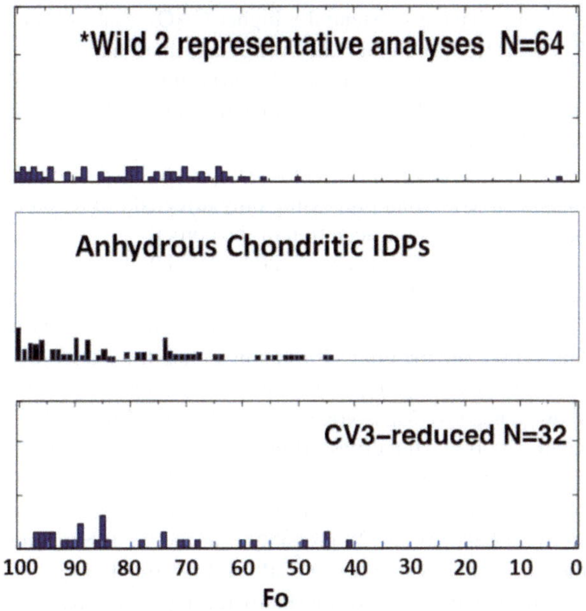

Fig. 4 Comparison of the range of Fo compositions of olivines from Wild 2 coma dust, anhydrous chondritic IDPs and reduced CV chondrite matrix. From Frank et al. (2014), Zolensky (2022)

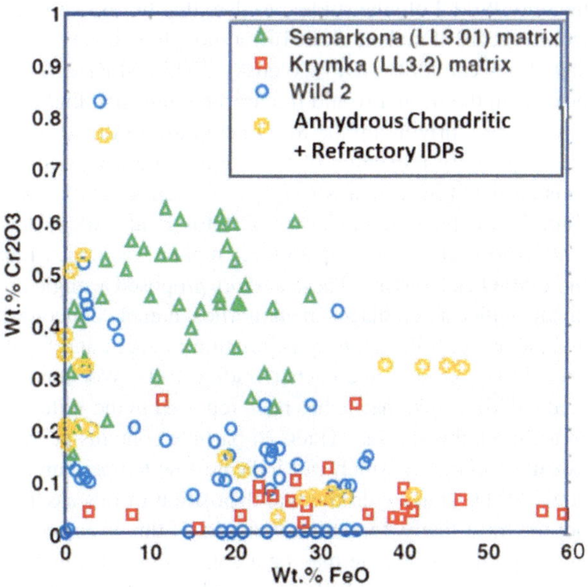

Fig. 5 Comparison of the Cr_2O_3 vs FeO content of olivine from Wild 2 coma dust, anhydrous chondritic and refractory IDPs, Semarkona matrix and Krymka matrix. Wild 2 olivine composition is closest to mildly-metamorphosed Krymka matrix olivine. From Frank et al. (2014), Zolensky (2022)

2002; Frank et al. 2014; Zolensky 2022; Mikouchi et al. 2022) and indicate an unmelted source (Ebel et al. 2012). In general, the FeO-rich olivine in Wild 2 has a wide range of MnO/FeO ratios, similar to CR/CH isolated matrix grains, but also includes other MnO-rich grains that are rare in chondrites (Frank et al. 2014). On the other hand, the Cr_2O_3 content of the FeO-rich Wild 2 olivine is distinctly lower than isolated matrix grains in the least equilibrated carbonaceous- (e.g., Acfer 094 (ungrouped C2) and ALH A77307 (CO3.0)) and ordinary chondrites (e.g., Semarkona (LL3.00) and QUE 97008 (L3.05)). However, many Wild

2 olivines have a substantially higher Cr_2O_3 content than those in slightly metamorphosed Krymka (LL3.2). In fact, the overall distribution of Cr_2O_3 content in Wild 2 olivine appears to be similar to that for isolated matrix grains in MET 96503 (L3.10). This result suggests that mild thermal metamorphism equivalent to that experienced by type 3.1 ordinary chondrites has modified some of the Wild 2 olivine, which would be a maximum temperature of ~400 °C, with large uncertainties (Huss and Lewis 1994; Huss et al. 2006). Of course, since this putative thermal processing only affected a fraction of the Wild 2 olivine it, obviously occurred before final accretion of the comet.

According to Fukuda et al. (2020) Wild 2 olivine grains from track 77 exhibit only small variations in δ^{25}Mg values from -1.0 to 0.6‰), which can be explained by mass-dependent fractionation from Mg isotopic compositions of the Earth and bulk meteorites, suggesting that Wild 2 olivine grains formed in a chondritic reservoir with respect to Mg isotope compositions. They also reported that one ^{16}O-rich Wild 2 LIME olivine grain in the same track shows a negative Mg isotope fractionation ($\delta^{25}Mg = -0.8 \pm 0.4$‰, $\delta^{26}Mg = -1.4 \pm 0.9$‰) relative to bulk chondrites. Minor element abundances of the olivine in this same grain are in good agreement with those of olivines from some carbonaceous chondrite AOAs. These observed similarities suggest that this particular grain formed in an environment similar to AOAs (Ebel et al. 2012), probably near the proto-Sun, and was then transported outward to the Kuiper belt where comet 81P/Wild 2 formed.

Unusual materials are constantly turning up among the Wild 2 coma particles. The olivine in the terminal particle of track 26 is nearly pure fayalite with up to 5.1 wt.% MnO and is the only Wild 2 olivine studied to date that has no known compositional analog in the meteorite record (Frank et al. 2014), although a single fayalite grain has been reported in an anhydrous chondritic IDP (Zolensky 2022). MnO-enriched olivine formed by an igneous process in the nebula would require very unusual, highly reducing conditions for Fe metal to condense, driving up the MnO/FeO ratio in the residual melt. This origin is incompatible with co-crystallization of fayalite and thus this grain must not have an igneous origin. This particle also contains tridymite (Zolensky et al. 2006; Joswiak et al. 2012). Equilibrium thermodynamic models by Zolotov et al. (2006) demonstrate that pure fayalite generally forms at lower temperatures, higher water/rock ratios, and at higher pressure than more MgO-rich olivine. These authors proposed a single process to explain the formation of fayalite/silica assemblages in chondrites, namely aqueous dissolution of amorphous silicate and kamacite, followed by precipitation of silica and fayalite. Amorphous silicates are expected to be present in comets (Bradley 1994; Wooden et al. 1999; Rietmeijer 2002; Lisse et al. 2006), and kamacite has been reported in the Wild 2 samples by Joswiak et al. (2012). In addition, this unusual Track 26 particle contains pentlandite and orthorhombic cubanite (ideally $CuFe_2S_3$), which can indicate low-temperature, hydrothermal processing (Berger et al. 2011). Thus, hydrothermal deposition of this assemblage appears to be a possible, if not the most likely, formation scenario. If this is correct, the extremely high MnO content in this particle's fayalite may be a compositional signature of a presumed precursor amorphous silicate unique to Wild 2 among known astromaterials. Of course, as described below, aqueous/hydrothermal activity in Wild 2 appears to have been very minor, localized, and/or heterogeneous. An alternative scenario is the collisional break-up of an early-forming KBO that underwent significant aqueous alteration. Material ejected from such KBO collisions could have been incorporated into Wild 2. Some evidence to support this may come from CI chondrites. For example, Gounelle et al. (2006) made trajectory calculations for the well-observed Orgueil (CI1) meteoroid fireball and inferred a probable KBO origin.

The major conclusions of the Frank et al. (2014) and Fukuda et al. (2020) studies of Wild 2 olivine were as follows: (1) Wild 2 accreted material from various carbonaceous and ordinary chondrite-forming regions; (2) fragments from amoeboid olivine aggregates (AOAs), type I chondrules, and type II chondrules were incorporated into Wild 2; (3) there is a striking deficiency of material from type I chondrules in Wild 2 relative to unmetamorphosed chondrites; (4) many Wild 2 grains appear to have experienced secondary thermal metamorphism. Zhang et al. (2023) suggested that the Wild 2 olivines with very low $\Delta^{17}O$ could all have formed as part of amoeboid olivine aggregates (AOA) which, like the CAI (see below), were transported outward from the inner disk of the solar system (Brownlee et al. 2006; Brownlee 2014); (5) relict grain compositions that are exceptionally rare in all chondrite chondrules (e.g. Marrocchi et al. 2019) appear to be quite abundant in Wild 2.

The latest, most precise, O isotope study of fragments (largely but not exclusively olivine) from a very large track (2 cm long), and consequently very large fragments, revealed a negative correlation between $\Delta^{17}O$ and Mg#, and that oxygen isotopes were clearly on the PCM line, supporting their formation in a very oxidized environment followed by migration to the comet accretion region (Zhang et al. 2024). These authors proposed that the migrating grains were well pulverized and mixed before final incorporation into Wild 2.

Wild 2's extraordinarily diverse mineral assemblage indicates that its components were forged over a wider range of formation times and/or positions in the early solar nebula than any known single chondrite parent body. Frank et al. (2014) concluded that Wild 2 had a late assembly date and included material from the break-up and dispersion of diverse bodies, including some that had been internally heated. This is consistent with the apparent absence of extinct ^{26}Al (Nakashima et al. 2015; Westphal et al. 2017) and generally longer accretion times for KBOs than for asteroids. Frank et al. further proposed that the olivine in the early solar system radial flow proposed by Brownlee et al. (2012) was dominated by ^{16}O-rich, refractory and AOA-like forsteritic olivine grains, rare relict grain compositions, and type II chondrule fragments. Scarcity of type I chondrule olivine would yield the flat distribution of fayalite contents for Wild 2. Therefore, the radial transport of both ^{16}O-rich forsterite and type II chondrule olivine was efficient, but transport of type I chondrule olivine was relatively inefficient. If ^{16}O-rich olivine formed prior to typical type I chondrules, then the efficiency of outward transport of inner solar system material to the Kuiper Belt may have been episodic. These results and observations provide valuable constraints to dynamical astrophysical models.

3.3 Pyroxene

After olivine, pyroxene is the most abundant mineral group found among the Wild 2 samples. Diffraction studies revealed that a significant fraction of the magnesium-rich Wild 2 low-calcium pyroxenes are orthoenstatite, requiring slow cooling, but the majority of Wild 2 pyroxenes have unknown crystal structures (Leroux et al. 2008a,b,c). Thus, we don't know the ratio of ortho- to clinopyroxene, which limits interpretation of formation scenarios. The compositional range displayed by the low-calcium pyroxene is also as extensive as the olivine, with En_{52-98} and a significant frequency peak centered at En_{95} (Zolensky et al. 2006). Low-calcium pyroxene usually coexists with olivine in the Wild 2 particles, but the Mg/Fe ratios for coexisting olivine and pyroxene are not always similar. For example, Track 17 contains olivine in the range Fo_{55-69}, while associated low-calcium pyroxene is a far less homogeneous En_{52-96}. The extreme compositional range of Wild 2 low-Ca pyroxene is again similar to the anhydrous chondritic IDPs, and significantly broader than that observed in most chondrites, including Murchison (CM2) and Orgueil (CI1). It should be possible to measure compositional profiles across the boundaries of adjacent olivine and low-calcium pyroxene crystals and between olivine and metal grains, to determine the extent of Mg and Fe zoning, which would yield the thermal histories of these assemblages, if they are in thermodynamic equilibrium. This is particularly important since the Wild 2 grains have apparently experienced such diverse formation histories. However, this project would require measurements of many grains to high compositional accuracy.

Fortunately, a few Wild 2 pyroxenes have been the subject of careful investigation. Both orthoenstatite and Ca-rich pyroxenes are frequently present, but the former predominates in abundance. Leroux et al. (2008a,b,c) reported pyroxene-bearing terminal particles which contain lamellar intergrowths of pigeonite and diopside on the (001) plane. This microstructure is typical for an igneous process and lamellae formation by exsolution during cooling. Width and wavelength of the lamellae indicate a cooling rate in the range 10-100 oC/h. This is within the range of experimentally-determined subsolidus cooling rates for chondrules (Schrader et al. 2018).

The abundance of Ca-rich pyroxene (mainly augite and diopside) in the Wild 2 particles appears to be beyond what is normally found in chondritic IDPs or unequilibrated chondrites (Leroux et al. 2008a,b,c; Joswiak et al. 2012). Some of the diopside could have formed during thermal metamorphism, but the available information is inconclusive. Joswiak et al. (2009, 2012) reported a high abundance of Na + Cr-rich pyroxene (kosmochlor) in half of the aerogel tracks they have investigated. These are part of an assemblage they named "Kool grains" (more properly called particles) (Kool = kosmochloric high-Ca pyroxene + FeO-rich olivine). The high Na, Cr and sometimes higher Mn contents of these pyroxenes appears to be in excess of what has been reported from most chondrites, although similar pyroxenes have been reported in the Orgueil CI1 chondrite (Greshake and Bischoff 1996) and in some anhydrous chondritic IDPs (Joswiak et al. 2009). Joswiak reported that fine-grained (submicron) Kool particles from Wild 2 have textures suggestive of crystallization from melts while coarse-grained ($>1\ \mu m$) Kool particles are often glass-free and could have formed by thermal metamorphism in the nebula. The major and minor element distributions between clinopyroxenes and coexisting FeO-rich olivines in the Kool particles lacking glass were reported as being consistent with formation at or near equilibrium. However, in glass-bearing fine-grained Kool particles, high concentrations of Na in the clinopyroxenes were inconsistent with experimentally-determined partition coefficients at equilibrium. Finally, these authors proposed that Kool particles could have been precursors to type II chondrules, though evidence for this proposal from actual chondrule studies was apparently lacking.

Frank et al. (2013) extended their study of pyroxene compositions in Wild 2 and in the matrices of unequilibrated chondrites (CM2, CR2, CH3, CB3, CO3, CV3-oxidized, CV3-reduced, L/LL3.0-4, H3.15, EH3, EL6, Acfer 094 (ungrouped C2), and K3) and reported that these confirmed the results from their study of olivine (Frank et al. 2014) in all essential respects. They reported pyroxene compositions in Wild 2 characteristic of L, LL, CH, and aubrite matrix materials. They reported a preponderance of low-Fe, Mn-rich pyroxene in Wild 2, a composition that is extremely rare in chondrite matrix. The chemical compositions presented in their study provide clear evidence that in addition to diverse nebular components associated with multiple chondrite types, Wild 2 incorporated materials that likely derived from evolved, internally heated asteroids. This conclusion obviously requires late formation of the Wild 2 nucleus.

Bridges et al. (2012) reported that one terminal particle from track 154 consists of an Al-rich diopside (9-11 wt% Al_2O_3) and pigeonite ($En_{80}Wo_{15}$) with considerable TiO_2 (0.5 wt%) and Al_2O_3 (5.2 wt%), all consistent with its being a fragment of an Al-rich chondrule typical of carbonaceous chondrites. The oxygen isotopic composition of the particle is also consistent with derivation from an Al-rich chondrule, but distinct from a typical CAI.

To summarize, Wild 2 contains anhydrous ferromagnesian silicates either from a wide range of nebular formation locations, some of which were apparently not sampled by known meteorites, or from a unique, previously unsampled, nebular reservoir. A significant number of these materials appear to have been affected by mild thermal metamorphism, with the environment (preaccretionary vs post-accretionary) being uncertain. Joswiak and coworkers have suggested that Kool pyroxenes formed through metamorphism at high pressure, which would require the rapid formation and early destruction of a large (500-1000 km diameter) parent body, an idea for which there is some support (e.g., Brown et al. 2007). However, Joswiak et al. (2012) point out that Kool pyroxenes could also have formed by condensation from a fractionated nebular gas reservoir which had been enriched in volatile components by partial evaporation of preexisting solids, or through magmatic processes at low pressures. Clearly, additional studies of Wild 2 pyroxene are desirable.

3.4 Sulfides

Sulfides are the only mineral group found in all extraterrestrial materials. Fe-Ni sulfides are also ubiquitous in the Wild 2 particles. Unfortunately sulfides were also one of the phases most affected by heating during capture in the aerogel (Barrett et al. 1992), making their study very difficult (Zolensky et al. 2006; Leroux et al. 2008a,b). Several coarse-grained Wild 2 particles are dominated by FeS or pentlandite. We refer here to troilite (stoichiometric FeS) and pyrrhotite ($Fe_{1-x}S$) collectively as "FeS" unless the exact phase was determined (e.g., Gainsforth et al. 2019). The vast majority of Wild 2 sulfides have compositions close to stoichiometric FeS, with less than 2 atom% Ni, although 7 atom% Ni has been reported (Gainsforth et al. 2019). The latter measurement is probably pentlandite exsolved within troilite or pyrrhotite. The general lack of compositions between FeS and pentlandite (with intermediate solid solution compositions) suggests (but does not require) that FeS and pentlandite formed directly as crystalline species, i.e., did not form as amorphous phases which later became annealed (Vaughan and Craig 1978). The remaining Wild 2 phases in the Fe-Ni-S ternary (approximately half analyzed so far) have compositions that trend from FeS directly towards the Fe apex of the Fe-Ni-S ternary diagram (e.g, Sack and Ebel 2006). These appear to be troilite with intermixed kamacite (Leroux et al. 2008a,b). These phases are almost certainly a result of capture heating in the aerogel, and their presence can be used to gauge the degree of capture modification of the associated Wild 2 grains.

The few verified pentlandite crystals in Wild 2 tracks are intriguing since this phase is frequently an indicator of low-temperature metamorphism under oxidizing conditions, and/or aqueous alteration (Zolensky and Thomas 1995; Bullock et al. 2005). Ni-rich sulfides are not found in totally anhydrous chondritic IDPs, but are found in IDPs that experienced aqueous alteration, and are abundant in carbonaceous chondrites that are aqueously altered. Therefore, at face value the presence of these few Ni-rich sulfides suggests that some Wild 2 particles could have experienced a limited degree of aqueous alteration, but does not require this. Gainsforth et al. (2019) described Ni-rich pyrrhotite exsolution in a large troilite crystal suggestive of heating to several hundred degrees C that was apparently not caused by sample capture.

Berger et al. (2011) reported a crystal of (Zn, Fe)S containing essentially equal amounts of zinc and iron in a Wild 2 particle. They called this phase sphalerite, although no diffraction data were provided to establish that it was not in fact the polytype wurtzite. The presence of true sphalerite in equilibrium with Fe and FeS (as was apparently the case here) could permit use of the sphalerite cosmobarometer (Hutchison and Scott 1983) to determine conditions of sulfide formation. Joswiak et al. (2012) also reported "sphalerite" in Wild 2 particles from two separate tracks, again solely on the basis of composition so again the phase could have been wurtzite. In this case the Zn sulfides were in contact with pyrrhotite, pentlandite and ferroan olivine. Since the ZnS polytypes can have different stability relations, with wurtzite probably being the high-temperature form (see Bowles et al. 2011), verification of the exact phase present in Wild 2, an ostensibly simple task, could help constrain formation conditions.

Berger et al. (2011) found submicron blebs of cubanite in a small particle from track 26 that also contained pentlandite, fayalite and tridymite. As mentioned earlier, this fragment has thus far proved to be unique.

3.5 High-Temperature Rocks – Calcium Aluminum-Rich Inclusions, Amoeboid Olivine Aggregates and Chondrules

A major surprise of the Stardust Mission was the rather abundant occurrence (~0.5 vol%) of high temperature mineral assemblages, most notably calcium aluminum rich (CAI-like) inclusions (Zolensky et al. 2006; McKeegan et al. 2006; Simon et al. 2008; Chi et al. 2009; Schmitz et al. 2009; Joswiak et al. 2012, 2017; Rotundi et al. 2014), amoeboid olivine aggregate (AOA)-like objects, and aggregates with obvious igneous origins which are very similar to chondrules from carbonaceous chondrites (Nakamura et al. 2008). In a survey of the Wild 2 results, Joswiak et al. (2017) reported that approximately 2–3% of investigated particles larger than 1 μm are CAI-like and nearly 50% of all bulbous Stardust tracks contain at least one CAI. Accordingly, they estimated that ~0.5 volume % of captured Wild 2 coma dust is CAI-like.

The Wild 2 CAIs described to date basically have concentric mineral structures, with spinel-anorthite cores surrounded by Al, Ti-rich clinopyroxene (Fig. 6). The first and still best studied Wild 2 CAI -"Inti", from Track 25- consists mainly of anorthite, gehlenite, spinel and diopside. Crystals of osbornite (TiN), (V, Ti)N, and Fe-Ni metal are present within all phases except diopside. The metal grains contain inclusions of platinum-group element alloy and Mo-, W-alloys. All these phases have high formation temperatures, especially the nitrides, which would have condensed in the nebula at 1600-2000 K, depending upon the local oxygen fugacity (Ebel 2006). As reported by Chi et al. (2009), Track 25 fassaite ($Ca(Mg, Ti, Al) (Si, Al)_2O_6$) was found to have a Ti^{3+}/Ti^{4+} ratio consistent with the fassaite in other solar system CAIs (Beckett 1986). The oxygen fugacity of formation estimated from this ratio (-19.4 ± 1.3), assuming equilibration among phases at ~1500 K, is near the canonical solar nebula value of -18.1 ± 0.3, and in close agreement with that reported for fassaite-bearing Allende CAIs (-19.8 ± 0.9).

It should be possible to measure compositional profiles across the boundaries of spinel within silicates to determine the extent of Mg zoning, which would elucidate one portion of the thermal history of the Wild 2 CAI. The oxygen isotope compositions of the Wild 2 CAIs are along the trend defined by carbonaceous chondrite CAIs (McKeegan et al. 2006; Simon et al. 2008). All of the cometary refractory objects are moderately refractory, fine grained, spinel-rich inclusions such as those found in primitive chondrites – none of the most refractory CAI types have been reported. This lack could indicate a higher abundance of small, moderately refractory CAI nodules in the nebula (Joswiak et al. 2017), but could also be misleading due to a still limited number of samples. The presence in Wild 2 of CAIs that are thought to have formed in the inner solar system was probably the first demonstration that there was transport of materials from the inner to the outer nebula, by processes still poorly understood (Brownlee et al. 2006).

Track 25 CAI osbornite and fassaite exhibit fine-scale variations in V and Ti, respectively (Chi et al. 2009). This lack of equilibration reveals that this CAI was never annealed at temperatures sufficient to cause chemical homogenization, and thus suggests relatively

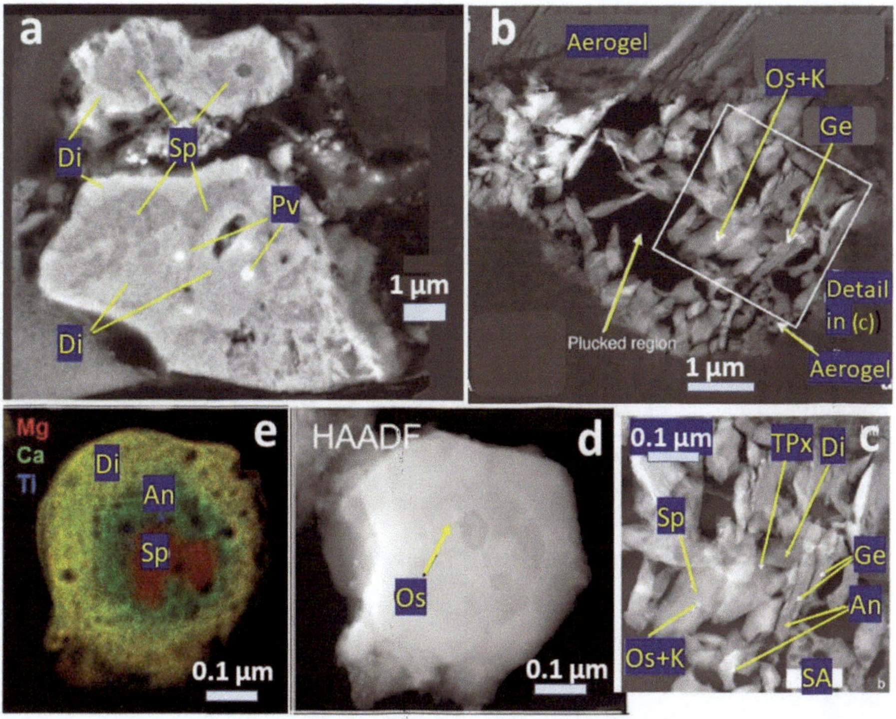

Fig. 6 CAI fragments recognized from among Wild 2 coma dust. (a) BSE image of a CAI fragment (called "Inti") from track 25. (b) Dark field TEM image of a microtomed slice of a subsample from (a). (c) closeup of (b) showing details of the refractory mineralogy. (d) High-angle annular dark field image and (e) composite RGB element maps of a microtomed slice of CAI fragment WF216 from track 172, consisting of concentric layers of spinel, anorthite, and Al-poor to Al-rich diopside. A V-bearing osbornite inclusion in anorthite is indicated in (d). Os = V-bearing osbornite, Di = diopside, An = anorthite, Sp = spinel, Pv = perovskite, TPx = Ti-bearing pyroxene, Ge = gehlenite, K = kamacite. Images after Simon et al. (2008), Brownlee et al. (2012), Joswiak et al. (2017)

early and rapid transport of this particular CAI from the inner to outer solar system. Chi et al. (2009) further found that Track 25 fassaite has a Ti^{3+}/Ti^{4+} ratio requiring formation at reducing conditions (see above). This low oxygen fugacity appeared to rule out formation by the X-wind model, which features highly oxidizing conditions in a reconnection region in the inner solar nebular disk (Shu et al. 2001). Their preferred scenario was CAI formation of the most refractory CAI components (e.g. osbornite) within 0.5 AU of the protosun, followed by outward transport and later incorporation with the remaining CAI phases (including fassaite and perovskite). The final CAI then continued to travel outward to the comet-forming region beyond 15AU. Wild CAIs are most similar to fine-grained meteoritic CAIs (FGIs) (Joswiak et al. 2017). These are porous aggregates of similarly rimmed nodules that are usually only a micron or so across. Some FGIs in meteorites contain thousands of nodules which therefore must have been a dominant solid in some local regions in the solar nebula.

Numerous Wild 2 particle fragments have now been found to be identical in many or all respects to chondrules from carbonaceous chondrites (e.g. Nakamura et al. 2008; Leroux et al. 2008c; Zhang et al. 2023, 2024; Joswiak et al. 2024) as described in part above (Fig. 7). These chondrule-like Wild 2 grains are characterized by (1) igneous porphyritic or

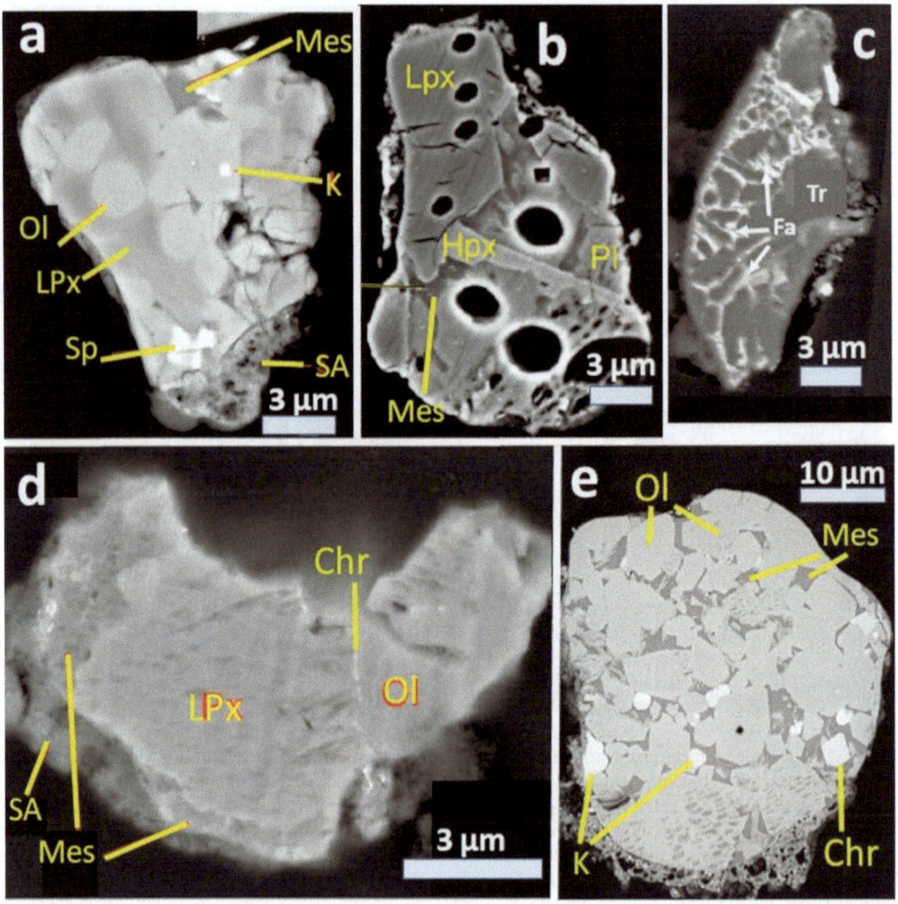

Fig. 7 Fragments of chondrules and a complete chondrule from among Wild 2 coma grains. (a) BSE image of a cross-section of track 35 fragment Torajiro, displaying a porphyritic texture. (b) BSE image of a cross section of a track 81 fragment (C2092,7,81,1,0) named "Pyxie", another porphyritic object; holes are from SIMS measurements; (c) BSE image of a cross section of a fragment from track 26 named "Ada", which contains crystals of tridymite with interstitial fayalite; (d) BSE image of a cross section of another fragment (C2054,0,35,4) from track 35 named "Lilly", which contains low-Ca pyroxene ($En_{89}Wo_4$) in contact with Fo_{91} olivine with 2.0 wt% MnO and Si- and Al-rich mesostasis; (e) Complete porphyritic chondrule from track 227 with olivine, chromite, kamacite and albitic glass mesostasis. Ol = olivine, Mes = mesostasis, LPx = low-Ca pyroxene, HPx = high-Ca pyroxene, Sp = spinel, Chr = chromite, Pl = plagioclase, Tr = tridymite, K-kamacite, Fa = fayalite, SA = silica aerogel. Images after Nakamura et al. (2008), Joswiak et al. (2012), Nakashima et al. (2015), Gainsforth et al. (2015), Joswiak et al. (2024)

poikilitic textures suggestive of partial melting at high temperatures; (2) presence of glass formed during rapid cooling, (3) dominant occurrence of Mg-rich olivine and low-Ca pyroxene with minor amounts of rounded Fe-metal inclusions suggestive of crystallization under reducing conditions, (4) the presence of ^{16}O-rich "relict" olivine grains that have equilibrated Mg/Fe ratios because Fe (being smaller) could more easily diffuse into the "relict" grain while oxygen did not - the POP Wild 2 olivines are equilibrated in Fe but some retain large ^{16}O effects, and (5) apparent absence of completely melted chondrules such as the barred olivine (BO) type. One notable chondrule-like object is a myrmekitic intergrowth of nepheline, olivine, augite, plagioclase, and a Ca phosphate (Zolensky et al. 2008). Some of these objects contain crystalline silica-rich phases including tridymite, cristobalite and roedderite, and other rare phases including schreibersite (Joswiak et al. 2012). Compared with typical porphyritic chondrules in chondrites, the crystal size of olivine in Wild 2 chondrule-like objects is much smaller – typically 2-10 μm. However, some chondrites, such as the ALH-85085 CH chondrite, have similarly fine-grained chondrules (Grossman et al. 1988). Major element abundances of olivine, pyroxene, and mesostasis glass in the chondrule-like particles are within the ranges of compositions defined by a majority of meteorite chondrules (Nakamura et al. 2008). However, the range of oxygen isotope ratios of these particles spans the entire range observed for chondrules in carbonaceous chondrites. None of the data plot on or above the terrestrial mass fractionation line, which clearly resolves these particles from ordinary, Rumuruti, and enstatite chondrite material (Nakamura et al. 2008). Therefore, the "chondrules" in Wild 2 are most similar to type I and II olivine-pyroxene chondrules in carbonaceous chondrites, which derive from C- and D-class asteroids (Tachibana et al. 2022), with type II being the most abundant. Therefore, Wild 2 chondrules appear to have a strong genetic relation with materials in the outer asteroid belt. The largest Wild 2 chondrule reported to date is 60 μm type II chondrule with Fo_{62} olivine from Track 227 (Joswiak et al. 2024) (Fig. 7e).

It is worth noting that despite the fact that no barred chondrules fragments have been reported from Wild 2 samples, there is reason to look for them. Zhang et al. 2021, show two barred olivine chondrule fragments found in a single giant cluster anhydrous chondritic IDP that is thought (by them) to be cometary and shown to have mineralogies similar to Wild 2.

As shown by Nakashima et al. (2012) the $\Delta^{17}O$ values vary inversely with Mg# for Wild 2 chondrule-like objects and carbonaceous chondrite chondrules, but not for chondrules from LL3 chondrites and enstatite type 3 chondrites. The ^{17}O compositional range indicates that Wild 2 chondrule-like objects are most similar to chondrules in the CR and Tagish Lake-like chondrites (such as WIS91600), which indicates that they likely come from a reservoir with similar dust/ice ratios as these chondrites (Defouilloy et al. 2017). However, the differences in the Mg# range between olivine in Wild 2 and carbonaceous chondrite chondrules imply that the 81P/Wild 2 reservoir was comparatively more oxidized, with a higher dust enrichment.

Since the oxygen isotope compositions of these Wild 2 CAI-like and chondrule-like objects are identical to some meteorite analogues, these objects probably all formed within the inner solar system as opposed to some other disk environment. The ^{16}O-richest reported composition of Wild 2 chondrule-like objects is $\sim -51‰$ $\delta^{18}O$ (Defouilloy et al. 2017; Zhang et al. 2023) and that of the Wild 2 "CAI" Inti is $\sim -41\delta^{18}O$ (Simon et al. 2008), suggesting that both sampled a common oxygen reservoir during formation. This is more ^{16}O-rich than most bulk meteoritic CAIs, but very close to the compositions of the most ^{16}O-rich, unaltered components of refractory inclusions from carbonaceous chondrites (Simon et al. 2008; Yurimoto et al. 2008). This implies that the Wild 2 chondrule-like objects and CAI formed in the same region, probably near the Sun. This in turn further supports Brownlee's proposal that a solar system wide dynamic process was operating to carry inner solar system solids (CAI and chondrules) out to the comet nucleus formation region (Brownlee et al. 2006; Brownlee 2014).

As previously mentioned, some Wild 2 olivine grains are proposed to have originated within AOAs (Joswiak et al. 2017; Zhang et al. 2023), although these particular olivines reportedly exhibit a wide range of Mg# (Fo_{0-97}) with high $\Delta^{17}O$ values largely between $\sim -2‰$ and $\sim 2‰$, suggesting formation in a very oxidized environment, which is unlike olivine from typical AOAs which are very ^{16}O rich and nearly pure forsterite. Iron-poor fragments have $\Delta^{17}O$ down to −6‰, and iron-rich ones tend to have higher $\Delta^{17}O$ (up to 7‰), displaying an overall negative correlation with their Mg#, which Zhang et al. (2023) suggest could be due to the addition or involvement of ^{16}O-poor ice during their formation, as proposed for CR chondrules (Tenner et al. 2015). Zhang et al. (2023) further speculated that the Wild 2 chondrule- and AOA-like objects could have formed late (>3 Ma after CAI) in a localized region outboard from the CR chondrule-forming region and therefore migrated only a short distance to the Wild 2 nucleus accretion location.

3.6 Non-crystalline Silicates

Since it is commonly proposed that the earliest nebular silicates (at least) were dominated by poorly-crystalline materials (Paquette et al. 2011), it is fundamentally important to determine the nature of any amorphous cometary silicates that may be present in the Wild 2 samples. We know from infrared spectroscopy (as much as we can "know" anything from such observations) that glassy or amorphous silicates are present in comets (Hanner and Zolensky 2010). In the anhydrous chondritic IDPs of proposed cometary origin, submicron-sized partially amorphous aggregates called GEMS (glass with embedded metal and sulfides) can comprise over 50% of the volume (Ishii et al. 2008). Thus, it was anticipated that GEMS would be prevalent in the Wild 2 samples. Unfortunately, distinguishing small (0.1-0.5 μm) GEMS from the silica aerogel that melted during particle impact has proven to be extremely difficult. Leroux et al. (2008a) and Ishii (2019) reported on detailed studies of the material along a number of track walls, describing a silica-rich matrix where cometary material has mixed with melted aerogel. Fe-Ni and Fe-Ni-S inclusions are present, as are localized regions of high Mg abundance, along with minor amounts of Ca, Al, Fe. In contrast to the IDPs, the GEMs-like material in aerogel has very low Mg abundance. Moreover, this material is generally found along tracks having a terminal FeS particle. Similar material, including reduced Fe-Ni inclusions, has been produced in laboratory experiments when FeS particles were accelerated into silica aerogel (Barrett et al. 1992), leading to the conclusion that most, if not all, of the GEMS-like material among the Wild 2 samples is a byproduct of the impact process. On the other hand, nanometer scale compositional mapping has identified Mg-rich domains within the molten aerogel, of a size and composition consistent with GEMS (Keller and Messenger 2008; Leroux et al. 2008a). Thus, there are hints that cometary GEMS are present in the sample. Hopefully this rather important issue will be resolved in the near future.

Some chondrule-like objects contain unaltered mesostasis glass. For example, the track 35 terminal grain named "Torajiro" (C2054,0,35,6) contains glass (see composition in Nakamura et al. 2008). The complete porphyritic chondrule from track 227 includes albitic glass mesostasis (Joswiak et al. 2024).

3.7 Fine-Grained Material

As mentioned above, the majority of fine-grained comet coma materials were apparently melted and intimately mixed with silica aerogel during sample capture, making it very challenging to characterize. Fortunately, several investigated Wild 2 terminal particles include well-preserved fine-grained material, including "Febo" (track 57), "Ada" (track 26), "Iris" (track 74), "Andromeda" (track 191), and track 227 (Brownlee et al. 2006; Matrajt et al. 2008; Stodolna et al. 2012b; Ishii et al. 2018; Gainsforth et al. 2019; Joswiak et al. 2024). In each case a large crystal, usually a sulfide and/or ferromagnesian silicate, and one case a large chondrule (track 227) shielded adjacent nanoscale material from significant heating

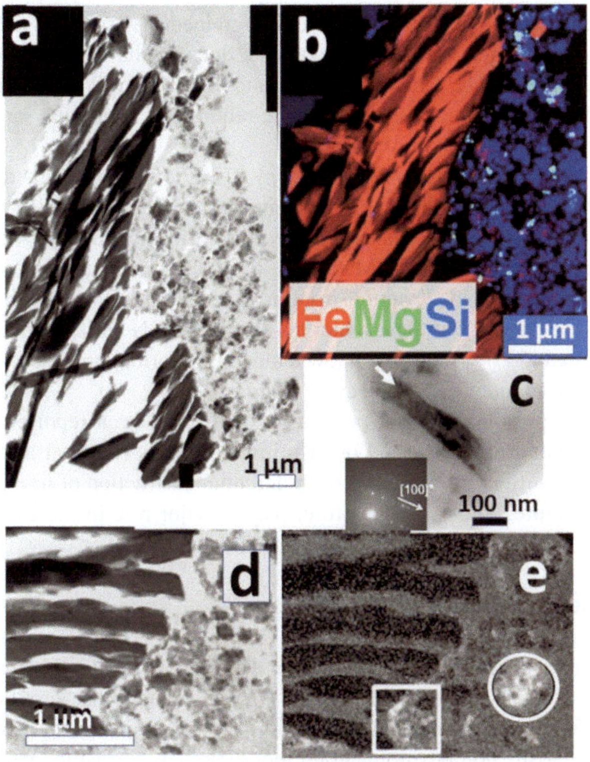

Fig. 8 Wild 2 particles exhibiting preserved fine-grained cometary material. (a) Bright field TEM image of a microtomed slice of a fragment from track 191 called "Andromeda", showing the sulfide (chattered from microtomy) on the left, and fine-grained material on the right. (b) Composite RGB element maps of a reduced area from (a); red is troilite, light blue is probably olivine and/or low-Ca pyroxene, dark blue are nanoscale aggregates of silicates. (c) Dark field TEM image of an enstatite whisker (arrowed) from a fragment from track 74 called "Iris", with an inset diffraction pattern showing that the elongation direction is roughly along the [100] direction. (d) TEM image of a microtomed slice of a terminal particle from track 57 called "Febo". This is a zero-loss image obtained with the zero-loss peak of the area of the particle with the highest carbon abundance. (e) C-EELS map of (d), with the white areas indicated by the square and circle exhibiting the highest carbon abundance. The doughnut shaped C material in the circle may be a hydrocarbon nanoglobule. Images after Matrajt et al. (2008), Stodolna et al. (2012b), Gainsforth et al. (2019)

during collection in aerogel (Fig. 8). The fine-grained material has been reported to include olivine, low-Ca pyroxene, kosmochloric diopside (Na-Cr coupled substitution), plagioclase-like material, Fe-Ni sulfides, Fe-Ni metal, probable chromite, poorly crystalline aluminosilicates, and C-bearing phases. All these phases are also found as coarser-grained material in the Wild 2 particles, with the possible exception of the carbonaceous material. The carbonaceous phases are systematically associated with ^{15}N enrichments, consistent with an organic, cometary material. In the particle Ada the carbonaceous phase is also associated with a D enrichment whose value is comparable to enrichments observed in IDPs (Matrajt et al. 2008). The C-rich areas are sometimes associated with O and show the presence of a carbonyl ($C = O$) feature at 288.5 eV, which also supports their organic nature. Although the size and level of the ^{15}N enrichment found in these C-rich areas is comparable with C-rich inclusions previously reported in IDPs, the C-XANES spectra of the organic materials in Febo and Ada is different from the ones observed in IDPs and carbonaceous chondrites.

Gainsforth et al. (2019) reported that the fine-grained material in "Andromeda", which appears to be the best-preserved known example, does not have a solar composition. It is depleted in Na, Mg, Al, Ca, Cr, Mn, and Ni, and enriched in P, S, Fe, and K relative to CI. Clearly additional, similar investigations are required as additional samples of the fine-grained materials are found and investigated.

Carbon Electron Energy Loss Spectroscopy (EELS) of fine-grained assemblages in "Ada" and "Febo" reveal round structures that in form and basic chemistry appear identical to the organic nanoglobules now known to be present in most organic-bearing chondrites (Nakamura-Messenger et al. 2006; Matrajt et al. 2008; De Gregorio et al. 2013) and recently reported from asteroids Ryugu (Stroud et al. 2024) and Bennu (Nguyen et al. 2024). Matrajt et al. (2008) noted that these materials all exhibit ^{15}N enrichments up to 200‰ and D enrichments up to 500‰, requiring a low temperature formation location, i.e., outer solar system or possibly an interstellar molecular cloud.

The amount of presolar material present in Wild 2 has been reported to be consistent with anhydrous chondritic IDPs, and other primitive bodies (Floss et al. 2013). However, this estimate again suffers from the fact of preferential destruction of fine-grained materials during sample capture. The foils, therefore, play a major role in the estimate of presolar grain abundance in Wild 2, which remains, at best, a lower limit.

3.8 Low-Temperature Inorganic Phases

No phyllosilicates or hydrous phases have been definitively identified in any Wild 2 sample (Zolensky et al. 2006; Westphal et al. 2017). Despite the fact that significant heating and structural modification accompanied collection of many grains in the aerogel, we should have seen characteristic compositions, grain morphologies, and lattice fringes of phyllosilicates had they been present, even if they had been thermally modified (Barrett et al. 1992; Hörz et al. 2000). During heating, phyllosilicates lose interlayer water and OH, and interlayer lattice fringes collapse in a characteristic manner. Eventually, a neoformed anhydrous silicate is formed, generally orthopyroxene or olivine. Fortunately, the neoformed silicates generally retain the flaky morphologies of the original phyllosilicates, however, and so are still recognizable as transformed phyllosilicates. We have observed this phenomenon in naturally- as well as artificially-heated (metamorphosed) carbonaceous chondrites and analogues (Akai 1988; Zolensky et al. 1991; Barrett et al. 1992; Okudaira et al. 2004; Tonui et al. 2014). Additional evidence that some phyllosilicates would have survived capture is presented by the partial survival of scattered, delicate, low-temperature and organic-rich Wild 2 phases exhibiting deuterium and ^{15}N excesses (Sandford et al. 2006; Cody et al. 2008). These fine-grained materials survived capture in the aerogel with minimal heating when attached to larger, more refractory grains. Phyllosilicates would have behaved similarly.

Berger et al. (2011) reported a tiny amount of cubanite in track 26. On Earth cubanite is generally indicative of low-temperature, hydrothermal processing in terrestrial environments. However, this cubanite has not been reported in any additional Wild 2 sample. Another possible alteration product is the magnetite that has been reported in a few tracks (Hicks et al. 2017). Once again, magnetite appears to be rare in Wild 2 samples, unlike the situation in thoroughly aqueously-altered chondrite meteorites, Ryugu and Bennu samples (Nakamura et al. 2022; King et al. 2024), and in any case can be formed by igneous processes (Keller et al. 2022).

Micron-sized crystals of calcite and Fe-Mg carbonate have been reported from three Wild 2 particles (Mikouchi et al. 2007; Wirick et al. 2007), but it is now clear that carbonates are exceedingly rare, if not mainly terrestrial contaminant grains. However, carbonates in metamorphosed carbonaceous chondrites have devolatilized to hydroxides or oxides (Tonui et al. 2014). Thus, it is possible that carbonates were more abundant in the Wild 2 particles before capture and capture heating, but this has not been established.

Thus, the lack of phyllosilicates, and rarity of carbonates among the examined Wild 2 grains suggests that they could not have composed more than a few percent of the more coarse-grained fraction of captured Wild 2 sample. Accordingly, Wild 2 seems to have not experienced any significant degree of aqueous alteration, or hosted liquid water at any time.

4 Organic Matter

One characteristic of comets is their high abundance of organics, which is already seen by remote sensing analyses in cometary atmospheres (Biver and Bockelée-Morvan 2019). These species are associated with sublimating ices when the comet gets closer to the sun, that release volatile organic compounds. These volatile species contain mainly CH(N)O bearing molecules, with a great variety of both saturated and unsaturated species containing CH-, CHN-, CHS-, CHO2-, CHNO- functional groups, and the CN radical and related molecules (Altwegg et al. 2017; Hänni et al. 2021, 2022). The aminoacid glycine, as well as methylamine and ethylamine were also found in the gas phase, originating probably preferentially from the sublimation of the dust particles rather than from ices from the nucleus (Altwegg et al. 2016; Hadraoui et al. 2019, 2021). Glycine was also found on Al foil exposed to the atmosphere of comet 81P/Wild 2 during the Stardust mission (Elsila et al. 2009). The ammonium salts stated earlier could be in the form of ammonium hydrosulphide and fluoride) (Altwegg et al. 2020, 2022; Poch et al. 2020). Organic matter was not very well preserved in the comet Wild2 samples brought back by Stardust, because of high-speed collection (Brownlee 2014; Keller and Messenger 2008; Sandford et al. 2006). In some place however, there were organic phases showing complex organic bonds, including mainly alkenes, aromatic $C = C$ and carboxyl $C = O$, in a variety of textures, including organic nanoglobules (De Gregorio et al. 2011). The organic matter in Wild 2 samples show spectral similarities with that of IOM from carbonaceous chondrites (De Gregorio et al. 2011).

COSIMA/Rosetta analyses also showed that 67P dust particles contain about 45 weight% of solid organic matter, bearing similarities with the insoluble organic matter extracted from carbonaceous chondrites (IOM) (Bardyn et al. 2017; Fray et al. 2016). The atomic H/C and N/C ratios of 67P dust measured in the organic matter (1.04 ± 0.16 and 0.035 ± 0.011, respectively (Isnard et al. 2019; Fray et al. 2017)) can be compared to that of insoluble organic matter extracted from carbonaceous chondrites and from the analyses of asteroid Ryugu. The H/C ratio is about twice that of carbonaceous chondrite IOMs (Alexander et al. 2007), and about 5 times that measured in asteroid Ryugu whole-rock samples (Yokoyama et al. 2023), but lower than the H/C ratio of volatile species measured in the coma by ROSINA on Rosetta ($H/C = 1.56 \pm 0.04$ (Hänni et al. 2022). This high H/C ratio in the coma species is in turn compatible with that of the soluble organic matter in the Murchison carbonaceous chondrite (Schmitt-Kopplin et al. 2010). The average N/C ratio of 67P dust particles is about one order of magnitude lower than the photosphere value ($N/C = 0.24 \pm 0.06$) (Lodders 2021), but it is compatible with that of primitive carbonaceous chondrite meteorites ($N/C \sim 0.03$) (Alexander et al. 2007) and of Ryugu IOM samples ($N/C \sim 0.15$) (Yabuta et al. 2023).

4.1 Combined Textural and Compositional Study

Each stardust track represents a unique hypervelocity impact event. Tracks were cut from aerogel tiles in thin, triangular "keystones" (Westphal et al. 2002) translucent to optical light. The nature of each track-forming event, including the original state of the impactor, is recorded in 3D track morphology. Material distribution has been recovered by synchrotron X-ray fluorescence (S-XRF) mapping of whole tracks (e.g., Flynn et al. 2006).

Ebel et al. (2009) used laser scanning confocal microscopy (LSCM), to acquire 3D imagery of whole tracks at high resolution (<80 nm/pixel in x and y, ~ 300 in z the optic axis). They developed 3D deconvolution methods to perform first-order corrections of optical aberrations induced by light diffraction and refractive index mismatches (Greenberg and Ebel 2010; White et al. 2012). S-XRF mapping of 3D-imaged tracks was done at 2 microns/pixel. For many tracks, the S-XRF maps were done at two angles, allowing elimination of aerogel surface contamination.

Greenberg and Ebel (2012) used these combined data to measure track volumes, entry hole size, and cross-sectional profiles of tracks. This allowed them to estimate the original impactor size, eccentricity, and volume for Tracks 152 and 163, for densities of 1.9 g/cc and 1.1 g/cc, respectively. These track data offer the ability to search track geometry and chemical maps to identify specific particles of interest for detailed chemical and isotopic analysis. These data are available through NASA JSC Curation and will enable future discoveries of interesting grains and physical properties of impactors in the Stardust sample suite.

5 Conclusions

Here we summarize the most important Wild 2 mineralogical and compositional results which provide constraints to early solar system evolution. As expected, the cometary solids were mineralogically similar to anhydrous chondritic IDPs with the exception that GEMS have not between reliably reported in the former. Important similarities include the very wide range of olivine and pyroxene bulk compositions (including minor elements), which probably reflects a wide range of sampled grain formation environments. The abundance of presolar materials in the Wild 2 samples is also approximately equal to that in anhydrous chondritic IDPs rather than being dramatically higher. High-temperature aggregates including chondrules (mainly type II), CAI and possibly AOAs are also present in Wild 2, but probably at less than 10% overall total abundance. There was a pre-mission expectation that comet solids would be dominated by either non-crystalline materials, nanoscale crystalline materials, or presolar grains. While it initially appeared that none of these expectations proved to be correct, it has become apparent that the preferential destruction of the finest-grained Wild 2 material during capture in the aerogel has largely erased its presence. Only further work on the rare occurrences of surviving fine-grained Wild 2 material will reveal its true nature. Organic matter was not very well preserved in the comet Wild2 samples because of the collection process. However, organic phases were preserved in some samples. These materials showing complex organic bonds, including mainly alkenes, aromatic $C=C$ and carboxyl $C=O$, in a variety of textures, including organic nanoglobules. The organic matter in Wild 2 samples show spectral similarities with that of IOM from carbonaceous chondrites.

Chronology of the collected solids has not been well established, but there is no evidence for captured live ^{26}Al, which could indicate comet accretion after this nuclide had effectively completely decayed. The implications of these major results are that Wild 2 sampled widely across the entire solar nebula, much wider than any other known astromaterials, potentially at a relatively late time compared to the oldest available astromaterials. However, a final caution is that the captured Wild 2 materials described here derived from a small number (possibly as low as 3) interior regions of a single Jupiter Family comet.

There is also limited mineralogical information from comet 67P provided by the Rosetta Mission. The investigated 67P dust particles also contained about 45 weight% of solid organic matter, bearing similarities with the IOM extracted from carbonaceous chondrites. The H/C ratio of 67P IOM is about twice that of most carbonaceous chondrite IOMs and about 5 times that measured in asteroid Ryugu whole-rock samples, but lower than the H/C ratio of volatile species measured in the 67P coma. This high H/C ratio in the coma species is in turn compatible with that of the soluble organic matter in investigated carbonaceous chondrites such as Murchison. The average N/C ratio of 67P dust particles is about one order of magnitude lower than the photosphere value, but it is compatible with that of primitive carbonaceous chondrite meteorites and of Ryugu IOM samples.

Only future work can establish how these two comets compare to the bulk of comets. Unfortunately, there are no additional comet sample analysis or return missions definitely planned at this time.

There is now sufficient data on the samples returned to Earth by the Hayabusa2 and OSIRIS-REx missions to C-complex asteroids Ryugu and Bennu, respectively, to permit comparisons to be made with comets Wild 2 and 67P. Ryugu and Bennu are both 2nd generation (at least) asteroids, formed from reaccretion of a small fraction of the fragments from much larger disaggregated earlier-generation bodies (Nakamura et al. 2022; Nakashima et al. 2023; King et al. 2024). While the precursor material for the Ryugu and Bennu asteroids may have been mineralogically similar to Wild 2 and 67P, both asteroids have experienced thorough, in fact almost complete, alteration by the activity of a brine – aqueous alteration, though not at a sufficiently high temperature for it to have been actual hydrothermal alteration. Amorphous and crystalline silicates, metal, oxides, sulfides, phosphides, etc., have been dissolved and subsequently replaced by an assemblage characteristic of the CI chondrites – namely saponite, serpentine, Fe-Ni sulfides, magnetite, Ca-Mg-Fe carbonates, Ca-Mg-Na phosphates and evolved organics, to name the most abundant phases. Only minor amounts of the pre-aqueous alteration mineralogical assemblage has survived in the sampled asteroids. Hydrous chondritic IDPs have suffered the same fate (Rietmeijer 1998). Wild 2 and, apparently, 67P escaped this processing, having as final bodies never having witnessed internal temperatures or pressures sufficiently high to melt water ice. It is interesting to note that the two sampled C-complex asteroids are effectively identical to a very rare class of meteorites (CI). We have no way, currently, of understanding how representative 81P/Wild 2 and 67P are of Jupiter Family comets.

Acknowledgements This paper benefitted greatly from detailed reviews by Don Brownlee and one anonymous individual.

Funding MZ was supported by the NASA Stardust and Hayabusa2 Missions Participating Scientist Programs.

Declarations

Competing Interests The authors declare that they have no conflict of interest.

References

Akai J (1988) Incompletely transformed serpentine-type phyllosilicates in the matrix of Antarctic CM chondrites. Geochim Cosmochim Acta 52:1593–1599

Alexander CMO, Fogel M, Yabuta H, Cody GD (2007) The origin and evolution of chondrites recorded in the elemental and isotopic compositions of their macromolecular organic matter. Geochim Cosmochim Acta 71:4380–4403

Altwegg K, Balsiger H, Bar-Nun A, Berthelier J-J, Bieler A, Bochsler P, Briois C, Calmonte U, Combi MR, Cottin H, De Keyser J, Dhooghe F, Fiethe B, Fuselier SA, Gasc S, Gombosi TI, Hansen KC, Haessig M, Jäckel A, Kopp E, Korth A, Le Roy L, Mall U, Marty B, Mousis O, Owen T, Rème H, Rubin M, Sémon T, Tzou C-Y, Hunter Waite J, Wurz P (2016) Prebiotic chemicals—amino acid and phosphorus—in the coma of comet 67P/Churyumov-Gerasimenko. Sci Adv 2:e1600285

Altwegg K, Balsiger H, Berthelier JJ, Bieler A, Calmonte U, Fuselier SA, Goesmann F, Gasc S, Gombosi TI, Le Roy L, de Keyser J, Morse A, Rubin M, Schuhmann M, Taylor MGGT, Tzou CY, Wright I (2017) Organics in comet 67P – a first comparative analysis of mass spectra from ROSINA–DFMS, COSAC and Ptolemy. Mon Not R Astron Soc 469:S130–S141

Altwegg K, Balsiger H, Hänni N, Rubin M, Schuhmann M, Schroeder I, Sémon T, Wampfler S, Berthelier J-J, Briois C, Combi M, Gombosi TI, Cottin H, De Keyser J, Dhooghe F, Fiethe B, Fuselier SA (2020) Evidence of ammonium salts in comet 67P as explanation for the nitrogen depletion in cometary comae. Nat Astron 4:533–540

Altwegg K, Combi M, Fuselier SA, Hänni N, De Keyser J, Mahjoub A, Müller DR, Pestoni B, Rubin M, Wampfler SF (2022) Abundant ammonium hydrosulphide embedded in cometary dust grains. Mon Not R Astron Soc 516:3900–3910

Anders E, Grevesse N (1989) Abundances of the elements: meteoritic and solar. Geochim Cosmochim Acta 53:197–214

Bardyn A, Baklouti D, Cottin H, Fray N, Briois C, Paquette J, Stenzel O, Engrand C, Fischer H, Hornung K, Isnard R, Langevin Y, Lehto H, Le Roy L, Ligier N, Merouane S, Modica P, Orthous-Daunay F-R, Rynö J, Schulz R, Silén J, Thirkell L, Varmuza K, Zaprudin B, Kissel J, Hilchenbach M (2017) Carbon-rich dust in comet 67P/Churyumov-Gerasimenko measured by COSIMA/Rosetta. Mon Not R Astron Soc 469:S712

Barrett RA, Zolensky ME, Horz F, Lindstrom DJ, Gibson EK (1992) Suitability of silica aerogel as a capture medium for interplanetary dust. In: Ryder G, Sharpton V (eds) Proceedings of the 19th Lunar and Planetary Science Conference, pp 203–212

Beckett JR (1986) The origin of calcium-, aluminum-rich inclusions from carbonaceous chondrites: an experimental study. PhD thesis. University of Chicago

Berger E, Zega T, Keller L, Lauretta D (2011) Evidence for aqueous activity on comet 81P/Wild 2 from sulfide mineral assemblages in Stardust samples and CI chondrites. Geochim Cosmochim Acta 75:3501–3515

Biver N, Bockelée-Morvan D (2019) Complex organic molecules in comets from remote-sensing observations at millimeter wavelengths. ACS Earth Space Chem 3:1550–1555

Bowles J, Howie R, Vaughan D, Zussman J (2011) Rock-forming minerals, volume 5A: non-silicates: oxides, hydroxides and sulphides, 2nd edn. The Geological Society, London, pp 817–819

Bradley J (1994) Chemically-anomolous, preaccretionally irradiated grains in interplanetary dust from comets. Science 265:925–929

Brearley A, Jones R (1998) Chondritic meteorites. In: Papike JJ (ed) Planetary materials. Reviews in Mineralogy, vol 36. MSA/De Gruyter, pp 3-01–398

Bridges J, Changela H, Nayakshin S, Starkey N, Franchi I (2012) Chondrule fragments from comet Wild2: evidence for high temperature processing in the outer Solar System. Earth Planet Sci Lett 341–344:186–194

Brown M, Barkume K, Ragozzine D, Schaller E (2007) A collisional family of icy objects in the Kuiper Belt. Nature 446:294–299

Brownlee D (2014) The Stardust mission: analyzing samples from the edge of the Solar System. Annu Rev Earth Planet Sci 42:179–205

Brownlee D, Joswiak D, Love S, Nier A, Schlutter D, Bradley J (1993) Identification of cometary and asteroidal particles in stratospheric IDP collections. In: Proceedings of the 24th Lunar and Planetary Science Conference. Lunar and Planetary Institute, Houston, pp 205–206

Brownlee D, Tsou P, Aléon J, Alexander C, Araki T, Bajt S, Baratta G, Bastien R, Bland P, Bleuet P, Borg J, Bradley J, Brearley A, Brenker F, Brennan S, Bridges J, Browning N, Brucato J, Brucato H, Bullock E, Burchell M, Busemann H, Butterworth A, Chaussidon M, Cheuvront A, Chi M, Cintala M, Clark B, Clemett S, Cody G, Colangeli L, Cooper G, Cordier P, Daghlian C, Dai Z, D'Hendecourt L, Djouadi Z, Dominguez G, Duxbury T, Dworkin J, Ebel D, Economou T, Fairey S, Fallon S, Ferrini G, Ferroir T, Fleckenstein H, Floss C, Flynn F, Franchi I, Fries M, Gainsforth Z, Gallien J-P, Genge M, Gilles

M, Gillet P, Gilmour J, Glavin D, Gounelle M, Grady M, Graham G, Grant P, Green S, Grossemy F, Grossman L, Grossman J, Guan Y, Hagiya K, Harvey R, Heck P, Herzog G, Hoppe P, Hörz F, Huth J, Hutcheon I, Ishii H, Ito M, Jacob D, Jacobsen C, Jacobsen S, Joswiak D, Kearsley A, Keller L, Khodja H, Kilcoyne D, Kissel J, Krot A, Langenhorst F, Lanzirotti A, Le L, Leshin L, Leitner J, Lemelle L, Leroux H, Liu M-C, Luening K, Lyon I, MacPherson G, Marcus M, Marhas K, Matrajt G, Meibom A, Mennella V, Messenger K, Mikouchi T, Mostefaoui S, Nakamura T, Nakano T, Newville M, Nittler L, Ohnishi I, Ohsumi K, Okudaira K, Papanastassiou D, Palma R, Palumbo M, Pepin R, Perkins D, Perronnet M, Pianetta P, Rao W, Rietmeijer F, Robert F, Rost D, Rotundi A, Ryan R, Sandford S, Schwandt C, See T, Schlutter D, Sheffield-Parker J, Simionovici A, Simon S, Sitnitsky I, Snead C, Spencer M, Stadermann F, Steele A, Stephan T, Stroud R, Susini J, Sutton S, Taheri M, Taylor S, Teslich N, Tomeoka K, Tomioka N, Toppani A, Trigo-Rodríguez J, Troadec D, Tsuchiyama A, Tuzolino A, Tyliszczak T, Uesugi K, Velbel M, Vellenga J, Vicenzi E, Vincze L, Warren J, Weber I, Weisberg M, Westphal A, Wirick S, Wooden D, Wopenka B, Wozniakiewicz P, Wright I, Yabuta H, Yano H, Young E, Zare R, Zega T, Ziegler K, Zimmerman L, Zinner E, Zolensky M (2006) Comet Wild 2 under a microscope. Science 314:1711–1716

Brownlee D, Joswiak D, Matrajt G (2012) Overview of the rocky component of Wild 2 comet samples: insight into the early Solar System, relationship with meteoritic materials and the differences between comets and asteroids. Meteorit Planet Sci 47:453–470

Bullock E, Gounelle M, Lauretta D, Grady M, Russell S (2005) Mineralogy and texture of Fe-Ni sulfides in CI1 chondrites: clues to the extent of aqueous alteration on the CI1 parent body. Geochim Cosmochim Acta 69:2687–2700

Chi M, Ishii H, Simon S, Bradley J, Dai Z, Joswiak D, Browning N, Matrajt D (2009) The origin of refractory minerals in comet 81P/Wild 2. Geochim Cosmochim Acta 73:7150–7161

Cody GD, Ade H, Alexander MO'D, Araki T, Butterworth A, Fleckenstein H, Flynn G, Gilles MK, Jacobsen C, Kilcoyne ALD, Messenger K, Sandford SA, Tyliszczak T, Westpha AJ, Wirick S, Yabuta H (2008) Quantitative organic and light-element analysis of comet 81P/Wild 2 particles using C-, N-, and O-μ-XANES. Meteorit Planet Sci 43:353–365

De Gregorio BT, Stroud RM, Nittler LR, Alexander CMO, Kilcoyne ALD, Zega TJ (2010) Isotopic anomalies in organic nanoglobules from comet 81P/Wild 2: comparison to Murchison nanoglobules and isotopic anomalies induced in terrestrial organics by electron irradiation. Geochim Cosmochim Acta 74:4454–4470

De Gregorio BT, Stroud RM, Cody GD, Nittler LR, David Kilcoyne AL, Wirick S (2011) Correlated microanalysis of cometary organic grains returned by Stardust. Meteorit Planet Sci 46:1376–1396

De Gregorio B, Stroud R, Nittler L, Alexander C, Bassim N, Cody G, Kilcoyne A, Sandford S, Milam S, Nuevo M, Zega T (2013) Isotopic and chemical variation of organic nanoglobules in primitive meteorites. Meteorit Planet Sci 48:904–928

Defouilloy C, Nakashima D, Joswiak D, Brownlee D, Tenner T, Kita N (2017) Origin of crystalline silicates from comet 81P/Wild 2: combined study on their oxygen isotopes and mineral chemistry. Earth Planet Sci Lett 465:145–154

Ebel DS (2006) Condensation of rocky material in astrophysical environments. In: Lauretta DS, McSween HY Jr (eds) Meteorites and the early Solar System II. University of Arizona Press, Tucson, pp 253–277

Ebel DS, Greenberg M, Rivers ML, Newville M (2009) 3-dimensional textural and compositional analysis of particle tracks and fragmentation history in aerogel. Meteorit Planet Sci 44:1445–1463

Ebel D, Weisberg M, Becket J (2012) Thermochemical stability of low-iron, manganese-enriched olivine in astrophysical environments. Meteorit Planet Sci 47:585–593

Ebel DS, Zolensky ME, Gonzalez C, Rodriguez M, Vu J (2023) Comet Wild 2 sample track 3D images and S-XRF stereo maps. In: 86th meteoritical society meeting. Abs. #6026

Elsila JE, Glavin DP, Dworkin JP (2009) Cometary glycine detected in samples returned by Stardust. Meteorit Planet Sci 44:1323–1330

Floss C, Stadermann FJ, Kearsley AT, Burchell MJ, Ong WJ (2013) The abundance of presolar grains in comet 81P/Wild 2. Astrophys J 763:140

Flynn GJ, Bajt S, Sutton SR, Zolensky M, Thomas KL, Keller LP (1996) The abundance pattern of elements having low nebular condensation temperatures in the IDPs: evidence for a new type of chondritic material. In: Gustafson B, Hanner M (eds) Physics, chemistry and dynamics of interplanetary dust. ASP Conference Series, vol 104. Astronomical Society of the Pacific, San Francisco, pp 291–294

Flynn GJ, Bleuet P, Borg J, Bradley J, Brenker F, Brennan S, Bridges J, Brownlee DE, Bullock E, Clark BC, Dai ZR, Daghlian C, Djouadi Z, Fakra S, Ferroir T, Floss C, Franchi IA, Gainsforth Z, Gallien J-P, Gillet Ph, Grant PG, Graham GA, Green SF, Grossemy F, Heck P, Herzog GF, Hoppe P, Hörz F, Huth J, Ignatyev KI, Ishii HA, Joswiak D, Kearsley AT, Khodja H, Lanzirotti A, Leitner J, Lemelle L, Leroux H, Luening K, MacPherson G, Marhas KK, Marcus MA, Matrajt G, Nakamura T, Nakano T, Newville M, Papanastassiou DA, Pianetta P, Rao W, Rietmeijer FJM, Rost D, Schwandt CS, See TH,

Sheffield-Parker J, Simionovic A, Sitnitsky I, Snead CJ, Stadermann FJ, Stephan T, Stroud RM, Susini J, Suzuki Y, Sutton SR, Taylor S, Teslich N, Troadec D, Tsou P, Tsuchiyama A, Uesugi K, Westphal A, Wozniakiewicz P, Vicenzi E, Vincze L, Zolensky ME (2006) Elemental compositions of comet 81P/Wild 2 samples collected by Stardust. Science 314:1730–1733

Frank D, Zolensky M, Le L, Weisberg M, Kimura M (2013) Highly reduced forsterite and enstatite from Stardust track 61: implications for radial transport of E asteroid material. In: 44th Lunar and Planetary Science Conference. Abstract 3082

Frank D, Zolensky M, Le L (2014) Olivine in terminal particles of Stardust aerogel tracks and analogous grains in chondrite matrix. Geochim Cosmochim Acta 142:240–259

Fray N, Bardyn A, Cottin H, Altwegg K, Baklouti D, Briois C, Colangeli L, Engrand C, Fischer H, Glasmachers A, Grün E, Haerendel G, Henkel H, Höfner H, Hornung K, Jessberger EK, Koch A, Krüger H, Langevin Y, Lehto H, Lehto K, Le Roy L, Merouan S, Modica P, Orthous-Daunay F-R, Paquette J, Raulin F, Rynö J, Schulz R, Silén J, Siljeström S, Steiger W, Stenzel O, Stephan T, Thirkell L, Thomas R, Torkar K, Varmuza K, Wanczek K-P, Zaprudin B, Kissel J, Hilchenbach M (2016) High-molecular-weight organic matter in the particles of comet 67P/Churyumov-Gerasimenko. Nature 538:72–74

Fray N, Bardyn A, Cottin H, Baklouti D, Briois C, Engran C, Fischer H, Hornung K, Isnard R, Langevin Y, Lehto H, Le Roy L, Mellado EM, Merouane S, Modica P, Orthous-Daunay F-R, Paquette J, Rynö J, Schulz R, Silén J, Siljeström S, Stenzel O, Thirkell L, Varmuza K, Zaprudin B, Kissel J, Hilchenbach M (2017) Nitrogen-to-carbon atomic ratio measured by COSIMA in the particles of comet 67P/Churyumov-Gerasimenko. Mon Not R Astron Soc 469:S506–S516

Fukuda K, Brownlee D, Joswiak D, Tenner T, Kimura M, Kita N (2020) Correlated isotopic and chemical evidence for condensation origins of olivine in comet 81P/Wild 2 and in AOAs from CV and CO chondrites. Geochim Cosmochim Acta 293:544–574

Gainsforth Z, Butterworth AL, Stodolna J, Westphal AJ, Huss GR, Nagashima K, Ogliore R, Brownlee DE, Joswiak D, Tyliszczak T, Simionovici AS (2015) Constraints on the formation environment of two chondrule-like igneous particles from comet 81P/Wild 2. Meteorit Planet Sci 50:976–1004

Gainsforth Z, Westphal A, Butterworth A, Jilly-Rehak C, Brownlee D, Joswiak D, Ogliore R, Zolensky M, Bechtel H, Ebel D, Huss G, Sandford S, White A (2019) Fine-grained material associated with a large sulfide returned from comet 81P/Wild 2. Meteorit Planet Sci 54:1069–1091

Gardner E, Lehto HJ, Lehto K, Fray N, Bardyn A, Lönnberg T, Merouane S, Isnard R, Cottin H, Hilchenbach M, the C. team (2020) The detection of solid phosphorus and fluorine in the dust from the coma of comet 67P/Churyumov-Gerasimenko. Mon Not R Astron Soc 499:1870–1873

Gounelle M, Devouard B, Engrand C, Genge M, Topani A, Leroux H (2002) TEM study of Antarctic micrometeorites: a preliminary report. Meteorit Planet Sci 37:A55

Gounelle M, Spurny P, Bland P (2006) The orbit and atmospheric trajectory of the Orgueil meteorite from historical records. Meteorit Planet Sci 41:135–150

Greenberg M, Ebel DS (2010) Laser scanning confocal microscopy of Stardust material. Geosphere 6:515–523

Greenberg M, Ebel DS (2012) Properties of original impactors estimated from three-dimensional analysis of whole Stardust tracks. Meteorit Planet Sci 47:634–648

Greshake A, Bischoff A (1996) Chromium-bearing phases in Orgueil (CI): discovery of magnesiochromite (MgCr2O4), ureyite (NaCrSi2O6), and chromium oxide (Cr2O3). Lunar Planet Sci 27:461

Grossman J, Rubin A, MacPherson G (1988) A unique volatile-poor carbonaceous chondrite with implications for nebular agglomeration and fractionation processes. Earth Planet Sci Lett 91:33–54

Hadraoui K, Cottin H, Ivanovski SL, Zapf P, Altwegg K, Benilan Y, Biver N, Della Corte V, Fray N, Lasue J, Merouane S, Rotundi A, Zakharov V (2019) Distributed glycine in comet 67P/Churyumov-Gerasimenko. Astron Astrophys 630:A32

Hadraoui K, Cottin H, Ivanovski SL, Zapf P, Altwegg K, Benilan Y, Biver N, Della Corte V, Fray N, Lasue J, Merouane S, Rotundi A, Zakharov V (2021) Distributed glycine in comet 67P/Churyumov-Gerasimenko (corrigendum). Astron Astrophys 651:C2

Hanner M, Zolensky M (2010) The mineralogy of comets. In: Henning T (ed) Astromineralogy. Lecture Notes in Physics, vol 815. Springer, Berlin, Heidelberg, pp 203–232

Hänni N, Altwegg K, Balsiger H, Combi M, Fuselier SA, De Keyser J, Pestoni B, Rubin M, Wampfler SF (2021) Cyanogen, cyanoacetylene, and acetonitrile in comet 67P and their relation to the cyano radical. Astron Astrophys 647:A22

Hänni N, Altwegg K, Combi M, Fuselier SA, De Keyser J, Rubin M, Wampfler SF (2022) Identification and characterization of a new ensemble of cometary organic molecules. Nat Commun 13:3639

Heck PR, Hoppe P, Huth J (2012) Sulfur four isotope NanoSIMS analysis of comet-81P/Wild 2 dust in impact craters on aluminum foil C2037N from NASA's Stardust mission. Meteorit Planet Sci 47:649–659

Hicks J, Macarthur J, Bridges J, Price M, Wickham-Eade J, Burchell M, Hansford G, Butterworth A, Gurman S, Baker S (2017) Magnetite in comet Wild 2: evidence for parent body aqueous alteration. Meteorit Planet Sci 52:2075–2096

Hörz F, Zolensky M, Bernhard R, See T, Warren J (2000) Impact features and projectile residues in aerogel exposed on Mir. Icarus 147:559–579

Huss GR, Lewis RS (1994) Noble gases in presolar diamonds II: component abundances reflect thermal processing. Meteoritics 29:811–829

Huss G, Rubin A, Grossman J (2006) Thermal metamorphism in chondrites. In: Lauretta DS, McSween HY Jr (eds) Meteorites and the early Solar System II. University of Arizona Press, Tucson, pp 567–586

Hutchison M, Scott S (1983) Experimental calibration of the sphalerite cosmobarometer. Geochim Cosmochim Acta 47:101–108

Ishii H (2019) Comparison of GEMS in interplanetary dust particles and GEMS-like objects in a Stardust impact track in aerogel. Meteorit Planet Sci 54:202–219

Ishii HA, Bradley JP, Dai ZR, Chi M, Kearsley AT, Burchell MJ et al (2008) Comparison of comet 81P/Wild 2 dust with interplanetary dust from comets. Science 319:447

Ishii H, Bradley J, Dai Z, Chi M, Kearsley A, Burchell M, Browning N, Molster F (2010) Comparison of comet 81P/Wild 2 dust with interplanetary dust from comets. Science 319:447–450

Ishii H, Bradley J, Bechtel H, Brownlee D, Bustillo K, Ciston J, Cuzzi J, Floss C, Joswiak D (2018) Multiple generations of grain aggregation in different environments preceded Solar System body formation. Proc Natl Acad Sci 115:6608–6613

Isnard R, Bardyn A, Fray N, Briois C, Cottin H, Paquette J, Stenzel O, Alexander C, Baklouti D, Engrand C, Orthous-Daunay FR, Siljeström S, Varmuza K, Hilchenbach M (2019) H/C elemental ratio of the refractory organic matter in cometary particles of 67P/Churyumov-Gerasimenko. Astron Astrophys 630:A27

Jessberger E (1999) Rocky cometary particulates: their elemental, isotopic and mineralogical ingredients. Space Sci Rev 90:91–97

Joswiak D, Brownlee D, Matrajt G, Westphal A, Snead C (2009) Kosmochloric Ca-rich pyroxenes and FeO-rich olivines (Kool grains) and associated phases in Stardust tracks and chondritic porous interplanetary dust particles: possible precursors to FeO-rich type II chondrules in ordinary chondrites. Meteorit Planet Sci 44:1561–1588

Joswiak D, Brownlee D, Matrajt G, Westphal A, Snead C, Gainsforth Z (2012) Comprehensive examination of large mineral and rock fragments in Stardust tracks: mineralogy, analogous extraterrestrial materials, and source regions. Meteorit Planet Sci 47:471–524

Joswiak D, Brownlee D, Nguyen A, Messenger S (2017) Refractory materials in comet samples. Meteorit Planet Sci 52:1612–1648

Joswiak D, Brownlee D, Zhang M, Kita N (2024) Microchondrules and lithic fragments in fine-grained chondrule rims and equivalent materials in comets. In: 55th Lunar and Planetary Science Conference. Abstract 1677

Kearsley A, Burchell M, Price M, Cole M, Wozniakiewicz P, Ishii H, Bradley J, Fries M, Foster N (2012) Experimental impact features in Stardust aerogel: how track morphology reflects particle structure, composition, and density. Meteorit Planet Sci 47:737–762

Keller L, Messenger S (2008) Relict amorphous silicates in Stardust samples. In: 71st annual meteoritical society meeting. Abstracts

Keller T, Tornos F, Hanchar J, Pietruszka D, Soldati A, Dingwell D, Suckale J (2022) Genetic model of the El Laco magnetite-apatite deposits by extrusion of iron-rich melt. Nat Commun 13:6114

King A, Tu V, Schofield P, Najorka J, Russell S, Bates H, Zega T, McCoy T, Keller L, Haenecour P, Thompson M, Thomas-Keprta K, Le L, Connolly H, Lauretta D (2024) The mineralogy of asteroid Bennu: first results from X-ray powder diffraction of returned samples. In: 55th Lunar and Planetary Science Conference. Abstract 1109

Kissel J (1986) Mass spectrometric studies of Halley's comet. In: Todd JFJ (ed) Advances in mass spectrometry 1985. Wiley, Chichester, pp 175–184

Klöck W, Thomas KL, McKay DS, Palme P (1989) Unusual olivine and pyroxene compositions in interplanetary dust and unequilibrated ordinary chondrites. Nature 339:126–128

Langevin Y, Hilchenbach M, Ligier N, Merouane S, Hornung K, Engrand C, Schulz R, Kissel J, Rynö J, Eng P (2016) Typology of dust particles collected by the COSIMA mass spectrometer in the inner coma of 67P/Churyumov Gerasimenko. Icarus 271:76–97

Lasue J, Maroger I, Botet R, Garnier P, Merouane S, Mannel T, Levasseur-Regourd AC, Bentley MS (2019) Flattened loose particles from numerical simulations compared to particles collected by Rosetta. Astron Astrophys 630:A28

Leroux H, Jacob D, Stodolna J, Nakamura-Messenger K, Zolensky M (2008c) Igneous Ca-rich pyroxene in comet 81P/Wild 2. Am Mineral 93:1933–1936

Leroux H, Rietmeijer FJM, Velbel MA, Brearley AJ, Jacob D, Langenhorst F, Bridges JC, Zega TJ, Stroud RM, Cordier P, Harvey RP, Lee M, Gounelle M, Zolensky ME (2008a) A TEM study of thermally modified comet 81P/Wild 2 dust particles by interactions with the aerogel matrix during the Stardust capture process. Meteorit Planet Sci 43:97–120

Leroux H, Stroud RM, Dai ZR, Graham G, Troadec D, Bradley J, Teslich N, Borg J, Kearsley A, Hörz F (2008b) Transmission electron microscopy of cometary residues from micron-sized craters in the Stardust Al-foils. Meteorit Planet Sci 43:143–160

Lisse CM, VanCleve J, Adams AC, A'Hearn MF, Fernandez YR, Farnham TL, Armus L, Grillmair CJ, Ingalls J, Belton MJS, Groussin O, McFadden LA, Meech KJ, Schultz PH, Clark BC, Feaga LM, Sunshine JM (2006) Spitzer spectral observations of the deep impact ejecta. Science 313:635–640

Lodders K (2020) Solar elemental abundances. In: The Oxford research encyclopedia of planetary science. Oxford University Press, London, p 68

Lodders K (2021) Relative atomic Solar System abundances, mass fractions, and atomic masses of the elements and their isotopes, composition of the solar photosphere, and compositions of the major chondritic meteorite groups. Space Sci Rev 217:44

Mannel T, Bentley MS, Boakes PD, Jeszenszky H, Ehrenfreund P, Engrand C, Koeberl C, Levasseur-Regourd AC, Romstedt J, Schmied R, Torkar K, Weber I (2019) Dust of comet 67P/Churyumov-Gerasimenko collected by Rosetta/MIDAS: classification and extension to the nanometre scale. Astron Astrophys 630:A26

Marrocchi Y, Euverte R, Villeneuve J, Batanova V, Welsch B, Ferrière L, Jacquet E (2019) Formation of CV chondrules by recycling of amoeboid olivine aggregate-like precursors. Geochim Cosmochim Acta 247:121–141

Matrajt G, Ito M, Wirick S, Messenger S, Brownlee D, Joswiak D, Flynn G, Sandford S, Snead C, Westphal A (2008) Carbon investigation of two Stardust particles: a TEM, nanoSIMS, and XANES study. Meteorit Planet Sci 43:315–334

McKeegan K, Aleon J, Bradley J, Brownlee D, Busemann H, Butterworth A, Chaussidon M, Fallon S, Floss C, Gilmour J, Gounelle M, Graham G, Guan Y, Heck P, Hoppe P, Hutcheon I, Huth J, Ishii H, Ito M, Jacobsen S, Kearsley A, Leshin L, Liu M, Lyon I, Marhas K, Marty B, Matrajt G, Meibom A, Messenger S, Mostefaoui S, Mukhopadhyay S, Nakamura-Messenger K, Nittler L, Palma R, Pepin R, Papanastassiou D, Robert F, Schlutter D, Snead C, Stadermann F, Stroud R, Tsou P, Westphal A, Young E, Ziegler K, Zimmermann L, Zinner E (2006) Isotopic compositions of cometary matter returned by Stardust. Science 314:1724–1728

Meier R, Owen TC, Matthews HE, Jewitts DC, Bockelée-Morvan D, Biver N, Crovisier J, Gautier D (1998) A determination of the HDO/H2O ratio in comet C/1995 O1 (Hale-Bopp). Science 279:842–844

Mikouchi T, Tachikawa O, Hagiya K, Ohsumi K, Zolensky M (2007) Mineralogy and crystallography of comet 81P/Wild 2 particles. In: 38th Lunar and Planetary Science Conference. #1946

Mikouchi T, Nakamura T, Zolensky M, Yoshida H, Nakashima D, Hagiya K, Morita T, Kikuiri M, Amano K, Kagawa E, Yurimoto H, Noguchi T, Okazaki R, Yabuta H, Naraoka H, Sakamoto K, Tachibana S, Watanabe S, Tsuda Y (2022) Olivine compositional variation of asteroid Ryugu samples: possible precursors of Ryugu's parent asteroid. In: 53rd Lunar and Planetary Science Conference. Abstract 1935

Müller DR, Altwegg K, Berthelier JJ, Combi M, De Keyser J, Fuselier SA, Hänni N, Pestoni B, Rubin M, Schroeder IRHG, Wampfler SF (2022) High D/H ratios in water and alkanes in comet 67P/Churyumov-Gerasimenko measured with Rosetta/ROSINA DFMS. Astron Astrophys 662:A69

Nakamura T, Noguchi T, Tsuchiyama A, Ushikubo T, Kita N, Valley J, Zolensky M, Kakazu Y, Sakamoto K, Mashio E, Uesugi K, Nakano T (2008) Chondrulelike objects in short-period comet 81P/Wild. Science 321:1664–1667

Nakamura T, Matsumoto M, Amano K, Enokido Y, Zolensky M, Mikouchi T, Genda H, Tanaka S, Zolotov M, Kurosawa K, Wakita S, Hyoudo R, Nagano H, Nakashima D, Takahashi Y, Fujioka Y, Kikuiri M, Kagawa E, Matsuoka M, Brearley A, Tsuchiyama A, Uesugi M, Matsuno J, Kimura Y, Sato M, Milliken R, Tatsumi E, Sugita S, Hiroi T, Kitazato K, Brownlee D, Joswiak D, Takahashi M, Ninomiya K, Takahashi T, Osawa T, Terada K, Brenker F, Tkalcec B, Vincze L, Brunetto R, Aléon-Toppani A, Chan Q, Roskosz M, Viennet J, Beck P, Alp E, Michikami T, Nagaashi Y, Tsuji T, Ino Y, Martinez J, Han J, Dolocan A, Bodnar R, Tanaka M, Yoshida H, Sugiyama K, King A, Fukushi K, Suga H, Yamashita S, Kawai T, Inoue K, Nakato A, Noguchi T, Vilas F, Hendrix A, Jaramillo C, Lorin D, Dominguez G, Gainsforth Z, Engrand C, Duprat J, Russell S, Bonato E, Ma C, Kawamoto T, Wada T, Watanabe S, Endo R, Enju S, Riu L, Rubino S, Tack P, Takeshita S, Takeichi Y, Takeuchi A, Takigawa A, Takir D, Tanigaki T, Taniguchi A, Tsukamoto K, Yagi T, Yamada S, Yamamoto K, Yamashita Y, Yasutake M, Uesugi K, Umegaki I, Huan CI, Ishizaki T, Okumura S, Palomba E, Pilorget C, Potin S, Abdulkareem A, Anada S, Araki Y, Sakatani N, Schultz C, Sekizawa O, Sitzman S, Sugiura K, Sun M, Dartois E, De Pauw E, Dionnet Z, Djouadi Z, Falkenberg G, Fujita R, Fukuma T, Gearba I, Hagiya K, Hu M, Kato T, Kawamura T, Kimura M, Kubo M, Langenhorst F, Lantz C, Lavina B, Lindner M, Vekemans B, Baklouti D, Bazi B, Borondics F, Nagasawa S, Nishiyama G, Nitta K, Mathurin J, Matsumoto T, Mitsukawa I, Miura H, Miyake A, Miyake Y, Yurimoto H, Okazaki R, Yabuta H, Naraoka H, Sakamoto K, Tachibana S, Connolly H, Lauretta D, Yoshitake M, Yoshikawa M, Yoshikawa K, Yoshihara K, Yokota Y, Yogata K, Yano H, Yamamoto Y, Yamamoto D, Yamada M, Yamada T, Yada T, Wada K, Usui T, Tsukizaki

R, Terui F, Takeuchi H, Takei Y, Suzuki A, Soejima H, Shirai K, Shimaki Y, Senshu H, Sawada H, Saiki T, Ozaki M, Ono G, Okada T, Ogawa N, Ogawa K, Noguchi R, Noda H, Nishimura M, Namiki N, Nakazawa S, Morota T, Miyazaki A, Miura A, Mimasu Y, Matsumoto K, Kumagai K, Kouyama T, Kikuchi S, Kawahara K, Kameda S, Iwata T, Ishihara Y, Ishiguro M, Ikeda H, Hosoda S, Honda R, Honda C, Hitomi Y, Hirata N, Hirata N, Hayashi T, Hayakawa M, Hatakeda K, Furuya S, Fukai R, Fujii A, Cho Y, Arakawa M, Abe M, Watanabe S, Tsuda Y (2022) Formation and evolution of carbonaceous asteroid Ryugu: direct evidence from returned samples. Science 377:eabn8671

Nakamura-Messenger K, Messenger S, Keller LP, Clemett SJ, Zolensky ME (2006) Organic globules in the Tagish Lake Meteorite: remnants of the protosolar disk. Science 314:1439–1442

Nakamura-Messenger K, Keller L, Clemett S, Messenger S, Ito M (2011) Nanometer-scale anatomy of entire Stardust tracks. Meteorit Planet Sci 46:1033–1051

Nakashima D, Ushikubo T, Joswiak DJ, Brownlee DE, Matrajt G, Weisberg MK, Zolensky ME, Kita NT (2012) Oxygen isotopes in crystalline silicates of comet Wild 2: a comparison of oxygen isotope systematics between Wild 2 particles and chondritic materials. Earth Planet Sci Lett 357–358:355–365

Nakashima D, Ushikubo T, Kita N, Weisberg M, Zolensky ZM, Ebel DS (2015) Late formation of a comet Wild 2 crystalline silicate particle, Pyxie, inferred from Al-Mg chronology of plagioclase. Earth Planet Sci Lett 410:54–61

Nakashima D, Nakamura T, Zhang M, Kita NT, Mikouchi T, Yoshida H, Enokido Y, Morita T, Kikuiri M, Amano K, Kagawa E, Yada T, Nishimura M, Nakato A, Miyazaki A, Yogata K, Abe M, Okada T, Usui T, Yoshikawa M, Saiki T, Tanaka S, Nakazawa S, Terui F, Yurimoto H, Noguchi T, Yabuta H, Naraoka H, Okazaki R, Sakamoto K, Watanabe S, Tachibana S, Tsuda Y (2023) Chondrule-like objects and Ca-Al-rich inclusions in Ryugu may potentially be the oldest Solar System materials. Nat Commun 14:532

Nguyen AN, Berger EL, Nakamura-Messenger K, Messenger S, Keller LP (2017) Coordinated mineralogical and isotopic analyses of a cosmic symplectite discovered in a comet 81P/Wild 2 sample. Meteorit Planet Sci 52:2004–2016

Nguyen A, Clemett S, Thomas-Keprta K, Keller L, Glavin D, Dworkin J, Barnes J, Zega T, McCoy T, Thompson M, Haenecour P, King A, Connolly H, Lauretta D (2024) Coordinated analysis of isotopically anomalous nanoglobules and insoluble organic matter in quick-look samples from asteroid Bennu. In: 55th Lunar and Planetary Science Conference. Abstract 2500

Ogliore RC, Butterworth A, Gainsforth Z, Huss GR, Nagashima K, Stodolna J, Westphal AJ (2012a) Sulfur isotope measurements of a Stardust fragment. In: Lunar and Planetary Science Conference. Abstract 1670

Ogliore RC, Huss GR, Nagashima K, Butterworth AL, Gainsforth Z, Stodolna J, Westphal AJ, Joswiak D, Tyliszczak T (2012b) Incorporation of a late-forming chondrule into comet Wild 2. Astrophys J Lett 745:L19

Ogliore R, Nagashima K, Huss G, Westphal A, Gainsforth Z, Buterworth A (2015) Oxygen isotopic composition of coarse- and fine-grained material from comet 81P/Wild 2. Geochim Cosmochim Acta 166:74–91

Okudaira K, Noguchi T, Nakamura T, Sugita S, Sekine Y, Yano H (2004) Evaluation of mineralogical alteration of micrometeoroid analog materials captured in aerogel. Adv Space Res 34:2299–2304

Paquette J, Ferguson F, Nuth J (2011) A model of silicate grain nucleation and growth in circumstellar outflows. Astrophys J 732:62

Paquette JA, Hornung K, Stenzel OJ, Rynö J, Silen J, Kissel J, Hilchenbach M, COSIMA Team (2017) The 34S/32S isotopic ratio measured in the dust of comet 67P/Churyumov-Gerasimenko by Rosetta/COSIMA. Mon Not R Astron Soc 469:S230–S237

Paquette JA, Engrand C, Hilchenbach M, Fray N, Stenzel OJ, Silen J, Rynö J, Kissel J, COSIMA Team (2018) The oxygen isotopic composition (18O/16O) in the dust of comet 67P/Churyumov-Gerasimenko measured by COSIMA on-board Rosetta. Mon Not R Astron Soc 477:3836–3844

Paquette JA, Fray N, Bardyn A, Engrand C, Alexander CMO, Siljeström S, Cottin H, Merouane S, Isnard R, Stenzel OJ, Fischer H, Rynö J, Kissel J, Hilchenbach M (2021) D/H in the refractory organics of comet 67P/Churyumov-Gerasimenko measured by Rosetta/COSIMA. Mon Not R Astron Soc 504:4940

Poch O, Istiqomah I, Quirico E, Beck P, Schmitt B, Theulé P, Faure A, Hily-Blant P, Bonal L, Raponi A, Ciarniello M, Rousseau B, Potin S, Brissaud O, Flandinet L, Filacchione G, Pommerol A, Thomas N, Kappel D, Mennella V, Moroz L, Vinogradoff V, Arnold G, Erard S, Bockelée-Morvan D, Leyrat C, Capaccioni F, De Sanctis MC, Longobardo A, Mancarella F, Palomba E, Tosi F (2020) Ammonium salts are a reservoir of nitrogen on a cometary nucleus and possibly on some asteroids. Science 367:eaaw7462

Rietmeijer F (1998) Interplanetary dust particles. In: Papike JJ (ed) Planetary materials. Reviews in Mineralogy, vol 36. MSA/De Gruyter, pp 2-1–2-95

Rietmeijer FJM (2002) The earliest chemical dust evolution in the solar nebula. Geochemistry 62:1–45

Rotundi A, Rietmeijer F, Ferrari M, Della Corte V, Baratta G, Brunetto R, Dartois E, Djouadi Z, Merouane S, Borg J, Brucato J, Le Sergeant D'Hendecourt L, Mennella V, Palumbo M, Palumbo P (2014) Two refractory Wild 2 terminal particles from a carrot-shaped track characterized combining MIR/FIR/Raman microspectroscopy and FE-SEM/EDS analyses. Meteorit Planet Sci 49:550–575

Rotundi A, Sierks H, Della Corte V, Fulle M, Gutierrez P, Lara L, Barbieri C, Lamy P, Rodrigo R, Koschny D, Rickman H, Keller H, López-Moreno J, Accolla M, Agarwal J, A'Hearn M, Altobelli N, Angrilli F, Barucci M, Bertaux J-L, Bertini I, Bodewits D, Bussoletti E, Colangeli L, Cosi M, Cremonese G, Crif J-F, Da Deppo V, Davidsson B, Debei S, De Cecco M, Esposito F, Ferrari M, Fornasier S, Giovane F, Gustafson B, Green S, Groussin O, Grün E, Güttler C, Herranz M, Hviid S, Ip W, Ivanovski S, Jerónimo J, Jorda L, Knollenberg J, Kramm R, Kührt E, Küppers M, Lazzarin M, Leese M, López-Jiménez A, Lucarelli F, Lowry C, Marzari F, Epifani E, McDonnell J, Mennella V, Michalik H, Molina A, Morales R, Moreno F, Mottola S, Naletto G, Oklay N, Ortiz J, Palomba E, Palumbo P, Perrin J-M, Rodríguez J, Sabau L, Snodgrass C, Sordini R, Thomas N, Tubiana C, Vincent J-B, Weissman P, Wenzel K-P, Zakharov V, Zarnecki J (2015) Dust measurements in the coma of comet 67P/Churyumov-Gerasimenko inbound to the Sun. Science 347:aaa3905

Rubin A, Ma C (2021) Meteorite mineralogy. Cambridge University Press

Rubin M, Altwegg K, Berthelier J-J, Combi MR, De Keyser J, Dhooghe F, Fuselier S, Gombosi TI, Hänni N, Müller D, Pestoni B, Wampfler SF, Wurz P (2022) Refractory elements in the gas phase for comet 67P/Churyumov-Gerasimenko. Astron Astrophys 658:A87

Sandford S, Aléon J, Alexander CMO'D, Araki T, Bajt S, Baratta GA, Borg J, Brucato JR, Burchell MJ, Busemann H, Butterworth A, Clemett SJ, Cody G, Colangeli L, Cooper C, D'Hendecourt L, Djouadi Z, Dworkin JP, Ferrini G, Fleckenstein H, Flynn GJ, Franchi IA, Fries M, Gilles MK, Glavin DP, Gounelle M, Grossemy F, Jacobsen C, Keller LP, Kilcoyne AL, Leitner J, Matrajt G, Meibom A, Mennella V, Mostefaoui S, Nittler LR, Palumbo ME, Robert F, Rotundi A, Snead CJ, Spencer MK, Steele A, Stephan T, Tyliszczak T, Westphal AJ, Wirick S, Wopenka B, Yabuta H, Zare RN, Zolensky M (2006) Organics captured from comet Wild 2 by the Stardust spacecraft. Science 314:1720–1724

Sansberro I, Fray N, Cottin H, Baklouti D, Bardyn A, Briois C, Engrand C, Paquette J, Silén J, Stenzel OJ, Hilchenbach M (2022) Variability in the elemental composition of the dust particles of 67P/Churyumov-Gerasimenko. In: 44th COSPAR scientific assembly. Held 16-24 July 44, p 175

Schmitt-Kopplin P, Gabelica Z, Gougeon RD, Fekete A, Kanawati B, Harir M, Gebefuegi I, Eckel G, Hertkorn N (2010) High molecular diversity of extraterrestrial organic matter in Murchison meteorite revealed 40 years after its fall. Proc Natl Acad Sci 107:2763–2768

Schmitz S, Brenker F, Schoonjans T, Vekemans B, Silversmit G, Vincze L, Burghammer M, Riekel C (2009) In situ identification of a CAI candidate in 81P/Wild 2 cometary dust by confocal high resolution synchrotron X-ray fluorescence. Geochim Cosmochim Acta 73:5483–5492

Schrader DL, Fu RR, Desch SJ, Davidson J (2018) The background temperature of the protoplanetary disk within the first four million years of the Solar System. Earth Planet Sci Lett 504:30–37

Shu F, Shang H, Gounelle M, Glassgold A, Lee T (2001) The origin of chondrules and refractory inclusions in chondritic meteorites. Astrophys J 548:1029–1050

Simon S, Joswiak D, Ishii H, Bradley J, Chi M, Grossman L, Aléon J, Brownlee D, Fallon S, Hutcheon I, Matrajt G, Mckeegan K (2008) A refractory inclusion returned by Stardust from comet 81P/Wild 2. 43:1861–1877

Stadermann FJ, Hoppe P, Floss C, Heck PR, Horz F, Huth J, Kearsley AT, Leitner J, Marhas KK, McKeegan KD, Stephan T (2008) Stardust in Stardust - the C, N, and O isotopic compositions of Wild 2 cometary matter in Al foil impacts. Meteorit Planet Sci 43:299–313

Stodolna J, Gainsforth Z, Butterworth A, Westphal A (2012b) Characterization of preserved primitive fine-grained material from the Jupiter family comet 81P/Wild 2 – a new link between comets and CP-IDPs. Earth Planet Sci Lett 388:367–373

Stodolna J, Jacob D, Leroux H (2012a) Mineralogy and petrology of Stardust particles encased in the bulb of track 80: TEM investigation of the Wild 2 fine-grained material. Geochim Cosmochim Acta 87:35–50

Stroud R, Barosch J, Bonal L, Burgess K, Cody G, De Gregorio B, Daly L, Dartois E, Dobrică E, Duprat J, Engrand C, Harries D, Hashiguchi M, Ishii KY, Kilcoyne A, Langenhorst F, Lee M, Nittler L, Quirico E, Okumura T, Remusat L, Sandford S, Yabuta H, Abe M, Abreu N, Bagot P, Beck P, Bejach L, Bland P, Bradley J, Bridges J, Cymes B, Dazzi A, de la Peña F, Deniset-Besseau A, Enju S, Enokido Y, Frank D, Gray J, Haruta M, Hata S, Hicks L, Igami Y, Jacob D, Kamide K, Komatsu M, Laforet S, Leroux H, Le Guillou C, Martins Z, Marinova M, Martinez M, Mathurin J, Matsumoto M, Matsumoto T, Matsuno J, McFadzean S, Michikami T, Mitsukawa I, Miyake A, Miyahara M, Miyazaki A, Montagnac G, Mostefaoui S, Nakamura T, Nakato A, Naraoka H, Nakauchi Y, Nakazawa S, Nishimura M, Noguchi T, Ohtaki K, Ohigashi T, Okada T, Okumura S, Okazaki R, Phan T, Rebois R, Sakamoto K, Saiki T, Saito H, Seto Y, Shigenaka M, Smith W, Suga H, Sun M, Tachibana S, Takahashi Y, Takeichi Y, Takeuchi A, Takigawa A, Tamenori Y, Tanaka S, Terui F, Thompson M, Tomioka N, Tsuchiyama A, Tsuda Y, Uesugi K, Uesugi M, Usui T, Maximilien Verdier-Paoletti M, Wakabayashi D, Watanabe S, Yada T, Yamashita T, Yasutake M, Yogata K, Yoshikawa M, Yurimoto H, Zanetta P-M, Zega T, Zolensky M (2024) Electron microscopy observations of the diversity of Ryugu organic matter and its relationship to minerals at the micro-to-nanoscale. Meteorit Planet Sci 59:1–21

Tachibana S, Sawada H, Okazaki R, Takano Y, Sakamoto K, Miura Y, Okamoto C, Yano H, Yamanouchi S, Michel P, Zhang Y, Schwartz S, Thuillet F, Yurimoto H, Nakamura T, Noguchi T, Yabuta H, Naraoka H, Tsuchiyama A, Imae N, Kurosawa K, Nakamura A, Ogawa K, Sugita S, Morota T, Honda R, Kameda S, Tatsumi E, Cho Y, Yoshioka K, Yokota Y, Hayakawa M, Matsuoka M, Sakatani N, Yamada M, Kouyama T, Suzuki H, Honda C, Yoshimitsu T, Kubota T, Demura H, Yada T, Nishimura M, Yogata K, Nakato A, Yoshitake M, Suzuki A, Furuya S, Hatakeda K, Miyazaki A, Kumagai K, Okada T, Abe M, Usui T, Ireland T, Fujimoto M, Yamada T, Arakawa M, Connolly H, Fujii A, Hasegawa S, Hirata N, Hirose C, Hosoda S, Iijima Y, Ikeda H, Ishiguro M, Ishihara Y, Iwata T, Kikuchi S, Kitazato K, Lauretta D, Libourel G, Marty B, Matsumoto K, Michikami T, Mimasu Y, Miura A, Mori O, Nakamura-Messenger K, Namiki N, Nguyen A, Nittler L, Noda H, Noguchi R, Ogawa N, Ono G, Ozaki M, Senshu H, Shimada T, Shimaki Y, Shirai K, Soldini S, Takahashi T, Takei Y, Takeuchi H, Tsukizaki R, Wada K, Yamamoto Y, Yoshikawa K, Yumoto K, Zolensky M, Nakazawa S, Terui F, Tanaka S, Saiki T, Yoshikawa M, Watanabe S, Tsuda Y (2022) Pebbles and sand on asteroid (162173) Ryugu: in situ observation and particles returned to. Earth Sci 375:1011–1016

Tenner T, Nakashima D, Ushikubo T, Kita N, Weisberg M (2015) Oxygen isotope ratios of FeO-poor chondrules in CR3 chondrites: influence of dust enrichment and H_2O during chondrule formation. Geochim Cosmochim Acta 148:228–250

Tonui E, Zolensky M, Hiroi T, Nakamura T, Lipschutz M, Wang M-S, Okudaira K (2014) Petrographic, chemical and spectroscopic evidence for thermal metamorphism in carbonaceous chondrites I: CI and CM chondrites. Geochim Cosmochim Acta 126:284–306

Vaughan D, Craig J (1978) Mineral chemistry of metal sulfides. Cambridge University Press, Cambridge

Westphal AJ, Snead C, Borg J, Quirico E, Raynal P-I, Zolensky ME, Ferrini G, Colangeli L, Palumbo P (2002) Small hypervelocity particles captured in aerogel collectors: location, extraction, handling and storage. Meteorit Planet Sci 37:855–865

Westphal A, Bridges J, Brownlee D, Butterworth A, De Gregorio B, Dominguez G, Gainsforth Z, Flynn G, Ishii H, Joswiak D, Nittler L, Ogliore R, Pepin R, Palma R, Stephan T, Zolensky M (2017) The future of Stardust science. Meteoritics and planetary. Science 52:1859–1898

White AJ, Ebel DS, Greenberg M (2012) A new deconvolution technique for 3-dimensional laser confocal microscopy of Stardust tracks in aerogel. In: Lunar and planetary science XLIII. Abs. #1542

Wirick S, Leroux H, Tomeoka K, Zolensky M, Flynn G, Tyliszczak T, Butterworth A, Tomioka N, Ohnishi I, Nakamura-Messenger K, Sandford S, Keller L, Jacobsen C (2007) Carbonates found in Stardust aerogel tracks. In: Lunar and planetary science XXXVIII

Wooden DH, Harker DE, Woodward CE, Butner HM, Koike C, Witteborn FC, McMurty CW (1999) Silicate mineralogy of the dust in the inner coma of comet C/1995 01 (Hale-Bopp) pre- and postperihelion. Astrophys J 517:1034–1058

Wozniakiewicz P, Ishii H, Kearsley A, Bradley J, Price M, Burchell M, Teslich N, Cole M (2015) The survivability of phyllosilicates and carbonates impacting Stardust Al foils: facilitating the search for cometary water. Meteorit Planet Sci 50:2003–2023

Wurz P, Rubin M, Altwegg K, Balsiger H, Berthelier J-J, Bieler A, Calmont U, De Keyser J, Fiethe B, Fuselier SA, Galli A, Gasc S, Gombosi TI, Jäckel A, Le Ro L, Mall UA, Rème H, Tenishev V, Tzou C-Y (2015) Solar wind sputtering of dust on the surface of 67P/Churyumov-Gerasimenko. Astron Astrophys 583:A22

Yabuta H, Cody GD, Engrand C, Kebukawa Y, De Gregorio B, Bonal L, Remusat L, Stroud R, Quirico E, Nittler L, Hashiguchi M, Komatsu M, Okumura T, Mathurin J, Dartois E, Duprat J, Takahashi Y, Takeichi Y, Kilcoyne D, Yamashita S, Dazzi A, Deniset-Besseau A, Sandford S, Martins Z, Tamenori Y, Ohigashi T, Suga H, Wakabayashi D, Verdier-Paoletti M, Mostefaoui S, Montagnac G, Barosch J, Kamide K, Shigenaka M, Bejach L, Matsumoto M, Enokido Y, Noguchi T, Yurimoto H, Nakamura T, Okazaki R, Naraoka H, Sakamoto K, Connolly HC, Lauretta DS, Abe M, Okada T, Yada T, Nishimura M, Yogata K, Nakato A, Yoshitake M, Iwamae A, Furuya S, Hatakeda K, Miyazaki A, Soejima H, Hitomi Y, Kumagai K, Usui T, Hayashi T, Yamamoto D, Fukai R, Sugita S, Kitazato K, Hirata N, Honda R, Morota T, Tatsumi E, Sakatani N, Namiki N, Matsumoto K, Noguchi R, Wada K, Senshu H, Ogawa K, Yokota Y, Ishihara Y, Shimaki Y, Yamada M, Honda C, Michikami T, Matsuoka M, Hirata N, Arakawa M, Okamoto C, Ishiguro M, Jaumann R, Bibring J-P, Grott M, Schröder S, Otto K, Pilorget C, Schmitz N, Biele J, Ho T-M, Moussi-Soffys A, Miura A, Noda H, Yamada T, Yoshihara K, Kawahara K, Ikeda H, Yamamoto Y, Shirai K, Kikuchi S, Ogawa N, Takeuchi H, Ono G, Mimasu Y, Yoshikawa K, Takei Y, Fujii A, Iijima Y, Nakazawa S, Hosoda S, Iwata T, Hayakawa M, Sawada H, Yano H, Tsukizaki R, Ozaki M, Terui F, Tanaka S, Fujimoto M, Yoshikawa M, Saiki T, Tachibana S, Watanabe S, Tsuda Y (2023) Macromolecular organic matter in samples of the asteroid (162173) Ryugu. Science 379:eabn9057

Yokoyama T, Wadhwa M, Lizuka T, Rai V, Gautam I, Hibiya Y, Masuda Y, Haba M, Fukai R, Hines R, Phelan N, Abe Y, Aléon J, Alexander C, Amari S, Amelin Y, Bajo K, Bizzarro M, Bouvier A, Carlson

R, Chaussidon M, Choi B, Dauphas N, Davis A, Di Rocco T, Fujiya W, Hidaka H, Homma H, Hoppe P, Huss G, Ichida K, Ireland T, Ishikawa A, Itoh S, Kawasaki N, Kita N, Kitajima K, Kleine T, Komatani S, Krot A, Liu M, McKeegan K, Morita M, Motomura K, Moynier F, Nakai I, Nagashima K, Nguyen A, Nittler L, Onose M, Pack A, Park C, Piani L, Qin L, Russell S, Sakamoto N, Schönbächler M, Tafla L, Tang H, Terada K, Terada Y, Usui T, Wada S, Walker R, Yamashita K, Yin Q, Yoneda S, Young E, Yui H, Zhang A, Nakamura T, Naraoka H, Noguchi T, Okazaki R, Sakamoto K, Yabuta H, Abe M, Miyazaki A, Nakato A, Nishimura M, Okada T, Yada T, Yogata K, Nakazawa S, Saiki T, Tanaka S, Terui F, Tsuda Y, Watanabe S, Yoshikawa M, Tachibana S, Yurimoto H (2023) Water circulation in Ryugu asteroid affected the distribution of nucleosynthetic isotope anomalies in returned sample. Sci Adv 9:eadi7048

Yurimoto H, Krot A, Choi B-G, Aleon J, Kunihiro T, Brearley A (2008) Oxygen isotopes of chondritic components. In: MacPherson GJ (ed) Oxygen in the Solar System. Reviews in mineralogy and geochemistry, vol 68. Mineralogical Society of America, Chantilly, pp 141–186

Zhang M, Defouilloy C, Joswiak D, Brownlee D, Nakashima D, Siron G, Kitajima K, Kita N (2021) Oxygen isotope systematics of crystalline silicates in a giant cluster IDP: a genetic link to Wild 2 particles and primitive chondrite chondrules. Earth Planet Sci Lett 564:116928

Zhang M, Brownlee D, Joswiak D, Kita N (2023) Rocky fragments in the largest track of comet 81P/Wild 2–chondrule debris formed in the icy outer Solar System? In: 54th Lunar and Planetary Science Conference. Abstract 1711

Zhang M, Brownlee D, Joswiak D, Kita N (2024) Petrologically diverse silicate fragments in the largest Stardust track of comet 81P/Wild 2. In: 55th Lunar and Planetary Science Conference. Abstract 1321

Zolensky M (2022) Understanding cometary olivine. In: 2022 Goldschmidt conference abstracts

Zolensky ME, Barrett R (1994) Compositional variations of olivines and pyroxenes in chondritic interplanetary dust particles. In Workshop on the Analysis of Interplanetary Dust Particles, LPI Technical Report 94-02, 54-56

Zolensky M, Thomas K (1995) Iron and iron-nickel sulfides in chondritic interplanetary dust particles. Geochim Cosmochim Acta 59:4707–4712

Zolensky M, Prinz M, Lipschutz M (1991) Mineralogy and thermal history of Y-86720, Y-86720 and B-7904. In: 16th symposium on Antarctic meteorites, 78-2

Zolensky M, Nakamura K, Weisberg MK, Prinz M, Nakamura T, Ohsumi K, Saitow A, Mukai M, Gounelle M (2003) A primitive dark inclusion with radiation-damaged silicates in the Ningqiang carbonaceous chondrite. Meteorit Planet Sci 38:305–322

Zolensky ME, Zega T, Yano H, Wirick S, Westphal A, Mike K, Iris Weber I, Warren JL, Velbel M, Tsuchiyama A, Tsou P, Toppani A, Tomioka N, Tomeoka K, Teslich N, Taheri M, Susini J, Stroud R, Stephan T, Stadermann F, Snead C, Simon S, Simionovici A, See T, Robert F, Rietmeijer F, Rao W, Perronnet M, Papanastassiou D, Okudaira K, Ohsumi K, Ohnishi I, Nakamura-Messenger K, Nakamura T, Mostefaoui S, Mikouchi T, Meibom A, Matrajt G, Marcus M, Leroux H, Lemelle L, Le L, Lanzirotti A, Langenhorst F, Krot A, Keller L, Kearsley A, Joswiak D, Jacob D, Ishii H, Harvey R, Hagiya K, Grossman L, Grossman J, Graham G, Gounelle M, Gillet P, Genge M, Flynn G, Ferroir T, Fallon S, Ebel D, Dai Z-R, Cordier P, Clark B, Chi M, Butterworth A, Brownlee D, Bridges J, Brennan S, Brearley A, Bradley J, Bleuet P, Bland P, Bastien R (2006) Mineralogy and petrology of comet Wild 2 nucleus samples. Science 314:1735–1739

Zolensky M, Nakamura-Messenger K, Rietmeijer F, Leroux H, Mikouchi T, Ohsumi K, Simon S, Grossman L, Stephan T, Weisberg M, Velbel M, Zega T, Stroud R, Tomeoka K, Ohnishi I, Tomioka N, Nakamura T, Matrajt G, Joswiak D, Brownlee D, Langenhorst F, Krot A, Kearsley A, Ishii H, Graham G, Dai Z-R, Chi M, Bradley J, Hagiya K, Gounelle M, Bridges J (2008) Comparing Wild-2 particles to chondrites and IDPs. Meteorit Planet Sci 43:261–272

Zolotov MY, Mironenko MV, Shock EL (2006) Thermodynamic constraints on fayalite formation on parent bodies of chondrites. Meteorit Planet Sci 41:1775–1796

Authors and Affiliations

Michael Zolensky[1] · Cecile Engrand[2] · Tomoki Nakamura[3] · Denton Ebel[4]

✉ M. Zolensky
michael.e.zolensky@nasa.gov

✉ T. Nakamura

1 ARES, NASA Johnson Space Center, Houston TX, USA

2 CNRS-University Paris Sud, 91405 Orsay Campus, Paris, France

3 Department of Earth Science, Tohoku University, Sendai, Japan

4 Division of Mineralogy, American Museum of Natural History, New York NY, USA

Space Science Reviews (2025) 221:118
https://doi.org/10.1007/s11214-025-01243-w

Physicochemical Controls on the Compositions of the Earth and Planets

Paolo A. Sossi[1] · **Remco C. Hin**[2] · **Thorsten Kleine**[3] · **Alessandro Morbidelli**[4,5] · **Francis Nimmo**[6]

Received: 4 January 2025 / Accepted: 1 November 2025 / Published online: 25 November 2025

Abstract
Despite the fact that the terrestrial planets all formed from the protoplanetary disk, their bulk compositions show marked departures from that of material condensing from a canonical H_2-rich solar nebula. Metallic cores fix the oxygen fugacities (fO_2s) of the planets to between ~5 (Mercury) and ~1 log units below the iron-wüstite (IW) buffer, orders of magnitude higher than that of the nebular gas. Their oxidised character is coupled with a lack of volatile elements with respect to the solar nebula. Here we show that condensates from a canonical solar gas at different temperatures (T_0) produce bulk compositions with Fe/O (by mass) ranging from ~0.93 ($T_0 = 1250$ K) to ~0.81 ($T_0 = 400$ K), far lower than that of Earth at 1.06. Because the reaction $Fe(s) + H_2O(g) = FeO(s) + H_2(g)$ proceeds only below ~600 K, temperatures at which most moderately volatile elements (MVEs) have already condensed, oxidised planets are expected to be rich in volatiles, and vice-versa. That this is not observed suggests that planets *i*) did not accrete from equilibrium nebular condensates and/or *ii*) underwent additional volatile depletion/fO_2 changes at conditions distinct from those of the solar nebula. Indeed, MVE abundances in small telluric bodies (Moon, Vesta) are consistent with evaporation/condensation at ΔIW-1 and ~1400–1800 K, while the extent of mass-dependent isotopic fractionation observed implies this occurred near- or at equilibrium. On the other hand, the volatile-depleted elemental- yet near-chondritic isotopic compositions of larger telluric bodies (Earth, Mars) reflect mixing of bodies that had themselves experienced different extents of volatile depletion, overprinted by accretion of volatile-undepleted material. On the basis of isotopic anomalies in Cr- and Ti in the BSE, such undepleted matter has been proposed to be CI chondrites, which would comprise 40% by mass if the proto-Earth were ureilite-like. However, this would result in an overabundance of volatile elements in the BSE, requiring significant loss thereafter, which has yet to be demonstrated. On the other hand, 6% CI material added late to an enstatite chondrite-like proto-Earth would broadly match the BSE composition. However, because the Earth is an end-member in isotopic anomalies of heavier elements, no combination of existing meteorites alone can account for its chemical- and isotopic composition. Instead, the Earth is most likely made partially or essentially entirely from an NC-like missing component. If so, the oxidised-, yet volatile-poor nature of differentiated bodies in the inner solar system, including Earth and Mars, is a property intrinsic to the NC reservoir.

Extended author information available on the last page of the article

1 Introduction

The present-day locations and properties of the terrestrial planets, Mercury, Venus, Earth and Mars (Sect. 2), are the culmination of chemical- and physical processes occurring through space and time that followed the collapse of the molecular cloud. Observations of young stars show that these systems are governed by the interaction of an H_2-rich gas with the central star (Sect. 3). Temporal evolution from a 'protostellar' or Class 0 stage in which molecular cloud material infalls to the central star occurs within $\sim 10^4$ to 10^5 yr (Class I) and forms envelopes extending from $\sim 10^3$ to 10^4 au. Over timescales of 10^6 to 10^7 yr, these envelopes develop into the more common T-Tauri stars, which are invariably surrounded by a circumstellar disk of 1 – 10 Jupiter masses (Montmerle et al. 2006). The lifetimes of such disks range from $\sim$1 – 10 Myr (Hartmann et al. 1998; Hillenbrand 2008), with that of our Sun having been inferred to have dissipated by 3.8 Myr after the condensation of the first solids by measurements of magnetic fields preserved in angrite meteorites (Wang et al. 2017).

The theory for the evolution of disks was initially developed assuming that turbulence is the main source of viscosity in disks, so that viscosity is proportional to the product of the square of scale height (the size of the largest turbulent eddies) and the orbital frequency (Lynden-Bell and Pringle 1974). However, it now appears that turbulent viscosity can be high only in the initial phase of the disk, due to ionising temperatures (Balbus and Hawley 1998), Reynolds stresses from gas infalling onto the disk (Kuznetsova et al. 2022) and/or gravitational instabilities (Rafikov 2016). Once this initial phase is over, turbulence becomes too weak in most parts of the disk to account for the observed gas accretion rates (e.g., Cassen 2006, see Sect. 3). This mismatch implies an alternative driver of accretion in protoplanetary disks. The strong X-ray emission of T-Tauri stars is testament to the magnetisation of both the star and disk (Feigelson and Montmerle 1999). It is now understood that the magnetic effects not only govern the collapse of gas from the initial cloud onto the protostar and its disk (Matsumoto and Tomisaka 2004; Joos et al. 2012; Krumholz and Federrath 2019), but also the transport of gas through the disk thanks to the removal of angular momentum in disk winds (Suzuki and Inutsuka 2009; Bai and Stone 2013; Gressel et al. 2015). Disk evolution with winds (Tabone et al. 2022) leads, on a typical timescale of $\sim$5 Myr, to accretion rates that are sufficiently low so as to result in the removal of the remaining gas by photoevaporation (Alexander et al. 2006).

Chemically, these protoplanetary disks are composed predominantly of H_2 ($\sim$91 mol.%) and He ($\sim$8 mol%), with diminishing quantities of other major components, namely, in order of decreasing abundance O, C, Ne, N, Mg, Si and Fe (Lodders 2010). Relative to the protoplanetary disk (and hence the Sun), the terrestrial planets are impoverished in H, C, N, He and other rare gases (Aston 1924) that attests to their accretion from predominantly condensed matter rather than from capture of large quantities of nebular gas. The composition of this material has typically been reconciled in the framework of equilibrium condensation (Larimer 1967; Grossman and Larimer 1974; Lodders 2003, see Sect. 4). Chemical equilibrium, though not mandated, is invoked for these calculations throughout the entire temperature range over which condensation occurs at an assumed, and constant, pressure of 10^{-4} bar, taken to be representative of the midplane for a Sun-like star.

The first solids predicted to have condensed in these calculations are analogous to calcium-aluminium-rich inclusions (CAIs) preserved in some meteorites (MacPherson 2006). They are also the oldest materials to have originated within the Solar System (establishing the t_0 age in relative chronology, Amelin et al. 2002; Connelly et al. 2012), lending credence to the concept of equilibrium condensation. The 'matrix' component of chondrites may also result from near-complete condensation of the nebular gas (save for the ice-forming elements; H, C, N, O and the noble gases) (Grossman and Larimer 1974). The third component, chondrules, may have originally condensed from the solar nebula, but subsequently experienced partial melting (e.g., Hewins et al. 2005; Ebel et al. 2018; Hellmann et al. 2020). Mixtures between these components likely gave rise to the chemical variability observed in chondrites (Humayun and Cassen 2000; Braukmüller et al. 2018; Alexander 2019), yet the nature of the processes producing such mixtures, and whether they reflect nebular-wide gradients, remains unclear (Li et al. 2021; Boyce et al. 2024).

The importance of chondrites lies in their adoption, to first order, as the building blocks of the planets (Ringwood 1966; Safronov 1969). While the constancy of the ratios of refractory lithophile elements (RLEs) to within $\sim$10% among chondrites substantiates this approximation, the abundances of the major planet-forming elements (Fe, Mg, Si, O) vary markedly (up to a factor $\sim$2, Yoshizaki and McDonough 2020). In detail, the core/mantle ratios of rocky bodies, which range from $\sim$0.65 for Mercury to 0.01 for the Moon by mass (Sect. 2, Ringwood 1966; Righter et al. 2006), contrasts with the range expected from chondrites, at roughly 0.2 – 0.3 (e.g., Bercovici et al. 2022). Indeed, no set of mixtures of chondrites is readily able to reproduce the chemical compositions of the planets (Drake and Righter 2002; Campbell and O'Neill 2012), which may instead reflect chemical transformations unique to growing planets (Sect. 5).

One such macroscopic difference is the differentiation of planetary bodies into a core and mantle, implying they experienced melting. Moreover, the terrestrial planets, as well as small telluric bodies (such as Vesta or the Moon), are depleted in volatile elements in a manner unlike that observed in chondritic meteorites (Witt-Eickschen et al. 2009; Mezger et al. 2021). Sossi et al. (2022) noted that the depletion of elements as a function of their volatilities approximates a cumulative normal distribution, suggesting that rocky bodies accreted by the stochastic combination of many smaller bodies that each experienced volatile depletion at different temperatures. Whether a heliocentric gradient in the temperatures, and hence compositions of bodies is required is unclear (see also Palme 2000). This observation only supports the notion that mixing of material with disparate thermal histories led to the formation of the planets, but does not speak to the locus or conditions at which volatile depletion occurred.

An enduring, though as yet inconclusive thesis states that the volatile depletion that affected small telluric bodies occurred under different physicochemical conditions to those prevailing during the formation of chondrites (Kreutzberger et al. 1986; Wulf et al. 1995; O'Neill and Palme 2008; Visscher and Fegley 2013; Norris and Wood 2017; Sossi et al. 2019, Sect. 6). The observation that at least the Earth, with a single-stage Hf-W age of 34$\pm$3 Myr (Kleine and Walker 2017) experienced protracted growth permits the possibility that melting and evaporation took place *after* dispersal of the nebular gas and hence under more oxidised (i.e., higher fO_2) conditions. Evidence for such 'post-nebular' evaporation may be found in the Mn/Na ratio, which is elevated in all small telluric bodies with respect to chondrites, though the bulk silicate Earth falls within the chondritic range (O'Neill and Palme 2008; Siebert et al. 2018).

The mass-dependent variations in the stable isotopic compositions of these elements provide another means with which to interrogate the origin of volatile depletion among planetary materials (Sect. 6). The coupled isotopic variations in Fe, Si and potentially Mg to superchondritic (i.e., heavy) values suggest that volatile depletion processes, rather than core formation, were most likely responsible for the range of isotopic compositions of planetary materials (Poitrasson et al. 2004; Pringle et al. 2014; Dauphas et al. 2015; Sossi et al. 2016; Hin et al. 2017; Bourdon et al. 2018; Sossi and Shahar 2021). On the other hand, the mass-dependent isotopic ratios of moderately volatile elements, such as K and Zn, are near-chondritic in the Earth and Mars (Wang et al. 2016; Sossi et al. 2018b; Paquet et al. 2023), whereas small telluric bodies, such as Vesta, are isotopically heavy (or extremely light in the case of the angrite parent body, Hu et al. 2022). These observations support the inference of Sossi et al. (2022), in which volatile element depletion affecting small telluric bodies generates isotopic fractionation, which is then progressively overprinted by mixing as the body accretes isotopically unfractionated, volatile-undepleted material.

Such undepleted material is often equated with CI chondrites, a hypothesis reinforced by the discovery, in 2011 (Warren 2011), of two discrete clusters of meteorites in their isotopic compositions of Cr and Ti (Trinquier et al. 2007, 2009), when normalised to mass-dependent variations. These observations prompted the proposal that two reservoirs existed in the early Solar System; the NC and CC groups (Sect. 7). That this dichotomy is likely spatial rather than temporal in nature comes from the overlap in accretion ages of both NC- and CC bodies (Kruijer et al. 2017; Bollard et al. 2017), with the former presumed to represent the inner- and the latter the outer solar system. Furthermore, within these groups, isotopically distinct reservoirs exist, implying that mixing, should it have taken place during the accretion of planetary bodies, was likely local in nature. Whether this decreases the likelihood of dynamical scenarios in which the bodies of the early solar system were agitated and resulted in eccentric, crossing orbits, such as in the Grand Tack (Mah and Brasser 2021; Mah et al. 2022; Woo et al. 2022), remains to be seen.

Earth and Mars fall close to- or within the NC group for these two elements (e.g., Warren 2011; Dauphas 2017). Some small, telluric bodies also occupy the NC group space, and, for Cr, Ti and other iron-peak elements, together define an 'NC trend' that passes linearly through or near to the composition of CI chondrites (Trinquier et al. 2009; Williams et al. 2020; Palme and Mezger 2024). On this basis, some models argue that the Earth is made predominantly ($\sim$95%) of EH-like material (Kleine et al. 2023; Dauphas et al. 2024), whereas others suggest a ureilite-like NC end-member and require up to $\sim$40% CC material (Schiller et al. 2018). However, the perceived necessity for the addition of carbonaceous chondrites may be circumstantial, since CCs lie on fractionation lines distinct from terrestrial in triple O isotope space (Clayton and Mayeda 1999), implying that they, in combination with ureilites or EH-like material, are unlikely to represent a significant portion of the Earth on a mass basis. Moreover, for the isotopes of the heavy elements, notably Mo and Zr, there is no suitable member of the extant CC group that would produce the composition of the Earth from an existing NC member. Possible solutions state that the Earth accreted some 'missing' material not represented in the meteorite collection for an EH-like proto-Earth (Burkhardt et al. 2021) or preferentially lost s-process-depleted material via envelope processing for a ureilite-like end-member (Onyett et al. 2023).

Here, we synthesise the major physical properties of the terrestrial planets and other differentiated telluric bodies and compare them with chondrites on an empirical basis (Sect. 2). Section 3 describes the thermophysical states of protoplanetary disks and their evolution in time so as to identify possible loci and times of planet formation. In Sect. 4, we characterise the chemistry of solids that condense from the solar nebula under the assumption of equilibrium, before exploring the effects of non-equilibrium condensation on the stability and compositions of condensates. Whether mixtures of such condensates are consistent with the bulk compositions of telluric bodies are evaluated in Sect. 5, and the influence of chemical equilibrium between core- and mantle at a range of pressure-temperature conditions on the gross structure of differentiated bodies is examined. The effect of planetary mass on the chemical and isotopic effects of volatile depletion processes are characterised in Sect. 6. Finally, a critical assessment of the distribution of nucleosynthetic anomalies, both across planetary materials and among different elements in the Earth and Mars is undertaken in Sect. 7, with a view to developing a coherent astrophysical model for planetary formation.

2 Physical- and Chemical Nature of Terrestrial Planets

2.1 Physical Structure and Distribution

The distribution of the mass of rocky material in the inner Solar System is concentrated in the terrestrial planets, with two, Earth and Venus, together comprising 91.8% (see Table 1). The integrated mean of the mass therefore is nearly symmetrically centred about 0.897 AU, an observation that is used to inform models of the initial mass distribution of the protoplanetary disk (Weidenschilling 1977). Despite the symmetry in mass distribution, the constitution of the planets is not uniform. The internal structures of the terrestrial planets are *differentiated* and can be approximated by silicate mantles and iron-rich cores, but the core/mantle ratio of the terrestrial planets broadly decreases with increasing semi-major axis (e.g., Ringwood 1966), ranging from 0.67±0.06 in Mercury (Hauck et al. 2013) to 0.204±0.015 in Mars (Khan et al. 2023) and 0.008–0.015 in the Moon (Sossi et al. 2025) In detail, however, Venus has a slightly lower uncompressed density (by 2%) than does the Earth, but its moment of inertia (0.337±0.024, Margot et al. 2021) overlaps with that of the Earth (0.3307, Williams 1994), meaning its core/mantle ratio is indistinguishable, within error, from that of the Earth (0.325).

2.2 Chemical Composition

2.2.1 Major Rock-Forming Elements

The core/mantle ratio, is, to first order, an indicator of the relative proportions of Fe/O accreted by the planetary body (as seen for chondritic meteorites; Urey and Craig 1953). There is no requirement for planets to adhere to the chondritic range, not only due to potential chemical gradients in the solar nebula (e.g., Larimer and Bartholomay 1979; Palme 2000) but also owing to giant impacts that can preferentially strip silicate material (e.g., Fegley and Cameron 1987; O'Neill and Palme 2008). Irrespective of the pathway via which differentiated bodies accreted, at equilibrium, the amount of oxygen that is available to oxidise metallic iron is described by;

$$\mathrm{Fe(core)} + \frac{1}{2}\mathrm{O}_2 = \mathrm{FeO(mantle)}, \tag{1}$$

in which FeO is dissolved in a multicomponent silicate (e.g., liquid phase), and metallic iron forms a discrete phase (including minor Ni, Co and light elements), that then constitutes the core of the body. Equation (1) represents the iron-wüstite (IW) buffer, and can be used to constrain the oxygen fugacity, fO_2, relative to IW, which is given by;

$$\begin{aligned}\Delta\mathrm{IW} &= \log_{10} f O_2(\mathrm{body}) - \log_{10} f O_2(\mathrm{IW}) \\ &= 2\log_{10}\left(\frac{a\mathrm{FeO}}{a\mathrm{Fe}.K_{(1)}}\right) - 2\log_{10}\left(\frac{1}{K_{(1)}}\right) = 2\log_{10}\left(\frac{a\mathrm{Fe}O}{a\mathrm{Fe}}\right),\end{aligned} \tag{2}$$

where a refers to the activity and $K_{(1)}$ the equilibrium constant of Eq. (1) (see, for example, O'Neill and Pownceby 1993). If the mantle FeO content of a given planetary body (see

Table 1 Selected physical and chemical properties of planetary bodies

	Semi-major axis (AU)	±	log(mass, kg)	±	CMF	±	FeO (wt.%)	±	Mg#	±	ΣFe	±	ΔIW	±
Mercury	0.387	-	23.519	-	0.67	0.06	0.21	0.10	0.98	0.02	0.604	0.054	−5.40	0.40
Venus	0.723	-	24.687	-	0.27	0.05	9.0	3.0	0.88	0.04	0.294	0.056	−2.11	0.35
Earth	1.000	-	24.776	-	0.325	-	8.1	0.2	0.890	0.005	0.335	-	−2.20	0.02
Moon	1.000	-	22.866	-	0.012	0.003	11	3	0.84	0.04	0.095	0.025	−1.58	0.26
Mars	1.524	-	23.807	-	0.204	0.015	13.8	1.0	0.81	0.02	0.298	0.029	−1.74	0.07
Vesta	2.36	-	20.413	-	0.15	0.07	24.4	3.0	0.68	0.03	0.296	0.068	−1.24	0.12
UPB	2.7*	0.5	21*	1	0.20	0.09	22.5	7.0	0.70	0.07	0.31	0.10	−1.36	0.25
APB	2.7*	0.5	21*	1	0.17	0.04	22	4	0.71	0.04	0.295	0.048	−1.33	0.11

CMF = core mass fraction. ΣFe = fraction of planetary mass comprised by total Fe (FeO+Fe). ΔIW = Oxygen fugacity with respect to the iron-wüstite buffer, $\Delta IW = \log fO_2 - \log fO_2(IW)$. References: Mercury - (Hauck et al. 2013; Namur et al. 2016), Venus (Basaltic Volcanism Study Project 1981; Helbert et al. 2021; Shah et al. 2022; Rodríguez-Mozos and Moya 2022), Earth (Frost and McCammon 2008; Palme and O'Neill 2014), Moon (Sossi et al. 2025), Mars (Khan et al. 2022, 2023), Vesta - (Zuber et al. 2011; Toplis et al. 2013), Ureilite Parent Body (UPB) - (Goodrich et al. 2007), Angrite Parent Body (APB) - (Longhi 1999; McKibbin and O'Neill 2018; Tissot et al. 2022). The semi-major axes and masses of the UPB and APB are unknown. They are thought to reside (or have resided) in the asteroid belt, and to have (or have had) masses similar to that of Vesta. Nominal ranges of 2.7±0.5 AU and $10^{21\pm1}$ kg for semi-major axis and mass, respectively, are given here.

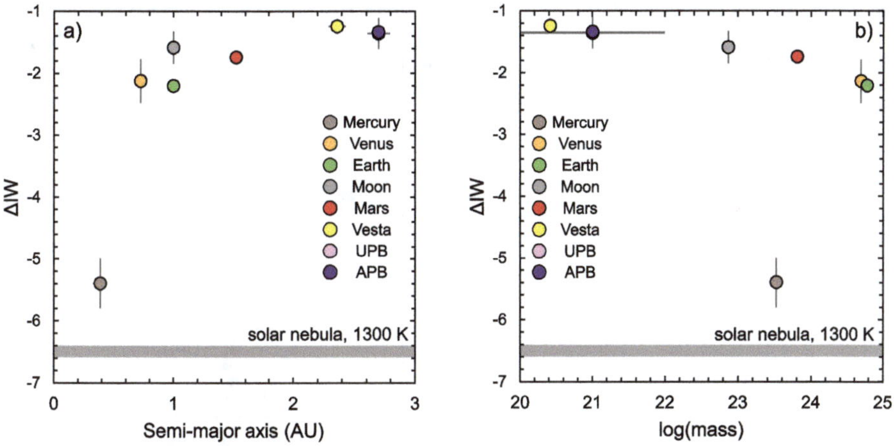

Fig. 1 Oxygen fugacities, with respect to the iron-wüstite (IW) buffer (O'Neill and Pownceby 1993) of the planets as a function of a) heliocentric distance (in AU) and b) the $\log_{10}$ of their masses. Also shown is the fO_2 defined by a solar gas at 1300 K and 10^{-4} bar (Grossman et al. 2008). Source data and abbreviations are shown in Table 1

Table 1) is assumed to be well-mixed and to represent a quantity determined by equilibrium with the core, there is a general tendency for the fO_2 to increase as a function of heliocentric distance (Fig. 1a), but also with decreasing mass (Fig. 1b). The small telluric bodies in the asteroid belt define values near ΔIW-1.3, Mars has an intermediate value of ΔIW-1.7, and the Earth the most reduced, ΔIW-2.2 (see also Wadhwa 2008; Frost and McCammon 2008). The Moon is displaced to higher values with respect to the Earth (see O'Neill 1991; Sossi et al. 2025), whereas Mercury has a conspicuously lower fO_2 (Namur et al. 2016). Even so, the fO_2 inferred for core formation on Mercury is close to that for a gas of solar composition at 1300 K, which has an H_2/H_2O ratio of 5×10^{-4}, corresponding to $\Delta IW = -6.5$ (e.g., Grossman et al. 2008). Thus, the fO_2s recorded by all differentiated planetary bodies, with the possible exception of Mercury, are more oxidised than that expected for equilibrium with the solar nebula.

Most bodies are consistent with an amount of total iron ($\sum$Fe ~ 28 - 33 wt.%, Table 1) that overlaps with those found in chondrites (18 - 30 wt.%, Wasson and Kallemeyn 1988), though it should be noted that estimates for core mass fractions for the small telluric bodies are made on the assumption of a chondritic bulk composition (e.g., Toplis et al. 2013). This notwithstanding, Mercury and the Moon are again exceptions and require roughly 60 wt.% and 10 wt.% total Fe, respectively (Table 1; Fig. 2). Comparisons of the oxygen fugacities of differentiated bodies with those of chondritic meteorites are hampered because the most primitive examples (i.e., type 3 chondrites) are a collection of unequilibrated grains and therefore the fO_2 is undefined. A relative scale of oxidation may be instead provided by the Mg# of silicates contained therein (calculated on a molar basis);

$$\mathrm{Mg\#} = \frac{\mathrm{Mg}}{[\mathrm{Mg} + \mathrm{Fe}^{2+}]}. \tag{3}$$

Because Mg is lithophile and Ni is siderophile, plots of the Fe/Mg and Ni/Mg ratios of groups of bulk chondrites can be used in conjunction with Eq. (1) to infer the oxygen fugacity at which their silicates formed (Larimer and Anders 1970). As Ni is almost entirely hosted in metal grains, the Fe/Mg ratio of the silicate fraction of a given chondrite group

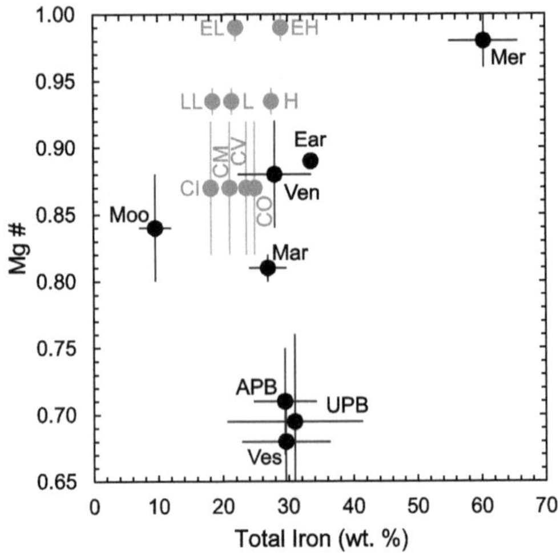

Fig. 2 The Mg # of differentiated (black) and undifferentiated (grey) bodies as a function of their total iron contents. Source data is from Wasson and Kallemeyn (1988) for chondritic meteorites. See Table 1 for abbreviations

is recovered at Fe/Ni = 0. Carbonaceous chondrites yield higher intercepts $(Fe/Mg)_0 = 0.35\pm0.16$ than do the ordinary chondrites $(Fe/Mg)_0 = 0.16\pm0.03$, while that for enstatite chondrites cannot be reliably assessed from only two groups, but their constituent silicates have Fe/Mg very near 0 (see also O'Neill and Palme 1998) and here we adopt 0.01 ± 0.01. These are compared to the range of Mg#s observed among planetary mantles, as a proxy for fO_2, in Fig. 2.

Chondritic meteorites occupy but a limited parameter space, down to Mg#s ~0.82, within uncertainty. Strikingly, all small telluric bodies (STBs) and Mars have mantle Mg#s far below even silicate phases in carbonaceous chondrites (CI, CM, CV, CO in Fig. 2). It should be noted that chemical equilibration in the presence of S could result in fO_2s higher than the IW buffer (Bercovici et al. 2022). Hence, the STBs must have experienced additional oxidation relative to that observed in at least the ordinary and enstatite meteorites, and potentially also the carbonaceous chondrites, considering some cores of STBs are thought to be S-poor (Steenstra et al. 2020), though others are inferred to contain up to 19 wt.% S (Hirschmann et al. 2021). To first order, the vertical trend in Fig. 2 implies differences in Fe/O ratios at constant total iron contents. This could not have been produced through physical, binary mixtures of metal and silicate, which would instead give rise to coupled changes in Fe/O and ΣFe, as vaguely defined by the trend passing through the Moon, Venus, Earth and Mercury. Therefore, and unlike refractory elements, there is no *a priori* reason to expect chondritic ratios (which differ even between themselves) among the major rock-forming elements (Fe, Mg, Si, O) in the terrestrial planets (see also Yoshizaki and McDonough 2020).

2.2.2 Volatile Elements

Unlike the major, rock-forming elements, abundances of the volatile elements vary by orders of magnitude among planetary materials. In order to derive their abundances in planetary mantles, they are typically referenced to a refractory element of similar geochemical character during partial melting of the mantle, such as La or U for the moderately volatile K, or Sr for Rb (e.g., Halliday and Porcelli 2001). Furthermore, both K and U are natural gamma-ray emitters and their abundances can thus be inferred via remote sensing (e.g., Prettyman et al. 2006; Peplowski et al. 2012). Ratios of these elements for the mantles of bodies in

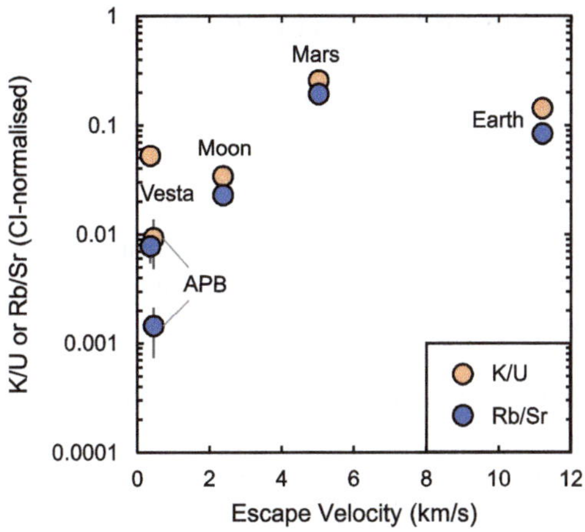

Fig. 3 The K/U and Rb/Sr ratios, normalised to those in CI chondrites, of differentiated planetary bodies. References: Angrite parent body - (Dauphas et al. 2022), Vesta - (Sossi et al. 2022), Moon - (Sossi et al. 2025), Mars and Earth - (Dauphas et al. 2022)

which they are well constrained, are plotted as a function of the escape velocity of the body, $v_e = \sqrt{\frac{2GM}{r}}$, in Fig. 3.

In broad terms, the planets, Mars and Earth, are more volatile-rich, as attested to by their markedly higher K/U and Rb/Sr ratios, than are the small telluric bodies, the Moon, Vesta and the APB (see also McCubbin and Barnes 2019). In detail, however, Mars appears to have higher quantities of volatile elements than does the Earth (Yoshizaki and McDonough 2020; Khan et al. 2022; Dauphas et al. 2022), a conclusion reinforced from higher S content of the Martian core (∼10 wt.% Khan et al. 2023) relative to that of Earth (∼2 wt.% S, Dreibus and Palme 1996; Suer et al. 2017). Moreover, despite the greater mass of the Moon relative to Vesta, both exhibit similar degrees of volatile depletion (Fig. 3). The exception is Mn, which is essentially undepleted with respect to Mg and CI chondrites in Vesta, but is lower, by nearly a factor of 4, in the Moon (Sossi et al. 2025). This observation was used as evidence to suggest that the Moon was derived from the Earth's mantle after core formation (Ringwood and Kesson 1977; Sossi et al. 2025). The APB remains the most highly volatile-depleted body known in the solar system, and thus defines a hierarchy from (most volatile-rich) Mars > Earth > Moon ∼ Vesta > APB (least volatile-rich). Therefore, differences in the present-day escape velocities of planetary bodies (proportional to mass for constant density) may only partially account for the observed range in their volatile element contents. Available observations do not support any trend in K/U with heliocentric distance (McCubbin and Barnes 2019), either, though volatile abundances in Mercury and Venus remain essentially unconstrained at the resolution required to draw such conclusions. The surface of Mercury, as revealed by MESSENGER, compared to the Earth and even Mars, is rich in C (1.4±0.9 wt.%, Peplowski et al. 2016; Nittler et al. 2018) and S (∼2.5 wt.%, Nittler et al. 2011; Weider et al. 2016), which, in the canonical nebular condensation sequence, are regarded as ice-forming and moderately volatile elements, respectively. This leaves at least three possibilities; i) the Hermean surface is richer in these elements than is its interior, ii) the region of the inner disk from which Mercury formed did not experience extensive volatile depletion and/or iii) C and S did not behave in a volatile manner as expected for a typical solar nebular composition.

3 Thermal States of Accreting Disks

The first steps of planet formation – condensation of the first solids, aggregation of dust into planetesimals, formation of protoplanets – occur in a protoplanetary disk. The temperature of the protoplanetary disk dictates which chemical species are in the solid or gaseous forms, and therefore influences the volatile element content (both moderately and highly) of the bodies depending on where and when they form. At a given time and location, gas in the disk can be moving either outwards or inwards. The following discussion details physics applicable to typical solar-mass (G-type) stars.

3.1 Disks Dominated by Internal Viscosity

When modelling viscous disks, following Shakura and Sunyaev (1973, 1976), it is customary to assume that the viscosity is

$$\nu = \alpha H^2 \Omega = \alpha c_s H, \tag{4}$$

where Ω is the orbital frequency at distance r from the central star, c_s is the sound speed and

$$H = \frac{c_s}{\Omega} = \sqrt{\frac{\gamma_{ad} R T r^3}{\mu G M_*}} \tag{5}$$

is the pressure scale height of the disk assumed to be vertically isothermal and in hydrostatic equilibrium, T is the temperature and γ_{ad}, R, μ are the adiabatic index-, constant- and mean molecular weight of an ideal gas, with G gravitational constant, and M_* the mass of the star. The coefficient α is a proportionality parameter with values between 0 and 1. Its precise value is not well known as it depends on the physical conditions inside the disk (degree of ionisation, coupling with the magnetic field, as discussed below in Sect. 3.3). A viscous disk spreads radially. Over most of the radial extension of the disk, the gas is transported towards the central star at a speed $v_r = -\frac{3}{2}\frac{\nu}{r}$, but near the disk's outer edge, the gas radial velocity becomes positive. This allows the disk to grow in radial extent over time, while transporting gas inwards (hence the name "accretion disk") and angular momentum outwards. The radius where the radial velocity of the gas changes sign increases over time (Lynden-Bell and Pringle 1974) (see Fig. 6); as a consequence, a fixed point in the disk (relative to the star) will see outwards-then-inwards gas motion.

The accretion rate on the star is given by:

$$\dot{M} = 2\pi r v_r \Sigma \tag{6}$$

where Σ is the surface density of the disk at distance r. The friction among different "rings" of the disk in differential rotation generates heat on the disk's midplane at a rate

$$Q_+^{\mathrm{visc}} = \frac{9}{4}\nu\Sigma\Omega^2 = \frac{3}{4\pi}\dot{M}\Omega^2 \tag{7}$$

The irradiation from the central star deposits heat at the disk's effective surface H_s, where it is absorbed at a rate

$$Q_+^{\mathrm{irr}} = \frac{\phi L_*}{2\pi r^2}, \tag{8}$$

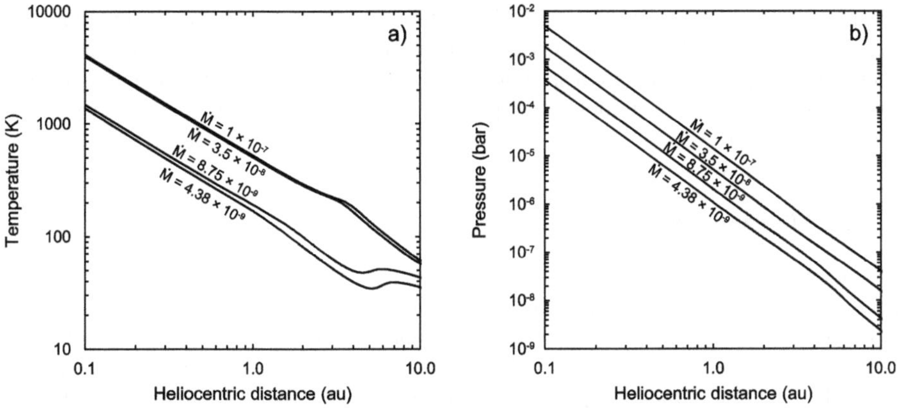

Fig. 4 Radial distribution of **a)** temperature in Kelvin and **b)** pressure in bar in the midplane of the disk for different stellar mass accretion rates ($\dot{M}$) in solar masses/yr. Here $\alpha = 0.0054$. The temperature of the snowline (i.e., the temperature at which the reaction $H_2O(g) = H_2O(s)$ proceeds to the right) is roughly between 220 – 150 K, depending on pressure. NB: Temperatures and pressures inward of 1 au are extrapolated. Constructed with data from Bitsch et al. (2015)

where L_* is the stellar luminosity and ϕ is the incidence angle between the H_s surface and the stellar rays. Notice that viscous heating is proportional to the accretion rate $\dot{M}$, hence to the product $\nu\Sigma$ while irradiation heating is independent of both. The disk cools at the H_s surface by radiating in space as a black body, at the rate

$$Q_- = 2\sigma_{\rm SB} T_{\rm eff}^4, \tag{9}$$

where $\sigma_{\rm SB}$ is the Stefan-Boltzmann constant and $T_{\rm eff}$ is the temperature of the disk at the surface H_s. The energy balance equation

$$Q_- = Q_+^{\rm visc} + Q_+^{\rm irr} \tag{10}$$

sets $T_{\rm eff}$ assuming steady state has been reached. The temperature on the midplane T is then related to $T_{\rm eff}$ via the equation

$$T = \left[\frac{(Q_+^{\rm irr} + \frac{3\tau}{4} Q_+^{\rm visc})}{Q_-}\right]^{1/4} T_{\rm eff} \tag{11}$$

where $\tau = \kappa\Sigma$ is the optical depth of the disk and κ its opacity (see Dullemond 2013 for more details). These simplified analytical calculations are useful to get a general view of the structure of the disk and its dependence on physical parameters. A more precise calculation can be obtained with hydrodynamical simulations, where $Q_+^{\rm visc}$ is imposed on each cell of the disk (substituting Σ with ρdz, where ρ is the local density and dz the vertical extent of the cell) and $Q_+^{\rm irr}$ and Q_- are computed for each cell accounting for the local gas optical depth (see for instance Bitsch et al. 2014, 2015).

Figure 4 shows the radial distribution of the midplane disk temperature as a function of the stellar accretion rate. The latter is constrained by observations Hartmann et al. (1998) to be dependent on the age of the star-disk system, as

$$\log\dot{M} = -8 \pm 0.1 - (1.4 \pm 0.3)\log t, \tag{12}$$

where $\dot{M}$ is measured in $M_\odot/yr$ and t is the time in Myr. The transition from a viscous-dominated to an irradiation-dominated disk is in the 5–10 au region. The calculation presented in the figure is made for a specific value of $\alpha = 5.4 \times 10^{-3}$. In the region dominated by the viscous heating the midplane temperature is proportional to $(\dot{M}^2/\alpha)^{1/5}$ (Batygin and Morbidelli 2022). Also, notice that the temperature at 1 au barely exceeds 500 K, even in the highest accretion case of $\dot{M} = 10^{-7} M_\odot/yr$. We return to this in Sect. 3.4.

3.2 Dust Dynamics and Disk Chemistry

By looking at Fig. 4 one could naïvely think that, as the disk cools over time (i.e. with decreasing $\dot{M}$) the gas condenses increasingly volatile species and that the chemical composition of the disk in its solid and gaseous parts simply depends on the local temperature. This is not true in general because (i) the radial motion of gas towards the central star is typically faster than the rate of radial displacement of a condensation line (i.e. the disk cooling rate) and (ii) the solid material is carried by dust particles that rapidly grow to sizes at which grain dynamics partially decouples from gas dynamics (Cassen 2001; Morbidelli et al. 2015).

The consequence of (i) is that "all [gas] trajectories pass from the cool side of the [condensation] front to the hot side; no condensation occurs" (Cassen 2001, their Fig. 13); the only exception being in the region where the gas radial motion is outward or near the radius where the radial velocity of the gas changes sign and so is small, as illustrated schematically in Fig. 5. Condensation can occur only if there is a continuous presence of the concerned species in gaseous form just inward of the condensation front, sustained by inward radial drift of dust and sublimation.

The consequence of (ii) is that, because, typically the inward radial flow of dust is faster than the flow of gas, the evaporated species has, in general, a higher concentration, relative to hydrogen, at the locus of evaporation than in the initial stellar composition (i.e., the mean solar nebula). This is illustrated in the top panel of Fig. 5.

On the other hand, if the radial drift of the dust is blocked by a pressure maximum appearing in the disk (for instance due to the formation of a gap-opening planet, but also to the formation of rings due to magnetic effects, Béthune et al. 2017), there is no evaporation at the condensation front and therefore no possible recondensation during disk cooling (bottom panel, Fig. 5). Morbidelli et al. (2015) proposed that the blockage of the radial drift of icy grains, possibly due to the formation of Jupiter, prevented the inner disk from becoming enriched in ice relative to the solar value, even when the snowline was eventually able to pass inward of 1 au, as expected from the decay of $\dot{M}$ over time and the corresponding decrease of the disk's temperature (Fig. 4, Eq. (12)). This scenario can explain the deficit in water (and other volatile elements) in bodies formed in the inner solar system (McCubbin and Barnes 2019, see also Sect. 7).

One can imagine also more complicated scenarios. For instance, if the dust trap is located between the H_2O and CO condensation lines, CO(g) would be released but water vapour would not. As a result, the inner disk would have an enhanced C/O ratio relative to the original ratio in the disk and the star. Thus, depending on the location, number and efficacy of dust traps, various scenarios can arise that influence the chemistry of the disk inward of the trap(s). Predicting the actual evolution of the dust, which, in turn, depends on dust growth, radial extension of the disk, formation of pressure maxima, and so on, is currently an active field of research.

3.3 Disks Dominated by Magnetised Winds

Up to this point, we have considered that the disk evolves and is heated under the effect of its own viscosity (cf. Eq. (4)). However, the origin of the disk's viscosity is elusive. In the

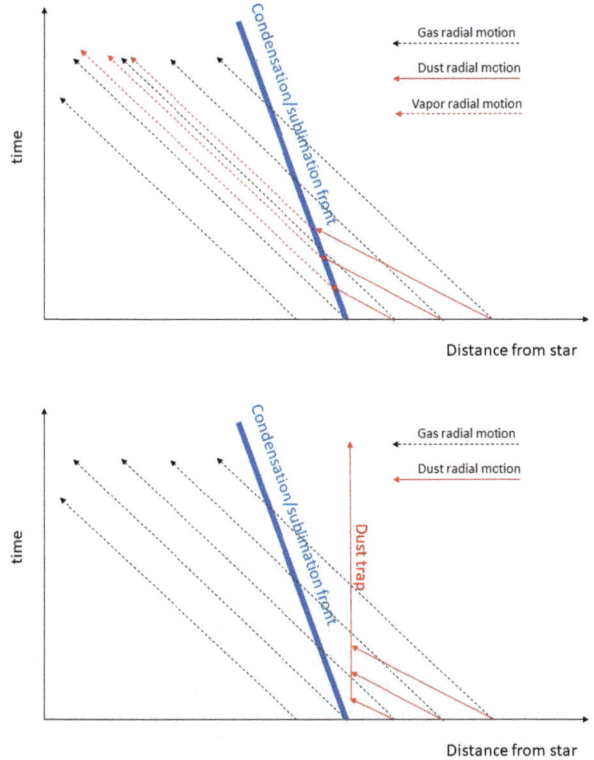

Fig. 5 A schematic illustration of the radial motion of the evaporation/condensation front of gas and dust. The top panel shows the situation prevailing in a continuous disk. The radial motion of the gas (mostly hydrogen) towards the central star (dashed black arrows) is faster than the radial displacement of the evaporation/condensation front (thick blue line) that is due to disk cooling over time (cf. Eq. (12)). The dust moves towards the star (solid red arrows) even faster than does the gas. When crossing the evaporation front, the released vapour then moves inward (red dashed arrows) at the same speed as the gas. As a consequence, the vapour/hydrogen ratio is increased with respect to the its initial ratio (i.e. the stellar abundance). Indeed, the red dashed arrows are denser in space than the black dashed lines, which indicates an enhanced density of the evaporated species. The bottom panel shows the case where the dust is trapped at some location beyond the evaporation front, due to the formation of a pressure maximum. In this case, the dust does not reach the evaporation front. Therefore, even if the disk cools, there is no further condensation possible

α-disk model expounded in Sect. 3.1, the disk evolution timescale is given (Hartmann et al. 1998; Cassen 2006):

$$t = \frac{{r_d}^2}{\nu} \tag{13}$$

where r_d is the characteristic length scale of the disk. For a typical disk extent of 100 AU at 1 Myr, as constrained by observations, ν is 10^{15} cm^2/s. Molecular viscosity would be several orders of magnitude smaller, $\sim 10^6$ cm^2/s (Cassen 2006), than the viscosity considered in the simulations of Bitsch et al. (2015), as shown in Fig. 4. Thus, an additional source of viscosity is needed to reconcile the viscous disk model with observed disk properties. For a decade or so, it was generally believed that this additional viscosity contribution arises from turbulence within the disk, itself generated by the so-called magneto-rotational instability (Balbus and Hawley 1998). However, this instability occurs only if there is a strong coupling between the magnetic field and the gas, which requires a large amount of gas ionisation. Ionisation is expected to occur only at locations in the disk where it is very hot and the alkali elements have sublimed, i.e. above $\sim$ 1000 K or so (Umebayashi 1983; Turner et al. 2007; Flock et al. 2017).

A strong effective viscosity can also be generated if there are hydro-dynamical stresses in the disk. These may be the consequence of the formation of spiral waves in the disk, such as those generated when the disk is so dense so as to become nearly gravitationally unstable (Rafikov 2016). The accretion of gas from the molecular cloud onto the disk can also generate large effective viscosities (Kuznetsova et al. 2022). In the absence of these non-generic mechanisms, the viscosity in the disk is now expected to be small, with a value α of the order of 10^{-4} or smaller, set by weak disk instabilities such as the vertical shear instability (Stoll and Kley 2014) or the convective overstability (Klahr and Hubbard 2014).

If the transport of gas towards the central star were solely due to disk's (i.e., molecular) viscosity, such low values of α would imply very massive disks (i.e. large Σ, large r_d, Eq. (13)) in order to account for the values of $\dot{M}$ that are observed. But the observed disk masses do not seem as high as required (Manara et al. 2016, 2018). This suggests that another mechanism than internal viscosity is responsible for the mass transport towards the central star.

The emerging view is that, in weakly ionised disks (i.e., beyond $\sim$1 au), the interaction with magnetic fields generates a magnetised wind, which ejects ions and extracts angular momentum from the disk, promoting the inward radial motion of the remaining gas (Béthune et al. 2017; Lesur et al. 2023). These magnetic fields are thought to have been inherited from the molecular cloud collapse stage, and subsequently lace the protoplanetary disk (Stephens et al. 2017; Fu et al. 2021). Though there is no consensus as to how such fields evolve (Lesur 2021), stellar accretion due to magnetised winds breaks the relationship between $\dot{M}$, α and Σ that is typical of viscous disks (Sect. 3.1). The main consequence, for what concerns this chapter, is that $\dot{M}$ is no longer related to the disk's temperature, unlike in the curves shown in Fig. 4. In absence of viscous heating, disks are solely heated by stellar irradiation and therefore are expected to be cold (Chiang and Youdin 2010), with a snowline ($\sim$170 K) inwards of 1 au. Because such temperatures would have resulted in water-rich bodies in the inner solar system, which are not observed, disks clearly must have been much warmer than this, at least initially, owing to the high viscosity induced by one of the sources of stress in the disk (see above). An additional source of heating could come from the existence of a current sheet at about $3H$ (Latter et al. 2010; Gressel et al. 2015), but the effect is expected to be small (Mori et al. 2019).

3.4 A View of Disk Evolution

In light of the foregoing discussion, we can envision a general view of disk evolution. Initially, when stellar collapse starts and the disk forms, the disk is small and dense. Its effective viscosity can be very high (even $\alpha = 0.01 - 0.1$) due to stresses exerted on it by the infalling gas and by the anisotropy in the gas distribution generated by conditions close to gravitational instability. The disk is therefore very hot (it is also strongly irradiated by the accretion shock at the stellar surface) and spreads very rapidly outwards (Hueso and Guillot 2005; Pignatale et al. 2017; Morbidelli et al. 2022; Marschall and Morbidelli 2023); see Fig. 6. At this time, the disk is at its hottest; significantly higher than that of a standard accretion disk (compare Fig. 6 with Fig. 4), reaching $\sim$ 1000 K at 1 au. During the radial expansion stage, the gas flows from the hot side of the condensation fronts towards the cold side (inward to outward), whereas this is reversed in the later evolutionary stage. Thus, the outward expansion stage is likely to result in direct condensation of gas, such as that leading

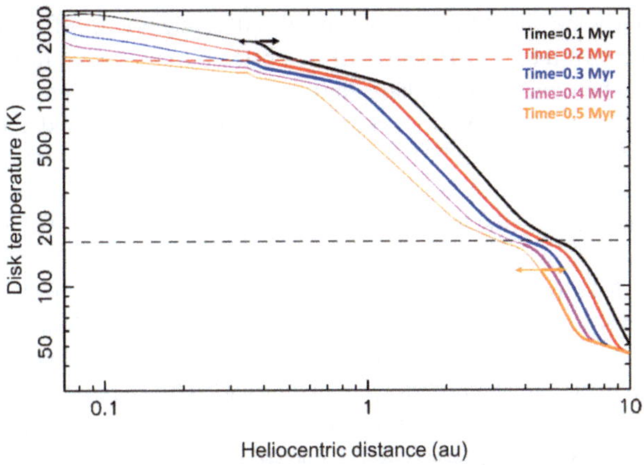

Fig. 6 Temperature in the disk as a function of time during the epoch when the disk is accreting material from the envelope surrounding the forming star. The thick part of each curve shows the region where the radial velocity of the gas is positive (outward), whereas the thin part depicts the accretion part of the disk (negative radial velocity), as also indicated by the black and orange arrows. The horizontal dashed lines mark the condensation temperature of water ($T = 170$ K, black), and rocks ($T = 1400$ K, red). The intersection of these lines with the various coloured curves identify the location of the condensation/sublimation fronts of these elements as a function of time. From Morbidelli et al. (2022)

to the formation of primitive grains, namely calcium-aluminium-rich inclusions, CAIs and amoeboid-olivine-aggregates, AOAs (Pignatale et al. 2017; Marschall and Morbidelli 2023). Notice that, from Fig. 6, the CAI condensation temperature is met at 0.3–0.6 au during the early stages of the disk, where these grains are thought to have formed (e.g., Gounelle et al. 2001). It should be noted that such refractory grains show evidence for having formed by very rapid cooling (within $\sim 10^{-1} - 10^{-2}$ yr, Marrocchi et al. 2019), and it is therefore unlikely that the thermal regime under which they condensed is representative of gross disk structure.

When the rate of infall of new gas onto the protoplanetary disk wanes, typically after a few 10^5 yr according to observations, and once the disk has spread sufficiently in the radial direction to be far from gravitationally unstable conditions, the viscosity rapidly drops to low values, corresponding to $\alpha < 10^{-4}$. Consequently, the temperature also drops very rapidly to approach that of a passively irradiated disk (Chiang and Youdin 2010). We remind readers that such a disk has a temperature of 120 K at 1 au and decaying as $1/r^{3/7}$. The transport of gas towards the central star is now dominated by the process of angular momentum removal in magnetised disk winds.

The relative timing of the radial spreading of the disk and subsequent accretion (inward transport) stages is unclear and the formation ages of planetary bodies are also used to inform models. For instance, Lichtenberg et al. (2021) assume an outflow phase that results in condensation from $\sim$0.2 Myr and persists until 0.7 Myr over 1.5 – 7 au, in order to explain the inferred iron meteorite (and other achondrite parent body) formation times. The disk cools and contracts thereafter, resulting in the formation of a second generation of planetesimals from 3 – 15 au at 0.7 – 5 Myr after t_0. By contrast, models which approximate α-disks with a decreasing value of α lead to the contemporaneous generation of two (Morbidelli et al. 2022) or three (Izidoro et al. 2022) planetesimal populations from $\sim$0.1 – 0.5 Myr at 1, 5 and, in Izidoro et al. (2022), 30 au for nominally 'rock', 'ice' and 'soot', respectively.

While the formation times and locations detailed above are based on the temperature in the disk (which evolves with time), they do not describe how these dust grains migrate in the disk thereafter. One should expect a generalised condensation of volatiles during the rapid cooling phase of the disk. However, if an obstacle to the radial drift of dust appears in the disk before the rapid cooling phase starts, as suggested by the preservation of the so-called isotopic dichotomy (see Sect. 7) established during the disk's radial expansion phase (Kruijer et al. 2017; Nanne et al. 2019; Brasser and Mojzsis 2020; Morbidelli et al. 2022), the inner disk may remain volatile depleted because volatile-rich dust cannot penetrate the inner disk to be accreted by the local planetesimals (Morbidelli et al. 2015).

The efficiency with which particles are able to traverse any such barrier between the outer- and inner disks is inversely proportional to their size, owing to the better coupling of smaller particles to the surrounding gas (Kalyaan et al. 2023; Stammler et al. 2023). Consequently, such barriers are expected to have been permeable to small grains. Empirical constraints on plausible threshold sizes were first provided by Haugbølle et al. (2019) on the basis of the size distribution of CAI grains in an ordinary chondrite, concluding that only grains smaller than 100 – 300 μm would be able to penetrate the barrier. This estimate was subsequently revised downward to 46±48 μm by Dunham et al. (2023) after having examined some 76 sections across a range of chondrites, both NC and CC. This figure overlaps with the 50μm size distribution peak observed by Hezel et al. (2008) in CCs. Furthermore, the region occupied by CAIs, on average, in NC chondrites is 0.0097±0.0002 %, compared to an average of 3-5 % for the most CAI-rich CCs (CK and CV, Hezel et al. 2008; Ebel et al. 2024). Therefore, CAIs are up to 500× less abundant in the NC-forming region than they were in the CC-forming region. Although these CAIs likely formed in close proximity to the Sun, their abundance in CCs implies they were rapidly transported to regions of the disk at which they formed, plausibly the outer solar system (Shu et al. 1997; Desch et al. 2010; Jacquet et al. 2024). If true and such CAIs resided in the outer solar system, then, at most, 1 % of these particles were able to (re-)enter the inner solar system. Interestingly, the NC chondritic planetesimals, which are thought to have formed in the inner solar system (parent bodies of ordinary and enstatite chondrites), are highly depleted in water (McCubbin and Barnes 2019), but only mildly depleted in more moderately volatile elements such as Na, Zn, K (Wasson and Kallemeyn 1988). This may suggest that the drift of dust was blocked after that the temperature at 1–2 au had decreased below $\sim$ 800 K, but before it decreased to 170 K.

4 Condensation of the Solar Nebula

4.1 Equilibrium Condensation

The discussion in Sect. 3 has highlighted the dynamic nature of the protoplanetary disk, both in time and in space. This contrasts with the 'static' picture afforded by the canonical cosmochemical understanding of disk evolution, which is frequently tied to nebular condensation temperatures of the elements, T_c, from a gas of fixed composition (the Sun) and pressure (10^{-4} bar; Lodders 2003; Wood et al. 2019). The utility of this volatility scale lies in its simplicity; it quantifies the temperature at which half of the mass of a given element exists in its condensed (liquid or solid) state (Larimer 1967). However, this simplicity comes at the expense of information as to the condensing mineral assemblage, and thus the temperature range over which condensation occurs. For oxides and metals, condensation stoichiometries take the general form:

$$M^{x+n}O_{\frac{x+n}{2}}(l,s) = M^{x}O_{\frac{x}{2}}(g) + \frac{n}{4}O_2(g) \tag{14}$$

where x is the oxidation state of the metal, M, in the gas phase and n the number of electrons exchanged. The condensation temperature of M in the above reaction can be described by (e.g., Sossi et al. 2019; Ebel 2021):

$$T = \frac{-\Delta H}{R\left(\frac{n}{4}\ln fO_2 + \ln P + \ln f_{\text{vap}}^{i,T}\right) - \Delta S} \tag{15}$$

where P is the total pressure, ΔH and ΔS are the enthalpy and entropy change, respectively, of Eq. (14) and $f_{vap}^{i,T}$ is the fraction of element i remaining in the vapour;

$$f_{\text{vap}}^{i,T} = N_{\text{vap}}^{i,T}/N_{\text{vap}}^{i,0} \tag{16}$$

where N is the number of moles of i in the gas phase at any given temperature, T, relative to the total abundance, 0. Values therefore range between $0 < f_{\text{vap}}^{i,T} < 1$. Condensation curves of the elements for any given reaction occur over a similar temperature interval, relating to the fact that gas-solid reactions have similar $\Delta G/RT$ among one another (cf. Eq. (15)), where ΔG is the change in Gibbs energy. Because ΔG is large relative to RT, the temperature intervals over which condensation occurs are small (~50 – 100 K) (Larimer 1967). This is typically the case for the major elements (Mg, Fe, and Si), whose condensation reactions are *not* limited by the availability of other elements in the gas phase. However, the 50% condensation temperature is not an intrinsic thermodynamic property of the element, but is related to the nature of the particular condensation reaction(s) for a given composition, pressure and temperature. When a single equilibrium predominates, a constant ΔH with respect to T is a good approximation, and the first derivative of Eq. (15) with respect to $f_{\text{vap}}^{i,T}$ yields the condensation rate;

$$dT = \frac{\Delta H R}{f_{\text{vap}}^{i,T}\left(R\left(\frac{n}{4}\ln fO_2 + \ln P + \ln f_{\text{vap}}^{i,T}\right) - \Delta S\right)^2} df_{\text{vap}}^{i} \tag{17}$$

The functional form of Eq. (17) indicates that the temperature interval is proportional to $1/[f_{\text{vap}}\ln f_{\text{vap}}^2]$, leading to decreasing condensation rates as temperature declines. These phenomena are best understood by examining the equilibrium condensation curves of the elements from a solar gas at 10^{-4} bar in Fig. 7, computed with FactSage 8.2 (Bale et al. 2016).

It is evident that some elements condense largely into a single phase, such as Al into Al_2O_3 (corundum) or S into FeS (troilite). The behaviour of others, however, is dictated by several condensation reactions occurring at different temperatures, the most prominent example being Si, which condenses into olivine at ~1345 K via the reaction;

$$2Mg(g) + SiO(g) + 3/2O_2(g) = Mg_2SiO_4(s), \tag{18}$$

which then reacts with remaining SiO(g) to form orthopyroxene at ~1275 K, according to:

$$Mg_2SiO_4(s) + SiO(g) = 2MgSiO_3(s). \tag{19}$$

This gives rise to a T_c^{50} of roughly 1320 K, which, thermodynamically, does not correspond to any particular reaction, but rather to the arithmetic average of the two major reactions. The root of the second derivative of the instantaneous fraction condensed with respect to temperature, d^2f/dT^2 gives the peak of the condensation curve, and each of the major reactions (contributing >5% to the total condensed fraction of an element, $f_c > 0.05$) are shown, together with f_c at that temperature, in Table 2.

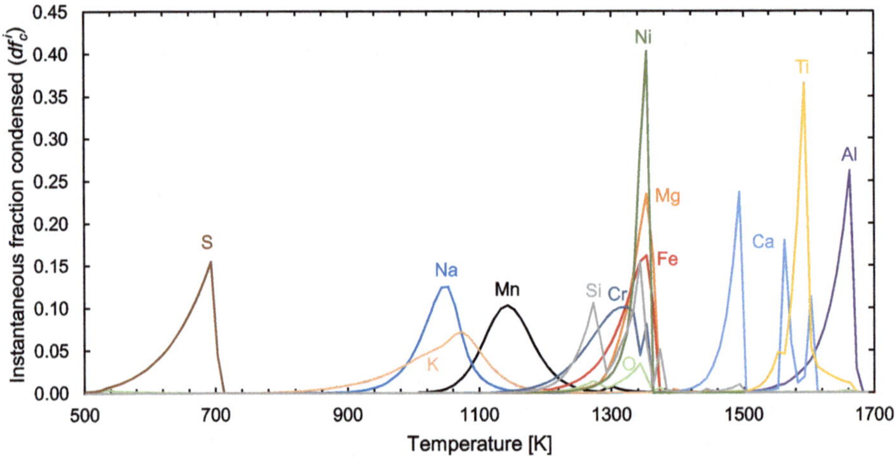

Fig. 7 Condensation curves of the elements from a gas of solar composition (Lodders 2010) at 10^{-4} bar total pressure as calculated with FactSage 8.2. It shows curves of the instantaneous fraction condensed (df_c^i) of a given element, i, as a function of temperature computed at 10 K intervals (dT). The quantity $df_c^i = f_c^{i,T_2} - f_c^{i,T_1}$, where $f_c^{i,T} = 1 - f_{\text{vap}}^{i,T}$. Elements with multiple peaks (notably Ca and Si) condense via a series of reactions. See text for details

Table 2 Major condensation reactions in the solar nebula

Element	Major gas(es)	Mineral	Mineral name	T [K]	$f_c^{i,T}$
Al	Al > AlOH > Al_2O	Al_2O_3	Corundum	1663	0.29
Ti	TiO ≫ Ti	$CaTiO_3$	Perovskite	1593	0.54
Ca	Ca ≫ CaOH	$CaAl_{12}O_{19}$	Hibonite	1603	0.11
Ca	Ca ≫ CaOH	$CaAl_4O_7$	Grossite	1563	0.38
Ca	Ca ≫ CaOH	$Ca_2Al_2SiO_7$	Melilite	1493	0.63
Ni	Ni ≫ NiH, NiS	Fe alloy	Fe alloy	1353	0.40
Mg	Mg	Mg_2SiO_4	Olivine	1343	0.24
Fe	Fe	Fe alloy	Fe alloy	1340	0.35
Cr	Cr ≫ $Cr(OH)_2$	Fe alloy	Fe alloy	1353	0.08
Cr	Cr ≫ $Cr(OH)_2$	$MgCr_2O_4$	Spinel	1318	0.37
Si	SiO > SiS	$CaAl_2Si_2O_8$	Feldspar	1373	0.09
Si	SiO > SiS	Mg_2SiO_4	Olivine	1343	0.31
Si	SiO > SiS	$MgSiO_3$	Orthopyroxene	1273	0.76
O	CO ~ H_2O ≫ SiO	Mg_2SiO_4	Olivine	1343	0.08
O	CO ~ H_2O ≫ SiO	$MgSiO_3$	Orthopyroxene	1273	0.17
Mn	Mn	Mg_2SiO_4	Olivine	1143	0.48
Na	Na > NaCl	$CaAl_2Si_2O_8$	Feldspar	1048	0.49
K	K > KCl	$CaAl_2Si_2O_8$	Feldspar	1083	0.40
S	H_2S	FeS	Troilite	693	0.20

T refers to the root of the second derivative of $f_c^{i,T}$ (the fraction condensed) with respect to temperature (i.e., the peak temperature in Fig. 7).

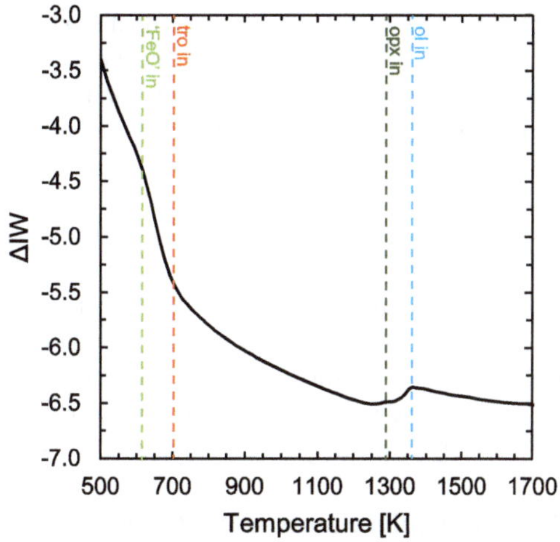

Fig. 8 Evolution of oxygen fugacity (fO_2) relative to the iron-wüstite buffer (O'Neill and Pownceby 1993) during cooling and equilibrium condensation of the solar nebula gas at 10^{-4} bar. Dashed vertical lines show the onset of condensation of major phases; ol = olivine (blue), opx = orthopyroxene (dark green), tro = troilite (orange) and FeO as a component in olivine, 'FeO' (light green)

Equation (18) proceeds through the consumption of O_2 gas. As a result, the fO_2 of the solar nebula decreases relative to those defined by solid-solid buffers as condensation of the major elements (notably Mg and Si) occur (Fig. 8). That of Fe (and Ni) is oxygen-neutral, as its reaction involves no electron exchange, yet would result in net increase in fO_2 of the gas phase by increasing its O/Fe ratio, given that fO_2 is proportional to xO_2P, and the total pressure is held constant. This does not arise, however, because this effect is not sufficient to offset the decrease in fO_2 caused by reaction (18) occurring at the same temperature (Fig. 8). Little condensation occurs thereafter, and oxygen fugacity then increases monotonically below 1250 K owing to internal gas phase reactions (chiefly $H_2O = 0.5O_2 + H_2$ and $CO_2 = 0.5O_2 + CO$) that favour the left-hand-side down-temperature. This increase in fO_2 relative to the iron-wüstite buffer is accelerated upon troilite precipitation (Fig. 8).

Mineralogically, oxidation is manifest in the redistribution of Fe from its host predominantly in Fe metal above ~700 K into increasingly fayalitic olivine and troilite (FeS) below this temperature threshold. The latter transformation is described by the well-known equilibrium (Lewis 1972):

$$Fe(s) + H_2S(g) = FeS(s) + H_2(g) \tag{20}$$

which, owing to the equal number of molecules of gas in the reactants and products, is independent of total pressure. Although O is not shown in Eq. (20), because the gas is in chemical equilibrium, it can equally be written;

$$Fe(s) + H_2S(g) + 1/2O_2(g) = FeS(s) + H_2O(g) \tag{21}$$

explicitly highlighting the increasing fO_2 that preempts FeS condensation. This reaction results in a reduction of the modal abundance of Fe metal by a factor 2, from ~27% (by mass) to 13% by 600 K, causing a concomitant increase in FeS to 15%. The consumption of the remaining condensed Fe^0 proceeds due to increasing fO_2 (fH_2O/fH_2) below ~600 K (Lewis 1972);

$$2Fe(s) + H_2O(g) = FeO(s) + H_2(g) \tag{22}$$

here, FeO is treated as a component in olivine, and reaction (22) tempers the fO_2 increase shown in Fig. 8. The Mg# (molar Mg/[Mg + Fe^{2+}]) in olivine decreases from ~0.98 at

700 K to 0.71 by 520 K in our computation (see also Grossman 1972), consuming almost the entirety of the remainder of the Fe^0 by this temperature. Like Eq. (20), Eq. (22) is pressure-independent. Consequently, and because reactions (20) and (22) occur at similar temperatures, the abundance of S and the FeO/Fe ratio of condensed material are correlated. The small fractions of remaining metal are oxidised below ~380 K by:

$$3Fe(s) + 4H_2O(g) = Fe_3O_4(s) + 4H_2(g) \tag{23}$$

though diffusion-controlled growth rates mean that reaction (23) is unlikely to proceed far (Hong and Fegley 1998). At lower temperatures still (below ~350 K), olivine and orthopyroxene may be consumed by the condensation of H_2O to form serpentine, talc and brucite, however, the low temperatures mean that reaction kinetics are sluggish, and such minerals are unlikely to form by direct condensation (Fegley 2000).

4.2 Non-equilibrium and Fractional Condensation

The idealised condensation sequence presented above is subject to the implicit assumption that all condensed material remains in contact with the same parcel of gas over its entire cooling history. This assumption characterises all such equilibrium models (e.g., Grossman 1972; Wood and Hashimoto 1993; Lodders 2003; Wood et al. 2019). For this condition to hold, it implies a static, slow-cooling solar nebula, in which the gas-solid equilibration timescale (t_{eq}) is at least as rapid as the cooling timescale (t_{cool}). Observations of T-Tauri stars indicate mean nebular lifetimes of ~1 – 10 Myr (e.g., Armitage et al. 2003). Kinetic considerations show that $t_{eq} < t_{cool}$ is not met, even in a static disk, below ~ 400 K owing to the exponential dependence of diffusion coefficients with reciprocal temperature, putting a lower limit on the equilibrium assumption (Fegley 2000).

Temperatures of 400 K represent a lower limit because the analysis in Sect. 3 emphasises the dynamic nature of disks. Gas and dust not only move radially inward or outward depending on their location in the disk, but also vertically as condensed grains coagulate and settle to the midplane (Cassen 2001). The degree to which settling occurs depends on the relative timescales of coagulation (t_{coag}) to the dynamical evolution of the disk (t_{dyn}), with the former of the order of 10^4 yr (Weidenschilling and Cuzzi 1993) and the latter ~10^5 yr (and increasing with time). Perfect equilibrium is therefore unlikely on dynamical grounds. Settling of coagulated dust to the midplane has several consequences; *i)* it reduces disk opacities, leading to enhanced cooling rates (see Sect. 3), *ii)* it fractionates volatile- from refractory elements and *iii)* it enhances the dust/gas ratio of the midplane.

The presence of CAIs in contact with chondrules and matrix is evidence of disequilibrium among chondritic meteorites; the condensation sequence illustrated in Fig. 7 predicts that refractory phases (corundum, perovskite, hibonite, melilite) are no longer present below ~1400 K, having been replaced by feldspar, pyroxenes and olivine (e.g., Grossman and Larimer 1974). Settling to the midplane and subsequent isolation from the evolving gas is one plausible mechanism for preserving CAIs. Alternatively, the cooling rate of the disk could have been so rapid (see Sect. 3.2) so as to prevent equilibration of newly condensed phases down-temperature. This occurs when the reciprocal cooling rate dt^{-1} is of the order of the condensation rate, dN_i/dt, which is in turn approximated by the Hertz-Knudsen equation:

$$\frac{dN_i}{dt} = A\frac{\gamma_{cond}(p_{i,eq} - p_i)}{\sqrt{2\pi m_i k_B T}}, \tag{24}$$

where N is the number of gaseous particles of i, t is the time, A is the surface area, γ_{cond} the dimensionless condensation coefficient with value $0 < \gamma_{\text{cond}} < 1$, m is the mass of i and k_B the Boltzmann constant. This process would engender supersaturation (i.e., $p_i > p_{i,\text{eq}}$) of the remaining gas phase components that are unable to condense at their equilibrium $P - T$ conditions, thereby resulting in condensation of new solids at higher-than-equilibrium temperatures (see Grossman et al. 2012). More sluggish are reactions that involve the conversion of an already condensed phase plus a gas to a new condensed phase;

$$\text{solid}(1) + \text{gas} = \text{solid}(2). \tag{25}$$

Equation (19) is an example of one such reaction. Simple collision theory (SCT) has been developed to predict the rates of these reactions (Fegley 1988, 2000), which scale the Hertz-Knudsen equation (Eq. (24)) by an additional activation energy, E_a,

$$\frac{dN_i}{dt}_{rxn} = \frac{dN_i}{dt} e^{-\frac{-E_a}{RT}}, \tag{26}$$

resulting in a linear scaling for evaporation/condensation kinetics with time. After some time, however, the growth rim of the new solid(2) becomes diffusion-limited, producing a parabolic growth curve where dN_i is proportional to $\sqrt{dt}$ (Fegley 2000). A detailed investigation of how cooling rates affect the mineralogy of the resulting phases is underway (Charnoz et al. 2025).

Chemical evidence for deviations from canonical solar nebula conditions and/or equilibrium comes from the low Mg# (0.86±0.003) of olivine and pyroxene in chondrules (Larimer and Anders 1970; Huang et al. 1996; O'Neill and Palme 1998, Fig. 2). For the canonical case of equilibrium condensation of a solar gas, these Mg#s would be achieved by $\sim$ 575 K. At these temperatures, chondrules should have their full complement of S, Zn and other moderately volatile elements, which is not observed (Wasson and Kallemeyn 1988). Ebel and Grossman (2000) suggest high FeO contents could arise from CI-like dust enrichment in the disk (either at the midplane or globally), resulting in fO_2 of ΔIW-3.1, ΔIW-1.7, and ΔIW-1.2 for dust enrichment factors of 100x, 500x and 1000x, respectively. Resulting olivine Mg#s at 1200 K are 0.96, 0.8 and 0.65-0.60, respectively (Ebel and Grossman 2000), suggesting $\sim 300 - 400$x dust enrichment is required. Using the FactSage model, Mokhtari and Bourdon (2025) find slightly lower Mg#, 0.90, at 1200 K and 100× CI dust enrichment, implying roughly 150× dust enrichment is required to achieve Mg # = 0.86. Enrichment factors above $\sim$100 at the midplane are difficult to obtain, however, in dynamical models of the solar nebula (Cassen 2001). Appealing to the accretion of ices to oxidise iron requires comparable degrees of enrichment ($\sim$250×) to achieve a similar H_2O/H_2 ratio (7.6×10^{-2}, Fegley and Palme 1985) which are, in turn, higher than thought plausible by transport of ice-rich planetesimals into the inner disk (5–100 ×) in the α-disk-based model of Ciesla and Cuzzi (2006).

5 Mixing and Differentiation During Planet Formation

5.1 Estimating Bulk Compositions

In predicting the bulk compositions of the planets, T_c is of limited use, as it is undefined for elements that do not condense to more than 50% relative to their initial mass. This is particularly relevant in the case of O, whose abundance in the planets is much higher than

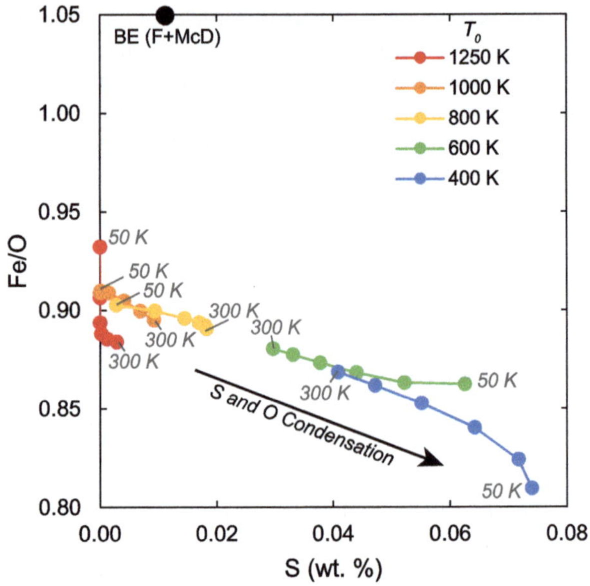

Fig. 9 Modelled bulk planetary compositions emerging from mixing of equilibrium nebular condensates from a Gaussian distribution as a function of temperature at 10^{-4} bar. The coloured lines correspond to different values of T_0 (red = 1250 K, orange = 1000 K, yellow = 800 K, green = 600 K, blue = 400 K) and grey numbers denote the standard deviation of the Gaussian distribution, σ. The black point illustrates the composition of the Bulk Earth (BE, Fischer and McDonough 2025)

its T_c (180 K) would suggest. Roughly 20% of the total nebular O budget is condensed by 1200 K, owing to the formation of solids in which it is bound with metals, yet only a further $\sim 3\%$ condenses by 500 K, associated with the oxidation of Fe to FeO (Eq. (22)). Hence, the remaining $\sim 77\%$ condenses as $H_2O(s)$ at much lower temperatures ($\sim$ 180 K; Lodders 2003).

Therefore, the use of nebular condensation temperatures alone is insufficient to define the bulk compositions of planetary bodies. To test the extent to which the composition of the Earth and planets can be reproduced through mixing of components formed by nebular condensation, a Gaussian distribution (Eq. (27)) is employed to sample the compositions of the equilibrium condensates over a range of temperatures. This model, shown to provide a sound match to the volatile element composition of the Earth (Sossi et al. 2022), takes the form here:

$$N_{i,\text{body}} = \sum_T \left[\frac{N^T_{i,\text{body}}}{\sigma\sqrt{2\pi}} \exp\left(-\frac{1}{2}\left[\frac{T - T_0}{\sigma}\right]^2 \right) \right] \tag{27}$$

where $N_{i,\text{body}}$ is the number of moles of element i in the mixture (planetary body), T_0 is the mean temperature of the Gaussian distribution and σ its standard deviation, and is therefore the sum across all temperatures (T) of the product of the number of moles in the body at each temperature step ($N^T_{i,\text{body}}$) and the Gaussian probability distribution calculated at T. The mole fraction is then given:

$$x_{i,\text{body}} = \frac{N_{i,\text{body}}}{\sum_i N_{i,\text{body}}} \tag{28}$$

The values of T_0 and σ are varied between ranges of 400 – 1250 K and 50 – 300 K, respectively and results shown in Fig. 9.

The Earth's composition (BE, Fischer and McDonough 2025) is not captured by such models; falling to higher Fe/O than any mixture of equilibrium nebular condensates. This results from both a high Fe abundance (nearly 32% by mass) *and* a low O abundance (∼30% by mass) in the BE compared to condensate mixtures; 27–29.5% for Fe and 32–33% for O. Such iron enrichment in the Earth has been cited as evidence of its 'non-chondritic' bulk composition (cf. O'Neill and Palme 2008). Mechanisms causing departures from chondritic compositions include, but are not limited to the physical 'sorting' of nebular condensates, such as the preferential incorporation of early-condensed forsterite into feedstocks for the Earth relative to later-condensing enstatite, as inferred on the basis of Si isotopes (Dauphas et al. 2015). Erosion of mantle- relative to core material is another plausible process, mooted to have given rise to the Earth's superchondritic Fe/Mg ratio (2.1 by mass, O'Neill and Palme 2008). Numerical simulations support the dynamical plausibility of collisional erosion in producing superchondritic Fe/Mg ratios (Carter et al. 2015, 2018). The potential for evaporative loss at a post-nebular stage to fractionate Fe/Mg is explored in Sect. 6.

Importantly, the Fe/O ratios and S contents of mixtures of nebular condensates are anti-correlated. This reflects the co-condensation of S, starting near 700 K at 10^{-4} bar, and additional O into the fayalite component of olivine near 550 K (Eq. (22), Fig. 7). At $T_0 > 700$ K, increasing σ at constant T_0 leads to increasing S contents, whereas the opposite is true for $T_0 < 700$ K. Thus, the more volatile-rich the body, the more oxidised it is expected to be. Concretely, values of T_0 around 900 K would be required (at constant $\sigma = 225\ K$, see Sossi et al. 2022) in order to condense quantities of S sufficient to match its estimated BE abundance (∼1 wt.% Fischer and McDonough 2025). However, at this T_0 value, the Earth would be too enriched in moderately volatile elements (Na, K, Zn, among others) relative to observations (which indicate T_0 for the Earth of ∼1150 K, Sossi et al. 2022). This problem is exacerbated in the small telluric bodies; Vesta, APB and the Moon are more oxidised than the Earth (Fig. 1) yet more depleted in volatile elements (Fig. 3). Together, these considerations are grounds to reject the hypothesis that the composition of the Earth (and, by extension, those of the other terrestrial planets) reflect mixing of components produced by equilibrium condensation of a canonical solar nebula alone.

5.2 Internal Differentiation

In order to translate planetary bulk compositions into those of their observable mantle and core, internal differentiation must be considered (e.g., Ringwood 1966). Core-mantle differentiation in growing planets is thought to have occurred during transient periods of (partial) melting at high pressures (several – to tens of GPa) and temperatures (>2000 K) informed by experiments and *ab-initio* simulations (e.g., Ringwood et al. 1991; Thibault and Walter 1995; O'Neill et al. 1998; Gessmann et al. 2001; Rubie et al. 2004; Wade and Wood 2005; Ricolleau et al. 2011; Fischer et al. 2015, 2020; Suer et al. 2017; Huang and Badro 2018; Li et al. 2020; Huang et al. 2021; Blanchard et al. 2022).

To leverage these constraints, here we develop a model that takes as input parameters the pressure and temperature of core-mantle equilibrium, and the bulk composition of the material (Sect. 5.1) in the system Si-Al-Ca-Mg-Fe-Ni-O, which, together, comprise ∼98.5 wt.% of the Earth (Palme and O'Neill 2014). This approach was first formalised by Rubie et al. (2004) and expanded by Frost et al. (2008) and Rubie et al. (2011). Our treatment is formally incomplete as it neglects other, minor core-forming elements, notably Co, S, C and H, which, together, likely comprise ∼2.5 wt.% of the core (Hirose et al. 2021). The partitioning of the alloying elements — Ni, O and Si — are described according to Eq. (29) and in Table 3 over a range from 1 bar to 100 GPa at a temperature corresponding to the peridotite liquidus of Andrault et al. (2011). Three compositions are taken from mixtures of nebular condensates at different T_0 and σ values from Sect. 5.1, and the fourth is the Bulk Earth composition (Table 4).

Table 3 Parameters for metal-silicate partitioning exchange reactions

	$\log K$	$\ln\gamma$	n	Reference
Si	0.52-13,000/T	0.5-5500/T	4	S13, F15
Fe	-	0	2	Ri11
O	0.7-5000/T+(22P)/T	0	-	Ri11, S13, F15
Ni	1.06+1553/T-(98P)/T	0	2	Ru11

Ri11 = Ricolleau et al. (2011), Ru11 = Rubie et al. (2011), S13 = Siebert et al. (2013), F15 = Fischer et al. (2015)

Table 4 Modelled bulk compositions using the parameters for Eq. (27) as described in Sect. 5.1. All elements in wt. fraction. Bulk Earth from Fischer and McDonough (2025)

	Fully reduced	Bulk Earth	Intermediate	Fully oxidised
Parameters				
T_0	1250	-	600	400
σ	50	-	50	50
Elements				
Al	0.0166	0.0159	0.0141	0.0138
Ca	0.0174	0.0167	0.0148	0.0145
Mg	0.1668	0.1578	0.1524	0.1491
Si	0.1636	0.1567	0.1721	0.1684
Fe	0.2971	0.3268	0.2913	0.2850
O	0.3187	0.3073	0.3377	0.3520
Ni	0.0198	0.0187	0.0176	0.0172
Fe/O	0.932	1.063	0.863	0.809

As iron is the dominant element in planetary cores, and is typically present at several weight percent in the coexisting mantles (cf. Table 1), element partitioning is described by an exchange coefficient with Fe (e.g., Wade and Wood 2005),

$$\frac{n}{2}\text{Fe} + M\text{O}_{\frac{n}{2}} = \frac{n}{2}\text{FeO} + M \tag{29}$$

which circumvents the definition of $f\text{O}_2$ but requires that n, the number of electrons transferred in the reaction, be known. The abundances of Ni, O and Si in the mantle are allowed to vary in order to satisfy the equation;

$$x(M)_{\text{core}} = \frac{K_{(29)}(a\text{Fe})^{\frac{n}{2}}aM\text{O}_{\frac{n}{2}}}{\gamma M(a\text{FeO})^{\frac{n}{2}}} = \frac{(xM)_T - (xM)_{\text{mantle}}(1-CMF)}{CMF} \tag{30}$$

where the subscript T is the total mass and CMF the core mass fraction. The values of $(xM)_{\text{mantle}}$ and CMF are iterated until the left-hand side is equivalent to the right-hand

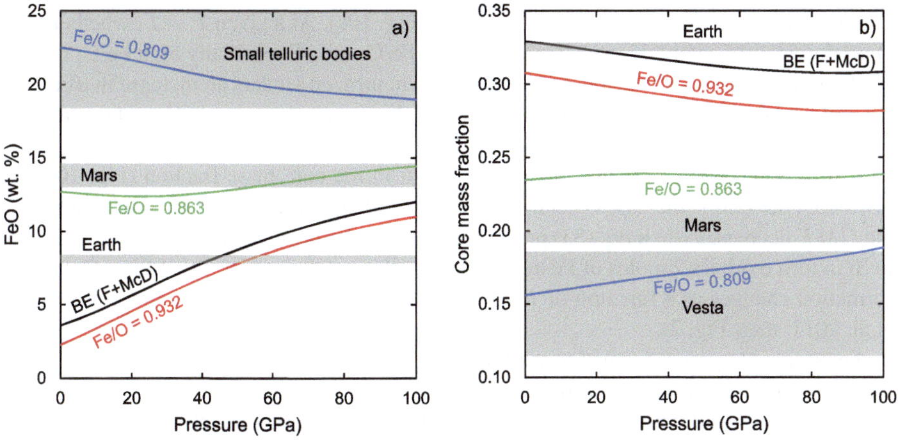

Fig. 10 The equilibrium **a)** FeO content in the mantle and **b)** core mass fraction of synthetic rocky planets calculated along the peridotite liquidus of Andrault et al. (2011) plotted as a function of pressure for different bulk compositions with Fe/O (by weight) ratios ranging from fully reduced (all Fe as Fe^0), 0.932 (red), 0.863 (green) and 0.809 (blue, fully oxidised, all Fe as FeO). These bulk compositions are derived from the model presented in Sect. 5.1 at different T_0 and σ values. The black curve has Fe/O = 1.063 and is the Bulk Earth composition of Fischer and McDonough (2025). All compositions are shown in Table 4. Observed FeO contents and core mass fractions of selected telluric bodies are shown as grey horizontal bars (see Table 1 for data sources)

side. The iteration is performed with the additional constraints that;

$$x(\mathrm{O})_{\mathrm{mantle}} = 1.5[x(\mathrm{Al})_{\mathrm{mantle}}] + x(\mathrm{Ca})_{\mathrm{mantle}} + x(\mathrm{Mg})_{\mathrm{mantle}} + 2[x(\mathrm{Si})_{\mathrm{mantle}}] + x(\mathrm{Fe})_{\mathrm{mantle}} + x(\mathrm{Ni})_{\mathrm{mantle}} \tag{31}$$

and

$$\sum_M x(M)_{\mathrm{mantle}} = 1 \tag{32}$$

The remaining elements, Al, Ca and Mg, are assumed to be perfectly lithophile, hence their mantle abundances are given analytically by $C_{\mathrm{mantle}} = C_T/(1 - CMF)$. This leaves *four unknowns* - $x(\mathrm{Fe})_{\mathrm{mantle}}$, $x(\mathrm{Ni})_{\mathrm{mantle}}$, $x(\mathrm{Si})_{\mathrm{mantle}}$ and CMF, which are soluble with four constraints (Eq. (30) for M = Ni, Si and O and Eq. (32)).

As first proposed by Javoy (1995), even for precursor materials that contain all Fe as Fe^0 (i.e., in reduced form), the process of core-mantle equilibration at increasing $P - T$ results in progressively higher FeO contents in the mantle (Fig. 10). This behaviour arises due to the competition between the incorporation of Si in the core via;

$$\mathrm{SiO_2(mantle)} + 2\mathrm{Fe(core)} = \mathrm{Si(core)} + 2\mathrm{FeO(mantle)} \tag{33}$$

and that of O by;

$$\mathrm{FeO(mantle)} = \mathrm{Fe(core)} + \mathrm{O(core)} \tag{34}$$

Both reactions are self-limiting (cf. O'Neill et al. 1998): FeO contents tend to converge as $P - T$ increase along the peridotite liquidus, which reflects the tendency for both Si and

O to partition into the core as $P-T$ increases (Fig. 10a). At a given $P-T$, whether Si or O predominantly enters the core depends on the Fe/O ratio. For initially high Fe/O, the low aFeO promotes Si dissolution in the core, which, in turn, engenders an increase in aFeO via Eq. (33); a negative feedback that results in a keeling over of mantle FeO content (red and black curves, Fig. 10). Because mass is conserved, any change in mantle FeO content must be compensated for by a concomitant adjustment in the core mass fraction (Fig. 10b). At low Fe/O, the CMF increases because both Fe and O enter the core (Eq. (34)). Conversely, the CMF decreases when Eq. (33) prevails over Eq. (34), as the reduction of 1 mole of SiO_2 to Si in turn oxidises 2 moles of Fe to FeO. A corollary of the model is that the fO_2 of core formation changes as a function of P and T at *constant* bulk composition (see also Rubie et al. 2011, their Fig. 2).

For single-stage core formation with the BE composition, we find FeO $= 8.33$ wt.% and Ni $= 1881$ ppm in the mantle with a CMF of 0.315 at $P = 40$ GPa and $T = 3150$ K. These values compare well with the 8.1±0.2 wt.% FeO and 1860±93 ppm Ni in the BSE (Palme and O'Neill 2014) and a CMF of ~0.325 (Table 1). At these conditions, our model predicts Si $= 3.6$ wt.% and O $= 1.2$ wt.% in the core, consistent with other estimates (O'Neill and Palme 1998; Badro et al. 2014; Dauphas et al. 2015; Hirose et al. 2021). That there is no single composition that fits the observed CMF and FeO at a given pressure should not be taken as evidence against single-stage core formation. At 40 GPa, this is likely remedied by the additional S (~2 wt.%), Co + P + Cr (~1.5 wt.%) in the core, which are not included in our model. Addition of these quantities to the CMF value of 0.315 at 40 GPa would result in the observed CMF of 0.325, meaning single-stage core formation is permitted for this set of elements.

Such estimates also match the 30–50 GPa and 3000 K inferred from D_{Ni} and D_{Co} (Thibault and Walter 1995; Li and Agee 1996; Bouhifd and Jephcoat 2003; Wade and Wood 2005; Kegler et al. 2008). Wade and Wood (2005) and Rubie et al. (2011, 2015) emphasised that, although homogeneous accretion models are readily able to satisfy the observed FeO, Ni and Co abundances in Earth's mantle (as also shown here), they lead to elevated mantle Cr and V contents, which would instead require temperatures ~650 K higher. A notable caveat to their models lies in the implicit assumption that the extant families of chondrites *de facto* formed Earth. Should this assumption be relaxed, it remains to be seen whether heterogeneous accretion is required from a chemical standpoint. Moreover, Siebert et al. (2013) demonstrated V and Cr abundances could be fit, independent of whether accretion were homogeneous or heterogeneous with respect to composition, at (mean) pressures of equilibrium between 40 and 60 GPa, by virtue of the negative interaction parameters between Cr-O and V-O in iron-rich alloys. In either case, as is evident from Fig. 10 and highlighted by Rubie et al. (2015), the composition of Earth-forming material would need to have been initially somewhat reducing (Fe/O > 0.9) and high in total Fe (~32 wt.%; Table 4), as lower Fe/O ratios lead to mantle FeO contents that are too high relative to that observed (8.1±0.02 wt.%; Table 1).

On the other hand, small telluric bodies have mantle FeO contents and core mass fractions (Table 1) that appear to require oxidised starting materials (Fe/O < 0.86; Fig. 10) because the low-pressure (< 5 GPa) core formation that must have occurred on these bodies would have led to temperatures insufficient (e.g., Cartier et al. 2024), for self-oxidation to operate to the extent to reach the ~20 wt.% FeO observed *via* Eq. (33) in their mantles. Mantle FeO contents inferred for the silicate mantles of iron meteorite parent bodies are almost invariably high, with all but the IIABs lying between 10–25 wt.% (Grewal et al. 2024). Rather, the volatile-poor nature of STBs, such as Vesta and the angrite parent body (e.g., O'Neill and Palme 2008) together with the wide range of volatile contents in iron meteorites (Scott and

Wasson 1975; Hirschmann et al. 2021) at near constant mantle FeO (Grewal et al. 2024) indicates that the relationship between fO_2 and volatile element content expected for nebular condensation (Figs. 8, 9) decoupled during the formation of STBs. Finally, even at the lowest Fe/O ratios considered here, FeO contents relevant to Mercury's mantle (0.2±0.1 wt.% Namur et al. 2016; Nittler et al. 2018, Table 1) are not obtained, suggesting it accreted material more reduced and with higher metal/silicate ratios than that produced through equilibrium condensation and/or contains significant S or C that are not considered in this model.

6 Physicochemical Conditions of Volatile Depletion

One possible explanation for the divergence in the compositions of the terrestrial planets and STBs relative to chondrites is that they (or their components) underwent chemical fractionation under conditions that differed from those set by the nebular gas (e.g., O'Neill and Palme 2008).

6.1 Chemical Fractionation

Here we examine the *relative* fractionation between two volatile elements, $M1$ and $M2$, such that the physical mechanism by which the two elements were lost cancels from the operation and their relative fractionation is dependent on the thermodynamics of the presumed vaporisation reaction alone.

For most metals, vaporisation reactions are simple because one condensed component (e.g., MnO) and one gas species (e.g., Mn) predominate over a range of planetary- and nebular conditions (Table 2, O'Neill 1991; Sossi et al. 2019). When comparing the fraction of an element vaporised ($fM1$) relative to that of a second comparator element ($fM2$), the total pressure of the system cancels, as does the total budget of the element in the system, to yield (Sossi et al. 2025):

$$\frac{f_{M1}^{\text{vap}}}{(1-f_{M1}^{\text{vap}})} = \frac{f_{M2}^{\text{vap}}}{(1-f_{M2}^{\text{vap}})}\left(\frac{\gamma_{M1}}{\gamma_{M2}}\right)(fO_2)^{\frac{\Delta n_{M2-M1}}{4}} \exp\left(\frac{(\Delta G^o_{M2}-\Delta G^o_{M1})}{RT}\right), \tag{35}$$

where $(\Delta G^o_{M2} - \Delta G^o_{M1})$ relates to the standard state free energy of the exchange reaction between the pure components vapour and condensed phase(s), such as

$$\text{Mn(g)} + \text{NaO}_{0.5}\text{(s, l)} + \frac{1}{4}\text{O}_2\text{(g)} = \text{Na(g)} + \text{MnO(s, l)}. \tag{36}$$

Because relative entropy changes for gas-liquid or gas-solid reactions are dominated by the (ideal) gas, the quantity $(\Delta G^o_{M2} - \Delta G^o_{M1})$ is nearly independent of temperature, and exchange reactions become more discriminating (i.e., larger fractionation of $M1$ from $M2$) proportional to exp($1/T$). While the free energy change $(\Delta G^o_{M2} - \Delta G^o_{M1})$ is readily tabulated from thermodynamic databases (e.g. JANAF), the activity coefficients for many components in silicate melts are uncertain (see Sossi and Fegley 2018; Fegley et al. 2023). However, divalent metal oxides mix near-ideally in silicate liquids (O'Neill and Eggins 2002; Wood and Wade 2013), while activity coefficients (γ) for the alkali metal oxides are typically very small but relatively well characterised (Charles 1967; Mathieu et al. 2011; Sossi et al. 2019). Here, measurements of activity coefficients at a given temperature are extrapolated according to:

$$\frac{\ln\gamma}{\ln\gamma_{ref}} = \frac{T_{\text{ref}}}{T}, \tag{37}$$

Table 5 Thermodynamic properties of evaporation reactions for selected moderately volatile elements of the form Eq. (14)

	ΔH^o (kJ/mol)	ΔS^o (kJ/mol.K)	n	γ_{ref}^*	Source
Li	419.0	0.141	1	0.2	(Sossi et al. 2019)
K	236.1	0.122	1	7.3×10^{-5}	(Sossi et al. 2019; Wolf et al. 2023)
Na	267.0	0.121	1	5.0×10^{-4}	(Sossi et al. 2019; Wolf et al. 2023)
Mn	606.3	0.173	2	1	(Kohn and Schofield 1994)
Mg	666.4	0.186	2	0.20	(Wolf et al. 2023)

$^*\gamma_{ref}$ calculated for the oxide in silicate liquid at a reference temperature of 1673 K.

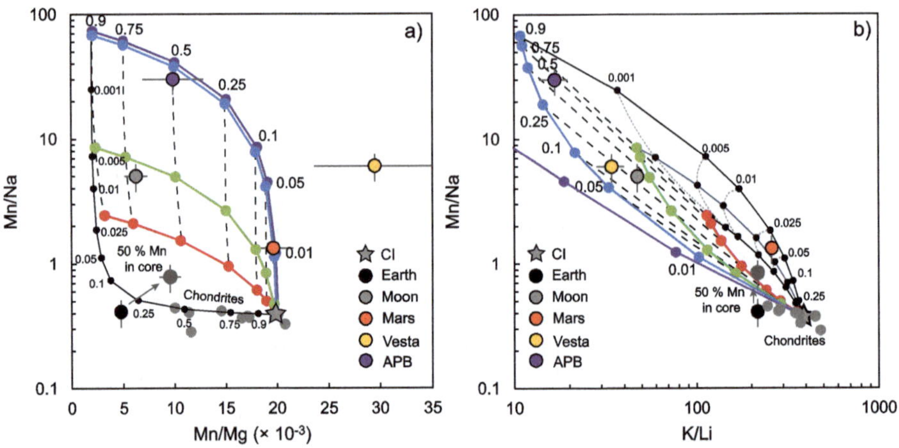

Fig. 11 **a)** Mn/Na vs. Mn/Mg and **b)** Mn/Na vs. K/Li ratios among planetary bodies. The coloured curves denote the evolution of the composition of an evaporation residue calculated using Eq. (35) at 1100 K (purple), 1500 K (blue), 2000 K (green) and 2500 K (red) at a fixed relative oxygen fugacity (ΔIW-1, except for 1100 K, which was calculated at ΔIW-6, see text), with numbers denoting the fraction of Mn loss undergone by the body. The black dashed lines connect equal degrees of Mn loss at different temperatures. The solid black curves trace a binary mixture of a volatile-depleted end-member with 90% Mn loss and a CI-like end-member. The small numbers denote the mass fraction of CI in the mixture. On a), only the volatile-depleted end-member at 1500 K is shown due to clarity, whereas in b) the end-members are taken at the three temperatures, and the dotted grey lines connect equal fractions of CI material. The grey point gives the composition of the Earth assuming 50% of its Mn budget resides in the core (see Siebert et al. 2018). The grey field denotes the compositional range of chondrites (Wasson and Kallemeyn 1988; Siebert et al. 2018)

where γ tends to 1 at infinite temperature. Values of γ_{ref} and T_{ref} are given in Table 5.

For a canonical solar nebula gas, the fO_2 is roughly ΔIW-6 during the condensation of most rock-forming elements (Fig. 8). The fO_2 during melting and vaporisation/condensation on planets *after* the dispersal of the nebular gas, reflects the composition of the evaporating/condensing material (e.g., Visscher and Fegley 2013). The FeO contents of small telluric bodies and oxybarometry of basaltic achondrites imply oxygen fugacities near ΔIW-1 (Fig. 1, Table 1, Wadhwa 2008).

The change in the Mn/Na (Fig. 11a) and K/Li (Fig. 11b) ratios of evaporation residues/partial condensates are computed as a function of the Mn/Mg ratio using Eq. (35). The Mn/Na ratio is not only sensitive to fO_2 (O'Neill and Palme 2008), but also to temperature at constant relative fO_2. For 25% Mn loss, 70%, 89% and 98.5% of Na is lost at

2500 K, 2000 K and 1500 K, respectively. Therefore, at higher temperatures, the fraction of Na lost approaches that of the Mn lost. That is, evaporation exchange reactions (Eq. (35)) become less discriminating. For 25% Mn loss under nebular conditions (1100 K, ΔIW-6), the Na loss is nearly identical to the 1500 K, ΔIW-1 case (98.6% lost), but the resulting K/Li is 10$\times$ lower (compare blue and purple curves). The APB has lost $\sim$50% of its Mn, and has Mn/Na of $\sim$30, achievable for single-stage evaporation at ΔIW-1 and 1500 K; conditions also consistent with its K/Li. By contrast, the Mn/Na and Mn/Mg range of chondrites is explained by mixtures of a volatile-depleted component and CI chondrites (thin black line, Fig. 11a), equivalent to chondrules (volatile-poor) and matrix (volatile-rich) (e.g., Alexander 2019; Hellmann et al. 2020).

The Earth has Mn/Na – Mn/Mg ratios lower than for any chondrite, its Mn/Na being too low to have been set by evaporation alone. The degree of Na depletion ($\sim$ 80% lost) in the Earth can be used to infer the maximum permitted Mn loss by volatility alone. Roughly 36% Mn would have been lost at 2500 K, decreasing to 2% for 1500 K or in the solar nebula. This implies between 1/2- to 2/3rds of the Earth's Mn budget is in the core (see also Drake et al. 1989; Ringwood et al. 1991; Siebert et al. 2018); a similar result arises when assuming Li is entirely lithophile and is as volatile as Mn.

Compositions of STBs can be reproduced by vapour loss at a single set of temperature-fO_2, favouring relatively 'high' temperatures of about 1400 – 1800 K and relatively high fO_2 at ΔIW-1 compared to conditions in the solar nebula. Vesta is a possible exception (Fig. 11a), suggesting estimates for its mantle Mg content are too low. The fact that the elemental fractionations observed differ empirically from those of chondrites indicates distinct processes were responsible in setting their bulk compositions; that is, evaporation (either in a single-stage or an average of multiple events) for STBs rather than mixing among components as is relevant for chondrites (Alexander 2019). This result, coupled with the ancient (< 5 Myr after CAIs) volatile depletion ages implied by the low $^{87}Sr/^{86}Sr_i$ (Hans et al. 2013) leads to the conclusion that there must have been localised, 'oxidised' domains at the loci of planetesimal formation, even in the presence of the nebular gas.

On the other hand, the compositions of Mars and Earth (see also Sect. 6.3.2 for Earth) cannot be reconciled with volatile loss at any given temperature. This is because the temperature inferred from Fig. 11a differs from that deduced from 11b. Implicit in these models is the absence of the major volatiles, such as H, C and S, which may modify the stable gas species considered in equations such as Eq. (36), that assume monatomic gases. Thermodynamic models indicate the stability of hydroxide- and sulfide-bearing metallic species, in particular (Fegley et al. 2016; Ivanov et al. 2022), yet a systematic investigation of how such differences influence the extent of volatile loss are yet to be undertaken. Nevertheless, within the framework explored here, mixing between volatile-depleted and volatile-rich end-members is required. It is also clear from Fig. 11, however, that these end-members cannot have been identical for Earth and Mars. These observations suggest that more complex, multi-component mixtures of volatile-poor and volatile-rich materials are likely needed to account for the compositions of the Earth and Mars.

6.2 Isotopic Fractionation

Partitioning of an element between two phases during cosmo-/geochemical processes is often accompanied by fractionation of its isotopes. The estimated mantle isotopic compositions of planetary objects are determined via analyses of (ultra-)mafic rocks. On Earth, this typically involves peridotites (e.g., Sossi et al. 2016; Savage et al. 2015), or, for (highly) incompatible elements, basalts/komatiites (e.g., Dauphas et al. 2010; Hibbert et al. 2012; Jerram et al. 2020). For other planetary bodies, we are limited to basaltic samples, as there are no mantle samples. Consequently, particularly for achondrites, the limited number of samples (cf. the angrite parent body, Keil 2012), and their genetic relationship with their source (cf. geochemical signatures in Martian shergottites, Borg et al. 2002) introduces uncertainty into the assignment of the measured isotopic signature to that of the body. Nonetheless, isotopic bulk compositions determined in this manner are compiled in Table 6.

6.2.1 The Major Elements Mg, Si and Fe

The estimated Mg, Si and Fe isotope ratios of bulk, differentiated planetary bodies are similar to- or higher than those of the (averages of) carbonaceous, ordinary and enstatite chondrites (Fig. 12a,b; Table 6). Taking hydrothermal alteration for type 1 and 2 carbonaceous chondrites into account (Young et al. 1999; Hin et al. 2017), estimated ratios of Mg, Si and Fe isotopes are higher in the BSE than they are in chondrites. The same applies to the Moon (e.g., Klaver et al. 2024; Armytage et al. 2012; Sossi et al. 2016), whose isotopic composition is indistinguishable from that of the Earth for these elements. Vesta (as estimated largely by eucrites) and Mars (via lherzolitic shergottites), on the other hand, have Fe and Si isotope ratios that overlap with carbonaceous and ordinary chondrites (though heavier than ECs; Fig. 12, Dauphas et al. 2015). Although these bodies may appear heavier in $\delta^{26/24}$Mg (light grey squares) the current estimates should be corrected downward by $\sim$0.04‰ to account for the effect of silicate differentiation (Liu et al. 2022, 2023), which also renders their values similar to chondrites (see also Young et al. 2019, black symbols in Fig. 12a).

The Mg, Si and Fe isotope ratios in bulk planetary objects correlate (Fig. 12a, b), implying a control by the same cosmo-/geochemical fractionation process. Because Mg, Si and Fe are major elements, only reservoir separation involving significant mass fractions of these elements or with large isotope fractionation factors, could have affected their mantle isotopic ratios. This leaves core-mantle-crust separation or vapour-condensed-phase fractionation (Poitrasson et al. 2004; Sossi et al. 2016; Hin et al. 2017; Young et al. 2019). Silicon and, in particular, Mg are sparingly soluble in metallic Fe during the differentiation of small telluric bodies. This factor, compounded with the fact that the Fe isotope fractionation factor between metal and silicate is small at low pressures (Hin et al. 2012) in the absence of S (Kubik et al. 2022), means that core formation is unlikely to have caused substantial isotopic fractionation of Fe, and, by extension, Mg and Si (although debate is ongoing, see Bourdon et al. 2018; Shahar and Young 2020). Despite the fact that the $\Delta Si_{met-sil}$ is considerably larger than that for Fe (Shahar et al. 2009; Hin et al. 2014), the observation that the angrite and HED parent bodies, despite both being too oxidised to host Si metal in their cores (Fig. 1), have significantly heavier Si (and Fe in the case of the APB) isotopes than chondrites (cf. Pringle et al. 2014; Liu et al. 2017), casts further doubt over the role of core formation, making vapour-condensed phase fractionation a more likely process.

6.2.2 Cr and the Moderately Volatile Elements

In marked contrast to Mg, Si and Fe, the isotope ratios of Cr in bulk, differentiated planetary bodies are similar to- or lower than those in chondrites. Earth and Mars have $\delta^{53/52}$Cr within error of all types of chondrites, while the Moon and Vesta have significantly lower values (Table 6, see Sect. 6.3.1). The isotope ratios of MVEs (the lithophiles Zn, K and Rb, and the siderophiles Cu and Ga) often vary widely compared to those of the major elements and Cr, even within samples from a single parent body, including chondrites (Paniello et al. 2012). The isotopic ratios of K, Zn, and Rb are similar to chondritic values in the Earth and Mars, but are enriched in heavy isotopes in the Moon and Vesta (Table 6; Fig. 12c,d).

Table 6 Estimated mean values of the mass-dependent, stable, isotopic compositions of planetary bodies. All values expressed in per mille deviations from the listed standard

	APB	EPB	Moon	Mars	Earth	CC	OC	EC	References
$\delta^{7/6}Li_{L\text{-}SVEC}$		3.7	3.8	4.4	3.5	3.2	2.7	1.9	1-3
95% c.i.		0.1	0.8	1.6	0.5	0.6	0.4	0.6	
$\delta^{25/24}Mg_{DSM\text{-}3}$	−0.079	−0.114	−0.130	−0.113	−0.121	−0.140		−0.134	4-6
95% c.i.	0.010	0.008	0.010	0.006	0.003	0.010		0.004	
$\delta^{30/28}Si_{NBS\text{-}28}$	−0.21	−0.42	−0.29	−0.48	−0.30	−0.47		−0.69	7,8
95% c.i.	0.03	0.03	0.08	0.03	0.03	0.03		0.05	
$\delta^{41/39}K_{SRM3141a}$	−1.70	0.41	−0.04	−0.30	−0.42	−0.25	−0.78	−0.19	9-17
95% c.i.		0.08	0.06	0.03	0.01	0.14	0.34*	0.39*	
$\delta^{53/52}Cr_{SRM979}$		−0.220	−0.216	−0.17	−0.124	−0.128	−0.102	−0.050	18-22
95% c.i.		0.030	0.020	0.08*	0.030	0.011	0.006	0.006	
$\delta^{57/54}Fe_{IRMM014}$	0.190	0.010	0.08	−0.010	0.050	−0.010			23,24
95% c.i.	0.020	0.010	0.03	0.020	0.010	0.010			
$\delta^{65/63}Cu_{JMC\text{-}Lyon}$		0.5	0.50		0.70	−0.7	−0.15	−0.25	25-27
95% c.i.		0.5**	0.10		0.010	1.3*	0.5*	0.08	
$\delta^{66/64}Zn_{JMC\text{-}Lyon}$		1.1	1.39	0.237	0.170	0.36	0.01	0.23	28-35
95% c.i.		2.2**	0.12	0.030	0.011	0.08	0.32	0.07	
$\delta^{71/69}Ga_{IPGP}$			0.14		0.000	−0.15	−0.6	−0.23	36-38
95% c.i.					0.016	0.11	0.8*	0.10	
$\delta^{87/85}Rb_{SRM984}$	−1.2	1.0	0.03	0.10	−0.13	0.11	−0.12	0.02	39-42
95% c.i.		1.4	0.03	0.03	0.06	0.10	0.20	0.27	

*Uncertainty quoted as 2 s instead of 95 percent confidence interval as the data in literature do not pass Wilk's test for a normal distribution around a single population mean.

**Uncertainty quoted as 2 s instead of 95 percent confidence interval because of the potential presence of a (unquantifiable) systematic uncertainty (see text).

For Mg, Si and Fe, no OC (and EC for Fe) data are shown, because they have been grouped with CC as their compositions are not statistically different. References: 1-3: Magna et al. (2006, 2014), von Strandmann et al. (2011)

4-6: Hin et al. (2017), Liu et al. (2023), Klaver et al. (2024)

7,8: Armytage et al. (2011), Dauphas et al. (2015)

9-17: Tian et al. (2019, 2020), Ku and Jacobsen (2020), Wang et al. (2016), Hu et al. (2022), Bloom et al. (2020), Jiang et al. (2021), Koefoed et al. (2020)

18-22: Schoenberg et al. (2008, 2016), Bonnand et al. (2016), Sossi et al. (2018a), Zhu et al. (2019, 2021)

23, 24: Sossi et al. (2016), Poitrasson et al. (2019)

25-27: Herzog et al. (2009), Savage et al. (2015), Dhaliwal et al. (2024)

28-35: Luck et al. (2005), Moynier et al. (2011), Paniello et al. (2012), Sossi et al. (2018b), Paquet et al. (2023), Fang et al. (2024)

36-38: Kato and Moynier (2017), Kato et al. (2017), Wimpenny et al. (2022)

39-42: Pringle and Moynier (2017), Nie et al. (2023), Wang et al. (2023, 2024)

K: chondrite data are from observed falls only to avoid potential effects of terrestrial weathering.

Zn: as for K, chondrite data are from observed falls only, with EC data deriving from EH only due to suspected metamorphism effects on the analysed EL6 falls.

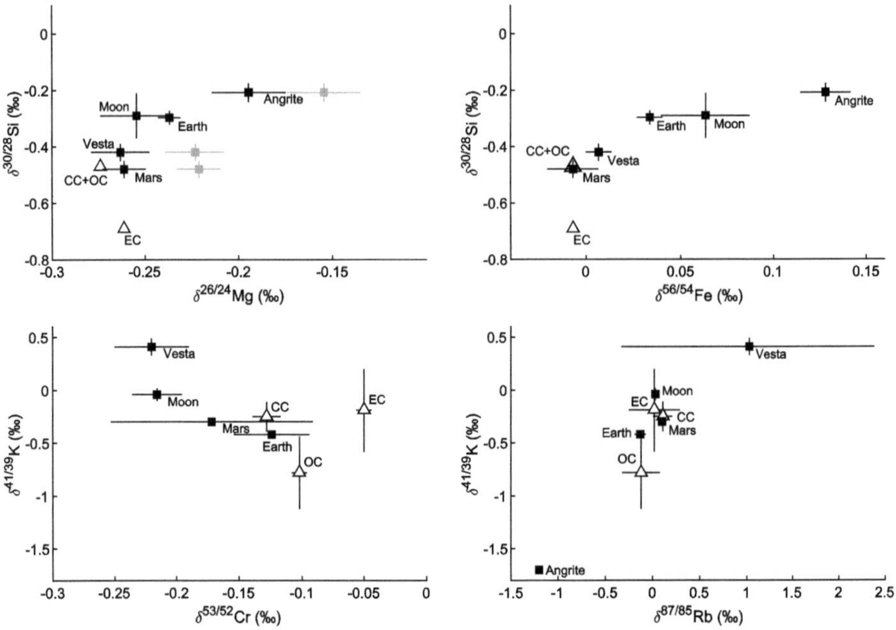

Fig. 12 Mass-dependent isotopic variations among bulk differentiated (squares) and undifferentiated (triangles) planetary objects. **a)** $\delta^{30/28}$Si vs. $\delta^{26/24}$Mg. The grey points correspond to measured values pre-correction (see main text for details). **b)** $\delta^{30/28}$Si vs. $\delta^{56/54}$Fe. **c)** $\delta^{41/39}$K vs. $\delta^{53/52}$Cr. **d)** $\delta^{41/39}$K vs. $\delta^{87/85}$Rb. See Table 6 for data sources

This tendency is also seen in the scarcer data on Cu and Ga isotopes, which are chondritic in the Earth, but significantly higher in lunar and vestan rocks (Table 6; Kato and Moynier 2017; Day et al. 2019; Wimpenny et al. 2022). The APB, instead, appears to be enriched in light K and Rb isotopes, interpreted as reflecting partial (re)condensation of these elements (Hu et al. 2022; Wang et al. 2024). Within current measurement precision, no significant variations occur among the Li isotope ratios of the various planetary objects (cf. Tomascak et al. 2016).

None of the isotopic variations in the aforementioned MVEs correlate with those in the major elements, Mg, Si and Fe. As shown in Fig. 12c,d, however, they correlate with one another (see also Wang et al. 2024), again implying that a common process controlled their observed variation. As first pointed out by Day and Moynier (2014) and extended by Tian et al. (2021), higher isotopic ratios are found on bodies with lower escape velocities. We show in Fig. 13a that such correlations among the Earth, Mars, the Moon and Vesta occur for the isotopes of K, Cr and Rb (of the MVE isotopic systems with data for these four planetary objects, only Zn does not yield a statistically significant correlation). In the case of K and Rb, data are also available for the APB (Hu et al. 2022; Wang et al. 2024), yet do not follow the trend defined by the other planetary bodies, irrespective of the assumed mass of the APB (Tissot et al. 2022).

Only three isotope systems currently display these correlations, which are based on only four data points per isotopic system, but they seem a worthy endeavour for future investigation. Following normalisation by $\Delta m/m^2$, the variations in Fig. 13a must be caused by either differences in the fractionation factor between vapour and condensed phase(s) or by the extent of mass loss, which is itself proportional to the relative volatilities of the elements.

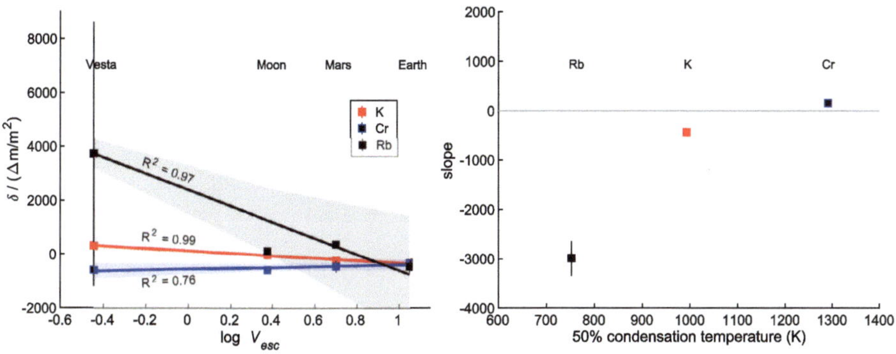

Fig. 13 Mass-dependent isotopic fractionation among bulk planetary objects and its relation to the escape velocity of planetary object **(a)** and their volatility as parameterised by their 50% condensation temperature **(b)**. Isotopic compositions have been normalised to the relative mass differences of the isotopes ($\Delta m/m^2$)

In Fig. 13b, we show that the slopes of the regressions presented in Fig. 13a become steeper for increasingly volatile elements (as quantified by T_c^{50}). As such, differences in K, Cr and Rb isotopic ratios among the different planetary bodies are controlled by the extent of mass loss experienced by the body in volatilisation/condensation event(s), that is, a Rayleigh process with a finite element budget. Because the isotopes of Cr and the MVEs in the Earth and Mars are not fractionated from the range of chondrites, these bodies show no isotopic evidence for vapour loss of these elements of the sort preserved in the STBs, despite the fact that they also have lower MVE abundances than in most chondrites.

6.3 Physical Causes of Volatile Loss

Planetary bodies with magma exposed at their surfaces will develop a vapour atmosphere, with the surface pressure determined by the magma temperature and its composition (e.g., Wolf et al. 2023). If the atmosphere is no longer hydrostatic, upwards motion can drive atmospheric loss, so-called hydrodynamic escape (Young et al. 2019; Chao et al. 2021). Whether hydrodynamic escape is efficient or not depends on the competition between thermal energy and gravity. More efficient escape occurs from small, high temperature bodies, as quantified in the parameter λ,

$$\lambda = \frac{GM}{r}\frac{\mu}{RT_s} \tag{38}$$

where G is the gravitational constant, M and r are the mass and radius of the body, μ is the atmospheric mean molar mass, T_s is the surface temperature and R is the gas constant. A large value of λ indicates inefficient escape (gravity wins). For a diatomic gas, the critical value of λ above which hydrodynamic escape shuts off is 2.4-3.6 (Volkov et al. 2011). For an adiabatic atmosphere, the critical value can be shown to be $\gamma_{\text{ad}}/(\gamma_{\text{ad}} - 1)$ (Zahnle and Kasting 1986), where γ_{ad} is the ratio of the specific heat at constant pressure over constant volume. For a diatomic gas this ratio is 7/5, so the critical value of λ in this case is 3.5.

Escape of atmospheres with a $\mu \geq 0.03$ kg/mol ($\geq$air) is only possible on <1000 km radii bodies (smaller than the Moon; Fig. 14). This limit increases to Mars-sized bodies for $\lambda = 3$ for $\mu = 0.004$ kg/mol at 2000 K. Hydrodynamic escape from the fully-formed Earth is not possible (except possibly for pure hydrogen, and even then only at very high temperatures,

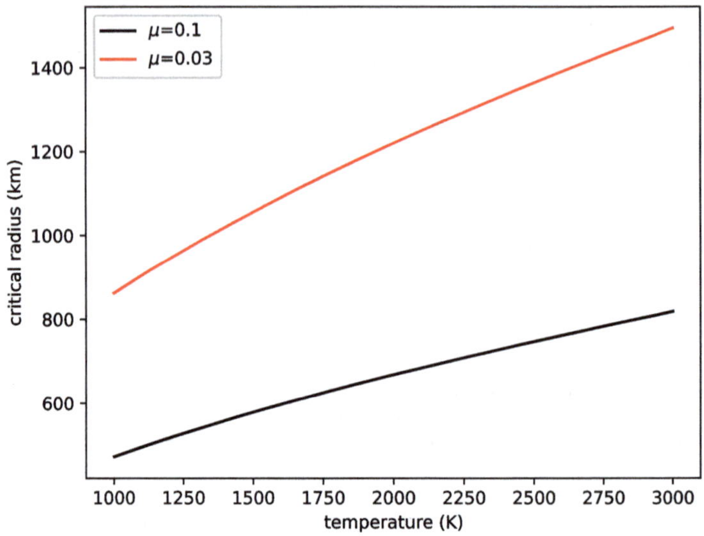

Fig. 14 Radius above which hydrodynamic escape is shut off, as a function of surface temperature, for two different molar masses μ in kg/mol. Here, we assume the critical value of λ is to be 3, and a bulk density of 4000 kg m^{-3}

≥3500 K). Hydrodynamic escape is impeded relative to that predicted from Eq. (38) if the surrounding disk has a non-zero pressure. It is also limited by the thermal energy available: heat is both radiated and advected away from the hot surface, and as the magma cools the vapour pressure drops, reducing the outwards flux of material. This limits the length of time available for escape to occur in the absence of external heat sources. There are two main sources of energy available to drive escape;

- Decay of ^{26}Al. Although this process likely caused melting and differentiation of early-formed solar system bodies, its ability to drive hydrodynamic escape is impeded by the rate at which it supplies thermal energy. A magma body radiating at 700 K (similar to terrestrial lava lakes; Patrick et al. 2016) loses energy at a rate of 14 kW m^{-2}. Conversely, ^{26}Al decay would produce a surface heat flux of 0.5 kW m^{-2} in a 1000 km radius body at t_0, and less thereafter. As a result, ^{26}Al heating cannot produce significant mass loss in bodies of such size (we note that the contrary finding expressed in Young et al. (2019) is due to a numerical error; Young, pers. comm.).
- Impacts. Sufficiently energetic impacts can cause wide-spread melting and vaporisation (Nakajima et al. 2021). Hydrodynamic loss is very sensitive to the melt temperature because it determines the surface vapour density, which in turn governs the mass loss rate through the equation:

$$\frac{dN_i}{dt} = 4\pi (r_B)^2 C_s n_i \tag{39}$$

where r_B is the critical radius (i.e. that at which the gas outflow speed equals the sound speed), C_s the sound speed at r_B, and n_i the number density of the vapour. During conventional accretion, impact velocities are typically comparable to the escape velocity of the target body (Safronov 1969; Agnor et al. 1999), which depends on its mass. Higher impact velocities produce more melting and higher temperatures (Nakajima et al. 2021). But if impact velocity is controlled by target size, then the high-velocity impacts will typically only take place on large bodies, from which hydrodynamic escape is prohibited (see above). There are several potential ways in which this trade-off can be overcome.

The first is to appeal to some event stirring up the planetesimals so that their impact velocities are higher than would be expected based on the arguments above. For instance, Hin et al. (2017) argued that high-velocity impacts driven by excited bodies on highly eccentric orbits, as induced by gas giant migration (Carter et al. 2015) yield sufficient energy to cause planetesimal melt production and vapour loss. Calogero et al. (2025) explored this hypothesis in more detail, focusing on potassium. A second mechanism states that planetary envelopes enriched in rock-forming elements can exchange mass with the ambient nebular gas. As long as the molecular mass is dominated by that of H_2 (0.002 kg/mol), escape from envelopes may be efficient (Steinmeyer et al. 2023). A third is to consider impact-produced melt that is not contained within the target planet, such as an impact-generated proto-lunar disk. Here, the liquid is hot but the effective gravity of the proto-lunar disk is low, facilitating escape from the Moon, but not from the Earth-Moon system (Charnoz et al. 2021).

While hydrodynamic escape is efficient in engendering mass loss, it does not itself produce large mass fractionation between isotopes in the atmosphere (Hunten et al. 1987). However, isotopic fractionation can occur at the melt-vapour interface, the magnitude and direction of which depends on two factors that trade off against each other;

- Temperature. Higher temperatures lead to smaller isotopic fractionation factors if occurring at equilibrium (proportional to $1/T^2$).
- Degree of equilibrium. If Langmuir (kinetic) fractionation prevails (i.e., $p_i < p_{i,\mathrm{eq}}$), then the fractionation factor, α is ${}^{\frac{i}{j}}\alpha_{\mathrm{kin}} = \sqrt{\frac{m_j}{m_i}}$, where m is the isotopic mass. If equilibrium prevails (i.e., $p_i = p_{i,\mathrm{eq}}$), then, in most instances, the gas phase is also enriched in lighter isotopes, but the fractionation factor is closer to unity (Young et al. 2019). This is because most rock-forming elements evaporate as monatomic gases (e.g., K^0, Na^0, Mg^0, see Sossi et al. 2019) or as oxide species with a *lower* mean oxidation state than in the condensed phase (e.g., SiO(g) vs. SiO_2(s,l), GeO(g) vs. GeO_2(s,l)).

6.3.1 Volatile Loss Due to Evaporation Only

If evaporation mimics a Rayleigh process (see Sect. 6.2.2), then the degree of isotopic fractionation follows:

$$\frac{{}^{i/j}R_{\mathrm{res}}}{{}^{i/j}R_0} = F^{\left({}^{i/j}\alpha_{\mathrm{gas/cond}}-1\right)}, \tag{40}$$

where R is the isotopic ratio of the residue (res) and the system (0), respectively and F is the fraction of isotope j remaining in the residue. Here, we examine the correlations between $\frac{{}^{i/j}R_{res}}{{}^{i/j}R_0}$ and F fit with values of α, and compare them to those expected from Langmuir (kinetic) and Knudsen (equilibrium) fractionation.

The prime case in which the expected correlation between F and $\frac{{}^{i/j}R_{\mathrm{res}}}{{}^{i/j}R_0}$ is observed is the Moon. Taking the BSE as ${}^{i/j}R_0$, $\delta_{\mathrm{Moon}} - \delta_{\mathrm{BSE}}$ correlates with the degree of depletion, F (Fig. 15). The fitted values of α (for Rb, K, Zn and Cu) are consistent with values of $p_i/p_{i,\mathrm{eq}}$ of 0.99, should evaporation have taken place at ~2500 K (Nie and Dauphas 2019), or 1 (i.e., equilibrium) for T ~1300 K (Tartèse et al. 2021; Dauphas et al. 2022). Hence, the fitted value of α is degenerate with respect to $p_i/p_{i,\mathrm{eq}}$ and temperature. However, Sn and Cr are anomalous, in that they preserve *lighter* isotopic ratios in the Moon than in the bulk silicate

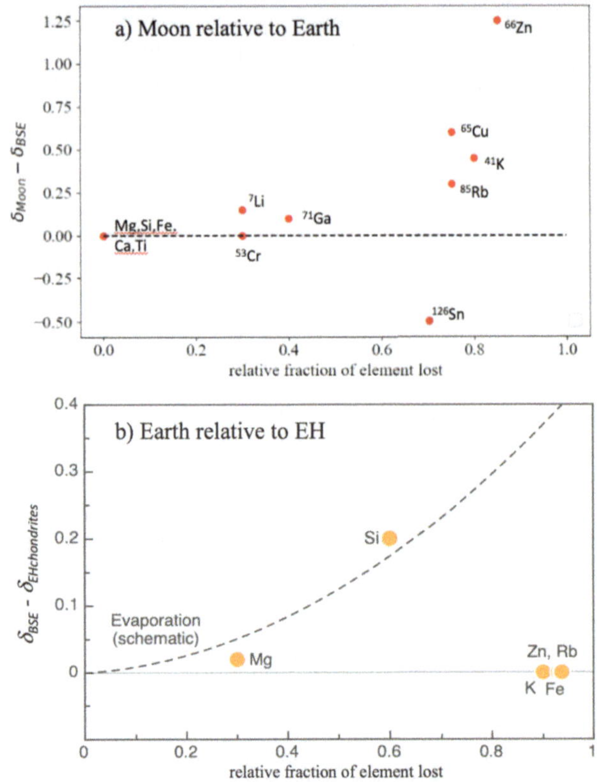

Fig. 15 a) The isotopic differences between the bulk silicate Earth and Moon for the isotopes of various elements, against the depletion of those elements in the Moon relative to the bulk silicate Earth. b) Mass-dependent isotopic fractionation in the bulk silicate Earth relative to EH chondrites for selected elements as a function of their elemental depletion (calculated as $X/\text{Ca}_{\text{BSE}}/X/\text{Ca}_{\text{EH}}$, where $X =$ Mg or Si). The dashed line shows the expected increase in isotopic fractionation for larger degrees of evaporative loss. Despite their greater degrees of depletion, the moderately volatile elements (K, Zn, Rb) do not show any isotopic deviation with respect to EH chondrites. Source data are listed Table 6

Earth (Sossi et al. 2018a; Wang et al. 2019). Because both Cr and Sn have gas species in which the metal is more strongly bound than in the condensed phase, fractionation at equilibrium can uniquely account for this observation in a vapour-condensed phase system. This interpretation, and the negative slope in Fig. 13 thus support loss of Cr and Sn on STBs at relatively oxidised conditions (∼IW) (see Sect. 6.1, Zhu et al. 2019).

The propensity for (near-) equilibrium isotopic fractionation to have prevailed reflects; *i)* the fact that vapour phase reactions reach equilibrium in 10^{-9} to 1 s (Fegley et al. 2020); much shorter than lifetimes of magma oceans (∼10^2–10^7 yr; Salvador and Samuel 2023) and *ii)* $p_i/p_{i,\text{eq}} > 0.95$ when the (far-field) total pressure (P) exceeds ∼10^{-8} bar (Young et al. 2019; Tang and Young 2020). In summary, evidence points to near- or at equilibrium values of $p_i/p_{i,\text{eq}}$ (> 0.99), at least for the Moon. Such conditions are more readily achieved on planetesimals that are Moon-mass or larger (Young et al. 2019), which indicates that volatile loss on the Moon likely occurred when it was close to its present-day mass, though does not require it.

6.3.2 Volatile Loss Due to Evaporation + Mixing

Correlations between isotopic fractionation in a given planetary body and its escape velocity (Fig. 13a) could imply that those bodies experienced volatilisation and vapour loss at their (near-)final masses, as posited by Tian et al. (2021). The calculations in Sect. 6.3 and the observations in Sect. 6.3.1 support the idea that Vesta and the Moon (see also Charnoz et al. 2021), were able to lose vapour by hydrodynamic escape, even at their present day masses.

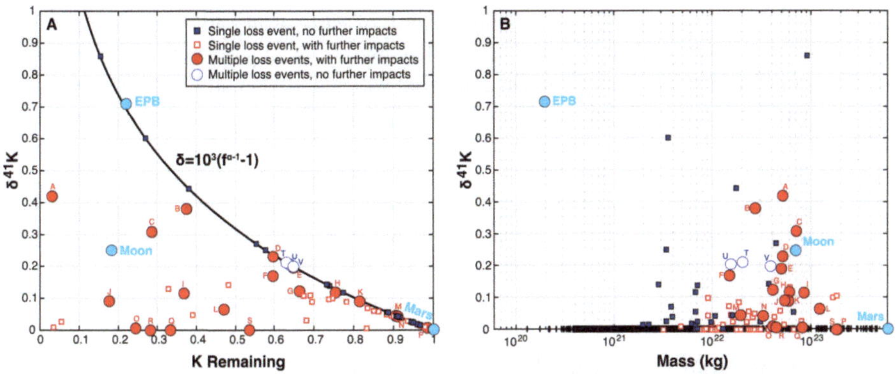

Fig. 16 Calculations using output from N-body simulations (Carter et al. 2015) showing the K isotope fractionation (expressed as δ^{41}K) with respect to the **a)** fraction of K remaining and **b)** the mass of individual planetary bodies, assuming a Mars-like starting composition. Different symbols show whether bodies experienced multiple impact-driven evaporative events. The solid line in a) is that for evaporative loss of K governed by the Rayleigh equation with a value of $\alpha = 0.999541$. See Calogero et al. (2025) for further details

Conversely, Mars, and in particular the Earth, were barely able to lose any vapour at their current masses. As opposed to the Moon and Vesta, Earth and Mars do not preserve any detectable mass-dependent isotopic fractionation relative to chondrites, yet they are clearly volatile-depleted with respect to CI chondrites (Fig. 15b).

Figure 15b shows that Mg and Si, relative to Ca and EH chondrites ($X/\mathrm{Ca}_{BSE}/X/\mathrm{Ca}_{EH}$), are slightly (32%) and moderately (59%) depleted, respectively, in the BSE and are also marginally isotopically heavy (Hin et al. 2017). These figures decrease to 20% for Mg and 29% for Si when normalised to Ca and CI chondrites. Therefore, the depletion of Si and Mg observed in the BSE relative to chondrites (both normalised to Ca) is coupled with heavy isotopic enrichment, whereas the elemental depletion in MVEs records no isotopic fractionation compared to chondrites. The simplest explanation for this duality is that the Earth (and Mars) accreted bodies in which MVEs underwent near-complete evaporative loss (or were hardly accreted at all), and mixed with bodies of undepleted, potentially chondritic material (Hin et al. 2017; Sossi et al. 2022). On the Earth's total mass-basis, this undepleted material is insignificant such that the isotope ratios of the major elements (Mg, Si and Fe) remained essentially unaffected, yet the budgets of the MVEs were entirely overprinted. This implies that there must be at least two populations of smaller bodies, one isotopically heavy and volatile-poor and the other volatile-undepleted and isotopically lighter (i.e., unfractionated relative to chondrites).

To test this idea, output from an N-body model (Carter et al. 2015) was analysed to examine the competition between the chemical and isotopic fractionation induced by impact-driven evaporation of K and the mixing of these bodies over time (Calogero et al. 2025). Bodies were assumed to have an initial bulk and isotopic composition identical to that of Mars. Subsequent impacts in some cases caused melting, magma ocean formation and hydrodynamic loss of K. Isotopic fractionation was assumed to occur in such cases with $\alpha = 0.999541$. Other collisions did not generate melting and were thus assumed to simply mix the compositions of the two colliding objects.

These simulations show that in general, as predicted by Eq. (38), smaller bodies are the most isotopically fractionated (Fig. 16b). However, they are not necessarily the most depleted in K. Indeed, other, larger bodies, which have experienced more impacts, can be *more* depleted in K, but have lower (i.e., more chondritic) δ^{41}K values (e.g. bodies I and

L in Fig. 16). This arises because the combination of early evaporative loss in small bodies followed by mixing results in larger bodies having isotopically unfractionated signatures (because more isotopically unfractionated material was added). For an Earth-sized body, this process naturally results in broadly chondritic δ^{41}K and δ^{66}Zn values, because the budget of these elements is entirely dominated by the addition of volatile-rich material that experienced little- to no evaporation. This is the scenario envisaged by Sossi et al. (2022), and states that, all else being equal, the compositions (both chemical and isotopic) of larger bodies will be determined to a greater extent by mixing, and those of smaller bodies by partial evaporation.

7 Provenance of Planet-Building Materials

The provenance of planetary building materials can be quantified using nucleosynthetic isotope anomalies. These arise through the heterogeneous distribution of presolar material in the solar protoplanetary disk and, as such, allow genetic links among and between meteorites and planets to be examined. The isotope anomalies in a given element are quantified by their part per ten-thousand deviation (ε) from a terrestrial reference standard (in most cases close to but often not identical with the BSE), following internal normalisation to a chosen ratio:

$$\varepsilon^i X = \left(\frac{({}^iX/{}^jX)_{\text{body}}}{({}^iX/{}^jX)_{\text{std}}} - 1 \right) \times 10{,}000 \tag{41}$$

where i and j are the numerator- and denominator isotope, respectively, and X is the element of interest. Implicit in computing isotope anomalies in this way is the assumption that the chosen internal normalising ratio of all samples is terrestrial. For samples having nucleosynthetic isotope anomalies this is almost certainly not true, and so calculated $\varepsilon^i X$ values may not reflect anomalies on a particular isotope, but rather the combined effects of abundance variations of several isotopes, including those used for the internal normalisation.

A key observation from the nucleosynthetic isotope anomalies, as well as Δ^{17}O (whose variations are not nucleosynthetic in origin), is the notion of an isotopic dichotomy between non-carbonaceous (NC) and carbonaceous chondrite (CC) type materials (see Bermingham et al. 2020; Kleine et al. 2020; Kruijer et al. 2020, for recent reviews). The first indication of what is now known as the NC-CC dichotomy dates back to Trinquier et al. (2007), who observed that, for nucleosynthetic isotope anomalies, differentiated meteorites and carbonaceous chondrites define two distinct fields. Warren (2011) then showed that isotope anomalies in O, Ti, Cr, and Ni, meteorites always fall into two distinct clusters, which he termed the non-carbonaceous (NC) and carbonaceous (CC) groups. Since then, the number of elements for which the NC-CC dichotomy has been identified has grown considerably, making it a fundamental characteristic of the early solar system. Importantly, the dichotomy was also identified for isotopes of Mo (Budde et al. 2016; Poole et al. 2017), in which it exists not only for chondrites but also for iron meteorites, which derive from parent bodies that accreted within the first 1 Myr of the solar system (Kruijer et al. 2017). Thus, together, these observations reveal that the dichotomy was established early, and was maintained for essentially the entire lifetime of the disk.

It is now widely thought that the NC reservoir represents the inner disk, and the CC reservoir the outer disk (Warren 2011; Budde et al. 2016; Kruijer et al. 2017). Below we assess whether or not CC material is the dominant source of *i)* volatile elements to the

Earth and *ii)* oxidation relative to the solar nebula among telluric bodies. Owing to their distinct nucleosynthetic heritage, we discuss the iron-peak (Sect. 7.1) and heavier elements (Sect. 7.2) separately.

7.1 Iron-Peak Elements

The 'iron-peak' elements have binding energies per nucleon that are higher than those of the surrounding nuclides, leading to enhanced abundances during stellar nucleosynthesis via the alpha process (Clayton 2003). Typically, these elements are Ti, V, Cr, Mn, Fe, Ni, Cu and Zn. Here, we also extend this classification to Si (produced by O burning) and Ca (produced by Si burning), owing to their similar masses and nucleosynthetic heritage.

In a diagram of ε^{50}Ti versus ε^{54}Cr, the composition of the BSE plots within the NC field, albeit as an endmember, with excesses in neutron-rich isotopes compared to other NCs, pointing towards the CC field (Trinquier et al. 2009). Thus, the composition of the BSE can be produced by mixtures of the extant groups of NC and CC meteorites for these two systems. However, the choice of end-members to represent each group is non-unique. One interpretation (hereinafter hypothesis A) first mooted by Warren (2011), and developed by Schiller et al. (2018) states that the isotopic composition of the Earth evolved over time from ureilite-like (i.e., the meteorites defining the lower end of the NC trend) to its present-day composition by the addition of CI-like material. Schiller et al. (2018) based their arguments on anomalies in ^{48}Ca, but the correlated nature of the isotope anomalies in Ca, Ti, and Cr among the NC meteorites means that the same argument holds for any of these elements. An alternative hypothesis (B), is built on the observation that, in many isotopic systems, including Cr and Ti, the Earth (and Moon) have compositions similar to those of enstatite chondrites (ECs) (e.g., Warren 2011; Dauphas 2017). In this scenario, the Earth accreted $\sim$95% EC-like material, which could reflect objects with a uniform, EC-like isotopic composition (Dauphas 2017; Dauphas et al. 2024), but also via accretion of a variety of NC materials with a range of isotopic compositions, whose average is that of the ECs (Burkhardt et al. 2021).

As pointed out by Warren (2011), hypotheses A and B lead to very different CC fractions inferred for the Earth. The CI-derived fraction for each element, X, in the BSE can be computed by mass balance:

$$f_{X,\mathrm{CI}} = \frac{\varepsilon^i X_{\mathrm{NC}} - \varepsilon^i X_{\mathrm{BSE}}}{\varepsilon^i X_{\mathrm{NC}} - \varepsilon^i X_{\mathrm{CI}}} \tag{42}$$

where NC denotes the isotopic composition of the NC material in Earth, which can either be ureilite- (hypothesis A) or EC-like (hypothesis B). Use of Eq. (42) implies identical element concentrations in each reservoir. For identical element concentrations in each reservoir, the CI-derived fraction for an element X is the same as the apparent mass fraction of CI material accreted to the Earth. This is likely to hold for RLEs (e.g., Ti) and potentially the main components (Cr, Si), but may not for volatile (e.g., Zn) and siderophile elements (e.g., Fe, Ni). This is because these elements are depleted in the BSE relative to chondrites as a result of volatile depletion (Steller et al. 2022; Savage et al. 2022) or core formation, respectively (Dauphas 2017).

In both hypotheses, the apparent fraction of CI-like material is near-constant for Ca, Ti, and Cr, at $\sim$5% for the BSE assuming an EC-like NC end-member, or $\sim$40% for a ureilite-like NC end-member (Fig. 17). As noted above, this constancy of inferred CI fractions in both hypotheses reflects the correlated nature of the isotope anomalies in these elements among the NC meteorites, and the position of Earth on the neutron-rich end of the NC trend.

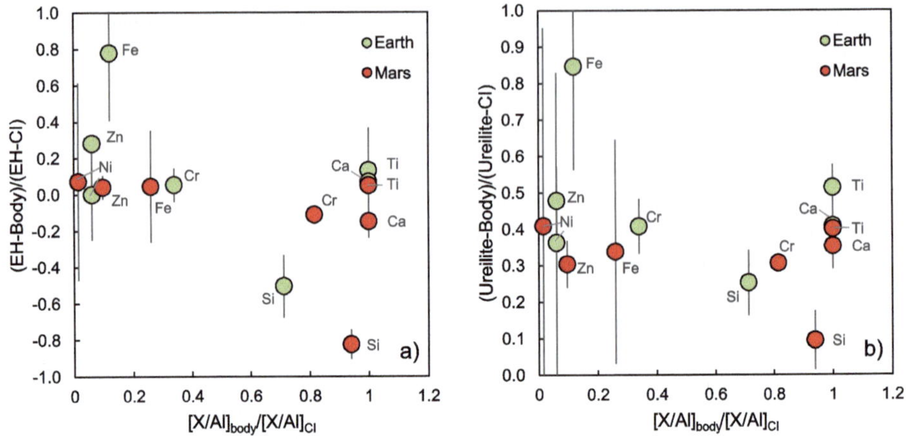

Fig. 17 The fraction of each element in the bulk silicate Earth (BSE, green) and bulk silicate Mars (BSM, red) brought by the CI component in mixtures between **a)** EH-like and CI-like and **b)** ureilite-like and CI-like isotopic end-members, each with identical concentrations, as a function of their Al, CI-normalised abundances in the BSE and BSM. Note the large uncertainties for Fe and Ni, owing to the relatively small degree of isotopic variation relative to the uncertainty on the analyses. Isotope ratios used are μ^{30}Si, ε^{48}Ca, ε^{50}Ti, ε^{54}Cr, ε^{54}Fe and ε^{62}Ni. Data from (Trinquier et al. 2007, 2009; Schiller et al. 2018, 2020; Burkhardt et al. 2021; Steller et al. 2022; Savage et al. 2022; Onyett et al. 2023)

For Ni, using ε^{62}Ni as in Fig. 17, the inferred CI fractions are similar to those obtained from the three aforementioned elements, but the small Ni isotope variations have larger uncertainties and would also be compatible with a wider range of CI fractions (Steele et al. 2012; Tang and Dauphas 2014; Nanne et al. 2019). For example, when using the ε^{60}Ni anomaly, the BSE and CI chondrites are indistinguishable within error (Spitzer et al. 2024). For Fe, the ε^{54}Fe values of CI chondrites and the BSE are indistinguishable, leading to high apparent CI fractions in the BSE (0.79±0.39 and 0.84±0.27 for an EC-CI mixture or Ureilite-CI mixture, respectively; Fig. 17) (Schiller et al. 2020; Hopp et al. 2022b). However, CI chondrites are unlikely to have significantly contributed to the Fe budget of the BSE because, as noted by Hopp et al. (2022b), the isotopic compositions of both Cr (less siderophile) and Ni (more siderophile) in CIs are distinct from the BSE. We note that the apparent contradiction can be resolved by considering that the BSE lies at the neutron-rich (i.e., ^{54}Fe-rich) end of the NC trend, indicating that Fe is entirely consistent with an NC origin (see also Hopp et al. 2022a).

For Si the apparent fractions of CI-like material are systematically lower than that from other elements (and extend to negative values), which indicates that the $\varepsilon^{30}Si_{BSE}$ is *lower* (=0) than both CI (ε^{30}Si = ~0.3) and EC (ε^{30}Si = ~0.1) (see Onyett et al. 2023). On this basis, Onyett et al. (2023) argued that EC-CI mixtures are precluded as proto-Earth source material. However, because ECs are fractionated mass-dependently in δ^{30}Si, correcting the ε^{30}Si for such fractionation yields a revised value of ε^{30}Si = 0.04±0.03 (Dauphas et al. 2024) which scarcely differs from that of the BSE and hence is broadly consistent with hypothesis B. Because such a correction is implicit on mass-dependent fractionation occurring at high temperatures, additional data to better resolve any potential differences between ECs and the BSE are essential in discerning between hypotheses A and B.

A key distinction between hypotheses A and B can instead be made using Zn isotopic composition of the BSE (Steller et al. 2022; Savage et al. 2022; Martins et al. 2023). In hypothesis A, the fraction of CI-derived Zn is 0.48±0.09 (Fig. 17b); indistinguishable from Cr

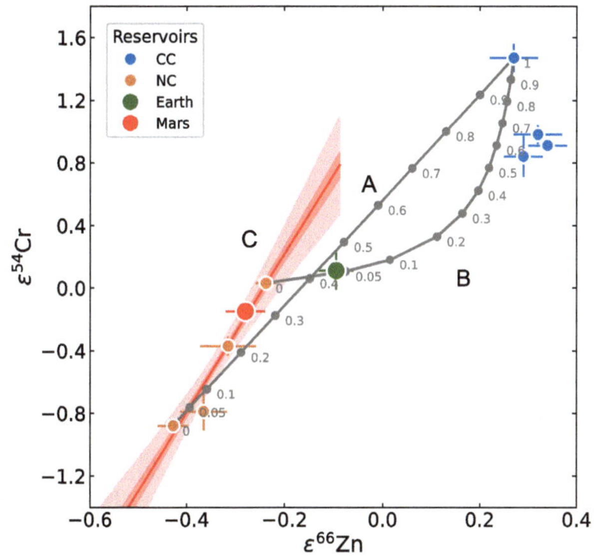

Fig. 18 The ε^{54}Cr-ε^{66}Zn variation among non-carbonaceous meteorites (NC, orange), carbonaceous meteorites (CC, blue), the bulk silicate Earth (Earth, green), and bulk silicate Mars (Mars, red). The grey curves denote mixing lines between CI chondrites and, A - a ureilite-like end-member with Zn/Cr ratios identical to CI chondrites (0.12) and B - an EC-like end-member with 35 ppm Zn ($Zn/Cr_{EC} = 0.014$). Tick marks show the mass fraction of CI-like material in the mixture. C denotes a fit according to the York method (York et al. 2004) assuming uncorrelated errors in x and y and its associated 1-σ (dark red) and 2-σ (light red) error envelope, yielding $\varepsilon^{54}\text{Cr} = (5.06\pm1.05)\varepsilon^{66}\text{Zn} + 1.23\pm0.33$. All data taken from Steller et al. (2022) for chondrites and the Earth, and Kleine et al. (2023), Paquet et al. (2023) for Mars using ^{67}Zn/^{64}Zn-normalised values

(0.41±0.07) and Ti (0.51±0.06). Hypothesis B yields a higher apparent CI-derived fraction computed from Zn (0.29±0.05) compared to Cr (0.07±0.03) and Ti (0.05±0.09; Fig. 17a). In ε^{66}Zn versus ε^{54}Cr space, using the Zn concentrations of CC meteorites and the BSE, any mixing line between the CC field that passes through the BSE intersects the NC field near the *isotopic* composition of the ECs with an average ∼35 ppm Zn for the NC material (Kleine et al. 2023; Kleine and Nimmo 2025, curve 'B', Fig. 18). This implies that the BSE contains ∼6% CI-like material by mass in hypothesis B. Notably, this ∼6% CI chondrite-like material is identical to the CC *mass* fractions inferred from the isotope anomalies in Ca, Ti, and Cr in Hypothesis B (Fig. 17), meaning Earth would have accreted predominantly from NC material with an EC-like isotopic composition, on average, with a 6% contribution of CI material, which delivered ∼ 30% of Earth's Zn budget. Hypothesis A, in contrast to Hypothesis B, would require that the NC (ureilite) and CC (CI) end-members had the same Zn/Cr ratios such that the mixing line passes through the BSE (curve 'A', Fig. 18). Indeed, Zn/Cr ratios in ureilites are similar, to within a factor ∼2, of those in CI chondrites (Brugier et al. 2019; Zhu et al. 2022). In this scenario, the apparent CI-like fraction in the BSE inferred from Zn matches those from Ca, Ti and Cr, but, in contrast to hypothesis B, CI chondrites must represent ∼40% of the mass of the Earth.

Variations in oxygen isotopes, although not nucleosynthetic in origin, nevertheless provide information on a sample's provenance. Ureilites are heterogeneous with respect to δ^{17}O and δ^{18}O (Clayton and Mayeda 1996; Kruttasch and Mezger 2025), yet, unlike other NC bodies, they lie on the Carbonaceous Chondrite Anhydrous Mineral Line (CCAM), which defines a slope of unity in a three-isotope plot (Clayton et al. 1977). Because CI chondrites

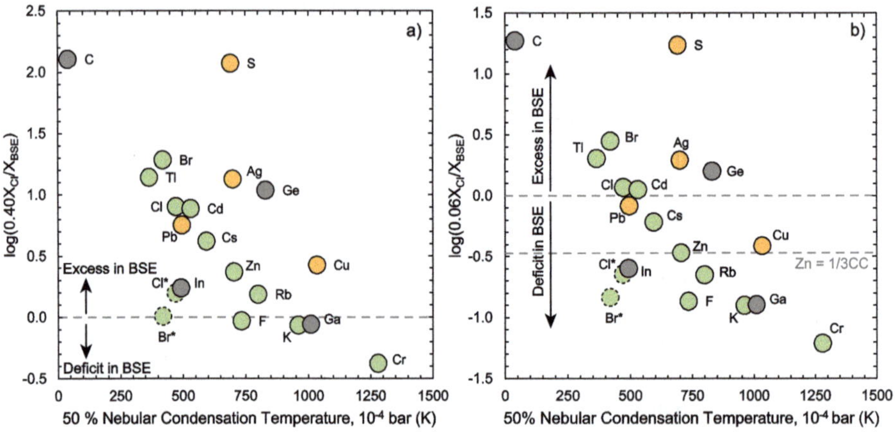

Fig. 19 The $\log_{10}$ of the ratio of the concentrations of a selection of moderately- and highly volatile elements for a) 40% and b) 6% (by weight) given by addition of CI chondrites to an otherwise volatile-free proto-Earth, normalised by the present-day bulk silicate Earth (BSE) abundances of these elements. Green = lithophile, grey = siderophile, orange = chalcophile. Cl* and Br* refer to the CI chondrite abundances of Clay et al. (2017). Values that exceed 0 indicate that these elements are more abundant in 0.40× (a) or 0.06× (b) CI chondrite than in the BSE

also lie on the CCAM with $\delta^{17}O \sim 9‰$ and $\delta^{18}O \sim 16‰$ (Clayton and Mayeda 1999), any mixture of ureilite-like and CI-like material, even if a specific proportion can produce the $\Delta^{17}O$ composition of the BSE, does not pass through the BSE composition ($\delta^{17}O = 2.7‰$, $\delta^{18}O = 5.2‰$, Eiler 2001). As such, ureilite-CI mixtures appear not to be compatible with the O isotope composition of the BSE.

These distinct interpretations provide a testable hypothesis, namely, which (if either) of the two scenarios is consistent with the volatile element budget of the BSE? Fig. 19 shows the abundances of a selection of volatile elements for mixing between a volatile-free proto-Earth and a) 40%- and b) 6% by mass of CI chondrites. In the latter case, this is equivalent to ~30% of the Zn budget of the BSE as explained above and posited by Steller et al. (2022), Savage et al. (2022) in the framework of hypothesis B. The choice of a volatile-free proto-Earth provides upper limits for the inferred volatile element abundances.

Delivery of 6% CI to an otherwise volatile-free proto-Earth (hypothesis B) would account for roughly the entire present-day budget of Pb, while the budgets of C, S, Ag, Ge, Tl and Br (±Cl, Cd) would be in excess of those observed. Because Ag, Ge, C and S are more siderophile than is Zn, their additional sequestration into the core *after* the delivery of CI material could reconcile their predicted excesses (Mann et al. 2009; Wood et al. 2014; Blanchard et al. 2022). The same argument cannot be made for Tl, Br, Cl and Cd, whose metal-silicate partition coefficients are similar to (Cd; Wang et al. 2016) or lower than (Br, Cl, Tl; Kuwahara et al. 2017; Yang et al. 2023) those for Zn. The excess halogen problem can be resolved if the CI abundances of (Clay et al. 2017) are adopted. Hence, hypothesis B remains plausible without significant volatile loss post-CI-addition (though additional core formation after CI addition is needed).

The addition of 40% CI chondrites by mass in hypothesis A, results in an overabundance of all volatile elements, including Zn. Therefore, the BSE must have to have undergone volatile loss *after* the delivery of CI material for hypothesis A to hold, which is very difficult because thermal loss of volatiles (except perhaps hydrogen) is not energetically feasible from a body >10% of Earth's mass (see Sect. 6.3). Alternatively, the material was only isotopically CI-like, but had similar and BSE-like elemental ratios (such as Cr/Zn), see above. A final possibility states that ureilite- and CI-like material was accreted to the Earth in constant 60:40 proportions, before mixing to form a uniform reservoir, from which a fraction of the Zn budget was lost (Bizzarro et al. 2025). To this end, sublimation of accreting pebbles in H_2-rich planetary envelopes was proposed in order to explain the depletion of S relative to more refractory elements (Steinmeyer et al. 2023), but is yet to be tested for a larger suite of elements. Moreover, this volatile loss must not have engendered any perceptible mass-dependent isotopic fractionation (Sect. 6.2.2, Sossi et al. 2022). Therefore, for hypothesis A to remain valid, it requires any elemental fractionation occurring in the H_2-rich envelope of the growing Earth to reproduce both its smooth decline in abundances of volatile elements with condensation temperature, and their lack of isotopic fractionation. No such results have yet been demonstrated.

Finally, owing to the relatively few studies conducted on Zn isotopic variations among meteorites (Steller et al. 2022; Savage et al. 2022; Martins et al. 2023), Fig. 18 shows that the BSE is displaced from trend defined by NC bodies and Mars. However, the 2-σ uncertainty is sufficiently large that the BSE falls close to its extension (envelope 'C'), though more data are required to provide a statistical assessment. Alternatively, the apparent deviation of the BSE from the ^{66}Zn-^{54}Cr trend might reflect the late-stage addition of volatile-rich NC material having a composition beyond the NC trend (i.e., unsampled NC material). Provided this material is volatile-rich compared to the composition of the proto-Earth, it would exert a stronger control on the BSE's isotope composition for Zn than for Cr, a potential explanation why the BSE appears to diverge from the NC trend for these two elements. This permits the possibility that the Zn and Cr isotopic compositions of the BSE are consistent with those of an NC body that lies to higher ε^{66}Zn and ε^{54}Cr than any existing NC meteorite group.

7.2 Heavy Elements

These elements, defined as those with $Z > 26$ (Fe), were synthesised by nuclear processes distinct from those of the lighter elements. In brief, nuclides form by neutron capture at different rates; the s-process (slow neutron capture) and r-process (rapid neutron capture), which produce roughly equal amounts of nuclides, and the less widespread p-process that gives rise to proton-rich nuclei. The rates of neutron capture (slow or rapid) are determined relative to the β^--decay rate of the target nuclide, and depend on the nuclide's neutron capture cross section, which is highly temperature-dependent (Schramm and Norman 1976). In general, r-process nucleosynthesis produces more neutron-rich nuclides of a given element relative to the s-process, as well as higher-Z elements. Isotopic variations among heavy elements have been reported for, for example, Sr (Moynier et al. 2012; Hans et al. 2013; Schneider et al. 2023), Zr (Akram et al. 2015; Render et al. 2022), Mo (Dauphas et al. 2002; Burkhardt et al. 2011), Ru (Chen et al. 2010; Fischer-Gödde et al. 2015), Nd and Sm (Qin et al. 2011; Boyet and Gannoun 2013; Burkhardt et al. 2016; Frossard et al. 2022).

The BSE is an end-member among *all* planetary materials in the isotopic compositions observed for heavy elements, as has been shown for Mo (Burkhardt et al. 2011; Budde et al. 2019), Zr (Akram et al. 2015; Render et al. 2022), Nd (Burkhardt et al. 2016), and Ru (Fischer-Gödde and Kleine 2017). Importantly, this implies that the BSE composition is inconsistent with any mixture of NC- and CC components currently available, an interpretation that is otherwise permitted from iron-peak elements alone (see Sect. 7.1). Specifically, for isotopic variations between an iron-peak element and a heavy element (e.g., ε^{50}Ti vs. ε^{96}Zr), the NC trend is transverse to variations among CCs (see Burkhardt et al. 2021; Render et al. 2022; Yap and Tissot 2023; Kleine and Nimmo 2025) such that no mixture of

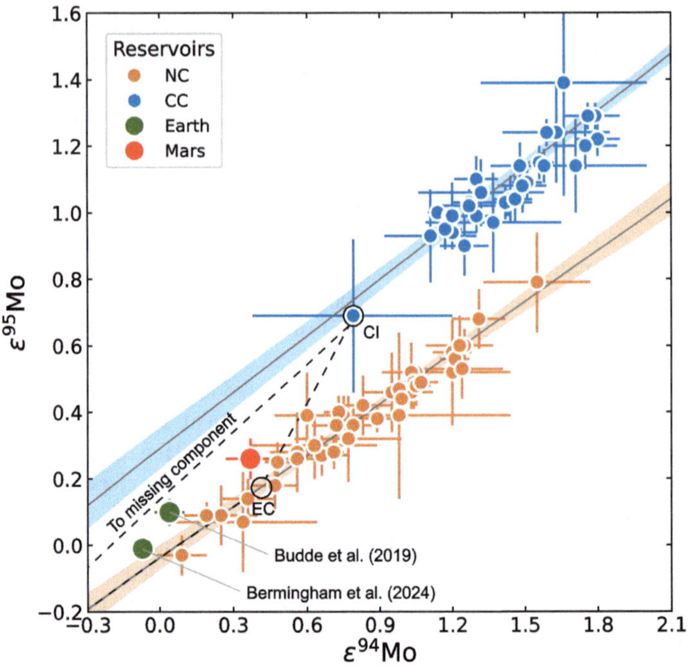

Fig. 20 Compositions of CC meteorites (blue circles), NC meteorites (red circles), the Earth (green circle) and Mars (red circle) expressed in terms of their ε^{94}Mo and ε^{95}Mo isotopic compositions. Data from Spitzer et al. (2025). The CC and NC trends were fit according to the York method (York et al. 2004) assuming uncorrelated errors in x and y, and yield $\varepsilon^{95}\text{Mo}_{NC} = (0.517\pm0.028)\times \varepsilon^{94}\text{Mo}_{NC}$ - 0.045 ± 0.026 and $\varepsilon^{95}\text{Mo}_{CC} = (0.560\pm0.023) \times \varepsilon^{94}\text{Mo}_{CC} + 0.298\pm0.037$ (all uncertainties 2-σ; red and blue envelopes). Black circles indicate compositions of end-members in a three-component mixture reproducing the composition of the BSE in hypothesis B, in which the 'missing component' has $\varepsilon^{94}\text{Mo} = -0.60$ and $\varepsilon^{95}\text{Mo} = -0.30$ (Burkhardt et al. 2021). In this scheme the contributions to the Mo budget of the BSE of Budde et al. (2019) can be reproduced by 50% missing component, 20% EC and 30% CI. The BSE value of Bermingham et al. (2024) does not require a CC contribution within uncertainty, but does not exclude it

a known NC and a (known) CC body passes through the terrestrial composition. Consequently, the processes that lead to variability among NC bodies were distinct from those that gave rise to variation among CC bodies, and the two populations cannot be considered as sampling different portions of a continuum, as in hypothesis A.

Isotopic variations induced by the r- and s-processes for a given isotopic ratio may often counteract one another, preventing an unequivocal assessment of their nucleosynthetic origin (Schneider et al. 2023). Molybdenum is an exception, because r- and s-process variations can be distinguished in ε-ε including one of the two p-process Mo nuclides (i.e., ^{92}Mo or ^{94}Mo) (Burkhardt et al. 2011). The CC and NC meteorite groups in ε^{94}Mo-ε^{95}Mo space (Budde et al. 2016), form two lines, which reflect s-process heterogeneities (Budde et al. 2016), while the offset between the two lines reflects a near-constant r-process-excess in CC over NC materials (Fig. 20).

Spitzer et al. (2020), based on Mo isotopic data representing 15 different parent bodies, including early- (i.e., iron meteorites, achondrites) and late-formed bodies (i.e., chondrites) found that the NC-line ($\varepsilon^{95}\text{Mo} = [0.528\pm0.045]\varepsilon^{94}\text{Mo} - 0.058\pm0.040$) is slightly shallower than the CC-line ($\varepsilon^{95}\text{Mo} = [0.596\pm0.006]\varepsilon^{94}\text{Mo} + 0.264\pm0.010$). More recently, Bermingham et al. (2024) argued that the slope of the NC-line is shallower again

(ε^{95}Mo = [0.517±0.042]ε^{94}Mo + 0.01±0.02); however, this result is not only based on a lower number of samples (six iron meteorites), but also is influenced by the Mo isotope data for two IAB irons, whose composition differs from that reported in an earlier study (Worsham et al. 2017). The latest compilation of Spitzer et al. (2025), yields $\varepsilon^{95}Mo_{NC}$ = (0.517±0.028)× $\varepsilon^{94}Mo_{NC}$ - 0.045±0.026 (95% confidence interval, Fig. 20) based on data for 41 distinct NC bodies and including available data for IAB irons from different studies with an indistinguishable slope but a lower intercept than that of Bermingham et al. (2024).

Based on Mo isotope data for several terrestrial igneous rocks, Budde et al. (2019) report ε^{94}Mo = 0.04±0.06 and ε^{95}Mo = 0.10±0.04 (95% confidence interval) for the BSE, indicating that, in the ε^{94}Mo–ε^{95}Mo diagram, the BSE plots between the NC- and CC-lines (Fig. 20). From the position of the BSE in this diagram, these authors calculated that 46±15% of the BSE's Mo derives from the CC reservoir. Budde et al. (2023) showed that there is some uncertainty on the BSE's Mo isotope composition owing to non-exponential isotope fractionation, either intrinsic to the sample, during the measurement, or both, but also argued that this does not affect the fraction of CC material in the BSE, though uncertainties are larger. Conversely, Bermingham et al. (2024), based largely on data for molybdenites, argued that the BSE composition is different from that reported by Budde et al. (2019), and quote ε^{94}Mo = −0.07±0.03 and ε^{94}Mo = −0.01±0.03. Because this composition overlaps with that of a single IAB iron meteorite and, hence, is on the limit of resolution at 1-σ from the NC line and within uncertainty at 2-σ (Fig. 20), these authors argue for a purely NC origin of the BSE's Mo.

In hypothesis B, the fraction of CC-derived Mo is similar to the 30% CC-derived Zn in the BSE deduced by Steller et al. (2022) and Savage et al. (2022). In this model, given that enstatite chondrites are taken as the NC end-member, the fraction of CC-derived Mo in the BSE can be calculated using Δ^{95}Mo, which represents a measure of the position of a sample relative to the NC- and CC-lines. The BSE composition of Bermingham et al. (2024) results in Δ^{95}Mo = 3±3 for the BSE; using Δ^{95}Mo = −7±4 for enstatite chondrites and Δ^{95}95Mo = 26±2 for the CC reservoir then results in 31±17% CC-derived Mo in the BSE for hypothesis B, whereas using instead the value of Budde et al. (2019) yields 41±17% CC-derived Mo in the BSE (see Nimmo et al. 2024). Therefore, hypothesis B indicates some of the BSE's Mo derived from CC material. In this model, because Mo (siderophile) and Zn (volatile) are depleted in the BSE with respect to RLEs, their isotopic compositions do not reflect the bulk CC fraction in the Earth, but rather the late-stage addition of volatile-rich, CC material to the Earth (Kleine and Nimmo 2025; Nimmo et al. 2024). In hypothesis B, the BSE's Mo isotopic composition is reconciled by a three-component mixture of CI, EC and s-process-enriched NC material (the missing component, see Fig. 20, Burkhardt et al. 2021; Budde et al. 2019).

On the other hand, if the Mo BSE interpretation of Bermingham et al. (2024) is correct, and if the isotopic composition of Zn is also taken to be consistent with that of an NC body, then this leads to a third hypothesis (C) in which the isotopic composition of the BSE is an entirely NC body whose composition is related to, but more extreme than other NC bodies. It should be noted that, converting the fraction of the CC-derived Mo in the BSE from hypothesis B into an actual mass of CC bodies accreted by Earth is difficult, mainly because the Mo isotopic composition of the missing component in hypothesis B is not known precisely, and, since almost all the Mo of a body might be expected to reside in its metallic core, only a fraction of this Mo may now be recorded in the BSE. Thus, hypotheses B and C may eventually be reconciled if the mass fraction of the CC contribution to the BSE is small (i.e., within uncertainty). Hypothesis C naturally resolves the apparent contradiction in the CI-like Fe isotopic composition of the BSE (see Sect. 7.1) as well as the s-process

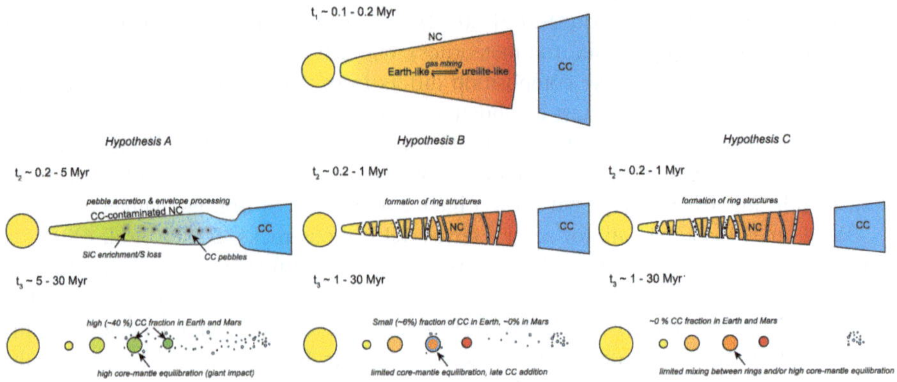

Fig. 21 Schematic illustration of the astrophysical setting of planetary accretion through time

excesses observed in nuclides of heavy elements such as Zr, because, in both cases, the BSE is consistent with a linear extension of the NC trend (see also Render et al. 2022). These ideas remain to be thoroughly tested.

7.3 Implications for the Astrophysical Setting and Redox State of the Earth and Planets

It is clear that hypothesis A requires considerable quantities of outer solar system material (~40%), hypothesis B requires a small fraction (~6%), while hypothesis C states that there is essentially none within some as yet undefined uncertainty (~0%). A schematic illustration as to how these scenarios could have played out spatio-temporally is shown in Fig. 21.

The initial conditions (t_1, Fig. 21) for all models invoke inner (NC) or outer (CC) solar system provenance, the approximate boundary between which is the present-day asteroid belt (Kruijer et al. 2017). Physical isolation of these two reservoirs is grounded in the observation that, isotopically, there is a dearth of planetary materials that show compositions intermediate to these groups. Isotopic variations in CIs, particularly in Fe as well as Cr (Schiller et al. 2020; Hopp et al. 2022b; van Kooten et al. 2024), have been used as a basis for the existence of a third reservoir (Yap and Tissot 2023; Dauphas et al. 2024). Nevertheless, the CI reservoir is thought, initially, to have resided in the outer solar system.

The interpretations diverge at t_2 (cf. Fig. 21), where hypothesis A indicates significant inflow of CC material to the inner disk, which was initially sporadic to explain the compositions of early-formed NC planetesimals (ureilite parent body, angrite parent body, Schiller et al. 2018) but ramped up over time, resulting in ~40% CC in the fully-formed Earth. In hypothesis A, this reflects the inward drift of CI-like pebbles (e.g., Johansen et al. 2023). In hypotheses B and C, the contribution of CC material was either limited and occurred late during the accretion of the Earth (t_3, B) or was absent altogether (t_3, C). One mechanism for arresting inward drift would have been upon cooling of the disk and the formation of rings associated with pressure bumps (Sect. 3, Whipple 1972; Carrera et al. 2021), compounded by the subsequent formation of giant planets (Lau et al. 2024) that hindered transport of condensing grains outside of heliocentrically restricted zones (Fig. 21, t_2). In this way, the rings could have evolved locally in terms of chemical compositions through ^{26}Al heating and collisions within the rings, but would have undergone little mixing outside of these zones, consistent with the preservation of bodies in the NC trend with distinct isotopic compositions.

Mars does not show strong evidence for heterogeneous accretion (in that its isotopic composition lies between EC and OC for most systems, see Dauphas et al. 2024; Bizzarro et al. 2025; Liebske et al. 2025) and also contains *less* CC material than does the Earth (see Paquet et al. 2023; Kleine et al. 2023, at least for hypotheses A and B), despite its more proximal location to the presumed CC reservoir in the outer solar system. This poses a dynamical challenge, particularly because Mars formed earlier than did the Earth (Dauphas and Pourmand 2011) and would have therefore been able to accrete any CC material scattered inwards during the early growth and migration of Jupiter (see Kleine and Nimmo 2025, for a discussion). As a result, pebble accretion is hard to reconcile with the CC-poor nature of Mars. The observations can, however, be explained if Earth but not Mars received a late CC-embryo impact, as discussed in Nimmo et al. (2024). Recent dynamical simulations (Branco et al. 2025) support this argument.

Hypothesis C states that Mars and Earth are equally (and essentially entirely) NC, and that the Earth either i) accreted material of, on average, uniform provenance through time and/or ii) any pre-existing signature of heterogeneous accretion in the BSE was reset by perfect core-mantle equilibration, in which the isotopic composition of the mantle exchanges completely with that of the underlying core at the completion of Earth's accretion. Homogeneous accretion with respect to chemical composition is unlikely on the basis that increasing fO_2 during terrestrial core formation appears to be required by Rubie et al. (2015, 2025), (see discussion in Sect. 5.2, Siebert et al. 2013). Although this does not *per se* mandate a difference in the nucleosynthetic origin of such material, it would seem logical that chemical differences, especially in oxidation state, are linked to nucleosynthetic variations. On the other hand, fO_2 changes accompany changing $P-T$ conditions of chemically identical material (Sect. 5.2), such that this process cannot be used as an argument for accretion of compositionally or isotopically heterogeneous material. Therefore, while hypothesis B is consistent with many observations, hypothesis C, though promising, will require further investigation, in particular for the Zn and Mo isotope signatures of the BSE, and its ability to reproduce the BSE's depletion in siderophile elements with homogeneous accretion.

Importantly, hypothesis A appeals to CI material to oxidise the inner disk. Taking the bulk composition of CI chondrites, and assuming that all H associated with H_2O is lost while O remains behind, mixtures between an NC body (here EH) and CI chondrites are used to compute Fe/O ratios. For an Fe/O of 1.04 for EH and 0.40 for CI by mass (Wasson and Kallemeyn 1988), addition of 6% CI material gives a bulk Earth Fe/O ratio of 0.97, whereas 40% results in 0.70. The best estimate for the bulk Earth yields 1.07 (Table 4), suggesting O must be lost, too, in hypothesis A to reproduce the composition of the Earth. This would imply mass loss of O-bearing rocky materials (e.g., the Earth's mantle), as the amount of O that binds with nominally volatile components (e.g., CO_2) is insignificant with respect to that stored in the mantle. Hypothesis C provides no information as to the oxidation state of the Earth, other than to imply that it must be intrinsic to the material of the inner solar system.

8 Summary, Future Directions and Extension to Other Planets

Here we summarise some of the key findings from this work, and highlight possible future directions to understand terrestrial planet formation in our Solar System.

- The major telluric bodies, in terms of their core-mantle ratios, are slightly- (Earth) to strongly (Moon, Mercury) non-chondritic, while small telluric bodies are more oxidised (i.e., have more Fe as FeO) than any chondrite. Consequently, the Mg and Fe concentrations of the Earth, Moon and Mercury cannot be explained by any mixture of condensed materials from a canonical solar nebula. Other processes, such as a non-solar bulk composition, physical sorting of metal/silicate grains and/or collisional erosion are required.

- Disk viscosities are much higher than those controlled by internal (molecular) viscosities in order to match the observed mass accretion rate of T-Tauri stars. The Sun must have lost angular momentum to the planets via early, likely magnetised outflows. The modern thinking on disks therefore states that they probably transitioned from hot and rapidly radially expanding ($>$ 1000 K inward of 1 AU) to cold and accreting very early on in their history (within a few 100 kyr). In the cold phase the snow line was inwards of 1 AU.
- Equilibrium and non-equilibrium condensation of the solar nebula effectively discriminates between elements according to their volatilities, owing to the limited range of temperatures over which the budget of a given element condenses (50–100 K). The compositions of small telluric bodies *do* adhere to this expectation. However, the terrestrial planets do not exhibit the step-pattern expected for condensation/evaporation at a single temperature, indicating they accreted from a mixture of materials that experienced different thermal histories.
- Equilibrium condensation of the solar nebula predicts that there should be a correlation between the moderately volatile element content of the condensed grains and their oxygen content. This results from the condensation of S into FeS and additional O into FeO at similar temperatures (starting at ~700 K and ~550 K, respectively, independent of pressure). Because O condenses after S, this predicts that small telluric bodies and many iron meteorite parent bodies, which are almost exclusively oxidised (i.e., inferred mantle FeO contents >10 wt.%), should have their full complement of moderately volatile elements, which is not observed. Instead, this suggests either i) alternative mechanisms for their oxidation and/or ii) subsequent loss of highly volatile (H, C) and moderately volatile (S, Zn, etc.) elements at conditions distinct from those of the solar nebula.
- The mantle FeO content of the Earth can be produced from initially reduced materials (all Fe as Fe^0), provided core-mantle equilibration occurred at mean conditions of ~40 GPa and ~3150 K, but requires higher Fe/O ratios (~1.06) than found in chondrites (0.4, CI to 1.04, EH) or in mixtures of equilibrium condensates (~0.81 to 0.93).
- Enrichment in heavy isotopes of moderately volatile elements (except for Cr, which tends to lighter values) on small telluric bodies is correlated to the extent of their elemental depletion, implying a Rayleigh process governing their loss. A correlation also exists between the isotopic fractionation and the present-day escape velocities of bodies over a range of masses (from Vesta to Earth). Evaporative loss is generally expected to result in the residues (planets) being enriched in heavier isotopes, but is not effective for bodies of Moon-mass or greater unless their atmospheres were H_2-dominated.
- The inferred fractionation factors of mass-dependent stable isotope fractionation, together with the light isotope enrichment of Cr in small telluric bodies (Moon, Vesta, but not the angrite parent body) are indicative of the attainment of near-equilibrium between vapour- and condensed phase(s) during their separation. This high degree of equilibration implies that there is no means of discriminating between evaporation or condensation.
- Volatile depletion from small telluric bodies (except the Moon) must have occurred within a few million years after the condensation of the first solids, and therefore likely in the presence of the nebular gas. Nevertheless, the ratios of moderately volatile elements (Mn/Na, K/Li, Mn/Mg) show evidence for evaporation under more oxidised conditions (ΔIW-1) and at higher temperatures (~1400–1800 K) than in a canonical solar nebula.

- The growth of planets is consistent with mixing of small bodies of variable composition through collisions to produce increasingly more massive bodies. This process reconciles the observation that small telluric bodies have step function-like abundance patterns together with isotopically heavy moderately volatile element stable isotope ratios, while more massive bodies have increasingly smooth abundance patterns, are more volatile-rich, and have chondritic stable isotope compositions.
- A ureilite-like proto-Earth for Cr- and Ti isotopes implies 40% by mass CI chondrites to match the composition of the Earth. This delivers too many volatile elements with respect to the present-day Earth's mantle and would therefore require volatile loss thereafter. Furthermore, this loss would have had to have happened without significant fractionation of mass-dependent stable isotope compositions and the preferential loss of s-process depleted material to satisfy the Mo isotope composition of the BSE. Further work on volatile loss mechanisms from a protoplanet embedded in an H_2-rich disk is required to assess this hypothesis.
- An enstatite chondrite-like proto-Earth requires only 6% by mass CI chondrite addition, broadly in-line with the present-day abundances of the volatile elements in the BSE. So that this 6% of CI material delivers the $\sim$30–40% of the Mo and Zn budgets, the proto-Earth's mantle would have had to have had sub-CI Mo and Zn abundances (e.g. by core formation and volatile loss, respectively). This hypothesis implies the Earth accreted heterogeneously with respect to early-accreted inner (NC) material and late-accreted (CC) material. Because the BSE is an end-member with respect to the compositions of heavy elements (Zr, Mo and Ru), the NC component must contain a mixture of on average enstatite chondrite-like material and a missing component not preserved in the meteorite collection.
- The evidence for a minor contribution of CC material to an otherwise NC Earth (hypothesis B) is based entirely on the BSE's Zn and Mo isotope composition. For both some uncertainty exists in the estimated CC fractions, which is mostly related to uncertainties in the isotopic composition of the missing component accreted by Earth. Provided the CC fractions for these two elements are even smaller than currently estimated, as permitted by their uncertainties, the Earth may be considered an essentially pure NC body whose composition is more extreme than other NC bodies uniformly across all elements (hypothesis C).
- For hypotheses B and C, where the Earth is essentially entirely NC, there must have been a population of oxidised, volatile-depleted bodies in the inner solar system, given that the Earth represents nearly 50% of its mass and is markedly poorer in both highly- and moderately volatile elements than chondrites. The oxidation states of these bodies must have been somewhat elevated (ΔIW-1 to -2) with respect to the solar nebula (ΔIW-6). The Earth and Mars obtained their moderately volatile element budgets largely ($\sim$60–100%, Earth) or almost entirely (Mars) from NC bodies, implying that oxidised, MVE-rich bodies were endemic to the inner disk.
- One mechanism to prevent ingress of CC material into the inner disk is through trapping at pressure bumps and/or giant planet growth. However, cooling of a solar composition gas to completion should still result in CI-like abundances of volatile elements, which are not observed in the terrestrial planets. As such, it may imply that relatively high temperatures were sustained in the inner disk until the dispersal of gas and/or this dispersal occurred earlier than it did in the CC-forming region.
- Evidence for a clear heliocentric gradient in composition preserved in the planets is incomplete, despite a decrease in core/mantle ratio with distance from the Sun. The composition of Mercury will be key to constraining the extent to which volatile elements were

able to condense at small heliocentric distances. Additional constraints on its moment of inertia and core density will be instrumental to facilitate better estimates of its composition.

Acknowledgements We thank the handling editor, Dominik Hezel, for his patience and persistence in helping us see this contribution over the finish line, and to two anonymous reviewers, who highlighted the role of dust transport, envelope processing, oxygen isotope constraints, and other factors in the cosmochemical evolution of the early solar system. This contribution was the fruit of an ISSI Meeting held in Bern in July, 2023, a follow-up held at ETH Zürich in October, 2023 and finally in IPG Paris in May, 2024.

Funding Information Open access funding provided by Swiss Federal Institute of Technology Zurich. PAS is grateful to the Swiss National Science Foundation (SNSF) via an Eccellenza Professorship (203668) and the Swiss State Secretariat for Education, Research and Innovation (SERI) under contract number MB22.00033, a SERI-funded ERC Starting Grant "2ATMO". FN acknowledges support from NSF-CSEDI-2054876 and NASA-EW-80NSSC18K0594. RCH acknowledges support from the European Research Council (ERC-STG-2020 grant 949417 - VapLoss) and the Italian Ministry of University and Research (NRRP M4C2 Investment 1.2 Young Researchers grant ERC-PI_0000002 - VapourTime).

Declarations

Competing Interests The authors declare no competing interests.

References

Agnor CB, Canup RM, Levison HF (1999) On the character and consequences of large impacts in the late stage of terrestrial planet formation. Icarus 142(1):219–237

Akram W, Schönbächler M, Bisterzo S, Gallino R (2015) Zirconium isotope evidence for the heterogeneous distribution of s-process materials in the Solar System. Geochim Cosmochim Acta 165:484–500

Alexander CM (2019) Quantitative models for the elemental and isotopic fractionations in chondrites: the carbonaceous chondrites. Geochim Cosmochim Acta 254:277–309

Alexander RD, Clarke C, Pringle J (2006) Photoevaporation of protoplanetary discs–II. Evolutionary models and observable properties. Mon Not R Astron Soc 369(1):229–239

Amelin Y, Krot AN, Hutcheon ID, Ulyanov AA (2002) Lead isotopic ages of chondrules and calcium-aluminum-rich inclusions. Science 297(5587):1678–1683

Andrault D, Bolfan-Casanova N, Nigro GL, Bouhifd MA, Garbarino G, Mezouar M (2011) Solidus and liquidus profiles of chondritic mantle: implication for melting of the Earth across its history. Earth Planet Sci Lett 304(1–2):251–259

Armitage PJ, Clarke CJ, Palla F (2003) Dispersion in the lifetime and accretion rate of T Tauri discs. Mon Not R Astron Soc 342(4):1139–1146

Armytage R, Georg R, Savage P, Williams H, Halliday A (2011) Silicon isotopes in meteorites and planetary core formation. Geochim Cosmochim Acta 75(13):3662–3676

Armytage R, Georg R, Williams H, Halliday A (2012) Silicon isotopes in lunar rocks: implications for the moon's formation and the early history of the Earth. Geochim Cosmochim Acta 77:504–514

Aston F (1924) The rarity of the inert gases on the Earth. Nature 114(2874):786–786

Badro J, Côté AS, Brodholt JP (2014) A seismologically consistent compositional model of Earth's core. Proc Natl Acad Sci USA 111(21):7542–7545

Bai X-N, Stone JM (2013) Local study of accretion disks with a strong vertical magnetic field: magnetorotational instability and disk outflow. Astrophys J 767(1):30

Balbus SA, Hawley JF (1998) Instability, turbulence, and enhanced transport in accretion disks. Rev Mod Phys 70:1–53

Bale CW, Bélisle E, Chartrand P, Decterov SA, Eriksson G, Gheribi AE, Hack K, Jung I-H, Kang Y-B, Melançon J, Pelton AD, Petersen S, Robelin C, Sangster J, Spencer P, Van Ende M-A (2016) FactSage thermochemical software and databases, 2010—2016. Calphad 54:35–53. https://doi.org/10.1016/j.calphad.2016.05.002

Basaltic Volcanism Study Project (1981) Basaltic volcanism on the terrestrial planets. Pergamon Press, Inc., New York

Batygin K, Morbidelli A (2022) Self-consistent model for dust-gas coupling in protoplanetary disks. Astron Astrophys 666:A19

Bercovici HL, Elkins-Tanton LT, O'Rourke JG, Schaefer L (2022) The effects of bulk composition on planetesimal core sulfur content and size. Icarus 380:114976

Bermingham KR, Füri E, Lodders K, Marty B (2020) The NC-CC isotope dichotomy: implications for the chemical and isotopic evolution of the early Solar System. Space Sci Rev 216:1–29

Bermingham K, Tornabene H, Walker R, Godfrey L, Meyer B, Piccoli P, Mojzsis S (2024) The non-carbonaceous nature of Earth's late-stage accretion. Geochim Cosmochim Acta

Béthune W, Lesur G, Ferreira J (2017) Global simulations of protoplanetary disks with net magnetic flux - I. Non-ideal MHD case. Astron Astrophys 600:A75

Bitsch B, Morbidelli A, Lega E, Crida A (2014) Stellar irradiated discs and implications on migration of embedded planets-II. Accreting-discs. Astron Astrophys 564:A135

Bitsch B, Johansen A, Lambrechts M, Morbidelli A (2015) The structure of protoplanetary discs around evolving young stars. Astron Astrophys 575:A28

Bizzarro M, Johansen A, Dorn C (2025) The cosmochemistry of planetary systems. Nat Rev Chem 9:378–396

Blanchard I, Rubie DC, Jennings ES, Franchi IA, Zhao X, Petitgirard S, Miyajima N, Jacobson SA, Morbidelli A (2022) The metal–silicate partitioning of carbon during Earth's accretion and its distribution in the early Solar System. Earth Planet Sci Lett 580:117374

Bloom H, Lodders K, Chen H, Zhao C, Tian Z, Koefoed P, Pető MK, Jiang Y, Wang K (2020) Potassium isotope compositions of carbonaceous and ordinary chondrites: implications on the origin of volatile depletion in the early Solar System. Geochim Cosmochim Acta 277:111–131

Bollard J, Connelly JN, Whitehouse MJ, Pringle EA, Bonal L, Jørgensen JK, Nordlund Å, Moynier F, Bizzarro M (2017) Early formation of planetary building blocks inferred from Pb isotopic ages of chondrules. Sci Adv 3(8):e1700407

Bonnand P, Williams H, Parkinson I, Wood B, Halliday A (2016) Stable chromium isotopic composition of meteorites and metal–silicate experiments: implications for fractionation during core formation. Earth Planet Sci Lett 435:14–21

Borg LE, Nyquist LE, Wiesmann H, Reese Y (2002) Constraints on the petrogenesis of Martian meteorites from the Rb-Sr and Sm-Nd isotopic systematics of the lherzolitic shergottites ALH77005 and LEW88516. Geochim Cosmochim Acta 66(11):2037–2053

Bouhifd MA, Jephcoat AP (2003) The effect of pressure on partitioning of ni and co between silicate and iron-rich metal liquids: a diamond-anvil cell study. Earth Planet Sci Lett 209(1–2):245–255

Bourdon B, Roskosz M, Hin RC (2018) Isotope tracers of core formation. Earth-Sci Rev 181:61–81

Boyce JW, McCubbin FM, Lunning N, Anderson T (2024) Large carbonaceous chondrite parent bodies favored by abundance–volatility modeling: a possible chemical signature of pebble accretion. Planet Sci J 5(2):53

Boyet M, Gannoun A (2013) Nucleosynthetic nd isotope anomalies in primitive enstatite chondrites. Geochim Cosmochim Acta 121:652–666

Branco D, Machado P, Raymond SN (2025) Dynamical origin of theia, the last giant impactor on Earth. Icarus:116724

Brasser R, Mojzsis SJ (2020) The partitioning of the inner and outer Solar System by a structured protoplanetary disk. Nat Astron 4:492–499

Braukmüller N, Wombacher F, Hezel DC, Escoube R, Münker C (2018) The chemical composition of carbonaceous chondrites: implications for volatile element depletion, complementarity and alteration. Geochim Cosmochim Acta 239:17–48

Brugier Y-A, Barrat J-A, Gueguen B, Agranier A, Yamaguchi A, Bischoff A (2019) Zinc isotopic variations in ureilites. Geochim Cosmochim Acta 246:450–460

Budde G, Burkhardt C, Brennecka GA, Fischer-Gödde M, Kruijer TS, Kleine T (2016) Molybdenum isotopic evidence for the origin of chondrules and a distinct genetic heritage of carbonaceous and non-carbonaceous meteorites. Earth Planet Sci Lett 454:293–303

Budde G, Burkhardt C, Kleine T (2019) Molybdenum isotopic evidence for the late accretion of outer Solar System material to Earth. Nat Astron 3(8):736–741

Budde G, Tissot FL, Kleine T, Marquez RT (2023) Spurious molybdenum isotope anomalies resulting from non-exponential mass fractionation. Geochemistry 83(3):126007
Burkhardt C, Kleine T, Oberli F, Pack A, Bourdon B, Wieler R (2011) Molybdenum isotope anomalies in meteorites: constraints on solar nebula evolution and origin of the Earth. Earth Planet Sci Lett 312(3–4):390–400
Burkhardt C, Borg L, Brennecka G, Shollenberger Q, Dauphas N, Kleine T (2016) A nucleosynthetic origin for the Earth's anomalous 142nd composition. Nature 537(7620):394–398
Burkhardt C, Spitzer F, Morbidelli A, Budde G, Render JH, Kruijer TS, Kleine T (2021) Terrestrial planet formation from lost inner Solar System material. Sci Adv 7(52):eabj7601
Calogero MA, Nimmo F, Hin RC (2025) Can impact-induced heating drive moderately volatile element loss? Earth Planet Sci Lett 669:119580
Campbell IH, O'Neill HSC (2012) Evidence against a chondritic Earth. Nature 483(7391):553–558
Carrera D, Simon JB, Li R, Kretke KA, Klahr H (2021) Protoplanetary disk rings as sites for planetesimal formation. Astron J 161(2):96
Carter PJ, Leinhardt ZM, Elliott T, Walter MJ, Stewart ST (2015) Compositional evolution during rocky protoplanet accretion. Astrophys J 813(1):72
Carter PJ, Leinhardt ZM, Elliott T, Stewart ST, Walter MJ (2018) Collisional stripping of planetary crusts. Earth Planet Sci Lett 484:276–286
Cartier C, Llado L, Pirotte H, Tissandier L, Namur O, Collinet M, Wang S-J, Charlier B (2024) Partitioning of nickel and cobalt between metal and silicate melts: expanding the oxy-barometer to reducing conditions. Geochim Cosmochim Acta 367:142–164
Cassen P (2001) Nebular thermal evolution and the properties of primitive planetary materials. Meteorit Planet Sci 36(5):671–700
Cassen P (2006) Protostellar disks and planet formation. In: Mayor M et al (eds) Extrasolar planets. Saas-fee adv courses, vol 31. Springer, Berlin, pp 369–448
Chao K-H, deGraffenried R, Lach M, Nelson W, Truax K, Gaidos E (2021) Lava worlds: from early Earth to exoplanets. Geochemistry 81(2):125735
Charles R (1967) Activities in Li_2O-, Na_2O, and K_2O-SiO_2 solutions. J Am Ceram Soc 50(12):631–641
Charnoz S, et al (2021) Tidal pull of the Earth strips the proto-moon of its volatiles. Icarus 364:114451
Charnoz S, Jérôme A, Chaussidon M, Sossi PA, Marrocchi Y, Franco P (2025) Forming the first solids precursors in the Solar System through kinetic condensation. Tech rep, Copernicus Meetings
Chen J, Papanastassiou D, Wasserburg G (2010) Ruthenium endemic isotope effects in chondrites and differentiated meteorites. Geochim Cosmochim Acta 74(13):3851–3862
Chiang E, Youdin A (2010) Forming planetesimals in solar and extrasolar nebulae. Annu Rev Earth Planet Sci 38:493–522
Ciesla FJ, Cuzzi JN (2006) The evolution of the water distribution in a viscous protoplanetary disk. Icarus 181(1):178–204
Clay PL, Burgess R, Busemann H, Ruzié-Hamilton L, Joachim B, Day JM, Ballentine CJ (2017) Halogens in chondritic meteorites and terrestrial accretion. Nature 551(7682):614–618
Clayton D (2003) Handbook of isotopes in the cosmos: hydrogen to gallium. Cambridge University Press
Clayton RN, Mayeda TK (1996) Oxygen isotope studies of achondrites. Geochim Cosmochim Acta 60(11):1999–2017
Clayton RN, Mayeda TK (1999) Oxygen isotope studies of carbonaceous chondrites. Geochim Cosmochim Acta 63(13–14):2089–2104
Clayton RN, Onuma N, Grossman L, Mayeda TK (1977) Distribution of the pre-solar component in Allende and other carbonaceous chondrites. Earth Planet Sci Lett 34(2):209–224
Connelly JN, Bizzarro M, Krot AN, Nordlund Å, Wielandt D, Ivanova MA (2012) The absolute chronology and thermal processing of solids in the solar protoplanetary disk. Science 338(6107):651–655
Dauphas N (2017) The isotopic nature of the Earth's accreting material through time. Nature 541(7638):521–524
Dauphas N, Pourmand A (2011) Hf–W–Th evidence for rapid growth of Mars and its status as a planetary embryo. Nature 473(7348):489–492
Dauphas N, Marty B, Reisberg L (2002) Molybdenum nucleosynthetic dichotomy revealed in primitive meteorites. Astrophys J 569(2):L139
Dauphas N, Teng F-Z, Arndt NT (2010) Magnesium and iron isotopes in 2.7 Ga Alexo komatiites: mantle signatures, no evidence for Soret diffusion, and identification of diffusive transport in zoned olivine. Geochim Cosmochim Acta 74(11):3274–3291
Dauphas N, Poitrasson F, Burkhardt C, Kobayashi H, Kurosawa K (2015) Planetary and meteoritic Mg/Si and δ^{30}Si variations inherited from solar nebula chemistry. Earth Planet Sci Lett 427:236–248
Dauphas N, et al (2022) The extent, nature, and origin of K and Rb depletions and isotopic fractionations in Earth, the Moon, and other planetary bodies. Planet Sci J 3(2):29

Dauphas N, Hopp T, Nesvorný D (2024) Bayesian inference on the isotopic building blocks of Mars and Earth. Icarus 408:115805
Day JM, Moynier F (2014) Evaporative fractionation of volatile stable isotopes and their bearing on the origin of the Moon. Philos Trans R Soc A, Math Phys Eng Sci 372:20130259
Day JM, Sossi PA, Shearer CK, Moynier F (2019) Volatile distributions in and on the Moon revealed by Cu and Fe isotopes in the 'Rusty Rock' 66095. Geochim Cosmochim Acta 266:131–143
Desch S, Morris M, Connolly H, Boss AP (2010) A critical examination of the x-wind model for chondrule and calcium-rich, aluminum-rich inclusion formation and radionuclide production. Astrophys J 725(1):692
Dhaliwal JK, Day JM, Creech JB, Moynier F (2024) Copper and zinc isotope compositions of pristine eucrites as analogues for differentiated planetary feedstocks. Earth Planet Sci Lett 637:118740
Drake MJ, Righter K (2002) Determining the composition of the Earth. Nature 416(6876):39–44
Drake MJ, Newsom HE, Capobianco CJ (1989) V, Cr, and Mn in the Earth, Moon, EPB, and SPB and the origin of the Moon: experimental studies. Geochim Cosmochim Acta 53(8):2101–2111
Dreibus G, Palme H (1996) Cosmochemical constraints on the sulfur content in the Earth's core. Geochim Cosmochim Acta 60(7):1125–1130
Dullemond CP (2013) Theoretical Models of the Structure of Protoplanetary Disks, ecole Doctorale des Houches. Chronology of Solar System formation V: the first solids and the first planetesimals, February 10, 2013 - February 15, 2013
Dunham E, Sheikh A, Opara D, Matsuda N, Liu M-C, McKeegan K (2023) Calcium–aluminum-rich inclusions in non-carbonaceous chondrites: abundances, sizes, and mineralogy. Meteorit Planet Sci 58(5):643–671
Ebel DS (2021) Condensation calculations in planetary ccience and cosmochemistry. In: Oxford research encyclopedia of planetary science. arXiv:2306.12645
Ebel DS, Grossman L (2000) Condensation in dust-enriched systems. Geochim Cosmochim Acta 64(2):339–366
Ebel DS, Alexander CMO, Libourel G (2018) Vapor–melt exchange: constraints on chondrite formation conditions and processes. In: Russell SS, Connolly Jr HC, Krot AN (eds) Chondrules: records of protoplanetary disk processes. Cambridge University Press, Cambridge, pp 151–174
Ebel DS, Gemma ME, Alpert SP, Bayron J, Lobo AH, Weisberg MK (2024) Abundance, sizes, and major element compositions of components in CR and LL chondrites: formation from single reservoirs. Meteorit Planet Sci 59(9):2276–2295
Eiler JM (2001) Oxygen isotope variations of basaltic lavas and upper mantle rocks. Rev Mineral Geochem 43(1):319–364
Fang L, Moynier F, Barrat J-A, Yamaguchi A, Paquet M, Chaussidon M (2024) The origin of 4-Vesta's volatile depletion revealed by the zinc isotopic composition of diogenites. Sci Adv 10(33):eadl1007
Fegley B Jr (1988) Cosmochemical trends of volatile elements in the Solar System. In: Nuth JA, Sylvester P (eds) Workshop on the origins of solar systems. LPI tech rep 88-04. Lunar and Planetary Institute, Houston, pp 51–60
Fegley B Jr, Palme H (1985) Evidence for oxidizing conditions in the solar nebula from Mo and W depletions in refractory inclusions in carbonaceous chondrites. Earth Planet Sci Lett 72(4):311–326
Fegley B Jr, Lodders K, Jacobson NS (2020) Volatile element chemistry during accretion of the Earth. Geochemistry 80(1):125594
Fegley B (2000) Kinetics of gas-grain reactions in the solar nebula. Space Sci Rev 92:177–200
Fegley B, Cameron A (1987) A vaporization model for iron/silicate fractionation in the Mercury protoplanet. Earth Planet Sci Lett 82(3–4):207–222
Fegley B, Jacobson NS, Williams K, Plane J, Schaefer L, Lodders K (2016) Solubility of rock in steam atmospheres of planets. Astrophys J 824(2):103
Fegley B, Lodders K, Jacobson NS (2023) Chemical equilibrium calculations for bulk silicate Earth material at high temperatures. Geochemistry 83(2):125961
Feigelson ED, Montmerle T (1999) High-energy processes in young stellar objects. Annu Rev Astron Astrophys 37(1):363–408
Fischer RA, McDonough WF (2025) Earth's core composition and core formation. In: Treatise of geochemistry, 3rd edn., vol 1, pp 17–71
Fischer RA, Nakajima Y, Campbell AJ, Frost DJ, Harries D, Langenhorst F, Miyajima N, Pollok K, Rubie DC (2015) High pressure metal–silicate partitioning of Ni, Co, V, Cr, Si, and O. Geochim Cosmochim Acta 167:177–194
Fischer RA, Cottrell E, Hauri E, Lee KK, Le Voyer M (2020) The carbon content of Earth and its core. Proc Natl Acad Sci USA 117(16):8743–8749
Fischer-Gödde M, Kleine T (2017) Ruthenium isotopic evidence for an inner Solar System origin of the late veneer. Nature 541(7638):525–527

Fischer-Gödde M, Burkhardt C, Kruijer TS, Kleine T (2015) Ru isotope heterogeneity in the solar protoplanetary disk. Geochim Cosmochim Acta 168:151–171
Flock M, Fromang S, Turner NJ, Benisty M (2017) 3d radiation nonideal magnetohydrodynamical simulations of the inner rim in protoplanetary disks. Astrophys J 835(2):230
Frossard P, Israel C, Bouvier A, Boyet M (2022) Earth's composition was modified by collisional erosion. Science 377(6614):1529–1532
Frost DJ, McCammon CA (2008) The redox state of Earth's mantle. Annu Rev Earth Planet Sci 36:389–420
Frost D, Mann U, Asahara Y, Rubie D (2008) The redox state of the mantle during and just after core formation. Philos Trans R Soc A, Math Phys Eng Sci 366(1883):4315–4337
Fu RR, Volk MW, Bilardello D, Libourel G, Lesur GR, Ben Dor O (2021) The fine-scale magnetic history of the Allende meteorite: implications for the structure of the solar nebula. AGU Adv 2(3):e2021AV000486
Gessmann C, Wood B, Rubie D, Kilburn M (2001) Solubility of silicon in liquid metal at high pressure: implications for the composition of the Earth's core. Earth Planet Sci Lett 184(2):367–376
Goodrich CA, Van Orman JA, Wilson L (2007) Fractional melting and smelting on the ureilite parent body. Geochim Cosmochim Acta 71(11):2876–2895
Gounelle M, Shu FH, Shang H, Glassgold A, Rehm K, Lee T (2001) Extinct radioactivities and protosolar cosmic rays: self-shielding and lightelements. Astrophys J 548(2):1051
Gressel O, Turner NJ, Nelson RP, McNally CP (2015) Global simulations of protoplanetary disks with Ohmic resistivity and amibpolar diffusion. Astrophys J 801(2):84
Grewal DS, Nie NX, Zhang B, Izidoro A, Asimow PD (2024) Accretion of the earliest inner Solar System planetesimals beyond the water snowline. Nat Astron 8(3):290–297
Grossman L (1972) Condensation in the primitive solar nebula. Geochim Cosmochim Acta 36(5):597–619
Grossman L, Larimer JW (1974) Early chemical history of the Solar System. Rev Geophys 12(1):71–101
Grossman L, Beckett JR, Fedkin AV, Simon SB, Ciesla FJ (2008) Redox conditions in the solar nebula: observational, experimental, and theoretical constraints. Rev Mineral Geochem 68(1):93–140
Grossman L, Fedkin AV, Simon SB (2012) Formation of the first oxidized iron in the Solar System. Meteorit Planet Sci 47(12):2160–2169
Halliday A, Porcelli D (2001) In search of lost planets–the paleocosmochemistry of the inner Solar System. Earth Planet Sci Lett 192(4):545–559
Hans U, Kleine T, Bourdon B (2013) Rb-sr chronology of volatile depletion in differentiated protoplanets: babi, ador and all revisited. Earth Planet Sci Lett 374:204–214
Hartmann L, Calvet N, Gullbring E, D'Alessio P (1998) Accretion and the evolution of T Tauri disks. Astrophys J 495(1):385
Hauck SA, et al (2013) The curious case of Mercury's internal structure. J Geophys Res Planets 118(6):1204–1220
Haugbølle T, Weber P, Wielandt DP, Benítez-Llambay P, Bizzarro M, Gressel O, Pessah ME (2019) Probing the protosolar disk using dust filtering at gaps in the early Solar System. Astron J 158(2):55
Helbert J, Maturilli A, Dyar M, Alemanno G (2021) Deriving iron contents from past and future Venus surface spectra with new high-temperature laboratory emissivity data. Sci Adv 7(3):eaba9428
Hellmann JL, Hopp T, Burkhardt C, Kleine T (2020) Origin of volatile element depletion among carbonaceous chondrites. Earth Planet Sci Lett 549:116508
Herzog G, Moynier F, Albarède F, Berezhnoy A (2009) Isotopic and elemental abundances of copper and zinc in lunar samples, Zagami, Pele's hairs, and a terrestrial basalt. Geochim Cosmochim Acta 73(19):5884–5904
Hewins R, Connolly H, Lofgren G Jr, Libourel G (2005) Experimental constraints on chondrule formation. In: Krot AN, Scott ERD, Reipurth B (eds) Chondrites and the protoplanetary disk. ASP Conference Series, vol 341. Astronomical Society of the Pacific, San Francisco, pp 286–316
Hezel DC, Russell SS, Ross AJ, Kearsley AT (2008) Modal abundances of CAIs: implications for bulk chondrite element abundances and fractionations. Meteorit Planet Sci 43(11):1879–1894
Hibbert KE, Williams H, Kerr AC, Puchtel I (2012) Iron isotopes in ancient and modern komatiites: evidence in support of an oxidised mantle from archean to present. Earth Planet Sci Lett 321:198–207
Hillenbrand LA (2008) Observational constraints on dust disk lifetimes: implications for planet formation. In: Livio M, Sahu K, Valenti J (eds) A decade of extrasolar planets around normal stars: proceedings of the space telescope science institute symposium. Cambridge University Press, Cambridge, pp 84–105
Hin RC, Schmidt MW, Bourdon B (2012) Experimental evidence for the absence of iron isotope fractionation between metal and silicate liquids at 1 GPa and 1250–1300 °C and its cosmochemical consequences. Geochim Cosmochim Acta 93:164–181
Hin RC, Fitoussi C, Schmidt MW, Bourdon B (2014) Experimental determination of the Si isotope fractionation factor between liquid metal and liquid silicate. Earth Planet Sci Lett 387:55–66

Hin RC, et al (2017) Magnesium isotope evidence that accretional vapour loss shapes planetary compositions. Nature 549(7673):511–515

Hirose K, Wood B, Vočadlo L (2021) Light elements in the Earth's core. Nat Rev Earth Environ 2(9):645–658

Hirschmann MM, Bergin EA, Blake GA, Ciesla FJ, Li J (2021) Early volatile depletion on planetesimals inferred from C–S systematics of iron meteorite parent bodies. Proc Natl Acad Sci USA 118(13):e2026779118

Hong Y, Fegley B Jr (1998) Experimental studies of magnetite formation in the solar nebula. Meteorit Planet Sci 33(5):1101–1112

Hopp T, et al (2022b) Ryugu's nucleosynthetic heritage from the outskirts of the Solar System. Sci Adv 8(46):eadd8141

Hopp T, Dauphas N, Spitzer F, Burkhardt C, Kleine T (2022a) Earth's accretion inferred from iron isotopic anomalies of supernova nuclear statistical equilibrium origin. Earth Planet Sci Lett 577:117245

Hu Y, Moynier F, Bizzarro M (2022) Potassium isotope heterogeneity in the early Solar System controlled by extensive evaporation and partial recondensation. Nat Commun 13(1):7669

Huang D, Badro J (2018) Fe-ni ideality during core formation on Earth. Am Mineral 103(10):1707–1710

Huang S, Lu J, Prinz M, Weisberg MK, Benoit PH, Sears DW (1996) Chondrules: their diversity and the role of open-system processes during their formation. Icarus 122(2):316–346

Huang D, Siebert J, Badro J (2021) High pressure partitioning behavior of Mo and W and late sulfur delivery during Earth's core formation. Geochim Cosmochim Acta 310:19–31

Hueso R, Guillot T (2005) Evolution of protoplanetary disks: constraints from DM Tauri and GM Aurigae. Astron Astrophys 442(2):703–725

Humayun M, Cassen P (2000) Processes determining the volatile abundances of the meteorites and terrestrial planets. In: Canup RM, Righter K (eds) Origin of the Earth and Moon. University of Arizona Press, Tucson, pp 3–23

Hunten DM, Pepin RO, Walker JC (1987) Mass fractionation in hydrodynamic escape. Icarus 69(3):532–549

Ivanov D, Fitoussi C, Bourdon B (2022) Trace element volatility and the conditions of liquid-vapor separation in the proto-lunar disk. Icarus 386:115143

Izidoro A, Dasgupta R, Raymond SN, Deienno R, Bitsch B, Isella A (2022) Planetesimal rings as the cause of the Solar System's planetary architecture. Nat Astron 6(3):357–366

Jacquet E, Dullemond C, Drążkowska J, Desch S (2024) The early Solar System and its meteoritical witnesses. Space Sci Rev 220(7):78

Javoy M (1995) The integral enstatite chondrite model of the Earth. Geophys Res Lett 22(16):2219–2222

Jerram M, Bonnand P, Kerr AC, Nisbet EG, Puchtel IS, Halliday AN (2020) The δ53cr isotope composition of komatiite flows and implications for the composition of the bulk silicate Earth. Chem Geol 551:119761

Jiang Y, Koefoed P, Pravdivtseva O, Chen H, Li C-H, Huang F, Qin L-P, Liu J, Wang K (2021) Early Solar System aqueous activity: K isotope evidence from Allende. Meteorit Planet Sci 56(1):61–76

Johansen A, Ronnet T, Schiller M, Deng Z, Bizzarro M (2023) Anatomy of rocky planets formed by rapid pebble accretion-I. How icy pebbles determine the core fraction and feo contents. Astron Astrophys 671:A74

Joos M, Hennebelle P, Ciardi A (2012) Protostellar disk formation and transport of angular momentum during magnetized core collapse. Astron Astrophys 543:A128

Kalyaan A, Pinilla P, Krijt S, Banzatti A, Rosotti G, Mulders GD, Lambrechts M, Long F, Herczeg GJ (2023) The effect of dust evolution and traps on inner disk water enrichment. Astrophys J 954(1):66

Kato C, Moynier F (2017) Gallium isotopic evidence for extensive volatile loss from the moon during its formation. Sci Adv 3(7):e1700571

Kato C, Moynier F, Foriel J, Teng F-Z, Puchtel IS (2017) The gallium isotopic composition of the bulk silicate Earth. Chem Geol 448:164–172

Kegler P, Holzheid A, Frost D, Rubie D, Dohmen R, Palme H (2008) New Ni and Co metal-silicate partitioning data and their relevance for an early terrestrial magma ocean. Earth Planet Sci Lett 268(1–2):28–40

Keil K (2012) Angrites, a small but diverse suite of ancient, silica-undersaturated volcanic-plutonic mafic meteorites, and the history of their parent asteroid. Geochemistry 72(3):191–218

Khan A, Sossi PA, Liebske C, Rivoldini A, Giardini D (2022) Geophysical and cosmochemical evidence for a volatile-rich Mars. Earth Planet Sci Lett 578:117330

Khan A, Huang D, Durán C, Sossi PA, Giardini D, Murakami M (2023) Evidence for a liquid silicate layer atop the Martian core. Nature 622(7984):718–723

Klahr H, Hubbard A (2014) Convective overstability in radially stratified accretion disks under thermal relaxation. Astrophys J 788(1):21

Klaver M, Klemme S, Liu X-N, Hin RC, Coath CD, Anand M, Lissenberg CJ, Berndt J, Elliott T (2024) Titanium-rich basaltic melts on the moon modulated by reactive flow processes. Nat Geosci:1–6

Kleine T, Nimmo F (2025) Origin of the Earth. In: Anbar A, Weis D (eds) Treatise on geochemistry, 3rd edn, vol 7. Elsevier, pp 325–381

Kleine T, Walker RJ (2017) Tungsten isotopes in planets. Annu Rev Earth Planet Sci 45:389–417

Kleine T, Budde G, Burkhardt C, Kruijer T, Worsham E, Morbidelli A, Nimmo F (2020) The non-carbonaceous–carbonaceous meteorite dichotomy. Space Sci Rev 216:1–27

Kleine T, Steller T, Burkhardt C, Nimmo F (2023) An inner Solar System origin of volatile elements in Mars. Icarus 397:115519

Koefoed P, Pravdivtseva O, Chen H, Gerritzen C, Thiemens MM, Wang K (2020) Potassium isotope systematics of the LL4 chondrite Hamlet: Implications for chondrule formation and alteration. Meteorit Planet Sci 55(8)

Kohn SC, Schofield PF (1994) The importance of melt composition in controlling trace-element behaviour: an experimental study of Mn and Zn partitioning between forsterite and silicate melts. Chem Geol 117(1–4):73–87

Kreutzberger ME, Drake MJ, Jones JH (1986) Origin of the Earth's moon: constraints from alkali volatile trace elements. Geochim Cosmochim Acta 50(1):91–98

Kruijer TS, Burkhardt C, Budde G, Kleine T (2017) Age of Jupiter inferred from the distinct genetics and formation times of meteorites. Proc Natl Acad Sci USA 114(26):6712–6716

Kruijer TS, Kleine T, Borg LE (2020) The great isotopic dichotomy of the early Solar System. Nat Astron 4(1):32–40

Krumholz MR, Federrath C (2019) The role of magnetic fields in setting the star formation rate and the initial mass function. Front Astron Space Sci 6:7

Kruttasch PM, Mezger K (2025) Time of proto-Earth reservoir formation and volatile element depletion from ^{53}Mn-^{53}Cr chronometry. Sci Adv 11(31):eadw1280

Ku Y, Jacobsen S (2020) Potassium isotope anomalies in meteorites inherited from the protosolar molecular cloud. Sci Adv 6(41):eabd0511

Kubik E, Sossi PA, Siebert J, Inglis E, Roskosz M, Rego ES, Wehr N, Moynier F (2022) The absence of an effect of nickel on iron isotope fractionation during core formation. Geochim Cosmochim Acta 327:186–199

Kuwahara H, Gotou H, Shinmei T, Ogawa N, Yamaguchi A, Takahata N, Sano Y, Yagi T, Sugita S (2017) High pressure experiments on metal-silicate partitioning of chlorine in a magma ocean: implications for terrestrial chlorine depletion. Geochem Geophys Geosyst 18(11):3929–3945

Kuznetsova A, Bae J, Hartmann L, Low M-MM (2022) Anisotropic infall and substructure formation in embedded disks. Astrophys J 928(1):92

Larimer JW (1967) Chemical fractionations in meteorites—I. Condensation of the elements. Geochim Cosmochim Acta 31(8):1215–1238

Larimer J, Anders E (1970) Chemical fractionations in meteorites—III. Major element fractionations in chondrites. Geochim Cosmochim Acta 34(3):367–387

Larimer JW, Bartholomay M (1979) The role of carbon and oxygen in cosmic gases: some applications to the chemistry and mineralogy of enstatite chondrites. Geochim Cosmochim Acta 43(9):1455–1466

Latter HN, Fromang S, Gressel O (2010) MRI channel flows in vertically stratified models of accretion discs. Mon Not R Astron Soc 406(2):848–862

Lau TCH, Birnstiel T, Drążkowska J, Stammler SM (2024) Sequential giant planet formation initiated by disc substructure. Astron Astrophys 688:A22

Lesur GRJ (2021) Magnetohydrodynamics of protoplanetary discs. J Plasma Phys 87(1):205870101

Lesur G, et al (2023) Hydro-, magnetohydro-, and dust-gas dynamics of protoplanetary disks. In: Inutsuka S, Aikawa Y, Muto T, Tomida K, Tamura M (eds) Protostars and planets VII, vol 534. Astronomical Society of the, Pacific, pp 465–500

Lewis JS (1972) Low temperature condensation from the solar nebula. Icarus 16(2):241–252

Li J, Agee CB (1996) Geochemistry of mantle–core differentiation at high pressure. Nature 381(6584):686–689

Li Y, Vočadlo L, Sun T, Brodholt JP (2020) The Earth's core as a reservoir of water. Nat Geosci 13(6):453–458

Li M, Huang S, Zhu Z, Petaev MI, Steffen JH (2021) Maximum temperatures in evolving protoplanetary discs and composition of planetary building blocks. Mon Not R Astron Soc 503(4):5254–5262

Lichtenberg T, Drażkowska J, Schönbächler M, Golabek GJ, Hands TO (2021) Bifurcation of planetary building blocks during Solar System formation. Science 371(6527):365–370

Liebske C, Khan A, McLennan SM, Sossi PA (2025) What is Mars (not) made of? A joint isotopic, geochemical and geophysical analysis. Icarus:116666

Liu J, et al (2017) Iron isotopic fractionation between silicate mantle and metallic core at high pressure. Nat Commun 8(1):14377

Liu X, Hin RC, Coath CD, van Soest M, Mekekhova E, Elliott T (2022) Equilibrium olivine-melt Mg isotopic fractionation explains high δ^{26}Mg values in arc lavas. Geochem Perspect Lett 22:42–47

Liu X-N, Hin RC, Coath CD, Bizimis M, Su L, Ionov DA, Takazawa E, Brooker R, Elliott T (2023) The magnesium isotopic composition of the mantle. Geochim Cosmochim Acta 358:12–26
Lodders K (2003) Solar System abundances and condensation temperatures of the elements. Astrophys J 591(2):1220
Lodders K (2010) Solar System abundances of the elements. In: Goswami A, Reddy B (eds) Principles and perspectives in cosmochemistry. Astrophysics and space science proceedings. Springer, Berlin, pp 379–417
Longhi J (1999) Phase equilibrium constraints on angrite petrogenesis. Geochim Cosmochim Acta 63(3–4):573–585
Luck J-M, Othman DB, Albarède F (2005) Zn and cu isotopic variations in chondrites and iron meteorites: early solar nebula reservoirs and parent-body processes. Geochim Cosmochim Acta 69(22):5351–5363
Lynden-Bell D, Pringle JE (1974) The evolution of viscous discs and the origin of the nebular variables. Mon Not R Astron Soc 168(3):603–637
MacPherson GJ (2006) Calcium-aluminum-rich inclusions in chondritic meteorites. In: Holland HD, Turekian KK (eds) Treatise on geochemistry, vol 1: meteorites and cosmochemical processes, 2nd edn., Elsevier, Oxford, pp 139–179
Magna T, Wiechert U, Halliday AN (2006) New constraints on the lithium isotope compositions of the moon and terrestrial planets. Earth Planet Sci Lett 243(3–4):336–353
Magna T, Šimčíková M, Moynier F (2014) Lithium systematics in howardite–eucrite–diogenite meteorites: implications for crust–mantle evolution of planetary embryos. Geochim Cosmochim Acta 125:131–145
Mah J, Brasser R (2021) Isotopically distinct terrestrial planets via local accretion. Icarus 354:114052
Mah J, Brasser R, Woo J, Bouvier A, Mojzsis S (2022) Evidence of a primordial isotopic gradient in the inner region of the solar protoplanetary disc. Astron Astrophys 660:A36
Manara CF, et al (2016) Evidence for a correlation between mass accretion rates onto young stars and the mass of their protoplanetary disks. Astron Astrophys 591:L3
Manara CF, Morbidelli A, Guillot T (2018) Why do protoplanetary disks appear not massive enough to form the known exoplanet population? Astron Astrophys 618:L3
Mann U, Frost DJ, Rubie DC (2009) Evidence for high-pressure core-mantle differentiation from the metal–silicate partitioning of lithophile and weakly-siderophile elements. Geochim Cosmochim Acta 73(24):7360–7386
Margot J-L, Campbell DB, Giorgini JD, Jao JS, Snedeker LG, Ghigo FD, Bonsall A (2021) Spin state and moment of inertia of Venus. Nat Astron 5(7):676–683
Marrocchi Y, Villeneuve J, Jacquet E, Piralla M, Chaussidon M (2019) Rapid condensation of the first Solar System solids. Proc Natl Acad Sci USA 116(47):23461–23466
Marschall R, Morbidelli A (2023) An inflationary disk phase to explain extended protoplanetary dust disks. Astron Astrophys 677:A136
Martins R, Kuthning S, Coles BJ, Kreissig K, Rehkämper M (2023) Nucleosynthetic isotope anomalies of zinc in meteorites constrain the origin of Earth's volatiles. Science 379(6630):369–372
Mathieu R, Libourel G, Deloule E, Tissandier L, Rapin C, Podor R (2011) Na_2O solubility in CaO–MgO–SiO_2 melts. Geochim Cosmochim Acta 75(2):608–628
Matsumoto T, Tomisaka K (2004) Directions of outflows, disks, magnetic fields, and rotation of young stellar objects in collapsing molecular cloud cores. Astrophys J 616(1):266
McCubbin FM, Barnes JJ (2019) Origin and abundances of h2o in the terrestrial planets, moon, and asteroids. Earth Planet Sci Lett 526:115771
McKibbin SJ, O'Neill HSC (2018) Petrogenesis of d'Orbigny-like angrite meteorites and the role of spinel in the angrite source. Meteorit Planet Sci 53(2):306–325
Mezger K, Maltese A, Vollstaedt H (2021) Accretion and differentiation of early planetary bodies as recorded in the composition of the silicate Earth. Icarus 365:114497
Mokhtari M, Bourdon B (2025) Condensation of major and trace elements in dust-rich environments. Icarus:116801
Montmerle T, Augereau J-C, Chaussidon M, Gounelle M, Marty B, Morbidelli A (2006) 3. Solar System formation and early evolution: the first 100 million years. Earth Moon Planets 98:39–95
Morbidelli A, Lambrechts M, Jacobson S, Bitsch B (2015) The great dichotomy of the Solar System: small terrestrial embryos and massive giant planet cores. Icarus 258:418–429
Morbidelli A, Baillié K, Batygin K, Charnoz S, Guillot T, Rubie D, Kleine T (2022) Contemporary formation of early Solar System planetesimals at two distinct radial locations. Nat Astron 6(1):72–79
Mori S, Bai X-N, Okuzumi S (2019) Temperature structure in the inner regions of protoplanetary disks: inefficient accretion heating controlled by nonideal magnetohydrodynamics. Astrophys J 872(1):98
Moynier F, Paniello RC, Gounelle M, Albarède F, Beck P, Podosek F, Zanda B (2011) Nature of volatile depletion and genetic relationships in enstatite chondrites and aubrites inferred from zn isotopes. Geochim Cosmochim Acta 75(1):297–307

Moynier F, Day JM, Okui W, Yokoyama T, Bouvier A, Walker RJ, Podosek FA (2012) Planetary-scale strontium isotopic heterogeneity and the age of volatile depletion of early Solar System materials. Astrophys J 758(1):45

Nakajima M, Golabek GJ, Wünnemann K, Rubie DC, Burger C, Melosh HJ, Jacobson SA, Manske L, Hull SD (2021) Scaling laws for the geometry of an impact-induced magma ocean. Earth Planet Sci Lett 568:116983

Namur O, Charlier B, Holtz F, Cartier C, McCammon C (2016) Sulfur solubility in reduced mafic silicate melts: implications for the speciation and distribution of sulfur on Mercury. Earth Planet Sci Lett 448:102–114

Nanne JA, Nimmo F, Cuzzi JN, Kleine T (2019) Origin of the non-carbonaceous–carbonaceous meteorite dichotomy. Earth Planet Sci Lett 511:44–54

Nie NX, Dauphas N (2019) Vapor drainage in the protolunar disk as the cause for the depletion in volatile elements of the moon. Astrophys J Lett 884(2):L48

Nie NX, Chen X-Y, Zhang ZJ, Hu JY, Liu W, Tissot FL, Teng F-Z, Shahar A, Dauphas N (2023) Rubidium and potassium isotopic variations in chondrites and Mars: accretion signatures and planetary overprints. Geochim Cosmochim Acta 344:207–229

Nimmo F, Kleine T, Morbidelli A, Nesvorny D (2024) Mechanisms and timing of carbonaceous chondrite delivery to the Earth. Earth Planet Sci Lett 648:119112

Nittler LR, et al (2011) The major-element composition of Mercury's surface from messenger X-ray spectrometry. Science 333(6051):1847–1850

Nittler LR, Chabot NL, Grove TL, Peplowski PN (2018) The chemical composition of Mercury. In: Solomon SC, Nittler LR, Anderson BJ (eds) Mercury: the view after MESSENGER. Cambridge University Press, Cambridge, pp 30–51

Norris CA, Wood BJ (2017) Earth's volatile contents established by melting and vaporization. Nature 549(7673):507–510

O'Neill HSC (1991) The origin of the moon and the early history of the Earth—a chemical model. Part 1: the moon. Geochim Cosmochim Acta 55(4):1135–1157

O'Neill HSC, Eggins SM (2002) The effect of melt composition on trace element partitioning: an experimental investigation of the activity coefficients of FeO, NiO, CoO, MoO_2 and MoO_3 in silicate melts. Chem Geol 186(1–2):151–181

O'Neill HS, Palme H (1998) Composition of the silicate Earth: implications for accretion and core formation. In: The Earth's mantle: composition, structure, and evolution, pp 3–126

O'Neill HSC, Palme H (2008) Collisional erosion and the non-chondritic composition of the terrestrial planets. Philos Trans R Soc A, Math Phys Eng Sci 366(1883):4205–4238

O'Neill HSC, Pownceby MI (1993) Thermodynamic data from redox reactions at high temperatures. I. An experimental and theoretical assessment of the electrochemical method using stabilized zirconia electrolytes, with revised values for the Fe-"FeO", Co-CoO, Ni-NiO and Cu-Cu_2O oxygen buffers, and new data for the W-WO_2 buffer. Contrib Mineral Petrol 114(3):296–314

O'Neill HSC, Canil D, Rubie DC (1998) Oxide-metal equilibria to 2500 °C and 25 gpa: implications for core formation and the light component in the Earth's core. J Geophys Res, Solid Earth 103(B6):12239–12260

Onyett IJ, Schiller M, Makhatadze GV, Deng Z, Johansen A, Bizzarro M (2023) Silicon isotope constraints on terrestrial planet accretion. Nature 619(7970):539–544

Palme H (2000) Are there chemical gradients in the inner Solar System? Space Sci Rev 92(1):237–262

Palme H, Mezger K (2024) Nucleosynthetic isotope variations in chondritic meteorites and their relationship to bulk chemistry. Meteorit Planet Sci 59(2):382–394

Palme H, O'Neill HSC (2014) Cosmochemical estimates of mantle composition. In: Holland HD, Turekian KK (eds) Treatise on geochemistry, vol 3. Elsevier, Amsterdam, pp 1–39

Paniello RC, Day JM, Moynier F (2012) Zinc isotopic evidence for the origin of the moon. Nature 490(7420):376–379

Paquet M, Sossi PA, Moynier F (2023) Origin and abundances of volatiles on Mars from the zinc isotopic composition of Martian meteorites. Earth Planet Sci Lett 611:118126

Patrick MR, Orr T, Sutton A, Lev E, Thelen W, Fee D (2016) Shallowly driven fluctuations in lava lake outgassing (gas pistoning), kīlauea volcano. Earth Planet Sci Lett 433:326–338

Peplowski PN, et al (2012) Variations in the abundances of potassium and thorium on the surface of Mercury: Results from the messenger gamma-ray spectrometer. J Geophys Res Planets 117(E12)

Peplowski PN, et al (2016) Remote sensing evidence for an ancient carbon-bearing crust on Mercury. Nat Geosci 9(4):273–276

Pignatale FC, Gonzalez J-F, Cuello N, Bourdon B, Fitoussi C (2017) Size and density sorting of dust grains in SPH simulations of protoplanetary discs. Mon Not R Astron Soc 469(1):237–254

Poitrasson F, Halliday AN, Lee D-C, Levasseur S, Teutsch N (2004) Iron isotope differences between Earth, moon, Mars and Vesta as possible records of contrasted accretion mechanisms. Earth Planet Sci Lett 223(3–4):253–266

Poitrasson F, Zambardi T, Magna T, Neal CR (2019) A reassessment of the iron isotope composition of the moon and its implications for the accretion and differentiation of terrestrial planets. Geochim Cosmochim Acta 267:257–274

Poole GM, Rehkämper M, Coles BJ, Goldberg T, Smith CL (2017) Nucleosynthetic molybdenum isotope anomalies in iron meteorites–new evidence for thermal processing of solar nebula material. Earth Planet Sci Lett 473:215–226

Prettyman TH, Hagerty J, Elphic R, Feldman W, Lawrence D, McKinney G, Vaniman D (2006) Elemental composition of the lunar surface: Analysis of gamma ray spectroscopy data from lunar prospector. J Geophys Res Planets 111(E12)

Pringle EA, Moynier F (2017) Rubidium isotopic composition of the Earth, meteorites, and the moon: evidence for the origin of volatile loss during planetary accretion. Earth Planet Sci Lett 473:62–70

Pringle EA, Moynier F, Savage PS, Badro J, Barrat J-A (2014) Silicon isotopes in angrites and volatile loss in planetesimals. Proc Natl Acad Sci USA 111(48):17029–17032

Qin L, Carlson RW, Alexander CM (2011) Correlated nucleosynthetic isotopic variability in cr, sr, ba, sm, nd and hf in Murchison and que 97008. Geochim Cosmochim Acta 75(24):7806–7828

Rafikov RR (2016) Protoplanetary disk heating and evolution driven by spiral density waves. Astrophys J 831(2):122

Render J, Brennecka GA, Burkhardt C, Kleine T (2022) Solar System evolution and terrestrial planet accretion determined by zr isotopic signatures of meteorites. Earth Planet Sci Lett 595:117748

Ricolleau A, Fei Y, Corgne A, Siebert J, Badro J (2011) Oxygen and silicon contents of Earth's core from high pressure metal–silicate partitioning experiments. Earth Planet Sci Lett 310(3–4):409–421

Righter K, Drake MJ, Scott E (2006) Compositional relationships between meteorites and terrestrial planets. In: Meteorites and the early Solar System II, p 803

Ringwood AE (1966) Chemical evolution of the terrestrial planets. Geochim Cosmochim Acta 30(1):41–104

Ringwood A, Kesson S (1977) Basaltic magmatism and the bulk composition of the moon: II. siderophile and volatile elements in moon, Earth and chondrites: implications for lunar origin. Moon 16(4):425–464

Ringwood A, Kato T, Hibberson W, Ware N (1991) Partitioning of cr, v, and mn between mantles and cores of differentiated planetesimals: implications for giant impact hypothesis of lunar origin. Icarus 89(1):122–128

Rodríguez-Mozos J-M, Moya A (2022) Internal structures and magnetic moments of rocky planets-application to the first exoplanets discovered by tess. Astron Astrophys 661:A101

Rubie DC, Gessmann CK, Frost DJ (2004) Partitioning of oxygen during core formation on the Earth and Mars. Nature 429(6987):58–61

Rubie DC, Frost DJ, Mann U, Asahara Y, Nimmo F, Tsuno K, Kegler P, Holzheid A, Palme H (2011) Heterogeneous accretion, composition and core–mantle differentiation of the Earth. Earth Planet Sci Lett 301(1–2):31–42

Rubie DC, Jacobson SA, Morbidelli A, O'Brien DP, Young ED, de Vries J, Nimmo F, Palme H, Frost DJ (2015) Accretion and differentiation of the terrestrial planets with implications for the compositions of early-formed Solar System bodies and accretion of water. Icarus 248:89–108

Rubie D, Dale K, Nathan G, Nakajima M, Jennings E, Golabek G, Jacobson S, Morbidelli A (2025) Tungsten isotope evolution during Earth's formation and new constraints on the viability of accretion simulations. Earth Planet Sci Lett 651:119139

Safronov V (1969) Evolution of protoplanetary cloud and formation of the Earth and planets. Nauka, Moscow. NASA Tech Trans F-677

Salvador A, Samuel H (2023) Convective outgassing efficiency in planetary magma oceans: insights from computational fluid dynamics. Icarus 390:115265

Savage PS, Moynier F, Chen H, Shofner G, Siebert J, Badro J, Puchtel I (2015) Copper isotope evidence for large-scale sulphide fractionation during Earth's differentiation. Geochem Perspect Lett

Savage PS, Moynier F, Boyet M (2022) Zinc isotope anomalies in primitive meteorites identify the outer Solar System as an important source of Earth's volatile inventory. Icarus 386:115172

Schiller M, Bizzarro M, Fernandes VA (2018) Isotopic evolution of the protoplanetary disk and the building blocks of Earth and the moon. Nature 555(7697):507–510

Schiller M, Bizzarro M, Siebert J (2020) Iron isotope evidence for very rapid accretion and differentiation of the proto-Earth. Sci Adv 6(7):eaay7604

Schneider JM, Burkhardt C, Kleine T (2023) Distribution of s-, r-, and p-process nuclides in the early Solar System inferred from sr isotope anomalies in meteorites. Astrophys J Lett 952(1):L25

Schoenberg R, Zink S, Staubwasser M, von Blanckenburg F (2008) The stable cr isotope inventory of solid Earth reservoirs determined by double spike mc-icp-ms. Chem Geol 249(3–4):294–306

Schoenberg R, Merdian A, Holmden C, Kleinhanns IC, Haßler K, Wille M, Reitter E (2016) The stable cr isotopic compositions of chondrites and silicate planetary reservoirs. Geochim Cosmochim Acta 183:14–30
Schramm D, Norman E (1976) Heavy element nucleosynthesis. In: Proceedings 3rd international conference on nuclei far from stability. CERN, Geneva, pp 570–579
Scott ER, Wasson JT (1975) Classification and properties of iron meteorites. Rev Geophys 13(4):527–546
Shah O, Helled R, Alibert Y, Mezger K (2022) Possible chemical composition and interior structure models of Venus inferred from numerical modelling. Astrophys J 926(2):217
Shahar A, Young ED (2020) An assessment of iron isotope fractionation during core formation. Chem Geol 554:119800
Shahar A, Ziegler K, Young ED, Ricolleau A, Schauble EA, Fei Y (2009) Experimentally determined si isotope fractionation between silicate and fe metal and implications for Earth's core formation. Earth Planet Sci Lett 288(1–2):228–234
Shakura NI, Sunyaev RA (1973) Black holes in binary systems. Observational appearance. Astron Astrophys 24:337–355
Shakura N, Sunyaev R (1976) A theory of the instability of disk accretion on to black holes and the variability of binary X-ray sources, galactic nuclei and quasars. Mon Not R Astron Soc 175(3):613–632
Shu FH, Shang H, Glassgold AE, Lee T (1997) X-rays and fluctuating x-winds from protostars. Science 277(5331):1475–1479
Siebert J, Badro J, Antonangeli D, Ryerson FJ (2013) Terrestrial accretion under oxidizing conditions. Science 339(6124):1194–1197
Siebert J, Sossi PA, Blanchard I, Mahan B, Badro J, Moynier F (2018) Chondritic mn/na ratio and limited post-nebular volatile loss of the Earth. Earth Planet Sci Lett 485:130–139
Sossi PA, Fegley B (2018) Thermodynamics of element volatility and its application to planetary processes. In: King P, Fegley Jr B, Seward T (eds) High temperature gas-solid reactions in Earth and planetary processes. De Gruyter, Berlin, pp 393–460
Sossi PA, Shahar A (2021) Ironing out isotopic differences among rocky bodies. Elements 17(6):407–412
Sossi PA, Nebel O, Foden J (2016) Iron isotope systematics in planetary reservoirs. Earth Planet Sci Lett 452:295–308
Sossi PA, Moynier F, Van Zuilen K (2018a) Volatile loss following cooling and accretion of the moon revealed by chromium isotopes. Proc Natl Acad Sci USA 115(43):10920–10925
Sossi PA, Nebel O, O'Neill HSC, Moynier F (2018b) Zinc isotope composition of the Earth and its behaviour during planetary accretion. Chem Geol 477:73–84
Sossi PA, Klemme S, O'Neill HSC, Berndt J, Moynier F (2019) Evaporation of moderately volatile elements from silicate melts: experiments and theory. Geochim Cosmochim Acta 260:204–231
Sossi PA, Stotz IL, Jacobson SA, Morbidelli A, O'Neill HSC (2022) Stochastic accretion of the Earth. Nat Astron 6(8):951–960
Sossi PA, Nakajima M, Khan A (2025) Composition, structure and origin of the moon. In: Treatise of geochemistry, 3rd edn., vol 7, pp 417–479
Spitzer F, Burkhardt C, Budde G, Kruijer TS, Morbidelli A, Kleine T (2020) Isotopic evolution of the inner Solar System inferred from molybdenum isotopes in meteorites. Astrophys J Lett 898(1):L2
Spitzer F, et al (2024) The ni isotopic composition of Ryugu reveals a common accretion region for carbonaceous chondrites. Sci Adv 10(39):eadp2426
Spitzer F, Burkhardt C, Kruijer TS, Kleine T (2025) Comparison of the earliest NC and CC planetesimals: Evidence from ungrouped iron meteorites. Geochim Cosmochim Acta 397:134–148
Stammler SM, Lichtenberg T, Drążkowska J, Birnstiel T (2023) Leaky dust traps: how fragmentation impacts dust filtering by planets. Astron Astrophys 670:L5
Steele RC, Coath CD, Regelous M, Russell S, Elliott T (2012) Neutron-poor nickel isotope anomalies in meteorites. Astrophys J 758(1):59
Steenstra E, Berndt J, Klemme S, Rohrbach A, Bullock E, Van Westrenen W (2020) An experimental assessment of the potential of sulfide saturation of the source regions of eucrites and angrites: implications for asteroidal models of core formation, late accretion and volatile element depletions. Geochim Cosmochim Acta 269:39–62
Steinmeyer M-L, Woitke P, Johansen A (2023) Sublimation of refractory minerals in the gas envelopes of accreting rocky planets. Astron Astrophys 677:A181
Steller T, Burkhardt C, Yang C, Kleine T (2022) Nucleosynthetic zinc isotope anomalies reveal a dual origin of terrestrial volatiles. Icarus 386:115171
Stephens IW, et al (2017) ALMA reveals transition of polarization pattern with wavelength in HL Tau's disk. Astrophys J 851(1):55
Stoll MHR, Kley W (2014) Vertical shear instability in accretion disc models with radiation transport. Astron Astrophys 572:A77

Suer T-A, Siebert J, Remusat L, Menguy N, Fiquet G (2017) A sulfur-poor terrestrial core inferred from metal–silicate partitioning experiments. Earth Planet Sci Lett 469:84–97
Suzuki TK, Inutsuka S-i (2009) Disk winds driven by magnetorotational instability and dispersal of protoplanetary disks. Astrophys J 691(1):L49
Tabone B, Rosotti GP, Cridland AJ, Armitage PJ, Lodato G (2022) Secular evolution of mhd wind-driven discs: analytical solutions in the expanded α-framework. Mon Not R Astron Soc 512(2):2290–2309
Tang H, Dauphas N (2014) ^{60}Fe–^{60}Ni chronology of core formation in Mars. Earth Planet Sci Lett 390:264–274
Tang H, Young E (2020) Evaporation from the lunar magma ocean was not the mechanism for fractionation of the Moon's moderately volatile elements. Planet Sci J 1(2):49
Tartèse R, Sossi PA, Moynier F (2021) Conditions and extent of volatile loss from the Moon during formation of the Procellarum basin. Proc Natl Acad Sci USA 118(12):e2023023118
Thibault Y, Walter MJ (1995) The influence of pressure and temperature on the metal-silicate partition coefficients of nickel and cobalt in a model C1 chondrite and implications for metal segregation in a deep magma ocean. Geochim Cosmochim Acta 59(5):991–1002
Tian Z, Chen H, Fegley B Jr, Lodders K, Barrat J-A, Day JM, Wang K (2019) Potassium isotopic compositions of howardite-eucrite-diogenite meteorites. Geochim Cosmochim Acta 266:611–632
Tian Z, Jolliff BL, Korotev RL, Fegley B Jr, Lodders K, Day JM, Chen H, Wang K (2020) Potassium isotopic composition of the Moon. Geochim Cosmochim Acta 280:263–280
Tian Z, et al (2021) Potassium isotope composition of Mars reveals a mechanism of planetary volatile retention. Proc Natl Acad Sci USA 118(39):e2101155118
Tissot FL, Collinet M, Namur O, Grove TL (2022) The case for the angrite parent body as the archetypal first-generation planetesimal: large, reduced and mg-enriched. Geochim Cosmochim Acta 338:278–301
Tomascak PB, Magna T, Dohmen R, et al (2016) Advances in lithium isotope geochemistry. Springer, Cham
Toplis M, et al (2013) Chondritic models of 4 Vesta: implications for geochemical and geophysical properties. Meteorit Planet Sci 48(11):2300–2315
Trinquier A, Birck J-L, Allegre CJ (2007) Widespread ^{54}Cr heterogeneity in the inner Solar System. Astrophys J 655(2):1179
Trinquier A, Elliott T, Ulfbeck D, Coath C, Krot AN, Bizzarro M (2009) Origin of nucleosynthetic isotope heterogeneity in the solar protoplanetary disk. Science 324(5925):374–376
Turner NJ, Sano T, Dziourkevitch N (2007) Turbulent mixing and the dead zone in protostellar disks. Astrophys J 659(1):729
Umebayashi T (1983) The densities of charged particles in very dense interstellar clouds. Prog Theor Phys 69(2):480–502
Urey HC, Craig H (1953) The composition of the stone meteorites and the origin of the meteorites. Geochim Cosmochim Acta 4(1–2):36–82
van Kooten E, Zhao X, Franchi I, Tung P-Y, Fairclough S, Walmsley J, Onyett I, Schiller M, Bizzarro M (2024) The nucleosynthetic fingerprint of the outermost protoplanetary disk and early Solar System dynamics. Sci Adv 10(24):eadp1613
Visscher C, Fegley B (2013) Chemistry of impact-generated silicate melt-vapor debris disks. Astrophys J Lett 767(1):L12
Volkov AN, Johnson RE, Tucker OJ, Erwin JT (2011) Thermally driven atmospheric escape: transition from hydrodynamic to Jeans escape. Astrophys J Lett 729(2):L24
von Strandmann PAP, Elliott T, Marschall HR, Coath C, Lai Y-J, Jeffcoate AB, Ionov DA (2011) Variations of Li and Mg isotope ratios in bulk chondrites and mantle xenoliths. Geochim Cosmochim Acta 75(18):5247–5268
Wade J, Wood B (2005) Core formation and the oxidation state of the Earth. Earth Planet Sci Lett 236(1–2):78–95
Wadhwa M (2008) Redox conditions on small bodies, the moon and Mars. Rev Mineral Geochem 68(1):493–510
Wang Z, Laurenz V, Petitgirard S, Becker H (2016) Earth's moderately volatile element composition may not be chondritic: evidence from In, Cd and Zn. Earth Planet Sci Lett 435:136–146
Wang H, Weiss BP, Bai X-N, Downey BG, Wang J, Wang J, Suavet C, Fu RR, Zucolotto ME (2017) Lifetime of the solar nebula constrained by meteorite paleomagnetism. Science 355(6325):623–627
Wang X, Fitoussi C, Bourdon B, Fegley B Jr, Charnoz S (2019) Tin isotopes indicative of liquid–vapour equilibration and separation in the Moon-forming disk. Nat Geosci 12(9):707–711
Wang B, Moynier F, Jackson MG, Huang F, Hu X, Halldórsson SA, Dai W, Devos G (2023) Rubidium isotopic fractionation during magmatic processes and the composition of the bulk silicate Earth. Geochim Cosmochim Acta 354:38–50
Wang B, Moynier F, Hu Y (2024) Rubidium isotopic compositions of angrites controlled by extensive evaporation and partial recondensation. Proc Natl Acad Sci USA 121(1):e2311402121

Warren PH (2011) Stable-isotopic anomalies and the accretionary assemblage of the Earth and Mars: a subordinate role for carbonaceous chondrites. Earth Planet Sci Lett 311(1–2):93–100
Wasson JT, Kallemeyn GW (1988) Compositions of chondrites. Philos Trans R Soc Lond Ser A, Math Phys Sci 325(1587):535–544
Weidenschilling S (1977) Aerodynamics of solid bodies in the solar nebula. Mon Not R Astron Soc 180(2):57–70
Weidenschilling S, Cuzzi JN (1993) Formation of planetesimals in the solar nebula. In: Levy EH, Lunine JI (eds) Protostars and planets III. University of Arizona Press, Tucson, pp 1031–1060
Weider SZ, et al (2016) Evidence from messenger for sulfur-and carbon-driven explosive volcanism on Mercury. Geophys Res Lett 43(8):3653–3661
Whipple FL (1972) On certain aerodynamic processes for asteroids and comets. In: Evlius A (ed) From plasma to planet. Proceedings of the twenty-first Nobel symposium. Wiley Interscience, New York, p 211
Williams JG (1994) Contributions to the Earth's obliquity rate, precession, and nutation. Astron J 108(2):711–724
Williams CD, Sanborn ME, Defouilloy C, Yin Q-Z, Kita NT, Ebel DS, Yamakawa A, Yamashita K (2020) Chondrules reveal large-scale outward transport of inner Solar System materials in the protoplanetary disk. Proc Natl Acad Sci USA 117(38):23426–23435
Wimpenny J, Borg L, Sio CKI (2022) The gallium isotopic composition of the moon. Earth Planet Sci Lett 578:117318
Witt-Eickschen G, Palme H, O'neill HSC, Allen CM (2009) The geochemistry of the volatile trace elements as, cd, ga, in and sn in the Earth's mantle: new evidence from in situ analyses of mantle xenoliths. Geochim Cosmochim Acta 73(6):1755–1778
Wolf AS, Jäggi N, Sossi PA, Bower DJ (2023) Vaporock: thermodynamics of vaporized silicate melts for modeling volcanic outgassing and magma ocean atmospheres. Astrophys J 947(2):64
Woo JMY, Brasser R, Grimm SL, Timpe ML, Stadel J (2022) The terrestrial planet formation paradox inferred from high-resolution n-body simulations. Icarus 371:114692
Wood JA, Hashimoto A (1993) Mineral equilibrium in fractionated nebular systems. Geochim Cosmochim Acta 57(10):2377–2388
Wood BJ, Wade J (2013) Activities and volatilities of trace components in silicate melts: a novel use of metal–silicate partitioning data. Contrib Mineral Petrol 166:911–921
Wood BJ, Kiseeva ES, Mirolo FJ (2014) Accretion and core formation: the effects of sulfur on metal–silicate partition coefficients. Geochim Cosmochim Acta 145:248–267
Wood BJ, Smythe DJ, Harrison T (2019) The condensation temperatures of the elements: a reappraisal. Am Mineral 104(6):844–856
Worsham EA, Bermingham KR, Walker RJ (2017) Characterizing cosmochemical materials with genetic affinities to the Earth: genetic and chronological diversity within the iab iron meteorite complex. Earth Planet Sci Lett 467:157–166
Wulf A, Palme H, Jochum K (1995) Fractionation of volatile elements in the early Solar System: evidence from heating experiments on primitive meteorites. Planet Space Sci 43(3–4):451–468
Yang X, Du Z, Li Y (2023) Experimental constraints on metal-silicate partitioning of chlorine and implications for planetary core formation. Geochim Cosmochim Acta 355:62–74
Yap TE, Tissot FL (2023) The NC-CC dichotomy explained by significant addition of CAI-like dust to the Bulk Molecular Cloud (BMC) composition. Icarus 405:115680
York D, Evensen NM, Martinez ML, De Basabe Delgado J (2004) Unified equations for the slope, intercept, and standard errors of the best straight line. Am J Phys 72(3):367–375
Yoshizaki T, McDonough WF (2020) The composition of Mars. Geochim Cosmochim Acta 273:137–162
Young ED, Ash RD, England P, Rumble D III (1999) Fluid flow in chondritic parent bodies: deciphering the compositions of planetesimals. Science 286(5443):1331–1335
Young E, Shahar A, Nimmo F, Schlichting H, Schauble E, Tang H, Labidi J (2019) Near-equilibrium isotope fractionation during planetesimal evaporation. Icarus 323:1–15
Zahnle KJ, Kasting JF (1986) Mass fractionation during transonic escape and implications for loss of water from Mars and Venus. Icarus 68(3):462–480
Zhu K, Sossi PA, Siebert J, Moynier F (2019) Tracking the volatile and magmatic history of Vesta from chromium stable isotope variations in eucrite and diogenite meteorites. Geochim Cosmochim Acta 266:598–610
Zhu K, et al (2021) Chromium stable isotope panorama of chondrites and implications for Earth early accretion. Astrophys J 923(1):94
Zhu K, Barrat J-A, Yamaguchi A, Rouxel O, Germain Y, Langlade J, Moynier F (2022) Nickel and chromium stable isotopic composition of ureilites: implications for the Earth's core formation and differentiation of the ureilite parent body. Geophys Res Lett 49(7):e2021GL095557

Zuber MT, McSween HY, Binzel RP, Elkins-Tanton LT, Konopliv AS, Pieters CM, Smith DE (2011) Origin, internal structure and evolution of 4 Vesta. Space Sci Rev 163:77–93

Publisher's Note Springer Nature remains neutral with regard to jurisdictional claims in published maps and institutional affiliations.

Authors and Affiliations

Paolo A. Sossi[1] · Remco C. Hin[2] · Thorsten Kleine[3] · Alessandro Morbidelli[4,5] · Francis Nimmo[6]

✉ P.A. Sossi
paolo.sossi@eaps.ethz.ch

1 Institute of Geochemistry and Petrology, ETH Zürich, Clausiusstrasse 25, CH-8092, Zürich, Switzerland

2 Institute of Environmental Geology and Geoengineering, Consiglio Nazionale delle Ricerche (CNR), Via Sandro Botticelli 23, 20133, Milan, Italy

3 Max Planck Institute for Solar System Research, Justus-von-Liebig-Weg 3, 37077, Göttingen, Germany

4 Laboratoire Lagrange, Université Côte d'Azur, Observatoire de la Côte d'Azur, CNRS, CS 34229, F-06304, Nice, France

5 Collège de France, CNRS, PSL Université, Sorbonne Université, F-75014, Paris, France

6 Department of Earth & Planetary Sciences, University of California Santa Cruz, CA 95064, Santa Cruz, USA

Space Science Reviews (2025) 221:54
https://doi.org/10.1007/s11214-025-01182-6

Condensation and the Volatility Trend of the Earth

Katharina Lodders[1] · Bruce Fegley[1] · Klaus Mezger[2] · Denton Ebel[3]

Received: 5 November 2024 / Accepted: 6 June 2025 / Published online: 17 June 2025

Abstract

This article describes condensation of the elements and use of condensation temperatures to plot and interpret Earth's apparent volatility trend. Major points covered include the following. (1) Updated 50% condensation temperatures (T_{50}) for all naturally occurring elements, Tc, and Pu are tabulated at 10^{-2} to 10^{-8} bar total pressure for solar composition material. (2) Condensation temperatures are mainly controlled by the Gibbs energy of condensation reactions and also by the Gibbs energy of ideal mixing if elements (compounds) condense in a solution. The additional Gibbs energy change due to non-ideal solution, i.e., activity coefficients $\neq 1$, is a secondary effect. (3) The theoretically correct relationship between condensation temperature and fraction condensed (α_M) is derived from mass balance and chemical thermodynamic considerations. For major elements the condensation temperature is inversely proportional to log (1-α_M). For trace elements dissolving in solid solution the condensation temperature is inversely proportional to log [(1-α_M)/α_M]. (4) The maximum amount of element condensed per K^{-1}, i.e., the maximum in [$d\alpha_M$/d(1/T)] is at the inflection point in the logistic (sigmoid) curve for an element, which is also at (or close to) the 50% condensation temperature. (5) Plots of normalized elemental abundances versus 50% condensation temperatures (volatility trends) are qualitative indicators of elemental fractionations due to volatility because they do not use the theoretically correct and quantitative relationship between condensation temperature and fraction condensed. (6) Volatility trend plots for average elemental abundances in CM, CO, CV, CR, H, L, LL, EH, EL chondrites show different "trends" for moderately and highly volatile elements, which may be linear, curved, a step function, or plateau. A comparison of three abundance sets for CM and CV chondrites shows trends depend on which elements are plotted, which data sources are used, and which temperature range is considered. (7) Proposed mechanisms for volatile element depletion in carbonaceous chondrites and the Earth are reviewed. (8) Some possible implications of volatile element abundances in the bulk silicate Earth are discussed.

1 Introduction

Chemical equilibrium calculations of element distribution between gas and solid (liquid) in a solar composition system – aka condensation calculations – have been done for over 70 years and are widely used in cosmochemistry and planetary science (e.g., Urey 1952; Lord 1965, Larimer 1967, 1973; Lewis 1972a,b; Grossman 1972, 1973; Grossman and Larimer 1974; Grossman and Olsen 1974; Wasson and Chou 1974, Boynton 1975, 1978; Palme and Wlotzka 1976; Wasson and Wai 1976; Kelly and Larimer 1977, Wai and Wasson 1977, 1979;

Extended author information available on the last page of the article

Sears 1978; Davis and Grossman 1979; Lewis et al. 1979; Fegley and Lewis 1980; Sears 1980; Fegley 1983, Saxena and Eriksson 1983a,b, Fegley and Palme 1985; Kornacki and Fegley 1986; Palme et al. 1988; Fegley and Prinn 1989; Lodders and Fegley 1993; Wood and Hashimoto 1993; Yoneda and Grossman 1995; Petaev and Wood 1998; Lauretta and Lodders 1997; Ebel and Grossman 2000; Ireland and Fegley 2000; Lodders 2003; Petaev et al. 2003; Ebel 2006; Fegley and Schaefer 2010; Wood et al. 2019; Lodders and Fegley 2023; Timmermann et al. 2023). Lodders and Fegley (1997a) and Ebel (2023) review the history of condensation calculations, and the papers cited above describe many details of the calculations. The major conclusion from this work is that gas – solid (liquid) condensation of the elements (compounds) was the major factor determining elemental abundances in the bodies of the solar system (Larimer 1988). This paper focuses on a subset of solar system bodies, namely chondritic meteorites and the Earth.

The condensation temperatures of elements are the basis for the cosmochemical classification of refractory, moderately volatile, and highly volatile elements (Larimer 1988). Refractory elements condense at higher temperatures than do the major rock-forming elements Mg, Si, and Fe. Moderately volatile elements (MVE) condense at temperatures between the condensation temperatures of the major elements and sulfur (as troilite FeS). Highly volatile elements condense at temperatures below that of sulfur. The exact demarcation temperatures are pressure dependent as shown in Table 1. At 10^{-4} bar total pressure, refractory elements condense at T > 1300–1360 K, moderately volatile elements condense between 1300 and 660 K, and highly volatile elements condense below 660 K. The cosmochemical classification also uses a fourth group, the siderophile elements, because all the metals are often either depleted (enriched) as a group (Larimer 1988). The different volatility of metals in solar composition gas leads to subdivision into refractory siderophile elements and volatile siderophile elements.

The cosmochemical classification of the elements is often combined with Goldschmidt's geochemical classification because with a few exceptions the elements have similar geochemical behavior in the early solar system as on Earth (Larimer 1988). Goldschmidt (1923) defined chalcophile, lithophile, and siderophile elements in terms of their affinity for sulfide, silicate, and metallic melts. However, the preference of an element for sulfide, silicate, or metallic host phases during gas – solid (melt) condensation (evaporation) under prevailing temperature, total pressure, oxygen fugacity fO_2, and sulfur fugacity fS_2 conditions is similar to its preference for these three phases in terrestrial and industrial settings under the same conditions. This point is illustrated by Ellingham diagrams, log fO_2 – 1/T, log fS_2 – 1/T plots, and predominance area diagrams (isothermal log plots of fO_2 – fS_2, or fO_2 – P_M, or fS_2 – P_M where P_M is monatomic metal gas pressure) (Ellingham 1944; Kellogg 1966; Ebel et al. 2012). Because the geochemical behavior of an element depends upon temperature, pressure, fO_2, and fS_2, changes in these variables during nebular condensation, parent body metamorphism, and planetary accretion can change an element's geochemical behavior.

Some elements, notably Fe, display chalcophile (FeS), lithophile (magnetite, ferromagnesian silicates), and siderophile (Fe metal alloy) tendencies in condensation calculations and in meteorites. Iron is the most abundant metal and is denoted as a siderophile element in the rest of this article. Chromium is another metal with chalcophile ($FeCr_2S_4$), lithophile ($FeCr_2O_4$), and siderophile (Cr in Fe alloy) behavior. It condenses dissolved in Fe metal but is also found in chromite in ordinary chondrites and in Earth's mantle and as daubréelite in enstatite chondrites. The fO_2 of solar composition gas intersects the fO_2 for the Cr – Cr_2O_3 buffer (Holzheid and O'Neill 1995) in the vicinity of the 50% condensation temperature of Cr in Fe alloy (1299 K) at 10^{-4} bar pressure. Other elements with mixed geochemical behavior in chondrites and condensation calculations include Ga, P, and Zn (siderophile and lithophile), and Ag, As, Cu, Pb, Sb, and Sn (siderophile and chalcophile). The phase into which an element condenses and/or occurs in Earth's mantle determines its geochemical classification on the figures in this paper.

Most refractory elements are lithophile elements condensing into oxide and/or silicate minerals. Geochemical analyses of meteorites show that to first approximation, these elements behave as a group with about a 10–20% deviation from the mean enrichment (depletion) relative to CI chondritic abundances (Vollstaedt et al. 2020 and this work). As mentioned later there are some exceptions. A few refractory elements are siderophile, notably W, Re, Os, Ir, Mo, Ru, Pt, and Rh that condense into refractory metal nuggets (Palme and Wlotzka 1976; Fegley and Palme 1985). The moderately volatile elements (MVE) are a mixture of chalcophile, lithophile, and siderophile elements plus one halogen (F), while the highly volatile elements (HVE) are mainly atmophile (H, C, N, O, noble gases), chalcophile (Cd, Hg, In, Tl) elements, and halogens (Cl, Br, I).

The abundances of MVE in carbonaceous chondrites and lithophile MVE in the accessible silicate portion of the Earth correlate qualitatively with condensation temperatures; more volatile elements with lower condensation temperatures are more depleted (relative to CI chondrites) than less volatile elements with higher condensation temperatures (e.g., as argued by Wai and Wasson 1977; Palme et al. 1988; Palme and O'Neill 2014; Alexander 2019b). The MVE abundances in non-carbonaceous chondrites such as ordinary and enstatite chondrites are less clearly correlated with condensation temperature. Lithophile and chalcophile MVE can be as abundant as more refractory lithophile and siderophile elements (Alexander 2019a).

The abundances of highly volatile elements (HVE) in chondrites are notoriously difficult to determine and vary with metamorphic grade in non-carbonaceous chondrites (e.g., Lipschutz and Woolum 1988). The elements H, C, N, and O are HVE according to Larimer's (1988) definition and we include them in subsequent plots. The correlation, if any, of highly volatile elemental abundances in meteorites and planetary bodies (such as Earth) with condensation temperatures is less clear. Some analytical data apparently show that subsets of highly volatile elements have constant depletions in CM, CO, CV, and CR carbonaceous chondrites, and in the bulk silicate Earth, relative to CI chondritic abundances, independent of their condensation temperatures (Krahenbühl et al. 1973; Takahashi et al. 1978, Braukmüller et al. 2018, 2019). The temperature dependence of trends for moderately and highly volatile elements (linear, curved, or step function) and the mechanism(s) for any trends remain active research areas, and are topics in this review. The focus is on Earth's apparent volatility trend, or more accurately the accessible portion of the Earth commonly known as the bulk silicate Earth (BSE).

The BSE is used in lieu of the bulk Earth composition because the latter is unknown without assumption-dependent modeling of elemental inventories of the Earth's core. However, the BSE composition is also an estimate because most mantle samples originate from the top 200 km of the upper mantle, which is about 27% of the mantle's total mass (Sect. 6.2, Fegley et al. 2020). The majority of the mantle is unsampled, but arguments can be made that the upper mantle composition is representative of the bulk mantle (e.g., see Palme and O'Neill 2014).

This paper is organized as follows. Section 2 presents 50% condensation temperatures for all natural elements at total nebular pressures of 10^{-2} to 10^{-8} bar. Section 3 mathematically describes the relationship between condensation temperatures and the fraction condensed of an element. Section 4 reviews figures showing volatility trends for CM, CO, CV, CR, H, L, LL, EH, and EL chondrites and the BSE. Section 5 discusses interpretations of the volatility trends given in Sect. 4 and related topics such as the mechanism(s) for volatile element depletion and isotopic constraints on the origin of Earth's volatile elements. Section 6 summarizes key points and suggests further work to resolve existing controversial issues.

2 Condensation Temperatures of the Elements

Table 1 gives 50% condensation temperatures for all natural elements in solar composition material (Lodders 2003; Lodders and Fegley 2023). The Table 1 values are used throughout this paper. The elements are in atomic number order starting with H (1) and ending at Pu (94). The radioactive elements Pm (61) and Po (84) onward are not included. Exceptions are short-lived Tc (43), long-lived Th (90), U (92), and short-lived Pu (94). Technetium and Pu are included because some Tc isotopes might have been present and ^{244}Pu (82 Ma half-life) was present in the early solar system and is important for Pu – I – Xe dating (Hudson et al. 1989; Avice and Marty 2014). Calculations are done as described earlier (e.g., Kornacki and Fegley 1986; Lodders 2003, and Lodders and Fegley 2023). The condensation temperatures of trace elements dissolved in more abundant host phases are independent of the assumed trace element abundances (see Kornacki and Fegley 1986 and equations below). The thermodynamic data sources are given in the Appendix and include the vapor pressure data used for highly volatile elements (H, C, N, noble gases) and a few updates for the alkalis (Na, K, Rb, Cs), halogens (Cl, Br, I), In, Tl, Tc, and Pu. The relative importance of activity coefficients for condensation calculations is discussed with mathematical examples in the Appendix. Table 1 deliberately covers a wide pressure range and interpolation of 50% condensation temperatures (T_{50}) between pressures shown can be done with linear fit equations as shown in the Appendix.

A few comments about Table 1 are in order. The 50% condensation temperatures for H_2 ice are fictive values at constant assumed total pressure, which drops to 43% of the initial value with half H_2 condensed (e.g., see Lewis 1972b). Because of the large amount of hydrogen in solar composition material (with atomic abundance ratios of H/Si $\simeq$ 24,000 and H/O $\simeq$ 2000) only a tiny fraction of hydrogen is condensed as hydrous minerals and/or water ice at higher temperature. The amount of oxygen in anhydrous rock is likewise insufficient for 50% condensation because the O/(2Si + Mg) atomic abundance ratio is about five. Lodders (2003) discusses the amount of oxygen retained in rock in Sect. 3.4.1 of her paper. Helium never condenses because the condensation temperature of liquid He is below the cosmic microwave background temperature of 3 K. The condensation temperatures for C and N in Table 1 assume complete chemical equilibrium.

Chemical kinetics may inhibit some gas phase and gas-solid reactions such as reduction of CO to CH_4, reduction of N_2 to NH_3, Fe_3O_4 formation, and hydrous silicate formation (e.g., Lewis and Prinn 1980; Fegley and Prinn 1989; Fegley 2000). It is also possible that condensation of highly volatile elements such as Cd, In, Tl, Br, Cl, and I may be affected by slow diffusion rates in host phases, but detailed consideration of kinetic effects is beyond the scope of this paper.

Table 1 50% condensation temperatures (K) at total nebular pressure (bar) of log P =

Element	−2	−4	−6	−8
H	10.6	7.4	5.7	4.6
He	<3	<3	<3	<3
Li	1319	1152	1003	891
Be	1566	1451	1349	1253
B	1027	945	875	814
C	48	40	34	30
N	139	124	111	101
O	210	182	160	143
F	775	717	699	697
Ne	10.8	9.1	7.7	6.8
Na	1077	978	895	826
Mg	1476	1330	1214	1115
Al	1795	1655	1533	1427
Si	1451	1313	1201	1111
P	1421	1273	1142	1029
S	661	661	661	660
Cl	460	418	383	354
Ar	54	46	41	36
K	1063	975	899	834
Ca	1683	1517	1382	1269
Sc	1704	1629	1505	1398
Ti	1742	1581	1449	1337
V	1711	1429	1371	1208
Cr	1490	1299	1152	1036
Mn	1290	1165	1049	966
Fe	1531	1334	1182	1063
Co	1552	1352	1200	1078
Ni	1552	1353	1200	1078
Cu	1200	1041	916	817
Zn	720	697	668	611
Ga	1206	965	730	665
Ge	1076	887	755	658
As	1230	1069	943	839
Se	693	693	693	693
Br	466	423	387	357
Kr	63	56	50	45
Rb	1161	1019	913	823
Sr	1682	1515	1379	1265
Y	1821	1607	1480	1378
Zr	1902	1742	1606	1489
Nb	1759	1559	1390	1276
Mo	1759	1592	1449	1324
Tc	1734	1570	1434	1320

Table 1 (*Continued*)

Element	−2	−4	−6	−8
Ru	1712	1551	1418	1306
Rh	1578	1390	1256	1153
Pd	1527	1324	1168	1045
Ag	1158	999	879	785
Cd	614	540	481	435
In	560	430	366	315
Sn	909	706	619	552
Sb	1148	983	859	764
Te	811	712	691	641
I	349	316	291	270
Xe	81	72	66	60
Cs	1103	857	729	641
Ba	1684	1490	1357	1252
La	1748	1583	1439	1326
Ce	1725	1538	1320	1266
Pr	1756	1583	1437	1321
Nd	1771	1613	1457	1327
Sm	1748	1590	1456	1328
Eu	1524	1366	1218	1133
Gd	1816	1619	1460	1346
Tb	1819	1619	1460	1346
Dy	1817	1619	1460	1346
Ho	1819	1619	1460	1346
Er	1821	1619	1460	1346
Tm	1815	1619	1460	1346
Yb	1699	1487	1393	1296
Lu	1821	1619	1460	1346
Hf	1827	1683	1572	1474
Ta	1774	1577	1451	1327
W	1988	1793	1626	1480
Re	2005	1821	1669	1539
Os	1993	1813	1662	1533
Ir	1770	1603	1464	1347
Pt	1585	1408	1277	1173
Au	1295	1065	902	783
Hg[a]	273	248	226	208
Tl	475	404	361	337
Pb	869	730	642	578
Bi	879	749	653	580
Th	1821	1619	1460	1346
U	1787	1610	1456	1327
Pu	1742	1580	1442	1329

[a]Ideal solution of HgS + HgSe + HgTe in FeS with HgTe being the most important

3 Relationship of Condensation Temperatures and Fraction Condensed

This section describes how the elemental fraction condensed is related to the condensation temperature. This relation is not linear (e.g., Larimer 1967, 1973, Kelly and Larimer 1977). The fraction condensed depends on solar elemental abundances, thermodynamic variables, and total pressure. This is illustrated for condensation of siderophile elements into a metal alloy, which is the easiest system to describe, but the general formalism also applies to oxide and sulfide condensation and to more complex gas phase speciation.

The condensation reaction of a monatomic gas into a metal alloy for any element M is

$$\mathrm{M\ (g) = M\ (alloy)} \tag{1}$$

This reaction applies to refractory siderophile elements such as Re, Os, Ir, Pt, Rh, Pd; to the most abundant siderophile elements Fe, Ni, Co; and to moderately and highly volatile siderophile elements such as Ag, As, Au, Bi, Cr, Cu, and Pb (e.g., see Table 2.3 of Lodders and Fegley 1998). The equilibrium constant for reaction (1) is K_1 and the equilibrium constant expression solved for the activity of M in the alloy is

$$a_{\mathrm{M,alloy}} = K_1 X_{\mathrm{M,gas}} P_{\mathrm{tot}} = X_{\mathrm{M,\ alloy}} \gamma_{\mathrm{M}} \tag{2}$$

where $X_{\mathrm{M,\ gas}}$ is the mole fraction of monatomic M gas times P_{tot}, the total pressure. The activity, $a_{\mathrm{M,\ alloy}}$ equals the mole fraction of M in the metal alloy, $X_{\mathrm{M\ alloy}}$, times the activity coefficient of M, γ_{M}, in the alloy. The condensation temperature enters via the equilibrium constant K_1.

The fraction condensed, α_{M}, is computed by relating the two mole fractions (one for the gas, one for the alloy) to the bulk composition (here solar). The total number of M atoms is N_{Mtot}, which is the sum of atoms in the gas, $N_{\mathrm{Mtot}}(1-\alpha_{\mathrm{M}})$, and $N_{\mathrm{Mtot}}(\alpha_{\mathrm{M}})$, atoms in condensate(s).

On the left-hand side (LHS) of Eq. (2), the mole fraction of monatomic gas X_{M} is the number of monatomic gas atoms divided by the sum of *all* other atomic and molecular gases:

$$X_{\mathrm{M,gas}} = \frac{\mathrm{M}}{\mathrm{H + H^+ + H_2 + He + CO + CH_4 + H_2O + N_2 + NH_3 + Si + SiO + Fe} + \cdots + \mathrm{M}} \approx \frac{M}{\mathrm{H_2 + H + He}} \tag{3}$$

For simplicity, Eq. (3) denotes numbers of atoms and molecules by the atomic and molecular formulas, e.g., N_{He} = He and N_{CO} = CO. The rest of this discussion uses the same convention, otherwise the equations appear even more complicated. Molecular and monatomic hydrogen and He dominate (>99%) the total pressure in a solar composition system (total pressure = the sum of all partial pressures). At temperatures where condensates become stable the denominator is usually largely H_2 and He but at higher temperatures and especially at higher temperatures and lower total pressures, the contributions of monatomic H, H^+, and electrons can become dominant and must be included. This is done by considering that the abundances of H_2, H, H^+, H^-, and electrons are coupled through their equilibrium reactions with preservation of mass and charge balance, and can be expressed in terms of each other, e.g., $p(\mathrm{H}) = K_{\mathrm{H}}\ p(\mathrm{H_2})^{0.5}$ for the (total pressure-dependent) equilibrium 1/2 H_2 = H. For reference, seven neutral and charged species of hydrogen plus the electron gas are fully considered. Many other ions of other gases are also considered but are not discussed further. Other gases (H_2O, CO, CH_4, N_2, noble gases, SiO, etc.) in the denominator can be neglected to a first approximation here but are fully included in the computations. Mass-balance, charge balance, and total pressure effects are constrained in the calculations for all H-bearing compounds and all compounds of all elements.

The abundance of the major gases H_2 and He can be written using their solar atomic elemental abundances as N_{Htot} and N_{Hetot} so that $H_2 \approx 0.5\ N_{Htot}$ and He = N_{Hetot}. At low temperatures, the condensation of hydrated silicates, H_2O, CH_4, and NH_3 ices reduces the amount of hydrogen in the gas by a factor of (1-α_H) from the total H abundance, where α_H stands for the fraction of H condensed. Full mass-balance for hydrogen at low temperatures requires solving coupled gas equilibria and solving the distribution of H between gases and condensates. For most practical purposes, the number of H_2 molecules is $H_2 \approx 0.5\ N_{Htot}$ (1-α_H) = 0.5 N_{Htot} for $H_2 \gg [H + H^+ + H^- + \ldots + 2\ H_2O + 3\ NH_3 + 4\ CH_4 + \ldots]$ and α_H is zero.

As noted earlier, the number of M atoms in the gas equals N_{Mtot}(1- α_M). Therefore Eq. (3) for the mole fraction for monatomic M gas $X_{M,\ gas}$ can be rewritten as

$$X_{M,gas} = \frac{M}{H_2 + He} = \frac{N_{Mtot}\,(1 - \alpha_M)}{0.5N_{Htot} + N_{Hetot}} \tag{4}$$

Using the astronomical epsilon notation for solar abundances for any element relative to H:

$$\varepsilon_M = N_{Mtot}/N_{Htot} \tag{5}$$

for He and M in Eq. (3) gives

$$X_{Mgas} = \frac{N_{Mtot}\,(1 - \alpha_M)}{(0.5N_{Htot} + N_{Htot}\varepsilon_{He})} = \frac{N_{Mtot}\,(1 - \alpha_M)}{N_{Htot}\,(0.5 + \varepsilon_{He})} = \varepsilon_M \frac{(1 - \alpha_M)}{(0.5 + \varepsilon He)} \tag{6}$$

For reference, solar composition has ε_{He} of about 1/10. The sum (0.5+ εHe) can be treated as a constant as long as H_2 is the major H-bearing gas, i.e., $H_2 \gg [H + H^+ + H^- + \ldots]$.

The right-hand side (RHS) of Eq. (2) gives the mole fraction of M in the alloy as the number of condensed M atoms, $\alpha_M N_{Mtot}$ over the sum of all condensed atoms (= $\alpha_M N_{Mtot}$ plus all other $\Sigma\ \alpha_i N_{itot}$) in the alloy:

$$X_{M\ alloy} = \frac{\alpha_M N_{Mtot}}{\alpha_M N_{Mtot} + \sum_{i\neq M}\alpha_i N_{itot}} = \frac{\varepsilon_M\alpha_M}{\varepsilon_M\alpha_M + \sum_{i\neq M}\varepsilon_i\alpha_i} \tag{7}$$

where the RHS follows from expanding with N_{Htot} and making use of Eq. (5). Now the gas and alloy mole fractions are inserted back into Eq. (2):

$$a_{M,alloy} = K_1 \frac{\varepsilon_M}{(0.5 + \varepsilon_{He})}(1 - \alpha_M)P_{tot} = \frac{\varepsilon_M\alpha_M}{\varepsilon_M\alpha_M + \sum_{i\neq M}\varepsilon_i\alpha_I}\gamma_M \tag{8}$$

The solar abundance ε_M cancels out from the numerators on the RHS and LHS of Eq. (8) but remains in the denominator on the RHS. Rearranging gives:

$$\frac{\alpha_M}{(1 - \alpha_M)\left(\varepsilon_M\alpha_M + \sum_{i\neq M}\varepsilon_i\alpha_i\right)} = K_1 \frac{1}{(0.5 + \varepsilon_{He})\gamma_M} P_{tot} \tag{9}$$

Equation (9) is the equivalent to Eq. (2) where activity and partial pressure were replaced by mole fractions which in turn were related to abundances and fraction condensed. Before introducing the temperature dependence through the equilibrium constant, it is useful to check the limiting cases for Eq. (9).

For *pure metal* condensation, the activity coefficient of pure M metal is unity ($\gamma_M = 1$) and there are no other elements in the alloy ($\sum_{i\neq M} \varepsilon_i \alpha_i = 0$), and then the term containing the fraction condensed for a major host phase element simplifies to

$$\frac{1}{(1-\alpha_M)} = K_1 \frac{\varepsilon_M}{(0.5+\ \varepsilon_{He})} P_{tot}$$

$$\text{for } M = \text{ major element in major phase and } \gamma_M = 1 \tag{10}$$

This remains a reasonable approximation for the major element during alloy formation as long as $\varepsilon_M\, \alpha_M \approx\ \varepsilon_M\, \alpha_M + \sum_{i\neq M} \varepsilon_i \alpha_I$, which is the case for Fe with much higher solar abundance than most of the other siderophile elements that do not contribute significantly to the total number of condensed atoms in the denominator in Eq. (9).

The other approximation for Eq. (9) is for siderophile *trace elements* that are condensing at temperatures where most of the abundant host elements Fe and Ni are already condensed. If the trace element contribution to the total number of atoms in the alloy is negligible such that $\varepsilon_M \alpha_M \ll \sum_{i\neq M} \varepsilon_i \alpha_I$, then $\sum_{i\neq M} \varepsilon_i \alpha_i \approx\ \varepsilon_M \alpha_M + \sum_{i\neq M} \varepsilon_i \alpha_I$ and Eq. (9) becomes

$$\frac{\alpha_M}{(1-\alpha_M)} \approx K_1 \frac{\sum_{i\neq M} \varepsilon_i \alpha_i}{(0.5+\varepsilon_{He})\gamma_M} P_{tot} \quad \text{for M} = \text{ trace element} \tag{11}$$

This simplification does not apply to condensation of highly refractory elements such as Os, Re, and Ir which are similar in abundance, but it works reasonably well for siderophile elements more volatile than Fe and Ni when $\alpha_{Fe} = \alpha_{Ni} = 1$ and $\sum_{i\neq M} \varepsilon_i \alpha_i \approx\ \varepsilon_{Fe} + \varepsilon_{Ni}$.

The *temperature dependence* enters through the equilibrium constant (obtained from thermodynamic data compilations) in the form:

$$\log K_1 = \Delta S_1/(R \ln 10) - \Delta H_1/(RT \ln 10) \tag{12}$$

where ΔH_1 and ΔS_1 are the enthalpy and entropy of the condensation reaction (1) respectively, for element M, and R is the gas constant. Taking the log of Eq. (9), inserting for log K_1 in Eq. (12), and re-arranging gives:

$$\frac{1}{T_{\alpha,M(alloy)}} = \frac{R \ln 10}{\Delta H_1}\left[\frac{\Delta S_1}{R \ln 10} - \log\left(0.5+\varepsilon_{He,sun}\right) - \log \frac{\alpha_M}{(1-\alpha_M)\left(\varepsilon_M \alpha_M + \sum_{i\neq M} \varepsilon_i \alpha_I\right)} + \log P_{tot} - \log \gamma_M\right] \tag{13}$$

If needed, the excess enthalpy and entropy of the activity coefficient ($\ln \gamma_M = H^E/RT - S^E/R$) can be combined with the enthalpy and entropy of the reaction.

Setting $\alpha_M = 0.5$ and solving for T gives the 50% condensation temperature of M into the alloy. The 50% condensation temperatures of the different elements into the alloy are interdependent through the overall alloy composition given in the term $\sum_{i\neq M} \varepsilon_i \alpha_I$ where each α_i

depends on temperature. For each element in the alloy there is an equation like Eq. (13) or similarly applicable ones that are coupled non-linearly and solved iteratively to obtain the fraction condensed for each element and the composition of the multi-component alloy.

With the approximations in Eq. (10) the condensation temperatures of pure metals from their monatomic gases are

$$\frac{1}{T_{\alpha,M(\text{pure})}} = \frac{R\ln 10}{\Delta H_1}\left[\frac{\Delta S_1}{R\ln 10} - \log(0.5+\varepsilon_{He}) - \log(1-\alpha_M) + \log(\varepsilon_M) + \log P_{tot}\right] \tag{14}$$

and for a trace element into an alloy:

$$\frac{1}{T_{\alpha,M(\text{trace in alloy})}} \approx \frac{R\ln 10}{\Delta H_1}\left[\frac{\Delta S_1}{R\ln 10} - \log(0.5+\varepsilon_{He}) - \log\frac{\alpha_M}{(1-\alpha_M)} + \log\frac{1}{\left(\sum_{i\neq M}\varepsilon_i\alpha_I\right)} + \log P_{tot} - \log\gamma_M\right] \tag{15}$$

For all cases the condensation temperature is *not* a linear function of the fraction condensed but instead is inversely proportional to

$$\frac{1}{T_{\alpha_M}} \propto -\log\frac{\alpha_M}{(1-\alpha_M)\left(\varepsilon_M\alpha_M + \sum_{i\neq M}\varepsilon_i\alpha_I\right)} = \log\frac{(1-\alpha_M)\left(\varepsilon_M\alpha_M + \sum_{i\neq M}\varepsilon_i\alpha_I\right)}{\alpha_M}$$

per Eq. (13); also note the minus sign in the inverse proportionality relationship. For the two limiting approximations above for major or trace elements the condensation temperature becomes inversely proportional to

$$\log\frac{(1-\alpha_M)}{\alpha_M}$$

for trace elements from Eq. (15) when using the approximation $\varepsilon_M\alpha_M \ll \sum_{i\neq M}\varepsilon_i\alpha_I$ as done in Eq. (11).

For major elements the condensation temperature is inversely proportional to

$$\log(1-\alpha_M)$$

as follows from Eq. (14) with the simplification as done in Eq. (10).

The general expression for the *fraction condensed with temperature* from Eqs. (9) or (13) is:

$$\log\frac{\alpha_M}{(1-\alpha_M)\left(\varepsilon_M\alpha_M + \sum_{i\neq M}\varepsilon_i\alpha_I\right)} = -\frac{\Delta H_1}{T_{\alpha,M(\text{alloy})}R\ln 10} + \frac{\Delta S_1}{R\ln 10} - \log(0.5+\varepsilon_{He}) + \log P_{tot} - \log\gamma_M \tag{16}$$

The conclusion is that the fraction condensed is not a simple function of $1/T_\alpha$.

Figure 1b shows the elemental fraction condensed as a function of $10^4/T_\alpha$ for elements condensing as minerals or dissolving in solid solution in a more abundant mineral. Major element condensation as silicate or another major phase, or trace element condensation into solid solution into silicates can be described by similar equations as illustrated for alloys above. Note the differences between curve shapes and slopes for major elements and trace elements.

The abundant elements (e.g., Fe, Al, Si, Mg, P, S, F, Cl) show steep curves and condense as more or less pure phases e.g., metal alloy, silicates, FeS, Fe_3P, F-apatite, and NaCl that also serve as host phases for other minor (trace) elements. In contrast, there are the sigmoid-like shapes for trace elements condensing into solid solutions (e.g., Na, K, B that condense into anorthite at lower temperatures, Ga condensing into Fe-Ni metal and anorthite, Br and I into NaCl). Figure 1a is similar to Fig. 6 of Larimer (1967) where fast cooling corresponds to condensation of pure elements and compounds and slow cooling allows formation of solid solutions. Next comes a description of why the curves for the fractions condensed for pure phases containing major elements differ in shape from those of trace elements in solid solutions.

For pure metals (or alloys very rich in element M), using the approximation in Eq. (10) gives a simplified expression for the fraction condensed from Eq. (16)

$$-\log(1-\alpha_M) = -\frac{\Delta H_1}{T_{\alpha,M(pure)} R \ln 10} + \frac{\Delta S_1}{R \ln 10} + \log\left(\frac{\varepsilon_M}{0.5+\varepsilon_{He}}\right) + \log P_{tot}$$

for pure metal (17a)

Equation (17a) can be further approximated for small fractions condensed of a pure phase ($\alpha_M < 0.25$) where, $-\ln(1-\alpha_M) \approx \alpha_M$ to give

$$\alpha_{M(pure)}/\ln 10 \approx -\frac{\Delta H_1}{T_{\alpha,M(pure)} R \ln 10} + \frac{\Delta S_1}{R \ln 10} - \log(0.5+\varepsilon_{He}) + \log(\varepsilon_M) + \log P_{tot} \quad (17b)$$

for small $\alpha_{M(pure)}$.

Equation (17b) is the only approximation that gives a linear relationship of the fraction condensed ($\alpha_{M(pure)}$) with $-(1/T)$. This approximation only applies to a limited amount of a major element condensed and does not apply to trace elements. But it illustrates why the fraction condensed for an element forming a pure phase shows a steep increase with T_α and $1/T_\alpha$ (Fig. 1a, 1b).

The circles with element symbols in Fig. 1a,b indicate the position of the BSE abundances normalized to CI-chondrites (standing in as proxy for solar) and refractory elements. There are steep increases in the fractions condensed over a very narrow temperature interval for many elements, especially for elements that form major or "pure" phases (e.g., Al, Mg, Si, P, but also F in F-apatite, Cl in NaCl). For these elements, the curves follow log ($1-\alpha_M$). The sigmoid curves are for elements going into solid solution (e.g., Mn, K, B, Cs, Br, but also Ga which dissolves in both metal and feldspar) – see Eq. (23). Thermodynamic data for Cs dissolution in feldspar are estimated (see the Appendix) and possibly affect the shape of its condensation curve. Indium has a shallow slope until Cl and Br condensation removes halogens from the gas phase. Then the α_{In} slope becomes steeper and all In condenses within a few K. The steep increase in the fraction condensed with decreasing T is the reason why

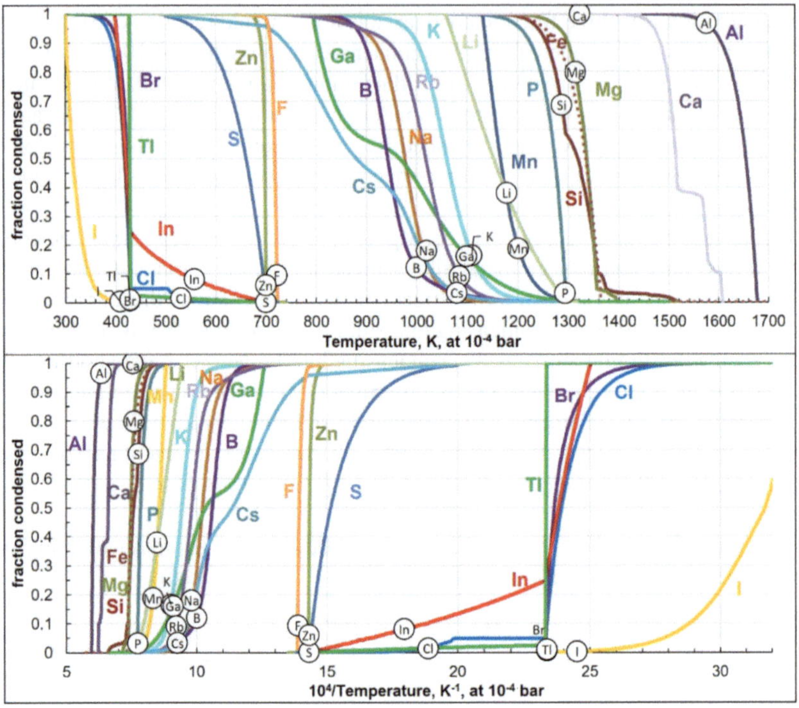

Fig. 1 Fraction condensed α_M at 10^{-4} bar total pressure. (a) on top: α_M plotted versus T. (b) on bottom: theoretically more applicable representation of logistic function of α_M vs inverse temperature (10^4/T scale)

50% condensation temperatures suggest an apparent trend with temperature. The bottom diagram has the proper logistic function display (see Eq. (21) below). The reader should not confuse this treatment with the averaging of T_{50} values done by Sossi et al. (2022) to deduce an average condensation temperature for the Earth.

For trace elements condensing into solid solution, the approximation in Eq. (11) gives from Eq. (12),

$$\log\frac{\alpha_M}{(1-\alpha_M)} = -\frac{\Delta H_1}{T_{\alpha,M(\text{alloy})} R \ln 10} + \frac{\Delta S_1}{R \ln 10} + \log\left(\frac{\sum_{i\neq M}\varepsilon_i\alpha_i}{0.5+\varepsilon_{He}}\right) - \log\gamma_M + \log P_{tot}$$

for trace elements (18)

The dependence on the trace element abundance cancels out in the solid solution approximation, and as noted at Eq. (11), when all Fe and Ni are condensed as metal (no Fe in oxides), $\sum_{i\neq M}\varepsilon_{i,sun}\alpha_i \approx \varepsilon_{Fe} + \varepsilon_{Ni}$.

Thus, for major element condensation, Eq. (17a)–(17b) gives the dependence with –(1/T) as log (1/(1-α)) but for trace element condensation into a solution, Eq. (18) shows the dependence is log (α /(1-α)) giving rise to the sigmoid curve shape in Fig. 1a,b. The curves for many trace elements in Fig. 1b often show inflection points near or at the 50% condensation temperatures. This is the outcome of Eq. (18), which is a logistic function when re-written for α. From Eq. (18) in exponential form:

$$\frac{\alpha_M}{(1-\alpha_M)} \approx \frac{P_{tot}}{\gamma_M} \frac{\sum_{i \neq M} \varepsilon_i \alpha_i}{(0.5 + \varepsilon_{He})} \exp\left(-\frac{\Delta H_1}{T_{\alpha,M} R} + \frac{\Delta S_1}{R}\right)$$
$$= \frac{1}{u} \exp\left(+\frac{\Delta S_1}{R}\right) \exp\left(-\frac{\Delta H_1}{T_{\alpha,M} R}\right) \tag{19}$$

Inverting Eq. (19) gives:

$$\frac{1-\alpha_M}{\alpha_M} = \frac{1}{\alpha_M} - 1 = \frac{\gamma_M}{P_{tot}} \frac{(0.5 + \varepsilon_{He})}{\sum_{i \neq M} \varepsilon_i \alpha_i} \exp\left(-\frac{\Delta S_1}{R}\right) \exp\left(+\frac{\Delta H_1}{T_{\alpha,M} R}\right)$$
$$= u \exp\left(-\frac{\Delta S_1}{R}\right) \exp\left(+\frac{\Delta H_1}{T_{\alpha,M} R}\right) \tag{20}$$

The u is shorthand for all constants and variables in the preexponential term. Solving for α gives the logistic equation for the cumulate distribution function of the fraction condensed as a function of 1/T:

$$\alpha_M = \frac{1}{1 + \frac{\gamma_M}{P_{tot}} \frac{(0.5+\varepsilon_{He})}{\sum_{i \neq M} \varepsilon_i \alpha_i} \exp\left(-\frac{\Delta S_1}{R}\right) \exp\left(+\frac{\Delta H_1}{T_{\alpha(M)} R}\right)}$$
$$= \frac{1}{1 + u \exp\left(-\frac{\Delta S_1}{R}\right) \exp\left(+\frac{\Delta H_1}{T_{\alpha(M)} R}\right)} \tag{21}$$

For $\alpha_M = 0.5$ at T_{50} in Eqs. (19) or (20) it follows that

$$\exp\left(+\frac{\Delta S_1}{R}\right) \exp\left(-\frac{\Delta H_1}{T_{50} R}\right) = \frac{\gamma_M}{P_{tot}} \frac{(0.5 + \varepsilon_{He})}{\sum_{i \neq M} \varepsilon_i \alpha_i} = u \tag{22}$$

(Note the multiple sign changes for the enthalpy and entropy terms from Eqs. (18) to (22)). Substituting Eq. (22) for u into Eq. (21) gives

$$\alpha_M = \frac{1}{1 + \exp\left(-\frac{\Delta H_1}{R}\left(\frac{1}{T_{50(M)}} - \frac{1}{T_{\alpha(M)}}\right)\right)} \tag{23}$$

Equation (23) illustrates that the 50% condensation temperature is not only a convenient volatility measure for a given element but also has practical use for describing the fraction of an element condensed if the 50% condensation temperature is known. The simple logistic function Eq. (23) from the approximation (!) of trace element condensation, Eqs. (18, 21), has limited use if the enthalpy and entropy of condensation and/or activity coefficient vary significantly with temperature and/or the alloy (or host phase) composition changes significantly with temperature. However, the approximation works fairly well near the 50% condensation temperature.

It is significant that the 50% condensation temperatures are often found around the inflection point of the Hill slope (slope of approximately linear portion of sigmoid curve), and for simple cases Eq. (23) already makes that point. The first derivative of Eq. (21) for the cumulate logistic function for α_M, $d\alpha_M/d(1/T)$ shows that the inflection point corresponds to a local maximum for the amount of an element condensed per inverse temperature step.

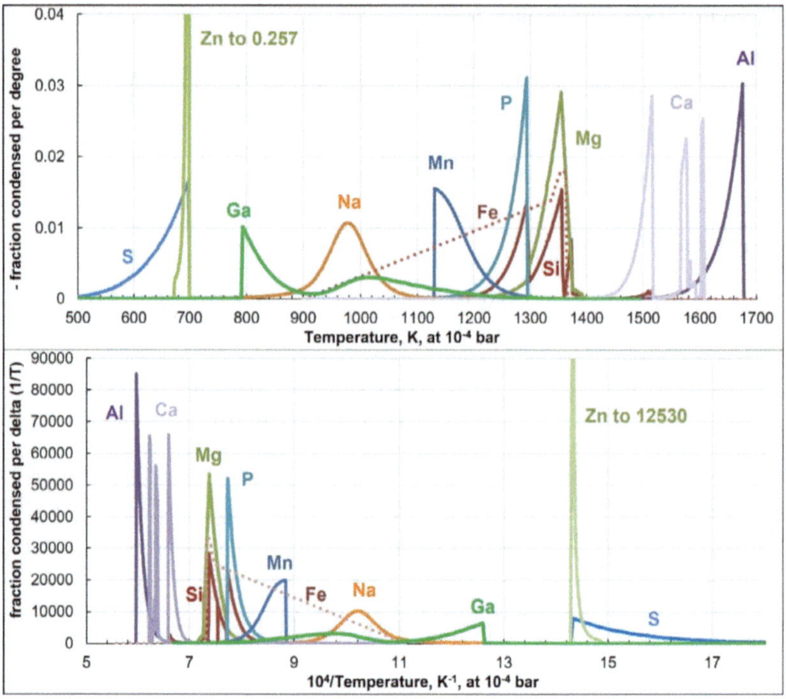

Fig. 2 (a) on top: The negative of the differential amount condensed per degree. (b) at bottom: The differentials of Eqs. (17a)–(17b) and (21) give the fraction condensed per (1/T) interval

This important point that the 50% condensation temperatures often correspond to the largest amount of an element condensing per inverse temperature step is shown in Fig. 2. Figure 2a plots the negative of the fraction condensed per degree, and Fig. 2b the proper form of differential amount condensed per (1/T). This is another good reason why 50% condensation temperatures are useful proxies for relative volatilities of the elements.

The logistic Eq. (21) is for trace elements and is easily differentiated to show the significance of the 50% condensation temperature and fraction condensed per (1/T). Proper calculations for every element make use of the general Eq. (16) which is difficult to differentiate. The differential of the re-arranged Eq. (17a)–(17b) is more practical for estimating the fractional condensation per 1/T interval for elements condensing as pure phases or elements that are major constituents of pure phases. This leads to somewhat different curvatures for α, which can be seen for the different elements illustrated in Fig. 2b.

The multiple local maxima for Ca in Fig. 2 illustrate the importance of multiple stable phases appearing in the condensation sequence for condensation efficiency. Calcium starts to condense as hibonite ($CaAl_{12}O_{19}$), then upon further cooling grossite ($CaAl_4O_7$), melilite etc. each boost the amount of Ca condensing per (1/T). So, each new stable phase is associated with at least one local maximum for differential condensation. A similar situation is shown in Fig. 2 for the trace element Ga, which condenses into both metal and feldspar.

An important feature in Fig. 1b is that the fractions condensed rise steeply along the Hill slope of the sigmoid curves over a narrow temperature (1/T) interval so that smaller and higher fractions condensed occur near the 50% condensation temperature, which means that the temperature (or 1/T) ranges over which most of an element condenses are relatively narrow. For example, at 10^{-4} bar, Li has a 50% condensation temperature T_{50}(Li) = 1152 K and within ±50 K of this temperature, 29 to 78% of all Li condenses. Similarly, for Na (T_{50} = 978), and within ±50 K of the 50% condensation temperature, 10–86% of all Na condenses. Fluorine is an extreme case as between the onset of F-apatite stability at 723 K and 700 K, more than 98% of F is condensed (the T_{50} (F) = 717 K). This may explain why the 50% condensation temperature sometimes describes "condensation trends" for fractions condensed other than 50%. But it only works if an element has a steep Hill slope. But to apply this to a set of elements, the elements in question also must condense within a narrow temperature interval with overlapping fraction condensed curves, as there can be only one temperature and pressure for an equilibrium condensation configuration for a given set of elemental abundances (see point (4) in Sect. 5). This is important for carbonaceous chondrites where observed fractionations often are within one order of magnitude only and could explain some (and only some) observed "trends" with inverse condensation temperatures in these cases.

However, for the BSE, the overall observed elemental fractionations span about four orders of magnitude (including C and N), so the 50% condensation temperatures are not necessarily close to the temperatures where less than 10% is condensed for elements such as the halogens. The circles in Fig. 1 show that the 50% condensation temperatures are far away from the temperatures that correspond to the observed fractions condensed for the CI-normalized BSE abundances. For example, the small amount of iodine in the BSE corresponds to a condensation temperature far from T_{50} for iodine. Similarly, there are two Cl-bearing condensates (chlorapatite and halite). While chlorapatite could explain a small amount of Cl in the BSE, chlorine is not 50% condensed until a much lower temperature when halite (NaCl) forms. Fegley and Schaefer (2010) first showed that the amount of P available for chlorapatite is insufficient to accommodate 50% of all chlorine and that halite condensation is needed to condense all the chlorine remaining in the nebular gas. Similarly, Ga condenses into metal alloy and anorthite, making Ga condensation dependent on the stability of two host phases, and the 50% condensation temperature is far from that for the Ga abundance in the BSE. The indium fraction condensed (Fig. 1a, 1b) increases sharply in the vicinity of the Br and Cl 50% condensation temperatures because their condensation decreases the partial pressures of the two major In-bearing gases, which are InBr (98.8%) and InCl (1.2%). These examples show there is no prescription that fits all elements.

To summarize this section, the popular volatility trend plots of normalized fraction condensed (α) versus 50% condensation temperature do not use the theoretically correct relationships discussed above (e.g., Sun 1982; McDonough and Sun 1995; Palme and O'Neill 2014; Wood et al. 2019). The fraction condensed for a particular element is not a linear function of temperature as originally shown by Larimer (1967) and also seen in Fig. 1. It is only because the 50% condensed point falls in a pseudo-linear region of the sigmoid-shaped condensation curve for an element that plots of normalized abundances versus T_{50} are qualitatively useful.

4 Volatility Trends for Chondrites and the Earth

Considering refractory elements first, Grossman (1973) and Wänke et al. (1974) measured that refractory lithophile and siderophile elements are enriched in CAIs from the Allende CV chondrite by about 20× CI chondritic abundances. Conversely, Grossman and Wänke et al. found that moderately volatile lithophile and siderophile elements, with lower condensation temperatures, are depleted in the CAIs they studied relative to their CI chondritic

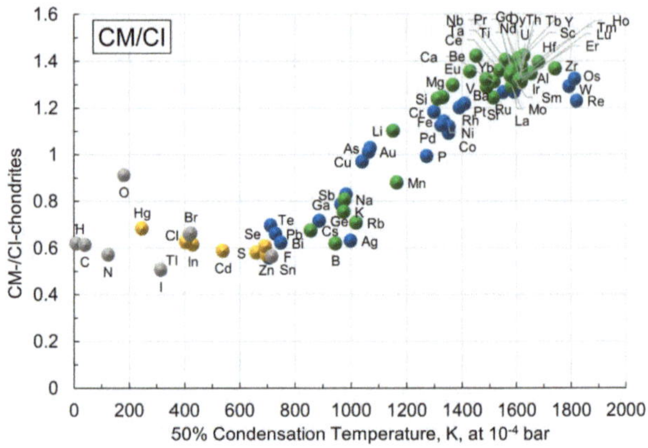

Fig. 3 CI-chondrite normalized concentrations in CM-chondrites. The Cl (418 K) and Br (423 K) points overlap in this and subsequent graphs. There is also some overlap among RLE. Color coding indicates chalcophile (yellow), lithophile (green), siderophile (blue), and halogens, H, C, N, O (grey) in this and subsequent figures

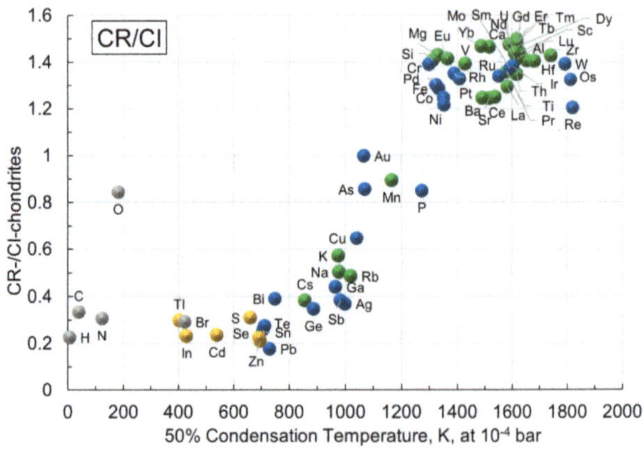

Fig. 4 CI-chondrite normalized concentrations in CR-chondrites. Note the incomplete coverage for known elemental composition of CR-chondrites

abundances. Wasson and Kallemeyn (1988) show refractory lithophile elements are about equally enriched (CM, CV, CO, CR chondrites) or depleted (H, L, LL, EH, EL chondrites) relative to CI, Si normalized abundances. The variable amounts of refractory lithophile elements (RLE) in chondrites can be explained by the addition (subtraction) of a CAI-like component from CI composition with the Mg/Si ratio balanced by olivine (see Figs. 7.3.5–7.3.6 and associated text in Larimer and Wasson 1988). However more recent high-resolution analyses of RLE show the RLE are not uniformly present in chondrites and planets (e.g., Dauphas and Pourmand 2015). The Allende CV chondrite (cf. Fig. 6) is the best example because bulk Allende displays a Group II rare earth element pattern (Tanaka and Masuda 1973; Fegley and Kornacki 1984; Stracke et al. 2012; Dauphas and Pourmand 2015).

In Figs. 3 to 13, the RLE show ca. 10-20% deviation from the mean enrichment (depletion) in the volatility trends for carbonaceous, ordinary, and enstatite chondrites. To first approximation, RLE are present in constant proportions, but in detail the different volatilities of the RLE are apparent. Figure 3 to 13 display CI chondrite normalized abundances for all natural chemically reactive elements (color coded as yellow – chalcophile, green – lithophile, blue – siderophile, gray - other) versus temperature to facilitate comparison with published volatility trend plots. The elemental concentration data for CI chondrites are from Lodders et al. (this volume) and from Lodders (2021) and Lodders and Fegley (2023) for the other chondrites.

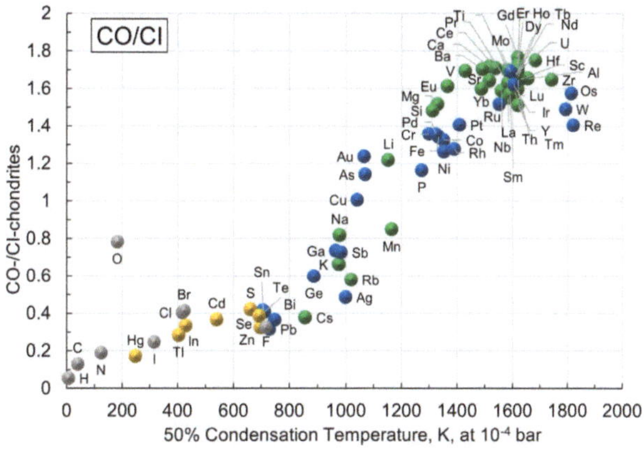

Fig. 5 CI-chondrite normalized concentrations in CO-chondrites

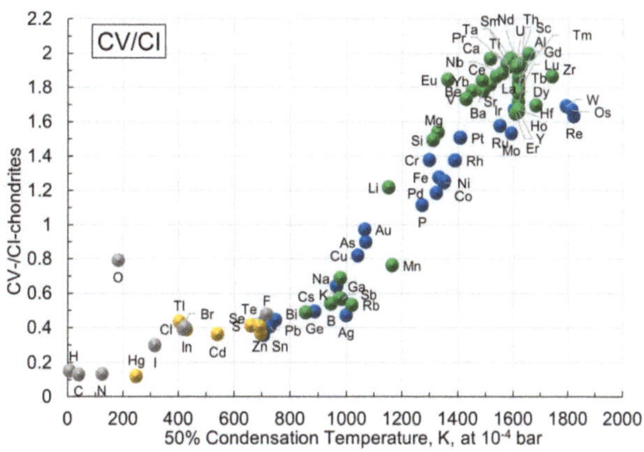

Fig. 6 CI-chondrite normalized concentrations in CV-chondrites

Wai and Wasson (1977) plot CI-chondrite normalized abundances of MVE in H and CM chondrites on a logarithmic scale versus 50% condensation temperatures at total pressures of 10^{-4} and 10^{-6} bar. Wasson and colleagues (Wasson and Chou 1974; Wai and Wasson 1977; Wasson 1977) proposed MVE abundances in CM and H chondrites followed a trend with a continuously negative slope. They suggested this trend was due to incomplete condensation of nebular gas as it cooled and was lost by some process. Wasson's proposed MVE volatility trend for ordinary chondrites was controversial because Anders (1977) interpreted MVE abundances in ordinary chondrites as "a terraced landscape where gently sloping plateaus alternate with steep declines at Cu and Te." Figures 3 to 6 and the volatility trends in Alexander (2019b) show MVE trends for carbonaceous chondrites like that proposed by Wasson and colleagues.

But volatility trend plots for ordinary chondrites in Alexander (2019a) and in Figs. 7 to 9 show different behavior for moderately volatile lithophile elements than seen in carbonaceous chondrites. With the exception of B (H, L, LL chondrites) and Cs (H, LL chondrites), moderately volatile lithophile elements plot together with RLE in the upper cluster of points. The steep break (or decline) at about 1000 K is seen best in Fig. 7 for H chondrites. Manganese, Li, Na, K, and Rb have higher abundances and Cs, Ga, B, and F have lower abundances. The high alkali abundances in ordinary and enstatite chondrites relative to CI chondrites are well known but not understood (e.g., Goles 1971a,b; Mason 1979). There is

Fig. 7 CI-chondrite normalized concentrations in H-chondrites

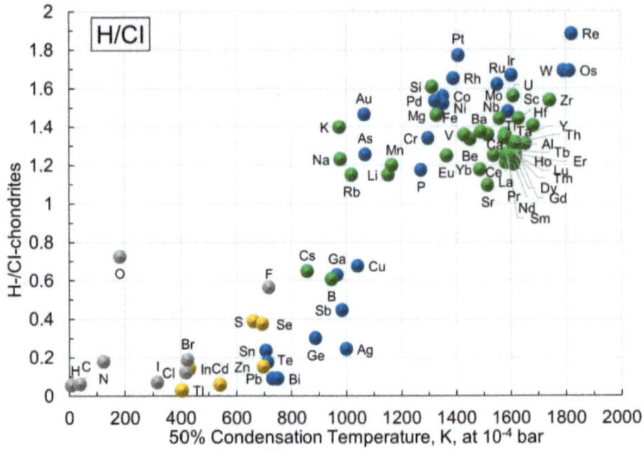

Fig. 8 CI-chondrite normalized concentrations in L-chondrites

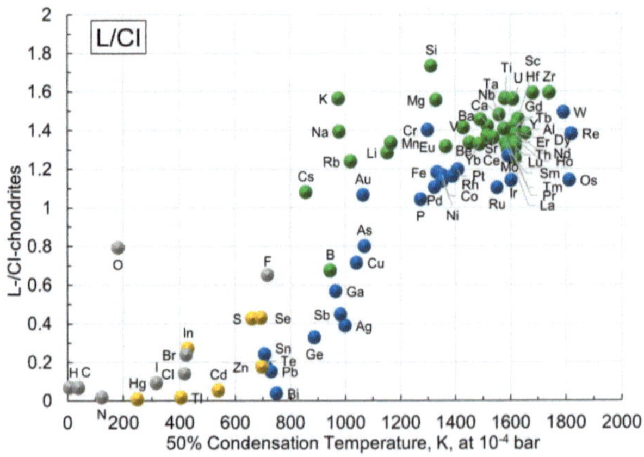

Fig. 9 CI-chondrite normalized concentrations in LL-chondrites

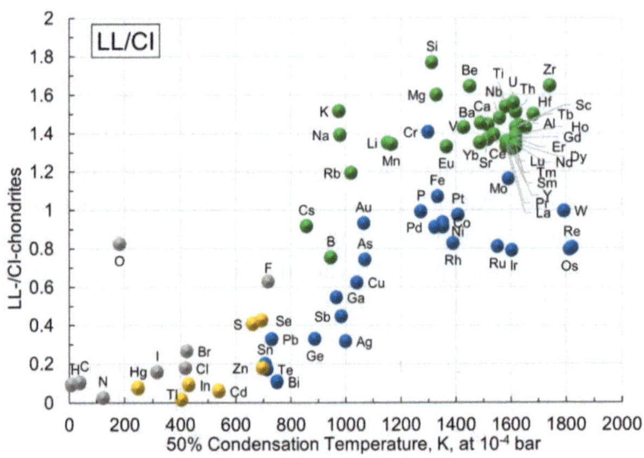

likely some connection with the solubility of alkalis in aqueous fluid and the presence of alkali halides in some ordinary chondrites, but further discussion is beyond the scope of this paper (Gast 1962; Lee et al. 2006; Zurfluh et al. 2013, Lodders et al. this volume).

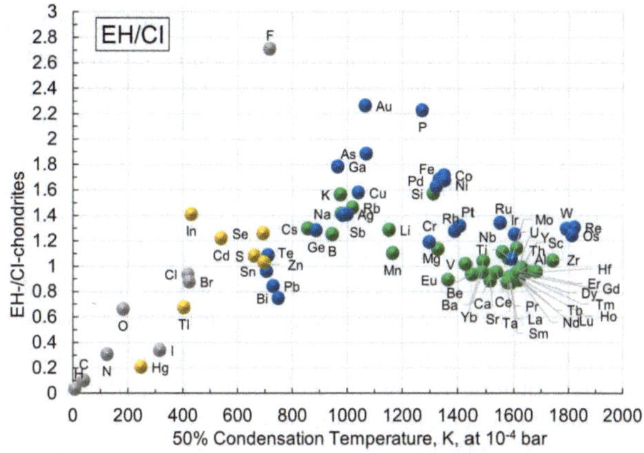

Fig. 10 CI-chondrite normalized concentrations in EH-chondrites

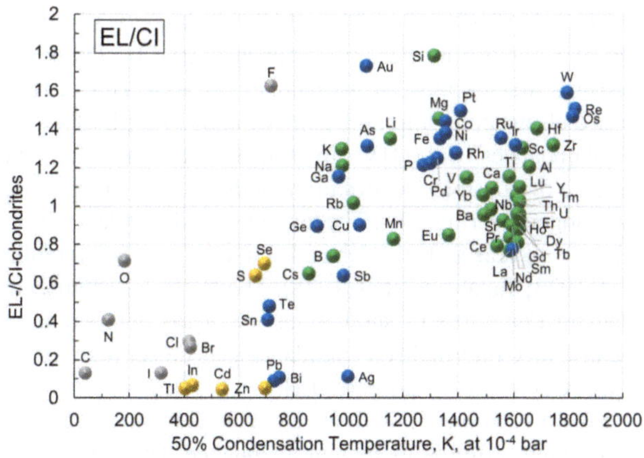

Fig. 11 CI-chondrite normalized concentrations in EL-chondrites

The siderophile elements in H chondrites cluster into two groups at higher and lower abundances with a steep break (decline) at about 1000 K. The L and LL chondrites apparently show a volatility trend for siderophile elements more volatile than Au (T_{50} = 1065 K), which is about as abundant as more refractory siderophile elements except the more abundant Cr, Re, and W (L chondrites), and Cr and Mo (LL chondrites).

The EH and EL chondrites in Figs. 10 and 11 display complicated trends with moderately volatile lithophile and siderophile elements more abundant than (or equally abundant with) the RLE and refractory siderophile elements. Figure 11 shows the MVE with T_{50} <1000 K (except F, B) are less abundant than the RLE and refractory siderophile elements in EL chondrites. The F abundance in EH and EL chondrites is high. Most of the fluorine analyses were done in the 1960s and on only seven different falls (see Table 2 of Lodders and Fegley 2023). It is important to reexamine the F abundance in enstatite chondrites to check the high F abundance. Alexander (2019a) gives a detailed discussion of volatility trends in ordinary, enstatite, and R chondrites.

Ringwood and Kesson (1977) plotted the increasing depletions of selected chalcophile, lithophile, and siderophile volatile elements in ordinary chondrites, Earth's mantle, and the Moon relative to their abundances in CI chondrites. They found step patterns for volatile depletions in all cases (their Figs. 1–3), but did not use condensation temperatures. Ringwood and Kesson concluded the observed depletions for ordinary chondrites and the BSE correlated at least qualitatively with the condensation temperatures of the elements in the solar nebula. Their depletion sequence for volatile elements in ordinary chondrites disagrees in detail with those plotted with more recent data in Figs. 7 to 9 for ordinary chondrites. Ringwood and Kesson (1977) emphasized the lunar depletions were different than those seen for ordinary chondrites and the BSE. Citing several prior papers by Ringwood, they concluded "... that the Moon was ultimately derived from material evaporated from the Earth's outer mantle and selectively recondensed at some distance from the Earth under unique conditions which permitted further loss of volatile components."

Palme et al. (1988) presented data showing that the abundances of moderately volatile elements (MVE) in carbonaceous chondrites, differentiated meteorites (e.g., eucrites, SNC meteorites, IIAB and IV irons), and the accessible silicate portions of the Earth and Moon correlate qualitatively with condensation temperatures; less abundant elements are those with lower condensation temperatures. Their review was the basis for subsequent modeling to explain the MVE depletions in terms of condensation during accretion of planetesimals in a cooling solar nebula (Cassen 2001; Ciesla 2008).

The behavior of the highly volatile elements (HVE) that condense at temperatures below that of S (as troilite FeS), was not discussed by Palme et al. (1988). The highly volatile elements include most of the halogens (Cl, Br, I), Cd, Hg, In, Tl, C, N, O, and H. As discussed elsewhere (Lipschutz and Woolum 1988; Lodders 2021; Lodders and Fegley 2023), the abundances of the HVE in chondrites are more uncertain than those of other elements because of possible redistribution and/or loss by metamorphism and/or aqueous alteration for halogen-bearing salts. The temperature dependence, if any, of HVE abundances in chondrites is unclear and still debated actively (e.g., Braukmüller et al. 2018, 2019, Krahenbühl et al. 1973; Takahashi et al. 1978). The HVE in carbonaceous, ordinary, and enstatite chondrites in Figs. 3 to 11 show a variety of low temperature "trends". Oxygen is interesting because its CI normalized abundance gets closer to the HVE trend in the sequence CM, CR $\approx$ LL, CO $\approx$ CV $\approx$ L, H, EL, EH.

Figure 12 compares volatile element abundance trends over the same temperature range using three different elemental abundance data sets for CM and CV chondrites: Alexander (2019b), Braukmüller et al. (2018, 2019), and Lodders (2021) and Lodders and Fegley (2023). The CI abundance data used for normalization are from Lodders et al. (this volume) in all cases. The plots show three different geometries for abundance "trends" that are due to the elements plotted and data used. Different low T cutoffs would further alter the apparent trends, which are not well defined and depend upon the data used.

Sun (1982) is apparently the first study to discuss and plot a volatility trend for the bulk silicate Earth using condensation temperatures at 10^{-4} atm total pressure. Sun (1982) reached several key conclusions, quoted here: "Above 700 K there are two parallel trends which are defined by lithophile elements (Al, Ca, REE, Ti, Mg, Si, Cr, Mn, Na, K, Rb, F, Zn etc.) and siderophile elements (W, Ni, Co, P, As, Ag, Sb and Ge) respectively. The depletion factor for the siderophile trend relative to the lithophile trend is about 0.085. Within each trend there is a continuous depletion towards lower temperature. A third trend is defined by noble metals (Ir, Os, Re, Pd, Pt and Au) with a depletion factor of about 0.003 relative to CI chondrites. These trends are interpreted in terms of core-mantle differentiation and volatility – controlled processes operating before and during earth accretion." Sun's Fig. 3 is the original volatility trend for the Earth and shows many of the same features as in volatility trends calculated with more recent elemental abundance data for the bulk silicate Earth (e.g., Kargel and Lewis 1993; McDonough and Sun 1995; Palme and O'Neill 2014; Wang et al. 2019). The siderophile element trend he identified involves several of the same elements that apparently show a volatility trend in L chondrites (Co, Ni, P, As, Sb, Ge, Ag). The common

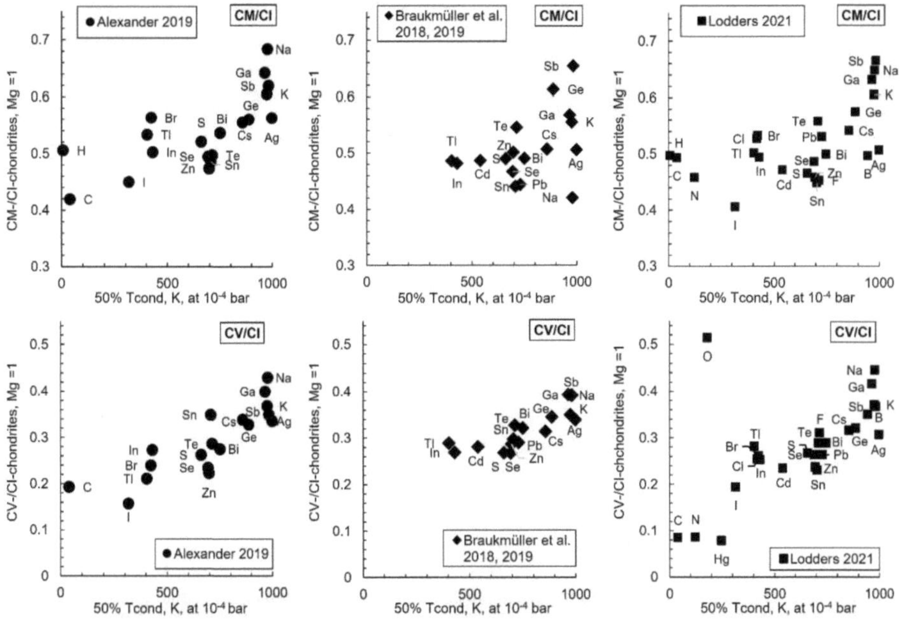

Fig. 12 Volatile element abundances of CM- and CV-chondrites relative to CI-chondrites and Mg versus condensation temperatures. Data from three references are shown. The values for the abscissa are the same in each graph. See text

siderophile elements in the two trends (L chondrites in Fig. 8 and in Sun's Fig. 3) are both depleted relative to lithophile elements of similar volatility, and for the same reason (but presumably different mechanisms) – metal loss from the sampled material.

With the exception of Wai and Wasson (1977), who focused on ordinary chondrites, all prior work presents volatility trends at 10^{-4} bar total pressure. Presumably this was done because no condensation temperatures were available for all elements at other pressures. But whatever the reason, the 10^{-4} bar – based volatility trend has become standard in cosmochemistry.

The four plots in Fig. 13 show the Earth's volatility trend computed with 50% condensation temperatures at four different total pressures from 10^{-2} to 10^{-8} bar to illustrate changes in condensation temperatures, the condensation sequence, and in chalcophile, lithophile, and siderophile element volatility trends.

The condensation temperatures are computed for a solar composition gas (Table 1). Thus, normalization of elemental concentrations in the BSE should be relative to the Sun for which atomic abundances are usually given relative to hydrogen. This normalization is impractical, and as generally done, normalization is to CI-chondrites, which are taken as representative for the condensable portion of the elements in the solar composition. The BSE composition is further normalized to the most abundant refractory elements in the BSE (Ca, Al) so the concentrations of refractory elements are at or near unity. This is equivalent to stating that the refractory lithophile elements are fully condensed and present in CI chondritic (i.e., solar) proportions within the bulk silicate Earth, which is the approach taken by Palme and O'Neill (2014). The quantitative relationship of condensation to elemental abundances requires normalizations at unity for the refractories because not more than 100% of an element can condense.

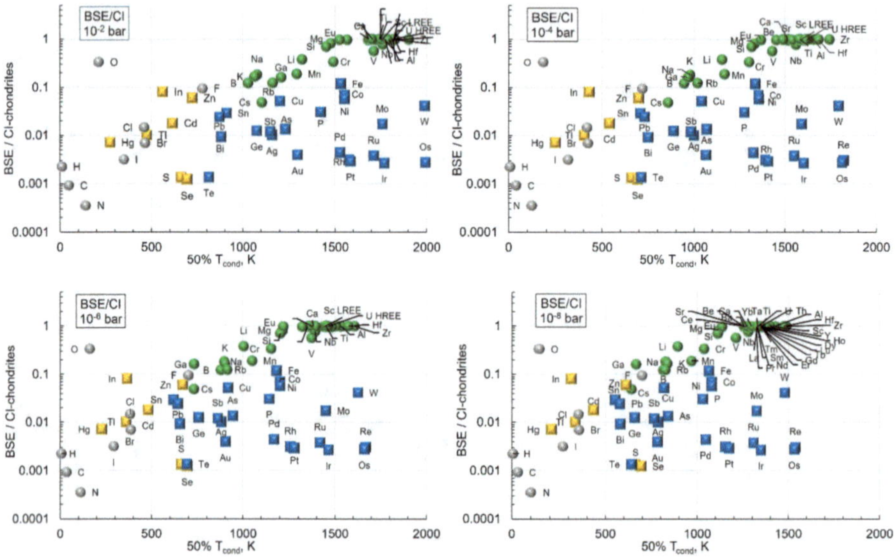

Fig. 13 BSE abundances normalized to CI-chondrites and refractory elements (Palme and O'Neill 2014, Lodders et al. this volume) versus 50% condensation temperatures from Table 1 at four different assumed total nebular pressures of 10^{-2}, 10^{-4}, 10^{-6}, and 10^{-8} bar. All plots are normalized to unity for the refractory lithophile elements. The abscissa is the same in all graphs

Figure 13a-d illustrate several important points. First, the normalized abundances vary by about four orders of magnitude (considering C and N). This is a much larger range than seen in chondrites. Second, the BSE volatility trend is more similar to that of the carbonaceous than the ordinary or enstatite chondrites (Figs. 3 to 11). Third, except for S and Se, all condensation temperatures decrease with decreasing total pressure. The sulfur condensation temperature is constant because its condensation reaction is independent of total pressure [H_2S (gas) + Fe (alloy) = FeS (troilite) + H_2 (gas)]. The selenium condensation temperature is constant because Se dissolves as $FeSe_{.96}$ in FeS (Lodders 2003).

The fourth important point is that the condensation sequence is different as a function of total pressure. The most well-known and important example is the reversal of the condensation temperatures of Fe versus Mg and Si at about 10^{-4} bar total pressure. Iron is more refractory than Mg and Si above this pressure and more volatile below 10^{-4} bar. The Ga and alkali (Na, K, Rb) condensation temperatures switch with decreasing total pressure with gallium being 130 K more refractory (at 10^{-2} bar) to 160 K more volatile (at 10^{-8} bar) than sodium. The deviation of In from other volatile chalcophile elements increases with decreasing pressure.

The fifth point is that Fig. 13 shows different trends for lithophile and siderophile elements, and noble metals as pointed out by Sun (1982). Sulfur, Se, and Te are grouped in CI-chondritic proportions with abundances lower than the highly siderophile elements Re, Os, Ru, Ir, Rh, Pt, Pd also with CI-chondritic proportions. With decreasing pressure, the sloped trend is better defined with more chalcophile and siderophile elements following the lithophile element trend.

One key question is which elements should be used to define any volatility trend. Should the BSE volatility trend be based only on lithophile elements (Mg, Si, Li, V, Mn, Cr, Ga, B, alkalis, and halogens) or also other elements that appear to follow the same trend such

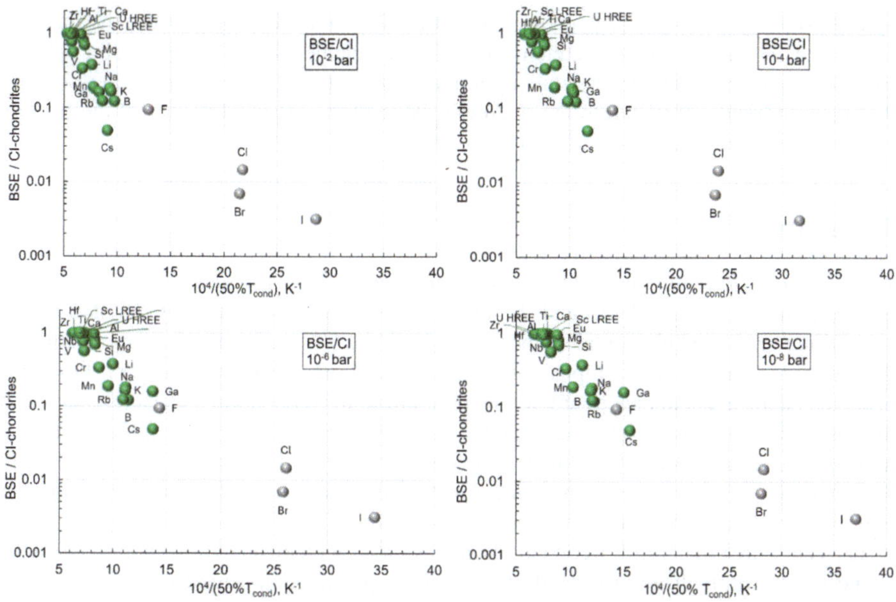

Fig. 14 The same data as in Fig. 13 but plotted versus inverse 50% condensation temperatures. Chalcophile and siderophile elements are not shown

as the atmophile elements and chalcophile elements (Kargel and Lewis 1993; Wang et al. 2019). The chalcophile elements are moderately or highly volatile elements that condense into troilite and/or as another sulfide (i.e., ZnS). Interpretation of the BSE abundances of Zn, Cd, In, Tl, S, Se, Te is complicated because they are affected by volatility, and core and sulfide matte formation. The two heavy Si-group elements Sn and Pb are MVE and depending upon T, fO_2, and fS_2 behave as chalcophile (PbS, SnS), lithophile (SnO_2, PbO) or siderophile elements. How to interpret their BSE abundances? Do Cr and the halogens behave as lithophiles or partly partition into the core (Ringwood 1979; Wänke 1981; Yuan and Steinle-Neumann 2023)? Fluorine in the BSE is mainly in the mantle and crust, Cl and Br are mainly in the oceans, and I is mainly in organic marine sediments.

Figure 13, which is widely used to discern linear trends with condensation temperatures, is plotted with the inappropriate functional relationship between fraction condensed and 50% condensation temperature. Inverse temperature, not temperature, is the correct independent variable to use. This was done by Humayun and Cassen (2006). The reason is the temperature dependence of the equilibrium constants used to compute condensation temperatures. Larimer (1967, 1973) was probably the first person to realize this. Each element follows a logistic condensation curve with temperature. But there is no common T (and P) where all elements can have the normalized abundances that they show in the commonly used volatility trend plot for the bulk silicate Earth. That is the root of the problem. If one looks at a given temperature in Fig. 1 one can see how much of an element should be condensed – for most elements, it is all or nothing.

When inverse temperature is used, the volatility trend looks completely different as shown in Fig. 14. The apparent linear diagonal "trend" is now a convex curve with decreasing temperature (or increasing 1/T). Thus, arguments based upon the shape of the volatility trends plotted as in Fig. 13 may not be meaningful.

But even Fig. 14 is formally incorrect for visualizing a linear trend because the fractions condensed (α values) for major elements and trace elements dissolved in solid solution have different functional dependence upon temperature (Eqs.(16), (23)). The reasons for the apparent trend are related to the steep Hill slope of the logistic function of the fraction condensed with inverse temperature, as described in Sect. 3.

5 Discussion

The premise of deriving a volatility trend and the nature of the problem. Figs. 3 to 13 show volatility trends for chondrites and the bulk silicate Earth (BSE) that are plotted as CI-chondritic normalized elemental abundances versus the 50% condensation temperatures of the elements from a solar composition gas, usually at 10^{-4} bar total pressure. Often the observed elemental abundances are doubly normalized to CI chondrites and a reference element (e.g., Si, Mg, Ca, Al, Yb), e.g., as done in Fig. 6-1 of Ebel et al. (2018). Unfortunately, in many studies that attempt to illustrate or to quantify volatility trends the observed abundances are not related to the correct functional dependence on condensation temperatures. Four underlying assumptions often made are as follows.

(1) The source material for condensation was *solar* in elemental composition. Using CI chondrites for normalization instead of solar photospheric composition is done for practical reasons because not all elements can be quantitatively measured in the sun, and the CI chondrites are the best available proxy for the solar condensable elements where comparison can be made (see e.g., Lodders (2020) and Lodders et al., this volume). The CI-normalized abundances are a measure of the elemental fractions condensed from an originally solar-composition-like gas. Normalizing to CI chondrites does not imply that the Earth or meteorite parent bodies formed from material of the same mineralogy or oxidation state as CI-chondrites.

(2) The linear or logarithmic elemental fractions condensed, α_M, are directly proportional to the 50% condensation temperatures. This assumption is incorrect. As explained above and in less detail by Larimer (1973), the correct proportionality of normalized abundances is log $(1-\alpha_M)$ for major elements or log $[(1-\alpha_M)/\alpha_M]$ for trace elements in solid solutions with *inverse* condensation temperatures. For several elements, a more complicated dependence of log $f(\alpha_M)$ with their inverse condensation temperatures applies.

(3) Although widely assumed, a total pressure of 10^{-4} bar is model specific. Temperature and pressure varied with radial distance from the proto-Sun, height above the nebular midplane, and with temporal evolution of the nebular accretion disk (e.g., Cameron and Fegley 1982; Ruden and Lin 1986; Cameron 1995; Cassen 2001; Ciesla 2008). Lewis (1974) derived a total nebular midplane pressure of 10^{-4} bar at 1 AU on the basis of planetary compositional data, an equilibrium condensation model, and particular adiabatic thermal profiles in the solar nebula. In general, lower total pressures are associated with lower temperatures in the nebular midplane.

As a consequence, more volatile elements condense at lower total pressures than more refractory elements. The net effect is a slightly lower condensation temperature than on an isobaric profile and a more shallow slope in conventional volatility plots like Figs. 3 to 13. For example, at 10^{-4} bar, the moderately and highly volatile elements condense from 1324 K (Pd T_{50}) to 248 K (Hg T_{50}). The change in pressure with temperature for an ideal gas adiabat with solar abundances of H_2 and He in this temperature region is given roughly by

$$\frac{d\log P}{d\log T} = \frac{(C_P/C_V)}{[(C_P/C_V)-1]} = \frac{\gamma}{(\gamma-1)} \sim 3.4$$

The change in pressure is about two orders of magnitude pressure drop from 1324 to 248 K. Interpolation of the condensation temperatures in Table 1 can be done to estimate condensation temperatures for thermal profile models of the solar nebula (see Appendix).

(4) It is often assumed that a physical basis exists for application of 50% condensation temperatures to define volatility trends. These element-specific temperatures describe the temperature, $T_{50}(M)$, where 50% of a given element M is condensed and the other half remains in the gas at constant total pressure. Thus, a relative volatility temperature scale can be derived for a given thermal profile such as an isobar or an adiabat. However, there is no inherent physical-chemical relationship between the observed normalized elemental abundances for any particular meteorite parent body (or the bulk silicate Earth) and the relative volatility of the elements. Inference of any such relationship must rely on imposition of a causal model. There is no fundamental reason to expect a relationship between 50% condensation temperatures at any single total pressure and elemental abundances in the BSE or any meteorite parent body.

Examining point (4) in more detail, if equilibrium condensation from a solar composition gas were the *sole* mechanism for establishing elemental abundances in chondrites or in the Earth, in principle there should only be one temperature and one total pressure for which the observed fractions condensed can be reproduced by full equilibrium condensation from a solar composition gas. This approach is analogous to finding a correspondence temperature that matches all gas equilibria in a volcanic gas (e.g., Symonds et al. 1994). But a basic argument indicates that the Earth's volatility trend cannot be matched at any single P, T point.

Consider the temperature where a refractory element is 50% condensed. Then at the same temperature, the fraction condensed of a volatile element with a lower condensation temperature obviously has to be less (or even be zero). Vice versa, at the temperature where the volatile element is 50% condensed, the refractory element would have a fraction condensed larger than 50%. Urey (1954) first identified this problem, which Larimer and Anders (1967) described succinctly: "Vapor pressure is a steep function of temperature; hence a partial depletion can be achieved only over a narrow temperature range, differing from element to element. Outside this range, all-or-nothing fractionations will be the rule, with complete retention or complete loss." Fig. 4 of Palme and Wlotzka (1976) shows an example of vapor pressure control; the more volatile siderophile elements such as Au are simply not present in refractory metal nugget grains condensed at high temperatures.

Accretion of the Earth. The chemical and isotope composition of the bulk Earth is distinctly different from all known meteorite compositions. The major differences are observed in the abundances of the moderately volatile elements relative to the refractory elements (e.g. Palme and O'Neill 2014). CI-chondrite normalized elemental abundances in the bulk silicate Earth (BSE) displayed versus condensation temperatures of the respective elements (Figs. 13 and 14), define a *rough* depletion trend for the volatile elements, e.g., Ebel et al. (2018) computed a r^2 coefficient of 0.33 for the BSE volatility trend. The apparent trend observed for the BSE composition is the result of several chemical processes starting from differentiation during the formation of solids in the solar nebula to planetesimal and planetary accretion to planet-wide differentiation. This path of chemical differentiation from solar nebula to a differentiated planet can be reconstructed from the elemental abundances in the BSE. Traditionally this volatile element depletion trend in the BSE is interpreted as indicating different extents of element condensation during cooling from the hot solar nebula (e.g. McDonough and Sun 1995). But these models do not consider secondary processes that modify elemental abundances in a planetary body after primary accretion, particularly core formation.

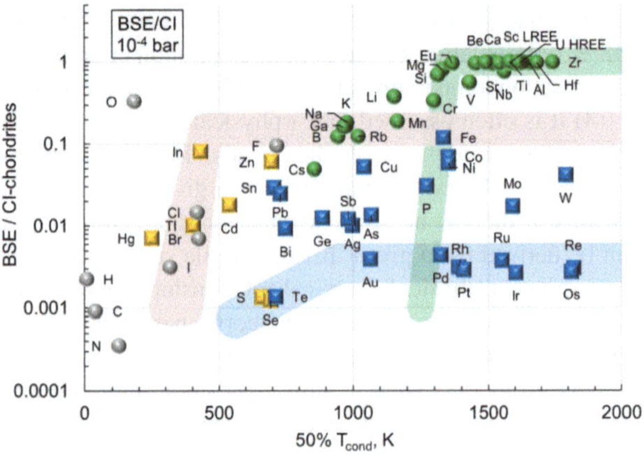

Fig. 15 Deconvolution of the apparent BSE volatility trend, modified after Mezger et al. (2021). The green curve is the volatility trend of the proto-Earth, about 85% of the BSE. The red curve is the volatility trend of the Moon-forming impactor Theia, about 15% of the BSE. The blue curve is the volatility trend of the late veneer, about 0.4% of the BSE

In order to explain the composition of Earth in comparison to different meteorites, qualitative and semi-quantitative accretion models have been presented (e.g., Ringwood 1979; Barshay 1981; Wänke 1981; Wänke et al. 1984; Wänke and Dreibus 1988; Lewis 1988, Lodders 1991, 2000, Lodders and Fegley 1997a; Albarède 2009; Schönbächler et al. 2010; Warren 2011; Liebske and Khan 2019; Sossi et al. 2022). These models use element abundances and/or isotope compositions to derive accretion mechanisms for the Earth and suggest chemical and isotope compositions for these components (e.g. Dauphas 2017). One key result is that the chemistry for the Earth is unlike the composition of any single meteorite group or a simple mixture of different known groups. The Earth has an end member composition for certain element abundances and isotope compositions (e.g., Dauphas et al. 2014; Fischer-Gödde et al., 2013; 2015, Mezger et al. 2020). The accretion models can be separated in two distinct groups. One type of model assumes continuous accretion of planetesimals to achieve Earth's current size (e.g. Sossi et al. 2022). The other model interprets the Earth as the result of planetesimal accretion until ca. 90% of its current size was achieved, and then the last 10% were added, possibly from a different feeding zone, as a giant impactor (e.g. Albarède 2009; Schönbächler et al. 2010; Mezger et al. 2021). Only the latter model accounts for the origin and peculiar composition of the Moon and is the preferred model.

The abundances of different chemical elements in BSE can be estimated with good precision for most elements (McDonough and Sun 1995; Palme and O'Neill 2014; Liebske and Khan 2019). There are somewhat larger uncertainties for the highly incompatible elements (e.g. Mezger et al. 2020) and the highly volatile elements, particularly halogens (e.g., Lodders and Fegley 2023). As shown in Fig. 15, the CI-chondrite normalized element abundances in the BSE define some moderately well-constrained trend when plotted versus condensation temperature. The moderately volatile elements scatter around a slope declining with condensation temperature. A common interpretation of this trend is that it represents different degrees of volatile element depletion; either due to incomplete condensation from the solar nebula (e.g., McDonough and Sun 1995) or later volatile loss during accretion (e.g., Halliday 2004). A consequence of incomplete condensation or partial evaporation is fractionation of stable metal isotopes. However, no such isotope variations are observed for moderately to strongly volatile elements that are depleted in BSE relative to the chondritic reservoir. Thus, based on the lack of stable isotope variations, the volatile element abundances in BSE are not the result of incomplete condensation nor partial evaporation (Mezger et al. 2020, 2021).

The slope of the apparent element depletion trend in the BSE shown by the moderately volatile elements is gentler than that derived from a cooling system where at a point in time the solids are removed from the gas phase, as expected during the initial stages of dust accretion and subsequent planetesimal formation. This relationship is demonstrated by the calculations in Sect. 3. In this case elements that condense above the temperature of dust removal will show chondritic abundances and elements that condense below this threshold temperature could not have condensed significantly. Only elements condensing within a narrow temperature range around the temperature interval where gas and solid fractionate, will condense to different degrees. As demonstrated by Figs. 1 and 2 the element pattern in rocky materials will show a steep depletion trend for volatile elements with the depletion starting around the temperature of gas-solid separation. Elements above this temperature interval are essentially completely condensed, while elements that condense below this interval are depleted (absent) in the solids that form planetesimals. The condensation accretion model of Barshay (1981) found this behavior for rocky elements and volatiles (S, H_2O, and C).

The element abundances in the BSE show this type of steep depletion pattern, if only the lithophile elements, which are unaffected by extraction of a metal or sulfide melt from the silicate reservoir, are considered. Many moderately volatile elements are siderophile and/or chalcophile and their abundances in the BSE are not *primary* signatures of accretion. If only the lithophile elements in the BSE are considered the element pattern shows two distinct groups with lithophile element abundances close to CI chondrite relative abundances (Fig. 15). The first group includes the refractory lithophile elements (green curve) and the second group is defined by the moderately volatile lithophile elements (red curve). The abundances of the MVE are ca. 15% that of the refractory lithophile elements. Thus, lithophile elements in the BSE define a "step" pattern with two distinct flat regions rather than continuously declining element abundances with decreasing condensation temperatures of the relevant elements. This relationship indicates that the BSE consists of two components. The first component, i.e. proto-Earth, had chondritic abundances of refractory elements and was strong depleted in moderately to highly volatile elements. The element pattern in Fig. 15 shows a depletion around the temperature interval where Mg, Si and Fe condensed. The second component that built the Earth was the Moon-forming impactor Theia. This smaller body (ca. 10% the mass of the current Earth, or 15% of the BSE mass) had chondritic abundances of the refractory elements and the moderately volatile elements but was depleted in highly volatile elements (see Fig. 15) (Maltese and Mezger 2020; Mezger et al. 2021). The combination of these two major planetary bodies and subsequent core formation accounts for the element patterns recorded in the present-day bulk silicate Earth. The abundances of the highly siderophile refractory elements as well as S, Se and Te are the result of a late veneer after core-mantle differentiation and account for ca 0.4% of the Earth.

A key feature of the BSE (Fig. 15) is the "step" pattern for elements, with high and chondritic abundances of refractory lithophile elements and low, but also chondritic relative abundances of moderately volatile lithophile elements. The chalcophile and siderophile elements fit these patterns, but their abundances were modified by core and sulfide matte formation. The "step" pattern results from accretion of material extremely depleted in volatile elements, with the end of efficient condensation during cooling of the solar nebula having occurred at different temperatures for different planetary bodies. The precursor material for the proto-Earth included refractory elements with 50% condensation temperatures $\geq T_{50}$ for Mg, Si, and Fe. The precursor material of Theia contained material with T_{50} values down to those of the halogens (Fig. 15). For Vesta such a steep pattern is also recorded with the transition region being around Mn (Fig. 1 in Sossi et al. 2022). Such element patterns are exactly as predicted for incomplete condensation. Due to the fact that the condensation interval for 95% condensation of an element is generally less than 50 K (Fig. 2), and distinct from other elements, flat chondrite normalized element patterns for refractory elements and near absence of volatile elements with a narrow temperature interval with moderate element depletion are expected (see Fig. 15 for the element patterns for proto-Earth and Theia). Mixing of different components with such element patterns gives an overall element depletion pattern with strongly different abundances, particularly of the more volatile elements. Such a mixture also accounts for the lack of stable isotope fractionation of volatile (i.e. depleted) elements in the BSE.

As a consequence, the element abundances in the BSE plotted against their respective condensation temperatures are not the sole result of processes in the solar nebula (i.e. incomplete condensation of the elements) nor planetary accretion processes (i.e. evaporation during accretion) but are a result of mixing of two planetary bodies with distinct element abundances that was followed by core formation. The gentle trend is the result of primary element abundances and the fact that many moderately volatile elements, that define the gentle depletion trend in BSE, were removed at least partially to the Earth`s core during early differentiation events.

The element patterns for the proto-Earth (Fig. 15) contrast with the element patterns of chondrites and in particular carbonaceous chondrites. This difference is possibly a result of dissimilar condensation pathways and processes at different locations in the solar nebula where the solids for these bodies formed. Based on its isotope composition, particularly s-process isotopes, Earth is an endmember among the rocky bodies of the solar system (e.g., Mezger et al. 2020) and represents the innermost material of the solar system available for study. The other planetary bodies, particularly the carbonaceous chondrite parent bodies, formed at greater heliocentric distances. These materials show a gentle decrease of element abundances, particularly for the moderately volatile elements, with decreasing condensation temperature and near chondritic relative abundances of highly volatile elements (Braukmüller et al. 2018, 2019). This gentle decrease in moderately volatile element depletion without concomitant stable isotope fractionation has been interpreted as the result of the change in element abundances in the solar nebula during cooling. The decreasing element abundance that correlates with the condensation temperature is likely due to a reduction of the surface density by photo-evaporation as a function of time and temperature (Vollstaedt et al. 2020). Sengupta et al. (2022) quantified the formation of the depletion patterns of chondrites by modelling open-system loss from high nebula altitudes caused by disk winds originating from the young Sun. This model reproduces the element abundances in differently depleted meteorite classes and accounts for the nearly constant stable isotope composition (no significant mass-dependent fractionation) of heavy elements as a function of the degree of depletion.

The comparison of the element abundances in proto-Earth with those in undifferentiated chondrites reveals two distinct depletion mechanisms for the inner and outer regions for the Solar System. Incomplete condensation of inner Solar System material resulted in strongly volatile element depleted solids; condensation of material at greater heliocentric distance than proto-Earth resulted in gradually depleted element patterns due to mechanical removal of gas from the surface of the cooling solar nebula.

Chalcophile elements in the BSE. The large depletion of the chalcogens S, Se, and Te, which exceeds that of highly siderophile elements (HSE), led to the concept of a Hadean sulfide matte that greatly depleted the abundances of S, Se, Te, and the HSE Re, Os, Ir, Ru, Pt, Rh, Pd, and Au (e.g., O'Neill 1991; Ballhaus et al. 2017). It is possible that the abundances of the volatile chalcophile elements Pb, Cd, In, Tl, and Hg were modified by formation of the sulfide matte and/or by evaporative loss during accretion. Fish et al. (1960) suggested that a natural analog of Soxhlet extraction involving a S-bearing vapor, perhaps S_2, another sulfur allotrope, SiS (g), or another S-bearing gas could transport the chalcophile volatiles Bi, In, Pb, Tl. More recently, Heck (2022) and Steenstra et al. (2023, 2024) studied melt – gas partitioning of moderately volatile chalcophile elements (Cu, Zn, Ge, As, Ag, Cd, In, Pb). But if sulfide – gas partitioning is involved an open system is required, i.e. gravitational escape of the volatilized species, otherwise no depletions are produced (Fegley et al. 2020). Kiseeva and Wood (2013, 2015) studied sulfide – silicate melt partitioning of volatile elements with chalcophile and lithophile behavior (including Cd, In, Pb, Tl) and found the partition coefficients are sensitive to prevailing oxygen and sulfur fugacities. Chalcophile behavior of halogens occurs under some conditions, e.g., iodine in meteoritic troilite (Goles and Anders 1961, 1962, Goswami et al. 1998), the chlorine-bearing mineral djerfisherite [K_6(Cu, Ni, Fe)(Fe, Ni, $Cu)_{24}S_{26}Cl$] in enstatite chondrites and Earth's mantle (Ebel and Sack 2013), partitioning of Cl, Br, and I into sulfide liquids (Lodders 1991; Steenstra et al. 2020). But the application of the melt – vapor and sulfide – silicate melt partitioning work to formation of a Hadean sulfide matte needs to be explored in more detail under appropriate (and possibly variable) fugacities (O_2, S_2, halogen), P, and T.

BSE volatile element inventory. Another question is the assessment of the volatile element budget of the BSE. If condensation (or evaporation) fractionation within a solar gas applies, there should not be a difference in relative abundances of lithophile, siderophile, and chalcophile elements of similar volatility within the *entire* Earth. However, we only have access to reliable estimates about the volatility-related fractionations of lithophile elements for the Earth's surface (atmosphere + hydrosphere + crust) and a part of the upper mantle, which are then used to give bulk silicate Earth abundances (e.g., see Sect. 6.2 of Fegley et al. 2020). But even these data are uncertain to some extent with abundances of the most volatile elements generally being the most uncertain. The elements with the largest ($\geq$40%, or unknown) uncertainties in Palme and O'Neill (2014) are B, F, S, Cl, Cu, Se, Br, Mo, Ag, Sb, Te, I, Cs, Hg, and Bi. With the exception of Mo these elements are moderately or highly volatile elements.

Finally, there is the conceptual issue of when the volatile element inventory of the Earth was established. Using solar condensation temperatures assumes that the building blocks of the Earth obtained their volatile complement by condensation (or evaporation) in equilibrium with the H and He-rich solar gas and not by evaporation from already accreted Earth materials after the moon-forming giant impact or other possible evaporation fractionations during accretion in the absence of H and He rich gas. Condensation (evaporation) temperatures of elements are significantly higher in anhydrous BSE vapor than in H and He-rich solar gas with the exception of some easily oxidized HSE (Ru, Os, Re) that can become much more volatile (Fegley et al. 2023).

Nebular vs. planetary origin of volatile element depletions. The origin of the volatile element depletion in meteorites and rocky planets is a long-standing question. Originally Anders (1964) proposed volatile element depletion resulted from variable mixtures of volatile-rich matrix and volatile-poor chondrules. Anders' two component model is generally discounted as it is now clear chondrules are not poor in all volatiles (cf. Palme et al. 1988). Wasson and colleagues (Wasson and Chou 1974; Wai and Wasson 1977; Wasson 1977) attributed volatile element depletion in chondrites to incomplete condensation as the solar nebula cooled and dissipated. Using potassium stable isotopes, Tian et al. (2021) showed a correlation between volatile loss and planetary size for Earth, Mars, the Moon, and asteroid 4 Vesta, the parent body of EHD meteorites: larger bodies have higher gravity and less volatile loss. This is an attractive model that needs further testing with other MVE isotopes.

Wasson's incomplete condensation model has stimulated the most theoretical work. Cassen (2001) and Humayun and Cassen 2006) showed condensation of MVE into coagulating solids in a cooling solar nebula could reproduce MVE depletions in CO and CV, but not CM, carbonaceous chondrites. Cassen used a relatively quiescent nebula with only radiative heat transport that did not include convective heat and mass transport (Cassen 2001, 1993). Ciesla (2008) reexamined nebular models of MVE depletions and showed that the MVE depletion patterns in carbonaceous chondrites are only reproduced by a narrow range of physical parameters that are inconsistent with astronomical observations of protoplanetary disks around young stars. Ciesla's models had more nebular turbulence than those of Cassen. Vollstaedt et al. (2020) involved photoevaporation of gas from the outer surface of the solar nebula to decrease nebular surface density (and hence opacity, temperature, and pressure) during MVE condensation into grains. They could match the MVE depletions in CM, CO, CV, and CK carbonaceous chondrites with reasonable physical parameters. Sengupta et al. (2022) quantified the formation of the depletion patterns of chondrites by modelling open-system loss from high nebula altitudes caused by disk winds originating from the young Sun. This model reproduces the element abundances in differently depleted meteorite classes and accounts for the nearly constant stable isotope composition (no significant mass-dependent fractionation) of heavy elements as a function of the degree of depletion. But none of the theoretical models explain why MVE depletions (volatility trends) are most evident in carbonaceous but not in ordinary or enstatite chondrites (cf. Figs. 3-11).

6 Summary

The following major points are covered in this paper.

(1) Table 1 gives 50% condensation temperatures (T_{50}) for all naturally occurring elements, Tc, and Pu at 10^{-2} to 10^{-8} bar total pressure for solar composition material.

(2) As shown in the Appendix, condensation temperatures are mainly controlled by the Gibbs energy of condensation reactions and also by the Gibbs energy of ideal mixing if elements (compounds) condense in a solution. The additional Gibbs energy change due to non-ideal solution, i.e., activity coefficients $\neq 1$, is a secondary effect.

(3) The theoretically correct relationships between condensation temperature and fraction condensed (α_M) is derived from mass balance and chemical thermodynamic considerations. These relationships are log $(1-\alpha_M)$ for major elements or log $[(1-\alpha_M)/\alpha_M]$ for trace elements in solid solutions with *inverse* condensation temperatures (K^{-1}).

(4) The maximum amount of element condensed per K^{-1}, i.e., the maximum in $[d\alpha_M/d(1/T)]$ is at the inflection point in the logistic (sigmoid) curve for an element, which is also at (or close to) the 50% condensation temperature.

(5) Plots of normalized elemental abundances versus 50% condensation temperatures (i.e., volatility trends) for meteorites and planets are qualitative indicators of elemental fractionations due to volatility. The popular volatility trend plots (α versus T_{50}) do not use the theoretically correct and quantitative relationship between condensation temperature and fraction condensed.

(6) Volatility trend plots for average elemental abundances in CM, CO, CV, CR, H, L, LL, EH, EL chondrites show different "trends" for moderately and highly volatile elements, which may be linear, curved, a step function, or plateau. A comparison of three abundance sets for CM and CV chondrites shows trends depend on which elements are plotted, which data sources are used, and which temperature range is considered.

(7) Proposed nebular mechanisms for the volatile element depletions in chondrites and the Earth are reviewed.

(8) Some possible implications of volatile element abundances in the BSE are reviewed and a preferred model for accretion of the Earth is presented.

Consideration of the volatility trends in chondrites and the bulk silicate Earth lead to the following questions:

Why do only carbonaceous chondrites show a clear volatility trend?

Why does the BSE apparently show a carbonaceous chondritic-like volatility trend if Zn and other volatile elements originated from non-carbonaceous chondritic material? Or do the trends look alike because terrestrial volatiles were sourced from carbonaceous chondritic material?

Further work is suggested to resolve two controversial issues:

The abundance of fluorine in enstatite chondrites, which looks anomalously high in Figs. 10 and 11, is based mainly on analyses done in the 1960s (Table 2 in Lodders and Fegley 2023). Their discussion of the analytical data suggests it may be useful to remeasure F abundances in EH and EL chondrites using modern methods.

If one assumes a monotonic decrease, indium appears anomalous in the BSE volatility trend (Fig. 13). This is not the case in the preferred model (Fig. 15) where Theia delivers indium. Witt-Eickschen et al. (2009) considered the possibility that indium is either too abundant in the BSE or erroneously too volatile. They propose the apparent problem is resolved if indium has about the same condensation temperature as zinc. This requires an activity coefficient of ca. 0.06 for InS dissolution in FeS at 697 K (10^{-4} bar) where FeS is stable, but S is not yet 50% condensed. As mentioned in the Appendix, compounds such as $FeIn_2S_4$ indicate nonideal dissolution with activity coefficients <1 for InS in FeS. But the absence of activity coefficient data meant ideal solution was assumed to compute T_{50} for indium. The InS activity coefficient in FeS can be measured by electromotive force measurements (e.g., see the discussion of emf measurements in Kubaschewski and Alcock 1979) and then used in condensation calculations.

Appendix

Interpolation of condensation temperatures. Interpolation of 50% condensation temperatures at intermediate pressures in the range of 10^{-2} to 10^{-8} bar is done with linear fit equations:

$$\frac{10^4}{T_{50}} = A + B \log P$$

For example, the 50% condensation temperature equation for oxygen is

$$\frac{10^4}{T_{50}} \text{(oxygen)} = 40.12654 - 3.72440 \log P$$

If desired, the fits can be done with inverse temperature and then the A and B coefficients are correspondingly smaller.

Thermodynamic data sources for pure compounds. Thermodynamic data for over 2000 compounds used in the calculations are listed in Fegley and Lodders (1994). Unless specified otherwise in the discussion, additional compounds and new (or revised) thermodynamic data are from the NIST-JANAF Tables (Chase 1998), the US Geological Survey tabulation of Robie and Hemingway (1995), Belov et al. (1999), Knacke et al. (1991), and the IVTAN code thermodynamic database described by Belov et al. (1999) and Gurvich et al. (1989–1996). As discussed in Fegley et al. (2023), the IVTAN thermodynamic database is the continually updated Russian counterpart to the JANAF Tables. Vapor pressure data for calculating the

50% condensation temperatures of H_2 are from Honig and Hook (1960), from Hultgren et al. (1973a) for liquid He and Ne ice, from Table 1.20 of Lodders and Fegley (1998) for C (as CH_4 ice), and Ar as argon clathrate hydrate ($Ar \cdot 6H_2O$ ice), Marboeuf et al. (2012) for Kr clathrate hydrate, Fray et al. (2010) for Xe clathrate hydrate, and from Haudenschild (1970) for N (as $NH_3 \cdot H_2O$ ice). The 50% condensation temperatures for H_2 are fictive values at constant assumed total pressure, which drops to 43% of the initial value with half H_2 condensed (e.g., see Lewis 1972). Helium never condenses because condensation temperatures for liquid He are below the 3 K cosmic background temperature.

Sodium and Potassium. Sodium and potassium condense into feldspar, which is modeled as a solution of anorthite, albite, and potassium feldspar using activity coefficients for the albite and potassium feldspar components computed from Elkins and Grove (1990). Above 1000 K, the activity coefficient equations for albite (Na) and potassium feldspar (K) are

$$\log \gamma_{Na} = 3.6 - \frac{3530}{T}$$

$$\log \gamma_{K} = 6.3 - \frac{4800}{T}$$

Ideal solution is assumed below 1000 K.

Rubidium. Twenty-three Rb-bearing gases and 18 Rb-bearing solids are included in the calculations. In the vicinity of its 50% condensation temperature as $RbAlSi_3O_8$ dissolved in feldspar, RbCl is the major Rb-bearing gas closely followed by monatomic Rb.

Thermodynamic data for the two Rb aluminosilicates included in the calculations are all estimates. The Rb analog of leucite is considered first. Hirao and Soga (1982) measured the absolute entropies of $MAlSi_2O_6$ where M = K (leucite), Rb, and Cs (pollucite) but their $S°_{298}$ values for leucite and pollucite disagree with those in Robie and Hemingway (1995) and Ogorodova et al. (2003) by large amounts and are incorrect. The $S°_{298}$ was estimated using ionic entropy values from Latimer (1951) and the assumption of $\Delta S_{298} = 0$ for the reaction

$$Rb^+ + KAlSi_2O_6 \text{ (leucite)} = K^+ + RbAlSi_2O_6 \text{(Rb leucite)}$$

The standard enthalpy of formation at 298.15 K was estimated as the mean of the values for leucite and pollucite. The heat capacity for T = 298 K to 1100 K (limited by the melting point of Rb_2SiO_3 at 1143 K) was estimated using a Neumann–Kopp approach ($\Delta C_P = 0$) for the reaction:

$$\begin{aligned} &1/2Rb_2SiO_3 \text{ (metasilicate)} + KAlSi_2O_6 \text{ (leucite)} \\ &\quad = 1/2\ K_2SiO_3 \text{ (metasilicate)} + RbAlSi_2O_6 \text{(Rb leucite)} \end{aligned}$$

The entropy and high temperature heat capacity of $RbAlSi_3O_8$ were estimated in a similar manner. The standard enthalpy of formation at 298.15 K was estimated from the reaction:

$$RbAlSi_2O_6 \text{(Rb leucite)} + SiO_2 \text{ (quartz)} = RbAlSi_3O_8 \text{ (feldspar)} \Delta H^\circ_{298} = -21.5 \text{ kJ mol}^{-1}$$

This $\Delta H°_{298}$ value is in trend with the analogous reactions for Na (dehydrated analcime – albite $\Delta H°_{298} = -29.81$ kJ mol^{-1}) and K (leucite – sanidine $\Delta H°_{298} = -25.67$ kJ mol^{-1}), i.e., about 4 kJ mol^{-1} smaller. For comparison at 1000 K, the estimated equilibrium constant (log K_f value) from this work for formation of $RbAlSi_3O_8$ from the elements is 168.05 versus 167.95 calculated by Wood et al. (2019). Ideal solution 50% condensation temperatures at 10^{-4} bar for Rb are 1033 K (this work), ~1080 K (Wasson 1985 citing Grossman and Larimer 1974), and 800 K (Lodders 2003).

There are no data for $RbAlSi_3O_8$ activity coefficients in feldspar and an estimate ($\log \gamma = 130/T$) was used from comparison of two approaches: (1) activity coefficients for Rb salts in the RbF – KF ($\log \gamma = 130/T$) and RbCl – KCl ($\log \gamma = 80/T$) systems (Sangster and Pelton 1987), and (2) from Beswick's (1973) distribution coefficient (F) measurements of Rb/K ratios in sanidine and chloride vapor assuming $\log \gamma = \log (1/F) = 1146/T - 0.37$. At 1000 K, the estimated values for the Rb feldspar activity coefficient are 1.4 (RbF – KF), 1.2 (RbF – KCl), and 6.0 (Beswick), respectively.

Cesium. Twenty-five Cs-bearing gases and twenty Cs-bearing solids are included in the calculations. In the vicinity of its 50% condensation temperature as $CsAlSi_3O_8$ dissolved in feldspar, CsCl is the major Cs-bearing gas followed by monatomic Cs.

Thermodynamic data for pollucite are from Xu et al. (2001) and Ogorodova et al. (2003). The pollucite thermodynamic data give a Cs vapor pressure in agreement with the Knudsen effusion mass spectrometry measurements of Odoj and Hilpert (1980). Taylor et al. (1989) measured a slightly more negative (by 1%) standard enthalpy of formation from equilibria in aqueous solution at 200 °C. The low temperature adiabatic calorimetry of Bennington et al. (1983) gives a standard entropy <0.5% larger than that of Ogorodova et al. (2003). In contrast to pollucite, thermodynamic data for the Cs analog of feldspar $CsAlSi_3O_8$ had to be estimated (as also done by Wood et al. 2019 but using different methods). The S°_{298} was estimated using ionic entropy values from Latimer (1951) and the assumption of $\Delta S_{298} = 0$ for the reaction

$$Cs^+ + KAlSi_3O_8 \text{ (sanidine)} = K^+ + CsAlSi_3O_8 \text{ (feldspar)}$$

The standard enthalpy of formation at 298.15 K was estimated from the reaction:

$$CsAlSi_2O_6 \text{ (pollucite)} + SiO_2 \text{ (quartz)} = CsAlSi_3O_8 \text{ (feldspar)} \Delta H^\circ_{298} = -21.5 \text{ kJ mol}^{-1}$$

The same ΔH°_{298} value was used for Rb and Cs feldspars for two reasons: (1) the ΔH°_f at 298 K for Rb leucite is an estimate and (2) trends in ΔH°_{298} values for formation of double oxides such as borates, carbonates, chromates, sulfates, and uranates from the constituent oxides show approximately the same values for Rb and Cs within a few kJ mol^{-1}. The heat capacity for T = 298 K to 1100 K (limited by the melting point of Cs_2SiO_3 at 1100 K) was estimated using a Neumann–Kopp approach ($\Delta C_P = 0$) for the reaction:

$$\begin{aligned} &1/2Cs_2SiO_3 \text{ (Cs metasilicate)} + KAlSi_3O_8 \text{ (sanidine)} \\ &= 1/2\, K_2SiO_3 \text{ (K metasilicate)} + CsAlSi_3O_8 \text{ (feldspar)} \end{aligned}$$

The metasilicates are preferable to the oxides for two reasons: (1) the low melting point of Cs_2O (768 K) limits the maximum temperature of the derived C_P equation of Cs feldspar to 768 K, which is lower than the condensation temperature of the Cs feldspar at 10^{-2} and 10^{-4} bar total pressure and (2) the enthalpies of fusion and heat capacities for Cs_2O and K_2O are unknown and are estimates. For comparison this work gives an estimated equilibrium constant ($\log K_f$ value) for formation of $CsAlSi_3O_8$ from the elements of 168.25 versus 168.09 calculated by Wood et al. (2019) at 1000 K.

There are no data for $CsAlSi_3O_8$ activity coefficients in feldspar and an estimate ($\log \gamma = 630/T$) was used from comparison of three approaches: (1) activity coefficients for Cs salts in the CsCl – KCl ($\log \gamma = 630/T$), CsBr – KBr ($\log \gamma = 470/T$), CsI – KI ($\log \gamma = 260/T$)

systems (Sangster and Pelton 1987), (2) from the Cs_2SO_4 activity coefficient (log γ = 650/T) derived from its solubility limit in K_2SO_4 (Levin et al. 1973), and (3) from Eugster's (1955) distribution coefficient (F) measurements of Cs/K ratios in sanidine and vapor assuming log γ = log(1/F) = 1108/T – 0.58. At 1000 K, the estimated values for the Cs feldspar activity coefficient are 4.3 (CsCl – KCl), 3.0 (CsBr – KBr), 1.8 (CsI – KI), 4.5 (Cs_2SO_4 – K_2SO_4), and 3.4 (Eugster), respectively.

Halogens. Fegley and Schaefer (2010) and Lodders and Fegley (2023) describe halogen condensation at 10^{-4} bar total pressure. This work extends calculations to other total pressures. The major F- and Cl-bearing gases at the 50% condensation temperatures are HF and HCl. The major Br-bearing gas at the Br 50% condensation temperature is HBr, which comprises 96.5% (10^{-8} bar) to 99.6% (10^{-2} bar) of all gaseous bromine. Indium monobromide InBr comes second and includes almost all the rest of gaseous bromine (3.5% to 0.4%) over the same pressure range. The major I-bearing gas at the I 50% condensation temperature is HI, which is almost 100% (10^{-2} bar) to 97.95% (10^{-8} bar) of all gaseous iodine. With decreasing pressure, the second most abundant I-bearing gas switches from SnI_2 (10^{-4}% at 10^{-2} bar) to GaI_3 (0.004% at 10^{-4} bar to 2.0% at 10^{-8} bar).

The activity coefficients for NaBr and NaI dissolution in NaCl are from Sangster and Pelton (1987) and are given by the equations:

$$\log \gamma_{\text{NaBr}} = \frac{287}{T} - 0.12$$

$$\log \gamma_{\text{NaI}} = \frac{1100}{T}$$

Indium and Thallium. Indium forms a thiospinel $FeIn_2S_4$ (known as indite) with iron sulfide (Mammedov et al. 2021). Thallium forms several ternary iron sulfides, e.g., $TlFeS_2$, $Tl_3Fe_2S_4$ (Zabel and Range 1979; Welz et al. 1989). Compound formation indicates indium and thallium sulfide dissolution in FeS is nonideal with activity coefficients less than unity. In the absence of any thermodynamic data ideal solution is assumed.

Technetium. Technetium is a siderophile element in group 7 (Mn, Tc, Re, Bh) with aqueous chemistry similar to Re but with a volatility intermediate between Ir (more refractory) and Mo (more volatile). Thermodynamic data for elemental Tc (s, l, g), Tc^+ (g), Tc_2 (g) TcO (g), Tc_2O_7 (g), TcC (s), TcO_2 (s), TcO_3 (s), and Tc_2O_7 (s, l) are from Powers (1991) and Rard et al. (1999). The Tc 50% condensation temperatures in Table 1 are computed for ideal solution in refractory metal alloy and agree within 10 K with those given earlier by Fegley et al. (1993). Monatomic Tc is the major gas and TcO is the second most abundant gas with a Tc/TcO ratio of 5.6×10^{-4} at the 50% condensation temperature of 1570 K at 10^{-4} bar total pressure. The Tc condensation temperature does not depend upon the assumed Tc abundance because the condensation temperatures of trace elements in solid solution do not depend upon their elemental abundances (Equation 8 of Kornacki and Fegley 1986).

Plutonium. Thermodynamic data for elemental Pu (s, l, g), Pu^+ (g), PuO (g), PuO_2 (g), PuO_2 (s), Pu_2O_3 (s), PuN (s), PuS (s), and $PuC_{0.88}$ (s) are from the IVTAN code database and Barin (1989). The Pu condensation temperatures in Table 1 are computed for ideal solution of Pu^{3+} ($PuO_{1.5}$) and Pu^{4+} (PuO^2) in hibonite and perovskite and are within ±2 K of those computed by Kornacki and Fegley (1986) using the METKON code (Palme and Wlotzka 1976; Fegley and Palme 1985).

Gibbs energy of mixing and activity coefficients. Many groups have used activity coefficients to model nonideal solutions of minor (trace) elements in more abundant host phases. Larimer (1973) considered nonideal solution of Pb, Bi, and Tl in Fe metal and of InS and

CdS in FeS. Kelly and Larimer (1977) computed 50% condensation temperatures for nonideal solution of Ga, Ge, and Pd in Fe alloy at high temperature. Wai and Wasson (1977, 1979) computed 50% condensation temperatures for nonideal solution of As, Au, Cu, Ga, Ge, and Sb in Fe alloy and ZnS in FeS. Sears (1978) considered several of the same elements. Boynton (1978) suggested Th, U, and Pu will have activity coefficients >1 in solid solution in perovskite. Davis and Grossman (1979) modeled ideal and nonideal condensation of REE oxides into perovskite and preferred ideal solution models to match observed REE patterns in Ca, Al-rich inclusions. Fegley (1980) calculated nonideal solution condensation of $BaTiO_3$ in perovskite from an assymetric solution Redlich–Kister analysis of the $BaTiO_3$ – $CaTiO_3$ phase diagram of DeVries and Roy (1955). At 5×10^{-3} bar pressure along the Lewis (1974) nebular adiabat, $\gamma(BaTiO_3) = 10$ at 1500 K, the Ba 50% condensation temperature. Fegley and Lewis (1980) calculated nonideal solution condensation temperatures for P as Fe_3P and in metal alloy and for Na and K in feldspar. Fraser and Rammensee (1982) measured activity coefficients in the Fe – Ni – Co system and used the METKON code (Palme and Wlotzka 1976) to compute alloy compositions in equilibrium with solar composition gas. Kornacki and Fegley (1986) computed the effect of variable activity coefficients on condensation of Ba, Sr, Pu, Th, and U dissolved as oxides in perovskite ($CaTiO_3$). Lodders (2003) and this work calculate nonideal solution condensation temperatures for As, Au, Cu, Ga, Ge, and Sb dissolution in Fe alloy using the same activity coefficient data as prior workers (Sears 1978, Wai and Wasson 1977, 1979). Wood et al. (2019) repeated the earlier calculations and also considered nonideal solution of several additional elements in metal alloy, oxide, and silicate host phases. Many of their activity coefficients were estimated from crystal/melt distribution coefficients, the semi-empirical Miedema model for metal alloys, or from a lattice strain model used for oxides, silicates, and sulfides. Before proceeding it is worth quantitatively considering the effects of activity coefficients for the Gibbs energy of condensation reactions.

The relationship between the activity a, activity coefficient γ, and mole fraction X is

$$a = \gamma X$$

The activity coefficient is unity for an ideal solution and is either <1 or >1 for nonideal solutions. Negative deviations from ideality are indicated by e.g., compound formation, mean γ <1 and higher condensation temperatures than ideal solution. Positive deviations from ideality are indicated by e.g., unmixing into two separate phases, mean γ >1 and lower condensation temperature than ideal solution. The size of the activity coefficient correction term (excess Gibbs energy)

$$RT \ln \gamma = \Delta \overline{\mathrm{G}}^E = \Delta \overline{\mathrm{H}}^E - T \Delta \overline{\mathrm{S}}^E$$

relative to the ΔG of condensation determines the importance of the activity coefficient term for the condensation temperature of a minor (trace) element (compound) dissolved in a host phase. For example, considering reaction (1) for M (gas) = M(alloy) we can write

$$\Delta \mathrm{G}_1 = \Delta \mathrm{H}_1 - T \Delta \mathrm{S}_1$$

Where the enthalpy and entropy of reaction are the same as previously defined in Eq. (12). With apologies to thermodynamic notation purists, we use simplified notation that is easier for most readers to understand. All quantities are on a molar basis. We can also write

$$\Delta \mathrm{G}_1 = \Delta \mathrm{G}^o_{\mathrm{pure}} + \Delta \overline{\mathrm{G}}^I + \Delta \overline{\mathrm{G}}^E$$

The Gibbs energy of condensation for reaction (1) is the sum of three Gibbs energy terms: that for the pure element (or compound) M, the ideal (I) solution of mixing term, and the non-ideal (E, excess) solution of mixing term. (The degree sign superscript denotes the pure element (or compound) at one bar pressure and is the value listed in the JANAF Tables or another thermodynamic data compilation. The overbars denote partial molal quantities, e.g., for element M in an alloy.) In general (e.g., see arguments by Ebel 2005; Ebel and Grossman 2005; Fegley et al. 2023 for silicate melts),

$$\Delta G_1 > \Delta \overline{G}^E$$

The ideal solution of mixing term is the partial molal ideal solution of mixing and is defined as

$$\Delta \overline{G}^I = RT \ln X_M$$

Taking Fe – a major element – as an example, at 10^{-4} bar total pressure, pure Fe metal is 50% condensed at 1331 K and Fe metal in ideal solution in metal alloy is 50% condensed at 1334 K, only 3 K higher. At 1334 K, $\Delta G^{\circ}_{pure} = -217{,}100$ J/mol for the reaction

$$\text{Fe (gas)} = \text{Fe (metal)}$$

The partial molal ideal solution of mixing of Fe in the metal alloy (with $X_{Fe} = 0.913$) is $\Delta \overline{G}^I = -1010$ J/mol for dissolution of pure Fe metal into the alloy

$$\text{Fe (metal)} = \text{Fe (alloy)}$$

and $RT\ln\gamma_{Fe} = -11$ J/mol from Fraser and Rammensee (1982) to account for

$$\text{Fe (ideal alloy)} = \text{Fe (nonideal alloy)}$$

(For reference, the other constituents of the alloy are primarily Ni and Co with mole fractions of $X_{Ni} = 0.083$ and $X_{Co} = 0.00396$.) This comparison shows that the ideal solution term is an insignificant part (0.5%) of the total Gibbs energy term. This result applies generally to chalcophile, lithophile, and siderophile elements (and compounds) that are major components of a solution, not only to iron metal. The 50% condensation temperature of a pure element (compound) is not greatly different from that of the 50% condensation temperature for the element (compound) that is a major component dissolved in an ideal solution. However, this is not the case for minor or trace elements dissolved in a host phase, which are considered next.

Palladium condensation illustrates the effect of ideal mixing and an activity coefficient <1 on condensation temperatures of a minor (trace) element dissolved in solution. Pure Pd metal is 50% condensed at 901 K at 10^{-4} bar total pressure. The 50% condensation temperature increases to 1321 K for ideal solution of Pd in Fe, Ni-rich metal alloy. Nonideal solution using the "most likely" $\gamma = 0.4$ increases the Pd condensation temperature to 1342 K (Palme et al. 1988). The Pd activity coefficient used by Palme et al. (1988) is the infinite dilution value for γ-Fe – Pd metal alloy from Hultgren et al. (1973a,b). At 1342 K, $\Delta G^{\circ}_{pure} = -209{,}720$ J/mol, the ideal solution term is $RT\ln X_{Pd} = -141{,}900$ J/mol, and the activity coefficient term $RT\ln\gamma = -10{,}220$ J/mol. The activity coefficient term is 2.8% of the total Gibbs energy term (−361,840 J/mol).

Thoria (ThO_2) condensation in a host phase such as hibonite or perovskite exemplifies the effect of an activity coefficient >1 on condensation temperatures. The results of Grossman

(1973) and Kornacki and Fegley (1986) are used for illustration. Grossman (1973) calculated pure ThO_2 to condense at 1496 K at 10^{-3} bar total pressure. CONDOR code calculations confirm that value, give a 50% condensation temperature of 1478 K, and show ThO_2 gas is 99.99% of total gaseous thorium. The condensation reaction is

$$ThO_2 \text{ (gas)} = ThO_2 \text{ (thoria)}$$

Kornacki and Fegley (1986) computed that ideal solution of ThO_2 in perovskite increases the ThO_2 50% condensation temperature to 1676 K at 10^{-3} bar, where perovskite first appears. Using ThO_2 activity coefficients >1, as argued by Boynton (1978), they computed lower 50% condensation temperatures (their Table 4). Taking $\gamma = 50$ for this example, ThO_2 is 50% condensed at 1675.4 K. The ΔG values for the condensation reaction of pure thoria, ideal solution of ThO_2 in perovskite, and RTlnγ for ThO_2 are below:

$$\Delta G_1 = \Delta G^o_{\text{pure}} + RT\,X_{ThO_2} + RT\ln\gamma_{ThO_2}$$

$$\Delta G_1 = -436{,}830 - 152{,}110 + 54{,}490$$

The activity coefficient term is about 10% of the total Gibbs energy (−534,450 J/mol) and opposite in sign to the other two terms. The positive RTlnγ decreases the 50% condensation temperature but the effect is tiny for ThO_2 because it is so refractory with a large negative ΔG of condensation. Uranium dioxide condensation is another example, and, in this case, the same size activity coefficient gives a larger decrease in the 50% condensation temperature because UO_2 is less refractory than thoria (Kornacki and Fegley 1986).

Acknowledgements Work at Washington University is supported by NSF Astronomy Program Grant AST-2108172 and by the McDonnell Center for the Space Sciences. All authors thank the International Space Science Institute for the invitation to the workshop, and Andy Davis, Dominik Hezel (editor), Jeff Cuzzi, and an anonymous referee for their constructive comments. BF thanks JS Lewis for introducing him to "condensation calculations" in the 1970s, and BF and KL also thank Herbert Palme for discussions about chemical thermodynamics and nebular condensation over the past 40+ years.

Funding Information Open access funding provided by University of Bern.

Declarations

Competing Interests The authors declare that they have no conflict of interest.

References

Albarède F (2009) Volatile accretion history of the terrestrial planets and dynamic implications. Nature 461:1227–1233

Alexander CMOD (2019a) Quantitative models for the elemental and isotopic fractionations in chondrites: the non-carbonaceous chondrites. Geochim Cosmochim Acta 254:246–276

Alexander CMOD (2019b) Quantitative models for the elemental and isotopic fractionations in chondrites: the carbonaceous chondrites. Geochim Cosmochim Acta 254:277–309
Anders E (1964) Origin, age, and composition of meteorites. Space Sci Rev 3:583–714
Anders E (1975) On the depletion of moderately volatile elements in chondrites. Meteoritics 10:283–286
Anders E (1977) Critique of "Nebular condensation of moderately volatile elements and their abundances in ordinary chondrites" by Chien M. Wai and John T. Wasson. Earth Planet Sci Lett 36:14–20
Avice G, Marty B (2014) The iodine – plutonium – xenon age of the Moon – Earth system revisited. Philos Trans R Soc Lond A 372:20130260
Ballhaus C, Fonseca ROC, Münker C, Rohrbach A, Nagel T, Speelmans IM, Helmy HM, Zirner A, Vogel AK, Heuser A (2017) The great sulfur depletion of Earth's mantle is not a signature of mantle – core equilibration. Contrib Mineral Petrol 172:68
Barin I (1989) Thermochemical Data of Pure Substances. 2 vols. VCH, Weinheim
Barshay SS (1981) Combined Condensation – Accretion Models of the Terrestrial Planets. PhD thesis, MIT, Cambridge MA
Belov GV, Iorish VS, Yungman VS (1999) IVTANTHERMO for Windows – Database on thermodynamic properties and related software. Calphad 23:173–180
Bennington KO, Beyer RP, Johnson GK (1983) Thermodynamic Properties of Pollucite (A Cesium – Aluminum Silicate). U.S. Bureau of Mines Report of Investigations 8779, 29pp
Beswick AE (1973) An experimental study of alkali metal distributions in feldspars and micas. Geochim Cosmochim Acta 37:183–208
Boynton WV (1975) Fractionation in the solar nebula: condensation of yttrium and the rare Earth elements. Geochim Cosmochim Acta 39:569–584
Boynton WV (1978) Fractionation in the solar nebula, II. Condensation of Th, U, Pu and Cm. Earth Planet Sci Lett 40:63–70
Braukmüller N, Wombacher F, Hezel DC, Escoube R, Münker C (2018) The chemical composition of carbonaceous chondrites: implications for volatile element depletion, complementarity and alteration. Geochim Cosmochim Acta 239:17–48
Braukmüller N, Wombacher F, Funk C, Münker C (2019) Earth's volatile element depletion pattern inherited from a carbonaceous chondrite-like source. Nat Geosci 12:564–569
Cameron AGW (1995) The first ten million years in the solar nebula. Meteorit Planet Sci 30:133–161
Cameron AGW, Fegley MB (1982) Nucleation and condensation in the primitive solar nebula. Icarus 52:1–13
Cassen P (1993) Why convective heat transport in the solar nebula was inefficient. Lunar Planet Sci XXIV:261–262
Cassen P (2001) Nebular thermal evolution and the properties of primitive planetary materials. Meteorit Planet Sci 36:671–700
Chase MW (1998) NIST-JANAF Thermochemical Tables, 4th ed., 2 vols., J Phys Chem Ref Data Monograph NIST and American Institute of Physics
Ciesla FJ (2008) Radial transport in the solar nebula: implications for moderately volatile element depletions in chondritic meteorites. Meteorit Planet Sci 43:639–655
Dauphas N (2017) The isotopic nature of the Earth's accreting material through time. Nature 541:521–524
Dauphas N, Pourmand A (2015) Thulium anomalies and rare Earth element patterns in meteorites and Earth: nebular fractionation and the nugget effect. Geochim Cosmochim Acta 163:234–261
Dauphas N, Burkhardt C, Warren PH, Teng F-Z (2014) Geochemical arguments for an Earth-like Moon-forming impactor. Philos Trans R Soc Lond A 372:20130244
Davis AM, Grossman L (1979) Condensation and fractionation of rare earths in the solar nebula. Geochim Cosmochim Acta 43:1611–1632
DeVries RC, Roy R (1955) Phase equilibria in the system $BaTiO_3$ – $CaTiO_3$. J Am Ceram Soc 38:142–146
Ebel DS (2005) Model evaporation of FeO-bearing liquids: applications to chondrules. Geochim Cosmochim Acta 69:3183–3193
Ebel DS (2006) Condensation of rocky material in astrophysical environments. In: Lauretta D, McSween HY Jr (eds) Meteorites and the Early Solar System II. University of Arizona Press, Tucson, pp 253–277
Ebel DS (2023) Condensation calculations in planetary science and cosmochemistry. In: The Oxford Research Encyclopedia of Planetary Science. Oxford University Press.
Ebel DS, Grossman L (2000) Condensation in dust-enriched systems. Geochim Cosmochim Acta 64:339–366
Ebel DS, Grossman L (2005) Spinel-bearing spherules condensed from the Chicxulub impact-vapor plume. Geology 33:293–296
Ebel DS, Sack RO (2013) Djerfisherite: nebular source of refractory potassium. Contrib Mineral Petrol 166:923–934
Ebel DS, Weisberg MK, Beckett JR (2012) Thermochemical stability of low-iron, manganese-enriched olivine in astrophysical environments. Meteorit Planet Sci 47:585–593

Ebel DS, Alexander CMOD, Libourel G (2018) Vapor-melt exchange – constraints on chondrite formation conditions and processes. In: Russell S, Connolly HC Jr, Krot AN (eds) Chap. 6 in chondrules: records of protoplanetary disk processes. Cambridge University Press, Cambridge, pp 151–174

Elkins LT, Grove TL (1990) Ternary feldspar experiments and thermodynamic models. Am Mineral 75:544–559

Ellingham HJT (1944) Reducibility of oxides and sulphides in metallurgical processes. J Soc Chem Ind, Lond 63:125–133

Eugster HP (1955) The cesium – potassium equilibrium in the system sanidine – water, Carnegie Inst. Washington Yearbook No. 54 (1954–1955), pp 112–113

Fegley MB Jr (1980) Chemical Fractionations in Solar Composition Material. PhD thesis, MIT. https://dspace.mit.edu/handle/1721.1/52853

Fegley B Jr (1983) Primordial retention of nitrogen by terrestrial planets and meteorites. Proc 13th Lunar Planet Sci Conf J Geophys Res 88:A853–A868

Fegley B Jr (2000) Kinetics of gas-grain reactions in the solar nebula. Space Sci Rev 92:177–200

Fegley B Jr, Kornacki AS (1984) The origin and mineral chemistry of Group II inclusions in carbonaceous chondrites. Lunar Planet Sci XV:262–263

Fegley B Jr, Lewis JS (1980) Volatile element chemistry in the solar nebula: Na, K, F, Cl, and P. Icarus 41:439–455

Fegley B Jr, Lodders K (1994) Chemical models of the deep atmospheres of Jupiter and Saturn. Icarus 110:117–154

Fegley B Jr, Palme H (1985) Evidence for oxidizing conditions in the solar nebula from Mo and W depletions in refractory inclusions in carbonaceous chondrites. Earth Planet Sci Lett 72:311–326

Fegley B Jr, Prinn RG (1989) Solar nebula chemistry: implications for volatiles in the Solar System. In: Weaver HA, Danly L (eds) The formation and evolution of planetary systems. Cambridge University Press, Cambridge, pp 171–211

Fegley B Jr, Schaefer L (2010) Cosmochemistry. In: Principles and perspectives in cosmochemistry. Springer, Berlin, pp 347–377

Fegley B Jr, Lodders K, Palme H (1993) Technetium chemistry in the solar nebula. Meteorit Planet Sci 28:346–347

Fegley B Jr, Lodders K, Jacobson NS (2020) Volatile element chemistry during accretion of the Earth. Geochemistry 80:125594

Fegley B Jr, Lodders K, Jacobson NS (2023) Chemical equilibrium calculations for bulk silicate Earth material at high temperatures. Geochemistry 83:125961

Fischer-Gödde M, Kleine T (2017) Ruthenium isotopic evidence for an inner Solar System origin of the late veneer. Nature 541:525–527

Fish RA, Goles GG, Anders E (1960) The record in the meteorites III. On the development of meteorites in asteroidal bodies. Astrophys J 132:243–258

Fraser DG, Rammensee W (1982) Activity measurements by Knudsen cell mass spectrometry – the system Fe-Co-Ni and implications for condensation processes in the solar nebula. Geochim Cosmochim Acta 46:549–556

Fray N, Marboeuf U, Brissaud O, Schmitt B (2010) Equilibrium data of methane, carbon dioxide, and xenon clathrate hydrates below the freezing point of water. Applications to astrophysical environments. J Chem Eng Data 55:5101–5108

Gast PW (1962) The isotopic composition of strontium and the age of stone meteorites – I. Geochim Cosmochim Acta 26:927–943

Goldschmidt VM (1923) Geochemische Verteilungsgesetze der Elemente. Skr Nor Vidensk-Akad Oslo, I Mat-Naturv Kl 3:3–17

Goles GG (1971a) Sodium. In: Mason B (ed) Handbook of elemental abundances in meteorites. Gordon & Breach, New York, pp 109–113

Goles GG (1971b) Potassium. In: Mason B (ed) Handbook of elemental abundances in meteorites. Gordon & Breach, New York, pp 149–169

Goles GG, Anders E (1961) On the geochemical character of iodine in meteorites. J Geophys Res 66:3075–3077

Goles GG, Anders E (1962) Abundances of iodine tellurium and uranium in meteorites. Geochim Cosmochim Acta 26:723–737

Goswami JN, Sahijpal S, Kehm K, Hohenberg CM, Swindle T, Grossman JN (1998) In situ determination of iodine content and iodine-xenon systematics in silicates and troilite phases in chondrules from the LL3 chondrite Semarkona. Meteorit Planet Sci 33:527–534

Grossman L (1972) Condensation in the primitive solar nebula. Geochim Cosmochim Acta 36:597–619

Grossman L (1973) Refractory trace elements in Ca-Al-rich inclusions in the Allende meteorite. Geochim Cosmochim Acta 37:1119–1140

Grossman L, Larimer JW (1974) Early chemical history of the Solar System. Rev Geophys Space Phys 12:71–101
Grossman L, Olsen E (1974) Origin of the high-temperature fraction of C2 chondrites. Geochim Cosmochim Acta 38:173–187
Gurvich LV, Veyts IV, Alcock CB (eds) (1989–1996) Thermodynamic Properties of Individual Substances. 4th ed., 3 parts in 6 vols. Hemisphere Publishing, New York
Halliday AN (2004) Mixing, volatile loss and compositional change during impact-driven accretion of the Earth. Nature 427:505–509
Haudenschild C (1970) Multi-phase ammonia-water system (Rev 1). JPL Space Programs Summary 37-66 vol III, pp 4–9
Heck W (2022) Experimental Degassing of Moderately Volatile Chalcophile Elements from Silicate Melt. MS Thesis, Tulane University
Hirao K, Soga N (1982) The heat capacity and a phase transition in leucite-type compounds. Yogyo Kyokai Shi 90:390–396
Holzheid A, O'Neill HStC (1995) The $Cr - Cr_2O_3$ oxygen buffer and the free energy of formation of Cr_2O_3 from high-temperature electrochemical measurements. Geochim Cosmochim Acta 59:475–479
Honig RE, Hook HO (1960) Vapor pressure data for some common gases. RCA Rev 21:360–368
Hudson GB, Kennedy BM, Podosek FA, Hohenberg CM (1989) The early Solar System abundance of ^{244}Pu as inferred from the St. Severin chondrite. Proc 19th Lunar Planet Sci Conf 19:547–557
Hultgren R, Desai PD, Hawkins DT, Gleiser M, Kelley KK (1973b) Selected values of the thermodynamic properties of binary alloys. American Society for Metals
Hultgren R, Desai PD, Hawkins DT, Gleiser M, Kelley KK, Wagman DD (1973a) Selected values of the thermodynamic properties of the elements. American Society for Metals
Humayun M, Cassen P (2006) Processes determining the volatile abundances of the meteorites and terrestrial planets. In: Canup RM, Righter K (eds) Origin of the Earth and Moon. University of Arizona Press, Tucson, pp 3–23
Ireland TR, Fegley B Jr (2000) The Solar System's earliest chemistry: systematics of refractory inclusions. Intl Geol Rev 42:865–894
Kargel JS, Lewis JS (1993) The composition and early evolution of Earth. Icarus 105:1–25
Kellogg HH (1966) Vaporization chemistry in extractive metallurgy. Trans Metall Soc AIME 236:602–615
Kelly WR, Larimer JW (1977) Chemical fractionations in meteorites – VIII. Iron meteorites and the cosmochemical history of the metal phase. Geochim Cosmochim Acta 41:93–111
Kiseeva ES, Wood BJ (2013) A simple model for chalcophile element partitioning between sulphide and silicate liquids with geochemical implications. Earth Planet Sci Lett 383:68–81
Kiseeva ES, Wood BJ (2015) The effects of composition and temperature on chalcophile and lithophile element partitioning into magmatic sulphides. Earth Planet Sci Lett 424:280–292
Knacke O, Kubaschewski O, Hesselmann K (1991) Thermochemical Data of Pure Substances, 2nd ed., 2 vols. Springer, New York
Kornacki AS, Fegley B Jr (1986) The abundance and relative volatility of refractory trace elements in Allende Ca, Al-rich inclusions: implications for chemical and physical processes in the solar nebula. Earth Planet Sci Lett 79:217–234
Krahenbühl U, Morgan JW, Ganapathy R, Anders E (1973) Abundance of 17 trace elements in carbonaceous chondrites. Geochim Cosmochim Acta 37:1353–1370
Kubaschewski O, Alcock CB (1979) Metallurgical thermochemistry, 5th edn. Pergamon, Elmsford
Larimer JW (1967) Chemical fractionations in meteorites – I. Condensation of the elements. Geochim Cosmochim Acta 31:1215–1238
Larimer JW (1973) Chemical fractionations in meteorites – VII. Cosmothermometry and cosmobarometry. Geochim Cosmochim Acta 37:1603–1623
Larimer JW (1988) The cosmochemical classification of the elements. In: Kerridge JF, Matthews MS (eds) Meteorites and the early Solar System. University of Arizona Press, Tucson, pp 375–389
Larimer JW, Anders E (1967) Chemical fractionations in meteorites – II. Abundance patterns and their interpretation. Geochim Cosmochim Acta 31:1239–1270
Larimer JW, Wasson JT (1988) Refractory lithophile elements. In: Kerridge JF, Matthews MS (eds) Meteorites and the early Solar System, pp 390–415
Latimer WM (1951) Methods of estimating the entropies of solid compounds. J Am Chem Soc 73:1480–1482
Lauretta DS, Lodders K (1997) The cosmochemical behavior of beryllium and boron. Earth Planet Sci Lett 146:315–327
Lee MR, Smith CL, Gordon SH, Hodson ME (2006) Laboratory simulation of terrestrial meteorite weathering using the Bensour (LL6) ordinary chondrite. Meteorit Planet Sci 41:1123–1138
Levin EM, Benedict JT, Sciarello JP, Monsour S (1973) The system $K_2SO_4 - Cs_2SO_4$. J Am Ceram Soc 56:427–430

Lewis JS (1972a) Metal/silicate fractionation in the Solar System. Earth Planet Sci Lett 15:286–290
Lewis JS (1972b) Low temperature condensation from the solar nebula. Icarus 16:241–252
Lewis JS (1974) The temperature gradient in the solar nebula. Science 186:440–443
Lewis JS (1988) Origin and composition of Mercury. In: Vilas F, Chapman CR, Matthews MS (eds) Mercury. Univ. of Arizona Press, Tucson, pp 651–666
Lewis JS, Prinn RG (1980) Kinetic inhibition of CO and N_2 reduction in the solar nebula. Astrophys J 238:357–364
Lewis JS, Prinn RG (1984) Planets and their atmospheres: origin and evolution. Academic Press, San Diego
Lewis JS, Barshay SS, Noyes B (1979) Primordial retention of carbon by the terrestrial planets. Icarus 37:190–206
Liebske C, Khan A (2019) On the principal building blocks of Earth and Mars. Icarus 322:121–134
Lipschutz ME, Woolum DS (1988) Highly labile elements. In: Kerridge JF, Matthews MS (eds) Meteorites and the early Solar System. University of Arizona Press, Tucson, pp 462–487
Lodders K (1991) Spurenelementverteilung zwischen Sulfid and Silikatschmelze und Kosmochemische Anwendungen. Ph.D. thesis, Univ. Mainz, Germany
Lodders K (2000) An oxygen isotope mixing model for the accretion and composition of rocky planets. Space Sci Rev 92:341–354
Lodders K (2003) Solar System abundances and condensation temperatures of the elements. Astrophys J 591:1220–1247
Lodders K (2020) Solar elemental abundances. In: The Oxford research encyclopedia of planetary science. Oxford University Press, London
Lodders K (2021) Relative atomic Solar System abundances, mass fractions, and atomic masses of the elements and their isotopes, composition of the solar photosphere, and compositions of the major chondrite groups. Space Sci Rev 217:44
Lodders K, Fegley B Jr (1993) Lanthanide and actinide chemistry at high C/O ratios in the solar nebula. Earth Planet Sci Lett 117:125–145
Lodders K, Fegley B Jr (1997a) An oxygen isotope model for the composition of Mars. Icarus 126:373–394
Lodders K, Fegley B Jr (1997b) Condensation chemistry of carbon stars. In: Bernatowicz TJ, Zinner EK (eds) Astrophysical implications of the laboratory study of presolar materials. AIP, New York, pp 391–423
Lodders K, Fegley B Jr (1998) The planetary scientist's companion. Oxford University Press, London
Lodders K, Fegley B Jr (2023) Solar System abundances and condensation temperatures of the halogens fluorine, chlorine, bromine, and iodine. Geochemistry 83:125957
Lord HC III (1965) Molecular equilibria and condensation in a solar nebula and cool stellar atmospheres. Icarus 4:279–288
Maltese A, Mezger K (2020) The Pb isotope evolution of Bulk Silicate Earth: constraints from its accretion and early differentiation history. Geochim Cosmochim Acta 271:179–193
Mammedov FM, Babanly DM, Amiraslanov IR, Tagiev DB, Babanly MB (2021) FeS – Ga_2S_3 – In_2S_3 system. Russ J Inorg Chem 66:1533–1543
Marboeuf U, Fray N, Brissaud O, Schmitt B, Bockelée-Morvan D, Gautier D (2012) Equilibrium pressure of ethane, acetylene, and krypton clathrate hydrates below the freezing point of water. J Chem Eng Data 57:3408–3415
Mason B (1979) Data of Geochemistry chapter B. Cosmochemistry, Part 1. Meteorites. U.S. Geological Survey Professional Paper 440-B-1, pp. B1-B132
McDonough WF, Sun S-S (1995) The composition of the Earth. Chem Geol 120:223–253
Mezger K, Schönbächler M, Bouvier A (2020) Accretion of the Earth—missing components? Space Sci Rev 216:1–24
Mezger K, Maltese A, Vollstaedt H (2021) Accretion and differentiation of early planetary bodies as recorded in the composition of the silicate Earth. Icarus 365:114497
Odoj R, Hilpert K (1980) Mass spectrometric study of the evaporation of crystalline compounds in the Cs_2O – Al_2O_3 – SiO_2 system. 1. The synthetic compound $CsAlSi_2O_6$ and the mineral pollucite. High Temp, High Press 12:93–98
Ogorodova LP, Melchakova LV, Kiseleva IA, Belitsky IA (2003) Thermochemical study of natural pollucite. Thermochim Acta 403:251–256
O'Neill HStC (1991) The origin of the Moon and the early history of the Earth – a chemical model. Part 2: the Earth. Geochim Cosmochim Acta 55:1159–1172
Palme H, O'Neill HStC (2014) Cosmochemical estimates of mantle composition. Treatise on geochemistry 2nd edn. Elsevier, Amsterdam
Palme H, Wlotzka F (1976) A metal particle from a Ca, Al-rich inclusion from the meteorite Allende, and the condensation of refractory siderophile elements. Earth Planet Sci Lett 33:45–60
Palme H, Larimer JW, Lipschutz ME (1988) Moderately volatile elements. In: Kerridge JF, Matthews MS (eds) Meteorites and the early Solar System, pp 436–461

Petaev MI, Wood JA (1998) The condensation with partial isolation (CWPI) model of condensation in the solar nebula. Meteorit Planet Sci 33:1123–1137

Petaev MI, Wood JA, Meibom A, Krot AN, Keil K (2003) The ZONMET thermodynamic and kinetic model of metal condensation. Geochim Cosmochim Acta 67:1737–1751

Powers DA (1991) The thermodynamic properties of technetium. High Temp Sci 31:105–120

Rard JA, Rand MH, Anderegg G, Wanner H, Sandino MCA, Östhols E (1999) Chemical thermodynamics of technetium. OECD Publishing, Paris

Ringwood AE (1979) The origin of the Earth and Moon. Springer, New York

Ringwood AE, Kesson SE (1977) Basaltic magmatism and the bulk composition of the Moon. II. Siderophile and volatile elements in Moon, Earth and chondrites: implications for lunar origin. Moon 16:425–464

Robie RA, Hemingway BS (1995) Thermodynamic Properties of Minerals and Related Substances at 298.15 K and 1 Bar (105 Pascals) Pressure and at Higher Temperatures. U.S. Geological Survey Bulletin No. 2131, U.S. Government Printing Office, Washington DC

Ruden SP, Lin DNC (1986) The global evolution of the primordial solar nebula. Astrophys J 308:883–901

Sangster J, Pelton AD (1987) Phase diagrams and thermodynamic properties of the 70 binary alkali halide systems having common ions. J Phys Chem Ref Data 16:509–561

Sarafian AR, Nielsen SG, Marschall HR, Gaetani GA, Hauri EH, Righter K, Sarafian E (2017) Angrite meteorites record the onset and flux of water to the inner Solar System. Geochim Cosmochim Acta 212:156–166

Saxena SK, Eriksson G (1983a) Low- to medium-temperature phase equilibria in a gas of solar composition. Earth Planet Sci Lett 65:7–16

Saxena SK, Eriksson G (1983b) High temperature phase equilibria in a solar-composition gas. Geochim Cosmochim Acta 47:1865–1874

Schönbächler M, Carlson R, Horan M, Mock T, Hauri EH (2010) Heterogeneous accretion and the moderately volatile element budget of the Earth. Science 328:884–887

Sears DW (1978) Condensation and the composition of iron meteorites. Earth Planet Sci Lett 41:128–138

Sears DW (1980) Formation of E chondrites and aubrites – a thermodynamic model. Icarus 43:184–202

Sengupta D, Estrada PR, Cuzzi JN, Humayun M (2022) Depletion of moderately volatile elements by open-system loss in the early solar nebula. Astrophys J 932:82

Sossi PA, Stotz IL, Jacobson SA, Morbidelli A, O'Neill HStC (2022) Stochastic accretion of the Earth. Nat Astron 6:951–960

Steenstra ES, van Haaster F, van Mulligan R, Flemetakis S, Berndt J, Klemme S, van Westrenen W (2020) An experimental assessment of the chalcophile behavior of F, Cl, Br and I: implications for the fate of halogens during planetary accretion and the formation of magmatic ore deposits. Geochim Cosmochim Acta 273:275–290

Steenstra ES, Renggli CJ, Berndt J, Klemme S (2023) Evaporation of moderately volatile elements from metal and sulfide melts: implications for volatile element abundances in magmatic iron meteorites. Earth Planet Sci Lett 622:118406

Steenstra ES, Renggli CJ, Berndt J, Klemme S (2024) Quantification of evaporative loss of volatile metals from planetary cores and metal-rich planetesimals. Geochim Cosmochim Acta 384:93–110

Stracke A, Palme H, Gellissen M, Münker C, Kleine T, Birbaum K, Günther D, Bourdon B, Zipfel J (2012) Refractory element fractionation in the Allende meteorite: implications for solar nebula condensation and the chondritic composition of parent bodies. Geochim Cosmochim Acta 85:114–141

Sun S-S (1982) Chemical composition and origin of the Earth's primitive mantle. Geochim Cosmochim Acta 46:179–192

Symonds RB, Rose WI, Bluth GJS, Gerlach TM (1994) Volcanic-gas studies: methods, results, and applications. In: Carroll MR, Holloway JR (eds) Volatiles in magmas. Mineralogical Society of America, Washington, pp 1–66

Takahashi H, Gros J, Higuchi H, Morgan JW, Anders E (1978) Volatile elements in chondrites: metamorphism or nebular fractionation? Geochim Cosmochim Acta 42:1859–1869

Tanaka T, Masuda A (1973) Rare-Earth elements in matrix, inclusions, and chondrules of the Allende meteorite. Icarus 19:523–530

Taylor P, DeVaal SD, Owen DG (1989) Stability relationships between solid cesium aluminosilicates in aqueous solutions at 200 °C. Can J Chem 67:76–81

Tian Z, Magna T, Day JMD, Mezger K, Scherer EE, Lodders K, Hin RC, Koefoed P, Bloom H, Wang K (2021) Potassium isotope composition of Mars reveals a mechanism of planetary volatile retention. Proc Natl Acad Sci USA 118:e2101155118

Timmermann A, Shan Y, Reiners A, Pack A (2023) Revisiting equilibrium condensation and rocky planet compositions. Astron Astrophys 676:A52

Urey HC (1952) The planets: their origin and development. Yale University Press, New Haven

Urey HC (1954) On the dissipation of gas and volatilized elements from protoplanets. Astrophys J Suppl Ser 1:147–173
Vollstaedt H, Mezger K, Allibert Y (2020) Carbonaceous chondrites and the condensation of elements from the solar nebula. Astrophys J 897:82
Wai CM, Wasson JT (1977) Nebular condensation of moderately volatile elements and their abundances in ordinary chondrites. Earth Planet Sci Lett 36:1–13
Wai CM, Wasson JT (1979) Nebular condensation of Ga, Ge and Sb and the chemical classification of iron meteorites. Nature 282:790–793
Wang HS, Lineweaver CH, Ireland TR (2019) The volatility trend of protosolar and terrestrial elemental abundances. Icarus 328:287–305
Wänke H (1981) Constitution of terrestrial planets. Philos Trans R Soc Lond 303A:287–302
Wänke H, Dreibus G (1988) Chemical composition and accretion history of terrestrial planets. Philos Trans R Soc Lond Ser A 325:545–557
Wänke H, Baddenhausen H, Palme H, Spettel B (1974) On the chemistry of the Allende inclusions and their origin as high temperature condensates. Earth Planet Sci Lett 23:1–7
Wänke H, Dreibus G, Jagoutz E (1984) Mantle chemistry and accretion history of the Earth. In: Kröner A (ed) Archean geochemistry. Springer, Berlin, pp 1–24
Warren PH (2011) Stable-isotopic anomalies and the accretionary assemblage of the Earth and Mars: A subordinate role for carbonaceous chondrites. Earth Planet Sci Lett 311:93–100
Wasson JT (1977) Reply to Edward Anders: a discussion of alternative models for explaining the distribution of moderately volatile elements in ordinary chondrites. Earth Planet Sci Lett 36:21–28
Wasson JT, Chou C-L (1974) Fractionation of moderately volatile elements in ordinary chondrites. Meteoritics 9:69–84
Wasson JT, Kallemeyn GW (1988) Composition of chondrites. Philos Trans R Soc Lond 325A:535–544
Wasson JT, Wai CM (1976) Explanation for the very low Ga and Ge concentrations in some iron meteorite groups. Nature 261:114–116
Welz D, Deppe P, Schaefer W, Sabrowsky H, Rosenberg M (1989) Magnetism of iron-sulfur tetrahedral frameworks in compounds with thallium. I. Chain structures. J Phys Chem Solids 50:297–308
Witt-Eickschen G, Palme H, O'Neill HStC, Allen CM (2009) The geochemistry of the volatile trace elements As, Cd, Ga, In and Sn in the Earth's mantle: new evidence from in situ analyses of mantle xenoliths. Geochim Cosmochim Acta 73:1755–1778
Wood JA, Hashimoto A (1993) Mineral equilibrium in fractionated nebular systems. Geochim Cosmochim Acta 57:2377–2388
Wood BJ, Smythe DJ, Harrison T (2019) The condensation temperatures of the elements: a reappraisal. Am Mineral 104:844–856
Xu H, Navrotsky A, Balmer ML, Su Y, Bitten ER (2001) Energetics of substituted pollucites along the $CsAlSi_2O_6 - CsTiSi_2O_{6.5}$ join: a high-temperature calorimetric study. J Am Ceram Soc 84:555–560
Yoneda S, Grossman L (1995) Condensation of CaO-MgO-Al_2O_3-SiO_2 liquids from cosmic gases. Geochim Cosmochim Acta 59:3413–3444
Yuan L, Steinle-Neumann G (2023) Earth's missing chlorine may be in the core. J Geophys Res 129:e2023JB027731
Zabel M, Range K-J (1979) Ternäre phase im system Eisen-Thallium-Schwefel. Z Naturforsch 34b:1–6
Zurfluh FJ, Hofmann BA, Gnos E, Eggenberger U (2013) "Sweating meteorites" – water-soluble salts and temperature variation in ordinary chondrites and soil from the hot desert of Oman. Meteorit Planet Sci 48:1958–1980

Publisher's Note Springer Nature remains neutral with regard to jurisdictional claims in published maps and institutional affiliations.

Authors and Affiliations

Katharina Lodders[1] · Bruce Fegley[1] · Klaus Mezger[2] · Denton Ebel[3]

✉ K. Mezger
klaus.mezger@unibe.ch

K. Lodders
lodders@wustl.edu

B. Fegley
bfegley@wustl.edu

D. Ebel
debel@amnh.org

1 McDonnell Center for Space Sciences and Department of Earth, Environmental, and Planetary Sciences, Washington University, St Louis, MO 63130, USA

2 Institut für Geologie and Center for Space and Habitability, Universität Bern, Baltzerstrasse 1+3, CH-3012 Bern, Switzerland

3 Department of Earth and Planetary Sciences, American Museum of Natural History, Central Park West at 79th Street, New York, NY 10024, USA